Identification and Ecology of Freshwater Arthropods in the Mediterranean Basin

Identification and Ecology of Freshwater Arthropods in the Mediterranean Basin

Edited by

Alain Maasri
Leibniz Institute of Freshwater Ecology and Inland Fisheries, Berlin, Germany
The Academy of Natural Sciences of Drexel University, Philadelphia, PA, United States

James H. Thorp
Kansas Biological Survey and Department of Ecology and Evolutionary Biology,
University of Kansas, Lawrence, KS, United States

ELSEVIER

Elsevier
Radarweg 29, PO Box 211, 1000 AE Amsterdam, Netherlands
125 London Wall, London EC2Y 5AS, United Kingdom
50 Hampshire Street, 5th Floor, Cambridge, MA 02139, United States

Notices
Knowledge and best practice in this field are constantly changing. As new research and experience broaden our understanding, changes in research methods, professional practices, or medical treatment may become necessary.

Practitioners and researchers must always rely on their own experience and knowledge in evaluating and using any information, methods, compounds, or experiments described herein. In using such information or methods they should be mindful of their own safety and the safety of others, including parties for whom they have a professional responsibility.

To the fullest extent of the law, neither the Publisher nor the authors, contributors, or editors, assume any liability for any injury and/or damage to persons or property as a matter of products liability, negligence or otherwise, or from any use or operation of any methods, products, instructions, or ideas contained in the material herein.

ISBN: 978-0-12-821844-0

For Information on all Elsevier publications
visit our website at https://www.elsevier.com/books-and-journals

Publisher: Candice G. Janco
Acquisitions Editor: Maria Elekidou
Editorial Project Manager: Sara Valentino
Production Project Manager: R. Vijay Bharath
Cover Designer: Vicky Pearson Esser

Typeset by MPS Limited, Chennai, India

Dedication

For all the amazing ecologists and taxonomists exploring and documenting the rich biodiversity of the Mediterranean Basin, without whom this book would have not been possible.

Amid a global biodiversity crisis where freshwater biodiversity is declining at unprecedented rates, we hope this book will provide additional tools to explore and study Mediterranean freshwater biodiversity and create an incentive to further protect it for generations to come.

—Alain Maasri and James H. Thorp

Contents

List of contributors

Pedro Abellán Department of Zoology, Faculty of Biology, University of Seville, Spain

Raúl Acosta FEHM-Lab (Freshwater Ecology, Hydrology and Management), Departament de Biologia Evolutiva, Ecologia i Ciències Ambientals, Facultat de Biologia, Universitat de Barcelona (UB), Barcelona, Catalonia/Spain; Institute of Environmental Assessment and Water Research (IDAEA), Spanish National Research Council (CSIC), Jordi Girona, Barcelona, Spain

Peter H. Adler Department of Plant and Environmental Sciences, Clemson University, Clemson, SC, United States

Enrique Baquero WasserCluster Lunz—Biologische Station GmbH, Lunz am See, Austria

Marta Sáinz Bariáin Spanish Institute of Oceanography, Oceanographic Research Center of Santander, Santander, Cantabria, Spain

Dani Boix GRECO, Institute of Aquatic Ecology, Faculty of Sciences, Universitat de Girona, Girona, Spain

Núria Bonada FEHM-Lab (Freshwater Ecology, Hydrology and Management), Departament de Biologia Evolutiva, Ecologia i Ciències Ambientals, Facultat de Biologia, Universitat de Barcelona (UB), Barcelona, Catalonia/Spain; Institut de Recerca de la Biodiversitat (IRBio), Universitat de Barcelona (UB), Barcelona, Catalonia/Spain

Maria Cristina Bruno Research and Innovation Centre, Fondazione Edmund Mach, San Michele all'Adige (Trento), Italy

Ana I. Camacho Museo Nacional de Ciencias Naturales (CSIC), Department of Biodiversity and Evolutionary Biology Department, Madrid, Spain

Gianmaria Carchini Società Italiana per lo Studio e la Conservazione delle Libellule ODV, Università degli Studi di Perugia, Perugia, Italy

Magdalini Christodoulou German Centre for Marine Biodiversity Research (DZMB), Senckenberg am Meer, Wilhelmshaven, Germany; Ö Landes-Kultur GmbH, Biology Centre, Linz, Austria

Fabio Cianferoni Research Institute on Terrestrial Ecosystems (IRET), National Research Council of Italy (CNR), Florence, Italy; Zoology, "La Specola," Natural History Museum of the University of Florence, Florence, Italy

Nicole Coineau Retired, Banyuls-sur-Mer, France

Gregory W. Courtney Department of Plant Pathology, Entomology, and Microbiology, Iowa State University, Ames, IA, United States

David Cunillera-Montcusí GRECO, Institute of Aquatic Ecology, Faculty of Sciences, Universitat de Girona, Girona, Spain; FEHM-Lab (Freshwater Ecology, Hydrology and Management), Departament de Biologia Evolutiva, Ecologia i Ciències Ambientals, Facultat de Biologia, Universitat de Barcelona (UB), Barcelona, Catalonia/Spain; Departamento de Ecología y Gestión Ambiental, Centro Universitario Regional del Este (CURE), Universidad de la República, Tacuarembó s/n, Maldonado, Uruguay

Romolo Fochetti Department for Innovation in Biological, Agro-food and Forest Systems, Tuscia University, Viterbo, Italy

Pau Fortuño FEHM-Lab (Freshwater Ecology, Hydrology and Management), Departament de Biologia Evolutiva, Ecologia i Ciències Ambientals, Facultat de Biologia, Universitat de Barcelona (UB), Barcelona, Catalonia/Spain; Institut de Recerca de la Biodiversitat (IRBio), Universitat de Barcelona (UB), Barcelona, Catalonia/Spain

Antonio J. García-Meseguer Department of Ecology and Hydrology, Faculty of Biology, University of Murcia, Spain

Stéphanie Gascón GRECO, Institute of Aquatic Ecology, Faculty of Sciences, Universitat de Girona, Girona, Spain

Jean-Luc Gattolliat State Museum of Natural Sciences, Department of Zoology, Lausanne, Switzerland; Department of Ecology and Evolution, Biophore, University of Lausanne, Lausanne, Switzerland

Sönke Hardersen Società Italiana per lo Studio e la Conservazione delle Libellule ODV, Università degli Studi di Perugia, Perugia, Italy; Reparto Carabinieri Biodiversità di Verona, Centro Nazionale Carabinieri Biodiversità "Bosco Fontana", Marmirolo, Italy

Ralph W. Holzenthal Department of Entomology, University of Minnesota, St. Paul, MN, United States

Rafael Jordana WasserCluster Lunz—Biologische Station GmbH, Lunz am See, Austria

Ioannis Karaouzas Institute of Marine Biological Resources and Inland Waters, Hellenic Centre for Marine Research, Anavyssos, Attica, Greece

Valeria Lencioni MUSE-Science Museum, Research and Museum Collections Office, Climate and Ecology Unit, Corso del Lavoro e della Scienza, Trento, Italy

Manuel Jesús López-Rodríguez Department of Ecology, University of Granada, Granada, Spain

Peter H. Langton 16 Irish Society Court, Coleraine, Co. Londonderry Northern Ireland, BT52 1GX

Alain Maasri Leibniz Institute of Freshwater Ecology and Inland Fisheries, Berlin, Germany; The Academy of Natural Sciences of Drexel University, Philadelphia, PA, United States

Federico Marrone Department of Biological, Chemical and Pharmaceutical Sciences and Technologies (STEBICEF), University of Palermo, Palermo, Italy

Claude Meisch National Natural History Museum, Luxembourg, Luxembourg

Francesc Mesquita-Joanes Cavanilles Institute of Biodiversity and Evolutionary Biology, University of València, Paterna, Spain

Giuseppe Messana Institute of Research on Terrestrial Ecosystems of the National Research Council (CNR-IRET), Firenze, Italy

Andrés Millán Department of Ecology and Hydrology, Faculty of Biology, University of Murcia, Spain

Joel Moubayed Freshwater and Marine biology, 10 rue des Fenouils, 34070 Montpellier, France

Isabel Muñoz Department of Evolutionary Biology, Ecology and Environmental Sciences, Universitat de Barcelona, Barcelona, Spain

Félix Picazo Department of Ecology/Research Unit Modeling Nature, Faculty of Sciences, University of Granada, Spain

Christophe Piscart French National Centre for Scientific Research (CNRS), University of Rennes, Research Unit ECOBIO, Rennes, France

Virginija Podeniene Institute of Biosciences, Life Sciences Center, Vilnius University, Vilnius, Lithuania

Michael Raupach Staatliche Naturwissenschaftliche Sammlungen Bayerns, Zoologische Staatssammlung München, München, Bayern, Germany

D. Christopher Rogers Kansas Biological Survey, and The Biodiversity Institute, The University of Kansas, Lawrence, KS, United States

Giampaolo Rossetti Department of Chemistry, Life Science and Environmental Sustainability, University of Parma, Parma, Italy

Jordi Sala GRECO, Institute of Aquatic Ecology, Faculty of Sciences, Universitat de Girona, Girona, Spain

David Sánchez-Fernández Department of Ecology and Hydrology, Faculty of Biology, University of Murcia, Spain

Michel Sartori State Museum of Natural Sciences, Department of Zoology, Lausanne, Switzerland; Department of Ecology and Evolution, Biophore, University of Lausanne, Lausanne, Switzerland

Fabio Stoch Department of Evolutionary Biology and Ecology, Free University of Brussels (ULB), Brussels, Belgium

Alain Thiéry Aix-Marseille Université, Avignon Université (CNRS, IRD, IMBE UMR), Institut Méditerranéen de Biodiversité et d'Écologie marine et continentale, Biomarqueurs, Environnement, Santé (BES), Avignon, France

James H. Thorp Kansas Biological Survey and Department of Ecology and Evolutionary Biology, University of Kansas, Lawrence, KS, United States

José Manuel Tierno de Figueroa Department of Zoology, University of Granada, Granada, Spain

Irene Tornero GRECO, Institute of Aquatic Ecology, Faculty of Sciences, Universitat de Girona, Girona, Spain

Kay Van Damme Centre for Academic Heritage and Archives and Ghent University Botanical Garden, Ghent University, Ghent, Belgium

Johann Waringer Department of Functional and Evolutionary Ecology, University of Vienna, Vienna, Austria

Karl J. Wittmann Department of Environmental Health, Medical University of Vienna, Vienna, Austria

Carmen Zamora-Muñoz Department of Zoology, University of Granada, Granada, Spain

Chapter 1

Introduction ☆

James H. Thorp[1] and Alain Maasri[2,3]

[1]*Kansas Biological Survey and Department of Ecology and Evolutionary Biology, University of Kansas, Lawrence, KS, United States,* [2]*Leibniz Institute of Freshwater Ecology and Inland Fisheries, Berlin, Germany,* [3]*The Academy of Natural Sciences of Drexel University, Philadelphia, PA, United States*

Introduction

Aside from being the cradle of western civilization and a region of high inherent cultural, geographical, and ecological importance, the Mediterranean Basin contains fascinating biodiversity in its freshwater lakes, streams, rivers, and wetlands. Mediterranean biomes occur primarily on five continents; however, this ecological area is by far most common in countries surrounding the Mediterranean Sea (Fig. 1.1).

The focal area of our book is the Mediterranean Basin defined by its bioclimatic region as shown in Fig. 1.1, and described more extensively in Chapter 2. The terrestrial component of this area consists primarily of dense scrubland adapted to drought, with a climate characterized by alternating cool-wet and hot-dry periods. Although natural lakes and wetlands exist, they are relatively scarce compared to those in more northern latitudes, especially in alpine regions. Rivers tend to be relatively small, with the largest from West to East being the Ebro in Spain, the Rhône in France, the Po in Italy, and the Maritsa (called also, Meriç or Evros) crossing Bulgaria, Greece, and Turkey.

Our book is focused on taxonomic keys that will help the user to identify inland freshwater arthropods at various classification levels from order to species, depending on the group and the needs of the individual researcher. The users of this book could be college students and professors as well as scientists and technicians in government agencies, private foundations, or environmental companies.

This book is limited to members of the Arthropoda, and within that phylum, insects (subphylum Crustacea, class Hexapoda) are covered in the greatest detail. They are generally keyed to the levels of genus or species (if the taxonomic knowledge in the literature is sufficient). Other crustacean members of this subphylum are also keyed to the genus or species level, while the Arachnida—a subphylum of Arthropoda containing spiders and mites—are discussed but only classified to the ordinal level.

Individual chapter authors were responsible for developing their chapters and taxonomic keys following chapter and classification rules set by the editors. These taxonomic keys are primarily based on microscopic identification, but occasional reference is made to genetic techniques to separate species—an increasing trend in some countries when primarily working on environmental assessment. These genetic techniques are discussed in Chapter 3.

Components of taxonomic chapters

Our primary focus in this book is providing the user with the information needed to identify aquatic arthropods at various taxonomic levels. To that end, we provide information on limits to identification for specific groups as well as methods to collect, preserve, and prepare specimens for identification. Each taxonomic chapter also contains information on the ecology of the group and relevant literature citations.

Chapter figures are mostly limited to photographs and line drawings needed to use the keys. In addition, some figures of habitats are present in some chapters to illustrate the diversity of Mediterranean freshwater ecosystems.

☆. This chapter was partially based in part on Chapter 1 (Thorp & Rogers, 2019) in a volume by Rogers and Thorp (2019) covering the entire Palearctic bioregion, but it was modified for the Mediterranean Basin with elimination of that chapter's taxonomic key and the addition or modification of other sections.

Identification and Ecology of Freshwater Arthropods in the Mediterranean Basin. DOI: https://doi.org/10.1016/B978-0-12-821844-0.00014-4

FIGURE 1.1 Map showing the Mediterranean Sea and surrounding countries. The Mediterranean Basin, as considered in this book, is the Mediterranean bioclimatic region delimited by the red line (see Médail, 2008, 2017). *Map produced by Joseph Shaybane.*

How to use this volume and the limits to identification[1]

Each chapter includes a basic introduction to the terms used in diagnosing the taxa of that section. Limitations to the current state of taxonomic knowledge may also be presented so that the reader may gage the reliability of the information presented. Only established peer-reviewed scientific literature is used to define the taxonomic categories and epithets included. All names should conform to the International Code of Zoological Nomenclature. All nomina and taxonomic arrangements used, as well as the rejection of old names, are based on peer-reviewed, validated, and published scientific literature. Provisional names and species designated "taxon 1" or "species 1" are not used unless they were previously recognized and accepted in the peer reviewed scientific literature (Richards & Rogers, 2011). No new species descriptions or previously unpublished taxonomic arrangements are presented. The keys are dichotomus and hierarchical, thus proceeding from higher to lower taxonomic levels.

Surprisingly, many users do not know how to interpret a dichotomus key, making the fundamental assumption that a correct identification answer is always present in the key. This assumption generally takes one of the following three forms.

1. All species are identifiable using a given key, which overlooks the fact that many new species have yet to be discovered and described. Generalized geographic ranges are provided for most taxa presented herein, yet species ranges naturally shrink, swell, and change elevation constantly, particularly as weather and climate patterns shift. In addition to these natural processes, some species are introduced intentionally or accidentally by humans.
2. All variation is accounted for in the key. As stated above, identification keys use specific, primary, diagnostic characters. Problems in identification are compounded by taxa that: (a) have different character states at different times; (b) only have diagnostic characters at certain life stages or in certain genders; and/or (c) have severely truncated morphology (often due to lack of sexual selection) and lack morphological characters to separate the species.
3. The key is a sufficient identification tool in and of itself. Identification keys are tools to aid in taxon identification and are primarily useful in eliminating incorrect taxa from the range of possible choices, narrowing the field to the

1. This section is derived extensively and occasionally word for word from that section present in Thorp & Rogers (2019) and largely written by D. Christopher Rogers.

names that may be applicable. The possibility that the specimen to be identified is new, a hybrid, anomalous, or a recent invasive colonist is always a possible answer.

Once one arrives at a name or group of possible names for a specimen in hand, the specimen should then be compared against descriptions, distribution maps, and figures of that and other taxa in that group. Direct comparison of the specimen at hand with identified museum material or using molecular comparisons is also sometimes necessary for correct identification.

Species are not immutable, fixed in location and form, thereby confounding keys and any other identification method, such as trait tables, character matrices, or even genetic analyses.

References

Médail, F. 2008. Ecosystems: Mediterranean. Pages 2296-2308 *in*: Jørgensen S.E. & Fath B. (eds) Encyclopedia of Ecology, Vol. 5. Elsevier, Oxford.

Médail, F. 2017. The specific vulnerability of plant biodiversity and vegetation on Mediterranean islands in the face of global change. Regional Environmental Change. 17: 1775-1790.

Richards, A.B. & Rogers, D.C. 2011. Southwest Association of Freshwater Invertebrate Taxonomists (SAFIT) list of freshwater macroinvertebrate taxa from California and adjacent states including standard taxonomic effort levels. 266 p.

Rogers, D.C. & Thorp, J.H. (eds.). 2019. Keys to Palaearctic Fauna. Vol. V *in*: Thorp and Covich's Freshwater Invertebrates. Academic Press, Elsevier.

Thorp, J.H. & D.C. Rogers. 2019. Introduction. Pages 1-4. D.C. Rogers & J.H. Thorp (eds.) Keys to Palaearctic Fauna, *in*: Thorp and Covich's Freshwater Invertebrates. Academic Press, Elsevier.

Chapter 2

Ecology of Mediterranean freshwater ecosystems

Alain Maasri[1,2] and Núria Bonada[3,4]
[1]Leibniz Institute of Freshwater Ecology and Inland Fisheries, Berlin, Germany, [2]The Academy of Natural Sciences of Drexel University, Philadelphia, PA, United States, [3]FEHM-Lab (Freshwater Ecology, Hydrology and Management), Departament de Biologia Evolutiva, Ecologia i Ciències Ambientals, Facultat de Biologia, Universitat de Barcelona (UB), Barcelona, Catalonia/Spain, [4]Institut de Recerca de la Biodiversitat (IRBio), Universitat de Barcelona (UB), Barcelona, Catalonia/Spain

Mediterranean climate

The Mediterranean climate is characterized by alternating cool-wet and hot-dry periods, and corresponds to one of the 31 climate classes initially described by Wladimir Köppen in 1900, and later updated by himself (Köppen, 1936) and others, notably by Rudolf Geiger (Kotteki et al., 2006). It occurs between 30 and 45 degrees latitude north and south of the equator, respectively, and is encountered mainly around the Mediterranean Basin (hence the name), and in other small parts of North and South America, South Africa, and Australia. For simplification throughout this book, the climate of the Mediterranean Basin will henceforth be referred to as the Mediterranean climate. Strong seasonality of precipitation and great variability on interannual and interdecadal time scales characterize the Mediterranean climate (Dünkeloh et al., 2003). This is supported by current observations and by an increasing number of palaeoenvironmental studies demonstrating significant historical climatic instability (Magny et al., 2013) accompanied by long-standing anthropogenic impacts (Vannière et al., 2013), which we will address later in this chapter.

The Mediterranean Basin, extending in the transitional zone between the temperate mid-latitude climate of Western and Central Europe, the Middle East, and the arid North African dry desert belt, is characterized by significant temporal and geographical climatic disparities. The climate is affected by interactions between mid-latitude and tropical climatic processes. Winters in the basin are influenced by the North Atlantic Oscillation over its western section, and the East Atlantic Oscillation and other patterns over its northern and eastern sections. Summers are affected by descending high-pressure areas due to downward motions leading to dry conditions, particularly over the southern part of the basin (Giorgi et al., 2008). In addition, the Mediterranean climate is responsive to subregional processes induced by a complex topography around the basin. The Mediterranean Sea constitutes an important source of moisture surrounded by major mountain ranges (Fig. 2.1). Among these mountain ranges from east to west are: the Mount Lebanon range (peak at 3088 m a.s.l.), the most eastern mountain range of the Mediterranean Basin; the Taurus Mountains (peak at 3756 m a.s.l.) separating the Mediterranean Sea from the central Anatolian Plateau; the Dinaric Alps (peak at 2694 m a.s.l.) separating the continental Balkan Peninsula from the Adriatic Sea; the Apennines (peak at 2912 m a.s.l.), a major mountain range along the length of the Italian Peninsula; the Alps (peak at 4809 m a.s.l.), a major crescent-shaped mountain range separating the Mediterranean Sea from Central Europe; the Pyrenees (peak at 3404 m a.s.l.), a major range of mountains in southwestern Europe that forms a natural border between Spain, Andorra, and France; the Baetic Mountains (peak at 3478 m a.s.l.) in the south of the Iberian Peninsula; and the Atlas Mountains (peak at 4167 m a.s.l.) across western North Africa. This complex physiography modulates climate at subregional spatial scales. Thus the combination of global and subregional scale climatic processes confers a high heterogeneity in climatic patterns to the Mediterranean Basin. Annual precipitations can range between 200 and 2000 mm, and it is not rare in some arid parts of the Mediterranean Basin (e.g., North Africa) that the total monthly precipitation falls in only a few hours. Absolute temperature maxima of 48°C and minima of −20°C have been recorded around the basin, with mean annual temperatures varying between 11°C and 19°C.

Identification and Ecology of Freshwater Arthropods in the Mediterranean Basin. DOI: https://doi.org/10.1016/B978-0-12-821844-0.00008-9

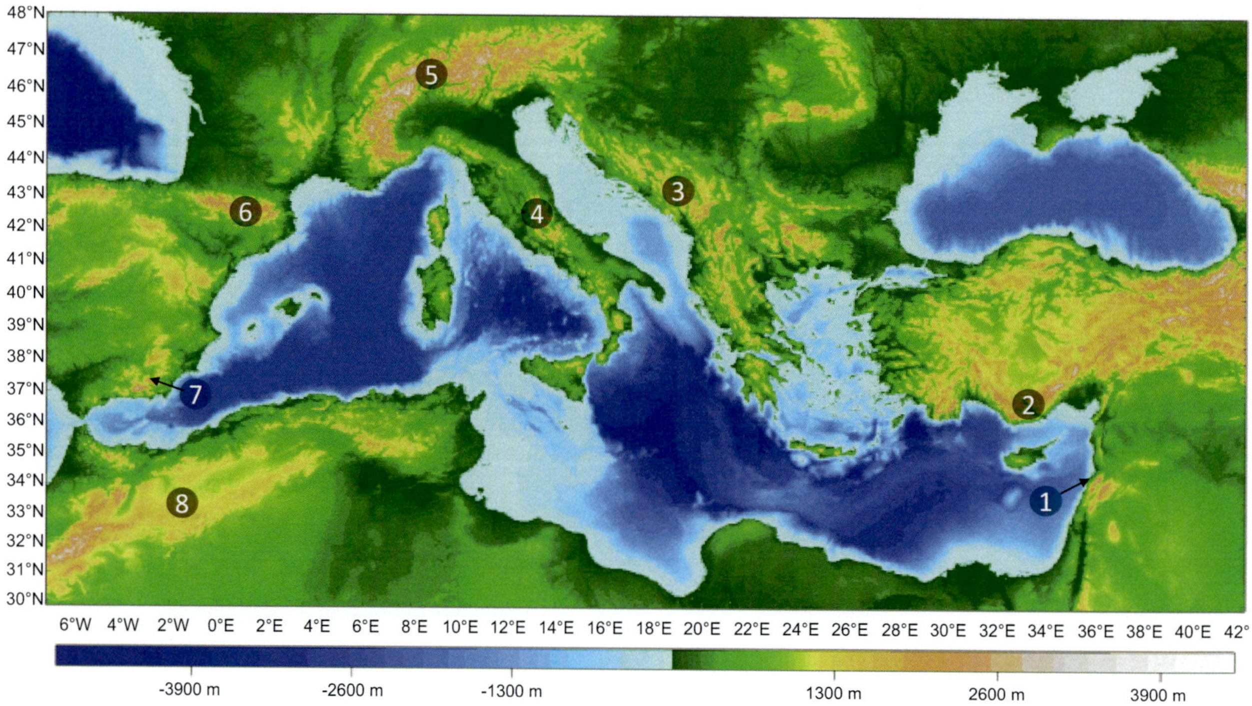

FIGURE 2.1 Topographic map showing the major mountain ranges of the Mediterranean Basin. From east to west (1) Mount Lebanon; (2) Taurus Mountains; (3) Dinaric Alps; (4) Apennines; (5) Alps; (6) Pyrenees; (7) Baetic Mountains; and (8) Atlas Mountains. *Modified after the original map available at www-3.unipv.it/cibra, used with permission from G. Pavan, University of Pavia, Italy.*

The Mediterranean Basin has been identified as among the most vulnerable regions facing climate change (Cramer et al., 2018). Observational records covering the period from 1860 to 2005 show that the decadal variability and the general tendency for mean annual climatic conditions are increasingly warmer and drier (Mariotti et al., 2015). Regional temperatures in the Mediterranean Basin are now around 1.3°C higher than temperatures from the late mid-19th-century, compared with an increase of only 0.85°C worldwide for the same time period (Guiot et al., 2016). Furthermore, an increase in temperature between 2.2°C and 5.1°C is expected before the end of this century (Nakiceovic et al., 2000), accompanied by a significant decrease in precipitation: between 4% and 27% (Christensen et al., 2008). Methodological dry spells, droughts, and heat waves are now more frequent, and their occurrence is expected to increase significantly primarily in the eastern Mediterranean (Zittis et al., 2015). Climate change is also heavily affecting the annual water cycle by increasing evaporation, reducing precipitation, and significantly altering the pattern and timing of rainfall. A global atmospheric temperature increase of 2°C will probably be accompanied by a reduction in precipitation of around 7% in the Mediterranean Basin, with a local reduction of winter precipitation (December through February) by 15%–20% in the coastal zones of Algeria and Tunisia, and by 10%–15% in Sicily, Greece, and Turkey (Vautard et al., 2014). Streamflow will generally be reduced. Many regions of the Mediterranean Basin like the Iberian Peninsula, Italy, and the Balkans are expected to experience continued drought intensifications, and stream flows are expected to decrease by up to 80% by the 2080s (Forzieri et al., 2014). Streamflow seasonality is also very likely to change with more irregularities in discharge and drought. An analysis of trends of river flood discharges from 1960 to 2010 showed that Mediterranean Europe (Spain, France, Italy, the Balkan peninsula, Greece, Turkey, and Cyprus) has witnessed a decrease in river discharge of 12%–24% over this time period (Blöschl et al., 2019). Water levels in lakes and wetlands will be similarly affected. Lefebvre et al. (2019) simulated the water balance of wetlands in 229 localities around the Mediterranean Basin and found that, while the majority of wetlands can persist with up to a 400 mm decrease in annual precipitations, the percentage of wetland habitats in a good state will decrease by 32% in 2050 and by 73% in 2100 (under the RCP8.5 scenario, a scenario that does not include any specific climate mitigation target). Overall, consistent studies show that progressive warming and drier conditions will greatly affect the already water-deficient and highly populated Mediterranean Basin. Climate change is therefore expected to drive major changes in freshwater ecosystems around the Mediterranean Basin in the current century.

Anthropogenic activities in the Mediterranean Basin

Human activities and anthropogenic disturbances around the Mediterranean Basin are millennia old. Paleoecological, historical, and archeological evidences have shown that since the onset of early civilizations, human activities have modified the Mediterranean landscape. Vannière et al. (2013), examining sedimentary charcoal records to estimate fire frequency and biomass burned for the last 16,000 years, showed a determinant anthropogenic driving force on fire regimes for 7000 years in Mediterranean Europe. Di Castri (1981) and later Blondel (2006) described complex coevolution that shaped the interactions between ecosystems and human societies around the Mediterranean Basin. Blondel (2006) reiterated this by stating that one cannot understand the structure and dynamics of current Mediterranean ecosystems and biodiversity without taking into account the history of the anthropogenic footprint in the Mediterranean Basin.

In a water-stressed region like the Mediterranean Basin, water continues to be at the center of many civilizations and societies. A great variety of ancient water supply techniques have been developed and documented around the basin exploiting fossil groundwater, major rivers, or intermittent sources of water, such as rainfall, runoff in intermittent streams, and shallow groundwater and wetlands. Modern-day anthropogenic activities have drastically altered many Mediterranean freshwater ecosystems. Although case studies are far too numerous to mention here, the following two examples illustrate the extent of observed alterations: Lake Ichkeul in northern Tunisia near the town of Bizerta, and the Durance River in southern France. Periodic change in freshwater supply modulated the functioning of Lake Ichkeul. During winter, freshwater inflow from six rivers predominated and guaranteed low salinity in the lake and associated wetlands, while in summer seawaters would enter from the adjacent Bizerta lagoon. This dynamic and exceptional natural system has been altered by damming and water abstraction from three of its six major inflowing rivers leading to drastic increases in salinity levels in the lake and the degradation of associated wetlands. The Durance River in Southern France, draining the French Alps down toward the Mediterranean Sea, was known for being highly dynamic, actively bringing about erosion and deposition, and often causing major floods. This 324 km long, braided Mediterranean river has been tamed by a series of 17 hydropower dams built over a 45-year timeframe, making it one of the most regulated rivers in the basin. This transformation was not without ecological consequences. The riverbed greatly narrowed, the braided structure of the river channel contracted, and the dynamic mosaic of aquatic and terrestrial habitats was heavily altered, allowing the expansion of agricultural exploitation of the nutrient-rich alluvium along the river floodplain (Maughna, 2015).

The Mediterranean Basin offers an exceptional diversity of natural and human-modified freshwater ecosystems. The complex historical processes, topography, and climate of the Mediterranean Basin provide a wide array of springs, streams, rivers, lakes, and wetlands home to exceptional aquatic biodiversity. However, the increasing anthropogenic pressures and the effects of climate change constitute undoubtedly the major threats of this century challenging the integrity of Mediterranean ecosystems and biodiversity.

Streams and rivers

Streams and rivers feeding the Mediterranean Basin have been used as sources of freshwater and energy, and transportation routes, thereby forming hubs for cultural, religious, and spiritual practices for many civilizations. For example, the gods of streams and rivers in Greek mythology were called Potamoi, a Greek term that means rivers. Thirteen major rivers contribute the most to the discharge in the Mediterranean Basin (Fig. 2.2): Ebro, Rhône, Po, Adige, Tiber, Neretva, Drin, Evros, Seyhan, Ceyhan, Nile, Shellif, and Moulouya. Their overall mean annual discharge is estimated to be more than 10,000 m^3/s, with those in Europe, in particular the Rhône and Po, contributing the most (Ludwig et al., 2009). This discharge has significantly decreased in the last decades in most Mediterranean catchments as a direct consequence of climate change and dam construction, with negative trends for half to two-thirds of the rivers (Skoulikidis et al., 1998; Ludwig et al., 2009). Hydrological alterations of natural river flows are very common around the Mediterranean Basin, especially in the European countries, where the disruption of natural flow dynamics has significantly impacted native biodiversity. For example, in Spain alone, there are more than 1000 large river reservoirs (Sabater et al., 2009). Hydrological alterations are still ongoing in many Mediterranean countries, such as those in the Balkans and the eastern Mediterranean, where numerous dams are either under construction or being planned (Zarfl et al., 2015).

The typology of streams and rivers in the Mediterranean Basin is very diverse (Fig. 2.3). Nine rivers are longer than 450 km and have basin orders between 3 and 5 (Vörösmarty et al., 2000). The Nile has the largest catchment but contributes less discharge to the Mediterranean Sea than other medium-size catchments in Europe (Ludwig et al., 2009). A large amount of its water is lost through evaporation but also through agricultural activities, which increased after the construction of the Aswan High Dam (Nixon, 2003). The central area of North Africa (i.e., Algeria, Tunisia, Libya)

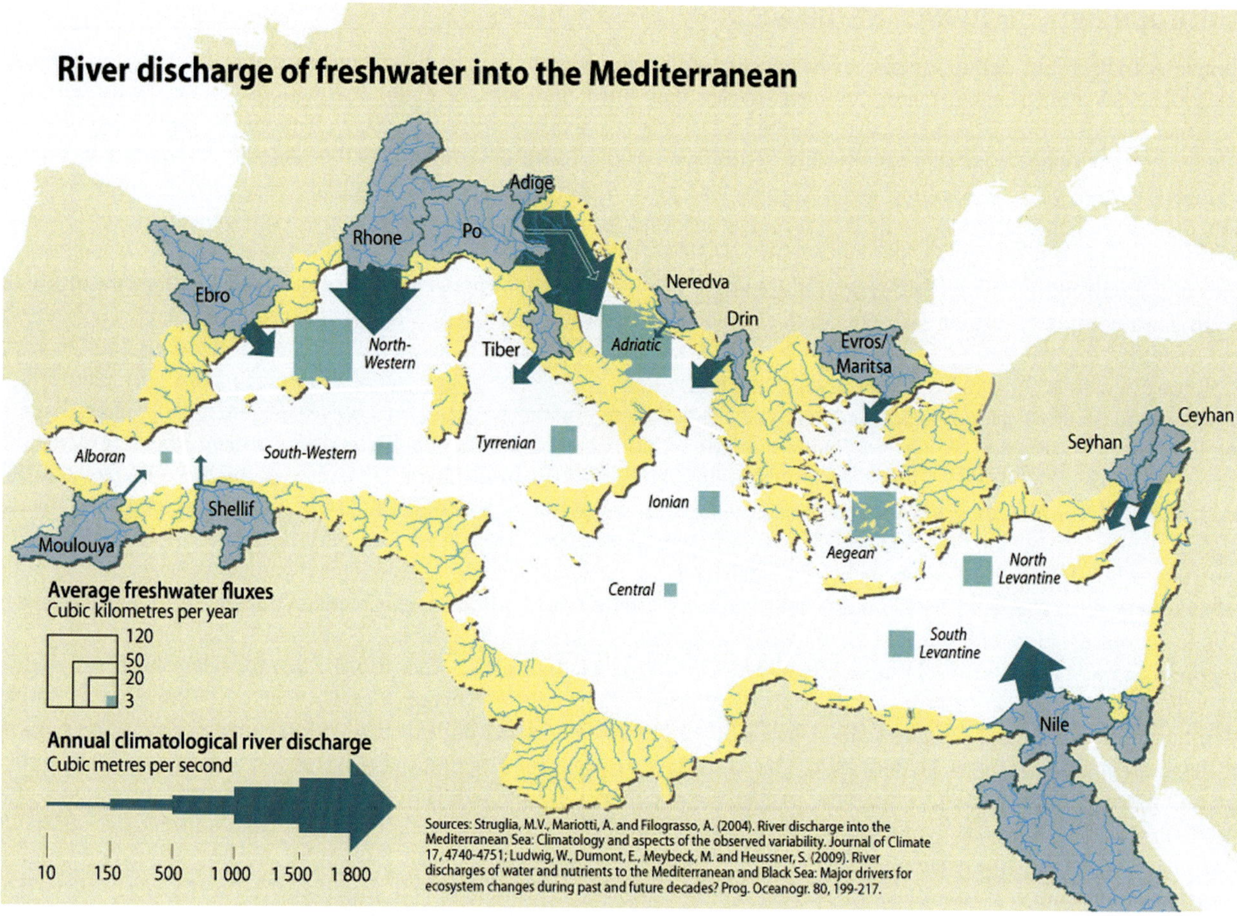

FIGURE 2.2 Annual freshwater inflow into the Mediterranean Sea from major rivers. *Used with permission from GRID-Arendal http://www.grida.no.*

also has large catchments, which are among the 50 largest in the world, but these only feed the Mediterranean Sea during floods (e.g., Araye and Qattara, Vörösmarty et al., 2000). The Ebro, Rhône, and Po in Europe are considered medium-size catchments on the global scale; however, they have a significant discharge to the Mediterranean Sea and vast deltas of high ecological interest. Few Mediterranean rivers are braided in their lower courses (e.g., Tagliamento and Vjosa), the vast majority being small coastal rivers with steep slopes when they drain through coastal ranges, such as in North–East Spain (Bonada et al., 2013). Indeed, many Mediterranean coastal streams and rivers have their sources in subalpine or alpine areas located near the coast (e.g., Rif, Sierra Nevada, Apennines, Dinaric Alps) or inland (e.g., Pyrenees, Alps, High Atlas), displaying, therefore, a variety of hydrological characteristics and seasonal patterns.

The Mediterranean Basin has a complex geological history. Paleogeographic reconstructions suggest that rivers' physiognomy has been mainly subjected to different geological events (Cavazza et al., 2003). For example, plate movements, collisions and splits, and the Alpine orogenesis *s.l.* defined the headwaters and shaped most of the river morphology currently observed in the Balkans area (Skoulikidis et al., 2009), and the larger rivers in Mediterranean Europe (e.g., Rhône and Po, Winterberg et al., 2019). Most likely, the Ebro was connected to the Atlantic Ocean until the end of the Eocene and connected to the Mediterranean Sea after the Messinian crisis (Babault et al., 2006), whereas the Rhône was already connected before the Messinian crisis. All these complex geological phenomena had consequences on the current aquatic biodiversity patterns. For example, the movement between the African and the Eurasian plates resulted in many species in common between south Spain and north of Morocco (Bonada et al., 2009). Quaternary events, including glaciations and the decrease of the level of the Mediterranean Sea by c. 120 m, also resulted in changes in river and lake morphology along with changes in their hydrological patterns due to the presence and retreat of glaciers (Skoulikidis et al., 2009). For instance, changes in the sea level modified the length and the morphology of the river mouth of the Po during the glaciations but also during the Messinian crisis, when some of the smaller rivers currently flowing directly to the Adriatic Sea (e.g., Adige and Tagliamento) were tributaries of the Po (Castiglioni et al., 2001). The geology of the vast majority of

FIGURE 2.3 Photos illustrating the variety of streams and rivers across the Mediterranean Basin. (A) Cetina River fed by karstic springs in Croatia; (B and C) Showing similarities between rivers in the Rif region of Morocco (B) and in the Andalusian region of Spain (C); (D) Ebro River in Spain; (E) Wadi Tafna draining a very arid area in Algeria; (F) Fuirosos Stream a coastal, temporary, high gradient stream in Spain; (G) A rambla in the Bárdenas Reales in Spain; (H) Mendavia a hypersaline river in Spain; (I) Tagliamento River a braided gravel-bed river in Italy; (J) Temporary river outflow to the Mediterranean Sea in the Samothraki Island in Greece; (K) Guarnón River a mountain river in Sierra Nevada, Spain; (L) Algars River forming calcareous tufa in Spain; (M) Rhône River in France; (N) Onyar River, an urban river in Girona, Spain; (O) Nile River in Egypt. *Photos courtesy Marko Miliša (A), Mouna Hafiani (E), Pau Fortuño (F), Andrés Millán (H), Núria Cid (I), Nikos Skoulikidis (J), Carmen Zamora (K), Sylvain Dolédec (M, N) and Vincent H. Resh (O).*

Mediterranean catchments is composed of calcareous, siliceous, or sedimentary rocks, or a combination of them. Siliceous rivers dominate the North African region and south of Spain, whereas calcareous rivers and karstic catchments are common in the western Mediterranean Basin, the Balkans, and the eastern Mediterranean. Natural saline rivers are also found in most arid regions of the Mediterranean Basin, with varying levels of salinity. In southeastern Spain and Sicily, several rivers are hypersaline with salinity values >100 g/L, whereas mesosaline rivers are also present in Algeria, Tunisia, and Israel (Millán et al., 2011; Arribas et al., 2009). These rivers have high conservation value and host many endemic species with particular adaptations.

All Mediterranean streams and rivers are characterized by two flow peaks, one in spring and another in autumn or winter. High precipitation in autumn characterizes most Mediterranean countries except those in the eastern Mediterranean, where winter precipitation is relatively more abundant. Low flow and flow intermittence are common in summer in all streams and rivers. Indeed, flow intermittence is the most recognized characteristic of Mediterranean rivers worldwide (Bonada et al., 2013). This is especially relevant in the Mediterranean Basin, where temporary rivers have been part of the daily landscapes of many civilizations, and received several popular names: "oueds" in North Africa, "arroyos" in Spain, "ravins" in France, and "rambles," "torrents," and "rieres" in Catalonia and Balearic Islands (Vidal-Abarca, 1990). The star example is perhaps the "Rambla de Barcelona," a street in Barcelona very popular among locals and tourists that was diverted outside the medieval wall during the 15th century, and was completely channelized during the beginning of the 20th century (Casassas-Simó et al., 1992). However, the occurrence of temporary rivers is not uniform around the Mediterranean Basin. Precipitation patterns decrease from north to south and from west to east, while the difference between summer-winter precipitation increases (Ludwig et al., 2009), with an expected high occurrence of temporary rivers in the southern and eastern parts of the basin. For example, France has c. 30% of streams and rivers with temporary flows, whereas Greece has c. 40% and Cyprus c. 80% (Skoulikidis et al., 2017). However, there are regional exceptions. The area of southeastern Spain has an arid climate similar to that found in several parts of North Africa; and in the eastern part of the basin, 98% of the streams and rivers are ephemeral (Gómez et al., 2005). The occurrence of temporary streams and rivers is expected to increase in the Mediterranean Basin due to climate change, with a higher frequency of extreme events such as floods, droughts, and wildfires (Filipe et al., 2013).

Lakes and wetlands

Lakes of the Mediterranean Basin are diverse and mirror the complex geology, topography, and tectonic processes that shaped this area. Tectonic movements and the convergence between three major tectonic plates, the African, Eurasian, and Arabian plates, and two minor plates, the Anatolian and Aegean plates, led to the formation of major lakes of tectonic and volcanic origins. Many of those are saline and will not be extensively addressed in this book. The most famous is the Dead Sea, a hypersaline endorheic lake bordered by Jordan, Israel, and Palestine. The Dead Sea lies in the upper part of the Great Rift Valley, a series of rifts and faults stretching from the Beqaa valley in Lebanon to Mozambique in Africa. The Dead Sea is currently receding at an alarming rate (estimated to be 1.2 m each year) due to the extreme reduction of its main inflow from the Jordan River. Other remarkable saline lakes are the series of shallow lakes south of the Atlas Mountains in North Africa. These saline shallow lakes stretch from the Gulf of Cabès in Tunisia to the west across Algeria. They are commonly called "chotts" and are characterized by their intermittence and their fluctuating water level. Many stay dry for much of the year until they receive rainfall in winter, or are reliant on groundwater and springs flowing down from the Atlas Mountains.

Most natural freshwater lakes of the Mediterranean Basin lie in the northern part of the basin where there is an overall water surplus, which contrasts with the water deficit and arid zone of North Africa, and the eastern Mediterranean. Lake Okrid, located in the mountain range of Sara-Pind at the border between North Macedonia and Albania, is one of the most remarkable tectonic lakes of the Mediterranean Basin. It is one of Europe's oldest lakes formed around 4 million years ago and covers a surface area of 358 km^2. Other major tectonic lakes are Lake Beyşehir in Turkey, considered to be the largest freshwater Mediterranean lake covering a surface area of 650 km^2. Volcanic crater lakes, or caldera lakes, are an important feature of the Italian landscape. Lake Bracciano and Lake Vico are two characteristic crater lakes of the Lazio region near Rome in central Italy (Fig. 2.4). Fewer natural lakes are of glacial origin. They occur mainly in high mountain ranges that feature the southern extent of the major ice sheets during the repeated glaciations of the Pleistocene (Roberts et al., 2009). These glacial lakes are in general small; however, few exceptions exist like Lake Como or Lake Maggiore in Italy. Most glacial lakes are oligotrophic, nutrient-limited, and have little to no emergent vascular vegetation. Lakes of karstic formations are common in the Balkan peninsula. A series of karst lakes can be observed in the Dinaric Alps (see Introduction), known to be the largest continuous karst area in southern Europe extending over 60,000 km^2 (Cerkvenik et al., 2018).

FIGURE 2.4 Two volcanic crater lakes of the Lazio region near Rome: Lake Bracciano (Lago Bracciano) to the left, and Lake Vico (Lago di Vico) to the right side of the photograph. *Photo originally published by NASA's Earth Observatory in September 3, 2013.*

Typical karst lakes of the Dinaric Alps are formed either in karst depressions or by natural calcareous tufa dams. The largest is Lake Shkodra between Albania and Montenegro, formed by a cryptodepression with an area that fluctuates seasonally between 370 and 530 km^2. Karst lakes tend to be distinctively alkaline (pH >8) due to the high proportion of carbonate rocks in their catchments, and are often connected to karst caves, ponors, and karst springs flowing in underground narrow channels. Freshwater coastal and lowland lakes are rare in the Mediterranean. The majority contain brackish water and are near major river deltas such as the deltas of the Po or Rhône. Most coastal lakes are shallow where the hydrological effects (e.g., water level fluctuation and water mixing by wind) have a determinant effect on their nutrient dynamics and biological processes. These lakes are often eutrophic and heavily impacted by human activities.

Deltaic wetlands and coastal lagoons remain the most extensive and varied wetlands around the Mediterranean Basin. From east to west, the major deltaic wetlands occur in the Evros (also called, Maritsa or Meriç) river delta in Greece, the Po river delta in Italy, the Rhône river delta in France, and the Ebro river delta in Spain. These wetlands are often a mosaic of salt and freshwater marshes. Salt marshes often occur immediately behind ridges surrounding coastal lagoons; however, they can extend far inland, maintained by saline water tables that bring salt to the surface driven by evaporation. The vegetation of salt marshes is often dominated by salt-tolerant plants belonging to the genera *Phragmites*, *Scirpus*, *Salicornia, Ruppia*, and *Juncus*. In river deltas, freshwater marshes occur mainly along river courses that often receive floodwaters. However, the majority of deltaic freshwater marshes around the Mediterranean Basin no longer receive floodwater naturally as the majority of large rivers have been diked. Freshwater marshes are often maintained artificially for biodiversity conservation purposes, birdwatching, and ecotourism. In reality, most deltas in the Mediterranean Basin have been extensively modified by human activities, and much of the wetlands were converted for agriculture or are heavily affected by transport infrastructures. For example, the airport of Barcelona is located in the middle of the Llobregat River delta and it is partially surrounded by a protected wetlands area. Fewer freshwater wetlands occur in river floodplains. They are now very rare as most have been drained for agriculture, and major rivers have been regulated by reservoirs to control flooding. The vegetation of freshwater wetlands is mainly composed of genera *Phragmites*, *Typha*, and *Scirpus*, which are replaced by *Carex* and *Cladium* in less wetted areas.

Mediterranean wetlands offer exceptional biodiversity. Many fish species occupy various habitats according to their tolerance of salinity. They also provide essential nesting and feeding areas for many species of birds. Extending over 1450 km^2 in southern France, the Camargue is among the most well-preserved wetlands in the Mediterranean Basin. It is home to thousands of flamingoes and waterfowls, attracting colonies up to several thousands of breeding or wintering birds. In the eastern Mediterranean, the Aammiq wetlands in Lebanon and the Hula wetlands in Israel are the largest remaining wetlands in the eastern Mediterranean. These wetlands are the most northern wetlands of the Great Rift Valley (see above) and are important resting areas for waterfowls and passerines migrating from Africa toward Asia and Europe. These wetlands have lost the majority of their surface areas to agriculture and water extraction and are critically endangered by the effects of climate change.

The role of disturbances in Mediterranean freshwater ecosystems

The effects of disturbance on biodiversity is a topic extensively studied in community ecology (e.g., Hughes et al., 2007). Freshwater ecosystems around the Mediterranean Basin are subjected to a combination of natural and anthropogenic disturbances. Natural disturbances include floods, droughts, seasonal drying and rewetting, wildfires, and salinity. They have been present during long geological periods (the Mediterranean climate is 3.2 myr old; Suc, 1984), which have resulted in many well-adapted and endemic species (Bonada & Resh, 2013). These natural disturbances influence communities, with stronger effects when natural and anthropogenic disturbances are present in combination (Bonada & Resh, 2013; Cid et al., 2017). In addition, the few studies comparing responses of communities along natural and anthropogenic disturbance gradients show comparable functional trends, with mean taxon functional richness increasing along the gradients (Gutiérrez-Cánovas et al., 2015). Assessing disturbances effects in Mediterranean freshwater ecosystems is, therefore a challenge, and new methods are required to disentangle the effects of natural and anthropogenic disturbances (Soria et al., 2020).

Floods and droughts are extreme events that disrupt arthropod communities in freshwater ecosystems (Bae & Park 2017; Piniewski et al., 2017; Lake, 2000). For example, severe floods lead to the reduction of arthropod abundances and changes in their feeding preferences (Pupilli et al., 2003). However, the high adaptability to drying and rewetting in the Mediterranean freshwater ecosystems, involving complex resilience and resistance strategies, confers some unique adaptations to these extreme events (a phenomenon named co-tolerance; Vinebrooke et al., 2004). This is especially relevant for droughts, given the numerous known species with life strategies adapted to the dry period (e.g., several species of Trichoptera and Plecoptera; Salavert et al., 2008; LóPez-RodríGuez et al., 2009).

Seasonal drying and rewetting are characteristic of temporary rivers and wetlands in the Mediterranean Basin. Drying usually occurs from late spring to later summer, whereas rewetting occurs from early autumn to early winter (Bonada & Resh, 2013). Because of the high seasonality and predictability of precipitation and temperature patterns in the Mediterranean climate (Tonkin et al., 2017), many species have synchronized their life cycles to these patterns or have strong abilities to disperse to or from refuges (Bonada et al., 2020). During the drying phase, organisms resist locally by migrating downwards into the hyporheic zone (Bruno et al., 2020), estivate in the seed bank, or stay in disconnected pools in the case of some temporary rivers. Others flee and search for perennial refuges (Bogan & Boersma, 2012; Bogan et al., 2017). During the rewetting phase, temporary habitats are recolonized from the sediments or organisms arriving from nearby refuges. Those can be upstream reaches in the case of temporary rivers, or nearby water bodies in the case of temporary rivers and wetlands (Bogan et al., 2017; Hershkovitz et al., 2013). In temporary rivers, aestivating in the seed bank is not the most common resistance strategy in contrast to regions with continental climates (Stubbington et al., 2013). Resting in disconnected pools or arriving from nearby refuges are the dominant resistance and resilience strategies for arthropod communities in Mediterranean freshwater ecosystems.

Wildfires are a common phenomenon in the Mediterranean Basin that substantially disrupt aquatic ecosystems. Surprisingly, in comparison with terrestrial vegetation, aquatic organisms do not have known strategies to resist and recover from fire (Verkaik et al., 2013). They rely on their intrinsic ability to recover from other severe disturbances, such as drying. Indeed, the recovery of aquatic communities after fire in Mediterranean ecosystems is twice as fast as in other climate regions (Verkaik et al., 2013).

Because of the presence of semiarid and arid regions, the Mediterranean Basin has several rivers and wetlands with significantly high salinity. In the case of rivers, they have been classified as hyposaline, meso-hypersaline, or hypersaline, depending on the degree of salinity concentration (Millán et al., 2011; Arribas et al., 2009). Aquatic communities change along salinity gradients, diversity often decreases, but species are replaced by others considered specialist and highly tolerant such as the Coleoptera *Ochthebius glaber* (Abellán et al., 2007). Saline aquatic ecosystems in the Mediterranean Basin are threatened by coastal developments and other anthropogenic disturbances. They are considered vulnerable and have a high conservation value (Millán et al., 2011; Gómez et al., 2005).

In contrast to natural disturbances, anthropogenic ones are very new on an evolutionary time scale and, therefore, few known species are adapted to them. As a result, under anthropogenic disturbances freshwater communities are often simplified and dominated by generalist highly tolerant species. Anthropogenic disturbances occur nowadays everywhere across the Mediterranean Basin (López-Doval et al., 2013; Cabrera, 2010). These include abiotic and biotic disturbances that produce sublethal or lethal effects on aquatic communities. Disturbances such as hydrological alterations, land-use changes, organic pollution, salinity, heavy metals, emergent pollutants, or biological invasions usually occur in combination. Their effects are often exacerbated by the seasonal hydrological variability in the Mediterranean Basin (Cid et al., 2017), and increasing water stresses due to climate change (Filipe et al., 2013). Coastal aquatic ecosystems and rivers in their lower parts are commonly affected by a combination of anthropogenic disturbances, with severe consequences on their communities that translate into a decreased biological diversity, and overall biological quality (Herrero et al., 2018). Urban rivers are another example of ecosystems affected by multiple disturbances around the Mediterranean Basin. Such rivers are often heavily impacted by pollution and channelization, despite some recent significant improvements due to the implementation of the Water Framework Directive in Europe. For instance, the Besòs River, north of Barcelona, was considered the most polluted river in Europe during the eighties. It has now significantly recovered even if its biological quality is still considered relatively poor (Fortuño et al., 2020). Mediterranean Europe often benefits from strong legislation and resources to address these disturbances, while Mediterranean countries in North Africa and the eastern Mediterranean Basin still suffer from the lack of proper legislation or insufficient implementation of existing ones.

Freshwater biodiversity and endemism

Freshwater biodiversity of the Mediterranean Basin is the result of a complex combination of geography, geological history, landscape ecology, and human history (Blondel et al., 2010; Tierno de Figueroa et al., 2013). Myers et al. (2000) described the Mediterranean Basin as one of the 25 global mega hotspots for biodiversity, while Blondel et al. (2010) described the basin as a zone where intricate interaction of taxa, hybridization, and speciation have been particularly favored and fostered, as compared to more homogeneous regions farther north and south. The biodiversity of the Mediterranean Basin is undoubtedly unique, threatened by ever-increasing anthropogenic and climatic disturbances, and shaped by major barriers for dispersal to the north, south, and east. For instance, the Mediterranean Basin includes a relatively low number of Afro-Tropical species compared to Euro-Asian species (Blondel et al., 2010). This has been associated with the Sahara Desert, a massive barrier to dispersal, whose aridity dates back to the shrinkage of the Tethys Sea during the Late Miocene (Zhang et al., 2014). Freshwater biodiversity of the Mediterranean Basin is estimated to be 6% of the global freshwater biodiversity, and 35% of the Palearctic biodiversity (Tierno de Figueroa et al., 2013). For example, more than 460 species of freshwater fishes are recorded in the Mediterranean Basin, accounting for 25% of the fishes known in the Palearctic (Lévêque et al., 2008). According to Freyhof et al. (2014), 322 species of freshwater fishes are present in the eastern Mediterranean region alone, with an estimated additional 84 "species" that remain undescribed. However, the Mediterranean region as defined by Freyhof et al. (2014) spreads beyond the area of the eastern Mediterranean Basin described in this book, to include, among others, Northern Anatolia, the upper and lower drainages of the Tigris and Euphrates, and south Caspian drainages.

Endemism is a common denominator of many taxonomic groups in the Mediterranean Basin. Tierno de Figueroa et al. (2013) estimated that 43% of 3551 legitimate species belonging to 22 animal and plant groups are endemic, while Fochetti (2012) estimated that 10% of the entire Italian freshwater fauna (5496 animal species) are endemic. Among the animal groups having high percentages of endemism are freshwater fishes. Çiçek et al. (2018), examining freshwater fish endemism in Turkey, found up to 40% and 34% of endemism in the Mediterranean drainages of Büyük Menderes and Antalya, respectively. This endemism is also documented in freshwater fishes around the Mediterranean Basin, where 64% (or 292 species) of identified freshwater fishes are considered endemics, the majority occurring in the Balkan Peninsula, western Turkey, and the Greek and Turkish islands (Tierno de Figueroa et al., 2013). High freshwater fish endemism in the Mediterranean Basin has been associated with having the Mediterranean drainages as a refuge for communities during the last glaciation and the presence of high mountain ranges, such as the Alps, that acted as a barrier to northern expansions of fishes and the recolonization of Mediterranean drainages by fishes coming from the north (Lévêque et al., 2008). Freshwater arthropod endemism is much more difficult to assess as many groups are still insufficiently studied and/or knowledge gaps are common in parts of the Mediterranean Basin, particularly in North Africa and the eastern Mediterranean Basin. However, in a comprehensive assessment carried out by Tierno de Figueroa et al. (2013) that accounted for 2402 species of arthropods, 40.4% (972 species) were found endemic to the Mediterranean Basin. Among the considered groups, the highest endemism was observed for Hydraneidae (Coleoptera, 57%), followed by Simuliidae (Diptera, 51%), and Trichoptera 45%. Ivković and Plant (2015), examining the biodiversity of Empididae (Diptera) in the Balkans. found a gradient of endemism, north-south, along the Dinaric Alps, and Fochetti (2020), examining Plecoptera endemicity in Italy, found that 30% of Italian stoneflies are endemic. This is comparable to the Plecoptera endemism observed in the entire Mediterranean Basin (43%, Tierno de Figueroa et al., 2013). Based on the current biodiversity knowledge, the endemism of freshwater Malacostraca is particularly high in karst systems (Väinölä et al., 2008) and the Mediterranean Basin islands (Christodoulou et al., 2016). Hupalo et al. (2021) estimated that more than half of the 181 species of Malacostraca identified on Mediterranean islands are endemic, even though the islands cover only 5% of the entire surface area of the basin. Sardinia, Crete, and the Balearic Islands have the highest percentage of endemism of Malacotraca (60%, or above, of the recorded species). Lastly, an initial examination of north African Malacostraca communities also suggests a high percentage of endemism of epigean freshwater Amphipoda (52%) in Algeria and Tunisia (Ayati et al., 2019).

Knowledge of freshwater biodiversity in the Mediterranean Basin is unquestionably constantly growing and evolving. In this book, experts in various taxonomic groups provide the most up-to-date comprehensive overview of freshwater arthropod biodiversity in the Mediterranean Basin.

References

Abellán, P., Gómez-Zurita, J., Millán, A., Sánchez-Fernández, D., Velasco, J., Galián, J. & Ribera, I. (2007) Conservation genetics in hypersaline inland waters: mitochondrial diversity and phylogeography of an endangered Iberian beetle (Coleoptera: Hydraenidae). *Conservation Genetics*, **8**, 79–88.

Arribas, P., Gutiérrez-Cánovas, C., Abellán, P., Sáncehz-Fernández, D., Picazo, F., Velasco, J. & Millán, A. (2009) Tipuficación de los ríos salinos ibéricos. *Ecosistemas*, **18**, 1–13.

Ayati, K., Hadjab, R., Khammar, H., Dhaouadi, S., Piscart, C. & Mahmoudi, E. (2019) Origin, diversity and distribution of freshwater epigean amphipods in Maghreb. *Annales de Limnologie - International Journal of Limnology*, **55**, 13.

Babault, J., Loget, N., Van Den Driessche, J., Castelltort, S., Bonnet, S. & Davy, P. (2006) Did the Ebro basin connect to the Mediterranean before the Messinian salinity crisis? *Geomorphology*, **81**, 155–165.

Bae, M.-J. & Park, Y.-S. (2017) Responses of the functional diversity of benthic macroinvertebrates to floods and droughts in small streams with different flow permanence. *Inland Waters*, **6**, 461–475.

Blondel, J. (2006) The 'Design' of Mediterranean Landscapes: A Millennial Story of Humans and Ecological Systems during the Historic Period. *Human Ecology*, **34**, 713–729.

Blondel, J., Aronson, J., Bodiou, J.-Y. & Boeuf, G. (2010) *The Mediterranean Region. Biological Diversity in Space and Time*, Oxford University Press Inc., New York.

Blöschl, G., Hall, J., Viglione, A., Perdigão, R.A.P., Parajka, J., Merz, B., Lun, D., Arheimer, B., Aronica, G.T., Bilibashi, A., Boháč, M., Bonacci, O., Borga, M., Čanjevac, I., Castellarin, A., Chirico, G.B., Claps, P., Frolova, N., Ganora, D., Gorbachova, L., Gül, A., Hannaford, J., Harrigan, S. & Kireeva, M. (2019) Changing climate both increases and decreases European river floods. *Nature*, **573**, 108–111.

Bogan, M.T. & Boersma, K.S. (2012) Aerial dispersal of aquatic invertebrates along and away from arid-land streams. *Freshwater Science*, **31**, 1131–1144.

Bogan, M.T., Chester, E.T., Datry, T., Murphy, A.L., Robson, B.J., Ruhí, A., Stubbington, R. & Whitney, J.E. (2017) Resistance, resilience, and community recovery in intermittent rivers and ephemeral streams. In: *Intermittent Rivers and Ephemeral Streams: Ecology and Management*. (Eds T. Datry & N. Bonada & A.J. Boulton), pp. 349–376. Elsevier, Inc., Cambridge, MA, USA.

Bonada, N. & Resh, V.H. (2013) Mediterranean-climate streams and rivers: geographically separated but ecological comparable freshwater systems. *Hydrobiologia*, **719**, 1–29.

Bonada, N., Cañedo-Argüelles, M., Gallart, F., Von Schiller, D., Fortuño, P., Latron, J., Llorens, P., Múrria, C., Soria, M., Vinyoles, D. & Cid, N. (2020) Conservation and management of isolated pools in temporary rivers. *Water*, **12**, 2870.

Bonada, N., Múrria, C., Zamora-Muñoz, C., El Alami, M., Poquet, J.M., Puntí, T., Moreno, J.L., Bennas, N., Alba-Tercedor, J., Ribera, C. & Prat, N. (2009) Using community and population approaches to understand how contemporary and historical factors have shaped species distribution in river ecosystems. *Global Ecology and Biogeography*, **18**, 201–213.

Bruno, M.C., Doretto, A., Boano, F., Ridolfi, L. & Fenoglio, S. (2020) Role of the hyporheic zone in increasing the resilience of mountain streams facing intermittency. *Water*, **12**, 2034.

Cabrera, E. (2010) *Water engineering and management through time: learning from history*, CRC Press.

Casassas-Simó, L. & Riba-Arderiu, O. (1992) Morfologia de la Rambla barcelonina. *Treballs de la Societat Catalana de Geografia*, **33-44**, 9–27.

Castiglioni, G.B., Bondesan, M., Elmi, C., Marchetti, E.G. & Pellegrini, L. (2001) Response of the fluvial system to environmental conditions. *Supplementi di Geograpfia Fisica e Dinamica Quaternaria*, **4**, 165–187.

Cavazza, W. & Wezel, C. (2003) The Mediterranean region – a geological primer. *Episodes*, **26**, 160–168.

Čerkvenik, R., Kranjc, A. & Mihevc, A. (2018) Karst wetlands in the Dinaric karst. In: *The Wetland Book*. (Eds Finlayson, C. & Milton, G. & Prentice, R. & Davidson, N.), pp. 1057–1065. Springer, Dordrecht.

Christensen, J.H., Boberg, F., Christensen, O.B. & Lucas-Pischer, P. (2008) On the need for bias correction of regional climate change projections of temperature and precipitation. *Geophysical Research Letters*, **35**, L20709.

Christodoulou, M., Anastasiadou, C., Jugovic, J. & Tzomos, T. (2016) Freshwater shrimps (Atyidae, Palaemonidae, Typhlocarididae) in the broader Mediterranean Region: distribution, life strategies, threats, conservation challenges and taxonomic issues. In: *A Global Overview of the Conservation of Freshwater Decapod Crustaceans*. (Eds T. Kawai & N. Cumberlidge), pp. 199–236. Springer, Cham.

Çiçek, E., Fricke, R., Sungur, S. & Eagderi, S. (2018) Endemic freshater fishes of Turkey. *FishTaxa*, **3**, 1–39.

Cid, N., Bonada, N., Carlson, S., Grantham, T.E., Gasith, A. & Resh, V.H. (2017) High variability is a defining component of Mediterranean-climate rivers and their biota. *Water*, **9**, 52.

Cramer, W., Guiot, J., Fader, M., Garrabou, J., Gattuso, J.-P., Iglesias, A., Lange, M., Lionello, P., Llasat, M.C., Paz, S., Peñuelas, J., Snoussi, M., Toreti, A., Tsimplis, M.N. & Xoplaki, E. (2018) Climate change and interconnected risks to sustainlable development in the Mediterranean. *Nature Climate Change*, **8**, 972–980.

Di Castri, F. (1981) Mediterranean-type shrublands of the world. In: *Mediterranean-Type Shrublands*. (Eds F. Di Castri & D.W. Goodall & R.L. Specht), pp. 1–52. Elsevier.
Dünkeloh, A. & Jacobeit, J. (2003) Circulation dynamics of Mediterranean precipitation variability 1948-98. *International Journal of Climatology*, **23**, 1843–1866.
Filipe, A.F., Lawrence, J.E. & Bonada, N. (2013) Vulnerability of stream biota to climate change in mediterranean climate regions: a synthesis of ecological responses and conservation challenges. *Hydrobiologia*, **719**, 331–351.
Fochetti, R. (2012) Italian freshwater biodiversity: status, threats and hints for its conservation. *Italian Journal of Zoology*, **79**, 2–8.
Fochetti, R. (2020) Endemism in the Italian stonefly-fauna. *Zootaxa*, **4**, 381–388.
Fortuño, P., Vinyoles, D., Fabre, N., Verkaik, I., Cid, N., Soria, M., Bonada, N., Gallart, F. & Prat, N. (2020) A polluted urban river as a resource to raise citizen awareness. The case of the Besòs river. In: *International Conference Daylighting Rivers: inquiry based learning for civic ecology, Florence*. pp. 130–135.
Forzieri, G., Feyen, L., Rojas, R., Flörke, M., Wimmer, F. & Bianchi, A. (2014) Ensemble projection of future streamflow droughts in Europe. *Hydrology and Earth System Sciences*, **18**, 85–108.
Freyhof, J., Ekmekçi, F.G., Ali, A., Khamees, N.R., Özuluğ, M., Hamidan, N., Küçük, H. & Smith, K.G. (2014) Freshwater Fishes, In: The status and distribution of freshwater biodiversity in the Eastern Mediterranean. K.G. Smith & V. Barrios & W.R.T. Darwall & C. Numa), pp. 19–42.
Giorgi, F. & Lionello, P. (2008) Climate change projections for the Mediterranean region. *Global and Planetary Change*, **63**, 90–104.
Gómez, R., Hurtado, I., Suarez, M.L. & Vidal-Abarca, M.R. (2005) Ramblas in south-east Spain: threatened and valuable ecosystems. *Aquatic Conservation: Marine and Freshwater Ecosystems*, **15**, 387–402.
Guiot, J. & Cramer, W. (2016) Climate change: The 2015 Paris agreement thresholds and Mediterranean basin ecosystems. *Science*, **354**, 465–468.
Gutiérrez-Cánovas, C., Sánchez-Fernández, D., Velasco, J., Millán, A. & Bonada, N. (2015) Similarity in the difference: changes in community functional features along natural and anthropogenic stress gradients. *Ecology*, **96**, 2458–2466.
Herrero, A., Gutiérrez-Cánovas, C., Vigiak, O., Lutz, S., Kumar, R., Gampe, D., Huber-García, V., Ludwig, R., Batalla, R. & Sabater, S. (2018) Multiple stressor effects on biological quality elements in the Ebro River: present diagnosis and predicted responses. *Science of the Total Environment*, **630**, 1608–1618.
Hershkovitz, Y. & Gasith, A. (2013) Resistance, resilience, and community dynamics in mediterranean-climate streams. *Hydrobiologia*, **719**, 59–75.
Hughes, A.R., Byrnes, J.E., Kimbro, D.L. & Stachowicz, J.J. (2007) Reciprocal relationships and potential feedbacks between biodiversity and disturbance. *Ecology Letters*, **10**, 849–864.
Hupalo, K., Stoch, F., Karaouzas, I., Wysocka, A., Rewicz, T., Mamos, T. & Grabowski, M. (2021) Freshwater Malacostraca of the Mediterranean Islands – diversity, origin and conservation perspectives. In: *Recent Advances in Freshwater Crustacean Biodiversity and Conservation*. (Eds T. Kawai & D.C. Rogers), p. 82. CRC Press, Boca Raton.
Ivković, M. & Plant, A. (2015) Aquatic insects in the Dinarides: identifying hotspots of endemism and species richness shaped by geological and hydrological history using Empididae (Diptera). *Insect Conservation and Diversity*, **8**, 302–312.
Köppen, W. (1936) Das geographische system der Klimate. In: *Handbuch der Klimatologie*. (Eds W. Köppen & R. Geiger), pp. 1–44. Gebrüder, Berlin.
Kotteki, M., Grieser, J., Beck, C., Rudolf, B. & Rubel, F. (2006) World map of the Köppen-Geiger climate classification updated. *Meteorologische Zeitschrift*, **15**, 259–263.
Lake, P.S. (2000) Disturbance, patchiness, and diversity in streams. *Journal of the North American Benthological Society*, **19**, 573–592.
Lefebvre, G., Redmond, L., Germain, C., Palazzi, E., Terzago, S., Willm, L. & Poulin, B. (2019) Predicting the vulnerability of seasonlly-flooded wetlands to climate change across the Mediterranean Basin. *Science of the Total Environment*, **692**, 546–555.
Lévêque, C., Oberdorff, T., Stiassny, M.L.J. & Tedesco, P.A. (2008) Global diversity of fish (Pisces) in freshwater. *Hydrobiologia*, **595**, 545–567.
López-Doval, J.C., Ginebreda, A., Caquet, T., Dahm, C.N., Petrovic, M., Barceló, D. & Muñoz, I. (2013) Pollution in mediterraneanclimate rivers. *Hydrobiologia*, **719**, 427–450.
LóPez-RodríGuez, M.J., De Figueroa, J.M.T., Fenoglio, S., Bo, T. & Alba-Tercedor, J. (2009) Life strategies of 3 Perlodidae species (Plecoptera) in a Mediterranean seasonal stream in southern Europe. *Journal of the North American Benthological Society*, **28**, 611–625.
Ludwig, W., Dumont, E., Meybeck, M. & Heussner, S. (2009) River discharges of water and nutrients to the Mediterranean and Black sea: major drivers for ecosystem changes during past and future decades? *Progress in Oceanography*, **80**, 199–217.
Magny, M., Combourieu-Nebout, N., De Beaulieu, J.L., Bout-Roumazeilles, V., Colombaroli, D., Desprat, S., Francke, A., Joannin, S., Ortu, E., Peyron, O., Revel, M., Sadori, L., Siani, G., Sicre, M.A., Samartin, S., Simonneau, A., Tinner, W., Vannière, B., Wagner, B., Zanchetta, G., Anselmetti, F., Brugiapaglia, E., Chapron, E., Debret, M., Desmet, M., Didier, J., Essallami, L., Galop, D., Gilli, A., Haas, J.N., Kallel, N., Millet, L., Stock, A., Turon, J.L. & Wirth, S. (2013) North–south palaeohydrological contrasts in the central Mediterranean during the Holocene: tentative synthesis and working hypotheses. *Climate of the Past*, **9**, 2043–2071.
Mariotti, A., Pan, Y., Zeng, N. & Alessandri, A. (2015) Long-term climate change in the Mediterranean region in the midst of decadal variability. *Climate Dynamics*, **44**, 1437–1456.
Maughna, N. (2015) The Serre-Ponçon Dam and the Durance River: The Founding Act towards the most Regulated French Waterway. *Arcadia*, **no. 16**.
Millán, A., Velasco, J., Gutiérrez-Cánovas, C., Arribas, P., Picazo, F., Sánchez-Fernández, D. & Abellán, P. (2011) Mediterranean saline streams in southeast Spain: What do we know?. *Journal of Arid Environments*, **75**, 1352–1359.
Myers, N., Mittermeier, R.A., Mittermeier, C.G., Da Fonseca, G.A.B. & Kent, J. (2000) Biodiversity hotspots for conservation priorities. *Nature*, **403**, 853–858.

Nakiceovic, N., Alcamo, J., Davis, G., De Vries, H.J.M., Fenhann, J., Gaffin, S., Gregory, K., Grubler, A., Jung, T.Y., Kram, T., La Rovere, E.L., Michaelis, L., Mori, S., Morita, T., Papper, W., Pitcher, H., Price, L., Riahi, K., Roehrl, A., Rogner, H.-H., Sankovski, A., Schlesinger, M., Shukla, P., Smith, S., Swart, R., Van Rooijen, S., Victor, N. & Dadi, Z. (2000) Emissions Scenarios. A Special Report of Working Group III of the Intergovernmental Panel on Climate Change. C.U. Press, p. 559, Cambridge.

Nixon, S.W. (2003) Replacing the Nile: Are Anthropogenic Nutrients Providing the Fertility Once Brought to the Mediterranean by a Great River? *Ambio*, **32**, 30–39.

Piniewski, M., Prudhomme, C., Acreman, M.C., Tylec, L., Oglecki, P. & Okruszko, T. (2017) Responses of fish and invertebrates to floods and drought in Europe. *Ecohydrology*, **10**, e1793.

Pupilli, E. & Puig, M.A. (2003) Effects of a major flood on the mayfly and stonefly populations in a Mediterranean stream (Matarranya stream, Ebro River basin, North East of Spain). In: *Research Update on Ephemeroptera and Plecoptera.* (Eds E. Gaino), pp. 381–389. University of Perugia, Perugia, Italy.

Roberts, N. & Reed, J. (2009) Lakes, wetlands, and Holocene environmental change. In: *The Physical Geography of the Mediterranean.* (Eds J.C. Woodward), pp. 255–286. Oxford University Press, Oxford.

Sabater, S., Feio, M.J., Graça, M., Muñoz, I. & Romaní, A.M. (2009) The Iberian Rivers. In: *Rivers of Europe.* (Eds K. Tockner & C. Robinson & U. Uhlinger), pp. 113–150. Academic Press.

Salavert, V., Zamora-MuñOz, C., Ruiz-RodríGuez, M., FernáNdez-CortéS, A. & Soler, J.J. (2008) Climatic conditions, diapause and migration in a troglophile caddisfly. *Freshwater Biology*, **53**, 1606–1617.

Skoulikidis, N.T. & Gritzalis, K. (1998) Greek river inputs in the Mediterranean. Their intra-annual an inter-annual variations. *Fresenius Environmental Bulletin*, **7**, 90–95.

Skoulikidis, N.T., Economou, A.N., Gritzalis, K.C. & Zogaris, S. (2009) Rivers of the Balkans. In: *Rivers of Europe.* (Eds K. Tockner & C. Robinson & U. Uhlinger), pp. 421–466. Academic Press.

Skoulikidis, N.T., Sabater, S., Datry, T., Morais, M.M., Buffagni, A., Dörflinger, G., Zogaris, S., Sánchez-Montoya, M.M., Bonada, N., Kalogianni, E., Rosado, J., Vardakas, L., De Girolamo, A.M. & Tockner, K. (2017) Non-perennial Mediterranean rivers in Europe: Status, pressures, and challenges for research and management. *Science of the Total Environment*, **577**, 1–18.

Soria, M., Gutiérrez-Cánovas, T., Bonada, N., Acosta, R., Rodríguez-Lozano, P., Fortuño, P., Burgazzi, G., Vinyoles, D., Gallart, F., Latron, J., Llorens, P., Prat, N. & Cid, N. (2020) Natural disturbances can produce misleading bioassessment results: Identifying metrics to detect anthropogenic impacts in intermittent rivers. *Journal of Applied Ecology*, **57**, 283–295.

Stubbington, R. & Datry, T. (2013) The macroinvertebrate seedbank promotes community persistence in temporary rivers across climate zones. *Freshwater Biology*, **58**, 1202–1220.

Suc, J.P. (1984) Origin and evolution of the Mediterranean vegetation and climate in Europe. *Nature*, **307**, 429–432.

Tierno De Figueroa, J.M., López-Rodríguez, M.J., Fenoglio, S., Sánchez-Castillo, P. & Fochetti, R. (2013) Freshwater biodiversity in the rivers of the Mediterranean basin. *Hydrobiologia*, **719**, 137–186.

Tonkin, J.D., Bogan, M.T., Bonada, N., Rios-Touma, B. & Lytle, D.A. (2017) Seasonality and predictability shape temporal species diversity. *Ecology*, **98**, 1201–1216.

Väinölä, R., Witt, J.D.S., Grabowski, M., Bradbury, J.H., Jazdzewski, K. & Sket, B. (2008) Global diversity of amphipods (Amphipoda; Crustacea) in freshwater. *Hydrobiologia*, **595**, 241–255.

Vannière, B., Blarquez, O., Rius, D., Doyen, E., Brücher, T., Colombaroli, D., Connor, S., Feurdean, A., Hickler, T., Kaltenrieder, P., Lemmen, C., Leys, B., Massa, C. & Olofsson, J. (2016) 7000-year human legacy of elevation-dependent European fire regimes. *Quaternary Science Reviews*, **132**, 206–212.

Vautard, R., Gobiet, A., Sobolowski, S., Kjellström, E., Stegehuis, A., Watkiss, P., Mendlik, T., Landgren, O., Nikulin, G., Teichmann, C. & Jacob, D. (2014) The European climate under a 2 °C global warming. *Environmental Research Letters*, **9**, 034006.

Verkaik, I., Rieradevall, M., Cooper, S.D., Melack, J.M., Dudley, T.L. & Prat, N. (2013) Fire as disturbance in Mediterranean climate streams. *Hydrobiologia*, **719**, 353–382.

Vidal-Abarca, M.R. (1990) Los ríos de las cuencas áridas y semiáridas: una perspectiva ecológica comparativa y de síntesis. *SCIENTIA gerundensis*, **16**, 219–228.

Vinebrooke, R.D., Cottingham, K.L., Norberg, J., Scheffer, M., Dodson, S.I., Maberly, S.C. & Sommer, U. (2004) Impacts of multiple stressors on biodiversity and ecosystem functioning: The role of species co-tolerance. *Oikos*, **104**, 451–457.

Vörösmarty, C.J., Fekete, B.M., Meybeck, M. & Lammers, R.B. (2000) Geomorphometric attributes of the global system of rivers at 30-minute spatial resolution. *Journal of Hydrology*, **237**, 17–39.

Winterberg, S. & Willett, S.D. (2019) Greater Alpine river network evolution, interpretations based on novel drainage analysis. *Swiss Journal of Geosciences*, **112**, 3–22.

Zarfl, C., Lumsdon, A.E., Berlekamp, J., Tydecks, L. & Tockner, K. (2015) A global boom in hydropower dam construction. *Aquatic Sciences*, **77**, 161–170.

Zhang, Z., Ramstein, G., Schuster, M., Li, C., Contoux, C. & Yan, Q. (2014) Aridification of the Sahara desert caused by Tethys Sea shrinkage during the Late Miocene. *Nature*, **513**, 401–404.

Zittis, G., Hadjinicolaou, P., Fnais, M. & Lelieveld, J. (2015) Projected changes in heat wave characteristics in the eastern Mediterranean and the Middle East. *Regional Environmental Change*, **16**, 1863–1876.

Chapter 3

Arthropoda

James H. Thorp[1] and Michael Raupach[2]
[1]Kansas Biological Survey and Department of Ecology and Evolutionary Biology, University of Kansas, Lawrence, KS, United States, [2]Staatliche Naturwissenschaftliche Sammlungen Bayerns, Zoologische Staatssammlung München, München, Bayern, Germany

Subchapter 3.1

Introduction to the Phylum Arthropoda

James H. Thorp
Kansas Biological Survey and Department of Ecology and Evolutionary Biology, University of Kansas, Lawrence, KS, United States

Overview

Based on their combined presence in terrestrial, marine, and freshwater habitats, no other metazoan taxon approaches the phylum Arthropoda in total species and ecological diversity. Although this is in great part due to the huge numbers of terrestrial insects and mites, members of the phylum are also quite abundant, diverse, and ecologically important in inland waters. In addition to having the greatest diversity of inland water species compared to other metazoan phyla, the Arthropoda are the most diverse anatomically. Arthropods are present in headwater streams to large rivers, wetlands and ephemeral pools, subterranean habitats, and small ponds to large freshwater lakes. Some groups also occur in saline lakes (e.g., some species of fairy shrimp in the class Branchiopoda) and subterranean environments (e.g., some isopod and decapod crustaceans). Their ecological roles span the gamut from herbivores and detritivores to invertebrate predators and some parasites (fish lice and mostly ectoparasitic mites).

Controversies over evolutionary kinships among invertebrates are a pervasive and nearly constant feature among taxonomically focused zoologists. Consequently, the lifespan of phylogenetic trees describing relationships among taxa from the species level to even kingdoms in rare cases is increasingly short as new molecular techniques reveal relationships not heretofore acknowledged. For example, one evolutionary theory (the "Articulata Hypothesis") places the phylum Arthropoda among other phyla with segmented bodies, in particular the Annelida, as discussed by Scholtz (2002). A contrasting theory (the "Ecdysozoan Hypothesis") unites Arthropoda with other phyla such as nemathelminth worms that periodically shed (molt) their outer layer (a three-layered cuticle) in a process called ecdysis (e.g., Daley & Drage, 2016).

Within the Arthropoda, a fairly dramatic taxonomic rearrangement occurred this century when Hexapoda (the taxon primarily encompassing the insects) was recognized as a clade nested within Crustacea (Regier et al., 2005, Regier et al., 2010; Tamone & Harrison, 2015)—a view that is widely accepted now. Aside from the more important taxonomic perspectives based on molecular data, this relationship makes evolutionary and ecological sense if we consider that ancient insects are likely to have evolved from traditional marine crustaceans that emerged from the ocean either directly on to land or first traveled on the pathway leading from an estuary to a stream and thence to land (and possibly in some cases back to freshwaters). The names for this united group of arthropods have included "Pancrustacea" and "Tetraconata" (Regier et al., 2005; Oakley et al., 2013), but we prefer to use the original name "Crustacea" and include Hexapoda as a major class within the subphylum Crustacea, as used in the Palearctic (Rogers & Thorp, 2019) and Neotropical volumes (Damborenea et al., 2020) in *Thorp and Covich's Freshwater Invertebrates* (Elsevier, Ltd.).

Identification and Ecology of Freshwater Arthropods in the Mediterranean Basin. DOI: https://doi.org/10.1016/B978-0-12-821844-0.00006-5

The phylum Arthropoda is now considered to contain three subphyla: Myriapoda, Chelicerata, and Crustacea (see keys to subphyla below). Myriapoda contains four classes, but only one of which contains any semiaquatic members: Diplopoda (millipedes) that has only a handful of semiaquatic species present on several continents. The only currently recognized species in Europe is *Serradium semiaquaticum*, a semiaquatic species reported from caves in northern Italy (Enghoff et al., 1997) but not currently in the Mediterranean Basin. It appears to enter the water to feed on fine-grained detritus in subterranean water films or pools, and it overlaps in distribution with its apparently terrestrial only species *S. hirsutipes*. In contrast, the subphylum Chelicerata has a very large number of freshwater representatives, with a few semiaquatic spiders and a vast number of mites. This chapter provides a relatively brief coverage of this huge group. More information on their biology can be found in Proctor et al. (2015) and Bartsch (2019).

The third subphylum of Arthropoda is Crustacea, which is briefly introduced in the current chapter. It is followed by separate chapters on four crustacean classes: Branchiopoda (Chapter 4), Ostracoda (Chapter 5), Copepoda (Chapter 6), and Malacostraca (Chapter 7). Two other classes of Crustacea are briefly discussed in this chapter: (a) class Ichthyostraca (also part of Chapter 6), which is discussed only in reference to the fish lice of the subclass Branchiura; and (2) Thecostraca, which includes the very few inland water barnacles. Following those separate crustacean classes is an introductory chapter on Hexapoda (Chapter 8) and eight chapters on various orders of insects (Chapters 9 to 15).

General ecology and distribution of Arthropods

Arthropods live almost everywhere from the deepest ocean trenches to the highest mountain ranges and can even be found carried by the winds as propagules and lightweight young stages. Among inland water species, the vast majority occupy surface freshwaters (lentic and lotic), but low diversities occur in subterranean habitats (limited by food supplies) and saline ponds and estuaries (osmotic conditions restrict most inland water taxa). The greatest diversity of species thrives as zoobenthos in streams and rivers and on the bottoms of ponds, lakes, and wetlands, but some groups (especially branchiopods, such as "water fleas") are mostly but not exclusively planktonic. Because they are usually susceptible to predation by aquatic and sometimes terrestrial vertebrates, many freshwater invertebrates live within the sediments or under rocks and plant material. For planktonic species where cover is largely unavailable, the common strategies are to be mostly transparent, have high rates of reproduction, and/or appear above the bottom only during darkness.

Although some benthic arthropods can thrive in low oxygen conditions, the majority require medium to high oxygen tensions. In a few cases (e.g., crayfishes), arthropods can crawl partially out of water when oxygen tensions are low; indeed, some crayfishes are known to migrate over land from one aquatic body to another on evenings when the humidity is high. Oxygen conditions are known to impose stream elevational limits on some aquatic insects.

Inland water arthropods are typically found in freshwater to slightly brackish waters. The exceptions are some large branchiopods (e.g., fairy shrimp), brine flies, and barnacles that can tolerate salinities approaching or even surpassing full-strength seawater.

The two most diverse arthropods found in inland waters are mites (subphylum Chelicerata, class Arachnida; this chapter) and insects (subphylum Crustacea, class Hexapoda; reviewed here but covered extensively in Chapters 8–15), followed in taxonomic richness by noninsect orders of crustaceans (discussed briefly in the present chapter and more extensively in Chapters 4–7).

Arthropod food webs

Arthropods occupy all trophic levels in freshwaters above primary producers. They can be the top predator in a wetland or headwater stream except where they cohabit with predatory fish. Arachnids (Chelicerata) include many predators (spiders in particular), but they also feature many ectoparasitic mites. Within the class Hexapoda, insects occupy a large range of ecological roles, whereas species of Entognatha (Collembola) are primarily algivores and detrivores. Members of the subclass Branchiura (fish lice in the class Ichthyostraca) are primarily ectoparasites, especially on fish but also on some amphibians. Suspension feeding on algae and dead organic matter as well as collector gathering on benthic substrates are especially common in seed shrimps (Ostracoda), water fleas and fairy shrimp (Branchiopoda), and barnacles (superclass Multicrustacea, class Thecostraca). Copepods (superclass Multicrustacea, class Copepoda) include many planktonic predators (principally cyclopoids), but also have abundant diversity of benthic (especially harpacticoids) and planktonic (primarily calanoids) herbivores and detrivores. Malacostracans encompass a huge diversity of both species and trophic positions. Despite including some of the largest freshwater invertebrates of the Mediterranean Basin, omnivory is more common among these crustaceans.

Biodiversity patterns, endemic nature, and conservation status

Biodiversity patterns

Freshwater biodiversity patterns of invertebrates (and other groups) in the rivers of the Mediterranean Basin were thoroughly reviewed by Tierno de Figueroa et al. (2013). They concluded that ~35% of known Palearctic freshwater species and more than 6% of the world's freshwater species are present in the Mediterranean Basin. Endemicity is high within this community, and many species are seriously threatened by normal urban, industrial, and agricultural development and by the effects of climate change on environmental conditions in rivers, wetlands, and lakes.

Nonindigenous species

The Mediterranean Basin, like much of the rest of the world, has been subject to intentional and accidental introduction of invertebrate species from other regions of the world via the pet trade, commercial shipping, and other avenues (see for example Darwall et al., 2014). For example, a detailed compendium of the distribution of the zebra mussel (*Dreissena polymorpha*) is given in one of the reports by the international, intergovernmental, not-for-profit organization known as CABI.org. Of arthropods, the intentional introduction and then accidental spread of the red swamp crayfish *Procambarus clarkii* throughout Europe has now led this North American species of commercial aquaculture value to reach even the islands of Malta and south Sicily. Another commercially important species, the giant crayfish *Cherax quadricarinatus* from Australia, has also apparently escaped confinement and could have a serious impact on native crustaceans and molluscs. Tiger mosquitoes, *Aedes albopictus*, invaded Europe from Asia in 1979 and is a major vector of blood-borne diseases. Freshwater scientists in the Mediterranean Basin need to be particularly aware of the presence of nonindigenous species and should report their occurrences promptly to local authorities.

Morphological characters needed in identification

Arthropods are inherently segmented animals in all life stages, though segmentation is more apparent in adults. Segmentation is present in both the main body and the appendages, though body segmentation is sometimes obscured by fusion of segments. Inland water species can be divided into those as adults with two or three body sections. Mites and spiders (subphylum Chelicerata, class Arachnida) have two body divisions: (1) an anterior cephalothorax in spiders (Fig. 3.1A and B) or a gnathosoma in mites (Fig. 3.2); and (2) a posterior abdomen in spiders (Fig. 3.1A and B) or an idiosoma in mites (Fig. 3.2). Three body divisions (head, thorax, and abdomen) are present in both the subphyla Myriapoda (millipedes; Fig. 3.3) and Crustacea—the latter including springtails (Fig. 3.4) and insects (Fig. 3.5A-D) as well as traditional crustaceans such as amphipods (Fig. 3.6), copepods, crayfish (Fig. 3.7A), and common shrimp or prawns (Fig. 3.7B). Antennae on the head region consist of one pair (millipedes, springtails, and insects) or two (crustaceans *sensu stricto*); chelicerates may have sensory oral palpi but lack true antennae. The number and function of appendages attached to the head, thorax, and abdomen (or from fused body segments) vary among and within the

(A) (B)

FIGURE 3.1 (A) Wolf spider *Paradosa californica*, which is semiaquatic and confined to walking on the surface of still water using hydrophobic hairs on its feet; (B) fishing spider *Dolomedes* carrying an egg sac with her chelicerae. *Photographs courtesy (A) Ken R. Schneider and (B) Heather Proctor (from Vol. I: Ecology and General Biology, Thorp and Rogers (ed.), 2015, in Thorp and Covich's Freshwater Invertebrates, 4th Edition, Elsevier).*

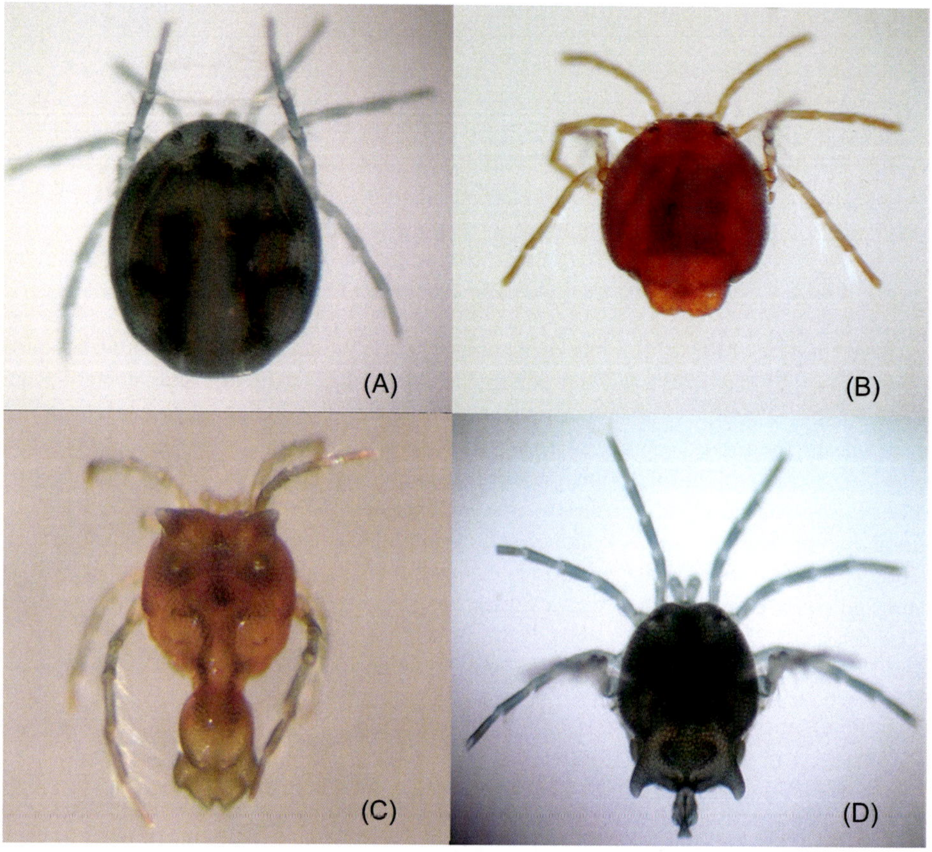

FIGURE 3.2 Diversity of shapes and colors in *Arrenurus* mites. *Photographs courtesy Bruce Smith. (from Vol. I: Ecology and General Biology, Thorp and Rogers (ed.), 2015, in Thorp and Covich's Freshwater Invertebrates, 4th Edition, Elsevier).*

FIGURE 3.3 *Serradium semiaquaticum* (Diplopoda, Polydesmidae). A cavernicolous, semiaquatic millipede from northern Italian caves. *Photograph courtesy Luca Cavallari (from Vol. I: Ecology and General Biology, Thorp and Rogers (ed.), 2015, in Thorp and Covich's Freshwater Invertebrates, 4th Edition, Elsevier).*

different subphyla, with the most prominent being those attached to the thoracic region, which are often ambulatory. Most thoracic segments in millipedes are fused (Fig. 3.3) with a pair of legs per diplosegment (giving the false impression of two pairs per segment) and more than 10 pairs total. Aquatic millipedes are extremely rare and mostly confined to the tropics, though first noted in Australia. However, the millipede *S. semiaquaticum* (Fig. 3.3) has been collected from a few alpine Italian caves just north of the Mediterranean Basin. It is considered semiaquatic because it submerges in cave streams to obtain food on the bottom (Enghoff et al., 1997). In contrast, larval arachnids have three pairs of legs in larvae, while adults have four pairs. Adult insects and springtails (Collembola) have three pairs of legs (hence the name Hexapoda). Traditional crustaceans, however, vary substantially in the number and function of thoracic limbs (e.g., locomotion, feeding, and defense). Three to seven pairs are generally present except in the crustacean order Branchiopoda (e.g., water fleas like *Daphnia*) where up to 19 pairs may occur. True walking legs in arthropods are restricted to thoracic limbs, but abdominal limbs (e.g., pleopods or "swimmerettes" in shrimp) may aid locomotion especially in benthic decapod crustaceans that periodically swim. They also may be employed in reproduction, respiration, and maternal care of eggs and young. Thoracic legs (or "thoracopods") are traditionally and still commonly referred to as being segmented, though some experts restrict this term to body segments.

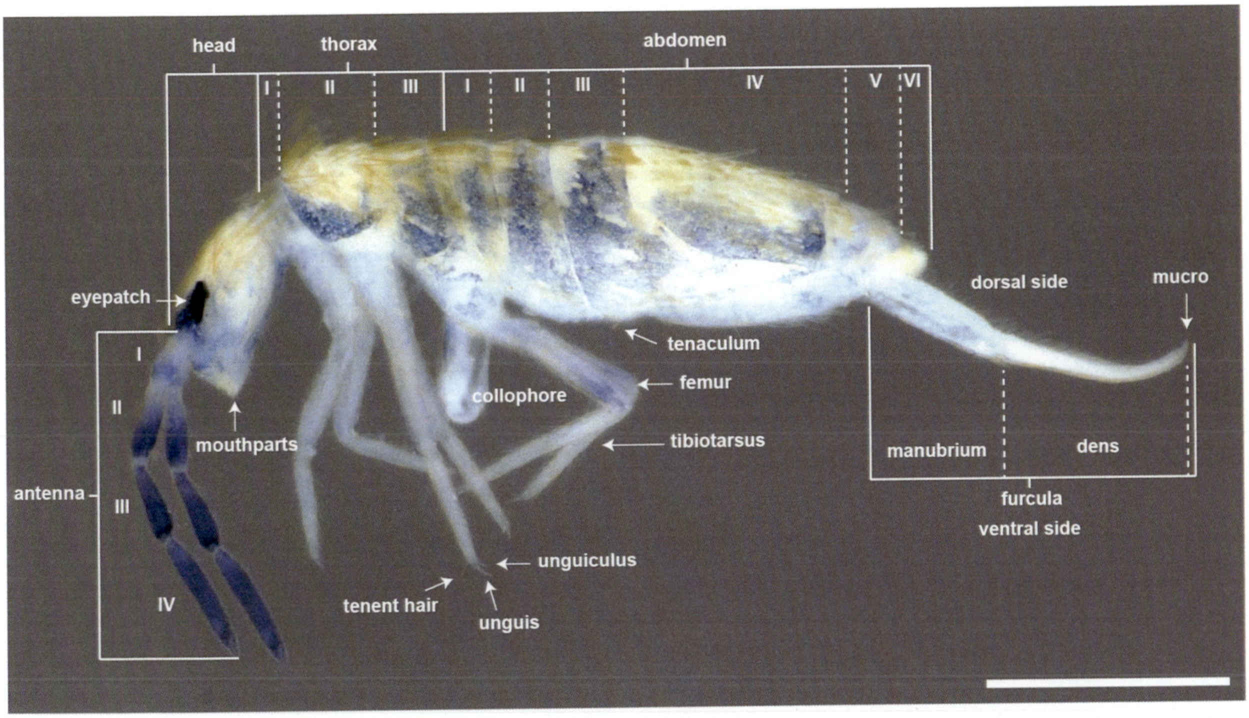

FIGURE 3.4 Lateral view of the external anatomy of a representative springtail (Hexapoda: Entognatha or Collembola). Scale bar = 0.5 mm. *Photograph courtesy Nikolas Cipola (from Vol. III: Keys to Neotropical Hexapoda, Hamada et al. (eds.), 2018, in Thorp and Covich's Freshwater Invertebrates, 4th Edition, Elsevier).*

FIGURE 3.5 Some examples of aquatic insects: (A) a water bug *Aphelocheirus* (Insecta, Heteroptera); (B) water beetle *Acilius*; (C) dragonfly nymph (Aeschnidae); and (D) alderfly larva *Sialis*. *Photographs courtesy (A) Stefan K. Hetz, (B-C) Matthew A. Hill, and (D) G. Kriska (from Vol. I: Ecology and General Biology, Thorp and Rogers (ed.), 2015, in Thorp and Covich's Freshwater Invertebrates, 4th Edition, Elsevier).*

FIGURE 3.6 *Crangonyx* sp., a crustacean amphipod found in surface and subsurface waters in the Mediterranean. *Photograph courtesy Matthew A. Hill from a Nearctic species (from Vol. I: Ecology and General Biology, Thorp and Rogers (ed.), 2015, in Thorp and Covich's Freshwater Invertebrates, 4th Edition, Elsevier).*

FIGURE 3.7 (A) *Procambarus clarkii*, a Nearctic crayfish (Crustacea, Decapoda) introduced in much of the world for aquaculture but which has often had a negative effect on local crustaceans; (B) another common decapod, the shrimp or prawn *Palaemon* sp. This genus occurs in freshwater, brackish, and nearshore marine waters. *Photograph courtesy Chris Lukhaup (from Vol. I: Ecology and General Biology, Thorp and Rogers (ed.), 2015, in Thorp and Covich's Freshwater Invertebrates, 4th Edition, Elsevier).*

Keys to freshwater Arthropoda

Additional information is given in the present chapter for those taxa shown in boldface.

Arthropoda: Subphyla

1 Less than 20 paired limbs, with 10 or fewer pairs for locomotion 2
1' Greater than 10 pairs of walking limbs; semiaquatic millipedes (Fig. 3.3) ....
.... Myriapoda, Diplopoda, one species .... *S. semiaquaticum* Enghoff et al., 1997
[Note: this species has not yet been collected in the Mediterranean Basin proper.]
2(1) Two body divisions: cephalothorax and abdomen in spiders (Aranae), or gnathosoma and idiosoma in mites (Acari); 3 (larval mites) or 4 pairs of legs; never with entirely enclosing, bivalved carapace ....
.... Chelicerata, class Arachnida
2' Three or more body divisions; if body is fused, then body is enclosed in a bivalved carapace; 1–2 pairs of antennae
.... Crustacea [Chapters 4–15]

Subphylum Chelicerata: Class Arachnida: Orders

The vast majority of arachnids are terrestrial, but a few spiders (Araneae) are semiaquatic, and many mites are aquatic (but still a very small minority). The taxonomy of the mites is still in flux, but the current divisions are based on Proctor et al. (2015).

1 Body prominently divided into cephalothorax and abdomen (Fig. 3.1); spinnerets present .... **Aranae**
1' Body with two, less conspicuous divisions: anterior gnathosoma (capitulum) and posterior idiosoma (Fig. 3.2); no spinnerets; mites .... **Acari**

Class Arachnida: Subclass Aranae: Families

1 Gnathosoma with 8 eyes separated equally into 2 distinct rows 2
1' Gnathosoma with 8 eyes in 2 rows but 2 lateral eyes of second row located somewhat posteriorly giving impression of 3 rows; wolf spiders or European "tarantulas" (Fig. 3.1A) .... Lycosidae
2 Mostly nondiving hunters on water surface; capture webs absent; fishing spiders (Fig. 3.1B and 3.8) .... Pisauridae, *Dolomedes*
2' Diving hunters with air-filled underwater webs; diving bell or water spider (Fig. 3.9) .... Cybaeidae, *Argyroneta aquatica* Clerck, 1757

Subclass Acari: Orders and Suborders

This key was based on those produced by a number of authors in three volumes in the series *Thorp and Covich's Freshwater Invertebrates* (Elsevier, Ltd), including contributions by Ian M. Smith in the Nearctic volume (Thorp & Rogers, 2015), Ilse Bartsch and D. Christopher Rogers in the Palearctic volume (Rogers & Thorp, 2019), and Natalia A. Fredes, Pablo A. Martínez, Almir Rogério Pepato, D. Christopher Rogers, and Pedro Henrique da Silva Conceição in the second Neotropical volume (Damborenea et al., 2020).

1 Four pairs of legs (Fig. 3.2); genital field well developed; pedipalp genu with 1 setae; idiosoma lacking series of paired glandularia .... adults 2
1' Three pairs of legs; genital field absent or not fully developed; pedipalp genu with 2 setae or idiosoma with series of paired glandularia .... larvae and nymphs
2 Chelicerae typically chelate and dentate; pedipalps simple; subcapitulum usually bearing rutellae or pseudorutellae; prodorsum usually with paired and well-defined bothridial setae bearing distinctively modified bothridial setae; opisthosoma usually with paired lateral glands (Fig. 3.10) .... order Sarcoptiformes, suborder Oribatida

FIGURE 3.8 Fishing spiders: (A) *Dolomedes plantarius*; and (B–D) *D. fimbriatus* as it dips below the water surface in pursuit of prey. *Photographs courtesy (A) Helen Smith; and (B-D) Stephen Barlow.*

FIGURE 3.9 Diving bell spider *Argyroneta aquatica* and its physical gill: (A) air clinging to abdominal hydrophobic hairs away from the diving bell; (B) a small diving bell, supported by an invisible web which is only large enough to hold the abdomen; (C) a large air bubble from the air–water interface that is held by the abdomen and carried down to the diving bell; and (D) a female in her diving bell, located below the cocoon. *Photographs courtesy Roger S. Seymour and Stefan K. Hetz (Seymour & Hetz, 2011).*

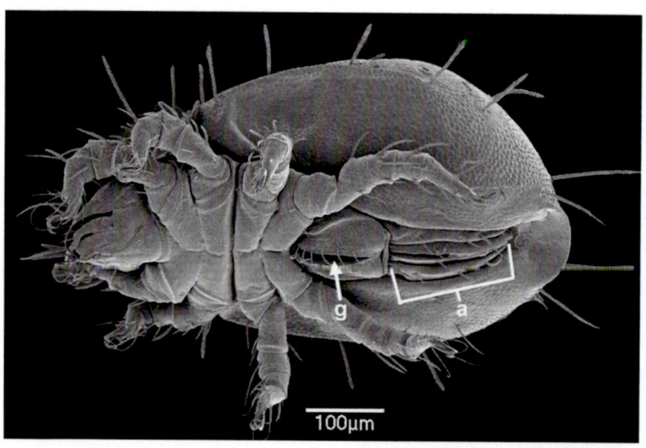

FIGURE 3.10 Scanning electronic micrograph (SEM) showing an oribatid mite from the ventral side. G = genital opening, and A = anal opening. *SEM courtesy Heather Proctor (from Vol. I: Ecology and General Biology, Thorp and Rogers (ed.), 2015, in Thorp and Covich's Freshwater Invertebrates, 4th Edition, Elsevier).*

2' Chelicerae with regressed and movable fixed digit as a hook or stylet-like structure; pedipalps either simple or with distal segments variously modified; subcapitulum lacking rutellae; prodorsum lacking well-defined bothridia and with sensillae setiform; opisthosoma lacking paired lateral glands (Figs 3.11–3.14) .... order Trombidiformes, suborder Prostigmata

FIGURE 3.11 SEM of the water mite *Eustigmaeus* (order Trombidiformes, suborder Prostigmata). This mite feeds on freshwater mosses (SEM by David Walter). *SEM photograph courtesy David Walter (from Vol. I: Ecology and General Biology, Thorp and Rogers (ed.), 2015, in Thorp and Covich's Freshwater Invertebrates, 4th Edition, Elsevier).*

Subphylum Crustacea: Classes

Individual couplets indicate the chapter where specific classes are discussed (or introduced in the case of Hexapoda). The only exceptions are the classes Ichthyostraca, subclass Branchiura (fish lice), and Thecostraca, subclass Cirripedia (barnacles) which are briefly discussed in the current chapter.

1 Genitalia at posterior end of thorax; more than 3 pairs of limbs; 2 pairs of antennae; wings absent 2
1' Genitalia not as above; larvae with or without limbs; but, if present, then less than 4 pairs (false legs present in some aquatic flies and butterflies); 1 pair of antennae; wings present or absent; springtails (Fig. 3.4) and insects (Fig. 3.5A–D and 3.15) .... class Hexapoda [Chapters 8–15]
2(1) Thoracopods (thoracic appendages) segmented 3
2' Thoracopods lamellar and not segmented; water fleas (Fig. 3.16), fairy and tadpole shrimps (Figs 3.17 and 3.18), etc .... class Branchiopoda [Chapter 4]
3(2) Carapace not bivalved 4
3' Carapace bivalved and enclosing entire animal; seed shrimps (Fig. 3.19) .... class Ostracoda [Chapter 5]
4(3) Adult not fused to the substrate 5
4' Adult fused to the substrate and present in inland and oceanic, saline waters; barnacles .... class Thecostraca, subclass Cirripedia [current chapter]
5(4) Adult with abdominal appendages present; telson or pleotelson present; crayfish, shrimps, crabs, etc. (Figs. 3.6 and 3.7) .... class Malacostraca [Chapter 7] (Fig. 3.20)
5' Adult with abdominal appendages absent 6
6(5) Carapace formed by a bilobed dorsal shield; paired compound eyes and naupliar eye on dorsal surface; abdomen rudimentary, unsegmented; parasitic; fish lice (Fig. 3.21) .... class Ichthyostraca, subclass Branchiura [current chapter and in more details in Chapter 6]
6' Carapace enveloping head; eyes, when present, represented by a naupliar eye; abdomen segmented (body of adult parasitic females strongly transformed and attached to the host) .... class Copepoda [Chapter 6]

Introduction to the subphylum Chelicerata, class Arachnida

Overview

The subphylum Chelicerata contains only one major group with freshwater representatives: the class Arachnida with its two subclasses—Aranae (spiders) and Acari (mites). The former is only marginally aquatic, as only a few species penetrate the surface of calm waters to prey on subsurface species, some other spider species prey on surface dwelling

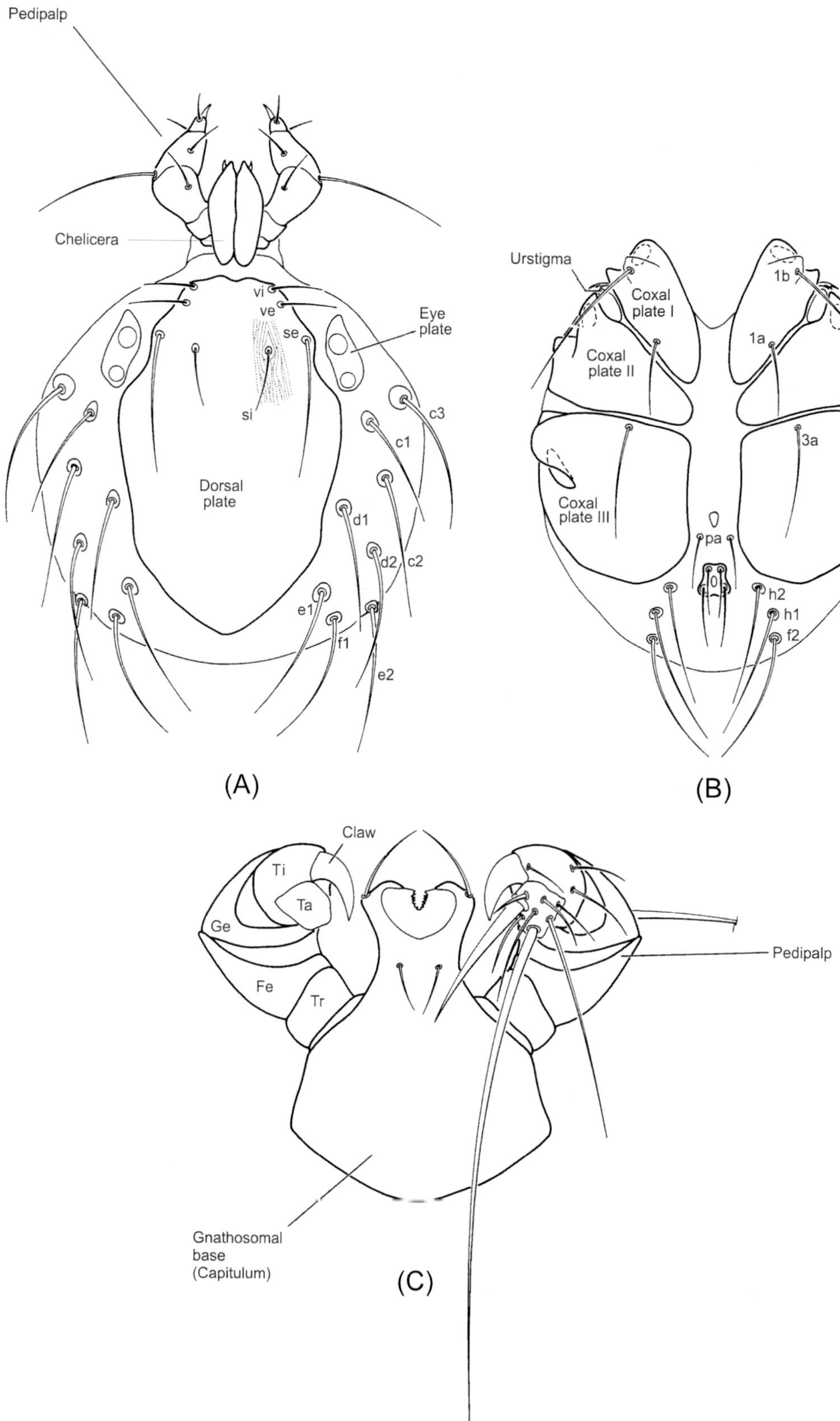

FIGURE 3.12 Labeled anatomy of water mite larva, *Sperchonopsis ecphyma* (order Trombidiformes, family Sperchontidae), shown without legs: (A) dorsal view, (B) ventral view excluding gnathosoma, and (c) ventral view of gnathosoma. *Drawings courtesy Heather Proctor (from Vol. I: Ecology and General Biology, Thorp and Rogers (ed.), 2015, in Thorp and Covich's Freshwater Invertebrates, 4th Edition, Elsevier).*

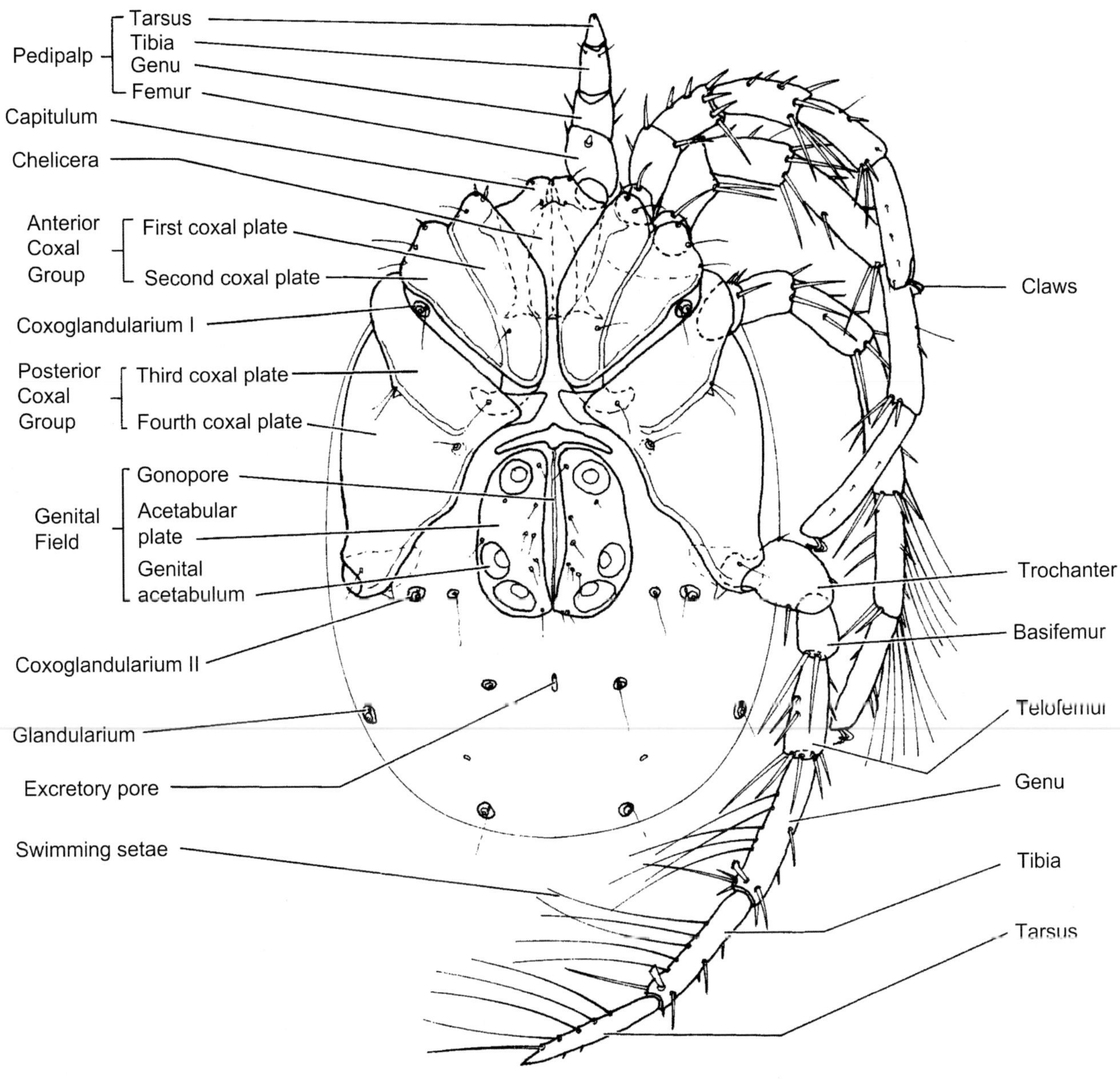

FIGURE 3.13 Drawing of the ventral view of an adult female mite of *Limnesia* sp. (order Trombidiformes). *Drawings courtesy Heather Proctor (from Vol. I: Ecology and General Biology, Thorp and Rogers (ed.), 2015, in Thorp and Covich's Freshwater Invertebrates, 4th Edition, Elsevier).*

aquatic invertebrates. In contrast, the latter include a large diversity of free-living and ectoparasitic freshwater mites. Three orders of the subclass Acari have freshwater representatives: Mesostigmata, Sarcoptiformes, and Trombidiformes, but only the latter two contain a few truly aquatic groups. Information on the ecology and diversity of members of this subphylum is only briefly covered in our book. For more information on the subphylum in general, consult the excellent discussion by Proctor et al. (2015). For information on the family Halacaridae of freshwater mites in the Palearctic, see Bartsch (2009, 2019).

Subclass Aranae: spiders associated with aquatic habitats

"Freshwater" spiders in the Mediterranean Basin include two ecological groups, both within the infraorder Araneomorphae: (1) those that hunt primarily on the surface of waters and usually extensively on land; they represent the vast majority of species; and (2) those that hunt and live mostly below the surface (still breathing atmospheric oxygen). The latter is primarily

FIGURE 3.14 Successively closer views of a raft of larval mite *Eylais* (order Trombidiformes) on East Mackey Lake, Alaska. *Photographs courtesy David Wartinbee (from Vol. I: Ecology and General Biology, Thorp and Rogers (ed.), 2015, in Thorp and Covich's Freshwater Invertebrates, 4th Edition, Elsevier).*

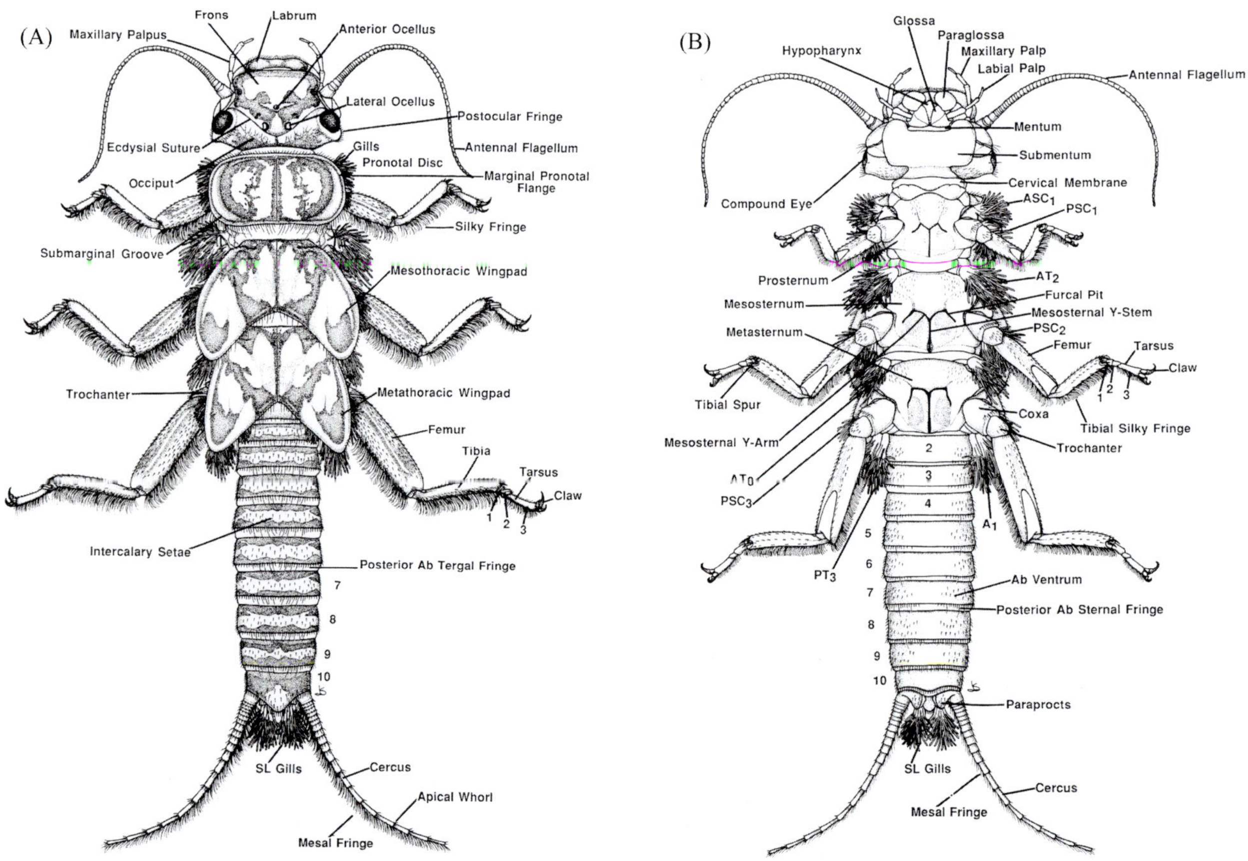

FIGURE 3.15 Dorsal (A) and ventral (B) anatomy of a larval stonefly (Insecta, Plecoptera), representing general structure of many aquatic insect larvae. *Drawings courtesy R. Edward DeWalt and Vincent H. Resh (from Vol. II: Keys to Nearctic Fauna, Thorp and Rogers (eds.), 2018, in Thorp and Covich's Freshwater Invertebrates, 4th Edition, Elsevier).*

FIGURE 3.16 Female *Daphnia* "water flea" (crustacean order Branchiopoda), with asexual eggs. *Photograph courtesy Dieter Ebert.*

FIGURE 3.17 Lateral view of the fairy shrimp *Artemia salina. Photograph courtesy Hans Hillewaert (© Hans Hillewaert).*

FIGURE 3.18 (A) Dorsal view of a tadpole shrimp *Lepidurus*; (B) ventral view of the tadpole shrimp *Triops. Photographs courtesy (A) Matthew A. Hill; (B) Brian K. Lang (both from Vol. I: Ecology and General Biology, Thorp and Rogers (ed.), 2015, in Thorp and Covich's Freshwater Invertebrates, 4th Edition, Elsevier).*

limited in the Palearctic to *Argyroneta aquatica*—the diving bell or water spider (family Dictynidae) (Fig. 3.9A–D). The former includes the European "tarantulas" or wolf spiders (family Lycosidae; Fig. 3.1A) and the fishing spiders [genus *Dolomedes*, family Pisauridae (Figs 3.1B and 3.8A–D)]. As is evident from Fig. 3.8A–D, fishing spiders can submerge in search of some prey or to escape predators, even though this behavior is less common than for *Argyroneta*.

Spiders are characterized by a well-developed cephalothorax and a reproductive abdomen (opisthosoma) separated from the cephalothorax by a slender pedicel (Proctor et al., 2015). The latter is reflective of the entirely liquid diet of

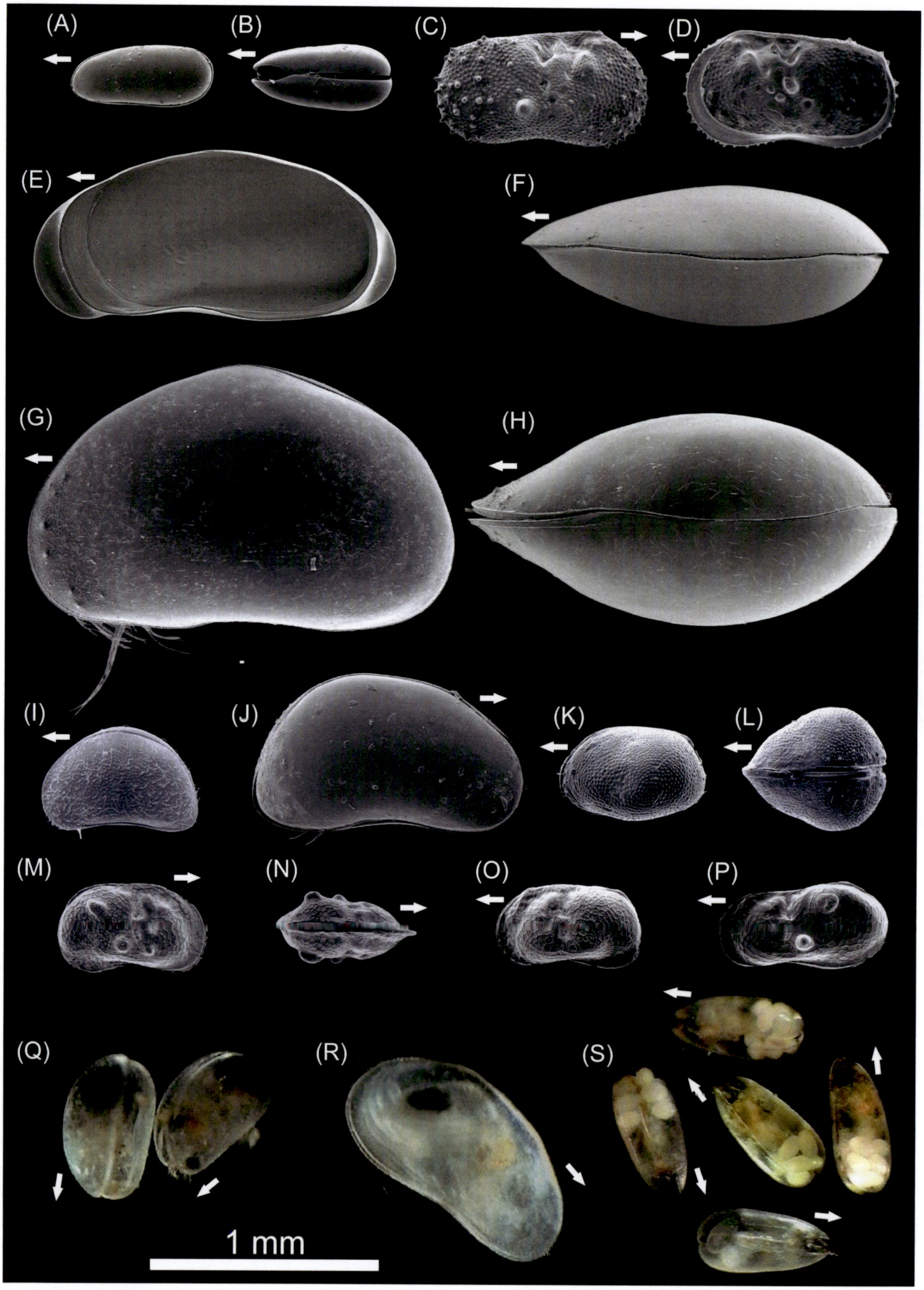

FIGURE 3.19 Scanning electron microscopy (A–P) and high resolution video microscopy (Q–S) images of freshwater ostracods. (A and B) *Darwinula*, adult female carapace left side (A) and ventral (B); (C and D) *Ilyocypris* adult female RV external: (C) and internal (D); (E and F) *Candonocypris*, adult female RV internal and carapace dorsal (F); (G and H) *Eucypris*, adult female carapace left side: (G) and dorsal (H); (I) *Potamocypris*, adult female carapace left side; (J) *Candona*, adult female carapace right side; (K and L) *Metacypris*, adult female carapace left side (K) and dorsal (L); (M–P) *Limnocythere*, tuberculate adult female RV external (M) and carapace dorsal (N), nontuberculate adult female LV external (O) and tuberculate male LV external (P); (Q) *Scottia*, two live adult females; R, *Candonopsis*, live adult male; S, *Darwinula*, five live adult females with eggs or juveniles in the brood chamber. *Photographs courtesy David J. Horne (from Rogers & Thorp, 2019).*

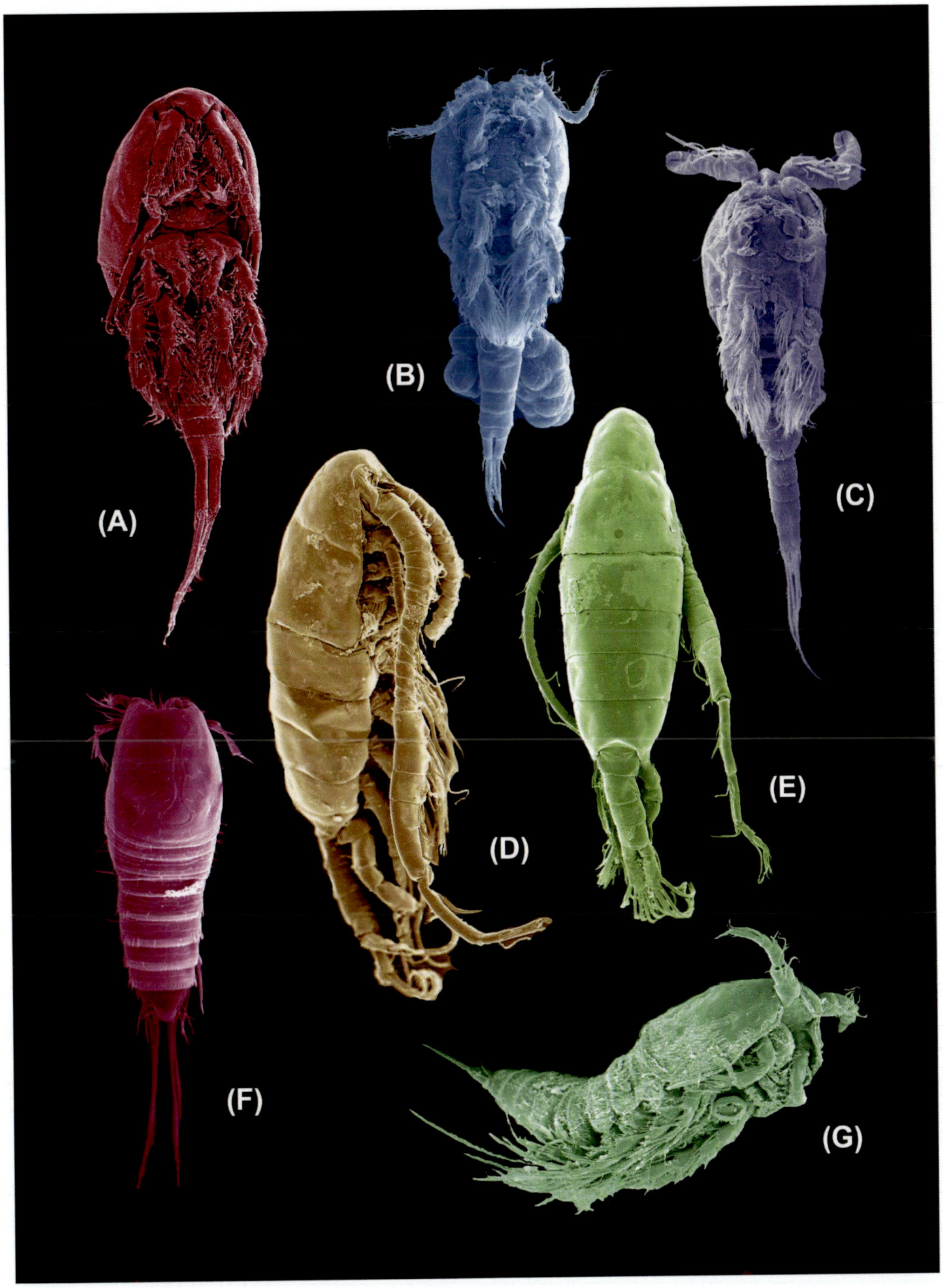

FIGURE 3.20 SEM colored specimens of three main groups of Copepoda in freshwater environments. (A–C) Cyclopoida, (A) female *Eucyclops* sp., ventral view; (B) egg-bearing female *Microcyclops* sp., ventral view; (C) male *Microcyclops* sp. (note two geniculate antennules); (D and E) Calanoida, Diaptomidae, (D) male *Leptodiaptomus* sp., lateral view; (E) male *Leptodiaptomus* sp., dorsal view (only the right antennule is geniculate); (F and G) Harpacticoida, (F) copepodite of *Elaphoidella* sp.; (G) female of *Cletocamptus* sp. *Courtesy Eduardo Suárez-Morales, with original SEM photos by Marcelo Silva-Briano (from Vol. I: Ecology and General Biology, Thorp and Rogers (ed.), 2015, in Thorp and Covich's Freshwater Invertebrates, 4th Edition, Elsevier).*

spiders. The cephalothorax contains four pairs of legs, up to eight pairs of single-lensed eyes, one pair of piercing chelicerae for killing their prey, and one pair of nonchelate palps for both handling prey and transferring sperm from males to females. The abdomen contains one to four pairs of spinnerets for building webs and binding prey, and also book lungs or tracheae and the reproductive organs.

FIGURE 3.21 Ventral view of the ectoparasitic fish lice *Argulus. Photograph courtesy John Pfieffer at EcoAnalyst, Inc. [From Vol. I: Ecology and General Biology, Thorp and Rogers (ed.), 2015.*

Freshwater spiders are limited ecologically to still or slow-moving waters of streams, wetlands, and the vegetated littoral zone of lakes. They feed primarily on aquatic insects (especially mosquito larvae) but will occasionally capture small fish or tadpoles. Surface-dwelling spiders can hide from their predators (especially birds but some fish) in surface vegetation and then rush out over the surface of the waters on hydrophobic legs to capture their aquatic prey or any small terrestrial prey that has been trapped by the water's surface tension. Subsurface spiders, primarily *Argyroneta aquatica*, build underwater, oxygenated nests where they wait for suitable prey as well as a mate (Fig. 3.9). They are unusual for spiders in that the males are larger than the females.

Subclass Acari: mites associated with aquatic habitats

Water mites are extremely abundant and diverse but easily overlooked members of the aquatic fauna of six continents where they occupy all portions of lake and stream habitats. They include both free-living and ectoparasitic forms. Fully and semiaquatic species have evolved primarily in the order Trombidiformes, suborder Prostigmata and secondarily in the order Sarcoptiformes, suborder Oribatida. Among the former, the most common mites are either in family Halacaridae or in the group Hydrachnidia. An excellent coverage of this group's morphology and ecology can be found in Proctor et al. (2015).

The mite body (Figs. 3.2 and 3.10–3.13) consists of an anterior gnathosoma with palps, chelicerae, and opening to the gut, and a posterior idiosoma with the legs and systems for reproduction, digestion, and excretion. Adults have four pairs of legs, while larvae have only three pairs.

Mites are extremely abundant in the benthos and on aquatic organisms. Proctor et al. (2015) indicated that the benthos of lakes and streams can contain 2000–5000 mites per square meter, with 50–75 species in 25–30 genera! An unusually high proportion of water mites (compared to other invertebrates) are bright red (Figs. 3.11 and 3.14). This would seemingly make them easy targets of predators, but the color suggests that mites are typically distasteful or even poisonous.

Freshwater mites can be divided ecologically into parasites and predators, with the larvae being markedly parasitic. Mite parasites are known to reduce egg production of adult insect hosts and can affect dispersal (e.g., in adult odonates). It is not unusual for some individuals of aquatic insects to have no mites, while others of the same species will be heavily invested. Nonparasitic mites can be voracious predators on insect eggs, chironomids, other small insects, ostracods, benthic cladocerans, and other aquatic species. Mites are eaten by some larger aquatic organisms, but much less commonly than would be suggested by their population sizes. This presumably relates to a distasteful or even poisonous nature.

Introduction to the subphylum Crustacea

Overview

This section is divided into a brief coverage of the crustacean class Hexapoda and a longer, but still modest coverage of the remaining crustaceans, which we will refer to here as "traditional" crustaceans, even though the name has absolutely no taxonomic significance. The difference in coverage reflects the presence of Chapter 8 in the current book which extensively describes the biology, ecology, and biodiversity of Hexapoda and summarizes the insect orders that are discussed more thoroughly in Chapters 9–15. The biology and ecology of these two groups are also discussed in a chapter on Hexapoda by Thorp and O'Neill (2015).

Subphylum Crustacea: class Hexapoda

The class Hexapoda consists of two subclasses: Entognatha (=Collembola or "springtails") and Insecta. Both subclasses are primarily terrestrial in terms of species richness, and the majority of orders in both subclasses have some species associated with aquatic habitats. However, only the insects represent major components of freshwater communities (for some examples, see Fig. 3.5A–D). Both can also be found in some oceanic environments, but the marine component is extremely small in species richness and spatial distribution in both cases. They share two distinguishing features within the phylum: (1) three pairs of jointed appendages are present in at least one life stage; and (2) their genitalia occur at the abdominal terminus. A general view of the anatomy of an aquatic insect larva—in this case a stonefly (Plecoptera)—is shown in Fig. 3.15.

Insecta

Insects are the only invertebrates where any species contains wings, and most adult insect species have them. Three body regions—or tagma—occur in this group: the head (with a single pair of antennae, complex mouthparts, paired compound eyes, and three or fewer ocelli), a thorax (with three segments, walking legs, and wings in adults), and the abdomen (Fig. 3.15).

Aquatic insects occur in all inland water habitats but are most common in stream and lake benthos, where their abundance and diversity are much greater in oxygenated habitats. They can be found in some oxygenated hyporheic environments but are rare in subterranean caves, possibly because of easy access. Some flies (Diptera) colonize the shallow edges of salt lakes and may occur in estuaries. A few taxa have been reported from islands near Antarctica (see Damborenea et al., 2020) and in northern polar regions. Aquatic insects have representatives in all trophic levels from detritivores to top invertebrate predators, and they occur in all countries within the Mediterranean Basin.

Collembola

Freshwater springtails are almost entirely associated with the surface of aquatic systems, and many of the few "freshwater species" seem also to occur in moist terrestrial environments. Springtails are rare in saline habitats, but *Anurida maritima* is a known maritime species in northern Europe which may be present in the Mediterranean Basin or is related to another regional springtail. One problem in studying these species is that it is not always clear whether you have collected true freshwater species or instead vagrants from the terrestrial environment.

Springtails are wingless hexapods in all life stages and are generally less than 6 mm long, with aquatic species usually under 3 mm. Locomotion is achieved by sudden release of a three-part abdominal structure known as a furcula (Fig. 3.4), whose release can propel the collembolan several centimeters through the air to escape a predator. The furcula of aquatic species is usually structured in a way that seems more adaptive to aquatic habitats because of its shape and greater surface area. Springtails feed primarily on algae and surface detritus.

Subphylum Crustacea: "Traditional" Crustaceans

The crustaceans that we are grouping here as "traditional" (i.e., the group predating addition of Hexapoda to other taxa within the subphylum) currently include six classes: Ichthyostraca (see below in section on Branchiura fish lice),

Thecostraca (see below on barnacles), Branchiopoda (Chapter 4), Ostracoda (Chapter 5), Copepoda (Chapter 6), and Malacostraca (Chapter 7). All these classes contain some marine species, but only Branchiura and Branchiopoda are almost exclusively freshwater denizens. The only group of these six subclasses that are extremely rare in inland waters is the class Thecostraca, subclass Cirripedia, which is represented by barnacles living in inland salt waters and fused to the substrate as adults.

These six classes of traditional crustaceans are characterized by: (1) more than three pairs of jointed appendages in at least one life stage; (2) genitalia at the juncture of the thorax and abdomen; (3) two pairs of antennae; and (4) the lack of wings in any life stage. The thoracopods (thoracic appendages) are lamellar and unsegmented in one class (Branchiopoda, like water fleas, fairy shrimp, etc.) but segmented in the other classes. The carapace is not bivalved except in seed shrimps (class Ostracoda), and either four (class Ichthyostraca, subclass Branchiura) or five or more pairs of thoracopods are present (e.g., class Malacostraca). Parasitism is rare among crustaceans except among Branchiura fish lice in the class Ichthyostraca.

Habitats

Some variation exists among crustacean classes in the typical habitats they occupy, but a large amount of overlap exists. Aside from ectoparasites of Branchiura and those in two families of copepods, only members of the class Thecostraca (barnacles cemented to rocks) are entirely benthic as adults, though most ostracods live on or slightly within the substrate. A more limited diversity of arthropods also colonize hyporheic and phreatic habitats, but these communities have been inadequately explored worldwide. Although malacostracans occupy all aquatic habitats, most families are associated with benthos. This is especially true of crustaceans characterized by a relatively large biomass, such as many decapod crustaceans (crabs, crayfishes, and shrimps). Malacostracans that live predominately in the water column are mostly "opossum shrimps" in the order Mysida; however, only a few species are found in the Mediterranean Basin, and those migrate at night from darker, deeper waters of a few lakes into surface layers. A sizeable diversity of crustaceans—primarily copepods and cladocerans—are primarily but not exclusively planktonic and represent substantial contributors to ecological processes in lotic and lentic habitats. Some crustaceans are frequently associated with subterranean habitats from the hyporheos to caverns, with the taxa most commonly limited to these habitats being in the malacostracan orders Thermosbaenacea and Bathynellacea. Crustaceans in the orders Amphipoda and Isopoda occupy both surface and subterranean habitats, but the total species diversity can be highest overall for all groundwater habitats combined because of the substantial effect of spatial isolation on evolution of new species. An extremely small numbers of inland water crustaceans live in saline environments. Other than the mostly marine barnacles, these include a few species of large branchiopods, such as fairy shrimp (Fig. 3.17A), including *Artemia franciscana* (Horváth et al., 2018). Branchiopoda also include species living in freshwater to slightly saline, ephemeral pools like tadpole shrimps (Fig. 3.17B), such as species of *Triops* and *Lepidurus* (Branchiopoda, Notostraca) in Cyprus (Tziortzis et al., 2014 and many other countries in the Mediterranean Basin).

The most common microhabitat for benthic crustaceans is the littoral zone of lentic and lotic habitats where oxygen and food are often in the greatest abundance and substrates are most complex, thereby providing some protection from predators. In rocky, grassy, and wood-laden systems, invertebrates can exploit these microhabitats for protection. However, in sand bed rivers, species are most successful if they live within the substrate (more common for insects than other crustaceans) or under riverbanks. In the rare Mediterranean rivers where natural hydrogeomorphic complexity still exists in part, benthic and planktonic crustaceans are often most abundant in low velocity side channels.

Food webs

Omnivory is characteristic of the greatest number of taxa in most groups of crustaceans because it allows the maximum flexibility for exploiting available food sources. The fish lice of Branchiura are an exception because as ectoparasites they are specialist in feeding on high energy tissue from their hosts. Some crustaceans, such as many planktivorous branchiopods, feed preferentially on algae but will consumer detritus if necessary. However, a few other members of this class are primarily predators or switch to becoming predators when they are larger (e.g., tadpole shrimp). Copepods include some taxa that are primarily predators, such as cyclopoids, but algivorous and detritivorous species are also common. Calanoid copepods are mostly filter feeders; but some species are predaceous, and others are generally benthic grazers which feed on a variety of living zoobenthos and particulate organic matter. Harpacticoid copepods also consume mostly algae, detritus, fungi, protozoa, and bacteria. Cannibalism (eating members of one's own species) and saprophagy (eating dead animals) are very common throughout the animal world, including among benthic crustaceans.

A brief introduction to the classes Branchiopoda, Ostracoda, Thecostraca, Copepoda, Ichthyostraca, and Malacostraca

Chapters 4–7 are focused on individual classes of the traditional Crustacea, and thus the current section is merely a prelude to more extensive information presented in those chapters. Other than the classes Thecostraca (barnacles) and Ichthyostraca (fish lice), these classes are moderately well represented in freshwaters. Additional information on the identification and distribution of these crustaceans can be found in Rogers and Thorp (2019) for the Palearctic.

Class Branchiopoda (Chapter 4)

Branchiopods are mostly small crustaceans found in lentic habitats (primarily) and streams. These small crustaceans (0.5–20 mm) are characterized by thoracic appendages (thoracopods) that are lamellar and unsegmented (Cáceres & Rogers, 2015). The three orders differ in the presence of a carapace and whether it is folded (Rogers et al., 2019). While most are confined to freshwaters, a few of the large branchiopods (e.g., fairy shrimp; Fig. 3.17A) thrive in saline lakes, ponds, and temporary wetlands around the world including in southern Europe, Anatolia, and northern Africa. Some of these lentic systems where they live are saltier than the Mediterranean Sea (e.g., the brine shrimp *Artemia salina* lives in the U.S. Great Salt Lake at 50 to >200 ppt salt compared to 35 ppt as the ocean average). The tadpole shrimps *Triops* and *Lepidurus* (order Notostraca; Fig. 3.18) are considered cosmopolitan in vernal pools of the basin from Spain to northern Africa. The more speciose order Diplostraca contains the diverse cladoceran "water fleas" including the cosmopolitan *Daphnia* (Fig. 3.16). Cladocera is a well known but no longer taxonomically acceptable name. These crustaceans occur in the pelagic and benthic zones of lentic systems and to some extent in rivers, especially in slow-flowing side channels. More information on the biology and ecology of this class is discussed in Cáceres and Rogers (2015), and taxonomic distributions are described for the Palearctic in Rogers et al. (2019) and in Chapter 4 of the current book.

Class Ostracoda (Chapter 5)

Ostracods (or ostracodes) are mostly microscopic crustaceans that are primarily marine but contain a large group of freshwater species in the order Podocopida, and some taxa occupy moist terrestrial soils. They are characterized by a distinctive, nonstriated, calcified bivalve shell (their carapace) which is periodically molted, and their thoracopods are segmented like other crustaceans (except the branchiopods) (Fig. 3.19). They range in size from 0.5 to 5 mm in most cases. These "seed shrimps" are mostly benthic in freshwaters, where they are primarily herbivores (eating true algae and sometimes cyanobacteria) and/or detritivores. Carnivory seems rare in this crustacean class but has been documented in some families. Ostracods occupy almost all aquatic habitats including subterranean waters and wet soil. One unique feature of this group is that ostracods are often important subjects of studies in paleobiology. More information on the biology and ecology of this class is discussed in Smith et al. (2015). Taxonomic distributions are described in Horne et al. (2019) for the Palearctic and in Chapter 5 of the current book.

Class Copepoda (Chapter 6)

This class contains animals (copepods) familiar to aquatic scientists working in freshwater and marine systems, but mostly under a different higher taxonomic name (previously class Maxillopoda, later Hexanauplia). Copepods are abundant in freshwaters (surface and subterranean), marine habitats, and even occasionally in humid terrestrial habitats. In lakes and wetlands, they occur mostly as zooplankton, but they are also well represented in or just above the bottom (especially the harpacticoid copepods). They are less common in the main channels of rivers, but they are often very abundant within side channels where current velocities are slow to absent. Some groups predominately consume phytoplankton, bacteria, and suspended organic matter (most calanoid copepods), where others, such as cyclopoid copepods, are often predators. Most freshwater copepods in the Mediterranean Basin are less than 3 mm long and feature a fusiform (especially planktonic forms) to cylindrical bodies. The body form features an anterior prosome (head and thoracic segments) and a posterior urosome with the genital and abdominal segments. Most have a single median eye, which can be bright red, especially in habitats lacking fish. Their life cycle includes multi-stage nauplius and copepodid larvae. More information on the biology and ecology of the copepods is discussed in Suárez-Morales (2015) for the copepods only, while the taxonomic distributions in the Palearctic are described for copepods in Lee and Lee (2019) for copepods and for Thecostraca in Rogers (2019) and by Van Syoc (2019). Chapter 6 of the current book describes the ecology and taxonomy of copepods in the Mediterranean Basin.

Class Thecostraca

For many readers one of the more surprising inland water crustaceans is found within the class Thecostraca. These inland water barnacles, especially *Amphibalanus improvisus* (the "bay barnacle"), tolerate salinities of 0.8–40 ppt (Suárez-Morales, 2015), which is much less than the 38–40 ppt found in the adjacent Mediterranean Sea. Another species reported from this region is *A. amphitrite*, but almost all other barnacles are confined to the oceans. Their life cycle includes free-living naupliar stages and a cypris larva which transforms into a calcified adult barnacle. The latter cements itself to a hard surface and feeds on plankton.

Class Ichthyostraca (Chapter 6)

This class currently contains two subclasses: (1) Pentastomida (tongue worms), which as adults parasitize terrestrial vertebrates; and (2) Branchiura, adults of which primarily obligate ectoparasites on fish (hence the name fish lice) and occasionally amphibians, but host specificity is not especially common (Møller, 2009). The estimated total diversity within the subclass Branchiura is perhaps 230 or more species (Møller, 2015), with the vast majority living in freshwaters. At least eight species of the largest family, Argulidae, have been reported from the Palearctic bioregion (Poly, 2008). However, only the most common worldwide genus, *Argulus*, has been collected from inland fresh waters of Europe (Fig. 3.21), with this genus represented by at least three inland water species: *A. coregoni*, *A. foliaceus*, and *A. japonicus*. The latter two are most common in the Mediterranean Basin, as *A. coregoni* has mostly a boreal distribution. *Argulus japonicus* species seems to have been introduced globally in association with stocking goldfish (*Carassius auratus*) and carp (*Cyprinus carpio*). In addition to these known freshwater species, some taxa have been collected from brackish and nearshore marine waters of the Mediterranean, including *A. arcassonensis* and *A. vittatus* from the waters off Algeria (Møller, 2015), and the latter also from the northern coast of Sicily (Bottari et al., 2017). Note, however, that it is possible that these species may still require freshwaters to reproduce successfully.

Fish lice occur primarily in lakes and wetlands. Although they can occasionally be collected from streams downstream of a lentic habitat, they require a lake habitat for successful reproduction via metanauplii larvae that hatch from eggs deposited on flat surfaces. Larval development is described for *A. foliaceus* in Rushton-Mellor and Boxshall (1994).

Argulids pierce the skin of their hosts, inject a toxin, and feed off the blood and other nearby tissues. They maintain their position or move about their host using hooks and suction. Mortality can result from heavy infestations in the host and from bacterial infections. Indeed, these ectoparasites can be a serious source of mortality and general debilitation of fish in aquacultural operations.

Their anatomy features four pairs of thoracopods, bilobed carapace (differing in shape among species), suction disks, paired compound eyes. Species and the sexes can be identified (especially in males) by the shape of the abdomen, its setae, and the depth of the abdominal incision.

Class Malacostraca (Chapter 7)

The crustaceans best known to society at large are in the class Malacostraca (Figs. 3.6 and 3.7). Among this class are amphipods and isopods (both in the superorder Peracarida) and the large order Decapoda with the familiar shrimps, crabs, and crayfishes. Entire orders (e.g., Thermosbaenacea and Bathynellacea) are subterranean, eyeless and depigmented; several subterranean families also occur in amphipods, isopods, mysids, and decapods. Four general body regions are present for members of this class: a head, pereon (or thorax), pleon (abdomen), and urosome (tail), each with 1–8 segments. The head and thorax may be fused into a cephalothorax. Each body region contains appendages modified for different purposes, including locomotion, respiration, feeding, sensory acquisition, and reproduction or care of young. Malacostracans contain the largest crustaceans in body size, and are economically the most important group to humans, with many species of decapods consumed by people. The most common ecological role they play in most types of freshwater habitats is as omnivores (e.g., combined detritivores, algivores, and predators).

Acknowledgments

A portion of this chapter generally followed the taxonomic approach and keys present in several chapters by R. Christopher Rogers and others in Vol. IV: Keys to Palearctic Fauna, in the series Thorp and Covich's Freshwater Invertebrates. We also benefitted from Christopher's general discussion of this phylum's biology. We are especially grateful to the useful comments from Dr. Fabio Stoch (Université libre de Bruxelles) on crustacean taxonomy.

Subchapter 3.2

Molecular tools for species identification

Michael J. Raupach
Staatliche Naturwissenschaftliche Sammlungen Bayerns, Zoologische Staatssammlung München Sektionsleitung Hemiptera, München, Germany

Introduction

Species identification represents a pivotal component for biodiversity studies and conservation planning but represents a challenge for many taxa when using morphological traits only (e.g., the correct identification of juveniles or larval stages). In the last few years, tremendous technological advances in molecular biology and other fields have significantly revolutionized biological sciences. New approaches and technologies have become available for the acquisition of molecular information, with a strong focus on the use of DNA sequence data (e.g., Raupach et al., 2016). In the following sections, I present the most prominent approaches and technologies of this highly dynamic field of science.

DNA barcoding

During the last few years, DNA barcoding has become the central component in the modern diagnostic toolbox of molecular biodiversity assessment studies (e.g., Hajibabaei et al., 2016; Lee et al., 2016) and taxonomic revisions (e.g., Huemer & Mutanen, 2015; Lin et al., 2018; Fernandez-Triana et al., 2019). DNA barcoding is based on the premise that a short, standardized sequence—for animals a 658 base pair fragment of the mitochondrial cytochrome *c* oxidase subunit I (COI) gene—can distinguish individuals of a species due to the fact that the genetic variation between species exceeds that within species (Hebert, Cywinska, et al., 2003; Hebert, Ratnasingham, et al., 2003; Hajibabaei et al., 2007). Barcode sequences are typically placed in the international Barcode of Life Data Systems database (BOLD; http://www.boldsystems.org). This database acts as the central core data interface and repository that allows researchers to collect, organize, and analyze DNA barcode data (Ratnasingham & Hebert, 2007). In addition to various analytical tools implemented in the BOLD workbench, DNA barcodes can be analyzed using the Barcode Index Number (BIN) system that clusters DNA barcodes in order to produce operational taxonomic units that closely correspond to species (Ratnasingham & Hebert, 2013).

In this context, the build-up of comprehensive DNA barcode libraries represents a pivotal task (e.g., Brandon-Mong et al., 2015; Curry et al., 2018). A recent survey showed that the DNA barcode coverage of freshwater taxa varies strongly among taxonomic groups and geographic regions in Europe (Weigand et al., 2019). Many species were actively targeted in specific barcode projects and are well represented at BOLD, such as amphibians (Hawlitschek et al., 2016, Zangl et al., 2020), fish (Geiger et al., 2014, Knebelsberger et al., 2015), mayflies, stoneflies, caddisflies (Morinière et al., 2017), chironomid diptera (Lin et al., 2015), and aquatic and semiaquatic true bugs (Raupach et al., 2014; Havemann et al., 2018). Other groups have much fewer records, however, and need to be updated, including crustaceans, dragonflies, mites, and molluscs.

Massive parallel sequencing: metabarcoding and the analysis of environmental DNA

To overcome the limitations of single-specimen identification methods for processing large numbers of specimens, recent studies have looked to high-throughput sequencing technologies to allow DNA barcode-based identification in a massively parallel manner (Piper et al., 2019). This approach, termed "metabarcoding" (Taberlet et al., 2012), produces many thousands of individual barcode sequences in a single reaction run and enables the parallel identification of specimens in large mixed communities (Yu et al., 2012; Porter & Hajibabaei, 2018; Elbrecht & Steinke, 2019; Piper et al., 2019). Parallel to this, environmental DNA (eDNA) studies which sequence DNA in soil or water without first isolating any organisms facilitate rapid biodiversity monitoring with only small sediment or water samples (Deiner et al., 2017; Lamb et al., 2019). While the perspectives of both approaches are promising and fascinating (Valentini et al., 2009;

Cristescu, 2014; Cristescu & Hebert, 2018), however, it should be noted that it is not possible to provide a valid quantitative estimation of individuals per analyzed species so far (e.g., Lamb et al., 2019).

Other approaches and technologies

In addition to the previously mentioned DNA sequence-based approaches, various other technologies can be used in modern species determination studies. In the case of freshwater taxa, the matrix-assisted laser desorption/ionization time-of flight mass spectroscopy has been successfully applied for the identification of freshwater copepods in the past (Riccardi et al., 2012). Moreover, the analysis of cuticular hydrocarbons by using coupled gas chromatography-mass spectrometry was employed to differentiate various water beetle species (Botella-Cruz et al., 2017). However, the identification of species has not to be restricted on molecular approaches only. For example, a recent study demonstrated the successful application of Convolutional Neural Networks for the identification of chironomid larvae in order to increase taxonomic resolution of biomonitoring data at minimal cost (Milošević et al., 2020).

The presented compilation does not cover all aspects, approaches, and technologies (see Raupach et al., 2016). However, it highlights that we are now able to document, describe, and identify species much more comprehensively than just a few years ago, and it is obvious that such approaches will become central in freshwater biodiversity research in the near future.

References

Bartsch, I. 2009. Checklist of marine and freshwater halacarid mite genera and species (Halacaridae, Acari) with notes on synonyms, habitats, distribution and descriptions of the taxa. Zootaxa 1998: 1–170.

Bartsch, I. 2019. Arachnida: Acari: Trombidiformes: Halacaridae. Chapter 16, pp. 554–563 *in*: D.C. Rogers and J.H. Thorp (eds.), Vol. IV: Keys to Palaearctic Fauna, *in*: Thorp and Covich's Freshwater Invertebrates, 4th Edition, Academic Press, Elsevier, Boston, MA.

Botella-Cruz, M., A. Villastrigo, S. Pallarés, E. López-Gallego, A. Millán, and J. Velasco. 2017. Cuticle hydrocarbons in saline aquatic beetles. PeerJ 5: e3562.

Bottari, T., A. Profeta, N. Spanò, F. Longo, and P. Rinelli. 2017. New host records for the marine fish ectoparasite *Argulus vittatus* (Crustacea: Branchiura: Argulidae). Comparative Parasitology 84: 64–66.

Brandon-Mong, G.J., H.M. Gan, K.W. Sing, P.S. Lee, P.E. Lim, and J.J. Wilson. 2015. DNA metabarcoding of insects and allies: an evaluation of primers and pipelines. Bulletin of Entomological Research 105: 717–727.

Cáceres, C.E. and D.C. Rogers. 2015. Class Branchiopoda. Chapter 28, pp. 687–708 *in*: J.H. Thorp and D.C. Rogers (eds.), Vol. I: Ecology and General Biology, *in*: Thorp and Covich's Freshwater Invertebrates, 4th Edition, Academic Press, Elsevier, Boston.

Cristescu, M.E. 2014. From barcoding single individuals to metabarcoding biological communities: towards an integrative approach to the study of global biodiversity. Trends in Ecology and Evolution 29: 566–571.

Cristescu, M.E. and P.D.N. Hebert. 2018. Uses and misuses of environmental DNA in biodiversity science. Annual Review of Ecology, Evolution, and Systematics 49: 209–230.

Curry, C.J., J.F. Gibson, S. Shokralla, M. Hajibabaei, and D.J. Baird. 2018. Identifying North American freshwater invertebrates using DNA barcodes: are existing COI sequence libraries fit for purpose? Freshwater Science 37: 178–189.

Daley, A.C. and H.B. Drage. 2016. The fossil record of ecdysis, and trends in the moulting behaviour of trilobites. Arthropod Structure and Development 45: 71–96.

Damborenea, C., D.C. Rogers, and J.H. Thorp (eds.). 2020. Keys to Neotropical and Antarctic Fauna. Vol. V *in*: Thorp and Covich's Freshwater Invertebrates, Academic Press, Elsevier, Boston.

Darwall,W., S. Carrizo, C. Numa, V. Barrios, J. Freyhof, and K. Smith. 2014. Freshwater key biodiversity areas in the Mediterranean Basin hotspot: Informing species conservation and development planning in freshwater ecosystems. Cambridge, UK and Malaga, Spain: IUCN. x + 86 pp.

Deiner, K., H.M. Bik, E. Mächler, M. Seymour, A. Lacoursière-Roussel, F. Altermatt, S. Creer, I. Bista, D.M. Lodge, N. de Vere, M.E. Pfrender, and L. Bernatchez. 2017. Environmental DNA metabarcoding: Transforming how we survey animal and plant communities. Molecular Ecology 26: 5872–5895.

Elbrecht, V. and D. Steinke. 2019. Scaling up DNA metabarcoding for freshwater macrozoobenthos monitoring. Freshwater Biology 64: 380–387.

Enghoff, H., G. Caoduro, J. Adis, and B. Messne. 1997. A new cavernicolous, semiaquatic species of Serradium (Diplopoda, Polydesmidae) and its terrestrial, sympatric congener. With notes on the genus *Serradium*. Zoologica Scripta 26:279–290.

Fernandez-Triana, J., C. Boudreault, T. Dapkey, M.A. Smith, W. Hallwachs, and D. Janzen. 2019. A revision of Dolichogenidea (Hymenoptera, Braconidae, Microgastrinae) with the second mediotergite broadly rectangular from Area de Conservación Guanacaste, Costa Rica. ZooKeys 835: 87–123.

Geiger, M.F., F. Herder, M.T. Monaghan,V. Almada, R. Barbieri, M. Bariche, P. Berrebi, J. Bohlen, M. Casal-Lopez, G.B. Delmastro, G.P. Denys, A. Dettai, I. Doadrio, E. Kalogianni, H. Kärst, M. Kottelat, M. Kovačić, M., Laporte, M., Lorenzoni, Z. Marĉi, M. Özuluğ, A. Perdices, S. Perea, H. Persat, S. Porcelotti, C. Puzzi, J. Robalo, R. Šanda, M. Schneider, V. Šlechtová, M. Stoumboudi, S. Walter, and J. Freyhof 2014. Spatial heterogeneity in the Mediterranean Biodiversity Hotspot affects barcoding accuracy of freshwater fishes. Molecular Ecology Resources 14: 1210–1221.

Hajibabaei, M., D.J. Baird, N.A. Fahner, R. Beiko, and G.B. Golding. 2016. A new way to contemplate Darwin's tangled bank: how DNA barcodes are reconnecting biodiversity science and biomonitoring. Philosophical Transactions of the Royal Society B: Biological Sciences 371: 20150330.

Hajibabaei, M., G.A.C. Singer, P.D.N. Hebert, and D.A. Hickey. 2007. DNA barcoding: how it can complements taxonomy, molecular phylogenetics and population genetics. Trends in Ecology and Evolution 23: 167–172.

Hamada, N., J.H. Thorp, and D.C. Rogers (eds.). 2018. Keys to Neotropical Hexapoda. Vol. III *in*: Thorp and Covich's Freshwater Invertebrates, 4th edition, Academic Press, Elsevier, Boston. 803 p.

Havemann, N., M.M. Gossner, L. Hendrich, J. Morinière, R. Niedringhaus, P. Schäfer, and M.J. Raupach. 2018. From water striders to water bugs: The molecular diversity of aquatic Heteroptera (Gerromorpha, Nepomorpha) of Germany based on DNA barcodes. PeerJ 6: e4577.

Hawlitschek, O., J. Morinière, A. Dunz, M. Franzen, D. Rödder, F. Glaw, and G. Haszprunar. 2016. Comprehensive DNA barcoding of the herpetofauna of Germany. Molecular Ecology Resources 16: 242–253.

Hebert, P.D.N., A. Cywinska, S.L. Ball, and J.R. deWaard 2003. Biological identifications through DNA barcodes. Proceedings of the Royal Society of London Series B: Biological Sciences 270: 313–321.

Hebert, P.D.N., S. Ratnasingham, and J.R. deWaard 2003. Barcoding animal life: cytochrome c oxidase subunit 1 divergences among closely related species. Proceedings of the Royal Society of London Series B: Biological Sciences 270: S96-S99.

Horváth, Z., C. Lejeusne, F. Amat, F.J. Sánchez-Fontenla, C.F. Vad, and A.J. Green. 2018. Eastern spread of the invasive *Artemia franciscana* in the Mediterranean Basin, with the first record from the Balkan Peninsula. Hydrobiologia 822: 229–235.

Huemer, P., and M. Mutanen. 2015. Alpha taxonomy of the genus *Kessleria* Nowicki, 1864, revisited in the light of DNA-barcoding. ZooKeys 503: 89–133.

Knebelsberger, T., A.R. Dunz, D. Neumann, and M.F. Geiger. 2015. Molecular diversity of Germany's freshwater fishes and lampreys assessed by DNA barcoding. Molecular Ecology Resources 15: 562–572.

Lamb, P.D., E. Hunter, J.K. Pinnegar, S. Creer, R.G. Davies, and M.I. Taylor. 2019. How quantitative is metabarcoding: A meta-analytical approach. Molecular Ecology 28: 420–430.

Lee, D.J. and W. Lee. 2019. Arthropoda: Copepoda. Chapter 16.4, pp. 761–781 *in*: D.C. Rogers and J.H. Thorp (eds.), Vol. IV: Keys to Palaearctic Fauna, *in*: Thorp and Covich's Freshwater Invertebrates, Academic Press, Elsevier, Boston.

Lee, P.S., H.M. Gan, G.R. Clements, and J.J. Wilson. 2016. Field calibration of blowfly-derived DNA against traditional methods for assessing mammal diversity in tropical forests. Genome 59: 1008–1022.

Lin, X., E. Stur, and T. Ekrem. 2015. Exploring genetic divergence in a species-rich insect genus using 2790 DNA barcodes. Public Library of Science ONE 10: e0138993.

Lin, X.-L., E. Stur, and T. Ekrem. 2018. DNA barcodes and morphology reveal unrecognized species in Chironomidae (Diptera). Insect Systematics & Evolution 49: 329–398.

Milošević, D., A. Milosavljević, B. Predić, A.S. Medeiros, D. Savić-Zdravković, M. Stojković Piperac, T. Kostić, F. Spasić, and F. Leese. 2020. Application of deep learning in aquatic bioassessment: Towards automated identification of non-biting midges. Science of the Total Environment 711: 135160.

Møller, O.M. 2009. Branchiura (Crustacea) - survey of historical literature and taxonomy. Arthropod Systematics and Phylogeny 67: 41–55.

Møller, O.M. 2015. Order Arguloida. Revista IDE@ - SEA, nº 103B: 1–8.

Morinière, J., L. Hendrich, M. Balke, A.J. Beermann, T. König, M. Hess, S. Koch, R. Müller, F. Leese, P.D.N. Hebert, A. Hausmann, C.D. Schubart, and G. Haszprunar. 2017. A DNA barcode library for Germany's mayflies, stoneflies and caddisflies (Ephemeroptera, Plecoptera and Trichoptera). Molecular Ecology Resources 17: 1293–1307.

Oakley, T.H., J.M. Wolfe, A.R. Lindgren, and A.K. Zaharoff. 2013. Phylotranscriptomics to bring the understudied into the fold: monophyletic Ostracoda, fossil placement, and pancrustacean phylogeny. Molecular Biology and Evolution 30: 215–233.

Piper, A.M., J. Batovska, N.O.I. Cogan, J. Weiss, J.P. Cunningham, B. Rodoni, and M.J. Blacket. 2019. Prospects and challenges of implementing DNA metabarcoding for high-throughput insect surveillance. GigaScience 8: 1–22.

Poly, W.J. 2008. Global diversity of fishlice (Crustacea: Branchiura: Argulidae) in freshwater. Hydrobiologia 595:209–212.

Porter, T.M. and M. Hajibabaei. 2018. Scaling up: A guide to high throughput genomic approaches for biodiversity analysis. Molecular Ecology 27: 313–338.

Proctor, H.C., I.M. Smith, D.R. Cook, and B.P. Smith. 2015. Subphylum Chelicerata, Class Arachnida. Chapter 25, pp. 599–660 *in*: J.H. Thorp and D.C. Rogers (eds.), Vol. I: Ecology and General Biology, in: Thorp and Covich's Freshwater Invertebrates, 4th Edition, Academic Press, Elsevier, Boston, MA.

Ratnasingham, S. and P.D.N. Hebert. 2007. BOLD: The Barcode of Life Data Systems. Molecular Ecology Notes 7: 355–364.

Ratnasingham, S. and P.D.N. Hebert. 2013. A DNA-based registry for all animal species: the Barcode Index Number (BIN) system. Public Library of Science ONE 8: e66213.

Raupach, M.J., Amann R., Q.D. Wheeler, and C. Roos. 2016. The application of "-omics" technologies for the classification and identification of animals. Organisms, Diversity & Evolution 16: 1–12.

Raupach, M.J., Hendrich L., S.M. Küchler, F. Deister, J. Morinière, and M.M. Gossner. 2014. Building-up of a DNA barcode library for true bugs (Insecta: Hemiptera: Heteroptera) of Germany reveals taxonomic uncertainties and surprises. Public Library of Science ONE 9: e106940.

Regier, J.C., Schultz J.W., and Kambic R.E.. 2005. Pancrustacean phylogeny: hexapods are terrestrial crustaceans and maxillopods are not monophyletic. Proceedings of the Royal Society B, 272: 395–401.

Regier, J.C., Shultz J.W., Zwick A., A. Hussey, B. Ball, R. Wetzer, J.W. Martin, and C.W. Cunningham. 2010. Arthropod relationships revealed by phylogenomic analysis of nuclear protein-coding sequences. Nature 463:1079–1083.

Riccardi, N., L. Lucini, C. Benagli, M. Welker, B. Wicht, and M. Tonolla. 2012. Potential of matrix-assisted laser desorption/ionization time-of-flight mass spectrometry (MALDI-TOF MS) for the identification of freshwater zooplankton: a pilot study with three *Eudiaptomus* (Copepoda: Diaptomidae) species. Journal of Plankton Research 34: 1–9.

Rogers, D.C. 2019. Arthropoda: Thecostraca. Chapter 16.5, pp. 781–787 in: D.C. Rogers and J.H. Thorp (eds.), Vol. IV: Keys to Palaearctic Fauna, *in*: Thorp and Covich's Freshwater Invertebrates, Academic Press, Elsevier, Boston.

Rogers, D.C. and Thorp J.H. (eds.). 2019. Keys to Palaearctic Fauna. Vol. IV *in*: Thorp and Covich's Freshwater Invertebrates, 4th edition, Academic Press, Elsevier, Boston. 920 p.

Rogers, D.C., Kotov A.A., A.Y. Sinev, S.M. Glagolev, N.M. Korovchinsky, N.N. Smirnov, and E.I. Bekker. 2019. Arthropoda, class Branchiopoda. Chapter 16.2, pp. 643–724 *in*: D.C. Rogers and J.H. Thorp (eds.), Vol. IV: Keys to Palaearctic Fauna, *in*: Thorp and Covich's Freshwater Invertebrates, Academic Press, Elsevier, Boston.

Rushton-Mellor, S.K. and G.A. Boxshall. 1994. The developmental sequence of *Argulus foliaceus* (Crustacea: Branchiura). Journal of Natural History 28: 763–785.

Scholtz, G. 2002. The Articulata hypothesis – or what is a segment? Organisms, Diversity & Evolution 2: 197–215.

Seymour, R.S. and S.K. Hetz. 2011. The diving bell and the spider: the physical gill of *Argyroneta aquatica*. Journal of Experimental Biology 214: 2175–2181; doi: 10.1242/jeb.056093

Smith, A.J., D.J. Horne, K. Martens, and I. Schön. 2015. Class Ostracoda. Chapter 30, pp. 757–780 *in*: J.H. Thorp and D.C. Rogers (eds.), Vol. I: Ecology and General Biology, of Thorp and Covich's Freshwater Invertebrates, 4th edition, Academic Press, Elsevier, Boston.

Suárez-Morales, E. 2015. Class Maxillopoda. Chapter 29, pp. 709–755 *in*: J.H. Thorp and D.C. Rogers (eds.), Vol. I: Ecology and General Biology, *in*: Thorp and Covich's Freshwater Invertebrates, 4th edition, Academic Press, Elsevier, Boston.

Taberlet, P., E. Coissac, F. Pompanon, C. Brochmann, and E. Willerslev. 2012. Towards nextgeneration biodiversity assessment using DNA metabarcoding. Molecular Ecology 21: 2045–2050.

Tamone, S.L. and J.F. Harrison 2015. Linking Insects with Crustacea: Physiology of the Pancrustacea: An Introduction to the Symposium. Integrative and Comparative Biology 55: 765–770.

Thorp, J.H. and B.J. O'Neill. 2015. Hexapoda - introduction to insects and Collembola. Chapter 33, pp. 849–871 *in*: J.H. Thorp and D.C. Rogers (eds.), Vol. I: Ecology and General Biology, *in*: Thorp and Covich's Freshwater Invertebrates, Academic Press, Elsevier, Boston.

Thorp, J.H. and D.C. Rogers (eds). 2015. Ecology and General Biology. Vol. I *in:* Thorp and Covich's Freshwater Invertebrates, 4th edition, Academic Press, Elsevier, Boston. 1118 p.

Tziortzis, I., S. Zogaris, A..Papatheodoulou, and F. Marrone. 2014. First record of the tadpole shrimp *Triops cancriformis* (Branchiopoda, Notostraca) in Cyprus. Limnetica 33: 341–348.

Valentini, A., F. Pompanon,, and P. Taberlet. 2009. DNA barcoding for ecologists. Trends in Ecology and Evolution 24: 110–117.

Van Syoc, R.J. 2019. Thecostraca: Subclass Sessilia. Chapter 16.5, pp. 782–787 in: D.C. Rogers and J.H. Thorp (eds.), Vol. IV: Keys to Palaearctic Fauna, *in*: Thorp and Covich's Freshwater Invertebrates, Academic Press, Elsevier, Boston.

Weigand, H., A.J. Beermann, F. Čiampor, F.O. Costa, Z. Csabai, S. Duarte, M.F. Geiger, M. Grabowski, F. Rimet, B. Rulik, M. Strand, N. Szucsich, A.M. Weigand, E. Willassen, S.A. Wyler, A. Bouchez, A. Borja, Z. Čiamporová-Zaťovičová, S. Ferreira, S.-D.B. Dijkstra, U. Eisendle, J. Freyhof, P. Gadawski, W. Graf, A. Haegerbaeumer, B.B. van der Hoorn, B. Japoshvili, L. Keresztes, E. Keskin, F. Leese, J.N. Macher, T. Mamos, G. Paz, V. Pešic, D.M. Pfannkuchen, M.A. Pfannkuchen, B.W. Price, B. Rinkevich, M.A.L. Teixeira, G. Várbíró, and T. Ekrem. 2019. DNA barcode reference libraries for the monitoring of aquatic biota in Europe: Gap-analysis and recommendations for future work. Science of the Total Environment 678: 499–524.

Yu, D.W., Y. Ji, Y., B.C. Emerson, X. Wang, C. Ye, C. Yang, and Z. Ding. 2012. Biodiversity soup: Metabarcoding of arthropods for rapid biodiversity assessment and biomonitoring. Methods in Ecology and Evolution 3: 613–623.

Zangl, L., D. Daill, S. Schweiger, G. Gassner, and S. Koblmüller. 2020. A reference DNA barcode library of Austrian amphibians and reptiles. Public Library of Science 15: e022935

Chapter 4

Class Branchiopoda

D. Christopher Rogers[1], Alain Thiéry[2] and Kay Van Damme[3]
[1]Kansas Biological Survey, and The Biodiversity Institute, The University of Kansas, Lawrence, KS, United States, [2]Aix-Marseille Université, Avignon Université (CNRS, IRD, IMBE UMR), Institut Méditerranéen de Biodiversité et d'Écologie marine et continentale, Biomarqueurs, Environnement, Santé (BES), Avignon, France, [3]Centre for Academic Heritage and Archives and Ghent University Botanical Garden, Ghent University, Ghent, Belgium

Introduction

Approximately 400 species of branchiopod crustaceans are found in Palearctic temporary and permanent aquatic habitats, from small seasonal wetlands to lakes, ponds and large rivers (Rogers et al., 2019) with a significant portion occurring in the Mediterranean Basin. Currently, 48 anostracans, 6 notostracans, and at least 8 Laevicaudata + Spinicaudata (groups that need revision) are reported from the Mediterranean Basin. Although cladocerans in the Mediterranean Basin have not been reviewed in the literature, approximately 145 taxa have been reported. The actual number is likely higher because of the presence of several unrevised species groups. Most branchiopods are planktonic or semibenthic and are often among the most abundant invertebrates in those habitats. Branchiopods are often used as indicators of aquatic ecosystem health, which ultimately reflects on land use practices and management (Rogers, 2009). Two invasive species of large branchiopods have been introduced (Muñoz et al., 2014) along with several cladocerans (Kotov et al., 2022). Much of this chapter is derivative of Rogers et al. (2019).

Limitations

Within the Mediterranean Basin, the large Branchiopoda comprises three orders: Anostraca with six families, Notostraca with one family, and Diplostraca. The specific taxonomic hierarchical terminology within the Diplostraca is unresolved. Workers on the higher systematics of the Branchiopoda recognize four primary clades: Laevicaudata, Spinicaudata, Cyclestherida, and Cladocera. These have been generally treated as suborders. However, within Cladocera are four clades that cladoceran workers treat as orders: Ctenopoda, Anomopoda, Haplopoda, and Onychopoda (Van Damme et al., 2022). The Diplostracan suborders Laevicaudata (one family), Spinicaudata (four families), and Cyclestherida (one family) are the clam shrimp. The remaining cladoceran groups are the water fleas. Cyclestherida, a circumtropical species complex, is not reported from the Mediterranean Basin.

The Mediterranean anostracans are reasonably well known in Southern Europe (Spain, Portugal, France, Italy, Balkans) and North Africa (Morocco, Algeria, Tunisia, Libya). The anostracan keys presented here are based primarily on male characters, but females can also be identified for some taxa. The anostracan genus *Chirocephalus* needs to be revised (Rogers & Soufi, 2013; Rogers et al., 2021). New species are discovered every few years and some older species may not be valid (Rogers, 2013).

The notostracans are represented by two genera, *Lepidurus* and *Triops*, both in need of revision; although great strides have been made in *Triops* (e.g., Korn et al., 2006; 2010; Korn & Hundsdoerfer, 2016), there are still too many undescribed taxa to make a useful identification key here. The laevicaudatan clam shrimp of the region are easily identified (Rogers & Olesen, 2016).

The spinicaudatan clam shrimp have been recently revised at higher taxonomic levels, but the species are more difficult to identify, and most genera need revision, especially in northern Africa and western Asia (Rogers, 2020; Schwentner et al., 2018, 2020).

Cladocera keys are primarily based upon adult females, which are readily identified to genus. Species level identifications typically require the use of a compound microscope, and many genera need to be revised or have incompletely

Identification and Ecology of Freshwater Arthropods in the Mediterranean Basin. DOI: https://doi.org/10.1016/B978-0-12-821844-0.00010-7

known species. The Cladocera are difficult to identify to species level, and there are no modern revisions for the Mediterranean Basin. The keys below are compiled from various sources; but for the cladoceran species complexes, which are under constant revision, we strongly recommend using specialized literature and the most recent taxonomical updates. The cladoceran keys are based on the morphology of adult parthenogenetic females (which are most common in nature).

Chydoridae is the most speciose branchiopod lineage, with many genera (e.g., *Alona*) still needing revision or are currently being revised. Several species groups harbor cryptic taxa, with some that cannot be distinguished without molecular data (e.g., *Chydorus sphaericus*-group). In such cases, the key indicates the species group as the identifiable unit, and specialized literature should be consulted for more recent (molecular) updates on their taxonomy. *Chydorus* species are difficult to identify, not well studied in the Mediterranean Basin, and more species are likely to be present. A few temperate Palearctic species are known here, with most being extremely rare. Only one Mediterranean species is endemic in Spain. For all *Chydorus*, the morphology and recent taxonomic updates much be checked in detail. Potential exotics (e.g., American *Chydorus brevilabris*) that have been reported from Europe can be expected here. *Chydorus* thermophilic species complexes that are widespread in (sub)tropical Africa and Asia could be also expected here (e.g., *C. eurynotus*-, *C. pubescens*-, and *C. parvus*-groups), but they have not yet been recorded (for morphology, see Smirnov, 1996). Records of *Chydorus latus* (Sars, 1862) in Mediterranean Basin (e.g., with one headpore in Italy) remain unresolved, this species was not included in our key (for drawings, see Flößner, 2000). *Alona* has been split into many genera, and substantial revision is continuing. The "core" lineage in *Alona* is the *Alona quadrangularis*-group. It is necessary to consult specialized and recent literature for all species that remain in this genus. A general checklist of names in *Alona* is provided by Van Damme et al. (2010), and several groups have since been revised. Because of the instability of the genus and no general consensus on some taxa, not all updates in *Alona* systematics are included in the key. For example, we did not include "*Flavalona*" and kept "*A.*" *costata* and "*A.*" *rustica* for now, as this move cannot be accepted until a more stable classification solution is found. Also "*Alona*" *intermedia* is kept separate in the key. Dissection of fine structures (limbs) is extremely important for separating all *Alona*-like genera; however, the following key is made as practical as possible.

Terminology and morphology

Palearctic branchiopods range in size from less than a millimeter to 10 cm. Adult branchiopods retain the reduced first antennae of the larvae. In notostracans the second antennae are greatly reduced. Diplostracan second antennae are typically branched, with anterior and posterior rami, and are used for propulsion. The anostracan second antenna is made of two antennomeres (proximal and distal). Anostracan second antennae are robust, and in males may have prehensile or cheliform antennal appendages (sometimes fused medially into a central cephalic appendage). Posteriolateral basal projections are called apophyses, while anteriomedial spiny patches are called pulvilli. In *Streptocephalus*, the antennal appendages are cheliform, divided into two rami which are referred to as a "thumb" and "finger" (Fig. 4.1K). The "thumb" is typically longer than the finger and may have a posterior spur. The "finger" may have one or more "teeth" on the anterior margin.

The head is proportioned to the body in anostracans, notostracans, and spinicaudatans, but is massive in the laevicaudatans and may be larger than the rest of the animal (Rogers & Olesen, 2016). A rostrum is absent in anostracans and notostracans.

The nauplii have a single median eye, which is reduced in most adult large branchiopods; but in the Notostraca, it becomes a large tubercle on the nuchal organ behind the eyes. The adult compound eyes are fused in laevicaudatans and diplostracans but are separate and borne on stalks in anostracans. In Cladocera, the naupliar eye may be posterior to the compound eye, and some cave-dwelling cladocerans lack eyes entirely. The nuchal organ lies on the dorsal surface of the head, although in most genera of the spinicaudatan family Limnadiidae it is borne on a short stalk.

Cladoceran sexes are morphologically similar, with small differences in appendages and body shape. Because most cladocerans reproduce asexually at least part of the time, most individuals will be females.

Sampling, preparation, and preservation

Detailed instructions regarding the collection and culture of branchiopods can be found in Martin et al. (2016) and Van Damme & Dumont (2010).

In the field, large branchiopods can be collected using hand nets, with a mesh of 1 mm for adults, and 250 μm for nauplii and juveniles. A net with 125 μm mesh will retain eggs and most identifiable juveniles. All branchiopods can be

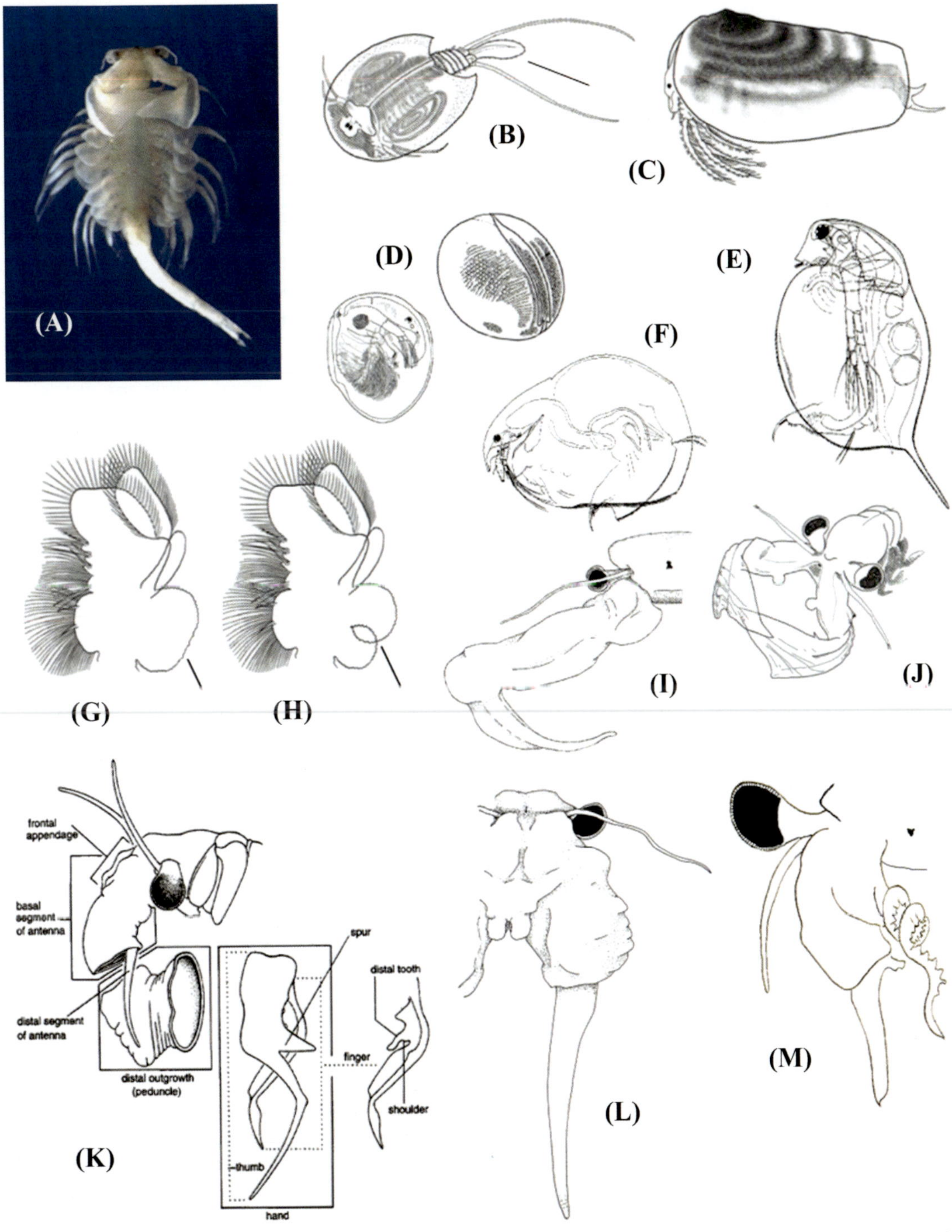

FIGURE 4.1 (A) *Artemia* sp. (B) *Lepidurus* sp. (line points to caudal lamina). (C) *Lepestheria* sp. (D) *Lynceus* sp. (E) *Daphnia* sp. (F) Euryceridae. (G) *Branchinecta* sp., left thoracopod V. (H) *Linderiella* sp., left thoracopod V. (I) *Branchinecta ferox*, male head, right half, anterior view. (J) *Artemia franciscana*, male head, left oblique view. (K) *Streptocephalus* sp. generalized male head, left lateral exploded view. (L) *Phallocryptus spinosus*, male head, right half, anterior view. (M) *Chirocephalus recticornis*, male head, right half, anterior view.

preserved in ethyl alcohol long term, and for purposes of identification can be manipulated in alcohol. Dissection is seldom necessary for large branchiopods, and it is recommended to have several animals available for identification. In the field, formaldehyde (4%–8% v/v) can be used to conserve the colors in living organisms, which will disappear in alcohol. Moreover, this formaldehyde solution prevents the loss of mass in the case of biomass studies.

Cladoceran collection mostly requires conical zooplankton dragnets (pulled by a rope) of 100–200 μm mesh size, which is necessary to collect the smallest species and all life stages. An extra, small hand net with the same mesh size is useful for going through plants (not practical with a conical dragnet) to study the diverse littoral taxa, and for (epi-)benthic species. Cladocerans sometimes require dissection for detailed species identification (in particular the Chydoridae), and a careful manipulation is needed to observe structures that are difficult to find (e.g., head pores, limbs, or fine structures on the postabdomen). The dissection is carried out under a stereomicroscope using tungsten needles and on a glass slide for detailed examination. For cladocerans in particular, it is useful to have many specimens because these small delicate animals are easily crushed. Cladocerans smaller than about 1 mm should be slide mounted for viewing under a compound microscope, preferably in a mixture of glycerol-ethanol and formaldehyde (4%). When transferring these animals from ethanol samples, it is important to avoid destroying the specimens by adding too much glycerol (which will shrivel them and structures will not be observable). The best cladoceran mounting medium is one which is more or less permanent and soluble in both water and alcohol solutions so specimens can be added directly to the medium without tedious dehydration. If you do not have access to chloral hydrate (currently a controlled substance), Hoyer's may be another option:

> Dissolve 30 g of ground Gum Arabic in 170 mL of hot distilled water. Add 200 g of chloral hydrate and 20 g of glycerine. Divide the Hoyer's into two batches. Let one sit in a warm place until it is as thick as honey, and keep the second batch in a closed jar so it remains thin. You can always add more water to thin either batch. Keep at least some of the thick and thin Hoyer's in screw-top eye dropper bottles.

To prepare a slide, use the eye dropper to put a small streak of dilute Hoyer's (about a quarter of a drop) toward one end of the slide. Arrange the specimens in this streak. It is wise to place four or so of what you think are the same kind of animal on a slide to allow for comparison and viewing of difficult characters. Delicate specimens can be protected from compression by putting a few pieces of broken coverslips around the specimens. Let the Hoyer's streak dry on a slide warmer or in a warm place (below 100°C). When the streak is dry, add a drop or two of the thick Hoyer's and gently lower a cover glass over the Hoyer's. Put the slide back into a warm place to dry again. Label the slide as to the date and location of the collection. It is important to use thickened Hoyer's in the last step; otherwise, the Hoyer's will shrink as it dries and produce large bubbles in the final preparation. However, if you use thickened Hoyer's, any small bubbles produced will disappear as the medium dries. These slides are semipermanent. If the climate is humid, the Hoyer's will thin and cover glasses will slip off the slide. If the climate is arid, the Hoyer's will desiccate enough so that the cover glasses pop off the slide. These problems may be avoided by storing the slides horizontally in a climate that has moderate humidity, and by ringing the slide with clear nail polish.

An alternate mounting method, especially useful for bosminids, chydorids, and macrothricids, is to use polyvinyl lactophenol stained with lignin pink. Temporary slides can be made using glycerol or glycerin jelly. Best of all for museum specimens is possibly Canada Balsam, although this requires the specimens to be dehydrated through an alcohol series. As cladocerans may become quite transparent, specimens or structures are sometimes colored with Bengal Red.

If specimens are required for DNA analysis, preservation in 95%–99% pure ethanol is necessary; high temperatures in the field can degrade the limited amounts of DNA in cladocerans rapidly, so keeping the samples in an ice chest upon collection is useful. The material should be stored as soon as possible at 4°C (if the DNA is not to be extracted within the next few months). It is highly advised to refresh 95%–99% ethanol within 24 h of sampling to ensure a good preservation of the DNA.

Branchiopods may be also reared from dried mud; resting eggs and ephippia can present species or genus characteristics. In the large branchiopods, only certain anostracan and spinicaudatans have useful characters on the egg shell (spines, ridges, other protrusions) for separating certain genera and species. For identification of cladoceran ephippia (which only occur in the Anomopoda) or other cladoceran resting eggs (e.g., Ctenopoda, Onychopoda), no specific literature is available for the Mediterranean Basin. Due to the importance of ephippia in paleolimnological reconstructions, some literature is available for several general Palearctic taxa (e.g., Late Caenozoic ephippia of Daphniidae; Kotov et al., 2019), yet identification is not straightforward and remains complex (Vandekerkhove et al., 2004). Hatching may be facilitated with distilled water and constant light in the first 24 h, as the mechanical aperture of the egg envelope is osmotically dependent. Detailed instructions regarding the collection and culture of branchiopods can be found in Martin et al. (2016) and Van Damme & Dumont (2010).

Keys to Branchiopoda

The keys to large branchiopods are derivative from Rogers et al. (2019).

Branchiopoda: Orders

1 Compound eyes sessile; carapace present or absent (Fig. 4.1B–F and K–M) 2
1' Compound eyes on stalks separated; carapace absent (Fig. 4.1A) .... **Anostraca**
2(1) Carapace broad, never folded or bivalved; eyes projected dorsally through carapace (Fig. 4.1B) .... Notostraca, one family: **Triopsidae**
2' Carapace bivalved, folded or reduced; eyes not projecting dorsally through carapace (Fig. 4.1C–F) .... **"Diplostraca"**

Branchiopoda: Anostraca: Families

1 Praeepipodites not cleft or divided (Fig. 4.1G) 2
1' Praeepipodites (Fig. 4.1H) obviously cleft, or entirely divided in two .... **Chirocephalidae**
2(1) Second antennal distal antennomere not broadly triangular; brood pouch without large pair of spines (Fig. 4.1I, L, and M) 3
2' Second antennal distal antennomere broadly triangular, apex acute (Fig. 4.1A and J); brood pouch with a ventral pair of large, recurved spines, halophiles, widespread .... Artemiidae, one genus: ***Artemia***
3(2) Gonopod rigid basal portions parallel, directed ventrally, with the bases closely united (Fig. 4.2B); male antennal and or frontal appendages present or not; female second antennae lamellar, two to three times as broad as thick 4
3' Gonopods directed ventrolaterally, widely separated at the base (Fig. 4.2A); male frontal and antennal appendages always absent, although rigid projections may be present; female second antennae subcylindrical in cross section; widespread .... Branchinectidae, one genus: ***Branchinecta***
4(3) Male second antennal proximal antennomere with or without appendages, if present, then appendages cylindrical or lamellar, but never cheliform; second antennae and appendages not laterally compressed (Figs. 4.1I, L, M, and 4.2F) 5
4' Male second antennal proximal antennomere terminating in a large, cheliform, multiramal appendage directed ventrally (Fig. 4.1K); second antennae and appendages laterally compressed; widespread .... Streptocephalidae, one genus: ***Streptocephalus***
5(4) Male second antennae with proximal antennomeres fused together into a rigid clypeus; clypeus soft proximally, and chitinized distally; *vas deferens* always without a dorsal loop 6
5' Male second antennae with proximal antennomeres fused basomedially, or medially coalescent, distal portions always free (Fig. 4.1L); *vas deferens* always with a dorsal loop (usually visible through the integument); male abdominal segments with long ventral spines; frontal appendage shorter than second antennae, bilobed (Fig. 4.1L) genus in need of revision.......... Thamnocephalidae, one genus: *Phallocryptus*
6(5) Rigid portion of gonopod with a basoventral, lamellar, subtriangular projection, with one or more spines along medial margin, shorter than apical spine; greatest width of eversible portion of gonopod 0.5–1.2 times the length; gonopod with one short, medial row of spines ending about one-third the length of the gonopod from the apex (Fig. 4.2B) .... **Tanymastigidae**
6' Rigid portion of gonopod without a basomedial projection, or if projection is present, then bearing one spine, shorter than the apical spine, on the lateral margin; greatest width of eversible portion of gonopod 0.25 or less the length; gonopod with one or more longitudinal rows of spines (Fig. 4.2C) .... Branchipodidae, one genus: ***Branchipus***

Branchiopoda: Anostraca: Chirocephalidae: Genera

1 Cercopods articulating at base, lateral, and medial margins edged in plumose setae (Fig. 4.3A); male second antennal proximal antennomere with anteriobasal appendages; everted portion of gonopod subcylindrical to spiniform (Fig. 4.2I) 2
1' Cercopods not articulating, rigid, short with a few ventrolateral separated setae (Fig. 4.2D); male second antennal proximal antennomere without anteriobasal appendages; everted portion of gonopods a broad, flattened process with distal margin crenulate .... *Branchinectella media* (Schmankewitsch, 1873)
[Halophilic; Mediterranean Basin, Palearctic]
2(1) Second antennal distal antennomere length $\leq$ proximal antennomere (Figs. 4.1M and 4.2H) .... ***Chirocephalus***
2' Second antennal appendage distal antennomere length $> 1.3 \times$ proximal antennomere (Fig. 4.2E and F) .... ***Linderiella***

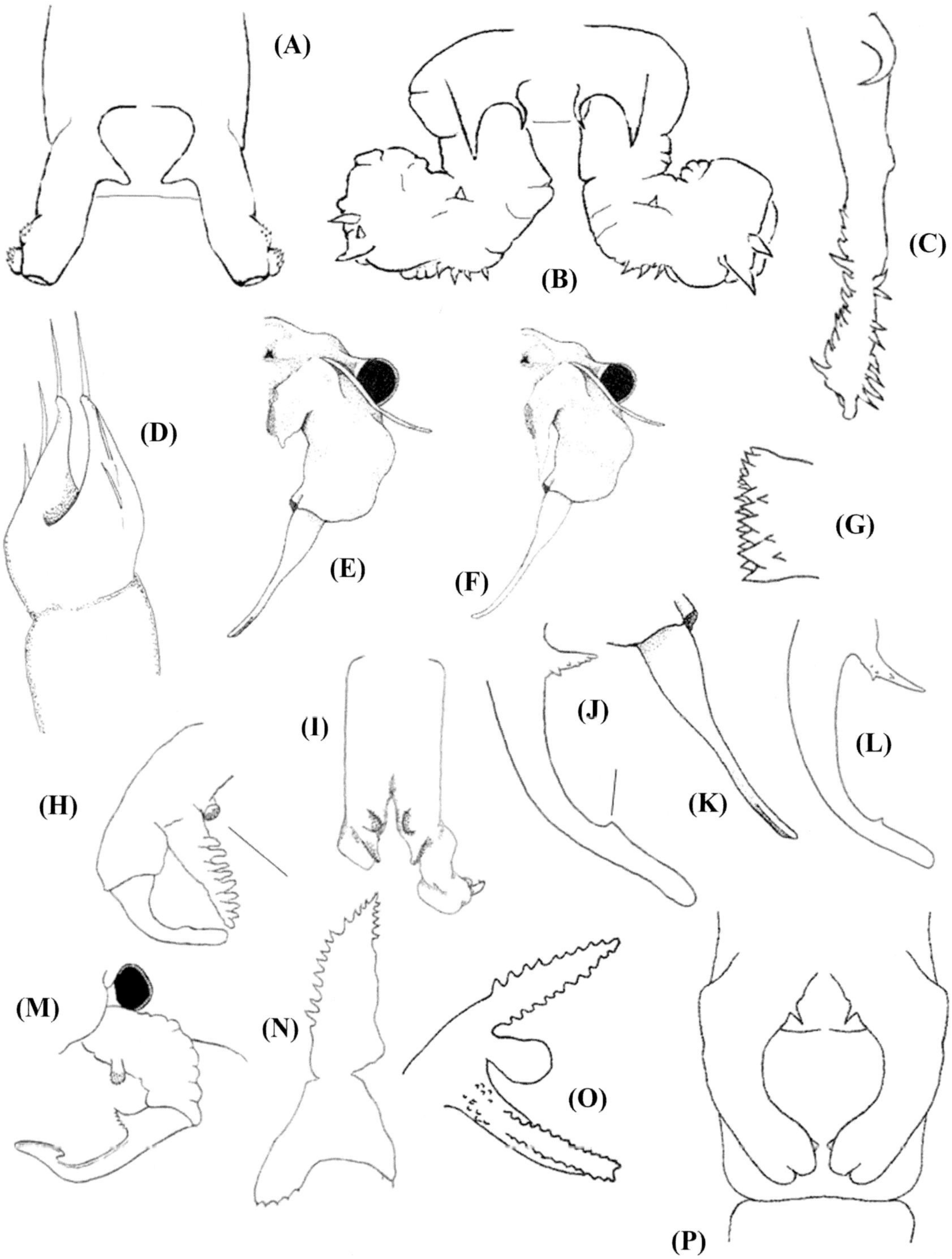

FIGURE 4.2 (A) *Branchinecta* sp. gonopods, ventral view. (B) *Tanymastix stagnalis* gonopods, ventral view. (C) *Branchipus schaefferi*, right gonopod, lateral view. (D) *Branchinectella media* cercopods, right oblique view. (E) *Linderiella massaliensis*, male head, left side, anterior view. (F) *Linderiella jebalae*, male head, left side, anterior view. (G) *Linderiella africana*, male left antennal appendage, anterior view. (H) *Chirocephalus shadini*, male right second antenna (line points to apophysis). (I) *Chirocephalus* sp. male genitalia, ventral view. (J) *Chirocephalus brteki*, male left second antenna, distal antennomere, anterior view (line points to subapical medial projection). (K) *Chirocephalus marchesonii*, male right second antenna, distal antennomere, anterior view. (L) *Chirocephalus vornatscheri*, male left second antenna, distal antennomere, anterior view. (M) *Chirocephalus bairdi*, male head, left side anterior view, antennal appendage removed. (N) *Chirocephalus bairdi*, right antennal appendage. (O) *Chirocephalus anatolicus*, right antennal appendage. (P) *Artemia franciscana*, gonopods, ventral view.

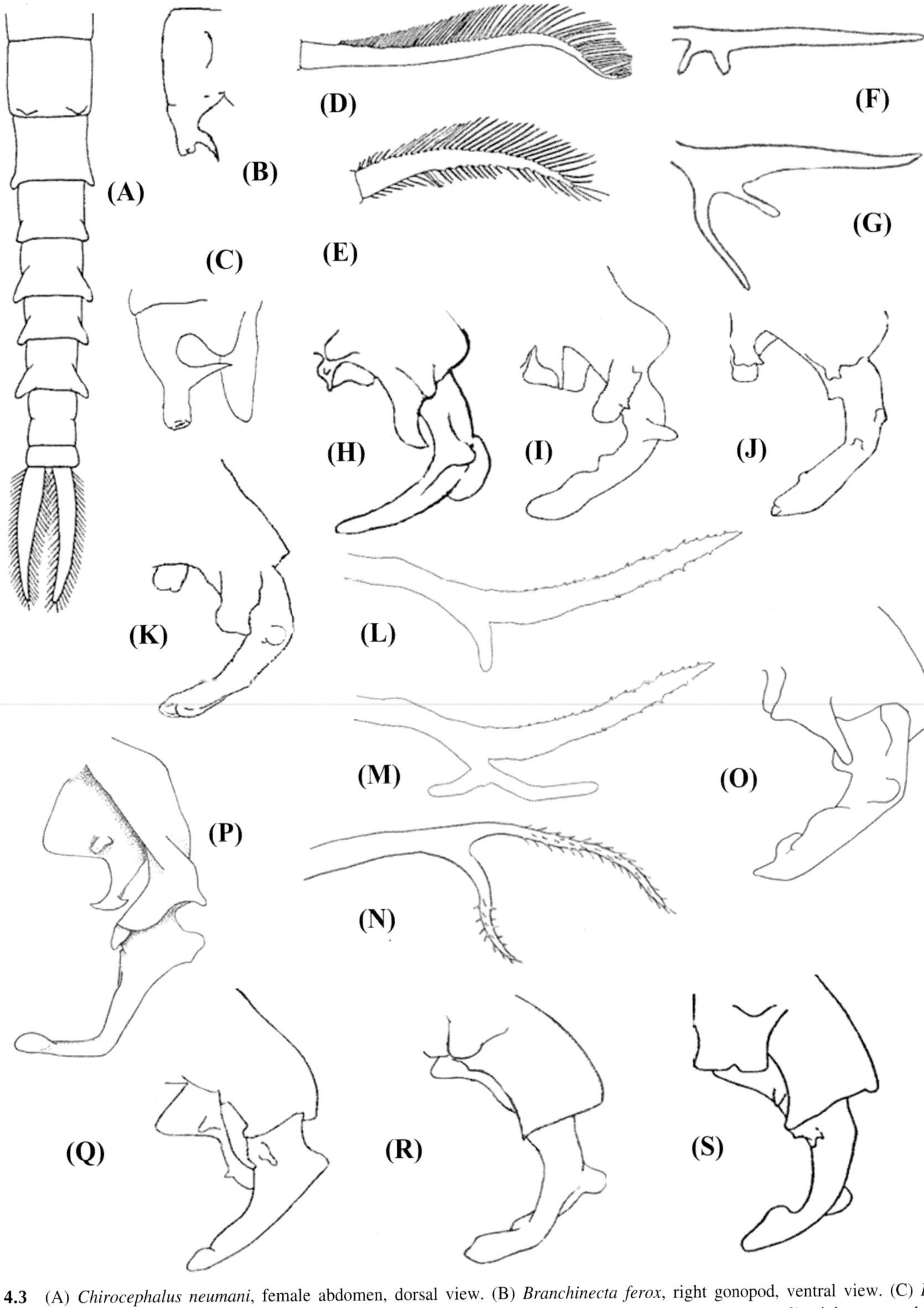

FIGURE 4.3 (A) *Chirocephalus neumani*, female abdomen, dorsal view. (B) *Branchinecta ferox*, right gonopod, ventral view. (C) *Branchinecta orientalis*, right gonopod, ventral view. (D) *Branchinecta ferox*, right cercopod, dorsal view. (E) *Branchinecta orientalis*, right cercopod, dorsal view. (F) *Tanymastix motasi*, antennal appendage. (G) *Tanymastix affinis*, antennal appendage. (H) *Tanymastix affinis*, left second antenna, anterior view. (I) *Tanymastix stagnalis*, left second antenna, anterior view. (J) *Tanymastix motasi*, left second antenna, anterior view. (K) *Tanymastix stellae*, left second antenna, anterior view. (L) *Tanymastigites lusitanica*, antennal appendage. (M) *Tanymastigites mzabica*, antennal appendage. (N) *Tanymastigites ajjeri*, antennal appendage. (O) *Tanymastigites lusitanica*, second antenna. (P) *Tanymastigites ajjeri*, second antenna. (Q) *Tanymastigites mzabica*, left second antenna, anterior view. (R) *Tanymastigites brteki*, left second antenna, anterior view. (S) *Tanymastigites cyrenaica*, left second antenna, anterior view.

Branchiopoda: Anostraca: Chirocephalidae: *Linderiella*: Species

1 Male antennal appendage apex pedunculate and asymmetrical, projecting distally (Fig. 4.2E and F) 2
1' Male antennal appendage pedunculate, symmetrical (Fig. 4.2G) .... *Linderiella africana* Thiéry, 1986 [Morocco]
2(1) Male second antenna, distal antennomere arcing in proximal third (Fig. 4.2E) 3
2' Male second antenna, distal antennomere arcing in distal third; antennal appendage distal projection long, tapering, without apical spines (Fig. 4.2F) .... *Linderiella jebalae* Boix et al., 2016 [Morocco]
3(2) Egg spines distally expanded .... *Linderiella baetica* Alonso & Garcia-de-Lomas, 2009 [Spain]
3' Egg spines acute .... *Linderiella massaliensis* Thiéry & Champeau, 1988 [France]

Branchiopoda: Anostraca: Chirocephalidae: *Chirocephalus*: Species

Note: The genus *Chirocephalus* needs to be revised. *Chirocephalus festae* Colosi, 1922 is known only from the type locality in Cyrenaika, Libya and is probably not a valid taxon.

1 Apophyses rounded apically (Fig. 4.2H) 2
1' Apophyses acute .... *Chirocephalus croaticus* Steuer, 1899 [Croatia]
2(1) Apophyses subconical, cylindrical, or broadly expanded (Fig. 4.2H and M) 3
2' Apophyses subglobular; antennal appendage with one lamella; antennal appendage longer than second antenna (Fig. 4.1M) .... *Chirocephalus recticornis* Brauer, 1877 [Tunisia]
3(2) Second antennae distal antennomere with a basal projection (Fig. 4.2J and L) 4
3' Second antennae distal antennomere without a basal projection (Fig. 4.2K) 20
4(3) Second antenna distal antennomere excised medially (Fig. 4.2J and M). 5
4' Second antenna distal antennomere not excised medially (Fig. 4.2L) 9
5(4) Second antenna distal antennomere arcing medially 70 degrees or more (Fig. 4.2J) 6
5' Second antennae distal antennomere arcing medially 45 degrees or less (Fig. 18.4J and M) 7
6(5) Second antenna distal antennomere with an acute or subacute, subapical, medial projection (Fig. 4.2J) .... *Chirocephalus brteki* Cottarelli, Aygen & Mura, 2010 [Turkey]
6' Second antennae distal antennomere lacking a subapical projection .... *Chirocephalus kerkyrensis* Pesta, 1936 [Greece, Italy]
7(5) Antennal appendage lower lamella divided (Fig. 4.2O) 8
7'Antennal appendage lower lamella entire (Fig. 4.2M and N) .... *Chirocephalus bairdi* (Brauer, 1877) [Israel, Turkey]
8(7) Second antenna distal antennomere proximomedial branch smooth, acute .... *Chirocephalus anatolicus* Cottarelli, Mura, & Özkütük, 2007 [Turkey]
8' Second antenna distal antennomere proximomedial branch spinose .... *Chirocephalus murae* Brtek & Cottarelli, 2006 [Turkey]
9(4) Second antenna distal antennomere with a short subapical projection 10
9' Second antenna distal antennomere lacking a short subapical projection (Fig. 4.2L) 11
10(9) Second antennae distal antennomere with basal projection spiniform, with several small conical projections along one margin .... *Chirocephalus vornatscheri* Brtek, 1968 [Bulgaria, Turkey]
10' Second antennae distal antennomere with a basal projection broad, terminating in 5–8 subequal spiniform projections .... *Chirocephalus sarpedonis* Cottarelli, Mura, Ippolito, & Marrone, 2017 [Turkey]
11(9) Second antenna distal antennomere longer than proximal antennomere 12

11' Second antenna distal antennomere subequal to or shorter than proximal antennomere 16
12(11) Female abdomen with lateral projections (Fig. 4.3A) 13
12' Female abdomen smooth; male antennal appendage upper lamella apex triangular, with subtending marginal papillae .... *Chirocephalus algidus* Cottarelli, Aygen & Mura, 2010
[Turkey]
13(12) Female abdominal lateral projections rounded (Fig. 4.3A) 14
13' Female abdominal lateral projections acute; male second antennal distal antennomere basomedial projection recurved apically .... *Chirocephalus paphlogonicus* Cottarelli, 1971
[Turkey]
14(13) Female without dorsal projection on genital segments 15
14' Female with dorsal projection on genital segments .... *Chirocephalus diaphanus* Prévost, 1803
[Western Palearctic]
15(14) Female with brood pouch smooth .... *Chirocephalus tauricus* Pesta, 1921
[Turkey]
15' brood pouch with lateral projections .... *Chirocephalus neumanni* Hartland-Rowe, 1967
[Israel, Iran]
16(11) Second antenna distal antennomere with a strong proximal bend medially, distally straight 17
16' Second antenna distal antennomere entirely straight or sinuate ....
.... *Chirocephalus sibyllae* Cottarelli & Mura, 1975
[Italy]
17(16) Second antenna distal antennomere subequal in length to proximal 18
17' Second antenna distal antennomere shorter than proximal 19
18(17) Antennal lower lamina with posterior margin broadly and evenly rounded ....
.... *Chirocephalus sanhadjaensis* Boumendjel, Rabet & Amarouayache, 2018
[Algeria]
18' Antennal lower lamina with posterior margin unevenly crenulate ....
.... *Chirocephalus cupreus* Cottarelli, Mura, & Özkütük, 2007
[Turkey]
19(17) Antennal appendage lower lamella broadly rounded medially ....
.... *Chirocephalus appendicularis* Vavra, 1905
[Lebanon, Syria, Turkey]
19' Antennal appendage lower lamella broadly triangular ....
.... *Chirocephalus salinus* Daday, 1910/*Chirocephalus reiseri* Marcus, 1913
[Bosnia & Herzegovina, France, Italy]
20(3) Antennal appendage length $\geq$ second antennae 21
20' Antennal appendage length $<$ second antennae .... *Chirocephalus ruffoi* Cottarelli & Mura, 1984
[Italy]
21(20) Second antennal distal antennomere with apex directed distally ....
.... *Chirocephalus ponticus* Beladjal & Mertens, 1997
[Turkey]
21' Second antennal distal antennomere with apex directed laterally (Fig. 4.2K); antennal appendage with marginal papillae longer on medial margin .... *Chirocephalus marchesonii* Ruffo & Vesentini, 1957
[Italy]

Branchiopoda: Anostraca: Artemiidae: *Artemia*: Species

Artemia contains the brine shrimps, all of which are halophiles (Fig. 4.1A). Most morphological characters are plastic, and wild collected animals may vary greatly, depending upon environmental conditions. Furthermore, there are several independent parthenogenic strains that have arisen from different species and are not generally separable without genetic analyses. Molecular analysis is the only really reliable tool for separating *Artemia* species. Two species are invasive in the Mediterranean Basin (Munoz et al., 2014).

1 Gonopods with basolateral spiniform projections (Fig. 4.2P) 2
1' Gonopods without basolateral spiniform projections; egg diameter less than 250 μm; male second antennae with apophyses subconical .... *Artemia salina* (L., 1758)
[Western and central Palearctic. Afrotropical]

2(1) Brood pouch with a ventral pair of large spines; male apophyses with microspinulae (need compound microscope) generally in pairs or triplets *Artemia franciscana* (Kellogg, 1906)
[Invasive Mediterranean Basin. Nearctic, Neotropical]
2' Brood pouch without a ventral pair of large spines; male apophyses with microspinulae (need compound microscope) single, well separated .... *Artemia persimilis* Piccinelli & Prosdocimi, 1968
[Invasive in Italy, but may no longer be extant. Neotropical]

Branchiopoda: Anostraca: Branchinectidae: *Branchinecta*: Species

1 Gonopod with proximal lobe not overreaching the ventrolateral spine (Fig. 4.3B); cercopods with plumose setae on medial margins only (Fig. 4.3D); labrum apex inerm .... *Branchinecta ferox* Milne-Edwards, 1840
[Central and western Palearctic]
1' Gonopod with proximal lobe overreaching the ventrolateral spine (Fig. 4.3C); cercopods with plumose setae on medial and lateral margins (Fig. 4.3E); labrum apex conical, covered with fine setulae ....
.... *Branchinecta orientalis* Sars, 1901
[Widespread]

Branchiopoda: Anostraca: Streptocephalidae: *Streptocephalus*: Species

The terminology of antennal appendages is presented in Fig. 4.1K.

1 Antennal appendage with "finger" roughly one third the length of the "thumb"; "thumb" with spines all along anterior margin .... *Streptocephalus rubricaudatus* (Klunzinger, 1867)
[Egypt]
1' Antennal appendage with "finger" and "thumb" subequal in length; "thumb" with few spines in a proximal group on anterior margin .... *Streptocephalus torvicornis* (Waga, 1842)
[Mediterranean Basin, Eastern Europe, Ponto Caspian Region]

Branchiopoda: Anostraca: Tanymastigidae: Genera

1 Gonopods each with a ventral lamellar projection with distomedial edge not margined with spines, but may have a single medial spine (Fig. 4.2B) .... ***Tanymastix***
1' Gonopods each with a ventral lamellar projection with distomedial edge margined with spines .... ***Tanymastigites***

Branchiopoda: Anostraca: *Tanymastix*: Species

1 Frontal appendage with first basolateral subbranch longer than second subbranch (Fig. 4.3G) 2
1' Frontal appendage with first basolateral subbranch shorter than or equal in length to second subbranch (Fig. 4.3F) 3
2(1) Second antennal proximal antennomere with ventrodistal projection falcate (Fig. 4.3H) ....
.... *Tanymastix affinis* Daday, 1910
[Morocco]
2' Second antennal proximal antennomere with ventrodistal projection broadly rounded with a digitiform lateral projection (Fig. 4.3I) .... *Tanymastix stagnalis* (L., 1758)
[Europe, North Africa]
3(1) Second antennal proximal antennomere with ventrodistal projection bearing angular, projecting lateral corners (Fig. 4.3J) .... *Tanymastix motasi* Orghidan, 1945
[Macedonia, Romania]
3' Second antennal proximal antennomere with ventrodistal projection smooth (Fig. 4.3K) ....
.... *Tanymastix stellae* Cottarelli, 1967
[France Corsica, Italy]

Branchiopoda: Anostraca: *Tanymastigites*: Species

From Thiéry & Rogers (2022).

1 Antennal appendage main branch, lateral ramus undivided (Fig. 4.3L) 2
1'Antennal appendage main branch, lateral ramus divided (Fig. 4.3M) 4
2(1) Antennal appendage main branch, lateral ramus length $\geq 0.4 \times$ medial ramus (Fig. 4.3N); second antenna distal antennomere without a ventrolateral ridge approximately at maximum point of curvature 3
2' Antennal appendage main branch, lateral ramus length $\leq 0.2 \times$ medial ramus (Fig. 4.3L); second antenna distal antennomere with a subquadrate ventrolateral ridge approximately at maximum point of curvature (Fig. 4.3O) .... *Tanymastigites lusitanica* Machado & Sala, 2013
[Portugal]
3(2) Second antenna proximal antennomere basally with an anterior lamellar process; second antenna proximal antennomere medial lamellar process lacking an anterior subconical projection; medial lamellar process distal lobe directed ventrally, narrowing and apically rounded; distal antennomere with a subapical, medial lobiform projection .... *Tanymastigites perrieri* (Daday, 1910)
[Northeast Africa]
3' Second antenna proximal antennomere basally without an anterior lamellar process; second antenna proximal antennomere medial lamellar process bearing a subconical projection; medial lamellar process distal lobe directed anterio-ventrally, apically spatulate; distal antennomere with rounded, lacking a projection (Fig. 4.3N and P) .... *Tanymastigites ajjeri* Thiéry & Rogers, 2022
[Libya]
4(1) Second antenna medially fused proximal antennomeres with anterior and distal projections (Fig. 4.3R and S) 5
4' Second antenna medially fused proximal antennomeres without anterior and distal projections (Fig. 4.3Q) .... *Tanymastigites mzabica* (Gauthier, 1928)
[Algeria]
5(4) Second antenna medially fused proximal antennomeres with anterior projections, low, inconspicuous, length < basal width; distal lamellar projections quadrate, with distal margin emarginate (Fig. 4.3S) .... *Tanymastigites cyrenaica* Brtek, 1972
[Libya, Saudi Arabia]
5' Second antenna medially fused proximal antennomeres with anterior projections prominent, conical, apically rounded, length $\geq$ basal width; distal lamellar projections with distal margins sinuate (Fig. 4.3R) .... *Tanymastigites brteki* Thiéry, 1986
[Morocco]

Branchiopoda: Anostraca: Branchipodidae: *Branchipus*: Species

1 Second antenna, distal antennomere lateral angle projecting (Fig. 4.4A, C, and D) 2
1' Second antenna, distal antennomere lateral angle rounded (Fig. 18.6B) .... *Branchipus laevicornis* Daday, 1912
[Turkey]
2(1) Second antenna, distal antennomere lateral projection flanged (Fig. 4.4C and D) 3
2' Second antenna, distal antennomere lateral projection digitiform (Fig. 4.4A) .... *Branchipus schaefferi* Fischer, 1834
[Northern Africa, temperate Europe, east to Pakistan, north to Russia]
3(2) Second antenna, distal antennomere lateral projection less than one-third the length of the antennomere (Fig. 4.4C) .... *Branchipus blanchardi* Daday, 1908
[France, Italy]
3' Second antenna, distal antennomere lateral projection one-half to one-third the length of the antennomere (Fig. 4.4D) .... *Branchipus cortesi* Alonso & Jaume, 1991
[Spain]

Branchiopoda: Notostraca: Triopsidae: Genera

One family: Triopsidae Kielhack, 1910. The notostracan genera both need revision. The presence of too many undescribed *Triops* species prevent us from presenting a useful identification key, although some good regional keys are available (e.g., Korn et al., 2006, 2010; Korn & Hundsdoerfer, 2016).

1 Telson without a caudal lamina projecting between cercopods .... *Triops*
1' Telson with a caudal lamina projecting between cercopods (Fig. 4.1 B) .... ***Lepidurus***

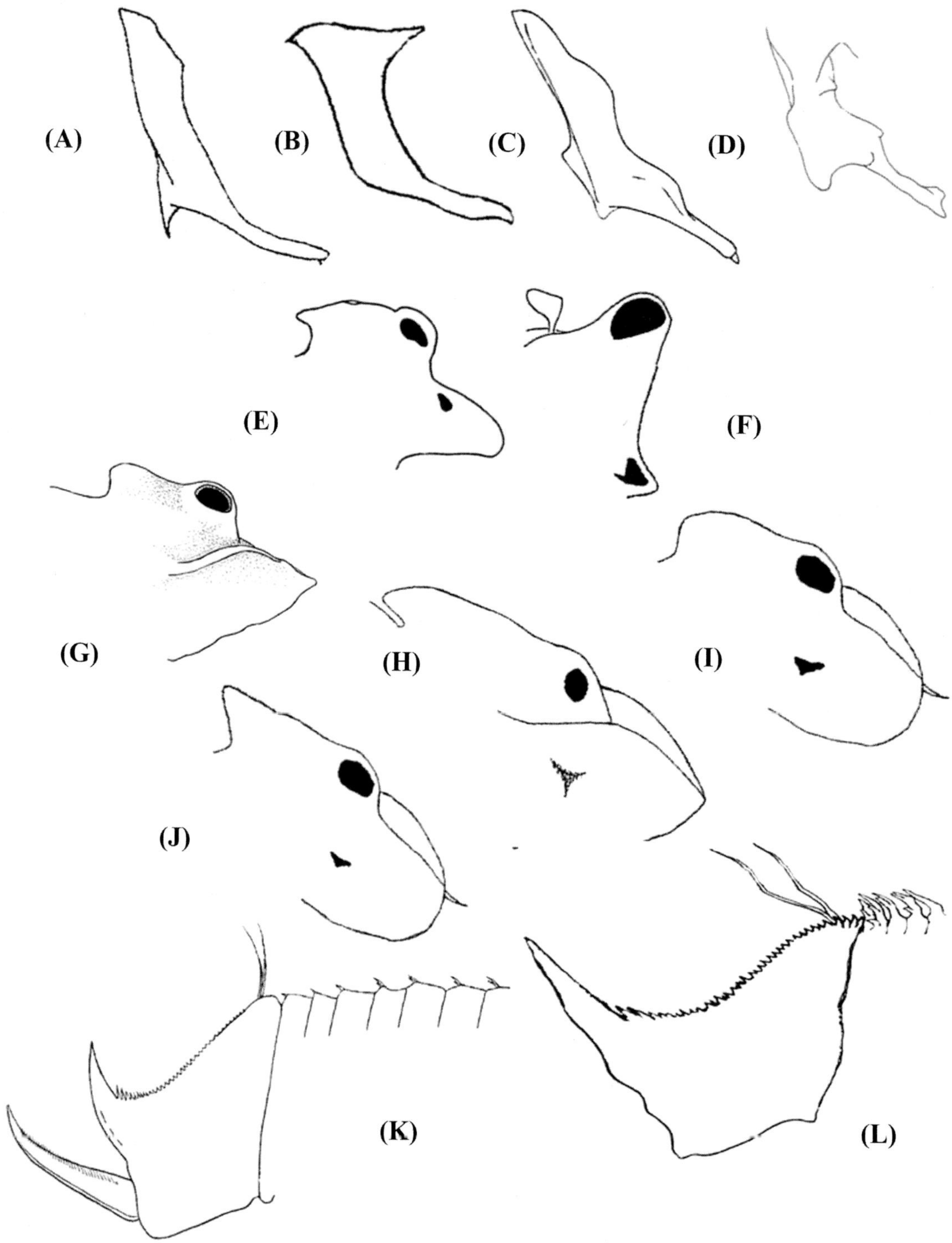

FIGURE 4.4 (A) *Branchipus schaefferi*, male right second antenna, distal antennomere, anterior view, antennal appendages absent. (B) *Branchipus laevicornis*, male right second antenna, distal antennomere, anterior view, antennal appendages absent. (C) *Branchipus blanchardi*, male right second antenna, distal antennomere, anterior view, antennal appendages absent. (D) *Branchipus cortesi*, male right second antenna, distal antennomere, anterior view, antennal appendages absent. (E) *Imnadia yeyetta*, head, lateral view. (F) *Limnadia lenticularis* head, lateral view. (G) *Eocyzicus* sp., head, lateral view. (H) *Cyzicus altus*, head, lateral view. (I) *Eoleptestheria* sp., head, lateral view. (J) *Leptestheria*, head, lateral view. (K) *Leptestheria*, telson and dorsal portion of abdomen, lateral view. (L) *Maghrebestheria maroccana*, telson and dorsal portion of abdomen, lateral view.

Branchiopoda: Notostraca: *Lepidurus*: Species

Adapted from Longhurst (1955).

1 Caudal lamina with 20–100 medial spines on a robust central keel .... *Lepidurus apus* L. 1758
[Europe]
1' Caudal lamina with three to 25 medial spines on a weak central keel .... *Lepidurus lubbocki* Brauer, 1873
[Mediterranean Basin]

Branchiopoda: "Diplostraca": Orders

1 Carapace folded, no true hinge present laterally flattened OR carapace reduced or absent (Fig. 4.1C, E, and F) 2
1' Carapace with a true, interlocking dorsal hinge; carapace subglobular, smooth, without growth lines, containing entire animal including head (Fig. 4.1D) .... Laevicaudata, one species: *Lynceus brachyurus* Müller, 1776
[Israel. Holarctic]
2(1) Carapace encompassing head, growth lines present, although may be faint (Fig. 4.1C) .... **Spinicaudata**
2' Carapace, if present, not encompassing head, growth lines absent (Fig. 4.1E and F) or rarely present "Cladocera" 3
3(2) Five or more thoracopods 4
3' Four thoracopods .... **Onychopoda**
4(3) Six similar thoracopods 5
4' Five or six differentiated thoracopods .... **Anomopoda**
5(4) Body not strongly elongate .... Ctenopoda, one family: **Sididae**
5' Body strongly elongate (Fig. 4.5A) .... Haplopoda, one species: *Leptodora kindtii* (Focke, 1844)
[Turkey. Palearctic]

Branchiopoda: Diplostraca: Spinicaudata: Families

Following Rogers (2020) and Schwentner et al. (2020).

1 Fornicies extend to rostrum (Fig. 4.4G–J) 2
1' Fornicies do not extend to rostrum (Fig. 4.4E and F) .... **Limnadiidae**
2(1) Swimming thoracopods lacking an anterior, medial, triangular lobe 3
2' Swimming thoracopods bearing an anterior, medial, triangular lobe .... **Leptestheriidae**
[This family is in need of revision]
3(2) Head with dorsoposterior deep cleft just anterad of carapace point of attachment (Fig. 4.4H) ....
.... Cyzicidae, one genus; *Cyzicus*
[Widely distributed, genus in need of revision]
3' Head lacking a dorsoposterior deep cleft (Fig. 4.4G) .... Eocyzicidae, one genus; *Eocyzicus*
[Widely distributed, genus in need of revision]

Branchiopoda: Diplostraca: Spinicaudata: Limnadiidae: Genera

Adapted from Rogers et al. (2012).

1 Frontal organ pedunculate (Fig. 4.4F) .... *Limnadia lenticularis* (L., 1761)
[Holarctic]
1' Frontal organ sessile (Fig. 4.4E) .... *Imnadia yeyetta* Hertzog, 1935
[Europe]

Branchiopoda: Diplostraca: Spinicaudata: Leptestheriidae: Genera

Adapted from Thiéry (1988).

1 Occiput acute (Fig. 4.4J) 2
1' Occiput rounded (Fig. 4.4I) .... *Eoleptestheria*
[Widely distributed, genus in need of revision]

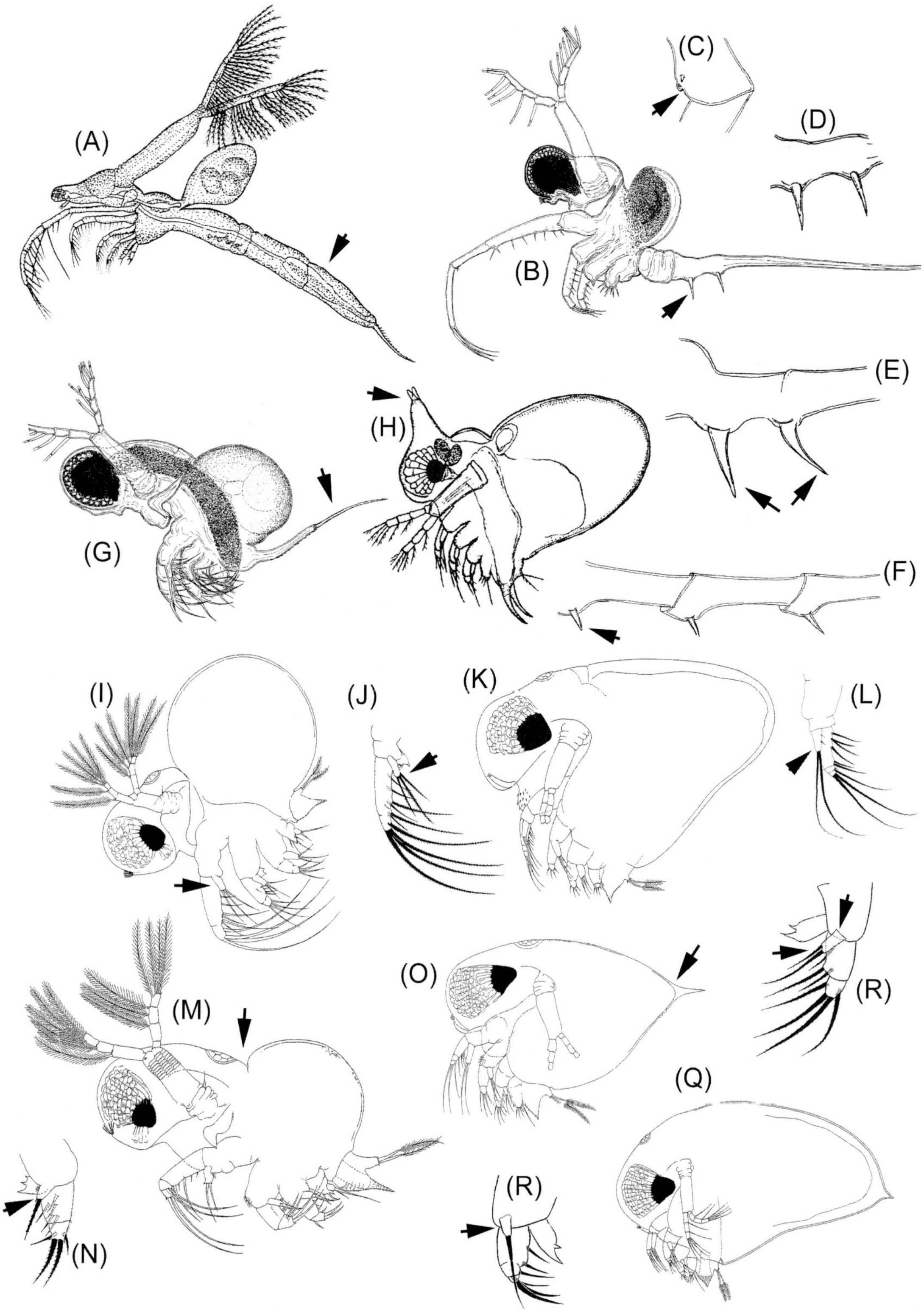

FIGURE 4.5 Representatives of Haplopoda and Onychopoda. (A) *Leptodora kindtii*, general view; (B–D) *Bythotrephes longimanus*, general view, claw of postabdomen and tail spine, and distal setae of second endopodal segment of second limb; (E) *Bythotrephes transcaucasicus*, claw of postabdomen and tail spine; (F) *Bythotrephes brevimanus*, claw of postabdomen and tail spines; (G) *Polyphemus pediculus*, general view; (H) *Cornigerius lacustris*, general view; (I and J) *Pleopis polyphemoides*, general view and first thoracic limb; (K and L) *Pseudoevadne tergestina*, general view and first thoracic limb; (M and N) *Podon intermedius*, general view and third thoracic limb; (O and P) *Evadne spinifera*, general view and third thoracic limb; (Q and R) *Evadne nordmanni*, general view and third thoracic limb. *(A) After Birge (1918); (B–G) After Rogers et al. (2019); (H) After Kotov et al. (2010); (I–R) After Alonso (1996).*

2(1) Posterior somites with dorsal spines simple, small, or absent (Fig. 4.4K) *Leptestheria*
[Widely distributed, genus in need of revision]
2' Posterior somites with dorsal spines large and bearing smaller spines on posterior surface (Fig. 4.4L) .... *Maghrebestheria maroccana* Thiéry, 1988
[Morocco]

Branchiopoda: Diplostraca: Cladocera: Onychopoda: Families

After Rogers et al. (2019)

1 Abdominal and postabdominal (caudal) body parts greatly reduced 2
1' Abdominal and postabdominal (caudal) body parts developed, the latter with long tail spine (Fig. 4.5B) .... Cercopagididae, one genus: ***Bythotrephes***
2(1) Postabdominal basal setae short .... **Podonidae**
2' Postabdominal basal setae long (Fig. 4.5G) .... Polyphemidae, one species: *Polyphemus pediculus* (Linnaeus, 1761) species group
[Holarctic. In need of revision]

Branchiopoda: Cladocera: Onychopoda: Cercopagididae: *Bythotrephes*: Species

The invasive *Cercopagis* (not included in the key below) could be expected in the future. After Rogers et al. (2019)

Adults with two or three claw pairs; limb I comparatively short, endopodite second article apical setae well developed 2
1' Adults with two pairs of straight claws situated closely on postabdomen and tail spine; limb I 80%–132% of body length, endopodite second article with apical setae reduced or absent (Fig. 4.5B–D) .... *Bythotrephes longimanus* Leydig, 1860
[Europe]
2(1) Adult body length 2.3–6.1 mm; tail spine 80%–216% body length, thick basally; postabdomen claws and tail spine long, stout, inserted closely, curved posteriorly (Fig. 4.5E) .... *Bythotrephes transcaucasicus* Behning, 1941
[Armenia, Georgia, Turkey]
2' Adult body length 1.2–2.6 mm; tail spine 120%–306% body length, thin basally; postabdomen claws and tail spine short, inserted rather sparsely, straight, directed either posteriorly or ventrally (Fig. 4.5F) .... *Bythotrephes brevimanus* Lilljeborg, 1901
[Europe]

Branchiopoda: Diplostraca: Cladocera: Onychopoda: Podonidae: Genera

The following are predatory species that are mainly marine though some venture into brackish water or lakes. Several are Ponto-Caspian invaders. Key after Alonso (1996) and Rogers et al. (2019). See also Rivier (1998).

1 Head rounded without outgrowths or horns 2
1' Head with dorsal outgrowths and/or horns (Fig. 4.5H) .... *Cornigerius lacustris* (Spandl, 1923)
[Turkey]
2(1) Limb II & III with exopodite with three to four setae 3
2' Limb II & III with exopodite with two setae 4
3(2) Limb I exopodite with three to four setae; carapace spherical (Fig. 4.5I and J) .... *Pleopis polyphemoides* (Leuckart, 1859)
[Widespread]
3' Limb I exopodite with two setae; carapace elliptical (Fig. 4.5K and L) .... *Pseudoevadne tergestina* (Claus, 1877)
[invasive Ponto-Caspian species]
4(3) Head separated from carapace by constriction; exopodite P2–P3 with one seta (Fig. 4.5M and N) .... *Podon intermedius* Lilljeborg, 1853
[Atlantic, North and Mediterranean Seas; Baltic]
4' Head and carapace continous; exopodite P2 and P3 both with two setae (Fig. 4.5O–R) .... ***Evadne***

Branchiopoda: Diplostraca: Cladocera: Onychopoda: *Evadne*: Species

Marine, may venture into coastal lagoons.

1 P3 exopodite with two setae (Fig. 4.5O and P) .... *Evadne spinifera* P.E. Müller, 1867 [Cosmopolitan]
1' P3 exopodite with one seta (Fig. 4.5Q and R) .... *Evadne nordmanni* Lovén, 1836 [Cosmopolitan]

Branchiopoda: Diplostraca: Cladocera: Anomopoda: Families

Keys adapted from Dumont & Negrea (2002) and Alonso (1996). See also Rogers et al. (2019) for other Palearctic families.

1 Second antenna with exopodite and endopodite each of three articles 2
1' Second antenna with exopodite of four articles, endopodite of three articles 3
2(1) Postbdomen with clearly defined postanal margin .... **Chydoridae**
2' Postabdomen with anus terminal (Fig. 4.6A and B) .... Eurycercidae, one species: *Eurycercus lamellatus* (O.F. Müller, 1776) [Palearctic]
3(1) Rostral apex not continuing into elongated first antennae 4
3' Rostral apex fused with elongated first antennae .... Bosminidae, one genus: ***Bosmina***
4(3) Head with first antennae ventral, not covered by the rostrum 5
4' Head with first antennae more posterior than ventral, partly or completely covered by rostrum .... **Daphniidae**
5(4) Valve ventral setae length $\geq 5 \times$ basal distance between setae 6
5' Valve ventral setae length $<3 \times$ basal distance between setae .... **Moinidae**
6(5) First antenna of one article; carapace without growth lines .... **Macrothricidae**
6' First antenna of two articles; carapace with or without growth lines .... Ilyocryptidae, one genus: ***Ilyocryptus***

Branchiopoda: Diplostraca: Cladocera: Anomopoda: Chydoridae: Subfamilies

Chydoridae is the most speciose branchiopod lineage, with many genera (e.g., *Alona*) still in need of revision. Cryptic taxa are common, with some indentifiable only using molecular techniques; in such cases, the key ends at species group.

1 Posteriodorsal carapace corner where the valves connect at least half as low as highest point of body; headshield with small head pores between the main head pores; postabdomen terminal claw with two basal spines .... **Chydorinae**
1' Posterodorsal carapace corner about as high as highest point of body; headshield small head pores if present lateral to main head pores; postabdomen terminal claw with a single basal spine .... **Aloninae**

Branchiopoda: Diplostraca: Cladocera: Anomopoda: Chydoridae: Chydorinae: Genera

1 Posterior ventral valve setae inserted marginally, most of length uncovered by the valve 2
1' Posterior ventral valve setae inserted inside valve margin, their length entirely covered by valve 3
2(1) Labrum with keel; postabdomen compressed and short; carapace without dark pigment spot .... ***Chydorus***
2' Labrum lacking a keel; postabdomen long and narrow; carapace center with large dark pigment spot (Fig. 4.6C and D) .... *Pseudochydorus globosus* (Baird, 1843) [Palearctic]

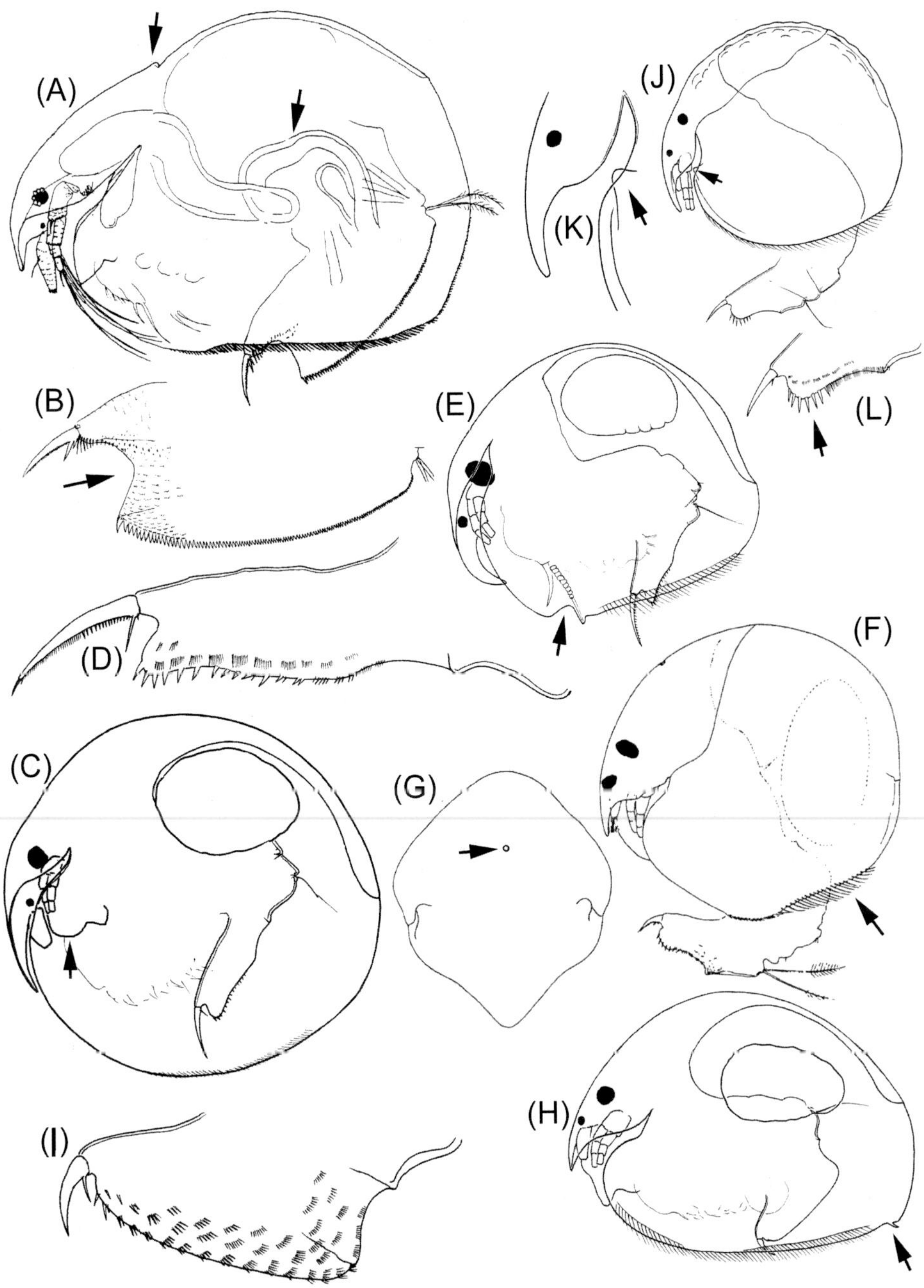

FIGURE 4.6 Representatives of Eurycercidae and Chydoridae (Chydorinae). (A and B) *Eurycercus lamellatus*, general view and postabdomen; (C and D) *Pseudochydorus globosus*, general view and postabdomen; (E) *Anchistropus emarginatus*, general view; (F and G) *Estateroporus gauthieri*, general view and headshield with head pore; (H and I) *Dunhevedia crassa*, general view and postabdomen; (J–L) *Paralona pigra*, general view, head and postabdomen. *After Rogers et al., 2019.*

3(1) Ventral valve margin entire 4

3' Ventral valve margin with robust triangular indentation; carapace with ventral gland (Fig. 4.6E) .... *Anchistropus emarginatus* Sars, 1862 [Palearctic]

4(3) Headshield with 0–1 major headpores; two small pores, if present, posterior to main pore 5

4' Headshield with two major headpores and two small pores in between 6

5(4) Labral keel with strong sharp denticles; adults headshield head pores absent .... ***Ephemeroporus***

5' Labral keel without strong sharp denticles, but may be serrated; adult headshield with one major head pore and two small lateral pores (Fig. 4.6F and G) .... *Estatheroporus gauthieri* Alonso, 1990 [Spain]

6(5) Headshield posterior to mandibular articulation point length $\leq$ headshield anterior portion 7
6' Headshield posterior to mandibular articulation point length $\geq 2\times$ anterior portion 8
7(6) Compound eye diameter $\leq$ ocellus diameter; postabdomen anal margin continuing almost straight into postanal margin 9
7' Compound eye diameter $\geq 3\times$ ocellus diameter; postabdomen with anal margin at nearly 90 degrees angle to postanal margin (Fig. 4.6H and I) .... *Dunhevedia crassa* King, 1853 complex [Widespread]
8(7) Postabdomen compressed, postanal margin length $<0.5\times$ anal margin (Fig. 4.6J–L) .... *Paralona pigra* (Sars, 1862) [Palearctic]
8' Postabdomen postanal margin length $\geq$ anal margin .... ***Pleuroxus***
9(7) Postabdomen length $3\times$ width; rostrum narrow, reaching ventral valve margin or beyond .... ***Disparalona***
9' Postabdomen length $\leq 2\times$ width; rostrum mostly blunt, not reaching ventral valve margin .... ***Alonella***

Branchiopoda: Cladocera: Anomopoda: Chydoridae: Chydorinae: *Chydorus*: Species

Chydorus is a difficult genus to identify, not well studied in the Mediterranean Basin, and more species are likely to be present. So far only a few temperate Palearctic species are known here, of which most are extremely rare, and one Mediterranean endemic is in Spain. However, *Chydorus sphaericus* forms a complex of species in the Palearctic that is impossible to distinguish using morphology of adult parthenogenetic females alone. For all *Chydorus*, it is necessary to examine morphological characters in detail and to check recent updates on taxonomy as well as molecular revisions while also considering potential exotics (e.g., American *Chydorus brevilabris*) that have been reported from Europe and can be expected here. *Chydorus* thermophilic species complexes that are widespread in (sub)tropical Africa and Asia could also be expected here (e.g., *C. eurynotus*-, *C. pubescens*-, and *C. parvus*-groups), though so far unrecorded in this basin (for morphology, see Smirnov, 1996). Because records of *Chydorus latus* (Sars, 1862) in the Mediterranean Basin (e.g., with one headpore in Italy) remain unresolved, this species was not included in the key below (for drawings see Flößner, 2000). Key after Rogers et al. (2019) with adaptations.

1 Postabdomen preanal margin smooth 2
1' Postabdomen preanal margin with small protrusions (Fig. 4.7A and B) .... *Chydorus pizarri* Alonso, 1988 [Spain]
2(1) Dorsal outline convex; no lateral expansions 3
2' Dorsal outline humpbacked; from anterior, bodyshape triangular, upper half with robust lateral expansions (Fig. 4.7C and D) .... *Chydorus gibbus* Sars, 1890 [Temperate Palearctic]
3(2) Second antenna exopod article I without seta; postabdomen postanal margin with many (11–15) small teeth of similar size (Fig. 4.7E and G) .... *Chydorus ovalis* Kurz, 1875 [Temperate Palearctic]
3' Second antenna exopod article I with seta; postabdomen postanal margin with fewer (5–9) long teeth of unequal sizes (Fig. 4.7H and J) .... *Chydorus sphaericus* (O.F. Müller, 1776)-group [Cosmopolitan]

Branchiopoda: Cladocera: Chydoridae: Chydorinae: *Ephemeroporus*: Species

After Alonso (1996) and Rogers et al. (2019). Check the general key to the subfamily for the very similar monotypic *Estatheroporus*, which retains the headpore in the adult stage (*Ephemeroporus* does not).

1 Carapace posteroventral corner with denticle(s) 2
1' Carapace posteroventral corner without denticle(s) (Fig. 4.8A–D) .... *Ephemeroporus phintonicus* (Margaritora, 1969) [Algeria, Baleares, Sardinia, Spain]
2(1) Labral keel with round apex 3
2' Labral keel with acute apex (Fig. 4.8E–G) .... *Ephemeroporus epiaphantoii* Alonso, 1987 [Spain]

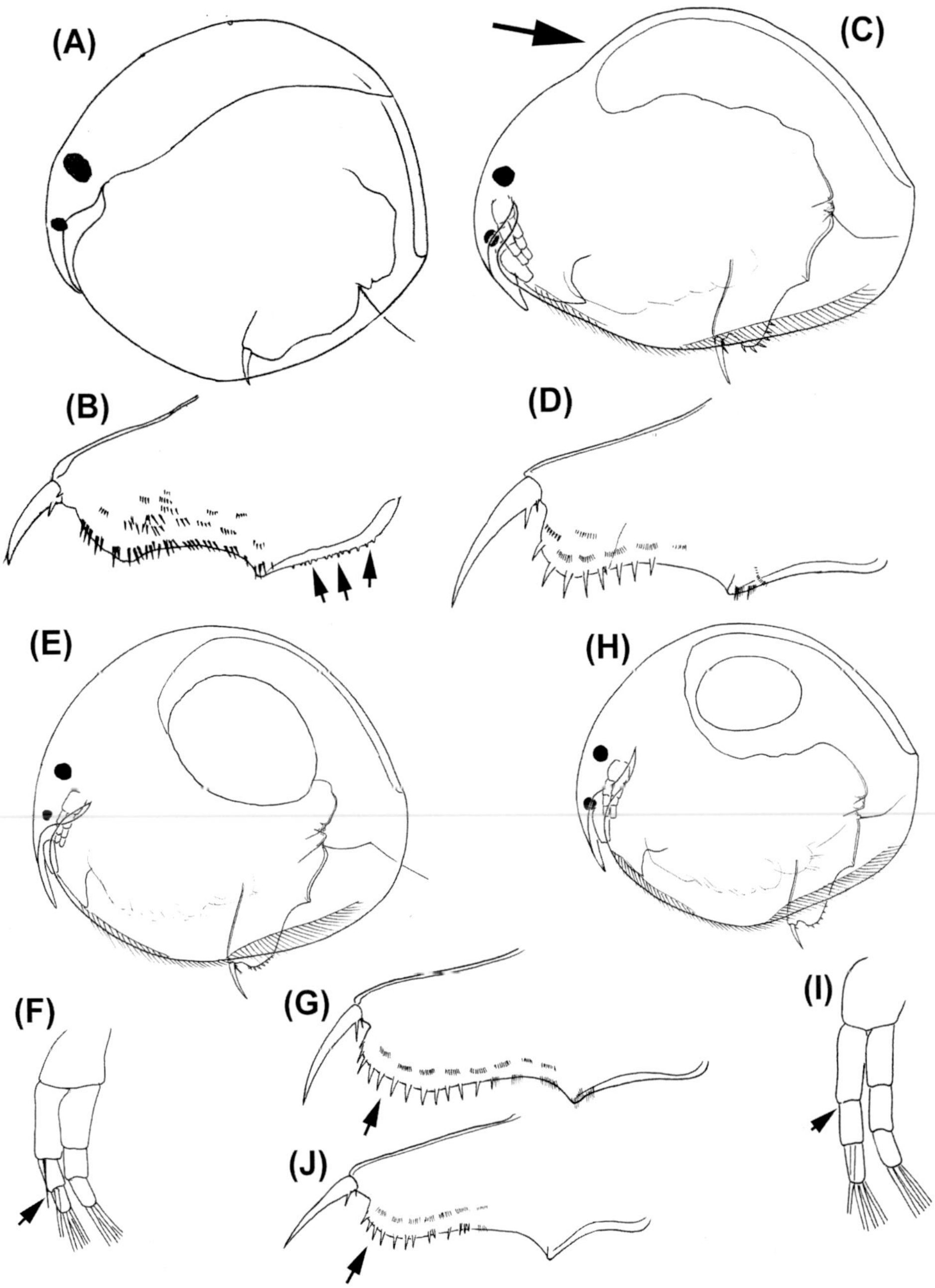

FIGURE 4.7 Representatives of Chydoridae (Chydorinae): (A and B) *Chydorus pizarri*, general view and postabdomen; (C and D) *Chydorus gibbus*, general view and postabdomen; (E–G) *Chydorus ovalis*, general view, second antenna and postabdomen; (H–J) *Chydorus sphaericus*, general view, second antenna and postabdomen. *(A and B) Redrawn from Alonso (1996) (C–J) After Rogers et al., 2019.*

3(2) Labral keel with round apex (Fig. 4.8H–J) .... *Ephemeroporus margalefi* Alonso, 1987 [Spain]

3' Labral keel with narrow rounded apex (Fig. 4.8K–M) .... *Ephemeroporus barroisi* (Richard, 1894) [Pantropical, southern Palearctic]

Branchiopoda: Cladocera: Anomopoda: Chydoridae: Chydorinae: *Pleuroxus*: Species

After Alonso (1996) and Rogers et al. (2019). Most *Pleuroxus* are part of unrevised (or partially revised) species complexes; please consult literature.

1 Bodyshape not globular; body tapering distally; rostrum long and pointed; postabdomen with dorsodistal marginal robust teeth 2

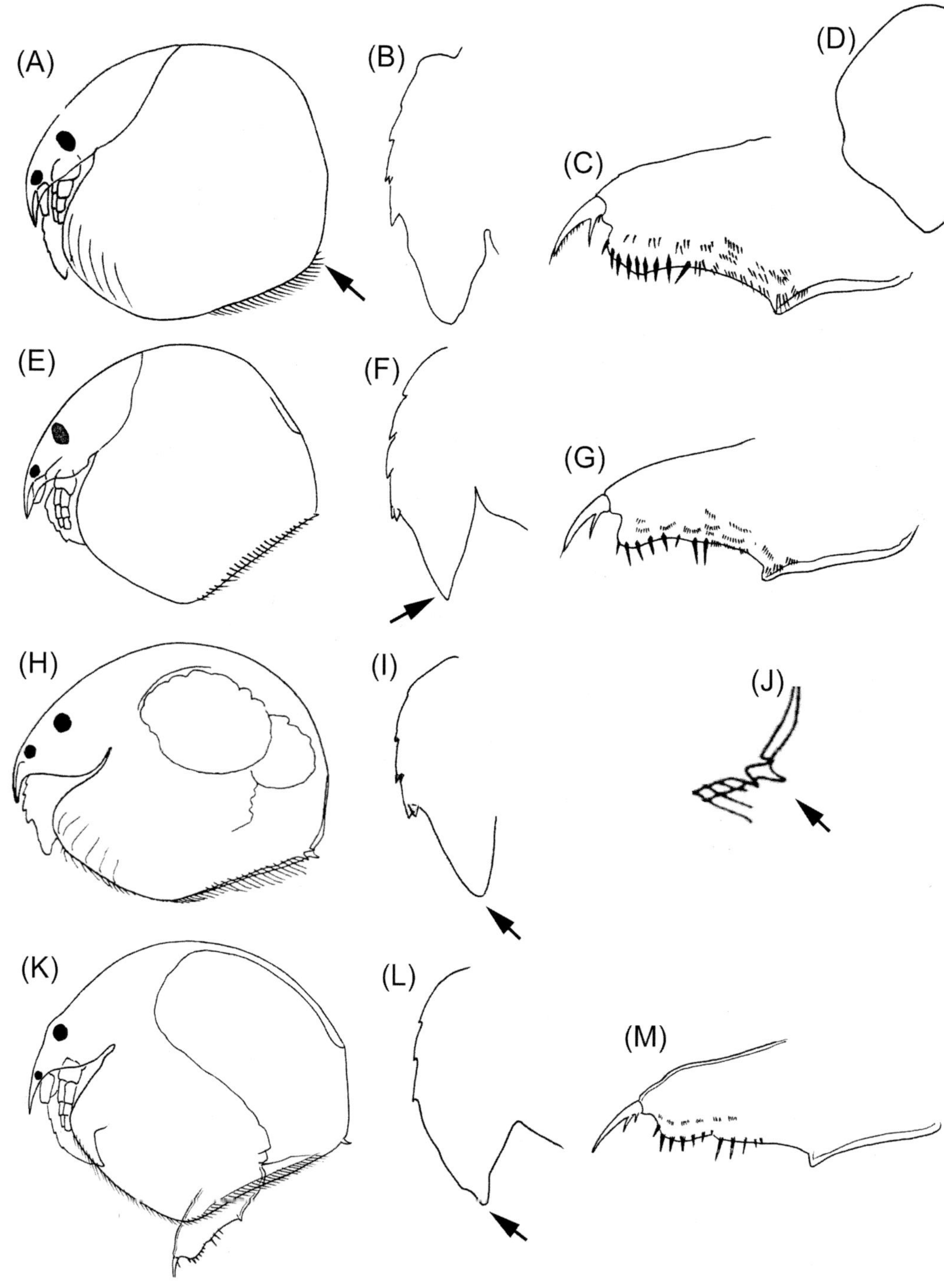

FIGURE 4.8 Representatives of Chydoridae (Chydorinae): (A–D) *Ephemeroporus phintonicus*, general view, labrum, postabdomen and headshield; (E–G) *Ephemeroporus epiaphantoii*, general view, labrum and postabdomen; (H–J) *Ephemeroporus margalefi*, general view, labrum and posteroventral corner of valves; (K–M) *Ephemeroporus barroisi*, general view, labrum and postabdomen. *(A–J)* Redrawn from *Alonso (1996)*. *(K–M) After Rogers et al., 2019.*

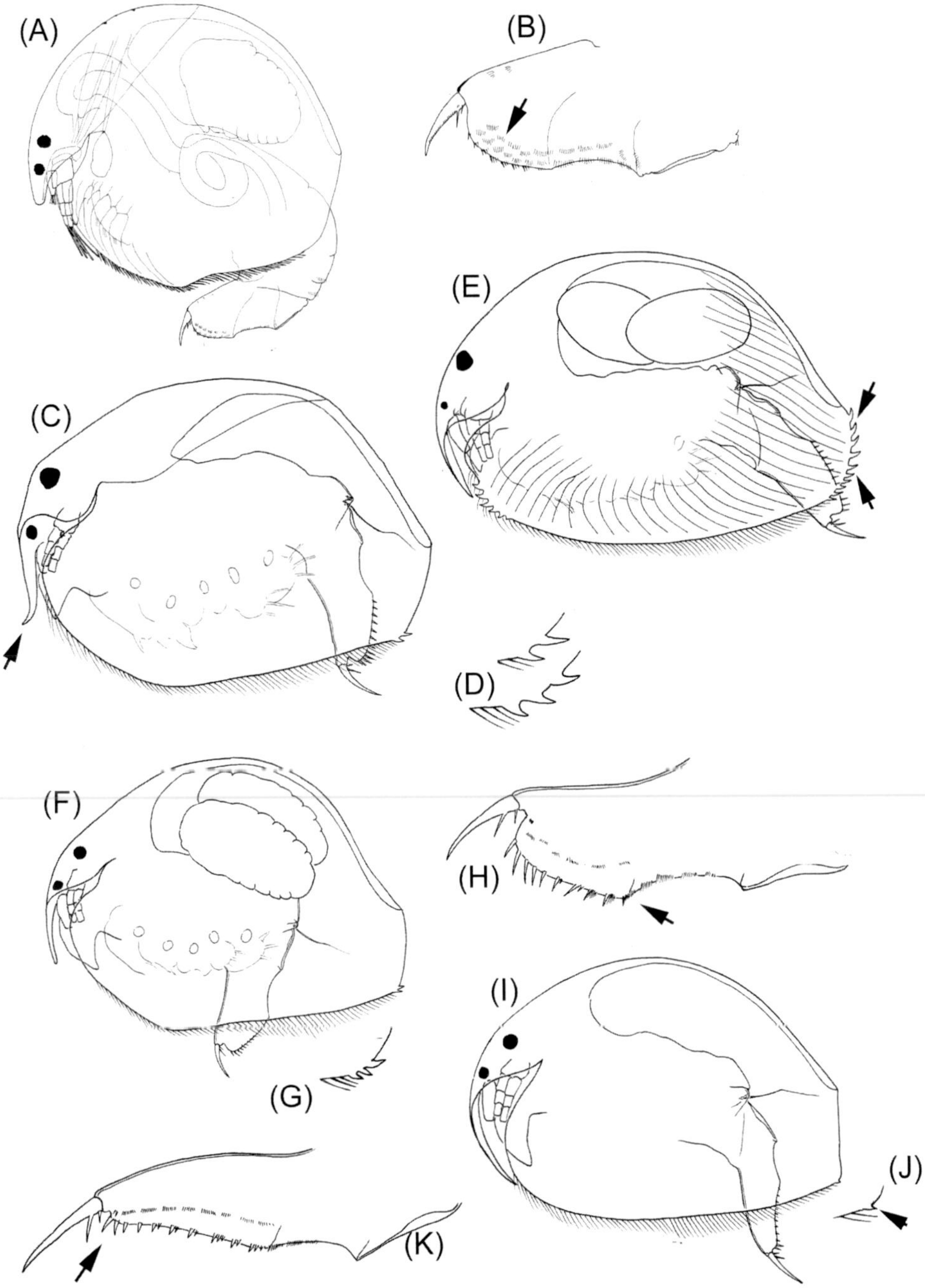

FIGURE 4.9 Representatives of Chydoridae (Chydorinae): (A and B) *Pleuroxus letourneuxi*, general view and postabdomen; (C and D) *Pleuroxus uncinatus*, general view and posteroventral valve corner; (E) *Pleuroxus truncatus*, general view; (F–H) *Pleuroxus aduncus*, general view, posteroventral valve corner and postabdomen; (I–K) *Pleuroxus laevis*, general view, posteroventral valve corner and postabdomen. *After Rogers et al., 2019.*

1' Bodyshape globular; rostrum short and blunt; postabdomen dorsal margins with small groups of denticles (Fig. 4.9A and B) .... *Pleuroxus letourneuxi* (Richard, 1888) [Mediterranean Basin]

2(1) Rostrum apex directed ventrally, not curved 3

2' Rostrum apex curved anteriorly (Fig. 4.9C and D) .... *Pleuroxus uncinatus* Baird, 1850 [Palearctic]

3(2) Carapace posterioventral corner with 0–4 denticles 4

3' Carapace posterior margins with >15 large denticles (Fig. 4.9E) .... *Pleuroxus truncatus* (O.F. Müller, 1785) [Palearctic]

4(3) Postabdomen postanal portion length ≥ 1.5 × anal portion 5

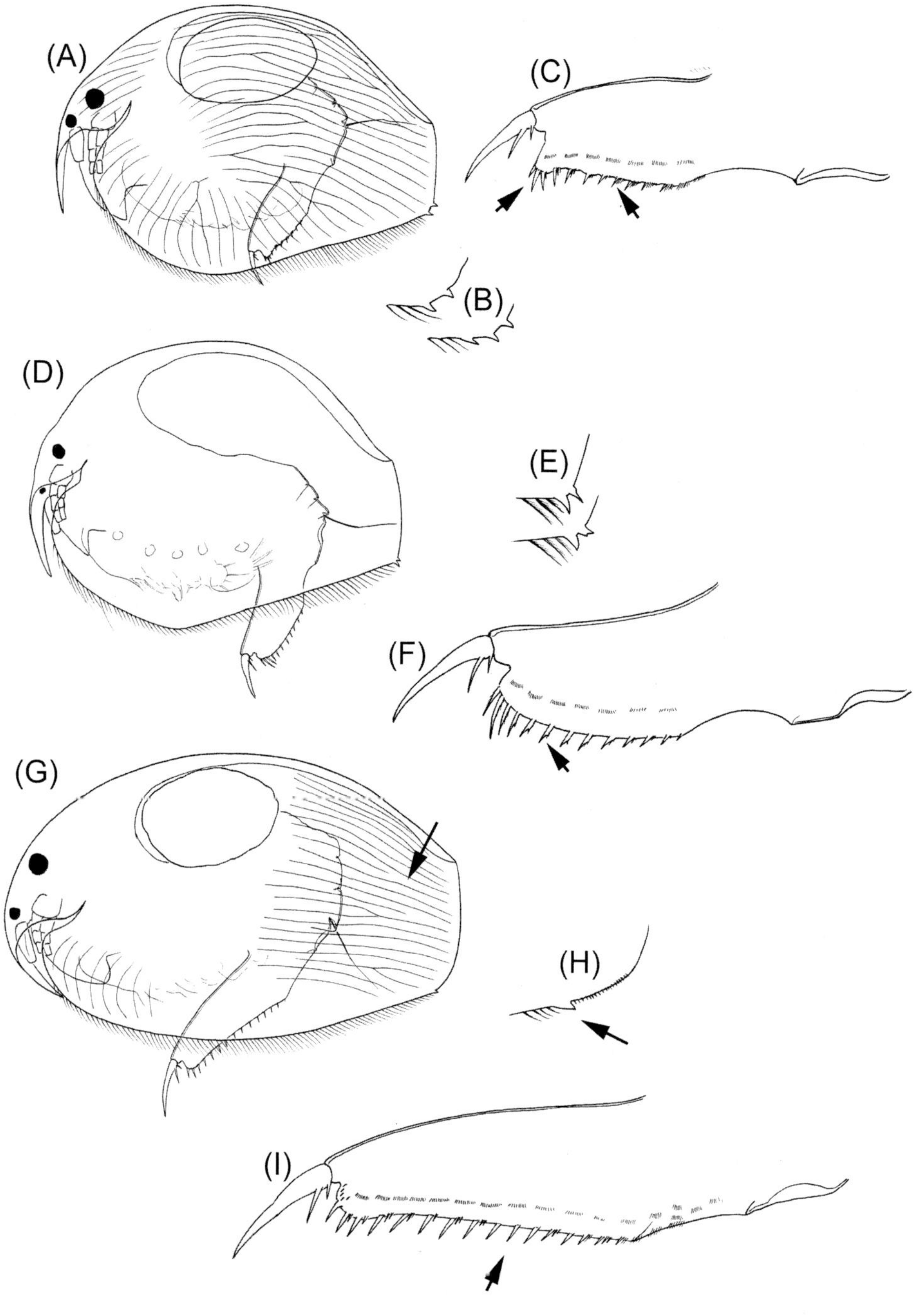

FIGURE 4.10 Representatives of Chydoridae (Chydorinae): (A–C) *Pleuroxus denticulatus*, general view, posteroventral valve corner and postabdomen; (E, F) *Pleuroxus trigonellus*, general view, posteroventral valve corner and postabdomen; (G–I) *Pleuroxus striatus*, general view, posteroventral valve corner and postabdomen. *After Rogers et al., 2019.*

4' Postabdomen postanal portion length subequal to anal portion (Fig. 4.9F–H) .... *Pleuroxus aduncus* (Jurine, 1820) [Palearctic]

5(4) Postabdomen with 2–3 distomarginal teeth markedly larger than others 6

5' Postabdomen with distomarginal teeth gradually increasing in size distally 7

6(5) Postabdomen postanal portion dorsal margin clearly concave; large distal marginal spines not on a projection (Fig. 4.10I–K) .... *Pleuroxus laevis* Sars, 1862 [Cosmopolitan]

6' Postabdomen postanal portion dorsal margin more straight; larger distal marginal spines on a projection (Fig. 4.10A–C) .... *Pleuroxus denticulatus* Birge, 1879 [Cosmopolitan]

7(5) Carapace surface without striae; postabdomen postanal portion dorsal margin length ~2.5 × anal portion (Fig. 4.10E and F) .... *Pleuroxus trigonellus* (O.F. Müller, 1776) [Palearctic]
7' Carapace with clear striae; postabdomen postanal portion dorsal margin length ~3.5 × anal portion (Fig. 4.10G–I) .... *Pleuroxus striatus* Schoedler, 1863 [Palearctic]

Branchiopoda: Cladocera: Anomopoda: Chydoridae: Chydorinae: *Disparalona*: Species

Sometimes confused with *Alonella*.

1 Carapace posterior valve corner without a denticle 2
1 Carapace posterior valve corner with a denticle (Fig. 4.11A–C) .... *Disparalona rostrata* (Koch, 1844) [Eurasia]
2(1) Postabdomen posteriodorsal margin next to terminal claw not protruding or angular; terminal claw with two short basal spines, with proximal one reduced, obscure (Fig. 4.11D–F) .... *Disparalona* cf. *leei* (Chien, 1970) [Holarctic]
2' Postabdomen posteriodorsal margin protruding and angular; terminal claw with two robust basal spines (Fig. 4.11G) .... *Disparalona hamata* (Birge, 1879) [Widespread]

Branchiopoda: Cladocera: Anomopoda: Chydoridae: Chydorinae: *Alonella*: Species

Modified after Rogers et al. (2019).

1 Body in lateral view elongate oval, ventral margin straight, valve ornamentation with polygons 2
1' Body globular, ventral valve margin strongly convex; valve ornamentation only nonanastomosing lines (Fig. 4.11H and I) .... *Alonella nana* (Baird, 1850) species complex [Palearctic]
2(1) Valves with small striae inside polygons (Fig. 4.11J–L) .... *Alonella excisa* (Fischer, 1854) species complex [Palearctic]
2' Valves without striae (Fig. 4.11M and N) .... *Alonella exigua* (Lilljeborg, 1901) species complex [Palearctic]

Branchiopoda: Diplostraca: Cladocera: Anomopoda: Chydoridae: Aloninae: Genera

Modified after Alonso (1996) and Rogers et al. (2019), with additions. *Alona* is unstable and unnatural, and it is slowly being revised. The *Alona* "core" lineage is the *Alona quadrangularis*-group. It is necessary to consult specialised and recent literature for each species. A checklist of *Alona* names is provided by Van Damme et al. (2010), and several groups have since been revised. Because of the instability of the genus and no general consensus on some taxa, not all updates in *Alona* systematics are included here. Dissection of limbs is extremely important for separating all *Alona*-like genera; however, the key is made as practical as possible. Furthermore, users must be aware that more than one *Karualona* species may be present, and more than one species of *Anthalona* may be expected, such as the widespread Afro-Asian *Anthalona harti* (Fig. 4.14H).

1 Compound eye and ocellus present; growth lines absent 2
1' Compound eye absent; growth lines present; only one major round headpore in headshield (Fig. 4.12A and B) .... *Monospilus dispar* Sars, 1862 [Palearctic]
2(1) Carapace dorsal midridge without dorsal keel, or keel weak 3
2' Carapace dorsal midridge with robust dorsal keel 17
3(2) Rostrum in lateral view with length ≤3 × width, directed ventrally 4
3' Rostrum in lateral view with length >5 × width, directed posteriorly (Fig. 4.12C and D) .... *Rhynchotalona falcata* (Sars, 1862) [Palearctic]

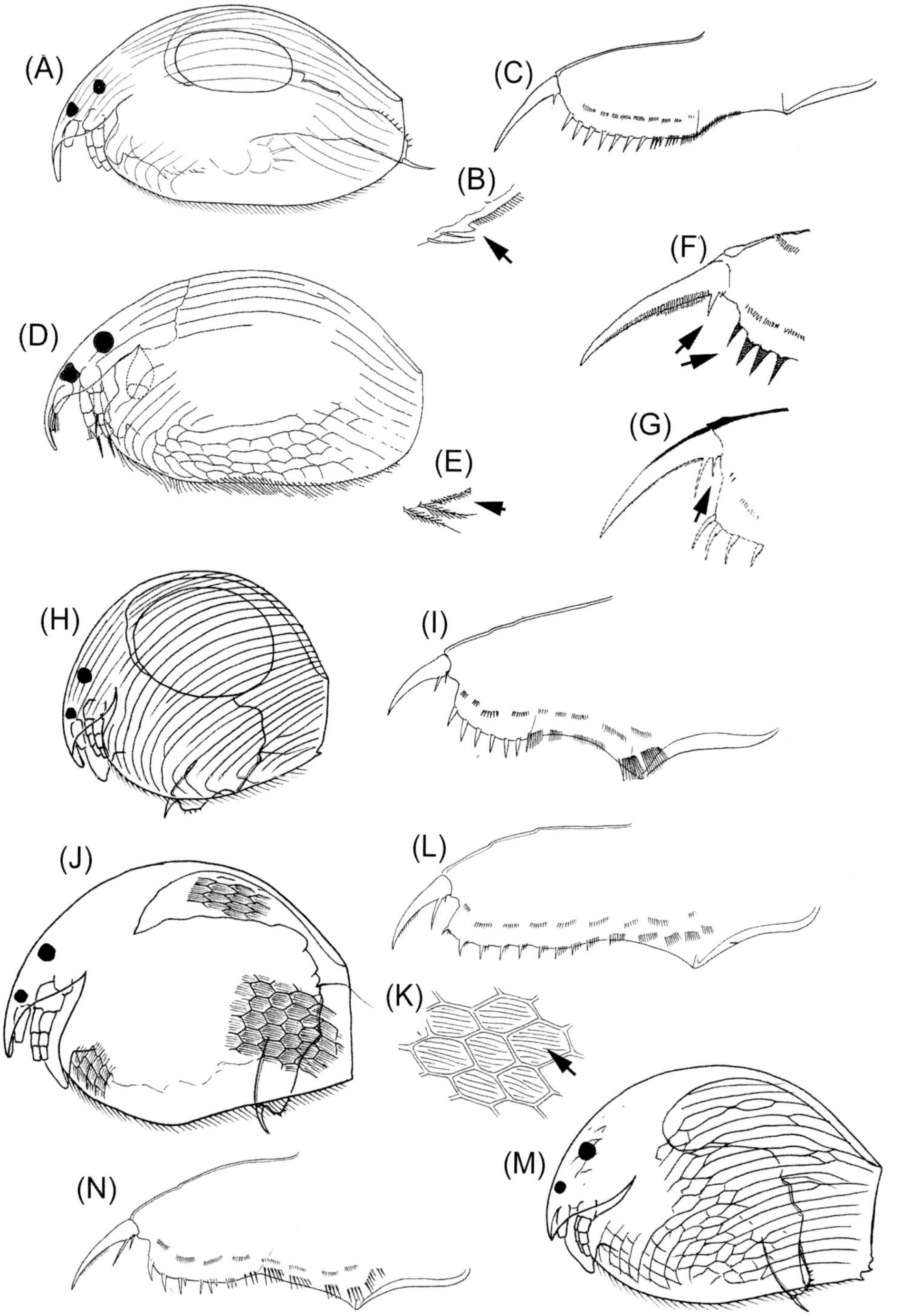

FIGURE 4.11 Representatives of Chydoridae (Chydorinae): (A–C) *Disparalona rostrata*, general view, posteroventral valve corner and postabdomen; (D–F) *Disparalona* cf. *leei*, general view, posteroventral valve corner and postabdomen (partim); (G) *Disparalona hamata*, postabdomen (partim); (H and I) *Alonella nana*, general view and postabdomen; (J–L) *Alonella excisa*, general view, valve ornamentation and postabdomen; (M and N) *Alonella exigua*, general view and postabdomen. *(A, C, H–N) After Rogers et al. (2019); (B, D-F) After Alonso (1996); (G) After Neritina et al. (2018).*

4(3) Postabdomen dorsal posteroventral portion with straight to slightly convex dorsal margin; lateral denticles short to moderate in length reaching maximally half their length over dorsal margin 5

4' Postabdomen strongly convex and D-shaped; lateral denticles long, reaching entirely over dorsal margin (Fig. 4.13A–J) .... ***Leydigia***

5(4) Postabdomen with marginal teeth gradually increasing in size distally, but no exceptional increase in length; two to three major headpores 6

5' Postabdomen with 2–3 larger distal marginal teeth, with length $\sim 0.5\times$ terminal claw (Fig. 4.12E and F); four major headpores .... *Oxyurella tenuicaudis* (Sars, 1862) [Palearctic]

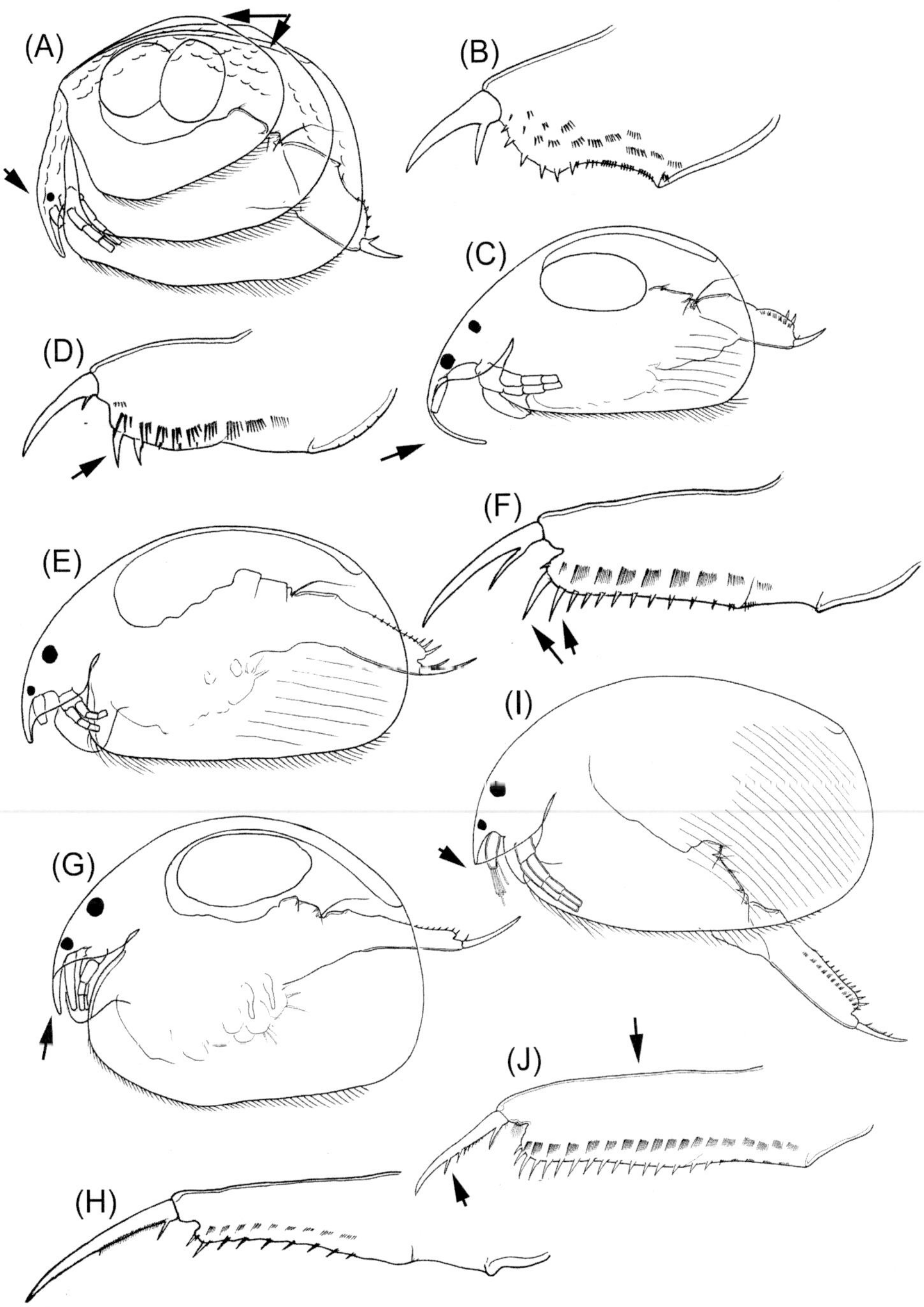

FIGURE 4.12 Representatives of Aloninae (Chydoridae): (A and B) *Monospilus dispar*, general view and postabdomen; (C and D) *Rhynchotalona falcata*, general view and postabdomen; (E and F) *Oxyurella tenuicaudis*, general view and postabdomen; (G and H) *Kurzia latissima*, general view and postabdomen; (I and J) *Alonopsis elongatus*, general view and postabdomen. *After Rogers et al., 2019.*

6(5) Three major headpores 7
6' Two major headpores may be broadly connected appearing as one long pore 13
7(6) Rostrum short, broad, and blunt (Fig. 4.12I) 8
7' Rostrum long and narrow with elongate apex (Fig. 4.12G and H) .... *Kurzia latissima* (Kurz, 1875) [Palearctic, Afrotropical]
8(7) Lateral headpores as round small holes 9
8' Lateral headpores as slits ("*Alona*" *costata*-group) .... ***"Alona"***
9(8) Postabomen length 2–2.5 × width; terminal claw with "normal" pectens, without extra robust dorsal spines (besides terminal spine) 10

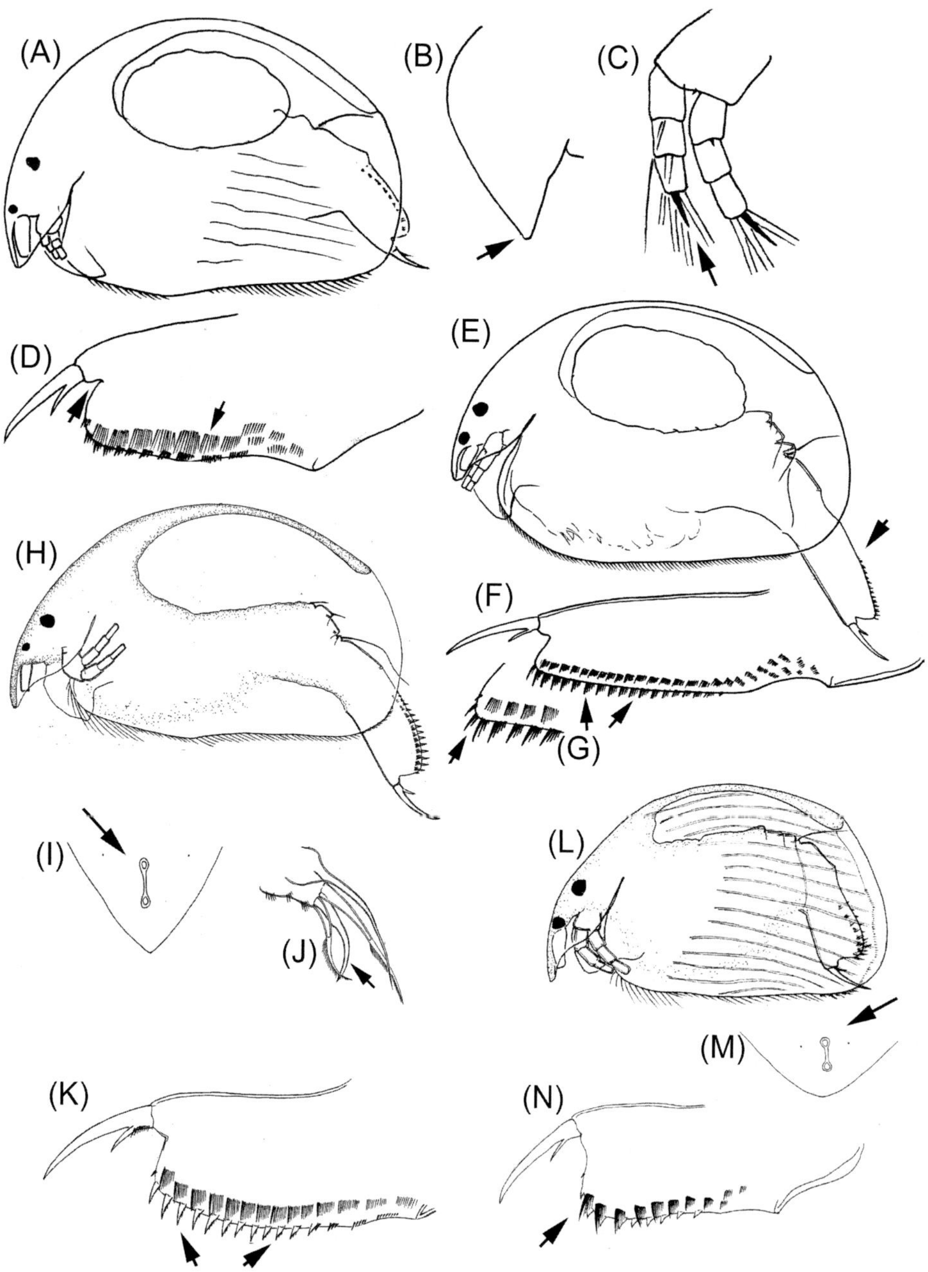

FIGURE 4.13 Representatives of Aloninae (Chydoridae): (A–D) *Phreatalona phreatica*, general view, labral keel, second antenna and postabdomen; (E–G) *Tretocephala ambigua*, general view, postabdomen and detail postabdomen; (H–K) *Biapertura affinis*, general view, posterior portion of headshield with headpores, detail of inner distal lobe of first limb and postabdomen; (L–N) *"Alona" intermedia*, general view, posterior portion of headshield with headpores and postabdomen. *After Rogers et al., 2019.*

9' Postabdomen length 3–4 × width; terminal claw with basal spine, pectens and two robust dorsal spines one half-way, one in distal third (Fig. 4.12I and J) .... *Alonopsis elongatus* (Sars, 1862) [Italy. Palearctic]

10(9) Labral keel without long narrowing apex; second antenna exopod proximal article with spine reaching at least to middle of second exopod article 11

10' Labral keel long with narrowing apex; second antenna exopod proximal article with spine never reaching middle of second exopod article (Fig. 4.13A–D) .... *Phreatalona phreatica* (Dumont, 1983) [France, Spain]

11(10) Thoracic limb III exopodite with seven setae 12

11' Thoracic limb III exopodite with six setae .... ***Coronatella* s. lat.**

12(11) At least one (but up to all three) major headpores disconnected .... ***Ovalona***

12' All three headpores clearly connected .... ***"Alona"***

13(6) Major headpores distinct; postabdomen length $\leq 3 \times$ width 14
13' Major headpores broadly connected, appearing as slit; postabdomen length $4-5 \times$ width (Fig. 4.13E–G) ....
.... *Tretocephala ambigua* (Lilljeborg, 1900)
[Palearctic]
14(13) Headshield posterior margin broadly round or obtuse; postabdomen dorsal margin with 8–11 marginal teeth (or groups of denticles) 15
14' Headshield posterior margin acute; postabdomen dorsal margin with 14–16 large teeth (Fig. 4.13H–K) ....
.... *Biapertura affinis* (Leydig, 1860)
[Palearctic]
15(14) Labral keel proximally with ≥ 1 small denticles; postabdomen dorsally with a deeply convex embayment next to terminal claw base (Fig. 4.14C and G) 16

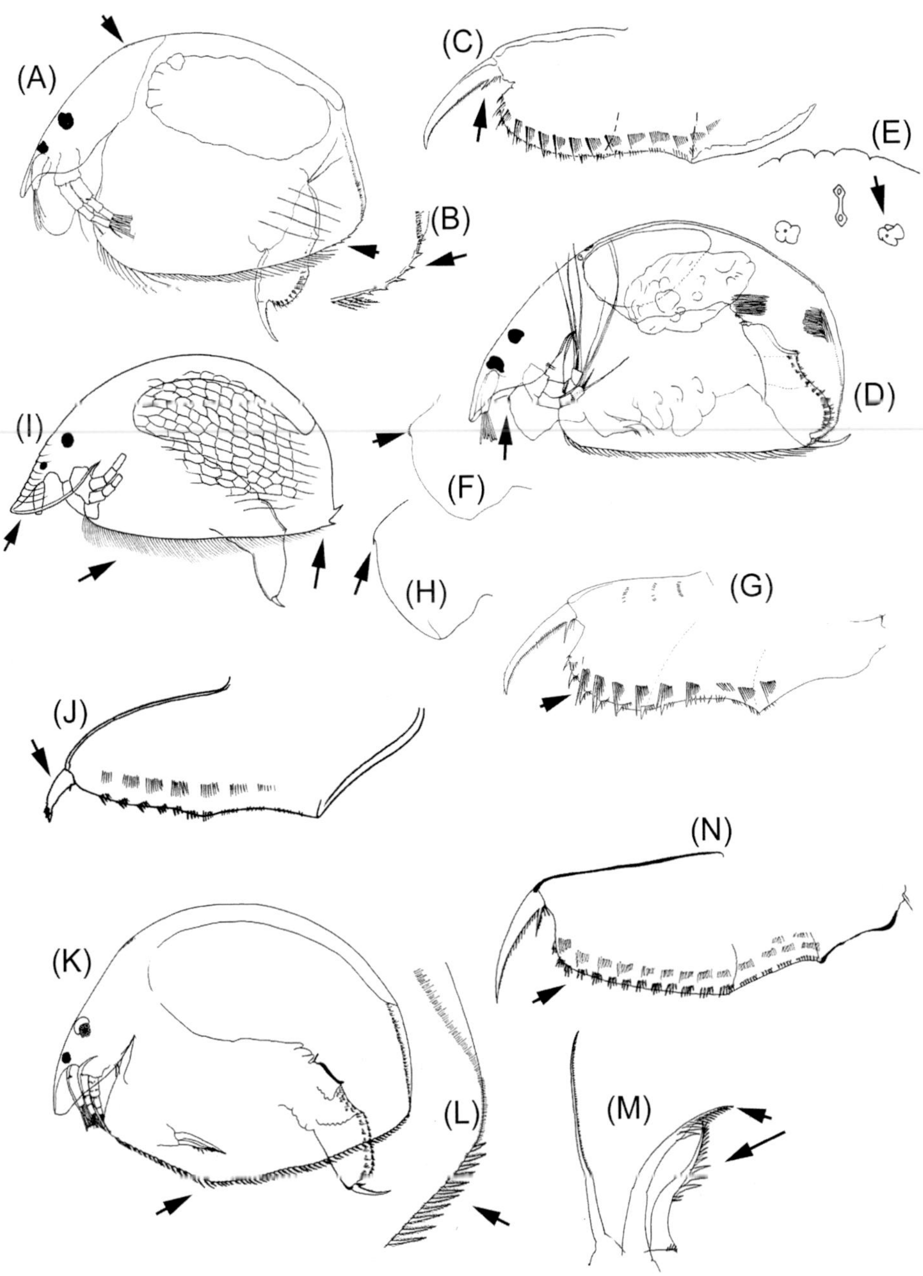

FIGURE 4.14 Representatives of Aloninae (Chydoridae): (A–C) *Karualona iberica*, general view, posteroventral valve corner and postabdomen; (D–G) *Anthalona mediterranea*, general view, posterior portion of headshield with headpores, labral keel and postabdomen; (H) *Anthalona harti*, labral keel; (I) *Graptoleberis testudinaria*, general view and postabdomen; (K–N) *Leberis punctatus*, general view, posteroventral corner of valves, outer and inner distal lobe of first limb and postabdomen. *After Rogers et al. (2019); (D–G) After Van Damme et al. (2011); (K–N) After Neritina & Sinev (2016).*

15' Labral keel proximally with a row of small setae; postabdomen dorsal portion without deeply convex embayment (Fig. 16.13L–N) .... *"Alona" intermedia* Sars, 1862
[Cosmopolitan species group in need of revision]
16(15) Valve posteroventral corner with 2–4 denticles; lateral headpores circular (Fig. 4.14A–C) .... *Karualona iberica* (Alonso & Pretus, 1989)
[Mediterranean Basin]
16' Valve posteroventral corner lacking denticles; lateral headpores with peculiar flower shaped structures (visible through the carapace) (Fig. 4.14D–G) .... *Anthalona mediterranea* (Yalim, 2005)
[Mediterranean Basin and Western Asia]
17(16) Postabdomen terminal claw length >4 × width; carapace ornamentation mainly lines; rostrum not directed anteriorly 4
17' Postabdomen terminal claw length ≤4 × width; carapace ornamentation a strong reticulate pattern; rostrum generally directed anteriorly (Fig. 4.14I and J) .... *Graptoleberis testudinaria* (Fischer, 1851)
[Cosmopolitan species group in need of revision]
18(17) Head in lateral view strongly convex; carapace ventral marginal setae filiform; thoracopod I distomedial lobe with three setae, one may be hamulate 19
18' Head not strongly convex; carapace ventral marginal setae short, spiniform; thoracopod I distomedial lobe with two setae (third reduced) (Fig. 4.14K–M) .... *Leberis punctatus* (Daday, 1898)
[Subtropical]
19(18) Postabdomen narrowing distally; postabdomen dorsomarginal teeth robust .... ***Camptocercus***
19' Postabdomen ventral and dorsal margins parallel; dorsomarginal teeth obscure .... ***Acroperus***

Branchiopoda: Cladocera: Anomopoda: Chydoridae: Aloninae: *Leydigia*: Species

For more information on this genus, see Kotov (2009). Key follows Rogers et al. (2019).

1 Terminal claw with robust basal denticle (Fig. 4.15E) 2
1' Terminal claw with basal denticle rudimentary or lacking (Fig. 4.15I) 3
2(1) Postabdomen distolaterally with 6–7 groups of long spines (Fig. 4.15A and B) .... *Leydigia korovchinskyi* Kotov & Alonso, 2010
[Spain]
2' Postabdomen distolaterally with 10–12 groups of long spines (Fig. 4.15C–F) .... *Leydigia leydigi* (Schoedler, 1863) species group
[Cosmopolitan]
3(1) Postabdomen with distolateral setal groups with distalmost setule length 2 × preceding one (Fig. 4.15G–I) .... *Leydigia acanthocercoides* (Fischer, 1854) species group
[Cosmopolitan]
3' Postabdomen with distalmost setule length <2 × preceding one (Fig. 4.15J) .... *Leydigia iberica* Kotov & Alonso, 2010
[Spain]

Branchiopoda: Cladocera: Anomopoda: Chydoridae: Aloninae: *Coronatella*: Species

These species groups are morphologically very close, characterized by five limb pairs and six setae on the thoracopod III exopodite (Van Damme & Dumont, 2008). The key includes the former *Alona rectangula*- and *Alona elegans*-groups. Users should be aware that *Coronatella* may comprise more than one species in the Mediterranean Basin.

1 Carapace with wide longitudinal striae, sometimes verrucae may be present, and about 10–15 transverse singule (never double) lines 2
1' Carapace with longitudinal striae often doubled, and >15 transverse lines (*Coronatella elegans*-group) 3
2(1) Postabdomen length ~2.5 × width; thoracopod I distal lobe with medial setae modified with two robust proximal spines (Fig. 4.16A–D) .... *Coronatella rectangula* (Sars, 1862) species group
[Cosmopolitan]
2' Postabdomen length 3 × width; thoracopod I distal lobe with medial denticles decreasing in size gradually distally (Fig. 4.16E–G) .... *Coronatella anemae* Van Damme & Dumont, 2008
[Palearctic]

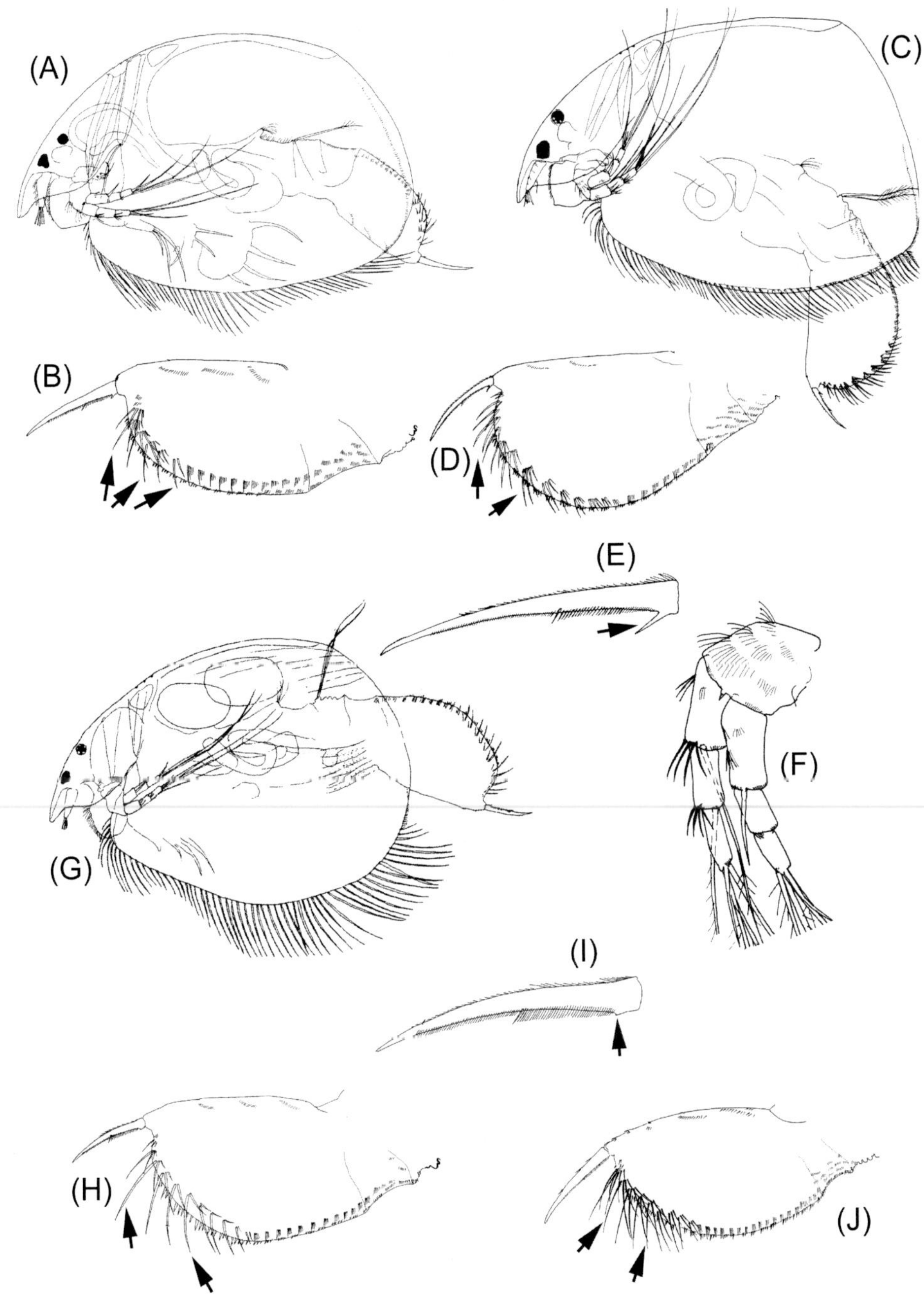

FIGURE 4.15 Representatives of Aloninae (Chydoridae): (A and B) *Leydigia korovchinskyi*, general view and postabdomen; (C–F) *Leydigia leydigi*, general view, postabdomen, terminal claw and second antenna; (G–I) *Leydigia acanthocercoides*, general view, postabdomen and terminal claw; (J) *Leydigia iberica*, postabdomen. *After Rogers et al., 2019.*

3(1) Postabdomen with single row of lateral fascicles 4
3' Postabdomen with double row of lateral fascicles (Fig. 4.16H) .... *Coronatella orellanai* (Alonso, 1996) [Spain]
4(3) Carapace striation with lines nearly equidistant; postabdominal preanal margin shorter than anal margin (Fig. 4.16I and J) .... *Coronatella elegans* (Kurz, 1875) [Palearctic]
4' Carapace striation with lines alternating in distance between further and closer, forming narrower double lines; postabdominal preanal margin subequal or longer than anal margin (Fig. 4.16K and L) .... *Coronatella salina* (Alonso, 1996) [Spain]

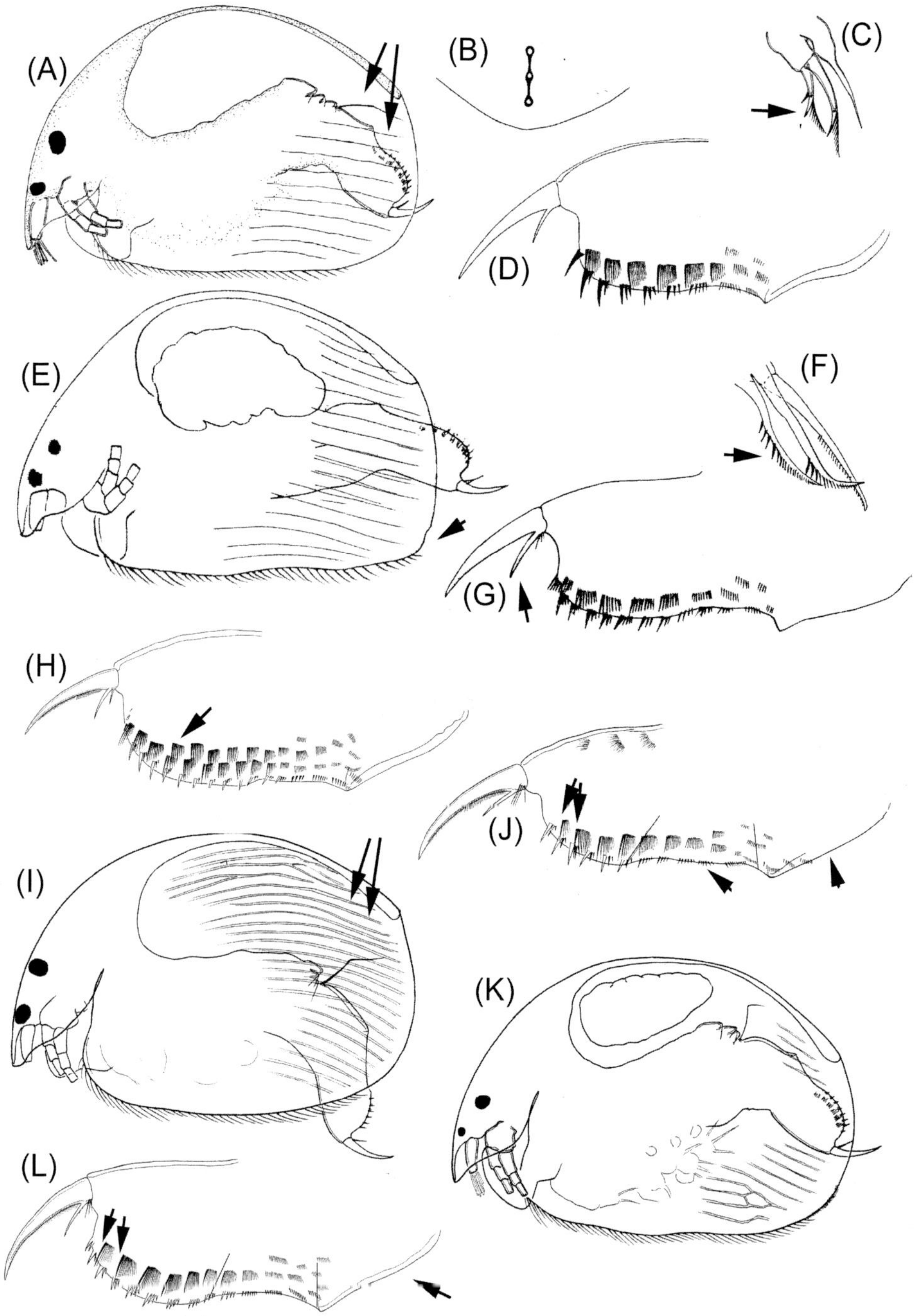

FIGURE 4.16 Representatives of Aloninae (Chydoridae): (A–D) *Coronatella rectangula*, general view, posterior portion of headshield with headpores, inner distal lobe of first limb and postabdomen; (E–G) *Coronatella anemae*, general view, inner distal lobe of first limb and postabdomen; (H) *Coronatella orellanai*, postabdomen; (I and J) *Coronatella elegans*, general view and postabdomen; (K and L) *Coronatella salina*, general view and postabdomen. *After Rogers et al. (2019)* and *Alonso (1996); (E–G) After Van Damme & Dumont (2008).*

Branchiopoda: Cladocera: Anomopoda: Chydoridae: Aloninae: *Ovalona*: Species

Follows Rogers et al. (2019); *Ovalona* comprises the former *Alona pulchella*-group.

1 Major headshield headpores not connected, all three equidistant 2
1' Major headpores with anterior and median pore connected; posterior headpore sometimes more distant than other two 3
2(1) Postabdomen narrowing slightly distally; postabdomen postanal marginal denticles subequal in size (Fig. 4.17A–C) .... *Ovalona nuragica* (Margaritora, 1971) [Western Mediterranean Basin]

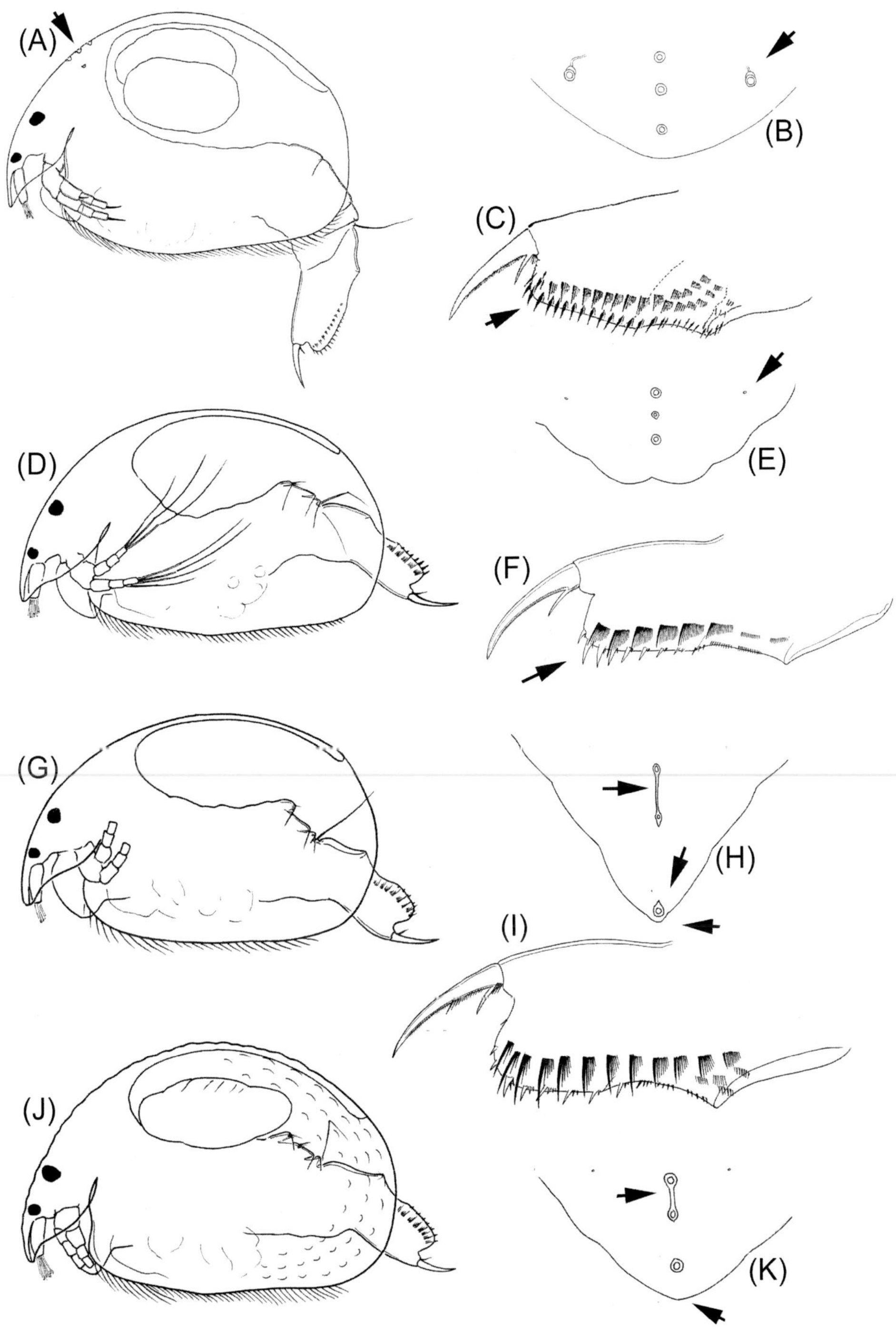

FIGURE 4.17 Representatives of Aloninae (Chydoridae): (A–C) *Ovalona nuragica*, general view, posterior portion of headshield with headpores and postabdomen; (D–F) *Ovalona cambouei*, general view, posterior portion of headshield and postabdomen; (G–I) *Ovalona azorica*, general view, posterior portion of headshield and postabdomen; (J and K) *Ovalona anastasia*, general view and posterior portion of headshield. *After Rogers et al., 2019. (C) After Alonso (1996).*

2' Postabdomen dorsal and ventral margins parallel; postabdomen postanal marginal denticles increasing in size distally (Fig. 4.17E and F) .... *Ovalona cambouei* (de Guerne & Richard, 1893) [Palearctic and Old World Subtropics]

3(1) Head shield distally acute (Fig. 4.17G–I) .... *Ovalona azorica* (Frenzel & Alonso, 1988) [Western Spain and Azores]

3' Head shield distal portion blunt (Fig. 4.17J and K) .... *Ovalona anastasia* Sinev, Alonso, Miracle & Sahuquillo, 2012 [Western Mediterranean Basin]

Branchiopoda: Cladocera: Anomopoda: Chydoridae: Aloninae: other *Alona*: Species

Alona is still under revision, with different species groups, which have been moved into separate genera (Van Damme et al., 2010). Strictly speaking, the only true *Alona* belong to the *A. quadrangularis*-group. The key below, partly after Rogers et al. (2019), does not include the *"Alona" elegans*-group, which belongs in *Coronatella*. Also "*Alona*" *intermedia* was included in the key to genera. The taxa below belong to very common, widespread species groups.

1 Lateral headpores simple, round, no structures underneath 2
1' Lateral headpores slits with vesicle-like structures underneath (*"Alona" costata*-group).. 3
2(1) Postabdomen postanal portion broadly rounded; valve with wide striae (Fig. 4.18A–D) .... *Alona quadrangularis* (O.F. Müller, 1776) species group [Cosmopolitan]
2' Postabdomen short, angular; valve striation smooth or with verruciform projections (Fig. 4.18E–G) .... "*Alona*" *guttata* Sars, 1862 species group [Cosmopolitan]
3(1) Lateral headpore slits length > distance between main headpores; postabdomen with posterodorsal angle acute (Fig. 4.18H–J) .... "*Alona*" *costata* Sars, 1862 species group [Cosmopolitan species group]
3' Lateral headpore slits length subequal to distance between main headpores; postabdomen with posterodorsal angle rounded (Fig. 4.18K–M) .... "*Alona*" *rustica* Scott, 1900 [Palearctic]

Branchiopoda: Cladocera: Anomopoda: Chydoridae: Aloninae: *Camptocercus*: Species

More species can be present in the Mediterranean Basin; please check revisions of this genus and Rogers et al. (2019) to confirm the identification.

1 Posteroventral valve corner with denticles (can be small!) (Fig. 4.19A–D) .... *Camptocercus rectirostris* Schödler, 1862 [Palearctic; rare in the Mediterranean Basin]
1' Posteroventral valve corner without denticles (Fig. 4.19E–G) .... *Camptocercus uncinatus* Smirnov, 1971 [Southern Europe including the Mediterranean Basin; also in parts of Asia and Africa]

Branchiopoda: Cladocera: Anomopoda: Chydoridae: Aloninae: *Acroperus*: Species

1 Second antenna endopod and exopod length subequal (Fig. 4.19H and I) .... *Acroperus angustatus* Sars, 1863 [Palearctic]
1' Second antenna endopod longer than exopod (Fig. 4.19J–L) .... *Acroperus harpae* (Baird, 1836) [Palearctic]

Branchiopoda: Diplostraca: Cladocera: Anomopoda: Bosminidae: Species Groups

Two subgenera occur in the Mediterranean Basin: *Bosmina (Bosmina)* and *Bosmina (Eubosmina)*. Please note that *B. coregoni* can be extremely variable in bodyshape. Following Rogers et al. (2019).

1 Lateral headpore situated along headshield lateral margin, close to base of second antenna (Fig. 4.20A–C) .... *Bosmina (Bosmina) coregoni* Baird, 1857 species group [Palearctic]
1' Lateral headpore situated far from the lateral margin of the headshield, close to mandibular articulation (Fig. 4.20D–I) .... *Bosmina (Eubosmina) longirostris* (O.F. Müller, 1776) species group [Cosmopolitan]

Branchiopoda: Diplostraca: Cladocera: Anomopoda: Daphniidae: Genera

Modified after Alonso (1996) and Flößner (2000).

1 Ventral carapace margin convex, not ending in a ventral spine 2

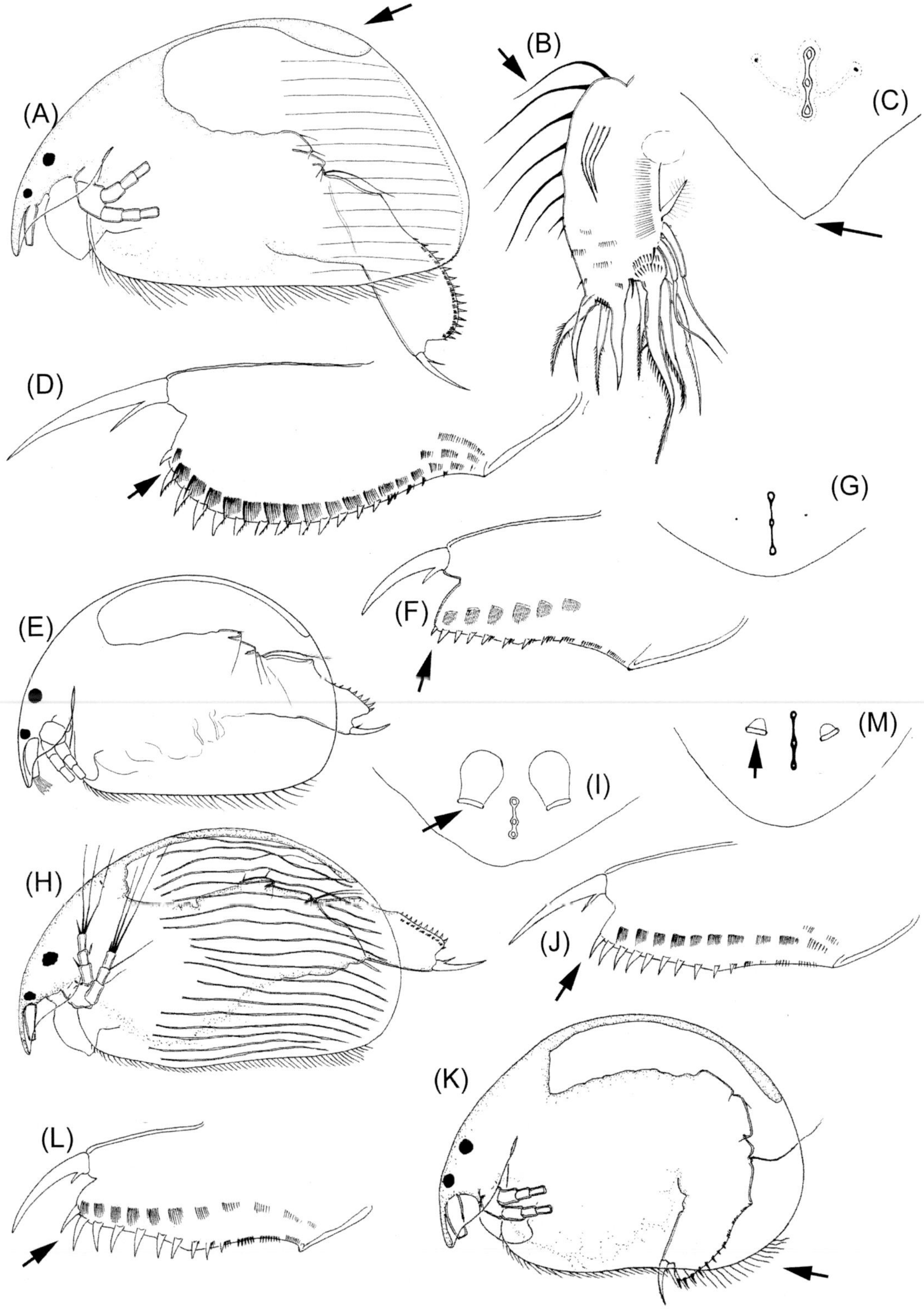

FIGURE 4.18 Representatives of Aloninae (Chydoridae): (A–D) *Alona quadrangularis*, general view, first limb, posterior margin headshield with headpores, and postabdomen; (E–G) *"Alona" guttata*, general view, postabdomen and posterior margin of headshield with headpores; (H–J) *"Alona" costata*, general view, posterior margin of headshield with head "slits," and postabdomen; (K–M) *"Alona" rustica*, general view, postabdomen and posterior margin of headshield with head "slits." *After Rogers et al., 2019.*

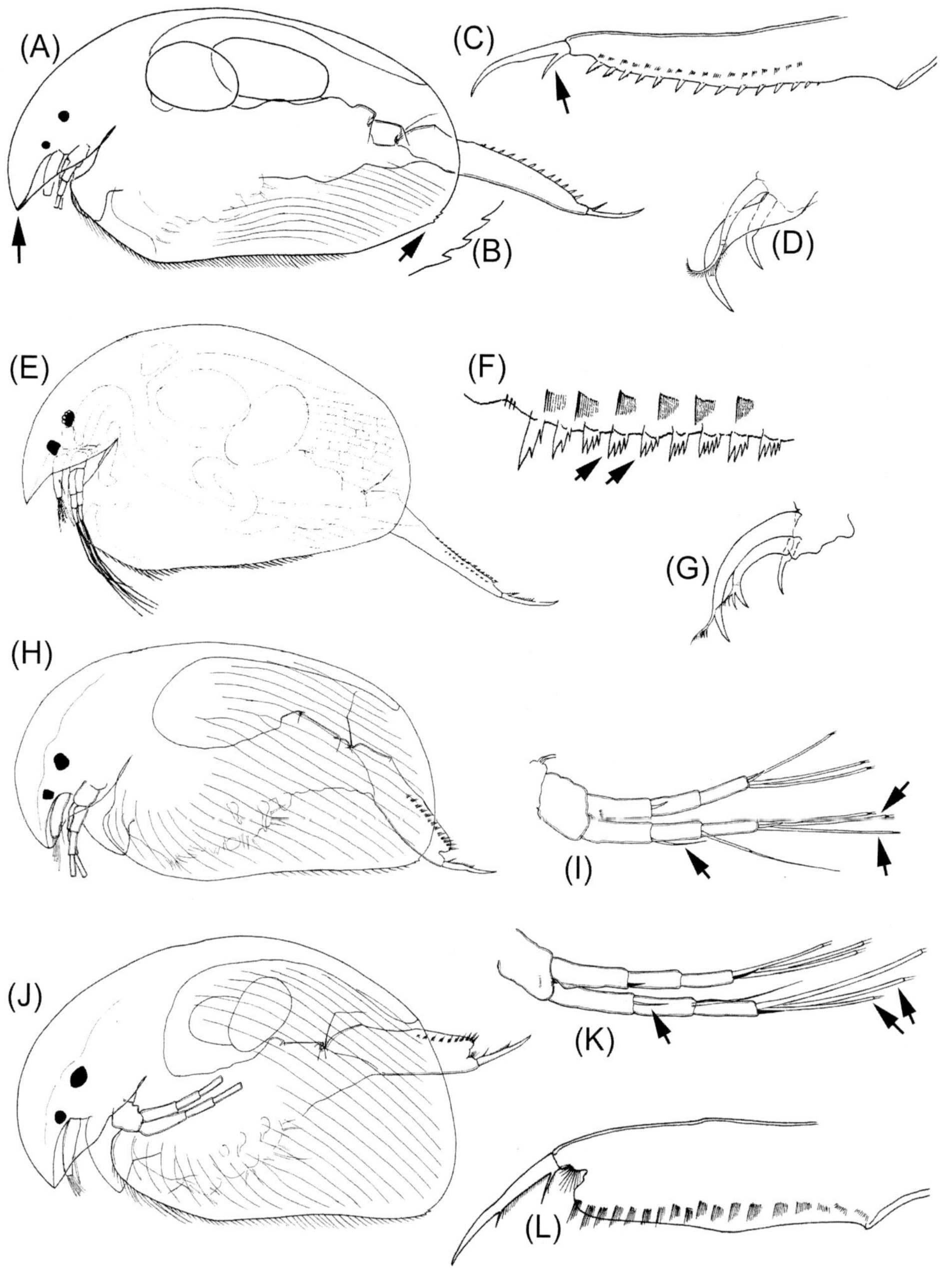

FIGURE 4.19 Representatives of Aloninae (Chydoridae): (A–D) *Camptocercus rectirostris*, general view, posteroventral valve corner, postabdomen and inner distal lobe of P1; (E–G) *Camptocercus uncinatus*, general view, marginal denticles on postabdomen and inner distal lobe of P1; (H and I) *Acroperus angstatus*, general view and second antenna; (J–L) *Acroperus harpae*, general view, second antenna and postabdomen. *After Rogers et al., 2019.*

1' Ventral carapace margin straight and ending in a ventral spine 4
2(1) Carapace dorsoposterior spine absent or sometimes a rounded bump 3
2' Carapace ending in a robust dorsoposterior spine (Figs. 4.21 and 4.22) .... ***Daphnia***
3(2) Postabdomen with anal margin strongly concave (Fig. 4.23) .... ***Simocephalus***
3' Postabdomen with anal margin straight (Figs. 4.24 and 4.25A–E) .... ***Ceriodaphnia***
4(3) Rostrum long, acute, tucked between carapace ventral margins; ocellus elongate and situated at least halfway between rostral apex and compound eye (Fig. 4.25L) .... *Megafenestra aurita* Dumont & Pensaert, 1983 [Palearctic, Afrotropical]
4' Rostrum blunt, free from carapace margins; ocellus small and immediately next to rostral apex (Fig. 4.25F–J) .... ***Scapholeberis***

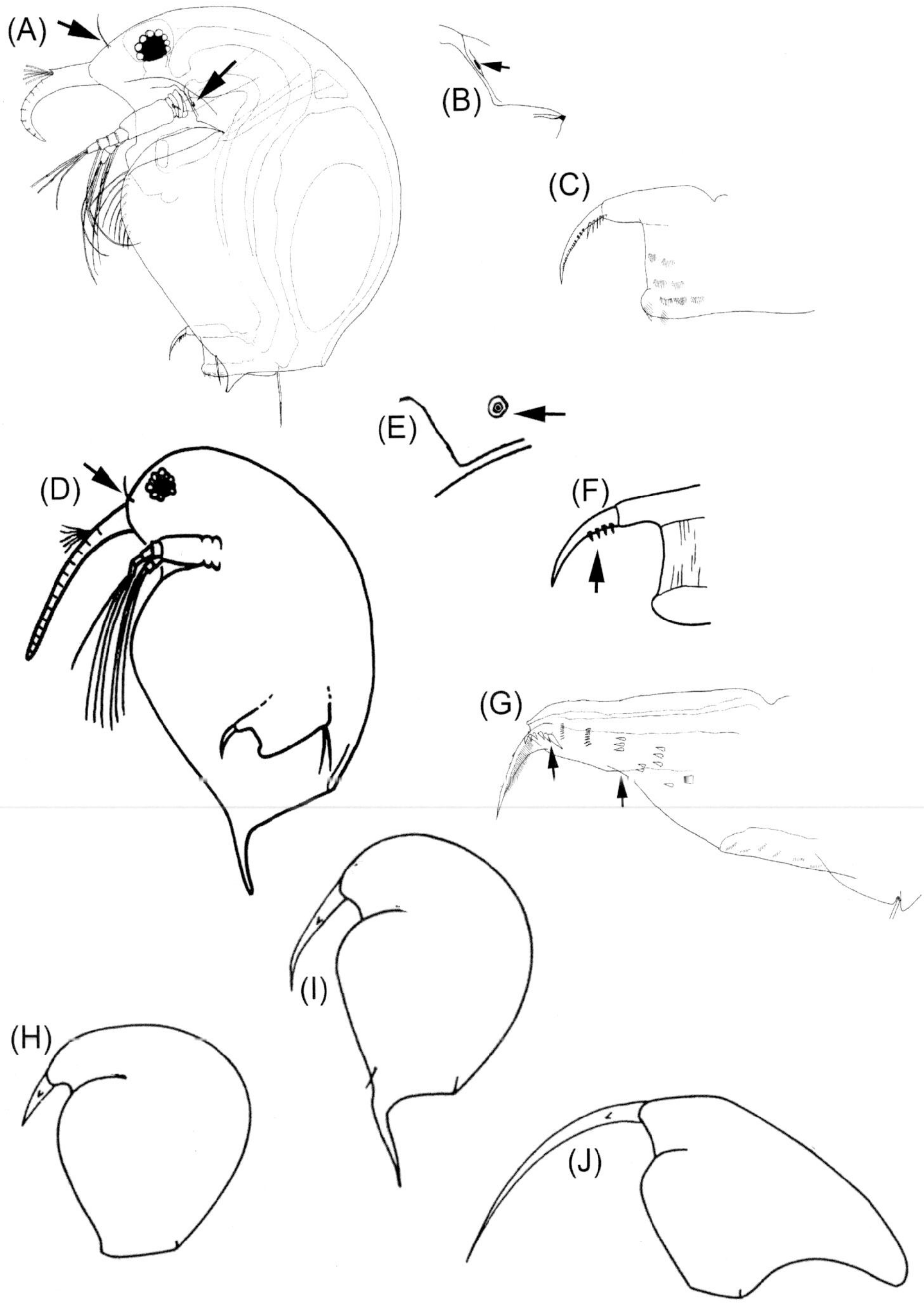

FIGURE 4.20 Representatives of Bosminidae: (A–C) *Bosmina longirostris*, general view, position lateral headpore and postabdomen; (D–J) *Bosmina coregoni*, general view, position lateral headpore, terminal claw of postabdomen, male postabdomen (Amoros, 1984); (G–I) *Bosmina coregoni*, different body shapes. *(A–C, J–I) After Rogers et al., 2019; (D, E) After Amoros (1984).*

Branchiopoda: Cladocera: Anomopoda: Daphniidae: *Daphnia*: Subgenera and Species

Follows Alonso (1996) and Rogers et al. (2019). Afrotropical populations can be expected in the Mediterranean Basin (e.g., populations of *D. carinata* were recorded from southern Algeria); refer to Benzie (2005) for more details.

1 Headshield posteriorly with deep embayment; carapace dorsal keel reaches head (subgenus *Ctenodaphnia*) 2
1' Headshield posteriorly acute; carapace dorsal keel not reaching head (subgenus *Daphnia*) 10
2(1) Postabdomen dorsal anal-postanal margin with concave embayment 3
2' Postabdomen dorsal anal-postanal margin lacking embayment 4
3(2) Body and head elongate; head midline without lateral grooves; postabdomen dorsal anal-postanal margin moderately concave (Fig. 4.21A–D) .... *Daphnia (Ctenodaphnia) deserti* Gauthier, 1937 [North Africa]

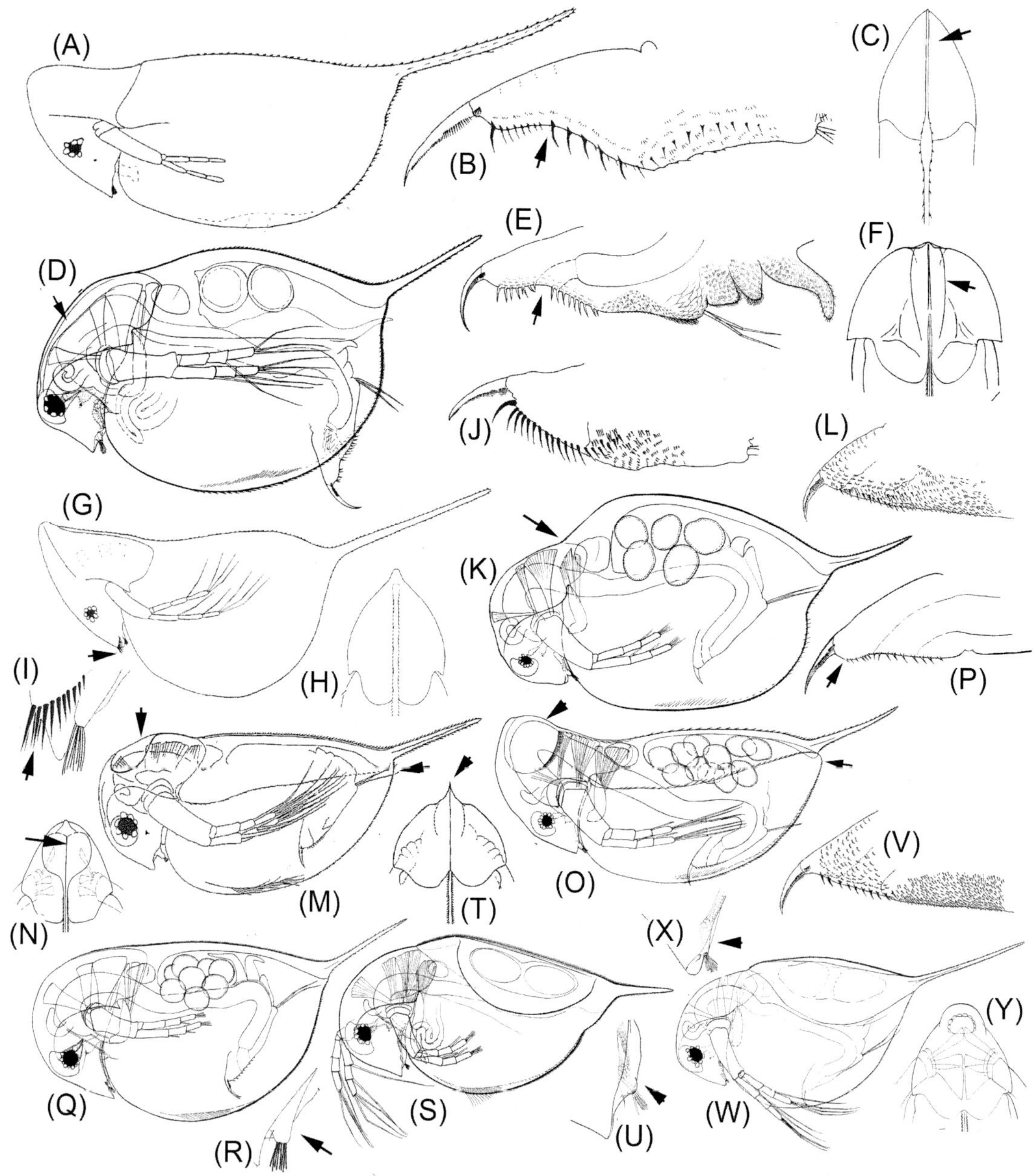

FIGURE 4.21 Representatives of Daphniidae – *Daphnia (Ctenodaphnia)*; (A–D) *Daphnia (Ctenodaphnia) deserti*, general view, postabdomen and head in dorsal view; (D–F) *Daphnia magna*, general view, postabdomen and head in dorsal view, (G–J) *Daphnia (Ctenodaphnia) barbata*, general view, head in dorsal view, rostrum and postabdomen; (K and L) *Daphnia (Ctenodaphnia) mediterranea*, general view and postabdomen; (M and N) *Daphnia (Ctenodaphnia) atkinsoni*, general view and head in dorsal view with head plate; (O and P) *Daphnia triquetra*, general view and postabdomen; (Q and R) *Daphnia similis*, general view and first antenna; (S–V) *Daphnia (Ctenodaphnia) hispanica*, general view (ephippial female), head in dorsal view, first antenna and postabdomen; (W–Y) *Daphnia (Ctenodaphnia) chevreuxi*, general view and head in dorsal view. *(A–C, G–J, R) After Benzie (2005); (L, T, and V) After Alonso (1996); Remaining images after Rogers et al. (2019).*

3' Body broad and robust; head midline with deep lateral grooves; postabdomen dorsal anal-postanal margin deeply concave, almost dividing postabdomen (Fig. 4.21D–F) .... *Daphnia* (*Ctenodaphnia) magna* Straus, 1820 [Palearctic]

4(2) Rostrum without spinule rows 5

4' Rostrum with transverse spinule rows (Fig. 4.21G–J) .... *Daphnia (Ctenodaphnia) barbata* Weltner, 1897 [Algeria, Afrotropics]

5(4) First antenna only fused at base, directed ventroposteriorly 6

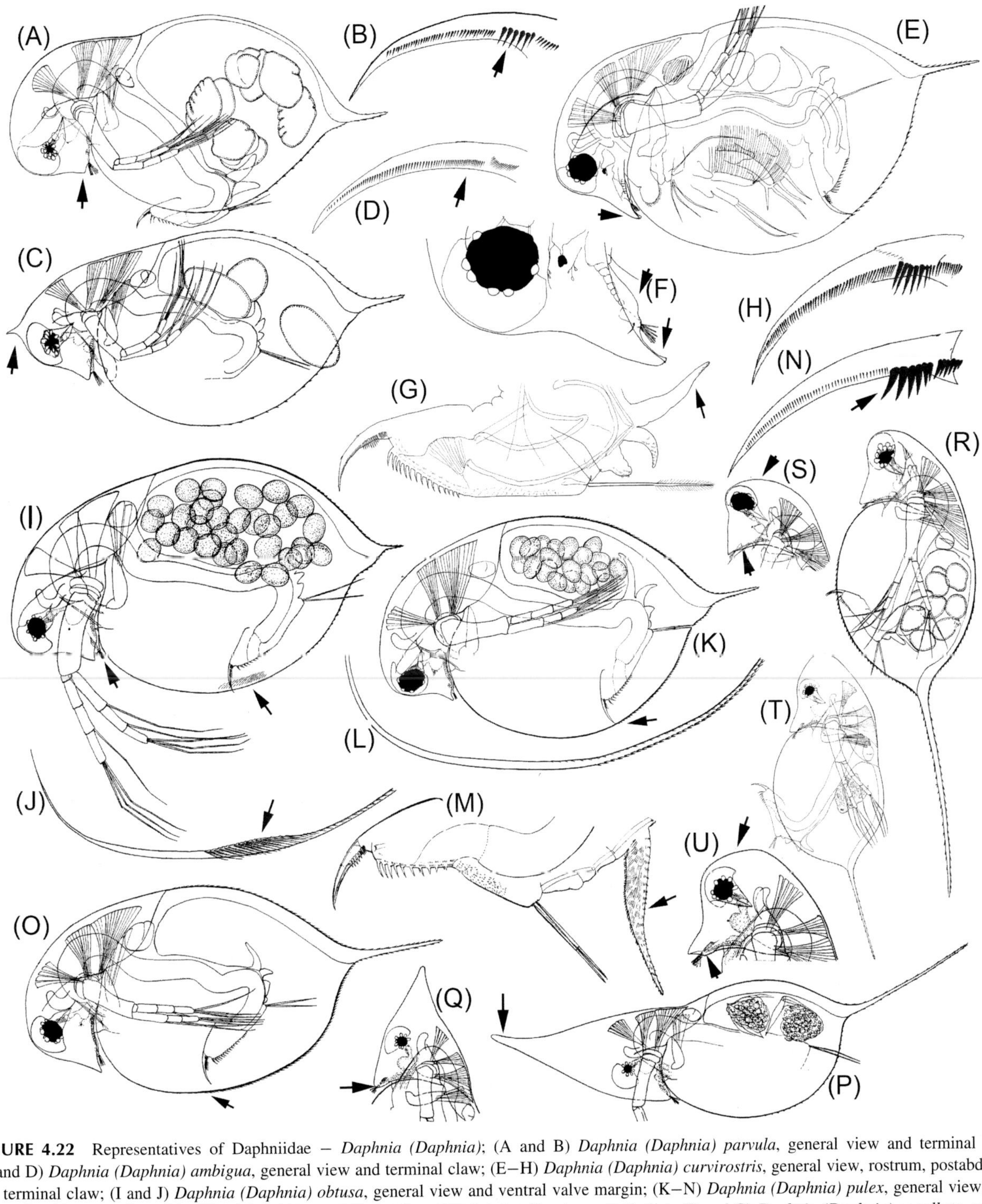

FIGURE 4.22 Representatives of Daphniidae – *Daphnia (Daphnia)*; (A and B) *Daphnia (Daphnia) parvula*, general view and terminal claw; (C and D) *Daphnia (Daphnia) ambigua*, general view and terminal claw; (E–H) *Daphnia (Daphnia) curvirostris*, general view, rostrum, postabdomen and terminal claw; (I and J) *Daphnia (Daphnia) obtusa*, general view and ventral valve margin; (K–N) *Daphnia (Daphnia) pulex*, general view, ventral valve margin, postabdomen and terminal claw; (O) *Daphnia (Daphnia) pulicaria*, general view; (P and Q) *Daphnia (Daphnia) cucullata*, general view and head; (R and S) *Daphnia (Daphnia) longispina*, general view and head; (T and U) *Daphnia (Daphnia) galeata*, general view and head. *After Rogers et al., 2019; (B, D, H, and N) After Benzie (2005).*

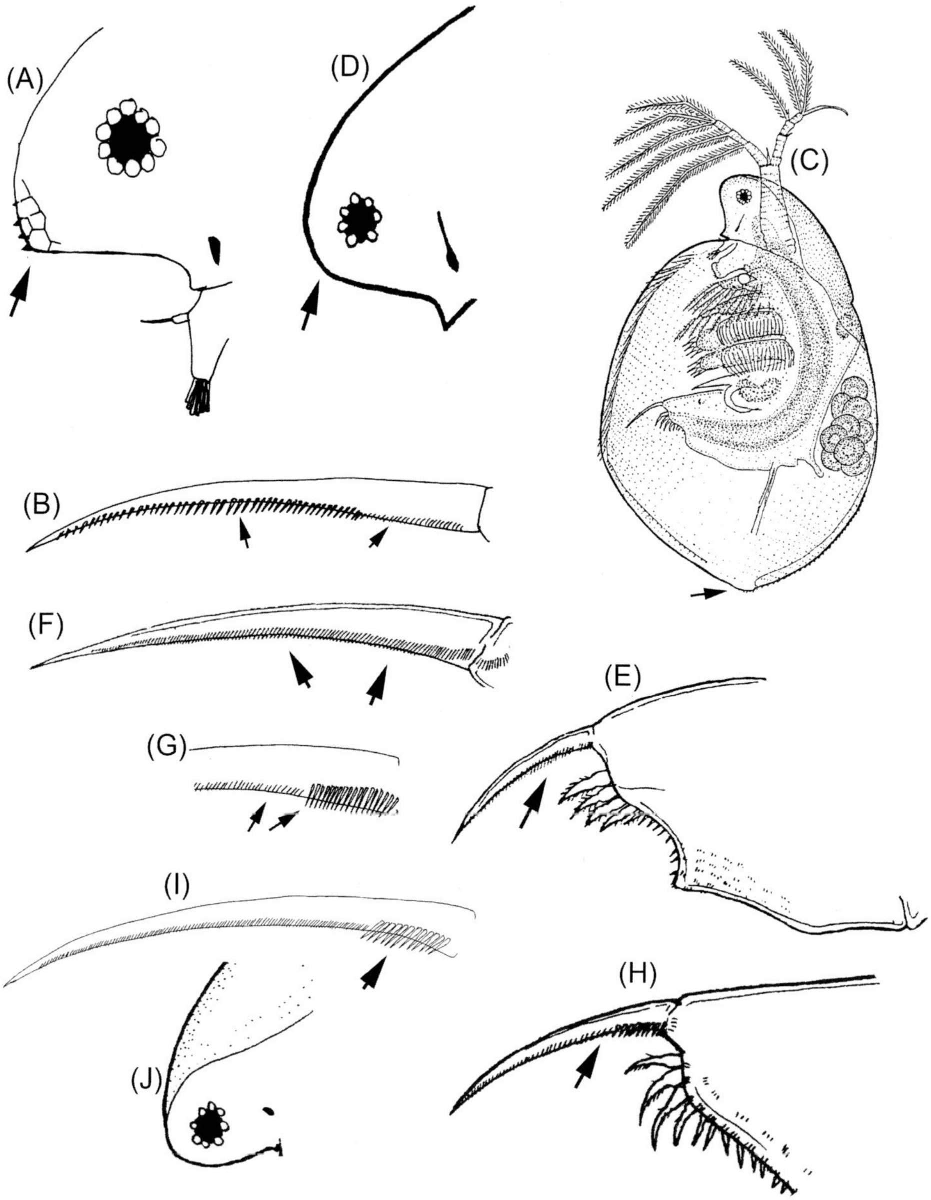

FIGURE 4.23 Representatives of Daphniidae (*Simocephalus*): (A and B) *Simocephalus serrulatus*, head and terminal claw; (C–F) *Simocephalus vetulus*, general view, head, postabdomen and terminal claw; (G) *Simocephalus congener*, terminal claw dorsal pecten; (H–J) *Simocephalus exspinosus*, postabdomen, terminal claw and head. *After Orlowa-Bienkowskaja (2001).*

5’ First antenna fused to rostrum for most of its length 8

6(5) Head with dorsal crown mostly of spinules; body with dorsolateral ridge, often with spines 7

6’ Head without dorsal crown; body without dorsolateral ridges on each side (Fig. 4.21K and L) .... *Daphnia (Ctenodaphnia) mediterranea* Alonso, 1985 [Mediterranean Basin]

7(6) Ventral valve area adjacent to caudal spine without a round projection (Fig. 4.21M and N) .... *Daphnia (Ctenodaphnia) atkinsoni* Baird, 1859 [Europe and circum-Mediterranean Basin]

7’ Ventral valve area adjacent to caudal spine with a round projection (Fig. 4.21O and P) .... *Daphnia (Ctenodaphnia) triquetra* Sars, 1903 [Southern Europe]

8(5) First antenna fused into an anntenal mound, aesthetascs not reaching rostral apex, directed away from rostral apex 9

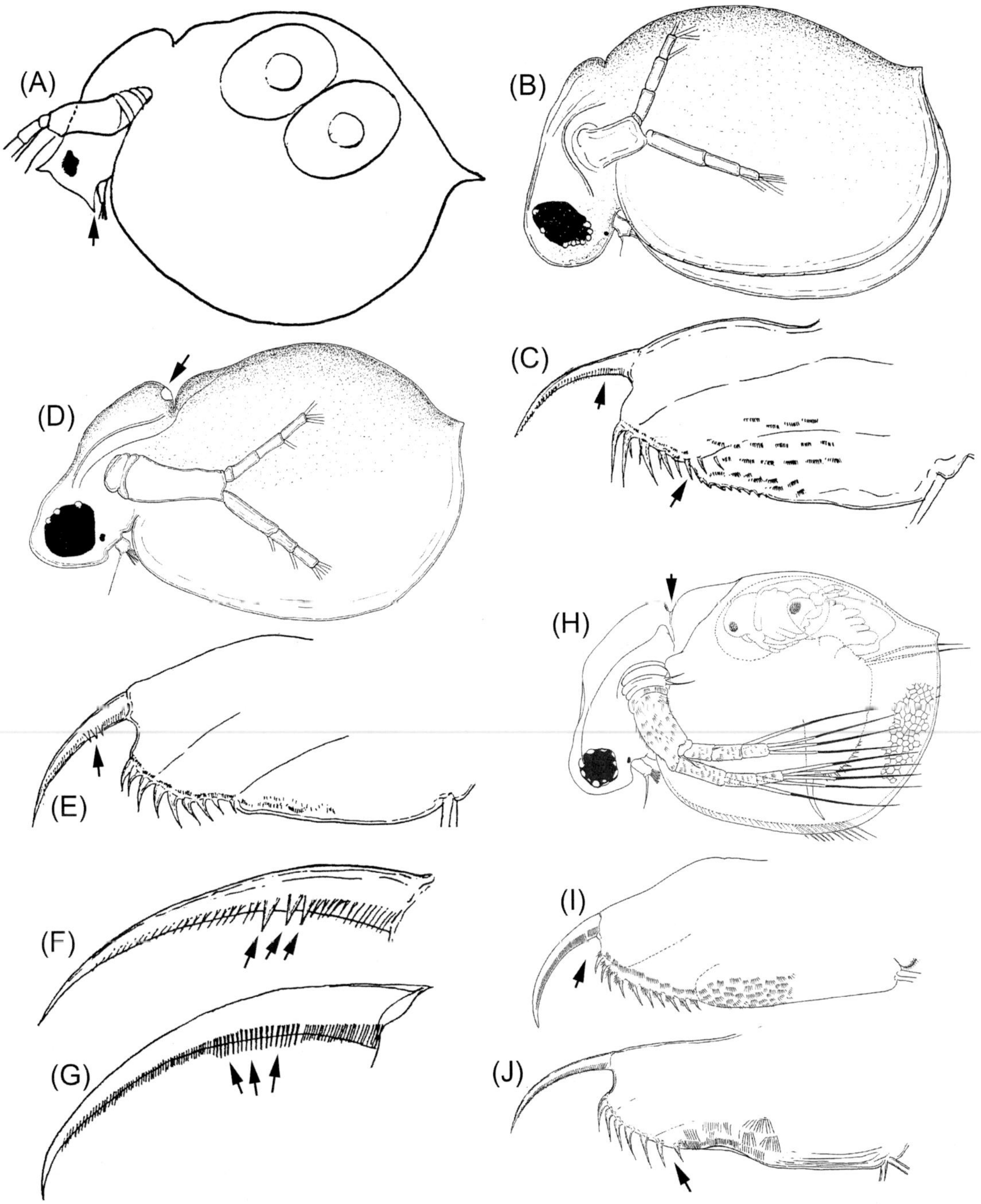

FIGURE 4.24 Representatives of Daphniidae (*Ceriodaphnia*); (A) *Ceriodaphnia cornuta*, general view; (B and C) *Ceriodaphnia megops*, general view and postabdomen; (D–F) *Ceriodaphnia reticulata*, general view, postabdomen and terminal claw; (G) *Ceriodaphnia dubia*, terminal claw; (H and I) *Ceriodaphnia smirnovi*, general view and postabdomen; (J) *Ceriodaphnia quadrangula*, postabdomen. *(A–G, J) After Rogers et al., 2019; (H–I) After Alonso et al., (2021).*

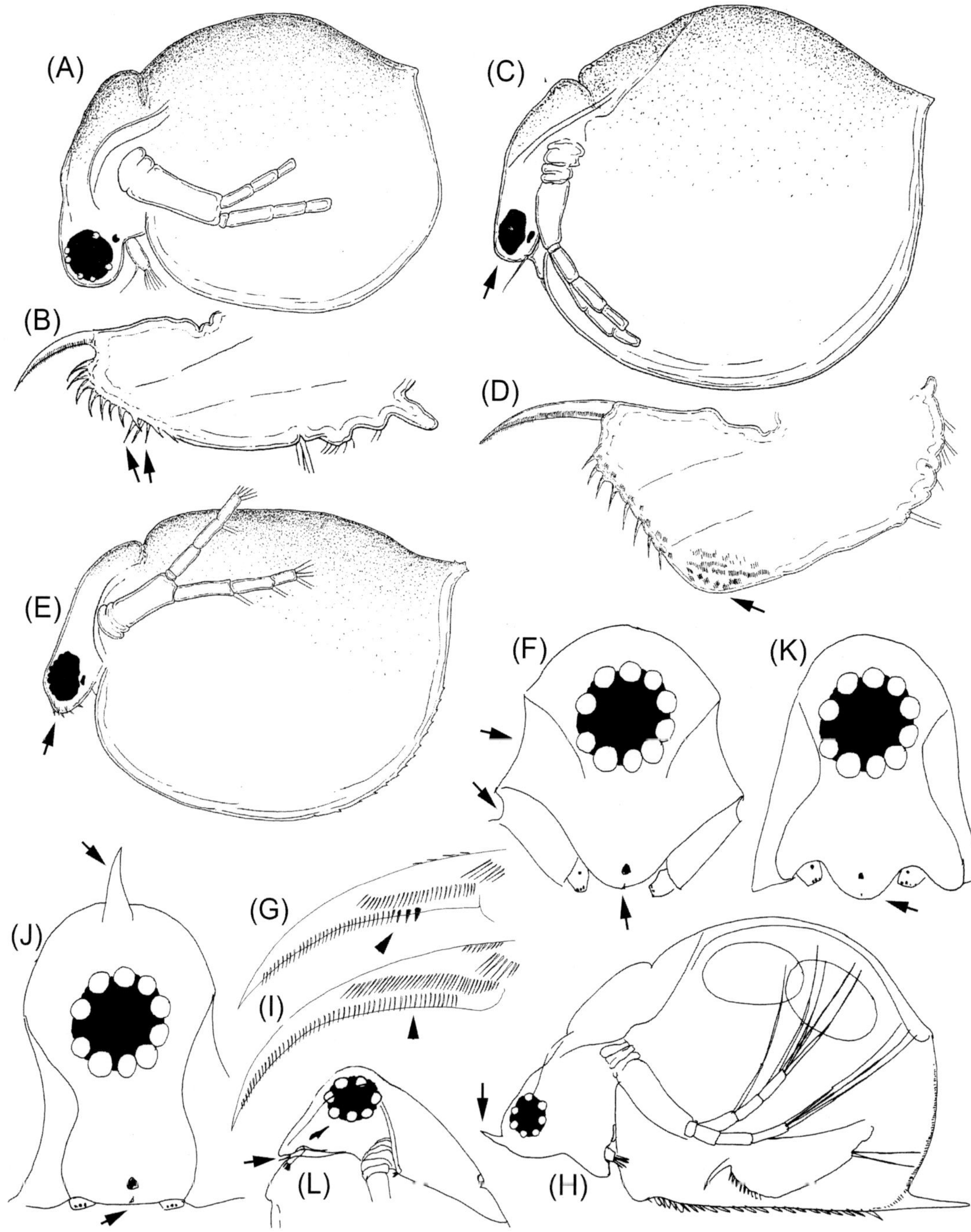

FIGURE 4.25 Representatives of Daphniidae (*Ceriodaphnia, Scapholeberis, Megafenestra*): (A and B) *Ceriodaphnia pulchella*, general view and postabdomen; (C and D) *Ceriodaphnia laticaudata*, general view and postabdomen; (E) *Ceriodaphnia rotunda*, general view; (F and G) *Scapholeberis rammneri*, head in frontal/ventral view and terminal claw; (H–J) *Scapholeberis mucronata*, general view, terminal claw and head in frontal/ventral view; (K) *Scapholeberis kingii*, head in frontal/ventral view. (L) *Megafenestra aurita*, head in lateral view. *After Rogers et al., 2019.*

8' First antenna partly fused, aesthetascs reaching rostral apex or beyond, directed distally (Fig. 4.21Q and R) .... *Daphnia (Ctenodaphnia) similis* Claus, 1876 species complex [Holarctic]
9(8) Head with large dorsomedial keel or crest (Fig. 4.21S–V) .... *Daphnia (Ctenodaphnia) hispanica* Glagolev & Alonso, 1990 [Spain]
9' Head without dorsomedial crest (Fig. 4.21W–Y) .... *Daphnia (Ctenodaphnia) chevreuxi* Richard, 1896 [Mediterranean Basin]
10(1) Adult females <1.5 mm; rostrum very short 11
10' Adult females larger than 1.5 mm; rostrum acute, not short 12
11(10) Head rounded; postabdomen terminal claw with second proximal pecten distinctly larger (Fig. 4.22A and B) .... *Daphnia (Daphnia) parvula* Fordyce, 1901 [Americas; invasive in Europe]
11' Head narrow, often acute; terminal claw second proximal pecten not distinctly larger (Fig. 4.22C and D) .... *Daphnia (Daphnia) ambigua* Scourfield, 1957 [Americans; invasive in Europe]
12(10) Postabdomen terminal claw with dorsal proximal pecten robust, consisting of robust teeth 13
12' Postabdomen terminal claw with dorsal pectens spinulae subequal in thickness 16
13(12) First antenna normal; rostrum not reaching, or reaching just beyond first antenna aesthetascs (at most 1x aesthetascs length beyond) 14
13' First antenna rudimentary or absent; rostrum reaching beyond aesthetascs by more than 1 × aesthetascs length (Fig. 4.22E–H) .... *Daphnia (Daphnia) curvirostris* Eylmann, 1887 [Palearctic]
14(13) Carapace ventral margin medially without long plumose setae 15
14' Carapace ventral margin medially with long plumose setae (Fig. 4.22I and J) .... *Daphnia (Daphnia) obtusa* Kurz, 1874 species group [Cosmopolitan]
15(14) Carapace valve ventral margin with less than half covered in denticles (Fig. 4.22K–N) .... *Daphnia (Daphnia) pulex* Leydig, 1860 species group [Cosmopolitan; hybrids are possible with *D. (D.) pulicaria*]
15' Carapace valve ventral margin with more than half covered in denticles (Fig. 4.22O) .... *Daphnia (Daphnia) pulicaria* Forbes, 1893 species group [Holarctic; hybrids are possible with *D. (D.) pulex*]
16(12) Ocellus present, small; head posterior margin with aesthetascs not protruding beyond rostral apex 17
16' Ocellus mostly absent; aesthetascs just reaching rostral apex or beyond (Fig. 4.22P and Q) .... *Daphnia (Daphnia) cucullata* Sars, 1862 [Palearctic]
17(16) Antennal mound between aesthetascs absent; helmet absent or rounded; compound eye in line with caudal spine (Fig. 4.22R and S) .... *Daphnia (Daphnia) longispina* O. F. Müller, 1785 species group [Cosmopolitan; hybrids known]
17' Antennal mound between aesthetascs present; helmet, if present, acute; compound eye more anterior to caudal spine axis (Fig. 4.22T and U) .... *Daphnia (Daphnia) galeata* Sars, 1864 species group [Palearctic; hybrids known]

Branchiopoda: Cladocera: Anomopoda: Daphniidae: *Simocephalus*: Species

Simocephalus in the Mediterranean Basin needs revision. The species below belong to four subgenera (*Aquipiculus, Coronocephalus, Echinocaudus* and *Simocephalus*); for more details, see Orlova-Bienkowskaja (2001). Following Rogers et al. (2019) and Flößner (2000).

1 Head without anterior denticles 2

1' Head with anterior denticles (Fig. 4.23A and B) .... *Simocephalus serrulatus* (Koch, 1841) species group [Cosmopolitan]
2(1) Terminal claw dorsally with distal pecten larger than proximal 3
2' Terminal claw with dorsal pecten one continuous row (Fig. 4.23C–F) .... *Simocephalus vetulus* (O.F. Müller, 1776) species group [Cosmopolitan]
3(2) Terminal claw proximally with c. 30 thin denticles (Fig. 4.23G) .... *Simocephalus congener* (Koch, 1841) [Palearctic]
3' Terminal claw proximally with ca. 15 large thick denticles (Fig. 4.23H–J) .... *Simocephalus exspinosus* (De Geer, 1778) species group [Palearctic]

Branchiopoda: Cladocera: Anomopoda: Daphniidae: *Ceriodaphnia*: Species

Ceriodaphnia is in need of revision. The taxa are hard to identify: morphologies are variable, there are many cryptic species, global keys are outdated, and there is potential hybridization. Identification to species groups is feasible, but it is necessary to check the recent specialized literature.

1 Rostrum rounded 2
1' Rostrum narrow and acute (plastic feature) (Fig. 4.24A) .... *Ceriodaphnia cornuta* Sars, 1885 species group [Subtropical]
2(1) Postabdomen dorsal margin without a deep depression near anus 3
2' Postabomen dorsal margin with a deep depression near anus (Fig. 4.24B and C) .... *Ceriodaphnia megops* Sars, 1862 [Eurasia]
3(2) Postabdominal claw with differentiated dorsal pecten, some spines thicker and larger than others 4
3' Postabdominal claw with all spines of similar thickness 5
4(3) Postabdomen terminal claw dorsally with medial pecten with 3–7 (mostly 4–5) strong spines (Fig. 4.24D–F) .... *Ceriodaphnia reticulata* (Jurine, 1820) species group [Cosmopolitan]
4' Postabdomen terminal claw dorsally with pecten with 20 long spines (Fig. 4.24G) .... *Ceriodaphnia dubia* Richard, 1894 species group [Cosmopolitan]
5(3') Postabdomen maximal width in proximal portion 6
5' Postabdomen maximal width in middle portion 7
6(5) Promixal anal denticles without smaller denticles in between (Fig. 4.24H and I) .... *Ceriodaphnia smirnovi* Alonso, Neretina & Ventura, 2021 [Mediterranean Basin; member of the *C. quadrangula*-group (Fig. 4.24J)]
6' Proximal anal denticles with smaller denticles in between (Fig. 4.25A and B) .... *Ceriodaphnia pulchella* (Sars, 1862) species group [Holarctic]
7(5) Head anteriorly without small denticles (Fig. 4.25C and D) .... *Ceriodaphnia laticaudata* P.E. Müller, 1867 [Holarctic]
7' Head anteriorly with small denticles (Fig. 4.25E) .... *Ceriodaphnia rotunda* Sars, 1862 [Palearctic]

Branchiopoda: Cladocera: Anomopoda: Daphniidae: *Scapholeberis*: Species

After Rogers et al. (2019); easily confused with *Megafenestra* (see Daphniidae key).

1 Head lacking deep grooves starting from the first antenna 2

1' Head with deep grooves starting from the first antenna base (Fig. 4.25F and G) .... *Scapholeberis rammneri* Dumont & Pensaert, 1983
[Holarctic]
2(1) Head in anterior view with rostrum broad and flat, first antenna not in deep depression; often head with dorsal spine (Fig. 4.25H–J) .... *Scapholeberis mucronata* (O.F. Müller, 1776)
[Palearctic]
2' Head in anterior view with rostrum convex, first antenna in deep depression; head rarely with dorsal spine (Fig. 4.25K) .... *Scapholeberis kingi* Sars, 1888
[Circumtropical species group in need of revision; also in southern Palearctic]

Branchiopoda: Diplostraca: Cladocera: Anomopoda: Moinidae: *Moina*: Species

Following Goulden (1968) and Rogers et al. (2019). Subgenera are not accepted here. North American-East Asian *M.* "*affinis*" reported from Italy (Goulden, 1968) and Sicily, potentially introduced, but relationships of these populations were not confirmed by molecular methods—the potentially exotic(?) taxon is included in the key below.

1 Postabdomen terminal claw with dorsal proximal pecten spines similar or 2–3 × larger than distal pecten 2
1' Postabdomen terminal claw with dorsal proximal pecten robust, with spines ≥5 × longer than distal pecten adjacent spines (Fig. 4.26A–C) .... *Moina brachiata* (Jurine, 1820) species group
[Palearctic]
2(1) Thoracopod I "penultimate" segment with anterior seta present 3
2' Thoracopod I without a seta in this position (Fig. 4.26D–F) .... *Moina salina* Daday, 1888
[Palearctic]
3(2) Thoracopod I "penultimate" segment with anterior seta with fine setulae (Fig. 4.26J) 4
3' Thoracopod I "penultimate" segment with anterior seta with spinulae (Fig. 4.26G) .... *Moina macrocopa* (Straus, 1820) species group
[Europe, Africa, and Southern Asia]
4(3) Head and valves with long setae (Fig. 4.26K) 5
4' Head and valves without long setae or only on head ventral surface (Fig. 4.26H–J) .... *Moina micrura* Kurz, 1874 species group
[Palearctic, Pantropics]
5(4) Body length <1.2 mm; head narrow, with robust supraocular depression (Fig. 4.26L) 6
5' Body length.2–1.6 mm; supraocular depression obscure (Fig. 4.26K) .... *Moina belli* Gurney, 1904
[Africa, western Asia]
6(5) Postabdomen with terminal claw pecten distinct; head and valve setae dense; posteroventral valve corner setules not in groups (Fig. 4.26L and M) .... *Moina affinis* Birge, 1893 species group
[North America-East Asia; potentially introduced]
6' Postabdomen with terminal claw pecten indistinct; head and valve setae sparse; posteroventral valve corner setules in groups (Fig. 4.26N and O) .... *Moina weismanni* Ishikawa, 1896
[Asia-Southern Europe]

Branchiopoda: Diplostraca: Cladocera: Anomopoda: Macrothricidae: Genera

Following Rogers et al. (2019). See also Smirnov (1992).

1 Intestine straight, without convolutions 2
1' Intestine with convolutions 5
2(1) Carapace ventral margin setae not lanceolate, not flattened 3
2' Carapace ventral margin setae lanceolate and flattened (Fig. 4.27A and B) .... *Lathonura rectirostris* (O.F. Muller, 1785)
[Holarctic]
3(2) Body moderately or not compressed laterally 4

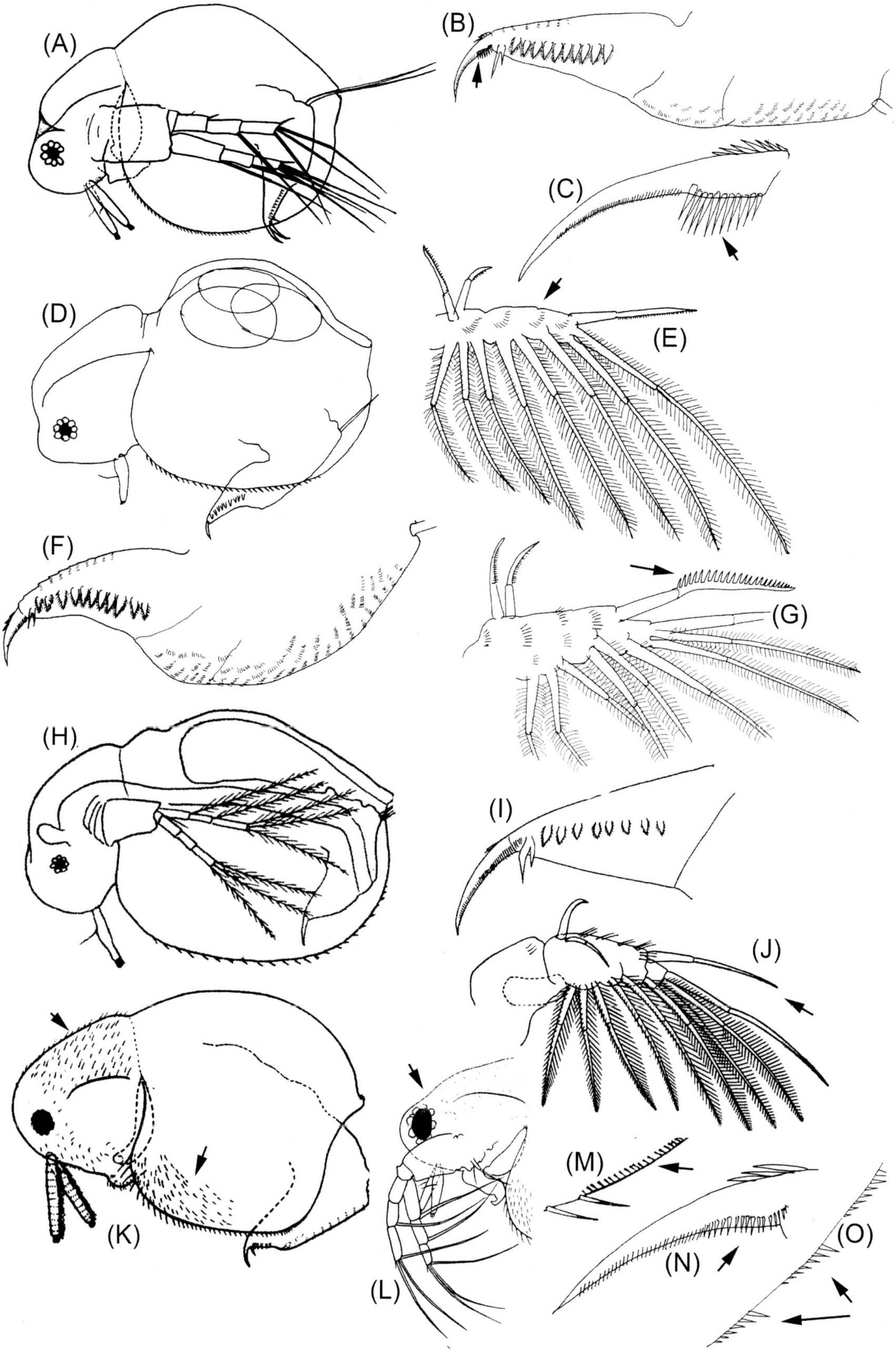

FIGURE 4.26 Representatives of Moinidae (*Moina*): (A–C) *Moina brachiata*, general view, postabdomen and terminal claw with large pecten; (D–F) *Moina salina*, general view, first limb and postabdomen; (G) *Moina macrocopa*, first limb; (H–J) *Moina micrura*, general view, postabdomen and first limb; (K) *Moina belli*, general view; (L and M) *Moina affinis*, head with supraocular depression and posteroventral valve corner setulation; (N and O) *Moina weismanni*, terminal claw of postabdomen and posteroventral valve corner setulation. *(A and J–M) After Goulden (1968). Other images after Rogers et al., 2019.*

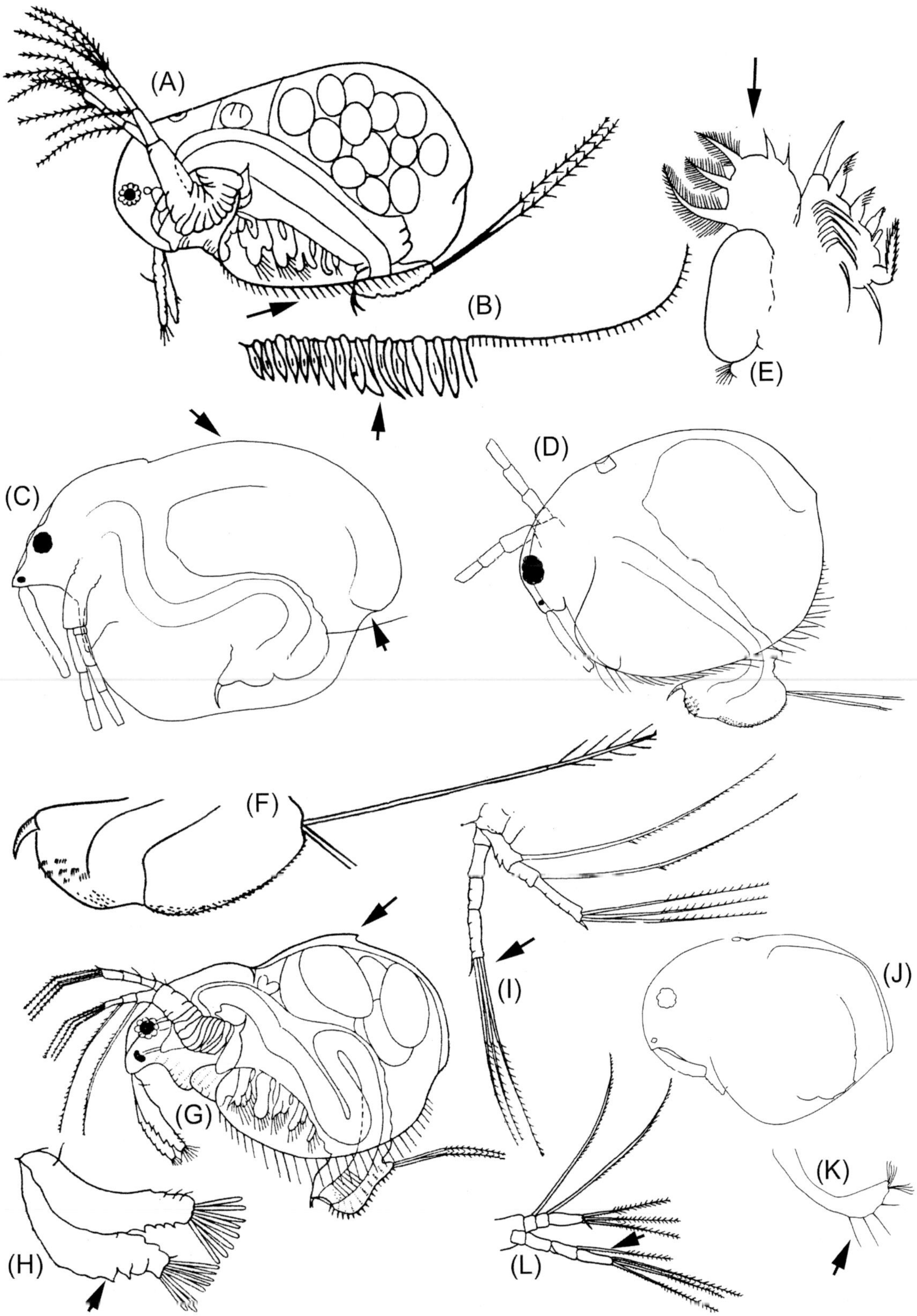

FIGURE 4.27 Representatives of Macrothricidae: (A and B) *Lathonura rectirostris*, general view, ventral valve margin; (C) *Bunops serricaudata*, general view; (D–F) *Wlassicsia pannonica*, general view, fourth thoracic limb (PIV) and postabdomen; (G–I) *Drepanothrix dentata*, general view, first antenna and second antenna; (J–L) *Streblocerus pygmaeus*, general view, first antenna and second antenna. *After Rogers et al., 2019. (G–I) After Lilljeborg (1900).*

3' Body strongly laterally compressed, with a high and acute dorsal keel (Fig. 4.27C) .... *Bunops serricaudata* (Daday, 1888)
[Palearctic]
4(3) Thoracopod IV exopodite with 2–3 setae .... ***Macrothrix***
4' Thoracopod IV exopodite with 5 setae (Fig. 4.27D–F) .... *Wlassicsia pannonica* Daday, 1904
[Palearctic]
5(1) Body with large dorsal spine; second antenna exopod (with four articles) with three apical setae and no lateral setae (Fig. 4.27G–I) .... *Drepanothrix dentata* (Eurén, 1861)
[Palearctic]
5' Body without large dorsal spine; second antenna exopod (with four articles) with three apical and one lateral seta (Fig. 4.27J–L) .... *Streblocerus serricaudatus* (Fischer, 1849)
[Palearctic]

Branchiopoda: Cladocera: Anomopoda: Macrothricidae: *Macrothrix*: Species

Following Rogers et al. (2019); *Macrothrix* needs revision; see also Smirnov (1992) and recent revisions per species group.

1 Head ventral margin straight to slightly convex, without a round projection; postabdomen without anal flaps 2
1' Head ventral margin with a round projection; postabdomen with anal flaps (Fig. 4.28A–C) .... *Macrothrix odiosa* Gurney, 1916
[Palearctic]
2(1) Carapace dorsal margin smooth or finely serrated 3
2' Carapace dorsal margin with a strongly serrated keel (Fig. 4.28D) .... *Macrothrix laticornis* (Jurine, 1820)
[Holarctic]
3(2) First antenna widening distally 4
3' First antenna not widening distally, evenly cylindrical 6
4(3) Carapace dorsal margin smooth or with minute folds 5
4' Carapace dorsal margin with distinct folds, appearing finely serrated (Fig. 4.28E–G) .... *Macrothrix spinosa* King, 1853
[Pantropical]
5(4) Postabdomen terminal claw length $< 0.5\times$ anal margin; first antenna aesthetascs strongly differ in size (Fig. 4.28H–J) .... *Macrothrix hirsuticornis* Norman & Brady, 1867
[Holarctic]
5' Postabdomen terminal claw length $> 0.5\times$ anal margin; first antenna aesthetascs similar (Fig. 4.28K and L) .... *Macrothrix dadayi* Behning, 1941
[Palearctic]
6(5) Postabdominal dorsal seta distal portion length ≥ 3 diameters of the seta; second antenna endopod proximal article seta with short spinules (Fig. 4.28M–P) .... *Macrothrix rosea* Jurine, 1820
[Holarctic]
6' Postabdominal dorsal seta distal portion length $2\times$ the diameter of the seta; second antenna endopod proximal article seta bearing 2–3 longer, robust spinules (Fig. 4.28Q and R) .... *Macrothrix triserialis* Brady, 1886
[Pantropical]

Branchiopoda: Diplostraca: Cladocera: Anomopoda: Ilyocryptidae: *Ilyocryptus*: Species

Follows Rogers et al. (2019) but is in need of revision for the Mediterranean Basin. See also Kotov & Štifter (2006).

1 Molting complete (no growth lines on carapace) 2
1' Molting incomplete (growth lines on carapace present) 3
2(1) Postabdomen with anus approximately midway along dorsal margin, preanal margin short (Fig. 4.29A and B) .. .. *Ilyocryptus agilis* Kurz, 1874
[Palearctic]
2' Postabdomen with anus subdistal, preanal margin long (Fig. 4.29C and D) .... *Ilyocryptus acutifrons* Sars, 1862
[Palearctic]

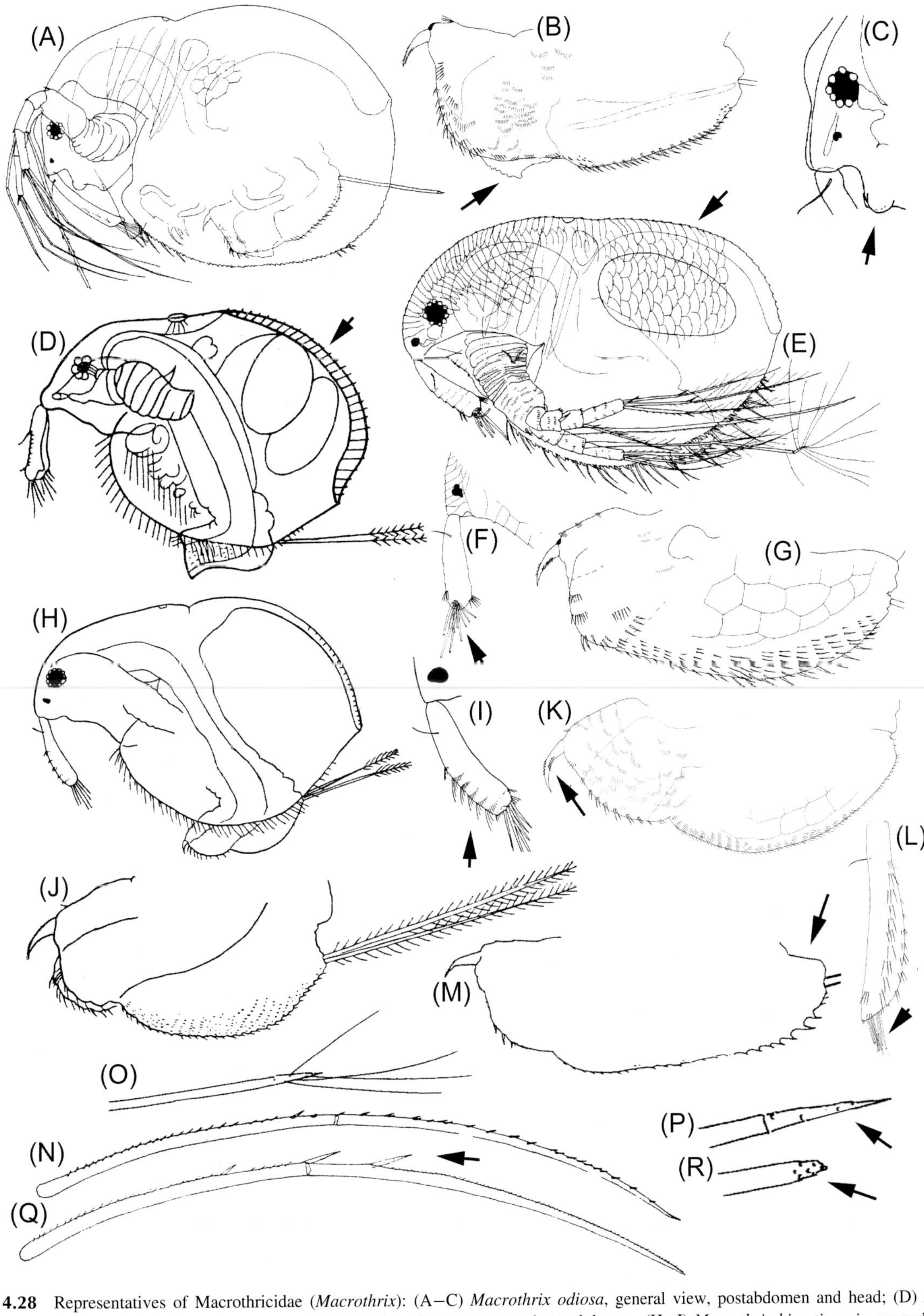

FIGURE 4.28 Representatives of Macrothricidae (*Macrothrix*): (A–C) *Macrothrix odiosa*, general view, postabdomen and head; (D) *Macrothrix laticornis*, general view; (E–G) *Macrothrix spinosa*, general view, first antenna and postabdomen; (H–J) *Macrothrix hirsuticornis*, general view, first antenna and postabdomen; (K and L) *Macrothrix dadayi*, postabdomen and first antenna; (M–P) *Macrothrix rosea*, postabdomen, seta on proximal segment of antennal endopod, postabdominal seta and distal segment (of postabdominal seta) without setules; (Q and R) *Macrothrix triserialis*, seta on proximal segment of antennal endopod and postabdominal seta distal segment without setules. *After Rogers et al., 2019; (D) After Lilljeborg (1900).*

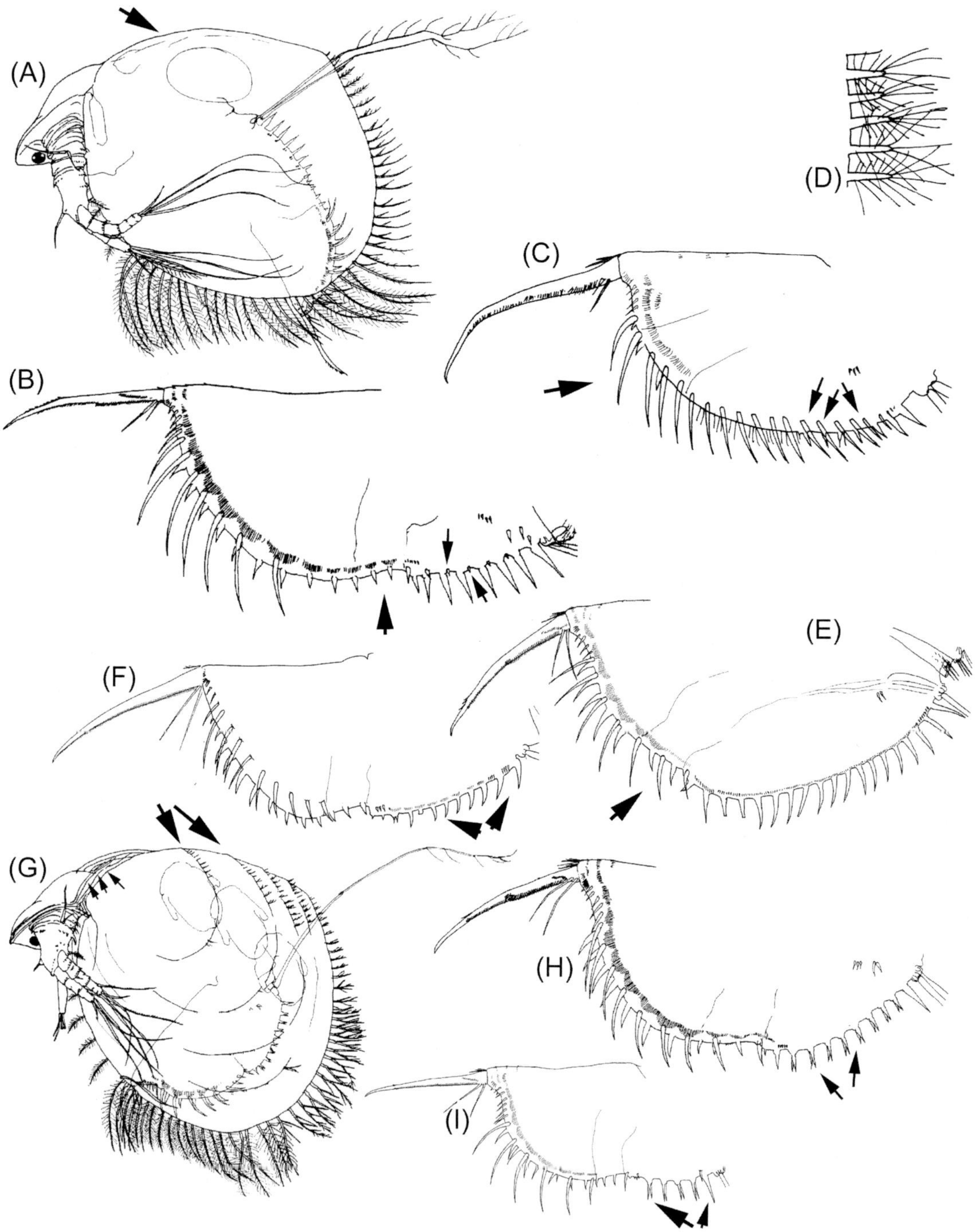

FIGURE 4.29 Representatives of Ilyocryptidae (*Ilyocryptus*): (A and B) *Ilyocryptus agilis*, general view and postabdomen; (C and D) *Ilyocryptus acutifrons*, postabdomen and armature of posterior valve margin; (E) *Ilyocryptus spinosus*, postabdomen; (F) *Ilyocryptus sordidus*, postabdomen; (G and H) *Ilyocryptus cuneatus*, general view and postabdomen; (I) *Ilyocryptus silvaeducensis*, postabdomen. *After Rogers et al., 2019.*

3(2) Postabdomen with anus approximately midway along dorsal margin, preanal margin short 4
3' Postabdomen with anus subdistal, preanal margin long (Fig. 4.29E) .... *Ilyocryptus spinosus* Stifter, 1988 [central and northern Europe]
4(3) Postabdomen with at least some preanal teeth paired 5
4' Postabdomen with all preanal teeth singular (Fig. 4.29F) .... *Ilyocryptus sordidus* (Liévin, 1848) species group [Holarctic]
5(4) Preanal margin long, with 9–12 teeth (Fig. 4.29G and H) .... *Ilyocryptus cuneatus* Stifter, 1988 [Palearctic]
5' Preanal margin short, with 5–8 teeth (Fig. 4.29I) .... *Ilyocryptus silvaeducensis* Romijn, 1919 [Western Europe]

Branchiopoda: Diplostraca: Cladocera: Ctenopoda: Sididae: Genera

1 Second antenna exopodite and endopodite each with different number of articles 2
1' Second antenna exopodite and endopodite each with two articles (Fig. 4.30A) .... *Penilia avirostris* Dana, 1849 [Marine planktonic species; may venture in lagoons]
2(1) Second antenna exopodite with two articles and endopodite with three articles 3
2' Second antenna exopodite with three articles and endopodite with two articles (Fig. 4.30B–D) .... *Sida crystallina* (O.F. Müller, 1776) [Palearctic]
3(2) Valve margins posteriorly and posterioventrally with long plumose setae; postabdomen with solitary lateral anal teeth (Fig. 4.30E) .... *Latonopsis australis* Sars, 1888 species group [Asia, Europe]
3' Valve margins posteriorly and posterioventrally without long plumose setae; postabdomen with clusters of lateral anal teeth (if present) .... ***Diaphanosoma***

Branchiopoda: Diplostraca: Cladocera: Ctenopoda: Sididae: *Diaphanosoma*: Species

This genus is very difficult to identify, and more species in the region may be present; see most recent identification keys for Sididae (Korovchinsky, 2018) and revisions per groups. Following Rogers et al. (2019).

1 Carapace ventral valve margins narrowly inflected, smoothly connecting with posterioventral margin 2
1' Carapace ventral valve margins with wide, flanged inflexion 6
2(1) Carapace posteroventral valve margin with denticles subequal 3
2' Carapace posteroventral valve margin with denticles unequal (Fig. 4.30F–J) .... *Diaphanosoma brachyurum* (Liévin, 1848) species group [Palearctic-Oriental]
3(2) Carapace valve margin with a posterodorsal spine (Fig. 4.30M) 4
3' Carapace valve margin without a posterodorsal spine 5
4(3) Head 40%–45% of body length; carapace valve posterioventral margins with 10–36 (usually 15–25) small denticles (Fig. 4.30K–M) .... *Diaphanosoma mongolianum* Ueno, 1938 [Palearctic, Afrotopical]
4' Head 34%–37% of body length; carapace valve posterioventral margins with 25–60 (usually 30–50) small denticles (Fig. 4.30N–P) .... *Diaphanosoma lacustris* Korinek, 1981 [Palearctic, Afrotopical]
5(4) Head dorsum slightly protruding; carapace posteroventral valve margin denticles relatively thin, not passing far along ventral margin (Fig. 4.31A and B) .... *Diaphanosoma macedonicum* Korovchinsky & Petkovski, 2014 [Macedonia]
5' Head dorsum not protruding, rectangular or roundish-rectangular; carapace posteroventral valve margin denticles relatively wide, passing far along ventral margin (Fig. 4.31C and D) .... *Diaphanosoma orghidani* Negrea, 1982 [Palearctic-Oriental region]

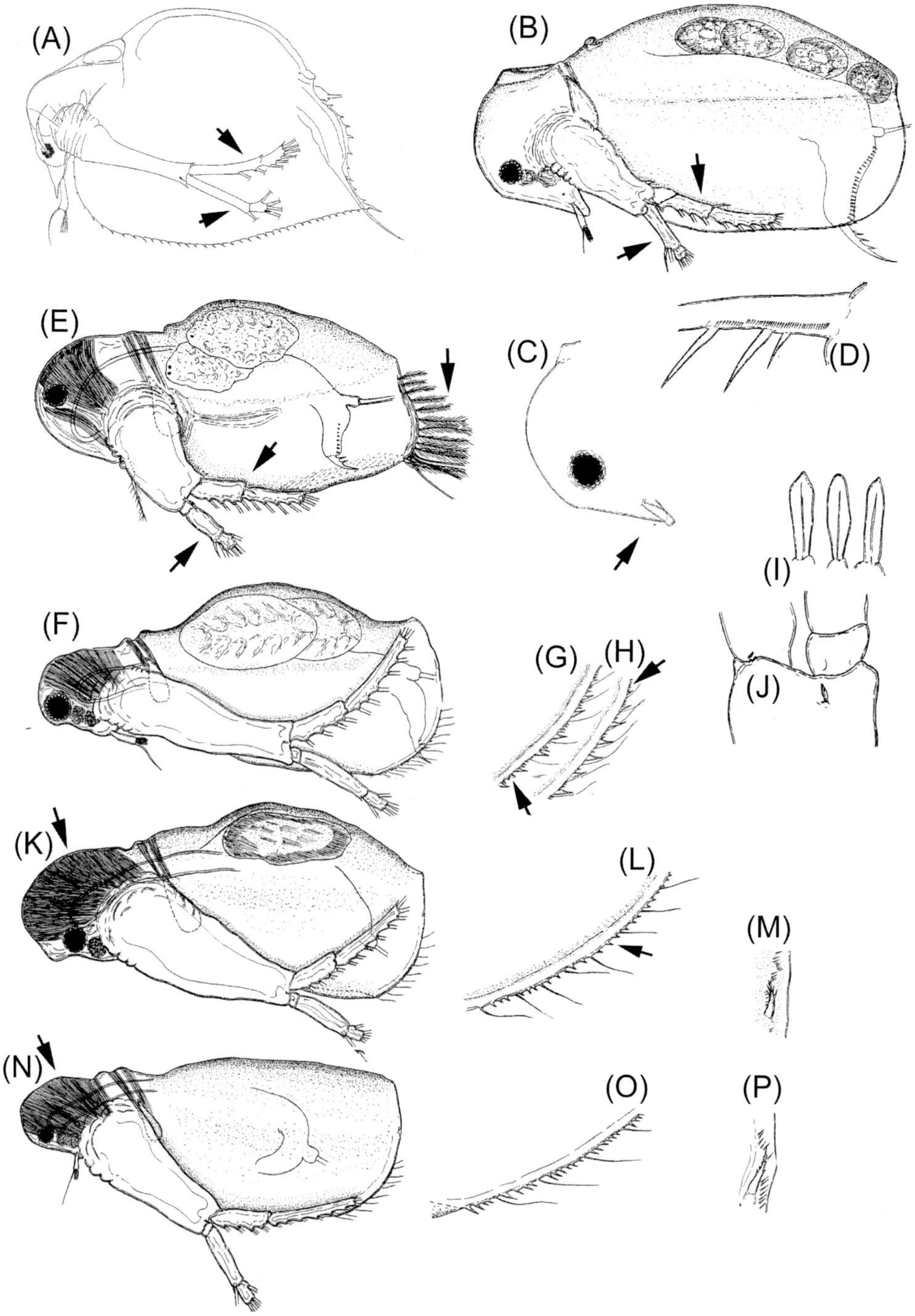

FIGURE 4.30 Representatives of Sididae: (A) *Penilia avirostris*, general view; (B–D) *Sida crystallina*, general view, head and basal part of terminal claw on postabdomen; (E) *Latonopsis australis*, general view; (F–J) *Diaphanosoma brachyurum*, general view, armature of postero-ventral valve portion, basal portion of antennal branches and spine in distal portion of basal segment of second antenna; (K–M) *Daphnia mongolianum*, general view, armature of posteroventral valve portion and spine on inner side of posteroventral valve portion; (N–P) *Diaphanosoma lacustris*, general view, armature of posteroventral valve portion and spine on inner side of posteroventral valve portion. *After Rogers et al., 2019.*

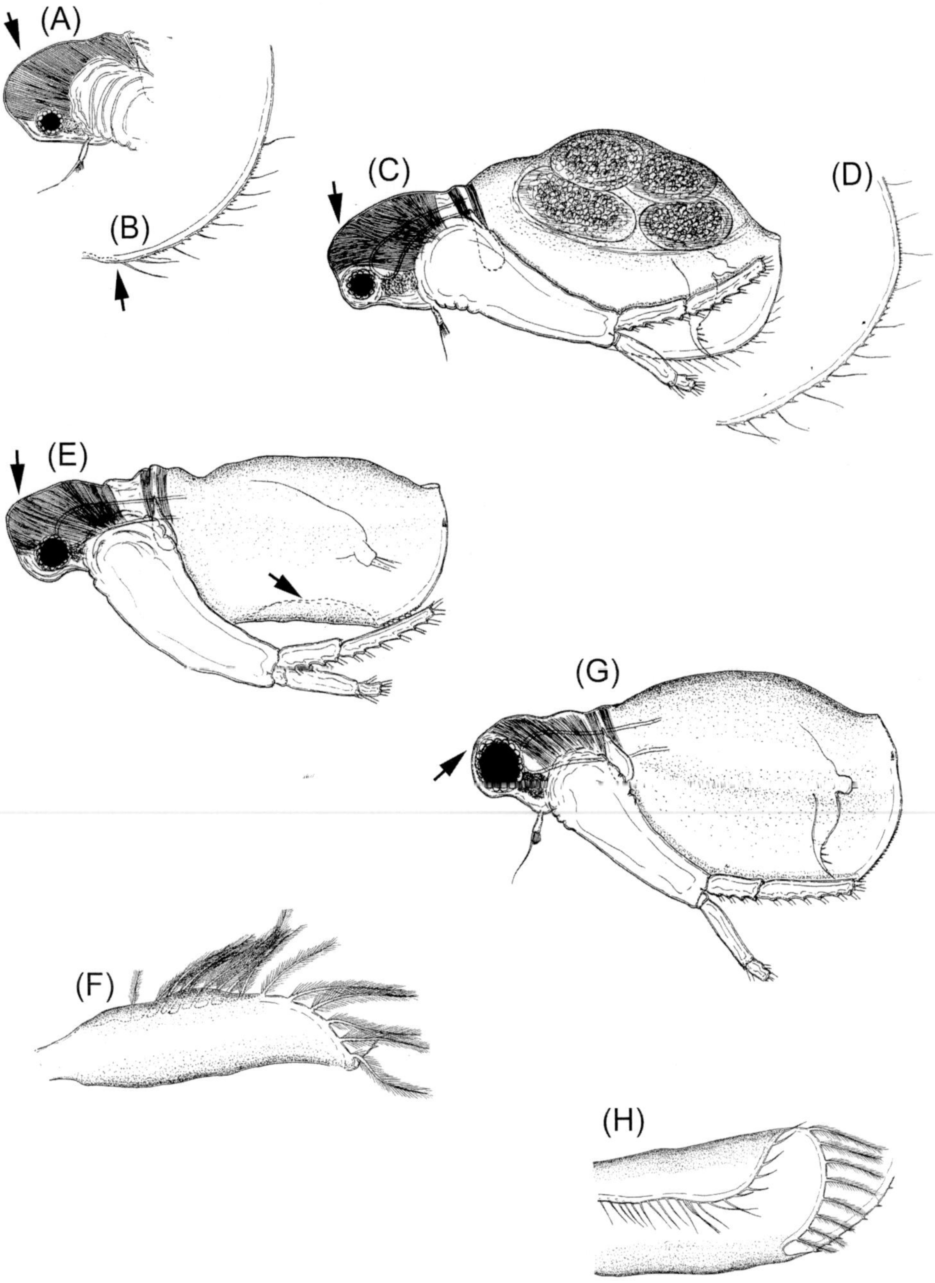

FIGURE 4.31 Representatives of Sididae (*Diaphanosoma*): (A and B) *Diaphanosoma macedonicum*, head and armature of posteroventral valve portion; (C and D) *Diaphanosoma orghidani*, general view and armature of posteroventral valve portion; (E and F) *Diaphanosoma excisum*, general view and ventral portion of valve; (G and H) *Diaphanosoma sarsi*, general view and ventral portion of valve. *After Rogers et al., 2019.*

6(5) Head rectangular with robust dorsum; carapace valves with ventral lamellar inflexion tapering distally and joining with posterioventral margin without depression (Fig. 4.31E and F) .... *Diaphanosoma excisum* Sars, 1885 [Afrotropical-Oriental]

6' Head roundish-rectangular with weakly developed dorsum; carapace valves with ventral lamelliform inflexion cleft near junction with posterioventral margin (Fig. 4.31G and H) .... *Diaphanosoma sarsi* Richard, 1894 [Afrotropical-Oriental]

Acknowledgments

We thank Alexey A. Kotov, Artem Y. Sinev, Nikolai M. Korovchinsky, and M. Alonso for the kind re-use of cladoceran drawings used in this chapter (after Rogers et al., 2019).

References

Alonso, M. 1996. Crustacea, Branchiopoda. In: Ramos, M.A. *et al.* (Eds), Fauna Ibérica. Vol. 7. Museo Nacional de Ciencias Naturales, CSIC, Madrid, 486 pp.

Alonso, M., Neretina, A.N. & Ventura, M. 2021. Ceriodaphnia smirnovi (Crustacea: Cladocera), a new species from the Mediterranean Region, and a phylogenetic analysis of the commonest species. Zootaxa, 2974: 1–46. https://doi.org/10.11646/zootaxa.4974.1.1.

Amoros, C. 1984. Introduction pratique à la systeématique des organismes des eaux continentales Françaises. 5. Crustacés Cladocères. Bulletin de la Société Linnéenne de Lyon, 53: 71–145.

Benzie, J.A.H. 2005. The genus Daphnia (including Daphniopsis) (Anomopoda: Daphniidae). In Dumont, H.J. (Ed.), Guide to the Identification of the Microinvertebrates of the Continental Waters of the World, Vol. 21, Backhuys Publishers, Leiden, 2005, pp. 1–376.

Birge, E.A. 1918. The water fleas (Cladocera). In: Ward & Wipple (Eds) Freshwater Biology. Wiley, New York, pp. 676–740.

Dumont H.J. & S.V. Negrea 2002. Introduction to the class Branchiopoda. In: Dumont H.J. (Ed). Guides to the identification of the microinvertebrates of the continental waters of the world, Vol. 21. Kenobi Productions, Belgium. 1–376 pp.

Flößner, D. 2000. Die Haplopoda und Cladocera (ohne Bosminidae) Mitteleuropas. Backhuys Publishers, Leiden,Netherlands, 428 pp.

Goulden, C.E. 1968. The Systematics and Evolution of the Moinidae. Transactions of the American Philosophical Society, 58: 1–101.

Korn, M. & A.K. Hundsdoerfer. 2016. Molecular phylogeny, morphology and taxonomy of Moroccan Triops granarius (Lucas, 1864) (Crustacea: Notostraca), with the description of two new species. Zootaxa, 4178: 328–346.

Korn, M., Green, A.J., Machado, M., García-de-Lomas, J., Cristo, M., Cancela da Fonseca, L., Frisch, D., Pérez-Bote, J.L. & Hundsdoerfer, A.K. 2010. Phylogeny, molecular ecology and taxonomy of southern Iberian lineages of Triops mauritanicus (Crustacea: Notostraca). Organisms Diversity & Evolution, 10: 409–440.

Korn, M., Marrone, F., Pérez-Bote, J.L., Machado, M., Cristo, M., Cancela da Fonseca, L. & Hundsdoerfer, A.K. 2006. Sister species within the Triops cancriformis lineage (Crustacea, Notostraca). Zoologica Scripta, 35: 301–322.

Korovchinsky N.M. 2018. Cladocera: Ctenopoda. Families Sididae, Holopediidae & Pseudopenilidae (Branchiopoda: Cladocera). In: Dumont, H.J. (Ed.) Identification Guides to the Plankton and Benthos of Inland Waters, 27. Backhuys Publ. & Margraf Publ. Weikerscheim (Germany).

Longhurst, A.R. 1955. A review of the Notostraca. Bulletin of the British Museum (Natural History), Zoology, 3: 1–57.

Kotov, A.A. 2009. A revision of Leydigia Kurz, 1875 (Anomopoda, Cladocera, Branchiopoda), and subgeneric differentiation within the genus. Zootaxa, 2028: 1–84.

Kotov A.A., A.Y. Sinev, S.M. Glagolev, & N.N. Smirnov. 2010. Water fleas (Cladocera). In: Alexeev V.R., & S.Y. Tsalolokhin (Eds.), Key book for zooplankton and zoobenthos of fresh waters of European Russia. KMK, Moscow. pp. 151–276. (In Russian)

Kotov A.A. & P. Štifter, 2006. Ilyocryptidae of the world. Kenobi Productions. In Dumont, H. J., Guides to the identification of the Microinvertebrates of the Continental Waters of the world, Vol. 22. Kenobi Productions, Ghent, Belgium & Backhuys Publishers, Leiden, The Netherlands,172 pp.

Kotov A.A., S.A. Kuzmina, L.A. Frolova, A.A. Zharov, A.N. Neretina, & N.N. Smirnov. 2019. Ephippia of the Daphniidae (Branchiopoda: Cladocera) in Late Caenozoic deposits: untapped source of information for palaeoenvironment reconstructions in the Northern Holarctic. Invertebrate Zoology, 16:183–199. doi: 10.15298/invertzool.16.2.06.

Kotov, A.A., D.P. Karabanov, & K. Van Damme. 2022. Non-Indigenous Cladocera (Crustacea: Branchiopoda): From a Few Notorious Cases to a Potential Global Faunal Mixing in Aquatic Ecosystems. Water, 14: 2806. https://doi.org/10.3390/w14182806.

Lilljeborg, W. 1900. Cladocera Sueciae. Nova Acta Reg. Soci. Sci. Uppsaliensis 19: 701 pp.

Martin, J., D.C. Rogers, & J. Olesen. 2016. Collecting and Processing Branchiopods. Journal of Crustacean Biology, 36: 396–401.

Muñoz, J., A. Gómez, J. Figuerola, F. Amat, C. Rico, & A.J. Green. 2014. Colonization and dispersal patterns of the invasive American brine shrimp Artemia franciscana (Branchiopoda: Anostraca) in the Mediterranean Region. Hydrobiologia, 726: 25–41 DOI 10.1007/s10750-013-1748-6.

Neretina, A.N., Garibian, P., Sinev, A.Y. & Kotov, A.A. 2018. Diversity of the subgenus Disparalona (Mixopleuroxus) Hudec, 2010 (Crustacea: Cladocera) in the New and Old World. Journal of Natural History 52: 1–51. DOI 10.1080/00222933.2017.1411987.

Neretina, A.N., Sinev, A.Y., 2016. A revision of the genus Leberis Smirnov, 1989 (Cladocera: Chydoridae) in the Old World and Australia. Zootaxa, 4079: 501–533. DOI: 10.11646/zootaxa.4079.5.1.

Orlova-Bienkowskaja, M.J. 2001. Cladocera: Anomopoda. Daphniidae: genus Simocephalus. In: H.J. Dumont (Ed.). Guides to the identification of the Microinvertebrates of the Continental Waters of the World, 17. Leiden, Backhuys Publishing.

Rivier, I.K., 1998. The predatory Cladocera (Onychopoda: Podonidae, Polyphemidae, Cercopagidae) and Leptodora of the world. In: Dumont, H.J. (Ed.), Guides to the Identification of the Microinvertebrates of the Continental Waters of the World, 13. Leiden, Backhuys Publishing, The Netherlands. 213 pp.

Rogers, D.C. 2009. Branchiopoda (Anostraca, Notostraca, Laevicaudata, Spinicaudata, Cyclestherida). In: Likens, G. F. (editor) Encyclopedia of Inland Waters, Vol. 2, pp. 242–249.

Rogers, D.C. 2013. Anostraca Catalogus. Raffles Bulletin of Zoology, 61: 525–546.

Rogers, D.C. 2020. Spinicaudata catalogus. Zoological Studies, 59: 45.

Rogers, D.C. & J. Olesen. 2016. Laevicaudata Catalogus (Crustacea: Branchiopoda). Arthropod Phylogeny & Systematics, 74: 221–240.

Rogers, D.C. & M. Soufi. 2013. A new species of Chirocephalus (Crustacea: Anostraca) from Iran. Zootaxa, 3609: 319–326.

Rogers, D.C., N. Rabet & S.C. Weeks. 2012. A revision of the extant genera of the clam shrimp family Limnadiidae. Journal of Crustacean Biology, 32: 827–842.

Rogers, D.C., A.A. Kotov, A.Y. Sinev, S.M. Glagolev, N.M. Korovchinsky, N.N. Smirnov & E.I. Bekker. 2019. Chapter 16.2. Arthropoda: Class Branchiopoda. In: D.C. Rogers & J.H. Thorp (eds.), Keys to Palaearctic Fauna; Thorp and Covich's Freshwater Invertebrates - Volume IV. Academic Press.

Rogers, D.C., F. Marrone & A. Hundsdoerfer. 2021. A new Chirocephalus gynandromorph (Branchiopoda: Anostraca: Chirocephalidae) from the Mediterranean Region and descriptions of associated specimens with deformities. Zoodiversity, 55: 459–466.

Schwentner, M., S. Richter, D.C. Rogers, & G. Giribet. 2018. Tetraconatan phylogenomics with special focus on Malacostraca and Branchiopoda. Proceedings of the Royal Society B, 285: 20181524.

Schwentner, M., N. Rabet, S. Richter, G. Giribet, S. Padhye, J.-F. Cart, C. Bonillo & D.C. Rogers. 2020. Phylogeny and Biogeography of Spinicaudata (Crustacea: Branchiopoda). Zoological Studies, 59: 44.

Smirnov, N.N. 1992. Cladocera: The Macrothricidae of the World. Guides to the identification of the Microinvertebrates of the Continental Waters of the World, 1, SPB Academic, Amsterdam.

Smirnov, N.N. 1996. Cladocera: The Chydorinae and Sayciinae of the World. Guides to the identification of the Microinvertebrates of the Continental Waters of the World, 11. Leiden, Backhuys Publishers.

Thiéry, A., 1988. Maghrebestheria maroccana n. gen., n. sp., nouveau représentant des Leptestheriidae au Maroc (Conchostraca). Crustaceana, 54: 43–56.

Thiéry, A. & D.C. Rogers. 2022. A new Tanymastigites species (Crustacea: Anostraca: Tanymastigidae) from Libya, with new large branchiopod records from Algeria. Journal of Crustacean Biology, 42: 1–10. https://doi.org/10.1093/jcbiol/ruac032.

Vandekerkhove, J., S. Declerck, M. Vanhove, L. Brendonck, E. Jeppesen, J.M. Conde Porcuna & L. De Meester. 2004. Use of ephippial morphology to assess richness of anomopods: potentials and pitfalls. Journal of Limnology, 63, Suppliment 1: 75–84.

Van Damme, K., Cornetti, L., Fields, P.D. & Ebert, D. 2022. Whole-genome phylogenetic reconstruction as a powerful tool to reveal homoplasy and ancient rapid radiation in waterflea evolution. Systematic Biology, 71: 777–787.

Van Damme, K., Kotov, A.A. & Dumont, H.J. 2010. A checklist of names in Alona Baird 1843 (Crustacea: Cladocera: Chydoridae) and their current status: an analysis of the taxonomy of a lump genus. Zootaxa, 2330: 1–63.

Van Damme, K. & Dumont, H.J. 2008. Further division of Alona Baird, 1843: separation and posiion of Coronatella Dybowski & Grochowski and Ovalona gen. n. (Crustacea: Cladocera). Zootaxa, 1960: 1–60.

Van Damme, K. & Dumont, H.J. 2010. Cladocera of the Lençóis Maranhenses (NE - Brazil): faunal composition and a reappraisal of Sars' Method. Brazilian Journal of Biology, 70: 755–779. doi:10.1590/S1519-69842010000400008.

Van Damme, K., Sinev, A.Y. & Dumont, H.J. 2011. Separation of Anthalona gen. n. from Alona Baird, 1843 (Branchiopoda: Cladocera: Anomopoda): morphology and evolution of scraping stenothermic alonines. Zootaxa. 2875: 1–64.

Chapter 5

Class Ostracoda

Francesc Mesquita-Joanes[1], Giampaolo Rossetti[2] and Claude Meisch[3]
[1]Cavanilles Institute of Biodiversity and Evolutionary Biology, University of València, Paterna, Spain, [2]Department of Chemistry, Life Science and Environmental Sustainability, University of Parma, Parma, Italy, [3]National Natural History Museum, Luxembourg, Luxembourg

Introduction

According to recent genomic data, the Ostracoda, together with Mystacocarida, Branchiura and Pentastomida, are considered to be members of the most basal group of the Pancrustacea: the Oligostraca (Lozano-Fernandez et al., 2019). Ostracods (Fig. 5.1) are amongst the oldest arthropods with living members, with a rich fossil record going back to the early Ordovician, although freshwater forms were not recorded until the Carboniferous (Rodriguez-Lazaro & Ruiz-Muñoz, 2012). Some of those early nonmarine ostracod groups, such as the superfamily Darwinuloidea, can still be found in freshwaters worldwide, although the superfamily Cypridoidea is at present the most diverse and abundant lineage in freshwaters (Martens et al., 2008).

The first living Ostracoda described using the Linnean system was arguably the freshwater species *Cypris pubera* (Mesquita-Joanes et al., 2020). Up to now, and according to the updated world list of freshwater ostracods by Meisch, Smith et al. (2019) there are 2330 subjective species in 270 genera, the Palearctic being the richest biogeographical region. However, ostracodology efforts have traditionally been biased towards central and northern European areas of the Palearctic, with less information available for the Mediterranean (e.g., Meisch, 2000; Sywula, 1974) and even less for other regions in Asia, Africa, or South America, notwithstanding more recent comprehensive works such as Karanovic (2012), Smith et al. (2015), Horne et al. (2019), and Higuti and Martens (2020). Nevertheless, during the past few decades, an increase in the sampling effort and more accessibility to the scientific literature has resulted in important amendments to the knowledge of the Mediterranean ostracod fauna, including checklists for European and African countries such as Spain (Baltanás et al., 1996; Castillo-Escrivà et al., 2023), Italy (Pieri et al., 2015, 2020), Slovenia (Griffiths & Brancelj, 1996), Turkey (Altınsaçlı & Griffiths, 2002; Külköylüoğlu et al., 2015; Özuluğ et al., 2018), Israel (Martens & Ortal, 1999), Algeria (Ghaouaci et al., 2017), and Tunisia (Marrone et al., 2020). In the present chapter, we briefly review the present knowledge on the classification, distribution, and ecology of freshwater

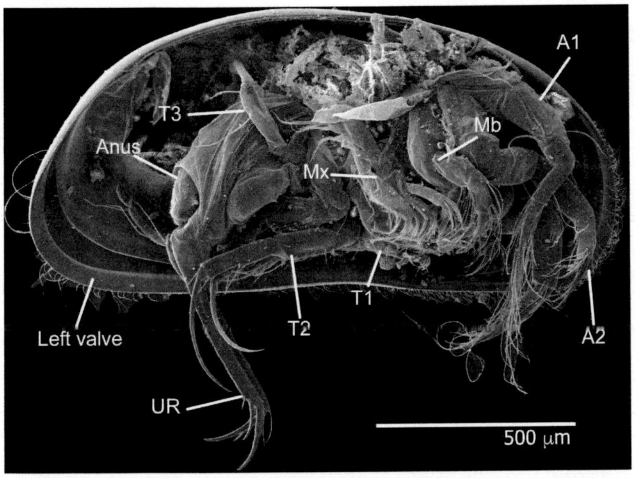

FIGURE 5.1 Scanning electron microscope (SEM) micrograph of a female specimen of *Herpetocypris brevicaudata* with the right valve removed to allow observation of soft parts (see text for abbreviations of appendage names).

Identification and Ecology of Freshwater Arthropods in the Mediterranean Basin. DOI: https://doi.org/10.1016/B978-0-12-821844-0.00013-2

ostracods in the Mediterranean Basin, aiming to serve as support for researchers working on the limnology of these little and poorly known, but diverse crustaceans in an area long and intensely affected by human impacts and that is expected to suffer the stressing effects of desertification and warming under the foreseen scenarios of global change.

General ecology and distribution

Understanding the ecology of Ostracoda has been instrumental for a plethora of studies focused on palaeoecology and environmental change, so that many previous works discussed the main factors affecting ostracod species ecology and distribution, mostly with the aim of using their remains for reconstructing palaeoenvironments. Some general reviews of freshwater ostracod biology and ecology include those by De Deckker (1983), Carbonel et al. (1988), De Deckker and Forester (1988), Meisch (2000), Martens and Horne (2009), Mesquita-Joanes, Smith, et al. (2012), and Smith et al. (2015). Considering such previous extensive body of literature on the topic, we will here focus mostly on aspects of the ecology and distribution of nonmarine ostracods in the Mediterranean Basin.

Environmental factors

The most important environmental factors affecting the distribution of ostracod species in continental waters do not differ much from those affecting many other aquatic invertebrates and include habitat hydroperiod (e.g., permanent vs. temporary waters), current velocity (lentic vs. lotic waters), water chemistry, temperature, food, and biotic interactions (Baltanás et al., 1990; Forester, 1986; Frenzel & Boomer, 2005; Horne et al., 2012; Mesquita-Joanes, Smith, et al., 2012; Neale, 1964). The availability of water either intermittently (e.g., temporary ponds and streams) or for the whole year in a continuous fashion (permanent lakes, springs, and rivers) is an essential habitat trait strongly determining which ostracod species are to be present. Only those that can produce diapausing eggs or other instars that can resist long dry periods, which is common in many Mediterranean water bodies, will persist in temporary waters. The ostracod fauna of Mediterranean temporary water bodies is dominated by Cyprididae species, particularly Cyprinotinae (*Heterocypris* spp.), Eucypridinae (*Eucypris* spp., *Trajancypris* spp.), Herpetocypridinae (*Herpetocypris* spp.), or Cypridopsinae (*Cypridopsis* spp., *Sarscypridopsis* spp., *Potamocypris* spp.), although it is not rare to find also some Ilyocyprididae (*Ilyocypris* spp.) or Limnocytheridae (*Paralimnocythere* spp., *Limnocythere* spp., *Leucocythere* spp.). The taxa most commonly found in permanent water bodies include Darwinulidae, Cytheroidea other than Limnocytheridae, and Candonidae, but also other members of the Cypridoidea. This duality comes apparent when analyzing the ostracod communities in a gradient of habitat types including temporary and permanent Mediterranean lakes or springs (Mezquita et al., 2005; Roca et al., 2000), or when considering the very low frequency and abundance of candonids and darwinulids in northern Africa, where Gauthier (1928) stressed the absence of precipitation in summer as one of his main conclusions: "J'ai constaté que le cachet très spécial de cette faune était dû à l'absence totale, durant l'été, de précipitations atmosphériques," a trait common to many areas of the Mediterranean Basin. Nevertheless, permanent waters are common in coastal wetlands around the Mediterranean, and also in freshwater (mountain) lakes and rivers of the northern shore and Anatolia, providing habitats for a high diversity of candonids and other species requiring long hydroperiods or without drought-resistant stages. Consequently, we could also consider the effect of the hydroperiod as a gradient rather than a duality. For instance, some species, even belonging to the same genus, prefer shorter while others prefer longer hydroperiods or even permanent waters, as in the case of *Potamocypris arcuata* vs. *P. villosa* (Meisch, 2000). Also, many permanent water bodies suffer changes in water level that facilitate the colonization of marginal habitats by species tolerating desiccation. In addition, even though diapausing eggs are a widespread strategy to withstand dry periods, some species, including some candonids, are known to tolerate desiccation periods as juveniles or adults, buried in the sediment (Aguilar-Alberola & Mesquita-Joanes, 2011; Horne, 1993; Tétart, 1974).

The importance of water flow velocity on ostracod species distribution has long been recognized. Freshwater ostracods are usually ecologically classified as either swimming or crawling species; the first usually have long swimming setae on the second antennae, while the second have short setae (Meisch, 2000). Crawling species are usually dominant in systems with high flow velocity such as rivers and streams, while swimming species, with long swimming setae, are most commonly found in lentic environments (e.g., Akdemir et al., 2016). For instance, we can usually find species with long swimming setae belonging to the genera *Potamocypris* (*P. villosa, P. arcuata*), *Ilyocypris* (*I. gibba*), or *Herpetocypris* (*H. chevreuxi, H. helenae*) in standing waters, while those with short swimming setae (e.g., *P. zschokkei, I. bradyi, I. inermis, H. brevicaudata*) seem to better stand drift risk under running water conditions. Yet when current velocity is too high, ostracods become very rare or absent in the epibenthic environment (Poquet & Mesquita-Joanes, 2011), although they might still colonize the interstitial habitat (Dole-Olivier et al., 2000).

The overwhelming effect of temperature on ostracod ecophysiology (e.g., Aguilar-Alberola & Mesquita-Joanes, 2014), as on many other organisms (Angilletta Jr., 2009), has been repeatedly highlighted when discussing their biogeography and distribution, as in Mediterranean epicontinental waters (Akdemir & Külköylüoğlu, 2014; Külköylüoğlu et al., 2016; Poquet & Mesquita-Joanes, 2011). This species—environment relationship has been considered a cornerstone when using ostracods as proxies for reconstructing palaeoenvironments (Horne & Mezquita, 2008; Horne et al., 2012). In the Mediterranean, the lower precipitation (and its effects on the hydroperiod), together with higher average temperatures compared to northern Europe, affects the biogeography of dominant species in a synergistic way. These climatic conditions result in high evaporation rates, short hydroperiods, and high environmental stochasticity, which can notably affect the faunal composition of similar environments, such as springs, in different regions (Rosati et al., 2017). And yet, the geographic heterogeneity provided by high mountain ranges in some Mediterranean countries allows species with preferences for colder environments and dominant in northern Europe to thrive in mountaintops with higher performance than typical circum-Mediterranean species (Poquet & Mesquita-Joanes, 2011).

Water chemistry, particularly ionic composition and dissolved salt content, has also been traditionally viewed as one of the most important local habitat traits affecting ostracod ecology and distribution. Indeed, ostracods seem to be excellent indicators of salinity and ionic composition in palaeolimnological research worldwide (De Deckker, 1983; Delorme, 1989; Forester, 1986), but also in the Mediterranean Basin, where these two factors appear repeatedly among the most important variables explaining ostracod species sorting (Gauthier, 1928; Külköylüoğlu et al., 2018; Külköylüoğlu, 2003; Mezquita et al., 2005; Mischke et al., 2012, 2014; Ramdani et al., 2001; Reed et al., 2012; Roca et al., 2000; Rossetti et al., 2005). Either when comparing different localities (Fig. 5.2) or even within the same locality, ostracod species are distributed depending on the salt content, which can be observed for instance within lakes with salinity gradients or in coastal wetlands with heterogeneous salinity distribution (Altınsaçlı & Mezquita, 2008; Altinsaçlı et al., 2018; Altinsaçlı, 2014; Marazanof, 1965; Perçin-Paçal et al., 2018; Valls, Zamora, et al., 2016).

Other limnological variables that can affect the occurrence of different species of Ostracoda in (Mediterranean) freshwaters are dissolved oxygen, pH, and trophic state. Some species such as *Cypria ophtalmica, Neglecandona neglecta, Darwinula stevensoni, Heterocypris incongruens*, or *Herpetocypris intermedia* are known to tolerate relatively high organic content and low oxygen concentration, although too high organic pollution leads to ostracod disappearance (Külköylüoğlu et al., 2007; Meisch, 2000; Mezquita et al., 1999; Poquet et al., 2008; Tétart, 1974). Such ecological differences in the distribution of species in relation to water chemistry can be observed even at the micro- and mesohabitat scales, with dominant species changing across a small distance of only a few meters (Mezquita et al., 2000).

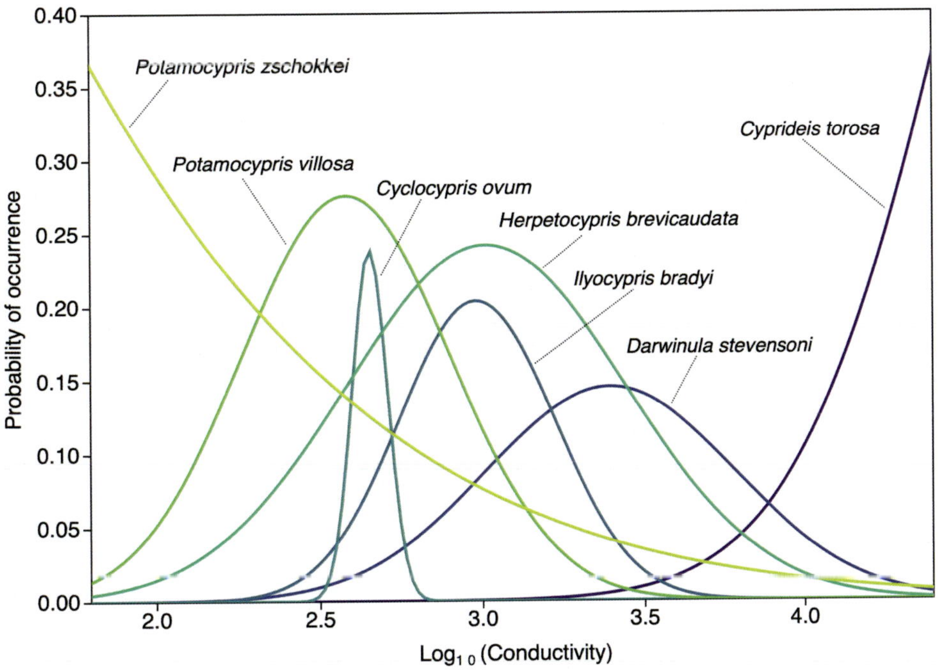

FIGURE 5.2 Example of probability of occurrences modeled as Gaussian responses to electric conductivity of several nonmarine ostracods from the Iberian Peninsula. Conductivity (Log_{10}-transformed) in $\mu S\ cm^{-1}$. *Data from Mezquita et al. (2005).*

We have discussed some important abiotic factors that influence ostracod distribution, but sometimes these interact with, or are modified by, biotic factors that also have a major role in determining which species are present or not in a particular site. One such interaction is the importance of temporary water bodies for some ostracods, particularly large-bodied ones (e.g., belonging to the genera *Cypris, Eucypris, Trajancypris, Heterocypris*, and *Bradleystrandesia*), because of not only their drought-resisting eggs but also the absence of effective predators. Fish are mostly absent from temporary water bodies unless these get connected to more permanent aquatic systems. Other vertebrates (e.g., amphibian larvae, waterfowl) and invertebrates (e.g., dragonflies, water bugs, dytiscid beetles, notostracans) can also influence the development of ostracod populations in Mediterranean permanent or temporary environments (Boix et al., 2006; Martins et al., 2009; Schmit et al., 2012; Vandekerkhove et al., 2012), which can even suffer from cannibalism (Rossi et al., 2011). Although most ostracods are considered herbivorous or detritivorous (Mesquita-Joanes, Smith, et al., 2012), they can also predate on other animals, even on amphibian larvae (Ottonello & Romano, 2011). In contrast to predation, very little is known on the parasites of freshwater (Mediterranean) ostracods (Mesquita-Joanes, Smith, et al., 2012), although there have been recent advances in the knowledge of their procariotic and eucaryotic symbionts in the past decade, partly thanks to the use of molecular techniques (Chatterjee et al., 2020; Mestre et al., 2019; Schön & Martens, 2020; Schön et al., 2019). However, direct evidence is largely lacking on the influence of these symbionts on their hosts and the potential competitive interactions among ostracods. It should be also noted that some ostracods belonging to the family Entocytheridae are strict commensals of other, larger crustaceans, such as decapods and isopods, with both native and exotic species present in Mediterranean environments (Mestre et al., 2013, 2014).

Adding dispersal and space to niche effects: ostracod metacommunities

The probability of finding an ostracod species in a particular site, as well as its abundance, does not depend uniquely on its ability to cope with local biotic or abiotic conditions. Besides these niche effects, all organisms are subjected to spatial effects (e.g., dispersal limitation and surplus), so that they can be found in sites with conditions unmatching their niche, or either not being able to access and colonize sites with appropriate conditions. The importance of these spatial effects, in combination with the niche-centered view, has promoted the research program in metacommunity ecology (Leibold et al., 2004). Empirical tests have shown the importance of both niche and neutral (spatial and stochastic) effects on aquatic metacommunities of many organisms worldwide (Cottenie, 2005; Soininen, 2014), including nonmarine ostracods (de Campos et al., 2018, 2019; Zhai et al., 2015). Disentangling spatial from environmental effects in Mediterranean freshwater ostracods has also been recently investigated for streams (Castillo-Escrivà, Rueda, et al., 2016; Escrivà et al., 2015), springs (Rosati et al., 2017), and lakes (Castillo-Escrivà et al., 2017; Castillo-Escrivà, Valls, et al., 2016) with variable results (Fig. 5.3): generally, a higher proportion of species variance is explained in lentic water bodies (and with relative higher importance of environmental factors) than in streams and springs, although the results might also depend on the spatial extent, environmental gradients, or analytical methods (Leibold & Chase, 2017). Accordingly, the composition of ostracod assemblages can be highly unpredictable, especially in small Mediterranean springs. However, it is important to note that (1) the wider the environmental gradient is (e.g., in terms of salinity or hydroperiod), the better we can predict changes in species assemblages in relation to environmental data; and (2) dispersal limitation and pathways (spatial effects) can often provide a better explanation of ostracod distributions than local environmental conditions.

Biogeography of the Mediterranean ostracod fauna

Metacommunity analyses have demonstrated the major role of space (e.g., through dispersal limitation) for ostracod species distribution at regional scales, which was already shown in biogeographic studies focusing on large spatial scales. In this sense, Martens et al. (2008) recognized the high degree of endemicity of Ostracoda at the continental scale compared with other aquatic invertebrates. Consequently, ostracods must have lower dispersal abilities than other crustaceans such as branchiopods or copepods, although the passive dispersal potential of ostracods via birds or mammals (including humans) has been widely demonstrated, also within the Mediterranean Basin (Frisch et al., 2007; Valls et al., 2017; Valls, Castillo-Escrivà, et al., 2016; Vanschoenwinkel et al., 2008; Waterkeyn et al., 2010). Indeed, with the help of humans, as for other organisms, some ostracods are being recorded as exotic and even invasive in the Mediterranean Basin and elsewhere (McKenzie & Moroni, 1986; Escrivà, et al., 2012; Mestre et al., 2013, 2016; Rossi et al., 2003; Smith et al., 2017; Valls et al., 2014). On the other hand, endemic species are relatively rare in freshwater bodies of the Mediterranean Basin; the exceptions are mostly the ancient Balkan lakes (some of which are included in the Mediterranean bioclimatic zone: Lake Dojran and L. Skadar) and the subterranean environment, which holds an

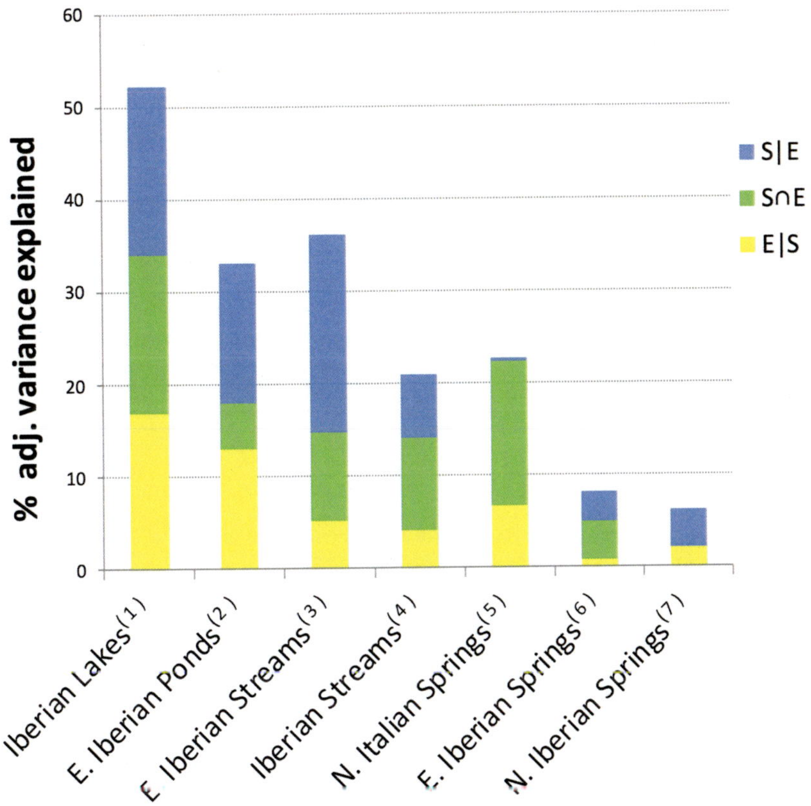

FIGURE 5.3 Comparison of pure environmental (E) and spatial (S) effects (as percentage of adjusted variance explained) and their overlap on ostracod metacommunities from the Iberian Peninsula and northern Italy. *References:* [(1)] *Castillo-Escrivà, Valls et al. (2016),* [(2)] *Gálvez et al. (2022),* [(3)] *Castillo-Escrivà, Rueda et al. (2016),* [(4)] *Escrivà et al. (2015),* [(5–7)] *Rosati et al. (2017).*

interesting ostracod fauna rich in endemisms, mostly belonging to the Candoninae and Sphaeromicolinae (Danielopol & Hart, 1985; Danielopol & Hartmann, 1986; Marmonier et al., 2005; Mazzini et al., 2017; Meisch, Scharf, et al., 2019).

The known species richness of nonmarine ostracods in Mediterranean countries is as high as, or even higher than, those of central or northern Europe (Özuluğ et al., 2018; Pieri et al., 2015), although the southern European areas have been less intensely explored than northern countries such as Germany or the UK. It would not be surprising that further effort would lead to significant increases in species richness in the Mediterranean Basin, considering its function as a large refuge area during glacial periods (Hewitt, 2004), in addition to the higher genetic variability in the widespread *Eucypris virens* species group at lower latitudes (Bode et al., 2010). Studies on the freshwater ostracods of northern Africa are even scarcer than in countries of the northern Mediterranean shore, and this, together with the lowest precipitation and larger desertic areas, reduces the expectation of a high diversity in northern Africa. Out of approximately 260 freshwater ostracod species known in the Mediterranean Basin, less than 60 are known from northern Africa. Nevertheless, taking this into account, the Mediterranean Basin is divided into six distinct regions and their similarities are analyzed regarding ostracod species occurrences (Fig. 5.4): the Iberian Peninsula (including the Balearic Islands), Italy and southern France (including Corsica, Sardinia, Sicily, and Malta), the Balkan Peninsula, Turkey and Cyprus, most of Northern Africa (Morocco, Algeria, Tunisia), and Northern Egypt together with the Middle East (Israel and the Mediterranean bioclimatic parts of Lebanon and Syria). Although these are very preliminary results, we found that the most similar areas are the Iberian Peninsula with France and Italy. However, the Balkan ostracod fauna is particularly distinct, not clustering with these two close areas of the northern Mediterranean shore and neither with Turkey, the latter sitting closer in the cluster to the western Mediterranean countries of the northern shore. The Middle East (plus Egypt), on the other hand, clusters with northern Africa. The separation of the northern and southern Mediterranean regions highlights the importance of the sea as a biogeographical barrier. The unexpected differences in the Balkan fauna, which hold even if we would exclude species unique for the area, and the higher similarity of the Turkish fauna to that of western Mediterranean regions require further investigation and might be related to the scarce knowledge of the Greek ostracod fauna and/or the major influence of central European ostracod fauna in the Balkan Peninsula. As in many other invertebrates, there is still a lack of

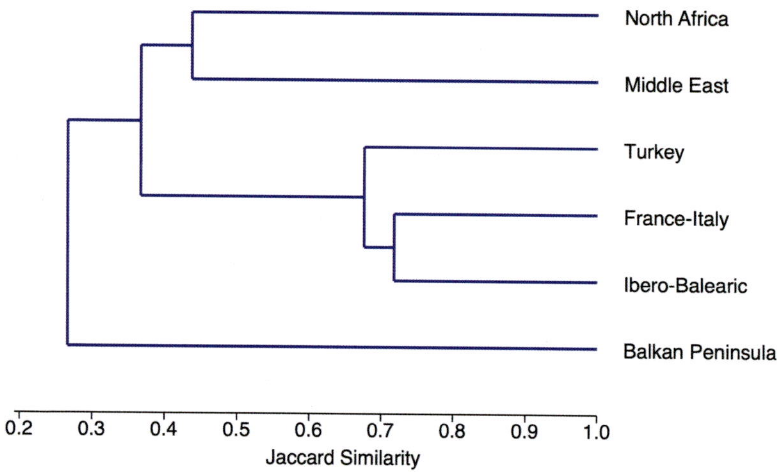

FIGURE 5.4 Dendrogram showing the clustering results of six major Mediterranean subregions according to their known ostracod fauna, based on occurrence data and using the UPGMA method and the Jaccard similarity measure. See text for further explanation.

information on the taxonomy, distribution, and ecology of Mediterranean ostracods; the Linnean, Wallacean, and other biodiversity shortfalls (Hortal et al., 2015) still need to be fully addressed for this group and biogeographic region if we want to appreciate and protect the high biodiversity of this area.

Morphological characters used in identification

We may separate two main groups of morphological traits widely used for the classification of ostracods: (1) those related to carapace morphology and ornamentation, and (2) those related to soft body parts, mostly focused on appendage morphology.

Carapace morphology

The ostracod shell morphology is one of the most important features commonly used for identification at different taxonomic levels, down to species in many cases. This reflects the variability of this most external aspect of ostracod morphology that is often the only remain found and used in paleontological studies. The ostracod shell is composed of two valves, namely left and right valves, according to the main body axis, which can be readily identified knowing the anterodorsal position of the medial naupliar eye in most podocopid ostracods.

The general aspect of an ostracod shell is commonly said to be similar to a seed (ostracods are usually known as "seed-shrimps"), or more particularly a bean, with the more convex part corresponding to the dorsal side and the most concave one to the ventral side. We can describe major shell differences between the three main groups of nonmarine ostracods (Fig. 5.5): the Darwinulidae, the only living family of the Darwinuloidea, generally present a more tubular shape (Fig. 5.5A), similar to a torpedo (except in the genus *Microdarwinula*). The ventral and dorsal margins are quite straight and parallel, only slightly curved (except at the frontal and posterior edges), and their middle transverse section is more circular than in other groups. Most members of the Cypridoidea (Fig. 5.5B–G) are usually more bean-shaped, with laterally compressed shells and, therefore, a more elliptical section, although there are also a number of globose species. This general bean shape is also common in the Cytheroidea, although their valves are usually more strongly calcified and more ornamented (Fig. 5.5H), while those of the Darwinuloidea and Cypridoidea are usually smoother and thinner.

There are different elements of the ostracod carapace (Fig. 5.6) that can be used for classification, in addition to its general shape. One of these traits is the hinge, which is very simple in Darwinuloidea and Cypridoidea, and more complex and variable in the Cytheroidea (Yamada, 2007). The two valves are not fully symmetrical, and the degree of asymmetry can be important for classification, so as the differences in valve closure, i.e., whether the right or left valve overlaps the other one when closed. Some examples of marked asymmetry include many members of Cypridoidea of the genera *Potamocypris* and *Heterocypris*. Another important trait for the classification of higher taxa is the set of adductor muscle scars, located near the center of each valve. These scars form a rosette in the Darwinuloidea, a vertical line of (usually four) scars in the Cytheroidea, and a more irregular and variable pawprint-like set in the Cypridoidea, which is quite conserved inside each family of this group (Fig. 5.7).

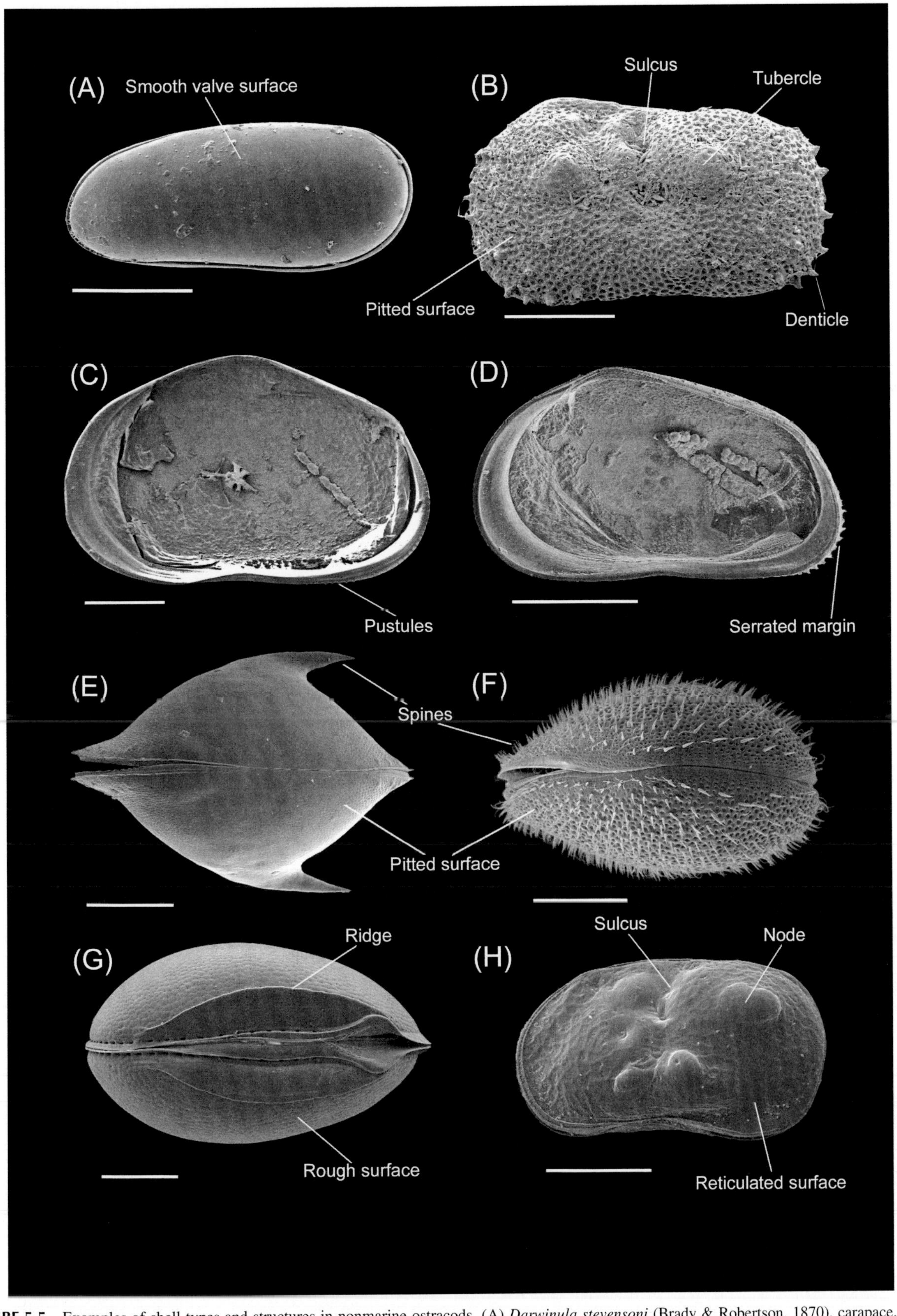

FIGURE 5.5 Examples of shell types and structures in nonmarine ostracods. (A) *Darwinula stevensoni* (Brady & Robertson, 1870), carapace, external view from left valve. Scale bar: 0.25 mm. (B) *Ilyocypris monstrifica* (Norman, 1862), left valve, external view. Scale bar: 0.25 mm. (C) *Heterocypris salina* (Brady, 1868), right valve, inner view. Scale bar: 0.20 mm. (D) *Prionocypris zenkeri* (Chyzer & Toth, 1858), right valve, inner view. Scale bar: 0.5 mm. (E) *Cypris bispinosa* Lucas, 1849, carapace, dorsal view. Scale bar: 0.5 mm. (F) *Sarscypridopsis aculeata* (Costa, 1847), carapace, dorsal view. Scale bar: 0.20 mm. (G) *Notodromas persica* (Gurney, 1921), carapace, ventral view. Scale bar: 0.2 mm. (H) *Limnocythere inopinata* (Baird, 1843), carapace, external view from left valve. Scale bar: 0.2 mm. *From (A, B) Meisch (1988); (C) Meisch and Broodbakker (1993).*

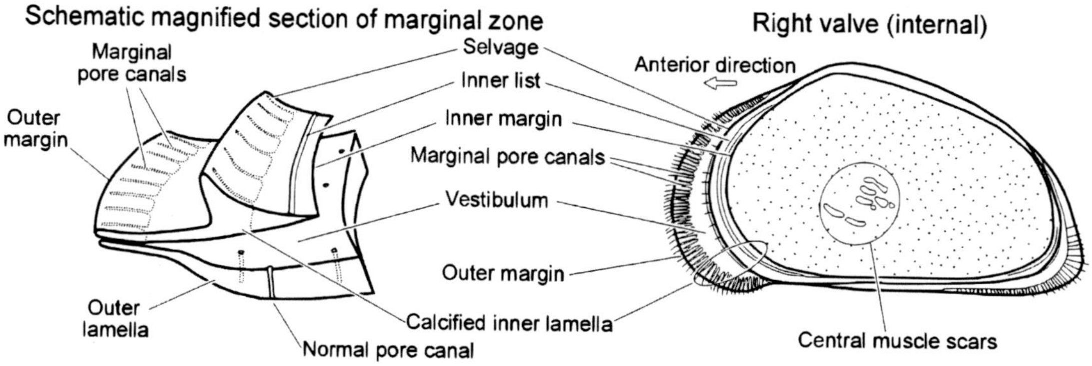

FIGURE 5.6 Main morphological features of a freshwater ostracod shell, with a detail of the marginal inner structures. *After Smith et al. (2015).*

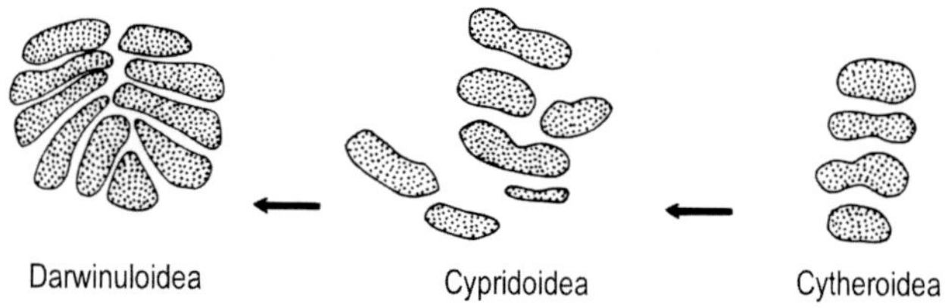

FIGURE 5.7 General aspect of muscle scars of the three superfamilies of nonmarine Ostracoda. Arrows point to anterior part of the shell. *From Meisch (2000).*

The external surface of the valves can be smooth, as in Darwinuloidea and most Cypridoidea, with the notable exception of the Ilyocyprididae, and scattered genera in other families, but always has at least some pores through which mechanosensory or chemosensory setae connect the external environment with the body. Many species also have a shell surface covered with numerous pits. This is common for instance in the *Ilyocypris* species, but also in some Cypridopsinae (Fig. 5.5). Some even have it covered in spines together with the hairs, as in *Sarscypridopsis aculeata*. The presence of small marginal spines or serrated margins is also common, as for instance in the cytheroid *Cyprideis torosa* or the cypridid *Cypris pubera*, or a row of pustules in some Cypridoidea as *Heterocypris* or *Physocypria*. We can even find larger lateral spines, as in *Cypris bispinosa*, but elements such as large spines are less common in the Mediterranean than in tropical species (Higuti & Martens, 2020). Anyway, the most ornamented ostracod shells are usually found in the cytheroids, such as species of the Limnocytheridae or Cytherideidae (but not the Entocytheridae, which have smooth carapaces).

The inner structure of the valves (Fig. 5.6) also allows distinction between ostracod taxa, and it is commonly used to differentiate similar species. Each valve has a calcified inner lamella (CIL), which may have different structures such as inner lists, teeth, selvage, grooves, pore canals, or septa. The width of this CIL, and also the number, shape, and position of these elements are essential for the description and differentiation of ostracod species.

Appendages and other chitinized structures

Besides carapace morphology, a number of chitinized structures are widely used for species identification of living ostracods. The main structures used in ostracod identification are the appendages (Baltanás & Mesquita-Joanes, 2015; Horne et al., 2019; Meisch, 2000; Smith et al., 2015). From front to rear, these are the antennulae (A1), antennae (A2), mandibulae (Mb), maxillulae (Mx), first thoracopods or fifth pair of limbs (T1/L5), second toracopods or sixth pair of limbs (T2/L6), third toracopods or seventh pair of limbs (T3/L7), and the uropods (UR) (Fig. 5.1). Around the mouth, there are also other chitinized parts, such as the upper lip, the hypostome, and the rake-like organs, although these have not been so much implemented in ostracod taxonomy (but see Smith, 2000). Other structures considered important for ostracod taxonomy when males are present are their paired penes and, in the Cyprididae, the Zenker organ (McGregor & Kesling, 1969). The copulatory apparatus of females, simpler than those of males, has been scarcely investigated (e.g., Matzke-Karasz & Martens, 2005; Smith & Kamiya, 2007) and is usually neglected.

The antennula (A1) (Fig. 5.8) has a variable number of segments (5–8) and is usually considered uniramous (Smith & Tsukagoshi, 2005). Although Karanovic (2005) suggests that the darwinuloid antennulae and the ancestral ostracod first appendage could be biramous, more recent works disagree with this interpretation (Boxshall & Jaume, 2013;

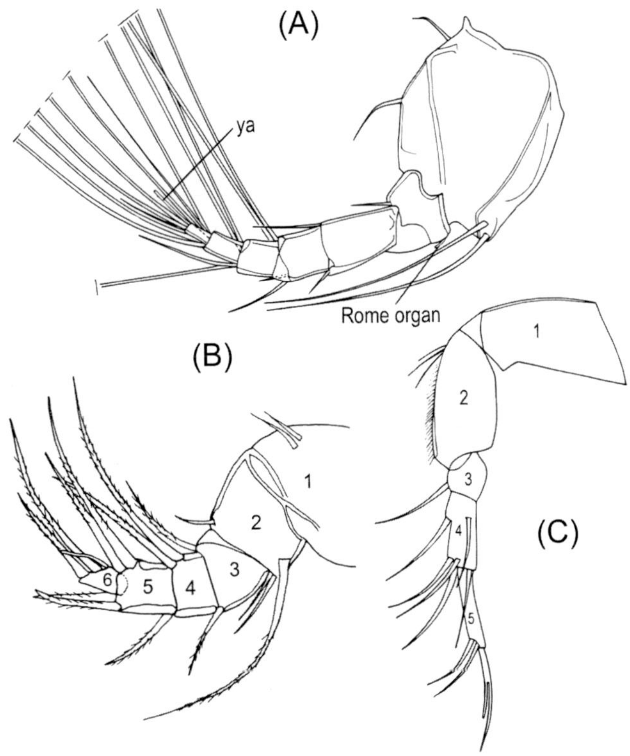

FIGURE 5.8 General aspect of the antennula (A1) of the three main superfamilies of nonmarine ostracods. (A) Cypridoidea; (B) Darwinuloidea; (C) Cytheroidea. *After Meisch (2000).*

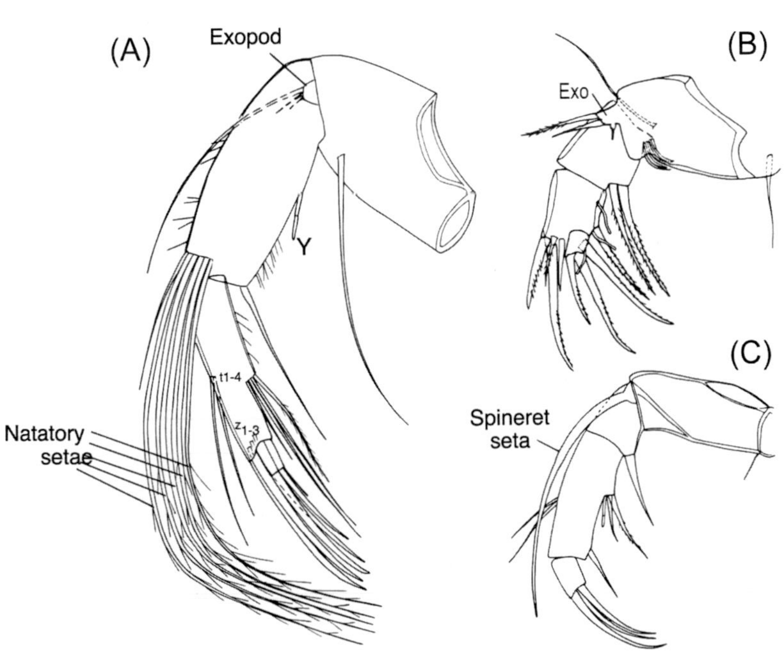

FIGURE 5.9 General aspect of the second antenna (A2) of the three main superfamilies of nonmarine ostracods. (A) Cypridoidea; (B) Darwinuloidea; (C) Cytheroidea. *After Meisch (2000).*

Boxshall et al., 2010). In Cypridoidea and Darwinuloidea, the A1 are usually curved upwards, but mostly downwards in the Cytheroidea. Most A1 segments have one or more setae, which can be very long in the Cypridoidea, and are then used for swimming. There are also some differentiated organs, such as aesthetascs, or the Rome and Wouters organs near the base of the A1 (Smith & Matzke-Karasz, 2008), all of these probably having a chemosensory function.

The antenna (A2) (Fig. 5.9) is biramous, but the exopod is reduced to a few setae in Cypridoidea and Darwinuloidea, and holds the large spinneret seta in the Cytheroidea. In cypridids, the first endopodal segment holds six setae near its distal part, five of which are very long in many species and are used for swimming. A chemosensory seta, i.e., an aesthetasc, is present on the ventral part of this segment in the Cypridoidea, and in the second segment in Darwinuloidea and Cytheroidea. Besides their use for swimming in some species, the A2 are mostly used for walking, digging, or grabbing food in most podocopids; that is probably why the last segment holds a number of modified strong setae in the form of (serrated) claws. The A2 have important sensory functions too, based on the presence of thin setae and aesthetascs. In the case of sexually reproducing species, it is common to observe a modification of some of these setae and claws in the male A2, probably related to mate search and recognition. This is the case for instance of the modified t- and z-setae in *Fabaeformiscandona* (Meisch, 2000; Smith & Kamiya, 2007) or the comb-like G_M in the A2 of the males of *Herpetocypris* (González-Mozo et al., 1996).

The mandibula (Mb) (Fig. 5.10) of nonmarine ostracods has a strongly calcified basal part with distal teeth: the coxa. This part is in general not used as a diagnostic character, but from the coxa arises the mandibular palp, which can be important for ostracod identification. This palp has a vibratory plate with a few filaments at its first segment. The chaetotaxy of the mandibular palp is used to distinguish between genera and species, in particular in the Candonidae and Darwinulidae.

The maxillula (Mx) (Fig. 5.11) is another appendage with masticatory and respiratory functions. But here, the vibratory plate is usually much larger than in the Mb, and with numerous filaments, except in the commensal Sphaeromicolinae, where the plate is absent. The Mx has a palp, usually two-segmented, and three endites or masticatory lobes, the latter being absent or vestigial in the Sphaeromicolinae. The palp and endites carry a number of setae,

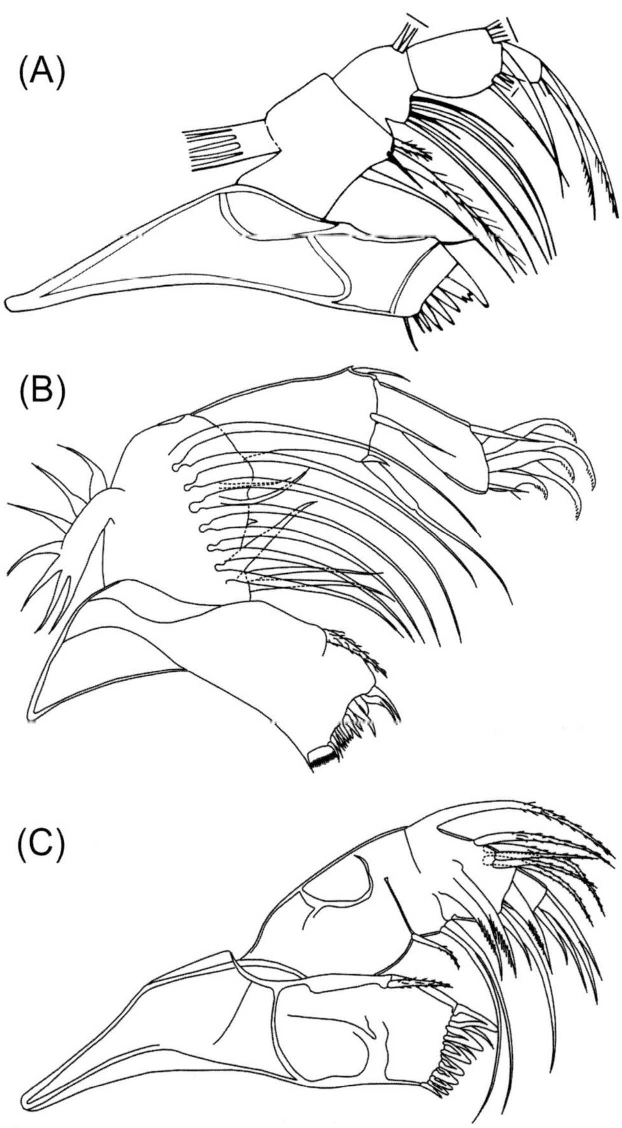

FIGURE 5.10 General aspect of the mandibula of the three main superfamilies of nonmarine ostracods. (A) Cypridoidea; (B) Darwinuloidea; (C) Cytheroidea. *After (A) Meisch (2000).*

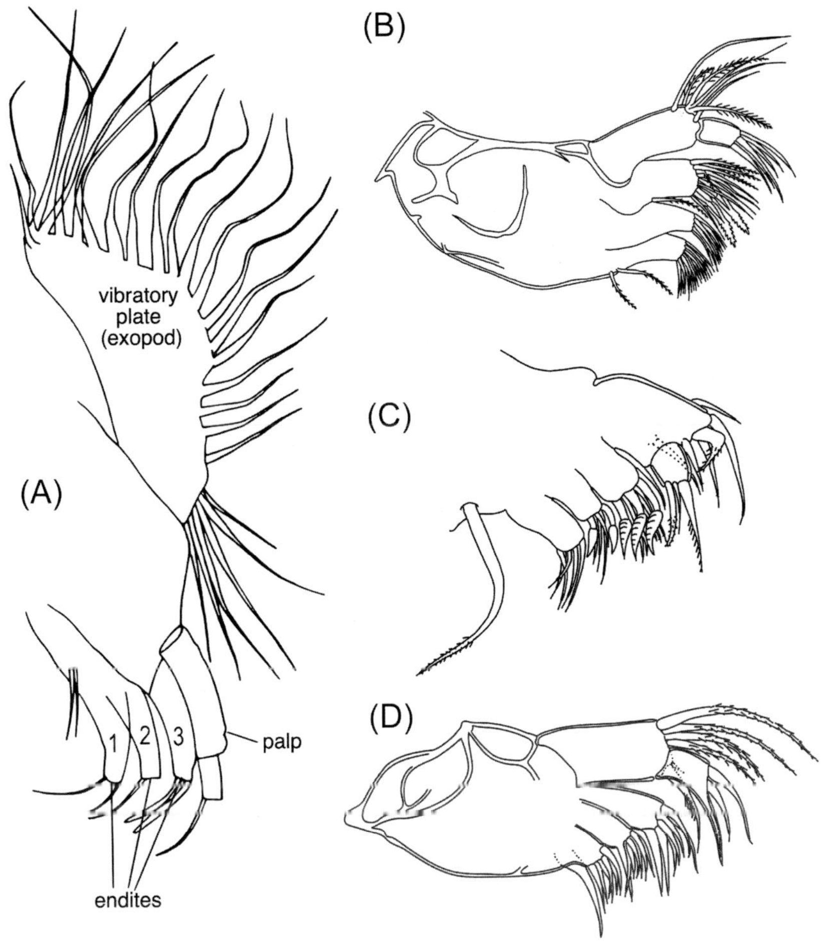

FIGURE 5.11 Maxillula of podocopid nonmarine ostracods. (A) General aspect including the endopod with endites and palp, and the exopod or vibratory plate. (B–D) Endopods of Cypridoidea (B), Darwinuloidea (C), and Cytheroidea (D). *After (A) Meisch (2000). Modified after (B) Aguilar-Alberola and Mesquita-Joanes (2011).*

and in the case of the third endite of the Cyprididae, two (rarely three) of them are usually enlarged to form special structures called *Zahnborsten*, or teeth bristles, which are either smooth or serrated.

The three thoracopods (T1–T3), i.e., appendages L5 to L7, are of similar shape and have a locomotory function (walking legs) in the Cytheroidea, but not so in the other nonmarine superfamilies (Fig. 5.12). In the Darwinuloidea, only T2–T3 share a similar form and locomotory function. T1 is somewhat reduced in the Darwinuloidea and Cypridoidea, and it has a distinct feeding and respiratory function (holding frontal setae and a posterior vibratory plate) in the females, but it is transformed into a clasping organ in the males. In the Cypridoidea, T3 is curved upwards forming a cleaning leg. The chaetotaxy and segmentation of all these appendages are of taxonomic importance.

The uropods (UR) (Fig. 5.13) are well developed in most Cypridoidea, but not so in the Darwinuloidea and Cytheroidea, where only a vestigial seta appears, along with a small vermiform, irregularly shaped structure in some members of the Darwinuloidea. In the Cypridoidea, the UR is remarkably reduced in the Cypridopsinae, consisting only of a small basis with a lateral small seta and a terminal flagellum. In most Cypridoidea, each UR is composed of a strong uropodal ramus (u-ramus) with (usually) two terminal claws and two setae. The uropodal rami are normally symmetrical, but in some groups they may differ, for instance in the serration of the posterior edge, as in some species of *Stenocypris* (Petkovski & Meisch, 1996). The structure of the chitinized uropodal attachment is also an important diagnostic character, particularly for the distinction of the Cypricercinae genera (Savatenalinton & Martens, 2009).

Other chitinized structures (Fig. 5.14) that are not commonly used in taxonomy are the rake-like organs (RLO), which help ostracods in feeding, being located in the mouth area between the upper lip and the hypostome, and the brush-like organs (BLO), probably a vestigial pair of appendages, present in males of the Cytheroidea and with unknown function (Horne et al., 2002; Meisch, 2000).

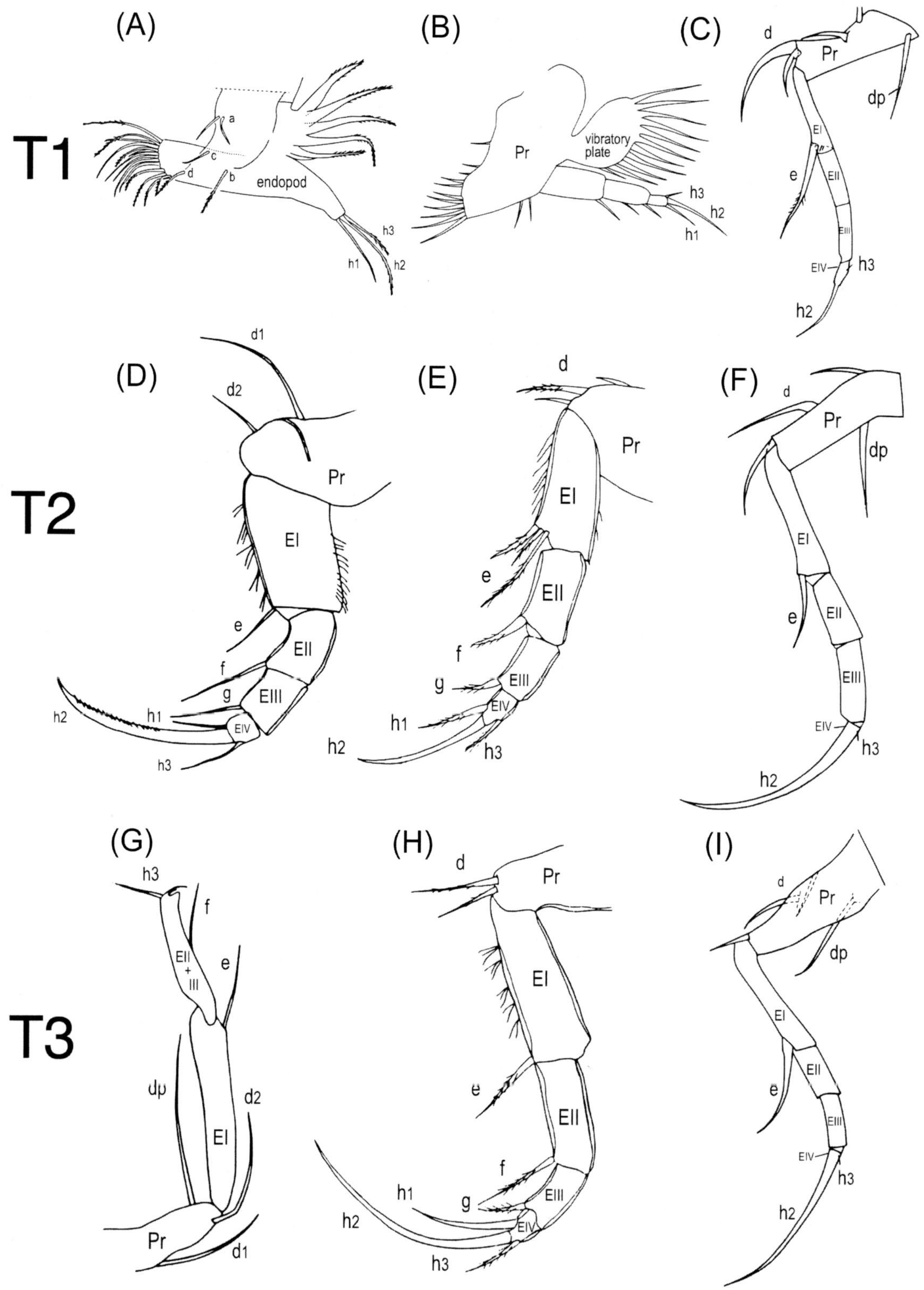

FIGURE 5.12 General aspect and parts of thoracopods T1–T3 in the three superfamilies of nonmarine ostracods. (A) T1 of Cypridoidea. (B) T1 of Darwinuloidea. (C) T1 of Cytheroidea. (D) T2 of Cypridoidea. (E) T2 of Darwinuloidea. (F) T2 of Cytheroidea. (G) T3 of Cypridoidea. (H) T3 of Darwinuloidea. (I) T3 of Cytheroidea. Pr: protopod. EI-EIV: endopod segments. a–d, d1, d2, dp, e–g, h1–h3: toracopod setae and claws. *After Meisch (2000).*

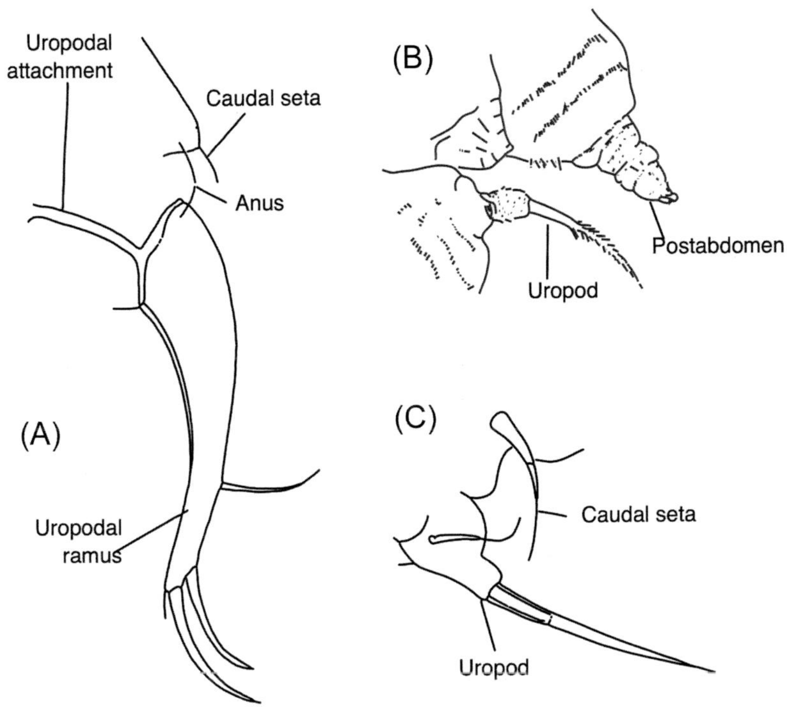

FIGURE 5.13 General morphology of the posterior part of the soft body of nonmarine ostracod, including the uropod. (A) Cypridoidea, (B) Darwinuloidea, (C) Cytheroidea. *After (A, C) Meisch (2000); (B) Rossetti and Martens (1998).*

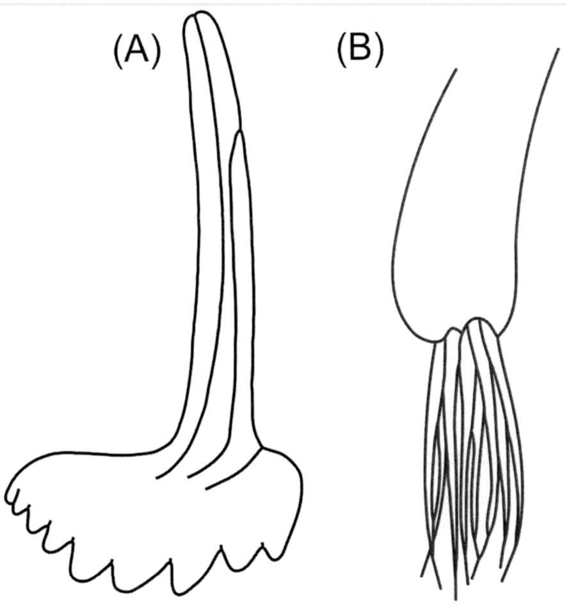

FIGURE 5.14 Morphology of (A) rake-like organ (RLO) and (B) brush-like organ (BLO). *From (A) Meisch (2000).*

Another important organ in ostracod classification is the male copulatory apparatus (Fig. 5.15) or penis, forming a double, symmetrical structure, so that ostracodologists refer to each half as the "hemipenis". The morphology of the hemipenis can help in distinguishing between close species, being particularly useful for the identification of the species of the Candonidae (Danielopol, 1969; Sywula, 1973). It is also the most important organ to distinguish among species of Entocytheridae (Hart & Hart, 1974). Although its inner structure and morphology can be complex, it has some distal lobes and inner processes that can be easily recognized and which are used in species identifications (Meisch, 2000).

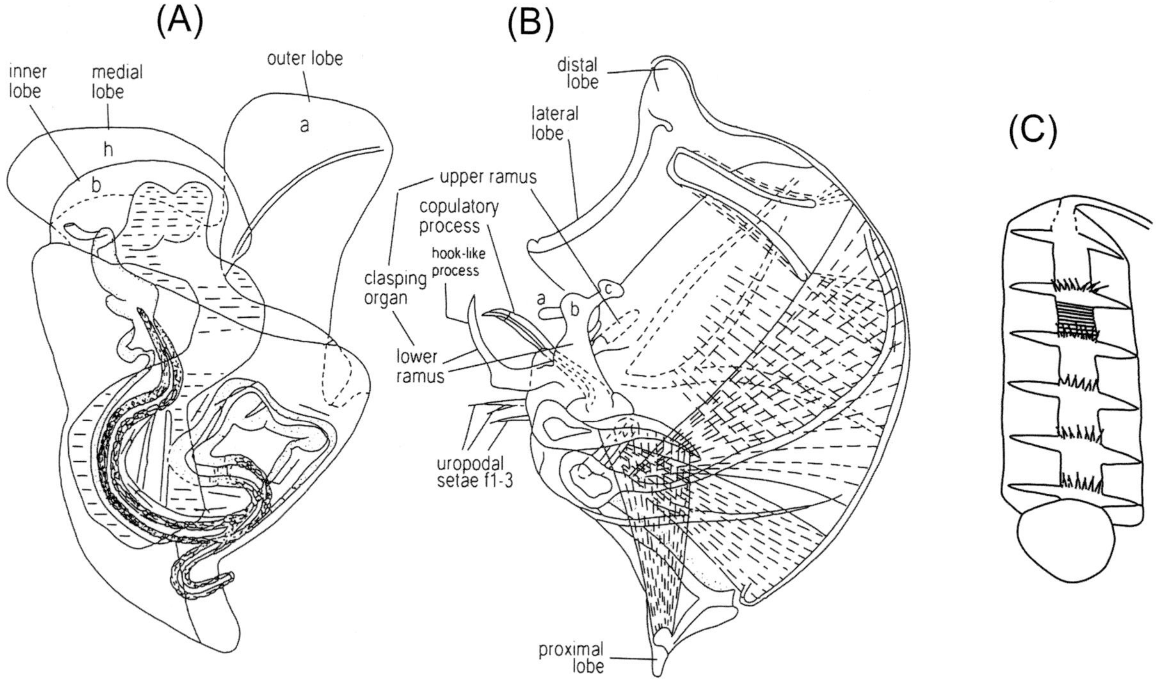

FIGURE 5.15 General morphology of nonmarine ostracod male sexual organs. (A) Hemipenis of Cypridoidea. (B) Hemipenis of Cytheroidea. (C) Zenker organ. *After Meisch (2000).*

The cypridoidean males present a Zenker organ (Fig. 5.15), which is a sperm pump that also has taxonomic potential despite the fact that it has usually been neglected (Matzke-Karasz, 1997).

Material preparation and preservation

Detailed information on the field and laboratory methods related to living ostracods has been reviewed and discussed in a number of papers and book chapters (e.g., Cohen & Oakley, 2017; Danielopol et al., 2002; Martens & Horne, 2016; Meisch, 2000; Namiotko et al., 2011), which we recommend for an exhaustive account on these subjects. Here we present a brief summary of the most common techniques that are routinely employed in the study of ostracods living in inland water bodies.

Nonmarine ostracods are commonly found in almost all types of aquatic ecosystems. Although most of them are prevalently benthic, many ostracod species are planktonic, neustonic, interstitial, periphytic, or subterranean. Only one semi-terrestrial species, *Scottia pseudobrowniana*, is so far known to occur in the Mediterranean Basin. Depending on the habitat characteristics, different sampling strategies are required, and the combination of different techniques is usually necessary to obtain a comprehensive assessment of the diversity of ostracod communities (Smith et al., 2015). For these reasons, it is difficult to establish standardized sampling protocols (Rossetti, 2007).

Sampling techniques mostly coincide with those used to collect other freshwater invertebrate groups, in particular, zooplankton and meio- and macrofauna. Performing quantitative samplings is complicated because of the high spatial heterogeneity usually observed in ostracod communities. Some sampling devices, for example bottom grab samplers and core samplers, allow to gather quantitative samples (Fig. 5.16), but their applicability and effectiveness strongly depend on sediment texture and the type of vegetation cover, and a large number of replicates are needed to obtain a reliable estimation of ostracod density. In addition, rare species may not be detected, and some ostracods may be able to escape from sampling gears by swimming or crawling. More frequently, collected samples are qualitative (Fig. 5.16). Semi-quantitative sampling methods based on a catch-per-unit effort can be used to assess the relative abundance of ostracod taxa. The choice among different sampling methods must also take into account the potential deleterious effects on particularly fragile or small environments, such as in the case of springs (Rosati et al., 2016), or the possibility of recovering passive sampling devices (e.g., traps, both baited and unbaited) within a few days from aquifers, surface sediments, hyporheic zone, etc. to avoid damage to the collected organisms. Many species of ostracods can colonize very small water bodies, both natural and artificial, in which standard sampling methods are not applicable; in

FIGURE 5.16 Collecting nonmarine ostracod samples. (A) Qualitative sample using a hand net from a Mediterranean wetland (Fondo d'Elx Natural Park), with permission from the sampler, Patricia Jerez. (B) Quantitative sample using an Ekman grab, lake Ojos de Villaverde (Spain).

these tiny systems, inexpensive and easily available tools such as cloth coffee strainers, plastic paint strainers, or modified entomologist exhausters can be used (Rossetti et al., 2006; Viehberg, 2002).

Many inland aquatic ecosystems in the Mediterranean Basin are temporary or ephemeral, often showing short filling phases. It is therefore not uncommon, when performing sampling campaigns, to visit completely dry water bodies. In this case, it is possible to use the so-called Sars' method, i.e., the collection of surface sediment samples which contain diapausing ostracod eggs and their rehydration in the laboratory to stimulate egg hatching and subsequent species identification based on late juvenile stages or adults (Marrone et al., 2019; Van Damme & Dumont, 2010). Usually, hatching does not require controlled conditions, although for some species different hydration/dehydration cycles and/or specific environmental conditions (e.g., photoperiod and temperature) may be necessary as cues for diapause termination. Nonmarine ostracods can be easily reared and maintained for long periods of time, given the generally low ecological requirements of these organisms and their high reproductive potential. A summary of rearing methods is reported by Schmit et al. (2007).

In the field, it is possible to pass the samples through a sieve or a conical funnel equipped with a net (in both cases with a mesh size of c. 0.5–1 cm) to retain the coarser material, which is then carefully rinsed to wash out all the remaining ostracods. The mesh pore size used to retain ostracods is usually between 100 and 250 μm. Whenever possible, a preliminary sorting of collected ostracods must be carried out on living material, pouring small amounts of water and sediment into white trays. After a while, once the lighter particles have been settled to the bottom and good water transparency has been obtained, the movement of the ostracods can be detected by the naked eye, and animals are collected using a pipette. Alternatively, sampled ostracods can be kept alive in plastic bottles and refrigerated until transported to the laboratory. This procedure usually does not harm the animals, provided that there is enough water in the bottles in relation to the sediment volume to ensure a good oxygen supply. If samples cannot be processed within

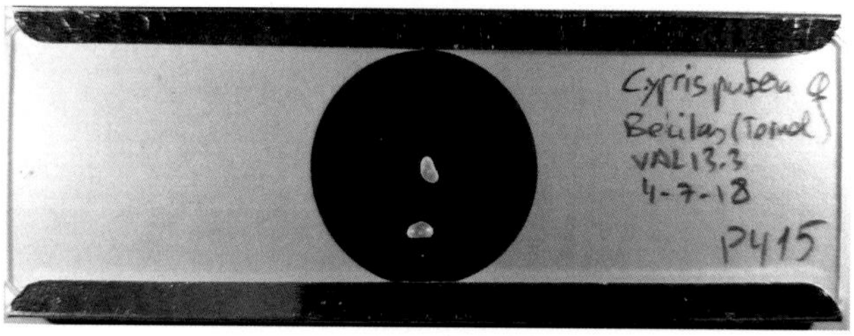

FIGURE 5.17 Ostracod valves stored dry in a micropalaeontological slide.

24–48 h after collection, it is strongly recommended to use a battery or AC-powered aerator equipped with a porous stone air diffuser (sold in aquarium stores).

Sorting operations are generally completed in the laboratory under a binocular microscope, both on living or preserved material. In order to increase the efficiency of extraction of living ostracods, a staining solution can be added to the samples together with the fixative (Danielopol et al., 2002). Quite effective methods have been proposed to shorten the sorting times for both preserved and unpreserved samples (Szlauer-Łukaszewska & Radziejewska, 2013). Once sorted from the collected material, specimens are usually stored in 70% ethanol in tubes sealed tightly to prevent evaporation. Alternatively, to avoid a tight closure of the ostracod valves which could make their dissection difficult, the specimens can be first fixed in 30% ethanol or less and afterward brought to the final concentration that guarantees their long-term preservation (Meisch, 2000).

In most cases, the examination of both soft parts and valves of adult specimens is needed for species identification. Methods for the gentle opening of closed carapaces with preservation of both valves have been suggested by Rossetti and Martens (1996) and Scharf et al. (2016). Valves are stored dry in micropalaeontological slides (Fig. 5.17), while dissected appendages are mounted in permanent glass slides, using glycerin or other mounting media. Soft parts can be stained with different chemicals to better show details of chaetotaxy, either before dissection or by adding the staining substance to the mounting medium (Namiotko et al., 2011). Some important taxonomic features of valves can only be revealed by scanning electron microscope observations.

Keys to Ostracoda

These keys use adult (mostly) female characters of freshwater ostracods in the class Podocopa, order Podocopida.

Ostracoda key: superfamilies, families, and subfamilies

1 Central muscle scars not arranged in a rosette; body with one or three pairs of walking legs 2

1' Central muscle scars arranged in a rosette; body with two pairs of walking legs (T2 and T3) (Figs. 5.7 and 5.12E and H) .... Darwinuloidea, one family (Fig. 5.18): **Darwinulidae**

2(1) Central muscle scars arranged like a "pawprint"; body with one pair of walking legs (T2), T3 an inverted cleaning limb (Figs. 5.7 and 5.12D and G) .... (Cypridoidea) 3

2' Central muscle scars arranged vertically in a row of four scars (some occasionally subdivided in two); body with three pairs of walking legs (T1, T2 and T3) (Figs. 5.7 and 5.12F and I) .... (Cytheroidea) 16

3(2) Valves without dorsomedian sulci and with short straight (less than 2/3 of valve length) or arched dorsal margins; T1 a maxilliped with a one-segmented endopodite (Fig. 5.12A); distal segment of cleaning leg (T3) with a pincer (Fig. 5.12G) or with three simple setae 4

3' Valves subrectangular with two dorsomedian sulci, a fairly straight dorsal margin (>3/4 of valve length) (Fig. 5.19A and B); T1 a maxilliped with a two-segmented endopodite (Fig. 5.19C); distal segment of cleaning leg (T3) with three simple setae (Fig. 5.19D) .... Ilyocyprididae, one genus: *Ilyocypris*

4(3) Carapace ovoid, elongate ovoid or reniform, rarely subrectangular, ventral margin convex, sinuous, or concave, rarely almost straight; A2 swimming setae well developed, reduced, or absent; maxillula third masticatory lobe with 2–3 teeth bristles (Fig. 5.11B) 5

4' Carapace ovoid with a fairly straight and flat ventral margin; A2 with well-developed swimming setae; maxillula third masticatory lobe with 6 teeth bristles (Fig. 5.20A) .... Notodromadidae, one genus in the Mediterranean Basin: *Notodromas*

5(4) Cleaning limb (T3) distally with three simple setae (Fig. 5.29J-M); carapace pigmented or unpigmented 6

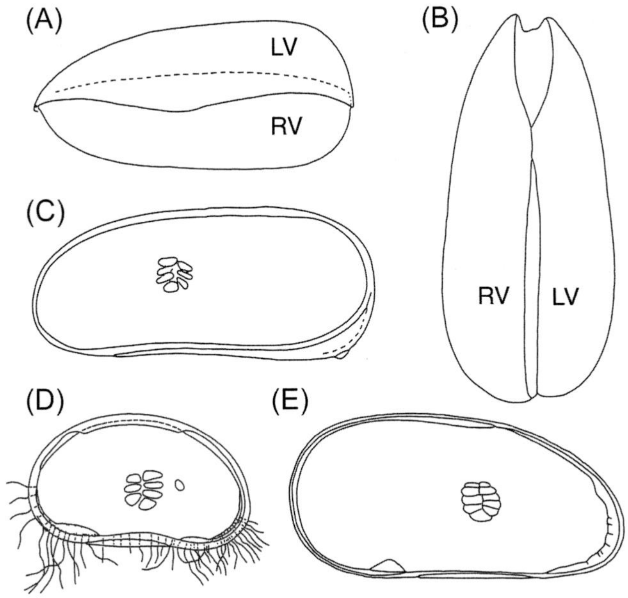

FIGURE 5.18 Valve morphology of some genera of Darwinuloidea. (A) *Vestalenula*, carapace, ventral view. (B) *Darwinula*, carapace, ventral view. (C) *Vestalenula*, right valve, inner view. (D) *Microdarwinula*, left valve, inner view. (E) *Penthesilenula*, left valve, inner view. *RV*, Right valve; *LV*, left valve. *After (A, C–E) Rossetti and Martens (1998); (B) Meisch (2000).*

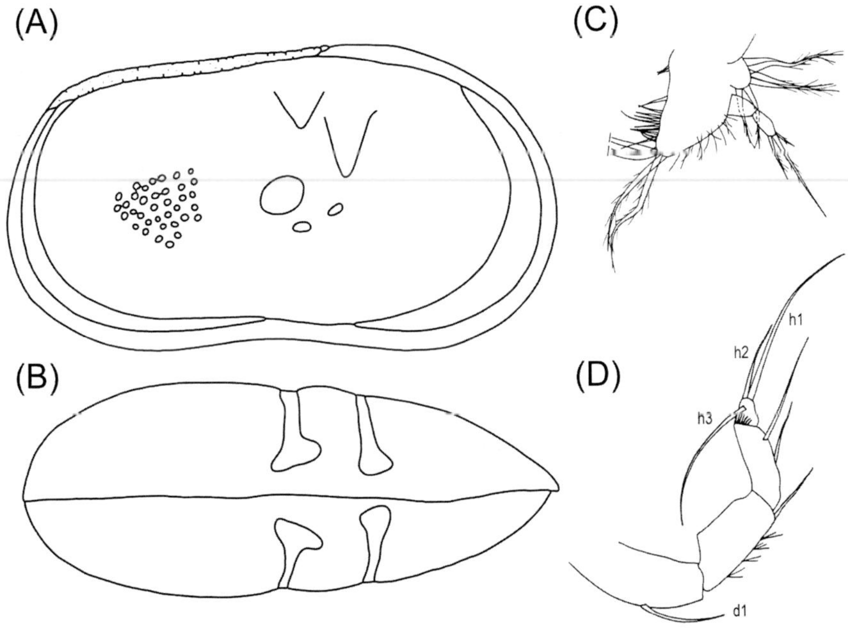

FIGURE 5.19 Distinctive characters of the genus *Ilyocypris*. (A) Left valve, inner view. (B) Carapace, dorsal view. (C) T1. (D) T3. *From Meisch (2000).*

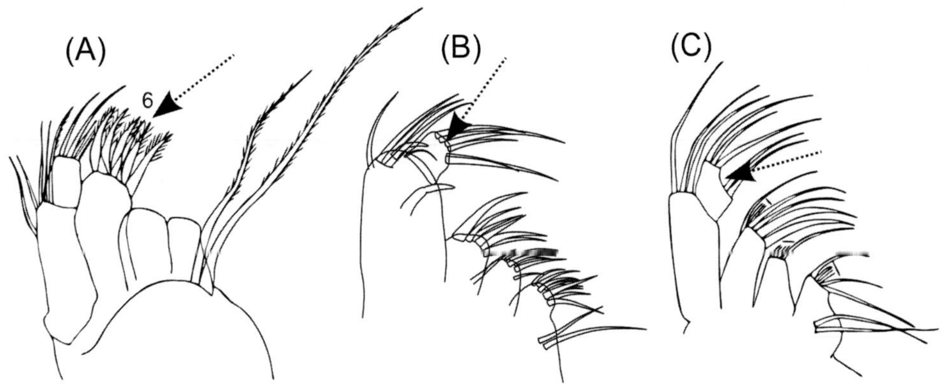

FIGURE 5.20 Schematic drawing of the maxillular endopod of some Cypridoidea ostracods, including three endites and palp. (A) Notodromadidae (setae of endites 1–2 not shown). (B) Cyclocyprididae. (C) Candonidae. Arrows point to third endite bristles in A and to last segment of palp in B–C. *From (A, C) Meisch (2000).*

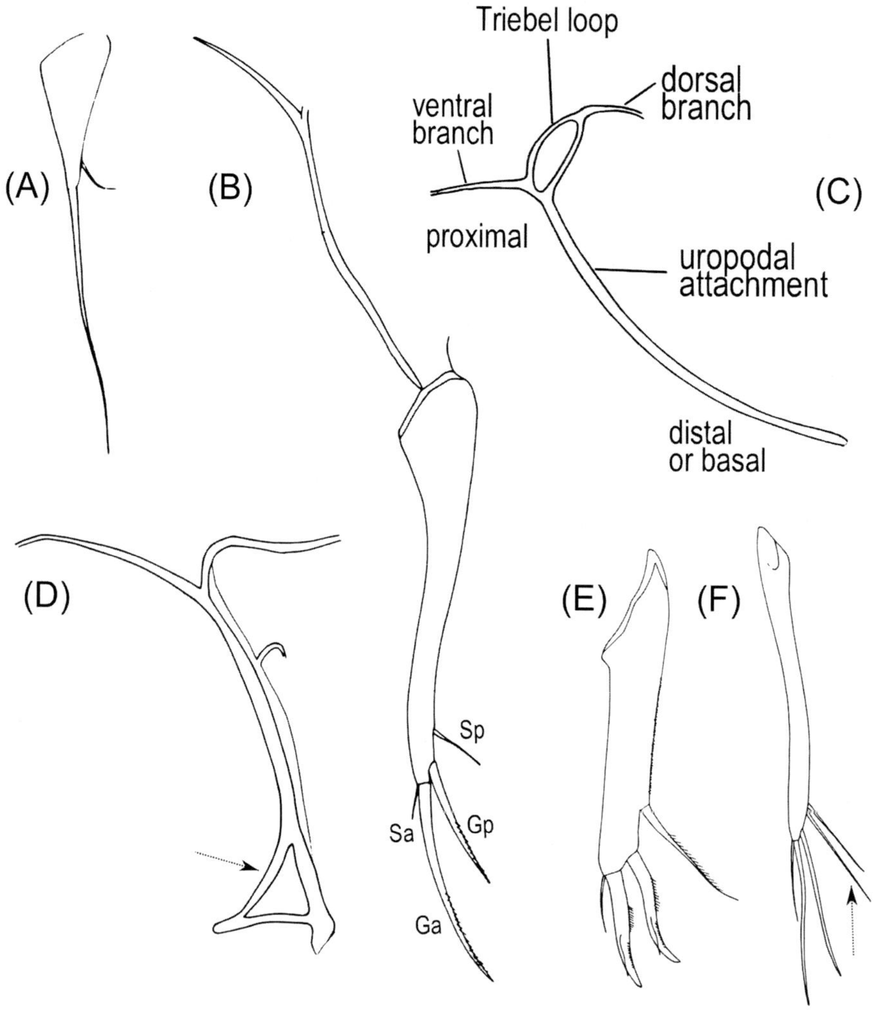

FIGURE 5.21 Uropod rami and attachments in Cypridoidea. (A) Uropod ramus of Cypridopsinae. (B) Uropod ramus and attachment of Cypridinae (Sp: posterior seta; Sa: anterior seta; Gp: posterior claw; Ga: anterior claw). (C) Uropod attachment of Cypricercinae, with nomenclature of different parts. (D) Uropod attachment of Herpetocypridinae. (E) Uropod ramus of Scottinae. (F) Uropod ramus of Hungarocypridinae. *After Meisch (2000).*

5' Cleaning limb (T3) distally with a pincer (Fig. 5.12G); carapace usually more or less conspicuously pigmented, rarely white .... (Cyprididae) 7

6(5) Carapace (sub)ovate in lateral view, more or less strongly colored, rarely white; A2 swimming setae well developed or more or less reduced in length, never completely missing; Mx1 palp terminal segment cylindrical, symmetrically shaped (Fig. 5.20B) .... **Cyclocyprididae**[1]

6' Carapace more elongate in lateral view, always unpigmented, pearly white, whitish or translucent; A2 swimming setae absent, totally reduced; Mx1 palp terminal segment asymmetrically shaped (Fig. 5.20C) .... **Candonidae**

7(5) U-ramus well developed, with two terminal claws (Fig. 5.13A) 8

7' U-ramus reduced, flagelliform, without terminal claws, or absent (Fig. 5.21A) .... **Cypridopsinae**

8(7) Endoskelatal attachment of u-ramus without a Triebel loop (Fig. 5.21B) 9

8' Endoskeletal attachment of u-ramus with a Triebel loop (Fig. 5.21C) .... **Cypricercinae**

9(8) Endoskeletal attachment of u-ramus without a distal triangular reinforcement 10

9' Endoskeletal attachment of u-ramus with a distal triangular reinforcement (weakly expressed in *Isocypris*) (Fig. 5.21D) .... **Herpetocypridinae**

1. According to recent molecular analysis by Hiruta et al. (2016), the subfamily Cyclocypridinae should be elevated to family level. We follow these and other previous authors and use here the family Cyclocyprididae.

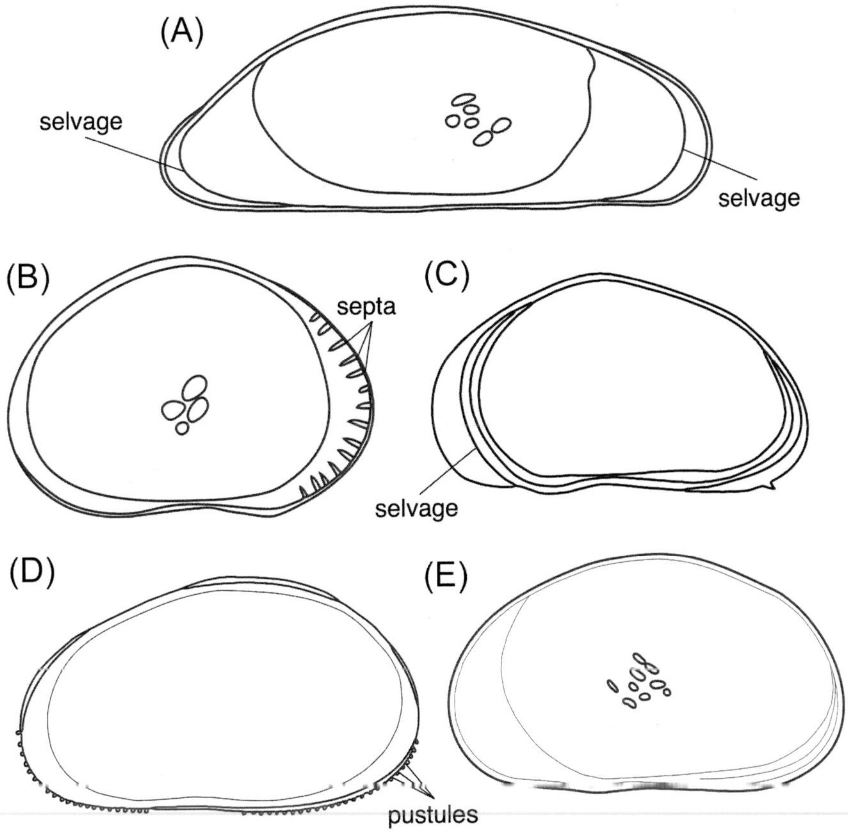

FIGURE 5.22 Carapace shape and structures in Cypridoidea. (A) Dolerocypridinae. (B) Cyprettinae. (C) Cypridinae. (D) Cyprinotinae. (E) Eucypridinae. *After Horne et al. (2019).*

10(9) U-ramus with one posterior seta (Fig. 5.21B and E); size < 4 mm 11
10' U-ramus with two posterior setae (Fig. 5.21F); size 4.5–5.0 mm .... Hungarocypridinae, one genus: *Hungarocypris*
11(10) Carapace ovoid or subtriangular in lateral view (Fig. 5.22B–E) 12
11' Carapace conspicuously elongate in lateral view (Fig. 5.22A) .... Dolerocypridinae, one genus: *Dolerocypris*
12(11) Marginal zone of valves without radial septa 13
12' Marginal zone of valves anteriorly with conspicuous radial septa (Fig. 5.22B) .... Cyprettinae, one genus in the Mediterranean Basin, *Cypretta*
13(12) U-ramus terminal claws long and slender (Fig. 5.21B) 14
13' U-ramus terminal claws short and thick (Fig. 5.21E) .... Scottiinae, one species in the Mediterranean Basin, *Scottia pseudobrowniana*
14(13) RV anteriorly without a conspicuous flat expansion, selvage therefore only slightly displaced inwards 15
14' RV anteriorly with a conspicuous flat expansion (flange), selvage therefore markedly displaced inwards (Fig. 5.22C) .... **Cypridinae**
15(14) Free margin of one valve (either RV or LV) anteriorly and posteriorly with rows of tiny pustules (sometimes missing) (Fig. 5.22D) .... **Cyprinotinae**
15' Free margin of valves smooth, without tiny pustules (Fig. 5.22E) .... **Eucypridinae**
16(2) All three walking legs (T1, T2, and T3) with long terminal claws, distinctly longer than distal segment (Fig. 5.12); free living species 17
16' All three walking legs with short, hook-shaped terminal claws, no longer than distal segment (Fig. 5.30A); commensal or ectoparasitic on larger crustaceans .... (Entocytheridae) 24
17(16) Carapace external surface smooth or ornamented; *Xestoleberis*-spot absent 18

17' Carapace external surface smooth; *Xestoleberis*-spot present (Fig. 5.30B) Xestoleberididae, one genus, ... *Xestoleberis*
18(17) Anterior marginal pore canals mostly simple (a few may be bifurcated) or not discernible 19
18' Anterior marginal pore canals conspicuously arranged in brush-like bunches (Fig. 5.30C) .... Hemicytheridae, one genus: *Tyrrhenocythere*
19(18) Mx1 branchial plate with > 10 filaments 20
19' Mx1 branchial plate with only 1–2 filaments (Fig. 5.30D) .... **Kliellidae**
20(19) T2 and T3 basal segment posterior setae slender or reduced, not annulated 21
20' T2 and T3 basal segment posterior setae stout, conspicuously annulated (Fig. 5.30E) .... **Cytherideidae**
21(20) A2 with three terminal claws (Fig. 5.31N). T1–T3 basal part of terminal claw roundly inflated, often with a short seta 22
21' A2 with two terminal claws. T1–T3 basal part of terminal claw not roundly inflated, that part without any seta . 23
22(21) Carapace not egg-shaped in dorsal view, with both ends pointed. Conspicuous lateral valve projections (tubercles) present in a number of species (Fig. 5.30F) .... **Limnocytheridae**
22' Carapace distinctly egg-shaped in dorsal view, with both ends rounded or roundly pointed (Fig. 5.30G). Valves without lateral projections (tubercles) .... **Timiriaseviidae**[2]
23(21) Hinge gongylodont (RV posterior tooth bilobate) (Fig. 5.30H) or henodont (RV posterior tooth simple) (Fig. 5.30I) .... **Loxoconchidae**
23' Hinge entomodont (RV posterior tooth with five cusps increasing in size toward posterior) (Fig. 5.30J) .... Leptocytheridae, one genus in the study area, *Leptocythere*
24(16) Mx1 with a branchial plate .... **Entocytherinae**
24' Mx1 without a branchial plate .... Sphaeromicolinae, one genus in the Mediterranean Basin, *Sphaeromicola*

Ostracoda: Darwinuloidea: Darwinulidae: genera

1 LV overlaps RV ventrally (Fig. 5.18A). Size < 0.60 mm (0.30–0.55 mm) 2
1' RV overlaps LV ventrally (Fig. 5.18B). Size > 0.60 mm (0.63–0.80 mm) .... *Darwinula*
2(1) Smaller valve (RV) without a posteroventral keel 3
2' Smaller valve (RV) with a posteroventral keel (Fig. 5.18C) .... *Vestalenula*
3(2) Central muscle scars situated at approximately mid-length of valves; valves ovate in lateral view (Fig. 5.18D) ... . *Microdarwinula*
3' Central muscle scars situated well in front of mid-length of valves (Fig. 5.18E); valves elongate in lateral view .... *Penthesilenula*

Ostracoda: Cypridoidea: Cyprididae: Cypridinae: genera

1 Walking leg 4-segmented (Fig. 5.23A). Third masticatory lobe of the maxillula with two barbed teeth bristles (Fig. 5.23C) .... *Cypris*
1' Walking leg 5-segmented. (Fig. 5.23B). Third masticatory lobe of the maxillula with two smooth teeth bristles and a reinforced, barbed bristle in between (Fig. 5.23D) .... *Chlamydotheca*

Ostracoda: Cypridoidea: Cyprididae: Cypridopsinae: genera

1 LV overlapping RV ventrally 2
1' RV overlapping LV ventrally 5
2(1) Valve surface smooth or pitted, without prominent concentric ridges 3
2' Valve surface ornamented with prominent concentric ridges (Fig. 5.24A) .... *Zonocypris*
3(2) LV posterior inner marginal zone without a conspicuous inner list; A2 swimming setae well developed or reduced 4

2. Relying on their analysis based on both morphological and genetic characters, Tanaka et al. (2021) recently suggested, but without defining it formally, the elevation of the subfamily Timiriaseviinae to the rank of a separate family from Cytheroidea, the Timiriaseviidae. Notwithstanding this formal shortcoming, the family Timiriaseviidae is accepted here as taxonomically valid.

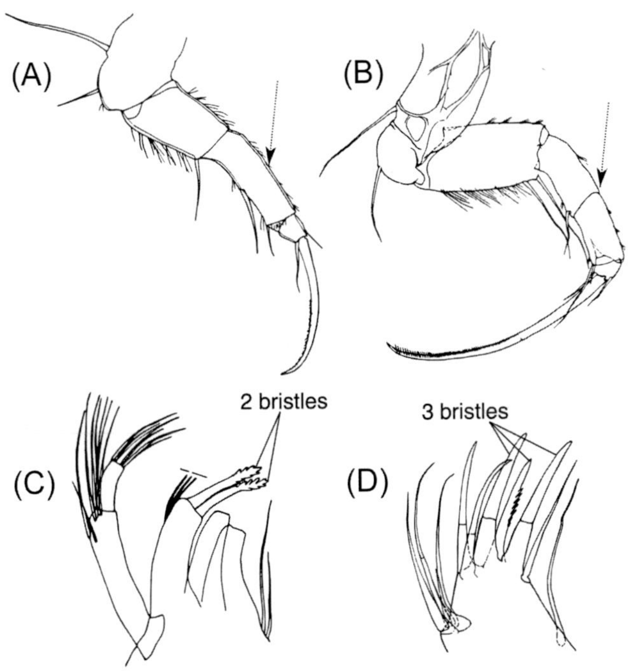

FIGURE 5.23 Second thoracopods (T2, walking legs) and details of maxillulae of Cypridinae genera. (A) T2 of the genus *Cypris*. (B) T2 of the genus *Chlamydotheca*. (C) Maxillula of the genus *Cypris*, with the presence of two barbed teeth bristles in the third endite. (D) Detail of the third endite of the maxillula of *Chlamydotheca*, with the presence of two smooth and one barbed teeth bristles. *After (A, C) Meisch (2000); (B) Roessler (1986a); (C) Roessler (1986b).*

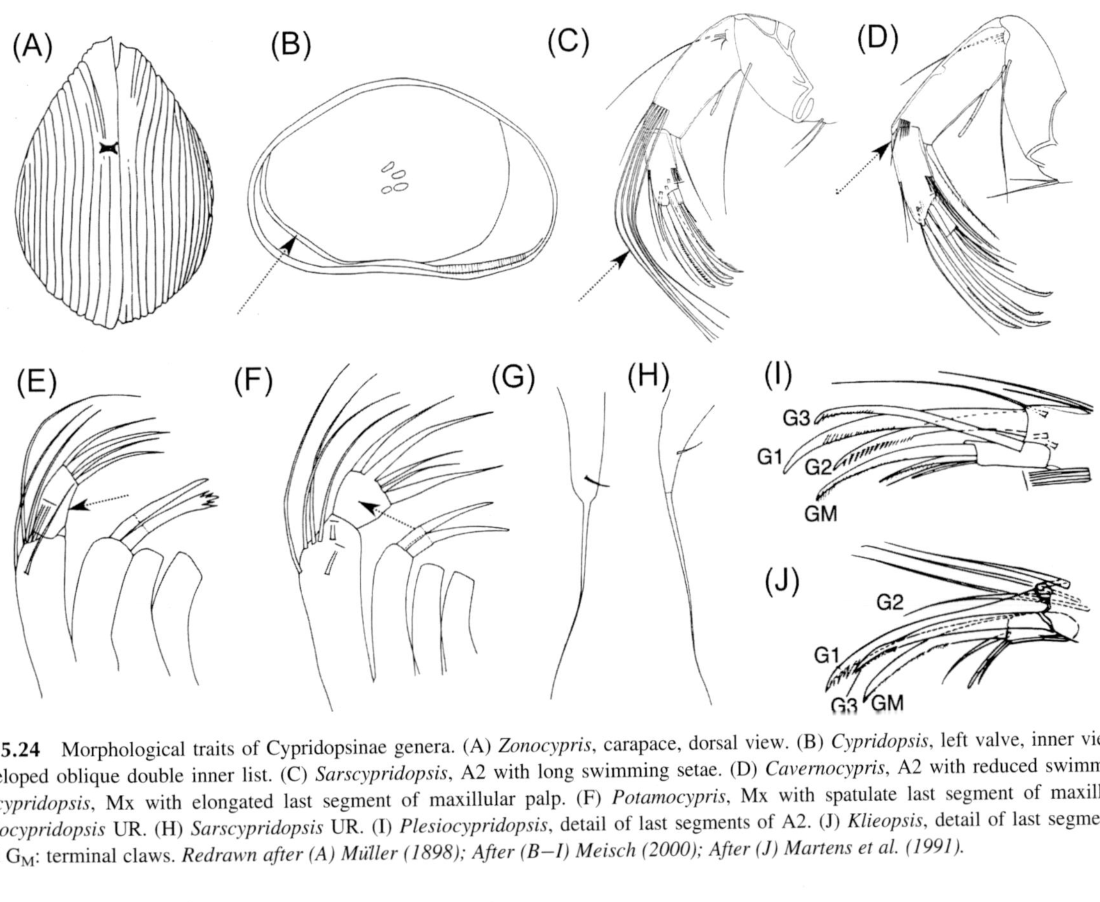

FIGURE 5.24 Morphological traits of Cypridopsinae genera. (A) *Zonocypris*, carapace, dorsal view. (B) *Cypridopsis*, left valve, inner view, with a well-developed oblique double inner list. (C) *Sarscypridopsis*, A2 with long swimming setae. (D) *Cavernocypris*, A2 with reduced swimming setae. (E) *Sarscypridopsis*, Mx with elongated last segment of maxillular palp. (F) *Potamocypris*, Mx with spatulate last segment of maxillular palp. (G) *Plesiocypridopsis* UR. (H) *Sarscypridopsis* UR. (I) *Plesiocypridopsis*, detail of last segments of A2. (J) *Klieopsis*, detail of last segments of A2. G_{1-3} and G_M: terminal claws. *Redrawn after (A) Müller (1898); After (B–I) Meisch (2000); After (J) Martens et al. (1991).*

3’ LV posterior inner marginal zone with a conspicuous, oblique double inner list; A2 swimming setae well developed (Fig. 5.24B and C) .... *Cypridopsis*
4(3) A2 swimming setae reduced (Fig. 5.24D); u-ramus present .... *Cavernocypris*
4’ A2 swimming setae reduced or well developed; u-ramus absent .... *Pseudocypridopsis*
5(1) Mx1 palp terminal segment elongated, with simple setae (Fig. 5.24E) 6
5’ Mx1 palp terminal segment spatulate, with 4–5 claw-like setae (Fig. 5.24F) .... *Potamocypris*
6 U-ramus present in both sexes 7
6’ U-ramus absent in both sexes .... *Martenscypridopsis*, one species in the study area, *Martenscypridopsis materia*
7(6) Trunk of u-ramus cylindrical, narrowing abruptly to distal flagellum (Fig. 5.24G); Mx1 third masticatory lobe with two serrated teeth bristles 8
7’ Trunk of u-ramus triangular, evenly tapering distally (Fig. 5.24H); Mx1 third masticatory lobe with one smooth and one serrated tooth bristle (Fig. 5.24E) .... *Sarscypridopsis*
8(7) A2 with four subequally long terminal claws (G1, G2, G3, and GM) (Fig. 5.24I) .... *Plesiocypridopsis*
8’ A2 with only three long terminal claws (G1, G3, and GM), G2 reduced to a long seta (Fig. 5.24J) .... *Klieopsis*, one species, *Klieopsis horai* (Klie, 1927)

Ostracoda: Cypridoidea: Cyprididae: Cypricercinae: genera

1 Carapace ovoid in lateral view, rounded in dorsal view (Fig. 5.25A and B) 2
1’ Carapace elongate in lateral view, laterally compressed in dorsal view (Fig. 5.25C and D) .... *Tanycypris*
2(1) Carapace approximately symmetrical in frontal or posterior view 3
2’ Carapace conspicuously asymmetrical in frontal or posterior view (Fig. 5.25E) .... *Bradleycypris*
3(2) Triebel loop situated centrally in the proximal fork of the u-ramus attachment (Fig. 5.25F) .... *Strandesia*
3’ Triebel loop situated in the dorsal branch of the u-ramus attachment (Fig. 5.25G) .... *Bradleystrandesia*

Ostracoda: Cypridoidea: Cyprididae: Herpetocypridinae: genera

1 Marginal zone of valves without radial septa 2
1’ Marginal zone of valves with conspicuous radial septa (Fig. 5.26A) .... *Stenocypris*

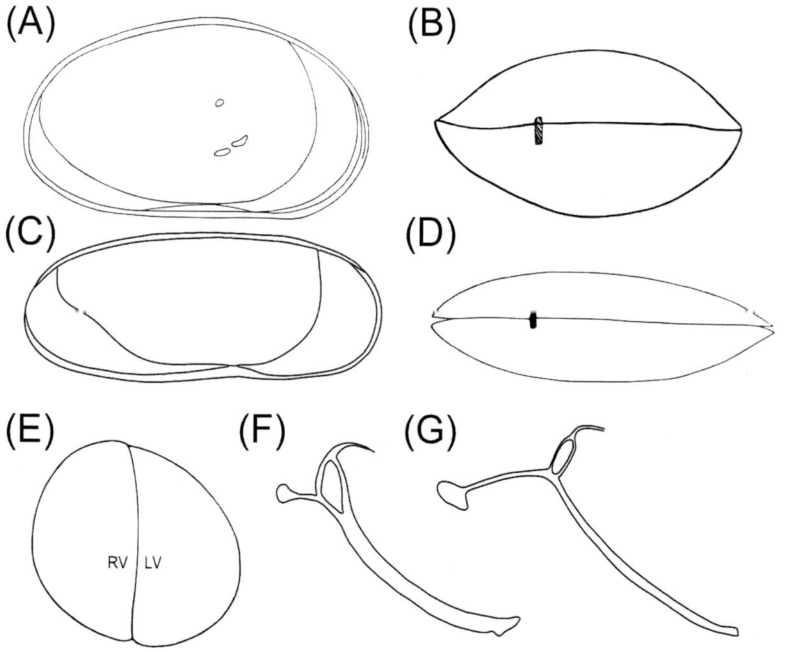

FIGURE 5.25 Morphological traits of Cypricercinae genera. (A) *Bradleycypris*, left valve, inner view. (B) *Bradleycypris*, carapace, dorsal view. (C) *Tanycypris*, left valve, inner view. (D) *Tanycypris*, carapace, dorsal view. (E) *Bradleycypris*, carapace, frontal view. (F) *Strandesia*, UR attachment. (G) *Bradleystrandesia*, UR attachment. *After (A, C, E, G) Meisch (2000); (B) Gauthier (1928); (D) Petkovski (1964); (F) Horne et al. (2019).*

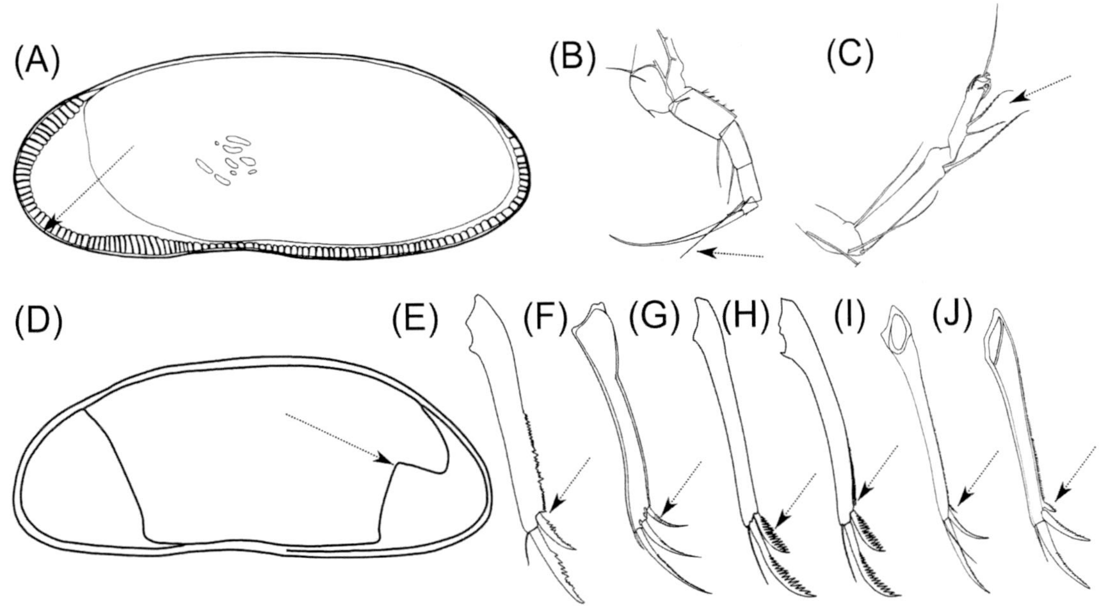

FIGURE 5.26 Morphological traits of Herpetocypridinae genera. (A) *Stenocypris*, right valve, inner view, with marginal radial septa. (B) *Isocypris*, T2. (C) *Candonocypris*, T3. (D) *Stenocypria*, right valve, inner view, with the posterior margin of the lamella conspicuously sinuous. (E) *Humphcypris*, UR. (F) *Ilyodromus*, UR. (G) *Chrissia*, left UR. (H) *Chrissia*, right UR. (I) *Herpetocypris*, UR. (J) *Psychrodromus*, UR. Arrows point to the posterior seta of the uropod in E,F,I,J, and to spines in claw or ramus in G and H. *After (A) Petkovski and Meisch (1996); (B, D, I, J) Meisch (2000); (C) Scharf et al. (2014); (E, G, H) Horne et al. (2019); (F) Higuti and Martens (2020).*

2(1) Walking leg, penultimate segment: g-seta of "normal" length (distinctly shorter than 1/2 L of the terminal claw) (Fig. 5.12D) 3
2' Walking leg, penultimate segment: g-seta unusually long (longer than 1/2 L of the terminal claw) (Fig. 5.26B) .... *Isocypris*, one species in the study area, *Isocypris beauchampi*
3(2) Cleaning leg with one f-seta 4
3' Cleaning leg with two f-setae (Fig. 5.26C) .... *Candonocypris*
4(3) Inner posterior margin simply rounded 5
4' Inner posterior margin conspicuously sinuous (Fig. 5.26D) .... *Stenocypria*
5(4) U-ramus: posterior seta small or either transformed into a small spine or a reinforced seta 6
5' U-ramus: posterior seta absent (Fig. 5.26E) .... *Humphcypris*
6(5) U-ramus: posterior seta small or transformed into a small spine 7
6' U-ramus: posterior seta reinforced, resembling a third distal claw (Fig. 5.26F) .... *Ilyodromus*
7(6) U-rami weakly asymmetrical or symmetrical, posterior margin smooth or with fine setules or spinules 8
7' U-rami conspicuously asymmetrical, posterior margin of at least one u-ramus with prominent spines (Fig. 5.26G and H) .... *Chrissia*
8(7) U-ramus: posterior seta small, untransformed (Fig. 5.26I) .... *Herpetocypris*
8' U-ramus: posterior seta transformed into a short spine (Fig. 5.26J) .... *Psychrodromus*

Ostracoda: Cypridoidea: Cyprididae: Cyprinotinae: genera

1 RV anteriorly and posteriorly with a row of tiny marginal pustules (Fig. 5.22D) 2
1' LV anteriorly and posteriorly with a row of tiny marginal pustules (Fig. 5.27A) .... *Hemicypris*
2(1) RV with a hump-like dorsal expansion (best seen in frontal or posterior view of the carapace) (Fig. 5.27B) .... *Cyprinotus*
2' RV without a hump-like dorsal expansion (although the LV may have one) (Fig. 5.22D) .... *Heterocypris*

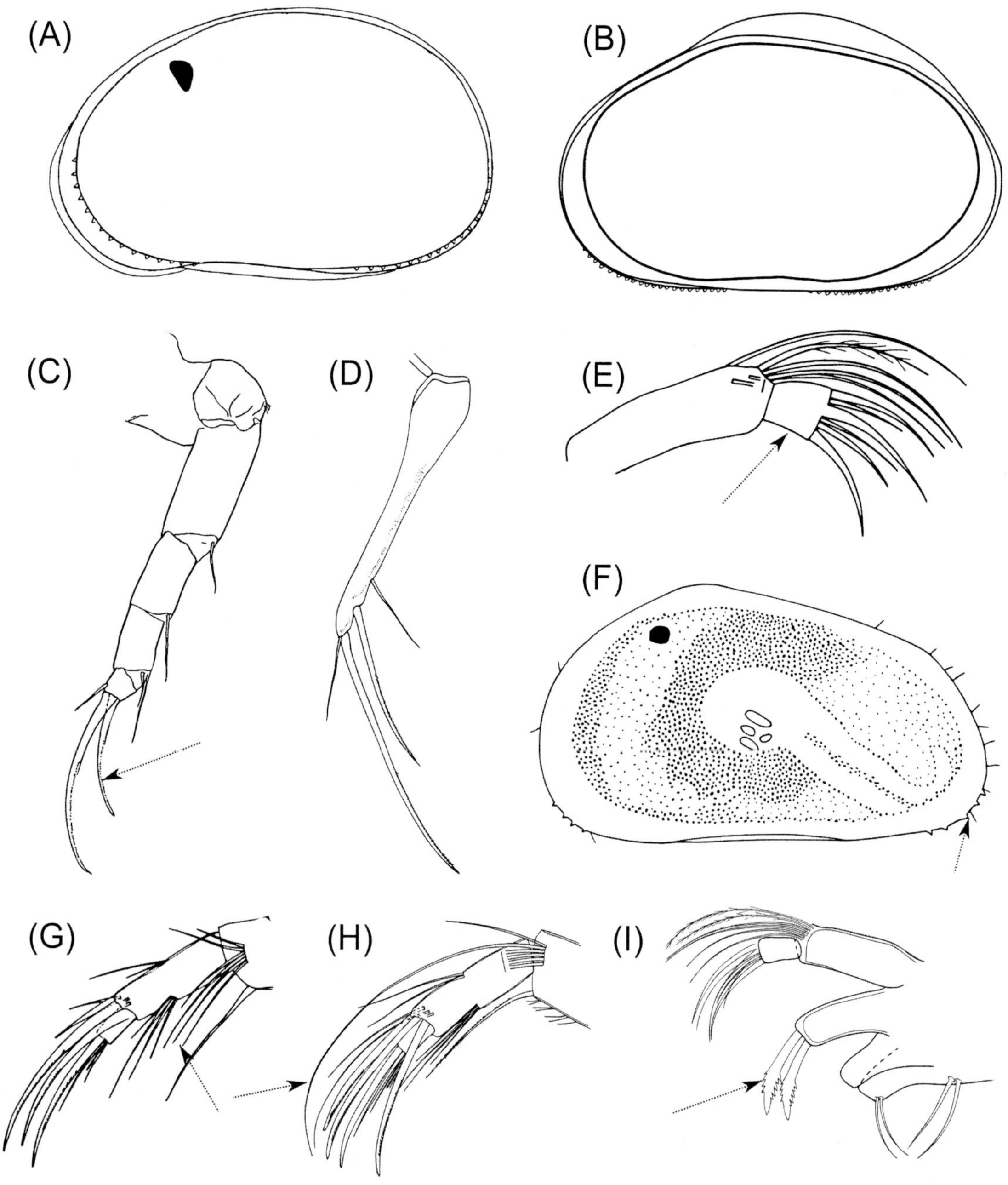

FIGURE 5.27 Morphological traits of Cyprinotinae and Eucypridinae genera. (A) *Hemicypris*, carapace, left valve lateral view. (B) *Cyprinotus*, right valve, inner view. (C) *Candelacypris*, T2 with two terminal claws. (D) *Koencypris*, UR. (E) *Prionocypris*, maxillular palp with a spatulate terminal segment. (F) *Prionocypris*, left valve, external view, showing a posteroventral row of spines. (G) *Tonnacypris*, distal part of A2 with reduced natatory setae. (H) *Trajancypris*, distal part of A2 with long natatory setae. (I) *Arctocypris*, maxillula with serrated teeth bristles. *Modified after (A) Gauthier (1933); After (B) Horne et al. (2019); After (C) Baltanás (2001); After (D–H) Meisch (2000); Modified after (I) Rasouli et al. (2016).*

Ostracoda: Cypridoidea: Cyprididae: Eucypridinae: genera

1 Walking leg with one terminal claw 2
1' Walking leg with two terminal claws (Fig. 5.27C) .... *Candelacypris*
2(1) U-ramus long terminal claw shorter than ramus 3
2' U-ramus long terminal claw conspicuously long, about as long as ramus (Fig. 5.27D) .... *Koencypris*, one species, *Koencypris ornata* (O.F. Müller, 1776)
3(2) Mx1 palp terminal segment cylindrical, longer than broad 6
3' Mx1 palp terminal segment short, spatulate (Fig. 5.27E) 4

4(3) Surface of valves smooth; valves posteroventrally with or without a row of spines 5
4' Surface of valves with pits; valves posteroventrally with a row of spines (Fig. 5.27F) .... *Prionocypris*
5(4) A2 swimming setae short (Fig. 5.27G) .... *Tonnacypris*
5' A2 swimming setae long, extending to tips of terminal claws or beyond (Fig. 5.27H) .... *Trajancypris*
6(3) Walking leg, setae d1 and d2 present, d2 well developed (Fig. 5.12D) .... *Eucypris*
6' Walking leg, seta d1 present, d2 tiny or absent 7
7(6) Mx1 third masticatory lobe, both teeth bristles smooth .... *Eucyprinotus*, one species, *Eucyprinotus rostratus* (Sywula, 1966)
7' Mx1 third masticatory lobe, both teeth bristles serrate (Fig. 5.27I) .... *Arctocypris*

Ostracoda: Cypridoidea: Cyclocyprididae: Cyclocypridinae: genera

1 Carapace laterally compressed or weakly inflated in dorsal view (Fig. 5.28A); cleaning leg penultimate segment distal seta (g) short or absent (Fig. 5.28C) 2
1' Carapace strongly inflated (tumid) in dorsal view (Fig. 5.28B); cleaning leg penultimate segment distal seta (g) well developed (Fig. 5.28D) .... *Cyclocypris*
2(1) RV with marginal pustules antero- and postero-ventrally (Fig. 5.28E) .... *Physocypria*
2' Both valve margins entirely smooth (Fig. 5.28F) .... *Cypria*

Ostracoda: Cypridoidea: Candonidae: Candoninae: genera

1 U-ramus posterior seta present (Fig. 5.13A) 3
1' U-ramus posterior seta absent (Fig. 5.29A) 2

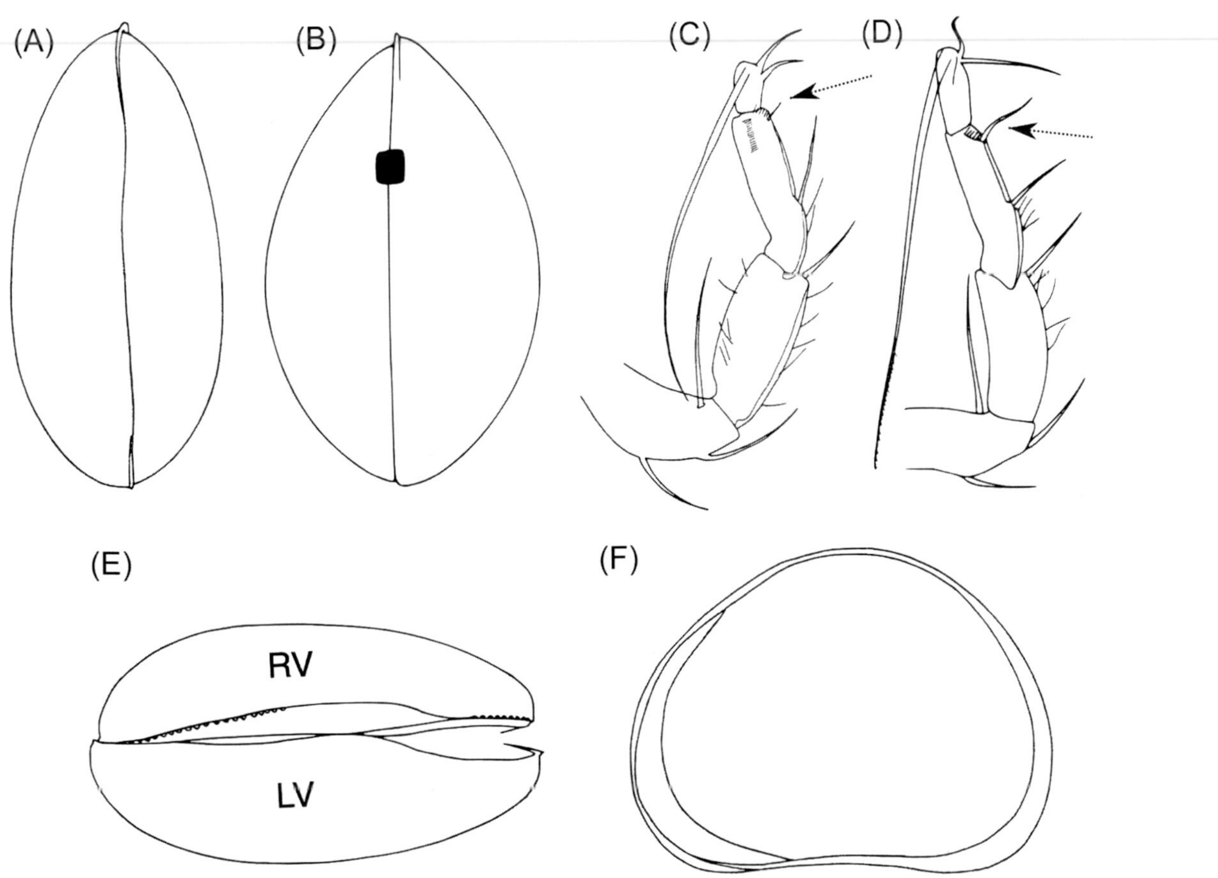

FIGURE 5.28 Morphological traits of Cyclocyprididae genera. (A) *Cypria*, carapace, dorsal view. (B) *Cyclocypris*, carapace, dorsal view. (C) *Cypria*, T3. (D) *Cyclocypris*, T3. (E) *Physocypria*, carapace, ventral view (RV: right valve, LV: left valve). (F) *Cypria*, right valve, inner view. *After Meisch (2000).*

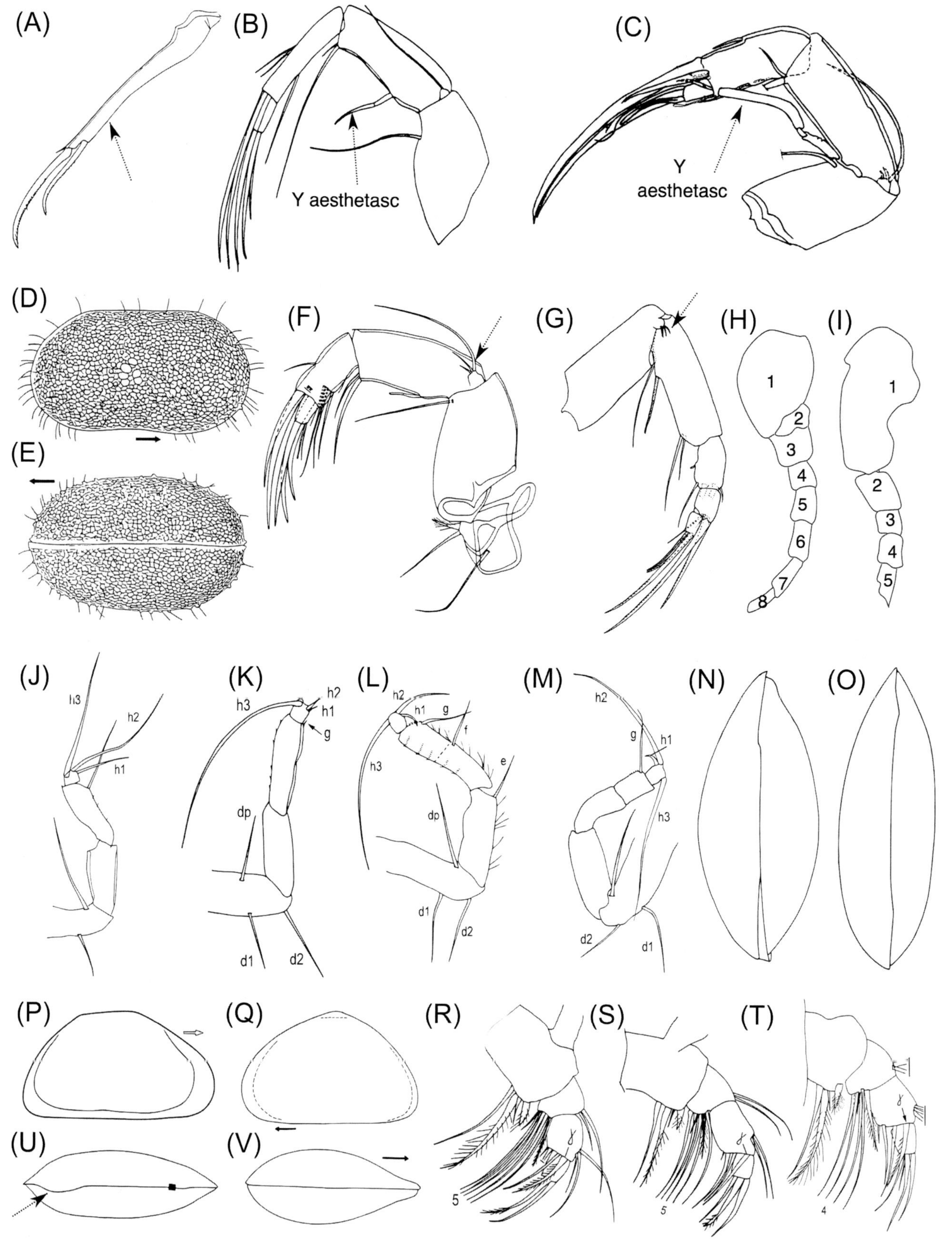

FIGURE 5.29 Morphological traits of Candonidae genera. (A) *Candonopsis*, UR without a posterior seta. (B) *Candonopsis*, A2 with a slender Y aesthetasc. (C) *Marococandona*, A2 with a large Y aesthetasc. (D) *Paracandona*, left valve, external view. (E) *Paracandona*, carapace, dorsal view. (F) *Candona*, A2. (G) *Trajancandona*, A2 with short exopod setae. (H) *Candona*, A1 (only segments). (I) *Nannocandona*, A1 (only segments). (J) *Candona*, T3. K: *Mixtacandona*, T3. (L) *Cryptocandona*, T3. (M) *Pseudocandona*, T3. J–M: d1, d2, dp, e, f, g, h1–3: specific setae. (N) *Pseudocandona*, carapace, dorsal view. (O) *Cryptocandona*, carapace, dorsal view. (P) *Marmocandona*, left valve, inner view. (Q) *Typhlocypris*, left valve, external view. (R) *Candona*, mandibular palp. (S) *Fabaeformiscandona*, mandibular palp. (T) *Neglecandona*, mandibular palp. (U) *Fabaeformiscandona*, carapace, dorsal view. (V) *Schellencandona*, carapace, dorsal view. *From (A, B, D–F, J–O, Q–V) Meisch (2000); Modified from (C) Marmonier et al. (2005); Modified after (G) Karanovic (1999); From (H, I, P) Horne et al. (2019).*

2(1) A2 aesthetasc Y slender, not conspicuously thick (Fig. 5.29B) .... *Candonopsis*
2' A2 aesthetasc Y huge, conspicuously thick (Fig. 5.29C) .... *Marococandona*
3(1) Surface of valves smooth or partially ornamented with pits 4
3' Surface of valves ornamented with a conspicuous net-like reticulation; carapace moderately inflated in dorsal view, with dorsal and ventral margins approximately parallel in lateral view (Fig. 5.29D and E) *Paracandona*
4(3) Carapace ovoid, subrectangular, subreniform or subtrapezoidal in lateral view; A2 exopodite with two short and one very long seta (Fig. 5.29F) 5
4' Carapace subrectangular in lateral view; A2 exopodite with three very short setae (Fig. 5.29G) *Trajancandona*
5(4) A1 with seven to eight articulated segments (Fig. 5.29H) 6
5' A1 with five articulated segments (Fig. 5.29I) .... *Nannocandona*
6(5) T3 cleaning leg distally with three long setae (Fig. 5.29J) 7
6' T3 cleaning leg distally with one long and two very short setae (Fig. 5.29K) .... *Mixtacandona*
7(6) T3 cleaning leg basal segment with three setae (d1, d2, and dp) (Fig. 5.29L) 8
7' T3 cleaning leg basal segment with two setae (d1 and dp) (Fig. 5.29J) 11
8(7) Carapace roundly broad in dorsal view (Fig. 5.29N); T3 cleaning leg seta f missing (Fig. 5.29M) 9
8' Carapace conspicuously narrow (laterally compressed) in dorsal view (Fig. 5.29O); T3 cleaning leg seta f present (Fig. 5.29L) .... *Cryptocandona*
9(8) Carapace subtriangular or subtrapezoidal in lateral view, with maximum height close to mid-length 10
9' Carapace subovate, subrectangular or subreniform in lateral view, with maximum height well behind mid-length .... *Pseudocandona*
10(9) Carapace subtrapezoidal in lateral view, dorsal margin forming a short straight or slightly concave section, approximately parallel to the ventral valve margin (Fig. 5.29P); valves conspicuously pitted in dorso-central valve area .... *Marmocandona*
10' Carapace subtriangular in lateral view (Fig. 5.29Q); valves weakly pitted in the area around the central muscle scars .... *Typhlocypris*
11(7) Md palp with plumose gamma-seta (Fig. 5.29R) 12
11' Md palp with smooth gamma-seta (Fig. 5.29S) 13
12(11) Md palp second segment with a bunch of five setae (Fig. 5.29R) .... *Candona*
12' Md palp second segment with a bunch of four setae (Fig. 5.29T) .... *Neglecandona*
13(11) Carapace in dorsal view: posterodorsal margin of one valve or both valves with a lobate expansion or flap which overlaps the opposing valve margin when closed (Fig. 5.29U) .... *Fabaeformiscandona*
13' Carapace in dorsal view: posterodorsal margins of valves simple, without lobate expansions (Fig. 5.29V) .... *Schellencandona*

Ostracoda: Cytheroidea: Entocytheridae: Entocytherinae: genera

1 External margin of horizontal ramus of clasping apparatus with talon or swelling (Fig. 5.31A) .... *Ankylocythere*
1' External margin of horizontal ramus of clasping apparatus without talon or swelling (Fig. 5.31B) .... *Uncinocythere*

Ostracoda: Cytheroidea: Kliellidae: genera

1 Carapace smooth (Fig. 5.31C), A1 with six articulated segments (Fig. 5.31D) .... *Kliella*, one species, *Kliella hyaloderma*
1' Carapace ornamented (Fig. 5.31E), A1 with seven articulated segments (Fig. 5.31F) .... *Nannokliella*, one species, *Nannokliella dictyoconcha*

Ostracoda: Cytheroidea: Cytherideidae: genera

1 RV with a small posteroventral spine (Fig. 5.31G); A2 with two terminal claws (Fig. 5.31H) .... *Cyprideis*
1' RV without a small posteroventral spine (Fig. 5.31I); A2 with three terminal claws (Fig. 5.31J) .... *Cytherissa*

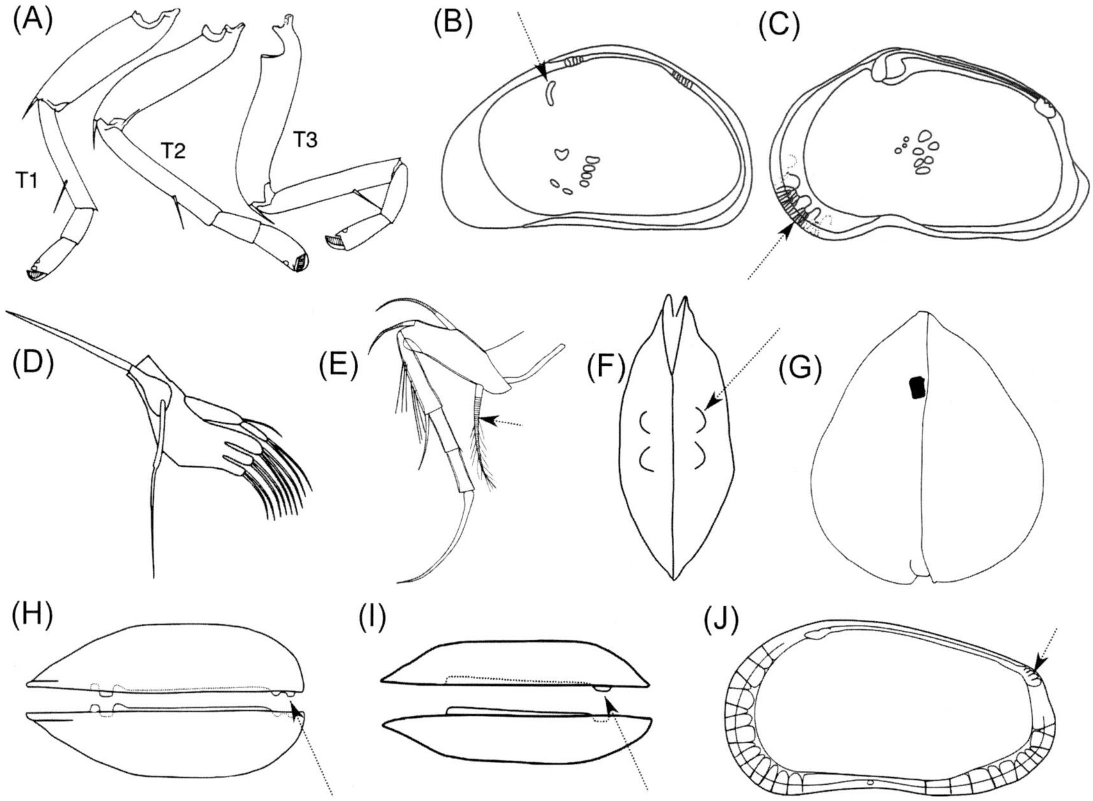

FIGURE 5.30 Morphological traits of Cytheroidea families. (A) Entocytheridae, thoracopods T1–T3. (B) Xestoleberidae, right valve, inner view, with a *Xestoleberis*-spot. (C) Hemicytheridae, right valve, inner view showing pore canals in brush-like bunches. (D) Kliellidae, maxillula. (E) Cytherideidae, T2 with a stout posterior seta. (F) Limnocytheridae, carapace, dorsal view with apparent tubercles. (G) Timiriaseviidae, carapace, dorsal view. (H) Loxoconchidae, species with gongylodont hinge, right (top) and left (bottom) valves, dorsal view. (I) Loxoconchidae, species with henodont hinge, right (top) and left (bottom) valves, dorsal view. (J) Leptocytheridae, right valve, inner view, showing posterior tooth with five cusps. *After (A, F, G) Meisch (2000); (B–D, H–J) Horne et al. (2019).*

Ostracoda: Cytheroidea: Limnocytheridae: genera

1 All marginal pore canals short and simple 2
1' Marginal pore canals long, some bifurcated (Fig. 5.31K) .... *Paralimnocythere*
2(1) A1 penultimate segment more than twice as long as broad (Fig. 5.8C) .... *Limnocythere*
2' A1 penultimate segment less than twice as long as broad (Fig. 5.31L) .... *Leucocythere*

Ostracoda: Cytheroidea: Timiriaseviidae: genera

1 Carapace very broad in dorsal view, with $W/L > 0.6$ (Fig. 5.30G) 2
1' Carapace slender in dorsal view, with $W/L < 0.5$ (Fig. 5.31M) .... *Gomphocythere*
2(1) A2 with three terminal claws (Fig. 5.31N) 3
2' A2 with one terminal claw (Fig. 5.31O) .... *Dolekiella*, one species, *Dolekiella europaea*
3(2) Carapace without dorsomedian sulci in each valve (Fig. 5.30G) .... *Metacypris*
3' Carapace with one dorsomedian sulcus in each valve (Fig. 5.31P) .... *Kovalevskiella*

Ostracoda: Cytheroidea: Loxoconchidae: genera

1 Hinge gongylodont (bilobate teeth anteriorly in LV and posteriorly in RV) 2
1' Hinge henodont (single posterior tooth in RV) (Figs. 5.30I and 5.31Q) ....
.... *Pseudolimnocythere*, one species, *Pseudolimnocythere hypogaea*
2(1) RV median hinge bar crenulate (Fig. 5.31R) .... *Loxoconcha*
2' RV median hinge bar smooth (Fig. 5.31S) .... *Cytheromorpha*, one species, *Cytheromorpha fuscata*

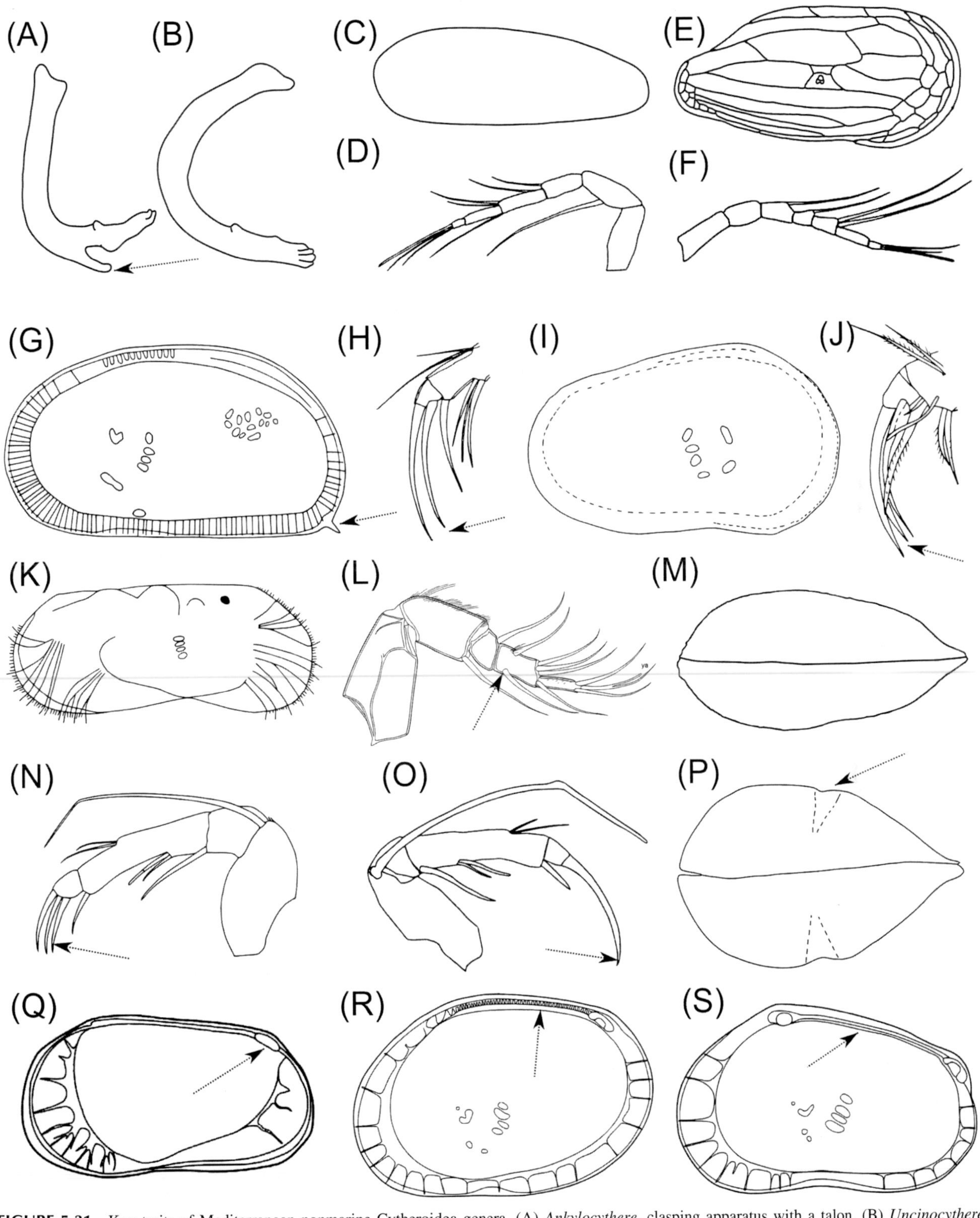

FIGURE 5.31 Key traits of Mediterranean nonmarine Cytheroidea genera. (A) *Ankylocythere*, clasping apparatus with a talon. (B) *Uncinocythere*, clasping apparatus. (C) *Kliella*, right valve, external view. (D) *Kliella*, A1. (E) *Nannokliella*, right valve, external view. (F) *Nannokliella*, A1. (G) *Cyprideis*, right valve, inner view, with posteroventral spine. (H) *Cyprideis*, distal part of A2 with two terminal claws. (I) *Cytherissa*, right valve, external view. (J) *Cytherissa*, distal part of A2 with three terminal claws. (K) *Paralimnocythere*, right valve, external view, showing branched pore canals. (L) *Leucocythere*, A1 with penultimate segment less than twice as long as broad. (M) *Gomphocythere*, carapace, dorsal view. (N) *Metacypris*, A2 with three terminal claws. (O) *Dolekiella*, A2 with one terminal claw. (P) *Kovalevskiella*, carapace, dorsal view showing dorsal sulci. (Q) *Pseudolimnocythere*, right valve, inner view, with a single posterior tooth. (R) *Loxoconcha*, right valve, inner view, with a crenulated hinge bar. (S) *Cytheromorpha*, right valve inner view, with a smooth hinge bar. *After (A, B) Mestre et al. (2013); (C, E) Schäfer (1945); (D, F, N, O, Q–S) Horne et al. (2019); (G–K, P) Meisch (2000).*

Acknowledgments

We are grateful to Alain Maasri and James H. Thorp for offering us the opportunity to contribute to this chapter, and to Koen Martens for suggesting it. Thanks to Ángel Baltanás for his initial contribution to this work, and to Dave Horne for his permission to use his figures. Thanks also to Josep A. Aguilar-Alberola and Maria Bisquert for allowing use of their unpublished drawings.

References

Aguilar-Alberola, J. A., & Mesquita-Joanes, F. (2011). Population dynamics and tolerance to desiccation in a crustacean ostracod adapted to life in small ephemeral water bodies. *Limnologica*, *41*, 348–355. Available from https://linkinghub.elsevier.com/retrieve/pii/S0075951111000168.

Aguilar-Alberola, J. A., & Mesquita-Joanes, F. (2014). Breaking the temperature-size rule: Thermal effects on growth, development and fecundity of a crustacean from temporary waters. *Journal of Thermal Biology*, *42*, 15–24. Available from https://linkinghub.elsevier.com/retrieve/pii/S0306456514000333.

Akdemir, D., & Külköylüoğlu, O. (2014). Preliminary study on distribution, diversity, and ecological characteristics of nonmarine Ostracoda (Crustacea) from the Erzincan region (Turkey). *Turkish Journal of Zoology*, *38*, 421–431. Available from https://doi.org/10.3906/zoo-1301-16.

Akdemir, D., Külköylüoğlu, O., Yavuzatmaca, M., & Sari, N. (2016). Freshwater ostracods (Crustacea) of Gaziantep (Turkey) and their habitat preferences according to movement ability. *Fundamental and Applied Limnology*, *187*, 307–314. Available from https://doi.org/10.1127/fal/2016/0665.

Altinsaçlı, S. (2014). Species diversity and distribution of Ostracoda (Crustacea) in mesosaline Lake Bafa (Aegean Region, Turkey). *Journal of Entomology and Zoology Studies*, *2*, 16–32. Available from https://www.entomoljournal.com/vol2Issue2/pdf/1.1.pdf.

Altınsaçlı, S., & Griffiths, H. I. (2002). A review of the occurrence and distribution of the recent non-marine Ostracoda (Crustacea) of Turkey. *Zoology in the Middle East*, *27*, 61–76. Available from http://www.tandfonline.com/doi/abs/10.1080/09397140.2002.10637941.

Altınsaçlı, S., & Mezquita, F. (2008). Ostracod fauna of salt Lake Acıgöl (Acı Tuz) (Turkey). *Journal of Natural History*, *42*, 1013–1025. Available from http://www.tandfonline.com/doi/abs/10.1080/00222930701851735.

Altinsaçlı, S., Perçin-Paçal, F., & Altinsaçlı, S. (2018). Assessments of environmental variables affecting the spatiotemporal distribution and habitat preferences of living Ostracoda (Crustacea) species in the Enez lagoon complex (Enez-Evros delta, Turkey). *Ecologica Montenegrina*, *19*, 130–151. Available from https://www.biotaxa.org/em/article/view/em.2018.19.14.

Angilletta, M. J., Jr. (2009). *Thermal adaptation: A theoretical and empirical synthesis* (p. 302) Oxford: Oxford University Press. Available from https://oxford.universitypressscholarship.com/view/10.1093/acprof:oso/9780198570875.001.1/acprof-9780198570875.

Baltanás, A. (2001). *Candelacypris* n. gen. (Crustacea, Ostracoda): A new genus from Iberian saline lakes, with a redescription of *Eucypris aragonica* Brehm & Margalef, 1948. *Bulletin de la Société des naturalistes luxembourgeois*, *101*, 183–192. Available from https://www.snl.lu/publications/bulletin/SNL_2001_101_183_192.pdf.

Baltanás, Á., Beroiz, B., & López, A. (1996). Lista faunistica y bibliografica de los ostracodos no-marinos (Crustacea, Ostracoda) de la Peninsula Ibérica, Islas Baleares e Islas Canarias. *Listas de la Flora y Fauna de las Aguas continentales de la Peninsula Iberica*, *12*, 1–71.

Baltanás, Á., & Mesquita-Joanes, F. (2015). Orden Podocopida. *Ide@-Sea*, *74*, 1–10. Available from http://www.sea-entomologia.org/IDE@/revista_74.pdf.

Baltanás, A., Montes, C., & Martino, P. (1990). Distribution patterns of ostracods in Iberian saline lakes. Influence of ecological factors. *Hydrobiologia*, *197*, 207–220. Available from https://link.springer.com/chapter/10.1007/978-94-009-0603-7_18.

Bode, S. N. S., Adolfsson, S., Lamatsch, D. K., Martins, M. J. F., Schmit, O., Vandekerkhove, J., Mezquita, F., Namiotko, T., Rossetti, G., Schön, I., Butlin, R. K., & Martens, K. (2010). Exceptional cryptic diversity and multiple origins of parthenogenesis in a freshwater ostracod. *Molecular Phylogenetics and Evolution*, *54*, 542–552. Available from https://linkinghub.elsevier.com/retrieve/pii/S1055790309003492.

Boix, D., Sala, J., Gascón, S., & Brucet, S. (2006). Predation in a temporary pond with special attention to the trophic role of *Triops cancriformis* (Crustacea: Branchiopoda: Notostraca). *Hydrobiologia*, *571*, 341–353. Available from https://link.springer.com/article/10.1007/s10750-006-0259-0.

Boxshall, G., & Jaume, D. (2013). Antennules and Antennae in the Crustacea. In L. Watling, & M. Thiel (Eds.), *The natural history of the Crustacea. Volume 1. Functional morphology and diversity* (pp. 199–236). Oxford: Oxford University Press. Available from https://oxford.universitypressscholarship.com/view/10.1093/acprof:osobl/9780195398038.001.0001/acprof-9780195398038-chapter-7.

Boxshall, G. A., Danielopol, D., Horne, D., Smith, R., & Tabacaru, I. (2010). A critique of biramous interpretations of the crustacean antennule. *Crustaceana*, *83*, 153–167. Available from https://brill.com/view/journals/cr/83/2/article-p153_3.xml.

Carbonel, P., Colin, J.-P., Danielopol, D. L., Löffler, H., & Neustrueva, I. (1988). Paleoecology of limnic ostracodes: A review of some major topics. *Palaeogeography, Palaeoclimatology, Palaeoecology*, *62*, 413–461. Available from https://www.sciencedirect.com/science/article/abs/pii/0031018288900661.

Castillo-Escrivà, A., Baltanás, Á., Camacho, A., Horne, D. J., Pretus, J. L., & Mesquita-Joanes, F. (2023). IMOST: a database for non-marine ostracods in the Iberian Peninsula, the Balearic Islands and Macaronesia. *Journal of Limnology*, *82*(s1), 2115. Available from https://doi.org/10.4081/jlimnol.2023.2115.

Castillo-Escrivà, A., Rueda, J., Zamora, L., Hernández, R., del Moral, M., & Mesquita-Joanes, F. (2016). The role of watercourse versus overland dispersal and niche effects on ostracod distribution in Mediterranean streams (eastern Iberian Peninsula). *Acta Oecologica*, *73*, 1–9. Available from https://linkinghub.elsevier.com/retrieve/pii/S1146609X16300224.

Castillo-Escrivà, A., Valls, L., Rochera, C., Camacho, A., & Mesquita-Joanes, F. (2016). Spatial and environmental analysis of an ostracod metacommunity from endorheic lakes. *Aquatic Sciences*, *78*, 707–716. Available from http://link.springer.com/10.1007/s00027-015-0462-z.

Castillo-Escrivà, A., Valls, L., Rochera, C., Camacho, A., & Mesquita-Joanes, F. (2017). Metacommunity dynamics of Ostracoda in temporary lakes: Overall strong niche effects except at the onset of the flooding period. *Limnologica, 62*, 104–110. Available from https://linkinghub.elsevier.com/retrieve/pii/S0075951116302092.

Chatterjee, T., Dovgal, I., Mayén-Estrada, R., & Fernandez-Leborans, G. (2020). A checklist of ciliates (Ciliophora) inhabiting on ostracods (Crustacea, Ostracoda). *Zootaxa, 4763*, 17–30. Available from https://www.biotaxa.org/Zootaxa/article/view/zootaxa.4763.1.2.

Cohen, A. C., & Oakley, T. H. (2017). Collecting and processing marine ostracods. *Journal of Crustacean Biology, 37*, 347–352. Available from https://academic.oup.com/jcb/article/37/3/347/3806681.

Cottenie, K. (2005). Integrating environmental and spatial processes in ecological community dynamics. *Ecology Letters, 8*, 1175–1182. Available from https://onlinelibrary.wiley.com/doi/full/10.1111/j.1461-0248.2005.00820.x.

Danielopol, D. L. (1969). Recherches sur la morphologie de l'organe copulateur male chez quelques ostracodes du genre *Candona* Baird. In J. W. Neale (Ed.), *The taxonomy, morphology and ecology of Recent Ostracoda* (pp. 136–153). Edinburgh: Oliver & Boyd.

Danielopol, D. L., & Hart, C. W., Jr. (1985). Notes of the center of origin and antiquity of the Spaeromicolinae, with a description of *Hobbsiella*, new genus (Ostracoda, Entocytheridae). *Stygologia, 1*, 54–70.

Danielopol, D. L., & Hartmann, G. (1986). Ostracoda. In L. Botosaneanu (Ed.), *Stygofauna Mundi. A Faunistic, Distributional, and Ecological Synthesis of the World Fauna inhabiting Subterranean Waters (including the Marine Interstitial)* (pp. 265–294). The Netherlands: Leiden.

Danielopol, D. L., Ito, E., Wansard, G., Kamiya, T., Cronin, T. M., & Baltanás, A. (2002). Techniques for collection and study of ostracoda. *Geophysical Monograph-American Geophysical Union, 131*, 65–97. Available from http://doi.wiley.com/10.1029/131GM04.

de Campos, R., Conceição, E. O., Martens, K., & Higuti, J. (2019). Extreme drought periods can change spatial effects on periphytic ostracod metacommunities in river-floodplain ecosystems. *Hydrobiologia, 828*, 369–381. Available from http://link.springer.com/10.1007/s10750-018-3825-3.

de Campos, R., Lansac-Tôha, F. M., Conceição, E. O., Martens, K., & Higuti, J. (2018). Factors affecting the metacommunity structure of periphytic ostracods (Crustacea, Ostracoda): A deconstruction approach based on biological traits. *Aquatic Sciences, 80*, 16. Available from http://doi.org/10.1007/s00027-018-0567-2.

De Deckker, P. (1983). Notes on the ecology and distribution of non-marine ostracods in Australia. *Hydrobiologia, 106*, 223–234. Available from https://link.springer.com/article/10.1007/BF00008120.

De Deckker, P., & Forester, R. M. (1988). The use of ostracods to reconstruct continental palaeoenvironmental records. In P. De Deckker, J.-P. Colin, & J.-P. Peypouquet (Eds.), *Ostracoda in the earth sciences* (pp. 175–199). Amsterdam: Elsevier.

Delorme, L. D. (1989). Methods in Quaternary ecology #7. Freshwater Ostracodes. *Geoscience Canada, 16*, 85–90.

Dole-Olivier, M. -J., Galassi, D. M. P., Marmonier, P., & Creuzé des Châtelliers, M. (2000). The biology and ecology of lotic microcrustaceans. *Freshwater Biology, 44*, 63–91. Available from https://onlinelibrary.wiley.com/doi/abs/10.1046/j.1365-2427.2000.00590.x.

Escrivà, A., Poquet, J., & Mesquita-Joanes, F. (2015). Effects of environmental and spatial variables on lotic ostracod metacommunity structure in the Iberian Peninsula. *Inland Waters, 5*, 283–294. Available from http://www.tandfonline.com/doi/full/10.5268/IW-5.3.771.

Escrivà, A., Smith, R., Aguilar-Alberola, J. A., Kamiya, T., Karanovic, I., Rueda, J., Schornikov, E. I., & Mesquita-Joanes, F. (2012). Global distribution of *Fabaeformiscandona subacuta*: an exotic invasive Ostracoda on the Iberian Peninsula? *Journal of Crustacean Biology, 32*, 949–961. Available from https://academic.oup.com/jcb/article-lookup/doi/10.1163/1937240X-00002096.

Forester, R. M. (1986). Determination of the dissolved anion composition of ancient lakes from fossil ostracodes. *Geology, 14*, 796–798. Available from https://pubs.geoscienceworld.org/gsa/geology/article/14/9/796/204169/Determination-of-the-dissolved-anion-composition.

Frenzel, P., & Boomer, I. (2005). The use of ostracods from marginal marine, brackish waters as bioindicators of modern and Quaternary environmental change. *Palaeogeography, Palaeoclimatology, Palaeoecology, 625*, 68–92. Available from https://www.sciencedirect.com/science/article/pii/S0031018205003330.

Frisch, D., Green, A. J., & Figuerola, J. (2007). High dispersal capacity of a broad spectrum of aquatic invertebrates via waterbirds. *Aquatic Sciences, 69*, 568–574. Available from http://link.springer.com/10.1007/s00027-007-0915-0.

Gálvez, Á., Magurran, A. E., Armengol, X., Savatenalinton, S., & Mesquita-Joanes, F. (2022). In T. Dalu, & R. J. Wasserman (Eds.), *Fundamentals of tropical freshwater wetlands* (pp. 549–586). London: Academic Press. Available from https://www.sciencedirect.com/science/article/pii/B978012822362800011 6?via%3Dihub.

Gauthier, H. (1928). *Recherches sur la faune des eaux continentales de l'Algérie et de la Tunisie* (p. 419) Alger: Minerva.

Gauthier, H. (1933). Entomostracés de Madagascar. 2e note. Description d'un nouveau *Cyprinotus* (Ostracodes). *Bulletin de la Societe Zoologique de France, 58*, 305–316.

Ghaouaci, S., Yavuzatmaca, M., Külköylüoğlu, O., & Amarouayache, M. (2017). An annotated checklist of the non-marine ostracods (Crustacea) of Algeria with some ecological notes. *Zootaxa, 4290*, 140–154. Available from https://www.mapress.com/zt/article/view/zootaxa.4290.1.8.

González-Mozo, M. E., Martens, K., & Baltanas, A. (1996). A taxonomic revision of European *Herpetocypris* Brady and Norman, 1889 (Crustacea, Ostracoda). *Bulletin de l'Institut Royal des Sciences Naturelles de Belgique, Biologie, 66*, 93–132.

Griffiths, H. I., & Brancelj, A. (1996). Preliminary list of freshwater Ostracoda (Crustacea) from Slovenia. *Annals for Istrian and Mediterranean Studies, 9*, 201–210.

Hart, D. G., & Hart, C. W., Jr. (1974). The ostracod family Entocytheridae. *Monographs of the Academy of Natural Sciences of Philadelphia, 18*, 1–239.

Hewitt, G. M. (2004). The structure of biodiversity—Insights from molecular phylogeography. *Frontiers in Zoology, 1*, 1–16. Available from https://frontiersinzoology.biomedcentral.com/articles/10.1186/1742-9994-1-4.

Higuti, J., & Martens, K. (2020). Chapter 20—Class Ostracoda. In (4th ed.D. C. Rogers, C. Damborenea, & J. Thorp (Eds.), *Thorp and Covich's freshwater invertebrates* (Vol. 5Amsterdam: Academic Press, Keys to Neotropical and Antarctic Fauna. Available from https://www.sciencedirect.com/science/article/pii/B9780128042250000204.

Hiruta, S., Kobayashi, N., Katoh, T., & Kajihara, H. (2016). Molecular phylogeny of cypridoid freshwater ostracods (Crustacea: Ostracoda), inferred from 18S and 28S rDNA sequences. *Zoological Science*, *33*(2), 179–185. Available from https://doi.org/10.2108/zs150103.

Horne, D. J., Cohen, A., & Martens, K. (2002). Taxonomy, morphology and biology of Quaternary and living ostracoda. In J. A. Holmes, & A. R. Chivas (Eds.), *The Ostracoda: Applications in Quaternary research* (pp. 5–36). Washington, DC: American Geophysical Union, Geophysical Monograph Series 131. Available from https://agupubs.onlinelibrary.wiley.com/doi/10.1029/131GM02, Geophysical Monograph Series 131.

Horne, D. J., Curry, B. B., & Mesquita-Joanes, F. (2012). Chapter five—Mutual climatic range methods for Quaternary ostracods. In D. J. Horne, J. A. Holmes, J. Rodriguez-Lazaro, & F. A. Viehberg (Eds.), *Developments in Quaternary Science* (Vol. 17, pp. 65–84). Elsevier. Available from https://linkinghub.elsevier.com/retrieve/pii/B9780444536365000056.

Horne, D. J., Meisch, C., & Martens, K. (2019). Chapter 16.3—Arthropoda: Ostracoda. In D. C. Rogers, C. Damborenea, & J. Thorp (Eds.), *Thorp and Covich's freshwater invertebrates,* Vol. 4, (4th ed., pp. 725–760). Amsterdam: Elsevier, Keys to Palaearctic Fauna. Available from https://linkinghub.elsevier.com/retrieve/pii/B9780123850249000198, Keys to Palaearctic Fauna.

Horne, D. J., & Mezquita, F. (2008). Palaeoclimatic applications of large databases: Developing and testing methods of palaeotemperature reconstruction using nonmarine ostracods. *Senckenbergiana lethaea*, *88*, 93–112. Available from http://link.springer.com/10.1007/BF03043981.

Horne, F. R. (1993). Survival strategy to escape desiccation in a freshwater ostracod. *Crustaceana*, *65*, 53–61. Available from https://brill.com/view/journals/cr/65/1/article-p53_8.xml.

Hortal, J., de Bello, F., Diniz-Filho, J. A. F., Lewinsohn, T. M., Lobo, J. M., & Ladle, R. J. (2015). Seven shortfalls that beset large-scale knowledge of biodiversity. *Annual Review of Ecology, Evolution, and Systematics*, *46*, 523–549. Available from http://www.annualreviews.org/doi/10.1146/annurev-ecolsys-112414-054400.

Karanovic, I. (1999). A new genus and two new species of Candoninae (Crustacea, Ostracoda) from Montenegro (SE Europe). *Mémoires de Biospéologie*, *26*, 47–57.

Karanovic, I. (2005). Comparative morphology of the Candoninae antennula, with remarks on the ancestral state in ostracods and a proposed new terminology. *Spixiana*, *28*, 141–160. Available from https://www.zobodat.at/pdf/Spixiana_028_0141-0160.pdf.

Karanovic, I. (2012). *Recent freshwater ostracods of the world* (p. 608) Heidelberg: Springer. Available from https://link.springer.com/book/10.1007/978-3-642-21810-1.

Külköylüoğlu, O. (2003). Ecology of freshwater Ostracoda (Crustacea) from lakes and reservoirs in Bolu, Turkey. *Journal of Freshwater Ecology*, *18*, 343–347. Available from http://www.tandfonline.com/doi/abs/10.1080/02705060.2003.9663968.

Külköylüoğlu, O., Akdemir, D., Yavuzatmaca, M., & Yilmaz, O. (2015). A checklist of recent non-marine Ostracoda (Crustacea) of Turkey with three new records. *Review of Hydrobiology*, *8*, 77–90. Available from http://www.reviewofhydrobiology.org/page/pdf.asp?pdf = 8-2/8-2-2-Full(1).pdf.

Külköylüoğlu, O., Dügel, M., & Kılıç, M. (2007). Ecological requirements of Ostracoda (Crustacea) in a heavily polluted shallow lake, Lake Yeniçağa (Bolu, Turkey). *Hydrobiologia*, *585*, 119–133. Available from http://link.springer.com/10.1007/s10750-007-0633-6.

Külköylüoğlu, O., Yavuzatmaca, M., Akdemir, D., Çelen, E., & Dalkıran, N. (2018). Ecological classification of the freshwater Ostracoda (Crustacea) based on physicochemical properties of waters and habitat preferences. *Annales de Limnologie—International Journal of Limnology*, *54*, 26. Available from https://www.limnology-journal.org/10.1051/limn/2018017.

Külköylüoğlu, O., Yavuzatmaca, M., Sarı, N., & Akdemir, D. (2016). Elevational distribution and species diversity of freshwater Ostracoda (Crustacea) in Çankırı region (Turkey). *Journal of Freshwater Ecology*, *31*, 219–230. Available from https://www.tandfonline.com/doi/full/10.1080/02705060.2015.1050467.

Leibold, M. A., & Chase, J. M. (2017). *Metacommunity ecology* (p. 504) Princeton, NJ: Princeton University Press. Available from https://www.jstor.org/stable/10.2307/j.ctt1wf4d24.

Leibold, M. A., Holyoak, M., Mouquet, N., Amarasekare, P., Chase, J. M., Hoopes, M. F., Holt, R. D., Shurin, J. B., Law, R., Tilman, D., Loreau, M., & Gonzalez, A. (2004). The metacommunity concept: A framework for multi-scale community ecology. *Ecology Letters*, *7*, 601–613. Available from https://onlinelibrary.wiley.com/doi/full/10.1111/j.1461-0248.2004.00608.x.

Lozano-Fernandez, J., Giacomelli, M., Fleming, J. F., Chen, A., Vinther, J., Thomsen, P. F., Glenner, H., Palero, F., Legg, D. A., Iliffe, T. M., Pisani, D., & Olesen, J. (2019). Pancrustacean evolution illuminated by taxon-rich genomic-scale data sets with an expanded remipede sampling. *Genome Biology and Evolution*, *11*, 2055–2070. Available from https://doi.org/10.1093/gbe/evz097.

Marazanof, F. (1965). Ostracodes de Camargue. *Annales de Limnologie*, *1*, 95–102. Available from http://www.limnology-journal.org/10.1051/limn/1965003.

Marmonier, P., Boulal, M., & Idbennacer, B. (2005). *Marococandona*, a new genus of Candonidae (Crustacea, Ostracoda) from Southern Morocco: Morphological characteristics and ecological requirements. *Annales de Limnologie—International Journal of Limnology*, *41*, 57–71. Available from http://www.limnology-journal.org/10.1051/limn/2005006.

Marrone, F., Alfonso, G., Stoch, F., Pieri, V., Alonso, M., Dretakis, M., & Naselli-Flores, L. (2019). An account on the non-malacostracan crustacean fauna from the inland waters of Crete, Greece, with the synonymization of *Arctodiaptomus piliger* Brehm, 1955 with *Arctodiaptomus alpinus* (Imhof, 1885)(Copepoda: Calanoida). *Limnetica*, *38*, 167–187. Available from http://limnetica.net/documentos/limnetica/limnetica-38-1-p-167.pdf.

Marrone, F., Pieri, V., Turki, S., & Rossetti, G. (2020). The recent non-marine ostracods of Tunisia: An updated checklist with remarks on their regional distribution patterns and ecological preferences. *Journal of Limnology*, *79*, 293–307. Available from https://jlimnol.it/index.php/jlimnol/article/view/jlimnol.2020.1982.

Martens, K., & Horne, D. J. (2009). Ostracoda. In G. Likens (Ed.), *Encyclopedia of inland waters* (pp. 405–414). Oxford: Elsevier.

Martens, K., & Horne, D. J. (2016). Collecting and processing living, non-marine ostracods. *Journal of Crustacean Biology*, *36*, 849–854. Available from https://academic.oup.com/jcb/article-lookup/doi/10.1163/1937240X-00002488.

Martens, K., Meisch, C., & Marmonier, P. (1991). On *Klieopsis* n. gen., with a redescription of *Cypridopsis horai* Klie, 1927 (Crustacea, Ostracoda). *Bulletin de l'Institut royal des Sciences naturelles de Belgique, Biologie*, *61*, 55–64.

Martens, K., & Ortal, R. (1999). Diversity and zoogeography of inland-water Ostracoda (Crustacea) in Israel (Levant). *Israel Journal of Zoology*, *45*, 159–173.

Martens, K., Schön, I., Meisch, C., & Horne, D. J. (2008). Global diversity of ostracods (Ostracoda, Crustacea) in freshwater. *Hydrobiologia*, *595*, 185–193. Available from https://link.springer.com/chapter/10.1007/978-1-4020-8259-7_20.

Martins, M. J. F., Vandekerkhove, J., Mezquita, F., Schmit, O., Rueda, J., Rossetti, G., & Namiotko, T. (2009). Dynamics of sexual and parthenogenetic populations of *Eucypris virens* (Crustacea: Ostracoda) in three temporary ponds. *Hydrobiologia*, *636*, 219–232. Available from http://link.springer.com/10.1007/s10750-009-9952-0.

Matzke-Karasz, R. (1997). Descriptive nomenclature and external morphology of the Zenker's organs of Cypridoidea (Crustacea, Ostracoda). *Sonderveröffentlichungen, Geologisches Institut der Universität zu Köln*, *114*, 295–315.

Matzke-Karasz, R., & Martens, K. (2005). The female reproductive organ in podocopid ostracods is homologous to five appendages: histological evidence from *Liocypris grandis* (Crustacea, Ostracoda). *Hydrobiologia*, *542*, 249–259. Available from https://link.springer.com/article/10.1007/s10750-004-3950-z.

Mazzini, I., Marrone, F., Arculeo, M., & Rossetti, G. (2017). Revision of Recent and fossil *Mixtacandona* Klie 1938 (Ostracoda, Candonidae) from Italy, with description of a new species. *Zootaxa*, *4221*, 323–340. Available from https://www.biotaxa.org/Zootaxa/article/view/zootaxa.4221.3.3.

McGregor, D. L., & Kesling, R. V. (1969). Copulatory adaptations in ostracods Part I. Hemipenes of *Candona*. *Contributions from The Museum of Paleontology, The University of Michigan*, *22*, 169–191.

McKenzie, K. G., & Moroni, A. (1986). Man as an agent of crustacean passive dispersal via useful plants—Exemplified by Ostracoda *ospiti esteri* of the Italian ricefields ecosystem—And implications arising therefrom. *Journal of Crustacean Biology*, *6*, 181–198. Available from https://academic.oup.com/jcb/article-abstract/6/2/181/2328012.

Meisch, C. (1988). Ostracodes récoltés à Paris. Avec une clef pour la détermination des espèces européennes du genre *Ilyocypris* (Crustacea, Ostracoda). *Bulletin de la Société des naturalistes luxembourgeois*, *88*, 145–163. Available from https://www.snl.lu/publications/bulletin/SNL_1988_088_145_163.pdf.

Meisch, C. (2000). *Freshwater Ostracoda of Western and Central Europe* (p. 552) Heidelberg: Spektrum Akademischer Verlag.

Meisch, C., & Broodbakker, N. W. (1993). Freshwater Ostracoda (Crustacea) collected by Prof. J.H. Stock on the Canary and Cape Verde islands. With an annotated checklist of the freshwater Ostracoda of the Azores, Madeira, the Canary, the Selvagens and Cape Verde islands. *Travaux scientifiques du musée national d'histoire naturelle de Luxembourg*, *19*, 3–47. Available from https://ps.mnhn.lu/ferrantia/publications/T.S.19.pdf.

Meisch, C., Scharf, B., Fuhrmann, R., & Thiéry, A. (2019). *Neglecandona altoides* (Petkovski, 1961) nov. comb. and the genus *Neglecandona* Krstić, 2006 (Crustacea, Ostracoda, Candonidae). *Bulletin de la Société des naturalistes luxembourgeois*, *121*, 237–264. Available from https://www.snl.lu/publications/bulletin/SNL_2019_121_237_264.pdf.

Meisch, C., Smith, R. J., & Martens, K. (2019). A subjective global checklist of the extant non-marine Ostracoda (Crustacea). *European Journal of Taxonomy*, *492*, 1–135. Available from https://europeanjournaloftaxonomy.eu/index.php/ejt/article/view/627.

Mesquita-Joanes, F., Aguilar-Alberola, J. A., Palero, F., & Rueda, J. (2020). A new species of *Cypris* (Crustacea: Ostracoda) from the Iberian Peninsula and the Balearic Islands, with comments on the first ostracod named using the Linnean system. *Zootaxa*, *4759*, 113–131. Available from https://www.biotaxa.org/Zootaxa/article/view/zootaxa.4759.1.8.

Mesquita-Joanes, F., Smith, A. J., & Viehberg, F. A. (2012). The ecology of Ostracoda across levels of biological organisation from individual to ecosystem. *Developments in Quaternary Science*, *17*, 15–35. Available from https://linkinghub.elsevier.com/retrieve/pii/B9780444536365000020.

Mestre, A., Aguilar-Alberola, J. A., Baldry, D., Balkis, H., Ellis, A., Gil-Delgado, J. A., Grabow, K., Klobučar, G., Kouba, A., Maguire, I., Martens, A., Mülayim, A., Rueda, J., Scharf, B., Soes, M., Monrós, J. S., & Mesquita-Joanes, F. (2013). Invasion biology in non-free-living species: Interactions between abiotic (climatic) and biotic (host availability) factors in geographical space in crayfish commensals (Ostracoda, Entocytheridae). *Ecology and Evolution*, *3*, 5237–5253. Available from http://doi.wiley.com/10.1002/ece3.897.

Mestre, A., Butlin, R. K., Kelso, W. E., Romaire, R., Bonvillain, C. P., Monrós, J. S., & Mesquita-Joanes, F. (2016). Contrasting patterns of genetic diversity and spatial structure in an invasive symbiont-host association. *Biological Invasions*, *18*, 3175–3191. Available from http://link.springer.com/10.1007/s10530-016-1207-1.

Mestre, A., Monrós, J. S., & Mesquita-Joanes, F. (2014). A review of the Entocytheridae (Ostracoda) of the world: Updated bibliographic and species checklists and global georeferenced database, with insights into host specificity and latitudinal patterns of species richness. *Crustaceana*, *87*, 923–951. Available from https://brill.com/view/journals/cr/87/8-9/article-p923_3.xml.

Mestre, A., Poulin, R., Holt, R. D., Barfield, M., Clamp, J. C., Fernandez-Leborans, G., & Mesquita-Joanes, F. (2019). The interplay of nested biotic interactions and the abiotic environment regulates populations of a hypersymbiont. *Journal of Animal Ecology*, *88*, 1998–2010. Available from https://besjournals.onlinelibrary.wiley.com/doi/10.1111/1365 2656.13091.

Mezquita, F., Hernández, R., & Rueda, J. (1999). Ecology and distribution of ostracods in a polluted Mediterranean river. *Palaeogeography, Palaeoclimatology, Palaeoecology*, *148*, 87–103. Available from https://linkinghub.elsevier.com/retrieve/pii/S0031018298001771.

Mezquita, F., Roca, J. R., Reed, J. M., & Wansard, G. (2005). Quantifying species-environment relationships in non-marine Ostracoda for ecological and palaeoecological studies: Examples using Iberian data. *Palaeogeography, Palaeoclimatology, Palaeoecology, 225*, 93–117. Available from https://linkinghub.elsevier.com/retrieve/pii/S0031018205003421.

Mezquita, F., Sanz-Brau, A., & Wansard, G. (2000). Habitat preferences and population dynamics of Ostracoda in a helocrene spring system. *Canadian Journal of Zoology, 78*, 840–847. Available from http://www.nrcresearchpress.com/doi/10.1139/z99-265.

Mischke, S., Almogi-Labin, A., Al-Saqarat, B., Rosenfeld, A., Elyashiv, H., Boomer, I., Stein, M., Lev, L., & Ito, E. (2014). An expanded ostracod-based conductivity transfer function for climate reconstruction in the Levant. *Quaternary Science Reviews, 93*, 91–105. Available from http://doi.org/10.1016/j.quascirev.2014.04.004.

Mischke, S., Ginat, H., Al-Saqarat, B., & Almogi-Labin, A. (2012). Ostracods from water bodies in hyperarid Israel and Jordan as habitat and water chemistry indicators. *Ecological Indicators, 14*, 87–99. Available from https://doi.org/10.1016/j.ecolind.2011.07.017.

Müller, G. W. (1898). Ergebnisse einer zoologischen Forschungsreise in Madagaskar und Ost-Afrika 1889–1895 von Dr. A. Voeltzkow: Die Ostracoden. *Abhandlungen der Senckenbergischen Naturforschenden Gesellschaft, 21*(2), 255–296.

Namiotko, T., Danielopol, D. L., & Baltanás, A. (2011). Soft body morphology, dissection and slide-preparation of Ostracoda: A primer. *Joannea—Geologie und Palaontologie*, 327–343. Available from https://www.zobodat.at/pdf/JoanGeo_011_0327-0343.pdf.

Neale, J. W. (1964). Some factors influencing the distribution of recent British Ostracoda. *Pubblicazioni della Stazione Zoologica di Napoli, 33*, 247–307.

Ottonello, D., & Romano, A. (2011). Ostracoda and Amphibia in temporary ponds: who is the prey? Unexpected trophic relation in a mediterranean freshwater habitat. *Aquatic Ecology, 45*, 55–62. Available from http://link.springer.com/10.1007/s10452-010-9323-y.

Özuluğ, O., Kubanç, S. N., Kubanç, C., & Demirci, G. İ. (2018). Checklist of Quaternary and Recent Ostracoda (Crustacea) species from Turkey with information on habitat preferences. *Turkish Journal of Bioscience and Collections, 2*, 51–100. Available from https://dergipark.org.tr/en/pub/tjbc/issue/35034/398747.

Perçin-Paçal, F., Altınsaçlı, S., Yardımcı, C. H., & Altınsaçlı, S. (2018). Assessment of factors influencing the spatio-temporal distribution, abundance, and diversity of Ostracoda (Crustacea) species in a coastal lagoon: Kamil Abdus lagoon (Istanbul, Turkey). *Fresenius Environmental Bulletin, 27*, 8392–8404.

Petkovski, T. (1964). Bemerkenswerte Entomostraken aus Jugoslavien. *Acta Musei Macedonici Scientiarum Naturalium, 9*(7), 147–182.

Petkovski, T., & Meisch, C. (1996). Species of the genus *Stenocypris* Sars, 1889 from the rice-fields of Macedonia (Crustacea, Ostracoda). *Travaux Scientifiques du Musée d'Histoire Naturelle de Luxembourg, 23*, 57–85. Available from https://ps.mnhn.lu/ferrantia/publications/T.S.23.pdf.

Pieri, V., Marrone, F., Martens, K., & Rossetti, G. (2020). An updated checklist of recent ostracods (Crustacea: Ostracoda) from inland waters of Sicily and adjacent small islands with notes on their distribution and ecology. *The European Zoological Journal, 87*, 714–740. Available from https://www.tandfonline.com/doi/full/10.1080/24750263.2020.1839581.

Pieri, V., Martens, K., Meisch, C., & Rossetti, G. (2015). An annotated checklist of the recent non-marine ostracods (Ostracoda: Crustacea) from Italy. *Zootaxa, 3919*, 271–305. Available from https://www.biotaxa.org/Zootaxa/article/view/zootaxa.3919.2.3.

Poquet, J. M., & Mesquita-Joanes, F. (2011). Combined effects of local environment and continental biogeography on the distribution of Ostracoda. *Freshwater Biology, 56*, 448–469. Available from http://doi.wiley.com/10.1111/j.1365-2427.2010.02511.x.

Poquet, J. M., Mezquita, F., Rueda, J., & Miracle, M. R. (2008). Loss of Ostracoda biodiversity in Western Mediterranean wetlands. *Aquatic Conservation: Marine and Freshwater Ecosystems, 18*, 280–296. Available from http://doi.wiley.com/10.1002/aqc.831.

Ramdani, M., Flower, R. J., Elkhiati, N., Birks, H. H., Kraïem, M. M., & Fathi, A. A. (2001). Zooplankton (Cladocera, Ostracoda), Chironomidae and other benthic faunal remains in sediment cores from nine North African wetland lakes: The CASSARINA project. *Aquatic Ecology, 35*, 389–403. Available from https://link.springer.com/article/10.1023/A:1011965226399.

Rasouli, H., Scharf, B., Meisch, C., & Aygen, C. (2016). An updated checklist of the Recent non-marine Ostracoda (Crustacea) of Iran, with a redescription of *Eucypris mareotica* (Fischer, 1855). *Zootaxa, 4254*(3), 273–292. Available from https://doi.org/10.11646/zootaxa.4154.3.3.

Reed, J. M., Mesquita-Joanes, F., & Griffiths, H. I. (2012). Multi-indicator conductivity transfer functions for Quaternary palaeoclimate reconstruction. *Journal of Paleolimnology, 47*, 251–275. Available from http://link.springer.com/10.1007/s10933-011-9574-1.

Roca, J. R., Mezquita, F., Rueda, J., Camacho, A., & Miracle, M. R. (2000). Endorheic versus karstic lakes: Patterns of ostracod distributions and lake typology in a Mediterranean landscape (Castilla—La Mancha, Spain). *Marine and Freshwater Research, 51*, 311–319. Available from http://www.publish.csiro.au/?paper = MF99103.

Rodriguez-Lazaro, J., & Ruiz-Muñoz, F. (2012). A general introduction to Ostracods. Morphology, distribution, fossil record and applications. *Developments in Quaternary Science, 17*, 1–14. Available from https://www.sciencedirect.com/science/article/pii/B9780444536365000019.

Roessler, E. W. (1986a). Estudios taxonómicos, ontogenéticos, ecológicos y etológicos sobre los ostrácodos de agua dulce en Colombia-V. Estudio taxonómico del género *Chlamydotheca* Saussure, 1858 (Ostracoda, Podocopida, Cyprididae). Parte II. El grupo *Chlamydotheca colombiensis* Roessler, 1985. *Caldasia, 14*(68–70), 585–616. Available from https://revistas.unal.edu.co/index.php/cal/article/view/34961/35221.

Roessler, E. W. (1986b). Estudios taxonómicos, ontogenéticos, ecológicos y etológicos sobre los ostrácodos de agua dulce en Colombia-V. Estudio taxonómico del género *Chlamydotheca* Saussure, 1858 (Ostracoda, Podocopida, Cyprididae). Parte III. El grupo *Chlamydotheca iheringi* (Sars, 1901). *Caldasia, 14*(68–70), 618–650. Available from https://revistas.unal.edu.co/index.php/cal/article/view/34965/35228.

Rosati, M., Cantonati, M., Fenoglio, S., Segadelli, S., Levati, G., & Rossetti, G. (2016). Is there an ideal protocol for sampling macroinvertebrates in springs? *Journal of Freshwater Ecology, 31*, 199–209. Available from http://www.tandfonline.com/doi/full/10.1080/02705060.2016.1149892.

Rosati, M., Rossetti, G., Cantonati, M., Pieri, V., Roca, J. R., & Mesquita-Joanes, F. (2017). Are aquatic assemblages from small water bodies more stochastic in dryer climates? An analysis of ostracod spring metacommunities. *Hydrobiologia, 793*, 199–212. Available from http://link.springer.com/10.1007/s10750-016-2938-9.

Rossetti, G. (2007). Ostracods: Sampling strategies in springs. *Monografie del Museo Tridentino di Scienze Naturali, 4*, 237–246.

Rossetti, G., & Martens, K. (1996). Redescription and morphological variability of *Darwinula stevensoni* (Brady & Robertson, 1870) (Crustacea, Ostracoda). *Bulletin de l'Institut Royal des Sciences Naturelles de Belgique, Biologie, 66*, 73–92.

Rossetti, G., & Martens, K. (1998). Taxonomic revision of the Recent and Holocene representatives of the Family Darwinulidae (Crustacea, Ostracoda), with a description of three new genera. *Bulletin de l'Institut Royal des Sciences Naturelles de Belgique, Biologie, 68*, 55–110.

Rossetti, G., Martens, K., Meisch, C., Tavernelli, S., & Pieri, V. (2006). Small is beautiful: diversity of freshwater ostracods (Crustacea, Ostracoda) in marginal habitats of the province of Parma (Northern Italy). *Journal of Limnology, 65*, 121–131. Available from http://jlimnol.it/index.php/jlimnol/article/view/jlimnol.2006.121.

Rossetti, G., Pieri, V., & Martens, K. (2005). Recent ostracods (Crustacea, Ostracoda) found in lowland springs of the provinces of Piacenza and Parma (Northern Italy). *Hydrobiologia, 542*, 287–296. Available from http://link.springer.com/10.1007/s10750-004-2566-7.

Rossi, V., Benassi, G., Belletti, F., & Menozzi, P. (2011). Colonization, population dynamics, predatory behaviour and cannibalism in *Heterocypris incongruens* (Crustacea: Ostracoda). *Journal of Limnology, 70*, 102–108. Available from http://jlimnol.it/index.php/jlimnol/article/view/jlimnol.2011.102.

Rossi, V., Benassi, G., Veneri, M., Bellavere, C., Menozzi, P., Moroni, A., & McKenzie, K. G. (2003). Ostracoda of the Italian ricefields thirty years on: New synthesis and hypothesis. *Journal of Limnology, 62*, 1–8. Available from http://jlimnol.it/index.php/jlimnol/article/view/jlimnol.2003.1.

Savatenalinton, S., & Martens, K. (2009). Generic revision of Cypricercinae McKenzie, 1971 (Crustacea, Ostracoda), with the description of three new genera and one new species and a phylogenetic analysis of the subfamily. *Hydrobiologia, 632*, 1–48. Available from https://link.springer.com/article/10.1007/s10750-009-9826-5.

Schäfer, H. W. (1945). Grundwasser-Ostracoden aus Griechenland. *Archiv für Hydrobiologie, 40*, 847–866.

Scharf, B., Meisch, C., Schön, I., & Martens, K. (2014). New records of *Candonocypris novaezelandiae* (Crustacea, Ostracoda) from Germany, Belgium, England and Tunisia. *Abhandlungen des Naturwissenschaftlichen Vereins zu Bremen, 47*(2), 337–344.

Scharf, B., Viehberg, F. A., & Meisch, C. (2016). Two new methods for opening closed carapaces of preserved Ostracoda (Crustacea). *Bulletin de la Société des naturalistes luxembourgeois, 118*, 149–154. Available from https://www.snl.lu/publications/bulletin/SNL_2016_118_149_153.pdf.

Schmit, O., Martens, K., & Mesquita-Joanes, F. (2012). Vulnerability of sexual and asexual *Eucypris virens* (Crustacea, Ostracoda) to predation: An experimental approach with dragonfly naiads. *Fundamental and Applied Limnology—Archiv für Hydrobiologie, 181*, 207–214. Available from https://www.schweizerbart.de/papers/fal/detail/181/79229/Vulnerability_of_sexual_and_asexual_Eucypris_virens_Crustacea_Ostracoda_to_predation_an_experimental_approach_with_dragonfly_naiads.

Schmit, O., Rossetti, G., Vandekerkhove, J., & Mezquita, F. (2007). Food selection in *Eucypris virens* (Crustacea: Ostracoda) under experimental conditions. *Hydrobiologia, 585*, 135–140. Available from http://link.springer.com/10.1007/s10750-007-0634-5.

Schön, I., Kamiya, T., Van den Berghe, T., Van den Broecke, L., & Martens, K. (2019). Novel *Cardinium* strains in non-marine ostracod (Crustacea) hosts from natural populations. *Molecular Phylogenetics and Evolution, 130*, 406–415. Available from https://linkinghub.elsevier.com/retrieve/pii/S1055790318303518.

Schön, I., & Martens, K. (2020). Are *Cardinium* infections causing asexuality in non-marine ostracods? *Hydrobiologia, 847*, 1651–1661. Available from http://link.springer.com/10.1007/s10750-019-04110-2.

Smith, A. J., Horne, D. J., Martens, K., & Schön, I. (2015). Class Ostracoda. In J. Thorp, & D. C. Rogers (Eds.), *Thorp and Covich's freshwater invertebrates* (4th ed., pp. 757–780). Amsterdam: Elsevier, Ecology and General Biology. Available from https://linkinghub.elsevier.com/retrieve/pii/B9780123850263000309.

Smith, R. J. (2000). The morphology of the upper lip of Cypridoidea ostracods: Taxonomic and phylogenetic significance. *Hydrobiologia, 418*, 169–184. Available from https://link.springer.com/article/10.1023/A:1003934528762.

Smith, R. J., & Kamiya, T. (2007). Copulatory behaviour and sexual morphology of three *Fabaeformiscandona* Krstić, 1972 (Candoninae, Ostracoda, Crustacea) species from Japan, including descriptions of two new species. *Hydrobiologia, 585*, 225–248. Available from http://link.springer.com/10.1007/s10750-007-0640-7.

Smith, R. J., & Matzke-Karasz, R. (2008). The organ on the first segment of the cypridoidean (Ostracoda, Crustacea) antennule: Morphology and phylogenetic significance. *Senckenbergiana lethaea, 88*, 127–140. Available from http://link.springer.com/10.1007/BF03043984.

Smith, R. J., & Tsukagoshi, A. (2005). The chaetotaxy, ontogeny and musculature of the antennule of podocopan ostracods (Crustacea). *Journal of Zoology, 265*, 157–177. Available from http://doi.wiley.com/10.1017/S095283690400617X.

Smith, R. J., Zhai, D., Savatenalinton, S., Kamiya, T., & Yu, N. (2017). A review of rice field ostracods (Crustacea) with a checklist of species. *Journal of Limnology, 77*, 1–16. Available from http://www.jlimnol.it/index.php/jlimnol/article/view/jlimnol.2017.1648.

Soininen, J. (2014). A quantitative analysis of species sorting across organisms and ecosystems. *Ecology, 95*, 3284–3292. Available from https://esajournals.onlinelibrary.wiley.com/doi/full/10.1890/13-2228.1.

Sywula, T. (1973). Notes on Ostracoda XIII: General plan of structure of the penis in subgenera *Candona* Baird, *Eucandona* Daday and *Typhlocypris* Vejd. of the genus *Candona* Baird. *Bulletin de l'Académie Polonaise des Sciences—Série des Sciences Biologiques, 21*, 123–125.

Sywula, T. (1974). Małzoraczki (Ostracoda). Fauna słodkowodna Polski. PWN. *Warszawa-Poznan, 24*, 1–315.

Szlauer-Łukaszewska, A., & Radziejewska, T. (2013). Two techniques of ostracod (Ostracoda, Crustacea) extraction from organic detritus-rich sediments. *Limnologica, 43*, 272–276. Available from https://linkinghub.elsevier.com/retrieve/pii/S0075951112000850.

Tanaka, H., Yoo, H., Pham, H. T. M., & Karanovic, I. (2021). Two new xylophile cytheroid ostracods (Crustacea) from Kuril-Kamchatka Trench, with remarks on the systematics and phylogeny of the family Keysercytheridae, Limnocytheridae, and Paradoxostomatidae. *Arthropod Systematics & Phylogeny, 79*, 171–188. Available from https://arthropod-systematics.arphahub.com/article/62282/.

Tétart, J. (1974). Les entomostracés des milieux peu profonds de la vallée du Rhône. Essai d'étude écologique: Composition des associations et répartition des espèces. *Travaux du Laboratoire d'hydrobiologie et de pisciculture de l'Université de Grenoble, 64-65*, 109–245.

Valls, L., Castillo-Escrivà, A., Barrera, L., Gómez, E., Gil-Delgado, J. A., Mesquita-Joanes, F., & Armengol, X. (2017). Differential endozoochory of aquatic invertebrates by two duck species in shallow lakes. *Acta Oecologica*, *80*, 39–46. Available from https://linkinghub.elsevier.com/retrieve/pii/S1146609X16301576.

Valls, L., Castillo-Escrivà, A., Mesquita-Joanes, F., & Armengol, X. (2016). Human-mediated dispersal of aquatic invertebrates with waterproof footwear. *Ambio*, *45*, 99–109. Available from http://link.springer.com/10.1007/s13280-015-0689-x.

Valls, L., Rueda, J., & Mesquita-Joanes, F. (2014). Rice fields as facilitators of freshwater invasions in protected wetlands: The case of Ostracoda (Crustacea) in the Albufera Natural Park (E Spain). *Zoological Studies*, *53*, 1–10. Available from https://zoologicalstudies.springeropen.com/articles/10.1186/s40555-014-0068-5.

Valls, L., Zamora, L., Rueda, J., & Mesquita-Joanes, F. (2016). Living and dead ostracod assemblages in a Coastal Mediterranean Wetland. *Wetlands*, *36*, 1–9. Available from http://link.springer.com/10.1007/s13157-015-0709-4.

Van Damme, K., & Dumont, H. (2010). Cladocera of the Lençóis Maranhenses (NE - Brazil): Faunal composition and a reappraisal of Sars' Method. *Brazilian Journal of Biology*, *70*, 755–779. Available from http://www.scielo.br/scielo.php?script = sci_arttext&pid = S1519-69842010000400008&lng = en&tlng = en.

Vandekerkhove, J., Namiotko, T., Hallmann, E., & Martens, K. (2012). Predation by macroinvertebrates on *Heterocypris incongruens* (Ostracoda) in temporary ponds: Impacts and responses. *Fundamental and Applied Limnology*, *181*, 39–47. Available from https://www.schweizerbart.de/papers/fal/detail/181/78404/Predation_by_macroinvertebrates_on_Heterocypris_in?af = crossref.

Vanschoenwinkel, B., Waterkeyn, A., Vandecasetsbeek, T., Pineau, O., Grillas, P., & Brendonck, L. (2008). Dispersal of freshwater invertebrates by large terrestrial mammals: A case study with wild boar (*Sus scrofa*) in Mediterranean wetlands. *Freshwater Biology*, *53*, 2264–2273. Available from http://doi.wiley.com/10.1111/j.1365-2427.2008.02071.x.

Viehberg, F. A. (2002). A new and simple method for qualitative sampling of meiobenthos-communities. *Limnologica*, *32*, 350–351. Available from https://linkinghub.elsevier.com/retrieve/pii/S0075951102800263.

Waterkeyn, A., Vanschoenwinkel, B., Elsen, S., Anton-Pardo, M., Grillas, P., & Brendonck, L. (2010). Unintentional dispersal of aquatic invertebrates via footwear and motor vehicles in a Mediterranean wetland area. *Aquatic Conservation: Marine and Freshwater Ecosystems*, *20*, 580–587. Available from http://doi.wiley.com/10.1002/aqc.1122.

Yamada, S. (2007). Structure and evolution of podocopan ostracod hinges. *Biological Journal of the Linnean Society*, *92*, 41–62. Available from https://academic.oup.com/biolinnean/article-lookup/doi/10.1111/j.1095-8312.2007.00870.x.

Zhai, M., Nováček, O., Výravský, D., Syrovátka, V., Bojková, J., & Helešic, J. (2015). Environmental and spatial control of ostracod assemblages in the Western Carpathian spring fens. *Hydrobiologia*, *745*, 225–239. Available from http://link.springer.com/10.1007/s10750-014-2104-1.

Chapter 6

Classes Copepoda and Ichthyostraca

Fabio Stoch[1], Federico Marrone[2] and Maria Cristina Bruno[3]

[1]Department of Evolutionary Biology and Ecology, Free University of Brussels (ULB), Brussels, Belgium, [2]Department of Biological, Chemical and Pharmaceutical Sciences and Technologies (STEBICEF), University of Palermo, Palermo, Italy, [3]Research and Innovation Centre, Fondazione Edmund Mach, San Michele all'Adige (Trento), Italy

Introduction

Copepoda are the largest microcrustacean class, with more than 14,000 described species (Horton et al., 2022). Of these, more than 2800 are reported from freshwaters. Copepods are defined as "one of the most successful groups of animals on earth" (Huys & Boxshall, 1991); this is true, especially for the oceans, but they are also abundant in all kinds of inland aquatic ecosystems. Indeed, they are common in aquatic and semiterrestrial habitats, ranging from the hypotelminorheic rivulets in soil (including moist soils and mosses) to ponds, lakes, running waters, and ground waters. Parasitic taxa include more than 6400 species and the highest number of families of all aquatic parasites. The name Copepoda derives from the Greek, meaning "oar-feet," and reflects the way they use their legs for swimming.

Four copepod orders have freshwater representatives (Calanoida, Cyclopoida, Harpacticoida, and Gelyelloida). The first three orders have representatives in the Mediterranean Basin and will be treated herein. Calanoids (Fig. 6.1A) are mostly planktonic in lakes, rivers, ponds, and pools, with a few species living exclusively in ground waters. Cyclopoids (Fig. 6.1B) are benthic, planktonic, as well as stygobitic (living in caves, interstitial, and even moist litter). Harpacticoids (Fig. 6.2) are generally benthic, with a few taxa living in wet mosses and wet soil; many species are stygobitic, with highly specialized forms living in phreatic, interstitial, and karstic habitats. True planktonic harpacticoids are not present in freshwaters, although they can be occasionally collected in open waters.

Adult copepods are sexually dimorphic; sexes can be easily distinguished: females are generally larger than males and have a genital somite (usually a double somite, formed by the fusion of the first two abdominal somites); moreover, spermatophores and/or external egg sacs are often visible. Males have the right (calanoids) or both (cyclopoids, harpacticoids) first antennae geniculate, that is, modified for grasping the female while mating. Other secondary characteristics are the transformation of the fifth pair of legs in a prehensile gripper in calanoids (used during mating), while harpacticoids may have modified second to fifth legs. Cyclopoids have usually identical legs in males and females, albeit in some species males have modified setae on swimming legs. In the parasitic cyclopoids treated herein, the males are usually "cyclopiform," while females show body modifications. In the most specialized species, females lose the "cyclopiform" body shape, which appears lobulated and elongated (Fig. 6.3). Most copepod species cannot be morphologically identified unless adult males and females are available.

During mating, the male copepod grips the female with his first pair of geniculate antennulae (Fig. 6.4) and with the fifth pair of legs, as in calanoids (Fig. 6.5). The male then transfers a spermatophore to the female genital opening using the thoracic legs. Eggs are sometimes laid directly into the water, but most species enclose them within egg sacs attached to the female's body until they hatch. Calanoids have one single egg sac (Fig. 6.1A), as in the harpacticoids, while cyclopoids bear two egg sacs (Fig. 6.1B). In temporary ponds, several species produce resistant eggs that can lie dormant in the sediments for long periods when the basin dries up and are easily transported by animals or wind. Eggs hatch into a first larval stage (nauplius: Fig. 6.6A), which then molts five or six times before a copepodid emerges (Fig. 6.6B). This second larval stage resembles the adult but has not a well-developed genital segment. After further five molts, the copepodid matures to the adult phase (Fig. 6.6C).

Identification and Ecology of Freshwater Arthropods in the Mediterranean Basin. DOI: https://doi.org/10.1016/B978-0-12-821844-0.00012-0

FIGURE 6.1 Habitus of ovigerous females of (A) *Diaptomus cyaneus* (Calanoida) and (B) *Cyclops divergens* (Cyclopoida). *Photos by J.-F. Cart.*

FIGURE 6.2 Habitus of ovigerous females of *Bryocamptus echinatus* (Harpacticoida). *Photo by J.-F. Cart.*

FIGURE 6.3 Habitus of ovigerous female of *Lernaea cyprinacea* (Cyclopoida). *Photo by J.-F. Cart.*

Copepod taxonomy (based almost exclusively on adult morphocharacters) is still developing. Copepoda, originally considered as Entomostraca (a polyphyletic group with no taxonomic value) and later allocated in different crustacean classes (Maxillopoda, Hexanauplia), is now considered a class. Following the most modern phylogenetic reconstructions, they are allocated in the superclass Multicrustacea (together with Tantulocarida, Thecostraca, and Malacostraca). This classification can be examined in detail in Horton et al. (2022). Furthermore, the classification of copepod orders is still under revision (but orders and families reported here are recognized in the most recent phylogenetic trees). Innovative morphotaxonomic techniques, based on the examination of microcharacteristics at high magnification, have led to the description of several new genera and species. Unfortunately, DNA taxonomy in this group is

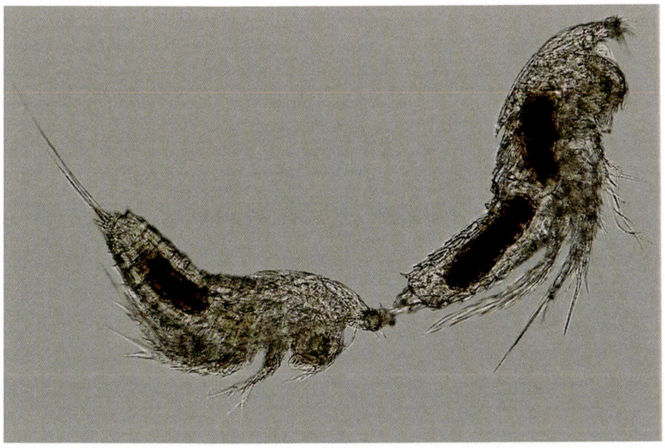

FIGURE 6.4 Mating couple of *Bryocamptus tatrensis* (Harpacticoida). *Photo by F. Stoch.*

FIGURE 6.5 Mating couple of *Diaptomus* sp. (Calanoida). *Photo by J.-F. Cart.*

still in its infancy, and several of these taxa (based on debatable morphocharacters whose affinity could be due to convergence or parallel evolution) are in urgent need of revision.

The class Ichthyostraca includes in freshwaters the subclass Branchiura (fish lice), with the single order Arguloida, represented in the Mediterranean Basin only by the family Argulidae (including four genera). Since accurate diversity and distribution data are lacking, the exact number of species occurring in inland waters of the Mediterranean Basin is not known. Apart from the widespread *Argulus foliaceus* (Linnaeus, 1758), at least one nonnative species, *Argulus japonicus* Thiele, 1900, has been introduced in freshwaters of the Mediterranean Basin, where it can reach high densities mostly in aquaculture plants and in aquaria.

General ecology and distribution

Copepoda

During their evolutionary history, copepods successfully colonized almost all the available water habitats. They can be found in waters of different salinity regimes, ranging from very soft freshwaters to hyperhaline inland, endorheic basins, and at all temperature regimes, including hot thermal springs (Huys & Boxshall, 1991). Freshwater habitats range from large lakes to ephemeral water bodies, including tree holes and other phytotelmata, mosses, and very small containers, as well as within benthic substrates (i.e., interstitial) and in every kind of groundwater, including karstic and epikarstic waters. Moreover, free-living cyclopoids and harpacticoids may represent an important component of the cryptozoic fauna in moist soils, especially in the hypothelminorheic habitat. Finally, some species are parasites of fish or symbionts of other invertebrates, such as crayfish.

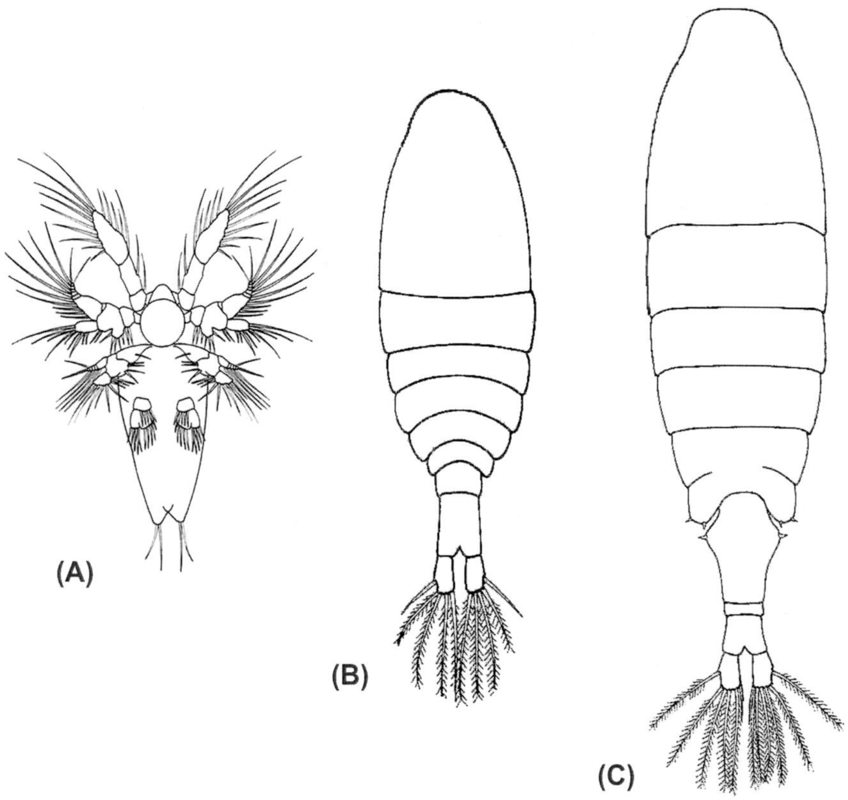

FIGURE 6.6 Developmental stages of a calanoid copepod: (A) nauplius (fourth stage of metanauplius); (B) copepodid (third stage); (C) adult female (not to scale).

Lakes and rivers. Copepods can be found in both the pelagic and benthic zones of lakes and large rivers. In pelagic environments, together with cladocerans and rotifers, they are the dominant group of zooplankton. Their feeding habits include filter-feeding, grazing on aquatic plants and substrates, as well as predation. When cooccurring with fishes, copepods are often small and diaphanous. In deeper lakes, they typically migrate vertically to optimize feeding opportunities while minimizing the risk of consumption by visual predators. In waterbodies where no or few visual predators occur, copepods have a larger size and intense pigmentation—a characteristic that protects them from UV radiation. The horizontal distribution of species in lakes and broad rivers is heterogeneous, with littoral species closely associated with vegetation alongside species strictly linked to open waters. However, occurrences in planktonic versus benthic habitats may be limited to certain life stages. Because of their abundance, copepod carcasses and fecal pellets play an important role in food webs and nutrient cycles.

Mediterranean ponds and pools. Water shortage is a limiting factor for Mediterranean freshwater habitats, and thus hydroperiod is considered the first structuring factor of pond and pool copepod communities (Sahuquillo & Miracle, 2013). Species richness increases for copepods with longer hydroperiods. Species usually display a temporal succession; therefore, longer hydroperiods allow the establishment of a higher number of species, with synchronically co-occurring species showing a considerable size differentiation to avoid or minimize niche overlap. Moreover, ponds with a shorter hydroperiod are usually smaller, thus reducing the likelihood of seeding by resting stages, and smaller ponds are less likely to be visited by waterfowls and other animals that play a major role in copepod dispersal. Dormancy in freshwater copepods is well studied (Dahms, 1995). The resting stages are eggs in calanoids and some harpacticoids, while some cyclopoids encyst mainly at the copepodid stage (IV and V stages). Other harpacticoids encyst at both copepodid and adult stages. Considering that temporary pond species hatch from resting stages, the premature drying up of a pond may prevent their populations from producing resistant eggs and cysts. Moreover, abrupt changes in hydroperiod can affect permanent pond dwellers as well, with the possible loss of some cyclopoids that lack resting stages (subfamily Eucyclopinae).

Springs and brooks. The small area around the spring mouth and associated network of branched rivulets and pools (eucrenal) harbors copepod assemblages consisting of at least three ecological groups: (1) benthic species that colonize springs from the brooks downstream (i.e., from the epirhithral), (2) stygobionts, colonizing from hypothelminorheic or ground waters feeding the spring and living in the interstitial habitat formed by gravel and sand near the spring mouth,

FIGURE 6.7 Habitus of the interstitial harpacticoids (females): *Stammericaris destillans* (top, dorsal view) and *S. diversitatis* (bottom, lateral view). *SEM photo by M. C. Bruno.*

and (3) species occurring only in springs (crenobionts). Few copepods found in springs are true crenobionts as most of these species also occur in the hypocrenal and epirhithral, in the littoral zone of lakes and marshes, in semiterrestrial habitats (like wet mosses and moist soil around springs), or in ground waters. In brooks and streams (epi- and mesorhithral), cyclopoid and harpacticoid copepods are abundant and diverse in the benthic and hyporheic habitats, where they feed as detritivores or, more rarely, as predators.

Subterranean waters. According to the degree of adaptation to groundwater life, copepods can be classified as stygobionts, stygophiles, and stygoxenes. Stygobionts are strictly linked to the groundwater environment during their entire life cycle, and frequently show adaptations to the biotic and abiotic conditions of subterranean waters (such as anophthalmy and cuticular depigmentation). Stygophiles can live and reproduce in subterranean habitats, as well as in some epigean marginal habitats, while stygoxenes are accidental or occasional in subterranean waters. A wide range of copepod body morphologies characterize stygobitic copepods because of different selective pressures and different evolutionary histories. For example, karstic pools and lakes in caves pose no relevant spatial constraint to the copepods living in them; some cyclopoids show elongation of antennae and swimming legs. In contrast, stygobiont calanoids, which are typically planktonic, show a tendency to have a reduced body size, a reduction in setation, and a lower number of segments (oligomerization) of swimming legs. A different condition can be found in the microcrevices of the epikarst and minute voids of unconsolidated sediments, colonized by cyclopoids and harpacticoids. In these small living spaces, copepods may be characterized by body miniaturization, together with an elongated, subcylindrical body form (Fig. 6.7). Swimming legs may be shortened, and undergo oligomerization (due to progenetic paedomorphosis); in this case, interstitial copepods move in the spaces among gravel or sand grains by pushing their body along the grains. Another common feature in interstitial copepods is the reduction in setation of cephalic appendages, which facilitates movement in sandy and muddy environments, avoiding the adhesion of particles on the body surface. As regards physiology, subterranean copepods produce fewer but larger eggs than surface ones, thereby enabling them to secure an endogenous food supply for the nauplii in the oligotrophic subterranean environment.

Parasitic species. Most of the described copepod species are free-living, some are commensal of invertebrates, while many parasitize invertebrates and fishes ($\sim$6500 described species, mostly marine). Some parasitic copepods resemble free-living forms in shape (Ergasilidae) while in other cases females have a highly modified habitus (Lernaeidae) and live attached to the gills of their hosts (Fig. 6.8), and can only be recognized as copepods by their larval forms (nauplius, copepodid) and the cyclopiform male. The most well-known lernaeid is the anchor worm (*Lernaea cyprinacea*), an economically important fish pest, which was globally spread with the introduction of goldfish and carp. The life cycle of anchor worms may be direct (with only one host), indirect (with an intermediate host), or with a transfer host. Species of the cyclopoid family Ergasilidae cause the most important damage in aquaculture and are widely distributed. Some of them (*Neoergasilus*) were introduced in fish farms and are now widespread in the Mediterranean Basin, and commonly found in zooplankton samples.

Interest as bioindicators. Given the high natural abundance, universal presence, and importance in the food web, understanding how chemical substances can impact this important class of crustaceans is central to ecotoxicology. The remarkable degree of habitat specialization suggests that copepod assemblages can be used as biological indicators of environmental quality and vulnerability. Finally, the use of copepods as bioindicators to monitor anthropogenic-dependent changes in the hydroperiod (due to land use and climate change) is feasible and highly recommended.

FIGURE 6.8 Parasitic female of *Lernaea cyprinacea* attached close to the fin of the host fish. *Photo by J.-F. Cart*

Geographical distribution. Surface cyclopoids and harpacticoids generally have wide ranges, but this may be partially due to the unresolved taxonomy of some taxa, as evidenced by recent molecular studies that revealed the presence of cryptic species (Kochanova et al., 2021). In contrast, calanoids, especially diaptomids, show reduced ranges despite their high dispersal potential (due to the production of resting eggs), which is probably related to the niche specialization of this group. Copepods from brackish waters, both coastal and inland ones, tend to have larger ranges than those from poorly mineralized waters, possibly due to a more frequent visitation by long-range dispersal vectors (especially birds) and to the harshness of environmental conditions, which reduce the number of potential colonizers. In contrast, groundwaters host a very high degree of endemics with copepod fauna often found exclusively in a single aquifer or karst massif (Galassi et al., 2013).

The rate of endemism and intraspecific genetic diversification is higher in areas with a semiarid climate, as species from decidedly steppe and desert climates have relatively wider distributions. This may result from the continuous nature of these areas in the absence of barriers and the scarcity of competitors, which effectively produces a biological filter to settlement in new sites. In the Mediterranean semiarid regions, we are witnessing the opposite situation to what occurs in areas affected by Quaternary glaciations, where copepods are genetically homogeneous over vast areas.

Ichthyostraca

All fish lice are ectoparasites on fishes, albeit some taxa also occur on amphibians or invertebrates. They feed on mucous, epithelial cells, and blood or hemolymph of their hosts. Notwithstanding their parasitic habits, they are also able to actively swim and can survive for long periods of their hosts; hence they can sometimes be collected from open waters. However, fish lice are usually found on the body surface, mouth, and gill chamber of fish hosts. No reliable data about the possible impact of argulids on native fish are available for the Mediterranean Basin.

Morphological characteristics used in identification

Copepoda

The morphologic terminology proposed by Huys and Boxshall (1991) is the most widely accepted among copepodologists, and it is used herein. Freshwater copepods are tiny crustaceans ranging between 0.3 and 5.0 mm, females of parasitic species may be longer. The body (Fig. 6.9) is divided into an anterior part (prosome) and a posterior part (urosome). In calanoids (superorder Gymnoplea), the single major articulation is between the fifth pedigerous somite and the genital segment, while in cyclopoids and harpacticoids (superorder Podoplea) it is located between the fourth and fifth pedigerous somites. The head is typically fused with the first thoracic somite (sometimes even with the second) to form the cephalosome. The thorax consists of seven postcephalic somites, including the one(s) fused with cephalosome and the genital segment. The number of urosomal segments is variable (from two to five, usually four with a fusion of two segments forming the double genital segment), the last one being the anal somite; the urosome ends with the caudal rami.

The head bears five pairs of cephalic appendages (Fig. 6.10), named first antennae or antennulae (A1), second antennae or antennae (A2), mandibles (Md), first maxillae (Mx1), and second maxillae (Mx2). The first antennae of male

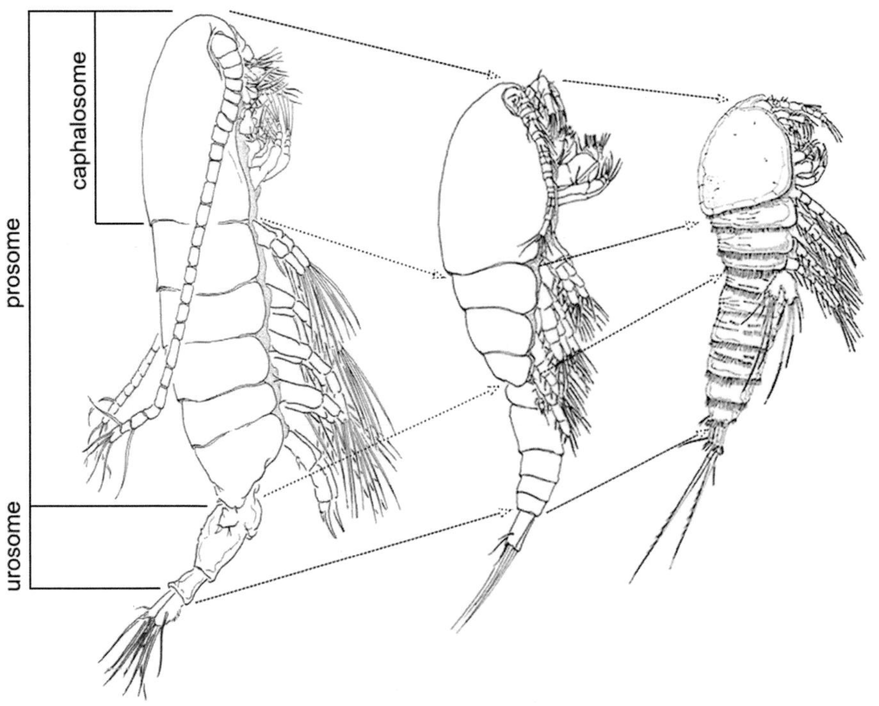

FIGURE 6.9 Body shape and organization of free-swimming calanoid, cyclopoid, and harpacticoid copepods. *Body shapes from Suárez-Morales et al. (2020), modified.*

copepods are geniculate (only the right one in calanoids). The A2 of calanoids and harpacticoids is biramous; in freshwater cyclopoids there is only an exopodal seta (families Halicyclopidae and Cyclopidae) that can be absent. The A2 is prehensile in Ergasilidae.

Six or seven pairs of thoracic appendages (maxillipeds and P1 to P5 or P6) are present in free-living copepods (Fig. 6.10). The first thoracic somite bears the maxillipeds, while the second through the fifth bear four pairs of biramous swimming legs (P1 to P4), each pair joined by an intercoxal sclerite. The swimming leg consists of two basal segments, that is, coxa (or coxopod) and basis (or basipod); the basis bears two rami, an inner one (endopod) and an outer one (exopod), both rami can be one to three segmented; the P4 endopod is often reduced in some cyclopoid and harpacticoid species. In calanoids, the fifth legs (P5) are well developed (symmetrical in the female, highly asymmetrical and modified and prehensile in the male), while in cyclopoids and harpacticoids they are reduced to a baseoendopod, bearing an endopod and an outer seta, and an exopod. In some cyclopid genera, the baseoendopod or the whole P5 are fused to somite; in this case, only two to three setae implanted on the somite are present. Cyclopoids and harpacticoids have vestigial sixth legs. Abdominal appendages are lacking in copepods; the urosome terminates in two caudal rami, typically bearing seven setae (for terminology, see Huys & Boxshall, 1991); in most cases, the number of setae is reduced.

Apart from the habitus, the most important taxonomic characteristics used in the key are segmentation and ornamentation of A1, shape of A2, and segmentation and setation of P1–P5. Other characteristics (shape and setation of P1–P6 and of caudal rami) are useful at the species level. Recent developments in copepod taxonomy require careful examination of minute characteristics, usually overlooked in the older keys, like tegument ornamentation, including the pattern of pores and sensillae, which require the use of Nomarski differential interference contrast (DIC) microscopy at 1000 × magnification or scanning electron microscope (SEM)/field emission scanning microscope (FESM) imaging. Finally, recent developments in molecular taxonomy are useful for species delimitation and detection of cryptic species, but these methods are outside the scope of this chapter.

A good, albeit outdated key to World copepod genera can be found in Dussart & Defaye (2001). Bledzki & Rybak (2016) reported pictorial keys and distribution maps for freshwater European zooplankton, while Lee and Lee (2019) reported a key for the freshwater copepods of the Palearctic region. There are no keys specific for the Mediterranean species; most available keys are useful for European copepods only, that is, Dussart (1967, 1969), Kiefer (1978), Fryer (1978a, 1978b), Einsle (1993), and Janetzky et al. (1996). These keys need to be integrated using monographs like

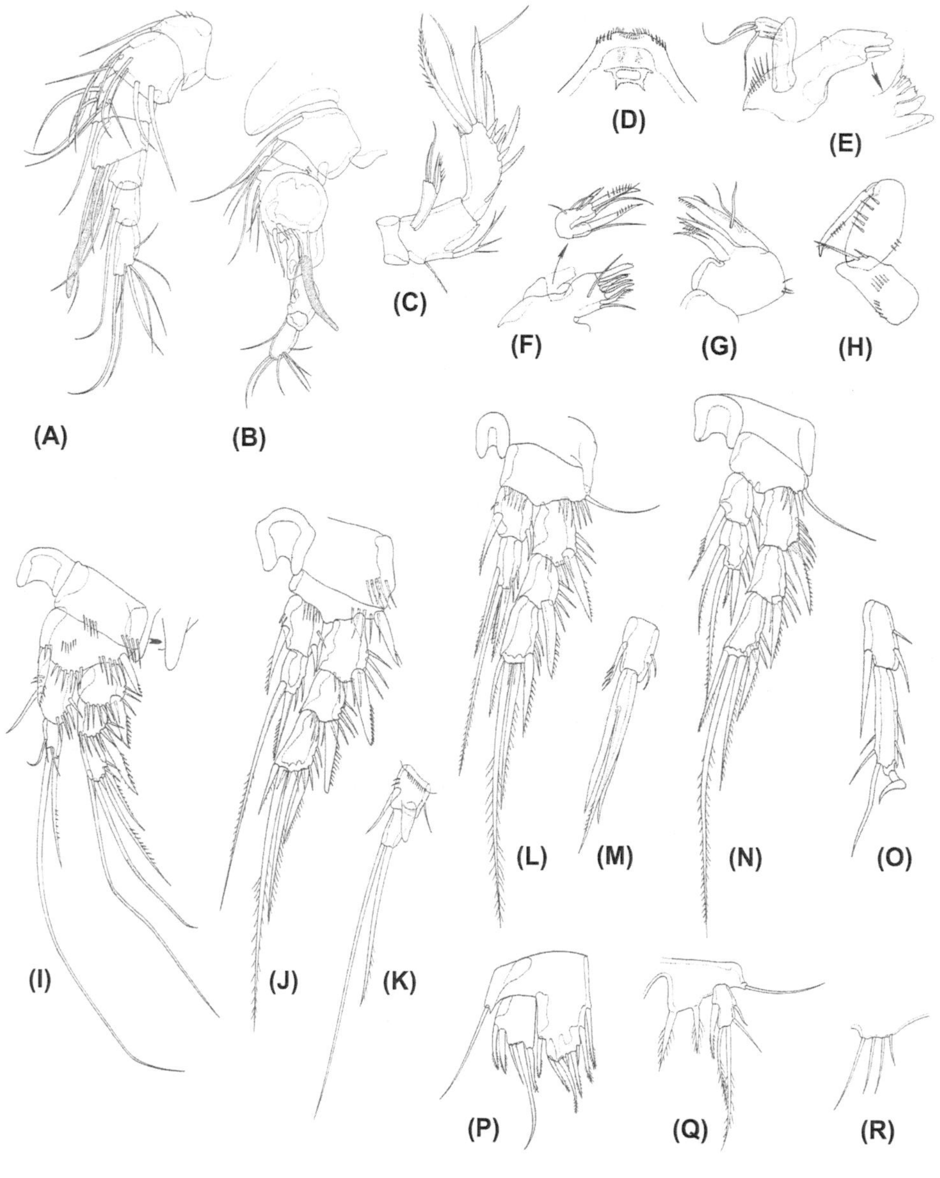

FIGURE 6.10 Appendages of a harpacticoid copepod of the genus *Moraria*. (A) Female antennula; (B) male antennula, geniculate; (C) antenna; (D) labium; (E) mandible; (F) maxillule; (G) maxilla; (H) maxilliped; (I) P1; (J) female P2; (K) male P2 endopod; (L) female P3; (M) male P3 endopod; (N) female P4; (O) male P4 endopod; (P) female P5; (Q) male P5; (R) P6. [Not to scale]. *From Stoch (1998).*

Bayly (1992): Centropagidae, Einsle (1996): *Cyclops*, *Megacyclops*, and *Acanthocyclops*, Hueda & Reid (2003): *Mesocyclops* and *Thermocyclops*, Karaytuğ (1999) *Paracyclops*, Ranga Reddy (1994), Rayner (1999): Paradiaptominae, Vives & Shmeleva (2007, 2010), Huys et al., 1996: marine species, sometimes present in brackish waters. The old monograph by Lang (1948) is still useful for harpacticoids, updated by the key from Wells (2007).

Ichthyostraca

Fish lice are characterized by a dorsoventrally flattened body (Fig. 6.11) covered by a shield-like carapace extending in two lobes, and a pair of compound eyes. Appendages are represented by the first antennae (antennulae) and second antennae, first maxillae (maxillulae) usually transformed in radially symmetrical suckers, four pairs of biramous thoracic legs, and an unsegmented abdomen, bearing tiny caudal rami (Fig. 6.11). Sexes are separate, with males showing elongated testes in the abdomen and secondary sexual characters on legs 2–4, and females bearing eggs in their thoracic region. Mating takes place on the host, which is later left by the females to attach the eggs to hard surfaces such as rocks or vegetation with an adhesive substance. Further information can be retrieved from Fryer (1978a, 1978b), Poly (2008), and Neethling & Avenant-Oldewage (2017).

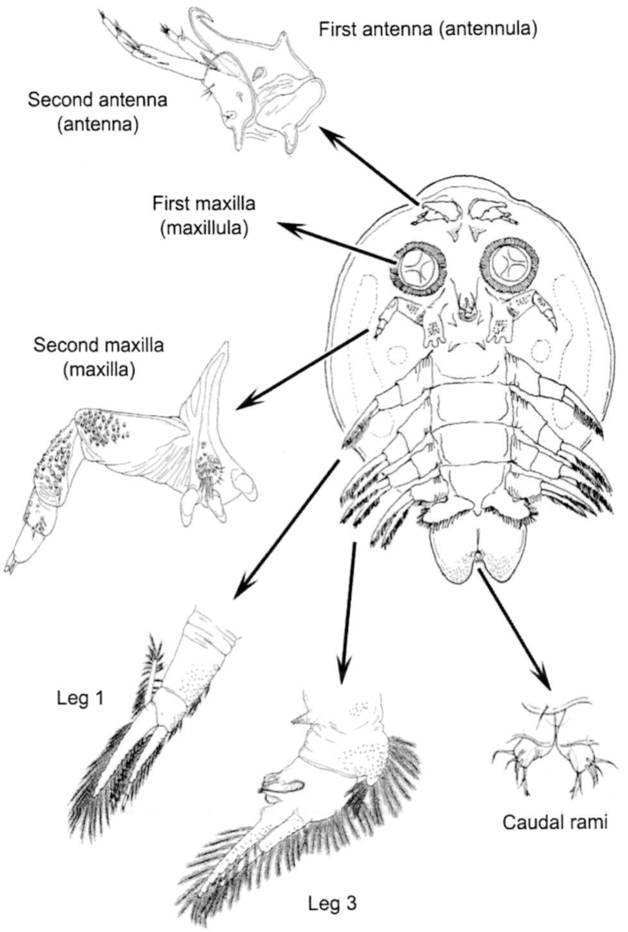

FIGURE 6.11 Habitus (ventral view) and appendages of *Argulus* (Branchiura). *From Suárez Morales (2020), modified.*

Material preparation and preservation

The collecting methods for copepods are described in detail by Huys & Boxshall (1991), and thus only a synthetic description will be provided herein. Zooplanktonic copepods can be collected following the usual protocols developed in limnology, by using water bottles, pumps, or plankton tow and throw nets (mesh size 60–100 μm). The use of nets is the most efficient method for fast-swimming calanoids and planktonic cyclopoids. The collection of meiofauna, including cyclopoids and harpacticoids, is exhaustively described by Stoch (2007) and is mainly based on the use of hand nets, benthic nets (kick sampling), drift nets at the spring mouth, and wet mosses squeezing. Malard et al. (2002) described the most suitable techniques to collect stygobitic species using the Karaman–Chappuis method, Bou–Rouch pump, Cvetkov nets, drift nets, and artificial substrates. Because copepods can be found in semiterrestrial microhabitats, they are sometimes collected in the pitfall traps used for soil Coleoptera. However, digging a hole in the litter and filtering the water using a fine mesh (60–100 μm) sieve may be the best way to obtain material from this cryptic environment. A Berlese device is useful for samples from soil and mosses. Free stages of parasitic copepods and branchiurans are collected using zooplankton methods; the collection of adults requires the careful examination of the fish host, usually its fins and gills.

Copepod and branchiuran samples should be fixed as soon as possible after collection. The fixing agent, most widely used until 20 years ago, is a 3%–5% formalin-buffered solution (Huys & Boxshall, 1991). Besides the caution required to use formaldehyde (this chemical is toxic and is classified as a possible human carcinogenetic agent), its use makes DNA extraction and sequencing more difficult. The best recommended fixative for this purpose is 96%–100% ethanol; filtering the sample through a sieve and pouring the residual (using a squeeze bottle with ethanol) into a collection bottle with the help of a funnel is the quickest way to fix fresh samples with detritus. Denatured ethanol must be avoided if the goal is DNA sequencing; moreover, it makes specimens brittle.

Sorting at the stereomicroscope (20 × −50 ×) is a long process, and alternative methods have been proposed (staining with Rose Bengal, density-gradient centrifugation); however, these methods may damage delicate specimens, hide minute morphocharacters, and prevent DNA sequencing. For this reason, manual extraction of copepods from the detritus to obtain clean specimens using a Pasteur pipette and/or needles is still the recommended method.

The preservation of samples depends on the study that will be performed. Specimens for DNA sequencing should be stored in 96%−100% ethanol and preserved at −20°C, while those used for morphological analysis are best preserved in 70% ethanol, which yields more relaxed specimens. Adding 5%−10% glycerin helps keeping specimens relaxed and prevents damage if alcohol evaporates. Glycerin is not suitable if specimens are to be examined under a SEM. Permanent storage in the ethanol/glycerin mix is recommended, using small polypropylene vials (polyethylene must be avoided because after a few years alcohol makes them fissured and brittle). Small vials, like Eppendorf's vials, must be stored in a larger jar, filled with the same preservative liquid of the vials, and closed with a hermetically sealed cap to prevent evaporation. Proper labeling is always essential.

Dissection is required for the identification of specimens. Before dissection, it is important to observe (and, when possible, photograph or draw) the cuticular ornamentation and length of antennules in relation to body length (which must be measured using an ocular micrometer). Individual specimens must be transferred to a slide without transferring unnecessary amounts of alcohol using bent stainless steel entomological micro-pins (minute pins, diameter 0.10 mm) mounted on a holder. Dissection is conducted in glycerin; lactic acid is used if specimens are to be mounted in polyvinyl lactophenol or in lactophenol as recommended by Huys & Boxshall (1991). Glycerin must be used if the permanent or semipermanent mounting medium is Faure's (or Hoyer's) liquid, glycerin jelly, or glycerin itself. Dissection is performed under a stereomicroscope (range of magnifications up to 60−100 × or more) using transmitted light. Dissection sequences for copepods have been described in detail by Huys & Boxshall (1991).

Glycerin is widely used as a mounting medium; up to four 10 × 10 mm coverslips can be used to mount separately on the same microscope slide antennae, mouthparts, P1−P4/P5, and the abdomen. Slides mounted using glycerin should be sealed using commercial epoxides to prevent evaporation and deterioration of the medium. Specimens mounted in glycerin have been studied after 50−80 years of their preparation (e.g., Kiefer's collection), and the specimens were still visible and in good condition. Larger specimens can be mounted in Faure's medium (which allows better positioning of appendages under the same coverslip, by letting the thin layers of liquid where the dissected parts are positioned to dry a bit, and then adding the coverslip with another drop of Faure's medium). Slides mounted using Faure's medium must be sealed as well, using for example Canada balsam. This medium may remain stable and provide good viewing conditions for 20 years or more; however, it must be protected from light and, in some cases, develops crystals damaging the specimens, or becomes opaque preventing the observation of small details. Transparent nail polish has been widely used as well to seal slides, but being its duration limited, it is recommended only for working (semipermanent) mounts. Polyvinyl lactophenol usually forms needle-like phenol crystals under the coverslip and obscures the material within about 10 years and, also in view of its toxicity, this mounting media is not recommended.

Specimen identification requires a compound microscope, possibly equipped with Nomarski DIC, and objectives up to 1000 × (oil immersion). When the examination of microcharacteristics is outside the observation range of light microscopy, specimens can be prepared for observation under a SEM, but this is rarely needed. Drawing can be obtained with the old-fashioned drawing tube (camera lucida), which gives the best results, or using multi-plane focused digital photographs assembled using dedicated software. Inking is best obtained using software producing vector images, which can be scaled as desired.

Keys to Copepoda and Ichthyostraca

Families and genera not represented in inland waters of the Mediterranean Basin but reported only from estuaries, lagoons, and marine coastal waters are not listed. As regards parasitic copepods, only ectoparasites whose adult stages are free swimming are considered herein.

Crustacea: Copepoda: orders

1 Major body constriction between pedigerous somite 5 and genital somite; male with one (typically the right) A1 geniculate; male P5 usually well-developed and prehensile .... **Calanoida**

1' Major body constriction between pedigerous somites 4 and 5; male with both A1 geniculate; male P5 small 2

2(1) P5 small, 1- to 3-semented or fused to somitae and reduced to one to three setae .... **Cyclopoida**

2' P5 broad, 1- to 2 segmented, basal thoracomere enlarged on medial margin .... **Harpacticoida**

Crustacea: Copepoda: Calanoida: families and subfamilies

1 P1 endopod 1-segmented 2
1' P1 endopod 2- or 3- segmented 3
2(1) Female P5 with reduced endopod (Fig. 6.12A) .... **Diaptomidae, Speodiaptominae**
2' Female P5 without endopod (Fig. 6.12B) .... **Temoridae**
3(1) P1 endopod 2-segmented 4
3' P1 endopod 3-segmented 6
4(3) P2–P4 endopods 2-segmented (Fig. 6.12C) .... **Acartiidae**
4' P2–P4 endopods 3-segmented (Fig. 6.12D) 5
5(4) P1 exopodal distal segment with two marginal, outer spines (Fig. 6.12E); right male A1 with three segments after the geniculation .... **Diaptomidae, Paradiaptominae**

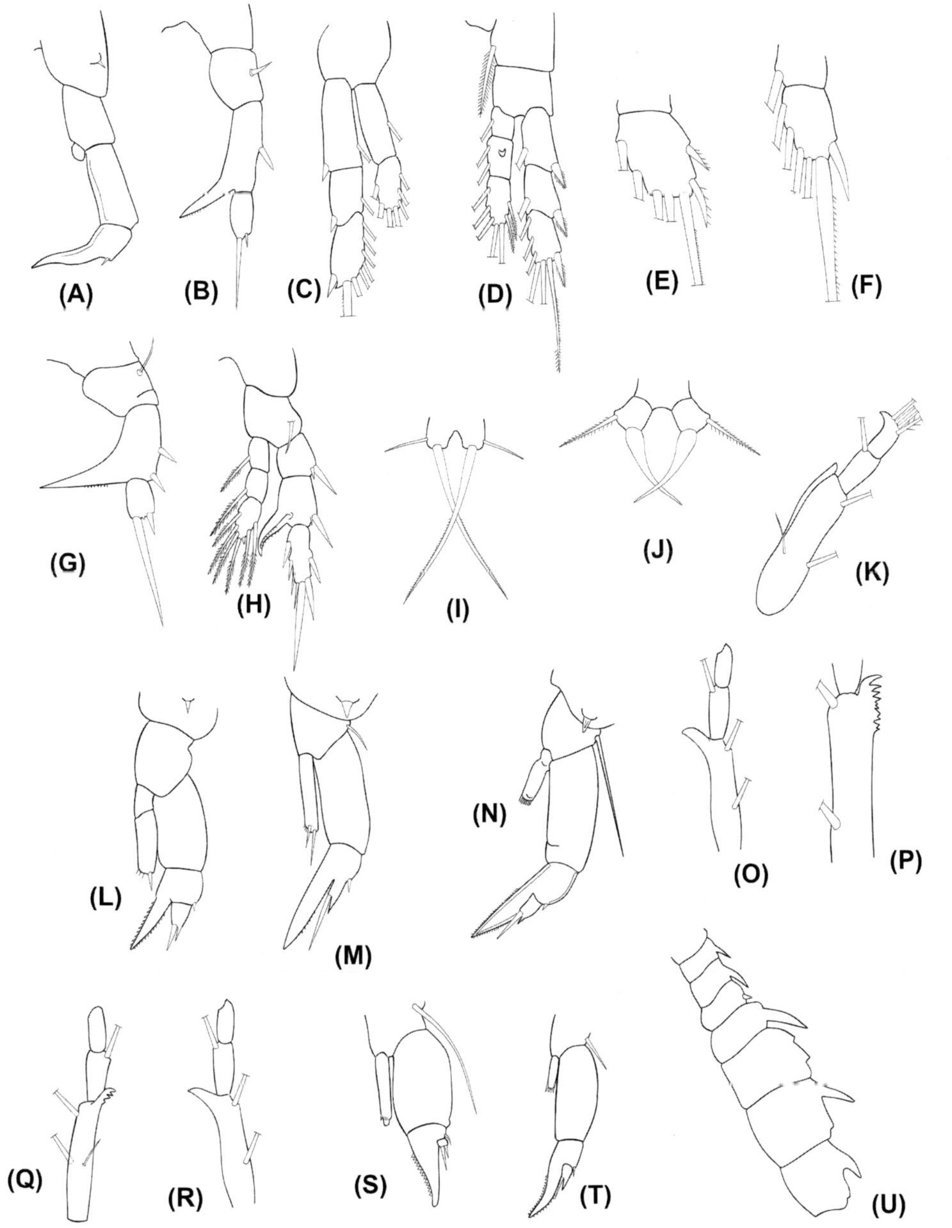

FIGURE 6.12 Representative structures of calanoid copepods: (A) Female P5 of *Troglodiaptomus sketi*; (B) female P5 of *Eurytemora grimmi*; (C) P2 of *Paracartia latisetosa*; (D) typical diaptomid P2; (E) P1 exopodal distal segment in Paradiaptominae; (F) P1 exopod in Diaptominae; (G) female P5 of *Calanipeda aquaedulcis*; (H) female P5 of *Boeckella triarticulata*; (I) female P5 of *P. latisetosa*; (J) female P5 of *Acartia clausii*; (K) last segments of male right A1 of *Acanthodiaptomus denticornis*; (L) female P5 of *Spelaeodiaptomus rouchi*; (M) female P5 of *Eudiaptomus vulgaris*; (N) female P5 of *Dussartius baeticus*; (O) last segments of male right A1 of *D. baeticus*; (P) antepenultimate segment of male right A1 of *Sinodiaptomus sarsi*; (Q) last segments of male right A1 of *Phyllodiaptomus blanci*; (R) last segments of male right A1 of *Copidodiaptomus numidicus*; (S) female P5 of *Neodiaptomus schmackeri*; (T) female P5 of *Copidodiaptomus steueri*; (U) segments 10–16 of the right male A1 of *Hemidiaptomus gurneyi*. [Not to scale]. *Figures redrawn, modified and corrected based on personal observations: from: (A) Petkvoski (1978), (C) Petkovski (1981), (B, E-G, I, K-R, U) Kiefer (1978), (D) Einsle (1996), (H) Alfonso & Belmonte (2008), (S) Alfonso et al. (2014), and (T) Stella (1982).*

5' P1 exopodal distal segment with one marginal, outer spine (Fig. 6.12F); right male A1 with four segments after the geniculation .... **Diaptomidae, Diaptominae**
6(3) Female P5 without endopod (Fig. 6.12G) .... **Pseudodiaptomidae**
6' Female P5 with a 3-segmented endopod (Fig. 6.12H) .... **Centropagidae**

Crustacea: Copepoda: Calanoida: Acartiidae: genera

1 Female A1 22-segmented .... *Pteriacartia*
1' Female A1 at most 19-segmented 2
2(1) Postero-lateral angles of prosome expanded in wing-like processes; female P5 with a plumose seta much shorter than terminal claw (Fig. 6.12I) .... *Paracartia*
2' Postero-lateral angles of prosome rounded or pointed, never expanded in wing-like processes; female P5 with a plumose seta usually longer than or subequal to terminal claw (Fig. 6.12J) .... *Acartia*
[The genus *Acartia* is divided by some authors into four or six subgenera, whose validity is doubtful]

Crustacea: Copepoda: Calanoida: Centropagidae: genera

Only the genus *Boeckella* occurs in the Mediterranean Basin, with the nonnative species *Boeckella triarticulata* Thomson, 1883.

Crustacea: Copepoda: Calanoida: Diaptomidae: Diaptominae: genera

1 Male A1 last segment with small claw-like process on its distal end (Fig. 6.12K) .... *Acanthodiaptomus*, one species: *A. denticornis* (Wierzejskii, 1887)
1' Male A1 last segment without such process 2
2(1') Female A1 and left male A1 with a single seta on the 11th segment 3
2' Female A1 and left male A1 with two setae on the 11th segment 10
3(2) Female P5 endopod with one or two apical spines (Fig. 6.12L and M) 4
3' Female P5 endopod with a row of hairs but without such spines (Fig. 6.12N) 5
4(3) Female P2 second exopod with an outer spine .... *Eudiaptomus*
4' Female P2 second exopod without such a spine .... *Spelaeodiaptomus*, one species, *S. rouchi* Dussart, 1970
[Stygobitic species]
5(3) Female prosome with a dorsal process on the penultimate pedigerous somite 6
5' Female prosome without such process 7
6(5) P2 second endopodal segment with a small dorsal lobus (Schmeil's lobus) (Fig. 6.12D). Right male A1 antepenultimate segment with a beak-like process (Fig. 6.12O) .... *Dussartius*, one species: *D. baeticus* Dussart, 1967
6' P2 second endopodal segment without such lobus. Right male A1 antepenultimate segment with a comb-like process (Fig. 6.12P) .... *Sinodiaptomus (Sinodiaptomus)*
7(5) Female genital somite 2-segmented .... *Thermodiaptomus*, one species: *T. galebi* (Barrois, 1891)
7' Female genital somite 3-segmented 8
8(7) Right male A1 antepenultimate segment with a comb-like process (Fig. 6.12Q) .... *Phyllodiaptomus*, one species: *P. blanci* (Guerne & Richard, 1896)
[Probably introduced]
8' Right male A1 antepenultimate segment with a beak-like process (Fig. 6.12R) 9
9(8) Female genital somite asymmetrical. Outer margins of the claws of female P5 second exopodal segment with few or no denticles (Fig. 6.12S) .... *Neodiaptomus*, one species: *N. schmackeri* (Poppe & Richard, 1892)
[Introduced]
9 Female genital somite symmetrical. Inner and outer margins of the claws of female P5 second exopodal segment with evenly distributed denticles (Fig. 6.12T) .... *Copidodiaptomus*
10(2) Female P5 endopod with one or two apical spines (Fig. 6.12M) 11
10' Female P5 endopod with a row of hairs but without such spines (Fig. 6.12N) 14
11(10) 16th segment of right male A1 with a tooth (Fig. 6.12U) 12
11' 16th segment of right male A1 without such a tooth 13

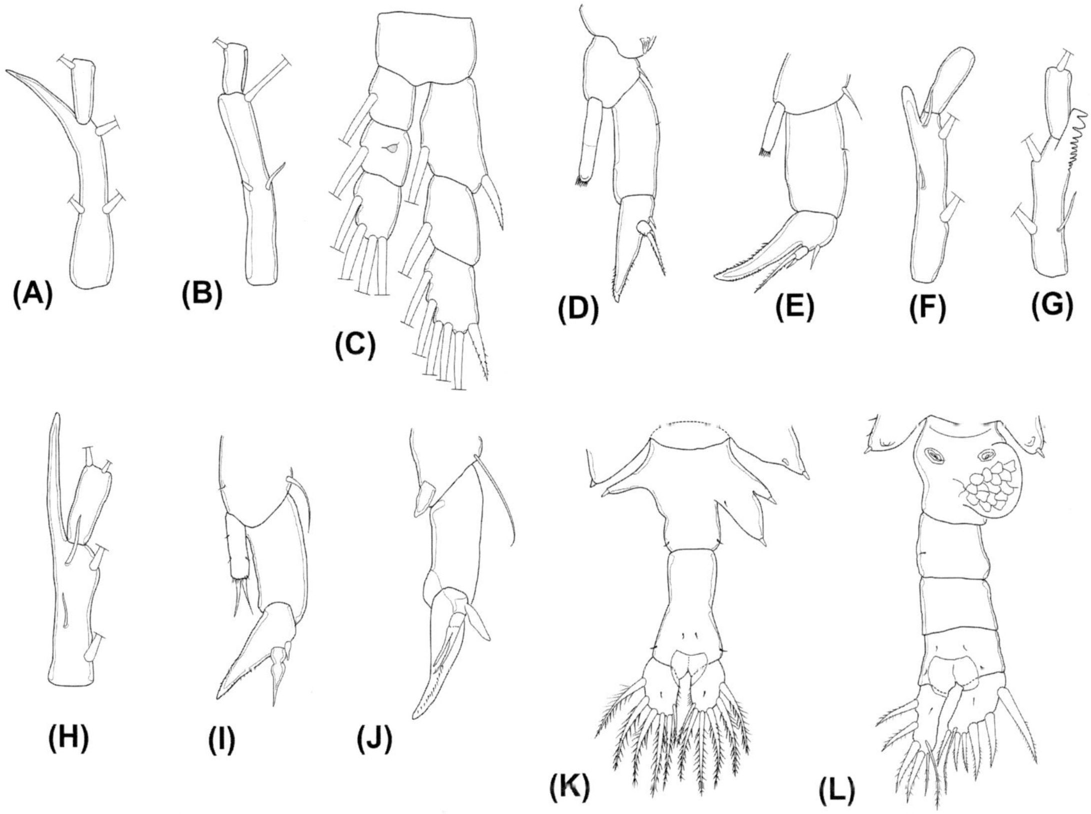

FIGURE 6.13 Representative structures of calanoid copepods: (A) Last segments of male right A1 of *Hemidiaptomus gurneyi*; (B) antepenultimate segment of right male A1 of *Hemidiaptomus roubaui*; (C) female P2 of *Stygodiaptomus kieferi*; (D) female P5 of *Mixodiaptomus kupelwieseri*; (E) female P5 of *Arctodiaptomus wierzejskii*; (F) antepenultimate segment of right male A1 of *Arctodiaptomus stephanidesi*; (G) last segments of right male A1 of *Arctodiaptomus wierzejskii*; (H) last segments of right male A1 of *Arctodiaptomus salinus*; (I) female P5 of *Metadiaptomus chevreuxi*; (J) female P5 of *Neolovenula alluaudi*; (K) female urosome of *Paradiaptomus similis*; (L) female urosome of *Metadiaptomus chevreuxi*. (Not to scale). *Figures redrawn, modified and corrected based on personal observations from (A, B) Stella (1982), and (D–L) Kiefer (1978).*

12(11) Antepenultimate segment of right male A1 with a well-developed tooth-shaped process (Fig. 6.13A) .... *Hemidiaptomus* (*Hemidiaptomus*)

12' Antepenultimate segment of right male A1 bordered by a hyaline lamella but without such process (Fig. 6.13B) .... *Hemidiaptomus* (*Occidodiaptomus*)

13(11) P2 second endopodal segment with a small dorsal lobus (Schmeil's lobus) (Fig. 6.12D) .... *Diaptomus* (*Diaptomus*)

13' P2 second endopodal segment without such lobus .... *Diaptomus* (*Chaetodiaptomus*)

14(10) P2–P4 endopods with four setae on their distal segments (Fig. 6.13C). Male right P5 with reduced or absent endopod .... *Stygodiaptomus*
[Stygobitic species]

14' P2–P4 endopods with seven setae on their distal segments (Fig. 6.12D). Male right P5 with well-developed endopodite 15

15(14) Female P5 endopod approximately of the same length of the first exopodal segment (Fig. 6.13D). Right male A1 antepenultimate segment with a hyaline lamella .... *Mixodiaptomus*

15' Female P5 endopod much shorter than the first exopodal segment (Fig. 6.13E). Right male A1 antepenultimate segment with a simple hyaline lamella or a distal, outer process 16

16(15) Right male A1 antepenultimate segment with a simple hyaline lamella .... *Arctodiaptomus* (*Mesodiaptomus*)

16' Right male A1 antepenultimate segment with a distal, outer process (Fig. 6.13F–H) 17

17(16) Right male A1 usually with a tooth on its 14th segment. Right male A1 with a well-developed stump-, claw-, hook-, or comb-like process on its antepenultimate segment (Fig. 6.13F and G) .... *Arctodiaptomus* (*Arctodiaptomus*)

17' Right male A1 without tooth on its 14th segment. Right male A1 with a slender, staff-like process on its antepenultimate segment (Fig. 6.13H) .... *Arctodiaptomus* (*Rhabdodiaptomus*)

Crustacea: Copepoda: Calanoida: Diaptomidae: Paradiaptominae: genera

1 Female P5 endopod with two terminal setae (Fig. 6.13I) 2
1' Female P5 endopod reduced and without terminal setae (Fig. 6.13J) .... *Neolovenula*, one species: *N. alluaudi* (Guerne & Richard, 1890)
2(1) Female urosome 2-segmented; female genital somite with asymmetrical pointed lobes (Fig. 6.13K) .... *Paradiaptomus*, one species: *P. similis* (Van Douwe, 1912)
2' Female urosome 3- or 4-segmented; female genital somite simple, or at most with an asymmetrical rounded lobe (Fig. 6.13L) .... *Metadiaptomus*

Crustacea: Copepoda: Calanoida: Diaptomidae: Speodiaptominae: genera

Only the stygobitic genus *Troglodiaptomus* occurs in the Mediterranean Basin, with a single species, *T. sketi* Petkvoski, 1978.

Crustacea: Copepoda: Calanoida: Pseudodiaptomidae: genera

1 Female A1 25-segmented .... *Calanipeda*, one species: *C. aquaedulcis* (Kritschagin, 1873)
1' Female A1 up to 23-segmented .... *Pseudodiaptomus*

Crustacea: Copepoda: Calanoida: Temoridae: genera

Only the genus *Eurytemora* occurs in the inland waters of the Mediterranean Basin.

Crustacea: Copepoda: Cyclopoida: families and subfamilies

1 Body of adult female transformed (Fig. 6.3), permanently attached to host, elongated, with no signs of segmentation; adult male cyclopiform, with long uniramous A1 .... **Lernaeidae**
[Ectoparasitic]
1' Body cyclopiform in both sexes 2
2(1) A2 prehensile and the most conspicuous appendage (Fig. 6.14A); body shape of female slightly transformed by swelling of prosome somites .... **Ergasilidae**
[Ectoparasitic]
2' A2 not prehensile; body shape of female not transformed 3
3(2) P5 exopod bearing four setae (Fig. 6.14B) .... **Halicyclopidae**
3' P5 exopod bearing two to three setae (Fig. 6.14C and D) 4
4(3) P5 exopod bearing three setae (Fig. 6.14C) .... **Cyclopidae, Eucyclopinae**
[Study area only; Stygocyclops bears two setae, Austriocyclops 1 seta, both living in Central Europe]
4' P5 exopod bearing two setae (Fig. 6.14D) .... **Cyclopidae, Cyclopinae**

Crustacea: Copepoda: Cyclopoida: Lernaeidae: genera

Lernaeidae are mesoparasites; females undergo a complete transformation in the modified parasitic adult stage, while copepodid stages and adult males maintain a cyclopiform body plan. The genus *Lernaea* is present in the Mediterranean Basin with the species *L. cyprinacea* (Linnaeus, 1758) (anchor worm), mainly parasitic on the gills of freshwater fishes. A second species, *L. phoxinacea* (Kollar in Krøyer, 1863) is present in Europe, but not known so far from the Mediterranean Basin, where its presence is likely.

Crustacea: Copepoda: Cyclopoida: Ergasilidae: genera

1 Cephalothorax triangular in dorsal view; coxobasis of prehensile A2 armed with a long spine; P1 basis with a stout tooth between exopod and endopod insertions .... *Neoergasilus*, one species: *N. japonicus* Harada, 1930
[usually on the fins on freshwater fishes; introduced]
1' Cephalothorax elongated in dorsal view; coxobasis of prehensile without spine; P1 basis without tooth between exopod and endopod insertions .... *Ergasilus*
[usually on the gills of freshwater fishes]

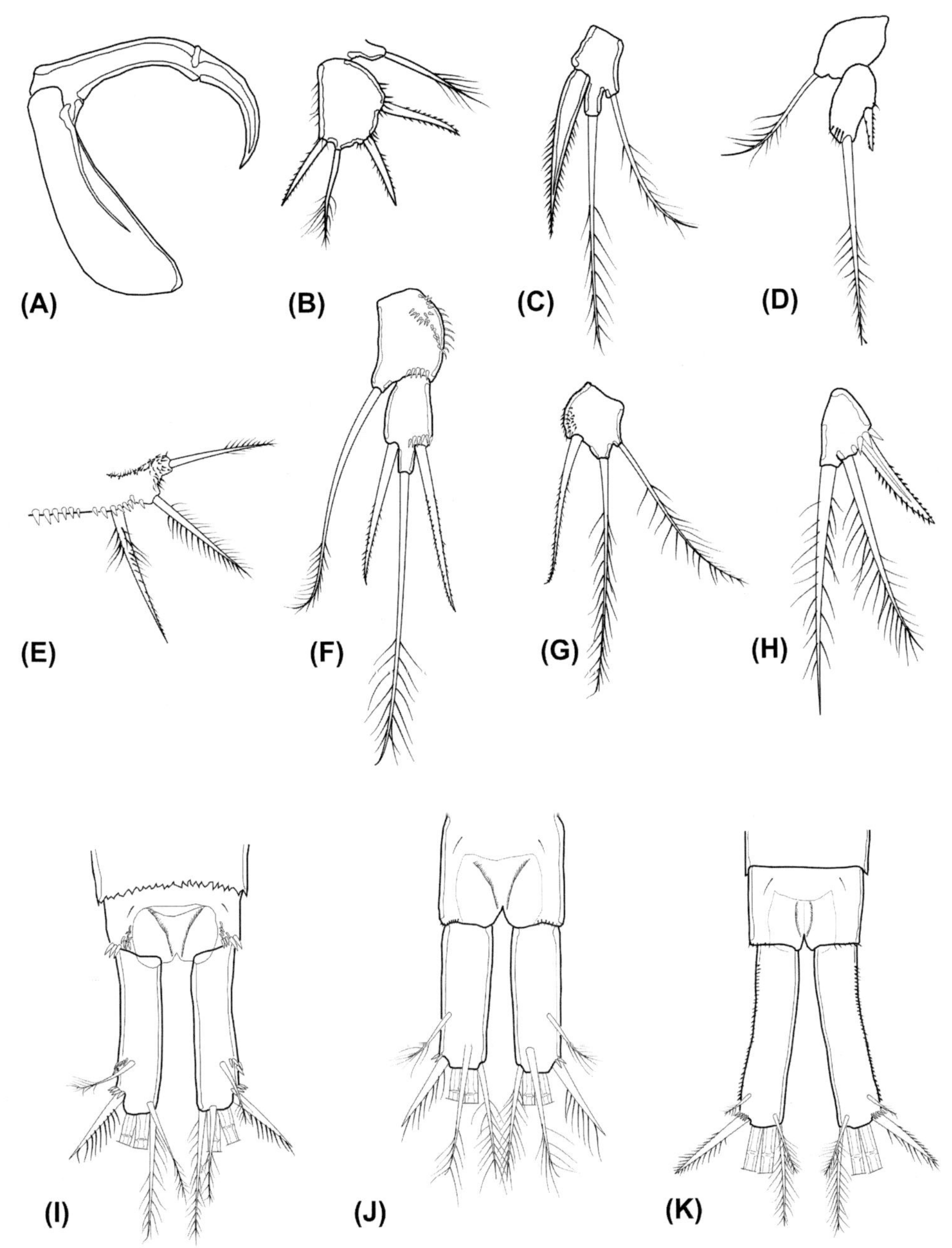

FIGURE 6.14 Representative structures of cyclopoid copepods: (A) Female A2 of *Neoergasilus japonicus*; (B) P5 exopod of *Halicyclops rotundipes*; (C) P5 of *Eucyclops serrulatus*; (D) P5 of *Cyclops divergens*; (E) P5 of *Ectocyclops phaleratus*; (F) P5 of *Macrocyclops albidus*; (G) P5 of *Tropocyclops prasinus*; (H) P5 of *Paracyclops imminutus*; (I) caudal rami of *Paracyclops imminutus*; (J) caudal rami of *Tropocyclops prasinus*; (K) Caudal rami of *Eucyclops serrulatus*. (Not to scale). *Figures redrawn, modified and corrected based on personal observations: from (A) Fryer (1978), (B) Dussart (1969), (C, D, F, G, and K) Kiefer (1978), and (H, I) Karaytuğ (1999).*

Crustacea: Copepoda: Cyclopoida: Halicyclopidae: genera

1. P5 3-segmented .... *Neocyclops*
1'. P5 2-segmented .... *Halicyclops*

Species mainly benthonic or interstitial in estuaries, lagoons, saltmarshes, anchialine caves, or freshwater interstitial.

Crustacea: Copepoda: Cyclopoida: Cyclopidae: Eucyclopinae: genera

1 P5 fused to somite, bearing two inner spines and one outer seta (Fig. 6.14E) .... *Ectocyclops*
1' P5 free, 1- or 2-segmented 2
2(1) P5 2-segmented, exopod bearing one inner spine, one medial seta, and one outer spine (Fig. 6.14F) .... *Macrocyclops*
2' P5 1-segmented (Fig. 6.14G and H) 3

3(2) A1 8- or 11-segmented in female, much shorter than cephalothorax; caudal rami of female with an oblique row of spinules on dorsal surface (Fig. 6.14I) .... *Paracyclops*
3' A1 11- or 12-segmented in female 4
4(3) A1 11-segmented in female, shorter than cephalothorax; caudal rami of female smooth .... *Ochridacyclops*, one species: *O. arndti* Kiefer, 1937 [Lakes Ochrid and Prespa]
4' A1 12-segmented in female 5
5(4) Anterior part of seminal receptacle tapering laterally into distorted horns; caudal rami short (2.5 times longer than wide), smooth (Fig. 6.14J) .... *Tropocyclops*, one species: *T. prasinus* (Fischer, 1860)
5' Anterior part of seminal receptacle not tapering laterally into horns; caudal rami usually with a longitudinal, outer row of spinules ("serra") of variable length (Fig. 6.14K), in some species reduced to few spinules close to the lateralmost terminal caudal seta .... *Eucyclops*
[The genus *Eucyclops* is divided by some authors in several subgenera, whose validity is doubtful]

Crustacea: Copepoda: Cyclopoida: Cyclopidae: Cyclopinae: genera

1 P5 fused to somite and reduced to three setae inserted on the posterior margin of pedigerous somite 5 or on a lobular expansion (Fig. 6.15A); P1–P4 endopod and exopod 2-segmented 2
1' P5 only partially fused with somite or 1–2-segmented; P1-P4 endopod and exopod 2- or 3-segmented 3
2(1) P5 fused to somite and reduced to three setae inserted on the posterior margin of pedigerous somite 5; P4 endopod 2 bearing three setae and a short outer spine; male P3 endopod 2 with a strongly modified medial spine (Fig. 6.15B) .... *Bryocyclops*, one species: *B. absalomi* Por, 1981
[Israel: Absalom (or Soreq) cave]

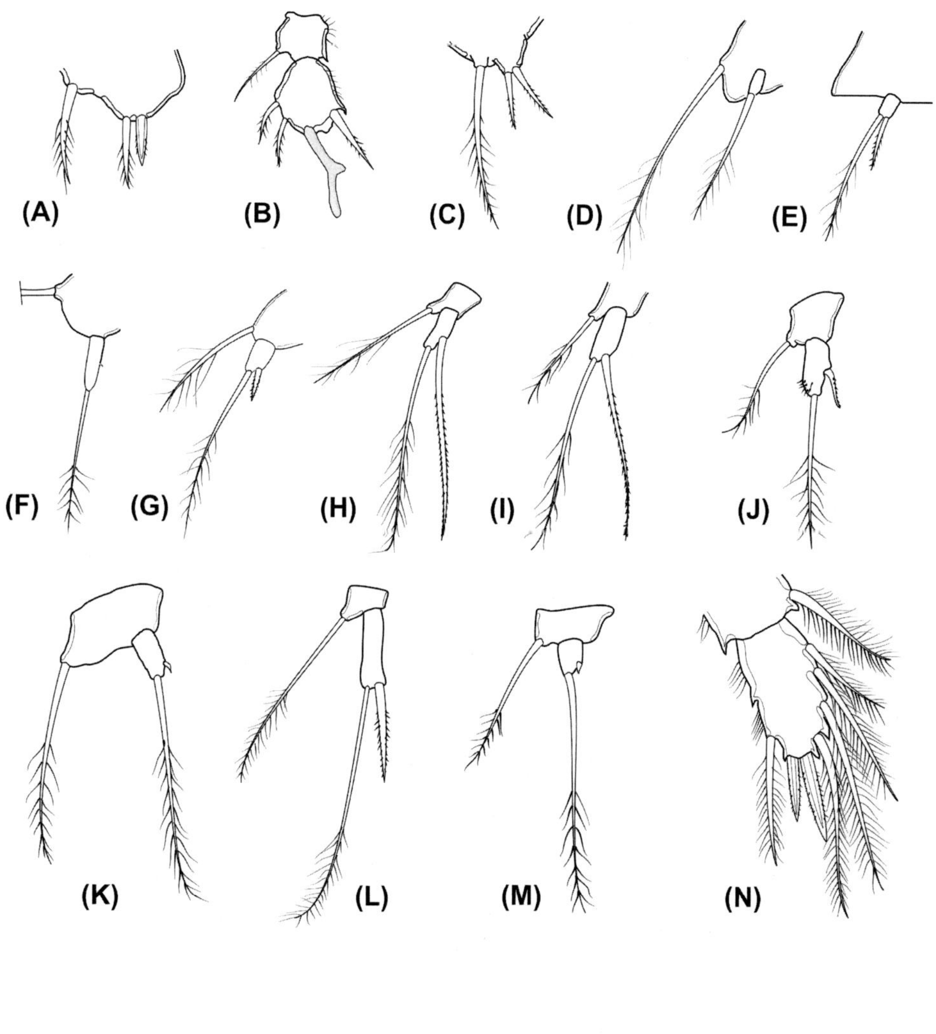

FIGURE 6.15 Representative structures of cyclopoid copepods: (A) P5 of *Hypocyclops kieferi*; (B) male P3 endopod 2 of *Bryocyclops absalomi*; (C) P5 of *Speocyclops italicus*; (D) P5 of *Cryptocyclops bicolor*; (E) P5 of *Graeteriella unisetigera*; (F) P5 of *Microcyclops varicans*; (G) P5 of *Metacyclops stammeri*; (H) P5 of *Mesocyclops leuckarti*; (I) P5 of *Thermocyclops crassus*; (J) P5 of *Cyclops strenuus*; (K) P5 of *Megacyclops viridis*; (L) P5 of *Diacyclops bicuspidatus*; (M) P5 of *Acanthocyclops gordani*; (N) P4 endopod 2 of *Monchenckocyclops mehmetadami*. [Not to scale]. *Figures redrawn, modified and corrected based on personal observations: from (A) Fiers (2012), (B) Por (1981), (D, H, I, L) Dussart (1969), (I, K) Einsle (1996), (N) Karaytuğ et al. (2018).*

2' P5 fused to somite and reduced to three setae inserted on a lobular expansion of pedigerous somite 5 (Fig. 6.15A); P4 endopod 2 bearing four setae and two apical spines of different length; male P3 endopod not sexually dimorphic .... *Hypocyclops*
[North Macedonia and Montenegro; stygobitic]
3(1) P5 partially fused to somite (Fig. 6.15C) and reduced to two contiguous segments (exopod bearing two short setae or spines and outer part of basis bearing a short seta); P4 endopod 2 bearing three setae and a long apical spine; male P3 endopod not sexually dimorphic .... *Speocyclops*
[Stygobitic]
3' P5 1- or 2-segmented 4
4(3) P5 1-segmented; basis fused to somite, exopod free (Fig. 6.15D and E) 5
4' P5 2-segmented; basis and exopod free 9
5(4) P5 outer seta absent on pedigerous somite 5 (Fig. 6.15E); P1–P4 endopods and exopods 2- or 3-segmented 6
5' P5 outer seta present on pedigerous somite 5 (Fig. 6.15D); P1–P4 endopods and exopods 2-segmented 7
6(5) P1–P4 2-segmented in female; male P4 exopod 3-segmented .... *Graeteriella* (*Graeteriella*)
[Stygobitic]
6' P1–P2 endopod and P1 exopod 3-segmented, P3 exopod and P3–P4 endopod and exopod 3-segmented .... *Graeteriella* (*Paragraeteriella*)
[Stygobitic]
7(5) P2–P4 basis with row of spinules on medial expansion; P4 endopod 2 with two apical spines, the outer one tiny, less than 1/6 length of inner apical spine; P5 exopod bearing a long distal seta and a tiny sublateral spine very difficult to discern (Fig. 6.15D) .... *Cryptocyclops*
7' P2–P4 basis with fine hairs on medial expansion 8
8(7) P4 endopod 2 with two well-developed apical spines of slightly different lengths; P5 exopod bearing a long distal seta and a tiny spine inserted at mid-segment length (Fig. 6.15F), missing in some species .... *Microcyclops*
8' P4 endopod 2 with only 1 well-developed apical spine; P5 exopod bearing a distal seta accompanied by a stout inner apical spine (Fig. 6.15G) .... *Metacyclops*
9(4) P5 exopod bearing an apical seta and an inner spiniform seta of similar length 10
9' P5 exopod bearing an apical seta and a short spine, of different length 12
10(9) P5 exopod bearing an apical seta and an inner spiniform seta inserted about mid-length of inner margin of segment (Fig. 6.15H); A1 17-segmented in female .... *Mesocyclops*
10' P5 exopod bearing an apical seta and a subapical spiniform seta of comparable length (Fig. 6.15I) 11
11(10) P1–P4 rami 3-segmented; A1 17-segmented in female .... *Thermocyclops*
11' P1–P4 rami 2-segmented; A1 11-segmented in female .... *Reidcyclops*, one species: *R. trajani* Reid & Strayer, 1994
[North Macedonia; stygobitic]
12(9) P5 exopod bearing an apical seta and a spine inserted about mid-length of inner margin of segment (Fig. 6.15J and K) 13
12' P5 exopod bearing an apical seta and a subapical or latero-distal spine (Fig. 6.15M) 14
13(12) P5 exopod inner spine stout and long, reaching or exceeding the distal margin of segment (Fig. 6.15J) .... *Cyclops*
13' P5 exopod inner spine tiny, sometimes fused to segment (Fig. 6.15K); P5 baseoendopod large, expanded towards the exopodal seta .... *Megacyclops*
14(12) P5 exopodal subapical or latero-distal spine short and stout, variable in length (Fig. 6.15M) 15
14' P5 exopodal subapical spine long (Fig. 6.15L), variable in length in stygobitic species 16
15(14) P1–P4 endopod 2-segmented, exopod 3-segmented; P1–P4 exopod with inner seta; P4 endopod 2 elongated, anterior side without spinule rows (Fig. 6.15N) .... *Monchenkocyclops*
[stygobitic; questionable genus]
15' P1–P4 differing in segmentation pattern; if with the same segmentation pattern, P1–P4 exopod without inner seta and with a spinular row on anterior side of P4 endopod 2 .... *Acanthocyclops*
16(14) A1 very long, reaching the genital double somite; P1–P4 rami, setae and spines elongated; seminal receptacle hammer-shaped .... *Kieferella*, a single species: *K. delamarei* Lescher-Moutoué, 1971
[Southern France; stygobitic]
16' A1 shorter, not reaching the posterior margin of prosome; seminal receptacle variable, but never hammer-shaped .... *Diacyclops*

Crustacea: Copepoda: Harpacticoida: families

1 Thoracic segment bearing P1 partially or totally fused to cephalosome; maxilliped prehensile or absent, never leaflike (Fig. 6.16A) 2
1' Thoracic segment bearing P1 not fused to the cephalosome; maxilliped flat, leaflike (Fig. 6.16B) .... **Phyllognathopodidae**
2(1) Body fusiform (spindle-shaped) or cylindrical; P5 usually 2-segmented (Fig. 6.16C); when fused into one plate, 2 lobes are still distinguishable (Fig. 6.16D) 3
2' Body elongated, vermiform, small body size (300–350 micron); P1 exopod 2 without spine on outer corner, P5 reduced to an unsegmented plate with setae and spines on its margin (Fig. 6.16E) .... **Parastenocarididae**
3(2) Body fusiform, caudal rami conical .... **Ectinosomatidae**
3' Body of another shape 4
4(3) P1 exopod 2-segmented; P1 endopod prehensile (Fig. 6.16F), segment 1 much longer than exopod; last segment ending in 1 stout claw .... **Laophontidae**
4' P1 exopod 3-segmented; P1 endopod prehensile (segment 1 longer than segments 2 and 3 combined), or not prehensile; last segment with terminal setae/spine 5
5 (4) P1 endopod 1 with inner seta .... **Arenopontiidae**
5' P1 endopod 1 without inner seta 6
6(5) A2 exopod absent. Maxilliped absent. P1 endopod not prehensile. Body cylindrical and elongate .... **Darcythompsoniidae**
6' A2 exopod present. Maxilliped well-developed, prehensile 7
7(6) P1 basis with inner spine, often transformed in male (Fig. 6.16G) 8
7' P1 basis without inner ornamentation 9
8(7) P2, last segment of endopod with long outer or apical spiniform process in male (Fig. 6.16H); female with paired lateral egg sacs .... **Miraciidae**
8' P2, last segment of endopod without apical processes in male (P2 endopod not sexually dimorphic); female with single egg sac .... **Ameiridae**
9(7) P5: a single a single fused plate bearing five setae in male (Fig. 6.16I); P5 not fused to baseoendopod in female (Fig. 6.16J). A1 very short: 4-segmented in female (Fig. 6.16K), 5-segmented in male .... **Nannopodidae**
9' P5 not fused to baseoendopod in both sexes (Fig. 6.16L and M). A1 long: 7–8 segmented in both sexes (Fig. 6.16N) .... **Canthocamptidae**

Crustacea: Copepoda: Harpacticoida: Ectinosomatidae: genera

1 P1–P4 endopods 2-segmented .... *Pseudectinosoma*
1' P1 endopod 2-segmented, P2–P4 endopods 3-segmented 2
2(1) Maxilla prehensile, with syncoxa and allobasis forming right angle (Fig. 6.16O); P5 exopod poorly developed, short, fused to baseoendopod in female (Fig. 6.16P), and distinct in male (Fig. 6.16Q), with three marginal and no surface setae; body very small (<300 μm) .... *Sigmatidium*, one species: *S. chappuisi* (Schäfer, 1951) [Stygobitic]
2 These characteristics not combined 3
3(2) Integument of somites with distinctive subrectangular pores; maxilla not prehensile, straight or with at most a slight angle between syncoxa and allobasis; P5 exopod well-developed, partly fused to baseoendopod in female (Fig. 6.16R) and distinct in male (Fig. 6.16S), with four marginal setae .... *Ectinosoma*
3' Integument of somites without distinctive subrectangular pores; maxilla not prehensile, with at most a slight angle between syncoxa and allobasis; P5 exopod well-developed, partly fused to baseoendopod in female (Fig. 6.16T) and distinct in male (Fig. 6.16U), with three marginal setae and one seta on anterior surface .... *Halectinosoma*

Crustacea: Copepoda: Harpacticoida: Laophontidae: genera

The family is mainly marine, one genus and one species recorded in freshwater of the Mediterranean Basin: *Onychocamptus mohammed* (Blanchard & Richard, 1891).

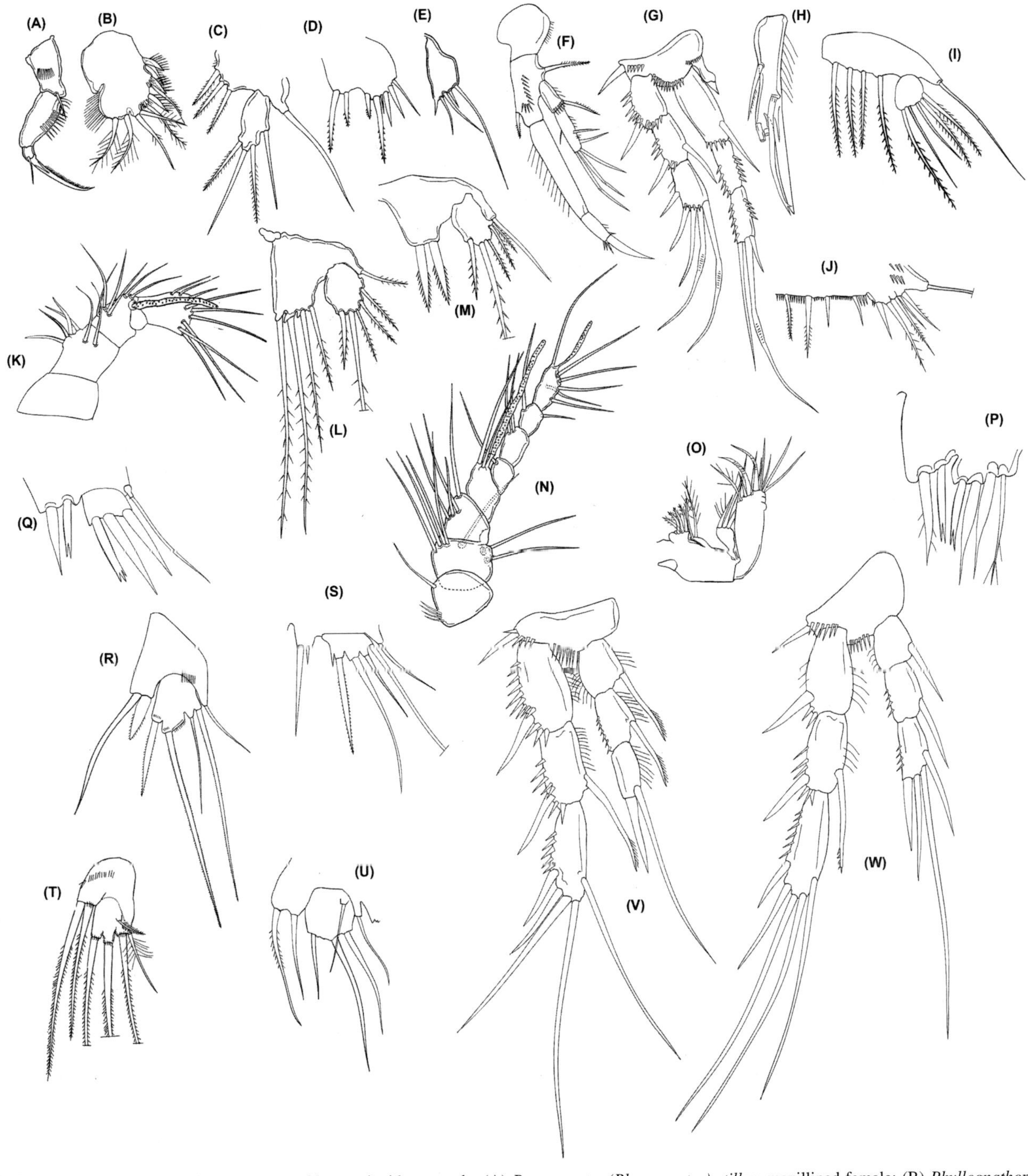

FIGURE 6.16 Representative structures of harpacticoid copepods: (A) *Bryocamptus (Rheocamptus) stillae*, maxilliped female; (B) *Phyllognathopus viguieri*, maxilliped female; (C) *Nitocrellopsis hellenica*, P5 female; (D) *Ceuthonectes pescei*, P5 male; (E) *Stammericaris vincentimariae*, P5 male; (F) *Onychocamptus mohammed*, P1 female; (G) *Parapseudoleptomesochra syriaca*, male P1; (H) *Schizopera samchunensis*, endopod P2 male; (I) *Nannopus palustris*, P5 female; (J) *N. palustris*, P5 male; (K) *N. palustris*, A1 female; (L) *Bryocamptus (R.) stillae*, P5 female; (M) *Bryocamptus (R.) stillae*, P5 male; (N) *B. (R.) stillae*, A1 female; (O) *Sigmatidium kunzi*, maxilla female; (P) *S. kunzi*, P5 female; (Q) *S. kunzi*, P5 male; (R) *Ectinosoma melaniceps*, P5 female; (S) *Ectinosoma dentatum*, P3 male; (T) *Halectinosoma parejae*, P5 female; (U) *Halectinosoma kliei*, P5 male; (V) *P. syriaca*, P2 male; (W) *P. syriaca*, P3 male. [Not to scale]. *Figures redrawn, modified and corrected based on personal observations from: (A, L, M, N) Cottarelli et al. (2012), (B) Galassi et al. (2011), (C) Cottarelli & Forniz (1993), (D) Cottarelli & Saporito (1985), (E) Bruno et al. (2020), (F, J) Gurney (1932), (G, V, W) Cottarelli, Puccetti & Saporito (1985), (H) Karaytuğ & Sak (2005), (I, K) Damian-Georgescu (1970), (O, P, Q, R) Huys et al. (1996), (S) Wells & Rao (1987), (T) Sciberras et al. (2018), and (U) Clément & Moore (2007).*

Crustacea: Copepoda: Harpacticoida: Arenopontiidae: genera

The family is mainly marine, only one genus (*Neoleptastacus*) recorded in the Mediterranean Basin (Italy), with two species, *N. phreaticus* (Cottarelli, Bruno & Venanzetti, 1994) and *N. speluncae* (Cottarelli, Bruno & Venanzetti, 1994).

Crustacea: Copepoda: Harpacticoida: Darcythompsoniidae: genera

The family is mainly marine, one genus and one species recorded in fresh and brackish waters of the Mediterranean Basin: *Leptocaris brevicornis* (Van Douwe, 1904).

Crustacea: Copepoda: Harpacticoida: Laophontidae: genera

The family is mainly marine, one genus (*Schizopera*) with two species recorded in freshwater of the Mediterranean Basin (Israel): *S. jugurtha* (Blanchard & Richard, 1891) and *S. samchunensis* (Karaytug & Sak, 2005).

Crustacea: Copepoda: Harpacticoida: Nannopodidae: genera

The family is mainly marine, one genus with one species recorded in fresh and brackish waters of the Mediterranean Basin: *Nannopus palustris* (Brady, 1880).

Crustacea: Copepoda: Harpacticoida: Ameiridae: genera and subgenera

1 P2–P4 endopods all 3-segmented 2
1' P2–P4 endopods not all 3-segmented 4
2(1) P3–P4 exopod 3 with two outer spines (Fig. 6.16V and W) .... *Parapseudoleptomesochra* [Stygobitic species]
2' P3–P4 exopod 3 with three outer spines 3
3(2) P1 endopod 3-segmented (Fig. 6.17A) .... *Nitokra*
3' P1 endopod 2-segmented .... *Psyllocamptus*, one species: *P. minutus* (Sars G.O., 1911)
4(1) P4 endopod 1-segmented 5
4' P4 endopod 2- or 3-segmented 6
5(4) P2 and P3 endopods 1- or 2-segmented, P2–P4 armature formula of the ultimate endopodal segment: 1.1.1 (Fig. 6.17B) .... *Stygonitocrella* (*Stygonitocrella*), one species: *S.* (*Stygonitocrella*) *guadalfensis* (Rouch, 1985) [France, stygobitic]
5' P2 and P3 endopods 1- or 2-segmented, P2–P4 armature formula of the last endopodal segment: 1.2.1 (Fig. 6.17C) .... *Megastygonitocrella*, one species: *M. petkovskii* (Pesce, 1985) [Greece, stygobitic]
6(4) P4 endopod 2-segmented 7
6' P4 endopod 3-segmented (Fig. 6.17D) .... *Praeleptomesochra*, one species: *Praeleptomesochra phreatica* (Pesce, 1981) [Morocco, Portugal]
7(6) P2–P3 endopods 3-segmented (Fig. 6.17E) .... *Nitocrellopsis* [Stygobitic species]
7' P2–P3 endopods 2-segmented (Fig. 6.17F) .... *Nitocrella* [Stygobitic species]

Crustacea: Copepoda: Harpacticoida: Canthocamptidae: genera and subgenera

1 P1 exopod 3-segmented 2
1' P1 exopod 2-segmented (Fig. 6.17G) .... *Maraenobiotus*
2 P5 exopod fused to baseoendopod 3
2' P5 exopod separated from baseoendopod 4
3 (2) Female: A1 7–8-segmented, P5 male exopod fused to baseoendopod to form two lobes, inner one with two setae, outer lobe with six setae (Pl. 6.16 D) .... *Ceuthonectes* [Stygobitic species]

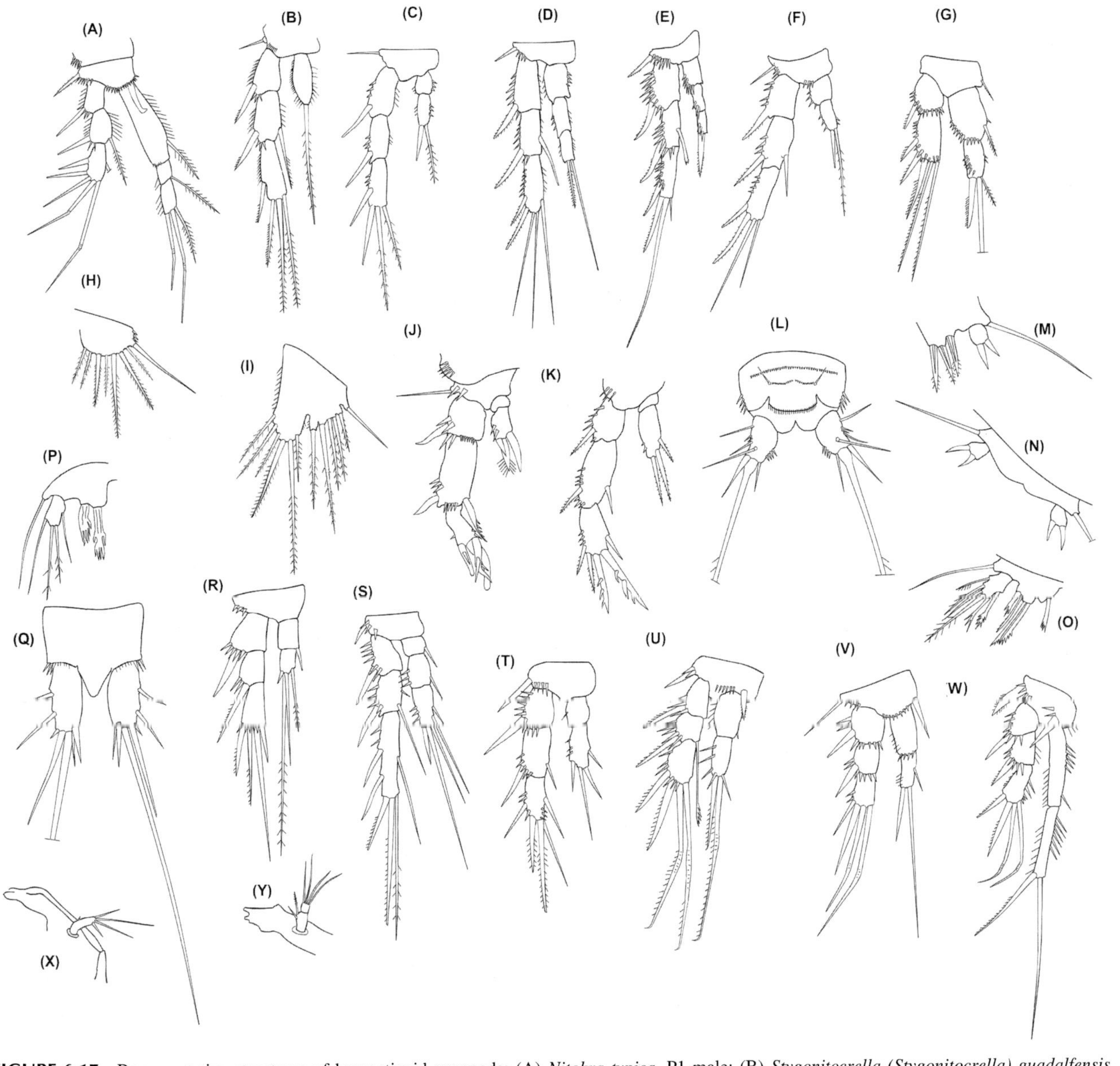

FIGURE 6.17 Representative structures of harpacticoid copepods: (A) *Nitokra typica*, P1 male; (B) *Stygonitocrella (Stygonitocrella) guadalfensis*, P3 female; (C) *Megastygonitocrella petkovski*, P3 female; (D) *Praeleptomesochra phreatica*, P4 female; (E) *Nitocrellopsis hellenica*, P2 female; (F) *Nitocrella achaie*, P2 female; (G) *Maraenobiotus vejdovski*, P1 female; (H) *Cletocamptus retrogressus*, P5 male; (I) *C. retrogressus*, P5 female; (J) *Paramorariopsis brigitae*, P4 male; (K) *Elaphoidella karamani*, P4 male; (L) *P. brigitae*, caudal ramus female, dorsal view; (M) *P. brigitae*, P5 female; (N) *P. brigitae*, P5 male (both P5s); (O) *Morariopsis dumonti*, P5 female; (P) *M. dumonti*, P5 male; (Q) *Moraria poppei*, caudal ramus female, dorsal view; (R) *M. poppei*, P2 female; (S) *Bryocamptus (Rheocamptus), zschokkei*, P3 female; (T) *Bryocamptus (Arcticocamptus) rhaeticus*, P2 female; (U) *Bryocamptus (Rheocamptus) stillae*, P1 male; (V) *Epactophanes richardi*, P1 female; (W) *Pesceus schmeili schmeili*, P1 female; (X) *Bryocamptus (B.) minutus*, mandible female; (Y) *Attheyella (Neomrazekiella) osmana*, mandible female. (Not to scale). *Figures redrawn, modified and corrected based on personal observations: from (A, H, I, Q, R, V, W) Dussart (1967), (B) Rouch (1985), (C) Pesce (1985), (D) Pesce (1981a), (E) Cottarelli & Forniz (1993), (F, K) Pesce (1981b), (G) Pesce et al. (1994), (J, L, M, N) Brancelj (2011), (O, P) Brancelj (2000), (S, T) Caramujo & Boavida (2009), (U) Cottarelli et al. (2012), (X) Gurney (1932), and (Y) Por (1983).*

3' Female: A1 6-segmented; P5 exopod (both sexes) fused to baseoendopod to form two lobes, inner lobe with six (female) or three (male) setae, outer lobe with six (female) or five (male) setae (Fig. 6.17H and I) .... *Cletocamptus*
4(2) Rostrum small, not extending past A1 segment 1 5
4' Rostrum large, extending past A1 segment 1, P1 endopod 2 -or 3-segmented .... *Mesochra*
5(4) P5 baseoendopod with at least 1 seta/spine in male or, if without setae/spines, P4 of male exopod 3 smaller than exopod 2 and with strong and antler-like transformed spines 6

5' baseoendopod without setae/spines in male, P4 of male exopod 3 approximately the same size of exopod 2 and with long and antler-like transformed spines (Fig. 6.17K) .... *Elaphoidella*
6(5) P1 endopod 2-segmented 7
6' P1 endopod 3-segmented 14
7(6) P1 endopod reaches from about midlength to slightly past end of exopod 3 8
7' P1 endopod much longer or shorter than exopod 13
8(7) P2–P3 endopods 1-segmented in female. Caudal branches distinctly divergent (Fig. 6.17L) 9
8' P2–P4 endopods 2- or 3-segmented in female 10
9(8) P4 endopod 2-segmented in female. P4, spines on exopod 3 antler-like transformed in male (Fig. 6.17J). P5: baseoendopod with three (female) or 0 spines, exopod short, with two spines (Fig. 6.17M and N) .... *Paramorariopsis* [Stygobitic species]
9' P4 endopod 2-segmented in female. P4, spines on exopod 3 not transformed in male. P5: baseoendopod with 5–6 (female) or 2 spines, exopod short, with 4–6 (female) or 4/5 (male) setae/ spines (Fig. 6.17O and P) .... *Morariopsis* [Stygobitic species]
10(8) Caudal rami straight or slightly divergent (Fig. 6.17Q). P2–P4 endopod 2-segmented in female (Fig. 6.17R); body cylindrical, elongated .... *Moraria*
10' These characteristics not combined 11
11(10) P3 exopod 3 with three outer spines (Fig. 6.17S) .... *Bryocamptus* (*Rheocamptus*), partim [*B.* (*Rheocamptus*) *zschokkei* group]
11' P3 exopod 3 with two outer spines 12
12(11) P1 exopod 2 inner seta small or missing (Fig. 6.17T) .... *Bryocamptus* (*Arcticocamptus*)
12 P1 exopod 2 inner seta long (Fig. 6.17U) .... *Bryocamptus* (*Rheocamptus*), partim [B. (*Rheocamptus*) *pygmaeus* group]
13(7) P1 endopod does not reach end of exopod 2 (Fig. 6.17V) .... *Epactophanes*, one species: *E. richardi* (Mrázek, 1893)
13' P1 endopod longer than exopod by half of endopod segment 3 or more (Fig. 6.17W) .... *Pesceus*, one species: *P. schmeili* (Mrázek, 1893)
14(6) P1 exopod 2 with inner seta 15
14' P1 exopod 2 without inner seta 17
15(14) Mandible without endopod (1-segmented) (Fig. 6.17X) .... *Bryocamptus* (*Bryocamptus*), partim [B. (*Bryocamptus*) minutus group]
15' Mandible with endopod (2-segmented) (Fig. 6.17Y) 16
16(15) P2 endopod 2 with five setae (Fig. 6.18A) .... *Attheyella* (*Attheyella*)
16' P2 endopod 2 with six setae (Fig. 6.18B) .... *Attheyella* (*Neomrazekiella*)
17(14) P5 baseoendopod next outermost seta not reduced, usually equal or longer than outermost seta in female (Fig. 6.18C); P4 endopod: outer corner of last segment with distinct spine or seta in male (Fig. 6.18D) 18
17' P5 baseoendopod next outermost seta tiny in female (Fig. 6.18E); P4 endopod: outer corner of last segment with spinous process in male (Fig. 6.18F) .... *Canthocamptus*
18(17) P5, baseoendopod with five to six (female) and two (male) setae, exopod with five (female) and six (male) setae in both sexes (Fig. 6.18G and H) .... *Bryocamptus* (*Echinocamptus*)
18' P5 baseoendopod with five (female) and one (male) setae, exopod with four setae in both sexes (Fig. 6.18C and I) .... *Pilocamptus* one species: *P. pilosus* (Douwe, 1910)

Crustacea: Copepoda: Harpacticoida: Parastenocarididae: genera

The identification of the genera and species of the family is usually based on males, as the taxonomy of females is very conservative. All Mediterranean species are stygobiotic, collected from hyporheic and psammic interstitial habitats, springs, cave, and epikarst. The validity of several genera are presently debated and under revision.

1 A1 coiled in male (Fig. 6.18J); P4 basis of male with large and slender hyaline process on anterior surface, between exopod and endopod, and no other chitinous structures on basis (Fig. 6.18K); genital field of female roundish and as broad as high .... *Proserpinicaris*

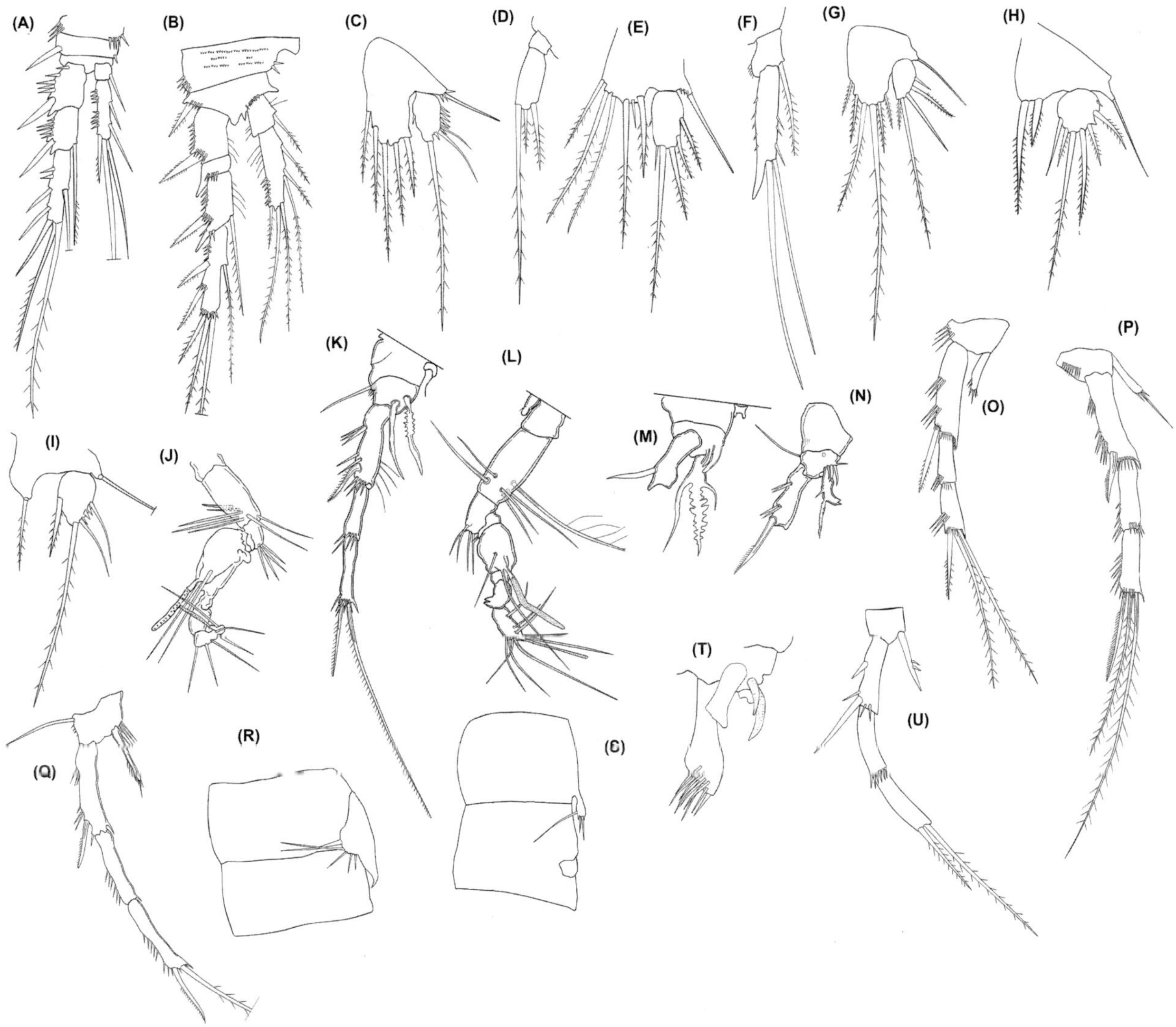

FIGURE 6.18 Representative structures of harpacticoid copepods: (A) *Attheyella (Attheyella) crassa*, P2 female; (B) *Attheyella (Neomrazekiella) osmana*, P2 female; (C) *Pilocamptus pilosus*, P5 female; (D) *Bryocamptus (Echinocamptus) echinatus*, P4 male; (E) *Canthocamptus staphylinus*, P5 female; (F) C. *staphylinus*, P4 endopod male; (G) *Bryocamptus (Echinocamptus) echinatus*, P5 female; (H) *B. (E.) echinatus*, P5 male; (I) *Pilocamptus pilosus*, P5 male; (J) *Proserpinicaris proserpina*, A1 male; (K) *Proserpinicaris specincola*, P4 male; (L) *Cottarellicaris sanctiangeli*, A1 male; (M) *C. sanctiangeli*, P4 endopod male; (N) *Stammericaris vincentimariae*, P4 endopod male; (O) *Simplicaris lethaea*, P2 female; (P) *Kinnecaris xanthi*, P2 female; (Q) *K. xanthi*, P4 male; (R) *Kinnecaris draconis*, first and second urosomite and P5 male, lateral view; (S) *Stammericaris vincentimariae*, first and second urosomite and P5 male, lateral view; (T) *Parastenocaris brevipes*, P4 endopod male; (U) *Italicocaris italica*, P4 male. [Not to scale]. *Figures redrawn, modified and corrected based on personal observations: from (A) Caramujo & Boavida (2009), (B) Por (1983), (C) Apostolov & Pesce (1989), (D, G, H) Kiefer (1959), (E, F) Dussart (1967), (I) Kiefer (1931), (J) Bruno & Cottarelli (1998), (K, L, M, N, S) Bruno et al. (2020), (O) Galassi & De Laurentiis (2004), (P, Q, R) Bruno & Cottarelli (2015), (T) Reid (1995), and (U) Chappuis (1953).*

1' A1 of pocket-knife shape in male (Fig. 6.18L); P4 basis of male with spinules (can be modified in claw- or tooth-like processes) medially of endopod, or at its base (Fig. 6.18M and N); genital field of female rectangular and band-like 2
2(1) P5 missing in both sexes, P1–P4 exopod 1 distinctly longer than exopod 2 or exopod 3, or as long as exopod 2 and exopod 3 combined (Fig. 6.18O) .... *Simplicaris*
2' P5 present in both sexes, P1–P4 exopod 1 as long as, or slightly longer than exopod 2 or exopod 3, or exopod 2 and exopod 3 combined (Fig. 6.18P) 3
3(2) P2 and P4 exopod 2 and 3 (both sexes) and P3 exopod 2 in female with longitudinal rows of spinules along outer margin (Fig. 6.18P and Q); P4 basis in male with a row of spinules above insertion of endopod, endopod not

modified, of cylindrical shape (Fig. 6.18Q); P5 formed by a single plate, enlarged, elongated, and trapezoid in both sexes, not adherent to the urosome but projecting outward to form a pronounced angle with the genital somite (Fig. 6.18R) .... *Kinnecaris*

3' P2 and P4 exopod 2 and 3 (both sexes) and P3 exopod 2 in female without longitudinal rows of spinules along outer margin; P4 basis in male with complex ornamentation, endopod often modified (Fig. 6.18M and N); P5 formed by a single plate, small, rectangular or trapezoidal, adherent to urosome in both sexes (Fig. 6.18S) 4

4(3) P4 basis of male with complex inner ornamentation: two large, claw-like processes and a larger hyaline claw, in some cases of transformed shape; endopod hyaline, with in most cases subterminal setae and/or spinules (Fig. 6.18T); P4 endopod of female long and spinulate; caudal rami tapering, with the dorsal seta and group of lateral setae inserted at midlength of rami in both sexes .... *Parastenocaris*

4' P4 basis of male with inner row of one to four curved spinules, or with no ornamentation; P4 endopod of female either spinulate or ending in a pinnate short seta; caudal rami cylindrical, with the dorsal seta and group of lateral setae inserted at end of rami in both sexes 5

5(4) P4 basis of male with inner row of one to four curved spinules, P4 endopod of male a curved plate with a pointed inner tip carrying two outgrowths at its outer border 6

5' P4 basis of male without ornamentation, P4 endopod of male a pointed, pinnate segment, shorter than exopod 1 (Fig. 6.18U); P4 endopod of female as in male, but almost as long as exopod 1 and exopod 2 .... *Italicocaris*, one species: *I. italica* (Chappuis, 1953) [Italy; genus in urgent need of revision]

6(5) P3 exopod of male with short, quandrangular apophysis; P4 of male: basis with inner row of one to four curved spinules decreasing in size laterally, endopod e a curved plate with a pointed inner tip carrying two outgrowths at its outer border, the distal one being an elongate lamella with undulating (crenulate) margins (Fig. 6.18M); P3 endopod of female from slightly shorter to slightly longer than P3 exopod 1 .... *Cottarellicaris*

6' P3 exopod of male with long, pointed apophysis; P4 of male: basis with an inner row of two to four curved spinules decreasing in size laterally, endopod a curved plate with a pointed inner tip carrying two outgrowths at its outer border, the distal one being a feathered or plain seta (Fig. 6.18N); P3 endopod of female half as long as or shorter than P3 exopod 1 .... *Stammericaris*

Crustacea: Ichthyostraca: order, family, and genus

The single subclass Branchiura, including the order Arguloida, comprises only the genus *Argulus* (family Argulidae) in the Mediterranean Basin; thus no identification key to argulid genera is provided herein. *Argulus* species can be distinguished mostly based on the shape of the carapace and its lobes, the size of the abdomen, the number of sclerites in the sucker rods, and the morphology of the second maxilla.

References

Alfonso, G. & G. Belmonte. 2008. Expanding distribution of *Boeckella triarticulata* (Thomson, 1883) (Copepoda: Calanoida: Centropagidae) in Southern Italy. Aquatic Invasions 3: 247–251.

Alfonso, G., Russo, R. & G. Belmonte. 2014. First record of the Asian diaptomid *Neodiaptomus schmackeri* (Poppe & Richard, 1892) (Crustacea: Copepoda: Calanoida) in Europe. Journal of Limnology 73: 584–592.

Apostolov, A. & G.L. Pesce. 1989. Copepodes harpacticoides stygobies de Bulgarie. Rivista di Idrobiologia 28:113–149.

Bayly, I.A.E. 1992. The non-marine Centropagidae (Copepoda: Calanoida) of the World. Guides to the Identification of the Microinvertebrates of the Continental Waters of the World. SPB Academic Publishing. 30 pp.

Bledzki, L.A. & J.I. Rybak. 2016. Freshwater Crustacean Zooplankton of Europe: Cladocera & Copepoda (Calanoida, Cyclopoida). Key to species identification, with notes on ecology, distribution, methods and introduction to data analysis. Springer International Publishing. Cham. 918 pp.

Brancelj, A. 2000. *Morariopsis dumonti* n. sp. (Crustacea: Copepoda: Harpacticoida) - a new species from an unsaturated karstic zone in Slovenia. Hydrobiologia 436: 73–80.

Brancelj, A. 2011. Copepoda from a deep-groundwater porous aquifer in contact with karst: description of a new species, *Paramoraropsis brigitae* n. sp. (Copepoda, Harpacticoida). Pages 85–104 *in*: D. Defaye, E. Suárez-Morales and J.C. Vaupel Klein (eds.), Crustaceana Monographs, Studies on Freshwater Copepoda: A Volume in Honour of Bernard Dussart. Brill.

Bruno, M.C. & V. Cottarelli. 1998. Description of *Parastenocaris amalasuntae* n. sp. and new data on *Parastenocaris proserpina* and *Parastenocaris pasquinii* from subterranean waters of central Italy (Copepoda, Harpacticoida). Italian Journal of Zoology 65: 121–136.

Bruno, M.C. & V. Cottarelli. 2015. First record of *Kinnecaris* (Copepoda: Harpacticoida: Parastenocarididae) from Turkey and Thailand; description of three new species and emended definition of the genus. Italian Journal of Zoology 82: 69–94.

Bruno, M.C., Cottarelli, V., Marrone, F., Grasso, R., Stefani, E., Vecchioni, L. & M.T. Spena. 2020. Morphological and molecular characterization of three new Parastenocarididae (Copepoda: Harpacticoida) from caves in Southern Italy. European Journal of Taxonomy 689: 1–46.

Caramujo, M.-J., & M.-J. Boavida, 2009. The Practical Identification of Harpacticoids (Copepoda, Harpacticoida) in Inland Waters of Central Portugal for Applied Studies. Crustaceana 82: 385–409.

Chappuis, P.A. 1953. Nouveaux Crustacés troglobies de l'Italie du Nord. Memorie del Museo Civico di Storia Naturale, Verona 4: 1–12.

Clément, M. & C. G. Moore 2007. Towards a revision of the genus *Halectinosoma* (Copepoda, Harpacticoida, Ectinosomatidae): new species from the North Atlantic and Arctic regions. Zoological Journal of the Linnean Society 149: 453–475.

Cottarelli, V. & P. E. Saporito. 1985. *Ceuthonectes pescei* n. sp., arpacticoide freatobio di Sardegna (Crustacea, Copepoda). Fragmenta Entomologica 18: 11–17.

Cottarelli, V. & C. Forniz. 1993. Due nuove specie di *Nitocrellopsis* Petkowski di acque freatiche delle isole di Kos e Tilos (Sporadi Meridionali) (Crustacea, Copepoda, Harpacticoida). Fragmenta Entomologica 24:131–145.

Cottarelli, V., A.C. Puccetti & P. E. Saporito. 1985. Una nuova *Parapseudoleptomesochra* di acque freatiche della Siria: *Parapseudoleptomesochra syriaca* n. sp. (Crustacea, Copepoda, Harpacticoida). Fragmenta Entomologica 18: 1–9.

Cottarelli V., M.C. Bruno, M.T. Spena & R. Grasso. 2012. Studies on subterranean copepods from Italy, with descriptions of two new epikarstic species from a cave in Sicily. Zoological Studies 51: 556–582.

Dahms, H.-U. 1995. Dormancy in the Copepoda - an overview. Hydrobiologia 306: 199–211.

Damian-Georgescu, A. 1970. Copepoda Harpacticoida (forme de apa dulce). Fauna Republicii Socialiste Romania, Crustacea 4(11): 1–252.

Dussart, B., 1967. Les Copépodes des eaux continentales d'Europe occidentale. Tome I: Calanoïdes et Harpacticoides, Faunes et Flores Actuelles. Editions N. Boubee & Cie, Paris. 500 pp.

Dussart, B., 1969. Les Copépodes des eaux continentales d'Europe occidentale. Tome II: Cyclopoïdes et Biologie, Faunes et Flores Actuelles. Editions N. Boubee & Cie, Paris. 292 pp.

Dussart, B.H. & D. Defaye. 2001. Introduction to the Copepoda, 2nd edition, revised and enlarged. Guides to the Identification of the Microinvertebrates of the Continental Waters of the World. Backhuys Publishers. 344 pp.

Einsle, U. 1996. Copepoda: Cyclopoida. Genera *Cyclops*, *Megacyclops*, *Acanthocyclops*. Guides to the Identification of the Microinvertebrates of the Continental Waters of the World. SPB Academic Publishing. 83 pp.

Einsle, U. 1993. Crustacea Copepoda Calanoida und Cyclopoida. Süßwasserfauna von Mitteleuropa, 8/4-1. Gustav Fischer Verlag. 209 pp.

Fiers, F. 2012. The generic concept of *Allocyclops* Kiefer, 1932: (Copepoda: Cyclopoida: Cyclopidae) an alternative view. Journal of Natural History 46: 175–247.

Fryer, G. 1978a. Free-living stages of freshwater parasitic Copepoda. Pages 344–367 *in*: F. Kiefer and G. Fryer, Das Zooplankton der Binnengewässer. 2. Teil, Die Binnengewässer. E. Schweizerbart'sche Verlagsbuchhandlung (Nägele u. Obermiller). Stuttgart.

Fryer, G. 1978b. Branchiura. Pages 368–374 *in:* F. Kiefer and G. Fryer, Das Zooplankton der Binnengewässer. 2. Teil, Die Binnengewässer. E. Schweizerbart'sche Verlagsbuchhandlung (Nägele u. Obermiller). Stuttgart.

Galassi, D. M. P. & P. De Laurentiis. 2004. Towards a revision of the genus Parastenocaris Kessler, 1913: establishment of *Simplicaris* gen. nov. grom groundwaters in central Italy and review of the *P. brevipes*-group (Copepoda, Harpacticoida, Parastenocarididae). Zoological Journal of the Linnean Society 140: 417–436.

Galassi, D. M.P., F. Stoch & A. Brancelj. 2013. Dissecting copepod diversity at different spatial scales in southern European groundwater. Journal of Natural History 47: 821–840.

Galassi, D.M.P., P. De Laurentiis & B. Fiasca. 2011. Systematics of the Phyllognathopodidae (Copepoda, Harpacticoida): re-examination of *Phyllognathopus vigueiri* (Maupas, 1892) and *Parbatocamptus jochenmartensi* Dumont and Maas, 1988, proposal of a new genus for *Phyllognathopus bassoti* Rouch, 1972, and description of a new species of *Phyllognathopus*. ZooKeys 104: 1–65.

Gurney, R. 1932. British Fresh-water Copepoda. II. Harpacticoida. Ray Society, London. 336 pp.

Horton, T., A. Kroh, S. Ahyong, N. Bailly, C. B. Boyko, et al. 2022. World Register of Marine Species (WoRMS). WoRMS Editorial Board.

Hueda, I. & J. W. Reid. 2003. Copepoda: Cyclopoida. Genera *Mesocyclops* and *Thermocyclops*. Guides to the Identification of the Microinvertebrates of the Continental Waters of the World. Backhuys Publishers. 318 pp.

Huys, R. & G. A. Boxshall. 1991. Copepod Evolution. The Ray Society, London. 468 pp.

Huys, R., J. M. Gee, C. Moore & R. Hamond. 1996. Marine and brackish water harpacticoid copepods. Part I. Field Studies Council, 1996. 352 pp.

Janetzky, W., R. Enderle & W. Noodt.,1996. Crustacea Copepoda Gelyelloida und Harpacticoida. Süßwasserfauna von Mitteleuropa. Gustav Fischer Verlag. 228 pp.

Karaytuğ, S. 1999. Genera *Paracyclops*, *Ochridacyclops* and key to the Eucyclopinae. Guides to the Identification of the Microinvertebrates of the Continental Waters of the World. Backhuys Publishers. 217 pp.

Karaytuğ, S., Bozkurt, A., Sönmez, S., 2018. A new hyporheic *Monchenkocyclops* Karanovic, Yoo & Lee, 2012 (Crustacea: Copepoda) from Turkey with special emphasis on antennulary homology. Zoosystema 40: 43–59.

Karaytug, S. & S. Sak. 2005. A new species of *Schizopera* Sars, 1906 (Copepoda: Harpacticoida) from Israel. Zoology in the Middle East 36: 33–42.

Kiefer, F. 1931. Kurze Diagnosen neuer Süsswasser Copepoden. Zoologischer Anzeiger 94: 219–224

Kiefer, F. 1959. Unterirdisch lebende Ruderfusskrebse von Hochrhein und Bodensee. Beiträge zur Naturkundlichen Forschung in Südwestdeutschland 18: 42–52.

Kiefer, F. 1978. Freilebende Copepoda, Pages 1–343 *in:* F. Kiefer and G. Fryer, Das Zooplankton der Binnengewässer. 2. Teil, Die Binnengewässer. E. Schweizerbart'sche Verlagsbuchhandlung (Nägele u. Obermiller). Stuttgart.

Kochanova, E., A. Nair, N. Sukhikh, R. Väinölä & A. Husby. 2021. Patterns of cryptic diversity and phylogeography in four freshwater copepod crustaceans in european lakes. Diversity 13: 448.

Lang, K., 1948. Monographie der Harpacticiden. Hakan Ohlsson's Boktryckeri, Lund. 2 vols., 1683 pp.

Lee, D. J. & W. Lee. 2019. Arthropoda: Copepoda, Pages 761–780 in: C. Rogers and J.H. Thorp (eds.), Keys to Palaearctic Fauna. Thorp and Covich's Freshwater Invertebrates. Elsevier/Academic Press.

Malard, F., M.-J. Dole-Olivier, J. Mathieu & F. Stoch. 2002. Sampling manual for the assessment of regional groundwater biodiversity. European project PASCALIS, 2.2.3 Assessing and conserving biodiversity, Technical Report. Lyon. 74 pp.

Neethling L. & A. Avenant-Oldewage (eds.). 2017. Branchiura. A compendium of the geographical distribution and a summary of their biology. Crustaceana Monographs 21: 1–212.

Pesce, G. L. 1981a. A new harpacticoid from phreatic waters of Morocco, and remarks on the genus *Praeleptomesochra* Lang (Crustacea Copepoda: Ameiridae). Zoological Bulletin Zoölogisch Museum, Universiteit van Amsterdam 8: 69–72.

Pesce, G. L. 1981b. Some harpaticoids from subterranean waters of Greece (Crustacea: Copepoda). Bollettino di Zoologia 48: 263–276.

Pesce, G. L. 1985. Stygobiological researches in subterranean waters of Lesbos (Greece) and description of *Stygonitocrella petkovskii* n. sp. (Crustacea Copepoda: Ameiridae). Fragmenta Balcanica, Musei Macedonici Scientiarium Naturalium 12: 125–13.

Pesce, G. L., D. M. P. Galassi & F. Stoch. 1994. Primo rinvenimento del genere *Maraenobiotus* Mrázek in Italia (Crustacea, Copepoda, Canthocamptidae). Fragmenta Entomologica 25: 161–173.

Petkvoski, T.K. 1978. *Troglodiaptomus sketi* n. gen., n. sp., ein neuer Hohlen-Calanoide vom Karstegelande Istriens (Crustacea, Copepoda). Acta Musei Macedonici Scientiarium Naturalium 15: 151–164.

Petkovski, T.K. 1981. *Stygodiaptomus kieferi* n. gen. et n. sp., Zweiter Hohlen-Calanoide vom Dinarischen Karstgebiet (Crustacea, Copepoda). Fragmenta Balcanica, Musei Macedonici Scientiarium Naturalium 11: 63–74.

Poly, W. J. 2008. Global diversity of fishlice (Crustacea: Branchiura: Argulidae) in freshwater. Hydrobiologia 595: 209–212.

Por, F.D. 1981. A new species of *Bryocyclops* (Copepoda: Cyclopoida) and of *Parastenocaris* (Copepoda: Harpacticoida) from a cave in Israel and some comments on the origin of the cavernicolous copepods. Israel Journal of Zoology 30: 35–46.

Por F. D. 1983 The freshwater Canthocamptidae (Copepoda: Harpacticoida) of Israel and Sinai, Israel Journal of Zoology 32: 113–134.

Ranga Reddy Y. 1994. Copepoda: Calanoida: Diaptomidae. Key to the genera *Heliodiaptomus, Allodiaptomus, Neodiaptomus, Phyllodiaptomus, Eodiaptomus, Arctodiaptomus* and *Sinodiaptomus*. Guides to the Identification of the Microinvertebrates of the Continental Waters of the World. SPB Academic Publishing. 221 pp.

Rayner, N.A. 1999. Copepoda: Calanoida. Diaptomidae: Paradiaptominae. Guides to the Identification of the Microinvertebrates of the Continental Waters of the World. Backhuys Publishers. 122 pp.

Reid, J. W. 1995. Redescription of *Parastenocaris brevipes* Kessler and description of a new species of *Parastenocaris* (Copepoda: Harpacticoida: Parastenocarididae) from the U.S.A. Canadian Journal of Zoology 73: 173–187.

Rouch, R. 1985. Une nouvelle *Stygonitocrella* (Copepoda, Harpacticoida) des eaux souterraines d'Andalousie, Espagne. Stygologia 1:118–127.

Sahuquillo, M. & M. R. Miracle. 2013. The role of historic and climatic factors in the distribution of crustacean communities in Iberian Mediterranean ponds. Freshwater Biology 58: 1251–1266.

Sciberras, M., R. Huys, V.N. Bulnes & N.J. Cazzaniga 2018. A new species of *Halectinosoma* Vervoort, 1962 (Copepoda: Harpacticoida) from Argentina, including a key to species with unusual leg armature patterns, notes on wrongly assigned taxa and an updated key to ectinosomatid genera. Marine Biodiversity 48: 407–422.

Stella, E. 1982. Calanoidi (Crustacea, Copepoda, Calanoida). Guide per il riconoscimento delle specie animali delle acque interne italiane. CNR AQ/1/140, 14: 1–67.

Stoch F., 1998. *Moraria alpina* n. sp. and redescription of *Moraria radovnae* Brancelj 1988, new rank, from Italian and Slovenian Alps (Crustacea, Copepoda, Harpacticoida). Studi Trentini di Scienze Naturali, Acta Biologica 73: 135–145.

Stoch, F. 2007. Copepods colonising italian springs, Pages 217–235 *in:* Cantonati, M., E. Bertuzzi and D. Spitale (eds.), The Spring Habitat: Biota and Sampling Methods, Monografie Del Museo Tridentino Di Scienze Naturali. Trento.

Suárez-Morales, E. 2020. Class Branchiura, Pages 663–796 in: C. Dambonarea, C. Rogers and J.H. Thorp (eds.), Keys to Neotropical and Antarctic Fauna. Thorp and Covich's Freshwater Invertebrates. Elsevier/Academic Press

Suárez-Morales, E., M.A. Gutiérrez-Aguirre, S., Gómez, G., Perbiche-Neves, D., Previattelli, E.N., dos Santos-Silva, C.E.F., da Rocha, N.F., Mercado-Salas, T.M., Marques, Y., Cruz-Quintana & A.M. Santana-Piñeros 2020. Class Copepoda, Pages 797–807 in: C. Dambonarea, C. Rogers and J.H. Thorp (eds.), Keys to Neotropical and Antarctic Fauna. Thorp and Covich's Freshwater Invertebrates. Elsevier/Academic Press.

Vives, F. & A.A. Shmeleva. 2007. Crustacea, Copépodos Marinos I: Calanoida. Fauna Ibérica. Museo Nacional de Ciencias Naturales, Consejo Superior de Investigaciones Científicas. Madrid. 1152 pp.

Vives, F. & A.A. Shmeleva. 2010. Crustacea, Copépodos Marinos II: Non Calanoida. Fauna Ibérica. Museo Nacional de Ciencias Naturales, Consejo Superior de Investigaciones Científicas. Madrid. 486 pp.

Wells, J.B.J. 2007. An annotated checklist and keys to the species of Copepoda Harpacticoida (Crustacea). Zootaxa 1568: 1–872.

Wells, J.B.J. & Rao G.C. 1987. Little harpacticoida (Crustacea, Copepoda) from the Andaman and Nicobar Islands. Memoirs of the Zoological Survey of India 16: 1–385.

Chapter 7

Class Malacostraca (subclass Eumalacostraca)

Christophe Piscart[1], Ana I. Camacho[2], Nicole Coineau[3], Magdalini Christodoulou[4,5], Giuseppe Messana[6] and Karl J. Wittmann[7]
[1]*French National Centre for Scientific Research (CNRS), University of Rennes, Research Unit ECOBIO, Rennes, France,* [2]*Museo Nacional de Ciencias Naturales (CSIC), Department of Biodiversity and Evolutionary Biology Department, Madrid, Spain,* [3]*Retired, Banyuls-sur-Mer, France,* [4]*German Centre for Marine Biodiversity Research (DZMB), Senckenberg am Meer, Wilhelmshaven, Germany,* [5]*Ö Landes-Kultur GmbH, Biology Centre, Linz, Austria,* [6]*Institute of Research on Terrestrial Ecosystems of the National Research Council (CNR-IRET), Firenze, Italy,* [7]*Department of Environmental Health, Medical University of Vienna, Vienna, Austria*

Subchapter 7.1

Introduction to Malacostraca

Christophe Piscart
French National Centre for Scientific Research (CNRS), University of Rennes 1, France

Introduction

Malacostraca are the most diversified class of Crustacea and represent more than half of total crustacean diversity. They are characterized by the development of ambulatory legs and specializations for benthic life, the absence of free-living larval development (brooding eggs instead), and the colonization of every kind of fresh and salted aquatic environments. Their body size range is very wide, from a few millimeters up to 3 m, even if the largest freshwater specimens (i.e., the Tasmanian giant freshwater crayfish) do not exceed 80 cm.

Malacostraca are composed of three subclasses (Eumalacostraca, Hoplocarida, Phyllocarida); two of them exclusively include marine species, and all the freshwater species of the class belong to the subclass Eumalacostraca. They constitute a very diversified group that includes around 75% of all freshwater crustacean species (Balian et al., 2008), and are represented by eight orders and at least 1086 species in the Mediterranean Basin (Table 7.1). The Mediterranean Basin can be considered as a hotspot for Malacostraca: it harbors 18% of the world freshwater species, up to one-third of the world known species of almost all orders, and up to more than 50% of Thermosbaenacea (Table 7.1). Only Decapoda are less diversified in this area, with only 2% of the world diversity of this order.

General ecology and distribution

Among freshwater Malacostraca of the Mediterranean Basin, most species ($\approx$73%) inhabit groundwaters, and the remaining species colonize running waters ($\approx$18%), lakes or ponds ($\approx$5%), or brackish waters (3.5%). The Mediterranean Basin is therefore characterized by a very high proportion of groundwater species, which represent around 24% of freshwater crustaceans in the world (Gibert & Culver, 2009). The proportion of obligate groundwater species in the Mediterranean Basin is likely related to two major historical events: (1) a strong vicariance related to the palaeological instability of the Mediterranean Sea and the sea level fluctuations (e.g., Messinian crisis) (Ayati et al., 2019), and (2) the relative long-term climatic stability observed throughout the Pleistocene period, when speciation events accumulated over time (Zagmajster et al., 2014).

Identification and Ecology of Freshwater Arthropods in the Mediterranean Basin. DOI: https://doi.org/10.1016/B978-0-12-821844-0.00018-1

TABLE 7.1 Mediterranean and global species diversity per order belonging to Malacostraca. The reference values of global diversity are those provided in (Balian et al., 2008) and updated with the World Register of Marine Species (Horton et al., 2020) and the present study.

	Mediterranean Basin	Global
Amphipoda	581	1866
Bathynellacea	71	219
Decapoda	63	2832
Ingolfiellida	9	20
Isopoda	336	994
Mysida	10	60
Stygiomysida	2	5
Thermosbaenacea	12	20
Total	1084	6014

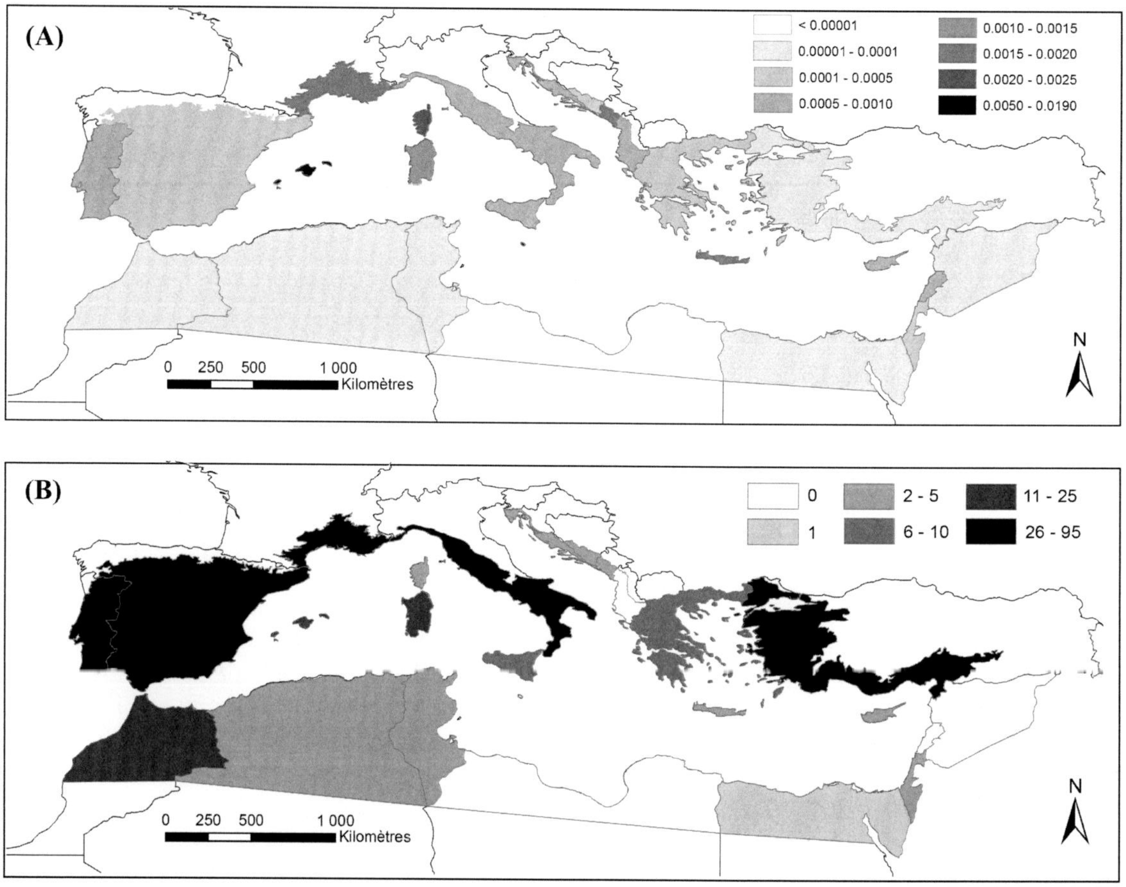

FIGURE 7.1 Geographic distribution of Malacostraca in the Mediterranean bioclimatic area. (A) Species richness was weighted by the country area (species.km-2); (B) the number of endemic species is given for each country.

Malacostraca are widely distributed throughout the Mediterranean Basin (Fig. 7.1A), but 80% of species are endemic (i.e., restricted to one country or one archipelago) (Fig. 7.1B). There also exists a clear positive correlation ($R^2 = 0.91$) between the total number of species in each country and the number of endemic ones. Groundwater species are generally

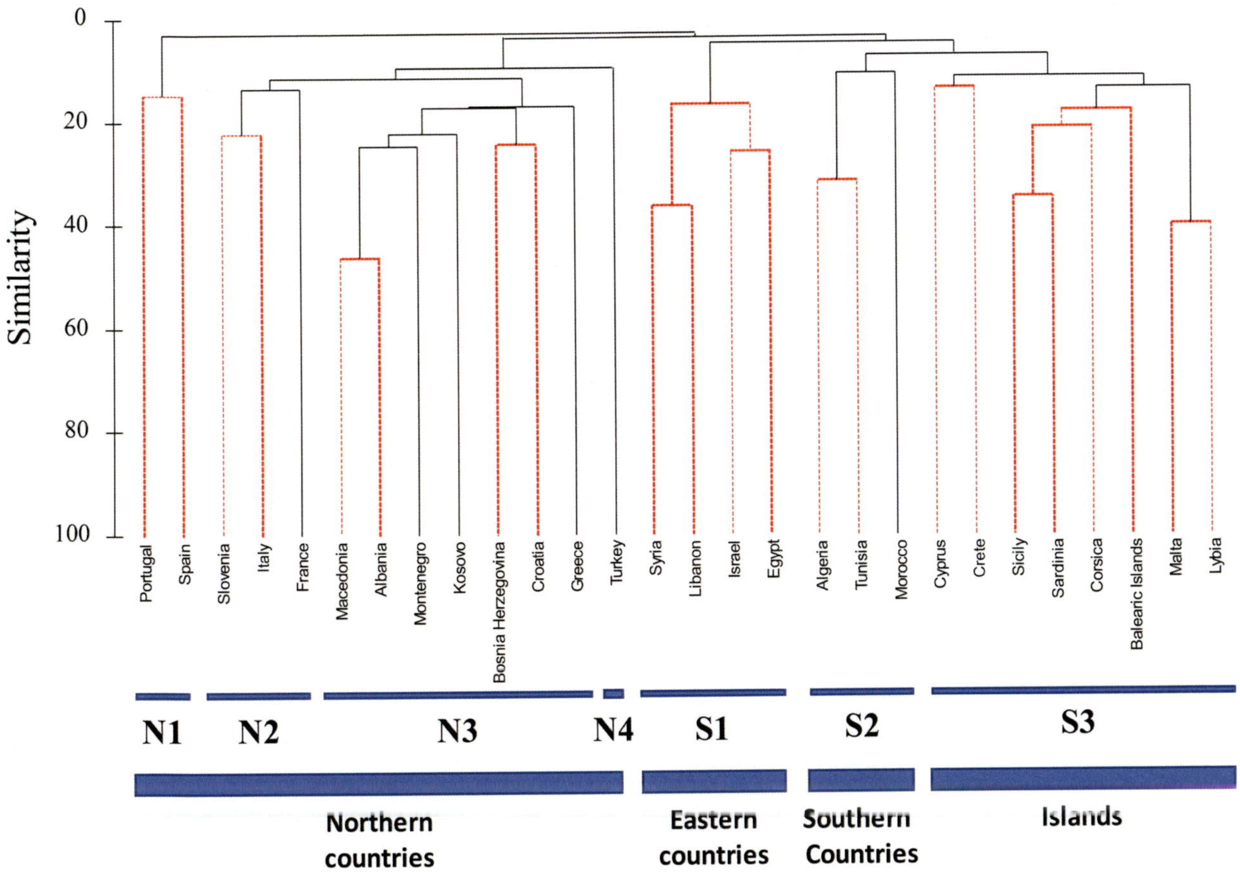

FIGURE 7.2 Results of the hierarchical cluster analysis of Bray Curtis similarity of faunal lists. Solid lines indicate significant differences ($P < .05$) using SIMPROF tests (Clarke et al., 2008) and dashed lines indicate groups of samples not significantly separated by the test.

restricted to less than two countries, whereas epigean species can be widely distributed, especially species able to survive in brackish waters. However, it is difficult to draw robust conclusions on the species distribution because information is lacking in many countries such as the Middle-East countries and about some groups (e.g., groundwater species). For example, countries in the north of the Mediterranean Basin harbor a higher diversity than southern countries do, but this difference might be related to poor knowledge about the southern part of the basin (Tuekam Kayo et al., 2012; Zagmajster et al., 2018).

General diversity patterns can be drawn despite this limitation. If we analyze diversity in the northern and in the southern parts of the basin separately, total diversity tends to decrease from west to east (Fig. 7.1A). In the Mediterranean islands, the Balearic Islands and Sardinia have the highest total species richness as well as the proportion of endemic species (48% and 37.5%, respectively). The analysis of presence/absence data for all Malacostraca clearly highlights four main groups of faunal assemblages (Fig. 7.2). One group includes countries from the northern countries of the Mediterranean Basin, and is divided into four sub-groups: Iberian countries (subgroup N1); France, Italy, and Slovenia (subgroup N2); Balkanian countries from Croatia to Greece (subgroup N3); and Turkey (subgroup N4). The second group includes the eastern countries and Egypt (group S1), the third one includes the southern countries (group S2), and finally the fourth one includes the Mediterranean islands and Lybia (group S3).

Key to Eumalacostraca

Cumcea and Tanaidacea are not included in this key

Eumalacostraca: orders

1 Carapace reduced or absent 2
1' Carapace well developed (Fig. 7.3B,F) 5

FIGURE 7.3 General aspect of Malacostraca. (A) From left to right *Dikerogammarus villosus* (Amphipoda, Gammaridae), *Chelicorophium robustum* (Amphipoda: Corophiidae), and *Niphargus schellenbergi* (Amphipoda: Nirphargidae). *Photogaph courtesy of JF Cart; (B), from left to right Pascifastacus leniusculus* (Decapoda: Astacidae), *photograph courtesy of J.F. Cart, Atyaephyra thyamisensis* (Decapoda: Atyidae) *and Potamon hippocratis* (Decapoda: Potamidae), *photograph courtesy of W. Klotz; (C), Asellus aquaticus* (Isopoda: Asellidae), *photograph credit of C. Piscart; (D), Hemimysis anomala* (Mysida: Mysidae), *photograph courtesy of J.F. Cart; (E), Iberobathynella andalusica* (Bathynellacea: Parabathynellidae), *photograph credit of* A.I. Camacho; *(F), Tethysbaena ledoyeri* (Thermaosbaenacea: Monodellidae), *photograph courtesy of P. Chevaldonné/CNRS.*

2 (1) Thoracopods uniramous, body vermiform or not 3
2’ Thorocapods biramous, body vermiform, less than 2 mm (Fig. 7.3E) .... **Bathynellacea**
3 (2) Body laterally compressed (Fig. 7.3A), telson and abdominal segments free 4
3’ Body dorsoventrally compressed (Fig. 7.3C), telson and abdominal segments fused into a pleotelson .... **Isopoda**
4 (3) Cephalothorax composed of head and first segment of the mesosome fused together (Fig. 7.10); presence or not of vestigial pedunculate eyes; eyes lacking; pleosome without epimera and with reduced pleopod/uropod appendages .... **Ingolfiellida**
4’ Head not fused with the first segment of the mesosome (Fig. 7.4); vestigial pedunculate not present; eyes present or lacking; pleosome with 3 epimera and normally developed appendages .... **Amphipoda**
5 (1’) Compounds eyes generally present, but absent in some groundwater genera, when present they are pedonculate; pleopods always developed 0.6
5’ Eyes absent, pleopods reduced and only on the two first abdominal segments, absent from remaining segments, eight pairs of thoracopods, first modified as a maxilliped .... **Thermosbaenacea**
6 (5) Seven pairs of thoracopods (Fig. 7.3D) 0.7
6’ Five pairs of walking legs, including chelae .... **Decapoda**
7 (6) Eyes normally developed, endopod of uropods with very large statocyst; sympod without ventral lobe; cornea mostly well-developed, rarely reduced .... **Mysida**
7’ Eyes reduced to flat ocular plate, endopod of uropods without statocyst; sympod with spiny ventral lobe; cornea reduced .... **Stygomysida**

Subchapter 7.2

Order Amphipoda

Christophe Piscart
French National Centre for Scientific Research (CNRS), University of Rennes 1, France

Introduction

Amphipoda Latreille 1816 are present in all types of continental waters (springs, rivers, and lakes) and in groundwaters (subterranean, interstitial, and cave systems) on all continents. Apart from a few exceptions, they are easily recognizable in the field from their shrimp-like habitus.

Biogeographic analysis indicates that amphipods were likely widespread in Pangea, Gondwana, and Laurentia (Väinölä et al., 2008). These studies were confirmed by the discovery of the oldest fossil *Rosagammarus minichiellus* in late Triassic limestone in Nevada (McMenamin et al., 2013). Marine ancestors subsequently colonized very early freshwaters during the Jurassic period before a probable massive diversification during the Cretaceous (Copilaş-Ciocianu et al., 2019) and Paleogene (Hou et al., 2014) periods.

The ancestors of Mediterranean amphipods may have an Atlantic origin around 40 Mya (Hou & Sket, 2016). They colonized its freshwater tributaries, like *Iberogammarus* and *Echinogammarus sensu stricto* did. Finally, some crossed the proto-Mediterranean sea to the lands located at its eastern edges, (the *Typhlogammarus* group and part of the *Sarothrogammarus* group), or also spread along the future Mediterranean coasts (*Rhipidogammarus*). Other groundwater groups (e.g., hadzioid-thalassoid Metacrangonyctidae amphipods) result from ancient continental-level vicariance (Bauzà-Ribot et al., 2012), or from more recent vicariance and long-distance dispersal (Pseudoniphargidae; Stokkan et al., 2018).

Among the 1870 amphipod species and subspecies recorded from fresh or inland waters worldwide (Horton et al., 2020), more than 31% are present in the Mediterranean Basin, where a total of 14 families, 50 genera, and 581 species are recorded; most of them are endemic to the basin.

General ecology and distribution

Amphipoda are widely distributed and colonize all types of marine and continental waters on all continents. They represent one of the highest invertebrate biomass values and are prey for many other organisms. They also support many functional processes in these ecosystems like organic matter recycling and bioturbation.

In the Mediterranean Basin, amphipods are present in all countries and big islands. We recorded 559 species (96%) restricted to only one type of environment (running water, lake, groundwater, brackish water) and only 22 euryecous species (i.e., present in a least two types of aquatic ecosystems). Euryecous species are present mainly in running waters, lakes, and brackish water, and only a few species (e.g., *Niphargus zagrebensis*) can live permanently either in groundwater or in surface waters.

The highest diversity is found in groundwaters, despite their very low density. They represent 74% of the total amphipod richness with 429 species, followed by running waters (127 species) and lakes (31 species). Only 17 species live solely in lakes and originally are found in Ohrid Lake, in the Balkan pininsula (8 species), or in Turkey (9 species); the other lacustrine species are euryecous species that also live in running waters (14 species) or in brackish environments (4 species).

Among the 14 families recorded so far, most of them are widespread worldwide, with members living in both fresh and marine waters. Others (Crangonyctidae, Niphargidae, Pontogammaridae, and Typhlogammaridae) have a Eurasian distribution and are only present in the central and eastern parts of the Mediterranean Basin, except *Crangonyx africanus* in Morocco. Metacrangonyctidae and Pseudoniphargidae are mainly distributed in the Mediterranean Basin but also have a few members in the Canary and Caribbean islands. Finally, only two families are endemic to the Mediterranean Basin: Salentinellidae with 14 species widely distributed around the Mediterranean Sea between Greece and Algeria, and Sensonatoridae with only one known species in Spain (*Sensonator valentienensis*; Notenboom, 1986).

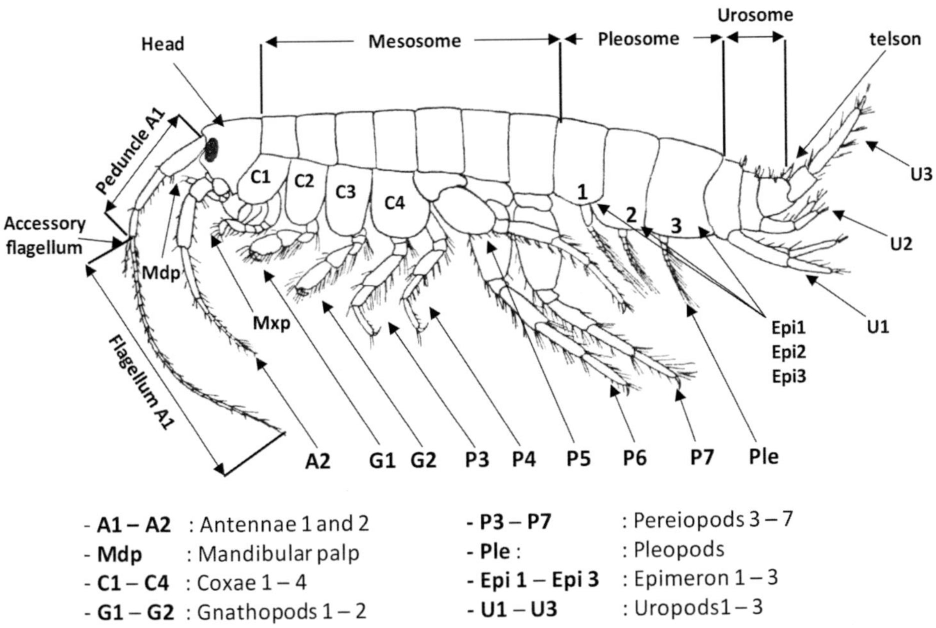

FIGURE 7.4 General morphology of Amphipoda.

Terminology and morphology

Amphipoda are medium/large-size Malacostraca (from 0.1 up to 35 mm in the Mediterranean Basin). Their body is typically compressed laterally, though less so for Corophiidae. It is divided into 13 segments that can be grouped into the head, the mesosome, the metasome, and the urosome (Fig. 7.4).

The head is fused to the thorax and bears two pairs of antennae. The upper pair is antennae 1, with a basal peduncle composed of three articles and a distal flagellum composed of many articles. The third peduncular article also bears an accessory flagellum on its distal end, usually composed of a reduced number (less than 10) of tiny cylindrical articles. The second pair of antennae (antennae 2) has a flagellum composed of several small articles and a peduncle composed of five articles; articles 4 and 5 are much longer than the first three articles (the first one is hidden by the head). Most epigean species (with few exceptions like epigean *N. zagrebensis;* S. Karaman, 1950) have one pair of more or less developed sessile compound eyes. Eyes are absent in obligate groundwater species. However, we often found eyed species in groundwaters, especially in the Mediterranean Basin, where epigean species can be present in a large proportion of groundwaters round the year. The mouthpart is on the ventral surface of the head and is composed of the labrum, labium, mandibles, maxillae I and II, and maxilliped.

The mesosome (or thorax or pereon) bears different kinds of legs, but lacks a carapace. It has eight pairs of uniramous appendages, the first of which are used as accessory mouthparts. Two pairs of gnathopods (modified pereopods 1 and 2) are used for grasping, usually with a last subchelate article; the next five pairs (pereopods 3–7) are walking legs. Gnathopods and pereopods 3 and 4 have modified coxae transformed into large coxal plates. Coxal gills are present on mesosome segments in various numbers according to species, and sternal gills are known only in Crangonyctidae.

The metasome (or abdomen or pleion) is divided into two parts: the pleosome and urosome segments. Each segment bears one pair of legs. The pleosome bears three pairs of swimming legs named pleopods, with multi-articulated rami. The urosome bears three pairs of uropods. The length and the shape of uropod 3 can be highly variable among genera. The basal part of the uropod is the peduncle that branches into an exopodite (with one or two articles, the second one can be short and hidden by the apical spines of article 1) and a medial branch named the endopodite (only one article or no article at all).

The telson—the final body segment—is attached above uropod 3.

Collection, preparation, and identification

Amphipoda are widely distributed in all aquatic environments, including groundwaters. Therefore, the sampling methods and devices are numerous and depend on the specific habitats where the fauna is found.

In surface waters, we can sample them using all kinds of sampling tools and methods (sweep net, kick sampling, grab sampler, dredges, and even artificial substrates) or just by taking stones, leaf litter, or macrophyte by hand or with a kitchen colander. There are rarely more than three species in any given site. However, it is strongly recommended to sample the different substrates because many species have different microdistribution patterns when they co-occur (Piscart et al., 2007).

In groundwaters, any kind of specific method for sampling this environment can be used (Malard et al., 2002). Baited bottle traps or nets exposed overnight are also very efficient for many groundwater organisms.

Amphipoda can be easily fixed and preserved in 70% ethanol or another fixative for morphological analysis and in 95% ethanol for molecular analysis.

Many species can be identified under a stereomicroscope, but identification at the species level may require dissection, mounting body parts and appendages on slides, and subsequent examination using a 400× or greater magnification microscope. Specimens in glycerol or ethanol can be handled under a low-power microscope with needles. Glycerol is used as an embedding medium for short-term slides. Baths in clearing solution (e.g., soft acids) or permanent mounting media are only required for depositing museum specimens. A same specimen can be used for both studies: the abdomen for DNA extraction and the other parts for dissection and mounting. This procedure ensures that the sequenced genes correspond to the same morphotype.

Taxonomic identification to the species level is generally only possible in males, but females can be used for higher taxonomic levels. The most important body parts for identification are the antennae, mouthpart, coxae, gnathopods, pereopods 5 and 7, epimera, telson, and uropods; but additional body parts need to be examined to identify certain species.

Limitations

Amphipoda are a highly diversified group with almost 600 known species, and even more remain undescribed. The keys below are designed to cover the 14 families and 49 genera; keys at the species level are provided only for the genera with a low number of species (usually less than 15). Rare records are included, unless they are insufficiently documented. Taxonomic synonymy has been updated, but taxonomy is not well established for some genera (e.g., *Echinogammarus, Laurogammarus*, and *Typhlogammarus*) and even families (e.g., Pontogammaridae), especially following recent molecular studies. For those genera and until a clear consensus is found, we assume the use of the most parsimonious point of view, even if alternate classifications may be found in other keys. As a consequence, some family, genus, and species names may change in the future.

Finally, even if our keys were designed to be as clear as possible, we should never forget that identifying certain groups of amphipods (especially groundwater species) can be very difficult and requires strong expertise. Do not hesitate to use alternative information or keys to consolidate your conclusions. We should keep in mind that keys, even at the family level, are designed for species of the Mediterranean Basin; they are not adapted to other amphipods worldwide.

Keys to Amphipoda

Amphipoda: families

1 Antenna 1 significantly shorter than antenna 2 2
1' Antenna 1 subequal or longer than antenna 2 4
2 (1) Antenna 1 accessory flagellum absent; eyes well developed 3
2' Antenna 1 accessory flagellum small, eyes absent .... Sensonatoridae (one genus: *Sensonator*) [Spain]
3 (2) Antenna 2 enlarged; mandibular palp present; urosomite dorsally flat .... **Corophiidae** [North-Eastern Mediterranean, Central Europe]
3' Urosomite enlarged, antenna 2 not enlarged, mandibular palp absent .... **Talitridae** [Peri-Mediterranean]
4 (1) Sternal gills absent 5
4' Sternal gills present .... **Crangonyctidae** [Europe, Morocco]
5 (4) Mandibular palp normally developed; eyes present or absent 6
5' Mandibular palp vestigial; eyes absent .... **Metacrangonyctidae** [Peri-Mediterranean]
6 (5) Coxae 1–3 subequal or size progressively increasing, coxae 4–5 of various size; uropod 1–2 peduncles with lateral spines; eyes present or absent 7
6' Coxae 1–3 small, size progressively decreasing, and coxa 4–5 larger; uropod 1–2 peduncles without lateral spines; eyes absent .... **Salentinellidae**

[Peri-Mediterranean]

7 (6) Antenna 1 longer than antenna 2; maxilla 1 inner plate with at least 1 apical setae and/or marginal setae, telson notched, emarginate or entire 8

7' Antenna 1 and 2 equal or subequal in length; maxilla 1 inner plate reduced with only 0 or 1 apical setae; telson entire .... Ischyroceridae (one genus *Jassa*)

[Sicily]

8 (7) Telson lobes almost completely separated, rami of uropod 3 lanceolate 9

8' Telson lobes partially fused, even sometime very slightly; uropod 3 rami straight more or less long 10

9 (8) Coxae 1–3 longer than wide or subrectangular; gnathopod 2 propodus elongated slightly stronger than gnathopod 1; labium inner lobe vestigial or absent; uropod 3 rami sub-equal .... **Hadziidae**

[Western Europe]

9' Coxae 1–3 small wider than long; gnathopod 2 propodus much stronger than gnathopod 1; labium inner lobe normally developed, except in the genus *Gammaropisa*; uropod 3 rami unequal, exopodite much longer than endopodite .... **Eriopisidae**

[Turkey, Tunisia]

10 (8') Telson entire or slightly emarginate (notch $< 30\%$ of the lobe), eyes absent 11

10' Telson deeply notched almost entirely, eyes present or absent 12

11 (10) Pleopods uniramous or inner ramus reduced or vestigial; uropod 3 rami subequal in length .... **Bogidiellidae**

[cosmopolitan]

11' Pleopods biramous, inner ramus well developed; uropod 3 exopodite much longer than endopodite .... **Pseudoniphargidae**

[Western Europe and North Africa]

12 (10') Inner lobe of maxilla 2 with oblique row of facial setae; eyes generally present of various sizes 13

12' Inner lobe of maxilla 2 without oblique row of facial setae; Labium without a more or less developed inner lobes, eye absent .... **Niphargidae**

[Southern and Eastern Europe, Minor Asia]

13 (12) Maxilla 1 outer plate with apical spines comb-like with few lateral teeth (except *Longigammarus*); pereopod 7 longer than 5; uropod 3 endopodite shorter than exopodite; eyes present or absent 14

13' Maxilla 1 outer plate with apical spines comb-like with numerous lateral teeth (> 10); pereopod 7 subequal in length than 5; uropod 3 subequally biramous; eyes absent .... **Typhlogammaridae**

[Balkan and Eastern Europe]

14 (13) Antenna 2 peduncle article 1 enlarged, bulbous; antenna 1 longer or subequal in length than antenna 2, antenna 1 slender, generally longer than a half of the body length, the diameter of the articles of the peduncle is not much different; uropod 1 with or without basofacial robust setae; urosomite 2 with multiple small robust setae across the somite .... **Gammaridae**

[Cosmopolitan]

14' Antenna 2 peduncle article 1 subequal than article 3; antennae 1–2 subequal in length and short ($\leq$ an half of the body length), antenna 1 stout, shorter than a half of the body length, the diameter of the articles of the peduncle strongly decrease in length and in width from article 1 to article 3; uropod 1 without basofacial robust setae; urosomite 2 without dorsal setae .... **Pontogammaridae**

[Western Europe (invasive), Eastern Europe, and Minor Asia]

Amphipoda: Bogidiellidae: genera

1 Molar nontriturative; maxilla 1 palp with 2 articles 2

1' Molar triturative; maxilla 1 palp with 1 or 2 articles 3

2 (1) Epistomal ridge with a sclerified subcircular area; pleopod rami very long with 9 articles .... *Maghrebidiella* (one species: *Maghrebidiella maroccana*; Diviacco & Ruffo, 1985)

[Morocco]

2' Epistomal ridge without slcerified area; pleopod rami with 3 articles .... *Hebraegidiella* (one species: *Hebraegidiella bromleyana*; G. Karaman, 1988)

[Israel]

3 (1) Pleopods 2, ramus article 2 of males with a modified seta .... *Stygogidiella* (one species: *Stygogidiella cypria*; G. Karaman, 1989)

[Cyprus]
3' Pleopods 2, ramus article 2 of males without modified seta 4
4 (3) Uropods not sexually dimorphous, uropods without transformed spines, pleopods endopodites small, vestigial or absent 5
4' Uropods sexually dimorphous, uropod 1 or 2 rami with transformed (rasp-like, spoon-like) spines; pleopods endopodites absent .... *Medigidiella*
[Western Mediterranean]
5 (4) Maxilla 1 outer lobe with 6 distal spines, telson deeply cleft ....
.... *Racovelia* (one species: *Racovelia birramea*; Jaume, Gràcia & Boxshall, 2007)
[Balearic Islands]
5'. Maxilla 1 outer lobe with 7 distal spines, telson entire or weakly cleft (<30%) .... *Bogidiella*
[cosmopolitan]

Amphipoda: Bogidiellidae: *Medigidiella*: species

1 Lenticular organ ≤ a half of the basis width or absent (not visible) 2
1' Lenticular organ very large, almost as large as the basis of pereopods 3–6, absent on pereopod 7 ....
.... *Medigidiella paolii* (Hovenkamp, 1983)
2 (1) Lenticular organ absent (not visible) 3
2' Lenticular organ visible on pereopod 3–7 7
3 (2) Pleopod inner rami absent; telson with apical spines only; basis of gnathopods 1 and 2 with only one posterior setae 4
3' Pleopod inner rami reduced but present; telson with apical and subapical spines; basis of gnathopods 1 and 2 with 2 posterior setae .... *Medigidiella hebraea* (Ruffo, 1963)
4 (3) Antenna 2 without groups of setae on peduncle articles; maxilliped outer lobe with 2 spines; gnathopod 1 propodus without short spines; uropod 1 peduncle with 1 spine 5
4' Antenna 2 with groups of setae on peduncle articles; maxilliped outer lobe with 3 spines; gnathopod 1 propodus with 2 or 3 short spines; uropod 1 peduncle without spine .... *Medigidiella paraichnusae* (Karaman, 1979)
5 (4) Gnathopod 2 propodus with 2 short spines; pereopod 7 propodus with 4–6 anterolateral setae; uropod 3 rami with 7–11 lateral spines 6
5' Gnathopod 2 propodus with 3 short spines; pereopod 7 propodus with 7–9 anterolateral setae; uropod 3 rami with 3–6 lateral spines .... *Medigidiella dalmatina* (S. Karaman, 1953)
6 (5) Gnathopods 1–2 propodi short; uropods 1–2 rami with short distal spines ....
.... *Medigidiella chappuisi chappuisi* (Ruffo, 1952)
6' Gnathopods 1–2 propodi longer; uropods 1–2 rami with relatively long distal spines ....
.... *Medigidiella chappuisi pescei* (G. Karaman 1989)
7 (2') Antenna 1 accessory flagellum ≤ 2 articles of the flagellum, peduncle articles with 2 spines; maxilliped outer lobe with bifid spines; gnathopod 1 propodus with 4–5 short spines 8
7' Antenna 1 accessory flagellum longer than the 3 articles of the flagellum, peduncle articles without spine; maxilliped outer lobe with simple spines; gnathopod 1 propodus with 7 short spines ....
.... *Medigidiella antennata* (Stock & Notenboom, 1988)
8 (7) Antenna 2 with aesthestascs; maxilla 2 without plumose setae; gnathopod 1 basis with only one long posterior seta; gnathopod 2 propodus with 4 short spines; telson with 4 plumose setae 9
8' Antenna 2 without aesthestascs; maxilla 2 without plumose setae; gnathopod 1 basis with 2 long posterior setae; gnathopod 2 propodus with 3 short spines; telson with at least 6 plumose setae ....
.... *Medigidiella aquatica* (Karaman, 1990)
9 (8) Lenticular organ small, about 1/3 of the basis width; labium outer lobe with spines; pereopod 7 propodus with less than 9 anterolateral setae, dactylus equal or longer than a half of the propodus; uropod 2 rami with modified distal spines in males .... *Medigidiella minautorus* (Ruffo & Schiecke, 1976)
9' Lenticular organ large, about 1/2 of the basis width; labium outer lobe with spines; pereopod 7 propodus with at least 9 anterolateral setae, dactylus shorter than a half of the propodus uropod 2 rami with unmodified distal spines in males .... *Medigidiella uncanata* (Stock & Notenboom, 1988)

Amphipoda: Corophiidae: Genera and species

1 Antenna 2 article 2 with gland very small .... 2 (genus: *Chelicorophium*) [invasive in Western Europe]
1' Antenna 2 article 2 with gland cone large, conspicuous .... *Corophium* (one species: *Corophium oriental* Schellenberg 1928) [Southern Europe]
2 (1) Antenna 2 peduncle article 5 with 1 distal teeth or without teeth 3
2' Antenna 2 peduncle article 5 with 2 distal teeth on the lower margin .... *Chelicorophium maeticum* (Sowinsky, 1898)
3 (2) Antenna 2 peduncle article 5 with 1 distal teeth (short or long) 4
3' Antenna 2 peduncle article 5 without distal teeth on the lower margin .... *Chelicorophium. sowinskyi* (Martynov, 1924)
4 (3) Antenna 2 peduncle article 5 with 1 long distal teeth; uropod 2 outer rami with 6 marginal spines .... *Chelicorophium robustum* (G.O. Sars, 1895)
4' Antenna 2 peduncle article 5 with 1 short distal teeth; uropod 2 outer rami with 3 marginal spines .... *Chelicorophium curvispinum* (Martynov, 1924)

Amphipoda: Crangonyctidae: genera and species

1 Urosomites coalesced; uropod 3 inner ramus absent 2
1' Urosomites free; uropod 3 inner ramus small or scale-like .... 6 (genus: *Crangonyx*) [Morocco, invasive in Western Europe]
2 (1) Body pigmented; pleopods with 2 coupling spines (retinacules); uropod 3 endopodite without terminal "squamous knob" .... 3 (genus: *Synurella*) [Balkan, Central Europe]
2' Body not pigmented; pleopods with 5–6 coupling spines (retinacules); uropod 3 endopodite with a terminal "squamous knob" .... *Pontonyx* (one species: *Pontonyx osellai;* Ruffo, 1974) [Turkey]
3 (2) Epimera 2 with spines at distal margin but never setae .... 3 (*Synurella ambulans*)
3' Epimera 2 with numerous setae at distal margin .... *Synurella longidactylus* (S. Karaman, 1929)
4 (3) Females's oostegite broader; eyes present or absent; pereopod dactyls in males longer and slenderer than in females 4
3' Female's oostegite narrow; eyes absent; pereopod dactyls subequal in shape between males and females .... *Synurella ambulans montenegrina* (G. Karaman, 1974)
5 (4) Eyes present; gnathopod 2 propodus in females is short and broad than in males .... *Synurella ambulans ambulans* (F. Müller, 1846)
5' Eyes absent; gnathopod 2 propodus in males is strongly dilated .... *Synurella ambulans subterranean* (S. Karaman, 1931)
6 (1') Eyes present, epigean species; gnathopod 2 palm with notched spines; pereopod 6 propodus much shorter than basis .... *Crangonyx pseudogracilis* (Bousfield, 1958)
6' Eyes absent, hypogean species; gnathopod 2 palm with simple spines; pereopod 6 propodus longer than basis and carpus .... *Crangonyx africanus* (Messouli, 2006)

Amphipoda: Eriopisidae: genera and species

1 Labium inner lobe absent; gnathopod 1 carpus slightly shorter than propodus; pleopod 3 greatly modified in males .... *Gammaropisa* (one species: *Gammaropisa argonoi;* Ruffo & Vigna-Taglianti, 1988)* [Turkey]
1' Labium inner lobe present; gnathopod 1 carpus longer than propodus; pleopod 3 not modified .... *Tunisopisa* (one species: *Tunisopisa seurati;* Gauthier, 1936) [Tunisia]

The position of *Gammaropisa* in Eriopisidae remains doubtful because it did not match with the revision of Lowry and Myers 2013, who defined Eriopisidae with an internal lobe in the labium but this lobe was not drawn by Ruffo 1987.

Amphipoda: Gammaridae: genera

1 Antenna 2 gland cone small, conical and straight 2
1' Antenna 2 gland cone extremely elongated and recurved going back toward coxae 1 .... *Laurogammarus* (one species: *Laurogammarus scutarensis*; Schaferna, 1922) [Skadar Lake drainage]
2 (1) Pleonites and pereonites without mediodorsal keels; pereopod 7 basal article with or without large posteriodistal lobe, urosome generally not carinate (except in some *Gammarus* and *Echinogammarus*) 3
2' Pleonites and some pereonites with mediodorsal keels; pereopod 7 basal article with a large posteriodistal lobe, urosome carinate .... *Amathillina* (one species: *Amathillina cristata*; Sars, 1894) [Turkey]
3 (2) Uropod 3 endopodite well developed, overreaching 25% of the exopodite 4
3' Uropod 3 endopodite very reduced, scale like, <25% of the exopodite 6
4 Maxilla 1 outer lobe with 17 curved spines; coxal gill 7 absent; eyes absent .... *Albanogammarus* (one species: *Albanogammarus inguscioi*; Ruffo, 1995) [Balkans]
4' Maxilla 1 outer lobe with straight simple or pectinate spines; coxal gill 7 present; eyes present or absent 5
5 Gnathopod 1 propodus in males larger than gnathopod 2; gnathopod 2 carpus elongated .... *Iberogammarus* [Spain, France]
5' Gnathopod 1 in male propodus smaller than gnathopod 2; uropod 3 endopodite ≥ 25% exopodite .... *Gammarus* [Cosmopolitan]
6 Uropod 1 reduced, not reaching apices of uropod 2, rami without lateral spines 7
6' Uropod 1 longer than uropod 2 8
7 Maxilla 1 outer lobe with many densely pectinate setae (like Typhlogammaridae), inner lobe wider than long .... *Longigammarus* [Peri-Mediterranean]
7' Maxilla 1 outer lobe with less densely pectinate setae (like Gammaridae), inner lobe longer than wide, pereopod 7 base elongated more than twice as long as wide .... *Rhipidogammarus* [Peri-Mediterranean]
8 (6') Uropod 1 longer than uropod 2 and reaching its apex, uropod 1 rami with several distal spines 9
8' Uropod 1 longer than uropod 2 but not reaching its apex, uropod 1 rami with only one strong distal spine .... *Tyrrhenogammarus* [Sardinia, Sicily]
9 (8) Urosomites 1 et 2 without dorsal elevations 10
9' Urosomites 1 et 2 with dorsal humps or cones with some apical spines .... *Dikerogammarus* [Turkey, invasive in Western Europe]
10 (9) Pereopod 7 ventroposterior lobe absent or small; uropod 1 outer ramus subequal to the inner ramus 11
10' Pereopod 7 ventroposterior lobe well developed, uropod 1 outer ramus reaching at least 60% of inner ramus 13
11 (10) Setae on at least either on pereopods 5–7, urosomite and/or epimera 12
11' Setae almost completely lacking on pereopods 5–7, urosomites and epimeron .... *Chaetogammarus* (one species: *Chaetogammarus saisensis*; Fadil et al., 2009) [Morocco]
12 (11) Gnathopod 1 propodus in males much larger than gnathopod 2, eyes elongate 2 × longer than wide .... *Echinogammarus sensu stricto* [Peri-Mediterranean]
12' Gnathopod 1 propodus in males equal of slightly longer than gnathopod 2, eyes ovoid or 1.5 × longer than wide .... *Echinogammarus/Homoeogammarus* [Peri-Mediterranean]
13 (10') Pereopod 7 ventroposterior lobe broad and rounded; uropod 1 outer ramus reaching 60% of inner ramus; uropod 3 with very long setae and marginal spines; telson with marginal spines .... *Ilvanella* (one species: *Ilvanella inexpectata*; Vigna-Taglianti, 1971) [Italia, Tuscan Archipelago]

13' Pereopod 7 ventroposterior lobe rounded-triangular; uropod 1 outer and inner rami subequal in length; uropod 3 with very few setae; telson without marginal spines .... *Jugogammarus* (one species: *Jugogammarus kusceri*; S. Karaman, 1931) [Slovenia]

Amphipoda: Gammaridae: *Dikerogammarus*: species

1 Antenna 1 peduncle segments long and slender 2
1' Antenna 1 peduncle segments short and swollen .... *Dikerogammarus gruberi* (Mateus & Mateus, 1990)
2 (1) Pereopod 7 basal segment with setae in the inner surface 3
2' Pereopod 7 basal segment without setae in the inner surface 4
3 (2) Antenna 1 peduncle segment 2 and antenna 2 segment 4–5 with few short setae (shorter than the diameter of the segment) on the inferior margin .... *Dikerogammarus istanbulensis* (Özbek and Özkan, 2011)
3' Antenna 1 peduncle segment 2 and antenna 2 segment 4–5 with many long setae (longer than the diameter of the segment) on the inferior margin .... *Dikerogammarus bispinosus* (Martynov, 1925)
4 (2') Antenna 2 flagellum with long setae forming a brush, urosomites 1 et 2 with dorsal humps well developed and armed with 3–6 upright apical spines and forming like a crater .... *Dikerogammarus villosus* (Sowinskyi, 1894)
4' Antenna 2 flagellum with short setae never forming a brush, urosomites 1 et 2 with small dorsal humps armed with 1–3 apical spines slightly oriented towards the telson .... *Dikerogammarus haemobaphes* (Eichwald, 1841)

Amphipoda: Gammaridae: *Iberogammarus*: species

1 Antenna 2 of male with many groups of long and densely implanted curled setae. Third segment of mandible palp with less than 6 terminal setae 2
1' Antenna 2 of male with straight setae. Third article of mandible palp with more than 6 terminal setae .... *Iberogammarus anisocheirus* (Ruffo, 1959)
2 (1) Gnathopod 1 propodus of males without the medial palmar spine. Basis of P7 with strong backward projecting ventroposterior corner .... *Iberogammarus macrocarpus* (Stock, 1969)
2' Gnathopod 1 propodus of males with a strong medial palmar spine. Basis of P7 with rounded, nonprojecting, ventroposterior corner .... *Iberogammarus toletanus* (Pinkster & Stock, 1970)

Amphipoda: Gammaridae: *Longigammarus*: species

1 Eyes present; maxilla 1 outer plate with 11 apical spines, slightly recurved .... *Longigammarus bruni* (G. Karaman, 1969)
1' Eyes present; maxilla 1 outer plate with 22 apical spines distally spatulate .... *Longigammarus planaisae* (Messana & Ruffo, 2001)

Amphipoda: Gammaridae: *Rhipidogammarus*: species

1 Antenna 1 peduncle article 2 with at least 4 groups of setae 2
1' Antenna 1 peduncle article 2 with 2–3 groups of setae 3
2 (1) Coxae 1–5 with long ventral setae, anterior margin of carpus of P7 with setae longer than spines .... *Rhipidogammarus triumvir* (Nottenboom, 1985)
2' Coxae 1–5 with short ventral setae only, anterior margin of carpus of P7 without long setae .... *Rhipidogammarus gordankaramani* (Özbek & Sket 2019)
3 (1') Antenna 1 flagellum with more than 35 articles; uropod 3 exopodite shorter (no more than 7× longer than endopodite) without seta or few long setae 4
3' Antenna 1 flagellum with more than 30 articles; uropod 3 exopodite long (10× longer than endopodite) with many long setae (2× longer than spines) .... *Rhipidogammarus variicauda* (Stock, 1978)

4 (3) Mandibular palp article 2 with long setae reaching the setae type D on the article 3; uropod 3 exopodite without seta .... *Rhipidogammarus karamani* (Stock, 1971)
4' Mandibular palp article 2 with setae never reaching the setae type D on the article 3; uropod 3 exopodite with few setae 2 × longer than spines .... *Rhipidogammarus rhipidiophorus* (Catta, 1878)

Amphipoda: Gammaridae: *Tyrrhenogammarus*: species

1 Pereopod 3 carpus with a fan of long setae; uropod 1 with subequal rami; epimera 3 with a slightly marked ventroposterior corner .... *Tyrrhenogammarus catacumbae* (G. Karaman & Ruffo, 1977)
1' Pereopod 3 carpus without long setae; uropod 1 inner ramus longer than outer ramus; epimera 3 without distinct ventroposterior corner .... *Tyrrhenogammarus sardous* (Karaman & Ruffo, 1989)

Amphipoda: Hadziidae: genera and species

1 Uropods 1 and 2 with row of 5 plumose setae, exopodite with dorsal setae; telson with only apical spines 2
1' Uropods 1 and 2 with row of 5 plumose setae, exopodite with dorsal setae; telson with marginal spines .... *Parhadizia* (one species: *Parhadizia sbordonii*; Vigna Taglianti, 1988) [Turkey]
2 Mandibular palp article 2 as long as article 1 .... 3 (genus: *Hadzia*)
2' Mandibular palp article 2 longer than article 1 .... 7 (genus: *Metahadzia*)
3 (2) Gnathopod 2 propod in females with entire convex palm; gnathopod 2 propod in males as wide as carpus; uropod 3 rami broad 0.4
3' Gnathopod 2 carpus in females with excavated, propodus palm concave; gnathopod 2 propodus in males wider than carpus; uropod 3 rami narrow 6
4 (3) Pleopods 1–3 each with 2 retinacula accompanied by one strong seta 5
4' Pleopods 1–3 each with 2 retinacula without strong seta .... *Hadzia fragilis stocki* (G. Karaman, 1989)
5 Pereopods 3–7 and antennae long and slender, including dactyl of pereopods 3–7; telson lobes with 2, occasionally 3 distal spines and without outer marginal spines .... *Hadzia fragilis fragilis* (S. Karaman, 1932)
5' Pereopods 3–7 and antennae shorter and stouter; telson lobes with 3–4 distal spines and 0–1 outer marginal spines .... *Hadzia fragilis drinensis* (G. Karaman, 1984)
6 (3') Pleopod 3 peduncle with 1–2 median spine on posterior margin .... *Hadzia gjorgjevici crispata* (G. Karaman, 1974)
6' Pleopod 3 peduncle without median spine on posterior margin .... *Hadzia gjorgjevici gjorgjevici* (S. Karaman, 1932)
7 (2') Uropod 2 peduncle distal end with a protuberance 8
7' Uropod 2 peduncle without protuberance 9
8 (7) Uropod 2 endopodite of males with transformed apical spine, peduncle with dorsomedial comb-like spines .... *Metahadzia uncispina* (Notenboom, 1988)
8'. Uropod 2 endopodite of males without transformed apical spine, peduncle without dorsomedial comb-like spine .... *Metahadzia tavaresi* (Mateus & Mateus, 1972)
9 (7') Lobe of telson with spines along both margins 10
9' Lobe of telson without spine along outer margin .... *Metahadzia minuta* (Ruffo, 1947)
10 (9) Maxilla 1 inner lobe long reaching the basis of spines of outer lobe, distally prominent .... *Metahadzia adriatica* (Pesce, 1979)
10' Maxilla 1 inner lobe short never reaching the basis of spines of outer lobe .... *Metahadzia helladis* (Pesce, 1980)

Amphipoda: Metacrangonyctidae: genera and species

1 Gnathopod 2 propodus not modified, palm slightly convex; gnathopod 1 merus without brush; coxae 1–3 subequal and longer than coxae 4–5 .... 2 (genus *Metacrangonyx*) [peri-Mediterranean]
1' Gnathopod 2 propodus modified in males, palm strongly concave; gnathopod 1 merus with many short setae forming a brush; coxae 1–5 decreasing progressively in length .... *Longipodacrangonyx* (one species: *Longipodacrangonyx maroccanus* Boutin & Messouli, 1988) [Morocco]
2 (1) Uropod 3 exopodite longer than peduncle, marginal spines present 3
2' Uropod 3 exopodite equal or shorter than peduncle, marginal spines absent 4
3 (2) Uropod 1 peduncle basofacial robust setae present .... *Metacrangonyx spinicaudatus* (Karaman & Pesce, 1980)
3' Uropod 1 peduncle basofacial robust setae absent .... *Metacrangonyx longicaudatus* (Ruffo, 1954)
4 (2') Uropod 3 exopodite with several terminal robust setae 5
4' Uropod 3 exopodite without one terminal robust seta 7
5 (4) Antenna 1 accessory flagellum with 2–3 articles 6
5' Antenna 1 accessory flagellum with 5 articles .... *Metacrangonyx longipes* (Chevreux, 1909)
6 (5) Telson with distal spines .... *Metacrangonyx remyi* (Balazuc & Ruffo, 1953)
6' Telson without apical spines .... *Metacrangonyx knidiiri* (Oulbaz et al., 1998)
7 (4') Uropod 1 peduncle basofacial robust setae present 8
7' Uropod 1 peduncle basofacial robust setae absent .... *Metacrangonyx ortali* (G. Karaman 1989)
8 (7) Mandibular palp with 2 or 3 articles 9
8' Mandibular palp with only 1 article 10
9 (8) Antenna 1 flagellum with only 10 articles in females (male unknown) .... *Metacrangonyx ilvanus* (Stoch, 1997)
9' Antenna 1 flagellum with 14 articles in females .... *Metacrangonyx sinaicus* (Ruffo, 1982)
10 (8') Antenna 1 flagellum with 11–14 articles 11
10' Antenna 1 flagellum with more than 15 articles 13
11 (10) Uropode 3 endopodite present but reduce, scale-like 12
11' Uropode 3 uniramous, endopodite absent .... *G. ruffoi* (Messouli, Boutin & Coineau, 1991)
12 (11) Antenna 1 article 1 with 2 fines facial spines; uropod 3 peduncle with 1 or 2 spines and one seta on ventral margin .... *Metacrangonyx aroudanensis* (Messouli, Boutin & Coineau, 1991)
12' Antenna 1 article 1 with 3 or 4 facial spines; uropod 3 peduncle with 2 spines and without seta on ventral margin .... *Metacrangonyx gineti* (Boutin & Messouli, 1988)
13 (10') Uropod 1 endopodite without baso-facial seta, telson with 2 apical setae clearly separated to each other, peduncle with one terminal seta 0.14
13' Uropod 1 endopodite with a slender baso-facial seta, telson with 2 apical setae very close to each other .... *Metacrangonyx panousei* (Ruffo, 1953)
14 (13) Habitus stocky; antenna 2 peduncle article 1 with 2 subterminal spines; uropod 3 peduncle with 1–3 dorso-apical spines; telson apex flat of slightly emarginate .... *Metacrangonyx delamarei* (Messouli, Boutin & Coineau, 1991)
14' Habitus slender; antenna 2 peduncle article 1 with 2 subterminal spines; uropod 3 peduncle with 1 dorso-apical spine; telson apex rounded .... *Metacrangonyx goulminensis* (Messouli, Boutin & Coineau, 1991)

Amphipoda: Niphargidae: genera

1 Urosomite not enlarged, not carminate; uropod 3 biramous, outer ramus with 1 or 2 articles 2
1' Urosomite enlarged, dorsally carinate, hiding urosomites 2 and 3; uropod 3 very short, outer ramus with 1 article, inner ramus absent .... *Carinurella* (one species: *Carinurella paradoxa*; Sket, 1964) [Italia, Slovenia]
2 (1) Uropod 3 exopodite of 1 article 3
2' Uropod 3 exopodite of 2 articles 6
3 (2) Uropod 3 exopodite very short, twice as long as its terminal spines; maxilla 1 palp with 1 article 4

3' Uropod 3 exopodite long, several times longer than peduncle; maxilla 1 palp with 2 articles .... *Haploginglymus* [Iberian Peninsula]

4 (3) Mandibular palp with 3 noticeable articles 5

4' Mandibular palp very short, reduced to only one small article .... *Chaetoniphargus* (one species: *Carinurella lubuskensis*; Karaman G.S. & Sket, 2019) [Croatia]

5 (4) Antenna 1 accessory flagellum with 1 article; mandibular palp article 3 with only terminal E-setae; telson slightly cleft .... *Niphargobates* (one species: *Niphargobates orophobata*; Sket, 1981) [Slovenia]

5' Antenna 1 accessory flagellum with two articles; mandibular palp article 3 with A, B, C, D and E-setae; telson deeply cleft .... *Niphargobatoides* (one species: *Niphargobatoides lefkodemonaki*; Sket, 1990) [Crete Island]

6 (2') Maxilliped palp article 4 with one, rarely 2 distal nails 7

6' Maxilliped palp article 4 with 3 distal nails .... *Exniphargus* (one species: *Exniphargus tzanisi*; G. Karaman, 2016) [Crete Island]

7. Maxilliped inner and outer plate short and broad; maxilla 2 both plates are short and broad; maxilla 1 inner plate very strong, broader than palpus, outer plate with 9–10 pectinate spines, palp short, not exceeding basis of outer plate spines .... *Foroniphargus* (one species: *Foroniphargus pori* G. Karaman 1985) [Israel]

7' Maxilliped inner and outer plate slender; maxilla 2 both plates are slender; maxilla 1 inner plate is slender, narrowed distally, outer plate with 7–9 simple to pectinate spines, palpus longer, exceeding basis of outer plate spines .... *Niphargus* [peri-Mediterranean except North Africa]

Amphipoda: Pontogammaridae: genera and species

1 Urosomites 1 et 2 with dorsal humps armed with apical spines (as in *Dikerogammarus*, but differing from it by strongly setose margins of coxae 1–4 and pereopods 5–7 basis) .... 2 (genus: *Turcogammarus*) [Greece, Turkey]

1' Urosomites 1 et 2 without dorsal elevations .... 3 (genus: *Pontogammarus*) [Central Europe, Turkey]

2 (1) Presence of a keel on the dorsum; telson lobes with setae longer than spines; antenna 1 peduncle with short setae, antenna 2 and pereopod 5–7 much less setose .... *Turcogammarus spandli* (S. Karaman, 1931)

2' No keel on the dorsum; telson lobes with setae shorter than spines; antenna 2 peduncle with some long setae; antenna 2 and pereopod 5–7 very setose .... *Turcogammarus turcarum* (Stock, 1974)

3 (1') Mandibular palp article 3 with 3 rows of setae type B and articles 2–3 with less and shorter setae; urosomite 2 with spines 4

3' Mandibular palp article 3 with 5 rows of setae type B and articles 2–3 with very long and numerous setae, much longer than the width of articles; urosomite 2 without spines .... *Pontogammarus maeticus* (Sowinsky, 1894)

4 (3) Uromosite 1 armed without setae only; coxae 1–4 and pereopods 3–5 bases with dense and numerous long setae .... *Pontogammarus aestuarius* (Derzhavin, 1924)

4' Urosomite 1 armed with spines and setae; coxae 1–4 and pereopods 3–5 bases with shorter and less numerous setae .... *Pontogammarus robustoïdes* (Sars, 1894)

Amphipoda: Pseudoniphargidae: genera

1 Body compressed and strongly curved; head anterior lobe strongly developed; coxa 1 ventrally wide; uropod 3 short, exopodite slightly (1.5 times) longer than peduncle .... *Parapseudoniphargus* (one species: *Parapseudoniphargus baetis*; Notenboom, 1988) [Spain]

1' Body outspread; head anterior lobe weakly developed; coxa 1 subrectangular; uropod 3 long, exopodite much longer than peduncle, at least 3 times .... *Pseudoniphargus* [Western Mediterranean Basin]

Amphipoda: Salentinellidae: genera and species

1 Uropod 1 longer than uropod 2; epimeral plates 2–3 as long as wide; telson cleft, elongated .... 2 (genus: *Salentinella*) [Western Mediterranean Basin]
1' Uropod 1 shorter than uropod 2; epimeral plates 2–3 elongated forming plates; telson uncleft with concave margins .... *Parasalentinella* (one species: *Parasalentinella rouchi*; Bou, 1971) [France]
2 (1) Posterodistal lobe of pereopod 7 merus almost reaching end of carpus 3
2' Posterodistal lobe of pereopod 7 merus not reaching beyond half the length of carpus 4
3 (2) Telson partially cleft .... *Salentinella seviliensis* (Platvoet, 1987)
3' Telson uncleft .... *Salentinella carracensis* (Platvoet, 1987)
4 (3') Antennae 1 and 2 peduncles without distal spines; pleopod peduncles with coupling spines; telson longer than wide 5
4' Antennae 1 and 2 peduncles with distal spines; pleopod peduncles without coupling spines; telson wider than long .... *Salentinella anae* (Messouli, Coineau & Boutin, 2002)
5 (4) Urosomite 3 lengthened, about 1.5 times as long as high 6
5' Urosomite 3 not lengthened, about as long as high 7
6 (5) Telson with dorsal setae implanted near apex of lobes .... *Salentinella petiti* (Coineau, 1968)
6' Telson with dorsal setae not implanted near apex of lobe .... *Salentinella longicaudata* (Platvoet, 1987)
7 (5') Telson deeply cleft over more than a half of lobes 8
7' Telson not cleft over 1/3 of lobes .... *Salentinella delamarei* (Coineau, 1962)
8 (7) Uropod 3 endopodite more than 15% of the exopodite 0.9
8' Uropod 3 endopodite less than 15% of the exopodite .... *Salentinella meijersae* (Platvoet, 1987)
9 (8) Uropod 1 peduncle without ventrodistal projection 10
9' Uropod 1 peduncle with a strong ventrodistal projection .... *Salentinella cazemierae* (Platvoet, 1987)
10 (9) Antenna 1 and 2 subequal in length, accessory flagellum shorter than the first peduncle article; uropod 3 endopodite longer than a half of the exopodite, exopodite article 2 less developed 11
10' Antenna 1 shorter than antenna 2, accessory flagellum longer than the first peduncle article; uropod 3 endopodite shorter than a half of the exopodite, exopodite article 2 well developed .... *Salentinella angelieri* (Ruffo & Delamare Deboutteville, 1952)
11 (10) Uropod 3 short, endopodite egal or less than 2/3 of exopodite; gnathopod 2 propodus palm strongly convex asymmetrically (largest part at 1/3 of the palm) 12
11' Uropod 3 slender, endopodite around 3/4 of exopodite; gnathopod 2 propodus palm convex symmetrically (largest part in the middle of the palm) .... *Salentinella gracillima* (Ruffo, 1947)
12 (11) Pereopod 4 dactylus shorter than a half of the propodus .... *Salentinella formenterae* (Platvoet, 1984)
12' Pereopod 4 dactylus equal to a half of the propodus .... *Salentinella gineti* (Balazuc, 1957)

Amphipoda: Typhlogammaridae: genera and species

1 Body not carinate; maxilla 1 palp symmetric; uropod 3 exopodite article 2 vestigial or absent 2
1' Body dorsally carinate, third mesosome segment and all metasome segments with one pair of dorso-lateral teeth each; maxilla 1 palp asymmetric (right maxilla different from the left one); uropod 3 exopodite article 2 ordinary .... *Metohia* (one species: *Metohia carinata*; Absolon, 1927) [Balkans]
2 (1) Antenna 2 gland cone shorter than article 3 of the peduncle; coxae 1–7 short diamond shaped or trapezoïdal; uropod 3 longer, much exceeding the top of uropods 1–2 .... *Typhlogammarus* (one species: *T. mrazeki*; Schäferna, 1907) [Balkans]
2' Antenna 2 gland cone very long overreaching article 3 of the peduncle; coxae 1–7 elongated subrectangular; uropod 3 short, not exceeding the top of uropods 1–2 .... *Accubogammarus* (one species: *Accubogammarus algor* (G. Karaman, 1973), subspecies not include in the key) [Balkans]

Subchapter 7.3

Order Bathynellacea

Ana I. Camacho[1] **and Nicole Coineau**[2]
[1]*Museo Nacional de Ciencias Naturales (CSIC), Dpto. Biodiversidad y Biología Evolutiva, Madrid, Spain*
[2]*67 Avenue du Puig del Mas, 66650 Banyuls-sur-Mer, France*

Introduction

Bathynellacea Chappuis, 1915 are obligate groundwater inhabitants (stygobionts) (limnic and oligo- to polyhaline waters) of all continents except Antarctica. Their interstitial life style would correlate with a decrease of the body size as well as with the reduction or loss of appendages and other organs (Coineau & Camacho, 2013), these features may be interpreted as evidence of an evolutionary process involving progenesis (Schminke, 1981; Schram, 1984).

The origin of Bathynellacea could be pre-Pangean. They might have been highly diversified in the Paleozoic seas, especially during the Carboniferous period. Some of these ancestors may have colonized the continental subterranean freshwaters of the Pangea as early as in the Permian and Triassic periods, and also part of the Gondwana (Camacho et al., 2018) before the Pangea break up. Thereafter, the fragmentation of Laurasia and Gondwana during the Mesozoic period (Golonka 2007) led to the current disjunct distribution of syncarids (Schram, 1977), and induced their independent evolution on each landmass. All the genera belonging to Parabathynellidae Noodt, 1965 from peninsular India show a Gondwanan heritage (Ranga Reddy 2011; Bandari et al., 2016). The most recent common ancestor of bathynellaceans can be dated back to the Triassic or Jurassic periods, as showed by our analysis (Camacho et al., 2021).

Their current distribution is important because they are restricted to freshwaters from remote geological times. They are one of the oldest groups of freshwater fauna and have no marine relatives for future colonizations, so they serve as indicators of ancient connections between the continents. They are considered a very suitable group for zoogeography studies, as they are old enough phylogenetically speaking, their capacity for dispersal is severely limited, and even today they are represented by a high number of species distributed worldwide. They can provide information on possible dispersal routes in past and present geological periods.

General ecology and distribution

Bathynellacea exclusively live in groundwater *sensu lato* (caves, springs, wells, and hyporrheic zones of rivers) as interstitial stygobionts (they inhabit the water filling the interstices between sand grains and gravels in sediments). Little is known about their life cycles. They live long, for example, *Paraiberobathynella fagei* (Delamare Deboutteville & Angelier, 1950) can live up to two and a half years. Fertilization is internal, but copulation has not been observed; they are K-strategists, that is, the female produces a single large egg that occupies almost its entire abdomen for almost 9 months (in the few cases observed to date), deposits it on the substrate, and a larva identical to the adult is immediately born but with only 3 or 4 pairs of legs, depending on the species. Metamorphosis takes place inside the egg. Due to the absence of a swimming larva, their ability to disperse is very low. The post-embryonic development or parazoeal phase (Schminke, 1981) (=larval phase 1, Serban, 1985), which occurs after birth, consists of 8–11 stages marked by molts, depending on the species.

They do not generally swim, except some species (e.g., Cho & Humphreys 2010 observed *Brevisomabathynella uramurdaliensis* swimming quite powerfully in an aquarium). They walk with their thoracopods. Some species show a swimming escape reaction. They touch the sand grains with their antennules and thus explore the interstices, resulting in discontinuous movements interrupted by frequent stops. Then they flex their pleon and leaning on their uropod and furca, they suddenly launch themselves forward, or change direction and move backwards. Thanks to their body flexibility, they can clean their antennae with the spines of their uropods and furca (personal observation). Their thigmotaxis will generally keep them in contact with the sandy sediment so that they can know their position (they are devoid of statocysts).

They graze and scrape deposits from sediment grains and also eat protists, fungi, and bacterial films. According to Serban (1980), Bathynellidae Grobben, 1905 comprise "filter and large food feeders," whereas Parabathynellidae would be "large food feeders." Some parabathynellids are carnivorous, eating harpacticoid copepods (*Iberobathynella* Schminke, 1973) or ostracods (*Brevisomabathynella* Cho, Park & Ranga Reddy, 2006). Cannibalism has been observed by Morimoto (1970), who reported large specimens of *Allobathynella* Morimoto & Miura, 1957 preying on young ones (Schminke & Cho 2013).

Bathynellacea are preyed upon by amphipods and subterranean fish.

They live in freshwater, although some species have been found in brackish groundwater, near the sea coastline. They can be considered cold stenotherms, and usually do not usually survive above 20°C. However, the genus *Thermobathynella* Capart, 1951 has been found in a hot spring (water at 55°C) and we found *Paraiberobathynella* cf *fagei* in the Sima de la Higuera (Murcia, Spain) living at 27°C.

Bathynellacea have been found from Alaska (Camacho et al., 2011) to New Zealand (Schminke, 1973), and their distribution is worldwide with high degrees of local endemic species (Camacho, 2019) and cryptic species (Camacho et al., 2011, 2018). The order Bathynellacea is composed of three families (338 species and 89 genera) (Camacho et al., 2021): Parabathynellidae, Bathynellidae, and Leptobathynellidae Noodt, 1965. Parabathynellidae (210 species, 45 genera) are currently known in temperate to tropical zones in both hemispheres; Bathynellidae (110 species, 37 genera) are known mainly in temperate zones of the northern hemisphere and some areas of the southern hemisphere (they are very frequent in Australia); while Leptobathynellidae (18 species, 7 genera) are only known in South America (Brazil, Argentina, Paraguay, and Chile), Africa (Ivory Coast and South Africa), California, and India. Parabathynellidae are more frequently found in samples than members of the other two families are, perhaps due to their larger size and consistency. Furthermore, their study is easier not only because of their size, but also because their morphological variability is greater and therefore it is easier to discriminate between genera and species. All Leptobathynellidae species are very thin and fragile, their dissection is very difficult, and their identification is almost impossible. To date, they are not known to occure in the Mediterranean Basin, but perhaps they have gone unnoticed in the samples, have escaped from the sampling nets, or they live in undersampled areas. However, we cannot rule out their presence in the Mditerranean Basin. The same happened in India: as soon as the sampling effort intensified, three *Parvulobathynella* Schminke (1973) species were found. The Bathynellidae family is characterized by its great homogeneity, so that it is very difficult to find characters that differentiate morphotypes. Large collections of specimens from this family collected all over the world have not yet been studied, because they are too difficult to identify. Morphological studies need to be supplemented with molecular data for researchers to discriminate species.

Following the emergence of molecular studies, cryptic species are being discovered in the group, both in Europe and Australia, and show that their diversity is even greater than suspected. To this we must add a significant unexplored and unsampled "underground aquatic world" (in caves and unconsolidated sediments), which is expected to yield many surprises.

In the Mediterranean Basin, 71 species belonging to 20 genera are formally described: 36 Parabathynellidae species (7 genera) and 35 Bathynellidae species (13 genera). The Iberian Peninsula is the area exhibiting the highest diversity: 41 species altogether, 30 Parabathynellidae species (5 genera), and 11 Bathynellidae species (5 genera). The *Iberobathynella* genus, with 22 species belonging to 3 subgenera from the Iberian Peninsula, is one of the most diverse in the world, behind *Allobathynella* with 25 species, and the cosmopolitan *Hexabathynella* Schminke, 1972 with 23 species distributed throughout the world (5 species in the area and 4 of them in the Iberian Peninsula) (Camacho, 2019). The greatest Bathynellidae diversity is found in France; thanks to the works of Serban, Coineau and Delamare Deboutteville (1971, 1972), 17 species (8 genera) are well documented; however, only 3 species of the family Parabathynellidae occur in France (several populations of *Parabathynella stygia* Chappuis, 1926 (=phreatica), *P. fagei*, and *Hexabathynella knoepfflery* (Coineau, 1964) from Corsica). The other findings are sporadic: one species in Israel, one in Morocco, and one in southern Tunisia.

All known species are short-range endemisms, as recent molecular studies revealed. Some species considered to be widely distributed in Spain, such as *Iberobathynella imuniensis* Camacho, 1987 (inhabitant of the Cantabrian coast, from the Pyrenees to Galicia) and *P. fagei* (described in France, in the Roussillon regin, and believed to live throughout eastern Spain, in Andalusia and Mallorca), only live in the type locality and areas nearby, while the rest of the populations is surely composed of different cryptic species, morphologically indistinguishable from the type species.

A small territory like the Iberian Peninsula is the most diverse area on earth in terms of bathynellids. The Ojo Guareña Natural Monument (Burgos) (a karst complex with more than 110 km of development), where eight species of Bathynellacea have been found (four are still undescribed: a new genus, two cryptic species of *Vejdovskybathynella edelweiss* and a cryptic species of *I. imuniensis*), is the most diverse place in the world and is considered a biodiversity hotspot of the group (Camacho et al., 2010).

Terminology and morphology

Microscopic Malacostraca are 0.5–3.0 mm in length, sometimes up to 6.3 mm, for example, *Billibathynella humphreysi* Cho, 2005. Eumalacostraca do not have a carapace and have biramous thoracopods. Their body is elongate and more or less straight (Fig. 7.5), worm-like and depigmented, female do not have a brood pouch. Eyes and statocysts are absent.

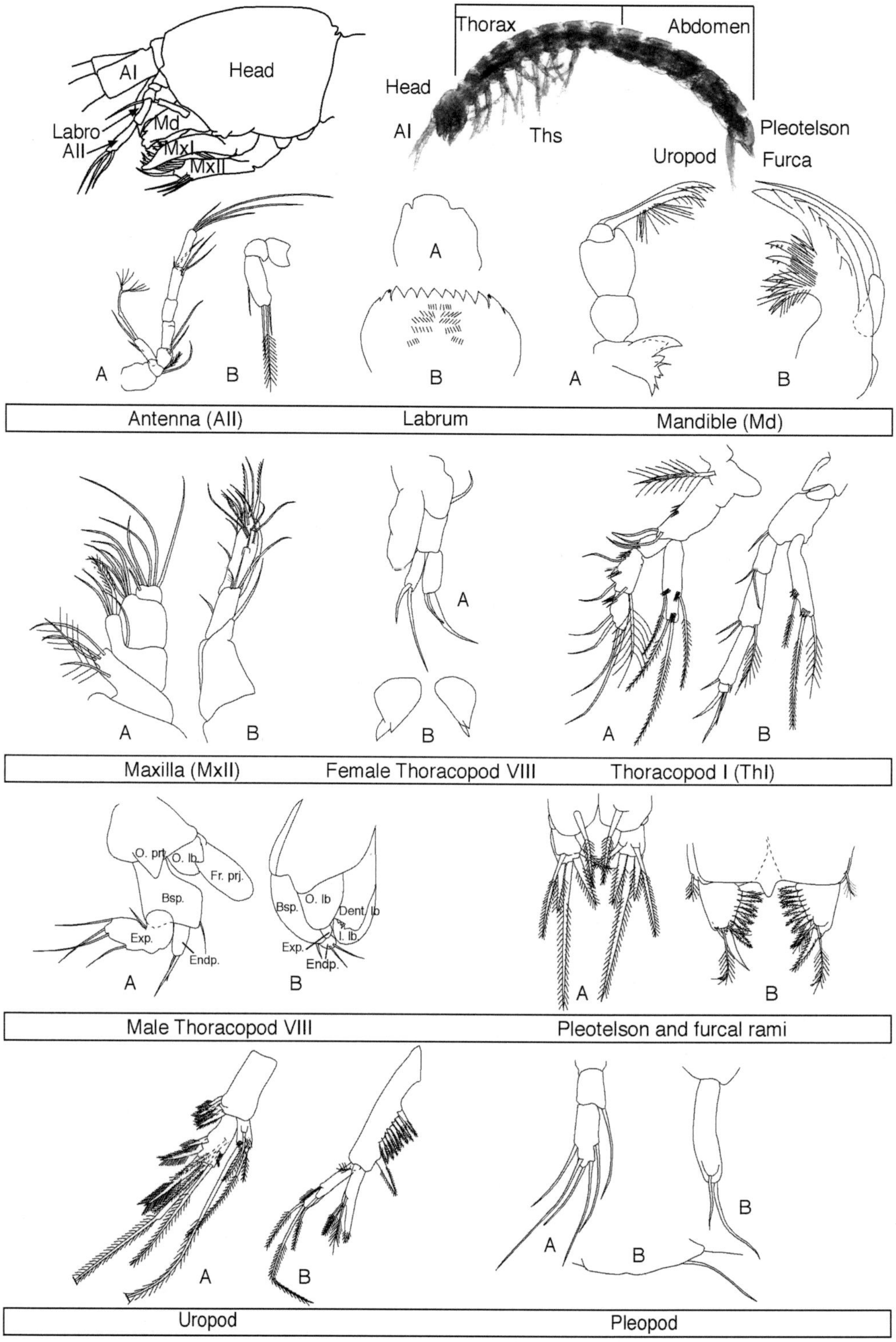

FIGURE 7.5 Head of Parabathynellidae and habitus of *Iberobathynella*. Schematic parts of Bathynellacea body. (A) Bathynellidae family and (B) Parabathynellidae family.

Their body is divided into the head, eight free thoracic segments with eight pairs of biramous thoracopods (thoracopod VIII is modified as a copulatory organ in the male and reduced in the female), five free abdominal segments without pleopods or with one or two uniramous pleopods, and a pleotelson with biramous uropods and furcal rami with spines (Fig. 7.5).

The head is longer than wide, with a fine cephalic capsule (Fig. 7.5). The anterior part has a couple of antennules (A.I) (article with sexual dimorphism in some genera) and a pair of antennae (A.II) with a variable number of articles. The five mouth pieces are located ventrally: labrum (single piece), a pair of paragnathes (present in all Bathynellidae species and a few species of genera belonging to Parabathynellidae and Leptobathynellidae) (Camacho et al., 2021), mandibles (Md), maxillules (Mx.I), and maxillae (Mx.II) (Fig. 7.5). The oral opening is located at the posterior end of the rostral third.

The thorax (Fig. 7.5) has eight well-defined segments (ancestral structure), a circular cross section of segments, with the diameter and length increasing towards the distal end. Eight pairs of thoracopods are present: the first seven ones are used for locomotion and the eighth one is transformed into a copulatory organ (penis) in males, and is greatly reduced (or even absent) in females. The demarcation between the thorax and abdomen is not very clear. Thoracopod I (Fig. 7.5) is the smallest, thoracopods II to IV or V increasing in length and the last ones of the same length. Biramous thoracopods (ThI, ThII, to ThVII), with exopod (1−12 articles in Parabathynellidae, 2 in Leptobathynellidae, and 1 in Bathynellidae) and endopod (4 articles in Parabathynellidae and Leptobathynellidae, and 3 or 4 articles in Bathynellidae) which can exhibit sexual dimorphism on the third article of Th VI (*Pacificabathynella* Schminke & Noodt, 1988 and *Paradoxibathynella* Serban, 2000) and on coxa of ThVII (Camacho et al., 2020); respiratory epipod (absent on ThI and sometimes on ThII and/or III), connected with coxa and basipod. Usually ctenidiae are present at the base of the setae on the exopod and along one of the articles on the endopod. ThVII is absent in *Hexabathynella* and *Hexaiberobathynella* Camacho & Serban, 1998.

Female ThVIII (Fig. 7.5) is smaller than other legs, very reduced in *Hexabathynella* and *Iberobathynella* (Fig. 7.5B), and the remains of articles are indistinguishable. It is still possible to identify the articles in some Bathynellidae (Fig. 7.5A) genera, although they are always very reduced.

Male ThVIII is massive, more or less rounded or curved; strongly modified but distinguishable articles (like lobes) of a typical leg. The terms used to designate each part are equivalent in the three families: (1) protopod, penial complex and basipod; basal region, outer lobe, inner lobe, and dentate lobe constitute the penial complex (these last three lobes are well distinguishable in Parabathynellidae (Fig. 7.5B); the genital opening is located between the lobes and some setae are usually found on the basipod; (2) endopod, only one article and setae; (3) exopod, only one article with some setae on the ventral edge. Some structures such as basipod and penial complex are complicated in the Bathynellidae family (Fig. 7.5A).

Five segments consitute the abdomen and the pleotelson is with furcal rami and uropods (Fig. 7.5 Habitus). Sometimes vestigial pleopods on the first and second segments. Setae smooth, barbed, or plumose basidorsally or basiventrally are present on the pleotelson.

The uropod lateral has sympod (with spines), exopod, and endopod, with specific setation. Furcal rami (with spines) is located on the pleotelson lobes.

Anus located in the terminal part of the pleotelson and covered by an operculum (anal operculum) in some Parabathynellidae species.

Collection, preparation, and identification

The methods for collecting samples depend on the specific habitat where the fauna is found in groundwater *sensu lato*. However, almost all of them are based on substrate removal and on filtration of the cloudy water with hand nets of *ad hoc* designs and a mesh size < 0.1 mm.

The aquatic environments sampled within caves are gours, puddles, pools, drips, lakes, and underground rivers. Outside, springs, wells, boreholes, permanent cores (for example in mining surveys) that intersect aquifers, and a hyporheic substratum (in lakes and rivers) should be sampled. Bathynelids have recently been found in samples from the Mesovoid Shallow Substratum (MSS) of the Sierra de Guadarrama (Madrid, Spain) arranged for the collection of terrestrial fauna (Camacho & Ortuño, 2019). Therefore sampling methods and devices are numerous and varied. A comprehensive compilation can be viewed at: *Sampling Manual for the Assessment of Regional Groundwater Biodiversity*, PASCALIS 2002 (Ed. by F. Malard, Asociate editors: Dole-Olivier, M.-J., J. Mathieu & F. Stoch).

Field samples can be kept alive in a refrigerator, frozen, or in absolute or 70% ethanol (this is a conservative liquid, but not a fixative one, so that the specimens deteriorate over time). In the case of living samples, it is convenient to wash and separate them as soon as possible. In this way, each type of fauna can be fixed in the most suitable fixative and according to the subsequent study you want to carry out. For morphological studies, it is appropriate to use 4%

formalin for 48 hours, while keeping the specimens in a refrigerator at 4°C. After washing, they can be immediately prepared for dissection and assembled in permanent preparations (Perina & Camacho, 2016), or the entire specimens can be preserved in 70% ethanol for subsequent studies. Bathynels have thin and fragile cuticles. If the specimens are to be preserved for long-term morphological studies, they must be fixed properly before being stored in the preservative liquid. When absolute or 70% ethanol is used for field samples, the specimens should also be washed and processed as soon as possible and thus be suitable for immediate molecular studies and short-term morphological studies as well. This way, the same specimen can and should be used for both studies: the abdomen for DNA extraction and the other parts for dissection and mounting in permanent preparations (after dipping in 4% formalin for 48 hours), with the same routine as in the previous case (Perina & Camacho, 2016). This procedure ensures that the sequenced genes correspond to a given morphotype. Samples frequently contain a mixture of species indistinguishable under the binocular microscope. If some specimens are used entirely for morphological studies and others are entirely destroyed by DNA extraction, it will be impossible to associate each gene sequence to a specific morphotype.

Taxonomic identification and the description of new species require a complete morphological study of specimens, males and females if possible. Anatomical observations can be performed using an oil immersion lens (100 × magnification) with an interference microscope equipped with a drawing tube. Photographs are not a substitute for drawings in taxonomic works but can complement them.

Well-prepared permanent preparations ensure good preservation of the specimens in scientific collections. They guarantee their survival and availability, at least on the medium term just as well-preserved DNA extracts and DNA collections of public institutions ensure access to them for future scientific studies.

Keys to Bathynellacea

Bathynellacea: families

1 Lengthened powerfully cephalic capsule; AI with 6 or 7 articles; 6–8 articles in AII, directed forward or perpendicular, with very well-developed exopod (Fig. 7.5A); labrum with smooth distal free edge (Fig. 7.6D); paragnathes present; Md, length/width = 3 (very long) (Fig. 7.6G), prehensile palp with 1–3 articles (Fig. 7.6H) in antero-ventral position, reduced *pars incisiva* (in antero medial part) and typical lobe (plate like) of *pars molaris* absent; MxII not prehensile (Fig. 7.6K); exopod of thoracopods with 1 article (Fig. 7.7A,B); male ThVIII remenbers "a leg" (Fig. 7.6D); pleopods of 2 articles; 3 or more oblique spine rows on sympod of uropod (Fig. 7.7F,H); dorsal seta on pleotelson .... **Bathynellidae** (Fig. 7.5)

2 Weakly elongated cephalic capsule; AI with 6–10 articles; 1–6 articles in AII, directed backwards, without exopod (Fig. 7.6B); labrum with dentate distal free edge (Fig. 10.2.E); paragnathes absent or rarely present; Md, length/width = 0.8 (high), with palp not prehensile of 1 article (Fig. 7.6I) in antero-dorsal position, large *pars incisiva* (occupying entire distal part) and typical lobe (plate like) of *pars molaris* present; MxII not prehensile (Fig. 7.6L); exopod of thoracopods multiarticled (Fig. 7.7C); male ThVIII different of "a leg" (Fig. 7.7E); when present (rarely) pleopod reduced; 5–30 inner spine row on sympod of uropod (Fig. 7.7G); ventral or dorso-lateral seta on pleotelson .... **Parabathynellidae** (Fig. 7.6)

3 Elongated cephalic capsule; AI with 6 articles; 5–7 articles in AII, directed backwards, with very rudimentary exopod (Fig. 7.6C); labrum with poory armed distal free edge (Fig. 7.6F); paragnathes present; Md, length/width = 1.5 (long), palp not prehensile with 1 or 2 articles (Fig. 7.6J), in anterodorsal position, reduced *pars incisiva* (occupying anterodistal part) and typical lobe (plate like) of *pars molaris* absent; MxII prehensile (Fig. 7.6M); exopod of thoracopods with 2 articles; male ThVIII not remenbering "a leg"; absent pleopods; 2 similar distal spines on sympod of uropod; no seta on pleotelson .... Leptobathynellidae (Fig. 7.7)

Bathynellacea: Bathynellidae: subfamilies

1 Penial region of male ThVIII with 3 lobes; AI with 7 articles; AI >, = or < AII; 3 articles in mandibular palp without sexual dimorphism; endopod of ThI to VII with 4 articles; female ThVIII with 1 article; 4 spines as minimum on sympod of uropod; 3–4 claws on endopod of uropod (Fig. 7.7H) .... **Bathynellinae**

1' Penial region of male ThVIII with 1 lobe; AI with 6 or 7 articles; AI > or = AII; 1–3 articles in mandibular palp with or no sexual dimorphism; endopod of ThI to VII with 3 or 4 articles; endopod of female ThVIII with 1 or 2 articles; 4 spines as maximum on sympod of uropod; 2–4 claws on endopod of uropod (Fig. 7.7F) .... **Gallobathynellinae**

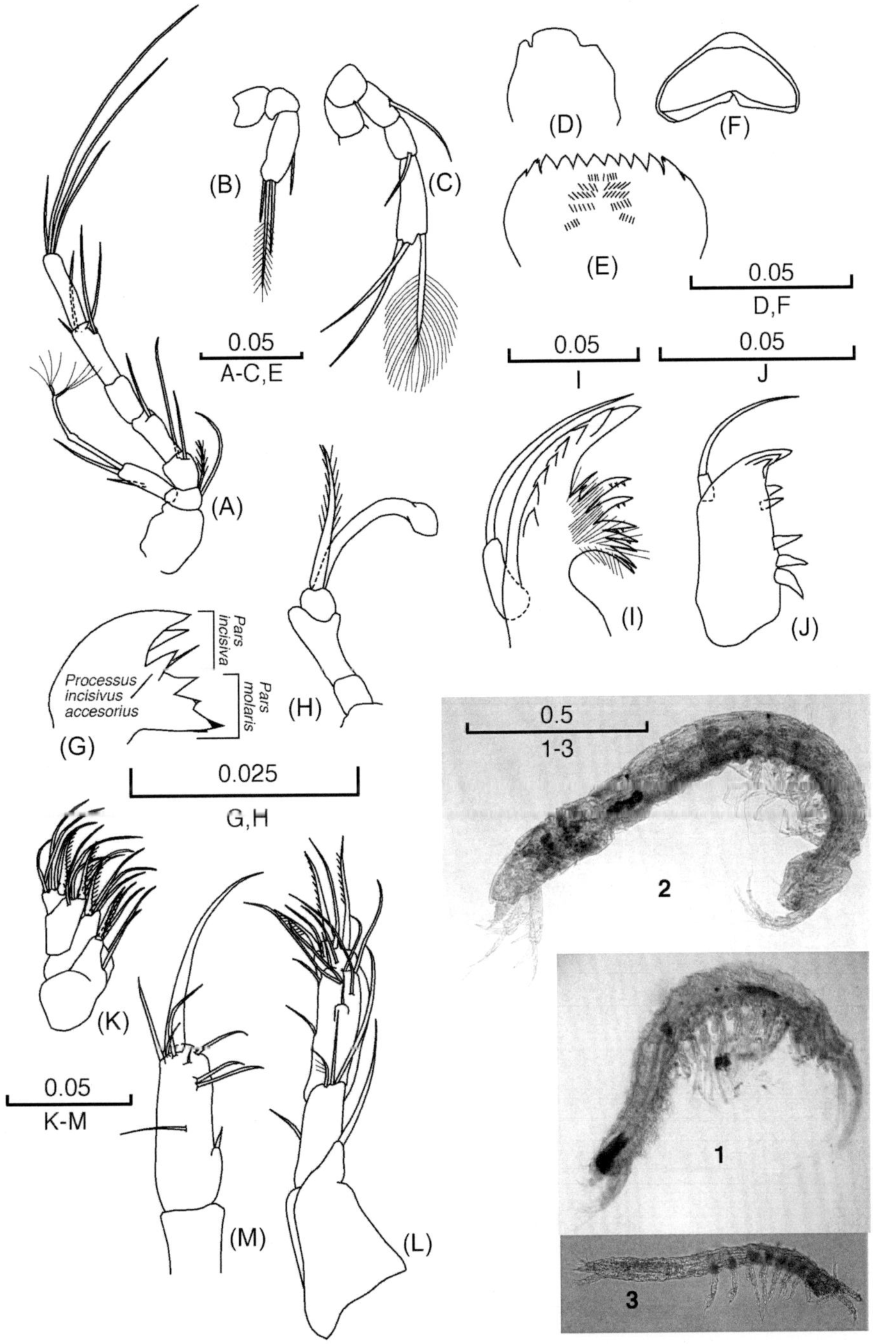

FIGURE 7.6 Habitus: 1, *Vejdovskybathynella edelweiss* (Camacho, 2007); 2, *Iberobathynella andalusica* (Camacho, 2007) and 3, *Parvulobathynella distincta* (Ranga Reddy, Bandari & Totakura, 2011). AII: (A) *Vejdovskybathynella edelweiss*, (B) *Iberobathynella andalusica* and (C) *Parvulobathynella distincta*. Labrum: (D) *Vejdovskybathynella edelweiss*, (E) *Iberobathynella andalusica*, and (F) *Parvulobathynella distincta*. Md: (D, E) *Vejdovskybathynella edelweiss*, (I) *Iberobathynella andalusica*, and (J) *Parvulobathynella distincta*. MxII: (K) *Vejdovskybathynella edelweiss*; (L) *Iberobathynella andalusica*, and (M) *Parvulobathynella distincta*.

Bathynellacea: Bathynellidae: Bathynellinae: genera

1 Six similar spines on sympod of uropod (Fig. 7.7I); 4 claws on endopod of uropod; first & second spines of furca the largest (Fig. 7.7J); mandibular palp without sexual dimorphism with 3 articles; 7 articles in AII; AI = AII; seta on medial segment of exopod of AII; 4 articles in endopod of ThI to VII .... *Bathynella*

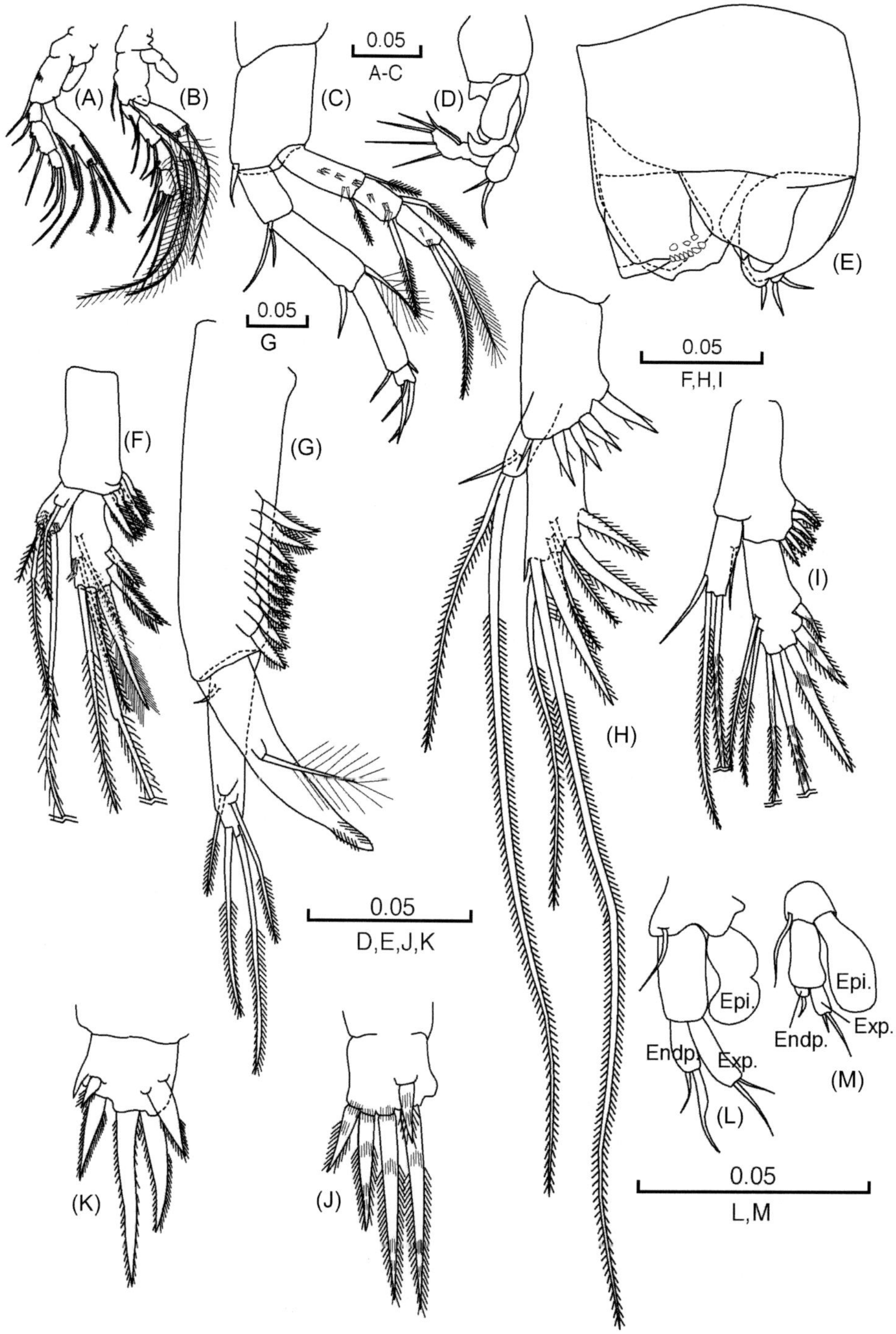

FIGURE 7.7 Thoracopods: (A) ThIII *Gallobathynella coiffaiti* (Delamare Deboutteville, 1961); (B) ThIII *Vejdovskybathynella edelweiss*; (C) ThI *Paraiberobathynella notenboomi* (Camacho, 1989). Male ThVIII: (D) *Vejdovskybathynella edelweiss*; (E) *Paraiberobathynella notenboomi*. Uropods: (F) *Vejdovskybathynella edelweiss*, (G) *Paraiberobathynella notenboomi*, (H) *Antrobathynella stammeri* (Jakobi, 1954), (I) *Bathynella paranatans* (Serban 1972). Furca: (J) *Bathynella paranatans*; (K) *Antrobathynella stammeri*. Female ThVIII: (L) *Paradoxiclamousella pirata* (Camacho, Dorda & Rey, 2013); (L) *Paradoxiclamousella fideli* (Camacho, Dorda & Rey, 2013).

1' Four different spines on sympod of uropod (Fig. 7.7H); 3 claws on endopod of uropod; second spines of furca the largest (1.1 times 3th) (Fig. 7.7K); mandibular palp with sexual dimorphism and 3 articles; 7 articles in AII; AI = AII; seta on medial segment of exopod of AII; 4 articles in endopod of ThI–VII .... *Antrobathynella* (one species: *Antrobathynella stammeri*; Jakobi, 1954) [Slovenia]

Bathynellacea: Bathynellidae: Gallobathynellinae: genera

1 AI with 6 articles; AII with 6 or 7 articles; AI > AII; Md palp with 3 articles without sexual dimorphism; ThI endopod with 4 articles 2
1' AI with 7 articles; 7 or 8 articles in AII; AI>, < or = AII; Md palp of 3 articles, with or without sexual dimorphism; ThI endopod with 3 or 4 articles 4
2 (1) AI with 6 articles; medial seta absent on exopod of AII; 3 articles on Md palp; ThI endopod of 4 articles; ThII to V endopod with 3 articles and ThVI–VII endopod with 3 or 4 articles; female ThVIII of 1 simplified article, without epipod 3
2' AII with 7 articles; medial seta present on AII exopod; Md palp of 3 articles; endopod ThI to VII with 4 articles; large female ThVIII epipod with 2 articles .... *Parameridiobathynella* (one species: *Parameridiobathynella gardensis* Serban & Leclerc, 1984) [France: Gard]
3 (2) ThII–VII with 3 articles; 5 spines on furca, the second longest .... *Meridiobathynella*
3' ThII to V endopod with 3 articles; ThVI–VII endopod with 4 articles; 5 spines on furca, the second very, very long .... *Hispanobathynella* (one species: *Hispanobathynella catalanensis* Serban, Coineau & Delamare Deboutteville, 1971) [Spain: Gerona]
4 (1) AI with 7 articles; AII with 7 articles; AI> or = AII; Md palp sexually dimorphic with 3 articles; ThI-V four-segmented endopod 5
4' AI with 7 articles; AII with 7 or 8 articles; AI>, < or = AII; Md palp sexually or not dimorphic with 1–3 articles; ThI–VII endopod with 4 articles or ThI–V endopod with 3 articles 6
5 (4) ThI–VII with 4 articles on endopod; epipod present on female ThVIII; 4 spines on sympod and 2 claws on endopod of uropod; 5 spines of different size on furca .... *Vejdovskybathynella*
5' ThI–V endopod with 4 articles and ThVI and VII endopod with 3 articles; female ThVIII without epipod; 3 spines on sympod and 4 claws on endopod of uropod; 5 spines of similar size on furca .... *Sardobathynella* (one species: *Sardobathynella cottarelli*; Serban, 1973) [Italy: Sardinia]
6 (4') AI with 7 articles; AII with 7 or 8 articles; ThI–V endopod with 3 articles; 1–3 articles on Md palp (dimorphic or not) 7
6' AI with 7 articles; AII with 8 articles; ThI–VII endopod with 4 articles; Md palp sexually dimorphic with 3 articles 10
7 (6) AI with 7 articles; AII with 7 articles; AI = AII; Md palp with 2 articles; well-developed endopod and exopod of female ThVIII; 4 spines on sympod and 2 claws on endopod of uropod; 5 spines of different size on furca .... *Clamousella* (one species: *Clamousella delayi* Serban, Coineau & Delamare Deboutteville, 1971) [France: Hérault]
7' AI with 7 articles; AII with 8 articles; AI> or < AII; Md palp with 1–3 articles; endopod and exopod of female ThVIII well developed or endopod absent; 3 or 4 spines on sympod and 2 or 3 claws on endopod of uropod; 5 spines of similar or different size on furca 8
8 (7) AI > AII; medial seta of AII exopod absent or present; Md palp with 2 or 3 articles; female ThVIII with well-developed endopod and exopod; 4 spines on sympod and two or three claws on endopod of uropod; 5 spines of similar or different size on furca 9

8' AI < AII; medial seta of exopod of AII present; Md palp of 1 article; endopod and exopod of female ThVIII absent; 3 spines on sympod and 2 claws on endopod of uropod; 5 spines of similar size on furca .... *Vandelibathynella* (one species: *Vandelibathynella vandeli*; Delamare Deboutteville & Chappuis, 1954) [France: Ariège]

9 (8) Medial seta of exopod of AII absent; Md palp not dimorphic with 2 or 3 articles; female ThVIII with large endopod and exopod; 4 spines on sympod and 2 or 3 claws on endopod of uropod; 5 spines of similar or different size on furca .... ***Gallobathynella***

9' Medial seta of exopod of AII present; Md palp dimorphic or not, with 3 articles; very small endopod of female ThVIII; 4 spines on sympod and 2 claws on endopod of uropod; 5 spines of similar size on furca .... ***Paradoxiclamousella***

10 (6) Medial seta of exopod of AII present; sexual dimorphism on mandibular palp; very short basipodal seta and well-developed endopod and exopod of female ThVIII; 3 or 4 spines on sympod and 4 claws on endopod of uropod .... *Pseudobathynella* (one species: *Pseudobathynella magniezi* Serban, Coineau & Delamare Deboutteville, 1972) [France: Allier]

10' Medial seta of exopod of AII absent; sexual dimorphism on labrum and mandibular palp; *pars molaris* of Md with complex morphology; very long basipodal seta and well-developed endopod and exopod on female ThVIII; 4 spines on sympod and 2 claws on endopod of uropod .... ***Delamareibathynella***

Bathynellacea: Bathynellidae: Gallobathynellinae: *Delamareibathynella:* species

1 One dentate lobe on *pars molaris* of Md .... *Delamareibathynella debouttevillei*; Serban, 1989 [France: Ardéche]

1' Two dentate lobes on *pars molaris* of Md .... *Delamareibathynella motasi*; Serban, 1992 [France: Allier]

Bathynellacea: Bathynellidae: Gallobathynellinae: *Gallobathynella:* species

1 Three or 4 teeth on *pars molaris* of Md 2

1' Two teeth on *pars molaris* of Md .... *Gallobathynella hispanica* (Delamare Deboutteville, 1954) [Spain: Tarragona]

2 (1) Pars molaris of Md with 3 teeth and 2 setae on mandibular palp of Md; 2 claws and 3 or 4 setae on endopod and 3 or 4 setae on exopod of uropod 3

2' *Pars molaris* with 4 teeth and 3 setae on mandibular palp of Md; 2 claws and 4 setae on endopod and 2 setae on exopod of uropod; second spine 3 times longer than first spine which is twice as long as dorsal spine and similar to fourth of furca .... *Gallobathynella juberthie* (Serban, Coineau & Delamare Deboutteville, 1971) [France: Pyrénées-Orientales]

3 (2) Similar 3 or 4 spines on sympod of uropod; 3 setae on exopod of uropod; epipod longer than basipod of female ThVIII 4

3' Uropod sympod with 4 spines, first the longest; 4 setae on exopod of uropod; epipod as long as basipod and exopod similar to endopod of female ThVIII; second spine 2.5 times longer than first spine which is as long as dorsal and fourth spines of furca .... *Gallobathynella boui* (Serban, Coineau & Delamare Deboutteville, 1971) [France: Hérault]

4 (3) Three spines on sympod of uropod; 4 setae on endopod of uropod; longer than basipod seta of the coxopod and exopod >> than endopod of female ThVIII; very small endopod of male ThVIII; second spine 4 times longer than first spine which is as long as dorsal spine of furca .... *Gallobathynella coiffaiti* (Delamare Deboutteville, 1961) [France: Hérault]

4' Four spines on sympod of uropod; 3 setae on endopod of uropod; no seta on coxopod and exopod > than endopod of female ThVIII; distal mamill on frontal projection of basipod and small endopod of male ThVIII; second spine almost as long as first spine which is longer than third and fourth, similar length of dorsal spine of furca .... *Gallobathynella tarissei* (Serban, Coineau & Delamare Deboutteville, 1971) [France: Aude]

Bathynellacea: Bathynellidae: Gallobathynellinae: *Paradoxiclamousella:* species

1 Barbed terminal claws of Md palp; two barbed setae on basipod of ThI–II; 12 setae on endopod of ThI; 4 setae on third article of endopod of ThII to IV; 5 setae on exopod of ThI to VII; almost square basipod of male ThVIII

without seta; epipod and coxal seta smaller than basipod of female ThVIII; exopod as long as basipod of female ThVIII; 2 setae on endopod of female ThVIII (Fig. 7.7L); 6 setae on pleopod; third spines twice as long as the first and second on furca .... *Paradoxiclamousella pirata* (Camacho, Dorda & Rey, 2013) [Spain: Cantabria]
1' Smooth terminal claws (very long on male) of Md palp; 2 smooth setae on basipod of ThI-II; 10 setae on endopod of ThI; ThII to IV endopod with 3 setae on third article; ThI to VII exopod with 4 setae; rectangular basipod of male ThVIII with 1 seta; epipod and coxal seta longer than basipod of female ThVIII; exopod as long half basipod of female ThVIII; 1 seta on very small endopod of female ThVIII (Fig. 7.7M); 5 setae on pleopod; third spines slightly longer than others of furca .... *Paradoxiclamousella fideli* (Camacho, Dorda & Rey, 2013) [Spain: Asturias]

Bathynellacea: Bathynellidae: Gallobathynellinae: *Vejdovskybathynella:* species

1 Fifth article without setae and 1 or 2 setae on second article of AII; medial seta of exopod of AII absent or present; no or 1 seta on first article and 1 or 2 setae on second article of ThIV endopod; very large, large or small frontal projection with spur on basipod of male ThVIII; 3 or 4 spines on sympod and 2 or 3 claws on endopod of uropod; furca with second spine longer than first 2
1' Second article of AII without setae and 1 seta on fifth article; medial seta of AII exopod absent; 2 setae on first and second articles of endopod of ThII to IV; very large, frontal projection without spur on basipod of male ThVIII; 4 spines on sympod and 2 claws on endopod of uropod; second spine twice as long as the first on furca ....
.... *Vejdovskybathynella leclerci* (Serban, 1989) [France: Ardèche]
2 (1) One seta on second article of AII; medial seta of exopod of AII absent or present; ThIII to V endopod with 3 setae on fourth article; dorsal spine as long as first spine and second spine 4 times as long as the first on furca 3
2' Two setae on second article of AII; medial seta of exopod of AII absent or present; ThI to IV endopod with 4 setae on fourth article; dorsal spine > or < than first spine and second spine 1.5, 2.0, or 2.5 times as long as the first on furca 5
3 (2) AII exopod with medial seta; exopod > endopod of female ThVIII; large endopod of male ThVIII; 5 or 7 setae on pleopod; 4 similar spines on sympod of uropod; 2 claws on endopod of uropod 4
3' AII exopod without medial seta; exopod >> endopod of female ThVIII; small endopod of male ThVIII; 6 setae on pleopod; uropod sympod with distal spines longer than other three spines; 3 claws on endopod of uropod ....
.... *Vejdovskybathynella caroloi* (Camacho, 2007) [Spain: Cantabria]
4 (3) Very large frontal projection of basipod of male ThVIII; exopod of male ThVIII four times longer than wide; 5 setae on pleopod .... *Vejdovskybathynella balazuci* (Serban & Leclerc, 1984) [France: Ardèche]
4' Male ThVIII basipod with large frontal projection; exopod of male ThVIII 3 times longer than wide; 7 setae on pleopod .... *Vejdovskybathynella espattyensis* (Serban & Leclerc, 1984) [France: Ardèche]
5 (2') AII exopod with or without medial seta; Md palp with sexual dimorphism; exopod = or >> endopod of female ThVIII; male ThVIII with small outer protuberance; male ThVIII exopod 2 or 3 times longer than wide; 4 spines on sympod and 2 or 3 claws on endopod of uropod; dorsal spine > than first spine of furca 6
5' AII exopod with medial seta; Md palp without sexual dimorphism; exopod < endopod of female ThVIII; male ThVIII with medium sized outer protuberance; exopod of male ThVIII 4 times longer than wide; 3 similar spines on sympod and 2 claws on endopod of uropod; dorsal spine < than first spine of furca ....
.... *Vejdovskybathynella vasconica* (Camacho, Dorda & Rey, 2013) [Spain: Vizcaya]
6 (5) Fourth article of AII with 1 seta and no seta on first article; AII exopod with medial seta; exopod = endopod of female ThVIII; male ThVIII with small frontal projection on basipod and large endopod; exopod of male ThVIII 2 times longer than wide; 4 similar spines on sympod and 3 claws on endopod of uropod ....
.... *Vejdovskybathynella edelweiss* (Camacho, 2007) [Spain: Burgos]
6' Fourth article of AII with 2 setae and 1 seta on first article; AII exopod without medial seta; exopod >> endopod of female ThVIII; male ThVIII with large frontal projection on basipod and very large endopod; exopod of male ThVIII 3 times longer than wide; 4 different spines (the basal smallest) on sympod and 2 claws on endopod of uropod .... *Vejdovskybathynella pascalis* (Camacho, 2007) [Spain: Cantabria]

Bathynellacea: Parabathynellidae: genera

1 Six pairs of thoracopods 2
1' Seven pairs of thoracopods 3

2 (1) AI with 6 articles; AII with 5 articles (Fig. 7.8A) .... ***Hexabathynella***
2' AI with 7 articles; AII with 3 articles (Fig. 7.8B) .... ***Hexaiberobathynella***
3 (1) AI with 6 or 7 articles; AII with 4 or 5 articles (Fig. 7.8C) 4
3' AI with 6 or 7 articles; AII with 3 articles (Fig. 7.8D) 5
4 (3) AI with 6 articles; AII with 5 articles (Fig. 7.8C); Furca with 3 spines (Fig. 7.8E) .... *Ctenobathynella* (one species: *Ctenobathynella calmani*; Por, 1968) [Israel: Tiberias Lake]
4' AI with 7 articles; AII with 4 or 5 articles; furca with 5 or 6 spines (Fig. 7.8F) .... ***Parabathynella***
5 (3') AI with 6 or 7 articles; 2 articles on exopod of ThII to V 6
5' AI with 7 articles; 3 or 4 articles on exopod of ThII to V .... ***Paraiberobathynella***
6 (5) AI with 6 or 7 articles; labrum with hook-like teeth (Fig. 7.8G); claws very curved on distal endite of MxI (Fig. 7.8I) .... *Guadalopebathynella* (one species: *Guadalopebathynella puchi*; Camacho & Serban, 1998) [Spain: Teruel]
6' AI with 7 articles; labrum with "normal teeth" (Fig. 7.8H); "normal claws" on distal endite of MxI (Fig. 7.8J)....... .... ***Iberobathynella***

Bathynellacea: Parabathynellidae: *Hexabathynella*: species

1 Md *pars molaris* with 5 or 6 teeth; 4 teeth on distal endite of Mx; 1 or 2 setae on first article of MxII; article fourth of ThI endopod with 2 or 3 setae; first pleopod absent or present; uropod inhomonomous or homonomous with 3–7 spines on sympod; uropod with 3 or 4 setae on exopod and 2 or 3 on endopod; 3 or 9–13 spines on furca 2
1' Md *pars molaris* with 4 teeth; distal endite of Mx with 5 teeth; 1 setae on first article of MxII; fourth article of ThI endopod with 3 setae; first pleopod present; 2 spines on sympod of inhomonomous uropod; uropod with 2 setae on exopod and 2 on endopod; 6–7 spines on furca .... *Hexabathynella knoepffleri* (Coineau, 1964) [France: Corsica]
2 (1) Md *pars molaris* with 5 or 6 teeth; 2 setae on first article of MxII; fourth article of ThI endopod with 2 setae; first pleopod absent or present; 6 or 7 spines on inhomonomous or homonomous sympod of uropod; 3 spines on furca 3
2' Md *pars molaris* with 5 teeth; 1 setae on first article of MxII; fourth article of ThI endopod with three setae; first pleopod present; 3 spines on inhomonomous sympod of uropod; 3 setae on exopod and 3 on endopod of uropod; 9–13 spines on furca .... *Hexabathynella minuta* (Noodt & Galhano, 1969) [Portugal: Barça. Spain: Sevilla]
3 (2) Fourth article of AII without seta or 1 seta; Md *pars molaris* with 5 teeth; first pleopod absent; sympod of uropod inhomonomous 4
3' Fourth article of AII with 2 setae; Md *pars molaris* with 6 teeth; first pleopod present; sympod of uropod homonomous; 3 setae on exopod and 2 on endopod of uropod .... *Hexabathynella sevillaensis* (Camacho, 2003) [Spain: Sevilla]
4 (3) Fourth article of AII with 1 seta; distal spine of Md not modified; second article of MxII with 3 setae; 4 setae on exopod and 3 on endopod of uropod .... *Hexabathynella valdecasasi* (Camacho, 2003) [Spain: Toledo]
4' Fourth segment of AII without seta; distal spine of Md modified; second segment of MxII with 4 setae; 3 setae on exopod and 2 on endopod of uropod .... *Hexabathynella nicoleiana* (Camacho, 1986) [Spain: Madrid, Guadalajara]

Bathynellacea: Parabathynellidae: *Hexaiberobathynella*: species

1 Mandibular palp not exceeding distal part of Md *pars incisiva*; 1 tooth on female ThVIII (Fig. 7.8K); 5–8 spines on sympod of uropod; 5–8 spines on furca .... *Hexaiberobathynella mateusi* (Galhano, 1967) [Portugal: Coimbra, Barcelos. Spain: Castellón, Granada, Guadalajara, Huesca, Jaén, Madrid, Soria, Teruel, Toledo]
1' Mandibular palp exceeding distal part of Md *pars incisiva*; 3 teeth on female ThVIII (Fig. 7.8L); 10–13 spines on sympod of the uropod; 8–10 spines on furca .... *Hexaiberobathynella hortezuelensis* (Camacho & Serban, 1998) [Spain: Soria]

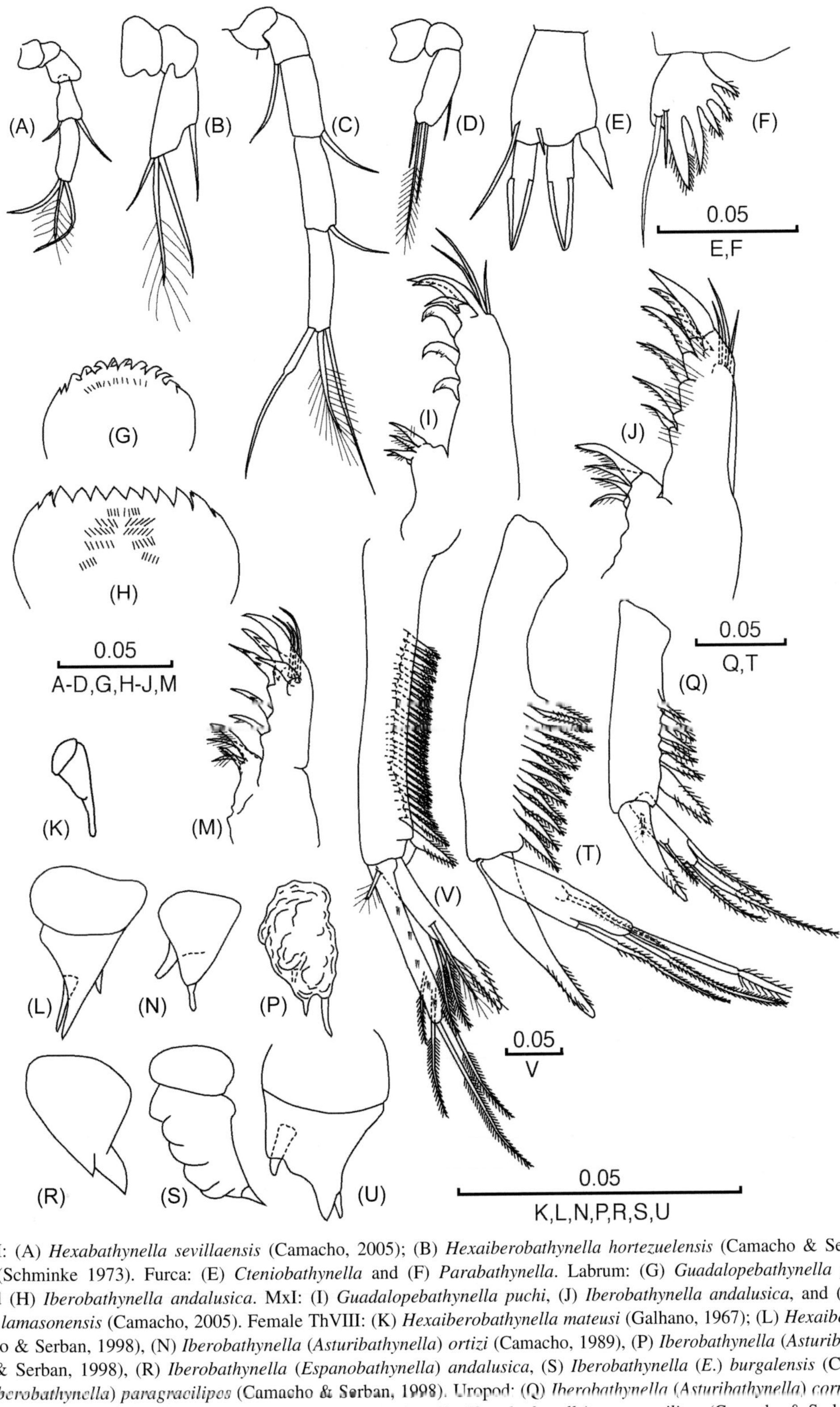

FIGURE 7.8 AII: (A) *Hexabathynella sevillaensis* (Camacho, 2005); (B) *Hexaiberobathynella hortezuelensis* (Camacho & Serban, 1998) and C, *Cteniobathynella* (Schminke 1973). Furca: (E) *Cteniobathynella* and (F) *Parabathynella*. Labrum: (G) *Guadalopebathynella puchi* (Camacho & Serban, 1998) and (H) *Iberobathynella andalusica*. MxI: (I) *Guadalopebathynella puchi*, (J) *Iberobathynella andalusica*, and (M) *Iberobathynella* (*Asturibathynella*) *lamasonensis* (Camacho, 2005). Female ThVIII: (K) *Hexaiberobathynella mateusi* (Galhano, 1967); (L) *Hexaiberobathynella hortezuelensis* (Camacho & Serban, 1998), (N) *Iberobathynella* (*Asturibathynella*) *ortizi* (Camacho, 1989), (P) *Iberobathynella* (*Asturibathynella*) *parasturiensis* (Camacho & Serban, 1998), (R) *Iberobathynella* (*Espanobathynella*) *andalusica*, (S) *Iberobathynella* (*E.*) *burgalensis* (Camacho, 2007), (U) *Iberobathynella* (*Iberobathynella*) *paragracilipes* (Camacho & Serban, 1998). Uropod: (Q) *Iberobathynella* (*Asturibathynella*) *cornejoensis* (Camacho, 2005), (T) *Iberobathynella* (*Asturibathynella*) *burgalensis*, (V) *Iberobathynella* (*Iberobathynella*) *paragracilipes* (Camacho & Serban, 1998).

Bathynellacea: Parabathynellidae: *Parabathynella*: species

1 AII with 5 articles; 9 teeth on labrum; 5 teeth on distal endite of MxI; 1 seta on first article of MxII; 2 setae on second and third articles of endopod of ThI; 5 spines on furca .... *Parabathynella motasi* (Dancau & Serban, 1973) [Slovenia]

1' AII with 4 articles; 12 teeth on labrum; 6 teeth on distal endite of MxI; 2 setae on first article of MxII; 3 setae on second and 1 on third article of ThI endopod; 6 spines on furca .... *Parabathynella stygia* (Chappuis, 1926) [Slovenia]

Bathynellacea: Parabathynellidae: *Paraiberobathynella*: species

1 Labrum with 8 main teeth; Md *pars incisiva* with 6–8 teeth; Md *pars molaris* with 6–11 teeth; inhomonomous sympod of the uropod; 7 or 8 spines on furca; exopod of ThI with 2 or 3 articles 2

1' Labrum with 9–11 teeth; Md *pars incisiva* with 5–9 teeth; Md *pars molaris* with 12 teeth; homonomous sympod of the uropod; 8–12 spines on furca; exopod of ThI, II, VI, and VII with 3 articles (Fig. 7.7C), and ThIII–V with 4 articles .... *Paraiberobathynella notenboomi* (Camacho, 1989) [Spain: Alicante]

2 (1) Exopod of ThI, II, and VI with 3 articles and 4 on ThIII to VI .... *Paraiberobathynella fagei* (Delamare Deboutteville & Angelier, 1950) [France: Le Boulou. Spain: Alicante, Almería, Asturias, Cádiz, Castellón, Gerona, Granada, Huesca, León, Lérida, Málaga, Mallorca, Murcia, Navarra, Orense, Teruel, Valencia]

2' Exopod of ThI with 2 articles and 3 on ThIII to VII .... *Paraiberobathynella maghrebensis* (Boutin & Coineau, 1987) [Morocco: Marrakech, Nador]

Bathynellacea: Parabathynellidae: *Iberobathynella*: species

1 Distal endite of MxI with 7 teeth (Fig. 7.8J); 0 or 1 seta on first article on MxII; 1 or 2 articles on exopod of ThI; inhomonomous or homonomous sympod of the uropod 2

1' Distal endite of MxI with 6 teeth (Fig. 7.8M); 0 or 1 seta on first article of MxII; 1 article on exopod of ThI; cuticle of female ThVIII smooth (Fig. 7.8N) or wrinkled (Fig. 7.8P); inhomonomous (Fig. 7.8Q) sympod of the uropod .... ***Iberobathynella (Asturibathynella)***

2 (1) One seta on first article on MxII; 1 or 2 articles on exopod of ThI; cuticle of female ThVIII smooth or wrinkled 3

2' Zero seta on first article of MxII; 1 article on exopod of ThI; cuticle of female ThVIII smooth .... *Iberobathynella pedroi* (Camacho, 2003) [Portugal: Coimbra]

3 (2) First article of ThII to VII endopod without dorsal seta; 2 setae on first article of ThII to VII exopod; 2 or 3 setae on second article of ThI endopod; 1 or 2 articles on ThI exopod; cuticle of female ThVIII smooth (Fig. 7.8R) or wrinkled (Fig. 7.8S); (Fig. 7.8T) uropod sympod inhomonomous or homonomous; 4–6 setae on exopod and 0, 1, or 3 on endopod of uropod .... ***Iberobathynella (Espanobathynella)***

3' First article of ThII to VII endopod with 1 dorsal seta; 3 setae on first article of ThII to VII exopod; 3 setae on second article of ThI endopod; ThI exopod with 2 articles, cuticle of female ThVIII smooth (Fig. 7.8U); uropod sympod homonomous (Fig. 7.8V); 4–8 setae on exopod and 3 on endopod of uropod .... ***Iberobathynella (Iberobathynella)***

Bathynellacea: Parabathynellidae: *Iberobathynella* (*Asturibathynella*): species

1 AI with 6 articles 2

1' AI with 7 articles 3

2 (1) First article of MxII with 1 seta; first article of ThI endopod with 2 setae; cuticle of female ThVIII smooth; 5 spines on sympod of uropod; 5 setae on exopod and 3 on endopod of uropod .... *Iberobathynella* (*Asturibathynella*) *celiana* (Camacho, 2003) [Spain: Sevilla]

2' First article of MxII without seta; first article of ThI endopod with 1 seta; cuticle of female ThVIII wrinkled; 9 spines on sympod of uropod; 4 setae on exopod and 2 on endopod of uropod .... *Iberobathynella* (*Asturibathynella*) *guarenensis* (Camacho, 2003) [Spain: Burgos]

3 (1') First article of MxII with 1 seta; first article of ThI endopod with 2 setae; cuticle of female ThVIII smooth or wrinkled; anal operculum protuded or not 4

3' First article of MxII without seta; first article of ThI endopod with 1 or 2 setae; cuticle of female ThVIII smooth or wrinkled; anal operculum protuded 6

4 (3) Md *pars molaris* with 7 teeth; Md *pars distales* with 6 or 7 teeth; cuticle of female ThVIII smooth or wrinkled; 7 or 8 spines on sympod of uropod; 4 or 5 setae on exopod and 1 or 3 on endopod of uropod; 6–10 spines on furca; anal operculum not protuded 5

4' Md *pars molaris* with 5 teeth; Md *pars distales* with 4 teeth; cuticle of female ThVIII smooth; 6 spines on sympod of uropod; 5 setae on exopod and no seta on endopod of uropod; 5 spines on furca; anal operculum protuded ... . *Iberobathynella* (*Asturibathynella*) *cornejoensis* (Camacho, 2005) [Spain: Burgos]

5 (4) Md *pars distalis* with 7 teeth; cuticle of female ThVIII wrinkled; 7 spines on sympod of uropod; 4 setae on exopod and 3 on endopod of uropod; 7–10 spines on furca .... *Iberobathynella* (*Asturibathynella*) *rouchi* (Camacho & Coineau, 1987) [Spain: Teruel, Huesca]

5' Md *pars distalis* with 6 teeth; cuticle of female ThVIII smooth; 8 spines on sympod of uropod; 5 setae on exopod and 1 on endopod of uropod; 6 or 7 spines on furca .... *Iberobathynella* (*Asturibathynella*) *imuniensis* (Camacho, 1987) [Spain: Álava, Asturias, Burgos, Cantabria, Huesca]

6 (3') Md *pars molaris* with 6–7 teeth, *pars distalis* with 5–7 teeth; first article of ThI endopod with 1 setae; cuticle of female ThVIII wrinkled or smooth; 5–9 spines on sympod of uropod; 3 or 4 setae on exopod and 1 or 2 on endopod of uropod 7

6' Md *pars molaris* with 5 teeth, *pars distales* with 4 teeth; first article of ThI endopod with 2 setae; cuticle of female ThVIII wrinkled; 9 or 10 spines on sympod of uropod; 4 setae on exopod and 2 on endopod of uropod .... *Iberobathynella* (*Asturibathynella*) *parasturiensis* (Camacho & Serban, 1998) [Spain: Asturias, Cantabria]

7 (6) Md *pars molaris* with 7 teeth, *pars distalis* with 5–7 teeth; cuticle of female ThVIII smooth or wrinkled; 5–7 spines on sympod of uropod; 3 or 4 setae on exopod and 2 on endopod of uropod 8

7' Md *pars molaris* with 6 teeth, *pars distalis* with 5 teeth; cuticle of female ThVIII smooth or wrinkled; 5–9 spines on sympod of uropod; 3 or 4 setae on exopod and 1 or 2 on endopod of uropod 9

8 (7) Cuticle of female ThVIII wrinkled; setules on distal end of inner lobe of male ThVIII present; 6–7 spines on sympod of uropod; 4 setae on exopod and 2 on endopod of uropod .... *Iberobathynella* (*Asturibathynella*) *asturiensis* (Serban & Comas, 1978) [Spain: Asturias, Cantabria, León]

8' Cuticle of female ThVIII smooth; distal part of inner lobe of male ThVIII without setules; 5–6 spines on sympod of uropod; 3 setae on exopod and 2 on endopod of uropod .... *Iberobathynella* (*Asturibathynella*) *cavadoensis* (Noodt & Galhano, 1969) [Portugal: Barcelos. Spain: León, Pontevedra]

9 (7') Cuticle of female ThVIII smooth; 5–7 spines on sympod of uropod; 3 or 4 setae on exopod and 1 or 2 on endopod of uropod 10

9' Cuticle of female ThVIII wrinkled; 9 spines on sympod of uropod; 4 setae on exopod and 2 on endopod of uropod .... *Iberobathynella* (*Asturibathynella*) *lamasonensis* (Camacho, 2005) [Spain: Cantabria]

10 (9) Five spines on sympod of uropod; 4 setae on exopod and 2 on endopod of uropod .... *Iberobathynella* (*Asturibathynella*) *ortizi* (Camacho, 1989) [Spain: Lugo]

10' Seven spines on sympod of uropod; 3 setae on exopod and 1 on endopod of uropod .... *Iberobathynella* (*Asturibathynella*) *serbani* (Camacho, 2003) [Portugal: Oporto]

Bathynellacea: Parabathynellidae: *Iberobathynella* (*Espanobathynella*): species

1 ThII to VII exopod with 2 articles; ThI with 1 or 2 articles; ThI endopod with 2 or 3 setae on second article; setules on distal part of inner lobe of male ThVIII present or absent; cuticle of female ThVIII smooth or wrinkled; sympod of uropod inhomonomous or homonomous 2

1' ThII, III, and VI, VII exopod with 2 articles and ThIV and V exopod with 3 articles; ThI exopod with 2 articles; second article of ThI endopod with 3 setae; no setules on distal part of male ThVIII inner lobe; cuticle of female ThVIII smooth; 12 or 13 spines on homonomous sympod, 6 setae on exopod and 1 seta on endopod of uropod .... *Iberobathynella* (*Espanobathynella*) *magna* (Camacho & Serban, 1998) [Spain: Asturias, Cantabria]

2 (1) Uropod with sympod inhomonomous; 1 or 2 articles on exopod of ThI; 3 setae on second article of ThI endopod; cuticle of female ThVIII smooth; anal operculum not pronounced 3
2' Sympod of the uropod homonomous; 1 or 2 articles on exopod of ThI; second article of ThI endopod with 2 setae; cuticle of female ThVIII smooth or wrinkled; anal operculum pronounced or not 4
3 (2) ThI exopod with 2 articles; setules on distal part of male ThVIII inner lobe present; 5 or 6 setae on exopod and 1 seta on endopod of uropod; anal operculum not pronounced ....
.... *Iberobathynella (Espanobathynella) espaniensis* (Serban & Comas, 1978) [Spain: Asturias, Guadalajara]
3' ThI exopod with 1 article; no setules on distal part of male ThVIII inner lobe; 5 setae on exopod and 3 setae on endopod of uropod; anal operculum pronounced ....
.... *Iberobathynella (Espanobathynella) andalusica* (Camacho, 2007) [Spain: Córdoba, Sevilla]
4 (2') ThI exopod with 1 article; cuticle of female ThVIII smooth; 7–9 spines on sympod, 5 setae on exopod and 1 seta on endopod of uropod ....
.... *Iberobathynella (Espanobathynella) cantabriensis* (Camacho & Serban, 1998) [Spain: Asturias, Cantabria]
4' ThI exopod with 2 articles; cuticle of female ThVIII wrinkled; 10–12 spines on sympod, 4 setae on exopod and no seta on endopod of uropod ....
.... *Iberobathynella (Espanobathynella) burgalensis* (Camacho, 2005) [Spain: Burgos]

Bathynellacea: Parabathynellidae: *Iberobathynella (Iberobathynella)*: species

1 Md *pars incisiva* with 9–12 teeth, *pars molaris* with 6–16 teeth; 2 or 3 setae on first article of ThI exopod; 9–27 spines on sympod, 5 setae on exopod and 3 seta on endopod of uropod 2
1' Md *pars incisiva* with 6 teeth, *pars molaris* with 10 teeth; 2 setae on first article of ThI exopod; 12 or 13 spines on sympod, 6–8 setae on exopod and 3 setae on endopod of uropod ....
.... *Iberobathynella (Iberobathynella) valbonensis* (Noodt & Galhano, 1969) [Portugal: Areinho de Valbom, Spain: Salamanca]
2 (1) Md *pars incisiva* with 7 or 8 teeth, *pars molaris* with 6–10 teeth; 2 or 3 setae on first article of ThI exopod; 9–12 spines on sympod of uropod 3
2' Md *pars incisiva* with 7–12 teeth, *pars molaris* with 8–16 teeth; 3 setae on first article of ThI exopod; 16–27 spines on sympod of uropod 4
3 (2) First article of ThI exopod with 2 setae; Md *pars incisiva* with 7 teeth, *pars molaris* with 6–8 teeth; 9–12 spines on sympod of uropod; 9–13 spines on furca ....
.... *Iberobathynella (Iberobathynella) lusitanica* (Braga, 1949) [Portugal: Porto]
3' First article of ThI exopod with 3 setae; Md *pars incisiva* with 7 or 8 teeth, *pars molaris* with 8–10 teeth; 9–11 spines on sympod of uropod; 10 spines on furca ....
.... *Iberobathynella (Iberobathynella) barcelensis* (Galhano, 1970) [Portugal: Barcelos]
4 (2) First article of ThVI exopod with 3 setae; Md *pars incisiva* with 10 teeth, *pars molaris* with 8–10 teeth; 16–19 spines on sympod of uropod; 6 or 7 spines on furca ....
.... *Iberobathynella (Iberobathynella) gracilipes* (Braga, 1960) [Portugal: Beira-Baixa]
4' First article of ThVI exopod with 2 setae; Md *pars incisiva* with 9–12 teeth, *pars molaris* with 8–16 teeth; 18–27 spines on sympod of uropod; 9–12 spines on furca ....
.... *Iberobathynella (Iberobathynella) paragracilipes* (Camacho & Serban, 1998) [Spain: Huelva]

Subchapter 7.4

Order Decapoda

Magdalini Christodoulou
German Centre for Marine Biodiversity Research (DZMB), Senckenberg am Meer, Südstrand 44, 26382 Wilhelmshaven, Germany

Introduction

Decapoda Latreille 1802 is the most species-rich order within Malacostraca with nearly 15,000 species (De Grave et al., 2009). The order includes many familiar groups such as crabs, shrimps, crayfish, lobsters, and hermit crabs that live in an impressive diversity of marine, freshwater, subterranean, and terrestrial habitats. The majority of decapod species are found in marine habitats, whereas approximately 3000 species live exclusively in freshwater habitats on almost all continents (Cumberlidge et al., 2015). These include over 1450 species of brachyuran crabs (Cumberlidge et al., 2015), about 800 species and subspecies of caridean and sergestoid shrimps (De Grave et al., 2015), around 669 species of crayfish (Crandall & De Grave, 2017), and 75 species of anomuran crabs (Cumberlidge et al., 2015; Bueno et al., 2016).

Freshwater crabs have a circumtropical distribution restricted to warm habitats covering five zoogeographical regions, from the Neotropical region to the Australasian region (Cumberlidge et al., 2015). Evolutionary studies suggest that freshwater crabs diverged early in the brachyuran phylogenetic tree. They are thought to have at least two independent origins, with four of the five freshwater crab families sharing common ancestry (Klaus et al., 2011; Tsang et al., 2014). Freshwater crabs are estimated to have diverged from their closest marine sister taxa after the break-up of Pangaea (~200 Mya), around 135 Mya (Tsang et al., 2014). Although the origin of Mediterranean crabs (*Potamon* spp.) remains unknown, their species diversity argues in favor of an East Asian origin (Jesse et al., 2011). Freshwater crayfish and anomuran crabs are distributed in temperate parts of the two hemispheres (Cumberlidge et al., 2015; Bueno de et al., 2016). The oldest crayfish fossils date back to the Permian and early Triassic periods, suggesting a widespread Pangaea distribution followed by a number of vicariance events that allowed the ancient crayfish population to diversify and spread throughout the continents of Laurasia and Gondwana (Crandall et al., 2000; Cumberlidge et al., 2015). Freshwater anomuran crabs, which mostly include *Aegla* species, seem to have a more recent origin—the early Tertiary period, about 60 Mya (Pérez-Losada et al., 2004). Finally, freshwater shrimps occur in both tropical and temperate freshwater habitats. Phylogenetic studies have shown a distant affinity among freshwater shrimp families implying that each family independently invaded continental and subterranean waters from an ancestral marine stock at different times (De Grave et al., 2008). The Mediterranean Basin was colonized through multiple ancient transoceanic dispersal or vicariance events (Von Rintelen et al., 2012; Carvalho et al., 2016).

Three groups of freshwater decapods are found in the Mediterranean Basin, namely crabs, shrimps, and crayfish. They belong to 6 families, 15 genera, and 63 species (1 family, 3 genera, and 3 species are non-native), representing about 2.1% of global diversity; most species are endemic to the region.

General ecology and distribution

Freshwater decapods successfully colonized a very diverse array of epigean and hypogean freshwater habitats reflected by a wide range of life histories as well as morphological and physiological adaptations. Because of their abundance and high biomass, they are integral components of food webs, represent a food source for a wide range of predators, and play an important role in nutrient recycling in freshwater ecosystems (Cumberlidge et al., 2015; De Grave et al., 2015).

A rich and highly endemic decapod fauna is found in the freshwater habitats of all countries and most islands of the Mediterranean Basin (Jesse et al., 2011; Kouba et al., 2014; Christodoulou et al., 2016). Their diversity comprises 6 families (15 genera, 63 species, and subspecies) in this region. The freshwater crab family Potamidae (1 genus, 15 species) is found in the surface waters of North Africa and the eastern Mediterranean Basin (Brandis et al., 2000; Jesse et al., 2011). Crayfish are represented in the Mediterranean Basin by one native family—Astacidae (4 genera, 7 species)—and one non-native family—Cambaridae (2 genera, 2 species). Native crayfish species are distributed in the northern Mediterranean Basin but are absent from continental North Africa (Kouba et al., 2014; Cumberlidge et al., 2015). The distribution of crayfish in the Mediterranean basin has been under great anthropogenic influence, resulting in three invasive species currently found in the western Mediterranean region, while numerous translocations of the native species took place across Europe (Kouba et al., 2014). The largest shrimp family is Atyidae (6 genera, 26 species, and subspecies) with representatives in epigean and subterranean systems throughout the Mediterranean Basin (Christodoulou et al., 2016), while Palaemonidae (1 genus, 9 species) are restricted to epigean systems throughout the basin (Christodoulou et al., 2016). Both families occur in all biogeographical regions, except Antarctica, but are most diverse in the Indo-Malayan region where the majority of palaemonid species live in marine and brackish waters whereas Atyidae almost exclusively live in freshwater (De Grave et al., 2015). The areas immediately surrounding the Mediterranean Sea harbor the world highest stygobiotic species diversity within Atyidae (16 species and subspecies) (Sket & Zakšek, 2009; Christodoulou et al., 2016). Finally, Typhlocarididae (1 genus, 4 species) are strictly subterranean and are the only endemic decapod family in the basin (Christodoulou et al., 2016), with species distributed in Israel, Italy, and Libya.

Terminology and mmorphology

Decapods exhibit a great variability in size and body form. Their size in freshwaters can vary from 1.0 cm for the bee shrimp (*Caridina cantonensis*) to more than 80 cm and 6 kg for some of the largest and heaviest freshwater invertebrates such as Tasmanian giant crayfish (*Astacopsis gouldi*). Mediterranean Basin decapods are small- to medium-sized, from about 2 cm (e.g., *Atyaephyra orientalis*) to about 20 cm (e.g., *Pontastacus leptodactylus*).

The general decapod body is composed of a head, thorax, and abdomen (Fig. 7.9). The body is compressed laterally in shrimps, compressed dorsoventrally in crayfish, but strongly depressed and flattened in brachyuran crabs. The first three of the eight thoracic segments of shrimps, crayfish, and crabs are fused with the head to form a cephalothorax covered by a shield—the carapace that extends over the whole head and thorax. The carapace extends laterally along the sides of the cephalothorax forming branchial chambers that enclose the gills.

The anterior part of the cephalothorax (head) carries a pair of compound stalked eyes, two pairs of antennae (antenna and antennula), and three pairs of mouth parts (mandible, maxillula, maxilla). The carapace of freshwater shrimps, crayfish, and aeglids (not present in the Mediterranean Basin) protrudes between the eyes and forms a rostrum, while it forms a front in freshwater crabs. The carapace can be ornamented with various spines, grooves, and depressions. The first three abdominal segments bear three pairs of maxillipeds—specialized appendages for feeding and locomotion. The next abdominal segments bear five pairs of pereiopods (ambulatory legs) modified either as walking legs or as chelate limbs (equipped with pinchers or setal brushes) termed chelipeds. In crayfish, the first three pairs of pereiopods are chelate, while in caridean shrimps only the first two pairs are chelate and in brachyuran crabs only the first pair is chelate. The name of the group Decapoda (=10 legs) derives from the presence of 10 pereiopods. The pereiopods have the typical biramous form of a crustacean limb composed from their proximal-most end to their distal-most end of a protopod with two podomeres (basis and coxa), an endopod with five podomeres (ischium, merus, carpus, propodus, and dactylus), and an exopod that can be fully developed, vestigial or absent. A small process—the epipod—and multidenticulate setae—setobranch setae—can be present on coxae.

The abdomen (pleon) follows the thorax. It is folded beneath the body in crabs, or extends posteriorly in shrimps and crayfish. Pleopods, modified for swimming, brooding eggs, ventilation, and burrowing appendages extend in pairs from the first five abdominal segments. The abdomen terminates in a telson, with a uropod on each side. Both the pleopods and the uropods are biramous, composed of a basal peduncle bearing two branches—the endopod and the exopod. The first two pleopods are sexually modified in male decapods.

Collection, preparation, and identification

Freshwater decapods are found in lotic, lentic, and subterranean environments. Different sampling methods are utilized depending on the species and the habitat. The freshwater shrimps that dwell in epigean waters are usually collected using strong, long-handled, fine-meshed dip nets, while dredges, seine nets, and sweep net electrofishing have also been reported as sampling techniques. In subterranean habitats, shrimps are caught with baited (e.g., with strips of raw beef) traps or by hand or with short hand nets and torchlight to illuminate the animals while scuba diving (Jaume & Brehier, 2005). Freshwater crayfish in shallow waters are collected by hand by turning rocks over. A dip net can be used in deeper waters while snorkeling or scuba diving, and seine nets or fyke nets can be used too. Finally, baited wire traps have proven especially efficient in deeper waters, especially at night. Crabs can be caught during the day by turning flat rocks over in rivers and streams. In deeper waters, baited traps or fyke nets can be used.

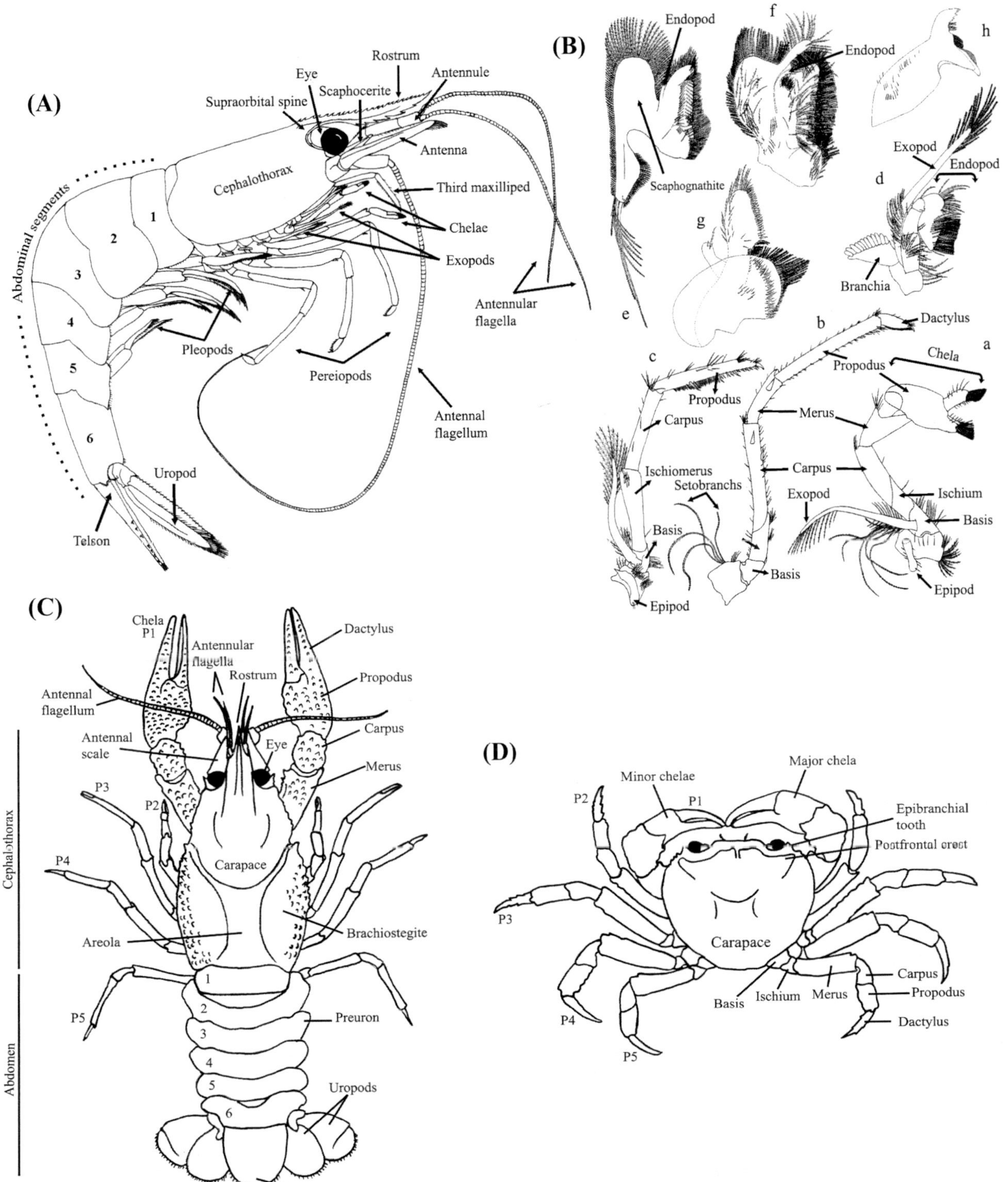

FIGURE 7.9 (A) General morphology of Caridea. (B) Mouthparts and Pereiopods of Caridea: a, first pereiopod (Chela); b, fifth pereiopod; c, third maxilliped; d, second maxilliped; e, maxilla; f, maxillula; h, mandible. (C) General morphology of Astacidea. (D) General morphology of Brachyura.

After collection, shrimps, crayfish, and crabs should be preserved in 70% ethanol for morphological analyses and in 95% ethanol for molecular analyses. Mediterranean decapods can be identified under a stereomicroscope down to the genus level. However, determination at the species level most often requires dissection and mounting of the mouthparts and appendages on slides and examination using a microscope (10 × –40 × magnification). Specimens or body parts can be dyed using different stains or cleared by submersion in weak acid (HCl, chlorine). The morphological identification of shrimps at the species level requires adult individuals; as for freshwater crabs, only males can be identified at the species level. Key morphological characteristics include among others rostrum armature, mouthparts, male pleopods, gonopods, telson armature, and pereiopod armature. Once the appendages are dissected, they are kept in smaller vials in the vial containing the specimen. It should be noted that many decapod species of the Mediterranean Basin can only be identified with certainty by DNA barcoding (Sket & Zakšek, 2009; Jesse et al., 2011; Christodoulou et al., 2016).

Limitations

The taxonomy of freshwater decapods in the Mediterranean Basin has been somewhat neglected in the past and has only recently grown more stable, as collections include more remote locations and molecular data are taken into account. Furthermore, high intra- and interspecific variability and sexual dimorphism complicate the morphological identification of decapods (Sket & Zakšek, 2009; Jugovic et al., 2010). Primary and secondary sexual characters in many decapods (Atyidae, Potamidae) are taxonomically important, but these characters are poorly developed in young animals (Sket & Zakšek, 2009; Jugovic et al., 2010; Christodoulou et al., 2016). Moreover, reliable identification of many species requires examining adult males because characters of taxonomic value have not been observed so far in females (Brandis et al., 2000; Jugovic et al., 2010). The current increased interest in Mediterranean epigean and hypogean decapods is based on more comprehensive sampling, molecular studies, and *a posteriori* examination of morphological characters that altogether improved taxonomy. Over the past two decades, the number of epigean freshwater shrimp species in the Mediterranean Basin has more than doubled, although controversies still remain about the precise number of species (e.g., *Atyaephyra* spp.), while new species may well still be awaiting discovery. Likewise, the number of troglobitic species has more than tripled, and more species are still awaiting description (Sket & Zakšek, 2009). Despite the increased interest in Mediterranean epigean and hypogean atyids and the new available methods, the taxonomy of these shrimps still remains unsettled. For example, the key to species of *Typhlocaris* was not included because it is in need of a revision as the descriptions of the genus species are based mainly on historical papers and the key characteristics are not well defined.

A number of Mediterranean decapods are cryptic species. Although they are genetically quite distinct, they appear morphologically identical to other species. Our keys can identify most species, but for some of them the key concludes to a species complex that molecular and geographical data can further distinguish. Finally, the keys are designed for Mediterranean freshwater decapods, they are not adapted for a global use.

Key to Decapoda

Decapoda: infraorders

1 Abdomen projecting posteriorly 2
1'Abdomen folded beneath body, tightly pressed to thorax .... Brachyura, **Potamidae (one genus: *Potamon*)**
[North Africa, Italian and Balkan Peninsulas, Middle East]
2 (1) Abdomen laterally compressed; pleuron of second abdominal somite overlapping pleura of second and third somite .... **Caridea**
[Mediterranean Basin]
2' Abdomen compressed dorsoventrally; pleuron of second somite not overlapping pleura of second and third somite .... **Astacidea**
[Europe]

Decapoda: Potamidae: *Potamon*: species

This *Potamon* key relies on the morphology of male first gonopod thus not suitable when only female specimens are present (Brandis et al., 2000). The key includes a total number of 15 species presented in Brandis et al. (2000) and Jesse et al. (2010, 2011). Four cryptic species exist within the genus where no morphological characters have been found yet to distinguish them, and COI gene nucleic acid sequence differences are required to distinguish them to species level (*).

1 Terminal article of first gonopod broad, more or less oval; mesial portion distinctly tumid; apex long and slightly curve 2
1' Terminal article of first gonopod conical or slender; mesial part not tumid 3

2 (1) Flexible zone of first gonopod V-shaped .... *Potamon fluviatile* (Herbst, 1785) species complex (*P. fluviatile*, *Potamon pelops**) [Albania, Croatia, Greece, Italy, Macedonia, Malta, Montenegro]
2' Flexible zone of first gonopod distinctly bilobed .... *Potamon algeriense* (Bott, 1967) [Algeria, Morocco, Tunisia]
3 (1') First gonopod with conical, cylindrical or triangular terminal article 4
3' First gonopod with very slender and elongated fusiform terminal article; flexible zone of first gonopod broad or only narrowed only in its lateral part .... *Potamon ibericum* (Bieberstein, 1809) species complex (*Potamon ibericum*, *Potamon kretaion**) [Greece, Turkey]
4 (3) Terminal article of first gonopod elongated and conical, mesial edge slightly bulging 5
4' Terminal article of first gonopod stout and triangular; subterminal part with projecting mesial edge 8
5 (4) Flexible zone of first gonopod flexible not symmetrically bilobed 6
5' Flexible zone of first gonopod symmetrically bilobed .... *Potamon potamios* (Olivier, 1804) species complex (*Potamon hippocratis*, *Potamon karpathos*, *Potamon potamios**) [Cyprus, Greece, Israel, Turkey]
6 (5) Flexible zone of first gonopod not distinctly V-shaped 7
6' Flexible zone of first gonopod distinctly V-shaped .... *Potamon rhodium* (Parisi, 1913) [Greece, Turkey]
7 (6) Flexible zone of first gonopod with only one linguiform elongate lobe .... *Potamon setigerum* (Rathbun, 1904) [Lebanon, Syria, Turkey]
7' Flexible zone of first gonopod slender, terminal article subcylindrical and narrow .... *Potamon bileki* (Pretzmann, 1971) [Turkey]
8 (4) Flexible zone of first gonopod V-shaped or with one mesial or lateral lobe 9
8' Flexible zone of first gonopod bilobed; terminal article of first gonopod asymmetric in shape, mesial margin bulging in its proximal half .... *Potamon magnum* (Pretzmann, 1962) [Turkey]
9 (8) Flexible zone of first gonopod distinctly V-shaped 10
9' Flexible zone of first gonopod with one lateral located linguiform elongated lobe; subterminal part of first gonopod with low mesial edge .... *Potamon hueceste* (Pretzmann, 1962) [Turkey]
10 (9) Terminal article of first gonopod almost triangular; anterolateral margin of carapace broad, serrated with unequally large teeth .... *Potamon mesopotamicum* (Brandis, Storch & Türkay, 1998) [Syria, Turkey]
10' Terminal article of first gonopod proximal half with straight mesial margins, in distal half becoming sharply deflected; anterolateral margin of carapace narrow serrated with very small regular teeth .... *Potamon persicum* (Pretzmann, 1962) [Turkey]

Decapoda: Caridea: families

1 First and second pereopod with chelae fingers lacking dense apical setal tufts 2
1' First and second pereopod with chelae bearing dense apical setal tufts .... **Atyidae** [peri-mediterranean]
2 (1) Carapace lacking longitudinal sutures; flagellum of first antenna with rami fused basally .... **Palaemonidae**, ***Palaemon*** [peri-mediterranean]
2' Carapace bearing pair of straight, longitudinal, dorsolateral sutures; cavernicolous .... Typhlocaridae, *Typhlocaris* [Israel, Italy, Tunisia]

Decapoda: Caridea: Atyidae: genera

1 At least first 3 pereiopods bearing exopods 2
1' Only first and second pereiopods bearing exopods .... ***Atyaephyra*** [Peri-Mediterranean]

2 (1) Appendix masculina of male second pleopod bearing spines longer than appendix diameter 3
2' Appendix masculina of male second pleopod bearing spines much shorter than appendix 5
3 (2) Supraorbital spines on carapace present; exopods of fifth pereiopods absent or reduced 4
3' Supraorbital spines on carapace absent; exopod of fifth pereiopods well developed .... ***Typhlatya*** [France, Spain]
4 (3) Eyes reduced, pigment absent; pterygostomian spine absent; rostrum short unarmed; dactylus of fifth pereiopod with less than 10 spiniform spines .... *Gallocaris* (one species: *Gallocaris inermis*; Fage, 1937) [France]
4' Eyes developed and pigmented; pterygostomian spine present; rostrum long and armed; dactylus of fifth pereiopod with more than 10 spiniform spines (comp-like) .... ***Dugastella*** [Morocco, Spain]
5 (2') Endopod of male first pleopod as a slightly bent plate, its appendix interna vestigial .... ***Troglocaris*** [Balkans, Italy]
5' Endopod of male first pleopod shaped as "9", that is, with a long, subapically laterally positioned appendix interna .... ***Spelaeocaris*** [Balkans]

Decapoda: Caridea: Atyidae: *Atyaephyra:* species

This *Atyaephyra* key is based on Christodoulou et al. (2012) and Jablonska et al. (2018), and it refers to individuals of carapace length $\geq$ 5.0 mm ($\geq$ 3.5 mm for *Atyaephyra orientalis*); however, the reader should keep in mind that this genus is still under taxonomic revision. Two cryptic species exist within the genus where no morphological characters have been found yet to distinguish them, and COI gene nucleic acid sequence differences are required to distinguish them to species level (*).

1 Propodite of third maxilliped bearing 8–38 (rarely 8) strong mesial spiniform setae 2
1' Propodite of third maxilliped bearing 0–8 (rarely 8) strong mesial spiniform setae 3
2 (1) Endopod terminal part of first male pleopod slender, tapering; merus of third pereiopod with 4–6 (rarely 3–9) spiniform setae; merus of fourth pereiopod with 3–5 (rarely 2–7) spiniform setae 4
2' Endopod terminal part of first male pleopod broad; merus of third pereiopod with 7–9 (rarely 6–10) spiniform setae; merus of fourth pereiopod with 6–8 (rarely 5–9) spiniform setae .... *A. orientalis* (Bouvier, 1913) [Israel, Syria, Turkey]
3 (1') Absence of post orbital teeth, leaving an unarmed proximal gap on dorsal surface of rostrum; basal endite of first maxilliped failing to reach or reaching distal end of exopod .... *Atyaephyra strymonensis* (Christodoulou, Antoniou, Magoulas & Koukouras, 2012) [Greece]
3' Post orbital teeth present without an unarmed proximal gap on dorsal surface of rostrum; basal endite of first maxilliped over-reaching distal end of exopod .... *Atyaephyra desmarestii* (Millet, 1831) species complex (*Atyaephyra acheronensis*, *Atyaephyra desmarestii*, *Atyaephyra tuerkayi*)
[Croatia, France, Italy, Montenegro, Portugal, Slovenia, Spain, Greece, Syria]
4 (2) Propodite of third maxilliped bearing 16–30 (rarely 11–38) strong mesial spiniform setae 5
4' Propodite of third maxilliped bearing 9–12 (rarely 8–20) strong mesial spiniform setae .... *Atyaephyra vladoi* (Jabłońska, Mamos, Zawal & Grabowski, 2018) [Albania, Montenegro]
5 (4) Stylocerite length 1.02–1.19 (rarely 0.97–1.19) times first antenullar segment length; fifth pleuron distal end pointed with acute angle .... *Atyaephyra thyamisensis* (Christodoulou, Antoniou, Magoulas & Koukouras, 2013) [Greece: Ipeiros, Ionian Islands]
5' Stylocerite length 0.82–1.0 (rarely 0.82–1.08) times first antenullar segment length; fifth pleuron distal end rounded, pointed with abtuse angle (rarely with acute) .... *Atyaephyra stankoi* (Karaman, 1972) [Greece]

Decapoda: Caridea: Atyidae: *Typhlatya:* species

1 Rostrum scarcely developed, shorter than reduced eye stalks; dactylus of fifth pereopod with row of 36 slender spiniform setae on flexor margin .... *Typhlatya miravetensis* (Sanz & Platvoet, 1995) [Spain]

1' Rostrum length subequal to reduced eye stalks; dactylus of fifth pereopod with row of 19–28 stout spiniform setae on flexor margin .... *Typhlatya arfaea* (Jaume & Bréhier, 2005) [France]

Decapoda: Caridea: Atyidae: *Dugastella:* species

1 Rivers and springs in Morocco .... *Dugastella marocana* (Bouvier, 1912)
1' Rivers, springs and lagoons in southern Spain .... *Dugastella valentina* (Ferrer Galdiano, 1924)

Decapoda: Caridea: Atyidae: *Troglocaris:* species

This *Atyaephyra* key is based on Sket & Zakšek (2009). Two cryptic species exist within the genus where no morphological characters have been found yet to distinguish them. COI gene nucleic acid sequence differences are required to distinguish them to species level (*).

1 Endopod of male first pleopod with more than 30, partially grouped, and long, inner marginal setae; rostrum with usually 15 ventral teeth .... *Troglocaris bosnica* (Sket & Zakšek, 2009) [Bosnia Herzegovina]
1' Endopod of male first pleopod with shorter and scarcer marginal setae; rostrum with fewer than 10 ventral teeth . ... *Troglocaris anophthalma* (Kollar, 1848) species complex (*T. anophthalma*, *Troglocaris planinensis**) [Bosnia Herzegovina, Croatia, Italy, Slovenia]

Decapoda: Caridea: Atyidae: *Spelaeocaris:* species

This *Spelaeocaris* key is based on Sket & Zakšek (2009) with the addition of one more species (*Spelaeocaris hercegovinensis*) according to the genetic study of Marin (2017).

1 Supraorbital spines on carapace absent; pereopod chelae and dactyli very stout 2
1' Supraorbital spines on carapace present; pereopod chelae and dactyli very narrow .... *S. hercegovinensis* (Babić, 1922) [Bosnia Herzegovina, Montenegro]
2 (1) Dactylus of fifth pereopod with more than 15 spines, comb-like; at least third and fourth mature male pereopods differentiated 3
2' Dactylus of fifth pereopod with fewer than 15 spiniform setae, not comb-like; mature male pereopods not differentiated; rostrum reduced, not serrated .... *Spelaeocaris neglecta* (Sket & Zakšek, 2009) [Croatia]
3 (2) Rostrum long or very short, but always with dorsal teeth; in mature males, only third and fourth pereopods differentiated; endopod lamina of first male pleopod apically strongly lobate 4
3' Rostrum very short and not serrated; in mature males, apical parts of third, fourth, and fifth pereopods widened; endopod lamina of male first pleopod apically and obliquely cut and not (or feebly) lobate .... *Spelaeocaris pretneri* (Matjašič, 1956) [Bosnia Herzegovina, Croatia]
4 (3) Appendix masculina of male second pleopod only a slightly longer than appendix interna; in adult males, propodi of third and fourth pereopod only slightly differentiated .... *Spelaeocaris kapelana* (Sket & Zakšek, 2009) [Croatia]
4' Appendix masculina of male second pleopod nearly twice as long as appendix interna; in adult males, propodi of third and fourth pereopod strongly differentiated .... *Spelaeocaris prasence* (Sket & Zakšek, 2009) [Bosnia Herzegovina, Montenegro]

Decapoda: Caridea: Palaemonidae: *Palaemon:* species

This *Palaemon* key is based on a recent revision (Tzomos & Koukouras, 2015) of the freshwater members of the genus in the Mediterranean Basin.

1 Fifth pleuron distal end not rounded 2
1' Fifth pleuron distal end rounded 7
2 (1) Fifth pleuron distal end strongly pointed 3
2' Fifth pleuron distal end subquadrate and not pointed 6

3 (2) Telson posterior margin with two plumose setae; third maxilliped exopod equal or longer than isciomerus; endopod of male first pleopod clearly shorter than exopod; appendix masculina shorter than the endopod of second pleopods 4

3' Telson posterior margin with 2–6 plumose setae; third maxilliped exopod shorter than isciomerus; endopod of male first pleopod slightly shorter than exopod; appendix masculina longer than the endopod of second pleopod .... *Palaemon mesogenitor* (Sollaud, 1912) [Algeria, Tunisia]

4 (3) Telson plumose setae fail to overpass the inner spiniform setae; first maxilliped endites separated with their inner edges forming an obtuse angle; branchiostegal tooth originating just before anterior margin of carapace 5

4' Telson plumose setae over pass the inner spiniform setae; first maxilliped endites arranged almost in a straight line forming just a notch between them; branchiostegal tooth marginal .... *P. varians* (Leach, 1814) [Algeria, France, Morocco, Portugal, Spain, Tunisia]

5 (4) Telson with 40–50 pairs of marginal lanceolate setae; endopod of male first pleopod deeply concave on the mesial portion of inner margin; endopod of female first pleopod with concave inner margin and rounded tip .... *Palaemon colossus* (Tzomos & Koukouras 2015) [Greece: Rhodos Island, Turkey]

5' Telson with 0–6 pairs of marginal lanceolate setae only on its proximal part; endopod of male first pleopod lightly concave on the mesial portion of inner margin; endopod of female first pleopod lanceolate .... *Palaemon antennarius* (H. Milne Edwards, 1837) [Albania, Croatia, Italy, Montenegro, Slovenia, Greece]

6 (2') Telson posterior margin with two plumose setae; endopod of male first pleopod almost as long as exopod; appendix masculina longer than the endopod of the male second pleopods .... *Palaemon turcorum* (Holthuis, 1961) [Turkey]

6' Telson posterior margin with 4–6 plumose setae; endopod of male first pleopod clearly shorter than the exopod; appendix masculina shorter than the endopod of male second pleopods .... *Palaemon zariquieyi* (Sollaud, 1938) [Spain]

7 (1') Telson posterior margin with 4–12 plumose setae 8

7' Telson posterior margin always with two plumose .... *Palaemon minos* (Tzomos & Koukouras, 2015) [Greece: Crete Island]

8 (7) Telson posterior margin with 10–12 plumose setae; third maxilliped exopod shorter than ischiomerus; outer margin of scaphocerite concave .... *Palaemon mesopotamicus* (Brandis, Storch & Türkay, 1998) [Syria, Turkey]

8' Telson posterior margin with 4–7 plumose setae; third maxilliped exopod equal or longer than ischiomerus; outer margin of scaphocerite straight .... *Palaemon migratorius* (Heller, 1862) [Egypt]

Decapoda: Astacidea: families

For the Astacidea we follow the classification of Crandall & De Grave (2017). The reader should keep in mind that some controversy exists on the status and validity of species which is further complicated with the presence of invasive species and the many human translocations of native species.

1 Ischium of male third and/or fourth pereiopods bearing conical hooks; females with annulus ventrali .... **Cambaridae**

1' Ischium of male pereiopods not bearing hooks; females lacking annulus ventralis .... **Astacidae**

Decapoda: Astacidea: Cambaridae: genera and species

1 First pleopod of mature males terminating in two rounded lobes and spiniform structure; females with movable annulus ventralis, not fused anteriorly to sternum .... *Procambarus clarkii* (Girard, 1852)
a) [Invasive: Cyprus, Egypt, France, Israel, Italy, Malta, Morocco, Spain, Tunisia]

1' First pleopod of mature males terminating in two triangular or elongate lobes; females with annulus ventralis fused anteriorly to sternum .... *Faxonius limosus* (Rafinesque, 1817) [Invasive: France, Italy, Spain]

Decapoda: Astacidea: Astacidae: genera

1 Telson with transverse suture 2
1' Telson without transverse suture ....
.... *Pacifastacus* (one species: *Pacifastacus leniusculus* (Dana, 1852)
[Invasive in Europe]
2 (1) Two pairs of postorbital ridges; first pleopod distal lobes unequal; second pleopod exopodite length subequal to endopod; third maxilliped merus with one medial spine, occasionally also with distal spine 3
2' Only one pair of postorbital ridges, or (rarely) with extremely reduced second pair; first pleopod distal lobes subequal; second pleopod exopod length distinctly smaller than endopod length; third maxilliped merus typically with medial spine row, rarely with one spine .... ***Austropotamobius***
[Europe]
3 (2) Second to fourth abdominal somites with pleura bearing acute spines; second pleopod with ventral process (talon) .... *Pontastacus* (one species: *Pontastacus Leptodactylus*; Eschscholtz, 1823)
[Bosnia Herzegovina, Croatia, France, Greece, Italy, Turkey]
3' Second to fourth abdominal somites with pleura rounded or angular, lacking spines, fourth somite with pleura bearing obscure spine or not; second pleopod without ventral process .... ***Astacus***
[Europe]

Decapoda: Astacidea: Astacidae: *Austropotamobius:* species

This *Austropotamobius* key is based on Souty-Grosset et al. (2006), but the reader should keep in mind that this genus is under constant revision. The separation of *Austropotamobius fulsianus* that was considered a subspecies of *Austropotamobius pallipes* is not clear. COI gene nucleic acid sequence differences are required to distinguish the two species (*).

1 Area behind cervical groove smooth, cervical spines absent; rostral borders triangle; chelae surface granulation very strong and rough; lower surface of antennal scale with denticulation ....
.... *Austropotamobius torrentium* (von Paula Schrank, 1803)
[Italian and Balkan Peninsulas]
1' Area behind cervical groove with cervical spines present; rostral borders triangle or trapezoid; chelae surface granulation small; lower surface of antennal scale with no denticulation ....
.... *A. pallipes* (Lereboullet, 1858) species complex (*A. pallipes*, *Austropotamobius fulsianus**)

Decapoda: Astacidea: Astacidae: *Astacus:* species

The key of *Astacus* is based on Storobogatov (1996). There is controversy whether *Astacus balcanicus* is a species or a subspecies of *Astacus astacus*.

1 Posterior part of telson completely separated from anterior part by a weak furrow; shape of posterior part roundly trapeziform; posterior pair of postorbital ridges with spinules at anterior ends; presence of few acute spines on each side immediately posterior to cervical groove .... *A. balcanicus* (Karaman, 1929)
1' Posterior part of telson incompletely separated from anterior one by a hardly distinct furrow; shape of posterior part semi-circular; posterior part of postorbital ridges devoid of spinules; presence of only one acute spine (usually) on each side immediately posterior to cervical groove .... *A. astacus* (Linnaeus, 1758)

Subchapter 7.5

Order Ingolfiellida

Christophe Piscart
French National Centre for Scientific Research (CNRS), University of Rennes 1, France

Introduction

Since their first discovery by Hansen at the beginning of the 1900s (Hansen, 1903) and until recently, ingolfiellids had been considered as a suborder of Amphipods. Several authors objected, and Lowry and Myers (2017) proposed to raise ingolfiellids to the order status only recently, after a large revision of Amphipoda. This Ingolfiellida order is a sister group of Amphipoda characterized by the lack of a carapace and coxal gills and the presence of vestigial pedunculate eyes ("eye-stalks" noticed by Hansen), a pleosome composed of six segments devoid of epimera, and reduced pleopod/uropod appendages.

This small order remains largely unknown and includes only two families—Ingolfiellidae (51 species) and the very rare Metaingolfiellidae (one species)—among which 20 are present in freshwaters (Horton et al., 2020). Species of this group are known on all continents and confirm that Ingolfiellida is a very old lineage that emerged around the Triassic Pangea (237 Mya) and diversified in interstitial freshwaters before further colonizing brackish and marine environments following evolutionary processes (Vonk and Schram, 2003).

The presence of the sole representative of Metaingolfiellidae—*Metaingolfiella mirabilis*—in Italy and species of the *Ingolfiella* genus with ancestral morphological characteristics (Vonk and Schram, 2003) suggests ancient colonization of the Mediterranean Basin by species of this order. The two Ingolfiellida families are present in this region and are represented by nine species belonging to two genera (*Metaingolfiella* and *Ingolfiella*) that represent 45% of the total number of known freshwater Ingolfiellida.

General ecology and distribution

Their ecology remains almost unknown, and most of the species are known from only one specimen (Vonk and Jaume, 2013). Despite a wide distribution in fresh and marine waters, a few species have a euryhaline distribution, and only *Ingolfiella manni* was observed in both fresh and brackish water environments in Chile.

Despite the small number of known species, Ingolfiellida are one of the most widespread groups in the crustacean world and have colonized an exceptional diversity of groundwaters and interstitial habitats ranging from deep sea water (−3521 m on the deep sea bottom for *Ingolfiella abyssi*) to the top of mountainous freshwater streams (2000 m asl. in the Andes Mountains for *Ingolfiella uspallatae*).

All Mediterranean species are distributed among the northern countries located between Spain and Greece. They have been found either in river sediments (*Ingolfiella catalanensis*, *Ingolfiella macedonica*, *Ingolfiella thibaudi*, and *Ingolfiella vandeli*), or in wells (*M. mirabilis*, *Ingolfiella acherontis*, and *Ingolfiella petkovskii*), and caves (*Ingolfiella cottarellii*, *Ingolfiella beatricis*). The distribution of many species is restricted to a very small area, and several species are known only from one river or one well.

Terminology and mmorphology

Ingolfiellida are small/medium-sized Malacostraca ranging from 1.5 to 23 mm in length worldwide but less than 2.5 mm in the Mediterranean Basin, except *M. mirabilis* which can reach 18.7 mm. Females are generally larger than males (when they are known). Their body is vermiform and adapted to life in interstitial environments such as those of bathnynellacean and bogidiellid amphipods. Their body is divided into 13 segments, which can be grouped into the head, the mesosome, the metasome, and the urosome (Fig. 7.10). The cephalothorax is composed of the head fused either with the first segment of the mesosome for Ingolfiellidae or with segments 1 and 2 of the mesosome for Metaingolfiellidae (Fig. 7.10). Their general morphology is similar to that of Amphipoda (cf. section Amphipoda in this chapter). However, Ingolfiellida differ from Amphipoda by the absence of eyes in all species. A vestigial

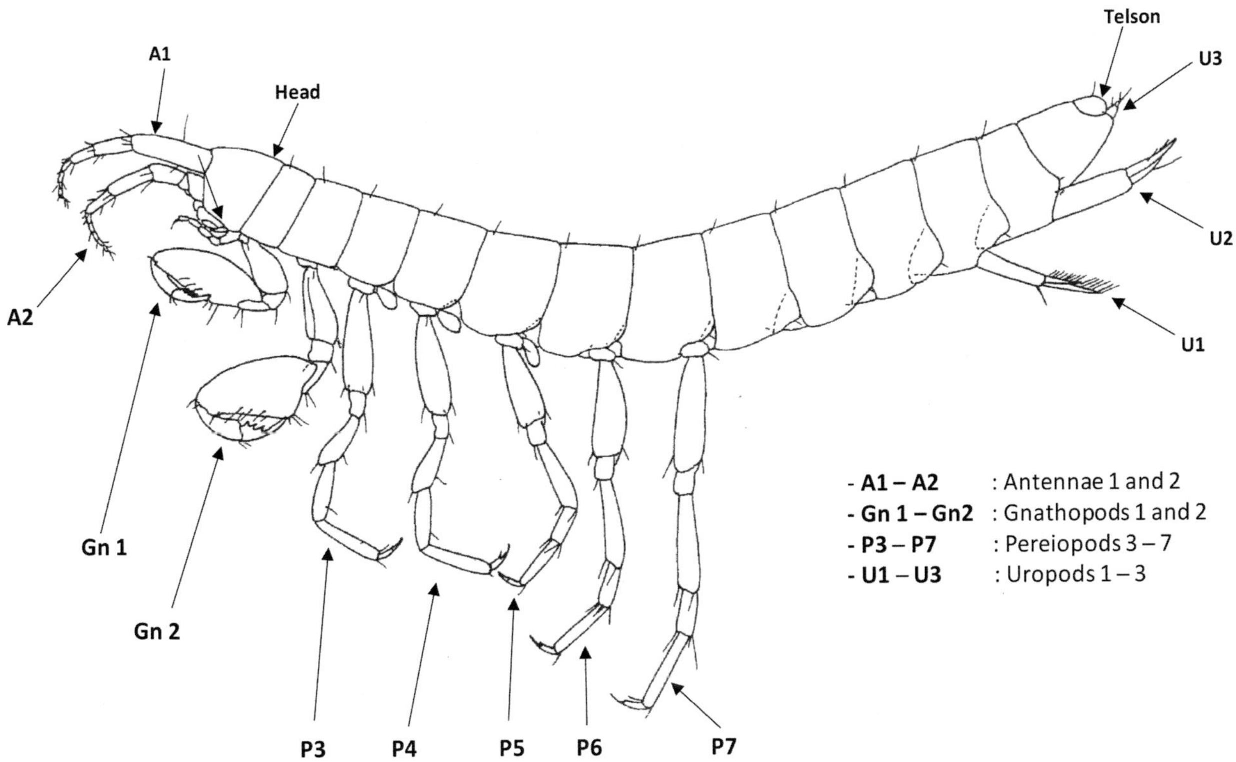

FIGURE 7.10 Habitus of *Ingolfiella cottarellii* (Ruffo & Vigna Taglianti, 1989). *Modified after Karaman 1993.*

pedunculate eye can be present is many Ingolfiellid species but is rarely observed in European species. Mouthparts are also similar, but the maxilliped does not have the ischial endite (outer plate). The shape of their gnathopods strongly differs from the shape of Amphipoda gnathopods: the "hand" of their gnathopods is formed by the fifth article (carpus) and the "finger" consists of two articles (propodus and dactyl) instead of one, whereas the "hand" of Amphipoda gnathopods consists of the sixth article (propodus) and the "finger" corresponds to the sole dactyl. Their mesosome also differs by the absence of coxal gills on their pereopods, unsegmented and uniramous pleopods, and the absence of epimera on their urosome.

Collection, preparation, and identification

Ingolfiellida are distributed worldwide but are very rare locally. In the Mediterranean Basin, all species are obligate groundwater species and can be sampled using any kind of sampling method adapted to hypogean species. Several kinds of specific materials (pumps, nets, traps) for this environment can be used (Malard et al., 2002).

Ingolfiellida can be easily fixed and preserved in 70% ethanol or another fixative for morphological analysis, and in 95% ethanol for genetic analysis.

The family and the genera can be identified under a stereomicroscope, but identification at the species level requires dissection, mounting of body parts and appendages on slides, and subsequent examination using a 400× or higher-magnification microscope. Specimens kept in glycerol or ethanol can be handled under a low-power microscope with needles. Glycerol is used as an embedding medium for short-term slides. Baths in clearing solution (e.g., soft acids) or permanent mounting media are only required when depositing museum specimens. The same specimen can be used for both studies: the abdomen for DNA extraction and the other parts for dissection and mounting. This procedure ensures that the sequenced genes correspond to the same morphotype.

Taxonomic identification at the species level is generally possible using males, but only females are known for some species (*I. acherontis*, *I. beatricis*, *I. macedonica*). The most important body parts for identification are the head, mouthparts, gnathopods, pleopods, and uropods. Additional body parts need to be examined to identify certain species.

Limitations

The key below is designed to cover all currently recognized freshwater ingolfiellid species of the Mediterranean Basin. Taxonomic synonymy has been updated. Rare records are included, unless they are insufficiently documented. For example, *Ingolfiella* (*Balchanella*) *acherontis* is included in the key, but it is not clearly described and its identification key should be used with caution. Species-specific secondary sexual characteristics (modified gnathopods or uropod 2, oostegites, etc.) are important for the taxonomy of Ingolfiellida but limited because several species are only known based on one sex (male or female). Another limitation of the key is that most species are only known from a small number of specimens, so that intraspecific variability generally remains unknown.

Key to Ingolfiellida

Ingolfiellida: families

1 Cephalothorax composed of head and first and second segment fused together; uropod 3 peduncles of right and left uropods fused together .... **Metaingolfiellidae** (one species: *M. mirabilis*; Ruffo, 1969)
1' Cephalothorax composed of head and first segment fused together; uropod 3 peduncles free .... **Ingolfiellidae** (one genus: ***Ingolfiella***)

Ingolfiellida: Ingolfiellidae: *Ingolfiella:* species

1 Cephalic ocular lobes developed 2
1' Cephalic ocular lobes absent 3
2 (1) Gnathopods 1 and 2 with 4 denticles on posterior margin of dactylus .... *I. beatricis* (Ruffo & Vonk, 2001)
2' Gnathopods 1 and 2 with 3 denticles on posterior margin of dactylus .... *I. acherontis* (S. Karaman, 1933)
3 (1') First pereonite (second thoracic segment) free 4
3' First pereonite (second thoracic segment) fused to cephalon .... *I. vandeli* (Bou, 1970)
4 (3) Maxilla 1 palp subequal to outer lobe, inner lobe with 1 or 2 setae 5
4'Maxilla 1 palp longer than outer lobe, inner lobe with 3 setae .... *I. petkovskii* (S. Karaman, 1957)
5 (4) Maxilla 1 inner lobe with 2 setae; gnathopods 1 and 2 with 3 denticles; pleopods 1–3 absent in females 6
5' Maxilla 1 inner lobe with one seta; gnathopods 1 and 2 with 4 denticles on posterior margin of dactylus, pleopods 1–3 present in females .... *I. macedonica* (S. Karaman, 1959)
6 (5) Uropod 1 outer ramus less than half of inner ramus length; uropod 2 longer than uropod 1; pereopod 3–4 claw simple 7
6' Uropod 1 outer ramus more than half of inner ramus length; uropod 1–2 subequal; pereopod 3–4 claw dentate or bifid .... *I. catalanensis* (Coineau, 1963)
7 (6) Antenna 1 accessory flagellum hardly longer than the first articles of the flagellum; peduncle uropod 2 with 6–7 rows of setae .... *I. cottarellii* (Ruffo & Vigna Taglianti, 1989)
7' Antenna 1 accessory flagellum longer than the two first articles of the flagellum; peduncle uropod 2 with 5 rows of setae .... *I. thibaudi* (Coineau, 1968)

Subchapter 7.6

Order Isopoda

Giuseppe Messana[1] **and Christophe Piscart**[2]
[1]*Institute of Research on Terrestrial Ecosystems (CNR-IRET), Sesto Fiorentino, Firenze, Italy*
[2]*French National Centre for Scientific Research (CNRS), University of Rennes 1, France*

Introduction

Isopoda Latreille, 1817 are a very diverse order both in the number of species and in their morphology. About 10% of all isopods—more than 10,000 described species—live in continental waters, mostly in groundwater (Wilson, 2008). They are easily recognizable in the field from their typical dorsally and ventrally flattened body shape, even if some groups of groundwater isopods (e.g., Microcerberidae) differ from this general habitus with a more or less thin and vermiform body.

Biogeographic analysis indicates that Isopoda, like Amphipoda, were likely widespread in Pangea, Gondwana, and Laurentia (Wilson, 2008). Freshwater colonization by isopods likely resulted from successive events, with the oldest events occurring during the Palaeozoic period (540–250 Mya) and involving ancestors of a few Asellota families and the entire suborder Phreatoicidea (not currently present in the Mediterranean Basin). The most ancient fossil records—*Hesslerella shermani*, a Phreatoicidea—dates back to 307 Mya (Schram, 1984). This hypothesis is congruent with molecular studies suggesting diversification events in freshwaters 400–500 Mya.

Among the 1000 species described in freshwaters, 336 are present in the Mediterranean Basin. The main taxonomic groups are Cirolanidae (nine genera), Asellidae (five genera), Stenasellidae (five genera), and Sphaeromatidae (five genera). A few taxa belonging to Janiridae, Lepidocharontidae, Microcerberidae, and Microparasellidae are also present (Boyko et al., 2020).

General ecology and distribution

Mediterranean freshwater isopods inhabit streams, rivers, ponds, lakes, and temporary pools where they can be found on the substrate or in the interstitial systems beneath the river bed (hyporheon). Most of the 299 described species are obligatory inhabitants of groundwaters (stygobionts), found in caves or other habitats where subterranean waters flow through; they are strictly endemic. Some taxa inhabit thermal waters (Rogers & Lewis, 2019). Only the epigean genus *Proasellus* (Asellidae) is widely distributed along the entire Mediterranean Basin; one species—*Proasellus coxalis*—is represented by various subspecies whose taxonomic rank should be accurately revised.

The suborder Asellota includes the highest number of described freshwater species, mostly in the Asellidae and Stenasellidae families. Out of the five genera composing Stenasellidae, which are all strictly subterranean without any epigean relative, *Stenasellus* has the widest distribution: it is present in the groundwaters of the Far East, Middle East, East Africa, and the western Mediterranean Basin, showing a biogeographic pattern linked to the ancient Thetys sea (Henry & Magniez, 1983; Messana et al., 2019). Some genera are much diversified but geographically restricted, for example, *Bragasellus* (20 species) and *Synasellus* (35 species), both restricted to the Iberian Peninsula. The janirid *Jaera sarsi*, originally from the Ponto-Caspian fauna but widely distributed in Western Europe, has been included even if it is invasive and present only in the largest rivers.

In the "Flabellifera" *sensu lato* group (e.g., Cymothoidea and Sphaeromatidea), the nine genera of Cirolanidae present in the Mediterranean Basin are the result of multiple colonizations of continental groundwater habitats (Thalassostygobionts) during several ancient marine ingressions (Baratti et al., 1999; Wilson, 2008). The genus *Sphaeromides* has a north Mediterranean distribution, *Typhlocirolana* a south-western one, while *Turcolana* has an eastern one. *Botolana* and *Moroccolana* are confined to Moroccan groundwaters with a strict relationship with *Typhlocirolana*, while *Faucheria* and *Kensleylana* are found in the groundwaters of southern France and the northern Iberian Peninsula, respectively. *Metacirolana* and *Saharolana* are endemic of the Baleric Islands and Tunisia, respectively.

The subterranean taxa are represented by several mostly scavenger or omnivorous taxa, with a carnivorous tendency. Epigean taxa feed on the periphyton or on detritus.

Terminology and morphology

Freshwater taxa include the typical dorso/ventral flattened isopods or narrow-bodied forms, and a few vermiform ones (Fig. 7.11). Adaptations mostly depend on colonized habitats. Epigean species are variously pigmented, while their subterranean counterparts are unpigmented and whitish. Likewise, eyes are present in the surface species and vary from reduced to completely absent in subterranean species.

The body is composed of three main sections (Fig. 7.12). In epigean taxa, the globular head (cephalon) has a couple of big composed dorsolateral sessile eyes and two pairs of uniramous antennae (A1 and A2); antenna 2 bears a rudimental exopodite (squama) in the more primitive taxa. The mouthpart is on the ventral surface of the head and is composed of the labrum, labium, mandibles, maxillae I and II, and maxilliped. The first thoracic segment is always fused to the cephalon and bears a couple of modified legs (maxillipeds).

The thorax (or pereon) is formed of seven very similar thoracic metamers (2–8) bearing seven pairs of walking legs (pereopods). The first pair is usually modified for grasping in most species. A couple of genital openings are present on the sternal face of pereionite 7 in males (genital papillae) and 6 in females. The abdomen (pleon) has six well-developed segments. The first five segments are usually free, the sixth one is fused to the telson to form a pleotelson. The five couples of appendices (pleopods) are biramous and mostly identical in the two sexes. Their rami have a respiratory and potential swimming role. The second pair of pleopods is modified into a copulatory organ in males (Fig. 7.13), to transfer sperm to the female genital opening. It has a unique construction in each species, and is frequently surrounded by accessory processes. The identification of most Isopoda, particularly Asellota, relies on the examination of this structure. The sixth pair (uropods) is variously modified in the diverse taxa.

Fertilized eggs develop in a ventral marsupium of the female, formed by the imbrication of lamellae originating from the coxopodites of pereopods: the oostegites. Isopoda have a typical biphasic molting: the rear part of the body molts before the anterior part.

Collection, preparation, and preservation

Collection methods vary depending upon the habitat. For surface and cave species, small hand nets with an appropriate mesh dimension or baited traps can be used. For interstitial hyporheic taxa, the most effective methods are the Bou-Rouch pump and the Karaman-Chappuis method (Malard et al., 2004). Isopods can be easily fixed and preserved in 70% ethanol or another fixative for morphological analysis and in 95% ethanol for molecular analysis. If many specimens are taken, ethanol should be replaced after a few days to prevent dilution by the animal's body water content. For conservation purposes, an ethanol (70/75°) plus a 5% glycerol mix is best suited to avoid articulations hardening, and also avoids accidental desiccation.

Specimens are best examined and viewed under low magnification for family or sometimes genus level identification. Species identification requires dissecting various parts of the body, which can be done by handling the specimen with fine forceps and pins. The simplest procedure consists in transferring the appendages into a drop of glycerin on a microscope slide. To see the fine details in heavily chitinized specimens, a semi-permanent slide with a clearing liquid such as Hoyers medium can be prepared (the slide should stay a few days at 35°C and then the cover should be sealed). Once dissected, the specimen is ready for examination, which may require a magnification of 100/1000 × for detailed examination depending on the size of the specimen. Scanning electron microscopy (SEM) observation is the best examination tool for fine details.

Limitations

Isopoda are a much diversified group of taxa with almost 1000 known freshwater species, and many more that remain undescribed. The keys proposed here are designed to cover 4 suborders, 8 families, and 33 genera. Oniscidea are not dealt with, although a few aquatic and several amphibious species are present in the Mediterranean Basin. Keys at the species level are only provided for genera with a low number of species (usually less than 15). Rare records are included, unless they are insufficiently documented. Taxonomic synonymy has been updated, but taxonomy remains unclear for some genera. However, several cirolanid, sphaeromatid, and asellid genera are badly described and need revision. As a consequence, some family, genus, and species names may change in the future.

Finally, even if our keys were designed to be as clear as possible, you should never forget that identifying certain groups, especially groundwater species, can be very difficult and requires strong expertise. Do not hesitate to use alternative information or keys to consolidate your conclusions. Keep in mind that keys, even at the family level, are designed for species of the Mediterranean Basin and are not adapted to other isopods worldwide.

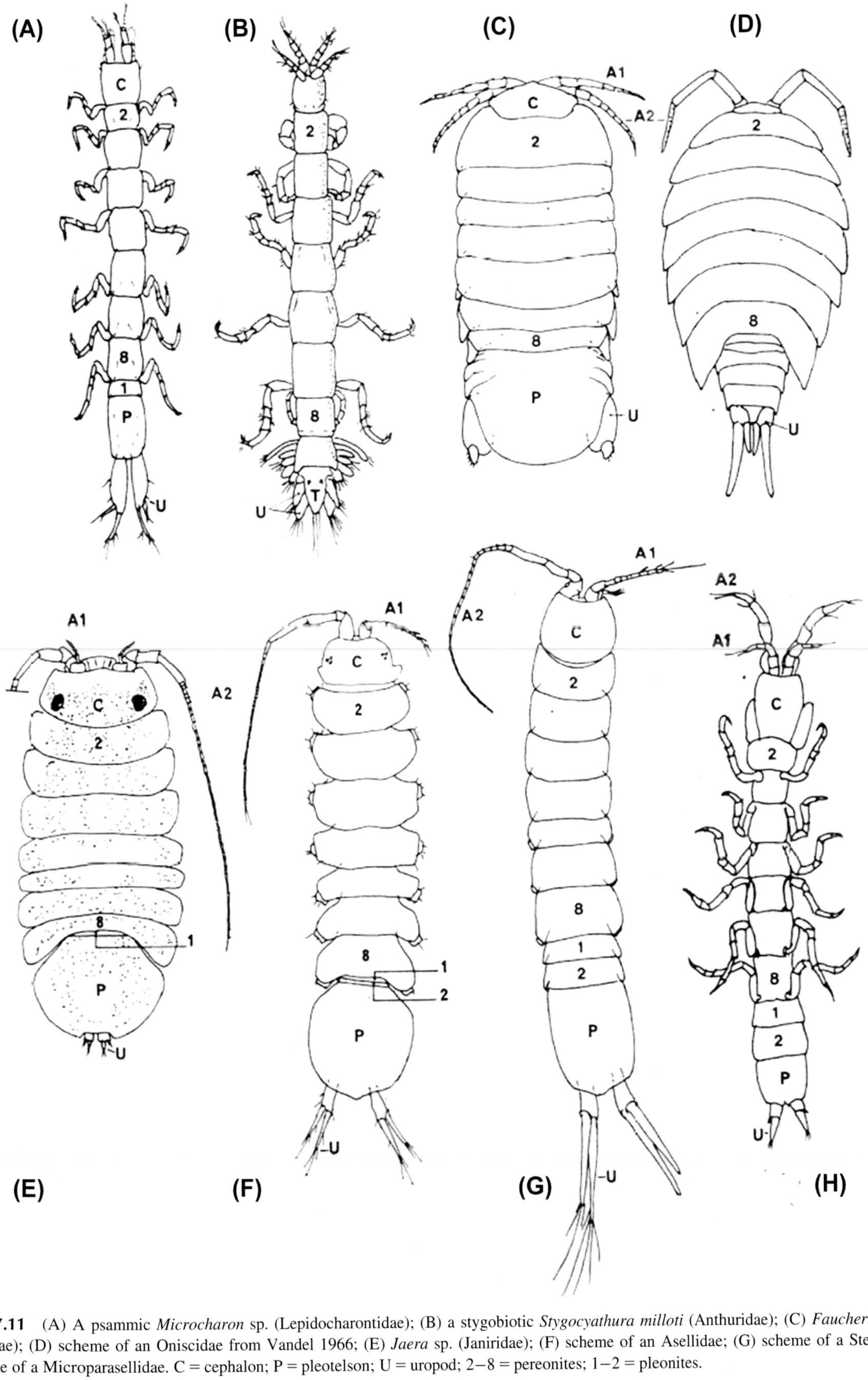

FIGURE 7.11 (A) A psammic *Microcharon* sp. (Lepidocharontidae); (B) a stygobiotic *Stygocyathura milloti* (Anthuridae); (C) *Faucheria faucheri* (Cirolanidae); (D) scheme of an Oniscidae from Vandel 1966; (E) *Jaera* sp. (Janiridae); (F) scheme of an Asellidae; (G) scheme of a Stenasellidae; (H) scheme of a Microparasellidae. C = cephalon; P = pleotelson; U = uropod; 2–8 = pereonites; 1–2 = pleonites.

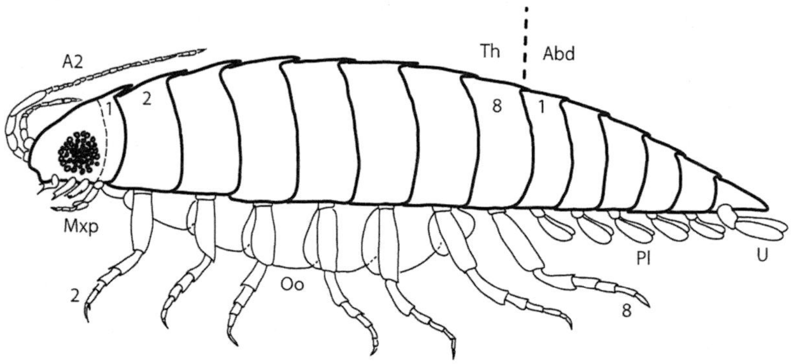

FIGURE 7.12 Schematic left lateral view of a female isopod. A2 = antennae; Abd = pleon; Mxp = maxilliped; Oo = oostegites; P = pleopods; U = uropods. 2–8 = free pereonites bearing pereopods (the first one fused to the cephalon bears the maxilliped); 1–5 = pleonites.

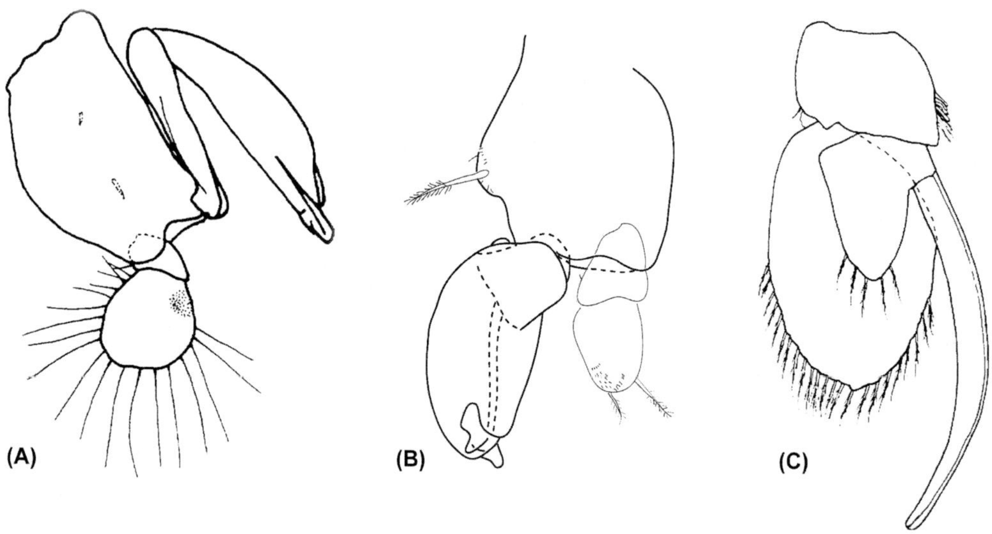

FIGURE 7.13 Second pleopod male of three different genera: 1. *Stenasellus*; 2. *Proasellus*; 2. *Typhlocirolana*.

Keys to Isopoda

Isopoda: suborders and families

1 Uropods styliform, inserted terminally on pleotelson 2
1' Uropods lamellar, inserted at pleotelson base 8
2 (1) Antennules normal, or if reduced not minute; pleon variable, with or without fused pleonites 3
2' Antennules vestigial, minute; pleon always of 5 free pleonites, plus the pleotelson .... **Oniscidea**
3 (2) One or two, free pleonites; body length variable; antenna peduncle usually with a scale; antennule rarely reduced, peduncle and flagellum distinct; maxilliped almost always with coupling setae on endite; female pleopod 2 uniramous; male pleopod 2 endopod large and geniculate .... 4 (**Asellota**)
3' Two long pleonites, and a pleotelson, all cylindrical; body length at least 6 times width; antenna peduncle without a scale; antennule reduced, peduncle indistinguishable from flagellum; maxilliped without coupling setae on endite; female pleopod 2 biramous; male pleopod 2 endopod not geniculate; interstitial .... **Microcerberidea (one family: Microcerberidae)**
4 (3) Two free pleonites 5
4' One or zero free pleonites 6
5 (4) Antenna 2 with an exopod; pleonite 1 and 2 wide, pronounced in dorsal view; endopodite of pleopod 2 ♂ biarticulated .... **Stenasellidae**
5' Antenna 2 without exopod; pleonite 1 and 2 short and narrow, not pronounced in dorsal view; endopodite of pleopod 2 ♂ with 1 article .... **Asellidae**

6 (4') Pleotelson lateral margins parallel; uropods longer than pleotelson; body elongate, length at least 6 times width 7
6' Pleotelson lateral margins converging; uropods shorter than pleotelson; body broad, length less than 4 times width .... **Janiridae (one genus: *Jaera*)**
7 (6) Uropods biramous, large; antennal podomeres shorter than flagellum; rostrum small or absent .... **Lepidocharontidae (one genus: *Microcharon*)**
7' Uropods uniramous; antennal podomeres longer than flagellym; head with prominent rostrum .... **Microparasellidae**
8 (1') Uropods not modified as a ventral operculum 9
8' Uropods modified as opercula, covering the entire pleopodal chamber; males with penes arising on sternum of pleonite 1, or on articulation between pereonite 7 and pleonite 1; mandibular molar process a stout, flattened grinding structure .... **Valvifera**
9 (8) Uropods articulating mediolaterally; pleon dorsally with 4 or 5 free pleonites plus the pleotelson .... **Cymothoida (one family: Cirolanidae)**
9' Uropods, particularly the peduncle/protopod, articulating dorsoventrally; pleon dorsally with four or fewer free pleonites plus the pleotelson .... **Sphaeromatidea (one family: Sphaeromatidae)**

Isopoda: Asellidae: genera

1 Coxae not obscured by pereonite, visible dorsally 2
1' Coxae reduced not visible dorsally .... *Bragasellus*
[Iberian Peninsula]
2 Pleopod 1 peduncles fused at base on at least 20% 3
2' Pleopod 1 peduncles free attached by at least 2 coupling hooks 4
3 Mandibular palp absent .... *Synasellus*
[Iberian Peninsula]
3' Mandibular palp present .... *Chthonasellus* (one species: *Chthonasellus bodoni* Argano & Messana, 1991)
[Italy]
4 Pleopod 2 endopod with basal spur and pronounced basal apophysis .... *Asellus*
[Europe]
4' Pleopod 2 endopod without basal spur and without pronounced basal apophysis .... *Proasellus*
[Peri-Mediterranean]

Isopoda: Janiridae: *Jaera:* species

1 Maxilliped palp article 2 wider than long; uropod exopodite and endopodite well developed, overlapping more than a half of the propodite 2
1' Maxilliped palp article 2 almost as long as wide; uropod exopodite and endopodite minute, not overlapping a half of the propodite width .... *Jaera sarsi* (Valkanov, 1936)
2 (1) Pleopod 1 ♂ praeoperculum narrow and pointed 3
2' Pleopod 1 ♂ praeoperculum T-shaped .... *Jaera italica* (Kesselyak, 1938)
3 (2) Pereopods 1 and 7 not sexually dimorphic; pleopods 1 praeoperculum wider at its distal end .... *Jaera schellenbergi* (Kesselyak, 1938)
3' Pereopod 1 and 7 sexually dimorphic; pleopods 1 praeoperculum restangular .... *Jaera nordmanni* (Rathke, 1837) (superspecies including 3 subspecies *Jaera n. nordmanni*, *Jaera n. balearica*, *Jaera n. brevicaudata* not included in this key)

Isopoda: Microparasellidae: genera and species

1 Somite lateral margins with denticles .... 2 (*Microparasellus*)
[Balkans, Greece, Middle East]
1' Somite lateral margins smooth .... *Angeliera* (one species: *Angeliera pheraticola* Chappuis & Delamare-Deboutteville, 1952)
[France, Italy]

2 (1) Uropods visible dorsally 3
2' Uropods not visible dorsally 4
3 (2) Pleopod 1 ♂ subtriangular distally; transverse section of the pleotelson ovoid .... *Microparasellus puteanus* (Karaman, 1933)
3' Pleopod 1 ♂ rounded distally; transverse section of the pleotelson sub-circular .... *Microparasellus hellenicus* (Argano & Pesce, 1979)
4 (3') Lateral expansion of the first thoracic segment lacking; uropods exopodite reduced, scale like .... *Microparasellus aloufi* (Coineau, 1968)
4'. Lateral expansion of the first thoracic segment present; uropods exopodite developed .... *Microparasellus libanicus* (Chappuis & Delamare-Deboutteville, 1954)

Isopoda: Stenasellidae: genera and species (except *Stenasellus*)

1 Body thinner than the preceding; uropods longer than 1/3 of Pleotelson 2
1' Body short, length around 2.5 width; uropods short, less than 1/5 of Pleotelson .... *Johanella* (one species: *Johanella purpurea* Monod, 1924) [Algeria]
2 (1) Pleopod 2 endopodite ♂ composed by 2 articles free, the proximal well developed, the distal never jointed on its internal margin 3
2' Pleopod 2 endopodite ♂ composed by 2 articles fused, proximal article very small, the distal furrowed .... *Metastenasellus* (one species: *Metastenasellus leysi* Magniez, 1986) [Algeria]
3 (2) Pleopod 1 ♂ propodite subquadrangular and well developed, exopodite flat 4
3' Pleopod 1 ♂ exopodite externally curved .... *Balkanostenasellus* (one species: *Balkanostenasellus skopljensis* (Karaman, 1937), superspecies including 4 subspecies: *Balkanostenasellus s. croaticus*, *Balkanostenasellus s. meridionalis*, *Balkanostenasellus s. skopljensis*, *Balkanostenasellus s. thermalis* not included in this key) [Balkan]
4 (3) Pleopod 2 endopodite ♂, distal article as a flat oval lamina, with a similar external appendix; pleopod 1 ♂ sympodite without coupling hooks .... *Magniezia* (one species: *Magniezia. gardei* Magniez (1975) [Morocco]
4' Pleopod 2 endopodite ♂, distal article long and thin, longitudinally folded with proximally divergent margins, sometimes distally fused; pleopod 1 ♂ sympodite with or without coupling hooks .... *Stenasellus* [Western Europe]

Isopoda: Stenasellidae: *Stenasellus*: species

1 Pleopod 2 ♂ endopodite, proximal article of the long and thin 2
1' Pleopod 2 ♂ endopodite, proximal article short and stubby 9
2 (1) Pleopod 1 ♂ protopodite without coupling hooks or with simple seta; pleopod 2 ♂ endopodite almost isodiametric with a large and denticulate terminal part; pleopods 4 and 5 exopodites ovoids, larger than endopodite .... 3 (*Stenasellus breuili*-group)
2' Pleopod 1 ♂ protopodite with coupling hooks; pleopod 2 ♂ endopodite, distal article fusoïd with acute apex; distal article of exopodite medially bented and bearing about 15 marginal setae; pleopods 4 and 5 exopodites varying from large and lamellar to thin or styliform .... 7 (*Stenasellus virei*-group)
3 (2) Pereopod 1 propodite robust 4
3' Pereopod 1 propodite thin and fragile .... *Stenasellus bragai* (Magniez, 1976)
4 (3) Pleopod 2 sympodite subquadrangular, endopodite distal article falciform with about 20 sternal setae 5
4' Pleopod 2 sympodite subpentagonal, endopodite distal article strong with a disto-externak expansion 6
5 (4) Pereopods 2–7 dactyli with 1 sternal spine .... *Stenasellus escolai* (Magniez, 1977)
5' Pereopods 2–7 dactyli with 2 sternal spines .... *Stenasellus. magniezi* (Escola, 1975)

6 (4') Pleopod 2 ♂ exopodite subovoid with 7 distal marginal plumose setae, endopodite distal article with a disto-external rounded expansion bearing 4 tiny spines .... *Stenasellus galhanoae* (Braga, 1962)
6' Pleopod 2 ♂ exopodite rounded with 18 distal marginal plumose setae, endopodite distal article with an external expansion bearing 8 spines and an internal lobe covered by tiny setae .... *Stenasellus breuili* (Racovitza, 1924)
7 (2') Pleopod 2 ♂ endopodite distal part stretched but not as an elongated beak 8
7' Pleopod 2 ♂ endopodite distal part stretched as an elongated beak .... *Stenasellus buili* (Remy, 1949)
8 (7) Pleopod 5 exopodite large almost totally covering endopodite and distally rounded .... *Stenasellus racovitzai* (Razzauti, 1925)
8' Pleopod 5 exopodite styliform, 5–6 times longer than large, always thinner than endopodite .... *Stenasellus virei* (Dollfus, 1897)
9 (1') Pleopod 2 ♂ endopodite distal article almost isodiametric, with an obtuse ending and a large terminal orifice, distal external part more or less pointed .... *Stenasellus nuragicus* (Argano, 1968)
9' Pleopod 2 ♂ endopodite distal article almost isodiametric, with an obtuse ending and a large terminal orifice, distal external part large .... *Stenasellus assorgiai* (Argano, 1968)

Isopoda: Cymothoida: Cirolanidae: genera

1 Eyes absent 2
1' Eyes minute; body length 5 mm .... *Saharolana* (one species: *Saharolana seurati* Monod, 1930) [Tunisia]
2 (1) Pleotelson completely fused to pleon; uropods with 1 terminal ramus; body length ≤ 5 mm 3
2' Pleotelson not as above; 5 free pleonite; uropods with 2 terminal rami; body length variable 4
3 (2) Pleotelson with 2 sutures visible laterally for the fused segments; uropod peduncle flattened dorsoventrally .... *Faucheria* (one species: *Faucheria faucheri* Dollfus & Viré, 1900) [France]
3' Pleotelson with only 1 suture visible laterally; uropod peduncle elongate and cylindrical, with a single, minute ramus .... *Kensleyana* (one species: *Kensleyana briani* Bruce & Herrando-Perez, 2005) [Spain]
4 (2') Pleonite 5 covered by pleonite 4; body length 1–10 mm 5
4' All 5 pleonite free; body length 10–33 mm .... *Sphaeromides* [France, Balkans]
5 (4) Not rolling into a ball; pleopods 1 not forming an operculum; pleotelson somewhat produced 6
5' Adapted to roll into a ball; pleopods 1 forming an operculum; pleotelson never produced .... *Turcolana* [East Mediterranean]
6 (5) Cephalon not laterally overlapped by pereonite 1 7
6' Cephalon posterior 50% embraced by pereonite 1; body length 7 mm .... *Marocolana* (one species *Marocolana delamarei;* Boutin, 1993) [Morocco]
7 (6) Clypeus not projecting; pleonite 5 overlapped by pleonite 4 8
7' Clypeus projecting; pleonite 5 overlapped by pleonite 4 .... *Metacirolana* (one species: *Metacirolana ponsi* Jaume & Garcia, 1992) [Spain]
8 (7) Uropod rami dorsoventrally flattened, triangular, subequal in length to peduncle, and shorter than pleotelson .. .. *Typhlocirolana* [South Europe, North Africa]
8' Uropod rami subcylindrical, styliform, length/width 1.5 × peduncle, longer than pleotelson .... *Botolana* (one species: *Botolana leptura* Botosaneanu, Boutin & Henry, 1985) [Morocco]

Isopoda: Cymothoida: Cirolanidae: *Sphaeromides*: species

1 Pleonite 3 wider than others 2
1' Pleonite 3 not wider than the others 3
2 (1) Fifth antennal article with straight margins .... *Sphaeromides bureschi* (Strouhal, 1963)
2' Fifth antennal article with inflate margins .... *Sphaeromides polateni* (Angelov, 1968)
3 (1') Pleotelson apex truncate .... *Sphaeromides virei* (Brian, 1923)
3' Pleotelson apex acute .... *Sphaeromides raymondi* (Dollfus, 1897)

Isopoda: Sphaeromatidea: Sphaeromatidae: genera and species (except *Monolistra*)

1 Pleopod 4 and 5 endopods smooth lacking pleats or fold 2
1' Pleopod 4 and 5 endopods with pleats or fold 5
2 (1) Pleotelson preceded by 2 or more pleonites fused at least dorsomedially 3
2' Pleotelson preceded by 2 free and articulated pleonites .... *Monolistra* [Italy, Balkans]
3 (2) Uropod short, around 0.25 × the pleotelson length .... *Merozoon vestigatum* (Sket, 2012) [Croatia]
3'Uropod length around 0.5 × pleotelson length .... 4 (*Caecosphaeroma*) [France]
4 (3) All pleonite fused with telson; pleopods short; uropod short, exopodite distinct .... *C. (Caecosphaeroma) virei* (Dolfus, 1896)
4' First pleonite partially free; pleopod well developed; uropod reduced scale like, exopodite not visible .... *Caecosphaeroma burgundum* (*Vireia*) *burgundum* (Dollfus, 1898)
5 (1') Maxilliped articles 2–4 lacking lobes, inferior margins with dense fringes of plumose setae .... *Sphaeroma* (one species: *Sphaeroma podicipitis* Monod, 1931) [South Europe, North Africa]
5' Maxilliped articles 2–4 more or less lobed, inferior margins with simple setae only .... 6 (*Lekanesphaera*) [Morocco]
6 (5) Pereopod 1 propodus with few to many distal setae next to rostrodistal spine 7
6' Pereopod 1 propodus without distal setae next to rostro-distal spine .... *Lekanesphaera hookeri* (Leach, 1814)
7 (6) Pereopod 1, setae on ischium and merus smooth or sparsely plumose only on their distal part 8
7' Pereopod 1, setae on ischium and merus entirely plumose, ending with long setules on their distal part .... *Lekanesphaera monodi* (Arcangeli, 1934)
8 (7) Pereopod 1 propodus with more than 3 setae inserted distally in transverse row next to rostro-distal spine. Antenna flagellum articles with fringe of long setae at distal interior angle, seta 2–3 times length of article 9
8' Pereopod 1 propodus with 2–3 setae inserted distally in transverse row next to rostro-distal spine. Antenna flagellum articles with fringe of few short setae, seta 1.5 times length of article .... *L. rugicauda* (Leach, 1814)
9 (8) Pleotelson subapically concave, slightly upcurved, dorsal surface slightly granular; uropod exopod, external margin with 5–7 little teeth, giving it a crenate appearance .... *Lekanesphaera hoestlandti* (Daguerre de Hureaux, Elkaim & Lejuez, 1965)
9' Pleotelson subapically concave but not upcurved, dorsal surface with row of 6–7 partly fused tubercles on each side of midline; Uropod exopod, external margin with 6–7 distinct teeth .... *Lekanesphaera panousei* (Daguerre de Hureaux, Elkaïm & Lejuez, 1964)

Isopoda: Microcerberidea: Microcerberidae: genera and species

1 Coxae of pereopods 2–4 ring-shaped; male pleopod 2 with saber-shaped endopod with pointed and laterally curved apex; pleopod 4 with elongate sympod, with 3 rami of similar size, only proximally covered by pleopod 3 *Microcerberus stygius* (Karaman, 1933)
1' Coxae of pereopods 2–4 more or less pointed; male pleopod 2 with elongate rectangular sympod, tiny rounded exopod, and apically bifid; Pleopod 4 well covered by pleopod 3, biramous, without articulations .... *Coxicerberus ruffoi* (Chappuis, 1953)

Subchapter 7.7

Orders Mysida and Stygiomysida

Karl J. Wittmann
Department of Environmental Health, Medical University of Vienna, Vienna. Austria

Introduction

Mysida

Peracarid crustaceans of the order Mysida are recognizable by their shrimp-like habitus, by separate, stalked eyes, fused cephalothorax, well-developed carapace and tail fan, by swimming and respiratory currents driven by rotating movements of their thoracic exopods, and by the absence of gills and of true chelae. The author's own global census (Feb 19, 2021) yielded a total of 178 genera and 1190 species considered valid. Doubtful cases excluded, an additional three genera and nine species are known from the Miocene (Voicu, 1974) down to the Triassic (San Vicente & Cartanyà, 2017). The main taxonomic groups are the family Petalophtalmidae with the subfamilies Petalophthalminae and Hansenomysinae, and the family Mysidae with 10 subfamilies, namely Boreomysinae, Erythropinae, Gastrosaccinae, Heteromysinae, Leptomysinae, Mysidellinae, Mysinae, Palaumysinae, Rhopalophthalminae, and Siriellinae.

Most extant species inhabit marine waters of all major sea basins from the shore down to the deep sea, usually with greatest species diversity in sublittoral habitats. Most marine and freshwater species live on or closely above the bottom, often in swarms; a few are pelagic. Some species burrow just below the sediment surface, others are associated with benthic invertebrates. Many species hide during the daytime and emerge at night to feed, mate, molt, and release young. Most species are omnivorous, some are microphytophagous, saprophagous, or carnivorous; none is parasitic. Dense populations can have marked effects on near-bottom plankton communities. High individual densities occur in turbid, meso- to oligohaline waters of lagoons and estuaries. Such habitats can provide main pathways for immigration and adaptation to freshwater. Quantitatively less important pathways involve marine and brackish groundwater. (Reddell, 1981) assumes a marine origin of the troglobitic mysids (i.e., Stygiomysida and Mysida) of Mexico, mainly found in brackish groundwater, less frequently in fresh groundwater. Such a distribution also holds true for the two Mediterranean Stygiomysida species (see below); however, the only troglobitic Mysida species (*Troglomysis vjetrenicensis*) of the Mediterranean lives exclusively in fresh groundwater.

A total of 26 genera and 75 species of Mysidae (and no Petalophthalmidae) are recorded in freshwater globally, in part based on single freshwater records of euryhaline species. Within Mysidae, as many as 71 species belong to the subfamily Mysinae, two to Leptomysinae (*Americamysis almyra* and *Tenagomysis chiltoni*), one to Heteromysinae (*Deltamysis holmquistae*), and one to Rhopalophthalminae (*Rhopalophthalmus chilkensis*). The remaining six subfamilies have not been recorded in freshwater.

The bulk of freshwater records includes 24 Ponto-Caspian endemics, followed by 12 species from Amazonia and adjacent freshwater systems, 9 Mexican, Central American and Caribbean endemics, 9 boreo-arctic species including extrazonal glacial relicts, and 6 Mediterranean endemics.

Including species of northeastern Atlantic and Ponto-Caspian origin, 10 species (adult body size 3–13 mm) have been recorded in Mediterranean freshwaters so far; all belong to the subfamily Mysinae within the family Mysidae. Among these, four Mediterranean endemics are restricted to freshwater, and two other endemics occur in fresh as well as brackish water. Among these six species, as many as four are stenoendemic in the Adriatic Sea, another one in the eastern Mediterranean including the Adriatic. The predominance of Adriatic endemics in Mediterranean freshwaters may have been favoured by alternating marine, lagoonary, and terrestrial phases (Correggiari et al., 1996) of the shallow northern Adriatic shelf during Pleistocene sea-level changes.

One brackish to marine northeastern Atlantic species of Mysidae (*Neomysis integer*) occurs marginally in freshwaters of the Mediterranean Basin. Two freshwater Ponto-Caspian endemics (*Hemimysis anomala* and *Limnomysis benedeni*) have recently expanded to the Rhône Delta on the Mediterranean coast of France. Finally, the details of when and how a third Ponto-Caspian endemic (*Paramysis lacustris*) arrived in an isolated freshwater lake without natural surface drainage in the highlands of Anatolia remain unknown.

Stygiomysida

Two families (Stygiomysidae and Lepidomysidae), two genera, and sixteen species of Stygiomysida are known globally. They show a Tethyan distribution, extending from Central America and the Caribbean to Mediterranean and Indian basins (Porter et al., 2008). Both known Mediterranean species (adult body size 6–13 mm) are stenoendemic in brackish groundwater in southeastern Italy and are rarely found there in freshwater.

General ecology and distribution

Stygiomysida

This order is represented by only two species in European waters. Both are stenoendemic of brackish groundwater in southeastern Italy and are rarely found in freshwater. Although their eyes are rudimentary, unpigmented, *Spelaeomysis bottazzii* and to a minor extent also *Stygiomysis hydruntina* are not strictly photophobic: the former species is regularly found in the dimly lit zone of wells, where it feeds on photoautotrophic organisms (Ariani & Wittmann, 2010).

Stygiomysis hydruntina has been reported from a few caves and wells near the Adriatic and Ionian coasts of the Salento Peninsula (southeastern Italy). It occurs mainly in deep brackish groundwater, rarely in near-surface low-salinity waters. Pesce et al. (1978) published three potential freshwater records without indication of salinity; however, a salinity of "0.3‰" is labeled on the slide containing parts of a male specimen from station PU/23 (own inspection of material kept at the Museo Civico di Storia Naturale, Verona). Besides, Pesce et al. (1978) indicated Monte Mario coordinates instead of Greenwich coordinates.

Spelaeomysis bottazzii has been reported from a number of caves, dolinas, wells, and springs near the Adriatic and Ionian coasts of the Salento Peninsula and from the Adriatic coast of Apulia (southeastern Italy). It occurs in deep brackish groundwater up to various near-surface mesohaline to anhaline waters (Inguscio, 1998). Breeding females are rare near the surface; most of them probably inhabit deep groundwater for more than 3 months of incubation. Laboratory females assume strongly reduced brood lamellae at the molt following incubation. This is interpreted as a strategy to replenish the trophic reserves required for sustaining a long incubation period in deep and trophically poor groundwater (Ariani & Wittmann, 2010).

Mysida

Only 10 species of Mysida have so far been found in true freshwater habitats (0.0–0.5 psu) of the Mediterranean Basin. The other two species given in the key below are expected but have not yet been recorded there in freshwater. Only four Mediterranean endemics are restricted to freshwater habitats:

Diamysis lacustris is stenoendemic in Lake Scutari on the southeastern coast of the Adriatic Sea, altitude 6 m; *T. vjetrenicensis* is stenoendemic in subterranean waters in the Vjetrenica cave system in Herzegovina, near the eastern Adriatic coast, altitude about 224 m; *Paramysis* (*Longidentia*) *adriatica* is known only from tributaries along the northern and eastern coasts of the Adriatic Sea, altitude 0–2 m. *Paramysis* (*Serrapalpisis*) *kosswigi* was reported from Lake Isikli and associated running waters of western Anatolia draining into the Aegean Basin, altitude 135–1007 m; previous records from tributaries of the Black Sea were referred to the sister species *P.* (*S.*) *lacustris* by (Wittmann et al., 2016).

An additional Mediterranean endemic, *D. fluviatilis*, occurs mainly in freshwater and to a lesser extent in brackish running waters, wells and estuaries along the northern coasts of the Adriatic Sea, altitude 0–16 m. Freshwater populations of *Diamysis mesohalobia heterandra* are confined to the Adriatic basin, altitude 0–3 m. Nonetheless, many more populations inhabit brackish waters of the eastern Mediterranean and Marmora basins.

The Mediterranean endemic *Diamysis hebraica* has not yet been reported from true freshwater. It has been found in three oligohaline streams and their springs on the coast of Israel (salinity 0.7–5 psu). An occurrence in the anhaline range appears plausible during potential periods of strong freshwater input.

P. (*S.*) *lacustris* is indigenous in anhaline to mesohaline waters of the Ponto-Caspian and Marmora basins. Within the boundaries of the Mediterranean Basin, it is found in the freshwater lakes Kus Gölü and Uluabat Gölü near the southern coast of the Marmora Sea. Material from the isolated freshwater Lake Beyşehir in the highlands of Anatolia, altitude 1115 m, was described as a separate subspecies, here not acknowledged *P. lacustris turcica* (Băcescu, 1948). The hydrological situation is quite complex: according to Günay (2010), at least part of the runoff enters the Mediterranean catchment through underground connections. This provides sufficient support for attributing this population to the Mediterranean fauna. Intensive intentional introductions of *P. lacustris* were made in the 1960s–70s in waters of eastern Europe, wherefrom it established in Lithuania and the Gulf of Finland (Audzijonytė et al., 2017). Nonintentional expansion occurred from the lower to the middle reach of the Danube River (Borza et al., 2019). If this expansion continues, this species may follow the below-discussed route to the Mediterranean coast of France.

Two other Ponto-Caspian endemics, *H. anomala* and *L. benedeni*, occur mainly in freshwater but have also been reported in brackish waters. They expanded from the Black Sea westward along the rivers Danube, Rhine, Rhône, and their associated artificial canals, and finally southward down to the delta of the Rhône on the Mediterranean coast of France, where they were first detected in 2007 and 2009, respectively (Wittmann & Ariani 2009; Wittmann et al., 2014).

Two indigenous northeastern Atlantic species inhabit brackish to marine waters, and marginally also freshwaters. *N. integer* inhabits the Mediterranean coasts of Spain and France, but not elsewhere in the Mediterranean Basin. It occurs marginally in freshwater of the Rhône River delta. *Mesopodopsis slabberi* is present in great numbers in brackish and marine, rarely in metahaline waters of the northeastern Atlantic, Baltic, Mediterranean, Marmora Sea, Black Sea, and the Sea of Azov. The lowest salinities for the Mediterranean records (0.9−1.4 psu) were reported in the estuaries of the Rhône and Po rivers and in a brackish drain in Corfu Island (Pesta 1935; Aguesse & Bigot 1960; Brun 1967). True freshwater records stem only from the Black Sea coast in waters of the delta and "Predelta" of the Danube (Dedju & Polischtuk 1968; Begun & Gomoiu 2007; Wittmann et al., 2016). Based on studies in Lake Sinoe, a barrier lagoon of the Predelta, Begun & Gomoiu (2007) concluded that this species is episodically found in freshwater but cannot form stable populations there. *M. slabberi* was recorded in this lagoon only under oligohaline conditions (1.1−2.1 psu) upon own sampling in 1985, 1998, and 2008.

Terminology and morphology

The terminology used here is according to (Wittmann et al., 2016). Only features using standard microscopy without dissection are considered below. Gross morphology and additional features are available in (Daneliya et al., 2019), with a focus on Ponto-Caspian species.

Among Malacostraca, the orders Stygiomysida and Mysida were formerly lumped together in the order Mysidacea based on their shared separate, stalked eyes at least in postnauplioid larvae (known so far), a respiratory carapace fused only with anterior thoracic somites, first and second antennae each with a three-segmented trunk, eight pairs of mostly biramous thoracopods (including maxillipeds), and the absence of gills. To our knowledge, the inner rami (endopods) are used for feeding and climbing, generally not for walking, and the outer rami (exopods) for swimming and driving respiratory currents through rotating movements. Both orders show five pairs of nonrespiratory, non-natatory pleopods. A marsupium is formed by one to seven pairs of oostegites (marsupial plates) arising as epipods in continuous series, starting with the ultimate thoracopod. The young pass through one embryonic (egg) and two nonfeeding larval stages in the marsupium, where they receive parental care rather than being merely stored, at least Mysida do so.

Based on genetic analyses, Meland & Willassen (2007) positioned the family Stygiomysidae closer to Mictacea than to Mysida, and proposed to establish Stygiomysida and Mysida as separate orders. The postlarval stages of Stygiomysida are distinguished from those of Mysida by reduced eyes, the presence of a spiny ventral lobe on the endopod of uropods (Fig. 7.16A,B; together with the telson forming a strong tail fan), by two laminar lobes flanking the male genital orifice instead of tubular penes, and by the absence of statocysts. All the Mysida dealt with here belong to the family Mysidae and are characterized by a large statocyst at the basis of the endopod of uropods. This organ contains one exceptionally large statolith (Fig. 7.16C) in the animal kingdom. The statoliths of most freshwater species, here represented by certain species of *Hemimysis*, *Limnomysis*, *Diamysis*, and *Paramysis*, are opaque because they are mineralized with vaterite, a metastable form of calcium carbonate. The statoliths of most marine and brackish-water species, and of just a few freshwater species (dealt with below: *Troglomysis*; rare in freshwater: *Mesopodopsis*, *Neomysis*) appear transparent and vitreous because they are mineralized with fluorite (CaF_2). However, statoliths can be damaged and may even vanish as fixation and preservation artefacts.

Both Stygiomysida species known in the Mediterranean Basin are readily distinguished by body shape (Figs 7.14 and 7.15) and size of the ventral lobe on the endopod of uropods (Fig. 7.16A,B). Careful inspection of antennal scale (Fig. 7.18) and telson (Fig. 7.19) is sufficient for identification of three Mysida species living in freshwaters of the Mediterranean Basin. Among additional features, the strongly reduced, obviously dysfunctional cornea (Fig. 7.20B) is a key feature to identify *Troglomysis*, whereas the structure of the carapace (Fig. 7.20) and of the thoracic exopods (Fig. 7.21) are important diagnostic features for *Diamysis*. For certain *Paramysis*, inspection of the paradactylary setae (Fig. 7.22; modified setae flanking the dactylus of thoracic endopods 3−8) is necessary for identification.

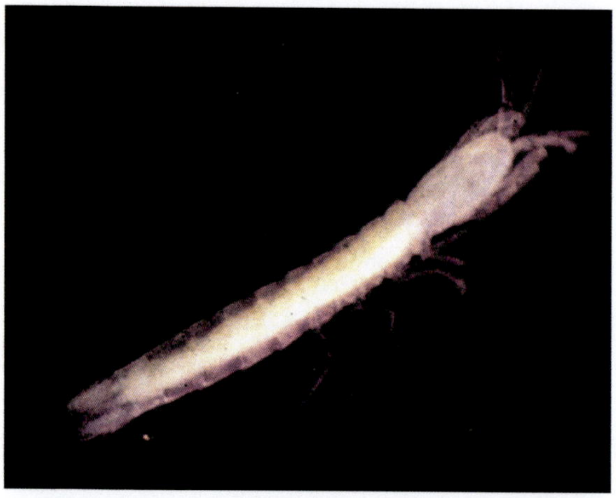

FIGURE 7.14 Stygiomysis hydruntina. *Photograph courtesy of Salvatore Inguscio.*

FIGURE 7.15 Spelaeomysis bottazzii. *Photograph courtesy of R. Pepe.*

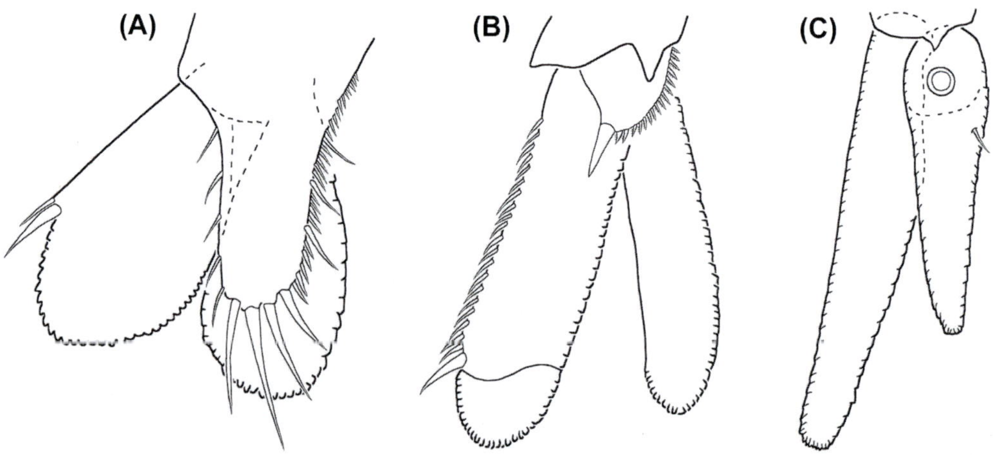

FIGURE 7.16 Uropods of (A) *Stygiomysis hydruntina*, (B) *Spelaeomysis bottazzii*, and (C) *Troglomysis vjetrenicensis*; (A–C) ventral view, setae omitted.

Collection, preparation, and preservation

Mysids are sampled during day and night with nets operated from the shore, with diver-operated nets, bottom nets, and epibenthic sledges. Baited bottle traps exposed overnight, and empty mollusc shells, stones with holes, discarded bottles, etc. collected during the day are treated *ex situ* with a few drops of clove oil to expel the animals. Fixation and preservation are performed with ethanol 80% for morphological analysis or 95% for genetic analysis. Identification at species level may require dissection, mounting of body parts and appendages on slides, and subsequent examination using a 400 × or higher-magnification microscope. Specimens in propylene glycol, glycerol, or ethanol can be handled under a low-power microscope with needles. Glycerol is used as an embedding medium for short-term slides, particularly for inspection of mineral statoliths that can be damaged by other media. The water-tolerant Swan-medium makes minute, superimposed objects well discernible owing to its strong bleaching power. After hardening at 60°C for about 15 hours, it requires very careful sealing within about 1 month to obtain multi-annual slides. Canada-Balsam requires dehydration in ascending ethanol series. It is recommended for permanent slides intended for deposition of museum specimens.

Limitations

The key below is designed to cover all currently recognized freshwater mysidacean species of the Mediterranean Basin. Rare records are included, unless they are insufficiently documented. Taxonomic synonymy is updated. Secondary sexual characteristics are crucial for the taxonomy of Mysida, but are restricted here in order to obtain a short-cut key applicable to specimens of any sex and stage of maturity (except juveniles in some cases).

Freshwater is defined as water with a salinity range of 0.0–0.5 practical salinity units (psu). Two species occurring in brackish waters of the Mediterranean (but so far recorded only in near but not pure freshwater) are included for potentially prolonging the validity of the key: *D. hebraica* from a documented lowest salinity of 0.7 psu, and *Mesopodopsis slabberi* found at a minimum of 0.9 psu. The latter species repeatedly found in true freshwaters of the Danube Delta along the Black Sea coast but not elsewhere in freshwater.

Key to Mysida and Stygiomysida

Mysida and Stygiomysida: orders

1 Endopod of uropods with very large statocyst; sympod without ventral lobe (Fig. 7.16C); cornea mostly well-developed (Figs 7.17 and Fig. 7.20A,C–F), rarely reduced (Fig. 7.20B) .... **Mysida (one family: Mysidae)**

1' Endopod of uropods without statocyst; sympod with spiny ventral lobe (Fig. 7.16A,B); cornea reduced (Figs 7.14 and 7.15) .... **Stygiomysida**

FIGURE 7.17 *Neomysis integer*, ventral.

Mysida and Stygiomysida: Stygiomysida: species

1 Habitus vermiform (Fig. 7.14), resembling the co-occurring thermosbaenacean *Monodella stygicola* (Ruffo, 1949); sympod of uropods with ventral lobe more than half length of unsegmented exopod (Fig. 7.16A) .... Stygiomysidae (one species: *Stygiomysis hydruntina;* Caroli, 1937) [South-Eastern Italy; mostly in brackish groundwater, occasionally also in freshwater]

1' Habitus moderately stout, shrimp-like (Fig. 7.15); sympod of uropods with ventral lobe less than 1/4 length of two-segmented exopod (Fig. 7.16B) .... Lepidomysidae (one species: *Spelaeomysis bottazii;* Caroli, 1924) [South-Eastern Italy; mostly in brackish groundwater, also in freshwater]

Mysida and Stygiomysida: Mysida: Mysidae: genera

1 Antennal scale setose all around (Fig. 7.18C–H) 3

1' Antennal scale with setose margins except for a proximal bare portion of outer margin (Fig. 7.18A,B) 2

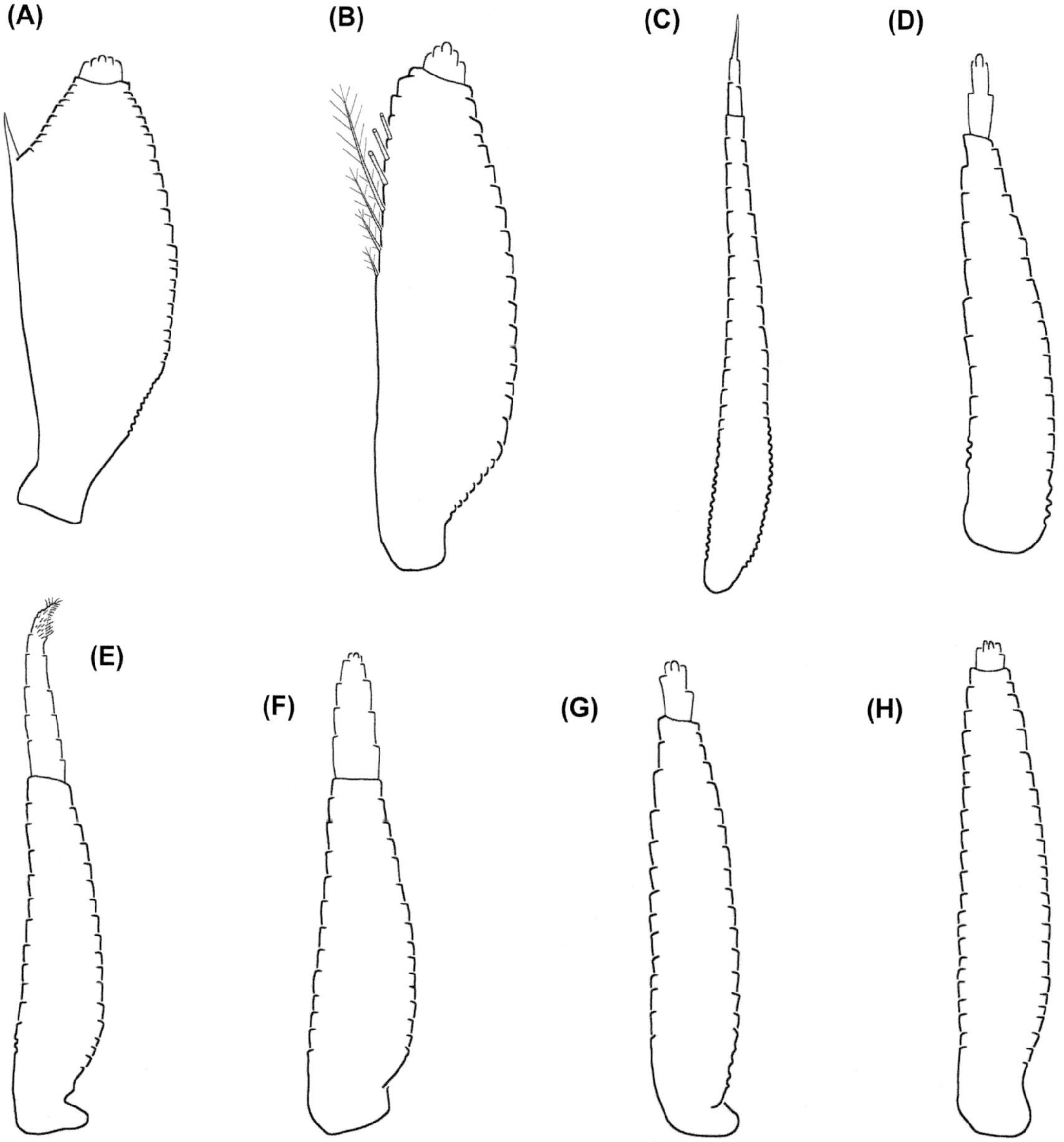

FIGURE 7.18 Antennal scale of (A) *Paramysis (Longidentia) adriatica*, (B) *Hemimysis anomala*, (C) *Neomysis integer*, (D) *Mesopodopsis slabberi*, (E) male *Limnomysis benedeni*, (F) female *L. benedeni*, (G) *Diamysis fluviatilis*, and (H) *Troglomysis vjetrenicensis*. Most (B and E) or all (A, C, D, and F–H) setae omitted, respectively.

2 (1') Outer margin of antennal scale with smooth portion ending in setae (Fig. 7.18B) .... *Hemimysis* (one species: *Hemimysis anomala;* G.O. Sars, 1907)
[Ponto-Caspian endemic; invasive in large parts of Europe and North America; expansion in the Rhône system down to the Mediterranean coast; mainly in freshwater]
2' Outer margin of antennal scale with smooth portion ending in a tooth (Fig. 7.18A) .... ***Paramysis***
3 (1) Telson entire, slender, subtriangular (Fig. 7.19A) .... *Neomysis* (one species: *N. integer;* Leach, 1814)
[North-Eastern Atlantic and North-Western Mediterranean; mainly in brackish and marine waters, marginally in freshwater]
3' Telson rhombohedral to subrectangular, with apical cleft (Fig. 7.19B,D–H) 4
4 (3) Telson without median lobe (Fig. 7.19D–H); eyestalks less than 3 times length of cornea (Fig. 7.20C–F) 5
4' Telson with large median lobe emerging from broad apical cleft (Fig. 7.19B); eyestalks 3–4 times as long as cornea (Fig. 7.20A) .... *Mesopodopsis* (one species: *M. slabberi;* Van Beneden, 1861)
[North-Eastern Atlantic, Baltic, Mediterranean, Marmora, and Ponto-Azov Seas; in oligo- to euhaline, rarely metahaline waters, episodically in freshwater of only the Danube Delta]
5 (4) Apical segment less than 20% antennal scale length; scale terminally broadly rounded in both sexes (Fig. 7.18G,H) 6

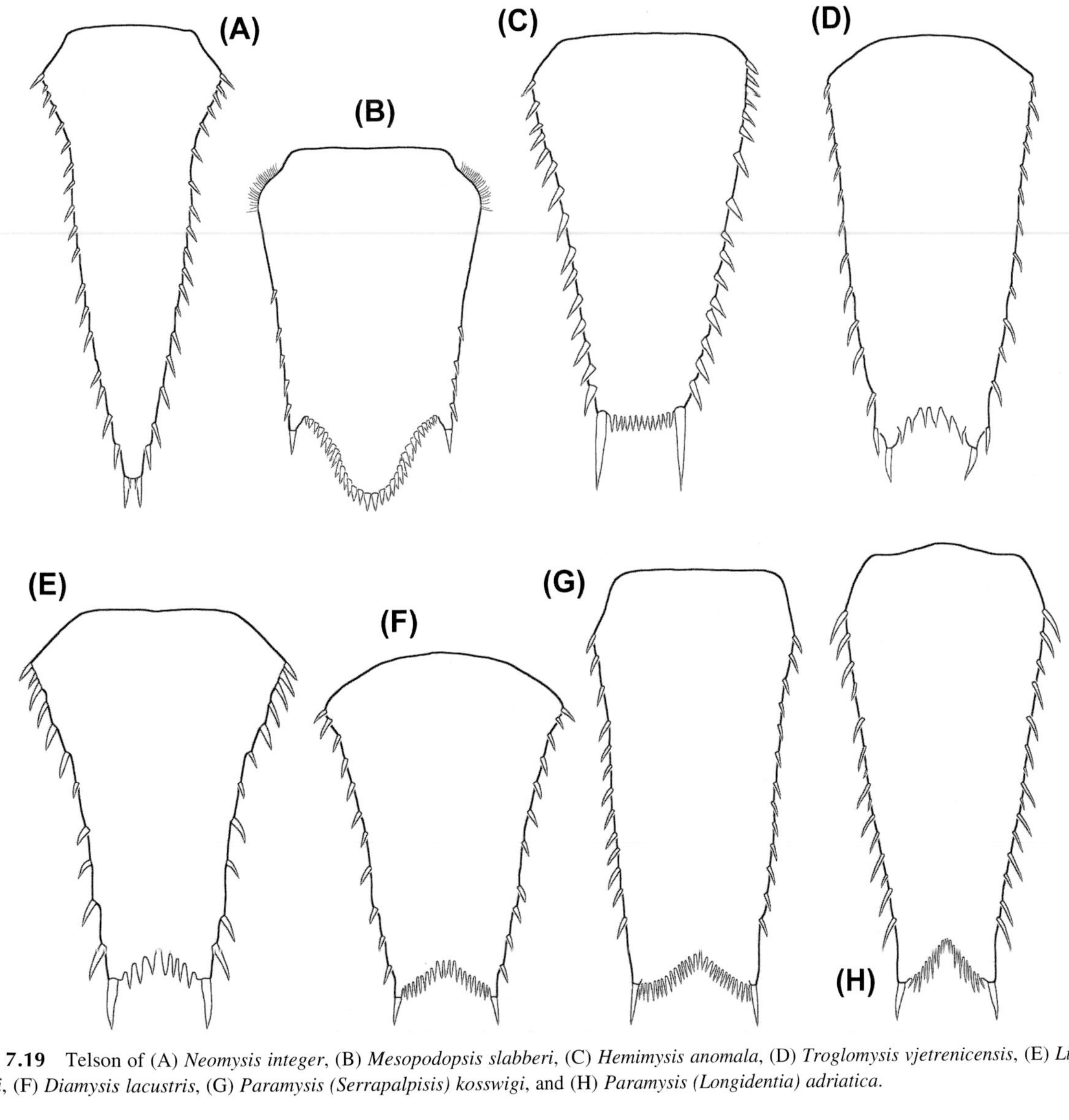

FIGURE 7.19 Telson of (A) *Neomysis integer*, (B) *Mesopodopsis slabberi*, (C) *Hemimysis anomala*, (D) *Troglomysis vjetrenicensis*, (E) *Limnomysis benedeni*, (F) *Diamysis lacustris*, (G) *Paramysis (Serrapalpisis) kosswigi*, and (H) *Paramysis (Longidentia) adriatica*.

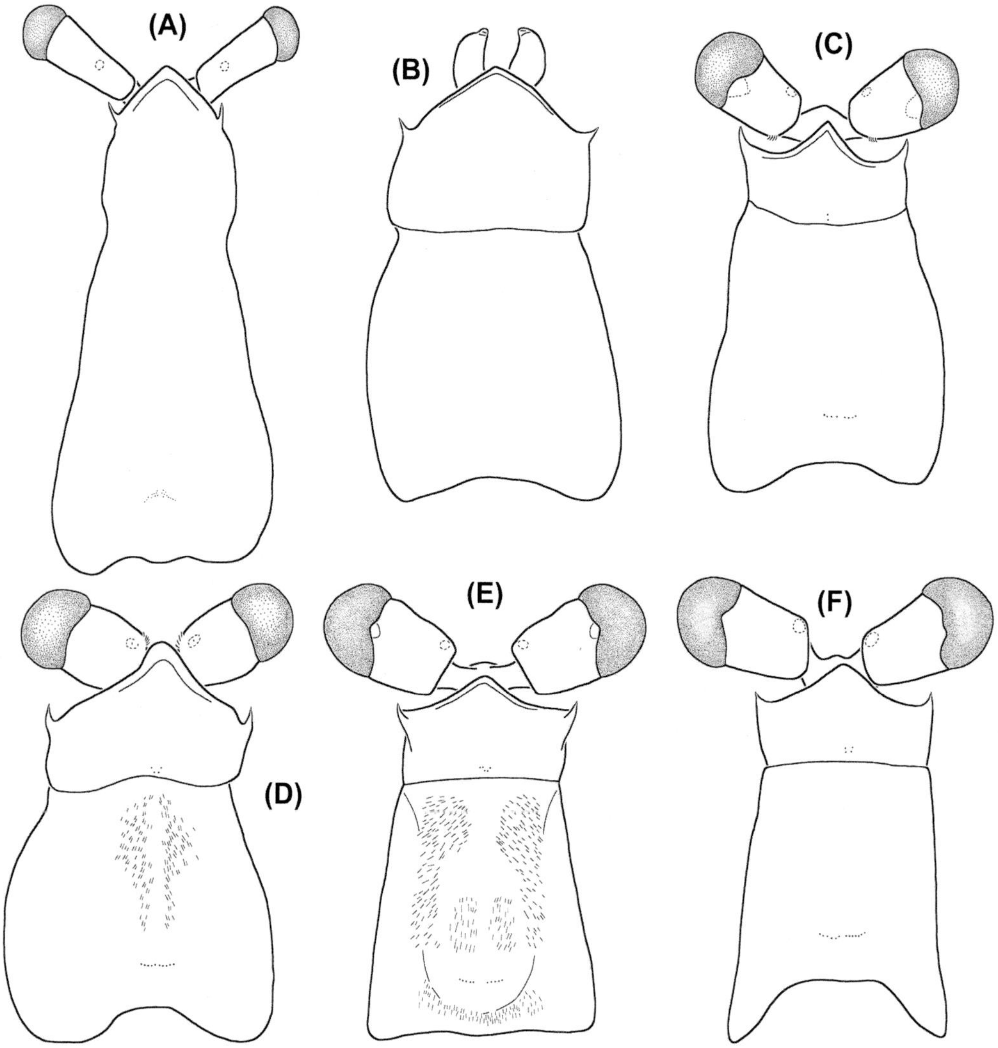

FIGURE 7.20 Eyes and carapace of (A) *Mesopodopsis slabberi*, (B) *Troglomysis vjetrenicensis*, (C) *Diamysis lacustris*, (D) male *Diamysis fluviatilis*, (E) male *Diamysis mesohalobia heterandra*, and (F) *Diamysis hebraica*.

5' Apical segment of antennal scale 25%–33% of scale length; scale terminally rounded in females (Fig. 7.18F), bluntly pointed in males (Fig. 7.18E) .... *Limnomysis* (one species: *Limnomysis benedeni;* Czerniavsky, 1882) [Ponto-Caspian endemic; introduced in Lake Balaton, Lake Aral, and waters of Turkmenistan; expansion to large parts of Europe, including the Rhône Delta at the Mediterranean coast; mainly in freshwater, also brackish]

6 (5) Cornea rudimentary, without pigment (Fig. 7.20B) .... *Troglomysis* (one species: *T. vjetrenicensis;* Stammer, 1933) [subterranean freshwaters in the cave Vjetrenica system, near the Eastern Adriatic coast]

6' Cornea normal, well pigmented (Fig. 7.20C–F) .... ***Diamysis***

Mysida and Stygiomysida: Mysida: Mysidae: *Diamysis:* species

1 Rostrum well rounded or forming a wide convex angle with rounded tip (Fig. 7.20D–F) 2

1' Rostrum angular, with acute or narrowly rounded tip (Fig. 7.20C) .... *D. lacustris;* (Băcescu, 1940) [freshwater Lake Scutari at the SE-coast of the Adriatic Sea]

2 (1) Basal plate of thoracic exopods 1–3 terminally ending in a spiniform projection (Fig. 7.21A) 3

2' Basal plate of thoracic exopods 1–3 terminally well rounded (Fig. 7.21B) .... *D. hebraica* (Almeida Prado-Por 1981) [oligohaline streams at the coast of Israel; not yet found in true freshwater]

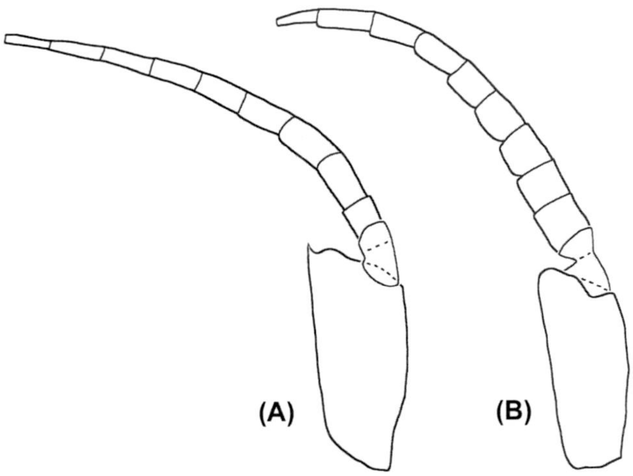

FIGURE 7.21 Third thoracic exopods of (A) *Diamysis fluviatilis* and (B) *Diamysis hebraica*; (A and B) setae omitted.

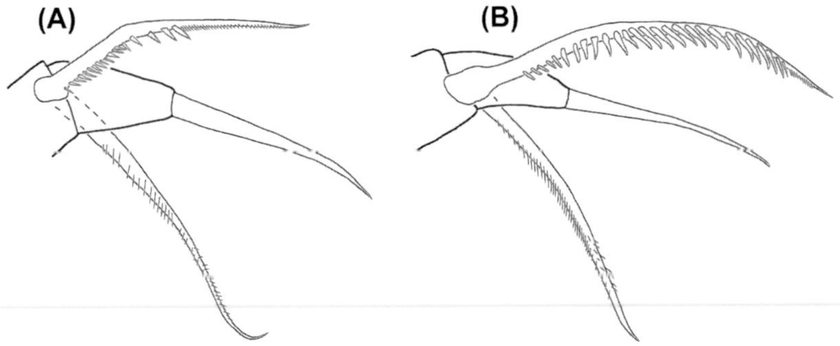

FIGURE 7.22 Dactylus with claw and paradactylary setae of fifth thoracic endopods of (A) *Paramysis (Serrapalpisis) kosswigi* and (B) *Paramysis (Serrapalpisis) lacustris*; (A and B) other types of setae omitted.

3 (2) Rostrum well rounded; male carapace with fringes arranged in only one pair of submedian stripes (Fig. 7.20D); female carapace without fringes .... *D. fluviatilis* (Wittmann & Ariani, 2012) [tributaries of the Northen Adriatic; mainly in freshwater]

3' Rostrum forms a wide convex angle with rounded tip; male carapace with fringes arranged in 1–2 pairs of submedian stripes plus a transverse row shortly in front of the posterior margin (Fig. 7.20E); ...... *D. mesohalobia heterandra* (Ariani & Wittmann 2000) [Adriatic, Ionian, and Marmora Seas; fresh and brackish waters]

Mysida and Stygiomysida: Mysida: Mysidae: *Paramysis:* species

1 Telson with shallow cleft forming an angle of more than 120° (Fig. 7.19G) 2

1' Telson with strong, subrectangular cleft (Fig. 7.19H) .... *P. (L.) adriatica* (Wittmann, Ariani & Daneliya, 2016) [freshwater tributaries of the Adriatic Sea]

2 (1) Denticulate paradactylary setae of thoracic endopods 5–8 errated' only along proximal 50%–65% length (Fig. 7.22A); cephalic region swollen (if well preserved) .... *P. (S.) kosswigi* (Băcescu, 1948) [freshwaters of W-Anatolia draining into the Aegean Basin]

2' Denticulate paradactylary setae of thoracic endopods 5–8 errated' over 70%–80% length (Fig. 7.22B); cephalic region not swollen .... *P. (S.) lacustris* (Czerniavsky, 1882) [expansive Ponto-Caspian endemic, also in (near)-Mediterranean freshwater lakes of Turkey]

Acknowledgements

We gratefully acknowledge C. Puch who helped us in different ways and Christophe Piscart for his valuable comments. This work was supported by PID2019-110243GB-I00 MICINN/FEDER project. The author is greatly indebted to Salvatore Inguscio (Nardò) for providing photos of Stygiomysida. Sincere thanks to Mikhail Daneliya (Helsinki) and to W. Wayne Price (Tampa) for important information about freshwater distribution of mysids.

Subchapter 7.8

Order Thermosbaenacea

Christophe Piscart
French National Centre for Scientific Research (CNRS), University of Rennes 1, France

Introduction

Thermosbaenacea Monod 1927 are an obscure group of small crustaceans living in groundwaters. Their taxonomic position was debated over for a long time before they were finally classified with Peracarida (see Wagner, 1994 for details). Their small size makes them difficult to identify in the field, and their elongate vermiform and somewhat dorsoventrally flattened body makes them look like microisopods or copepods. However, they are easily recognizable under a stereomicroscope from the presence of an obvious dorsal carapace. They differ from Mysidacea and Decapoda in that they are eyeless and have reduced pleopods present only on their first two abdominal segments.

The biogeographic origin of Thermosbaenacea still remains to be elucidated because available information is very scarce. We already know that their distribution area falls within the boundaries of the ancient Tethys belt, indicating that they may have had a very old ancestor in the Mesozoic period (252–66 Mya), close to Halosbaenidae that are the only known family distributed worldwide.

However, the diversification process likely occurred in the Mediterranean Basin between the marine regression of the late Miocene period and the Messinian salinity crisis (≅ 5.5 Mya) for the *Tethysbaena atlantomaroccana* group and the marine transgression during the Pliocene period (2.5–5 Mya) for the *Tethysbaena argentarii* group (Wagner, 1994).

Among the 40 Thermosbaenacea species recorded worldwide (Horton et al., 2020), less than 20 occur in freshwaters, among which 12 are present in the Mediterranean Basin. A total of 3 families, 4 genera, and 12 species are recorded in the Mediterranean Basin, all of them endemic to the region.

General ecology and distribution

Thermosbaenacea are all obligate groundwater species and colonize all types of marine and continental waters, even if they are always located close to the sea and are more frequent in marine and brackish waters. Some species can colonize hot water springs; for example, the type locality of *Thermosbaena mirabilis* has a temperature ranging from 44.5°C to 48°C. They have a strong affinity to low oxygen concentrations and are resistant to high temperatures and the presence of hydrogen sulfide (H_2S); near-anoxic conditions are even more favorable to them (Wagner, 2012). They are mainly found is detritus, but likely feed on bacteria and algae growing on detritus.

Terminology and morphology

Thermosbaenacea are small-sized Malacostraca, generally less than 5 mm long. Their body is elongate, vermiform and slightly flattened dorsoventrally, except *T. mirabilis* whose body is wider and looks like a flabelliform isopod.

Their body consists of a more or less developed carapace, a thorax formed of eight segments (thoracomeres 1–8), a pleon with generally six segments (except Thermosbaenidae whose sixth pleonite is fused with the telson forming a pleotelson) (Fig. 7.23).

The cephalic capsule is fused with the thoracic tergum through the maxillipedal metamere, and partly fused with the second thoracomere to form the carapace. The upper pair is antennae 1, with a basal peduncle of three articles and a distal flagellum of several articles with aesthetascs. The third peduncular article also bears an accessory flagellum on its distal end, usually composed of a reduced number of tiny cylindrical articles. The second pair of antennae is shorter than the first one; it has a flagellum of 8–11 small articles and a peduncle of 5 articles. Eyes are absent, but ocular scales can be present in some genera (e.g., *Limnosbaena*). A dorsal brood pouch is formed in ovigerous females, as an extension of the carapace, wherein eggs and juveniles are incubated. Their mouthpart is on the ventral surface below their carapace and is composed of the labrum, labium, mandibles, maxillae I and II, and maxilliped, all laterally connected to the carapace (see Wagner, 1994 for details).

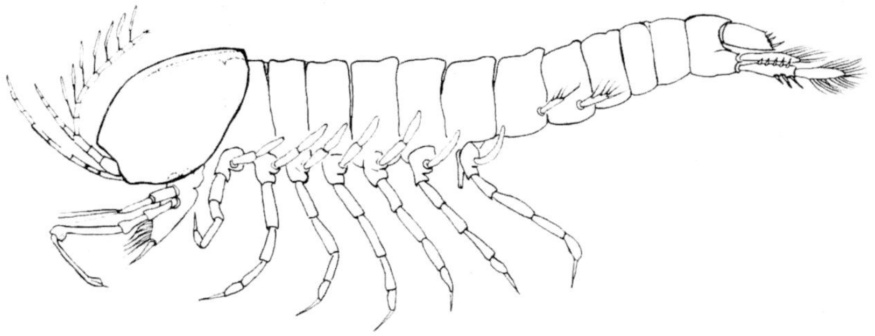

FIGURE 7.23 General morphology of Monodellidae. *Modified after Monod and Cals (1988).*

The thorax bears different kinds of legs generally eight pairs of biramous appendages, except for *Limnosbaena* (uniramous), the first of which are used as accessory mouthparts. One pair of gnathopods (modified pereopod 1) is possibly used for grasping detritus, the next six pairs are walking legs (pereopods 2–7), except in *Thermosbaena* only has five pereopods.

The pleon is composed of six segments (only five in *Thermosbaena*), but only three pairs of appendages. The first two pleonites bear a pair of reduced pleopods (pleopods 1–2); the first pair is scale-like in Halosbaenidae, and the last pleonite has a well-developed pair of biramous uropods. In *Thermosbaena*, the uropod is attached to the pleotelson and the endopodite is strongly reduced. Finally, a free telson (or pleotelson) is attached to the last pleonite and corresponds to the last part of the body. The shape and armature of the telson are typical of each genus and of each species to a lesser extent.

Collection, preparation, and identification

Thermosbaenacea are very rare and present only in groundwaters (aquifers, caves or springs). Therefore, any kind of specific material for this environment (Malard et al., 2002) can be used to collect them. Some species were even collected using a suction bottle or jars operated by divers in caves. Thermosbaenacea can be easily fixed and preserved in 2% buffered glutaraldehyde or in 95% ethanol for genetic analysis.

Determination at the species level requires dissection, mounting of body parts and appendages on slides, and subsequent examination using a 400 × or higher-magnification microscope or even observation in SEM (e.g., *Tethysbaena argentarii* group). Specimens in glycerol or ethanol can be handled under a low-power microscope with needles. Baths in clearing solution (e.g., soft acids) and cuticular staining (e.g., black chlorazol B) can be helpful. A same specimen can be used for both studies: the abdomen for DNA extraction and the other parts for dissection and mounting. This procedure ensures that the sequenced genes correspond to the same morphotype.

Taxonomic identification to the species level is generally based on males (if available), but females can be used too. Chaetotaxy is complex but very important for identification. The most important body parts for identification are the antennae, mouthparts, pereopod, pleopods, uropods and telson, but additional body parts need to be examined to identify certain species.

Limitations

Thermosbaenacea remain a poorly known order, with only 38 species described worldwide and hardly any molecular study to confirm their taxonomy. As a consequence, some families, genera, and species names may change in the future, and many species are still to be found. The keys below are designed to cover the 12 species inhabiting fresh and brackish waters of the Mediterranean Basin. Rare records are included, unless they are insufficiently documented.

Even if our keys were designed to be as clear as possible, we should keep in mind that identifying certain groups (e.g., the *Tethysbaena argentarii* group) is difficult and requires strong expertise. Do not hesitate to use alternative information or keys (e.g., Wagner 1994) to consolidate your conclusions. We should remain aware that the keys are designed for species of the Mediterranean basin; they are not adapted to other amphipods worldwide, even at the family level.

Keys to Thermosbaenacea

Adapted from Wagner (1994)

Thermosbaenacea: families

1 Body elongate; 7 pairs of pereopods; telson separate from pleonite 6 2
1' Body distinctly wider than high; 5 pairs of pereopods; telson fused with pleonite .... **Thermosbaenidae** (one species: *Thermosbaena mirabilis*; Monod, 1924) [Tunisia]
2 (1) Ocular scales absent; gnathopods with one strong serrate seta forming the unguis; propodus of gnathopods with spur; pleopod 1 with basal articulation .... **Monodellidae** [Peri-Mediterranean]
2' Ocular scales present; gnathopods with two or three (more or less modified) strong serrate seta forming a claw; propodus of gnathopods lacking spur; pleopod 1 with a small unarticulated lobe .... **Halosbaenidae** (one species: *Limnosbaena finki;* Mestrov & Lattinger-Penko, 1969) [Balkan]

Thermosbaenacea: Monodellidae: genera and species

1 Antenna 1 with 6–12 articles on the main flagellum; carpus of gnathopods with three long teasel setae; second segment of exopodite of pereopod 5 with medial plumose setae and with or without vestigial lateral seta; telson with group of three cuspidate setae on the distal corners .... 2 (genus *Tethysbaena*) [Peri-Mediterranean]
1' Antenna 1 with 10 articles on the main flagellum; carpus of gnathopods with four long strong teazel setae; second segment of exopodite of pereopod 5 without medial and lateral plumose setae; telson with group of two cuspidate setae on the distal corners .... *Monodella* (one species *Monodella stygicola;* Ruffo, 1949) [Italy]
2 (1) Telson wider than long; anal lobes protruding beyond the terminal stretch of the telson 3
2' Telson longer than wide; anal lobes not protruding beyond the terminal stretch of the telson 4
3 (2) Antenna 1 main flagellum with 7 articles; Maxilla 1 palp second article with 9 setules; Uropod endopodite with 18–22 plumose setae .... *Tethysbaena ophelicola* (Wagner, 2012)
3' Antenna 1 main flagellum with 8 articles; Maxilla 1 palp second article with 4 setules; Uropod endopodite with 18–22 plumose setae .... *Tethysbaena relicta* (Pór, 1962)
4 (2') Terminal stretch of the telson with one or more protuberances 5
4' Terminal stretch of the telson with one or more protuberances .... *Tethysbaena argentarii*-group (*T. argentarii* (Stella, 1951), *Tethysbaena aiakos* Wagner (1994), *Tethysbaena halophile* (S. Karaman, 1953), *Tethysbaena scabra* (Pretus, 1991) and *Tethysbaena siracusae* Wagner (1994))*
*The accurate identification of each species of the group needs SEM microscopy.
5 Article 2 of the mandibular palp with 6 plumidenticulate setae; Maxilla 2 outer lobe with 17 rake-like terminal setae; the presence of an additional dorsolateral subplumose seta on pleopod 2; Medial margin of the first segment of the exopodite of uropod with 5 stout plumose setae .... *Tethysbaena atlantomaroccana* (Boutin & Cals, 1985)
5'. Article 2 of the mandibular palp with 5 plumidenticulate setae; Maxilla 2 outer lobe with 25 rake-like terminal setae; the absence of an additional dorsolateral subplumose seta on pleopod 2; Medial margin of the first segment of the exopodite of uropod with 4 stout plumose setae .... *Tethysbaena tarsiensis* (Wagner 1994).

References

Aguesse, P. & L. Bigot. 1960. Observations floristiques et faunistiques sur un étang de Moyenne Camargue: la Baisse Salée de la Tour du Valat. Vie et Milieu 11: 284–307.

Almeida Prado-Por, M.S.D. 1981. Two new subspecies of the *Diamysis bahirensis* Sars species group (Crustacea: Mysidacea) from extreme salinity environments on the Israel and Sinai coasts. Israel Journal of Zoology 30: 161–175.

Ariani, A.P. & K.J. Wittmann. 2010. Feeding, reproduction, and development of the subterranean peracarid shrimp *Spelaeomysis bottazzii* (Lepidomysidae) from a brackish well in Apulia (southeastern Italy). Journal of Crustacean Biology 30: 384–392.

Audzijonytė, A., L. Baltrūnaitė, R. Väinölä & K. Arbačiauskas. 2017. Human-mediated lineage admixture in an expanding Ponto-Caspian crustacean species *Paramysis lacustris* created a novel genetic stock that now occupies European waters. Biological Invasions 19: 2443–2457.

Balian, E.V., H. Segers, K. Martens & C. Lévêque. 2008. An introduction to the Freshwater Animal Diversity Assessment (FADA) project. Pages 3–8 *in*: E.V. Balian, C. Lévêque, H. Segers, & K. Martens (eds), Freshwater Animal Diversity Assessment. Springer Netherlands, Dordrecht, Nehterland. https://doi.org/10.1007/978-1-4020-8259-7_1.

Baratti, M., Burchi, A., Messana, G., & Yacoubi-Khebiza, M. (1999). Inferring Phylogenetic History on five taxa of the genus Typhlocirolana (Isopoda Cirolanidae) by 12s sequence. Preliminary data. *Mémoires de Biospéologie*, *26*, 59–64.

Bauzà-Ribot, M.M., C. Juan, F. Nardi, P. Oromí, J. Pons & D. Jaume. 2012. Mitogenomic Phylogenetic Analysis Supports Continental-Scale Vicariance in Subterranean Thalassoid Crustaceans. Current Biology 22: 2069–2074.

Begun, T. & M.-T. Gomoiu. 2007. New data on cumacean and mysid populations (Crustacea – Peracarida) in Razelm – Sinoie lagoon complex. Page 69 *in:* Aquatic Biodiversity International Conference, Sibiu, Romania, 04–07, October 2007.

Borza, P., K. Kovács, A. György, J.K. Török & Á. Egri. 2019. The Ponto-Caspian mysid *Paramysis lacustris* (Czerniavsky, 1882) has colonized the Middle Danube. Knowledge & Management of Aquatic Ecosystems 420: 4 pp.

Boyko C.B., N.L. Bruce, K.A. Hadfield, K.L. Merrin, Y. Ota, G.C.B. Poore, S. Taiti, M. Schotte & G.D.F. Wilson. 2020. World Marine, Freshwater and Terrestrial Isopod Crustaceans database. http://www.marinespecies.org [Accessed 2020 December 31].

Brandis D., V. Storch & M. Türkay. 2000. Taxonomy and zoogeography of the freshwater crabs of Europe, North Africa, and the Middle East (Crustacea, Decapoda, Potamidae). Senckenbergiana Biologica 80(1): 5–56.

Brun, G. 1967. Etude écologique de l'estuaire du Grand Rhône. Bulletin de l'Institut Océanographique de Monaco 66 (1371): 1–46.

Bueno de S.L.S., R.M. Shimizu & J.C.B. Moraes. 2016. A remarkable anomuran: The taxon Aegla Leach, 1820. Taxonomic remarks, distribution, biology, diversity and conservation. Pages 23–64 *in:* T. Kawai & N. Cumberlidge (eds), A Global Overview of the Conservation of Freshwater Decapod Crustaceans. Springer, Cham. https://doi.org/10.1007/978-3-319-42527-62.

Camacho, A.I. 2019. Diversity, morphological homogeneity and genetic divergence in a taxonomically complex group of groundwater crustaceans: the little known Bathynellacea (Malacostraca). Bulletin de la Société d'Histoire Naturelle de Toulouse 154: 105–160.

Camacho, A.I., B.A. Dorda & I. Rey. 2011. Identifying cryptic speciation across groundwater populations: first COI sequences of Bathynellidae (Crustacea, Syncarida). Graellsia 67: 7–12.

Camacho, A.I., P. Mas-Peinado, S. Watiroyram, A. Brancelj, E. Bandari, B.A. Dorda, A. Casado & I. Rey. 2018. Molecular phylogeny of Parabathynellidae (Crustacea, Bathynellacea) and three new species from Thai caves. Contributions to Zoology 87 (4): 227–260.

Camacho, A.I., P. Mas-Peinado, S. Iepure, G. Perina, B.A. Dorda, A. Casado & I. Rey. 2020. Novel sexual dimorphism in a new genus of Bathynellidae from Russia, with a revision of phylogenetic relationships. Zoologica Scripta 49: 47–63.

Camacho, A.I., P. Mas-Peinado, Y. Ranga Reddy, E. Bandari, S. Shaik, G. Perina, B.A. Dorda, A. Casado & I. Rey. 2021. An integrated approach to re-evaluate the validity of the family Leptobathynellidae (Crustacea, Bathynellacea). Zoological Journal of the Linnean Society XX: 1–43. https://doi.org/10.1093/zoolinnean/zlaa121/6144087.

Carvalho F., S. De Grave & F.L. Mantelatto. 2016. An integrative approach to the evolution of shrimps of the genus *Palaemon* (Decapoda, Palaemonidae). Zoologica Scripta 46(4): 473–485. https://doi.org/10.1111/zsc.12228.

Christodoulou M, C. Anastasiadou, J. Jugovic & T. Tzomos. 2016. Freshwater shrimps (Atyidae, Palaemonidae, Typhlocarididae) in the broader Mediterranean region: Distribution, life strategies, threats, conservation, challenges and taxonomic issues. Pages 199–236 *in*: T. Kawai & N. Cumberlidge (eds), A Global Overview of the Conservation of Freshwater Decapod Crustaceans. Springer, Cham. https://doi.org/10.1007/978-3-319-42527-67.

Clarke, K.R., P.J. Somerfield & R.N. Gorley. 2008. Testing of null hypotheses in exploratory community analyses: similarity profiles and biota-environment linkage. Journal of Experimental Marine Biology and Ecology 366: 56–69.

Coineau, N. & A.I. Camacho. 2013. *Superorder Syncarida Packard, 1885*. Pages 357–449 in: J.C. von Vaupel Klein, M. Charmantier-Daures & F.R. Schram, (eds), Treatise on Zoology-Anatomy, Taxonomy, Biology. The Crustacea. Traité de Zoologie, Vol. 4 part A. Koninklijke Brill, Leiden, Netherland.

Copilaş-Ciocianu, D., D. Sidorov & A. Gontcharov. 2019. Adrift across tectonic plates: molecular phylogenetics supports the ancient Laurasian origin of old limnic crangonyctid amphipods. Organisms Diversity & Evolution 19: 191–207.

Correggiari, A., M. Roveri & F. Trincardi. 1996. Late Pleistocene and Holocene evolution of the North Adriatic Sea. Il Quaternario - Italian Journal of Quartenary Sciences 9: 697–704.

Crandall K.A. & S. De Grave. 2017. An updated classification of the freshwater crayfishes (Decapoda: Astacidea) of the world, with a complete species list. Journal of Crustacean Biology 37(5): 615–653. https://doi.org/10.1093/jcbiol/rux070.

Crandall K.A., D.J. Harris & J.W. Fetzner Jr. 2000. The monophyletic origin of freshwater crayfish estimated from nuclear and mitochondrial DNA sequences. Proceedings of the Royal Society of London 267: 1679–1686. https://doi.org/10.1098/rspb.2000.1195.

Cumberlidge N., H.H. Hobbs & D.M. Lodge. 2015. Chapter 32. Class Malacostraca, Order Decapoda. pp. 797–847 *in*: J. Thorp & D.C. Rogers (eds), Ecology and general Biology: Thorp and Covich's freshwater invertebrates. Academic Press, San Diego, U.S.A. https://doi.org/10.1016/B978-0-12-385026-3.00032-2.

Daneliya, M. E., Petryashov, V. V., & Väinölä, R. (2019). *Malacostraca: Mysida and Stygiomysida. Thorp and Covich's Freshwater Invertebrates. Keys to Palaearctic Fauna: Phylum Arthropoda: Malacostraca*, (4th, pp. 866–889). (vol. 4, pp. 866–889). Elsevier Inc.

Dedju, I.I. & W. Polischtuk. 1968. Fauna der höheren Krebsartigen im sowjetischen Donauabschnitt und ihre Rolle bei der Bildung der Bodenfauna des Pruth. Pages 277–283 in: A. Valkanov, B. Russev, and W. Naidenow (eds.), Limnologische Berichte der X. Jubiläumstagung der Arbeitsgemeinschaft Donauforschung. Bulgarien 10.-20. Oktober 1966. Verlag der Bulgarischen Akademie der Wissenschaften, Sofia.

De Grave S., Y. Cai & A. Anker 2008. Global diversity of shrimps (Crustacea: Decapoda: Caridea) in freshwater. Hydrobiologia 595: 287–293. https://doi.org/10.1007/s10750-007-9024-2.

De Grave S., N.D. Pentcheff, S.T. Ahyong, T.-Y. Chan, K.A. Crandall, P.C. Dworschak, D.L. Felder, R.M. Feldmann, C.H.J.M. Fransen, L.Y.D. Goulding, R. Lemaitre, M.E.Y. Low, J.W. Martin, P.K.L. Ng, C.E. Schweitzer, S.H. Tan, D. Tshudy & R. Wetzer. 2009. A classification of Recent and fossil genera of decapod crustaceans. Raffles Bulletin of Zoology 21: 1–109.

De Grave S., K.G. Smith, N.A. Adeler, D.J. Allen, F. Alvarez, A. Anker, Y. Cai, S.F. Carrizo, W. Klotz, F.L. Mantelatto, T.J. Page, J.-Y. Shy, J.L. Villalobos & D. Wowo. 2015. Dead Shrimp Blues: A Global Assessment of Extinction Risk in Freshwater Shrimps (Crustacea: Decapoda: Caridea). PLoS ONE 10: e0120198. https://doi.org/10.1371/journal.pone.0120198.

Gibert, J. & D.C. Culver. 2009. Assessing and conserving groundwater biodiversity: an introduction. Freshwater Biology 54: 639–648.

Günay, G. 2010. Chapter 10.6 - Geological and hydrogeological properties of Turkish karst and major karst springs. pp. 479-497 *in*: N. Kresic and Z. Stevanovic (eds.), Groundwater Hydrology of Springs: Engineering, Theory, Management and Sustainability. Butterworth-Heinemann: Elsevier, Oxford, UK.

Hansen, H.J. 1903. The Ingolfiellidæ, fam. n., a new Type of Amphipoda. Journal of the Linnean Society of London, Zoology 29: 117–133.

Henry J. & G. Magniez. 1983. Introduction pratique à la systématique des organismes des eaux continentales françaises - 4. Crustacés Isopodes (principalement Asellotes). Bullettin Mensuel de la Société Linnéenne de Lyon 10: 319–357.

Horton, T., A. Kroh, S. Ahyong, N. Bailly, & Z. Zhao. 2020. World Register of Marine Species (WoRMS). WoRMS Editorial Board, http://www.marinespecies.org. [Accessed 2020 December 31].

Hou, Z. & B. Sket. 2016. A review of Gammaridae (Crustacea: Amphipoda): the family extent, its evolutionary history, and taxonomic redefinition of genera. Zoological Journal of the Linnean Society 176: 323–348.

Hou, Z., B. Sket & S. Li 2014. Phylogenetic analyses of Gammaridae crustacean reveal different diversification patterns among sister lineages in the Tethyan region. Cladistics 30: 352–365.

Inguscio, S. 1998. Misidacei stigiobionti di Puglia. ideemultimediali, Nardò, Italia. 95 pp.

Jablonska A, T. Mamos, A. Zawal & M. Grabowski. 2018. Morphological and molecular evidence for a new shrimp species, *Atyaephyra vladoi* sp. nov. (Decapoda, Atyidae) in the ancient Skadar Lake system, Balkan Peninsula - its evolutionary relationships and demographic history. Zoologischer Anzeiger - A Journal of Comparative Zoology, 275: 66–79. https://doi.org/10.1016/j.jcz.2018.05.004.

Jaume D. & F. Bréhier. 2005. A new species of *Typhlatya* (Crustacea: Decapoda: Atyidae) from anchialine caves on the French Mediterranean coast. Zoological Journal of the Linnean Society 144: 387–414. https://doi.org/10.1111/j.1096-3642.2005.00175.x.

Jesse R., M. Grudinski, S. Klaus, B. Streit & M. Pfenninger. 2011. Evolution of freshwater crab diversity in the Aegean region (Crustacea: Brachyura: Potamidae). Molecular Phylogenetics and Evolution 59: 23–33. https://doi.org/10.1016/j.ympev.2010.12.011.

Jugovic J., S. Prevorčnik & B. Sket. 2010. Development of sexual characters in the cave shrimp genus *Troglocaris* (Crustacea: Decapoda: Atyidae) and their applicability in taxonomy. Zootaxa 2488: 1–21. https://doi.org/10.5281/zenodo.195554.

Klaus S., D.C.J. Yeo & S.T. Ahyong. 2011. Freshwater crab origins - laying Gondwana to rest. Zoologischer Anzeiger 250: 449–456. https://doi.org/10.1016/j.jcz.2011.07.001.

Kouba A., A. Petrusek, P. Kozák. 2014. Continental-wide distribution of crayfish species in Europe: update and maps. Knowledge and Management of Aquatic Ecosystems 413: 05. https://doi.org/10.1051/kmae/2014007.

Lowry, J.K. & A.A. Myers. 2017. A Phylogeny and Classification of the Amphipoda with the establishment of the new order Ingolfiellida (Crustacea: Peracarida). Zootaxa 4265: 1–89.

Magniez G. 1975. Observations sur la Biologie de *Stenasellus virei* (Crustacea Isopoda Asellota des eaux souterraines). International Journal of Speleology 7: 79–228.

Malard, F., M.-J. Dole-Olivier, J. Mathieu & F. Stoch. 2002. Sampling Manual for the Assessment of Regional Groundwater Biodiversity. Report of the European Project Pascalis. 110 pp.

Malard, F., Dole-Olivier, M.-J., Mathieu, J., Stoch, F., Boutin, C., Brancelj, A., Camacho, A.I., Fiers, F., Galassi, D., Gibert, J., Lefebure, T., Martin, P., Sket, B., & Valdecasas, A.G. 2004. Sampling Manual for the Assessment of Regional Groundwater Biodiversity - Figures PASCALIS Proyect Report.: 74.

McMenamin, M.A.S., L.P. Zapata & M.C. Hussey. 2013. A Triassic Giant Amphipod from Nevada, USA. Journal of Crustacean Biology Oxford Academic 33: 751–759.

Meland, K. & E. Willassen. 2007. The disunity of "Mysidacea" (Crustacea). Molecular Phylogenetics and Evolution 44: 1083–1104.

Messana G., K. Van Damme & R. Argano 2019. A new stygobiotic *Stenasellus* Dollfus, 1897 (Asellota: Stenasellidae) from Socotra Island, Yemen. Zootaxa 4683: 552–562.

Pérez-Losada M., Bond-Buckup G., Jara C.G. & Crandall K.A. (2004). Molecular systematics and biogreography of the southern South American freshwater "crabs" *Aegla* (Decapoda, Anomura, Aeglidae) using multiple heuristic tree search approaches. Systematic Biology, 53(5), 767–780. https://doi.org/10.1080/10635150490522331.

Perina, G. & A.I. Camacho, 2016. Permanent slides for morphological studies of small crustaceans: Serban's method and its variation applied on Bathynellacea (Malacostraca). Crustaceana 89(10): 1161–1173.

Pesce, G.L., G. Fusacchia, D. Maggi & P. Tetè. 1978. Ricerche faunistiche in acque freatiche del Salento. Thalassia Salentina 8: 1–51.

Pesta, O. 1935. Ein Mysidaceen-Nachweis auf der Insel Korfu (Griechenland). Zoologischer Anzeiger 111: 332–333.

Piscart, C., A. Manach, G.H. Copp & P. Marmonier. 2007. Distribution and microhabitats of native and non-native gammarids (Amphipoda, Crustacea) in Brittany, with particular reference to the endangered endemic sub-species *Gammarus duebeni celticus*. Journal of Biogeography 34: 524–533.

Porter, M.L., K. Meland & W. Price. 2008. Global diversity of mysids (Crustacea-Mysida) in freshwater. Hydrobiologia 595: 213–218, suppl.

Reddell, J. R. (1981). A review of the cavernicole fauna of Mexico, Guatemala, and Belize. *Bulletin of the Texas Memorial Museum*, *27*, 1–327.

Rogers C., J.J. Lewis. 2019. Malacostraca: Isopoda. Pages 837–844 in: D.C. Rogers & J.H. Thorp (eds), Freshwater Invertebrates. Vol. 4. Keys to Palearctic fauna. Academic Press. https://doi:0.1016/B978-0-12-385024-9.00022-8.

San Vicente, C., & Cartanyà, J. (2017). A new mysid (Crustacea, Mysida) from the Ladinian Stage (Middle Triassic) of Conca de Barberà (Catalonia, NE Iberian Peninsula). *Journal of Paleontology, 91*, 968–980. Available from http://doi.org/10.1017/jpa.2017.24.

Schminke, H.K. 1973. Evolution, System und Verbreitungsgeschichte der Familie Parabathynellidae (Bathynellacea, Malacostraca). Akademie der Wissenschaften und der Literatur Mainz, Mathematisch-Naturwissenschaftliche Klasse, Mikrofauna des Meeresbodens 24: 1–192.

Schminke, H.K. 1981. Adaptation of Bathynellacea (Crustacea, Syncarida) to life in the Interstitial ('Zoea Theory'). International Revue Ges. Hydrobiologie Hydrographie 66: 575–637.

Schram, F. 1984. Fossil Syncarida. Transaction San Diego Society Natural History 20: 189–246.

Serban, E. 1972. Bathynella (Podophallocarida, Bathynellacea). Travaux de l'Institut de Spéologie "Émile Racovitza" 11: 11–225.

Serban, E., N. Coineau & C. Delamare Deboutteville. 1972. Recherches sur les Crustacés souterrains et mésopsammiques. I. Les Bathynellacés (Malacostraca) des régions méridionales de l'Europe occidentale. La sous-famille des Gallobathynellinae. Mémoires du Muséum National d'Histoire Naturelle, sér. A, zoologie 75: 1–107.

Sket B. & V. Zakšek. 2009. European cave shrimp species (Decapoda: Caridea: Atyidae), redefined after a phylogenetic study; redefinition of some taxa, a new genus and four new *Troglocaris* species. Zoological Journal of the Linnean Society 155: 786–818. https://doi.org/10.1111/j.1096-3642.2008.00473.x.

Souty-Grosset C., D.M. Holdich, P.Y. Noël, J.D. Reynolds & P. Haffner. 2006. Atlas of Crayfish in Europe, Muséum national d'Histoire naturelle, Paris. Patrimoines naturels 64, 187 pp.

Stokkan, M., J.A. Jurado-Rivera, P. Oromí, C. Juan, D. Jaume & J. Pons. 2018. Species delimitation and mitogenome phylogenetics in the subterranean genus *Pseudoniphargus* (Crustacea: Amphipoda). Molecular Phylogenetics and Evolution 127: 988–999.

Storobogatov Y.I. 1996. Taxonomy and geographical distribution of crayfishes of Asia and East Europe (Crustacea: Decapoda: Astacoidei). Arthropoda Selecta 4 (3/4): 3–25.

Tsang L.M., C.D. Schubart, S.T. Ahyong, J.C.Y. Lai, E.Y.C. Au, T.-Y. Chan, K.L.P. Ng & K.H. Chu. 2014. Evolutionary History of True Crabs (Crustacea: Decapoda: Brachyura) and the Origin of Freshwater Crabs. Molecular Biology and Evolution 31 (5): 1173–1187. https://doi.org/10.1093/molbev/msu068.

Tuekam Kayo, R.P., P. Marmonier, S.H.Z. Togouet, M. Nola & C. Piscart. 2012. An Annotated Checklist of Freshwater Stygobiotic Crustaceans of Africa and Madagascar. Crustaceana 85: 1613–1631.

Tzomos T. & A. Koukouras. 2015. Redescription of *Palaemon antennarius* H. Milne Edwards, 1837 and *Palaemon migratorius* (Heller, 1862) (Crustacea, Decapoda, Palaemonidae) and description of two new species of the genus from the circum-Mediterranean area. Zootaxa 3905(1): 27–51. https://doi.org/10.11646/zootaxa.3905.1.2.

Väinölä, R., J.D.S. Witt, M. Grabowski, J.H. Bradbury, K. Jazdzewski & B. Sket. 2008. Global diversity of amphipods (Amphipoda; Crustacea) in freshwater. Hydrobiologia 595: 241–255.

Voicu, Gh. (1974). Identification des mysidés fossiles dans les dépôts du Miocene Superieur de la Paratethys Centrale et Orientale et leur importance paléontologique, stratigraphique et paléogéographique. *Geologicky Zbornik - Geologica Carpathica, 25*, 231–239.

Vonk, R., & D. Jaume. 2013. A new ingolfiellid amphipod from sandy beaches of the Gura Ici Islands, Western Halmahera (North Moluccas). The Raffles Bulletin of Zoology 62: 547–560.

Vonk, R., & F.R. Schram. 2003. Ingolfiellidea (Crustacea, Malacostraca, Amphipoda): a phylogenetic and biogeographic analysis. Contributions to Zoology 72: 39–72.

Von Rintelen K., Page T.J., Cai Y., Roe K., Stelbrink B., Kuhajda B.R., Iliffe T.M., Hughes J. & von Rintelen T. (2012). Drawn to the dark side: a molecular phylogeny of freshwater shrimps (Crustacea, Decapoda, Caridea, Atyidae) reveals frequent cave invasions and challenges current taxonomic hypotheses. Molecular Phylogenetics and Evolution 63(1): 82–96. https://doi.org/10.1016/j.ympev.2011.12.015.

Wagner, H. 1994. A Monographic review of the Thermosbaenacea (Crustacea: Peracarida). A study on their Morphology, Taxonomy, Phylogeny and Biogeography. 291 pp.

Wagner, H.P. 2012. *Tethysbaena ophelicola* n. sp. (Thermosbaenacea), a new prime consumer in the Ophel biome of the Ayyalon Cave, Israel. Crustaceana 85: 1571–1587. https://doi.org/10.1163/156854012X651646.

Wilson G.D.F. 2008. Global diversity of Isopod crustaceans (Crustacea; Isopoda) in freshwater. Hydrobiologia 595: 231–240. https://doi:10.1007/s10750-007-9019-z.

Wittmann, K. J. & A. P. Ariani, 2009. Reappraisal and range extension of non-indigenous Mysidae (Crustacea, Mysida) in continental and coastal waters of eastern France. Biological Invasions 11: 401–407.

Wittmann, K. J., Ariani, A. P., & Daneliya, M. (2016). The Mysidae (Crustacea: Peracarida: Mysida) in fresh and oligohaline waters of the Mediterranean. Taxonomy, biogeography, and bioinvasion. *Zootaxa, 4142*(1), 1–70. Available from http://doi.org/10.11646/zootaxa.4142.1.1.

Wittmann, K. J., Ariani A. P. & Lagardère J. -P. (2014). Chapter 54. Orders Lophogastrida Boas, 1883, Stygiomysida Tchindonova, 1981, and Mysida Boas, 1883 (also known collectively as Mysidacea). pp. 189-396, 404-406 *in:* J.C. von Vaupel Klein, M. Charmantier-Daures, and F.R. Schram (eds.), Treatise on Zoology - Anatomy, Taxonomy, Biology. The Crustacea. Revised and updated, as well as extended from the Traité de Zoologie, Vol. 4 Part B. Koninklijke Brill NV, Leiden.

Zagmajster, M., D. Eme, C. Fišer, D. Galassi, P. Marmonier, F. Stoch, J.-F. Cornu & F. Malard. 2014. Geographic variation in range size and beta diversity of groundwater crustaceans: insights from habitats with low thermal seasonality. Global Ecology and Biogeography 23: 1135–1145.

Zagmajster, M., F. Malard, D. Eme, & D.C. Culver, 2018. Subterranean Biodiversity Patterns from Global to Regional Scales. Pages 195–227 *in:* O.T. Moldovan, L. Kovac & S. Halse (eds), Cave Ecology. Springer, Cham.

Chapter 8

Class Hexapoda: general introduction

Dani Boix[1], Núria Bonada[2,3], Isabel Muñoz[4], Enrique Baquero[5], Rafael Jordana[5], David Cunillera-Montcusí[1,2,6], Irene Tornero[1], Pau Fortuño[2,3], Raúl Acosta[2,7], Stéphanie Gascón[1] and Jordi Sala[1]

[1]*GRECO, Institute of Aquatic Ecology, Faculty of Sciences, Universitat de Girona, Girona, Spain,* [2]*FEHM-Lab (Freshwater Ecology, Hydrology and Management), Departament de Biologia Evolutiva, Ecologia i Ciències Ambientals, Facultat de Biologia, Universitat de Barcelona (UB), Barcelona, Catalonia/Spain,* [3]*Institut de Recerca de la Biodiversitat (IRBio), Universitat de Barcelona (UB), Barcelona, Catalonia/Spain,* [4]*Department of Evolutionary Biology, Ecology and Environmental Sciences, Universitat de Barcelona, Barcelona, Spain,* [5]*WasserCluster Lunz—Biologische Station GmbH, Lunz am See, Austria,* [6]*Departamento de Ecología y Gestión Ambiental, Centro Universitario Regional del Este (CURE), Universidad de la República, Tacuarembó s/n, Maldonado, Uruguay,* [7]*Institute of Environmental Assessment and Water Research (IDAEA), Spanish National Research Council (CSIC), Jordi Girona, Barcelona, Spain*

Introduction to aquatic Hexapoda

In his classical paper on biodiversity, Robert May (1992) recognized that Hexapoda included the greatest proportion of global biodiversity. Nearly three-quarters of all named living animal species are insects (Mayhew, 2007). While the current insect decline worldwide is unquestionable (Cardoso et al., 2020; Hallmann et al., 2017; Wagner et al., 2021), data to assess large-scale spatial patterns in the severity of insect trends are not yet available (Montgomery et al., 2020) and veracity of alarmist insect decline statements is under revision due recent weak studies (Didham et al., 2020). Moreover, this biodiversity decline is only well documented for a few aquatic groups (Odonata, Ephemeroptera, and Plecoptera; Sánchez-Bayo & Wyckhuys, 2019) and some trends in declining abundance seem to be more related to terrestrial insects than aquatic ones (Van Klink et al., 2020 but see Desquilbet et al., 2020 and Jähnig et al., 2021). Similarly, the traditional view of the temporal trend of their biodiversity is that insect richness increased continuously over the evolutionary history of the group, but new approaches now recognize a Cretaceous peak in family richness and negligible net richness change over the past 125 Myr (Clapham et al., 2016).

In freshwater ecosystems, Hexapoda are a hyperdiverse group since the majority of the freshwater animal species are insects (60.4%; Balian et al., 2008). Moreover, some hundreds of water-dependent Collembola have also been identified (approximately 0.5% of the total diversity; Deharveng et al., 2008), although this group needs a more accurate estimate (Balian et al., 2008). Knowledge of the evolutionary patterns of freshwater hexapods is less developed than for terrestrial insects (but see Múrria et al., 2018), and this lack of knowledge contrasts with other aspects that have been widely studied, such as their ecology and habitat preferences (Dijkstra et al., 2014). Moreover, several ecological patterns of terrestrial hexapods do not seem to match the aquatic ones, showing differences in population and community dynamics (e.g., Fenoglio et al., 2016; Lancaster & Downes, 2018; Van Klink et al., 2020).

The unity of the morphological characters of hexapod organisms was already evident in the pre-Linnean taxonomies, and it was from the Linnaeus proposal that the main groups (i.e., orders such as Coleoptera, Hemiptera, Lepidoptera, Neuroptera, Hymenoptera and Diptera) were identified. In fact, Linnaeus' *Systema Naturae* named Insecta as the fifth Class of his proposal, but the term was used in a very different way than our modern concept, since it included all arthropods. The term "hexapods" appeared in some proposals of the 19th century accompanying the term Insecta to refer to the class in which insects and springtails were included (i.e., "Hexapod metamorphotic insects" in John O. Westwood's two-volumes *An Introduction to the Modern Classification of Insects* (1839–1840)). Springtails were included within insects until second half of the 20th century, when cladistic analyses favoured the split of Collembola from Insecta. Some 19th century authors, such as F. M. Brauer, pointed out the differences between Collembola and wingless insects (those that lost wings secondarily). According to the morphological and phylogenetic affinities between

Identification and Ecology of Freshwater Arthropods in the Mediterranean Basin. DOI: https://doi.org/10.1016/B978-0-12-821844-0.00019-3

both groups, they were assigned to the same class (Hexapoda). A comprehensive history of Hexapoda classification was published by Engel and Kristensen (2013). Therefore, two classes of freshwater hexapods can be identified, Collembola and Insecta. The former are organisms of <6 mm long and can be distinguished from insects by the presence of ventral tube or "collophore" on the first abdominal segment and in some species by a furca (see section *Keys to the Subclass Entognatha (Collembola)*) on the fourth segment (Hilsenhoff, 2001).

Subclass Collembola—aquatic taxa

Identification and sampling

Collembola is currently considered as a subclass of Hexapoda with four orders, two of them with elongate or cylindrical shape: Poduromorpha, with the three thoracic segments evident in dorsal view (all with chaetae); and Entomobryomorpha, with prothorax tergite reduced and without chaetae. The other two orders, Symphypleona and Neelipleona, are globular in shape, with abdominal segments indistinctly divided, or sometimes into two parts by a constriction. Neelipleona are diminute, without eyes, and with special sensilla and sensillar areas on the head and body. Many species (all groups) have a "furca," an unpaired ventral organ on the fourth segment, that allows them to spring considerable distances. The most characteristic feature of the group is the ventral tube (or collophore), with many functions under debate but surely used for gas, water, and ion exchange with the environment, and in some groups it is a cleaning organ. The antennae have four segments (sometimes subdivided or annulated) and always possess a sensillar area at the end of the third segment. The eyes are simple, with a maximum of eight ocelli, reduced in many species and, sometimes, there is a chemical receptor near them called the PAO (post antennal organ).

Aquatic Collembola groups

In the order Entomobryomorpha, the Isotomidae is the family that can be considered more related to freshwater aquatic environments, while Entomobryidae, Tomoceridae, and Oncopoduridae accumulate many stygofauna species (living on a water film or in the surface of the water). For example, the Isotomidae species *Isotomurus palustris* is frequently cited throughout the whole Holarctic Region living at riverbanks and other freshwater habitats, probably not neustonic. It may actually be a cluster of species (Carapelli et al., 2001).

Among Symphypleona, only Bourletiellidae, Sminthurididae, Dicyrtomidae, and Katiannidae have species associated with above-ground aquatic environments, while Arrhopalitidae includes many species specialized in cave habitats (water film on surfaces or neustonic on pools). The Neelipleona, with a few species, are also present in aquatic environments.

Sampling

With low-density populations and cryptic habitats, the only efficient method is the hand collection onto a white surface or directly from the water surface, capturing the specimens with a small aspirator, using an aquarium net, or taking them directly with a paintbrush. Small pitfall traps can be used at the edges of water bodies. From sandy-river banks and beaches the specimens can be retrieved with an aquarium net after making a hole (a meter above the water limit) allowing water poured into it.

Collembola ecology

Habitats and distribution

Most collembolans found on a water surface are not actually aquatic and may have just fallen into that habitat. Instead, springtails are predominantly soil, litter, and arboreal dwellers, with many tropical species occurring on vegetation and tree canopies (Palacios-Vargas, 2013). The relationship between springtails and water was known before the description of the *Podura aquatica* Linnaeus, 1758, when in 1740 De Geer observed large groups on the water in Holland (De Geer, 1740).

Due to its way of obtaining oxygen, through its cuticle, it can be said that all springtails are dependent on high ambient humidity. In some studies, it has been demonstrated that eggs laid into the soil may hatch after a long period of flooding (Tamm, 1984). Besides the species *P. aquatica*, only a limited number of species belonging to a few genera can be considered strictly aquatic: *Isotomurus*, *Hypogastrura, Axelsonia, Anurida*, and *Sminthurides* (Yue & Fu, 2000), especially the last one, that lives on water surfaces all its life cycle (Greenslade, n.d.). The "aquatic" springtails are specialized for various damp microhabitats (Waltz & McCafferty, 1979). The first group, cryophilic species, are primary aquatic-associated springtails, living permanently on the snow surface. This group will not be considered in this publication. The second group, epigean hydrophilic species, can live on low vegetation close to water, on other organisms, or directly on the surfaces of freshwater or marine habitats. The last is found on the surface tension layer. Only a few species can live submerged, such as *P. aquatica*. The third group, cave hydrophilic species, need a water-saturated atmosphere. Some are dependent on the water film on rocks and have specialized claws and empodia, other are frequent on the water of gours or pools in caves (Deharveng et al., 2008), again on the surface tension layer such as some *Arrhopalites* and Neelidae. The adaptation in this group, especially for the claws and empodia, will be treated later. Some species have been also considered aquatic when found into the interstices of soil components, for example in sandy soils near rivers and sea, or on the riverbanks and pond littoral. They are also found in soils that often be flooded. The impossibility of separating the species that should be considered water dependent from those that are never found in this environment makes us discard the numerous citations that exist in the bibliography for this group (i.e., Arbea & Ariza, 2012).

Only *P. aquatica* has a widespread distribution. *Isotomurus* is the dominant genera in nontropical regions, with *I. palustris* (probably a cluster of species, not all aquatic) (Carapelli et al., 2001) are along with all Holarctic region. *Sminthurides*, although present in the northern hemisphere, is more diversified in the tropics (Deharveng et al., 2008). In many cases, the hydrophilic species have a wide range of endemism, being higher in caves. The marine species have a greater range distribution because the capacity to stay at the water surface allows them to move along long coast distances.

Physiology and morphology

The cuticle of Collembola is composed of structures that give it stiffness. The composition, including wax, also makes it hydrophobic, enabling the animal to float on water. However, only the tubercles and joints have wax, leaving the depressed areas permeable to air that is trapped among the tubercles and acts as a physical gill when submerged (Noble-Nesbitt, 1963). If Collembola eggs hatch underwater the juveniles can be totally submerged, but when the cuticle is exposed to air, it changes and starts to repel water avoiding submersion (Greenslade, n.d.). Especially among neustonic species, the claws suffer an evident elongation, and the mucro acquires a wider shape. Adaptation to aquatic or semiaquatic habitats is a derived condition that has evolved independently several times in Collembola (D'Haese, 2002).

Neanuridae: *Anurida maritima* Guerin, 1839 is frequent on seaweed in marine waters.

Poduridae: *P. aquatica* (peat bog; Lek-Ang et al., 2007) is usually observed on the freshwater surface, deposits its eggs on water plants, and is believed to hibernate under water.

Onychiuridae: *Ongulonychiurus colpus* (Thibaud & Massoud, 1986) is a stalactite water film dweller based on its peculiar morphology.

Tullbergiidae: Many of the genera have species present on marine littoral sands, and it seems that they can survive in immersion during the tides. However, it is not clear if they really are aquatic.

Hypogastruridae: *Ongulogastrura longisensilla* (Thibaud & Massoud, 1983) lives on the water film on the stones of a cave from the French Basque Country (Oyanbeltza, Urkullu range). Some species have been found near the sea, such as *Xenilla maritima*, but the reality is that it is a species with wide distribution and edaphic.

Isotomidae: *I. palustris* was also observed by De Geer (1740). Some species of the genus are riverbank species that live near freshwater and are usually caught with phytoplankton or neuston nets.

Actaletidae: *Actaletes neptuni* (Giard, 1889) has been cited on the coast of the Iberian Peninsula in Galicia (Selga, 1971) and Coimbra on *Mytilus* (Gama, 1988).

Orchesellidae: Only *Orchesella quinquefasciata* has been cited in relation to the aquatic environment, on the northern beaches of the Mediterranean coast in Spain, on the Costa Brava (Arbea & Ariza, 2012), but it may not be justified to consider it aquatic as it is present in dry environments in many parts of Europe.

Entomobryidae: Some *Entomobrya* are halophiles, but it appears that they are not strictly aquatic, although they live in riverside vegetation.

Bourletiellidae: This family seems mostly associated with dry environments, although some species are present in wet meadows or littorals. *Heterosminthurus insignis* is a species of wide distribution in northern Europe and other parts of the world; it is present on floating plants and could be present in the Mediterranean Basin.

Sminthurididae: *Sminthurides aquaticus* has mentioned feeding on *Lemna* (Klugkist, 1907), and other *Sminthurides* species have been cited in relation to freshwater courses (Bretfeld, 1999).

Dicyrtomidae: Although many of the species in the family are found in wet habitats (forests or high-altitude areas), only three species have been found clearly related to the aquatic environment. *Dicyrtomina minuta* has been cited as thalassophile and present in the littoral by several authors (Strenzke, 1955; Weigmann, 1973), although it is abundant in dry environments. *Ptenothrix cavicola* is described from a jetty in a cave, making evident its relationship with water

(Cassagnau & Delamare-Deboutteville, 1955). Lastly, *Jordanathrix articulata articulata*, although present in dry environments at various points in the Mediterranean, has been found on aquatic vegetation on the Levantine coast (Rueda & Jordana, 2020).

Katiannidae: Only some species of the genus *Sminthurinus* have been related with aquatic environment, and among them, only *Sminthurinus concolor* is present in the Mediterranean Basin, where it is found in Spain (Minorca) on *Phragmites* near the seashore (Bretfeld, 1999).

Arrhopalitidae: Many are aquatic, practically all of which live in caves. Since they feed diatoms on *neuston*, and it is not uncommon to find carbonate crystals in its gut, which indicates that they can live on the water pools in caves.

Neelidae: It seems that its presence in the water is accidental, because aquatic organisms, such as diatoms, that are common in the diet of aquatic collembolans are not present in the gut contents of this family.

Subclass Insecta

Some biological notes on the subclass Insecta

The subclass Insecta belongs to the class Hexapoda and includes a large group of arthropods characterized by having three pairs of locomotory legs and external mouthparts, at least in their adult stage (Johnson & Triplehorn, 2004). Aquatic insects spend one or more life cycle stages in the freshwater habitat, with the majority moving to terrestrial habitats as adults. They play important ecological roles in both habitats as primary consumers, detritivores, predators, and pollinators (Lancaster & Downes, 2013). Primary aquatic insects are archaic elements of the fauna of continents. There are fossils that date more than 200 million years and are amongst the oldest members of the aquatic organisms (Illies, 1969; Sinitshenkova, 2003). In evolutionary terms, aquatic insects are essentially terrestrial insects that have found the way to live in freshwaters habitats (Marden, 2008). Due to this, they developed many morphological, physiological, and ecological adaptations that made them extremely diverse and abundant, providing excellent models for research on diversification (Múrria et al., 2018; Wootton, 1988). Their habitats exhibit marked spatial and temporal gradients in stability, and their amphibiotic lifestyles link strong habitat dependence with a response to change via dispersal (Dijkstra et al., 2014).

Most aquatic insect species are included in 10 taxonomic orders and may represent in abundance and richness terms the most important community in freshwater habitats (Wichard et al., 2002). In five orders, all their species are entirely aquatic in some part of their life cycle, usually, the immature stages, whereas the adult stages are terrestrial. These orders include insects such as mayflies (order Ephemeroptera), stoneflies (order Plecoptera), dragonflies (order Odonata), caddisflies (order Trichoptera), and alderflies (order Megaloptera). The other five orders are partially aquatic and often have a minority of species with aquatic immatures. The adults of these species are mostly terrestrial, although some species also have aquatic adults. These orders contain insects such as true bugs (order Hemiptera), sponge flies and lance lacewings (order Neuroptera), beetles (order Coleoptera), moths (order Lepidoptera), and true flies (order Diptera).

Insect life cycle

All insects begin their lives as eggs. After hatching, insects pass through various development stages until they become adults. This process is called metamorphosis (McGavin, 2001). A minority of the insects are ametabolous, i.e. the immature stages are very similar to adults, but have not developed the reproductive system (external genitalia). No aquatic insects have ametabolous development. Most of them alternate between the water and the terrestrial habitats and live temporarily on land or in the air. All are grouped into one of two development cycles. In the "hemimetabolous life cycle," a gradual metamorphosis occurs. The immature insect (larva or nymph) resembles the adult in form, eating habits, and body proportions. Nevertheless, it is different in body size and genitalia and wings development. Rudimentary wings are externally visible and develop gradually inside wing pads. The larvae or nymphaea grow gradually through a succession of molts. The term "nymph" is traditionally referred to larvae with hemimetabolous metamorphosis living in different habitats than the adults (Ephemeroptera, Plecoptera, and Odonata). However, in these cases, the term larva is also commonly accepted. In this section, we will use the simplest term and will call to all immature stages as "larvae." In the "holometabolous life cycle" (or complete metamorphosis) the larva also grows gradually, but the developmental stages differ greatly in morphology and habits from the adult. Larvae are wingless, and the form and habits are suited for growth and development rather than reproduction. Additionally, the change to the adult occurs during the inactive nonfeeding stage called a "pupa." Sometimes the eggs, larvae, pupae, and imagos are not adapted to live in the water at all (Wichard et al., 2002). Aquatic insects with holometabolous metamorphosis belong to Coleoptera, Trichoptera, Lepidoptera, Megaloptera, Neuroptera, and Diptera orders.

Insect body structure

An insect's body has an exoskeleton composed of chitin, a long-chain polymer derived from glucose. In most insects, especially the adults, the exoskeleton is heavily sclerotized and forms a series of dorsal, ventral, and lateral plates (Johnson & Triplehorn, 2004). However, in many insects, mainly larvae, the entire cuticle is flexible, although clearly segmented (McGavin, 2001). The body of the adult insect consists of three parts: head, thorax, and abdomen. The head is formed from several plates that join through sutures lines. Also, it has a pair of multisegmented antennae, a complex set of mouthparts, a pair of compound eyes, and up to three simple eyes (ocelli). The thorax contains most of the organ systems. It is also the center of insect locomotion. Thus, it has three pairs of segmented legs and in most adults, one or two pairs of functional wings. The abdomen usually consists of a relatively simple structure with up to 10 visible segments. Most lack appendages, and the external genitalia occur on the last segment.

The larval body contains the same parts such as the adults, although in some cases these may be undifferentiated or reduced (e.g., in some Diptera larvae) (Johnson & Triplehorn, 2004). Larvae of the holometabolous orders lack compound eyes and ocelli but have special eyes called stemmata. The larvae thorax may contain true legs (jointed), prolegs (no jointed), or lack of them altogether. All larvae lack wings, although in hemimetabolous orders these may develop gradually inside wing pads. Also, many aquatic larvae bear gills placed between the legs or in the cervical zone (Wichard et al., 2002). The abdomen may have appendages (prolegs), filaments, or suckers. Overall, the prolegs (thoracic or abdominal) usually have a locomotory function, and neither are true legs nor lack sclerotized segments. Additionally, lateral gills are generally very common in the aquatic larvae abdomen. The last abdominal segment may bear appendages, lobes, cercus, hairs, gills, caudal prolegs, or also may lack any structure.

Endemicity of aquatic insects and singular habitats in the Mediterranean Basin

Worldwide Mediterranean regions are characterized by high levels of endemicity of several biotic groups, including aquatic insects (Table 8.1). The high taxonomic and genetic diversity observed in the aquatic insects in the Mediterranean Basin is partly explained by the complex geographical and historical events that have affected the region through geological times. Moreover, the faunal elements are mainly of Palearctic origin; but due to the location of the Mediterranean Basin close to the southern edge of the Palearctic realm, there is an influence of the Ethiopian and Oriental realms, especially in the southeastern part (Botosaneanu & Gasith, 1971; Boudot et al., 2009; Por, 1975). This diversity, as in many other groups, has been shaped by the intrinsic role of the Mediterranean Sea as a dispersal barrier, together with the complicated palaeo-geographical and palaeo-climatic history (including microplate movements, orogens formations, the salinity crisis, intermittent connections between shores, and Quaternary climatic oscillations, among others), establishing several foci of biodiversity all across the Mediterranean Basin, especially in the peninsulas and mountain systems (Balletto & Casale, 1991). Few taxa with reduced populations and with disjunct distributions among different faunistic realms are considered to be pre-Quaternary relicts from an ancient biota that was highly transformed due to the climatic oscillations of the Pleistocene (Tierno de Figueroa et al., 2013), such as the ephemeropteran *Prosopistoma* spp. (with three species in the Mediterranean Basin, and one widely distributed in the West Palaearctic; Barber-James, 2009; Bojková & Soldán, 2015), the neuropteran genus *Nevrorthus* spp. (with five species in the Mediterranean Basin; Aspöck et al., 2017), and the trichopterans *Nyctiophylax gaditana*, *Pseudoneureclipsis* spp., *Calamoceras* spp., or *Larcasia* spp., among others (Malicky, 2014; Malicky, 2020; Martínez, 2014; Ruiz-García et al., 2013; Tachet et al., 2001). Other aquatic insects radiated in pre-Pleistocene times (e.g., the hemipteran *Velia* spp., or the coleopterans *Deronectes* spp., *Hydrochus* spp., *Hydraena* spp. ("*Haenydra*" lineage), *Ochthebius exsculptus* group; Berchi et al., 2018; García-Vázquez et al., 2016; Hidalgo-Galiana & Ribera, 2011; Ribera et al., 2010; Trizzino et al., 2011), some of them showing geographical distributions linked to the different geological evolutions of the western and eastern areas of the northern Peri-Tethys platforms or northern shore of the Mediterranean (Meulenkamp & Sissingh, 2003). However, climatic oscillations during the Pleistocene were very important in establishing modern biodiversity patterns in the Mediterranean, affecting all pre-Quaternary faunistic elements, as well as those aquatic insects that diversified during the Pleistocene (e.g., the odonates *Calopteryx* spp., *Aeshna cyanea*, the coleopterans *Meladema* spp., *Graptodytes* spp., *Hydraena gracilis* complex, or the trichopterans *Drusus* spp., *Annitella* spp.; García-Vázquez et al., 2017; Múrria et al., 2020; Previšić et al., 2009; Ribera & Faille, 2010; Simonsen et al., 2020; Sýkora et al., 2017; Weekers et al., 2001). Thus, the role of the Mediterranean peninsulas as refugia during this period allowed the survival of populations which accumulated higher levels of genetic diversity due to long continuities in the same areas (Hewitt, 2004), and some of these lineages were the sources of the recolonization of northern latitudes during interglacial periods. However, the roles of the Iberian, Italian, and Balkan peninsulas, as well as North Africa, were probably not

TABLE 8.1 Some data available of the endemicity (percentage of endemic species) in the worldwide Mediterranean regions.

Mediterranean area	Biotic group (% of endemic species)
California	Fish (63%)
Mediterranean Basin	Crustaceans (55%) Coleopterans–Hydraenidae (57%) Dipterans–Simuliidae (51%) Amphibians (60%) Fish (63%)
South America (Chile)	Aquatic plants (80%) Trichopterans (70%) Amphibians (61%)
South Africa	Aquatic plants (86%) Crustaceans (72%) Water mites (52%) Plecopterans (92%) Trichopterans (71%) Anurans (69%) Fish (52%)
Southwestern Australia	Oligochaetes (59%) Water mites (56%) Ephemeropterans (58%) Trichopterans (82%) Plecopterans (75%) Amphibians (81%)

Data sources: Ball et al. (2013), Davies and Stewart (2013), de Moor and Day (2013), Figueroa et al. (2013), Tierno de Figueroa et al. (2013).

similar in the evolution of the fauna due to the different origins, palaeo-geographic components, and landscape elements (Hewitt, 2011; Husemann et al., 2014). Moreover, the presence of diverging lineages within these highly heterogeneous refugial regions pointed to past isolated populations linked to high climatic stability, allowing to the existence of multiple (long-term or temporary) glacial refugia-within-refugia all across the Mediterranean Basin (Abellán & Svenning, 2014; Gómez & Lunt, 2007; Migliore et al., 2018).

The mechanisms of diversification in aquatic insects are influenced by their capacity to adapt to the varied environmental characteristics of the aquatic habitats. The habitat type can be broadly divided in lotic (running) and lentic (standing) waters, with contrasting characteristics related to the habitat stability and physical and chemical gradients. Lotic habitats are characterized by a high stability of the environment, from the geological and ecological perspectives, which results in a lower need to disperse (Ribera, 2008). In both habitat types, two main environmental gradients which create community structure have been identified: temporality and salinity (Brucet et al., 2012; Datry et al., 2014; Gasith & Resh, 1999; Leigh & Datry, 2017; Marchant et al., 2006; Wellborn et al., 1996). In this sense, temporary and saline waterbodies are not exclusive of the Mediterranean Basin, but they are not rare as in other regions, and they harbor singular communities (Álvarez-Cobelas et al., 2005; Millán et al., 2011). Temporary waterbodies had been considered as secondary habitats with respect to permanent ones; however, this point of view has changed considerably and nowadays they support significant biodiversity and provide valuable goods and services, especially in arid and semi-arid landscapes as in the Mediterranean Basin (Acuña et al., 2017; Boix et al., 2020; Calhoun et al., 2017; Vander Vorste et al., 2020). In some Mediterranean regions, comparable macroinvertebrate richness has been recorded at permanent and temporary sites in both lotic and lentic habitats (Boix et al., 2008; Bonada, Rieradevall, et al., 2007), but not in all the studies (Della Bella et al., 2005). Moreover, macroinvertebrate composition of permanent and temporary ponds showed differences and the latter support relevant biodiversity of rare and threatened species (Boix et al., 2008; Della Bella et al., 2005). Species richness in Mediterranean temporary ponds is dominated by arthropods: primarily insects and secondarily crustaceans (Boix et al., 2016 and references therein). The absence of fishes in Mediterranean temporary ponds gives to insects a relevant ecological role as main predators, but compared to other latitudes the predation pressure is not directly related to hydroperiod length (Boix et al., 2011; Schneider & Frost, 1996), since seasonal timing is also a relevant factor (Kneitel, 2014). In colder temperate regions longer hydroperiod implies more time with optimum weather for aerial dispersal, since ponds fill after snow melt. However, the Mediterranean autumnal-winter hydroperiods can be longer than spring ones, but the latter have better weather conditions for aerial dispersion (Boix et al., 2016). Moreover, Mediterranean temporary rivers harbor greater regional diversity, beta diversity, rarity, and endemicity levels of macroinvertebrates than temperate rivers (Bonada et al., 2017). This high biodiversity could be the result of anagenetic speciation due to isolation and species range contractions. On the other hand, in saline inland and coastal wetlands environmental factors, such as hydrology, have a crucial and different role with respect to freshwater in ecosystem functioning (e.g., food webs structure, eutrophication process and nutrient effects, confinement-flooding dynamics, etc.) (Bas-Silvestre et al., 2020; Beklioglu et al., 2007; Comín et al., 1992; Díaz et al., 1998). It is important to note that inland saline wetlands are mainly located in the south Mediterranean countries, with the exception of the Iberian Peninsula (Britton & Crivelli, 1993). Insect fauna of these habitats are characterized by a low richness and euryhaline species (Alcorlo et al., 2001; García et al., 1997; Gascón et al., 2008; Quintana et al., 1998). In coastal ponds, a decrease in the relative importance of the richness ratio of insects to crustaceans with increases in salinity has been reported (Boix et al., 2007). Finally, one of the most singular habitats are saline streams, which have received less attention due to their scarcity and scattered geographical distribution, and their low economic value as a water resource (Moreno et al., 2010). The harsh environmental conditions of high salinities imply an environmental filter for many species and shape the kind of organisms that inhabit them (Millán et al., 2011). These organisms showed adaptations to the extreme conditions (not only high salinity, but extreme temperatures and marked hydrological fluctuations of severe dry periods and floods). Insects are the best represented macroinvertebrates especially Diptera (mainly the families Chrironomidae, Ceratopogonidae, Ephydridae, Stratiomyidae, and Syrphidae), Heteroptera (Corixidae), and Coleoptera (Hydraenidae, Hydrophilidae, and Dytiscidae) (Millán et al., 2009; Moreno & De las Heras, 2009; Velasco et al., 2006). Although these habitats have a high conservation interest and a limited distribution (especially in the European context), the low social awareness of these habitats potentially exposes them to intense anthropogenic pressures (Millán et al., 2011).

Biological traits of the aquatic insects in Mediterranean climate

Biological traits are morphological, physiological, behavioral, or life history attributes that characterize organisms (Violle et al., 2007). They can be applied to individuals or to species (or other taxonomic levels) and have been evolutionarily acquired through the interaction of these organisms with the environment (Menezes et al., 2010; Statzner, Hildrew, et al., 2001). Therefore, according to the theory of the habitat template (based on the niche concept) (Southwood, 1977), the presence or absence of a species in a community depends on the match between their traits and the environment, acting at multiple spatial scales through a niche filtering process. More recently, local community composition is being also explained by dispersal processes of these species (another type of trait) or the landscape configuration favoring or limiting species' dispersal (Leibold et al., 2004). Trait analysis has thus become a central part of community ecology in both fundamental and applied research.

Besides their use in community ecology, traits have also been considered in other fields, such as molecular phylogenetics and biodiversity conservation (Cavender-Bares et al., 2004; Statzner, Hildrew, et al., 2001; Miatta et al., 2021). Their popular use in the last decades has resulted in a wide variety of trait databases (e.g., CESTES, Jeliazkov et al., 2020) that cover multiple taxonomic groups, with plants being the most popular organisms in trait studies. There are also several trait databases that include information on aquatic insects, and some of them have been widely used in Mediterranean regions worldwide. The Tachet database and subsequent additions (Bonada & Dolédec, 2011; Tachet et al., 2010), for example, have been applied to environmental gradients in the Mediterranean Basin countries. These studies have considered large-scale traits patterns (Bonada, Dolédec, et al., 2007; Morais et al., 2009; Statzner et al., 2007) or the response of traits to specific environmental gradients related to hydrology, habitat, or pollution (García-Roger et al., 2013; Mellado-Díaz et al., 2008; Vidal-Abarca et al., 2013). More recently, the DISPERSE database compiles trait information related to dispersal of aquatic insects and its potential use in fundamental and applied research (Sarremejane et al., 2020) (Table 8.2).

Trait databases for aquatic insects consider a wide variety of traits (with several trait categories) and include information for family, genus, or species level in form of presence/absence or some type of quantification, such as the fuzzy coding approach (Chevenet et al., 1994). All these databases consider potential trait data (i.e., not real trait measured values), and therefore their use is usually limited to large-scale studies or along strong environmental gradients (Bonada & Dolédec, 2018). In particular, the Tachet database and subsequent additions include information for 473 macroinvertebrate taxa (mostly genus level), with 331 aquatic insects. It comprises 11 biological traits and 63 categories coded

TABLE 8.2 List of general biological traits and trait categories describing aquatic insects according to Tachet et al. (2010) and list of dispersal-related traits according to Sarremejane et al. (2020).

References	Trait	Category
Tachet et al. (2010)	Maximal size	≤0.25 cm
		>0.25–0.5 cm
		>0.5–1 cm
		>1–2 cm
		>2–4 cm
		>4–8 cm
		>8 cm
	Life cycle duration	≤1 year
		>1 year
	Potential number of	<1
	Reproduction cycles	1
	Per year	>1
	Aquatic stages	Egg
		Larva
		Nymph
		Imago
	Reproduction	Ovoviviparity
		Isolated eggs, free
		Isolated eggs, cemented
		Clutches, cemented or fixed
		Clutches, free
		Clutches in vegetation (endophytic)
		Clutches, terrestrial
		Asexual reproduction
	Dissemination	Aquatic passive
		Aquatic active
		Aerial passive
		Aerial active
	Resistance form	Eggs, statoblasts, gemmules
		Cocoons

References	Trait	Category
Sarremejane et al. (2020)	Maximum body size (cm)	<0.25
		≥0.25–0.5
		≥0.5–1
		≥1–2
		≥2–4
		≥4–8
		≥8
	Female wing length (mm)	<5
		≥5–10
		≥10–15
		≥15–20
		≥20–30
		≥30–40
		≥40–50
		≥50
	Wing pair type	1 pair + halters
		1 pair + elytra or hemielytra
		1 pair + small hind wings
		2 similar-sized pairs
	Life-cycle duration	≤1 year
		>1 year
	Adult life span	<1 week
		≥1 week to 1 month
		≥ 1 month to 1 year
		≥1 year
	Lifelong fecundity	<100
	(number of eggs per female)	≥ 100–1000
		≥1000–3000
		≥3000
	Potential number of	<1

(Continued)

TABLE 8.2 (Continued)

References	Trait	Category	References	Trait	Category
		Cells against desiccation		Reproduction cycles	1
		Diapause or dormancy		Per year	>1
		None		Dispersal strategy	Aquatic active
	Respiration	Tegument			Aquatic passive
		Gill			Aerial active
		Plastron			Aerial passive
		Spiracle (aerial)		Propensity to drift	Rare/catastrophic
		Hydrostatic vesicle (aerial)			Occasional
	Locomotion and	Flier			Frequent
	Substrate relation	Surface swimmer			
		Swimmer			
		Crawler			
		Burrower (epibenthic)			
		Interstitial (endobenthic)			
		Temporarily attached			
		Permanently attached			
	Food	Fine sediment + microrganisms			
		Detritus < 1 mm			
		Plant detritus ≥ 1 mm			
		Living microphytes			
		Living macrophytes			
		Dead animal > 1 mm			
		Living microinvertebrates			
		Living macroinvertebrates			
		Vertebrates			
	Feeding habits	Absorber			
		Deposit feeder			
		Shredder			
		Scraper			
		Filter feeder			
		Piercer (plant or animal)			
		Predator (carver/engulfer/swallower)			
		Parasite, parasitoid			

using a fuzzy coding approach that consider within genera and across latitude variability (Table 8.2). It was built with information from c. 6000 published and unpublished studies and a time span of more than one century (the oldest publication used dated back from 1802). These studies were all performed in Europe and therefore, the Tachet database is supposed to be representative of the whole continent (Bonada & Dolédec, 2011). Despite this, its application to non-European Mediterranean Basin countries is potentially possible because most aquatic insect genera in these countries are also present in Europe and the use of the fuzzy coding provides some flexibility on its use. However, some genera found in these countries are not coded in the Tachet, and updates and adaptations of this database to all Mediterranean Basin countries are needed. For example, Bonada and Dolédec (2011) found 44 new genera present in Mediterranean Basin but not included in the Tachet database.

The particular characteristics of rivers in the Mediterranean Basin impose the presence of particular traits (Bonada & Resh, 2013; Cid et al., 2017). Mediterranean rivers are overall characterized by high seasonal and predictable flow patterns, with temporary rivers being a common river type. In comparison to other climate regions in the world, the seasonality and predictability of Mediterranean rivers (Tonkin et al., 2017) should have resulted in particular trait adaptations to the drying period. According to Lytle and Poff (2004), under these conditions, aquatic organisms should have acquired synchronized life history traits to the drying period instead of other strategies to escape the unfavorable moment. This is actually a common pattern in plants and terrestrial arthropods (Milla et al., 2010; Stamou, 1998) but also for aquatic insects. For example, the Mediterranean trichopteran *Mesophylax aspersus* have life cycles synchronized with the drying period. The pupae of this species emerge before the river dries up and adults estivate in nearby cave during the summer period and until the rewetting, when mating and egg laying take place (Salavert et al., 2008).

The characteristic hydrological variability in Mediterranean rivers has resulted in species with particular traits. In comparison to rivers in temperate Europe, aquatic insects in Mediterranean Basin have traits to cope with or recover from the dry period, along with dominant traits from pool habitats during summer (Hershkovitz & Gasith, 2013). For example, resistance forms, such as diapause or terrestrial egg laying, or resilient traits such as smaller sizes, aerial active dispersal, or multivoltinism were significantly more frequent in Mediterranean than in temperate rivers (Bonada & Dolédec, 2018; Bonada, Dolédec, et al., 2007). Many aquatic insects in the Mediterranean Basin have also traits adapted to the pool conditions, less frequent in the temperate rivers. Aquatic insects have usually larger sizes and imagos, and aerial respiration is common. This is because these pools habitats are usually characterized by Odonata, Coleoptera, and Heteroptera, whose richness values have also been considered as a measure of the temporary nature of rivers (Bonada et al., 2020; Bonada, Rieradevall, et al., 2006).

Resistance and resilience are the two response mechanisms to drying (Fig. 8.1). Several studies have found that dormancy is a common resistance mechanism in temporary rivers located in several climate regions of the world (Bogan et al., 2017; Strachan et al., 2015). Stubbington and Datry (2013) found a large proportion of aquatic insects in dry riverbeds in English temporary rivers, with dormancy stages such as eggs, larvae, pupae, or adults. Although the survival of this "seedbank" depends on the duration of the drying period (Storey & Quinn, 2013; Stubbington & Datry, 2013), some aquatic insects, such as the North American megalopteran *Neohermes filicornis*, can survive multiple periods of drying and rewetting (Cover et al., 2015). However, unless species are able to resist in disconnected pools, resilience strategies seem to be more common than resistance ones in Mediterranean Basin. The traits of most aquatic insects in these countries (smaller sizes, shorter life cycles, high dispersal ability) and the few existing rewetting experiments (Folch de la Iglesia, 2020) reveal that resilience plays a more important role than resistance when rivers completely dry up (Fig. 8.1). More studies on the potential resistance strategies of aquatic insects are, however, still needed. For example, there is evidence of the existence of aquatic insects with semivoltine life cycles in temporary rivers of the Mediterranean Basin, so specific strategies to survive the dry period in the sediments are required (e.g., the Plecoptera *Guadalgenus franzi*, López-Rodríguez et al., 2009).

Aquatic insects in the Mediterranean Basin are also supposed to be characterized by strong dispersal abilities. The alternative strategy to survive the dry period with a resistance stage is to leave the site before the river dries. This could be done by the aerial active dispersal of aquatic flying adults (i.e., Coleoptera or Heteroptera) or by synchronizing the life cycle with drying, with larvae or pupae emerging as adults and dispersing to other water bodies (Fig. 8.1). "To stay or leave" is considered an evolutionary trade-off (Bonada et al., 2017), and both strategies are related to traits allowing species to spread risk through space and time. Whereas spatial dispersal refers to the ability of species to leave a site and move to another one, temporal dispersal relies on the ability of species to enter into a dormancy stage and to remain in the site during the unfavorable conditions. Different species of aquatic insects are placed along the spatial and temporal dispersal continuum (Bonada et al., 2017). For example, the North American plecopteran *Mesocapnia arizonenis* has apterous males with aestivating nymphs (Gray, 1981), and the Iberian plecopteran *Guadalgenus franzi* most likely has similar dormancy strategies (López-Rodríguez et al., 2009), suggesting high temporal dispersal. In contrast, most beetles of the family Dytiscidae have strong dispersal abilities and unknown resistance strategies (Schäfer et al., 2006).

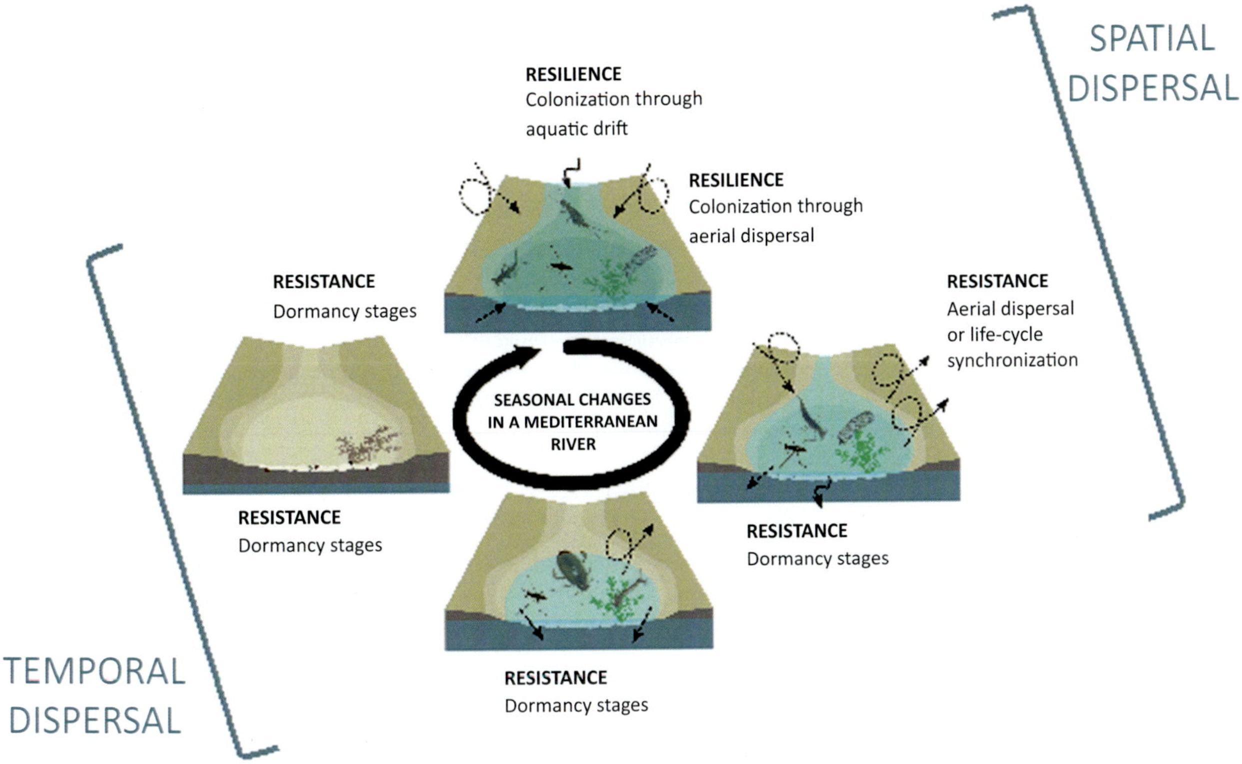

FIGURE 8.1 Resistance and resilience strategies of aquatic insects in Mediterranean rivers to drying. Spatial dispersal refers to the ability of species to leave a site and move to another one. Temporal dispersal relies on the ability of species to enter into a resistance stage and to remain in the site during the unfavorable conditions.

Most likely, as stated above, the aquatic insects in the Mediterranean Basin have a dominance of spatial dispersal but also may possess intermediate strategies (e.g., having both strategies, spatial and temporal dispersal), as in the case of *M. aspersus*. In this species adults flight dozens of kilometers to find caves where they estivate (Salavert et al., 2008). Actually, high spatial dispersal abilities are one of the most significant traits of insects in rivers of the Mediterranean Basin as a result of the higher dominance of Odonata, Coleoptera, and Heteroptera adapted to temporary rivers, as these three orders contain species known to be higher dispersers as adults (Bonada & Resh, 2013; Bonada, Dolédec, et al., 2007; Bonada et al., 2020).

Analyzing community composition by using a trait-based approach also provides information on functional processes (Violle et al., 2007). Most commonly used traits in aquatic insects are related to ecosystem functions such as body size, feeding habits, dispersal abilities, life history, and behavior characteristics (Bonada et al., 2017). Large aquatic insects, for example, contribute to increased biomass and ecosystem stability and decreased productivity, whereas shredders significantly contribute to the processing of organic matter. Species with high dispersal ability, short life cycles, and large offspring significantly increase ecosystem resilience, and some aquatic insects, such as cased-building Trichoptera, contribute to the retention of solids (Statzner, 2012). The understanding of the relationship between biodiversity and ecosystem functions is thus mediated by these traits related to functional processes. In a Mediterranean context, the high hydrological variability imposes seasonal biodiversity changes in taxonomic and trait richness and composition (e.g., García-Roger et al., 2013; Hershkovitz & Gasith, 2013) that can potentially compromise ecosystem functions. Temporal stability for functional diversity is, however, higher than for taxonomic diversity in the Mediterranean Basin (Dolédec et al., 2017), suggesting that there is a high trait redundancy and that ecosystem functions can remain relatively stable. In addition, in comparison to taxonomic diversity and composition, traits recover quickly after the dry period (Dolédec et al., 2017).

Beyond the use in fundamental research in the Mediterranean Basin, traits have been also used in applied research regarding bioassessment. The use of traits in bioassessment is considered a good approach and fulfils most criteria on rationale, implementation, and performance identified as the ideal bioassessment approach by Bonada, Prat, et al. (2006). They are able to respond to particular disturbances and, as mentioned before, to link disturbance effects to functional processes. For example, increases in organic matter would favor aquatic insects with aerial respiration by decreasing dissolved oxygen, whereas pollution by heavy metals would favor those with smaller sizes that bioaccumulate less (Statzner, Bis, et al., 2001). This ability to respond to different disturbances has been very useful in disentangling natural from human disturbances. This is of great importance in rivers in the Mediterranean Basin where the effects of human disturbances are exacerbated by the low or no-flow periods of temporary rivers. Functional metrics seem to work well in these conditions and respond to human disturbances regardless of hydrological variability (Soria et al., 2020). In addition, traits related to dispersal have also the potential to increase the reliability of the bioassessment of these temporary rivers (Cid et al., 2020). The recurrent shifts among flow, disconnected pools, and dry riverbeds in temporary rivers imply the need for recolonization of aquatic insects and thus the arrival of aquatic insects from other water bodies. If this arrival is in addition constrained by landscape configuration, the recovery of biodiversity during the rewetting in temporary rivers may be limited. In these conditions, the use of traditional biological indices that rely on species richness is not reliable because low richness does not imply low biological quality but rather high isolation (spatially and temporally). Therefore, the incorporation of the dispersal ability of aquatic insects and spatial constraints for it (i.e., considering the metacommunity approach) would contribute to improving the bioassessment of rivers in Mediterranean Basin and in other regions worldwide (Cid et al., 2020).

Dispersal and metacommunity dynamics

Hexapods are a group that show diverse dispersal modes among taxa and development stages, a strategy that may help increasing their success in dispersing effectively depending on the environmental conditions and the inhabited system characteristics. Thus, active dispersal by flight is the usual but not exclusive mode of dispersion for adult insects (Bilton et al., 2001). Some winged adults also present passive aerial dispersal since their movement is strongly affected by winds (e.g., Ceratopogonidae, Chironomidae, Culicidae; Grönroos et al., 2013), and some larvae might passively disperse by migratory birds (Green & Figuerola, 2005). In lotic systems the main dispersal modes include drift (downstream, active or passive), active movement of aquatic larvae (up- or downstream), and aerial dispersal of adult winged stages (active, passive) (Sondermann et al., 2015). In contrast, dispersal in lentic systems exclusively relies on aerial modes (both passive and active). However, dispersal processes are highly influenced by environmental conditions. For example, insect flight may be influenced by landscape, atmospheric, and habitat conditions and biological interactions (Boix et al., 2016), and drift highly depends on river flow. In fact, some winged adults such as damselflies, stoneflies, and caddisflies have some preferred conditions (microclimatic and light conditions) for dispersal and can even stop dispersing if conditions change (Briers et al., 2002; Collier & Smith, 1998). Thus, this emphasizes their ability to modify dispersal based on environmental conditions (Bilton et al., 2001; De Bie et al., 2012). Moreover, within lotic systems, the hydrology characteristics generate distinctive features at dispersal level. In Mediterranean systems the flow cessation or bed dryness favors flying to other less-dry sites, and therefore it promotes the aerial active dispersal. By contrast, in temperate streams, the permanent action of flow increases downstream drift, and, consequently, promotes aquatic passive dispersal (Bonada, Dolédec, et al., 2007). Interestingly, in pond systems, in which drift is not possible, and so aquatic dispersion among ponds is not relevant, there are also differences between pond typologies. Thus, Mediterranean ponds are more dominated by organisms showing a passive aerial dispersal mode than continental ponds, in which active aerial dispersal is enhanced (Céréghino et al., 2012).

But why is dispersal so important? Communities are not isolated entities but connected to other communities through dispersal movements. In fact, a set of local communities that are linked by dispersal of multiple potentially interacting species is the definition of a metacommunity (Hanski & Gilpin, 1991; Wilson, 1992). In the metacommunity context, communities are structured by a combination of local processes like environmental filtering and species interactions, and regional processes driven by the dispersal of organisms among localities on a landscape (Holyoak et al., 2005; Leibold et al., 2004). Therefore, inherent in the metacommunity perspective is the comprehension of how organisms move on landscapes (Brown et al., 2018). In the case of aquatic systems, an organisms' movement over the landscape, i.e., its dispersal, greatly relies on the connectivity among the communities that form the network (Altermatt, 2013; Brown et al., 2018; Liu et al., 2013). However, this connectivity is not static, and in some cases could change along and between years, mainly due to seasonal and interannual variability. This is the case of Mediterranean rivers that present sequential seasonal flooding and drying periods, with increasing loss of habitat connectivity over an annual cycle that can result in temporary isolated habitats especially during severe droughts. This temporal pattern generates different levels of hydrological connectivity among seasons, with an expansion phase in the wet period (i.e., autumn-winter) and a contraction phase in the dry period (i.e., spring-summer) (Bernal et al., 2013). This seasonal hydrological variability affects biological communities (Robson et al., 2013) and by extension, the metacommunity dynamics (Cañedo-Argüelles et al., 2020; Sarremejane et al., 2017). Then, during the wet period, communities are mainly directionally connected due to the water flux (functioning more like temperate rivers; Carrara et al., 2014; Tonkin et al., 2018), from headwaters to downstream since downstream reaches are connected to headwaters through drift of upstream fauna (Fig. 8.2). In contrast, during the dry period, with the increase of habitat fragmentation, this directional connectivity is broken and communities in isolated pools are randomly connected to other pools (functioning more like pond networks; Brown & Swan, 2010; Sarremejane et al., 2017). During this period, disconnected pools act as islands and drying impose strong environmental filtering constraining metacommunities (Sarremejane et al., 2017). This variability implies that the fauna present must adapt to unique stresses that require multiple adaptive mechanisms (Bonada & Resh, 2013). During the rewetting, metacommunities are re-established again and mainly explained by stochastic and dispersal-related processes favoring recolonization from permanent rivers (Sarremejane et al., 2017). Thus, organisms have life history, behavioral or morphological adaptations to resist floods and droughts through endurance or avoidance strategies (Robson et al., 2013). In addition to the seasonal changes, biological communities of Mediterranean rivers also experience long-term changes mainly linked to environmental conditions (Cañedo-Argüelles et al., 2020). Communities are dominated by pool-like taxa (e.g., Gerridae and Notonectidae; Bonada et al., 2008) in dry years (i.e., a year with values lower than the annual mean rainfall minus the standard deviation; Scian & Donnari, 1997). In contrast, in wet years (i.e., a year with values higher than the annual mean rainfall plus the standard deviation; Scian & Donnari, 1997) riffle-like taxa (e.g., Simuliidae, Hydropsychidae, Empididae; Bonada et al., 2008) are more abundant (Pace et al., 2013). According to literature (Razeng et al., 2016; Tonkin et al., 2018) strong aerial dispersers tend to show little spatial structure in most stream networks. However, the flow intermittence during dry years in Mediterranean rivers can significantly reduce the dispersal capacity of aerial dispersers (i.e., dry reaches act as barriers for the dispersal of flying insects that use water for laying their eggs), leading to spatially structured metacommunities (Cañedo-Argüelles et al., 2020). In contrast, strong drift dispersers generally best reflect environmental filters, even during dry years, suggesting that drift dispersal is so local that it does not influence assembly at metacommunity level (Cañedo-Argüelles et al., 2020).

Ponds and lakes metacommunity functioning is slightly different since communities are usually not directionally connected (headstream to downstream), and so their connectivity mainly relies on network structure (position, density, and configuration of the communities/patches in the network; Brown & Swan, 2010; Economo & Keitt, 2008) and the organisms' dispersal ability that ultimately configure their landscape perception (Borthagaray et al., 2014). Specifically, the organisms' dispersal ability determines the interaction between dispersal and landscape structure (Borthagaray et al., 2014; Cunillera-Montcusí, Boix, et al., 2020) and thus, has effects on the spatial processes and environmental control important in metacommunity organization (Heino, 2013). Several studies in the Mediterranean Basin have detected an important role of the species sorting processes (i.e., to select for species whose traits allow them to most

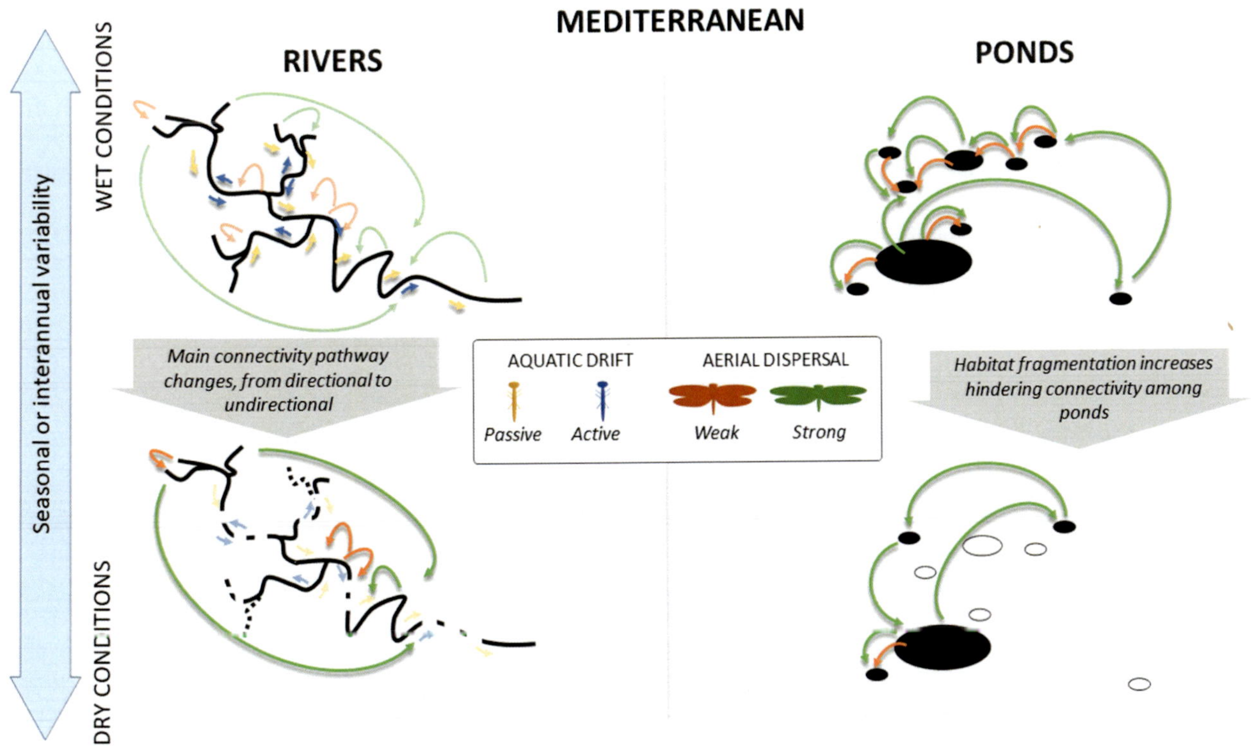

FIGURE 8.2 Main differences on dispersal dynamics between Mediterranean rivers and ponds in relation to seasonal or interannual variability.

effectively function in a given habitat; Leibold et al., 2017) over other mechanisms as a determinant of community assembly patterns in different organisms including hexapods: macrophytes (in this case together with dispersal limitation; García-Girón et al., 2019), ostracods (Castillo-Escrivà et al., 2017), macrofauna (i.e., invertebrates and amphibians; Tornero et al., 2018), and invertebrates (including cladoceran and macroinvertebrates; Waterkeyn et al., 2008). However, it is also true that the detection of some metacommunity archetypes, like the mass effects (i.e., a mechanism in which there is net flow of individuals created by differences in population size or density in different patches; Schmida & Wilson, 1985), is highly scale dependent, being more probably detected in smaller geographical ranges than in larger ones (Heino et al., 2014; Tornero et al., 2018). Besides this scale dependency, the type of organisms' dispersal is also relevant, since some types are more prone to track environmental conditions and so to show the species sorting signal (i.e., active aerial dispersers, like many insects).

On the other hand, the location of a pond within a pond network with a centrality-isolation gradient can influence insects' colonization, with the central ponds often having the highest abundance and richness of insects. In this sense, Cunillera-Montcusí, Boix, et al. (2020) found clearly differentiated abundance and richness patterns between central and isolated ponds for strong dispersers (i.e., Coleoptera: Dytiscidae, Helophoridae, and Hydrophilidae). Hence, when the neighboring waterbodies were drying, the central ponds experienced an increase in individual arrivals. However, weak dispersers (i.e., Diptera: Ceratopogonidae, Chironomidae, and Ephydridae) followed more or less the same pattern of increasing abundance during the hydroperiod independently of pond location. Moreover, in highly connected systems, in which the centrality-isolation gradient is blurred, species (mainly insect species) demonstrated weak dispersal limitations (Florencio et al., 2014; Tornero et al., 2016). Finally, the fact of being an active or passive disperser can also create patterns of nestedness within metacommunities (i.e., when species-poor sites contain subsets of assemblages found in species-rich sites; McAbendroth et al., 2005). In this sense, Florencio et al. (2011) observed that winged insects were more highly nested than nondispersing taxa, because the dispersal movements of insects after the dry period would help maintain the nested structure of the macroinvertebrate assemblages across the pond network.

In summary, aquatic insects often have adaptive strategies for effective dispersal, and especially in the case of those inhabiting the Mediterranean Basin, these type of strategies gain importance due to the recurrent drying or filling of both lotic and lentic ecosystems. Overall, the diverse literature including both lotic and lentic environments from the Mediterranean Basin has evidenced that both the dispersal mode (active or passive) and dispersal ability (weak or strong) of aquatic hexapods may have important effects on metacommunity structuring and dynamics.

The role of aquatic insects in food webs

Trophic interactions determine the structure of a community and underly crucial functions in ecosystems. Trophic relationships control energy flow from primary resources to consumers (bottom–up interactions), while, at the same time, some consumers potentially limit the biomass of some taxa through a top–down control. The strength of these consumer–resource interactions has large effects on community dynamics. The study of food webs enables the extension of the attributes of individuals and populations to ecosystem properties such as production and element cycling (Thompson et al., 2012). In this way, a food web perspective can help us better understand the links between biodiversity and ecosystem functioning and how to advance in the conservation and management of functional ecosystems.

In freshwater habitats under the Mediterranean climate, seasonality affects food source availability from terrestrial and aquatic primary producers. In streams and during the wet period (autumn-winter), longitudinal, lateral, and vertical flow connectivity is restored and the inputs of allochthonous organic matter (terrestrial origin) are distributed and processed in the river network. During the dry period (spring-summer), high temperature and flow contraction favor autochthonous (river origin) primary production although there is a reduction of river connectivity (Gasith & Resh, 1999). Ponds fed by fluvial drainage or directly by rainfall also reflect fluctuations in food availability related to the hydroperiod and interannual climate variations. In addition, the Mediterranean Basin has many karstic areas with spring-fed ponds that are highly stable (Sahuquillo & Miracle, 2013) as well as permanent streams. Thus, depending on the degree of flow intermittence, the Mediterranean inland waters span a wide range of hydrological conditions that can determine the quality and quantity of the food resources conditioning trophic relationships. The characteristics of the food resources, however, are also affected by factors other than the hydroperiod (e.g., water chemical characteristics and land uses). Since these factors interact to determine the biotic and abiotic conditions in the system, the food web in a particular freshwater system will reflect the consequences of the effects of all these factors.

Essentially, all aquatic insects are omnivorous. For example, shredders ingest not only leaf tissue and the associated microbiota, but also diatoms and other algae attached to the leaf surface, as well as small invertebrates. First described 47 years ago, the functional feeding group (FFG) approach (Cummins, 1973) is based on a correspondence between the categories of nutritional resources present in the environment and the analysis of insect feeding based on the morpho-behavioral mechanisms of food acquisition. Six FFGs were initially proposed and linked to six food resource categories. Among the FFGs there are obligate and facultative (i.e., flexible requirements for more than one food resource) members. There are few specialists feeding on a specific food resource during the entire life cycle. Despite this, the FFG designation is based on the preponderance of a food item and on the most probable feeding mode (Merritt et al., 2019). Among the insect orders, we find representatives of all the different FFGs. Instead of only classifying a taxon into its main FFG, a more accurate approach was adopted by Tachet et al. (2010) for European freshwater macroinvertebrates that involved using a fuzzy system for the proportional assignment of a taxon (at the family, genus, or species level) to different FFGs. Nevertheless, the values given to each taxon, which result from known variations in size, season, and place of origin, cannot always be extrapolated to the same taxon in a specific place and season (a similar pattern that also occurs with other labile traits such as voltinism, Bonada & Dolédec, 2018). This is especially important in Mediterranean freshwater systems, where the source, identity, and quality of the food available for consumers change over the course of the year, requiring long-term studies to characterize the consumer diet well (Tierno de Figueroa et al., 2019).

Insects can be top predators in a food web or be part of the diet of other top predators (other invertebrates, amphibians, and fish). The most common insect predators include those belonging to Odonata, Plecoptera, Megaloptera, Trichoptera, and Diptera. Mayflies and Diptera are the common prey taxa for invertebrate predators (Peckarsky & Lamberti, 2017).

How to analyze a food web

The most common approach for the identification of trophic relationships is the analysis of gut contents. This approach provides a minimum estimate of the biomass and diversity of the diet, although these can be underestimated in most aquatic consumers since some food may be unrecognizable or the predator may ingest only fluids or soft parts. Gut analysis provides information about what is ingested, but not all ingested material is necessarily assimilated. Analyses of stable isotope ratios of C ($\delta^{13}C$) and N ($\delta^{15}N$) have contributed to food web research over the last 30 years. Values of these isotopes in animal tissues have been used as indicators of food sources and trophic level, since $\delta^{13}C$ values can distinguish primary producers and increases in the $\delta^{15}N$ value can indicate a move to a higher trophic level (e.g., Muñoz et al., 2009; Vander Zanden & Rasmussen, 2001). Therefore, biplots for C and N isotopes reveal the food web structure. However, in aquatic ecosystems, the $\delta^{13}C$ of algae in periphyton is sometimes too variable and detrital sources from different origins (terrestrial or aquatic) are mixed, making assessment difficult. Similarly, for $\delta^{15}N$, the isotope enrichment factor per trophic level of freshwater invertebrates is likely to be smaller and more variable than that of other animals. Other isotopes that are used in diet analysis are $\delta^{2}H$ and $\delta^{34}S$, the latter is especially useful for discriminating between marine and terrestrial nutrient sources and may have utility in wetlands and saltmarshes.

The use of fatty acids as trophic markers has provided information on trophic links in open-ocean, estuarine, lake, and river food webs. Some polyunsaturated fatty acids are good markers for diatom, bacteria, or bryophytes and have been proposed as markers for a terrestrial versus aquatic matter origin in streams (Torres-Ruiz et al., 2007). The measurement of the stable N isotopic composition of amino acids (SIAA) has recently been applied to estimate the trophic position of consumers in freshwater systems. In amino acid metabolism, glutamic acid is subjected to deamination and transamination, which lead to increased isotope enrichment per trophic level. In contrast, phenylalanine retains its amino group during metabolism as animals cannot synthesize phenylalanine themselves, resulting in little isotope enrichment in the food chain. Thus, in a single food chain, the trophic level of an animal can be estimated only from its $\delta^{15}N$Glu and $\delta^{15}N$Phe values, in a similar way to that when using C and N isotopes (Ishikawa et al., 2014).

Recent advances in DNA-based approaches have led to more accurate methods for identifying prey species from gut or fecal samples. These techniques can be used in the cases where it is difficult to obtain good results with other methods, such as for insect predators that feed on fluids or when the morphological identification of remnants is impossible. A standardized DNA region (DNA barcode) is amplified by PCR, with the amplicons sequenced and then compared to a reference database for the identification of prey. Good results have been obtained with a wide taxonomic range of predators, fish, and insect adults like the members of Odonata. However, more detailed work is needed, particularly in freshwater systems (Kaunisto et al., 2017; Pompanon et al., 2012).

Food web structure

Most undisturbed stream food webs have approximately three or four trophic levels and are characterized by a high degree of connectivity. In general, disturbed aquatic systems or those under extreme conditions (e.g., long dry periods) typically have simplified food webs (Townsend et al., 1998). Reductions of food chain length linked to drying are associated with the loss of large top predators (fish or invertebrates) (Sabo et al., 2010). Detritus and primary producers, including algae, bryophytes, and vascular macrophytes, are the food sources that occupy the basal level. A similar structure is found in lentic systems, although the primary producers can sustain preferent and long food chains in some deep and permanent water bodies. Grazers (direct consumers of primary production) and detritivores (consumers of detritus coming from algal and terrestrial organic matter) occupy the primary consumer trophic level. As producers and detritus are associated with heterotrophic microbes and microorganisms (e.g., ciliates), these consumers function as both primary and secondary consumers. Predators often have mixed diets that include some combinations of detritus, diatoms, animal prey, and even other predators, placing them between trophic levels 3, 4, and, in some cases, 5. As in other systems, omnivores (consumers that feed on different trophic levels) often dominate communities in Mediterranean inland waters and it is common that the prey−predator interactions form a complex food web rather than a simple and linear food chain. High levels of omnivory buffer food web dynamics from fluctuations in water flow. Food depletion may increase cannibalism and intraguild predation in aquatic insects, which are common among the larvae of beetles and odonates and in omnivorous predators such as the caddisfly *Hydropsyche*.

Among detritivores, insect shredders are critical for mobilizing energy from leaf litter to consumers in streams and may play a similarly important role in small ponds (Holgerson et al., 2016). In Mediterranean rivers, the low abundance or the lack of shedders has been linked to the low quality of the organic matter when flow returns after drying (Álvarez & Pardo, 2009; López-Rodríguez et al., 2012). For example, in the permanent river Riera Major, an alder-shaded, second-order stream (NE Spain), the shredding caddisfly *Halesus radiatus* is dominant in autumn, but a high shredder biomass is present all year round because conditioned leaves are available for the insect. In contrast, in the nearby temporary river La Solana, a second-order tributary in the same basin, shredders are mainly present in autumn and significantly decline thereafter, with decreases in discharge and benthic organic matter (Muñoz, 2003).

The quantitative study of trophic link strengths in a permanent Mediterranean river food web in the south of Spain (Peralta-Maraver et al., 2017) revealed the presence of small groups of strong links and many other weak interactions and although the node with a high number of interactions was similar over time, the strength of the links differed seasonally. This structure contributed to a certain degree of stability against seasonal environmental variability in these systems.

Temporal variability of resource−consumer interactions

In most freshwater systems, detritus predominating as food source favors the heterotrophic pathway. Allochthonous matter and its associated microbiota are an important food source in many low-order streams. Small ponds are fed by abundant detritus originating from adjacent terrestrial systems. However, even in heavily shaded streams with low algal standing crops, algae support freshwater organisms with their rapid turnover and high nutritional value (low C:N ratios) compared to detritus and can strongly influence the structure of entire food webs. Similarly, most pond consumers feed on both algae and detritus coming from leaves, but algal-based pathways are prevalent, challenging the common assumption that small net-heterotrophic low-light ponds have detritus-based food webs.

In temporary Mediterranean streams, seasonality and its associated changes in hydrological conditions determine the relative importance of allochthonous versus autochthonous food sources over time. Pulses of riparian litterfall occur mostly in summer with water stress, with a second peak occurring in autumn due to leaf abscission (e.g., Sanpera-Calbet et al., 2016). When flow resumes in dry stream channels, high levels of light and temperature promote a rapid increase in periphyton biomass (Acuña et al., 2007; Álvarez & Pardo, 2009), thus increasing the importance of benthic primary producers. In Fuirosos, a third-order temporary river (NE Spain), the contribution of algal biofilm to the diet of the consumers is relevant during the entire hydrological cycle. The highest contribution of biofilm to collectors, shredders, and predators has been observed in the drying phase, but additionally for the last two groups, biofilms is also relevant when water flow is disconnected (fragmentation), coinciding with the highest abundance of autochthonous resources. The contribution of detritus is the greatest in the late recovery phase, following leaf autumn abscission (Fig. 8.3; Mas-Martí, 2014). The assimilation of resources is related more to their overall abundance than to their quality. Interestingly, the increase in the importance of autochthonous resources to the biomass of consumers is transferred up in the food web in the dry period.

The dependence of the relative dominance of a basal resource on water flow dynamics and hence, energy pathways through consumers has been observed in other Mediterranean streams with interannual differences. In the Eel River (California; Power et al., 2013), algal (*Cladophora*) growth is limited in dry years due to the control exerted by a grazer case-living caddisfly (predator resistant). During wet winters, floods kill or export these grazers and large *Cladophora* blooms proliferate in the following

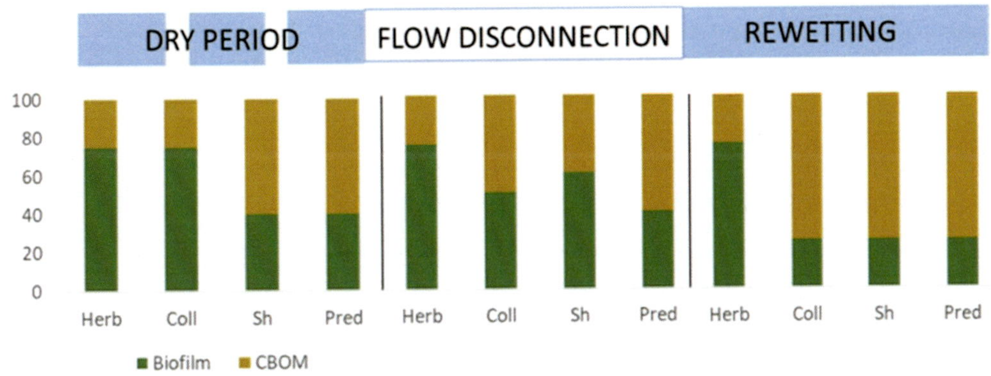

FIGURE 8.3 Mean percent contribution of epilithic biofilm and coarse benthic organic matter (CBOM) to consumers (analyzed by SIAR mixing model, using stable isotopic analysis) in the different hydrologic phases in Fuirosos stream: dry period, flow disconnection, and rewetting (flow recovery). *Herb*, Herbivores; *Coll*, collectors; *Sh*, shredders; *Pred*, predators. *Adapted from Mas-Martí, E. (2014).* Climate induced changes in headwater streams: Effects of warming and drought on resource-consumer trophic interactions *(Ph.D. thesis). Spain: University of Barcelona, Barcelona. 162 pp.*

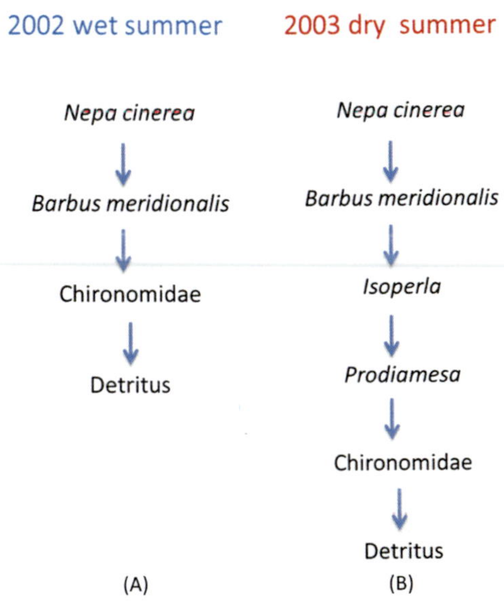

FIGURE 8.4 Example of a food chain in Fuirosos stream in two consecutive summers with different hydrological characteristics. (A) Summer 2002 (wet period preceded by high precipitations in winter-spring); (B) summer 2003 (dry summer with stream flow fragmentation).

spring and summer, which sustain other grazers like mayflies and free-living midges. In Fuirosos, algal sources are more important to food webs after wet years with severe winter floods. Winter floods remove the accumulated detritus and organic matter-rich sediments (Acuña et al., 2007). With lower storage of organic matter, the overall detritivore biomass decreases. These interannual hydrological differences also lead to differences in the food web structure between years in Fuirosos. For example, over two consecutive summers characterized by different winter-spring hydroperiods, there are clear differences in the properties of the food web (Fig. 8.4, Table 8.3). A drier winter-spring period favors higher diversity in the habitat patches (pools, riffles, and isolated pools) and food sources (algae, detritus and preys), producing longer average food chain length. The proportion of intermediate consumers is higher and the number of omnivorous species also increases.

In floodplains where water dynamics are governed by a seasonal inundation regime and lateral hydrological connectivity, a rich array of temporary water bodies occurs. Complexity of the trophic structure is governed by the persistence of water (O'Neill & Thorp, 2014). Food chain length and trophic complexity are low at the beginning of an inundation period due to high trophic redundancy. Later, insects colonize the ecosystem, bringing new trophic species that can increase the food chain length and trophic structural complexity. The longer the hydroperiod, the more diverse the trophic niches are and higher the colonization and insect oviposition.

Drying also determines spatial variation in the density and composition of communities at small spatial scales. For example, drying increases the density-dependence processes that also influence trophic interactions. In isolated pools after flow

TABLE 8.3 Food web properties in two consecutive years during summer.

	Summer 2002	Summer 2003
Average long chain	1.39	1.86
Maximum trophic levels	4	6
Ratio basal species/total	0.13	0.12
Ratio intermediate species/total	0.22	0.33
Ratio top species/total	0.65	0.55
Omnivorous species	7	9

Summer 2002 was a wet period preceded by high precipitations in winter-spring that maintained connected the water flow during all summer. Summer 2003 was drier with stream flow fragmentation.

fragmentation, the density of invertebrates increases, causing changes in predation intensity (Bonada et al., 2020; Gasith & Resh, 1999). The abundance of large-bodied insect predators (e.g., Dytiscidae and Odonata) is low during high flow, but becomes dominant during low flow (Acuña et al., 2005). In Fuirosos, for example, changes in habitat dynamics induce changes in food source availability for consumers. The odonate *Onychogomphus* is a predator, consuming midges and mayflies (mainly *Baetis* and *Ephemerella*), in summer and in reaches that maintain flow connectivity. However, it is omnivorous in disconnected pools consuming diatoms and detritus in addition to *Baetis* and worms (Naididae).

In standing ponds, the diets of predatory insects vary from highly specialized to broadly general. Species with similar sizes that forage in the same microhabitat (e.g., bottom or water column) have widely overlapping diets, but separable from those insect predators with different body sizes or habitat use, although individuals can shift between food webs according to food availability (Klecka & Boukal, 2012). Spatial connectivity between temporary ponds (Tornero et al., 2016) or disconnected pools in temporary rivers facilitates dispersion and colonization of the macrofauna from one pond to another, determining their trophic structure (McIntosh et al., 2017).

Ontogeny

Ontogenetic diet shifts associated with individual growth in size are common across all major groups of predatory aquatic insects. They occur in *Notonecta* bugs and in larval odonates in both running and standing waters (see references in Klecka & Boukal, 2012). Odonate larvae begin to feed on rotifers and even protozoans after hatching, later switching to larger benthic prey. Some predatory Plecoptera and Trichoptera are herbivorous during the first larval stages. In a seasonal stream, López-Rodriguez et al. (2018) described that the stonefly *Isoperla morenica* reduces its intake of animal matter with size, whereas the dragonfly *Onychogomphus forcipatus* increases it. Plasticity in food habits could drive the success in persisting in or colonizing ecosystems. If animals must adapt to the fluctuations in their habitat to complete their life cycle (Townsend et al., 1998), they may thus adapt their food preferences during their developmental period according to the quantity and quality of the food resources available.

Donor aquatic ecosystems to terrestrial ecosystems

Freshwater and terrestrial food webs are connected at multiple trophic levels by cross-habitat flows of organic matter and organisms (Larsen et al., 2016). The importance of allochthonous terrestrial organic matter to aquatic food webs was recognized decades ago with more recent studies describing reciprocal aquatic-terrestrial subsidies that strongly affect consumers and food webs in both habitats.

Aquatic ecosystems may be important sources of energy and nutrients for adjacent terrestrial systems. A large part of aquatic productivity is exported to land consumers (spiders, ground beetles, lizards, birds, and bats) via insect emergence or even larvae (some terrestrial birds, such as *Cinclus cinclus*, can dive in pursuit of aquatic insects). Most of the data about the importance of this process come from streams in which more than 60% of the emergent insect biomass are dipteran adults, followed by adults from Ephemeroptera, Plecoptera, Trichoptera, and Odonata (Popova et al., 2017).

Aquatic algae can also be consumed directly by specialist terrestrial algivores (e.g., grasshoppers), in which nearly 90% of the $\delta^{13}C$ signature indicates that the carbon derives from epilithic algae rather than terrestrial vegetation (Power et al., 2013). This aquatic source of energy is more available to terrestrial animals when water flow decreases in temporary rivers. The drying of lentic systems or the disconnected pools in temporary rivers can also accumulate aquatic insects in the shoreline, making them available for terrestrial predators. Aquatic insects may be a superior food quality source for terrestrial consumers, not only as a source of C, but because they contain high concentrations of essential polyunsaturated fatty acids obtained from aquatic food webs (Martin-Creuzburg et al., 2017).

Disturbance effects on aquatic insects

Aquatic insects have been acknowledged as quality indicators of freshwater systems that have been affected acute or chronically by a disturbance (Armitage et al., 1983; European Commission, 2000; Verdonschot & Nijboer, 2004). Their ubiquitous presence, high abundance, intermediate lifespan, and position in the trophic chain (i.e., between primary producers and secondary consumers) has made aquatic insects a more suitable candidate to indicate long-term environmental conditions and quality than for example chemical or physical properties of aquatic systems (see the section on "Biological assessment of water quality"). Consequently, literature focusing on aquatic insects and disturbances is abundant even within a restricted region such as the Mediterranean Basin. Thus, in this section we have focused on four main disturbances for aquatic systems: changes in water quality, habitat loss and fragmentation, climate change consequences, and invasive species (Fig. 8.5) that exemplify how aquatic insects group respond to a change in habitat conditions.

A disturbance could be defined as "(. . .) any relatively discrete event in the time that disrupts ecosystem, community or population structure and changes resources, substrate availability or the physical environment" (White & Pickett, 1985). This definition gives a wide window in which to include what is a disturbance, ranging from the direct change in environmental conditions due to human intervention (e.g., heavy metals, salinization, eutrophication), the landscape structure alteration due to the direct elimination of habitats (e.g., habitat loss, damming, habitat fragmentation), the change in the frequency or harshness of already natural processes (e.g., drought, floods, wildfires) or the arrival of another species (e.g., fish introduction or insect invasive species). All these disturbances are directly or indirectly human-mediated and therefore, caused by human activity either through specific punctual activities such as pesticides

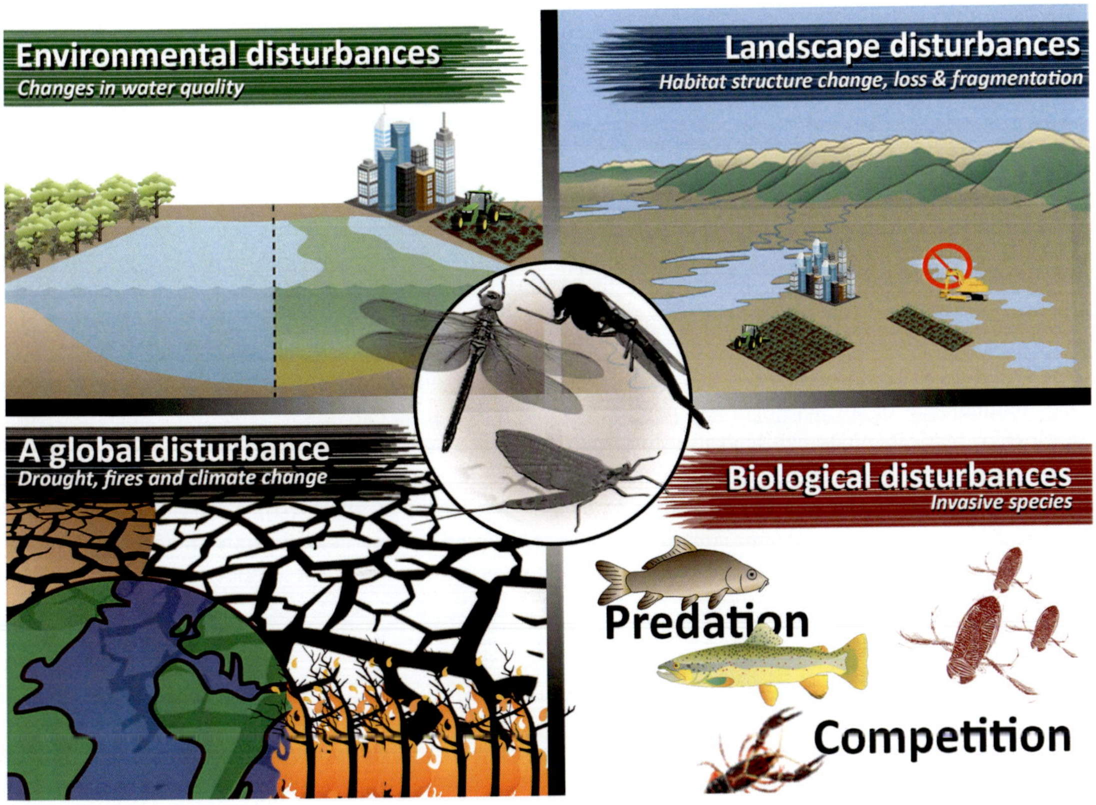

FIGURE 8.5 Schematic illustration of the four main disturbances affecting aquatic insects discussed in this chapter. Central invertebrate images (Odonata, Ephemeroptera, and Diptera) have been ceded by Jesús Ortiz and David Cunillera-Montcusí. Smaller drawings and symbols have been downloaded from IAN Image Library. Finally, small corixids, bottom-right section, are original drawings ceded by Rita Montcusí Rovira.

usage (Topaz et al., 2020) or mining activities (Solà et al., 2004) or by more diffuse and bigger-scale activities such as climate change (Alba-Tercedor et al., 2017) or land-use changes (Bruno et al., 2014). Overall, all these impacts are affecting insect abundances and diversity across the globe in what has been said as a death by a thousand cuts (Wagner et al., 2021). In contrast, some reports have detected an increase in aquatic insects' diversity but not in their abundance across Europe (Pilotto et al., 2020). Nevertheless, we still need better resolution and datasets that can provide a clearer view in this extremely complex problematic (Wagner et al., 2021).

Environmental disturbances: changes in water quality

In general, changes in water characteristics lead to a change in species presence or abundance as a consequence of species fitness limitations or direct toxicity. Nutrient increases primary production, and the consequent eutrophication can lead to dissolved oxygen depletion to which only few insect species can survive (e.g., Syrphidae dipterans which have a respiratory conduct to breath aerial oxygen). Therefore, lowland aquatic systems, where agricultural activity and human settlements (e.g., wastewaters) are more frequent, present depleted and less singular insect communities due to eutrophication (e.g., Bazzanti et al., 2017; Bruno et al., 2014; Fenoy & Casas, 2015; Mor et al., 2019; Pilotto et al., 2015; Sánchez-Morales et al., 2018), where only high tolerant taxa (e.g., Syrphidae or Ephydridae) can survive. Furthermore, pesticides usage, directly targeted to eliminate insects, can reach high toxicity values in flooding periods causing high stress situations in aquatic habitats (Topaz et al., 2020). However, increase in nutrients, as well as entrance of dissolved salts, can lead to changes on system salinity, which is one of the main aquatic insect community determinants (Brucet et al., 2012; Damásio et al., 2011; Pallarés et al., 2016, 2017; Picazo et al., 2012; Touaylia et al., 2013). In fact, even salinity decrease due to dilution can impact aquatic insects and loss of their singularity (Boix et al., 2008; Cañedo-Argüelles & Rieradevall, 2010; Gascón et al., 2009; Gutiérrez-Cánovas et al., 2012). Salinization impacts on insects imply physiological stress due to osmoregulation, which require previous adaptations in order to tolerate them (Cañedo-Argüelles, 2020; Pallarés et al., 2016, 2017). Finally, while eutrophication and salinization might be understood as natural processes that are fostered by human activities, heavy metal inputs in aquatic systems are less frequent and their direct toxicity and consequences can erode insect diversity from aquatic habitats and remain with toxic values for years if no intervention is carried (Damásio et al., 2011; Solà et al., 2004). In particular for temporary rivers, the impacts of all these environmental disturbances are exacerbated during the drying period, where pollutants become more concentrated in low or no flows (i.e., disconnected pools; Bonada et al., 2020).

Landscape disturbances: habitat structure change, loss and fragmentation

At the regional level, changes in habitat structure (alteration of river beds, damming, canopy loss) can cause changes in habitat characteristics (e.g., available resources, habitat conditions, populations disconnection) which can lead to a loss of insect richness and impoverishment of communities (Lobera et al., 2017). Such alterations of system structure also influence hydrological connectivity which can affect dispersal dynamics between communities, and consequently generate a loss of biodiversity due to a decrease in new organism arrivals (Gallardo et al., 2014). The impact of such alterations on the loss of biodiversity can become even greater than the impact of eutrophication (Bruno et al., 2014; Pilotto et al., 2015). In lotic systems connectivity relevance is assumed, but in lentic systems it is rarely acknowledged although the maintenance of an heterogenous pondscape is key for insect diversity (Jooste et al., 2020) and, consequently, the loss of lentic habitats can lead to a loss of invertebrate species diversity (Downing, 2010; Horváth et al., 2019; Pekel et al., 2016). In temporary rivers in the Mediterranean Basin, however, these landscape disturbances are also considered natural disturbances when linked to natural drying (Soria et al., 2020). As drying in the Mediterranean Basin has been a long-term phenomenon, aquatic insects are adapted to fragmentation and thus can be more resistant and resilient to human-driven landscape disturbances, a phenomenon known as co-tolerance (Soria et al., 2020).

Global disturbances: drought, fires, and climate change

On a wider global scale, climatic alterations can foster the previously commented consequences both at the local and the regional scales. The case of the Mediterranean Basin is especially threatening as the situation and the pressures under which the aquatic communities are subject have increased in the recent years compared to central European regions (Kalkman et al., 2008). In the Mediterranean Basin an increase in temperature and a decrease in water availability during the year (i.e., increase in extreme events) are expected (Kovats et al., 2014). Drought frequency and duration are strong determinants of insect communities (Boix et al., 2016; Hershkovitz & Gasith, 2013; Williams, 2006). An increase in drought harshness will probably led to the loss of some currently temporary habitats and the change from permanent to temporary classification of many of the Mediterranean waterbodies. One of the consequences of drought will be the increase in salinity values in water bodies which, as indicated, will constrain insect communities to those only adapted to them (Arribas et al., 2015; Boix et al., 2008; Brucet et al., 2012; Gascón et al., 2009; Picazo et al., 2012). Insect communities will therefore be forced to cope with such harsher conditions (Pace et al., 2013), with survival of only those specialized species (e.g., Coleptera species adapted to high salinity) that are better adapted to such new conditions (Pallarés et al., 2016, 2017). Water scarcity can also foster dissolved O_2 depletion, which can negatively affect sensitive species, thereby increasing species drift (Calapez et al., 2017). Furthermore, in lotic systems, stronger drought periods added to more frequent flood events can also decrease species diversity (López-Rodríguez et al., 2012). Overall, rises in temperature and water scarcity should increase Mediterranean aquatic systems harshness, leaving only the few species adapted to those conditions (Kefford et al., 2016). One example would be coleopterans adapted to high salinities (Pallarés et al., 2016), which would be expected to find more habitat available, unless other negative, human-induced degradations of this habitat are present (Arribas et al., 2015). However, some species, from more wet environments (e.g., high-altitude regions), which are endemic to the Mediterranean Basin might be completely lost due to the disappearance of their natural habitat (Guareschi et al., 2018; Kroll et al., 2017; Múrria et al., 2020). Therefore, future predictions based on climate change impacts in Mediterranean countries clearly forecast a generalized loss of insect species richness (Alba-Tercedor et al., 2017).

While climatic alterations due to global change can have a direct effect (i.e., average temperature increase and water scarcity), these conditions can also foster more punctual disturbances such as wildfires (Pausas & Fernández-Muñoz, 2012; Turco et al., 2018). As an example of catastrophic disturbances induced by global change, wildfires can impact aquatic systems indirectly while also fostering previously described impacts such as nutrient entrance, eutrophication, and structural changes to the habitat (McCullough et al., 2019; Minshall et al., 1989). In this sense, Mediterranean aquatic systems rapidly recover after these disturbances although some species can be impacted in the short term (Cunillera-Montcusí et al., 2019; Verkaik et al., 2013). However, at the same time, drought plays a key role in determining the normal community structure as well as their responses to wildfires (Cunillera-Montcusí, Arim, et al., 2020; Verkaik et al., 2015). Although such communities might be considered as "resilient," future scenarios might compromise this response (Cunillera-Montcusí et al., 2020).

Biological disturbances: invasive species

All the previously described disturbances could have short-term impacts or could produce more extensive impacts on environmental conditions and thus affect habitat structure for resident biota. However, changes in the biological interactions through the introduction of a predator (i.e., fish), another competitor (i.e., invasive insect or other invertebrate species), or a primary producer (i.e., algae) can also be considered as a disturbance and generate an impact on insect communities (e.g., Céspedes, Coccia, et al., 2019, Ferreras-Romero et al., 2016, Ladrera et al., 2018). Fish predation is another main determinant of aquatic systems insect diversity, especially for traditionally fishless habitats (e.g., temporary ponds; Boix et al., 2016), but even in aquatic systems with fish presence, introductions of fish invasive species (e.g., *Gambusia spp.*) can disrupt and change insect community characteristics (Anton-Pardo & Armengol, 2014) or decrease specific groups diversity (Odonata; Ferreras-Romero et al., 2016). Moreover, although less studied for insects, the arrival of introduced competitors may impact native species, such as the introduction of *Trichocorixa verticalis* (Hemiptera, Corixidae), which are competitively stronger in saline waters than native corixid species (e.g., *Sigara* spp.) (see the section on "Alien Aquatic Hexapods").

Use of aquatic insects in biological assessment of water quality

The pioneer studies on the relationship between environmental variables and the presence of specific species date back from the beginning of the 20th century (Elton, 1927; Grinnell, 1917; Hutchinson, 1957). This knowledge became one of the central core aspects of modern ecology and has been well stablished around the ecological niche concept, which is in continuous evolution and re-analysis (Chase, 2011; Leibold, 1995; Pocheville, 2015). This concept is also the fundamental background of several applied issues, such as the assessment of the ecological integrity of ecosystems based on changes on community composition in relation to disturbances. The need for the development of tools for the assessment of aquatic ecosystems favored the research on both the identification of most sensitive aquatic organisms (i.e., bioindicators) and the study of their relationship to disturbance (e.g., Liebmann, 1962; Margalef, 1955;

Woodiwiss, 1964). These tools were elaborated and validated during the second half of 20th century in many countries worldwide, including those in the Mediterranean Basin (Dallas, 2013), and many of them focused on the response of aquatic insects as part of the larger macroinvertebrate community (Birk et al., 2012; Rosenberg & Resh, 1993). In fact, the ecology and habitat preferences of aquatic insects have been intensively studied, in part because of their widespread use as bioindicators (Dijkstra et al., 2014).

Bioindicators are organisms that respond to an environmental stress or disturbance and, therefore, inform about the quality of the environment (Adams, 2002; Li et al., 2010). Diatoms, macrophytes, macroinvertebrates, and fish are the most commonly used bioindicators in aquatic ecosystems (e.g., Bonada, Prat, et al., 2006; Menetrey et al., 2011; Wright, 2000). In the case of macroinvertebrates, they have been used to evaluate pollution, heavy metals, nutrient enrichment, habitat loss, overexplotation, hydromorphological degradation, hydrology alteration, acidification, and general stressors (Li et al., 2010 and references therein). Moreover, they have been also used to assess ecological integrity (Birk et al., 2012; Rosenberg & Resh, 1993) or ecological risk (King & Richardson, 2003), and as a management tool useful in environment restoration projects (Kenney et al., 2009). Comparisons of the results obtained among bioindicators based in different biotic groups (Birk et al., 2012; Hering et al., 2006; Marzin et al., 2012) put in evidence that they have different sensitivities to human disturbances. For example, diatoms, macrophytes, and macroinvertebrates respond more quickly to nutrient enrichment or eutrophication than fish, whereas macroinvertebrates and fish are more sensitive to hydrological and morphological alteration.

Aquatic insects constitute the main bulk of macroinvertebrates used in bioassessment. Rosenberg & Resh (1993) and Bonada, Prat, et al. (2006), summarized the advantages of using macroinvertebrates for bioassessment purposes, all applicable to aquatic insects. Basically, they stated that macroinvertebrates are ubiquitous and sedentary, respond to a wide spectrum of environmental responses, can be used as "sentinel" organisms in experimental conditions, sampling is simple and inexpensive, and the taxonomy of many groups as well as their response to different types of pollution is relatively well-known (at least at family and genera level). Despite these advantages, Rosenberg and Resh (1993) also listed several methodological difficulties in using macroinvertebrates in bioassessment: invertebrates apparently do not respond to all impacts or they respond to other factors than water quality; quantitative data require exhaustive sampling and time-consuming sample processing; seasonal variation in abundance and distribution, specially of insects, could difficult comparisons among different seasons; and in lotic habitats the presence of some invertebrate taxa could be explained by the drift from other habitats.

A wide range of approaches have been developed in bioassessment using aquatic insects (Bonada, Prat, et al., 2006; Li et al., 2010; Beauger & Lair, 2014) and other macroinvertebrates. Four types of bioassessment methods or procedures based on macroinvertebrates have been identified according to Beauger and Lair (2014): two of them derived from the pioneering proposals (Trent Biotic Index (Woodiwiss, 1964) and Chandler's Biotic Index (Chandler, 1970), respectively), predictive methods and rapid bioassessment methods. The first method assigns a water quality category according to a double entry table consisting of (1) the more sensible taxa to pollution (identifying several assemblage groups from very sensible to high tolerant) and (2) macroinvertebrate richness (often family level). In contrast, the second method assigns a coefficient to each taxa. These coefficients indicate the level of pollution sensibility of each taxon (often coefficient take values from 1 to 10, being 1 the most tolerant and 10 the most sensitive). Then, the water quality category is obtained taking into account the coefficients of all taxa present in the sample. Another differences could also exist for example in the field work: in first type sampling is habitat-specific, while in the second one sampling is conducted in all habitats of the sampled point. The third type, predictive methods predict site-specific fauna composition and compare the observed and expected fauna (Dolédec et al., 1999; Wright, 2000). Finally, the fourth approach consists of rapid bioassessment methods. To clarify what a rapid bioassessment is, Fennessy et al. (2004) proposed the following four criteria that must charactertize the method: (1) it must measure condition/integrity; (2) it should be rapid (i.e., no more than two people a half day in the field and no more than a half day of office preparation and data analysis); (3) include an on-site assessment; and (4) results must be verifiable. Rapid bioassessment is a very useful tool for managers needing easily understandable results in a short time period (Lenat & Barbour, 1994; Resh & Jackson, 1993). Such methods using aquatic insects have been developed for streams and wetlands (Barbour et al., 1999; Fennessy et al., 2004; Growns et al., 1997).

Metrics used in Mediterranean bioassessment indices include taxa richness (usually at family level), diversity indices (included those based not only on abundance distribution but also on size classes), or metrics based on pollution tolerance. In EU Mediterranean countries following the implementation of the Water Framework Directive (Directive 2000/60/EC), these metrics are assessed in relation to a particular typology of water body and its reference condition (the value of the metric in absence of disturbances). This can be done by comparing the real values of the metrics with the reference condition of the typology, or by modeling the expected community composition in reference conditions, and thus comparing the real metric value with the expected one (i.e., using a multivariate approach). While the most common approaches used in Mediterranean countries consider the first option and combine unimetric (e.g., BMWP', EQAT, or HES; Alba-Tercedor & Sánchez-Ortega, 1988; Artemiadou & Lazaridou, 2005; Cañedo-Argüelles et al., 2012) or multimetric indices (e.g., AQEM, QAELS, or IMMi-T; Buffagni et al., 2004; Munné & Prat, 2009; Solimini et al., 2008), some attempts in predictive models have been also developed (i.e., MEDPACS; Poquet et al., 2009), or are being considered, even using the metacommunity framework (Cid et al., 2020). In addition to these bioassessment methods based on taxonomic composition, recent methods that consider functional approaches are becoming useful and are consided more reliable under particular conditions. For example, Soria et al. (2020) found that metrics based on biological traits responded better in temporary rivers than traditional taxonomic metrics. This is because they appear to better disentangle the effect of increasing human disturbance from the increasing flow intermittence. Despite their potential good performance in Mediterranean rivers and the currently extensive knowledge of the traits of aquatic insects (and of macroinvertebrates in general) in the EU Mediterranean countries, their use is not yet routine, most likely because of the difficulty of comparing their output to those derived from other metrics. Another challenge for the use of aquatic insects in Mediterranean countries is related to taxonomic issues. Despite most methods based on family level data, there are many taxonomic gaps and some regions have been more explored than others. For example, the family of aquatic insect Nevrorthidae (Neuroptera) was cited for the first time in the Iberian Peninsula in 2012, and it is not included in the biotic indices currently used in the area (Gavira et al., 2012). Recent advances in molecular techniques may overcome some of these taxonomic limitations, although this method has not been adopted often yet for macroinvertebrates (Blackman et al., 2019).

The long tradition of using aquatic insects in bioassessment varies considerably among countries and environments (i.e., lotic and lentic). For instance, central and north Europe have a longer tradition than in Mediterranean regions. For this reason, some existing indexes in Mediterranean countries represent a transposition of existing tools elaborated in other countries. In some cases, the use of indexes needs to be adapted, since differences in fauna and community structure exist among the country where the bioindicator was created and the country in which they are to be applied. One good example is the index IBMWP, created from BMWP British index (Hellawell, 1978) and first adapted only by means of expert criteria (Alba-Tercedor & Sánchez-Ortega, 1988) and later validated in a research project (i.e., GUADALMED; Prat & Bonada, 2002). However, the bioassessment tools created in other countries have to be used with caution, since they are usually region-specific. For example, even for the IBWMP, the scores of particular Trichoptera families are not still well adapted, and they did not consider highly tolerant species endemic of the Iberian Peninsula (Bonada et al., 2004). A similar situation currently exists between northern and southern Mediterranean countries and some European tools, and even USA methods are used in other regions such as Turkey and North African countries (e.g., Girgin, 2010; Karrouch & Chahlaoui, 2009; Sellam et al., 2016; Souilmi et al., 2017). In addition to concerns on the application of methods derived from other countries, a bigger effort is needed in the research on the ecology and impact-fauna responses in North African countries in order to obtain the essential knowledge to adapt adequately existing methodologies or to create new bioassessment tools (e.g., Benzina et al., 2018; Djitli et al., in press; Haggag et al., 2018; Korbaa et al., 2018; Souilmi et al., 2019). Similarly, the use of aquatic insects in bioassessments in lotic environments (streams and rivers) has a longer tradition than in lentic ones (wetlands, ponds, and lakes). In this sense, it is important to add that aquatic insects have a clear prevalence in the biological indicators used in lotic environments, while macroinvertebrate groups other than aquatic insects (i.e., molluscs, oligochaetes, or crustaceans; Boix et al., 2005; Mazzella et al., 2009; Mouthon, 1993) have more importance in lentic indexes.

Alien aquatic Hexapods

The impacts of biological invasions are considered a main component of global change and they are a leading cause of animal extinctions (Clavero & García-Berthou, 2006; Early et al., 2016; Simberloff et al., 2013). However, the number of aquatic alien insect species in the continental Mediterranean waters, and in world inland waters, is surprisingly low. Besides, as far as we know, no aquatic alien Collembola has been reported. Although in other world regions (i.e., South Africa and Australia) the number of exotic species could achieve the 20% of the Collembola fauna, few of these alien species are aquatic or inhabitant of wet habitats (Greenslade, 2018; Janion-Scheepers et al., 2015).

Several points make to this low number of alien insects astounding: (1) the high biodiversity of this group, being the richness group in continental waters (Balian et al., 2008); (2) the high number of aquatic alien species in other faunal groups (e.g., invertebrates such as crustaceans or vertebrates as fish) reported in some Mediterranean countries (Gherardi et al., 2008; Muñoz-Mas & García-Berthou, 2020) or in some waterbodies (Ricciardi, 2015); (3) the high proportion of terrestrial alien insects known in Mediterranean countries (e.g., in Italy insects represent between 80% and 90% of the introduced species in terrestrial habitats; Zapparoli, 2006), in fact insects are one of the groups with the world highest number of invasive species (Kenis et al., 2009); (4) compared to terrestrial systems, estuarine and inland waters are highly vulnerable to either unintentional or deliberate introductions of taxa and to their spread consequences (Dudgeon et al., 2006; Gherardi, 2007); and (5) freshwater habitats seemed to be more susceptible to invasion—and experienced more impacts—than coastal habitats (Tricarico et al., 2016). Therefore, although inland waters are among the most invaded ecosystems on our planet, the number of aquatic alien insects remains very low. Fenoglio et al. (2016) proposed explanations and summarized some biological, ecological, and anthropogenic factors that could explain, at least partially, this fact. The explanations identify some characteristics of aquatic insects that dimmish their potential invasive capacity: (1) economic interest in moving aquatic insects is currently limited; (2) associations between aquatic insects and host plants are extremely rare; (3) aquatic insects usually lack adaptations for overland or maritime transport; (4) aquatic insects seem to have less diverse reproductive strategies than terrestrial ones; (5) aquatic insects usually have an aquatic and a terrestrial stage; (6) aquatic insects usually have an aquatic and a terrestrial stage; (7) many aquatic insects live in running water environments. In fact, the few aquatic alien insects are exemptions of these general characteristics. For example, the hemipteran *T. verticalis*, and the dipterans *Aedes albopictus* and *Ae. japonicus* have biological adaptations that allow them to disperse by means of overland and marine transport. The water boatmen *Trichocorixa* tolerates waters with high conductivity and even marine waters (Gunter & Christmas, 1959; Hutchinson, 1931), while the *Aedes* mosquitoes can survive in small amounts of water, and have high reproductive rates and desiccation-resistant eggs that allow them spread worldwide via the trade of used tyres (Benedict et al., 2007; Kaufman & Fonseca, 2013). Within continental waters, temporary ones showed less susceptibility to the establishment of nonindigenous species (Naselli-Flores & Marrone, 2019); however, several of the alien insects listed below have been located in temporary waters (i.e., *T. verticalis*, *Aedes* spp., and *Stenopelmus rufinasus*).

The lack of knowledge of the past and present distribution of aquatic insects makes it difficult to identify exotic species and, therefore, could also explain the low number of the reported alien aquatic insects. Thus, some cosmopolitan aquatic insect species could be undocumented introductions due to the difficulties to have verifiable historical data which allow us to confirm as exotic one species (de Moor, 1992). Even some species could have described in a continent, but they are native from a different one, such as the case of gastropod *Haitia acuta* (Draparnaud) which has been described in France and it inhabits nowadays all types of freshwaters in Europe, but later it was identified as an American species (García-Berthou et al., 2007 and references therein). A similar case for aquatic insects could be *Culicoides paolae* (see below). In other situations, one species can be identified as alien because it does not have previously recorded (groups poor studied or changes in environmental conditions). One example are species that recovery their native distribution ranges after the environmental constraint that had initially caused their disappearance is reversed, but historical records documenting its disappearance do not exist (this could be the case of *Ametropus fragilis* in Germany; Berger & Rothe, 1999).

The low number of alien insect species does not imply that biological invasion is not a relevant subject for aquatic insects, since some of these few species can affect human health (i.e., some Culicidae and Ceratopogonidae). Moreover, some of the most dangerous invasive species have been spread by humans as biological control of insects (e.g., *Gambusia* spp. as control of mosquitoes), and aquatic and semiaquatic insects are used to control invasive species including alien insects (Hoddle, 2004; Shaw et al., 2018). Besides, insects could be affected by invasive species in several ways. Some impacts of invasive species on aquatic insects include (1) substitution of autochthonous plants by alien ones, which are food resource for insects in riparian habitats, imply changes in nutritional value of food that in turn affect growth rates and insect production (e.g., Canhoto & Graça, 1995; Going & Dudley, 2008); (2) fish introductions is one of the main threats of insects conservation (Polhemus, 1993), since it is long known and well stablished that some introductions imply a high predation impact on insects (e.g., Macan, 1965; Merkley et al., 2015; Pope et al., 2009); (3) similarly, other invasive species such as birds can increase predation pressure on insects (e.g., Marion, 2013); and (4) some invasive ecosystem engineer species, including many species of crayfish, significantly reduce the insect biodiversity and alter aquatic insect composition (e.g., Freeland-Riggert et al., 2016; Geiger et al., 2005; Rodríguez et al., 2005).

In this section, we will depict the few known cases of alien aquatic insects reported in Mediterranean inland waters and we also included doubtful cases for different reasons:

(1) *Ametropus fragilis* Albarda, 1878 (Ephemeroptera: Ametropodidae)

This species is known in Italy and Croatia (Ćuk et al., 2015; Turin et al., 1997). Its presence in Italy was considered an introduction because of the known distribution of Ametropodidae in Europe (and its absence in some adjacent countries with a well-known Ephemeroptera fauna, such as France, Austria, Switzerland, and Slovenia). It seems that it was the result of an accidental introduction related to fish restocking from eastern Europe (Turin et al., 1997). In contrast, when this species was located in Germany in 1998, the researchers considered it as a native species (taking into account its global distribution); and the absence of previous records was explained because it had been extinct, but they did not have knowledge of historic proof that supported this hypothesis (Berger & Rothe, 1999). This species has been extinct in several European countries, such as The Netherlands and Czech Republic, probably as a result of human activity, linked to fish restocking in rivers (Ćuk et al., 2015). The finding in Croatia in 2014 was located near the considered native area (near Hungary border) and far away of the Mediterranean coast (Ćuk et al., 2015).

(2) *Electrogena zebrata* (Hagen, 1864) (Ephemeroptera: Heptageniidae)

The distribution of this species constitutes a doubt of its origin. Although it is considered a Sardo-Corsican endemic, the only similar species is found in Lebanon and Israel. The morphology between them is very similar, so distinction of the two is impossible. This disjunct distribution cannot be explained by geological events, and some authors formulated the hypothesis that its presence in Sardinia was an involuntary introduction by the Phoenicians (Belfiore, 2006).

(3) *Trichocorixa verticalis verticalis* (Fieber, 1851) (Hemiptera: Corixidae)

This species, native to the Atlantic coast of America, is distributed throughout the Atlantic coast from Labrador to the north of Mexico, and the Caribbean islands (Jansson, 2002; Kment, 2006; Sailer, 1948). Its presence in the south western Iberian Peninsula has been documented from the last decade of 20th century (Sala & Boix, 2005), but it was in the first years of 21st century that its presence in this region was regularly documented and monitored in some areas as Doñana Natural Park where reproduction has been observed (Günther, 2004; Kment, 2006; Millán et al., 2005; Rodríguez-Pérez et al., 2009; Sala & Boix, 2005; Tornero et al., 2014). Its presence was also reported in Morocco (L'Mohdi et al., 2010), where it has significantly increased its distribution range in 2010s decade (Fouzi et al., 2020). Outside of its native range, it has also reported from New Caledonia (Jansson, 1982) and several sites in the KwaZulu-Natal region in South Africa (Nzimane River, Umhlatuze River, and Charter's Creek; Jansson & Reavell, 1999). Predictions of the potential new zones of invasion based on both climatic data and thermal physiological data (Guareschi et al., 2013) indicate that *T. v. verticalis* may expand well beyond its current distribution and find new habitable conditions in temperate areas, with an emphasis on coastal areas (including Europe, North and South America, Asia (mainly Arabian Peninsula, India and Myanmar), Australia, and New Zealand).

Fortunately, this species has attracted the attention of Mediterranean researchers and a significant number of studies on its biology in the invaded area and its interaction with native species have been done. It is a well-known halobiont species and usually inhabits brackish and saline waterbodies, even occurring in the open sea (Gunter & Christmas, 1959; Hutchinson, 1931; Sailer, 1948). Therefore, it is not surprising that in the invaded area it dominates permanent saline waters (Coccia et al., 2013). However, when its ecophysiology was compared with native Corixidae, its wide saline tolerance was lower than the tolerance range of the native species *Sigara selecta* (Carbonell et al., 2016). In contrast, *T. v. verticalis* took advantage of other environmental stress such as temperature and shows higher thermal plasticity than native *Sigara* spp. (Carbonell et al., 2016; Coccia et al., 2013). Comparison of biological traits was also done between *T. v. verticalis* and the native *Sigara* spp. (Carbonell et al., 2016; Céspedes, Coccia, et al., 2019; Coccia et al., 2016). The main conclusions were as follows: (1) according to wing morphometry, *T. v. verticalis* could be a stronger flier; (2) it shows higher fecundity, especially in saline waters; (3) it performs a continuous reproduction through the year, while native spp. mainly reproduce from spring to autumn; and (4) it changes diet preferences relying more on herbivory under the presence of native competitors. All these reasons explain that this species is a successful invader in saline waters, but they do not justify why it fails to invade freshwaters, although in experimental conditions it can reproduce and tolerate low salinities, even in the presence of native species (Carbonell et al., 2020). One explanation could be the higher infection levels by mite larvae in *T. v. verticalis* than in native corixids, and the fact that these infections are especially abundant in low saline waters (Céspedes, Stoks, et al., 2019; Sánchez et al., 2015). A complementary explanation could be predation, since differential predation by Odonata has been observed between *T. v. verticalis* and native corixids (Coccia et al., 2014), the latter being less predated (Odonata larvae achieved higher densities in low saline waters). Moreover, *T. v. verticalis* can also affect the community structure by means of predation over keystone species of saline wetlands, since its predation over *Artemia partenogenetica* implies a trophic cascade resulting a higher phytoplankton increase (Céspedes et al., 2017).

(4) *Stenopelmus rufinasus* Gyllenhal, 1836 (Coleoptera: Erirhinidae)

This aquatic weevil is native to North America (Southern and Western Unites States) (Richerson & Grigarick, 1967). It has been first recorded in France, a Mediterranean country, in 1898 (Bedel, 1901), from where it reached several European countries (it was recorded in United Kingdom, and the Netherlands, Ireland, Germany, Belgium, and Ukraine) including some Mediterranean ones as Portugal, Spain, and Italy (Carrapiço et al., 2011; Dana & Viva, 2006; Florencio et al., 2015; Gherardi et al., 2008). The few observations in the Iberian Peninsula indicate that this species is present in southern, northeastern, and western locations (Fernández et al., 2005; Mor et al., 2010; Tornero et al., 2014).

The current distribution of this species is closely related to the expansion of the aquatic fern *Azolla filiculoides*, which constitutes its main food source. *Azolla filiculoides* is native to the Americas but has become naturalized throughout the world by a variety of mechanisms, with humans as its main dispersal factor (Lumpkin & Plucknet, 1980). One of the strategies to mitigate the negative impacts of this aquatic fern is the use of *S. rufinasus* as a biocontrol agent in some countries as South Africa, but its present in Mediterranean Basin seems to be unintentional (Shaw et al., 2018).

(5) *Lissorhoptrus oryzophilus* Kuschel, 1951 (Coleoptera: Erirhinidae)

L. oryzophilus is a semi aquatic beetle native to North America (Canada, United States, and Mexico; Kuschel, 1951). In 1976, it was accidentally introduced into Japan (Tsuzuki and Isogawa, 1976) and from there the distribution range of this pest has expanded to several Asian countries (i.e., China, North and South Korea, Taiwan and India). It is also present in Caribbean and South American countries (i.e., Cuba, Dominican Republic, Colombia, Suriname and Venezuela) (CABI, 2023). In Europe, its presence was detected more recently in four Mediterranean countries: Italy, France, Greece and Spain in 2004, 2014, 2016 and 2018, respectively (Caldara et al., 2004; Ferrand, 2017; Giantsis et al., 2017; Montauban et al., 2021). The rice water weevil *L. oryzophilus* is considered one of the most important rice pests globally and in North America it causes annual losses of up to 25% of rice crops (Aghaee and Godfrey, 2014).

(6) *Sternolophus solieri* Laporte, 1840 (Coleoptera: Hydrophilidae)

This species is native to Asia and Australia, but it is currently widespread in western, central, and eastern Africa, Madagascar, the Cape Verdes, and the Comoros. It has reached the Mediterranean Basin in Algeria, Egypt, Israel, and Greece (Bird et al., 2017; Nasserzadeh et al., 2017). In Italy its presence is also documented; however, there are no reports of well established (i.e., self-sustaining) populations (Gherardi et al., 2008).

(7) *Pelosoma lafertei* Mulsant, 1844 (Coleoptera: Sphaeridiidae)

Species of *Pelosoma* are native to southern and central America and are currently present in France and Italy (Bameul, 1992; Costa et al., 2017 and references therein). This species and the previous one, *Sternolophus solieri*, possibly arrived by ship transport and seem not to adapt well to the Mediterranean climate, having never been found again since their introduction (Rocchi, 2006). This species is a typical inhabitant of phytotelmata in its native range (de Oliveira et al., 2018).

(8) *Cercyon* (*Paracycreon*) *laminatus* Sharp, 1873 (Coleoptera: Sphaeridiidae)

Its native distribution includes China, Taiwan, Japan, and the Russian Far East (Jia et al., 2011). The introduction of this species has been reported in Europe and the Hawaiian Islands (Hansen, 1999), where it is well-established, and it was also reported in Chile and Australia (Fikáček, 2009). At the moment, its sole presence in the Mediterranean Basin is in Italy (Gherardi et al., 2008). Although this species is included in xenodiversity inventories of inland waters (Gherardi et al., 2008), it is described as a terrestrial species, which inhabits various kinds of decaying organic matter and excrements of various mammals (Fikáček, 2009).

(9) *Cryptopleurum subtile* Sharp, 1884 (Coleoptera: Sphaeridiidae)

The native range of this beetle is East Asia, and it was probably introduced in Europe, northern Asia (excluding China), North America, and southern Asia (Lobl & Smetana, 2015). In the Mediterranean Basin, it has been reported in France, Italy, and Albania (CABI, 2021; Gherardi et al., 2008). Similar to the previous species, *C. laminatus* it is considered terrestrial although it is also included in alien species inventories of inland waters (i.e., Gherardi et al., 2008) and it was observed in woody debris in river banks (Dalley, 2014).

(10–13) *Aedes* spp. (Diptera: Culicidae): *Aedes aegypti* (Linnaeus, 1762); *Aedes albopictus* (Skuse, 1894); *Aedes japonicus* (Theobald, 1901); *Aedes koreicus* (Edwards, 1917).

At least four nonnative *Aedes* species have been recorded in the Mediterranean Basin; however, their distribution is very different. For example, *Ae. albopictus* was first introduced in 1970s in Albania and was later found in many Mediterranean countries (i.e., Syria, Lebanon, Palestine, Israel, Jordan, Turkey, Greece, Albania, Montenegro, Bosnia-Herzegovina, Croatia, Slovenia, Italy, Malta, San Marino, Monaco, France, Algeria, Spain, and Portugal). In contrast, *Ae. japonicus* and *Ae. koreicus* have been only reported in few countries: *Ae. japonicus* in Bosnia-Herzegovina, Croatia, Slovenia, and Italy (it is also established in France and Spain, but not in the Mediterranean climate region); and *Ae. koreicus* in Italy. Finally, *Ae. aegypti* is present in Egypt (far from the coast) and Turkey (Black Sea coast), but not in the Mediterranean climate region. This species was present in Mediterraean basin (i.e., Turkey, Greece, Italy, Spain), but it was eradicated. However, sporadic records have been recorded in Turkey, Israel, and Italy (Aranda et al., 2006; Ballardini et al., 2019; Schaffner & Mathis, 2014; ECDC, 2021; Eritja et al., 2019).

Three of these *Aedes* species are native to Asia, being the origin of *Ae. albopictus* and *Ae. koreicus* (southeast Asia), while *Ae. japonicus* is distributed in China, Japan, Korea, south-eastern Russia, and Taiwan (Ballardini et al., 2019; Eritja et al., 2019). In contrast, *Ae. aegypti* has its origin in sub-Saharan Africa, where the ancestor of the domestic form of *Ae. aegypti* is native (the domestic form is nowadays distributed around the world; Powell & Tabachnick, 2013). The distribution of these species is largely driven by both human movement and the presence of suitable climate; marine transport of tires, plants, and other goods seems a common pathway for several of these species (Eritja et al., 2005; Kraemer et al., 2019; Schaffner et al., 2009). However, different dispersion patterns have been reported, and some species spread by long distance importations (i.e., *Ae. aegypti*), while others (i.e., *Ae. albopictus*) expanded mainly along the fringes of its distribution (Kraemer et al., 2019). Future scenarios predict an increase of the distribution of these species related to accelerating urbanization, connectivity, and climate change conditions (Kraemer et al., 2019; Schaffner & Mathis, 2014). All these species are extensively studied because they are dispersal vectors for several viral diseases. Thus, (1) *Ae. aegypti* is vector for viral diseases such as yellow fever, dengue fever, and chikungunya (due to an effective vaccine, yellow fever is of less concern worldwide, although cases still occur; Powell & Tabachnick, 2013); (2) *Ae. albopictus* is vector of major human diseases such as chikungunya, dengue, yellow fever, West Nile virus, and encephalitis, but it is also a vector of the dog heartworm *Dirofilaria* spp. (Aranda et al., 2006; Ballardini et al., 2019; Cancrini et al., 2007); (3) *Ae. japonicus* is vector for several mosquito-borne pathogens, to date only the West Nile virus is a concern based on field evidence (Eritja et al., 2019); and (4) *Ae. koreicus* specimens were found infected by encephalitis virus and the heartworm *Dirofilaria* spp. and it is vector for the chikungunya virus (Ballardini et al., 2019 and references therein).

(14) *Culicoides paolae* Boorman, 1996 (Diptera: Ceratopogonidae)

It is a *Culicoides* species found in southern Italy (Boorman et al., 1996) that was described in 1996 and named *C. paolae*. In the following years, this species was found in several countries of the Mediterraean basin including Malta, Croatia, Tunisia, France, Argelia, Greece, and Spain (Estrada et al., 2011). However, some doubts exist on the validity of this species, since several specialists considered that it is a synonym of *C. jamaicensis* Edwards, 1922 because morphology of both species is very similar, but unusual differences were noted (Meiswinkel et al., 2004). At the moment, it has been placed in the subgenus *Drymodesmyia* which includes more than 20 endemic species to the tropical regions of the New World and, where known, breed in vegetative materials including the decaying leaves and fruits of Central American cacti. These points are used to propose the exotic origin of this species and its introduction related to the introduction of American cacti in Europe (in a similar way of the introduction of the related species *C. loughnani* (Edwards, 1922) in Australia from America; Dyce, 1969). Even it has been proposed that *C. paolae* was brought to Europe with the introduction of *Opuntia ficus-indica* by Christopher Columbus (Meiswinkel et al., 2004). However, the fact that this species was not discovered until 500 years later and that efforts to rear *C. paolae* from *Opuntia* have this far failed make this hypothesis doubtful.

The distribution knowledge of this species in the Mediterranean Basin has been improved by national surveys (e.g., in Spain and Italy), since *Culicoides* species (i.e., *C. imicola*) are known vectors of bluetongue and deer virus (Estrada et al., 2011; Meiswinkel et al., 2004). *C. paolae* was formerly included in *Remmia* subgenus, and this fact implied that it was considered a potential new vector. However, taxonomic revisions consider that this species has differences in morphology and biology (it feeds preferentially on birds). In fact, *C. paolae* is considered a potential vector of avian haemosporidians (Veiga et al., 2018). Similarly to two coleopterans previously commented above, this species is perhaps linked to terrestrial habitats, since some *Culicoides* species select rotting parts of vegetation or even wet manure to breed over aquatic habitats, and the breeding habitats of this species are not yet well known (Meiswinkel et al., 2004).

Taxonomic keys to the Subphylum Crustacea, Class Hexapoda

How to use these keys

The designed keys in this chapter have been written to identify aquatic springtails (Collembola) and insects that can generally be found in all types of Mediterranean freshwater habitats: streams, rivers, wetlands, ponds, and lakes. It does not include communities found in brackish water nor marine estuarine habitats. The key to Entognatha (Collembola) is to species, while the key to Insecta is limited to orders knowing that more detailed keys are provided in later chapters of this book. Only the aquatic stages of insects living on or below the water surface are treated here knowing that many insects migrate to terrestrial habitats as adults. This key should only be used for adults and last stadium larvae (mature larvae). Juvenile or first stadium larvae should not be used because their morphological characteristics are not fully developed, and that can lead to mistaken identifications. Only a minute proportion of Mediterranean freshwater insects have been illustrated, and identified specimens may differ from those examples drawn and shown here. However, if the identified and illustrated specimens contrast considerably, then the identification should be checked in the later chapters of this book.

The keys do not cover insects living at the shoreline. However, because these usually are found in the field samples and often confused with aquatic stages, a list of them is also included at the end of the Insecta key.

Key to the subclass Entognatha (Collembola)

This key is a modification or compilation of others, such as those included in Jordana and Arbea (1989), Bretfeld (1999), Potapov (2001), Thibaud et al. (2004), Christiansen and Snider (2008), Jordana (2012), Bellinger et al. (1996–2020), with some own contributions.

1 Body elongate and segmented (Figs. 8.6, 8.8, and 8.35A) 2
1' Body subglobular, not or partially segmented (Figs. 8.7 and 8.9). 3
2(1) Th I with tergal chaetae (Fig. 8.6) .... Poduromorpha
2' Th I without tergal chaetae (Fig. 8.8) .... Entomobryomorpha
3(1') Antennae as long or longer than head (Fig. 8.7) .... Symphypleona
3' Antennae shorter than head (Fig. 8.9) .... Neelipleona

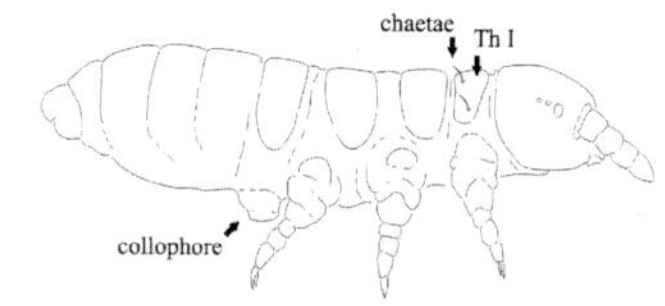

FIGURE 8.6 Habitus of a Poduromorpha showing chaetae on protoracic tergum (Th I).

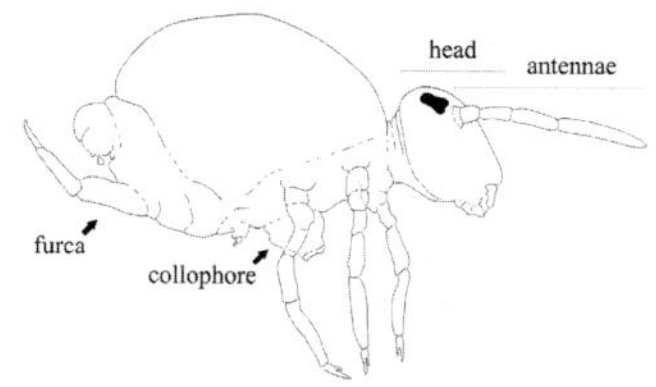

FIGURE 8.7 Habitus of a Symphypleona.

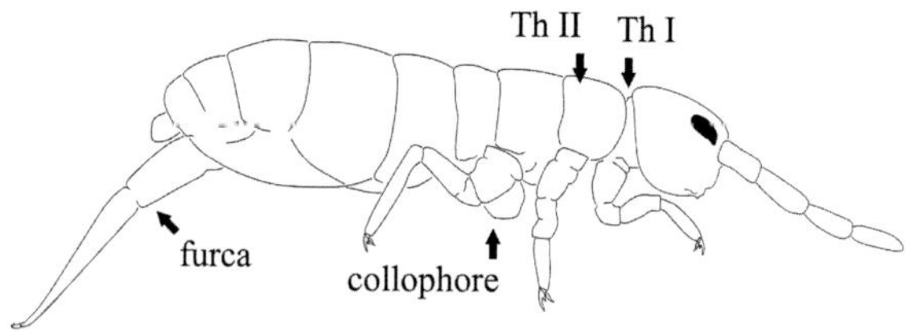

FIGURE 8.8 Habitus of a Isotomidae showing the absence of protoracic tergum.

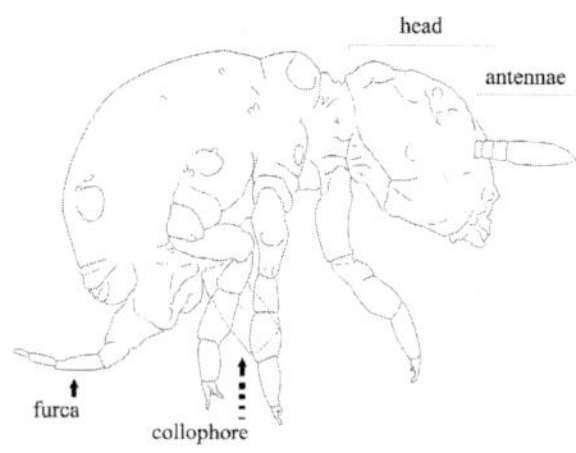

FIGURE 8.9 Habitus of a Neelipleona.

Poduromorpha

1 Mouthparts modified (Fig. 8.10) .... (Neanuridae) 2
1' Mouthparts with maxillae, and mandibles with a molar plate (Fig. 8.11) 7

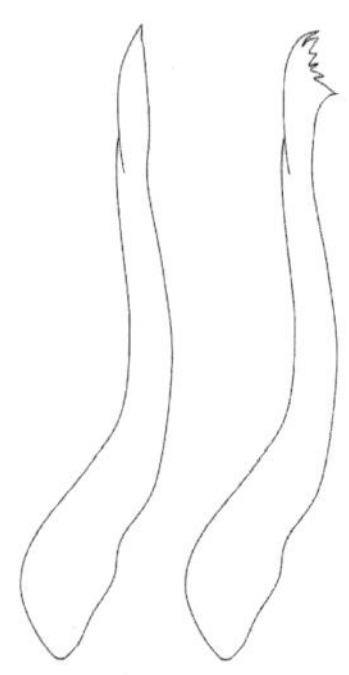

FIGURE 8.10 An example of modified mouthparts.

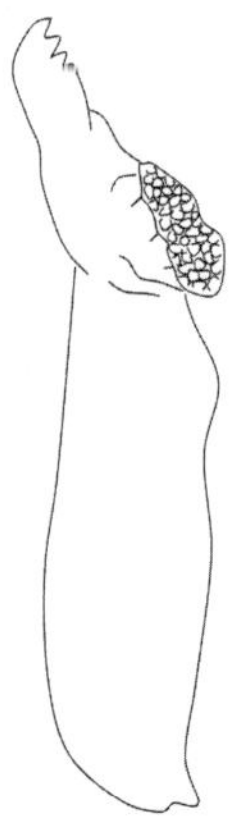

FIGURE 8.11 Maxylla with denticulate molar plate.

2(1) Furca absent; eight “s” chaetae on Ant IV (Fig. 8.12); PAO absent .... (Neanurinae) *Bilobella aurantiaca*
2’ Furca present or absent; fourth antennal segment different of eight “s” chaetae; PAO absent or present 3
3(2) Maxylla head triangular (Fig. 8.13) .... (Frieseinae) *Friesea acuminata*
3’ Maxylla head not triangular (Fig. 8.14) .... (Pseudachorutinae) 4

FIGURE 8.12 Tip of the antenna of *Bilobella aurantiaca* showing the eight “s” chaetae (Jordana et al., 1997). *Modified from Jordana, R., Arbea J. I., Simón C., & Luciáñez, M. J. (1997). Collembola Poduromorpha. In M. A. Ramos, et al. (Eds.),* Fauna Iberica *(Vol. 8, p. 807). Madrid, Spain: Museo Nacional de Ciencias Naturales, CSIC. Class Hexapoda: general introduction*

FIGURE 8.13 Maxylla of a *Friesea.*

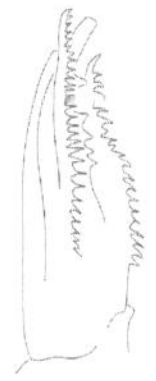

FIGURE 8.14 Maxylla with dentate lamellae of a Pseudachorutinae.

4(3) Maxylla with toothed or flabellated lamellae (Fig. 8.15) 5
4’ Simple and stylized maxilla; 0–2 + 0–2 eyes. Ant IV with “s” chaetae in candle-shaped flame (Fig. 8.16) .... *Micranurida pygmaea*
5(4) Mandible with normal teeth (Fig. 8.17) 6

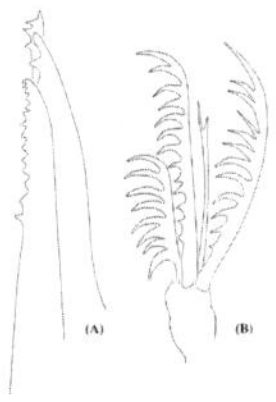

FIGURE 8.15 Maxylla dentate (A) and flabellated (B).

FIGURE 8.16 Tip of the antenna of *Micranurida pygmaea.*

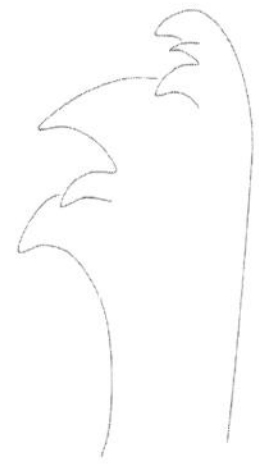

FIGURE 8.17 Mandible of *Anurida.*

5’ Mandible with two teeth and between them a finely serrated semicircular lamella (Fig. 8.18) .... *Anuridella calcarata*
6(5) Terga without additional chaetae .... *Anurida granaria*
6’ Terga plurichaetotic .... *A. maritima*
7(1) Dentes more than three times as long as manubrium, with distal rings of granules (Fig. 8.19) .... (Poduridae) *P. aquatica*
7’ Dentes with other shape 8
8(7) Body with pseudocelli; Ant III sensory organ as in Fig. 8.20; without eyes; body without pigmentation .... (Onychiuridae) 9
8’ Body without pseudocelli; Ant III sensory organ as in Fig. 8.21; 0–8 eyes; body with or without pigment .... (Hypogastruridae) 11
9(8) PAO with 3–5 lobes of one vesicle (Fig. 8.22). .... *Oligaphorura absoloni*
9’ PAO with 8 or more vesicles (Fig. 8.23) 10
10(9) Claw as long as antenna; Ant III sensory organ with two rows of guard papillae .... *O. colpus*
10’ Claw as long as first proximal antennal segment; Ant III sensory organ with one row of guard papillae .... *Thalassaphorura debilis*

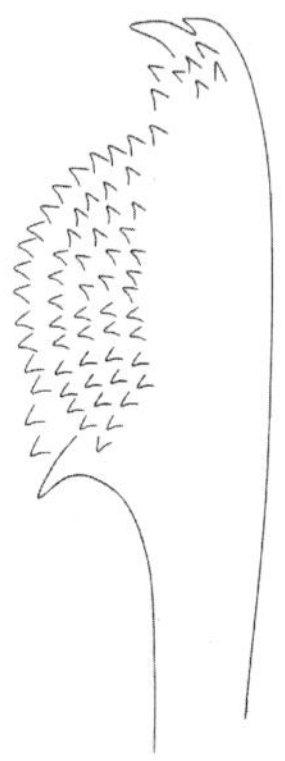

FIGURE 8.18 Mandible of *Anuridella calcarata*.

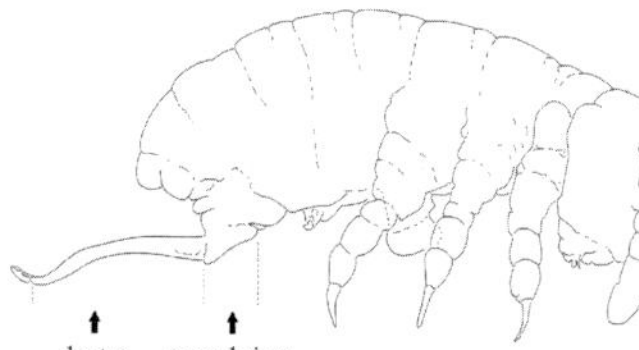

FIGURE 8.19 Habitus of *Podura aquatica*.

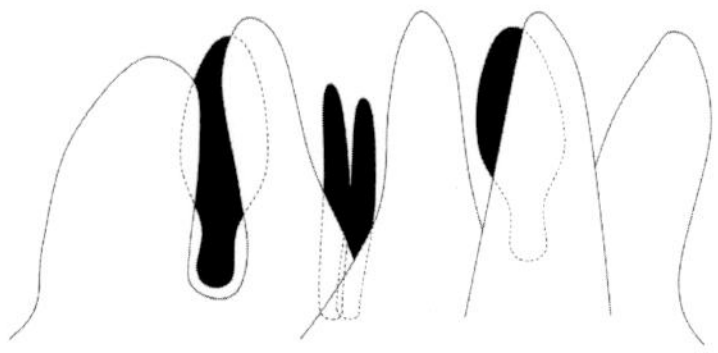

FIGURE 8.20 Ant III sensory organ of an Onychiuridae.

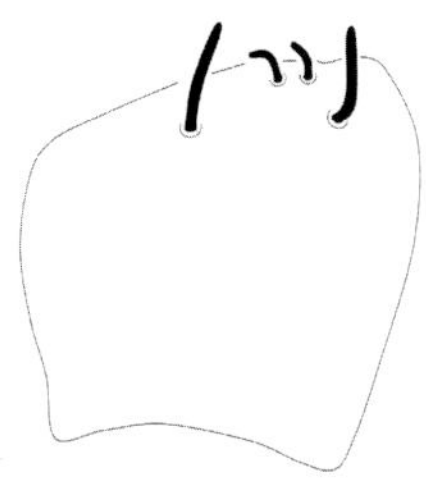

FIGURE 8.21 Ant III sensory organ of a Hypogastruridae.

FIGURE 8.22 PAO of *Oligaphorura absoloni*.

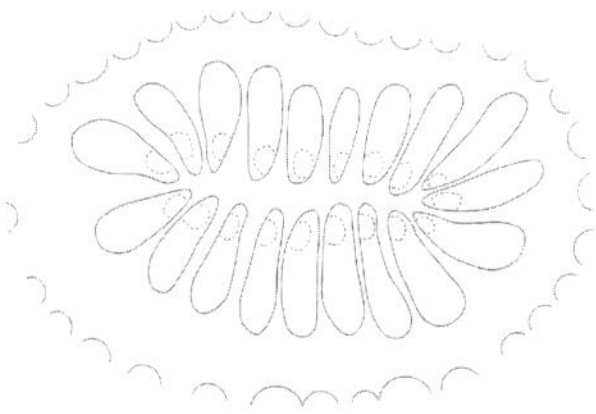

FIGURE 8.23 PAO of *Thalassaphorura debilis*.

11(8) Head with 8 + 8 eyes 12
11' Head with fewer than 8 + 8 eyes 15
12(11) Empodium 1/7th or more as long as inner claw 13
12' Empodium less than 1/9th as long as inner claw .... *Schoettella ununguiculata*
13(12) Tibiotarsus I–III with three knobbed tenent hairs (Fig. 8.24) .... *Hypogastrura viatica*
13' Tibiotarsus I–III with one weakly knobbed tenent hair 14
14(13) Retinaculum with 3 + 3 teeth .... *H. gisini*
14' Retinaculum with 4 + 4 teeth (Fig. 8.25) .... *H. vernalis*
15(11) PAO absent 16
15' PAO present (Fig. 8.26) 20
16(15) Head with 2 + 2 eyes or fewer per side .... *Acherontiella carusoi*
16' Head with 4 + 4–5 + 5 eyes per side 17

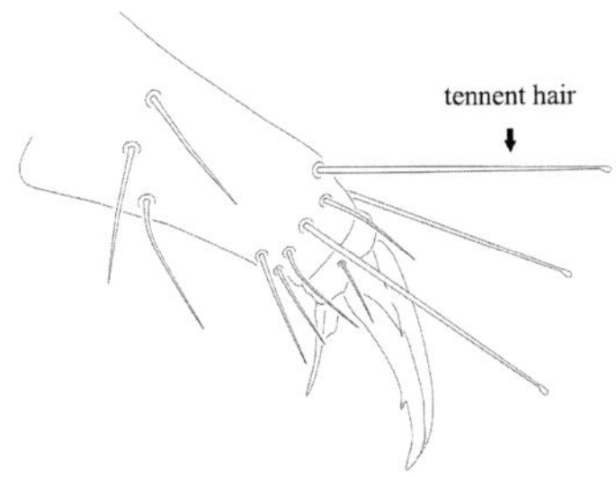

FIGURE 8.24 Tip of the leg of a *Hypogastrura viatica*.

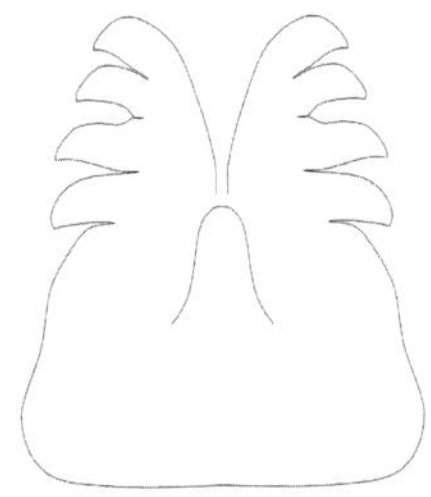

FIGURE 8.25 Retinaculum of *Hypogastrura vernalis*.

FIGURE 8.26 PAO of different species.

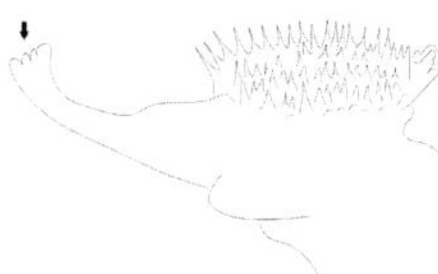

FIGURE 8.27 Mandible of *Paraxenylla affiniformis*.

17(16) Mandible with long thin apex and small apical teeth (Fig. 8.27, arrow .... *Paraxenylla affiniformis*
17' Mandible short without long apex, apical teeth large (Fig. 8.28, arrow) 18
18(17) Mucro separated from dens, and with two chaetae (Fig. 8.29); Th II–III with dorsal central chaetae in three rows (Fig. 8.30) or Th III with dorsal a_2 chaeta displaced distally relative to a_1 .... *Xenylla humicola*
18' Furca diversely structured; Th II with dorsal a_2 chaeta displaced distally relative to a_1 and dorsal p_2 displaced apically relative to p_1 19
19(18) Mucrodens more than twice as long as claw III (Fig. 8.31) .... *X. maritima*
19' Mucrodens less than twice as long as claw III, and with inner semicircular lamella (Fig. 8.32) .... *X. grisea*

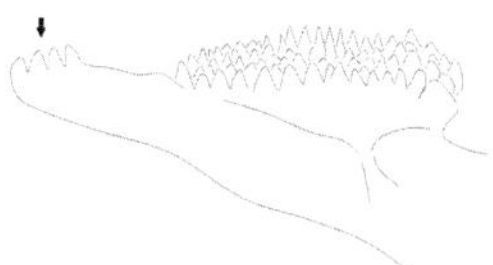

FIGURE 8.28 Mandible of *Xenylla*.

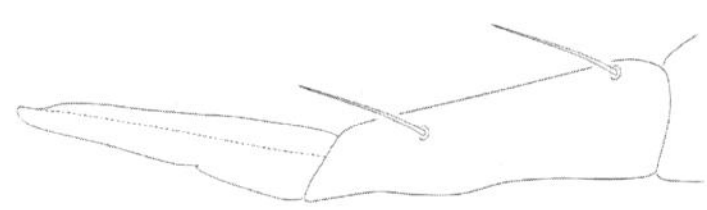

FIGURE 8.29 Dens and mucro of *Xenylla humicola*.

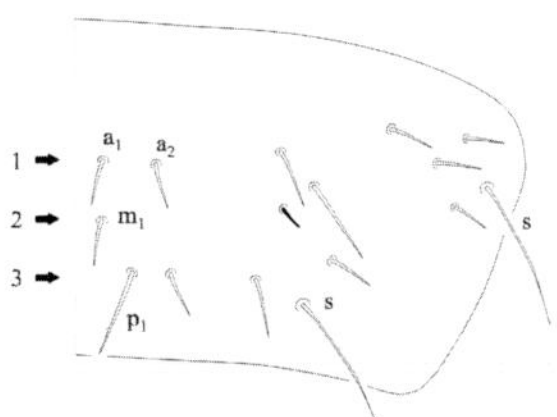

FIGURE 8.30 Th II–III of *Xenylla humicola*.

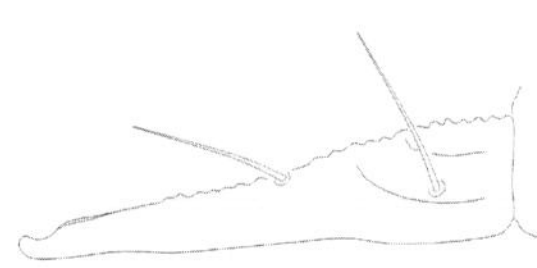

FIGURE 8.31 Mucrodens of *Xenylla maritima*.

20(15) Claw eight or more times as long as basal width (Fig. 8.33) .... *O. longisensilla*
20' Claw six or less times as long as basal width 21
21(20) 2–3 + 2–3 eyes .... *Typhlogastrura breuili*
21' 0 + 0 eyes .... *T. mendizabali*

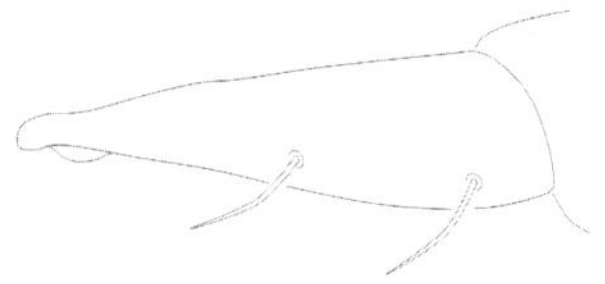

FIGURE 8.32 Mucrodens of *Xenylla grisea*.

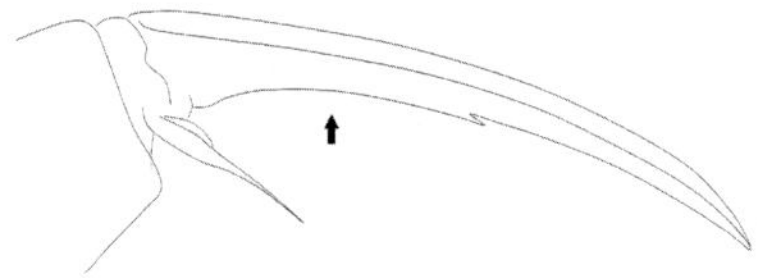

FIGURE 8.33 Claw of *Ongulogastrura longisensilla*.

Entomobryomorpha

1 Body chaetae smooth (Fig. 8.34A) or unilaterally ciliate (Fig. 8.34B); multilaterally ciliate macrochaetae always acuminate (Fig. 8.34C) 2

1' Body chaetae mostly ciliate; multilaterally ciliate macrochaetae sometimes cylindrical (Fig. 8.35B) usually truncate or broadened at the tip (Fig. 8.35C) .... (Entomobryidae) 20

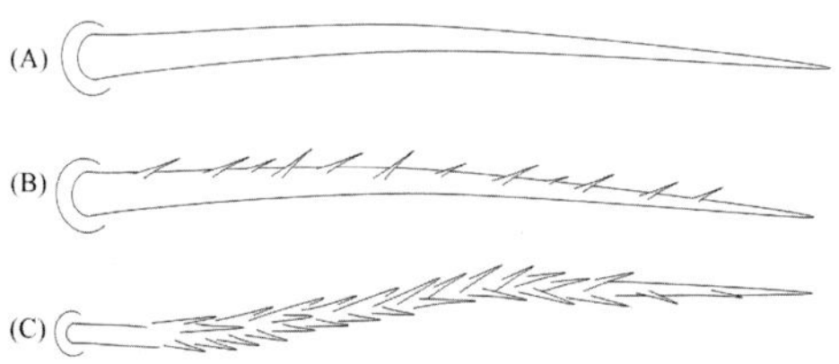

FIGURE 8.34 Some examples of chaetae of Isotomidae: (A), smooth chaeta; (B), unilaterally ciliate chaeta; (C), multilaterally ciliate macrochaetae.

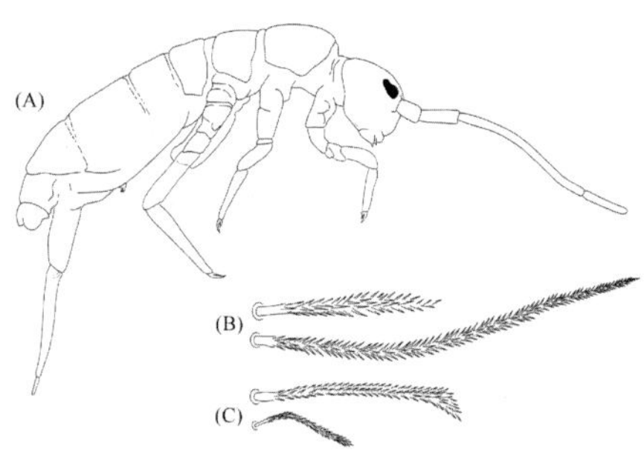

FIGURE 8.35 Habitus of Tomoceridae (A); some examples of chaetae of Entomobryomorpha except Isotomidae and Actaletidae: pointed (B) and capitate (C).

2(1) Abd IV–VI fused and 1.4 times the length of mesothorax to Abd III (Fig. 8.36); tenent hair leaf-shaped (Fig. 8.37, arrow) .... (Actaletidae) *A. neptuni*

2' Abd IV–VI not fused (Fig. 8.38), but if fused then shorter than the length of mesothorax to Abd III; tenent hair never leaf-like .... (Isotomidae) 3

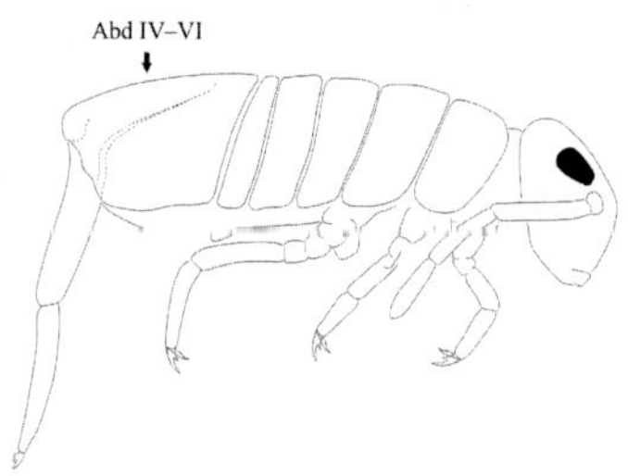

FIGURE 8.36 Habitus of *Actaletes neptuni*.

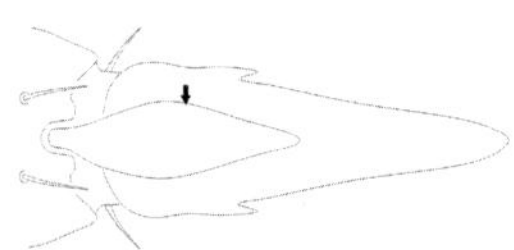

FIGURE 8.37 Tenent hair leaf-shaped of *Actaletes neptuni*.

3(2) Abd III larger than other abdominal segments 4

3' Abd III not larger than other abdominal segments 11

(3) Dens cylindrical, distally not narrowed (Fig. 8.39), normally multilaterally chaetaceous .... *Hydroisotoma schaefferi*

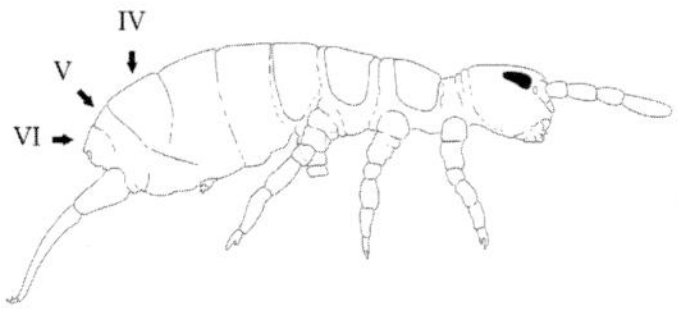

FIGURE 8.38 Habitus of Isotomidae.

FIGURE 8.39 Dens and mucro of *Hydroisotoma schaefferi*.

4' Dens clearly narrowed distally, dorsally crenulate, or knobbed; dorsal and ventral surfaces of dens clearly distinguished by the number and thickness of their chaetae (Fig. 8.40) 5
5(4) Abd II–IV without bothriotricha 6
5' Abd II–IV with bothriotricha (Fig. 8.41) 7

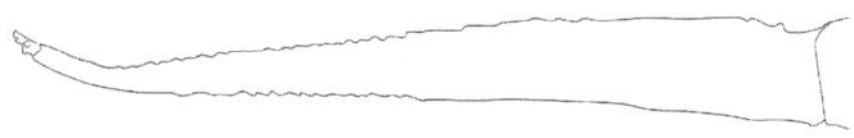

FIGURE 8.40 Dens of normal shape.

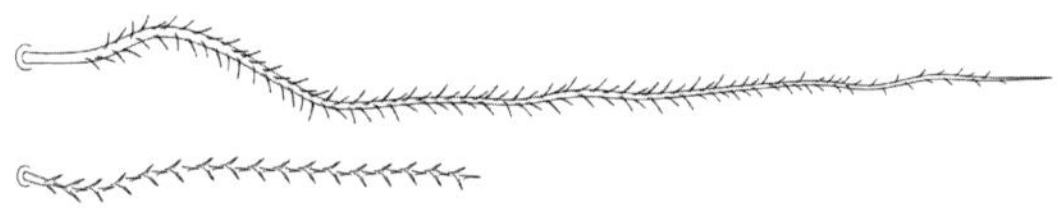

FIGURE 8.41 Two examples of bothriotricha.

6(5) One feathered chaeta on tibiotarsus II .... *Halisotoma boneti*
6' Two feathered chaeta on tibiotarsus II .... *H. maritima*
7(5) Ant III sensory organ with 10 or more thickened sensilla (Fig. 8.42) .... *Axelsonia litoralis*
7' Ant III sensory organ without that sensilla, bothriotricha ciliate 8
8(7) Pigment absent; claws long and slender; found in caves .... *Isotomurus subterraneus*
8' Pigment present, at least in eye region; claws variable in shape 9
9(8) Abd II–IV with 0, 1, 1 bothriotricha; Th II–Abd IV with transverse broad dark bands .... *Isotomurus gallicus*
9' Abd II–IV with 3, 3, 1 bothriotricha (absent in juvenile specimens); pigmentation variable, often with longitudinal stripes 10
10(9) Longitudinal median band interrupted or fainter on last abdominal segments, especially on Abd III .... *I. palustris* species complex
10' Longitudinal median band continuous, broad and dark, sometimes interrupted on Abd IV and V .... *I. unifasciatus*
11(3) Abd IV–VI dorsally fused (Fig. 8.43), often with a nonchaetaceous band between Abd IV and V, where this occurs the band is narrower than that among other segments 12

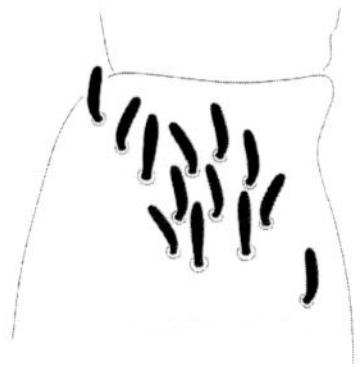

FIGURE 8.42 Ant III sensory organ of *Axelsonia litoralis*.

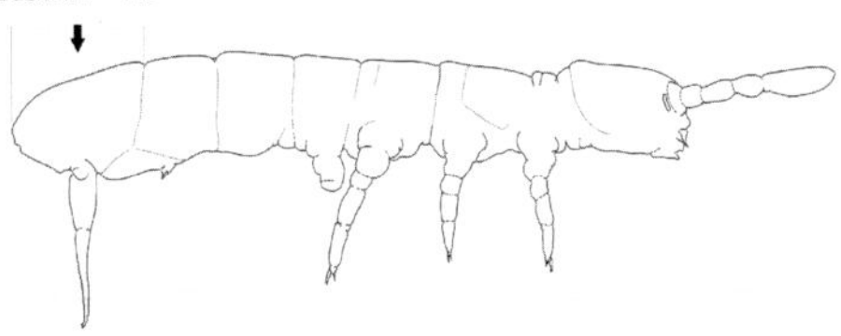

FIGURE 8.43 Habitus of *Folsomia*.

11' Abd IV–V not fused, if they seem fused, then the nonchaetaceous band is located with the same width as those among other segments 13
12(11) Head with 2 + 2 eyes .... *Folsomia quadrioculata*
12' Head with 3 + 3 eyes .... *F. sexoculata*

13(11) Cuticle granulate; dorsal surface of dens granulate, short, with ventral chaetae; mucro without chaetae, often with lamellae (Fig. 8.44) 14
13' Cuticle not granulate; dorsal surface of dens smooth, annulate, or with large knobs (Fig. 8.45); mucro may be chaetaceous 15
14(13) 2 + 2 anterior chaetae on manubrium .... *Pachyotoma crassicauda*
14' Without anterior chaetae on manubrium .... *P. levantina*

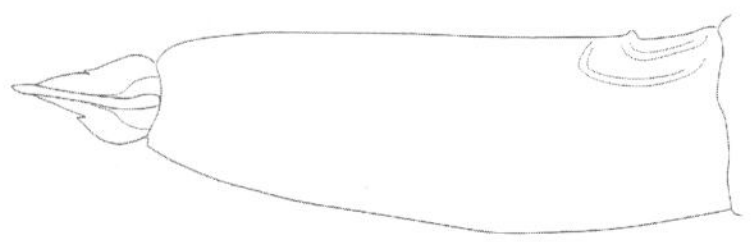

FIGURE 8.44 Dens and mucro of *Pachyotoma*.

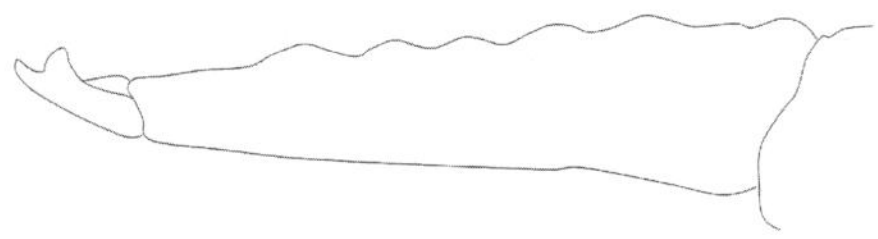

FIGURE 8.45 Dens and mucro of *Ballistura*.

15(13) Manubrium with two or more ventral chaetae and/or spines 16
15' Manubrium without ventral chaetae and/or spines 18
16(15) Only males with dorsal broad and numerous spines on body (Fig. 8.46) .... *Dimorphotoma porcellus*
16' Dorsal broad spines absent or present on both sexes 17
17(16) Abd V–VI not fused .... *Proisotoma minuta*
17' Abd V–VI fused .... *Cryptopygus thermophilus*
18(15) Dens smooth and cylindrical; mucro characteristic (Fig. 8.47), often with chaetae .... *Archisotoma interstitialis*
18' Dens annulated or knobbed; mucro different 19

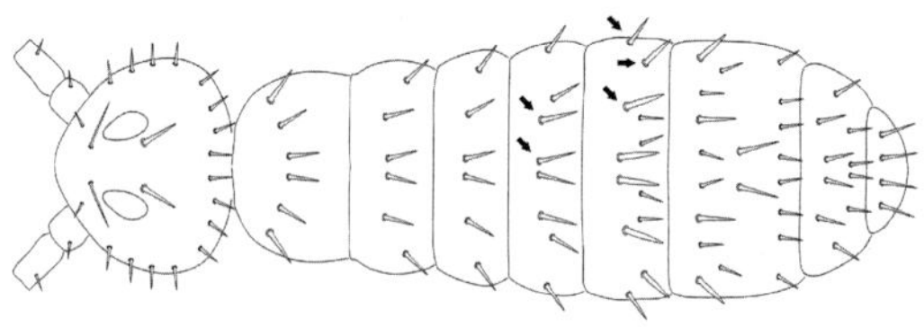

FIGURE 8.46 Habitus of *Dimorphotoma porcellus*.

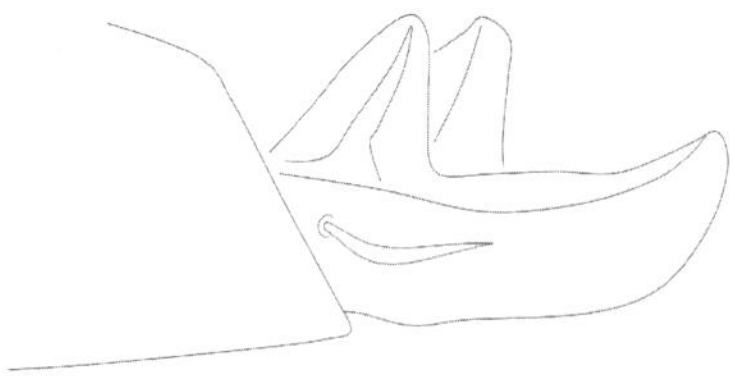

FIGURE 8.47 Mucro of *Archisotoma interstitialis*.

19(18) At least with the distal 2/3th of the anterior side of dens with chaetae; posterior side of dens without notches (Fig. 8.48) .... *Ballistura schoetti*
19' At most, the distal half of the anterior side of dens with chaetae; posterior side of dens with distinct notches (Fig. 8.49) .... *B. albertinae*

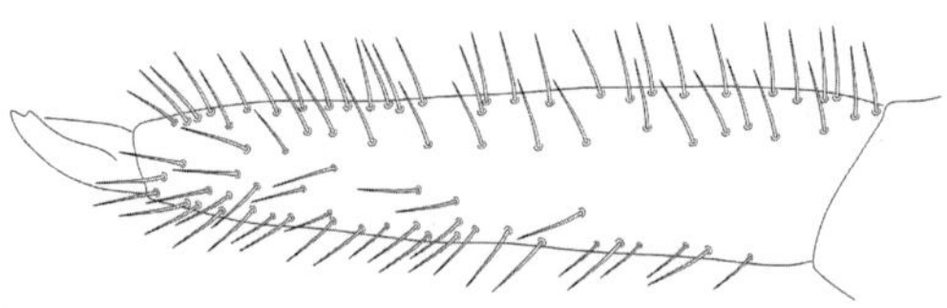

FIGURE 8.48 Dens and mucro of *Ballistura schoetti*.

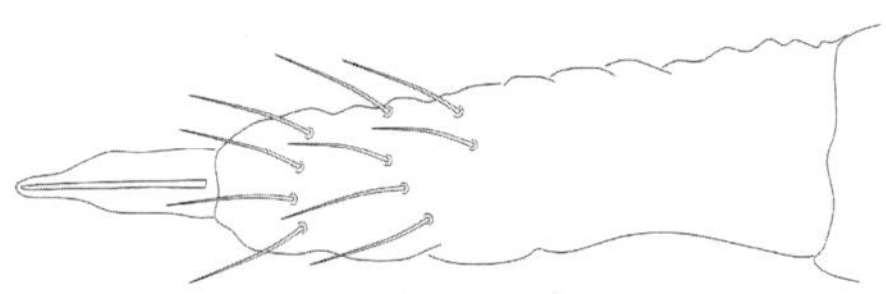

FIGURE 8.49 Dens and mucro of *Ballistura albertinae*.

20(1) Ventral face of dens with scales 21
20' Ventral face of dens without scales 24
21(20) Scales with coarse ribs, some pointed (Fig. 8.50) .... *Seira ferrarii*
21' Scales without ribs, finely denticulate, apical rounded (Fig. 8.51) 22

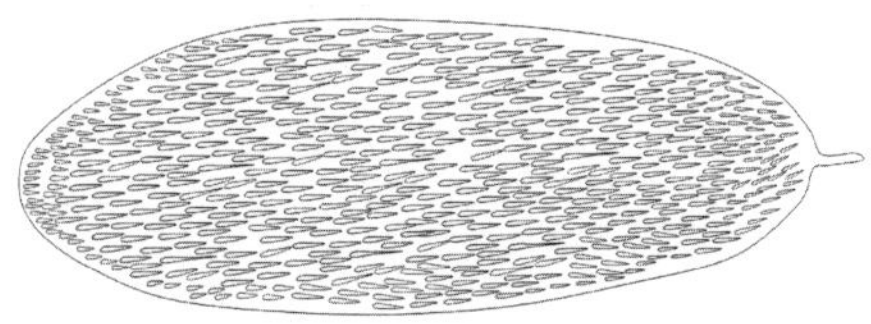

FIGURE 8.50 Scale of *Seira*.

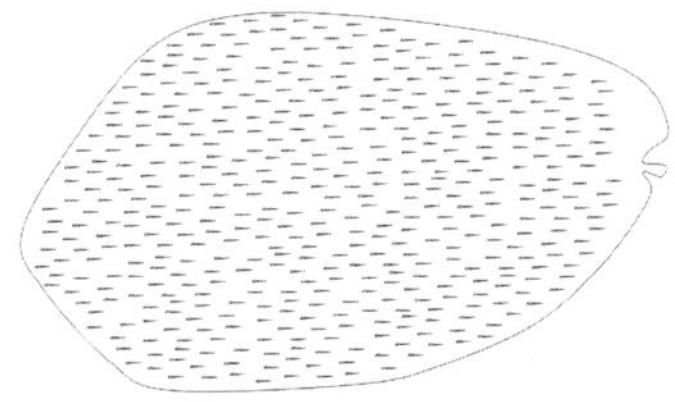

FIGURE 8.51 Scale of *Pseudosinella* or *Lepidocyrtus*.

22(21) Head with 0–6 + 0–6 eyes; claws with lateral teeth; first internal teeth (paired) well developed (like-wings) (Fig. 8.52) .... *Pseudosinella* spp.
22' Head with 8 + 8 eyes; claws with normal teeth (Fig. 8.53) 23
23(22) Body with blue color .... *Lepidocyrtus cyaneus*

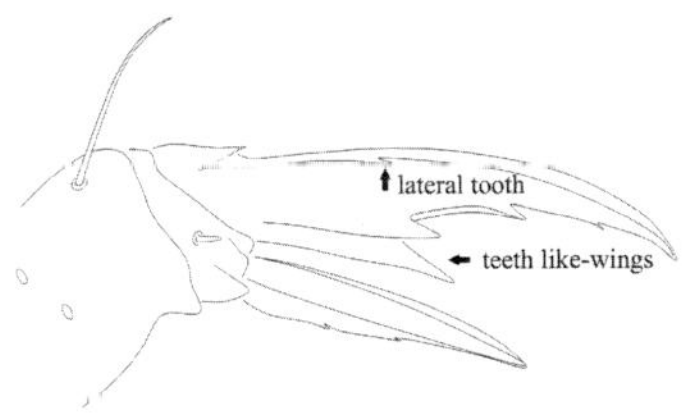

FIGURE 8.52 Claw of *Pseudosinella*.

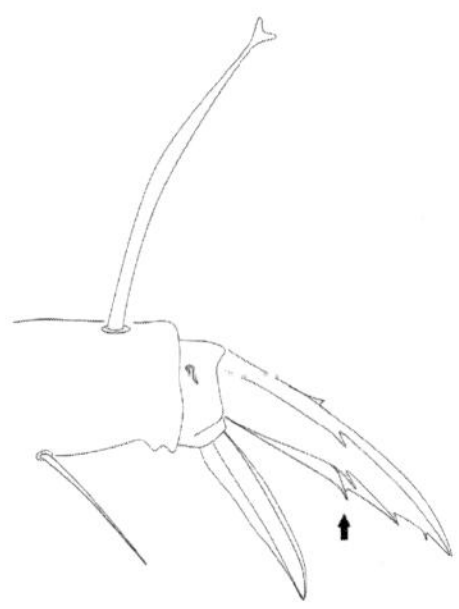

FIGURE 8.53 Claw of *Entomobrya*.

23' Body without color, or yellowish .... *L. lignorum*
24(20) Mucronal basal spine present (Fig. 8.54) .... *Entomobrya benaventi*
24' Mucronal basal spine absent 25
25(24) Nine dorsal macrochaetae on Abd II .... *Mesentotoma dollfusi*
25 Seven dorsal macrochaetae on Abd II .... *M. hispanica*

Symphypleona

1 Tenaculum with three teeth from juvenile phase II to adult (really with two teeth and a basal tubercle) (Fig. 8.55) .... (Bourletiellidae) *H. insignis*
1' Tenaculum with four teeth from juvenile phase II to adult (really with three teeth and a basal tubercle) 2

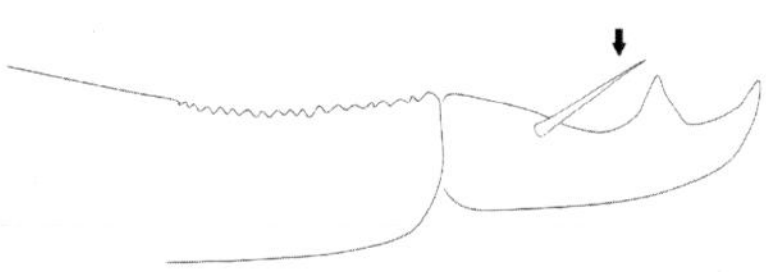

FIGURE 8.54 Mucro of *Entomobrya*.

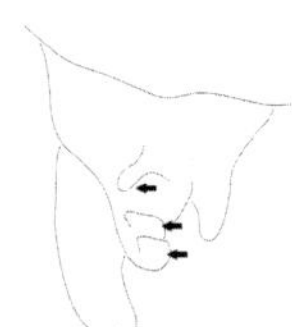

FIGURE 8.55 Tenaculum of Bourletiellidae.

2(1) Females without anal appendages. Males with Ant II–III modified as a fixation organ. Two pairs of bothriotricha in the Abd V. Large abdominal bothriotricha A, B, and C equidistant and open to back or forward (Fig. 8.56) 3
2' Female with anal appendages. Male with unmodified antennae. At most one pair of bothriotricha on Abd V. Large abdominal bothriotricha A, B, and C in another arrangement 4

3(2) Tibiotarsal III organ absent; mucro without a chaeta .... *Sphaeridia pumilis*
3' Tibiotarsal III organ present (Fig. 8.57) .... *Sminthurides* spp.

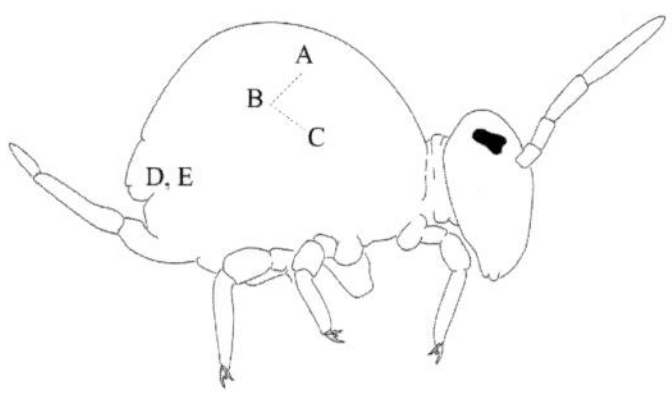

FIGURE 8.56 Habitus of Sminthuridae.

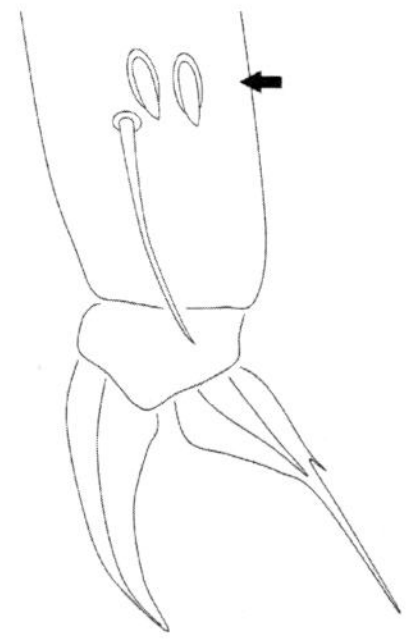

FIGURE 8.57 Tibiotarsal organ of *Sphaeridia pumilis*.

4(2) Ant IV shorter than III. Anal appendages clearly directed towards the anus (Fig. 8.58). Bothriotricha of the large abdominal at a backward angle, bothriotricum A apparently on a papilla (Fig. 8.59) ....
.... (Dicyrtomidae) *Dicyrtomina* spp., *Jordanathrix* spp.

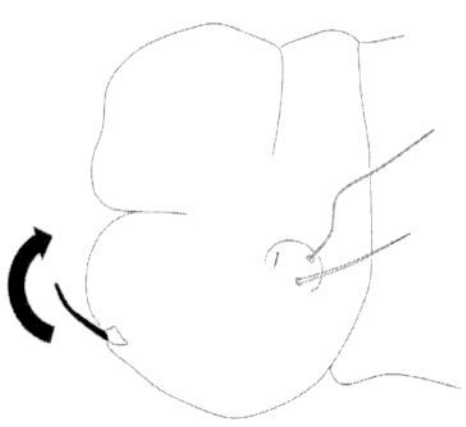

FIGURE 8.58 Small abdominal (lateral view) of a female Dicyrtomidae.

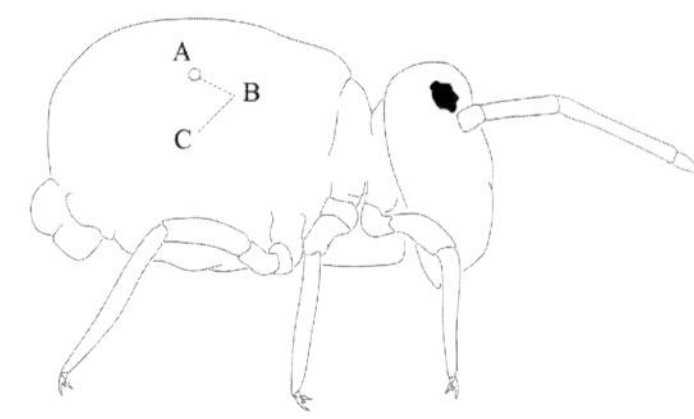

FIGURE 8.59 Habitus of a Dicyrtomidae. A, B, and C are the large abdominal bothriotricha.

4' Ant IV longer than Ant III. Bothriotricha form an angle open forward, with different angle (A and B or B and C near each other, and the other (Fig. 8.60). Anal appendages not clearly directed towards the anus, or directed towards genital opening (Fig. 8.61) 5

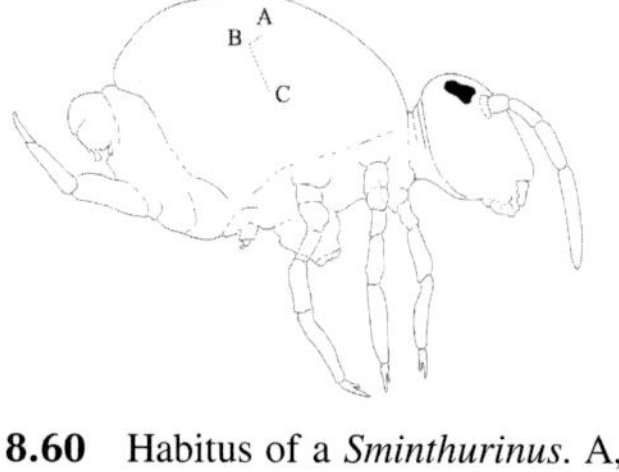

FIGURE 8.60 Habitus of a *Sminthurinus*. A, B, and C are the large abdominal bothriotricha.

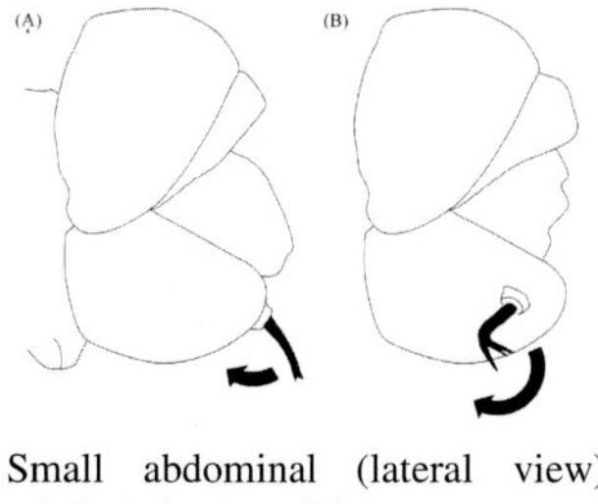

FIGURE 8.61 Small abdominal (lateral view) of a female *Arrhopalites* (A) and *Sminthurinus* (B).

5(4) Head with 8 + 8 eyes on each side. Tibiotarsi with erect and capitated chaetae (Fig. 8.62). Dentes without spines .... (Katiannidae) *Sminthurinus* spp.
5' Head with 0–2 eyes on each side. Body without pigment or very pale; tibiotarsus without capitated chaetae. Adults with spines on dens (Fig. 8.63) .... (Arrhopalitidae) *Arrhopalites* spp. (Fig. 8.64)

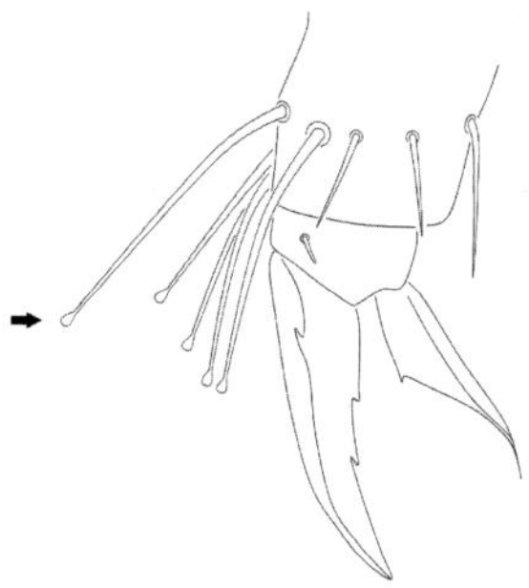

FIGURE 8.62 Tip of leg, claw, and empodium of *Sminthurinus*.

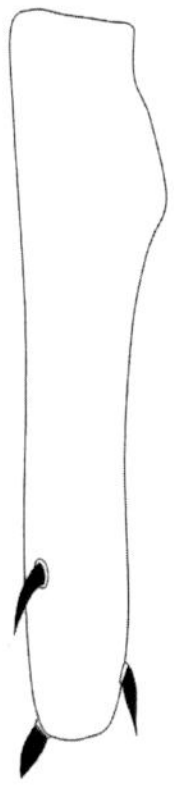

FIGURE 8.63 Dens of *Arrhopalites*.

Neelipleona (Neelidae)

1 Sensory fields of posterior large abdomen with two marginal chaetae (Fig. 8.65) .... *Neelus murinus* (Fig. 8.66)
1' Sensory fields of posterior large abdomen with five marginal chaetae (Fig. 8.67) .... *Megalothorax* spp. (Fig. 8.68)

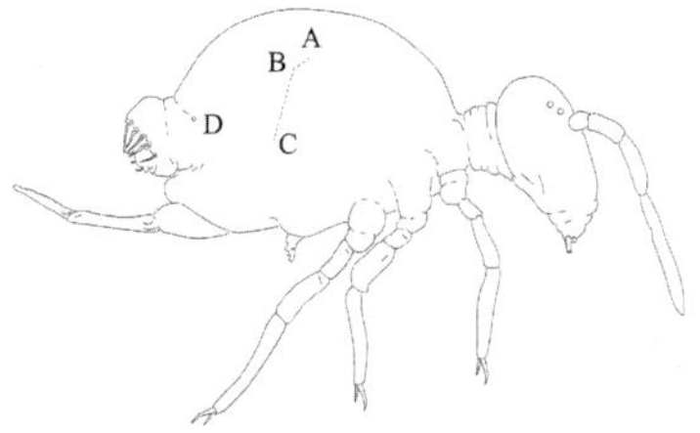

FIGURE 8.64 Habitus of a *Arrhopalites*. A, B, and C are the large abdominal bothriotricha, and D is the small abdominal bothriotricha.

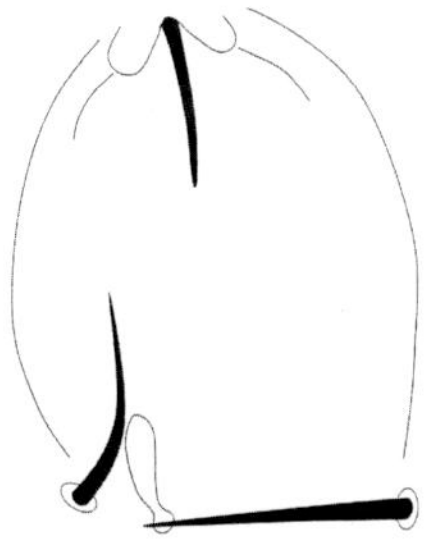

FIGURE 8.65 Sensory field of *Neelus*.

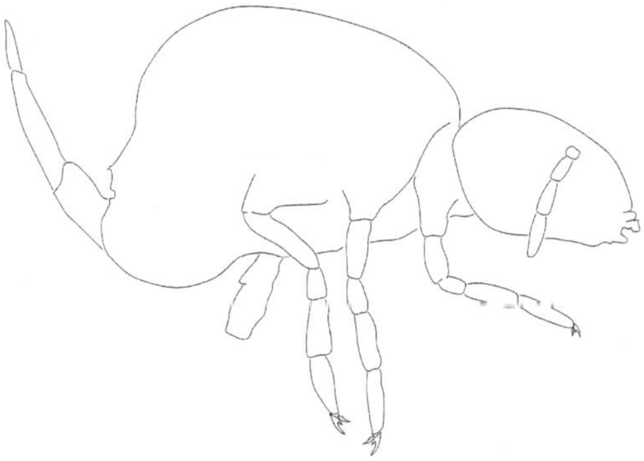

FIGURE 8.66 Habitus of *Neelus murinus*.

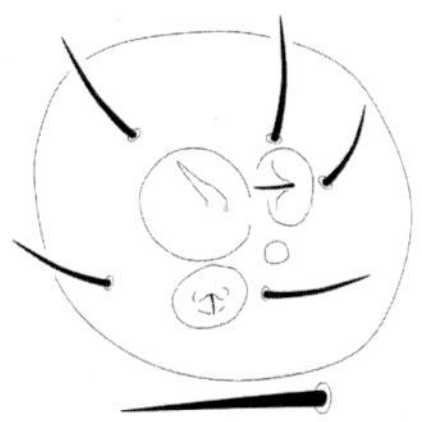

FIGURE 8.67 Sensory field of *Megalothorax*.

FIGURE 8.68 Habitus of *Megalothorax*.

Key to the subclass Insecta

1 Adults and larvae with sucking mouthparts forming a more-or-less prominent narrow tube. Thorax with three pairs of jointed legs (true legs). Forewings (if present) partly chitinized in the basal half and membranous in the distal half. Larvae similar to adults, but without developed wings. Last larval stadium usually must present wing pads .... **Hemiptera** (Fig. 8.69)

1' Adults and larvae usually without sucking mouthparts. But, if these are present, in larvae do not make up a single narrow tube and in adults have a pair of antenna with small clubs 2

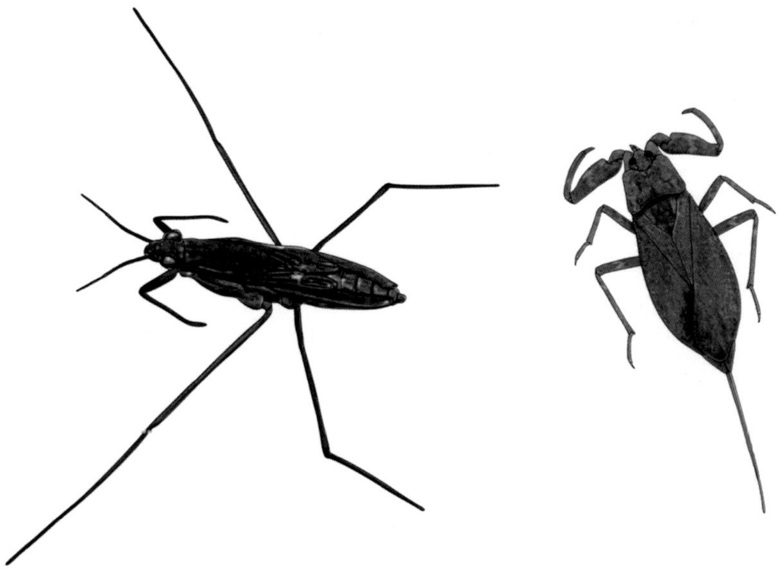

FIGURE 8.69 Two examples of Hemiptera: Gerridae (left) and Nepidae (right).

2(1') Adults with forewings rigid and strongly chitinized to form wing cases (elytra) .... **Coleoptera (Adults)** (Fig. 8.70)

2' Without wings. All insect immature (larvae) 3

3(2') Body curved to straight with short setae and usually white or yellow. Head heavily sclerotized and often brown or orange. First thoracic segment with a pair of dorsal sclerotized plates. Without thoracic legs. Abdominal segments

FIGURE 8.70 Two examples of Coleoptera (Adults): Dytiscidae (left) and Dryopidae (right).

1–8 with 2–5 dorsal transverse folds, segment 9 with 1 dorsal fold and segment 10 reduced to anal lobe .... **Coleoptera** (Curculionidae larvae) (Fig. 8.71)
3' Larvae without that combination of features 4
4(3') Body shape highly variable, but all without true legs. Usually with thoracic, abdominal, or caudal no jointed legs (i.e., false legs or prolegs). Although them must also be entirely absent. Filaments, suckers, extensions, setae, creeping welts, or gills may also be present throughout the body .... **Diptera** (Fig. 8.72)
4') All with true legs on each thoracic segment 5

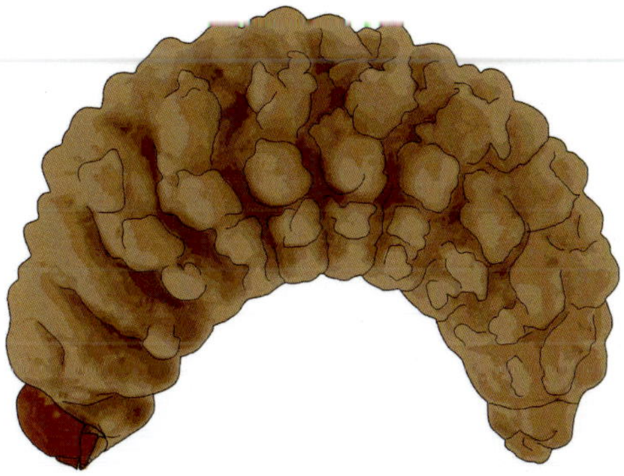

FIGURE 8.71 Larvae of Curculionidae (Coleoptera).

FIGURE 8.72 Two examples of Diptera: Chironomidae (left) and Limoniidae (right).

5(4') Larvae with caterpillar shape. Head heavily defined and sclerotized. Abdomen with ventral prolegs on segments 3–6 and the caudal segment, each with a ring of apical hooklets fine hooks (crochets) .... **Lepidoptera** (Fig. 8.73)
5' Larvae are not caterpillar-like 6
6(5') Mandibles are usually elongate, slender, and modified in two long sucking tubes on the head .... **Neuroptera** (Fig. 8.74)
6' Short and robust mandibles, no modified in sucking tubes 7
7(6') Abdomen with a single long caudal filament and with seven pairs of lateral filamentous gills .... **Megaloptera** (Fig. 8.75)
7' Abdomen without this combination of characters 8
8(7') Abdomen with one pair of anal prolegs, each ending in a sclerotized hook; larvae often live inside cases or nets .... **Trichoptera** (Fig. 8.76)
8' Abdomen without anal prolegs. Occasionally, two pairs of sclerotized hooks may be present. Larvae never live inside cases 9
9(8') Last larval stadium without wing pads developed. Thorax and abdomen highly variable. Abdominal segments may present dorsal projections, lateral gills, prolegs. Caudal segment may present gills, hooks, long filaments (urogomphi) .... **Coleoptera** (Larvae) (Fig. 8.77)
9' Last larval stadium with wing pads developed 10
10(9') Modified mouthparts into a mask-like structure, hinged basally and toothed apically, that can extend forward and retract rapidly. Slender species with three leaf-like, caudal gills. Stout species without caudal gills, but with spine-like processes at the body end .... **Odonata** (Fig. 8.78)
10' Mouthparts no modified into a mask 11
11 (10') Abdomen with three (less usually two) multisegmented caudal filaments and paired lateral gills. Legs with one tarsal claw .... **Ephemeroptera** (Fig. 8.79)
11' Abdomen with two multisegmented caudal. Without lateral abdominal gills, but cervical, thoracic, or caudal gills may be present. Legs with two tarsal claws .... **Plecoptera** (Fig. 8.80)

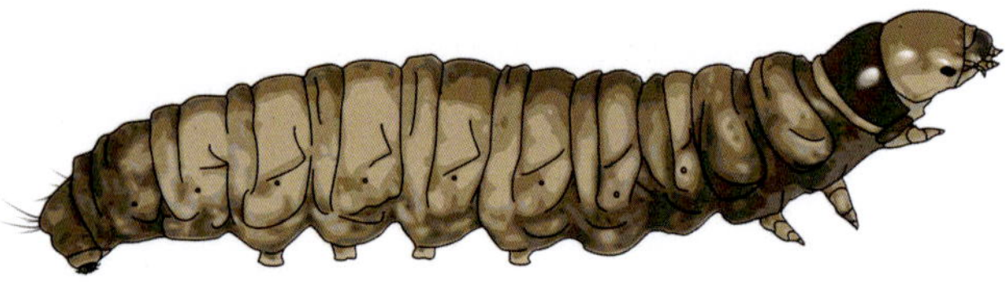

FIGURE 8.73 Example of Lepidoptera: Crambidae.

FIGURE 8.74 Example of Neuroptera: Osmylidae.

FIGURE 8.75 Example of Megaloptera: Sialidae.

FIGURE 8.76 Two examples of Trichoptera: Limnephilidae (left) and Hydropsychidae (right).

FIGURE 8.77 Two examples of Coleoptera (Larvae): Elmidae (left) and Hydrophilidae (right).

FIGURE 8.78 Two examples of Odonata: Aeshnidae (left) and Lestidae (right).

FIGURE 8.79 Two examples of Ephemeroptera: Baetidae (left) and Heptageniidae (right).

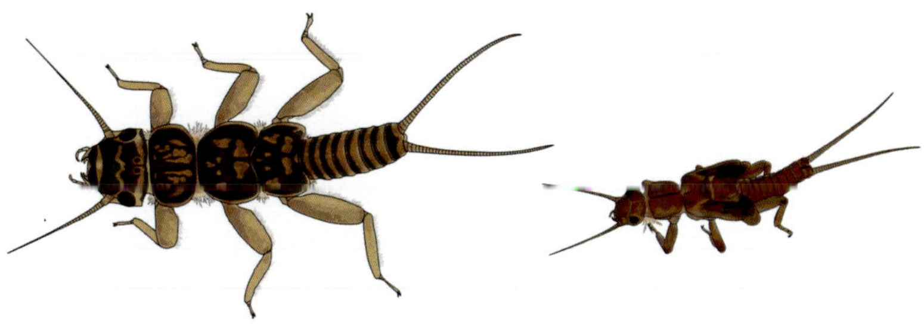

FIGURE 8.80 Two examples of Plecoptera: Perlidae (left) and Nemouridae (right).

Insect taxa marginally associated with the aquatic environment

Hymenoptera: This is a very large order of holometabolous insects that includes bees, wasps, and ants. Some species of wasps have terrestrial adults and larvae that are parasitic upon other aquatic organisms, often other insects. Most aquatic Hymenoptera are endoparasites of larvae that occur in aquatic plants (Lancaster & Downes, 2013). However, some are ectoparasitic and oviposit on larvae, prepupae, or pupae. Larvae are generally worm-like, and their morphology is the typical of internal parasites. These do not have legs, and the head is often very reduced (without ocelli and absent or reduced antennae). Hymenoptera will be treated more extensively in Chapter 16 of this book.

Shoreline insects often found in aquatic samples

The following three insect orders include insects that are primarily terrestrial, but because they live in the shoreline of rivers or lakes, are often found in aquatic samples, and thus could easily be confused with aquatic taxa:

Suborder Homoptera: The suborder Homoptera are grouped within Hemiptera. Consequently, they also have sucking mouthparts forming a prominent narrow tube, and their first pair of wings are membranous. Two families frequently found in the shoreline are Cicadellidae (leafhoppers) and Aphididae (aphids). The Cicadellidae are easily recognized by having the hind tibiae with one or more distinct keels and a row of movable spines on each side. Most Aphididae are wingless and have long antennae. Their main characteristic is the presence of a pair of abdominal tubes on the dorsal surface of their fifth abdominal segment.

Order Coleoptera: Excluding the aquatic species, most Coleoptera have terrestrial habits. But some species live in the shoreline. Two families can be found frequently in the shoreline. The Carabidae are predatory ground beetles with long and filamentous antennae, shiny black or metallic elytra, and long and run-adapted legs. The Staphylinidae (or rove beetles) are primarily distinguished by their short elytra typically leaving more than half of their abdominal segments exposed. Larvae of Carabidae and Staphylinidae are also frequently found in the shoreline and confused with Hydrophylidae or Dytiscidae. Both can be recognized by having in the last abdominal segment, two long cerci and a ventral tube or supporting structure.

Order Thysanoptera: Commonly called thrips, Thysanoptera are hemimetabolous minute insects whose distinctive characteristic is the presence of two pairs of wings fringed with hairs and asymmetrical rasping-sucking mouthparts.

References

Abellán, P., & Svenning, J.-C. (2014). Refugia within refugia – Patterns in endemism and genetic divergence are linked to Late Quaternary climate stability in the Iberian Peninsula. *Biological Journal of the Linnean Society*, *113*, 13–28.

Acuña, V., Giorgi, A., Muñoz, I., Sabater, F., & Sabater, S. (2007). Meteorological and riparian influences on organic matter dynamics in a forested Mediterranean stream. *Journal of the North American Benthological Society*, *26*, 54–69.

Acuña, V., Hunter, M., & Ruhí, A. (2017). Managing temporary streams and rivers as unique rather than second-class ecosystems. *Biological Conservation*, *211*, 12–19.

Acuña, V., Muñoz, I., Giorgi, A., Omella, M., Sabater, F., & Sabater, S. (2005). Drought and postdrought recovery cycles in an intermittent Mediterranean stream: Structural and functional aspects. *Journal of the North American Benthological Society*, *24*, 919–933.

Adams, S. M. (2002). *Biological indicators of aquatic ecosystem stress*. Bethesda: American Fisheries Society, 644 pp.

Aghaee, M., & Godfrey, L. D. (2014). A century of rice water weevil (Coleoptera: Curculionidae): a history of research and management with an emphasis on the United States. *Journal of Integrated Pest Management*, *5*, 1–14.

Alba-Tercedor, J., Sáinz-Bariáin, M., Poquet, J. M., & Rodríguez-López, R. (2017). Predicting River macroinvertebrate communities distributional shifts under future global change scenarios in the spanish Mediterranean area. *PLoS One*, *12*, 1–21.

Alba-Tercedor, J., & Sánchez-Ortega, A. (1988). Un método rápido y simple para evaluar la calidad biológica de las aguas corrientes basado en el de Hellawell (1978). *Limnetica*, *4*, 1–56.

Alcorlo, P., Baltanás, A., & Montes, C. (2001). Food-web structure in two shallow salt lakes in Los Monegros (NE Spain): Energetic vs dynamic constraints. *Hydrobiologia*, *466*, 307–316.

Altermatt, F. (2013). Diversity in riverine metacommunities: A network perspective. *Aquatic Ecology*, *47*, 365–377.

Álvarez, M., & Pardo, I. (2009). Dynamics in the trophic structure of the macroinvertebrate community in a Mediterranean, temporary stream. *Aquatic Sciences*, *71*, 202–213.

Álvarez-Cobelas, M., Rojo, C., & Angeler, D. G. (2005). Mediterranean limnology: Current status, gaps and the future. *Journal of Limnology*, *64*, 13–29.

Anton-Pardo, M., & Armengol, X. (2014). Aquatic invertebrate assemblages in ponds from coastal Mediterranean wetlands. *Annales de Limnologie*, *50*, 217–230.

Aranda, C., Eritja, R., & Roiz, D. (2006). First record and establishment of the mosquito *Aedes albopictus* in Spain. *Medical and Veterinary Entomology*, *20*, 150–152.

Arbea, J., & Ariza, E. (2012). Seasonal dynamics and population characteristics of Collembola communities on beaches of the Costa Brava (Girona, Spain). *Boletín de la Sociedad Entomológica Aragonesa (S.E.A.), 51*, 203–210.

Armitage, P. D., Moss, D., Wright, J. F., & Furse, M. T. (1983). The performance of a new biological water quality score system based on macroinvertebrates over a wide range of unpolluted running-water sites. *Water Research, 17*, 333–347.

Arribas, P., Abellán, P., Velasco, J., & Millán, A. (2015). Evolutionary ecology, biogeography and conservation of water beetles in Mediterranean saline ecosystems. *Limnetica, 34*, 481–494.

Artemiadou, V., & Lazaridou, M. (2005). Evaluation Score and Interpretation Index for the ecological quality of running waters in Central and Northern Hellas. *Environmental Monitoring and Assessment, 10*, 1–40.

Aspöck, U., Aspöck, H., & Liu, X. (2017). The Nevrorthidae, mistaken at all times: Phylogeny and review of present knowledge (Holometabola, Neuropterida, Neuroptera). *Deutsche Entomologische Zeitschrift, 64*, 77–110.

Balian, E. V., Segers, H., Martens, K., & Lévéque, C. (2008). The freshwater animal diversity assessment: An overview of the results. *Hydrobiologia, 595*, 627–637.

Ball, J. E., Bêche, L. A., Mendez, P. K., & Resh, V. H. (2013). Biodiversity in Mediterranean-climate streams of California. *Hydrobiologia, 719*, 187–213.

Ballardini, M., Ferretti, S., Chiaranz, G., Pautasso, A., Riina, M. V., Triglia, G., Verna, F., Bellavia, V., Radaelli, M. C., Berio, E., Accorsi, A., De Camilli, M., Cardellino, U., Fiorino, N., Acutis, P. L., Casalone, C., & Mignone, W. (2019). First report of the invasive mosquito *Aedes koreicus* (Diptera: Culicidae) and of its establishment in Liguria, northwest Italy. *Parasites & Vectors, 12*, 1–13.

Balletto, E., & Casale, A. (1991). Mediterranean insect conservation. In N. M. Collins, & J. A. Thomas (Eds.), *The conservation of insects and their habitats* (pp. 121–142). London: Academic Press.

Bameul, F. (1992). Première découverte d'un *Omicrus* Sharp dans la région éthiopienne: *O. hebaueri* n. sp.(Coleoptera, Hydrophilidae). *Bulletin de la Société entomologique de France, 97*, 373–379.

Barber-James, H. M. (2009). A preliminary phylogeny of Prosopistomatidae (Ephemeroptera) based on morphological characters of the larvae, and an assessment of their distribution. *Aquatic Insects, 31*, 149–166.

Barbour, M. T., Gerritsen, J., Snyder, B. D., & Stribling, J. B. (1999). *Rapid bioassessment protocols for use in streams and wadeable rivers: Periphyton, benthic macroinvertebrates and fish* (2nd ed.). Washington, DC: U.S. Environmental Protection Agency, Office of Water, EPA 841-B-99-002.

Bas-Silvestre, M., Quintana, X. D., Compte, J., Gascón, S., Boix, D., Antón-Pardo, M., & Obrador, B. (2020). Ecosystem metabolism dynamics and environmental drivers in Mediterranean confined coastal lagoons. *Estuarine, Coastal and Shelf Science, 245*, 106989.

Bazzanti, M., Mastrantuono, L., & Pilotto, F. (2017). Depth-related response of macroinvertebrates to the reversal of eutrophication in a Mediterranean lake: Implications for ecological assessment. *Science of the Total Environment, 579*, 456–465.

Beauger, A., & Lair, N. (2014). Analyse des principales méthodes de bio-évaluation basées sur les macroinvertébrés benthiques. *Bulletin de la Société Linnéenne de Lyon, 4*, 15–33.

Bedel, L. (1901). Description et moeurs d'un nouveau genre de curculionides de France. *Bulletin de la Société entomologique de France, 6*, 358–359.

Beklioglu, M., Romo, S., Kagalou, I., Quintana, X. D., & Bécares, E. (2007). State of the art in the functioning of shallow Mediterranean lakes: Workshop conclusions. *Hydrobiologia, 584*, 317–326.

Belfiore, C. (2006). Insecta Ephemeroptera. In S. Ruffo & F. Stoch (Eds.), Checklist and distribution of the Italian fauna. Memorie del Museo Civico di Storia Naturale di Verona, 2. Serie, *Sezione Scienze della Vita, 17*, 127–129.

Bellinger, P. F., Christiansen, K. A., & Janssens, F. (1996–2020). *Checklist of the Collembola of the World*. http://www.collembola.org. Last access: 12 May 2020.

Benedict, M. Q., Levine, R. S., Hawley, W. A., & Lounibos, L. P. (2007). Spread of the tiger: Global risk of invasion by the mosquito *Aedes albopictus*. *Vector Borne Zoonotic Diseases, 7*, 76–85.

Benzina, I., Si Bachir, A., Saheb, M., Santoul, F., & Céréghino, R. (2018). Macroinvertebrate communities respond to anthropogenic pressure in arid-land streams of north East Algeria. *Vie et milieu-Life and Environment, 68*, 271–280.

Berchi, G. M., Copilaş-Ciocianu, D., Kment, P., Buzzetti, F. M., Petrusek, A., Rákosy, L., Cianferoni, F., & Damgaard, J. (2018). Molecular phylogeny and biogeography of the West-Palaearctic Velia (Heteroptera: Gerromorpha: Veliidae). *Systematic Entomology, 43*, 262–276.

Berger, T., & Rothe, U. (1999). *Ametropus fragilis* Albarda 1878 (Insecta: Ephemeroptera) neu für Deutschland mit Anmerkungen zu Verbreitung. *Biologie und Status der Art. Lauterbornia, 37*, 177–197.

Bernal, S., von Schiller, D., Sabater, F., & Martí, E. (2013). Hydrological extremes modulate nutrient dynamics in mediterranean climate streams across different spatial scales. *Hydrobiologia, 719*, 31–42.

Bilton, D. T., Freeland, J. R., & Okamura, B. (2001). Dispersal in freshwater invertebrates. *Annual Review of Ecology and Systematics, 32*, 159–181.

Bird, M. S., Bilton, D. T., & Perissinotto, R. (2017). Diversity and distribution of polyphagan water beetles (Coleoptera) in the Lake St Lucia system, South Africa. *ZooKeys, 656*, 51–84.

Birk, S., Bonne, W., Borja, A., Brucet, S., Courrat, A., Poikane, S., Solimini, A., van de Bund, W., Zampoukas, N., & Hering, D. (2012). Three hundred ways to assess Europe's surface waters: an almost complete overview of biological methods to implement the Water Framework Directive. *Ecological indicators, 18*, 31–41.

Blackman, R. C., Mächler, E., Altermatt, F., Arnold, A., Beja, P., Boets, P., Egeter, B., Elbrecht, V., Filipe, A. F., Jones, J. I., Macher, J., Majaneva, M., Martins, F. M. S., Murria, C., Meissner, K., Pawlowski, J., Schmidt, Y. P. L., Zizka, V. M. A., Leese, F., ... Macher, J. (2019). Advancing the use of molecular methods for routine freshwater macroinvertebrate biomonitoring—the need for calibration. *Metabarcoding and Metagenomics, 3*, 49–457.

Bogan, M. T., Chester, E. T., Datry, T., Murphy, A. L., Robson, B. J., Ruhí, A., Stubbington, R., & Whitney, J. E. (2017). Resistance, resilience, and community recovery in intermittent rivers and ephemeral streams. In T. Datry, N. Bonada, & A. J. Boulton (Eds.), *Intermittent rivers and ephemeral streams: Ecology and management* (pp. 349–376). Cambridge, MA: Elsevier, Inc.

Boix, D., Calhoun, A. J. K., Mushet, D. M., Bell, K. P., Fitzsimons, J. A., & Isselin-Nondedeu, F. (2020). Conservation of temporary wetlands. In M. I. Goldstein, & D. A. DellaSala (Eds.), *Encyclopedia of the world's biomes* (pp. 279–294). Boston: Elsevier.

Boix, D., Gascón, S., Sala, J., Badosa, A., Brucet, S., López-Flores, R., Martinoy, M., Guifre, J., & Quintana, X. D. (2008). Patterns of composition and species richness of crustaceans and aquatic insects along environmental gradients in Mediterranean water bodies. *Hydrobiologia, 597*, 53–69.

Boix, D., Gascón, S., Sala, J., Martinoy, M., Gifre, J., & Quintana, X. D. (2005). A new index of water quality assessment in Mediterranean wetlands based on crustacean and insect assemblages: The case of Catalunya (NE Iberian Peninsula). *Aquatic Conservation: Marine and Freshwater Ecosystems, 15*, 635–651.

Boix, D., Kneitel, J., Robson, B. J., Duchet, C., Zúñiga, L., Day, J., Gascón, S., Sala, J., Quintana, X. D., & Blaustein, L. (2016). Invertebrates of freshwater temporary ponds in Mediterranean climates. In D. P. Batzer, & D. Boix (Eds.), *Invertebrates in freshwater wetlands. An international perspective on their ecology* (pp. 141–190). Cham: Springer International Publishing.

Boix, D., Magnusson, A. K., Gascón, S., Sala, J., & Williams, D. D. (2011). Environmental influence on flight activity and arrival paterns of aerial colonizers of temporary ponds. *Wetlands, 31*, 1227–1240.

Boix, D., Sala, J., Gascon, S., Martinoy, M., Gifre, J., Brucet, S., Badosa, A., López-Flores, R., & Quintana, X. D. (2007). Comparative biodiversity of crustaceans and aquatic insects from various water body types in coastal Mediterranean wetlands. *Hydrobiologia, 584*, 347–359.

Bojková, J., & Soldán, T. (2015). Two new species of the genus Prosopistoma (Ephemeroptera: Prosopistomatidae) from Iraq and Algeria. *Zootaxa, 4018*, 109–123.

Bonada, N., & Dolédec, S. (2011). Do mediterranean genera not included in Tachet et al. (2002) have mediterranean trait characteristics? *Limnetica, 30*, 129–142.

Bonada, N., & Dolédec, S. (2018). Does the Tachet trait database report voltinism variability of aquatic insects between Mediterranean and Scandinavian regions? *Aquatic Sciences, 80*, 7.

Bonada, N., & Resh, V. H. (2013). Mediterranean-climate streams and rivers: geographically separated but ecologically comparable freshwater system. *Hydrobiologia, 719*, 1–29.

Bonada, N., Cañedo-Argüelles, M., Gallart, F., Von Schiller, D., Fortuño, P., Latron, J., Llorens, P., Múrria, C., Soria, M., Vinyoles, D., & Cid, N. (2020). Conservation and management of isolated pools in temporary rivers. *Water, 12*, 2870.

Bonada, N., Carlson, S. M., Datry, T., Finn, D. S., Leigh, C., Lytle, D. A., Monaghan, M. T., & Tedesco, P. A. (2017). Genetic, evolutionary, and biogeographical processes in intermittent rivers and ephemeral streams. In T. Datry, N. Bonada, & A. J. Boulton (Eds.), *Intermittent rivers and ephemeral streams: Ecology and management* (pp. 405–431). Cambridge, MA: Elsevier, Inc.

Bonada, N., Dolédec, S., & Statzner, B. (2007). Taxonomic and biological trait differences of stream macroinvertebrate communities between mediterranean and temperate regions: Implications for future climatic scenarios. *Global Change Biology, 13*, 1658–1671.

Bonada, N., Prat, N., Resh, V. H., & Statzner, B. (2006). Developments in aquatic insect biomonitoring: A comparative analysis of recent approaches. *Annual Review of Entomology, 51*, 495–523.

Bonada, N., Rieradevall, M., & Prat, N. (2007). Macroinvertebrate community structure and biological traits related to flow permanence in a Mediterranean river network. *Hydrobiologia, 589*, 91–106.

Bonada, N., Rieradevall, M., Dallas, H., Davis, J., Day, J., Figueroa, R., Resh, V. H., & Prat, N. (2008). Multi-scale assessment of macroinvertebrate richness and composition in Mediterranean-climate rivers. *Freshwater Biology, 53*, 772–788.

Bonada, N., Rieradevall, M., Prat, N., & Resh, V. H. (2006). Benthic macroinvertebrate assemblages and macrohabitat connectivity in Mediterranean-climate streams of northern California. *Journal of the North American Benthological Society, 25*, 32–43.

Bonada, N., Zamora-Muñoz, C., Rieradevall, M., & Prat, N. (2004). Ecological profiles of caddisfly larvae in mediterranean streams: Implications for bioassessment methods. *Environmental Pollution, 132*, 509–521.

Boorman, J., Mellor, P. S., & Scaramozzino, P. (1996). A new species of *Culicoides* (Diptera, Ceratopogonidae) from southern Italy. *Parassitologia, 38*, 501–503.

Borthagaray, A. I., Barreneche, J. M., Abades, S., & Arim, M. (2014). Modularity along organism dispersal gradients challenges a prevailing view of abrupt transitions in animal landscape perception. *Ecography, 37*, 564–571.

Botosaneanu, L., & Gasith, A. (1971). Contributions taxonomiques et écologiques à la connaissance des Trichoptères (Insecta) d'Israel. *Israel Journal of Zoology, 20*, 89–129.

Boudot, J.-P., Kalkman, V. J., Azpilicueta, M., Bogdanović, T., Cordero, A., Degabriele, G., Dommanget, J.-L., Ferreira, S., Garrigós, B., Jović, M., Kotarac, M., Lopau, W., Marinov, M., Mihoković, N., Riservato, E., Samraoui, B., & Schneider, W. (2009). Atlas of the Odonata of the Mediterranean and North Africa. *Libellula* (Suppl. 9), 1–256.

Bretfeld, G. (1999). Synopses on Palaearctic Collembola, Volume 2. Symphypleona. *Abhandlungen und Berichte des Naturkundemuseums Görlitz, 71*, 1–318.

Briers, R. A., Cariss, H. M., & Gee, J. H. R. (2002). Dispersal of adult stoneflies (Plecoptera) from upland streams draining catchments with contrasting land-use. *Archiv Für Hydrobiologie, 155*, 627–644.

Britton, R. H., & Crivelli, A. J. (1993). Wetlands of southern Europe and North Africa: mediterranean wetlands. In D. F. Whigham, D. Dykyjová, & S. Hejný (Eds.), *Wetlands of the world: Inventory, ecology and management* (pp. 129–194). Dordrecht: Springer.

Brown, B. L., & Swan, C. M. (2010). Dendritic network structure constrains metacommunity properties in riverine ecosystems. *Journal of Animal Ecology, 79*, 571–580.

Brown, L. E., Khamis, K., Wilkes, M., Blaen, P., Brittain, J. E., Carrivick, J. L., Fell, S., Friberg, N., Füreder, L., Gislason, G. M., Hainie, S., Hannah, D. M., James, W. H. M., Lencioni, V., Olafsson, J. S., Robinson, C. T., Saltveit, S. J., Thompson, C., & Milner, A. M. (2018). Functional diversity and community assembly of river invertebrates show globally consistent responses to decreasing glacier cover. *Nature Ecology and Evolution, 2*, 325–333.

Brucet, S., Boix, D., Nathansen, L. W., Quintana, X. D., Jensen, E., Balayla, D., Meerhoff, M., & Jeppesen, E. (2012). Effects of temperature, salinity and fish in structuring the macroinvertebrate community in shallow lakes: Implications for effects of climate change. *PLoS One, 7*, e30877.

Bruno, D., Belmar, O., Sánchez-Fernández, D., Guareschi, S., Millán, A., & Velasco, J. (2014). Responses of mediterranean aquatic and riparian communities to human pressures at different spatial scales. *Ecological Indicators, 45*, 456–464.

Buffagni, A., Erba, S., Cazzola, M., & Kemp, J. L. (2004). The AQEM multimetric system for the southern Italian Apennines: Assessing the impact of water quality and habitat degradation on pool macroinvertebrates in Mediterranean rivers. *Hydrobiologia, 516*, 313–329.

CABI. (2021). *Cryptopleurum subtile. Invasive species compendium.* Wallingford, UK: CAB International. Available from https://www.cabi.org/isc/datasheet/113681.

CABI. (2023). *Lissorhoptrus oryzophilus* (rice water weevil). In *Invasive species compendium.* Wallingford, UK: CAB International. Available from https://doi.org/10.1079/cabicompendium.30992.

Calapez, A. R., Branco, P., Santos, J. M., Ferreira, T., Hein, T., Brito, A. G., & Feio, M. J. (2017). Macroinvertebrate short-term responses to flow variation and oxygen depletion: A mesocosm approach. *Science of the Total Environment, 599–600*, 1202–1212.

Calhoun, A. J. K., Mushet, D. M., Bell, K. P., Boix, D., Fizsimons, J. A., & Isselin-Nondedeu, F. (2017). Temporary wetlands: Challenges and solutions for protecting a 'disappearing' ecosystem. *Biological Conservation, 211*, 3–11.

Caldara, R., Diotti, L., & Regalin, R. (2004). First record for Europe of the rice water weevil, *Lissorhoptrus oryzophilus* Kuschel (Coleoptera, Curculionoidea, Erirhinidae). *Bollettino di Zoologia Agraria e di Bachicoltura, Serie II, 36*, 165–171.

Cancrini, G., Scaramozzino, P., Gabrielli, S., Paolo, M. D., Toma, L., & Romi, R. (2007). *Aedes albopictus* and *Culex pipiens* implicated as natural vectors of *Dirofilaria repens* in central Italy. *Journal of Medical Entomology, 44*, 1064–1066.

Cañedo-Argüelles, M. (2020). A review of recent advances and future challenges in freshwater salinization. *Limnetica, 39*, 185–211.

Cañedo-Argüelles, M., & Rieradevall, M. (2010). Disturbance caused by freshwater releases of different magnitude on the aquatic macroinvertebrate communities of two coastal lagoons. *Estuarine, Coastal and Shelf Science, 88*, 190–198.

Cañedo-Argüelles, M., Boix, D., Sánchez-Millaruelo, N., Sala, J., Caiola, N., Nebra, A., & Rieradevall, M. (2012). A rapid bioassessment tool for the evaluation of the water quality of transitional waters. *Estuarine, Coastal and Shelf Science, 111*, 129–138.

Cañedo-Argüelles, M., Gutiérrez-Cánovas, C., Acosta, R., Castro-López, D., Cid, N., Fortuño, P., Munné, A., Múrria, C., Pimentão, A. R., Sarremejane, R., Soria, M., Tarrats, P., Verkaik, I., Prat, N., & Bonada, N. (2020). As time goes by: 20 years of changes in the aquatic macroinvertebrate metacommunity of Mediterranean river networks. *Journal of Biogeography, 47*, 1861–1874.

Canhoto, C., & Graça, M. A. S. (1995). Food value of introduced eucalypt leaves for a Mediterranean stream detritivore: *Tipula lateralis. Freshwater Biology, 34*, 209–214.

Carapelli, A., Frati, F., Fanciulli, P. P., & Dallai, R. (2001). Taxonomic Revision of 14 South-western European Species of *Isotomurus* (Collembola, Isotomidae), with description of four new species and the designation of the neotype for I. palustris. *Zoologica Scripta, 30*, 115–143.

Carbonell, J. A., Céspedes, V., Coccia, C., & Green, A. J. (2020). An experimental test of interspecific competition between the alien boatman *Trichocorixa verticalis* and the native corixid *Sigara lateralis* (Hemiptera, Corixidae). *Aquatic Invasions, 15*, 318–334.

Carbonell, J. A., Céspedes, V., & Green, A. J. (2021). Is the spread of the alien water boatman *Trichocorixa verticalis verticalis* (Hemiptera, Corixidae) aided by zoochory and drought resistant eggs? *Freshwater Biology, 66*, 409–420.

Carbonell, J. A., Millán, A., Green, A. J., Céspedes, V., Coccia, C., & Velasco, J. (2016). What traits underpin the successful establishment and spread of the invasive water bug *Trichocorixa verticalis verticalis*? *Hydrobiologia, 768*, 273–286.

Cardoso, P., Barton, P. S., Birkhofer, K., Chichorro, F., Deacon, C., Fartmann, T., Fukushima, C. S., Gaigher, R., Habel, J. C., Hallmanng, C. A., Hill, M. J., Hochkirch, A., Kwak, M. L., Mammol, S., Noriega, J. A., Orfinger, A. B., Pedraza, F., Pryke, J. S., Roque, F. O., … Samways, M. J. (2020). Scientists' warning to humanity on insect extinctions. *Biological Conservation, 242*, 108426.

Carrapiço, F., Santos, R., & Serrano, A. (2011). First occurrence of *Stenopelmus rufinasus* Gyllenhal, 1835 (Coleoptera: Erirhinidae) in Portugal. *The Coleopterists Bulletin, 65*, 436–443.

Carrara, F., Rinaldo, A., Giometto, A., & Altermatt, F. (2014). Complex interaction of dendritic connectivity and hierarchical patch size on biodiversity in river-like landscapes. *The American Naturalist, 183*, 13–25.

Cassagnau, P., & Delamare-Deboutteville, C. (1955). Mission Henri Coiffait au Liban (1951). 3. Collemboles. *Archives de Zoologie Expérimentale et Générale, 91*(4), 365–395.

Castillo-Escrivà, A., Valls, L., Rochera, C., Camacho, A., & Mesquita-Joanes, F. (2017). Metacommunity dynamics of Ostracoda in temporary lakes: Overall strong niche effects except at the onset of the flooding period. *Limnologica, 62*, 104–110.

Cavender-Bares, J., Ackerly, D. D., Baum, D. A., & Bazzaz, F. A. (2004). Phylogenetic overdispersion in Floridian oak communities. *The American Naturalist, 163*, 823–843.

Céréghino, R., Oertli, B., Bazzanti, M., Coccia, C., Compin, A., Biggs, J., Bressi, N., Grillas, P., Hull, A., Kalettka, T., & Scher, O. (2012). Biological traits of European pond macroinvertebrates. *Hydrobiologia, 689*, 51–61.

Céspedes, V., Coccia, C., Carbonell, J. A., Sánchez, M. I., & Green, A. J. (2019). The life cycle of the alien boatman *Trichocorixa verticalis* (Hemiptera, Corixidae) in saline and hypersaline wetlands of south-west Spain. *Hydrobiologia, 827*, 309–324.

Céspedes, V., Sánchez, M. I., & Green, A. J. (2017). Predator–prey interactions between native brine shrimp Artemia parthenogenetica and the alien boatman *Trichocorixa verticalis*: influence of salinity, predator sex, and size, abundance and parasitic status of prey. *PeerJ*, *5*, e3554.

Céspedes, V., Stoks, R., Green, A. J., & Sánchez, M. I. (2019). Eco-immunology of native and invasive water bugs in response to water mite parasites: insights from phenoloxidase activity. *Biological Invasions*, *21*, 2431–2445.

Chandler, J. R. (1970). A biological approach to water quality management. *Water Pollution and Control*, *69*, 415–422.

Chase, J. M. (2011). Chapter 5: Ecological niche theory. In S. M. Scheiner, & M. R. Willig (Eds.), *The theory of ecology* (pp. 93–107). Chicago: The University Chicago Press.

Chevenet, F., Dolédec, S., & Chessel, D. (1994). A fuzzy coding approach for the analysis of long-term ecological data. *Freshwater Biology*, *31*, 295–309.

Christiansen, K., & Snider, R. J. (2008). Aquatic Collembola. In R. W. Merritt, K. W. Cummins, & M. B. Berg (Eds.), *An introduction to the aquatic insects of North America* (4th ed., pp. 165–179). Dubuque, IA: Kendall/Hunt Publishing Company.

Cid, N., Bonada, N., Carlson, S. M., Grantham, T. E., Gasith, A., & Resh, V. H. (2017). High variability is a defining component of Mediterranean-climate rivers and their biota. *Water*, *9*, 52.

Cid, N., Bonada, N., Heino, J., Cañedo-Argüelles, M., Crabot, J., Sarremejane, R., Soininen, J., Stubbington, R., & Datry, T. (2020). A metacommunity approach to improve biological assessments in highly dynamic freshwater ecosystems. *Bioscience*, *70*, 427–438.

Clapham, M. E., Karr, J. A., Nicholson, D. B., Ross, A. J., & Mayhe, P. J. (2016). Ancient origin of high taxonomic richness among insects. *Proceedings of the Royal Society B: Biological Sciences*, *283*, 20152476.

Clavero, M., & García-Berthou, E. (2006). Homogenization dynamics and introduction routes of invasive freshwater fish in the Iberian Peninsula. *Ecological Applications*, *16*, 2313–2324.

Coccia, C., Boyero, L., & Green, A. J. (2014). Can differential predation of native and alien corixids explain the success of *Trichocorixa verticalis verticalis* (Hemiptera, Corixidae) in the Iberian Peninsula? *Hydrobiologia*, *734*, 115–123.

Coccia, C., Calosi, P., Boyero, L., Green, A. J., & Bilton, D. T. (2013). Does ecophysiology determine invasion success? A comparison between the invasive boatman *Trichocorixa verticalis verticalis* and the native *Sigara lateralis* (Hemiptera, Corixidae) in South-West Spain. *PLoS One*, *8*, e63105.

Coccia, C., Fry, B., Ramírez, F., Boyero, L., Bunn, S. E., Diz-Salgado, C., Walton, M., Le Vay, L., & Green, A. J. (2016). Niche partitioning between invasive and native corixids (Hemiptera, Corixidae) in south-west Spain. *Aquatic sciences*, *78*, 779–791.

Collier, K. J., & Smith, B. J. (1998). Dispersal of adult caddisflies (Trichoptera) into forests alongside three New Zealand streams. *Hydrobiologia*, *361*, 53–65.

Comín, F. A., Rodó, X., & Comín, F. A. (1992). Lake Gallocanta (Aragón, NE Spain): a paradigm of fluctuations at different scales of time. *Limnetica*, *8*, 79–86.

Costa, S., Morchio, F., & Bodon, M. (2017). Macrobenthos alieno in Liguria: Stato attuale ed evoluzione del fenomeno. *Biologia Ambientale*, *31*, 183–190.

Cover, M. R., Seo, J. H., & Resh, V. H. (2015). Life history, burrowing behavior, and distribution of *Neohermes filicornis* (Megaloptera: Corydalidae), a long-lived aquatic insect in intermittent streams. *Western North American Naturalist*, *75*, 474–490.

Ćuk, R., Čmrlec, K., & Belfiore, C. (2015). The first record of *Ametropus fragilis* Albarda, 1878 (Insecta: Ephemeroptera) from Croatia. *Natura Croatica*, *24*, 151–157.

Cummins, K. W. (1973). Trophic relations of aquatic insects. *Annual Review of Entomology*, *18*, 183–206.

Cunillera-Montcusí, D., Arim, M., Gascón, S., Tornero, I., Sala, J., Boix, D., & Borthagaray, A. I. (2020). Addressing trait selection patterns in temporary ponds in response to wildfire disturbance and seasonal succession. *Journal of Animal Ecology*, *89*, 2134–2144.

Cunillera-Montcusí, D., Boix, D., Sala, J., Compte, J., Tornero, I., Quintana, X. D., & Gascón, S. (2020). Large- and small-regional-scale variables interact in the dispersal patterns of aquatic macroinvertebrates from temporary ponds. *Aquatic Ecology*, *54*, 1041–1058.

Cunillera-Montcusí, D., Gascón, S., Tornero, I., Sala, J., Àvila, N., Quintana, X. D., & Boix, D. (2019). Direct and indirect impacts of wildfire on faunal communities of Mediterranean temporary ponds. *Freshwater Biology*, *64*, 323–334.

Cunillera-Montcusí, D. (2020). *Resilience of aquatic metacommunities: implications for disturbance recovery* (Ph.D. thesis). Catalonia, Spain: Institute of Aquatic Ecology, Universitat de Girona. 317 pp.

D'Haese, C. (2002). Were the first springtails semi-aquatic? A phylogenetic approach by means of 28S rDNA and optimization alignment. *Proceedings of the Royal Society of London*, *269*, 1143–1151.

Dallas, H. F. (2013). Ecological status assessment in mediterranean rivers: Complexities and challanges in developing tools for assessing ecological status and defining reference conditions. *Hydrobiologia*, *719*, 483–507.

Dalley, G. (2014). Cryptopleurum subtile Sharp (Coleoptera, Hydrophilidae) in Scotland. *The Coleopterist*, *23*, 134.

Damásio, J., Fernández-Sanjuan, M., Sánchez-Avila, J., Lacorte, S., Prat, N., Rieradevall, M., Soares, A. M. V. M., & Barata, C. (2011). Multi-biochemical responses of benthic macroinvertebrate species as a complementary tool to diagnose the cause of community impairment in polluted rivers. *Water Research*, *45*, 3599–3613.

Dana, E. D., & Viva, S. (2006). *Stenopelmus rufinasus* Gyllenhal 1836 (Coleoptera: Erirhinidae) naturalized in Spain. *The Coleopterists Bulletin*, *60*, 41–42.

Datry, T., Larned, S. T., Fritz, K., Bogan, M. T., Wood, P. J., Meyer, E. I., & Santos, A. N. (2014). Broad-scale patterns of invertebrate richness and community composition in temporary rivers: Effects of flow intermittence. *Ecography*, *37*, 94–104.

Davies, P. M., & Stewart, B. A. (2013). Freshwater faunal biodiversity in the mediterranean climate waterways of southwestern Australia. *Hydrobiologia, 719*, 215–235.

De Bie, T., De Meester, L., Brendonck, L., Martens, K., Goddeeris, B., Ercken, D., Hampel, H., Denys, L., Vanhecke, L., Van der Gucht, K., Van Wilchelen, J., Vyverman, W., & Declerck, S. A. J. (2012). Body size and dispersal mode as key traits determining metacommunity structure of aquatic organisms. *Ecology Letters, 15*, 740–747.

De Geer, C. (1740). Experimenta et Observationes de parvulis insectis, agili saltu corpuscula sua in altum levantibus, quibus Poduræ nomen est. *Acta Societatis Regiae Scientiarum Upsaliensis*, 48–67.

de Moor, F. C. (1992). Factors influencing the establishment of aquatic insect invaders. *Transactions of the Royal Society of South Africa, 48*, 141–158.

de Moor, F. C., & Day, J. A. (2013). Aquatic biodiversity in the mediterranean region of South Africa. *Hydrobiologia, 719*, 237–268.

de Oliveira, T. R. S., de Moura, S. R., da Cruz, D. D., Loges, V., & Martins, C. F. (2018). Interaction and distribution of beetles (Insecta: Coleoptera) associated with *Heliconia bihai* (Heliconiaceae) inflorescences. *Florida Entomologist, 101*, 160–165.

Deharveng, L., D'Haese, C. A., & Bedos, A. (2007). Global diversity of springtails (Collembola; Hexapoda) in freshwater. *Hydrobiologia, 595*, 329–338.

Deharveng, L., D'Haese, C. A., & Bedos, A. (2008). Global diversity of springtails (Collembola; Hexapoda) in freshwater. *Hydrobiologia, 595*, 329–338.

Della Bella, V., Bazzanti, M., & Chiarotti, F. (2005). Macroinvertebrate diversity and conservation status of Mediterranean ponds in Italy: Water permanence and mesohabitat influence. *Aquatic conservation: marine and freshwater ecosystems, 15*, 583–600.

Desquilbet, M., Gaume, L., Grippa, M., Céréghino, R., Humbert, J. F., Bonmatin, J. M., Cornillon, P. A., Maes, D., Van Dyck, H., & Goulson, D. (2020). Comment on "Meta-analysis reveals declines in terrestrial but increases in freshwater insect abundances.". *Science (New York, N.Y.), 370*, eabd8947.

Díaz, P., Gerrero, M. C., Alcorlo, P., Baltanás, A., Florin, M., & Montes, C. (1998). Anthropogenic perturbations to the trophic structure in a permanent hypersaline shallow lake: La Salada de Chiprana (north-eastern Spain). *International Journal of Salt Lake Research, 7*, 187–210.

Didham, R. K., Basset, Y., Collins, C. M., Leather, S. R., Littlewood, N. A., Menz, M. H. M., Müller, J., Packer, L., Saunders, M. E., Schönrogge, K., Stewart, A. J. A., Yanoviak, S. P., & Hassall, C. (2020). Interpreting insect declines: Seven challenges and a way forward. *Insect Conservation and Diversity, 13*, 103–114.

Dijkstra, K. B., Monaghan, M. T., & Pauls, S. U. (2014). Freshwater biodiversity and insect diversification. *Annual Review of Entomology, 59*, 143–163.

Djitli, Y., Boix, D., Milla, A., Marniche, F., Tornero, I., Cunillera-Montcusí, D., Sala, J., Gascón, S., Quintana, X.D., & Daoudi-Hacini, S. (in press). Annual cycle of water quality and macroinvertebrate composition in Algerian wetlands: A case study of Lake Réghaïa (Algeria). Limnetica.

Dolédec, S., Statzner, B., & Bournard, M. (1999). Species traits for future biomonitoring across ecoregions: Patterns along a human-impacted river. *Freshwater Biology, 42*, 737–758.

Dolédec, S., Tilbian, J., & Bonada, N. (2017). Temporal variability in taxonomic and trait compositions of invertebrate assemblages in two climatic regions with contrasting flow regimes. *Science of the Total Environment, 599–600*, 1912–1921.

Downing, J. A. (2010). Emerging global role of small lakes and ponds: little things mean a lot. *Limnetica, 29*, 9–24.

Dudgeon, D., Arthington, A. H., Gessner, M. O., Kawabata, Z., Knowler, D. J., Lévêque, C., Naiman, R. J., Prieur-Richard, A. H., Soto, D., Stiassny, M. L. J., & Sullivan, C. A. (2006). Freshwater biodiversity: importance, threats, status and conservation challenges. *Biological Reviews, 81*, 163–182.

Dyce, A. L. (1969). Biting midges (Diptera: Ceratopogonidae) reared from rotting cactus in Australia. *Mosquito News, 29*, 644–649.

Early, R., Bradley, B. A., Dukes, J. S., Lawler, J. J., Olden, J. D., Blumenthal, D. M., Gonzalez, P., Grosholz, E. D., Ibañez, I., Miller, L. P., & Sorte, C. J. (2016). Global threats from invasive alien species in the twenty-first century and national response capacities. *Nature Communications, 7*, 12485.

ECDC. (2021). *European Centre for Disease CPrevention and Control and European Food Safety Authority*. Mosquito maps. Stockholm. Available from: https://ecdc.europa.eu/en/disease-vectors/surveillance-and-disease-data/mosquito-maps.

Economo, E. P., & Keitt, T. H. (2008). Species diversity in neutral metacommunities: A network approach. *Ecology Letters, 11*, 52–62.

Elton, C. S. (1927). *Animal ecology*. London: Sidgwick and Jackson, 296 p.

Engel, M. S., & Kristensen, N. P. (2013). A history of entomological classification. *Annual Review of Entomology, 58*, 585–607.

Eritja, R., Escosa, R., Lucientes, J., Marques, E., Roiz, D., & Ruiz, S. (2005). Worldwide invasion of vector mosquitoes: present European distribution and challenges for Spain. *Biological Invasions, 7*, 87–97.

Eritja, R., Ruiz-Arrondo, I., Delacour-Estrella, S., Schaffner, F., Álvarez-Chachero, J., Bengoa, M., Puig, M. A., Melero-Alcíbar, R., Oltra, A., & Bartumeus, F. (2019). First detection of *Aedes japonicus* in Spain: An unexpected finding triggered by citizen science. *Parasites & Vectors, 12*, 1–9.

Estrada, R., Carmona, V. J., Alarcón-Elbal, P. M., Miranda, M. A., Borrás, D., Roche, M. L., Tamarit, A., Navarro, J., & Lucientes, J. (2011). Primera cita de *Culicoides paolae* Boorman et al., 1996 (Diptera, Ceratopogonidae) para la Península Ibérica y aportaciones sobre su distribución. *Boletín de la Sociedad Entomológica Aragonesa, 49*, 217–221.

European Commission. (2000). *Directive 2000/60/EC. Establishing a framework for community action in the field of water policy*. European Commission PE-CONS 3639/1/100 Rev 1, Luxemburg. 327 pp.

Fennessy, M. S., Jacobs, A. D., & Kentula, M. E. (2004). *Review of rapid methods for assessing wetland condition. EPA/620/R-04/009*. Washington, DC: U.S. Environmental Protection Agency.

Fenoglio, S., Bonada, N., Guareschi, S., López-Rodríguez, M. J., Millán, A., & Tierno de Figueroa, J. M. (2016). Freshwater ecosystems and aquatic insects: A paradox in biological invasions. *Biology Letters*, *12*, 20151075.

Fenoy, E., & Casas, J. (2015). Two faces of agricultural intensification hanging over aquatic biodiversity: The case of chironomid diversity from farm ponds vs. natural wetlands in a coastal region. *Estuarine, Coastal and Shelf Science*, *157*, 99–108.

Fernández, J. L., Fernández, E., & Alonso-Zarazaga, M. A. (2005). Primera cita de *Stenopelmus rufinasus* Gyllenhal, 1835 en la Península Ibérica (Coleoptera, Erirhinidae). *Graellsia*, *61*, 139–140.

Ferrand, M. (2017). Présence en France du charançon aquatique du riz *Lissorhoptrus oryzophilus* Kuschel, 1952 [Coleoptera, Erirhinidae]. *Ephemera*, *18*, 31–39.

Ferreras-Romero, M., Márquez-Rodríguez, J., & Fernández-Delgado, C. (2016). Long-time effect of an invasive fish on the odonata assemblage in a Mediterranean lake and early response after rotenone treatment. *Odonatologica*, *45*, 7–21.

Figueroa, R., Bonada, N., Guevara, M., Pedreros, P., Correa-Araneda, F., Díaz, M. E., & Ruiz, V. H. (2013). Freshwater biodiversity and conservation in mediterranean climate streams of Chile. *Hydrobiologia*, *719*, 269–289.

Fikáček, M. (2009). Occurrence of introduced species of the genus *Cercyon* (Coleoptera: Hydrophilidae) in the Neotropical Region. *Revista de la Sociedad Entomológica Argentina*, *68*, 351–357.

Florencio, M., Díaz-Paniagua, C., Gómez-Rodríguez, C., & Serrano, L. (2014). Biodiversity patterns in a macroinvertebrate community of a temporary pond network. *Insect Conservation and Diversity*, *7*, 4–21.

Florencio, M., Díaz-Paniagua, C., Serrano, L., & Bilton, D. T. (2011). Spatio-temporal nested patterns in macroinvertebrate assemblages across a pond network with a wide hydroperiod range. *Oecologia*, *166*, 469–483.

Florencio, M., Fernández-Zamudio, R., Bilton, D. T., & Díaz-Paniagua, C. (2015). The exotic weevil *Stenopelmus rufinasus* Gyllenhal, 1835 (Coleoptera: Curculionidae) across a "host-free" pond network. *Limnetica*, *34*, 79–84.

Folch de la Iglesia, G. (2020). *Time to leave or to stay: Responses of aquatic invertebrates to flow intermittence* (Master thesis). Barcelona, Spain: Department of Evolutionary Biology, Ecology and Environmental Sciences, University of Barcelona, 35 pp.

Furse, M., Hering, D., Moog, O., Verdonschot, P., Johnson, R. K., Brabec, K., Gritzalis, K., Buffagni, A., Pinto, P., Friberg, N., Murray-Bligh, J., Kokes, J., Alber, R., Usseglio-Polatera, P., Haase, P., Sweeting, R., Bis, B., Szoszkiewicz, K., Soszka, H., ... Krno, I. (2006). The STAR project: Context, objectives and approaches. *Hydrobiologia*, *566*, 3–29.

Fouzi, T. A., Youness, M., Guy, C., Ali, B., & Andrés, M. (2020). The alien boatman *Trichocorixa verticalis verticalis* (Hemiptera: Corixidae) is expanding in Morocco. *Limnetica*, *39*, 49–59.

Freeland-Riggert, B. T., Cairns, S. H., Poulton, B. C., & Riggert, C. M. (2016). Differences found in the macroinvertebrate community composition in the presence or absence of the invasive alien crayfish, *Orconectes hylas*. *PLoS One*, *11*, e0150199.

Gallardo, B., Dolédec, S., Paillex, A., Arscott, D. B., Sheldon, F., Zilli, F., Mérigoux, S., Castella, E., & Comín, F. A. (2014). Response of benthic macroinvertebrates to gradients in hydrological connectivity: A comparison of temperate, subtropical, Mediterranean and semiarid river floodplains. *Freshwater Biology*, *59*, 630–648.

Gama, M. M. (1988). Note à propos de deux espèces de collemboles trouvées dans les cotes maritimes portugaises pour la premire fois. *Ciencia Biologica Ecology and Systematics (Portugal)*, *6*, 123–126.

García, C. M., García-Ruiz, R., Rendón, M., Niell, F. X., & Lucena, J. (1997). Hydrological cycle and interannual variability of the aquatic community in a temporary saline lake (Fuente de Piedra, Southern Spain). *Hydrobiologia*, *345*, 131–141.

García-Berthou, E., Boix, D., & Clavero, M. (2007). Non-indigenous animal species naturalized in Iberian inland wàters. In F. Gherardi (Ed.), *Biological invaders in inland waters: Profiles, distribution, and threats* (pp. 123–140). Dordrecht: Springer.

García-Girón, J., Wilkes, M., Fernández-Aláez, M., & Fernández-Aláez, C. (2019). Processes structuring macrophyte metacommunities in Mediterranean ponds: Combining novel methods to disentangle the role of dispersal limitation, species sorting and spatial scales. *Journal of Biogeography*, *46*, 646–656.

García-Roger, E. M., Sánchez-Montoya, M. M., Cid, N., Erba, S., Karaouzas, I., Verkaik, I., Rieradevall, M., Gómez, R., Suárez, M. L., Vidal-Abarca, M. R., Demartini, D., Buffagni, A., Skoulikidis, N., Bonada, N., & Prat, N. (2013). Spatial scale effects on taxonomic and biological trait diversity of aquatic macroinvertebrates in Mediterranean streams. *Fundamental and Applied Limnology*, *183*, 89–105

García-Vázquez, D., Bilton, D. T., Foster, G. N., & Ribera, I. (2017). Pleistocene range shifts, refugia and the origin of widespread species in western Palaearctic water beetles. *Molecular Phylogenetics and Evolution*, *114*, 122–136.

García-Vázquez, D., Bilton, D. T., Alonso, R., Benetti, C. J., Garrido, J., Valladares, L. F., & Ribera, I. (2016). Reconstructing ancient Mediterranean crossroads in *Deronectes* diving beetles. *Journal of Biogeography*, *43*, 1533–1545.

Gascón, S., Boix, D., & Sala, J. (2009). Are different biodiversity metrics related to the same factors? A case study from Mediterranean wetlands. *Biological Conservation*, *142*, 2602–2612.

Gascón, S., Boix, D., Sala, J., & Quintana, X. D. (2008). Relation between macroinvertebrate life strategies and habitat traits in Mediterranean salt marsh ponds (Empordà wetlands, NE Iberian Peninsula). *Hydrobiologia*, *597*, 71–83.

Gasith, A., & Resh, V. H. (1999). Streams in Mediterranean climate regions: abiotic influences and biotic responses to predictable seasonal events. *Annual Review of Ecology and Systematics*, *30*, 51–81.

Gavira, O., Sánchez, S., Carrasco, P., Ripoll, J., & Solís, S. (2012). Presence of the family Nevrorthidae (Neuroiptera) in the Iberian Peninsula. *Boletín de la Sociedad Entomológica Aragonesa*, *51*, 217–220.

Geiger, W., Alcorlo, P., Baltanas, A., & Montes, C. (2005). Impact of an introduced Crustacean on the trophic webs of Mediterranean wetlands. *Biological Invasions*, *7*, 49–73.

Gherardi, F. (2007). Biological invasions in inland waters: An overview. In F. Gherardi (Ed.), *Biological invaders in inland waters: Profiles, distribution, and threats. Invading Nature* (pp. 3–25). Dordrecht: Springer.

Gherardi, F., Bertolino, S., Bodon, M., Casellato, S., Cianfanelli, S., Ferraguti, M., Lori, E., Mura, G., Nocita, A., Riccardi, N., Rossetti, G., Rota, E., Scalera, R., Zerunian, S., & Tricarico, E. (2008). Animal xenodiversity in Italian inland waters: Distribution, modes of arrival, and pathways. *Biological Invasions*, *10*(4), 435–454.

Giantsis, I. A., Castells Sierra, J., & Chaskopoulou, A. (2017). The distribution of the invasive pest, rice water weevil *Lissorhoptrus oryzophilus* Kuschel (Coleoptera: Curculionidae), is expanding in Europe: First record in the Balkans, confirmed by CO1 DNA barcoding. *Phytoparasitica*, *45*, 147–149.

Girgin, S. (2010). Evaluation of the benthic macroinvertebrate distribution in a stream environment during summer using biotic index. *International Journal of Environmental Science & Technology*, *7*, 11–16.

Going, B. M., & Dudley, T. L. (2008). Invasive riparian plant litter alters aquatic insect growth. *Biological Invasions*, *10*, 1041–1051.

Gómez, A., & Lunt, D. H. (2007). Refugia within refugia: Patterns of phylogeographic concordance in the Iberian Peninsula. In S. Weiss, & N. Ferrand (Eds.), *Phylogeography of Southern European Refugia* (pp. 155–188). Dordrecht: Springer.

Gray, L. (1981). Species composition and life histories of aquatic insects in a lowland Sonoran Desert stream. *American Midland Naturalist*, *106*, 229–242.

Green, A. J., & Figuerola, J. (2005). Recent advances in the study of long- distance dispersal of aquatic invertebrates via birds. *Diversity and Distributions*, *11*, 149–156.

Greenslade, P. (n.d.). *The Aquatic Springtails (Insecta: Collembola) of South Africa*. http://www.ru.ac.za/static/departments/zoo/Martin/acollembola.html. Accessed 12 April 2020.

Greenslade, P. (2018). Why are there so many exotic Springtails in Australia? A review. *Soil Organisms*, *90*, 141–156.

Grinnell, J. (1917). The niche-relationships of the California Thrasher. *The Auk*, *34*, 427–433.

Grönroos, M., Heino, J., Siqueira, T., Landeiro, V. L., Kotanen, J., & Bini, L. M. (2013). Metacommunity structuring in stream networks: Roles of dispersal mode, distance type, and regional environmental context. *Ecology and Evolution*, *3*, 4473–4487.

Growns, J. E., Chessman, B. C., Jackson, J. E., & Ross, D. G. (1997). Rapid assessment of Australian rivers using macroinvertebrates: Cost and efficiency of 6 methods of sample processing. *Journal of the North American Benthological Society*, *16*, 682–693.

Guareschi, S., Coccia, C., Sánchez-Fernández, D., Carbonell, J. A., Velasco, J., Boyero, L., Green, A. J., & Millán, A. (2013). How far could the alien boatman *Trichocorixa verticalis verticalis* spread? Worldwide estimation of its current and future potential distribution. *PLoS One*, *8*, e59757.

Guareschi, S., Mellado-Díaz, A., Puig, M. Á., & Sánchez-Fernández, D. (2018). On the Iberian endemism *Eurylophella iberica* Keffermuller and Da Terra 1978 (Ephemeroptera, Ephemerellidae): current and future potential distributions, and assessment of the effectiveness of the natura 2000 network on its protection. *Journal of Insect Conservation*, *22*, 127–134.

Günther, H. (2004). *Trichocorixa verticalis verticalis* (Fieber), eine nearktische Ruderwanze in Europa. *Mitteilungen des Internationalen Entomologischen Vereins*, *29*, 45–49.

Gunter, G., & Christmas, J. Y. (1959). Corixid insects as part of the offshore fauna of the sea. *Ecology*, *40*, 724–725.

Gutiérrez-Cánovas, C., Hernández, J., Millán, A., & Velasco, J. (2012). Impact of chronic and pulse dilution disturbances on metabolism and trophic structure in a saline Mediterranean stream. *Hydrobiologia*, *686*, 225–239.

Haggag, A. A., Mahmoud, M. A., Bream, A. S., & Amer, M. S. (2018). Family variation of aquatic insects and water properties to assess freshwater quality in El-Mansouriya stream, Egypt. *African Entomology*, *26*, 162–173.

Hallmann, C. A., Sorg, M., Jongejans, E., Siepel, H., Hofland, N., Schwan, H., Stenmans, W., Müller, A., Sumser, H., Hörren, T., Goulson, D., & de Kroon, H. (2017). More than 75 percent decline over 27 years in total flying insect biomass in protected areas. *PLoS One*, *12*, e0185809.

Hansen, M. (1999). *World Catalogue of Insects 2: Hydrophiloidea (s. str.) (Coleoptera)*. Apollo Books, Stenstrup. 416 pp.

Hanski, I., & Gilpin, M. (1991). Metapopulation dynamics: Brief history and conceptual domain. *Biological Journal of the Linnean Society*, *42*, 3–16.

Heino, J. (2013). The importance of metacommunity ecology for environmental assessment research in the freshwater realm. *Biological Reviews*, *88*, 166–178.

Heino, J., Melo, A. S., Siqueira, T., Soininen, J., Valanko, S., & Bini, L. M. (2014). Metacommunity organisation, spatial extent and dispersal in aquatic systems: Patterns, processes and prospects. *Freshwater Biology*, *60*, 845–869.

Hellawell, J. M. (1978). *Biological surveillance of rivers, a biological monitoring hand-book*. UK: Water Research Centre, Medmenham and Stevenage. 331 pp.

Hering, D., Johnson, R. K., Kramm, S., Schmutz, S., Szoszkiewicz, K., & Verdonschot, P. F. (2006). Assessment of European streams with diatoms, macrophytes, macroinvertebrates and fish: a comparative metric-based analysis of organism response to stress. *Freshwater Biology*, *51*, 1757–1785.

Hershkovitz, Y., & Gasith, A. (2013). Resistance, resilience, and community dynamics in Mediterranean-climate streams. *Hydrobiologia*, *719*, 59–75.

Hewitt, G. M. (2004). The structure of biodiversity –insights from molecular phylogeography. *Frontiers in Zoology*, *1*, 4.

Hewitt, G. M. (2011). Mediterranean Peninsulas: The evolution of hotspots. In F. E. Zachos, & J. C. Habel (Eds.), *Biodiversity hotspots. Distribution and protection of conservation priority areas* (pp. 123–147). Heidelberg: Springer.

Hidalgo-Galiana, A., & Ribera, I. (2011). Late Miocene diversification of the genus Hydrochus (Coleoptera, Hydrochidae) in the west Mediterranean area. *Molecular Phylogenetics and Evolution*, *59*, 377–385.

Hilsenhoff, W. (2001). Diversity and classification of insects and collembola. In J. H. Thorp, & A. P. Covich (Eds.), *Ecology and classification of North American freshwater invertebrates* (pp. 661–731). New York: Academic press.

Hoddle, M. S. (2004). Restoring balance: Using exotic species to control invasive exotic species. *Conservation Biology*, *18*, 38–49.

Holgerson, M. A., Post, D. M., & Skelly, D. K. (2016). Reconciling the role of terrestrial leaves in pond food webs: A whole-ecosystem experiment. *Ecology*, *97*, 1771–1782.

Holyoak, M., Leibold, M. A., & Holt, R. D. (2005). *Metacommunities: Spatial dynamics and ecological communities* (p. 526) University of Chicago Press.

Horváth, Z., Ptacnik, R., Vad, C. F., & Chase, J. M. (2019). Habitat loss over six decades accelerates regional and local biodiversity loss via changing landscape connectance. *Ecology Letters*, *22*, 1019–1027.

Husemann, M., Schmitt, T., Zachos, F. E., Ulrich, W., & Habel, J. C. (2014). Palaearctic biogeography revisited: Evidence for the existence of a North African refugium for Western Palaearctic biota. *Journal of Biogeography*, *41*, 81–94.

Hutchinson, G. E. (1931). On the occurrence of *Trichocorixa* Kirkaldy (Corixidae, Hemiptera- Heteroptera) in salt water and its zoo-geographical significance. *The American Naturalist*, *65*, 573–574.

Hutchinson, G. E. (1957). Concluding remarks. *Cold Spring Harbor Symposia on Quantitative Biology*, *22*, 415–427.

Illies, J. (1969). Biogeography and ecology of Neotropical freshwater insects, especially those from running waters. In E. J. Fittkau, J. Illies, H. Klinge, G. H. Schwabe, & H. Sioli (Eds.), *Biogeography and ecology in South America* (Vol. 2, pp. 685–708). Monogr. Biol. Dr. W. Junk, The Hague, NL.

Ishikawa, N. F., Kato, Y., Togashi, H., Yoshimura, M., Yoshimizu, C., Okuda, N., & Tayasu, I. (2014). Stable nitrogen isotopic composition of amino acids reveals food web structure in stream ecosystems. *Oecologia*, *175*, 911–922.

Jähnig, S. C., Baranow, V., Altermatt, F., Cranston, P., Friedrichs-Manthey, M., Geist, J., He, F., Heino, J., Hering, D., Hölker, F., Jourdan, J., Kalinkat, G., Kiesel, J., Leese, F., Maasri, A., Monaghan, M., Schäfer, R., Tockner, K., Tonkin, J. D., & Domisch, S. (2021). Revisiting global trends in freshwater insects. *WIREs Water*, *8*, e1506.

Janion-Scheepers, C., Deharveng, L., Bedos, A., & Chown, S. L. (2015). Updated list of Collembola species currently recorded from South Africa. *ZooKeys*, *503*, 55–88.

Jansson, A. (1982). Notes on some Corixidae (Heteroptera) from New Guinea and New Caledonia. *Pacific Insects*, *24*, 95–103.

Jansson, A. (2002). New records of Corixidae (Heteroptera) from northeastern USA and eastern Canada, with one new synonymy. *Entomologica Fennica*, *13*, 85–88.

Jansson, A., & Reavell, P. E. (1999). North American species of *Trichocorixa* (Heteroptera: Corixidae) introduced into Africa. *African Entomology*, *7*, 295–297.

Jeliazkov, A., Mijatovic, D., Chantepie, S., Arlettaz, R., Andrew, N., Barbaro, L., Barsoum, N., Bartonova, A., Belskaya, E., Bonada, N., Brind'amour, A., Carvalho, R., Castro, H., Chmura, D., Choler, P., Chong-Seng, K., Cleary, D., Cormont, A., Cornwell, W., . . . Chase, J. (2020). A global database for metacommunity ecology, integrating species, traits, environment and space. *Scientific Data*, *7*, 6.

Jia, F.-L., Fikáček, M., & Ryndevich, S. K. (2011). Taxonomic notes on Chinese *Cercyon*: Description of a new species, new synonyms, and additional faunistic records (Coleoptera: Hydrophilidae: Sphaeridiinae). *Zootaxa*, *3090*, 41–56.

Johnson, N. F., & Triplehorn, C. A. (2004). *Borror and DeLong's introduction to the study of insects*. Brooks/Cole, 888 pp.

Jooste, M. L., Samways, M. J., & Deacon, C. (2020). Fluctuating pond water levels and aquatic insect persistence in a drought-prone Mediterranean-type climate. *Hydrobiologia*, *847*, 1315–1326.

Jordana, R. (2012). Capbryinae and Entomobryini. Synopses on Palaeartic Collembola volume 7/1. In W. Dunger & U. Burkhardt (Eds.), *Senckenberg Museum of Natural History Görlitz*. 391 pp.

Jordana, R., & Arbea, J. I. (1989). *Clave de identificación de los géneros de colémbolos de España (Insecta: Collembola). Navarra, Servicio de Publicaciones de la Universidad de Navarra S. A, Serie Zoológica* (19, pp. 1–16).

Jordana, R., Arbea, J. I., Simón, C., & Luciáñez, M. J. (1997). Collembola Poduromorpha. In M. A. Ramos, et al. (Eds.), *Fauna Iberica* (Vol. 8, p. 807). Madrid, Spain: Museo Nacional de Ciencias Naturales, CSIC.

Kalkman, V. J., Clausnitzer, V., Dijkstra, K. D. B., Orr, A. G., Paulson, D. R., & Tol, J. V. (2008). Global diversity of dragonflies (Odonata) in freshwater. *Hydrobiologia*, *595*, 351–363.

Karrouch, L., & Chahlaoui, A. (2009). Bio-évaluation de la qualité des eaux de l'oued Boufekrane (Meknès, Maroc). *Biomatec Echo*, *3*, 6–17.

Kaufman, M. G., & Fonseca, D. M. (2013). Invasion biology of *Aedes japonicus japonicus* (Diptera: Culicidae). *Annual Review of Entomology*, *59*, 31–49.

Kaunisto, K. M., Roslin, T., Sääksjärvi, I. E., & Vesterinen, E. J. (2017). Pellets of proof: First glimpse of the dietary composition of adult odonates as revealed by metabarcoding of feces. *Ecology and Evolution*, *7*, 8588–8598.

Kefford, B. J., Buchwalter, D., Cañedo-Argüelles, M., Davis, J., Duncan, R. P., Hoffman, A., & Thompson, R. (2016). Salinized rivers: Degraded systems or new habitats for salt-tolerant faunas? *Biology Letters*, *12*, 20151072.

Kenis, M., Auger-Rozenberg, M. A., Roques, A., Timms, L., Péré, C., Cock, M. J. W., Settele, J., Augustin, S., & Lopez-Vaamonde, C. (2009). Ecological effects of invasive alien insects. In D. Langor, & J. Sweeney (Eds.), *Ecological impacts of non-native invertebrates and fungi on terrestrial ecosystems* (pp. 21–45). Dordrecht: Springer.

Kenney, M. A., Sutton-Grier, A., Smith, R., & Gresens, S. (2009). Benthic macroinvertebrates as indicators of water quality: The intersection of science and policy. *Terrestrial Arthropod Reviews*, *2*, 99–128.

King, R. S., & Richardson, C. J. (2003). Integrating bioassessment and ecological risk assessment: An approach to developing numerical water-quality criteria. *Environmental Management*, *31*, 795–809.

Klecka, J., & Boukal, D. S. (2012). Who eats whom in a pool? A comparative study of prey selectivity by predatory aquatic insects. *PLoS One*, *7*, e37741.

Klugkist, C. E. (1907). Ein Parasit der *Lemna minor* L. *Sminthurus aquaticus* Bourl. *Abhandlungen herausgegeben vom Naturwissenschaftlichen Verein zu Bremen*, *19*, 45–46.

Kment, P. (2006). A contribution to the faunistics of aquatic and semiaquatic Bugs (Heteroptera: Nepomorpha, Gerromorpha) in Portugal, with the review of biology of the Neartic corixid *Trichocorixa verticalis* (Fieber, 1851). *Boletín de la Sociedad Entomológica Aragonesa*, *38*, 359–361.

Kneitel, J. M. (2014). Inundation timing, more than duration, affects the community structure of California vernal pool mesocosms. *Hydrobiologia*, *732*, 71–83.

Korbaa, M., Ferreras-Romero, M., Ruiz-García, A., & Boumaiza, M., M. (2018). TSOI – A new index based on Odonata populations to assess the conservation relevance of watercourses in Tunisia. *Odonatologica*, *47*, 43–72.

Kovats, R. S., Valentini, R., Bouwer, L. M., Georgopoulou, E., Jacob, D., Martin, E., Rounsevell, M., & Soussana, J. (2014). Europe. In V. R. Barros, C. B. Field, D. J. Dokken, M. D. Mastrandrea, K. J. Mach, T. E. Bilir, M. Chatterjee, K. L. Ebi, Y. O. Estrada, R. C. Genova, B. Girma, E. S. Kissel, A. N. Levy, S. MacCracken, P. R. Mastrandrea, & L. L. White (Eds.), *Climate change 2014: Impacts, adaptation, and vulnerability. Part B: Regional aspects. Contribution of working group II to the fifth assessment report of the intergovernmental panel on climate change* (pp. 1267–1326). Cambridge, United Kingdom and New York, NY: Cambridge University Press.

Kraemer, M. U., Reiner, R. C., Brady, O. J., Messina, J. P., Gilbert, M., Pigott, D. M., Yi, D., Johnson, K., Earl, L., Marczak, L. B., Shirude, S., Weaver, N. D., Bisanzio, D., Perkins, T. A., Lai, S., Lu, X., Jones, P., Coelho, G. E., Carvalho, R. G., ... Golding, N. (2019). Past and future spread of the arbovirus vectors *Aedes aegypti* and *Aedes albopictus*. *Nature microbiology*, *4*, 854–863.

Kroll, S. A., Ringler, N. H., de la Cruz Cano Costa, M., & De las Heras Ibañez, J. (2017). Macroinvertebrates on the front lines: Projected community response to temperature and precipitation changes in Mediterranean streams. *Journal of Freshwater Ecology*, *32*, 513–528.

Kuschel, G. S. V. D. (1951). Revisión de *Lissorhoptrus* LeConte y géneros vecinos de América. *Revista Chilena de Entomologia*, *1*, 23–74.

Ladrera, R., Gomà, J., & Prat, N. (2018). Effects of *Didymosphenia geminata* massive growth on stream communities: Smaller organisms and simplified food web structure. *PLoS One*, *13*, e0193545.

Lancaster, J., & Downes, B. J. (2013). *Aquatic entomology*. United Kingdom: Oxford University Press, 285 pp.

Lancaster, J., & Downes, B. J. (2018). Aquatic versus terrestrial insects: Real or presumed differences in population dynamics? *Insects*, *9*, 157.

Larsen, S., Muehlbauer, J. D., & Marti, E. (2016). Resource subsidies between stream and terrestrial ecosystems under global change. *Global Change Biology*, *22*, 2489–2504.

Leibold, M. A. (1995). The niche concept revisited: Mechanistic models and community context. *Ecology*, *76*, 1371–1382.

Leibold, M. A., Chase, J. M., & Ernest, S. K. M. (2017). Community assembly and the functioning of ecosystems: How metacommunity processes alter ecosystems attributes. *Ecology*, *98*, 909–919.

Leibold, M. A., Holyoak, M., Mouquet, N., Amarasekare, P., Chase, J. M., Hoopes, M. F., Holt, R. D., Shurin, J. B., Law, R., Tilman, D., Loreau, M., & Gonzalez, A. (2004). The metacommunity concept: A framework for multi-scale community ecology. *Ecology Letters*, *7*, 601–613.

Leigh, C., & Datry, T. (2017). Drying as a primary hydrological determinant of biodiversity in river systems: A broad-scale analysis. *Ecography*, *40*, 487–499.

Lek-Ang, S., Park, Y.-S., Ait-Mouloud, S., & Deharveng, L. (2007). Collembolan communities in a peat bog versus surrounding forest analyzed by using self-organizing map. *Ecological Modelling*, *203*, 9–17. Available from https://doi.org/10.1016/j.ecolmodel.2006.01.007.

Lenat, D. R., & Barbour, M. T. (1994). Using benthic macroinvertebrate community structure for rapid, cost effective, water quality monitoring: rapid bioassessment. In S. L. Loeb, & A. Spacie (Eds.), *Biological monitoring of aquatic systems* (pp. 187–215). Boca Raton, FL: Lewis Publishers.

Li, L., Zheng, B., & Liu, L. (2010). Biomonitoring and bioindicators used for river ecosystems: Definitions, approaches and trends. *Procedia environmental sciences*, *2*, 1510–1524.

Liebmann, H. (1962). *Handbuch der Frischwasser-und Abwasserbiologie. Band I, 2 Auflag. Verlag R.* München: Oldenbourg.

Liu, J., Soininen, J., Han, B. P., & Declerck, S. A. J. (2013). Effects of connectivity, dispersal directionality and functional traits on the metacommunity structure of river benthic diatoms. *Journal of Biogeography*, *40*, 2238–2248.

L'Mohdi, O., Bennas, N., Himmi, O., Hajji, K., El-Haissoufi, M., Hernando, C., Carbonell, J. A., & Millán, A. (2010). *Trichocorixa verticalis verticalis* (Fieber, 1851) (Hemiptera, Corixidae): A new exotic species in Morocco. *Boletín de la Sociedad Entomológica Aragonesa*, *46*, 395–400.

Lobera, G., Muñoz, I., López-Taranzón, D., & Batalla, R. J. (2017). Effects of flow regulation on river bed dynamics and invertebrate communities in a Mediterranean river. *Hydrobiologia*, *784*, 283–304.

Lobl, I., & Smetana, A. (2015). *Catalogue of Palaearctic Coleoptera, Volume 2: Hydrophiloidea—Staphylinoidea* (p. 942) Apollo Books.

López-Rodríguez, M. J., Peralta-Maraver, I., Gaetani, B., Sainz-Cantero, C. E., Fochetti, R., & Tierno de Figueroa, J. M. (2012). Diversity patterns and food web structure in a Mediterranean intermittent stream. *International Review of Hydrobiology*, *97*, 485–496.

López-Rodríguez, M. J., Tierno de Figueroa, J. M., Fenoglio, S., Bo, T., & Alba-Tercedor, J. (2009). Life strategies of 3 Perlodidae species (Plecoptera) in a Mediterranean seasonal stream in southern Europe. *Journal of the North American Benthological Society*, *28*, 611–625.

López Rodriguez, N., Luzón Ortega, J. M., & Tierno De Figueroa, J. M. (2018). A trophic approach to the study of the coexistence of several macroinvertebrate predators in a seasonal stream. *Vie et Milieu*, *68*, 263–270.

Lumpkin, T. A., & Plucknet, D. L. (1980). *Azolla*: Botany, physiology, and use as a green manure. *Economic Botany*, *34*, 111–153.

Lytle, D. A., & Poff, N. L. (2004). Adaptation to natural flow regimes. *Trends in Ecology and Evolution*, *19*, 94–100.

Macan, T. T. (1965). The influence of predation on the composition of fresh-water communities. In C. M. Tarzwell (Ed.), *Biological problems in water solution* (pp. 141–144). Cincinnati, OH: U.S. Department of Health, Education and Welfare.

Malicky, H. (2014). Neue Beiträge zur Kenntnis asiatischer und mediterraner Köcherfliegen (Trichoptera). *Braueria, 41*, 43–50.

Malicky, H. (2020). Expected and unexpected areas of distribution of caddisflies (Trichoptera). *Zoosymposia, 18*, 135–142.

Marchant, R., Ryan, D., & Metzeling, L. (2006). Regional and local species diversity patterns for lotic invertebrates across multiple drainage basins in Victoria. *Marine and Freshwater Research, 57*, 675–684.

Marden, J. (2008). Evolution and physiology of flight in aquatic insects. In J. Lancaster, & R. A. Briers (Eds.), *Aquatic insects: Challenges to populations* (pp. 230–249). CABI Publishing.

Margalef, R. (1955). *Los organismos indicadores en limnología*. Biología de las aguas continentales, 12. Ministerio de Agricultura, Madrid, 300 pp.

Marion, L. (2013). Is the Sacred ibis a real threat to biodiversity? Long-term study of its diet in non-native areas compared to native areas. *Comptes Rendus Biologies, 336*, 207–220.

Martin-Creuzburg, D., Kowarik, C., & Straile, D. (2017). Cross-ecosystem fluxes: Export of polyunsaturated fatty acids from aquatic to terrestrial ecosystems via emerging insects. *Science of the Total Environment, 577*, 174–182.

Martínez, J. (2014). *Biodiversidad de los tricópteros (Insecta: Trichoptera) de la península ibérica: estudio faunístico y biogeográfico* (Ph.D. thesis). Spain: Departamento de Zoología y Antropología Física, Universidade de Santiago de Compostela, Santiago de Compostela. 326 pp.

Marzin, A., Archaimbault, V., Belliard, J., Chauvin, C., Delmas, F., & Pont, D. (2012). Ecological assessment of running waters: Do macrophytes, macroinvertebrates, diatoms and fish show similar responses to human pressures? *Ecological Indicators, 23*, 56–65.

Mas-Martí, E. (2014). *Climate induced changes in headwater streams: Effects of warming and drought on resource-consumer trophic interactions* (Ph.D. thesis). Spain: University of Barcelona, Barcelona. 162 pp.

May, R. M. (1992). How many species inhabit the earth? *Scientific American, 267*, 42–49.

Mayhew, P. J. (2007). Why are there so many insect species? Perspectives from fossils and phylogenies. *Biological Reviews, 82*, 425–454.

Mazzella, L., de Bortoli, J., & Argillier, C. (2009). *Création d'un nouvel outil de bioindication basé sur les communautés d'invertébrés benthiques lacustres: méthodes d'échantillonnage et métriques candidates*. IRSTEA. 25 pp.

McAbendroth, L., Foggo, A., Rundle, S. D., & Bilton, D. T. (2005). Unravelling nestedness and spatial pattern in pond assemblages. *Journal of Animal Ecology, 74*, 41–49.

McCullough, I. M., King, K. B. S., Stachelek, J., Diaz, J., Soranno, P. A., & Cheruvelil, K. S. (2019). Applying the patch-matrix model to makes: A connectivity-based conservation framework. *Landscape Ecology, 34*, 2703–2718.

McGavin, G. C. (2001). *Essential entomology: An order-by-order introduction* (p. 328) United Kingdom: Oxford University Press.

McIntosh, A. R., Leigh, C., Boersma, K. S., McHugh, P. A., Febria, C., & García-Berthou, E. (2017). Food webs and trophic interactions in intermittent rivers and ephemeral streams. In T. Datry, N. Bonada, & A. Boulton (Eds.), *Intermittent rivers and ephemeral streams: ecology and management* (pp. 323–347). Elsevier.

Meiswinkel, R., Labuschagne, K., & Goffredo, M. (2004). Christopher Columbus and *Culicoides*: was *C. jamaicensis* Edwards, 1922 introduced into the Mediterranean 500 years ago and later re-named *C. paolae* Boorman 1996? *Veterinaria Italiana, 40*, 340–344.

Mellado-Díaz, A., Suarez-Alonso, M. L., & Vidal-Abarca, M. R. (2008). Biological traits of stream macroinvertebrates from a semi-arid catchment: Patterns along complex environmental gradients. *Freshwater Biology, 53*, 1–21.

Menetrey, N., Oertli, B., & Lachavanne, J. B. (2011). The CIEPT: A macroinvertebrate-based multimetric index for assessing the ecological quality of Swiss lowland ponds. *Ecological Indicators, 11*, 590–600.

Menezes, S., Baird, D. J., & Soares, A. M. V. M. (2010). Beyond taxonomy: A review of macroinvertebrate trait-based community descriptors as tools for freshwater biomonitoring. *Journal of Applied Ecology, 47*, 711–719.

Merkley, S. S., Rader, R. B., & Schaalje, G. B. (2015). Introduced Western Mosquitofish (Gambusia *affinis*) reduce the emergence of aquatic insects in a desert spring. *Freshwater Science (New York, N.Y.), 34*, 564–573.

Merritt, R. W., Cummins K. W. & Berg M. B. (Eds.) (2019). An Introduction to the Aquatic Insects of North America. Kendall/Hunt Publishing Company.

Metcalfe, J. L. (1989). Biological water quality assessment of running waters based on macroinvertebrate communities: History and present status in europe. *Environmental Pollution, 60*, 101–139.

Meulenkamp, J. E., & Sissingh, W. (2003). Tertiary palaeogeography and tectonostratigraphic evolution of the Northern and Southern Peri-Tethys platforms and the intermediate domains of the African–Eurasian convergent plate boundary zone. *Palaeogeography, Palaeoclimatology, Palaeoecology, 196*, 209–228.

Miatta, M., Bates, A. E., & Snelgrove, P. V. R. (2021). Incorporating biological traits into conservation strategies. *Annual Review of Marine Science, 13*, 421–443.

Migliore, J., Baumel, A., Leriche, A., Juin, M., & Médail, F. (2018). Surviving glaciations in the Mediterranean region: An alternative to the long-term refugia hypothesis. *Botanical Journal of the Linnean Society, 187*, 537–549.

Milla, R., Castro-Díez, P., & Montserrat-Martí, G. (2010). Phenology of Mediterranean woody plants from NE Spain: Synchrony, seasonality, and relationships among phenophases. *Flora, 205*, 190–199.

Millán, A., Abellán, P., Sánchez-Fernández, D., Gutiérrez-Cánovas, C., Picazo, F., Arribas, P., Belmar, O., & Velasco, J. (2009). Biodiversidad de macroinvertebrados en los ecosistemas acuáticos salinos en la Región de Murcia. In J. F. Carrasco, & K. Hueso (Eds.), *Los Paisajes Ibéricos de la Sal: 2. Humedales salinos de interior* (pp. 11–126). Guadalajara: Asociación de amigos de las salinas de interior.

Millán, A., Hernando, C., Aguileira, P., Castro, A., & Ribera, I. (2005). Los coleópteros acuáticos y semiacuáticos de Doñana: Reconocimiento de su biodiversidad y prioridades de conservación. *Boletín de la Sociedad Entomológica Aragonesa*, *36*, 157–164.

Millán, A., Velasco, J., Gutiérrez-Cánovas, C., Arribas, P., Picazo, F., Sánchez-Fernández, D., & Abellán, P. (2011). Mediterranean saline streams in southeast Spain: What do we know? *Journal of Arid Environments*, *75*, 1352–1359.

Minshall, G. W., Brock, J. T., & Varley, J. D. (1989). Wildfires and Yellowstone's stream ecosystems. *Bioscience*, *39*, 707–715.

Montgomery, G. A., Dunn, R. R., Fox, R., Jongejans, E., Leather, S. R., Saunders, M. E., Shortall, C. R., Tingley, M., & Wagner, D. L. (2020). Is the insect apocalypse upon us? How to find out. *Biological Conservation*, *241*, 108327.

Mor, J. R., Dolédec, S., Acuña, V., Sabater, S., & Muñoz, I. (2019). Invertebrate community responses to urban wastewater effluent pollution under different hydro-morphological conditions. *Environmental pollution*, *252*, 483–492.

Mor, J. R., Sabater, L. C., Masferrer, J., Sala, J., Font, J., & Boix, D. (2010). Presence of the exotic weevil *Stenopelmus rufinasus* Gyllenhal, 1836 (Coleoptera: Erirhinidae) in Ter River (NE Iberian Peninsula). *Boletín de la Sociedad Entomológica Aragonesa*, *46*, 367–372.

Morais, M., Pinto, P., Pedro, A., Battin, T., Gafny, S., Gerino, M., Marti, E., Puig, M., Pusch, A., Solimini, A., Voreadou, C., Sabater, F., & Usseglio-Polatera, P. (2009). Relationships among macroinvertebrate community structure, bio/ecological trait profiles, and environmental descriptors in European human-altered streams. *Verhandlungen des Internationalen Verein Limnologie*, *30*, 1234–1238.

Moreno, J. L., & De las Heras, J. (2009). Habitat selection and life cycle of aquatic Diptera in a semiarid saline stream in Spain—An approximation. *Lauterbornia*, *68*, 71–81.

Moreno, J. L., Angeler, D. J., & De las Heras, J. (2010). Seasonal dynamics of macroinvertebrate communities in a semiarid saline spring stream with contrasting environmental conditions. *Aquatic Ecology*, *44*, 177–193.

Montauban, C., Mas, M., Wangensteen, O. S., Sarto, I., Monteys, V., Fornós, D. G., Mola, X. F., & López-Baucells, A. (2021). Bats as natural samplers: First record of the invasive pest rice water weevil *Lissorhoptrus oryzophilus* in the Iberian Peninsula. *Crop Protection*, *141*, 105427.

Mouthon, J. (1993). Un indice biologique lacustre basé sur l'examen des peuplements de mollusques. *Bulletin Français de la Pêche et de la Pisciculture*, *331*, 397–406.

Munné, A., & Prat, N. (2009). Use of macroinvertebrate-based multimetric indices for water quality evaluation in Spanish Mediterranean rivers: An intercalibration approach with the IBMWP index. *Hydrobiologia*, *628*, 203–225.

Muñoz, I. (2003). Macroinvertebrate community structure in an intermittent and a permanent Mediterranean streams (NE Spain). *Limnetica*, *22*, 107–116.

Muñoz, I., Romaní, A. M., Rodriguez-Capitulo, A., González-Esteban, J., & Garcia-Berthou, E. (2009). Capítulo 19: Relaciones tróficas en el ecosistema fluvial. In S. Sabater, & A. Elosegi (Eds.), *Conceptos y Técnicas en ecología fluvial, Fundación BBVA, Bilbao, España* (pp. 347–366).

Muñoz-Mas, R., & García-Berthou, E. (2020). Alien animal introductions in Iberian inland waters: An update and analysis. *Science of the Total Environment*, *703*, 134505.

Múrria, C., Dolédec, S., Papadopoulou, A., Vogler, A. P., & Bonada, N. (2018). Ecological constraints from incumbent clades drive trait evolution across the tree-of-life of freshwater macroinvertebrates. *Ecography*, *41*, 1049–1063.

Múrria, C., Sáinz-Bariáin, M., Vogler, A. P., Viza, A., González, M., & Zamora-Muñoz, C. (2020). Vulnerability to climate change for two endemic high-elevation, low-dispersive *Annitella* species (Trichoptera) in Sierra Nevada, the southernmost high mountain in Europe. *Insect Conservation and Diversity*, *13*, 283–295.

Naselli-Flores, L., & Marrone, F. (2019). Different invasibility of permanent and temporary waterbodies in a semiarid Mediterranean Island. *Inland Waters*, *9*, 411–421.

Nasserzadeh, H., Alipanah, H., & Gilasian, E. (2017). Phylogenetic study of the genus *Sternolophus* Solier (Coleoptera, Hydrophilidae) based on adult morphology. *ZooKeys*, *712*, 69–85.

Noble-Nesbitt, J. (1963). Transpiration in *Podura aquatica* L. and the wetting properties of its cuticle. *The Journal of Experimental Biology*, *40*, 681–700.

O'Neill, B. J., & Thorp, J. H. (2014). Untangling food-web structure in an ephemeral ecosystem. *Freshwater Biology*, *59*, 1462–1473.

Pace, G., Bonada, N., & Prat, N. (2013). Long-term effects of climatic-hydrological drivers on macroinvertebrate richness and composition in two Mediterranean streams. *Freshwater Biology*, *58*, 1313–1328.

Palacios-Vargas, J. G. (2013). Biodiversidad de Collembola (Hexapoda: Entognatha) en México. *Revista Mexicana de Biodiversidad*, *84*, 220–231.

Pallarés, S., Botella-Cruz, M., Arribas, P., Millán, A., & Velasco, J. (2017). Aquatic insects in a multistress environment: Cross-tolerance to salinity and desiccation. *Journal of Experimental Biology*, *220*, 1277–1286.

Pallarés, S., Velasco, J., Millán, A., Bilton, D. T., & Arribas, P. (2016). Aquatic insects dealing with dehydration: Do desiccation resistance traits differ in species with contrasting habitat preferences? *PeerJ*, *2016*, 1–19.

Pausas, J. G., & Fernández-Muñoz, S. (2012). Fire regime changes in the western Mediterranean basin: From fuel-limited to drought-driven fire regime. *Climatic Change*, *110*, 215–226.

Peckarsky, B. L., & Lamberti, G. A. (2017). Invertebrate consumer-resource interactions. In F. R. Hauer, & G. Lamberti (Eds.), *Methods in stream ecology, Ecosystem structure* (3rd edition, pp. 379–398). Academic Press.

Pekel, J. F., Cottam, A., Gorelick, N., & Belward, A. S. (2016). High resolution mapping of global surface water and its long-term changes. *Nature*, *540*, 418–422.

Peralta-Maraver, I., López-Rodríguez, M. J., & Tierno de Figueroa., J. M. (2017). Structure, dynamics and stability of a Mediterranean river food web. *Marine and Freshwater Research*, *68*, 484–495.

Picazo, F., Bilton, D. T., Moreno, J. L., Sánchez-Fernández, D., & Millán, A. (2012). Water beetle biodiversity in Mediterranean standing waters: Assemblage composition, environmental drivers and nestedness patterns. *Insect Conservation and Diversity*, *5*, 146–158.

Pilotto, F., Bazzanti, M., Di Vito, V., Frosali, D., Livretti, F., Mastrantuono, L., Pusch, M. T., Sena, F., & Solimini, A. G. (2015). Relative impacts of morphological alteration to shorelines and eutrophication on littoral macroinvertebrates in Mediterranean lakes. *Freshwater Science*, *34*, 410–422.

Pilotto, F., Kühn, I., Adrian, R., Alber, R., Alignier, A., Andrews, C., Bäck, J., Barbaro, L., Beaumont, D., Beenaerts, N., Benham, S., Boukal, D. S., Bretagnolle, V., Camatti, E., Canullo, R., Cardoso, P. G., Ens, B. J., Everaert, G., Evtimova, V., . . . Haase, P. (2020). Meta-analysis of multidecadal biodiversity trends in Europe. *Nature Communications*, *11*, 3486.

Pocheville, A. (2015). The ecological niche: History and recent controversies. In T. Heams, P. Huneman, G. Lecointre, & M. Silberstein (Eds.), *Handbook of evolutionary thinking in the sciences* (pp. 547–586). Dordrecht: Springer.

Polhemus, D. A. (1993). Conservation of aquatic insects: Worldwide crisis or localized threats? *American Zoologist*, *33*, 588–598.

Pompanon, F., Deagle, B. E., Symondson, W. O. C., Brown, D. S., Jarman, S. N., & Taberlet, P. (2012). Who is eating what: Diet assessment using next generation sequencing. *Molecular Ecology*, *21*, 1931–1950.

Pope, K. L., Piovia-Scott, J., & Lawler, S. P. (2009). Changes in aquatic insect emergence in response to whole-lake experimental manipulations of introduced trout. *Freshwater Biology*, *54*, 982–993.

Popova, O. N., Haritonov, A. Y., Sushchik, N. N., Makhutova, O. N., Kalachova, G. S., Kolmakova, A. A., & Gladyshev, M. I. (2017). Export of aquatic productivity, including highly unsaturated fatty acids, to terrestrial ecosystems via Odonata. *Science of the Total Environment*, *581–582*, 40–48.

Poquet, J. M., Alba-Tercedor, J., Puntí, T., Sánchez-Montoya, M. M., Robles, S., Álvarez, M., Zamora-Muñoz, C., Sáinz-Cantero, C. E., Vidal-Abarca, M. R., Suárez, M. L., Toro, M., Pujante, A. M., Rieradevall, M., & Prat, N. (2009). The MEDiterranean Prediction and Classification System (MEDPACS): An implementation of the RIVPACS/AUSRIVAS predictive approach for assessing Mediterranean aquatic macroinvertebrate communities. *Hydrobiologia*, *623*, 153–171.

Por, F. D. (1975). An outline of the zoogeography of the Levant. *Zoologica Scripta*, *4*, 5–20.

Potapov, M. (2001). Isotomidae. Synopses on Palaearctic Collembola volume 3. In W. Dunger (Ed.), *Staaliches Museum für Naturkunde Görlitz*. 603 pp.

Powell, J. R., & Tabachnick, W. J. (2013). History of domestication and spread of *Aedes aegypti*—A review. *Memórias do Instituto Oswaldo Cruz*, *108*, 11–17.

Power, M. E., Holomuzki, J. R., & Lowe, R. L. (2013). Food webs in Mediterranean rivers. *Hydrobiologia*, *719*, 119–136.

Prat, N., & Bonada, N. (2002). Resultados del proyecto GUADALMED sobre el Estado Ecológico de los ríos Mediterráneos. *Limnetica*, *21*, 1–204.

Previšić, A., Walton, C., Kučinić, M., Mitrikeski, P. T., & Kerovec, M. (2009). Pleistocene divergence of Dinaric Drusus endemics (Trichoptera, Limnephilidae) in multiple microrefugia within the Balkan Peninsula. *Molecular Ecology*, *18*, 634–647.

Quintana, X. D., Comín, F. A., & Moreno-Amich, R. (1998). Nutrient and plankton dynamics in a Mediterranean salt marsh dominated by incidents of flooding. Part 2: Response of the zooplankton community to disturbances. *Journal of Plankton Research*, *20*, 2109–2127.

Razeng, E., Morán-Ordóñez, A., Box, J. B., Thompson, R., Davis, J., & Sunnucks, P. (2016). A potential role for overland dispersal in shaping aquatic invertebrate communities in arid regions. *Freshwater Biology*, *61*, 745–757.

Resh, V. H., & Jackson, J. K. (1993). Rapid assessment approaches to biomonitoring using benthic macroinvertebrates. In V. H. Resh, & D. M. Rosenberg (Eds.), *Freshwater biomonitoring and benthic macroinvertebrates* (pp. 195–233). New York: Chapman & Hall.

Ribera, I. (2008). Habitat constraints and the generation of diversity in freshwater macroinvertebrates. In J. Lancaster, & R. A. Briers (Eds.), *Aquatic insects: Challenges to populations* (pp. 289–311). Wallingford: CAB International.

Ribera, I., & Faille, A. (2010). A new microphthalmic stygobitic *Graptodytes* Seidlitz from Morocco, with a molecular phylogeny of the genus (Coleoptera, Dytiscidae). *Zootaxa*, *2641*, 1–14.

Ribera, I., Castro, A., & Hernando, C. (2010). *Ochthebius* (*Enicocerus*) *aguilerai* sp.n. from central Spain, with a molecular phylogeny of the Western Palaearctic species of *Enicocerus* (Coleoptera, Hydraenidae). *Zootaxa*, *2351*, 1–13.

Ricciardi, A. (2015). Ecology of invasive alien invertebrates. In J. H. Thorp, & D. C. Rogers (Eds.), *Thorp and Covich's freshwater invertebrates: Ecology and general biology* (pp. 83–91). Amsterdam: Academic Press.

Richerson, P. J., & Grigarick, A. A. (1967). The life history of *Stenopelmus rufinasus* (Coleoptera: Curculionidae). *Annals of the Entomological Society of America*, *60*, 351–354.

Robson, B. J., Chester, E. T., Mitchell, B. D., & Matthews, T. G. (2013). Disturbance and the role of refuges in mediterranean climate streams. *Hydrobiologia*, *719*, 77–91.

Rocchi, S. (2006). Insecta Coleoptera Hydrophiloidea. In S. Ruffo & F. Stoch (Eds.), *Checklist and Distribution of the Italian fauna* (pp. 167–168). Verona: Memorie del Museo Civico di Storia Naturale di Verona.

Rodríguez, C. F., Bécares, E., Fernández-Aláez, M., & Fernández-Aláez, C. (2005). Loss of diversity and degradation of wetlands as a result of introducing exotic crayfish. *Biological Invasions*, *7*, 75–85.

Rodríguez-Pérez, H., Florencio, M., Gómez-Rodríguez, C., Green, A. J., Díaz-Paniagua, C., & Serrano, L. (2009). Monitoring the invasion of the aquatic bug *Trichocorixa verticalis verticalis* (Hemiptera: Corixidae) in the wetlands of Doñana National Park (SW Spain). *Hydrobiologia*, *634*, 209–217.

Rosenberg, D. M., & Resh, V. H. (Eds.), (1993). *Freshwater biomonitoring and benthic macroinvertebrates*. New York: Chapman & Hall.

Rueda, J., & Jordana, R. (2020). Checklist of Collembola (Hexapoda: Entognatha) from "malladas" of the Devesa and Racó de l'Olla (Albufera Natural Park, Valencia, Spain) with a description of a sp. nov. *Limnetica*, *39*(1), 93–111.

Ruiz-García, A., Márquez-Rodríguez, J., & Ferreras-Romero, M. (2013). Discovery of *Nyctiophylax* (Trichoptera: Polycentropodidae) in Europe, with the description of a new species. *Freshwater Science, 32*, 169–175.

Sabo, J. L., Finlay, J. C., Kennedy, T., & Post, D. M. (2010). Food chain length in rivers. *Science (New York, N.Y.), 203*, 2005–2007.

Sahuquillo, M., & Miracle, M. R. (2013). The role of historic and climatic factors in the distribution of crustacean communities in Iberian Mediterranean ponds. *Freshwater Biology, 58*, 1251–1266.

Sailer, R. I. (1948). The genus *Trichocorixa* (Corixidae, Hemiptera). *In*: H. B. Hungerford, The Corixidae of the Western Hemisphere (Hemiptera). *The University of Kansas Science Bulletin, 32*, 289–407.

Sala, J., & Boix, D. (2005). Presence of the Nearctic water boatman *Trichocorixa verticalis verticalis* (Fieber, 1851) (Heteroptera, Corixidae) in the Algarve region (S Portugal). *Graellsia, 61*, 31–36.

Salavert, V., Zamora-Muñoz, C., Ruiz-Rodríguez, M., Fernández-Cortés, A., & Soler, J. J. (2008). Climatic conditions, diapause and migration in a troglophile caddisfly. *Freshwater Biology, 53*, 1606–1617.

Sánchez, M. I., Coccia, C., Valdecasas, A. G., Boyero, L., & Green, A. J. (2015). Parasitism by water mites in native and exotic Corixidae: Are mites limiting the invasion of the water boatman *Trichocorixa verticalis* (Fieber, 1851)? *Journal of Insect Conservation, 19*, 433–447.

Sánchez-Bayo, F., & Wyckhuys, K. A. (2019). Worldwide decline of the entomofauna: A review of its drivers. *Biological Conservation, 232*, 8–27.

Sánchez-Morales, M., Sabater, F., & Muñoz, I. (2018). Effects of urban wastewater on hyporheic habitat and invertebrates in Mediterranean streams. *Science of the Total Environment, 642*, 937–945.

Sanpera-Calbet, I., Acuña, V., Butturini, A., Marcé, R., & Muñoz, I. (2016). El Niño southern oscillation and seasonal drought drive riparian input dynamics in a Mediterranean stream. *Limnology and Oceanography, 61*, 214–226.

Sarremejane, R., Cañedo-Argüelles, M., Prat, N., Mykrä, H., Muatka, T., & Bonada, N. (2017). Do metacommunities vary through time? Intermittent rivers as model systems. *Journal of Biogeography, 44*, 2752–2763.

Sarremejane, R., Cid, N., Stubbington, R., Datry, T., Alp, M., Cañedo-Argüelles, M., Cordero-Rivera, A., Csabai, Z., Gutiérrez-Cánovas, C., Heino, J., Forcellini, M., Millán, A., Paillex, A., Paril, P., Polásek, M., Tierno De Figueroa, J. M., Usseglio-Polatera, P., Zamora-Muñoz, C., & Bonada, N. (2020). DISPERSE, a trait database to assess the dispersal potential of European aquatic macroinvertebrates. *Scientific Data, 7*, 386.

Schäfer, M. L., Lundkvist, E., Landin, J., Persson, T. Z., & Lundström, J. O. (2006). Influence of landscape structure on mosquitoes (Diptera: Culicidae) and dytiscids (Coleoptera: Dytiscidae) at five spatial scales in Swedish wetlands. *Wetlands, 26*, 57–68.

Schaffner, F., & Mathis, A. (2014). Dengue and dengue vectors in the WHO European region: Past, present, and scenarios for the future. *The Lancet Infectious Diseases, 14*, 1271–1280.

Schaffner, F., Bellini, R., Petric, D., & Scholte, E. J. (2009). The invasive mosquito *Aedes japonicus* in central Europe. *Medical and Veterinary Entomology, 23*, 448–451.

Schmida, A., & Wilson, M. V. (1985). Biological determinants of species diversity. *Journal of Biogeography, 12*, 1–20.

Schneider, D. W., & Frost, T. M. (1996). Habitat duration and community structure in temporary ponds. *Journal of the North American Benthological Society, 15*, 64–86.

Scian, B., & Donnari, M. (1997). Retrospective analysis of the palmer drought severity index in the semi-arid Pampas Region, Argentina. *International Journal of Climatology, 17*, 313–322.

Selga, D. (1971). Catálogo de los colémbolos de la Península Ibérica. *Graellsia, 24*, 133–283.

Sellam, N., Viñolas, A., Fatah, Z., & Moulai, R. (2016). L'utilisation des Coleoptera, Ephemeroptera et Diptera comme bioindicateurs de la qualite des eaux de quelques Oueds en Algérie. *Butlletí de la Institució Catalana d'Història Natural, 80*, 47–56.

Shaw, R. H., Ellison, C. A., Marchante, H., Pratt, C. F., Schaffner, U., Sforza, R. F., & Deltoro, V. (2018). Weed biological control in the European Union: From serendipity to strategy. *BioControl, 63*, 333–347.

Simberloff, D., Martin, J.-L., Genovesi, P., Maris, V., Wardle, D. A., Aronson, J., Courchamp, F., Galil, B., García-Berthou, E., Pascal, M., Pyšek, P., Sousa, R., Tabacchi, E., & Vila, M. (2013). Impacts of biological invasions: What's what and the way forward. *Trends in Ecology & Evolution, 28*, 58–66.

Simonsen, T. J., Olsen, K., & Djernæs, M. (2020). The African-Iberian connection in Odonata: mtDNA and ncDNA based phylogeography of *Aeshna cyanea* (Müller, 1764) (Odonata: Aeshnidae) in Western Palaearctic. *Arthropod Systematics & Phylogeny, 78*, 309–320.

Sinitshenkova, N. (2003). Main ecological events in aquatic insects history. *Acta zoologica cracoviensia, 46*(suppl. – Fossil Insects), 381–392.

Solà, C., Burgos, M., Plazuelo, Á., Toja, J., Plans, M., & Prat, N. (2004). Heavy metal bioaccumulation and macroinvertebrate community changes in a Mediterranean ctream affected by acid mine drainage and an accidental spill (Guadiamar river, SW Spain). *Science of the Total Environment, 333*, 109–126.

Solimini, A. G., Bazzanti, M., Ruggiero, A., & Carchini, G. (2008). Developing a multimetric index of ecological integrity based on macroinvertebrates of mountain ponds in central Italy. *Hydrobiologia, 597*, 109–123.

Sondermann, M., Gies, M., Hering, D., Schröder, M., & Feld, C. K. (2015). Modelling the effect of in-stream and terrestrial barriers on the dispersal of aquatic insect species: A case study from a Central European mountain catchment. *Fundamental and Applied Limnology, 186*, 99–115.

Soria, M., Gutiérrez-Cánovas, C., Bonada, N., Acosta, R., Rodríguez-Lozano, P., Fortuño, P., Burgazzi, G., Vinyoles, D., Gallart, F., Latron, J., Llorens, P., Prat, N., & Cid, N. (2020). Natural disturbances can produce misleading bioassessment results: Identifying metrics to detect anthropogenic impacts in intermittent rivers. *Journal of Applied Ecology, 57*, 283–295.

Souilmi, F., Ghedda, K., Fahde, A., Bellali, F., Tahraoui, S., & Malki, M. (2017). Pollution evaluation in the Oum Er Rbia River (Morocco) using macroinvertebrate-based indices andphysico-chemical parameters. *Journal of Materials and Environmental Sciences, 8*, 4840–4845.

Souilmi, F., Ghedda, K., Fahde, A., Fihri, F. E., Tahraoui, S., Elasri, F., & Malki, M. (2019). Taxonomic diversity of benthic macroinvertebrates along the Oum Er Rbia River (Morocco): Implications for water quality bio-monitoring using indicator species. *West African Journal of Applied Ecology*, *27*, 137–149.

Southwood, T. R. E. (1977). Habitat, the templet for ecological strategies? *Journal of Animal Ecology*, *46*, 337–365.

Stamou, G. P. (1998). *Arthropods of Mediterranean-Type Ecosystems* (p. 141) Berlin: Springer.

Statzner, B. (2012). Geomorphological implications of engineering bed sediments by lotic animals. *Geomorphology*, *157–158*, 49–65.

Statzner, B., Bis, B., Dolédec, S., & Usseglio-Polatera, P. (2001). Perspectives for biomonitoring at large spatial scales: A unified measure for the functional composition of invertebrate communities in European running waters. *Basic and Applied Ecology*, *2*, 73–85.

Statzner, B., Bonada, N., & Dolédec, S. (2007). Conservation of taxonomic and biological trait diversity of European stream macroinvertebrate communities: A case for a collective public database. *Biodiversity and Conservation*, *16*, 3609–3632.

Statzner, B., Hildrew, A. G., & Resh, V. H. (2001). Species traits and environmental constraints: Entomological research and the history of ecological theory. *Annual Review of Entomology*, *46*, 291–316.

Storey, R. G., & Quinn, J. M. (2013). Survival of aquatic invertebrates in dry bed sediments of intermittent streams: Temperature tolerances and implications for riparian management. *Freshwater Science*, *32*, 250–266.

Strachan, S. R., Chester, E. T., & Robson, B. J. (2015). Freshwater invertebrate life history strategies for surviving desiccation. *Springer Science Reviews*, *3*, 57–75.

Strenzke, K. (1955). Thalassobionte und thalassophile Collembola. *Die Tierwelt der Nord-und Ostsee*, *36*(2), 1–52.

Stubbington, R., & Datry, T. (2013). The macroinvertebrate seedbank promotes community persistence in temporary rivers across climate zones. *Freshwater Biology*, *58*, 1202–1220.

Sýkora, V., García-Vázquez, D., Sánchez-Fernández, D., & Ribera, I. (2017). Range expansion and ancestral niche reconstruction in the Mediterranean diving beetle genus *Meladema* (Coleoptera, Dytiscidae). *Zoologica Scripta*, *46*, 445–458.

Tachet, H., Morse, J. C., & Berly, A. (2001). The larva and pupa of *Pseudoneureclipsis lusitanicus* Malicky, 1980 (Trichoptera: Hydropsychoidea): description, ecological data and taxonomical considerations. *Aquatic Insects*, *23*, 93–106.

Tachet, H., Richoux, P., Bournaud, M., & Usseglio-Polatera, P. (2010). *Invertébrés d'eau douce: systématique, biologie, écologie* (3rd ed). Paris: CNRS Éditions, 588 pp.

Tamm, J. C. (1984). Surviving long submergence in the egg stage—A successful strategy of terrestrial arthropods living on flood plains (Collembola, Acari, Diptera). *Oecologia*, *61*, 417–419.

Thibaud, J. M., & Massoud, Z. (1983). Un nouveau genre d'insectes collemboles Hypogastruridae cavernicole du Pays Basque. *Mémoires de Biospéologie*, *10*, 317–319.

Thibaud, J. M., & Massoud, Z. (1986). Un nouveau genre d'Insectes Collemboles Onychiuridae cavernicoles des Picos de Europa (Espagne). *Bulletin du Muséum National d'Histoire Naturelle, Paris*, *8*(2), 327–331.

Thibaud, J. M., Schulz, H. J., & da Gama, M. M. (2004). Hypogastruridae. Synopses on Palaeartic Collembola volume 4. In W. Dunger (Eds.), *Staaliches Museum für Naturkunde Görlitz*. 287 pp.

Thompson, R. M., Brose, U., Dunne, J. A., Hall, R. O., Hladyz, S., Kitching, R. L., Martinez, N. D., Rantala, H., Romanuk, T. N., Stouffer, D. B., & Tylianakis, J. M. (2012). Food webs: Reconciling the structure and function of biodiversity. *Trends in Ecology and Evolution*, *27*, 689–697.

Tierno de Figueroa, J. M., López-Rodríguez, M. J., Fenoglio, S., Sánchez-Castillo, P., & Fochetti, R. (2013). Freshwater biodiversity in the rivers of the Mediterranean Basin. *Hydrobiologia*, *719*, 137–186.

Tierno de Figueroa, J. M., López-Rodríguez, M. J., & Villar-Argaiz, M. (2019). Spatial and seasonal variability in the trophic role of aquatic insects: An assessment of functional feeding group applicability. *Freshwater Biology*, *64*, 954–966.

Tonkin, J. D., Bogan, M. T., Bonada, N., Rios-Touma, B., & Lytle, D. A. (2017). Seasonality and predictability shape temporal species diversity. *Ecology*, *98*, 1201–1216.

Tonkin, J. D., Florian, A., Finn, D. S., Heino, J., Olden, J. D., Pauls, S. U., & Lytle, D. A. (2018). The role of dispersal in river network metacommunities: Patterns, processes, and pathways. *Freshwater Biology*, *63*, 141–163.

Topaz, T., Egozi, R., Suari, Y., Ben-Ari, J., Sade, T., Chefetz, B., & Yahel, G. (2020). Environmental risk dynamics of pesticides toxicity in a Mediterranean micro-estuary. *Environmental Pollution*, *265*, 114941.

Tornero, I., Boix, D., Bagella, S., Pinto-Cruz, C., Caria, M. C., Belo, A., Lumbreras, A., Sala, J., Compte, J., & Gascón, S. (2018). Dispersal mode and spatial extent influence distance-decay patterns in pond metacommunities. *PLoS One*, *13*, e0203119.

Tornero, I., Sala, J., Gascón, S., Avila, N., Quintana Pou, X., & Boix, D. (2016). Pond size effect on macrofauna community structure in a highly connected pond network. *Limnetica*, *35*, 337–354.

Tornero, I., Sala, J., Gascón, S., Àvila, N., Quintana, X. D., & Boix, D. (2014). Aquatic macrofauna of Vila Nova de Milfontes temporary ponds, with the first record of *Cyphon hilaris* Nyholm, 1944 (Coleoptera: Scirtidae) from Portugal. *Boletín de la Sociedad Entomológica Aragonesa*, *55*, 326–330.

Torres-Ruiz, M., Wehr, J. D., & Perrone, A. A. (2007). Trophic relations in a stream food web: Importance of fatty acids for macroinvertebrate consumers. *Journal of the North American Benthological Society*, *26*, 509–522.

Touaylia, S., Garrido, J., & Boumaiza, M. (2013). Abundance and diversity of the aquatic beetles in a Mediterranean stream system (northern Tunisia). *Annales de La Societe Entomologique de France*, *49*, 172–180.

Townsend, C. R., Thompson, R. M., McIntosh, A. R., Kilroy, C., Edwards, E., & Scarsbrook, M. R. (1998). Disturbance, resource supply, and food-web architecture in streams. *Ecology Letters*, *1*, 200–209.

Tricarico, E., Junqueira, A. O., & Dudgeon, D. (2016). Alien species in aquatic environments: a selective comparison of coastal and inland waters in tropical and temperate latitudes. *Aquatic Conservation: Marine and Freshwater Ecosystems, 26*, 872–891.

Trizzino, M., Audisio, P. A., Antonini, G., Mancini, E., & Ribera, I. (2011). Molecular phylogeny and diversification of the "*Haenydra*" lineage (Hydraenidae, genus *Hydraena*), a north-Mediterranean endemic-rich group of rheophilic Coleoptera. *Molecular Phylogenetics and Evolution, 61*, 772–783.

Tsuzuki, H., & Isogawa, Y. (1976). The occurrence of a new insect pest, the rice water weevil in Aichi prefecture. *Plant Protection, 30*, 341.

Turco, M., José Rosa-Cánovas, J., Bedia, J., Jerez, S., Pedro Montávez, J., Carmen Llasat, M., & Provenzale, A. (2018). Exacerbated fires in Mediterranean Europe due to anthropogenic warming projected with non-stationary climate-fire models. *Nature Communications, 3821*, 1–9.

Turin, P., Bilò, M. F., & Belfiore, C. (1997). Primo rinvenimento in Italia di *Ametropus fragilis* Albarda 1878 (Ephemeroptera: Ametropodidae). *Lavori della Societa venetiana di Scienza Naturale, 22*, 7–14.

Van Klink, R., Bowler, D. E., Gongalsky, K. B., Swengel, A. B., Gentile, A., & Chase, J. M. (2020). Meta-analysis reveals declines in terrestrial but increases in freshwater insect abundances. *Science (New York, N.Y.), 368*, 417–420.

Vander Vorste, R., Sarremejane, R., & Datry, T. (2020). Intermittent rivers and ephemeral streams: A unique biome with important contributions to biodiversity and ecosystem services. In M. I. Goldstein, & D. A. DellaSala (Eds.), *Encyclopedia of the world's biomes* (pp. 419–429). Boston: Elsevier.

Vander Zanden, M. J., & Rasmussen, J. B. (2001). Variation in dl5N and d13C trophic fractionation: Implications for aquatic food web studies. *Limnology and Oceanography, 46*, 2061–2066.

Veiga, J., Martínez-de la Puente, J., Václav, R., Figuerola, J., & Valera, F. (2018). *Culicoides paolae* and *C. circumscriptus* as potential vectors of avian haemosporidians in an arid ecosystem. *Parasites & Vectors, 11*, 1–10.

Velasco, J., Millán, A., Hernández, J., Gutiérrez, C., Abellán, P., Sánchez, D., & Ruiz, M. (2006). Response of biotic communities to salinity changes in a Mediterranean hypersaline stream. *Saline Systems, 2*, 1–15.

Verdonschot, P. F. M., & Nijboer, R. C. (2004). Testing the European stream typology of the water framework directive for macroinvertebrates. *Hydrobiologia, 516*, 35–54.

Verkaik, I., Reich, P., Prat, N., Rieradevall, M., & Baxter, C. V. (2015). Stream macroinvertebrate community responses to fire: Are they the same in different fire-prone biogeographic regions? *Freshwater Science, 34*, 1527–1541.

Verkaik, I., Rieradevall, M., Cooper, S. D., Melack, J. M., Dudley, T. J., & Prat, N. (2013). Fire as a disturbance in Mediterranean climate streams. *Hydrobiologia, 719*, 353–382.

Vidal-Abarca, M. R., Sánchez-Montoya, M. M., Guerrero, C., Gómez, R., Arce, M. I., García-García, V., & Suárez, M. L. (2013). Effects of intermittent stream flow on macroinvertebrate community composition and biological traits in a naturally saline Mediterranean stream. *Journal of Arid Environments, 99*, 28–40.

Violle, C., Navas, M.-L., Vile, D., Kazakou, E., Fortunel, C., Hummel, I., & Garnier, E. (2007). Let the concept of trait be functional!. *Oikos, 116*, 882–892.

Wagner, D. L., Grames, E. M., Forister, M. L., Berenbaum, M. R., & Stopak, D. (2021). Insect decline in the anthropocene: Death by a thousand cuts. *Proceedings of the National Academy of Sciences of the United States of America, 118*, 1–10.

Waltz, R. D., & McCafferty, W. P. (1979). *Freshwater springtails (Hexapoda: Collembola) of North America.* West Lafayette, IN: Purdue University Agricultural Experiment Station. 32 pp.

Waterkeyn, A., Grillas, P., Vanschoenwinkel, B., & Brendonck, L. (2008). Invertebrate community patterns in Mediterranean temporary wetlands along hydroperiod and salinity gradients. *Freshwater Biology, 53*, 1808–1822.

Weekers, P. H. H., De Jonckheere, J. F., & Dumont, H. J. (2001). Phylogenetic relationships inferred from ribosomal ITS sequences and biogeographic patterns in representatives of the genus Calopteryx (Insecta: Odonata) of the West Mediterranean and adjacent West European zone. *Molecular Phylogenetics and Evolution, 20*, 89–99.

Weigmann, G. (1973). Zur ökologie der Collembolen und Oribatiden im Grezbereich land-Meer (Collembola, Insecta-Oribatei, Acari) Zeitschrift für Wissenschaft. *Zoologie, 186*(3/4), 295–391.

Wellborn, G. A., Skelly, D. K., & Werner, E. E. (1996). Mechanisms creating community structure across a freshwater habitat gradient. *Annual Review of Ecology, Evolution, and Systematics, 27*, 337–363.

White, P. S., & Pickett, S. T. A. (1985). *The ecology of natural disturbance and patch dynamics* (p. 472) Orlando: Academic Press.

Wichard, W., Arens, W., & Eisenbeis, G. (2002). *Biological atlas of aquatic insects. Apollo Books* (p. 339) Denmark: Stenstrup.

Williams, D. D. (2006). *The biology of temporary waters* (p. 352) Oxford University Press.

Wilson, D. (1992). Complex interactions in metacommunities, with implications for biodiversity and higher levels of selection. *Ecology, 73*, 1984–2000.

Woodiwiss, F. S. (1964). The biological system of stream classification used by the River Trent Board. *Chemistry & Industry, 14*, 443–447.

Wootton, R. J. (1988). The historical ecology of aquatic insects: An overview. *Palaeogeography, Palaeoclimatology, Palaeoecology, 62*, 477–492.

Wright, J. F. (2000). An introduction to RIVPACS. In J. F. Wright, D. W. Sutcliffe, & M. T. Furse (Eds.), *Assessing the biological quality of fresh waters* (pp. 1–24). Ambleside, Cumbria: Freshwater Biological Association.

Yue, Q.-Y., & Fu, R.-S. (2000). New records and new species of freshwater springtails from China (Collembola). *Acta Entomologica Sinica, 43*(4), 394–402.

Zapparoli, M. (2006). The exotic species of the Italian fauna. In S. Ruffo & F. Stoch (Eds.), *Checklist and distribution of the Italian fauna* (pp. 57–61). Verona: Memorie del Museo Civico di Storia Naturale di Verona.

Chapter 9

Order Ephemeroptera

Michel Sartori[1,2] and Jean-Luc Gattolliat[1,2]

[1]*State Museum of Natural Sciences, Department of Zoology, Lausanne, Switzerland,* [2]*Department of Ecology and Evolution, Biophore, University of Lausanne, Lausanne, Switzerland*

Introduction

Ephemeroptera (so called mayflies) is a small order of primitive winged insects with an aquatic nymphal stage, which may last for some months to up to 3 years, and two aerial stages. When ready to emerge, the full grown nymph molts into a subimago, which is a sexually immature winged stage, a unique case among Hexapoda. After a period of a couple of minutes to a couple of days, this subimago molts into the imago reproductive stage. These adults do not feed and only fly for mating, followed by the female's oviposition. This period can be extremely brief (a couple of minutes), but in some species this can last a couple of days. Mayflies are strictly dependent of freshwaters for their life cycle, but a few species can be found in brackish waters.

Because of their sensitivity to environmental changes (both chemical and physical), Ephemeroptera belong to the orders of aquatic insects commonly used in biomonitoring programs. Currently, there are a little bit less than 4000 species described worldwide in more than 500 genera and 42 families (personal database of the authors, Kluge, 2021). The Mediterranean Basin, as defined in this book, harbors more than 300 species distributed in 51 genera, and 16 families.

General ecology and distribution

Ephemeroptera colonize all types of freshwater habitats, from small, rainfall created pools to large rivers, lakes, and springs. Their general biology and ecology have been treated by Sartori and Brittain (2015). Here, we focus on those mayflies inhabiting the Mediterranean Basin. A recent study (Buffagni, 2021) stressed the importance of the lentic–lotic characteristics of the habitat for mayfly (and other invertebrates) distribution in Sardinia; the main factor governing the distribution and optima of species along the lentic–lotic gradient was the range of flow types at a site. Indeed, there is a small proportion of mayfly genera which can be found in standing waters. They generally have a streamlined body for swimming (Baetidae, Siphlonuridae) or a flat body with some adaptations, such as the Caenidae which walk onto the sediment and possess an operculate second pair of gills for protecting its respiratory gills below. Some burrowers, especially the genus *Ephemera*, may also colonize standing waters where they burrow into sediment, and use their gills to actively circulate oxygenated water in their burrowing holes. Among Baetidae found in standing waters, the most common genus is by far *Cloeon* which possesses a number of remarkable adaptations, such as tolerance to oxygen depletion and ovoviviparity, thus allowing some of its species to colonize rather unique habitats.

The vast majority of mayflies, including those mentioned in standing waters, are found in running waters with both lotic and lentic habitat conditions. Compared to Plecoptera and Trichoptera, few mayfly species are restricted to the spring zone (crenal), mainly among the genera *Baetis* and *Electrogena*. Most mayfly genera can be found all along the rhithral zone, generally with a shift in species composition going downstream. The hyporhithral zone probably exhibits the richest mayfly communities, but due to severe anthropogenic impacts since the second half of the 20th century, these zones unfortunately lost many of their exclusive taxa, such as numerous species of the genera *Rhithrogena, Ecdyonurus, Raptobaetopus, Pseudocentroptilum*, etc. The largest river zones, especially the epipotamal, also suffered from human impacts, and some of its most important representative genera, such as *Kageronia*, *Dacnogenia*, and *Neoephemera*, are rare nowadays. In running waters in general, some genera are strictly petricolous and found on or

Identification and Ecology of Freshwater Arthropods in the Mediterranean Basin. DOI: https://doi.org/10.1016/B978-0-12-821844-0.00011-9

under cobbles and stones in the riverbed (e.g., Heptageniidae, Leptophlebiidae), whereas others colonize the lentic zones and the vegetation along the river banks (e.g., Ephemerellidae, Siphlonuridae). Some nymphs possess adaptation for burrowing into the sediment, such as mandibular tusks and strong forelegs. Among them, *Ephoron* and *Palingenia* are known to emerge in masse, forming huge swarms. *Palingenia longicauda* (Olivier, 1791), the largest European mayfly, was on the verge of extinction in the second half of the 20th century. Almost restricted to the Tisza watershed (sub-basin of the Danube River Basin), it has now recolonized some stretches of the Lower Danube River, and could perhaps reach Macedonia where it was recorded in the Vardar River (Ikonomov, 1964). The hyporheic habitat may be colonized by some mayfly genera, although to a lesser extent than Plecoptera. Among them one can mention *Acentrella* and *Thraulus*. To our knowledge, there is a single species which can be found only in temporary streams, *Metreletus balcanicus* (Ulmer, 1920). Other species, especially in semi-arid environments of North Africa and Spain, are also able to complete their life cycle during the period when the water flows; the resting stage is generally the eggs, which will hatch when the water comes back. Due to global warming and water abstraction for domestic or industrial uses, these drought periods tend to increase, and some species cannot tolerate these new conditions and are extirpated, such as among the genera *Oligoneuriopsis, Ephoron*, or *Potamanthus* in North Africa. Some genera such as *Oligoneuriella, Rhithrogena*, or *Epeorus* are very sensitive to oxygen depletion. They are, therefore, mainly found in higher elevation streams where the current is turbulent, and the water saturated in oxygen.

Endemism is generally high in Ephemeroptera, at least at the species level. Certain species can be found only on some islands, such as Corsica, Sardinia, Cyprus, or Aegean islands (e.g., Gattolliat et al., 2015; Soldán & Godunko, 2008, 2009). The three peninsulas, Iberian, Italian, and Balkan, possess an important percentage of endemic taxa, as also occurs in North Africa and the Levant. In general, endemic species can be mainly found in the rhitral zone above the hyporhitral, whereas nonendemic species are widespread downstream and in lentic habitats (Tierno de Figueroa et al., 2013). On the other hand, endemism at the generic level is low. To our knowledge, two genera are only found in the Mediterranean Basin, the ephemerellid *Quatica*, with two species, and the heptageniid *Anapos*, also known by two species, one in Corsica and Sardinia, and the other in the Levant. The once monospecific genus *Calliarcys* (Leptophlebiidae) was for a long time only known from the Iberian Peninsula, until another species was discovered from eastern Turkey (Godunko et al., 2015). Most genera have a broad Palearctic distribution. A couple of them present a distinct Afrotropical origin, having the Mediterranean Basin as their northern distribution limit. This is the case for *Cheleocloeon* (Baetidae, North Africa, and Levant) and *Oligoneuriopsis* (Oligoneuriidae, North Africa, Iberian Peninsula, Levant), which are represented in the Mediterranean Basin by two species each, whereas they are much more diversified in sub-Saharan areas (Barber-James et al., 2020).

Some genera are usually represented by a single species in the basin and are extremely rare. *Ametropus fragilis* Albarda, 1878 (Ametropodidae) is only known from a couple of localities in Italy and Croatia (Cuk et al., 2015; Sartori & Bauernfeind, 2020); its nymph is carnivorous and psammophilous, a habitat that has been highly disturbed. The North African endemics *Sparbarus kabyliensis* (Soldán, 1986) (Caenidae) and *Prosopistoma alaini* (Bojkova & Soldán, 2015) (Prosopistomatidae) have not been found for more than 30 years in well-studied regions, and may be locally extirpated (Benhadji et al., 2019). *Neoephemera maxima* (Joly, 1871) (Neoephemeridae) is reported from few localities in Macedonia and Albania (Bauernfeind, 2018). *Palingenia anatolica* (Jacob, 1977) from Turkey has not been reported since its description in the 1970s (Jacob, 1977), whereas *Palingenia orientalis* (Chopra, 1927) originally reported from the Jordan River in the Levant is certainly extinct in the area since the late 1940s (Sartori, 1992). *Isonychia ignota* (Walker, 1853) reaches its southern limit in the Mediterranean Basin. It is rare throughout Europe and is only known by a single locality in Spain from 1930. It was mentioned from Albania in the 1980s, but it seems more common in Turkey (Salur et al., 2016).

Some areas such as southern France, the Italian Peninsula, and Turkey were the subjects of extensive surveys and important systematic works. Thus, their fauna can be considered as well known, and updated checklists are available (see Buffagni et al., 2003; Salur et al., 2016; Tenchini et al., 2018). Other regions such as Maghreb and Israel have been recently the subject of important investigations, including both ecological and taxonomical works, improving and often challenging our knowledge (Benhadji et al., 2018, 2019; Kechemir et al., 2020; Yanai et al., 2018, 2020). Iberian and Balkan Peninsulas were subject to important surveys (Alba-Tercedor & Jaimez-Cuellar, 2003; Bauernfeind, 2003, 2018; Vilenica et al., 2015; Xerxa et al., 2019); unfortunately, specific identification remains uncertain in several cases. Moreover, regional endemism was not taken into account, and specimens were sometimes wrongly attributed to wide-spread Central European species.

Morphological characters needed in identification

For the identification of Ephemeroptera at the generic level, a good stereomicroscope with a magnification up to 80–100× is needed; additionally, some small and difficult characters can only be observed at higher magnification under a compound microscope. Species identification cannot be generally performed without making slide preparation of the mouthparts, legs, and abdominal tergites. Morphology terminology is explained in Figs. 9.1 and 9.2.

Barcoding, using a part of CO1 gene, can be considered in some case as a good option for species identification, especially when reasonably exhaustive libraries are available (Moriniere et al., 2017; Tenchini et al., 2018). It may also be of great help for associating ontogenetic stages or identifying poorly known stages. Therefore, barcoding must be considered as complementary to traditional morphological approaches.

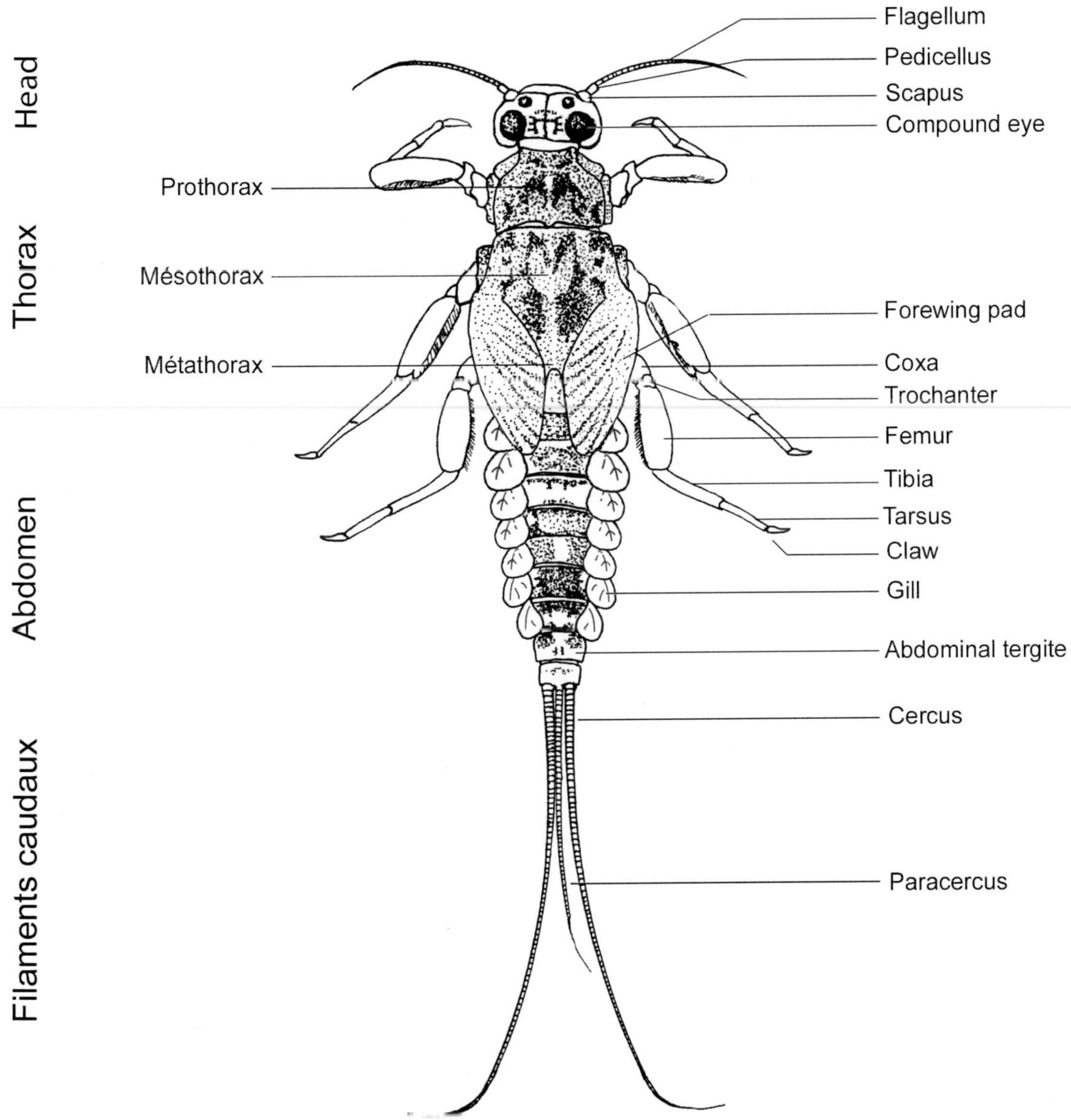

FIGURE 9.1 Morphological terms used in this chapter; body structures.

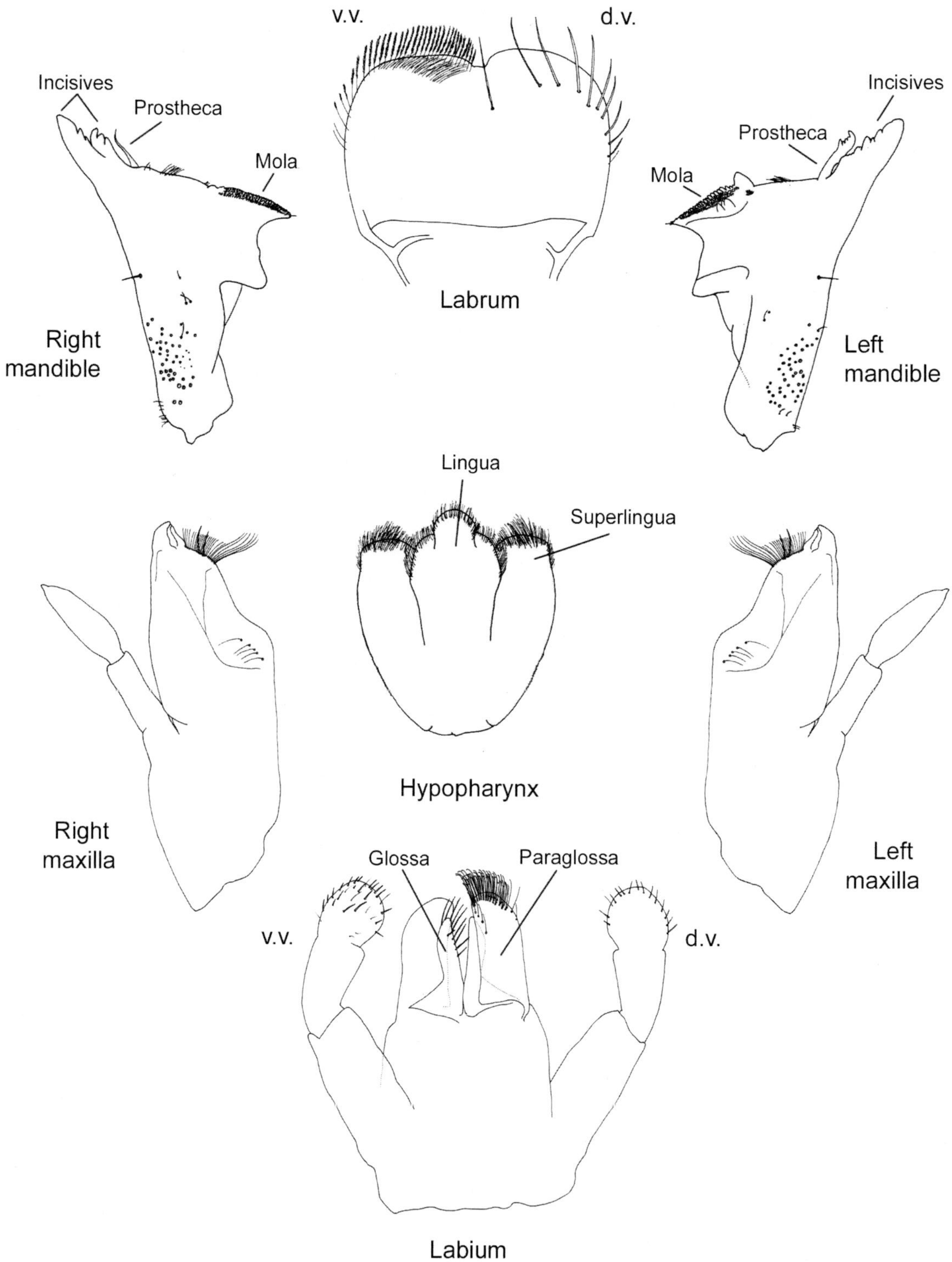

FIGURE 9.2 Morphological terms used in this chapter; mouthparts (d.v.: dorsal view; v.v.: ventral view).

Material preparation and preservation

All mayfly nymphs should always be preserved in fluid, preferably in pure ethanol. For morphological studies, a concentration of 75%–80% is recommended. If the material is designated for genetic analyses, then a minimum concentration of 95% should be privileged. In all cases, formalin should be avoided since it destroys the DNA and makes the cuticle very breakable, and hence renders identification much more difficult. Preservation in ethanol makes the color patterns fade with time. It is of the upmost importance that the preserved material be stored in the dark and if possible at cool temperatures (<6°C) or even in a freezer.

Keys to Ephemeroptera

Here, we present identification keys at the family level, then at the genus level for the families which are not monogeneric. These keys are based on different published works, mainly Bauernfeind and Soldán (2012), and Barber-James et al. (in prep) for the key to genera also found in the Afrotropical realm. The circumscription of some genera is still debated among taxonomists. We tried to have a pragmatic approach, allowing the reader to reach the most detailed level, whenever it is considered as a genus or a subgenus by different authors (see for instance for *Baetis*, *Choroterpes*, *Heptagenia*, or *Procloeon*).

Insecta: Ephemeroptera: Families

1 Nota fused, forming carapace-like structure encasing thorax and abdominal gills, giving nymphs an ellipsoid appearance (Fig. 9.3A), tarsal claws of all legs without denticles; size range of mature nymphs 3 to 9 mm; retractile cerci and paracercus of up to 1 mm .... Prosopistomatidae, one genus: *Prosopistoma*

1' Nota not fused as above, legs and gills at least partially visible from the dorsal view, claws with or without denticles, cerci (and paracercus) not retractile, longer than 1 mm 2

2(1) Nymphs of a burrowing habit, head with long mandibular tusks projecting forward (Fig. 9.4C), or if tusks short, gill I reduced, not feather-like or fringed; gills II–VII each consisting of two elongate lamellae with fringed margins (Fig. 9.4A) 3

2' Nymphs not of a burrowing habit; head without prominent mandibular tusks; gills II–VII not as above (e.g., Figs. 9.5A and 9.6B) 6

3(2) Abdominal gills arranged laterally (Fig. 9.4A), fore legs not modified for burrowing, tibiae cylindrical, mandibular tusks very reduced (Fig. 9.4B) .... Potamanthidae, one species: *Potamanthus luteus (Linnaeus, 1767)* [Algeria, Bosnia-Herzegovina, Croatia, France, Greece, Italy, Macedonia, Morocco, Slovenia, Spain, Tunisia, Turkey]

3' Abdominal gills folded dorsally over the top of the abdomen (Fig. 9.4E), fore legs modified for burrowing, tibiae flattened, mandibular tusks well developed 4

FIGURE 9.3 Larva of *Prosopistima pennigerum* (A), historical specimen collected in the Rhone River near Valence in May 1952—coll. C. Degrange); head and mandibular tusks of *Palingenia longicauda* (B) (scale bar: 1 mm).

FIGURE 9.4 Habitus (A, E) and mandibular tusks (B–D) of *Potamanthus luteus* (A, B), *Ephoron virgo* (C), and *Ephemera glaucops* (D, E) (scale bars: 1 mm).

4(3) Mandibular tusks stocky, with large irregular denticles laterally (Fig. 9.3B) .... Palingeniidae, one genus: *Palingenia*
4' Mandibular tusks, edentate, often slender, if stocky, then only with teeth apically 5
5(4) Mandibular tusks apically divergent, pointing upwards, frontal process present (Fig. 9.4E); hind legs with pointed tibial process (Fig. 9.4E) .... Ephemeridae, one genus: *Ephemera*
5' Mandibular tusks apically convergent (Fig. 9.4C); frontal process absent, or considerably reduced; hind legs without pointed tibial process .... Polymitarcyidae, one species: *Ephoron virgo (Olivier, 1791)* [Albania, Algeria, Croatia, France, Italy, Macedonia, Morocco, Portugal, Spain, Tunisia, Turkey]
6(2') Gill II operculate (Fig. 9.7A and B), covering the remaining gills III–VI which are highly pectinated, gill VII absent 7

FIGURE 9.5 Habitus of *Cloeon dipterum* (A), *Procloeon* sp (B), and *Bungona (Chopralla) pontica* (C, D) (scale bars: 1 mm).

6' Gill II not operculate, similar to the following ones, which may have different forms, but never pectinated; gill VII generally present 8
7(6) Operculate gills rounded or subquadrate (Fig. 9.7B), generally overlapping (Fig. 9.7D), with posterior fringe of setae or spines; operculate gills never coupled with rows of setae; metathoracic wingpads absent .... **Caenidae**
7' Operculate gills subquadrate, meeting medially without overlapping (Fig. 9.7A) but with dense row of setae which couple the opeculate gills; metathoracic wing pads present .... Neoephemeridae, one species *Neoephemera maxima (Joly, 1871)*
[Albania, France, Macedonia]
8(6') Head with lateral ocelli located posterior to (above) lateral branches of epicranial suture (Fig. 9.8E); hind wingpads small or absent; gills always ovoid with simple lamella (double in some genera), usually 7 pairs, occasionally 6 pairs (Fig. 9.9D–H) .... **Baetidae**

FIGURE 9.6 Habitus of *Ametropus fragilis*, a rare carnivorous and psammophilous species (scale bar: 1 mm).

FIGURE 9.7 Habitus of *Neoephemera maxima* (A), *Caenis pusilla* (B), and *Brachycercus harisella* (C); abdomen of *Sparbarus kabiliensis* (D); head of *Brachycercus harisella* (E) (scale bars: 1 mm).

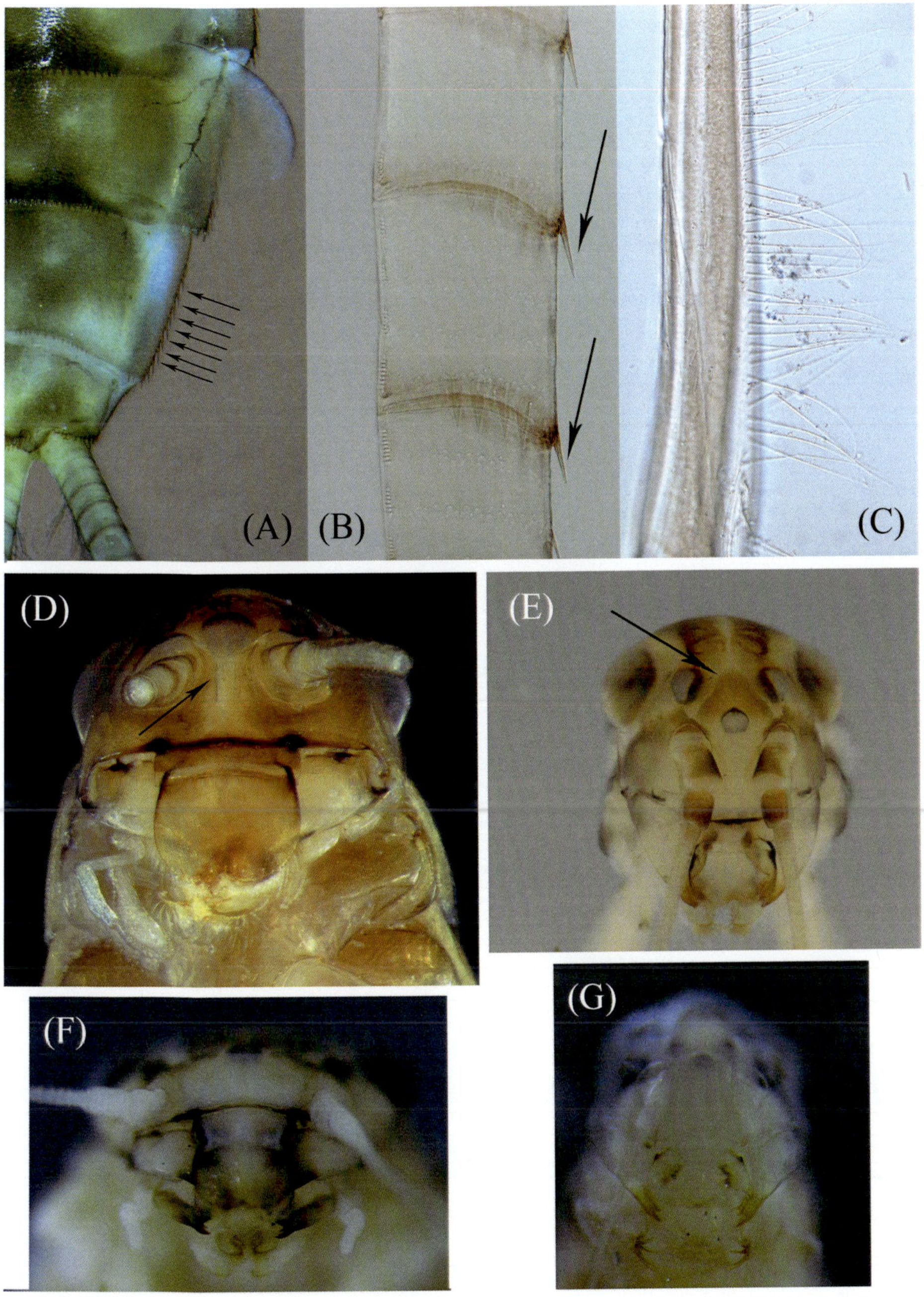

FIGURE 9.8 Lateral margin of abdominal segments VIII and IX of *Cloeon dipterum* (A); lateral margin of cercus of *Procloeon bifidum* (B); dorsal margin of tibia of *Acentrella sinaica* (C); head in frontal view of *Alainites sadati* (D), *Labiobaetis atrebatinus* (E), *Acentrella sinaica* (F), and *Raptobaetopus tenellus* (G).

8' Head with lateral ocelli located anterior to (below) lateral branches of epicranial suture (Fig. 9.10B); hind wing-pads usually present and well developed; gills with various modifications (e.g., Fig. 9.11B) 9

9(8') Forelegs with rows of long setae along inner margin of femur and tibia (Fig. 9.12A and B); base of maxillae or labium with accessory gills (Fig. 9.12C) 10

9' Forelegs without a row of long setae along inner margin of femur and tibia; base of mouthparts without accessory gills 11

10(9) Body flattened dorsoventrally (Fig. 9.12B); foretarsus without filtering setae (Fig. 9.12F); gill I ventrally oriented, lamellate section reduced, with pronounced fibrilliform section; gills II–VII small, in dorsal position .... **Oligoneuriidae**

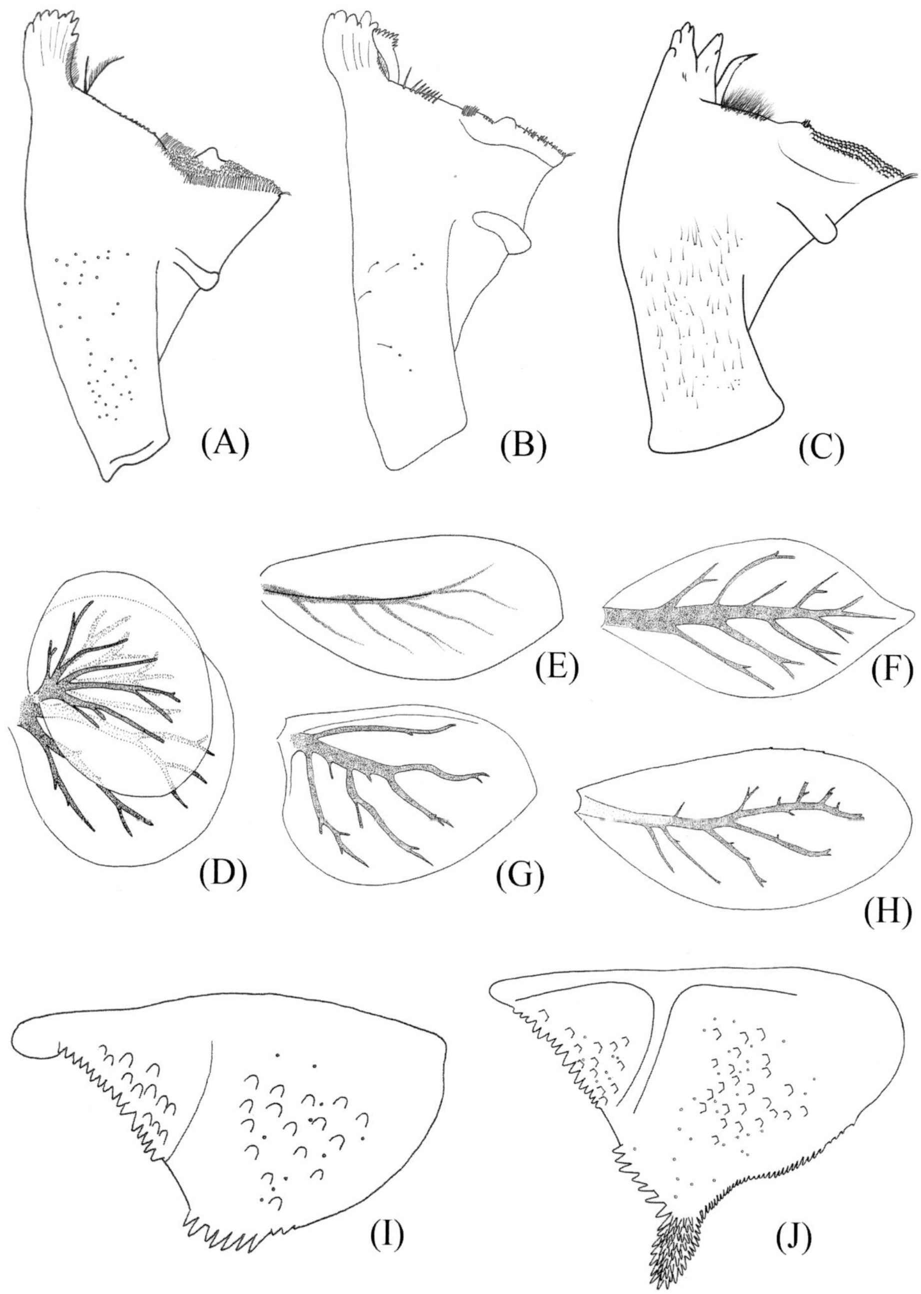

FIGURE 9.9 Right mandible of *Alainites sadati* (A), *Nigrobaetis* sp. (B), and *Cloeon perkinsi* (C); gill IV of *Cloeon dipterum* (D), *Bungona (Chopralla) pontica* (E), *Centroptilum luteolum* (F), *Procloeon bifidum* (G), and *Cheleocloeon soldani* (H); paraproct of *Nigrobaetis* sp. (I) and *Alainites sadati* (J).

10' Body cylindrical or laterally compressed (Fig. 9.12A), frons between the antennae projected forwards; foretarsus with filtering setae on inner margin (Fig. 9.12G) .... Isonychiidae, one species *Isonychia ignota (Walker, 1853)* [France, Greece, Macedonia, Spain, Turkey]

11(9') Body flattened dorsoventrally, eyes in dorsal position (e.g., Fig. 9.6A), gill I–VI composed of a lamellate dorsal portion more or less rounded and most often with fibrilliform ventral portion ventrally 12

11' Body not flattened dorsoventrally, eyes in lateral position, or if somewhat flattened, gill I–VI not as above 13

12(11')All legs stout, with femur always much broader than tibia; gills I–VI consisting of lamellate dorsal portion and fibrilliform ventral portion (Fig. 9.10D) .... **Heptageniidae**

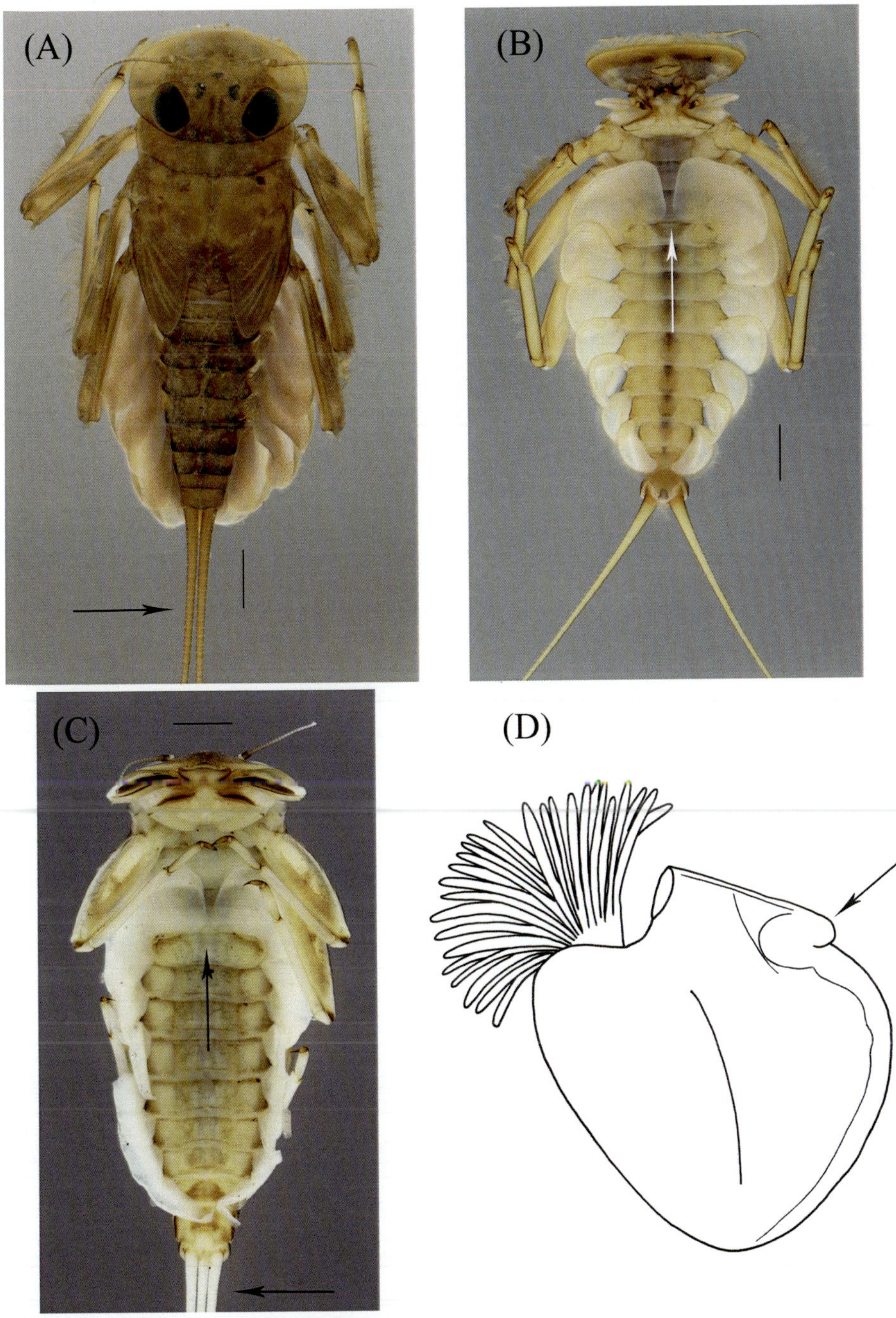

FIGURE 9.10 Habitus of *Epeorus (Epeorus) assimilis* in dorsal view (A); habitus of *Epeorus (Alpiron) alpicola* (B) and *Rhithrogena sartorii* (C) in ventral view; gill IV of *Epeorus (Caucasiron) insularis* (D) (scale bars: 1 mm) (Hrivniak et al., 2020). *Redrawn from Hrivniak, L., Sroka, P., Bojkova, J., Godunko, R. J. (2020). Identification guide to larvae of Caucasian Epeorus (Caucasiron) (Ephemeroptera, Heptageniidae).* ZooKeys, 986, *1–53.*

12' Mid- and hind legs very slender and long, with femur about the same width than tibia; forelegs reduced, similar to palpi (Fig. 9.6); gills only with dorsal lamellate portion, without fibrilliform ventral portion .... Ametropodidae, one species: *Ametropus fragilis Albarda, 1878* [Croatia, Italy]

13(11') Abdominal segment II without gills, gills IV–VI with a lamellate dorsal portion and a ventral portion bifurcate in two branches (Fig. 9.13C); forewing pads fused on more than half their length .... **Ephemerellidae**

13' Gill on segment II always present; gills IV–VI with different shape, either plate-like or composed of two plates which are divided in several branches; forewing pads separated on more than half their length 14

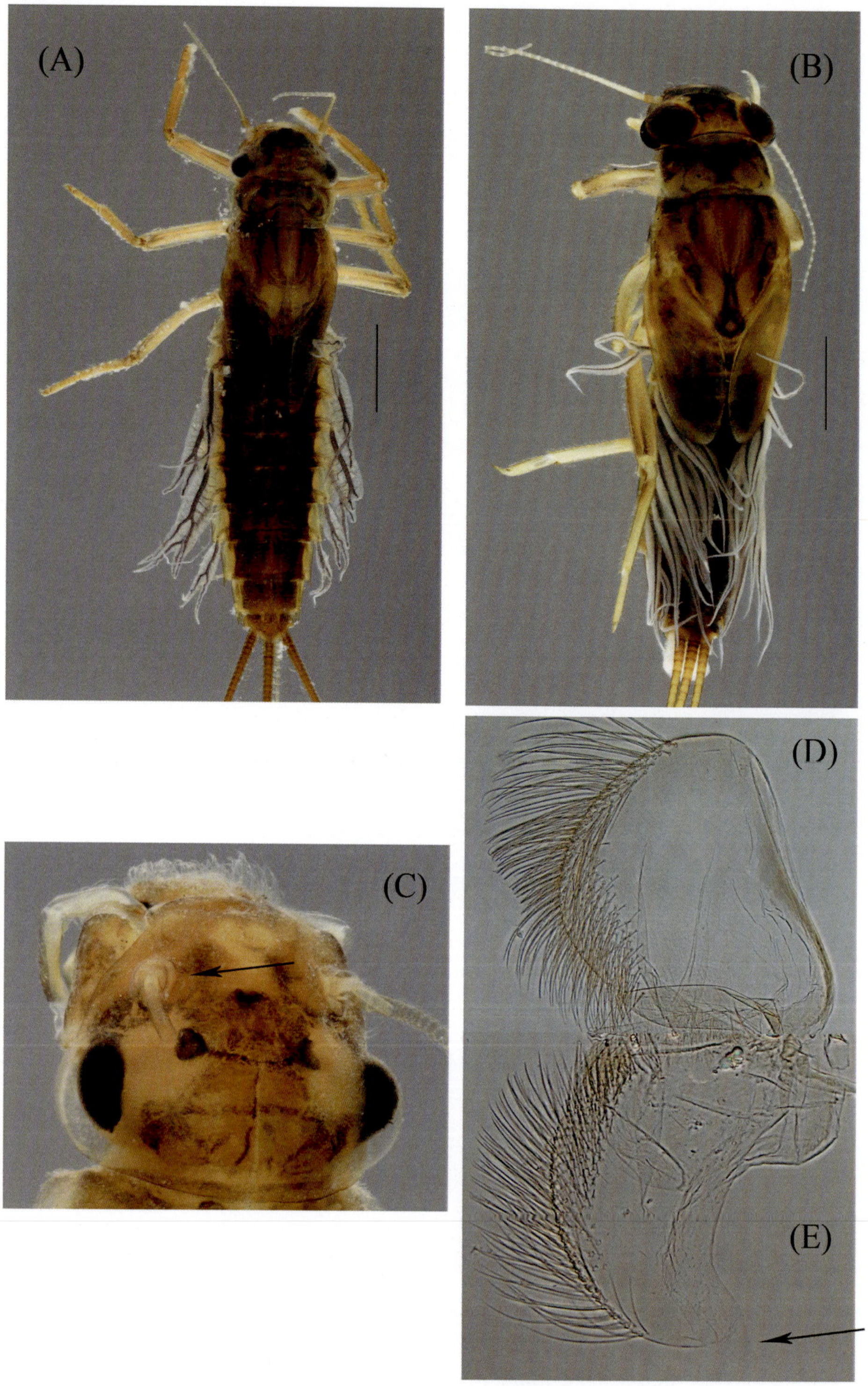

FIGURE 9.11 Habitus of *Calliarcys humilis* (A) and *Habroleptoides confusa* (B); head of *Paraleptophlebia submarginata* (C); superlingua of *Paraleptophlebia submarginata* (D) and *Habroleptoides umbratilis* (E) (scale bars: 1 mm).

FIGURE 9.12 Habitus of *Isonychia ignota* (A) and *Oligoneuriella rhenana* (B); accessory gills of *Oligoneuriella rhenana* (C); hind femur of *Oligoneuriella skoura* (D) and *Oligoneuriopsis skhounate* (E); fore tarsus of *Oligoneuriella rhenana* (F) and *Isonychia ignota* (G) (scale bars: 1 mm).

14(13') Gills plate-like, large and entire, with a single or double lamellae (Fig. 9.14A) 15

14' Gills never plate-like, forked (Fig. 9.11A), with two fringed lamellae (Fig. 9.15B) or with upper and lower lamellae ending in three processes (Fig. 9.15G) .... **Leptophlebiidae**

15(14) Posterolateral expansions on the abdomen well marked (Fig. 9.14A); gills simple or double, without outer reinforcement; mouthparts not elongated .... **Siphlonuridae**

15' Posterolateral expansions on the abdomen weakly developed; gills always simple, with an outer reinforcement (Fig. 9.16C); mouthparts elongated (Fig. 9.16C) .... **Ameletidae**

Keys to Genera

It is important to insist that these keys are only accurate for the Mediterranean Basin.

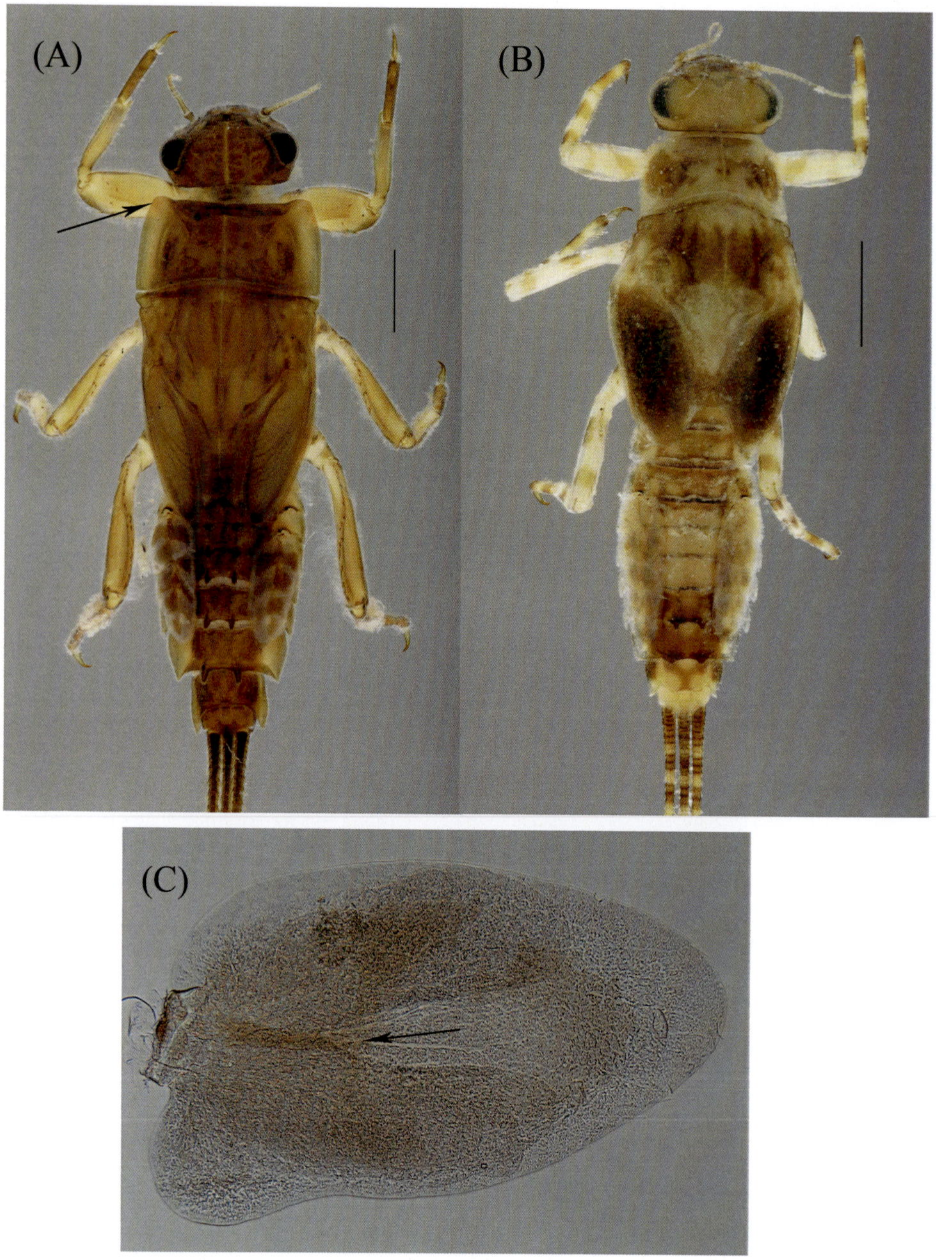

FIGURE 9.13 Habitus of *Quatica paradinasi* (A) and *Serratella ignita* (B); gill VI of *Serratella ignita* (C).

Insecta: Ephemeroptera: Baetidae: Genera

Baetidae is the most diversified family in the world and the second richest after Heptageniidae in the west Palearctic. *Baetis* is by far the most species rich genus in this area. Generic delimitation is still problematic in different lineages. Generic attribution of several species is highly problematic, and species are still regularly transferred from one genus to the other. For the tribe Cloeonini, we mostly follow Kluge's (2016) systematic with the following generic attributions: *Procloeon bifidum* (Bengtsson, 1912); *Procloeon calabrum* (Belfiore & D'Antonio, 1990); *Procloeon concinnum* (Eaton, 1885); *Procloeon fascicaudale* (Sowa, 1985); *Procloeon nana* (Bogoescu, 1951); *Procloeon pulchrum* (Eaton, 1885); *Procloeon stagnicola* Soldán & Thomas, 1983; *Pseudocentroptiloides shadini* (Kazlauskas, 1964); *Pseudocentroptilum unguiculatum* (Tshernova 1941) (=*Pseudocentroptilum motasi* Bogoescu 1947). The wide concept

FIGURE 9.14 Habitus (A) and lateral view (B) of *Siphlonurus (Siphlonurus) aestivalis*; lateral view of *Siphlonurus (Siphlurella) alternatus* (C) (scale bars: 1 mm).

of *Baetis* with several subgenera used by various authors (see Bauernfeind & Soldán, 2012 for a review) is highly polyphyletic and must be avoided. *Alainites* was tentatively considered as junior synonym of *Takobia* (Kluge & Novikova, 2014), but this synonymy was not followed in subsequent studies. *Bungona (Chopralla) pontica* (Sroka et al., 2019) is the single species of the genus in this area (only report from Turkey); this species was tentatively assigned to *Centroptella* by Kluge et al. (2020).

1 Highly modified mouthparts adapted to carnivorous behavior (Fig. 9.8G): maxillary palp much longer than galealacinia (segment II 3x segment I), long, slender apical denticles of maxilla, both canines with pointed, slender denticles, margin between prostheca and mola incurved, labrum subovoid without median emargination, labial palp 2-segmented, segment II enlarged, hooked, covered with numerous long thin setae .... One species: *Raptobaetopus tenellus*
[Bosnia-Herzegovina, France, Italy, Spain]
1' Mouthparts not modified for carnivorous behavior (Fig. 9.8D–F) 2

FIGURE 9.15 Habitus of *Habrophlebia hassainae* (A), *Leptophlebia marginata* (B), *Thraulus bellus* (C), *Choroterpes (Choroterpes) picteti* (D), and *Choroterpes (Euthraulus) ortali* (E); gill IV of *Choroterpes (Choroterpes) picteti* (F) and *Choroterpes (Euthraulus) ortali* (G) (scale bars: 1 mm).

2(1') Paracercus egal or subegal to cerci (Figs. 9.5A–D, and 9.17A and B); margin between prostheca and mola of right and generally left mandibles with very abundant setae (Fig. 9.9C); legs generally elongated: femur with dorsal and ventral margins subparallel, claws elongated with two rows of numerous denticles, being reduced, vestigial, or absent in some genera 3

2' Paracercus at most 2/3 of cerci (Figs. 9.17C–D and 9.18A–D); margin between prostheca and mola of right and left mandibles bare or at most with a few spine-like setae (Fig. 9.9A); legs with ovoid femur, claws short with one row of less than 20 denticles, never reduced, vestigial, or absent 10

3(2) Lateral margin of abdominal segments VIII and IX (sometimes also VI and VII) with strong spines (Fig. 9.8A) 4

FIGURE 9.16 Habitus (A) and lateral view (B) of *Ameletus inopinatus*; lateral view of *Metreletus balcanicus* (C) (scale bar: 1 mm).

3' Lateral margin of abdominal segments VI to IX without spines (or spines limited to the latero-apical corner) 8

4(3) Labrum pentagonal with distal margin straight and concave; glossae quadrangular, clearly shorter than paraglossae; segment III of the labial palp more than twice broader than long; maxilla stocky, apically truncated with short teeth. .... One species: *Pseudocentroptiloides shadini (Kazlauskas, 1964)* [France, Turkey]

4' Labrum rounded to sub-quadrangular with a central emargination; hooked glossae, at most slightly shorter than paraglossae; segment III of the labial palp as broad as long; maxilla elongated, with hooked teeth 5

5(4') Upper lamellae of gills II to VI symmetrical, lower lamellae with palmate tracheation (Fig. 9.9D) .... *Cloeon*

5' Upper lamellae of gills II to VI asymmetrical, lower lamellae with a central tracheation (Fig. 9.9E–H) 6

6 Cerci and paracercus with long bare last segments (as in Fig. 9.5A); hindwing pads always absent ...... *Similicloeon*

6' Cerci and paracercus without long bare last segments (Fig. 9.5B); hindwing pads present or absent 7

FIGURE 9.17 Habitus of *Centroptilum luteolum* (A), *Cheleocloeon soldani* (B), *Acentrella sinaica* (C), and *Labiobaetis neglectus* (D) (scale bars: 1 mm).

7(6') Lateral side of distal part of cercus, each segment with a single greatly enlarged spine (Fig. 9.8B) .... *Procloeon*
7' Lateral side of distal part of cercus, each segment without a single greatly enlarged spine .... One species: *Pseudocentroptilum unguiculatum (Tshernova, 1941)* [Greece, Romania]
8(3') Canine of both mandibles divided in two completely separated sets (Fig. 9.19B); gills apically acuminated (Fig. 9.9F) .... *Centroptilum*
8' Canine of both mandibles with two sets of incisors at least partially fused (Fig. 9.19A); gills apically rounded to slightly pointed (Fig. 9.9E and H) 9
9(8') Segment II of labial palp with a thumb like projection (Fig. 9.19C); legs with a short arc of thin setae on tibia only .... *Cheleocloeon*

FIGURE 9.18 Habitus of *Baetis noa* (A), *Baetis* gr. *alpinus* (B), *Alainites sadati* in lateral view (C), and *Nigrobaetis rhithralis* (D) (scale bar: 1 mm).

9' Segment II of labial palp without a distolateral projection (Fig. 9.19D), legs with long arcs of thin setae on femora, tibia and tarsi .... One species: *Bungona* (*Chopralla*) *pontica Sroka, Godunko & Gattolliat, 2019* [Turkey: Sinop Province]

10(2') Antenna located close together with a well-developed carina in between (Fig. 9.8D); body compressed laterally, mouthparts generally in a hypognathous position (Fig. 9.18C); segment III of the labial palp quadrangular 11

10' Antenna well separated without a carina in between (Fig. 9.8E and F); body not compressed laterally, mouthparts not in a hypognathous position; segment III of the labial palp ovoid or conical 12

11(10) Right mandible with a bifid prostheca (Fig. 9.9A); paraproct with an elongate prolongation on distal margin (Fig. 9.9J) .... *Alainites*

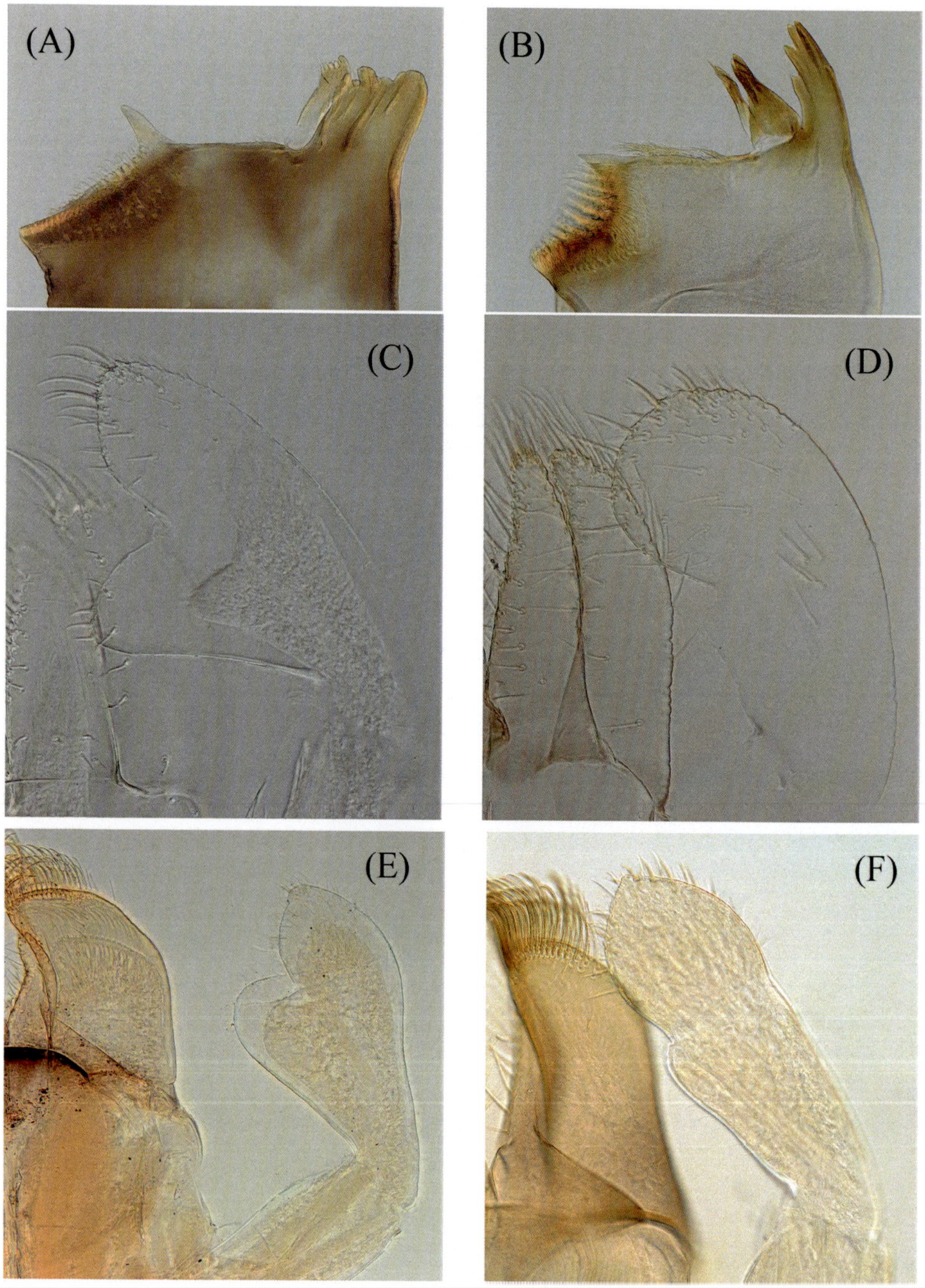

FIGURE 9.19 Left mandible of *Bungona (Chopralla) pontica* (A) and *Centroptilum luteolum* (B); labial palp of *Cheleocloeon soldani* (C), *Bungona (Chopralla) pontica* (D), *Labiobaetis neglectus* (E), and *Baetis gomerensis* (F).

11' Right mandible with a normal prostheca (Fig. 9.9B); paraproct without an elongate prolongation on distal margin (Fig. 9.9I) .... *Nigrobaetis*

12(10') Second segment of labial palp with a well-developed thumb like projection (Fig. 9.19E); hypopharynx with an apical tuft of strong dark setae; scapus with or without distolateral lobe; apex of maxillary palp with or without excavation .... *Labiobaetis*

12' Second segment of labial without or with a moderately developed thumb like projection (Fig. 9.19F); hypopharynx apically with numerous transparent setae; scapus without distolateral lobe; apex of maxillary palp without excavation 13

13(12') Legs elongated with a row of very thin setae on the dorsal margin of tibia (Fig. 9.8C); paracercus reduced to a few segments (Fig. 9.17C); mouthparts short and stocky (Fig. 9.8F) .... *Acentrella*
13' Legs without or with a restricted number of thin setae on the dorsal margin of tibia; paracercus generally equal to 2/3 of cerci (Fig. 9.18A), sometimes reduced to highly reduced (Fig. 9.18B); mouthpart not stocky (as in Fig. 9.8E) .... *Baetis*

Insecta: Ephemeroptera: Caenidae: Genera

A medium size family (less than 300 species worldwide, less than 50 in the Palearctic) mainly represented in the Mediterranean Basin by the genus *Caenis* with ca 20 species. The genera *Brachycercus* and *Sparbarus* are more rarely found.

1 Head with ocellar tubercles (Fig. 9.7E); legs long and narrow, all segments more or less of the same width (Fig. 9.7C); posterolateral expansions of the abdomen well expressed; maxillary and labial palps two-segmented 2
1' Head without ocellar tubercles; femora clearly broader than tibiae (Fig. 9.7B); posterolateral expansions of the abdomen moderately or weakly expressed; maxillary and labial palps three-segmented .... *Caenis*
2(1) Middle ocellar tubercle short and weakly developed; posterolateral expansion of tergite VI strongly bent upwards (Fig. 9.7D) .... One species: *Sparbarus kabyliensis (Soldán, 1986)* [Algeria, Tunisia]
2' Middle ocellar tubercles well developed (Fig. 9.7E); posterolateral expansion of tergite VI not bent upwards .... One species: *Brachycercus harrisella Harris, 1834* [France, Italy, Spain]

Insecta: Ephemeroptera: Oligoneuriidae: Genera

A small family (less than 70 species known worldwide), mainly represented in the area by the genus *Oligoneuriella* with less than 6 species. The Afrotropical genus *Oligoneuriopsis* reaches North Africa and the Iberian Peninsula with *O. skhounate* (Dakki & Giudicelli, 1980) and the Levant with *O. orontensis* (Koch, 1980).

1 Hind femora with few or with a small bunch of setae at base (Fig. 9.12D); gill lamellae without or with short and thick setae on the outer margin .... *Oligoneuriella*
1' Hind femora with a row of long setae at least on the proximal part (Fig. 9.12E); gill lamellae with long and thin setae on the outer margin .... *Oligoneuriopsis*

Insecta: Ephemeroptera: Heptageniidae: Genera and Subgenera

With the Baetidae, the most diversified family in the area, with more than 120 species. Most species rich genera are *Rhithrogena* (ca 60 species), *Ecdyonurus* (ca 30 species), and *Electrogena* (ca 25 species). Some taxa exhibit high endemism, especially on islands. The genus *Anapos* is only known by one Corso-Sardinian species, *A. zebratus* (Hagen, 1864) and one species in the Levant, *A. kugleri* (Demoulin, 1973).

1 Paracercus completely reduced, almost not visible (Fig. 9.10A and B) .... *Epeorus* 2
1' Paracercus always well developed (Figs. 9.10C and 9.20A) 4
2(1) First pair of gill not enlarged, in lateral position as the others .... *Epeorus (Epeorus)*
2' First pair of gills greatly enlarged in ventral position (Fig. 9.10B) 3
........3 (2') Gills II–VI with an antero-dorsal projection on the costal rib (Fig. 9.10D) .... One species: *Epeorus (Caucasiron) insularis (Braasch, 1983)*
[Greece: Samos Island]
3' Gills II–VI without an antero-dorsal projection on the costal rib *Epeorus (Alpiron)*
4(1') First pair of gills kidney-shaped, greatly enlarged and meeting ventrally to form a friction disk (Fig. 9.10C) .... *Rhithrogena*
4' First pair of gills laterally expanded, not larger than gill II 5
5(4') Pronotum with lateral expansions extending posteriorly (Fig. 9.21A) .... *Ecdyonurus* 6
5'Pronotum without lateral expansions extending posteriorly (as in Fig. 9.20A) 7

FIGURE 9.20 Habitus of *Heptagenia (Dacnogenia) coerulans* (A); lateral view of *Heptagenia (Kageronia) fuscogrisea* (B); maxilla of *Heptagenia (Dacnogenia) coerulans* (C); superlingua of *Heptagenia (Dacnogenia) coerulans* (D) and *Heptagenia (Heptagenia) flava* (E). (scale bars: 1 mm).

6 (5) Superlingua of hypopharynx with a row of long and thin setae, replaced by scattered short setae at the tip (Fig. 9.21C); apex of maxilla with more than 20 comb-shaped bristles .... *Ecdyonurus (Helvetoraeticus)*
6' Superlingua of hypopharynx with a row of long and thin setae including to the tip (Fig. 9.21D); apex of maxilla with less than 20 comb-shaped bristle .... *Ecdyonurus (Ecdyonurus)*
7(5') Outer margin of femora with irregular short or long setae; ventral surface of the maxillae with a row of setae parallel to inner margin (Fig. 9.20C); labial glossae elongated; apex of maxilla with less than 13 comb-shaped bristles; body coloration generally well contrasted .... *Heptagenia* 8

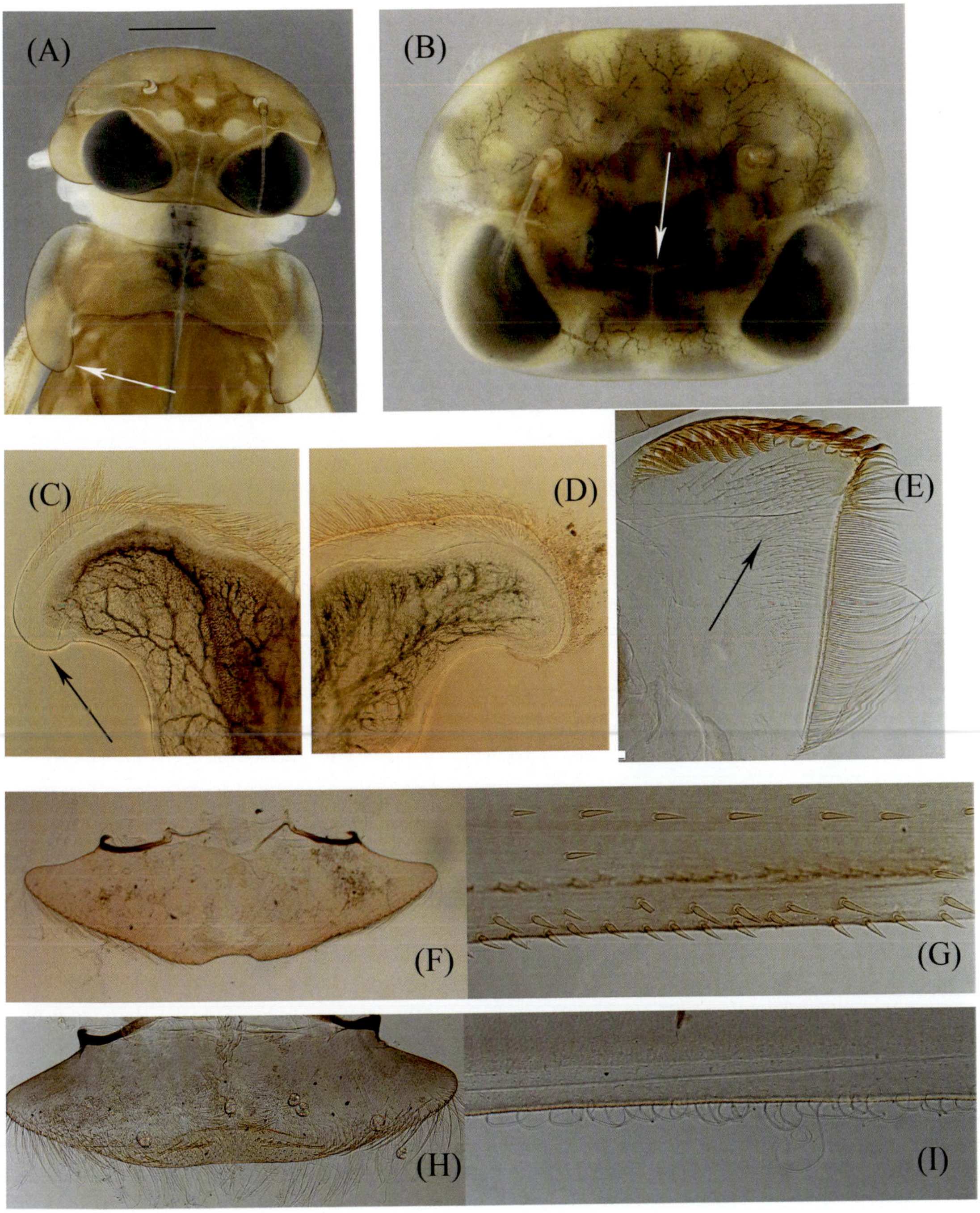

FIGURE 9.21 Head and prothorax *of Ecdyonurus (Helvetoraeticus) helveticus* (A); head of *Electrogena brulini* (B); superlingua of *Ecdyonurus (Helvetoraeticus) helveticus* (C) and *Ecdyonurus (Ecdyonurus) venosus* (D); maxilla of *Anapos kugleri* (E); labrum (F, H) and hind tibia outer margin (G, I) of *Anapos kugleri* (F), *Anapos zebratus* (G), *Electrogena lateralis* (H), and *Electrogena grandiae* (I).

7' Outer margin of femora with a regular row of long setae; ventral surface of the maxillae with scattered setae (Fig. 9.21E); labial glossae quadrangular; apex of maxilla with more than 13 comb-shaped bristles; body coloration generally less contrasted 10
8(7) Gill VII without a bunch of fibrillae; gill I broad, heart-shaped, with long and narrow tip (Fig. 9.20B); outer margin of femora with short, stout bristles .... One species: *Heptagenia (Kageronia) fuscogrisea (Retzius, 1783)* [Croatia, France, Macedonia]

8' Gill VII with a bunch of fibrillae (Fig. 9.20A); gill I slender, tongue-shaped or lanceolate; outer margin of femora with long, stout bristles and thin setae 9
9(8') Hypopharynx with superlingua short and broad (Fig. 9.20D) .... One species: *Heptagenia (Dacnogenia) coerulans Rostock, 1878* [Bosnia-Herzegovina, Croatia, France, Italy, Macedonia, Turkey]
9' Hypopharynx with superlingua long and with tip retrograde (Fig. 9.20E) .... *Heptagenia (Heptagenia)*
10(7') Labrum with distinct antero-median notch (Fig. 9.21F); outer margin of hind tibiae with one longitudinal row of stout spines (Fig. 9.21G) .... *Anapos*
10' Labrum without distinct antero-median notch (Fig. 9.21H); outer margin of hind tibiae without rows of stout spines (Fig. 9.21I) .... *Electrogena*

Insecta: Ephemeroptera: Ephemerellidae: Genera

Medium size family represented in the area by less than 15 species. The by far most abundant one is *Serratella ignita* (Poda, 1761), the only ephemerellid species reported from North Africa. Recent generic reorganization is not completely satisfactory, especially regarding the genus *Quatica*; two species are known, *Q. paradinasi* (Gonzalez del Tanago & Garcia de Jalon, 1981) in the Iberian Peninsula and *Q. ikonomovi* (Puthz, 1971) in Southern Balkans; however, they differ in many aspects.

1 Abdominal segment I with bristle-like gill; gills present on segments IV–VII (Fig. 9.22A) .... One species: *Eurylophella iberica Keffermüller & Da-Terra, 1978* [Portugal, Spain]
1' Abdominal segment I without gill; gills present on segments III–VII (Fig. 9.22E) 2
2(1') Abdomen wide and densely covered with long, hair-like setae (Fig. 9.22B); gill on segment III semi-opeculate, covering gills on segments IV–V (Fig. 9.22D); gill lamella on segment VI not visible in dorsal view .... *Torleya*
2' Abdomen elongated, sometimes covered with sparse and short setae (Fig. 9.13A and B); gill on segment III normal, slightly overlapping gills on segments IV–V (Fig. 9.22E); gill lamella on segment VI visible in dorsal view 3
3(2') Claw with a large, preapical stout denticle (Fig. 9.22C) .... *Teloganopsis*
3'Claw with evenly arranged denticles, or increasing or decreasing in size, but never with a stout preapical denticle 4
4(3') Ventral lamella of gills VI without deep cleft .... *Ephemerella*
4' Ventral lamella of gills VI with deep cleft (Fig. 9.13C) 5
5(4') Prothorax with anterolateral projections (Fig. 9.13A); abdominal terga with or without greatly elongate paired spines .... *Quatica*
5' Prothorax with anterolateral projections absent or only slightly developed (Fig. 9.13B); abdominal tergal spines moderately developed .... *Serratella*

Insecta: Ephemeroptera: Leptophlebiidae: Genera

A large family especially diversified in the tropical regions. Only 60 species are recorded in the Palearctic area, among them about 40 inhabit the Mediterranean Basin. The genera *Habrophlebia* and *Habroleptoides* are the most diversified.

1 Gills II–VII slender and bifurcate (Fig. 9.11A and B) 2
1' Gills II–VII broader, consisting of a dorsal and a ventral lamella (Fig. 9.15F–G) 4
2(1) Gills divided only in apical half (Fig. 9.11A) .... One species: *Calliarcys humilis Eaton, 1881* [Portugal, Spain]
2' Gills divided almost at the base (Fig. 9.11B) 3
3(2') Hypopharynx with superlinguae laterally well developed and pointed (Fig. 9.11E); antennae inserted on the margin of the cephalic capsule .... *Habroleptoides*
3' Hypopharynx with superlinguae laterally weakly developed and rounded (Fig. 9.11D); antennae inserted on the dorsal surface of the cephalic capsule (Fig. 9.11C) .... *Paraleptophlebia*

FIGURE 9.22 Habitus of *Eurylophella iberica* (A) and *Torleya major* (B); tarsal claw of *Teloganopsis mesoleuca* (C); gills arrangement of *Torleya major* (D) and *Ephemerella mucronata* (E).

4(1') Gills I–VII similar, dorsal and ventral lamellae with several long and slender processes (Fig. 9.15A) .... *Habrophlebia*
4' Gill I different in shape than gills II–VII; gills II–VII with different processes 5
5(4') Gill I bifurcate; gills II-VII with entire margin, dorsal and ventral lamellae more or less lanceolate (Fig. 9.11B) .... *Leptophlebia*
5' Gill I simple or bifurcate; gills II–VII with incisions, dorsal and ventral lamellae with three apical processes, or margin entirely fringed 6
6(5') Gill I bifurcate; gills II–VII with dorsal and ventral lamellae oval, with entire margin heavily fringed (Fig. 9.15C) .... *Thraulus*

6' Gill I simple; gills II–VII with dorsal and ventral lamellae quadrangular, with three apical processes .... *Choroterpes* 7
7(6') Gill apical processes of the same shape, slender, middle process narrow (Fig. 9.15E and G) .... *Choroterpes (Euthraulus)*
7' Gill apical processes of different shape, middle process lanceolate and much longer than the lateral ones (Fig. 9.15D and F) .... *Choroterpes (Choroterpes)*

Insecta: Ephemeroptera: Siphlonuridae: *Siphlonurus*: Subgenera

A small Holarctic family known by 50 species, less than 12 of them are present in the Mediterranean Basin. Iberian Peninsula is especially rich with 8 species.

1 Gill I and II constituted by two lamellae; gills III–VII simple (Fig. 9.14A and B) .... *Siphlonurus (Siphlonurus)*
1' Gills I–VII constituted by two lamellae (Fig. 9.14C) .... One species: *Siphlonurus (Siphlurella) alternatus (Say, 1824)* [Spain]

Insecta: Ephemeroptera: Ameletidae: Genera and Species

A small family especially diversified in the Nearctic region and in the East Palearctic. Only two species are known in Europe.

1 Cerci and paracercus shorter than body length (Fig. 9.16A); tibiae proximally and distally with a broad black transverse band (Fig. 9.16B); gills with a sclerotized line along the margin .... One species: *Ameletus inopinatus Eaton, 1887* [Bosnia-Herzegovina, Croatia, France, Greece, Macedonia, Slovenia, Turkey]
1' Cerci and paracercus longer than body length; tibiae uniformly colored; gills with a sclerotized line in submarginal position (Fig. 9.16C) .... One species: *Metreletus balcanicus (Ulmer, 1920)* [Bosnia-Herzegovina, Croatia, France, Kosovo, Turkey]

Acknowledgments

The authors want to thank the editors, Alain Maasri and James H. Thorp, for inviting us to write this chapter. We express our gratitude to Luke Jacobus (Purdue University Columbus) for fruitful discussions about the generic composition of the family Ephemerellidae, and to André Wagner for providing some rare specimens.

References

Alba-Tercedor, J., & Jaimez-Cuellar, P. (2003). Checklist and historical evolution of the knowledge of Ephemeroptera in the Iberian Peninsula, Balearic and Canary Islands. In E. Gaino (Ed.), *Research update on Ephemeroptera and Plecoptera* (pp. 91–97). Perugia, Italy: University of Perugia.

Barber-James, H. M., Gattolliat, J.-L., Pereira-da-Conceicoa, L. L., & Sartori, M. (In prep). Chapter 18.1. Ephemeroptera. In J. Day, D. C. Rodgers, & J. H. Thorp (Eds.), *Thorp & Covich's freshwater invertebrates* (4th ed., Vol. VII: Key to Afrotropical Fauna). New York: Academic Press.

Barber-James, H. M., Zrelli, S., Yanai, Z., & Sartori, M. (2020). A reassessment of the genus *Oligoneuriopsis* Crass, 1947 (Ephemeroptera, Oligoneuriidae, Oligoneuriellini). *ZooKeys*, *985*, 15–47.

Bauernfeind, E. (2003). The mayflies of Greece (Insecta: Ephemeroptera)—A provisional check-list. In E. Gaino (Ed.), *Research update on Ephemeroptera and Plecoptera* (pp. 99–107). Perugia, Italy: University of Perugia.

Bauernfeind, E. (2018). Mayflies (Ephemeroptera) of the River Vjosa. *Albania. Acta ZooBot Austria*, *155*, 155–162.

Bauernfeind, E., & Soldán, T. (2012). The Mayflies of Europe. Apollo Books. *Ollerup*, 781 pp.

Benhadji, N., Hassaine, K. A., & Sartori, M. (2018). *Habrophlebia hassainae*, a new mayfly species (Ephemeroptera: Leptophlebiidae) from North Africa. *Zootaxa*, *4403*, 557–569.

Benhadji, N., Hassaine, K. A., Gattolliat, J.-L., & Sartori, M. (2019). Thirty years after: An update to the mayflies composition in the Tafna basin (Algeria). *Zoosymposia*, *16*, 22–35.

Buffagni, A. (2021). The lentic and lotic characteristics of habitats determine the distribution of benthic macroinvertebrates in Mediterranean rivers. *Freshwater Biology*, *66*, 13–34.

Buffagni, A., Belfiore, C., Erba, S., Kemp, J. L., & Cazzola, M. (2003). A review of Ephemeroptera species distribution in Italy: Gains from recent studies and areas for future focus. In E. Gaino (Ed.), *Research update on Ephemeroptera and Plecoptera* (pp. 279–290). Perugia, Italy: University of Perugia.

Cuk, R., Cmrlec, K., & Belfiore, C. (2015). The first record of *Ametropus fragilis* Albarda, 1878 (Insecta: Ephemeroptera) from Croatia. *Natura Croatica, 24*, 151–157.

Gattolliat, J.-L., Cavallo, E., Vuataz, L., & Sartori, M. (2015). DNA barcoding of Corsican mayflies (Ephemeroptera) with implications on biogeography, systematics and biodiversity. *Arthropod Systematics & Phylogeny, 73*, 3–18.

Godunko, R. J., Sroka, P., Soldán, T., & Bojkova, J. (2015). The higher phylogeny of Leptophlebiidae (Insecta: Ephemeroptera), with description of a new species of *Calliarcys* Eaton, 1881. *Arthropod Systematics & Phylogeny, 73*, 259–280.

Hrivniak, L., Sroka, P., Bojkova, J., & Godunko, R. J. (2020). Identification guide to larvae of Caucasian *Epeorus (Caucasiron)* (Ephemeroptera, Heptageniidae). *ZooKeys, 986*, 1–53.

Ikonomov, P. (1964). Die Eintagsfliegen (Ephemeroptera) des Flusses Vardar. *Annuaire de la faculté des Sciences de l'Université de Skopje, 15*, 191–198.

Jacob, U. (1977). *Palingenia anatolica* n.sp. (Ephemeroptera, Palingeniidae) aus der Türkei. *Entomologische Nachrichten, 21*, 177–182.

Kechemir, L. H., Sartori, M., & Lounaci, A. (2020). An unexpected new species of *Habrophlebia* from Algeria (Ephemeroptera, Leptophlebiidae). *ZooKeys, 953*, 31–47.

Kluge, N. J. (2016). A new subgenus *Oculogaster* subgen. n. for viviparous representatives of *Procloeon* s. l., with discussion about status of the generic name *Austrocloeon* Barnard 1932 and the species name *africanum* Esben-Petersen 1913 *Cloeon* (Ephemeroptera, Baetidae). *Zootaxa, 4107*, 491–516.

Kluge, N. J. (2021). *Ephemeroptera of the world*. http://www.insecta.bio.spbu.ru. Last accessed January 29, 2021.

Kluge, N. J., & Novikova, E. A. (2014). Systematics of *Indobaetis* Muller-Liebenau & Morihara 1982, and related implications for some other Baetidae genera (Ephemeroptera). *Zootaxa, 3835*, 209–236.

Kluge, N. J., Godunko, R. J., & Svitok, M. (2020). Nomenclatural changes in *Centroptella* Braasch & Soldan, 1980 (Ephemeroptera, Baetidae). *ZooKeys, 914*, 81–125.

Moriniere, J., Hendrich, L., Balke, M., Beermann, A. J., Konig, T., Hess, M., Koch, S., Muller, R., Leese, F., Hebert, P. D. N., Hausmann, A., Schubart, C. D., & Haszprunar, G. (2017). A DNA barcode library for Germany's mayflies, stoneflies and caddisflies (Ephemeroptera, Plecoptera and Trichoptera). *Molecular Ecology Resources, 17*, 1293–1307.

Salur, A., Darilmaz, M. C., & Bauernfeind, E. (2016). An annotated catalogue of the mayfly fauna of Turkey (Insecta, Ephemeroptera). *ZooKeys, 620*, 67–118.

Sartori, M., & Brittain, J. E. (2015). Order Ephemeroptera. In J. H. Thorp, & D. C. Rogers (Eds.), *Ecology and general biology, Vol I: Thorp and Covich's freshwater invertebrates* (4th ed., pp. 873–891). New York: Academic Press.

Sartori, M., & Bauernfeind, E. (2020). Mayfly types and additional material (Insecta: Ephemeroptera) examined by F.-J. Pictet and A.-E. Pictet, housed in the Museums of Natural History of Geneva and Vienna. *Revue Suisse de zoologie, 127*(2), 315–339.

Sartori, M. (1992). Mayflies from Israel (Insecta; Ephemeroptera). I. Heptageniidae, Ephemerellidae, Leptophlebiidae, Palingeniidae. *Revue Suisse de Zoologie, 99*, 835–858.

Soldán, T., & Godunko, R. J. (2008). Two new species of the genus *Baetis* Leach, 1815 (Ephemeroptera: Baetidae) from Cyprus. *Annales Zoologici, 58*, 79–104.

Soldán, T., & Godunko, R. J. (2009). *Baetis zdenkae* sp nov., a new representative of the *Baetis buceratus* species-group (Ephemeroptera: Baetidae) from Rhodos (Greece) with notes to species-grouping of the subgenus *Baetis* Leach, 1815 s. str. *Zootaxa, 1972*, 1–19.

Tenchini, R., Cardoni, S., Piredda, R., Simeone, M. C., & Belfiore, C. (2018). DNA barcoding and faunistic criteria for a revised taxonomy of Italian Ephemeroptera. *European Zoological Journal, 85*, 254–267.

Tierno de Figueroa, J. M., López-Rodríguez, M. J., Fenoglio, S., Sánchez-Castillo, P., & Fochetti, R. (2013). Freshwater biodiversity in the rivers of the Mediterranean Basin. *Hydrobiologia, 719*, 137–186.

Vilenica, M., Gattolliat, J.-L., Mihaljevic, Z., & Sartori, M. (2015). Croatian mayflies (Insecta, Ephemeroptera): Species diversity and distribution patterns. *ZooKeys, 523*, 99–127.

Xerxa, B. L., Sartori, M., Gashi, A., & Gattolliat, J.-L. (2019). First checklist of mayflies (Insecta, Ephemeroptera) from Kosovo. *ZooKeys*, 69–82.

Yanai, Z., Gattolliat, J.-L., & Dorchin, N. (2018). Taxonomy of *Baetis* Leach in Israel (Ephemeroptera, Baetidae). *ZooKeys, 794*, 45–84.

Yanai, Z., Sartori, M., & Gattolliat, J.-L. (2020). Contribution to the mayflies (Insecta, Ephemeroptera) of Israel and the Palestinian Authority. *Check List, 16*, 229–236.

Chapter 10

Order Plecoptera

José Manuel Tierno de Figueroa[1], Manuel Jesús López-Rodríguez[2] and Romolo Fochetti[3]
[1]Department of Zoology, University of Granada, Granada, Spain, [2]Department of Ecology, University of Granada, Granada, Spain, [3]Department for Innovation in Biological, Agro-food and Forest Systems, Tuscia University, Viterbo, Italy

Introduction

Plecoptera is an amphibiotic insect order with a life cycle consisting of three stages: egg, nymph, and imago or adult. Eggs and nymphs are found in freshwater ecosystems, while adults are found in riparian or shoreline terrestrial ecosystems (with few exceptions not present in the Mediterranean Basin). This order is widely distributed on all continents except Antarctica and is present on the main continental islands as well as in Iceland (Hynes, 1988; Fochetti & Tierno de Figueroa, 2008a). It is composed of two suborders, Arctoperlaria and Antarctoperlaria, 17 families (Zwick, 2000; South et al., 2021), and approximately 3800 described species (DeWalt & Ower, 2019; DeWalt et al., 2020). Stoneflies mainly inhabit streams and rivers, but some species can also live in lentic systems (Hynes, 1976). They are very sensitive to physical, chemical, and biological alterations of their habitats, which, together with their low dispersal capacity, make them particularly vulnerable to global change (Fochetti & Tierno de Figueroa, 2006, 2008a; Tierno de Figueroa et al., 2010). Both adults and nymphs contribute significantly to the food webs of freshwater ecosystems (Tierno de Figueroa & López-Rodríguez, 2019), and they exhibit complex and diverse behaviors (Hynes, 1976; Zwick, 1980; Stewart, 1994, 2009; DeWalt et al., 2015).

In their review of the freshwater biodiversity of rivers from the Mediterranean Basin, Tierno de Figueroa et al. (2013) reported 340 stonefly species belonging to 32 genera, seven families, and two superfamilies (Perloidea and Nemouroidea). According to these authors, a high percentage of species, that is, 40.3% of the total (137 species), could be considered endemic to the Mediterranean Basin and that percentage could be significantly higher if species that only slightly exceed the geographical boundaries considered in that study were also included as endemic. Moreover, considering that there are some differences in the boundaries of the Mediterranean Basin between Tierno de Figueroa et al. (2013) and those considered in this book and that there have been many taxonomic and faunal studies on this order of insects in the area, the data presented here differ slightly. Therefore 419 species of Plecoptera can be considered Mediterranean according to the geographic boundaries accounted for in this book. Those species belong to 34 genera (*Agnetina* was not included in the previous paper because only the most western part of Turkey was considered there, and *Zwicknia* was not described as a different genus until 2014). The genera *Afroperlodes*, *Guadalgenus*, *Helenoperla*, and *Tyrrhenoleuctra* are endemic, and *Hemimelaena* and *Eoperla* are nearly endemic to the considered Mediterranean Basin. *Bulgaroperla* has a restricted global distribution, including part of the eastern Mediterranean, and the genera *Besdolus*, *Marthamea*, *Brachyptera*, and *Capnioneura* display the greatest specific diversity within the Mediterranean Basin.

General ecology and distribution

Nymphs of stoneflies from the Mediterranean Basin are found in streams from almost sea level [e.g., *Tyrrhenoleuctra tangerina* (Navás, 1922)] to nearly 3000 m a.s.l. (e.g., *Perlodes microcephalus* (Pictet, 1833) or *Isoperla nevada* Aubert, 1952). This suggests that populations are found under a wide array of environmental conditions, which sometimes originate from clearly segregated Plecoptera biocenosis. Mid to lowland biocenosis or those present in temporary streams are mainly characterized by euritherm species and would correspond to the thermophilous association defined by Aubert (1963), while those of higher elevation reaches are dominated by rheophilic, cold-stenotherm species,

Identification and Ecology of Freshwater Arthropods in the Mediterranean Basin. DOI: https://doi.org/10.1016/B978-0-12-821844-0.00004-1

corresponding to the orophilous association of Aubert (1963). Due to the specific tolerance ranges of stoneflies to some ecological factors, such as temperature, dissolved oxygen, and salinity, species diversity usually increases with elevation. Within particular stream reaches, nymphs seem to be found more frequently in certain microhabitats, such as those where pebbles or gravels predominate, and less frequently in muddy and sandy substrates (Tierno de Figueroa et al., 2003). This is probably because in muddy and sandy substrates the very small interstices between particles reduce the water circulation and therefore the available dissolved oxygen.

In their life cycle, stoneflies spend more time in the nymphal stage than in other stages, so factors related to the biotic and abiotic aquatic environment are important for determining the survival and, in general, the fitness of the individuals in each population. Most species are univoltine in the Mediterranean Basin, with some species being semivoltine/merovoltine [e.g., Perlidae as *Perla* spp. or *Dinocras cephalotes* (Curtis, 1827) or Perlodidae as *Guadalgenus franzi* (Aubert, 1963)]. Multivoltinism has not been recorded in any population in the Mediterranean Basin. Nonetheless, because multivoltinism is difficult to distinguish from the coexistence of several cohorts, it is not clear if some species are multivoltine or if different univoltine but temporally delayed cohorts are present in the same stream. This is typical of some species of the *Protonemura* genus, as in *Protonemura gevi* Tierno de Figueroa & López-Rodríguez, 2010, given the particular conditions of the cave this species inhabits (with a constant temperature throughout the year and an absence of light) may play an important role in its developmental pattern (López-Rodríguez & Tierno de Figueroa, 2012).

To cope with the drought period typical of many temporary streams in the Mediterranean Basin, some strategies have been selected in populations of particular species (see López-Rodríguez et al., 2017). One of these strategies is the existence of a resting stage in the middle of nymphal development, either diapause or quiescence, usually synchronized with the beginning and the end of the dry period. This allows semivoltine species, such as *G. franzi*, to survive this adverse phase and is usually accompanied by a previous displacement of nymphs to deeper substrates where water remains (Agüero-Pelegrín & Ferreras-Romero, 2002; López-Rodríguez et al., 2009b). Another relatively common strategy in these environments is the rapid growth rate that can be observed in the nymphs of several univoltine species, mainly in those that develop during the end of winter and/or the spring, that is, in the months prior to the estival drought. This is the case, for instance, for *Brachyptera vera cordubensis* Berthélemy & Baena, 1984 and *Capnioneura gelesae* Berthélemy & Baena, 1984 (López-Rodríguez et al., 2009a; López-Rodríguez et al., 2017). This fast growth ensures that the nymphs have reached the mature stage before conditions become harsh due to the absence of water. In some genera, such as *Capnioneura*, development even occurs sooner in species typically found in temporary streams than in those living in permanent aquatic environments, probably as the result of natural selection favoring nymphs that developed earlier within a population and eggs that hatched before the onset of adverse conditions. Finally, another strategy that has been favored by natural selection under the particular pressures that exist in these streams is the existence of asynchrony in the development of nymphs from the same population. This is reflected in the presence of several overlapping cohorts coexisting in the same stream reach that grow and develop in different months of the year, as occurs in *Tyrrhenoleuctra* cf. *minuta* (Klapálek, 1901) (López-Rodríguez et al., 2009a). This asynchronic development is the effect of an extended flight period combined with ovoviviparism (at least in the studied population). The presence of several cohorts in the same population would provide insurance should the drought period start earlier in a particular year. Moreover, other strategies in the adult or egg stages would contribute to the success of Mediterranean stoneflies.

In a review of the biodiversity of Mediterranean rivers, Tierno de Figueroa et al. (2013) found that most species of EPT (Ephemeroptera, Plecoptera, and Trichoptera), both endemic and nonendemic, were univoltine and that several endemic species used diapause as a form of resistance to droughts. Moreover, the largest proportion of both endemic and nonendemic species were *r*-strategists, which could be easily assumed to occur considering stoneflies exclusively. Additionally, Bonada and Dolédec (2011), working on traits of Mediterranean-exclusive genera of macroinvertebrates, found that these taxa exhibited specific traits to cope with the particular flow and climatic conditions occurring in the Mediterranean Basin.

The life cycle and the length and timing of nymphal development depend on the available trophic resources in a particular environment. Species with longer nymphal development may exploit different trophic resources throughout the year, which may differ both qualitatively and quantitatively, while those with a shorter development may have a narrower array of food items to choose from. Overall, the nymphs of Mediterranean stoneflies feed upon a wide assemblage of trophic resources and are fed upon by many other animals. Traditionally, Perloidea have been considered predators, while Nemouroidea have been considered phytophagous—detritivorous, but a wide array of exceptions exists worldwide (Tierno de Figueroa & López-Rodríguez, 2019). In the Mediterranean Basin, there are also particular records of trophic plasticity, on most occasions related to the resource availability associated with the natural variability in Mediterranean streams and rivers. For instance, *Isoperla morenica* Tierno de Figueroa & Luzón-Ortega, 2011 (formerly included in *I. curtata* Navás, 1924), belonging to a typical predator family (Perlodidae), behaves as a scraper-grazer in temporary streams when the availability of certain algae is very high; other Perloidea species also incorporate vegetal matter into their diets under these conditions (López-Rodríguez et al., 2009b).

Due to their wide trophic spectrum, both interspecific and intraspecific stonefly nymphs play different relevant functions in their ecosystems. They are found at different trophic levels and are considered both primary and secondary consumers. As mentioned before, some species in the Mediterranean Basin can be simultaneously predators and primary consumers in the same food web (e.g., López-Rodríguez et al., 2009b). Some species may be found in the detrital-based pathway, consuming leaves and other coarse particulate organic matter, while others are predominantly found in the producer-based pathways, feeding mainly on diatoms or other primary producers. The relative importance of stonefly nymphs as primary consumers in relation to other organisms is thought to be moderate to low, as other organisms have more specialized mouthparts or feeding structures that make them more efficient in that role (e.g., Gastropoda feeding on biofilm with their radula or Gammaridae feeding on leaves with their adapted mouthparts). Nonetheless, stonefly nymphs contribute to maintaining the flux of matter throughout these pathways, and their secondary production is comparable to that of other macroinvertebrates of the same functional feeding group (López-Rodríguez et al., 2009a). On the other hand, nymphs in the upper levels of the food web play a more important role due to their large size and the absence of many other organisms that may compete with them for the same function. Most likely, among lotic macroinvertebrates, only Odonata nymphs may be comparable in terms of size and voracity, but there are few Odonata species that share habitat with stoneflies. In many streams of the Mediterranean Basin, predatory stonefly nymphs are numerically the most dominant macroinvertebrates in the upper levels of the food web (Peralta-Maraver et al., 2017). As such, they may exert an important role as top-down control in these ecosystems, though this has not always been supported, at least when studying experimental macroinvertebrate biocoenoses (López-Rodríguez et al., 2018). Furthermore, stonefly nymphs are an important part of the diet of top-level predators, such as fish; therefore, they contribute to the flow of energy to the uppermost trophic levels in lotic ecosystem food webs.

As mentioned previously, nymphs are sensitive to changes in environmental factors and many species have narrow ranges of tolerance to those factors. This makes stoneflies particularly vulnerable to anthropogenic disturbances. In the framework of the current global change, some of the more important threats stoneflies face are the contamination of middle and lowland reaches due to human activities (agricultural, industrial, etc.), habitat destruction, stream continuity fragmentation, alteration of the environmental conditions both upstream and downstream of barriers (such as dams or hydroelectric power stations), water abstraction for irrigation or human supply, and changes in water temperature due to climate change. The latter causes even high-mountain species in pristine habitats to be endangered as their populations are directly affected by climate warming. In fact, a study assessing the vulnerability of stonefly species across Europe reported that most vulnerable species are found in the Alps, in the Pyrenees, and in the Iberian Peninsula, that is, near the limits of or within the Mediterranean Basin (Tierno de Figueroa et al., 2010). Species with small populations may be facing decline. As pointed out by Fochetti & Tierno de Figueroa (2008a), stoneflies have a high percentage of endemism globally and in the Mediterranean Basin specifically (Tierno de Figueroa et al., 2013). Several species have restricted distribution (Fochetti & Tierno de Figueroa, 2006) and therefore are very exposed to climate change. Indeed, some species are currently known only from their typical localities. Thus the overall vulnerability of this group makes it especially interesting to study the effects of anthropogenic disturbances and, particularly, the consequences of global change. In fact, the monitoring of some population-level parameters or autoecological traits of these organisms may provide a useful tool for managing lotic ecosystems in future climatic and socioeconomic scenarios.

Data on the distribution of the Plecoptera genera in the Mediterranean Basin are reported in Table 10.1.

Morphological characteristics needed for identification

Mediterranean Plecoptera nymphs show dull colors, with dark, gray or yellowish tones. The body can be hairless or have bristles of varying lengths and densities. Stonefly nymphs generally have a slender and flattened body, ending in two multiarticulate cerci at the tip of the abdomen. The first stages without wingpads are usually called neanids; individuals are called nymphs with the appearance of wingpads, which become gradually more developed with each molt. Nevertheless, the terms nymph or naiad (or even larvae) are usually used to refer to the complete preimaginal stage.

Head

Nymph head can be prognathous (Perlidae, Perlodidae, and Chloroperlidae), hypognathous (Nemouridae), or somewhat intermediate between the two (Taeniopterygidae, Capniidae, and Leuctridae). Starting from the posterior margin, from a

TABLE 10.1 Distribution of the Plecoptera genera in the Mediterranean area. NW: Mediterranean area of the Iberian Peninsula, France, Italy, and the northwestern Mediterranean islands; NE: Mediterranean areas of the Balkan countries and Greece, including their islands; TC: Mediterranean areas of Turkey and Cyprus; LSIP: Mediterranean area of Lebanon, Syria, Israel and Palestine; and M: Mediterranean area of Maghreb (Morocco, Algeria and Tunisia).

Taxa	NW	NE	TC	LSIP	M
Perlodidae					
Afroperlodes Miron & Zwick, 1973					●
Arcynopteryx Klapálek, 1904	●	●			
Besdolus Ricker, 1952	●	●			
Bulgaroperla Raušer, 1966		●	●		
Dictyogenus Klapálek, 1904	●	●			
Guadalgenus Stark & González del Tánago, 1986	●				
Hemimelaena Klapálek, 1907	●				●
Isogenus Newman, 1833	●				
Perlodes Banks, 1903	●	●	●	●	
Isoperla Banks, 1906	●	●	●	●	●
Perlidae					
Agnetina Klapálek, 1907			●		
Dinocras Klapálek, 1907	●	●			
Eoperla Illies, 1956	●	●			●
Helenoperla Sivec, 1997		●			
Marthamea Klapálek, 1907	●	●	●	●	●
Perla Geoffroy, 1762	●	●	●		●
Chloroperlidae					
Chloroperla Newman, 1836	●	●	●		
Pontoperla Zwick, 1967			●		
Siphonoperla Zwick, 1967	●	●	●	●	●
Xanthoperla Zwick, 1967	●	●	●		
Taeniopterygidae					
Taeniopteryx Pictet, 1842	●	●			
Brachyptera Newport, 1848	●	●	●	●	●
Rhabdiopteryx Klapálek, 1902	●	●	●		
Nemouridae					
Amphinemura Ris, 1902	●	●	●		●
Protonemura Kempny, 1898	●	●	●	●	●
Nemoura Latreille, 1796	●	●	●		●
Nemurella Kempny, 1898	●	●			
Capniidae					
Capnia Pictet, 1841	●	●	●	●	●
Capnioneura Ris, 1905	●	●	●		●

(Continued)

TABLE 10.1 (Continued)

Taxa	NW	NE	TC	LSIP	M
Capnopsis Morton, 1896	●	●	●		●
Zwicknia Murányi, 2014	●	●	●	●	
Leuctridae					
Leuctra Stephens, 1836	●	●	●	●	●
Pachyleuctra Despax, 1929	●				
Tyrrhenoleuctra Consiglio, 1957	●				●

dorsal view, the head shows a metopic suture that originates two postfrontal sutures, which extend up to the base of the antennae, forming the ecdysial or epicranial suture. An occipital fold with taxonomic interest can also be observed in some taxa. Neanids show isolated ommatidia, while nymphs present well-developed compound eyes and (in all the Mediterranean species) three ocelli: one central anterior and two lateral posterior. The antennae are threadlike, formed by a variable number of antennomers. The labrum is a broadly flattened plate hinged to the clypeus or frontoclypeus. Nymphs have chewing mouthparts, with slight differences in morphology inside the order, sometimes reflecting their feeding biology. The mandibles are heavily sclerotized, with both cutting and grinding edges. In the largely carnivorous Perloidea, the mandibles are enlarged and lack a grinding edge, while the cutting edge is formed by two groups of teeth. In contrast, in Nemouroidea (shredders, collectors-gatherers, or scrapers), the mandibles are short and strong and have a well-developed grinding edge. The maxilla has a five-segmented maxillary palp; the lacinia shows a maximum of two apical teeth. The labium has a labial palp consisting of three articles; the glossae and paraglossae are of the same dimension in the Nemouroidea, while the glossae are smaller in the Perloidea.

Thorax

The pro-, meso-, and metanotum are clearly differentiated and longitudinally divided by a suture. The pronotum can bear bristles of taxonomical value on the margin. The legs are long and stout. Both the tibiae and femora can bear fringes of swimming hairs or bristles. The tarsi are three-segmented, with the last segment bearing two terminal claws.

Abdomen

Cylindrical in shape, this structure is formed by 11 metameres, with the first 10 being well developed and the 11th being reduced and modified into two triangular ventral paraprocts and a dorsal lamina, joined to the 10th tergite. Tergites 1 and 2 are separated from their sternites by a membranous area. The remaining tergites (either all or some of them) are joint to the sternites, forming a complete ring. The cerci are always long and filiform.

Gills

The nymphal tracheal system is apneustic, and breathing takes place through the body surface or through gills, which can be differently shaped (finger-like, filamentous, or telescopic). Gill structures may be present in different body regions: on the head (in the postmentum of some Perlodidae) or on the thorax, either in the prosternite (the so-called cervical gills or prosternal gills of the *Protonemura* and *Amphinemura*) or in the pleura (thoracic pleural gills of Perlidae), on the coxae (only in *Taeniopteryx* in the Mediterranean Basin) or on the abdomen (in the anal region in some Perlidae).

Material preparation and preservation

Stonefly nymphs can be collected all year long using different sampling methods. For qualitative studies, a kick sampler can be used, while a Surber or Hess sampler, depending on the type of substrate, is required for quantitative research. These samplers must be placed on the stream bottom, moving the stones upstream (kick) or inside (Surber and Hess) to allow the nymphs to drift and enter the net. For collecting nymphs from the hyporheic zone, a Bou-Rouch pump can be employed. The collected material can be sorted in the field using a white plastic tray and must then be stored in plastic vials containing preserving liquids, usually 70% alcohol. For better fixation of internal organs, a mixture of 90% alcohol with 2% formaldehyde (40% by volume) can be used. Regarding molecular research, stonefly nymphs must be preserved in 95% to 99% alcohol or, if dealing with RNA, in RNA*later*™ (see, for instance, Tierno de Figueroa & Fochetti, 2014).

The morphological study of neanids and nymphs usually requires observation under a stereomicroscope. Specimens must often be dissected for the study of internal organs or structures. Organs can be mounted on microscope slides using a preserving (fixative) compound (e.g., Hoyer or Faure fixatives) prior to examination under a transmission microscope. Morphological and ultrastructural investigations of anatomic structures can be carried out by scanning and transmission electron microscopy. The use of these approaches requires specific preparation methods (see, for instance, Fausto et al., 2002).

Stonefly nymphs can be reared in the laboratory. In this regard, simple artificial water courses can be built, consisting of a continuous flow of water in a container, with a refrigerating system to keep the water at a constant or controlled temperature and a pump with an incorporated filter.

A more thorough and detailed treatment of collecting, rearing, preserving, and labeling Plecoptera nymphs can be found in DeWalt et al. (2015).

The study of nymphal feeding can be carried out by analyzing gut contents, dissecting the gut, or clearing the integument with suitable liquids (e.g., Hertwig's liquid) prior to the observation of ingested material through a microscope (for details, see Tierno de Figueroa & Fochetti, 2001). Data on nymphal trophic ecology can also be based on direct observations in nature or in the laboratory, experimental studies and stable isotope analyses (see, for instance, Hershey et al., 2006), or studies on digestive enzymatic activity (see, for instance, López-Rodríguez et al., 2012).

In some studies, such as those dealing with the secondary production of nymphs, dry mass may be necessary. As in other animal groups, dry mass is preferentially obtained from individuals that are preserved in 4% formaldehyde, dried at 60°C for 24 h, desiccated for 1 h and posteriorly weighed with a microbalance (see, for instance, Benke & Huryn, 2006).

For biometric studies (including those aimed at analyzing the life cycles or the growth rates of nymphs), preservation in 4% formaldehyde is preferred, but individuals preserved in 70% alcohol could also be used.

Keys

The keys presented here are based on those of Hynes (1977), Consiglio (1980), Sivec et al. (1988), Tierno de Figueroa et al. (2003), Fochetti & Tierno de Figueroa (2008b), Tachet et al. (2010) and, mainly, Zwick (2004) as well as the direct observation and study of specimens and original descriptions of some genera. Because the nymphs of many species have not been described yet or, in other cases, no valuable characters allow species differentiation, we have only included keys for families and genera.

Plecoptera: Families

1 Glossae reduced, considerably shorter than paraglossae (Fig. 10.1A); last segment of maxillary palp narrower than previous segments (Fig. 10.1B) .... (Superfamily Perloidea) 2

1' Glossae and paraglossae with similar lengths (Fig. 10.1C); last segment of maxillary palp as wide as previous segments (Fig. 10.1D) .... (Superfamily Nemouroidea) 4

2(1) Thoracic pleural gills absent 3

2' Thoracic pleural gills present (Fig. 10.2) .... **Perlidae**

3(2) Maxillary palpus with apical segment distinctly smaller than the previous segment (approximately 1/3–1/4 wide), needle-like, and set asymmetrically on the previous segment (Fig. 10.3A); outer margins of the wingpads rounded (Fig. 10.3B) .... **Chloroperlidae**

3' Maxillary palpus with apical segment only slightly smaller than the previous segment, not needle-like, and set symmetrically on the previous segment (Fig. 10.1B); outer margins of the wingpads straight (Fig. 10.3C) .... **Perlodidae**

4(1') Tarsal segment 2 clearly shorter than segment 1 (Fig. 10.4A) 5

4' Tarsal segments progressively longer from 1st to 3rd (Fig. 10.4B) .... **Taeniopterygidae**

5(4) Body streamlined and cylindrical; extended metathoracic legs not reaching the tip of the abdomen* (Fig. 10.5A) 6

5' Body stout; extended metathoracic legs reaching or surpassing the tip of the abdomen (Fig. 10.5B) .... **Nemouridae**

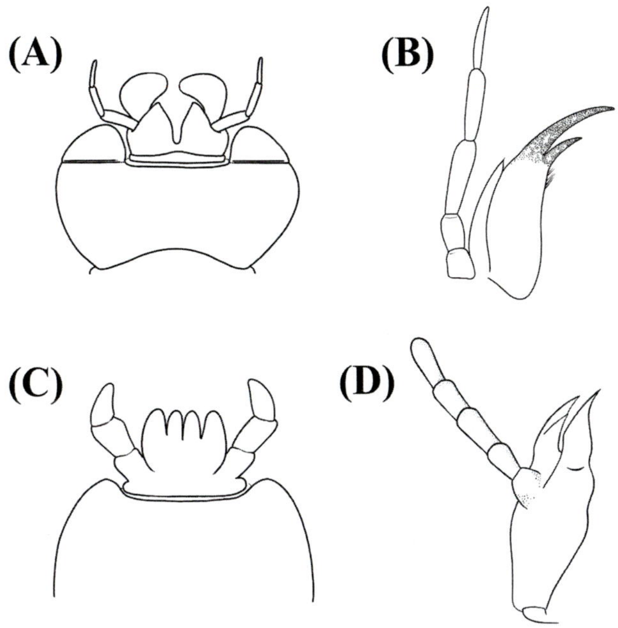

FIGURE 10.1 (A) Labium of a Perloidea; (B) maxilla of a Perloidea; (C) labium of a Nemouroidea; (D) maxilla of a Nemouroidea.

FIGURE 10.2 Photograph of a *Perla* showing its thoracic pleural gills.

6(5) Mentum small, not covering the basal part of the maxillae (Fig. 10.6A) .... **Capniidae**
6' Mentum very large, covering the basal part of the maxillae (Fig. 10.6B) .... **Leuctridae**

*Note: The disposition of mature nymph wingpads is usually employed for distinguishing Nemouridae (with wingpads divergent from the body axis) from Capniidae/Leuctridae (wingpads parallel or subparallel to the body axis). Nevertheless, within Capniidae, the genus *Capnioneura* has slightly divergent wingpads (Fig. 10.17C). However, *Capnioneura* can be easily distinguished from Nemouridae by its glabrous body and cerci, together with the thick basal parts of its cerci (Fig. 10.17E) (as well as the other characters reported in the keys).

Plecoptera: Perlodidae: Genera

(Ten genera and approximately 60 species in the Mediterranean Basin)

1 Paraprocts blunt (Fig. 10.7A) .... (Subfamily Perlodinae) 2
1' Paraprocts pointed (Fig. 10.7B) .... (Subfamily Isoperlinae) ***Isoperla***
2(1) Lacinia unidentate (Fig. 10.8A) 3

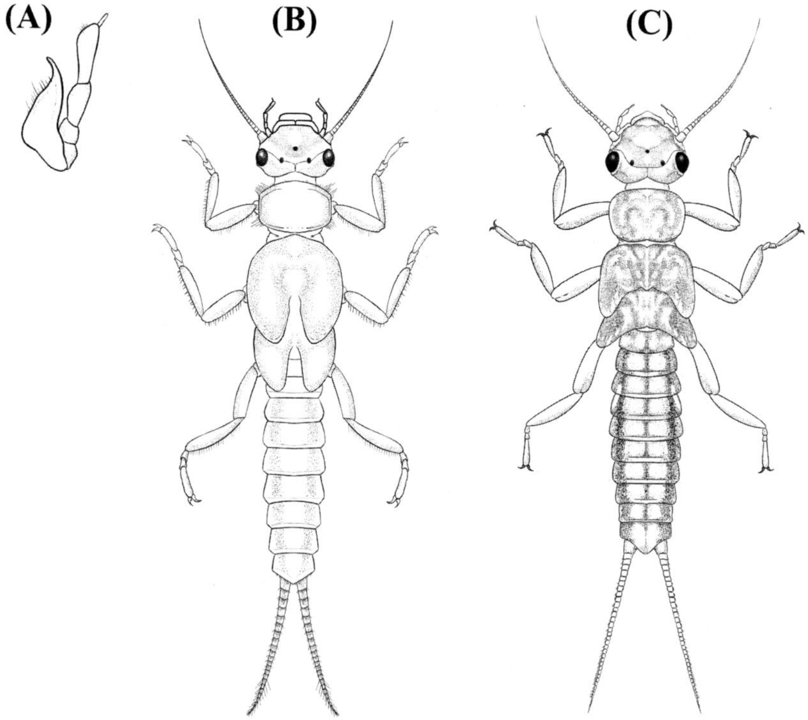

FIGURE 10.3 (A) Maxilla of a Chloroperlidae; (B) habitus of a Chloroperlidae; (C) habitus of a Perlodidae.

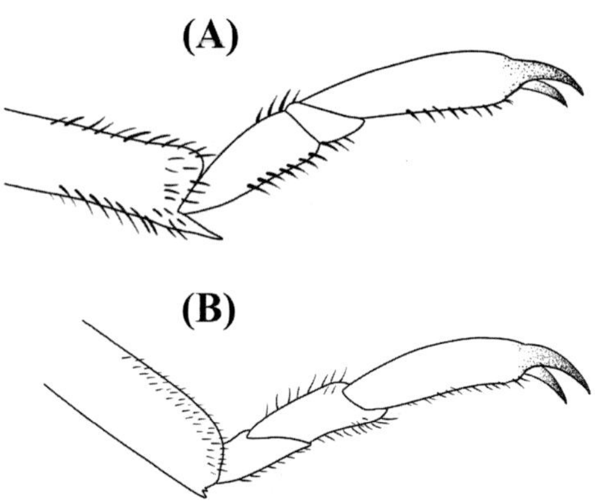

FIGURE 10.4 Distal part of the leg of (A) a Nemouridae; (B) a Taeniopterygidae.

2' Lacinia bidentate (with terminal and subterminal teeth) (Fig. 10.8B–F) 4
3(2) Head and pronotum hairless .... ***Afroperlodes***
3' Head and pronotum with hairs .... ***Hemimelaena***
4(2') Short finger-shaped gill on either side of the base of the postmentum absent. Abdominal segments 1 to 4 or 1 to 2 divided into tergite and sternite by a membranous pleura (Fig. 10.9A, B) 5
4' Short finger-shaped gill on either side of the base of the postmentum (Fig. 10.9D). Abdominal segments 1 to 3 divided into tergite and sternite by a membranous pleura (Fig. 10.9C) .... ***Arcynopteryx***
5(4) Abdominal segments 1 to 4 divided into tergite and sternite (Fig. 10.9A) .... ***Perlodes***
5' Abdominal segments 1 to 2 divided into tergite and sternite (Fig. 10.9B) 6
6(5') Lacinia with setal fringe along the inner edge beginning immediately next to the apical teeth (Fig. 10.8C–F) 7
6' Lacinia with setal fringe along the inner edge clearly separated from the apical teeth (Fig. 10.8B) .... ***Bulgaroperla***
7(6) Apical tooth of the lacinia short (approximately 1/3 of the lacinia length); inner margin of the lacinia with many setae (more than 9) (Fig. 10.8D–F) 8
7' Apical tooth of the lacinia long (approximately 1/2 of the lacinia length); inner margin of the lacinia with only few setae (approximately 3–4) (Fig. 10.8C) .... ***Guadalgenus***

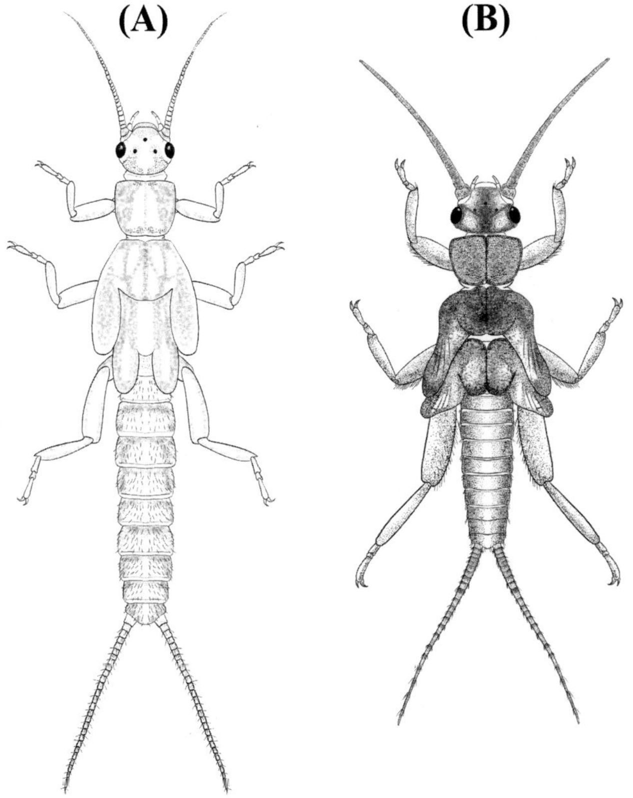

FIGURE 10.5 Habitus of (A) a Leuctridae; (B) a Nemouridae.

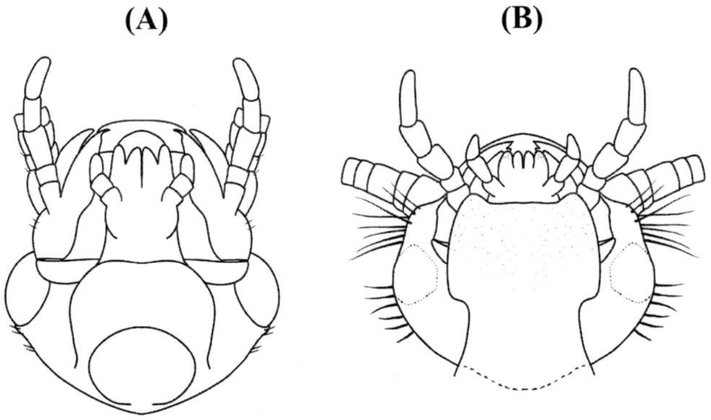

FIGURE 10.6 Head (in ventral view) of (A) a Capniidae; (B) a Leuctridae. *Figures redrawn from Zwick (2004).*

8(7) Inner edge of the lacinia below the subapical tooth with setae in a single long row (Fig. 10.8E, F) 9
8' Inner edge of the lacinia below the subapical tooth with setae forming a small patch (Fig. 10.8D) .... ***Besdolus***
9(8) Lacinia slender and only slightly notched below the subapical tooth (Fig. 10.8E) .... ***Isogenus***
9' Lacinia broad with a distinctive bulge below the subapical tooth (Fig. 10.8F) .... ***Dictyogenus***

Plecoptera: Perlidae: Genera

(Six genera and approximately 21 species in the Mediterranean Basin)

1 Postmentum with anterior lobes separated by sutures (Fig. 10.10A) 2
1' Postmentum with anterior lobes not separated by sutures (Fig. 10.10B) 3
2(1) Occipital fold angled forward (Fig. 10.11A); abdomen without pale bands, with only paired pale spots ....
.... ***Dinocras***
2' Occipital fold regularly curved (Fig. 10.11B); abdomen with two more or less complete pale bands .... ***Eoperla***

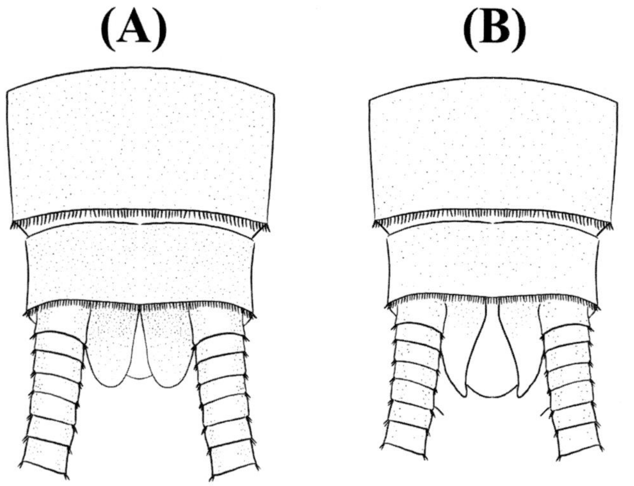

FIGURE 10.7 Abdomen tip of (A) a Perlodinae; (B) an Isoperlinae.

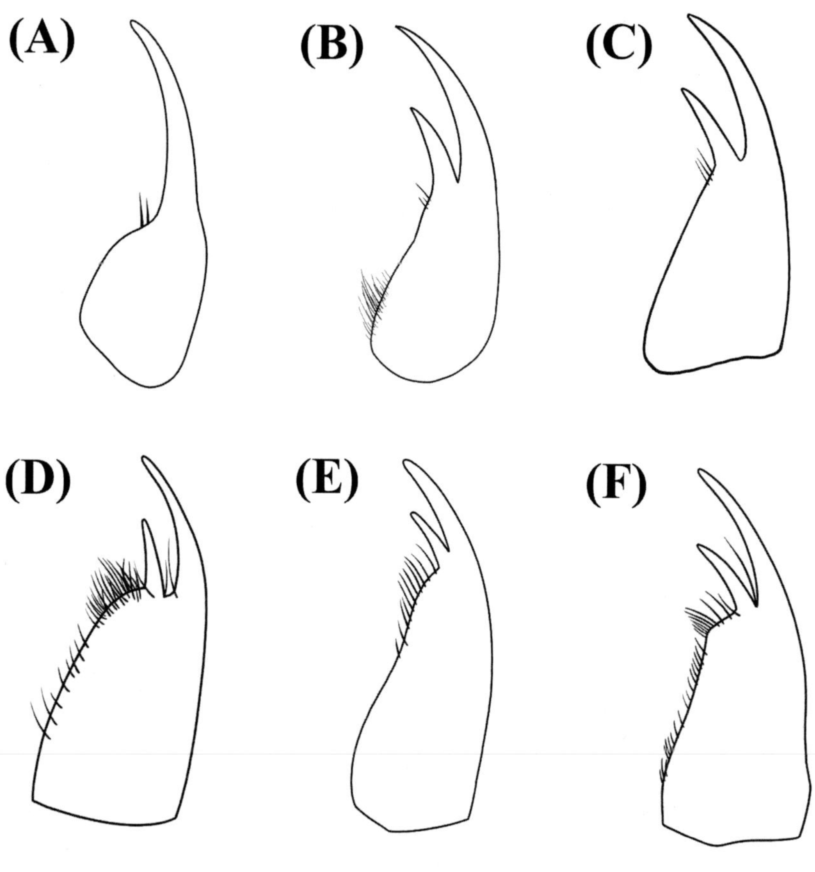

FIGURE 10.8 Lacinia of (A) *Hemimelaena* (similar in *Afroperlodes*); (B) *Bulgaroperla*; (C) *Guadalgenus*; (D) *Besdolus*; (E) *Isogenus*; (F) *Dictyogenus*. *Figures B-F redrawn from Zwick (2004).*

3(1') Occipital fold bent angularly forward behind inner edge of compound eye (Fig. 10.12A, B) 4
3' Occipital fold not bent angularly forward behind inner edge of compound eye (Fig. 10.13A) 5
4(3) Head with a dark band between the bases of the antennae leaving a pale area in front of the anterior ocellus (Fig. 10.12A) .... ***Marthamea***
4' Head with a dark, wide band between the bases of the antennae interrupted in the middle by three light areas (Figs. 10.12B) .... ***Agnetina***

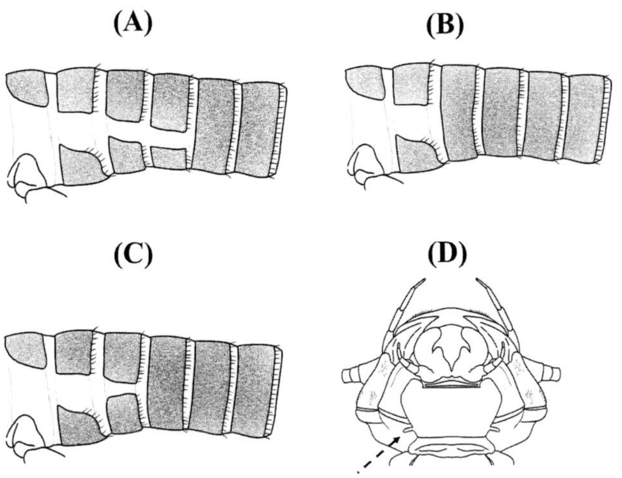

FIGURE 10.9 Proximal part (in lateral view) of the abdomen of different Perlodidae genera showing the abdominal segments: (A) first to fourth segments divided into tergites and sternites; (B) first and second segments divided into tergites and sternites; (C) first to third segments divided into tergites and sternites. (D) Head (in ventral view) of *Arcynopteryx* (finger-shaped gills shown with the arrow).

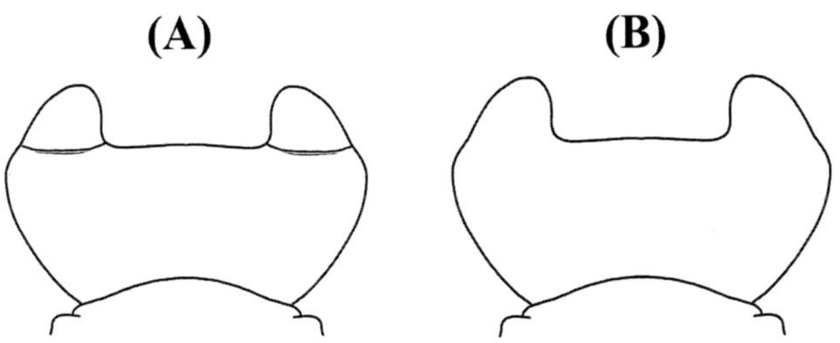

FIGURE 10.10 Postmentum with (A) anterior lobes separated by sutures; (B) anterior lobes not separated by sutures.

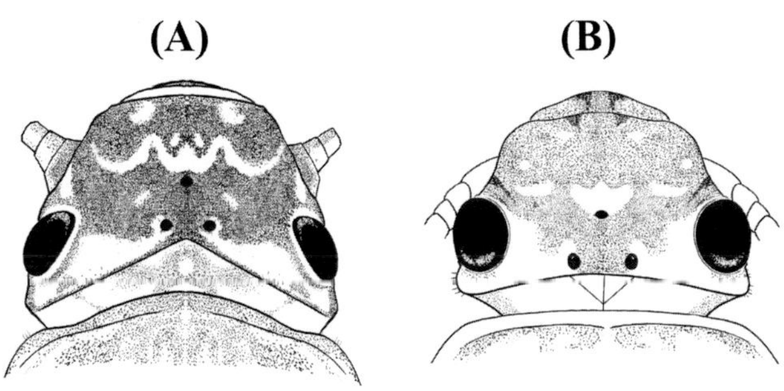

FIGURE 10.11 Head (in dorsal view) of (A) *Dinocras*; (B) *Eoperla*.

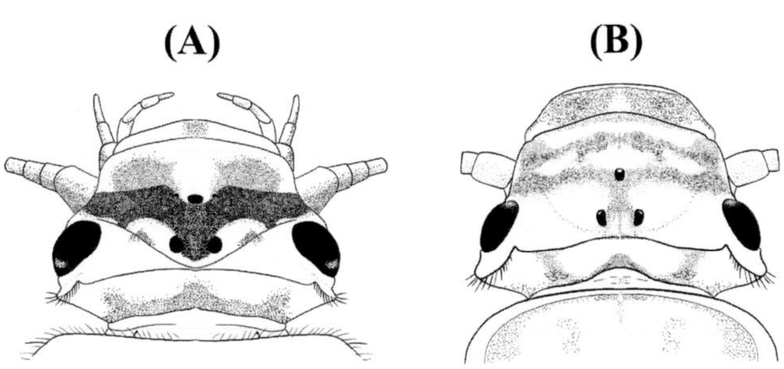

FIGURE 10.12 Head (in dorsal view) of (A) *Marthamea*; (B) *Agnetina*.

5(3') Basal cercus segments with rings of dark spines; pilosity around the cercus segments usually short, half of the segment length (Fig. 10.13B) (without considering those silky, long hairs forming a dorsal fringe, especially in the cercus basal part, exceptionally absent in *Perla bipunctata* Pictet, 1833) .... ***Perla***

5' Basal cercus segments without rings of dark spines; cercus surrounded by pilosity, longer than the segment; cerci without dorsal hair fringe (Fig. 10.13C) .... ***Helenoperla***

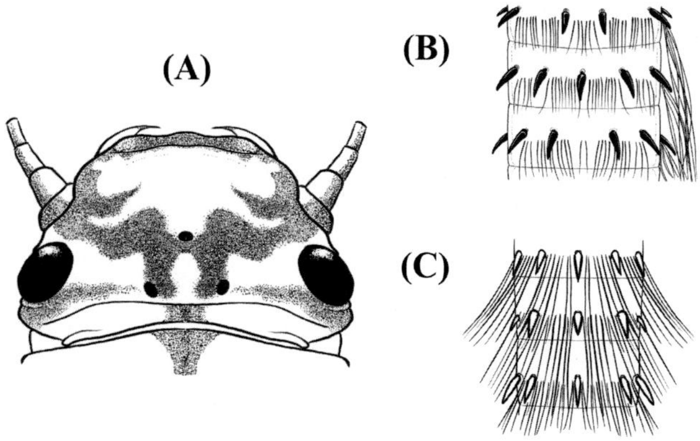

FIGURE 10.13 (A) Head (in dorsal view) of *Perla*. Basal cercus segments of (B) *Perla*; (C) *Helenoperla*. *Figures B and C redrawn from Zwick (2004).*

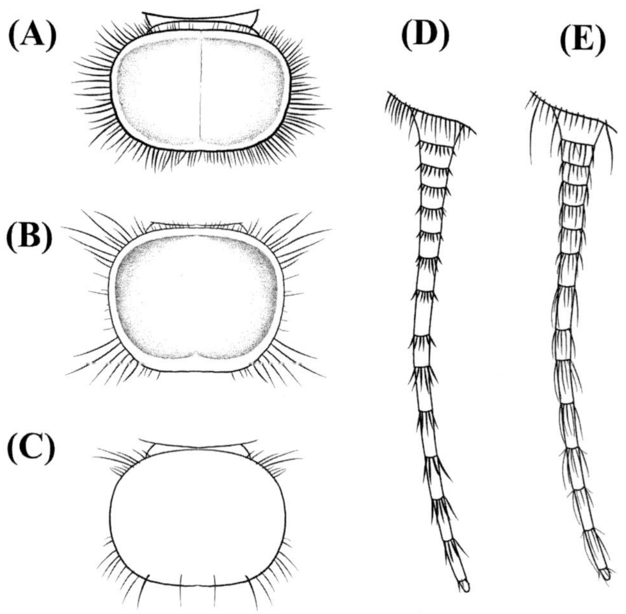

FIGURE 10.14 Pronotum of (A) *Siphonoperla*; (B) *Chloroperla*; (C) *Xanthoperla*. Cercus of (D) *Pontoperla*; (E) *Siphonoperla*. *Figures D and E were redrawn and modified from Zwick (2004).*

Plecoptera: Chloroperlidae: Genera

(Four genera and approximately 19 species in the Mediterranean Basin)

1 Elliptic pronotum, approximately 1.7 times wider than long, and with abundant marginal hairs all around (Fig. 10.14A) 2
1' Rectangular pronotum, approximately 1.5 times wider than long, with abundant marginal hairs mainly concentrated on corners, lateral portions hairless (Fig. 10.14B, C) 3
2(1) Cercus with setae shorter than the segment over which they extend (Fig. 10.14D) .... ***Pontoperla***
2' Cercus with some setae as long as or longer than the segment over which they extend (Fig. 10.14E) .... ***Siphonoperla***
3(1') Pronotum with delimiting folds at its front and rear margins and with hairs along them (Fig. 10.14B) .... ***Chloroperla***
3' Pronotum without delimiting folds and with only a few hairs along its front and rear margins (Fig. 10.14C) .... ***Xanthoperla***

Plecoptera: Taeniopterygidae: Genera

(Three genera and approximately 36 species in the Mediterranean Basin)

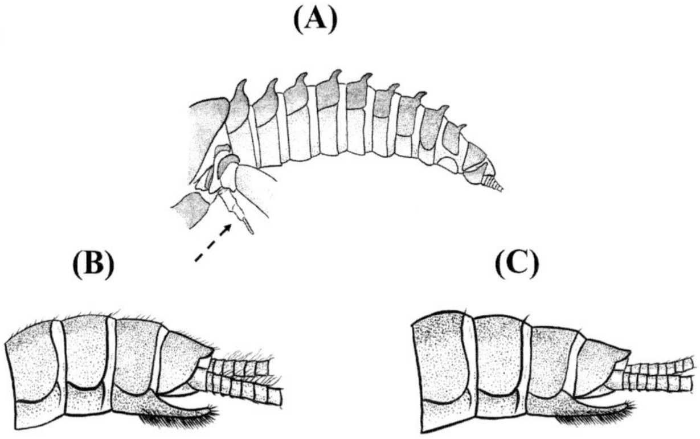

FIGURE 10.15 (A) Abdomen and posterior part of the metathorax (in lateral view) of *Taeniopteryx* (three-segmented coxal gill shown by the arrow). Distal part of the abdomen (in lateral view) of (B) *Brachyptera*; (C) *Rhabdiopteryx*.

1 Coxae without gills; abdominal terga without thorn-like processes 2
1' Three-segmented coxal gills present; several abdominal terga with thorn-like sclerotized processes (Fig. 10.15A) .... ***Taeniopteryx***
2(1) Abdominal tergites with setae along their distal margins not visible in lateral view; upper side of the cerci with one or more long hairs (Fig. 10.15B); head with a rounded epicranial suture* .... ***Brachyptera***
2' Abdominal tergites with setae along their distal margins visible in lateral view; upper side of the cerci without long hairs (Fig. 10.15C); head with a pointed epicranial suture .... ***Rhabdiopteryx***

Notes: nymphs of some *Brachyptera* and *Rhabdiopteryx* species have not yet been described, or their characteristics have not been confirmed (see Zwick, 2004). *This last character is not informative for *B. trifasciata* (Pictet, 1832), which has an epicranial suture with an intermediate shape between both genera (Consiglio, 1980; Zwick, 2004).

Plecoptera: Nemouridae: Genera

(Four genera and approximately 131 species in the Mediterranean Basin)

1 Cervical gills present (Fig. 10.16A, B) .... (Subfamily Amphinemurinae) 2
1' Cervical gills absent .... (Subfamily Nemourinae) 3
2(1) Cervical gills in two groups (one on each side), each composed of three finger-shaped gills (Fig. 10.16A) .... ***Protonemura***
2' Cervical gills in four groups (two on each side), composed of many filamentous gills (Fig. 10.16B) .... ***Amphinemura***
3(1') Metathoracic legs with 1st tarsal segment clearly shorter than the 3rd in late instars (Fig. 10.16C) .... ***Nemoura***
3' Very long metathoracic legs, with 1st tarsal segment as long as the 3rd in late instars (Fig. 10.16D) .... ***Nemurella***

Plecoptera: Capniidae: Genera

(Four genera and approximately 23 species in the Mediterranean Basin)

1 Body with hairs; wingpads parallel or subparallel to each other (Fig. 10.17A, B); cerci not particularly thick at their base and with bristles (Fig. 10.17D) 2
1' Body hairless; wingpads long, slender, and slightly divergent (Fig. 10.17C); cerci very thick at their base and hairless (Fig. 10.17E) .... ***Capnioneura***
2(1) Body covered by long hairs; eyes with a fringe of long hairs (Fig. 10.17F); wingpads stocky (Fig. 10.17A, F) .... ***Capnopsis***
2' Body with short hairs; eyes without a fringe of long hairs; wingpads thin (Fig. 10.17B) .... ***Capnia/Zwicknia****

*Note: The genus *Zwicknia* has been recently erected (Murányi et al., 2014) from a taxon previously belonging to the genus *Capnia* [*C. bifrons* (Newman, 1838)], supported by molecular data and adult morphological characteristics. Nevertheless, no general nymphal characters have been proposed for distinguishing *Zwicknia* from *Capnia*, and those characters employed previously to separate *Capnia bifrons* (currently *Zwicknia bifrons*) from some other *Capnia* species cannot be generally applied to other recently described species of *Zwicknia*, the nymphs of which are unknown in most cases.

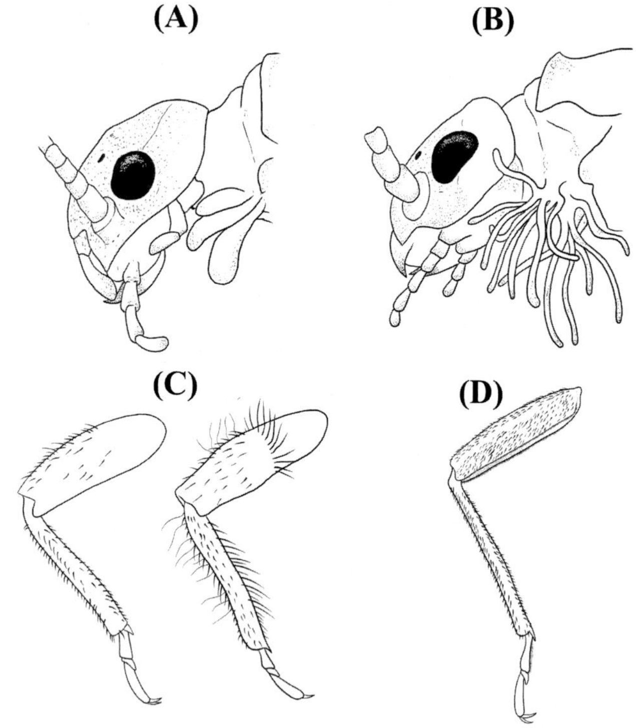

FIGURE 10.16 Head and cervical zone (in lateral view) of (A) *Protonemura*; (B) *Amphinemura*. Metathoracic leg of (C) two different species of *Nemoura*; (D) *Nemurella* in late instars.

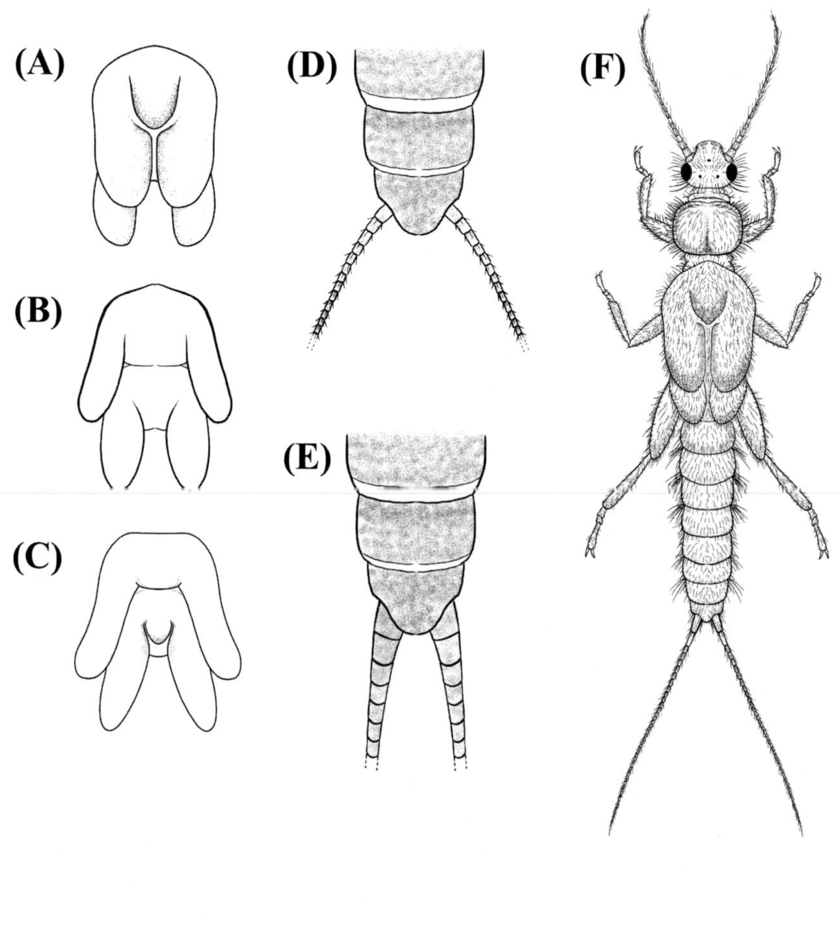

FIGURE 10.17 Wingpad silhouette (not showing hairs when present) of (A) *Capnopsis*; (B) *Capnia/Zwicknia*; (C) *Capnioneura*. Distal part of the abdomen showing the basal part of the cerci (in dorsal view) of (D) *Capnia/ Zwicknia*; (E) *Capnioneura*. (F) Habitus of *Capnopsis*.

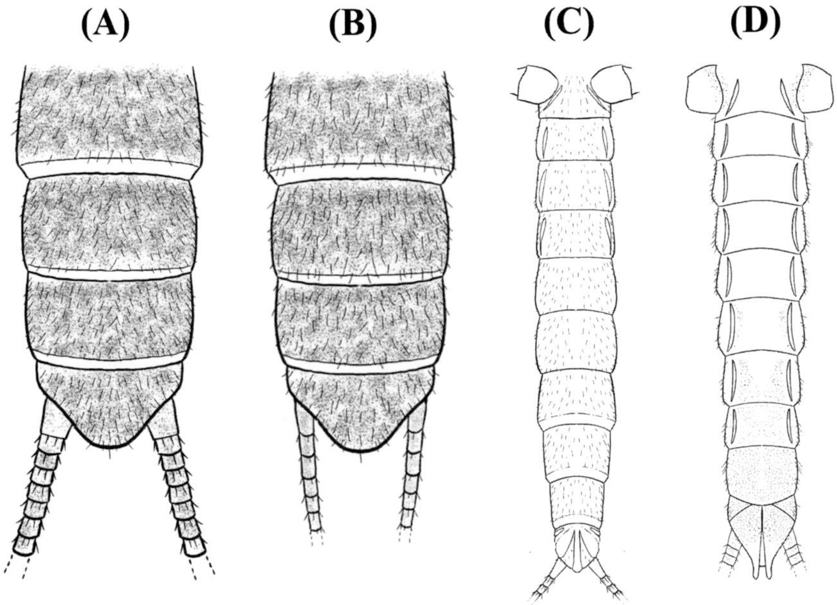

FIGURE 10.18 Distal part of the abdomen (in dorsal view) showing the proximal part of the cerci of (A) *Leuctra*; (B) *Pachyleuctra*. Abdomen (in ventral view) of (C) *Leuctra*; (D) *Tyrrhenoleuctra*. *Figures C and D redrawn from Consiglio (1980).*

Plecoptera: Leuctridae: Genera

(Three genera and approximately 129 species in the Mediterranean Basin)

1 Cercus approximately 1/5 the width of tergite 10 (Fig. 10.18A). Very slender body 2
1' Cercus slender (thread-like), less than 1/8 the width of tergite 10 (Fig. 10.18B). Relatively stout body .... ***Pachyleuctra***
2(1) Abdominal segments 1 to 4 divided into tergite and sternite by a membranous pleura (Fig. 10.18C) .... ***Leuctra***
2' Abdominal segments 1 to 8 divided into tergite and sternite by a membranous pleura (Fig. 10.18D) .... ***Tyrrhenoleuctra***

Acknowledgments

We want to thank to our colleagues Drs Dávid Murányi, Jean-Paul Reding, and Julio M. Luzón-Ortega for resolving our questions and, mainly, Dr Ignacio Peralta-Maraver for drawing the figures. We are particularly grateful to the editors, Drs Alain Maasri and James Thorp, for inviting us to participate in this book and for their support during the chapter development.

References

Agüero-Pelegrín, M. & M. Ferreras-Romero. 2002. The cycle of *Guadalgenus franzi* (Aubert, 1963) (Plecoptera: Perlodidae) in the Sierra Morena Mountains (southern Spain): semi-voltinism in seasonal streams of the Mediterranean Basin. Aquatic Insects 24: 237–245.

Aubert, J. 1963. Les Plécoptères de la péninsule ibérique. Eos 39: 23–107.

Benke, A.C. & A.D. Huryn. 2006. Chapter 29: Secondary production of macroinvertebrates. Pages 691–710 *in*: F.R. Hauer and G.A. Lamberti (eds.), Methods in Stream Ecology (Second edition). Academic Press, Burlington, MA.

Bonada, N. & S. Dolédec. 2011. Do Mediterranean genera not included in Tachet et al. 2002 have Mediterranean trait characteristics? Limnetica 30: 129–142.

Consiglio, C. 1980. Plecotteri (Plecoptera). Guide per il riconoscimento delle specie animali delle acque interne italiane. N. 9. Ed. C.N.R., Verona. 68 pp.

DeWalt, R.E. & G.D. Ower. 2019. Ecosystem services, global diversity, and rate of stonefly species descriptions (Insecta: Plecoptera). Insects 10: 99.

DeWalt, R.E., B.C. Kondratieff & J.B. Sandberg. 2015. Chapter 36: Order Plecoptera. Pages 933-949 *in*: J. Thorp and D.C. Rogers (eds.), Thorp and Covich's Freshwater Invertebrates (Fourth Edition, Volume I), Ecology and General Biology. Academic Press, Cambridge, MA.

DeWalt, R.E., M.D. Maehr, H. Hopkins, U. Neu-Becker & G. Stueber. 2020. Plecoptera Species File Online. Version 5.0/5.0. [March 20th 2020]. <http://Plecoptera.SpeciesFile.org>

Fausto A.M., M. Belardinelli, R. Fochetti, J.M. Tierno de Figueroa & M. Mazzini. 2002. Comparative spermatology in Plecoptera (Insecta). II. An ultrastructural investigation on four species of Systellognatha. Arthropod Structure & Development 31: 147–156.

Fochetti, R. & J.M. Tierno de Figueroa. 2006. Notes on diversity and conservation of the European fauna of Plecoptera (Insecta). Journal of Natural History 40: 2361–2369.

Fochetti, R. & J.M. Tierno de Figueroa. 2008a. Global diversity of stoneflies (Plecoptera; Insecta) in freshwater. *in*: Balian, E., K. Martens, C. Lévêque and H. Segers (eds.), A Global Assessment of Animal Diversity in Freshwater. Hydrobiologia 595: 365–377.

Fochetti, R. & J.M. Tierno de Figueroa. 2008b. Plecoptera. Fauna d'Italia, vol. 43, Ed. Calderini de Il Sole 24 ore, Milan, 339 pp.

Hershey, A.E., K. Fortino, B.J. Peterson & A.J. Ulseth. 2006. Chapter 27: Stream Food Webs. Pages 637-659 *in*: F.R. Hauer and G.A. Lamberti (eds.), Methods in Stream Ecology (Second edition). Academic Press, Burlington, MA.

Hynes, H.B.N. 1976. Biology of Plecoptera. Annual Review of Entomology 21: 135–153.

Hynes, H.B.N. 1977. A key to the adults and nymphs of the British stoneflies. Freshwater Biological Association, No 17, Ambleside, Cumbria. 90 pp.

Hynes, H.B.N. 1988. Biogeography and origins of the North American stoneflies (Plecoptera). The Memoirs of the Entomological Society of Canada 120(S144): 31–37.

López-Rodríguez, M.J. & J.M. Tierno de Figueroa. 2012. Life in the dark: on the biology of the cavernicolous stonefly *Protonemura gevi* (Insecta, Plecoptera). The American Naturalist 180: 684–691.

López-Rodríguez, M.J., J.M. Tierno de Figueroa & J. Alba-Tercedor. 2009a. Life history, feeding and secondary production of two Nemouroidea species (Plecoptera, Insecta) in a temporary stream of the Southern Iberian Peninsula. Fundamental and Applied Limnology-Archiv für Hydrobiologie 175: 161–170.

López-Rodríguez, M.J., J.M. Tierno de Figueroa, S. Fenoglio, T. Bo & J. Alba-Tercedor. 2009b. Life strategies of 3 Perlodidae species (Plecoptera) in a Mediterranean seasonal stream in southern Europe. Journal of the North American Benthological Society 28: 611–625.

López-Rodríguez, M.J., C.E. Trenzado, J.M. Tierno de Figueroa & A. Sanz. 2012. Digestive enzyme activity and trophic behavior in two predator aquatic insects (Plecoptera, Perlidae). A comparative study. Comparative Biochemistry and Physiology Part A: Molecular & Integrative Physiology 162: 31–35.

López-Rodríguez, M.J., P. Delgado-Juan, J.M. Luzón-Ortega & J.M. Tierno de Figueroa. 2017. Nymphal biology of *Capnioneura gelesae* Berthélemy & Baena, 1984 (Plecoptera, Capniidae) in temporary streams of Sierra Morena (Southern Spain). Limnetica 36: 45–53.

López-Rodríguez, M.J., C. Martínez-Megías, A.C. Salgado-Charrao, J.P. Cámara-Castro & J.M. Tierno de Figueroa. 2018. The effect of large predators on the decomposition rate and the macroinvertebrate colonization pattern of leaves in a Mediterranean stream. International Review of Hydrobiology 103: 90–98.

Murányi, D., M. Gamboa & K.M. Orci. 2014. *Zwicknia* gen. n., a new genus for the *Capnia bifrons* species group, with descriptions of three new species based on morphology, drumming signals and molecular genetics, and a synopsis of the West Palaearctic and Nearctic genera of Capniidae (Plecoptera). Zootaxa 3812: 1–82.

Peralta-Maraver, I., M.J. López-Rodríguez & J.M. Tierno de Figueroa. 2017. Structure, dynamics and stability of a Mediterranean river food web. Marine and Freshwater Research 68: 484–495.

Sivec, I., B.P. Stark & S. Uchida. 1988. Synopsis of the world genera of Perlinae (Plecoptera: Perlidae). Scopolia 16: 1–66.

South, E.J., R.K. Skinner, E.R. DeWalt, M.A. Davis, K.P. Johnson, V.A. Teslenko, J.J. Lee, R.L. Malison, J.M. Hwang, Y.J. Bae & L.W. Myers. 2021. A new family of stoneflies (Insecta: Plecoptera), Kathroperlidae, fam. n., with a phylogenomic analysis of the Paraperlinae (Plecoptera: Chloroperlidae). Insect Systematics and Diversity 5(4): 1–27.

Stewart, K.W. 1994. Theoretical considerations of mate finding and other adult behaviors of Plecoptera. Aquatic Insects 16: 95–104.

Stewart, K.W. 2009. Plecoptera: Stoneflies. Pages 810-813 *in*: V.H. Resh and R.T. Cardé (eds.), Encyclopedia of Insects. Academic Press, Burlington, MA.

Tachet, H., P. Richoux, M. Bournaud & P. Usseglio-Polatera. 2010. Invertébrés d'eau douce: Systématique, biologie, écologie. CNRS Editions, Paris. 608 pp.

Tierno de Figueroa J.M. & R. Fochetti. 2001. On the adult feeding of several European stoneflies. Entomological news 112: 128–132.

Tierno de Figueroa J.M. & R. Fochetti. 2014. A second new species of *Tyrrhenoleuctra* discovered by means of molecular data: *Tyrrhenoleuctra lusohispanica* n. sp. (Insecta: Plecoptera). Zootaxa 3764: 587–593.

Tierno de Figueroa, J.M. & J.M. López-Rodríguez. 2019. Trophic ecology of Plecoptera (Insecta): a review. The European Zoological Journal 86: 79–102.

Tierno de Figueroa, J.M., A. Sánchez-Ortega, P. Membiela-Iglesia & J.M. Luzón-Ortega. 2003. Plecoptera. Fauna Ibérica, vol. 22, Ramos, M.A. et al. (eds.). Museo Nacional de Ciencias Naturales, CSIC, Madrid. 404 pp.

Tierno de Figueroa, J.M., J.M. López-Rodríguez, A. Lorenz, W. Graf, A. Schmidt-Kloiber & D. Hering. 2010. Vulnerable taxa of European Plecoptera (Insecta) in the context of climate change. Biodiversity and Conservation 19: 1269–1277.

Tierno de Figueroa, J.M., J.M. López-Rodríguez, S. Fenoglio, P. Sánchez-Castillo & R. Fochetti. 2013. Freshwater biodiversity in the rivers of the Mediterranean Basin. *in*: N. Bonada and V. H. Resh (eds.), Streams in Mediterranean climate regions: lessons learned from the last decade. Hydrobiologia 719: 137–186.

Zwick, P. 1980. Plecoptera (Steinfliegen). *in*: Handbuch der Zoologie. Walter de Gruyter, Berlin. 115 pp.

Zwick, P. 2000. Phylogenetic system and zoogeography of the Plecoptera. Annual Review of Entomology 45: 709–746.

Zwick, P. 2004. Key to the West Paleartic genera of stoneflies (Plecoptera) in the larval stage. Limnologica 34: 315–348.

Chapter 11

Order Odonata

Gianmaria Carchini[1] and Sönke Hardersen[1,2]
[1]*Società Italiana per lo Studio e la Conservazione delle Libellule ODV, Università degli Studi di Perugia, Perugia, Italy,* [2]*Reparto Carabinieri Biodiversità di Verona, Centro Nazionale Carabinieri Biodiversità "Bosco Fontana", Marmirolo, Italy*

Introduction

Odonates or dragonflies are hemimetabolous insects, and their life cycle includes three stages: egg, aquatic larva, and terrestrial imago or adult. As arthropods, odonates shed their exoskeleton in order to grow in a process called molting. Some authors refer to the abandoned cuticles as "larval skins" if they derive from molts preceding the last one and call the skin that is left after the last molt "exuvia." To avoid confusion, we will only use the term "exuvia" (plural exuviae) in this key. The first larval instars lack the external wing sheaths and historically were called the "neanidae" stage or "naiades." The subsequent instars have wing sheaths and were often termed nymphs. However, in the recent scientific literature, to avoid confusion, it is generally preferred to use the term "larvae" to indicate all aquatic instars, with no further distinction. There are 6317 named odonate species in the world (Paulson & Schorr, 2021), all included in the order Odonata. They are divided into three suborders: the Zygoptera and Anisoptera with about 3000 species each, and the Anisozygoptera (not present in the Mediterranean basin) with only four species.

Morphological characters

The larva body (Fig. 11.1A and B) is organized in a similar way to that of the adult and consists of the head, the pro-, meso-, and metathorax (the last two being fused into a pterothorax), and the abdomen which is composed of 10 visible segments plus an 11th which is reduced and transformed. Larvae always have dull colors (green, yellowish, and light or dark brown), which is often variable among larvae of the same species. Furthermore, the cuticle has a brown color, with patterns that might be characteristics for certain species and can therefore be important for species identification. However, the larvae may show wide variations in the intensity of the cuticle's pigmentation. For example, immediately after molting, larvae will be almost colorless and later will darken until they become completely brown. Sometimes the patterns of the cuticle are almost invisible, and this should be considered when using these as taxonomical characters. The largest part of the body is the abdomen; it can be conical and elongated (suborder Zygoptera) or ovoid in shape, shorter, and more or less flattened in the dorsal-ventral direction (suborder Anisoptera). The maximum body length when larval development is complete is approximately 1.5–6 cm. The larvae of odonates are distinguished from all other aquatic insects by the presence of a prey catching organ, resulting from the modification of the labium; it is sometimes called a "mask." It consists of rather conspicuous labial segments, which at rest are kept folded under the head and thorax but which can be extended very rapidly to catch prey.

Head

The head is approximately pentagonal or rectangular when seen from above, with slightly marked sutures. Compound eyes (Fig. 11.1A and B) are always well developed, placed laterally on the head and forming two protuberances in the profile of the head. The number of ommatidia is small in the first larval instars (as few as 7), but increases with each molt, reaching a relatively high number before metamorphosis. Ocelli are sometimes visible in the last instar, but they are never functional. The antennae are inserted approximately on a line between the anterior edges of the eyes, with a minimum of three segments in the first larval instar. This number increases with age to a maximum of seven (rarely eight) segments. Post-ocular lobes (termed occiput by some authors) are well developed, generally expanding backwards beyond the joint of the head and the prothorax.

Identification and Ecology of Freshwater Arthropods in the Mediterranean Basin. DOI: https://doi.org/10.1016/B978-0-12-821844-0.00007-7

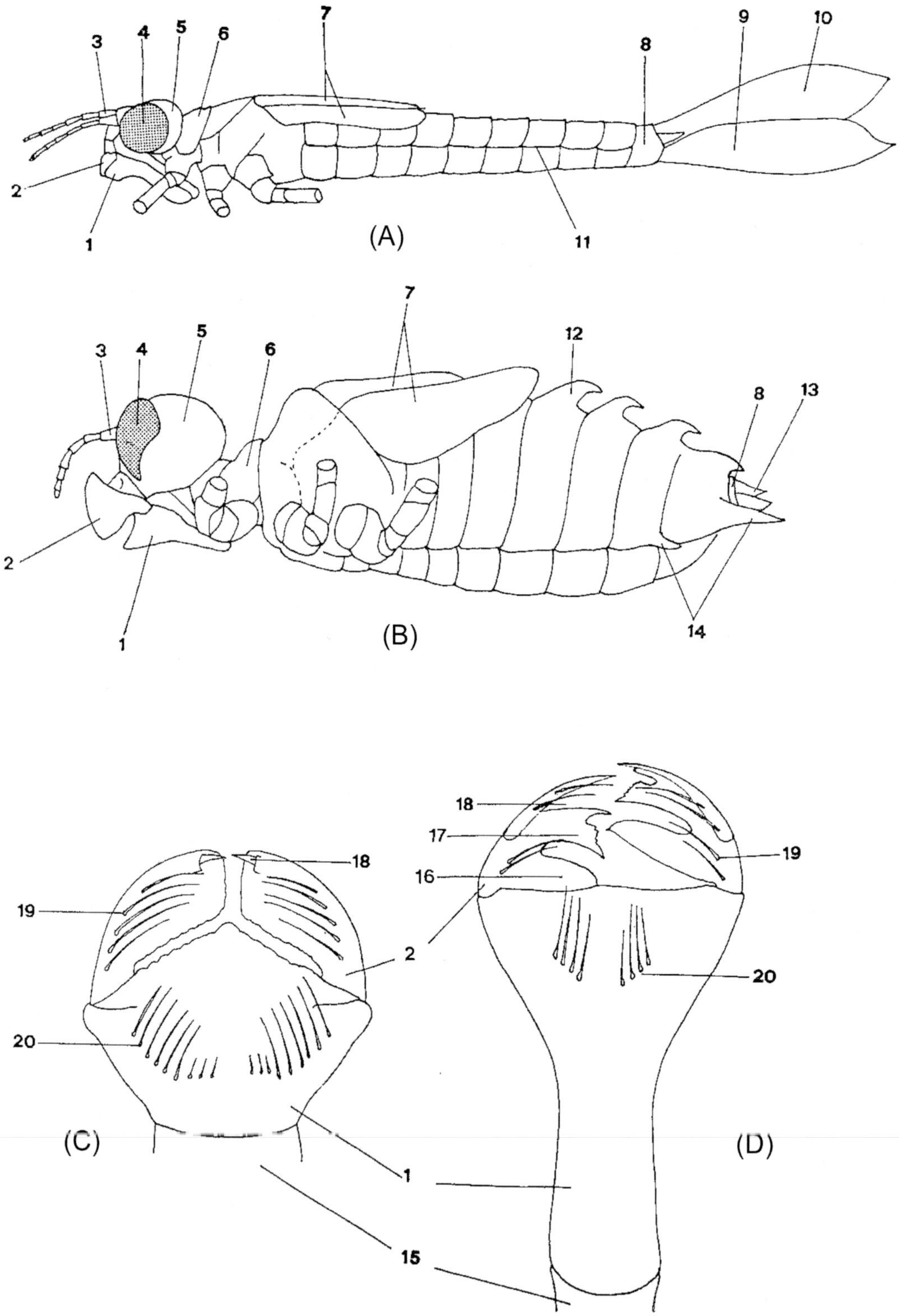

FIGURE 11.1 Schematic diagram of the morphology of odonate larvae. (A, B) Body in side view of Zygoptera and Anisoptera respectively; (C, D) inner surface of the labium of Anisoptera and Zygoptera respectively. Numbers indicate: 1, prementum; 2, palpus; 3, antennae; 4, compound eye; 5, postocular lobes; 6, prothorax; 7, wing sheaths; 8, 10 th abdominal segment or S10; 9, lateral caudal lamella; 10, median caudal lamella; 11, lateral carinae; 12, middorsal spines; 13, anal pyramid; 14, lateral spines; 15, postmentum; 16, adoral (or medial) branch of palpus; 17, aboral (or lateral) branch of palpus; 18, movable hook; 19, setae of the palpus; 20, setae of the prementum.

Mouth structure

The labrum is movable and partially covers the very strong mandibles; maxillae are thinner than mandibles and contain sharp teeth. The labium is present as a mask, which can be extended rapidly to catch prey, and consists of the postmentum, prementum, and labial palps. The backwardly directed postmentum is rectangular and longitudinally elongated and hinged onto the ventral side of the head, behind the opening of the mouth. The prementum is also approximately rectangular in shape but enlarged distally. In some cases, it has the shape of a racket, hinged on the postmentum and pointing forwards. In cross section the distal portion of the prementum can be flat (Gomphidae, Aeshnidae), but it can also be clearly concave in the distal part (other families of Anisoptera) or in a less pronounced concave (Zygoptera). In these two latter cases, two series of long spiniform setae are inserted on the dorsal surface of the prementum, which becomes the inside surface of the mask when at rest. These are aligned in two series and slanted with respect to the midline. In some species, other setae are also present but are shorter than the former and either scattered on the surface or aligned along the margins (marginal setae). Labial palpi consist of two segments, inserted at the lateral distal corners of the prementum, with the axis of rotation parallel to the dorsal-ventral axis. The first segment is more pronounced than the second, and in general it has a row of long spiniform setae on the dorsal margin. In some species, a second series of setae is present on the outside (marginal setae) but shorter than the former. The distal margin can be either smooth, dentate, or split into two branches. The second segment is always spiniform, pointed and can also have a row of long setae (Lestidae). In this chapter, the term palpi is exclusively used for the labial palpi (the maxillae have none). The morphology of the mask changes little during the development of the larvae, but the number of setae keeps increasing until the late instars. For the labium, we employ the terminology of Corbet (1953) in the simplified version used by Gardner (1954) and as reported in the captions of Fig. 11.1. Accordingly, the term "palpus" (plural "palpi") is used here to indicate exclusively the first segment, while the second will be called a "movable hook."

Thorax

The prothorax is not fused but is almost immovable with respect to the mesothorax. At the joint of the coxae, two supracoxal apophysis (Aeshnidae) can be present. Meso- and metathorax are fused, forming a synthorax or pterothorax, which is not inclined backwards as in the adults and bears wing sheaths. These normally become visible midway through the larval developmental period and begin as small protuberance and becoming quickly larger. In larger larvae these extend backwards, generally covering the tergites of the first abdominal segments. In late instars, the wing venation of the adult is visible in the wing sheaths. The legs are formed by coxa, femur (with fused trochanter), tibia, tarsus, and pretarsus. They are used for walking and are always well developed; femurs and tibias can have long or spiniform setae but are not used for swimming or catching prey. In some cases (Gomphidae, some Libellulidae, Cordulegastridae), the legs are robust and short and allow the larvae to burrow into the sand or mud. The tarsi consist of three segments (reduced to two in Gomphidae); but in very young larvae, these consist of only one segment. The pretarsus always has two claws.

Abdomen

The abdomen is always composed of 10 distinct segments, plus a much reduced eleventh. In this chapter, the first 10 complete segments are indicated as S1–S10. The shape of the abdomen is rather different in the two suborders, as described below.

In the Zygoptera, the abdomen is long and slender (Fig. 11.1A). In cross section, the abdominal segments are almost circular with a dorsal sclerite (tergite) extending downwards. A longitudinal ridge is present slightly below the middle; hereafter, it is termed the "lateral carina." A row of short, backward pointing spiniform setae can be present on this ridge. Below the lateral carina, the cuticle is thin and flexible, thus allowing the movement of the ventral sclerite (sternite). In the distal-most abdominal segments, the lateral carina is more clearly visible, but it can also be reduced in S9 and is always absent in S10. The latter segment consists of one single cylindrical sclerite, without a thin cuticle. The distal end of abdominal segment S10 carries the reduced S11 in the form of three short appendages: a medial-dorsal epiproct (or supra-anal lamina) and two ventral and lateral paraprocts (or subanal lamina). All these bear elongated branchial gills which are tubular in shape in the early instars and leaf-shaped and expanded in the sagittal plane in the subsequent instars. These gills are termed "lamellae" in the following text. In the lamellae two large tracheal trunks are present which divide into more or less dense branches or ramifications that are not always visible. On the ventral and dorsal edges of the lamellae, long hair-like setae and short spiniform setae are inserted. In many species, the lamellae are divided by a so-called nodal line into a rigid proximal portion, which has short spiniform setae on the edges, and a flexible distal portion, which has long hair-like setae along the edges. In a few species this division is absent, while in others it is not very evident and almost invisible in the larvae. It is more evident in the exuviae when prepared on a slide and observed in transmitted light. Form, color, number, and layout of the tracheal ramifications and setae are highly variable among the various species. At about the fifth instar, two short and pointy elements called "cerci" appear between epiproct and paraprocts.

In the Anisoptera the diameters of the central abdominal segments are larger than those of the first and the last ones, resulting in an abdomen with a spindle-like shape (Fig. 11.1B). When seen in cross section, the abdominal segments appear flattened in the dorsal-ventral direction, to a greater or lesser extent. The tergites can have a more or less (Gomphidae) sharp projection in the middle that will be referred to as "dorsal spine" in this text. The posterior-inferior edges of the tergites can be extended backwards, forming two spines, here referred to as "lateral spines." The tergite forms a movable joint with two ventral lateral sclerites or pleurae which in turn form movable joints with a ventral element or sternite. Middorsal and lateral spines are present to a different extent in the various abdominal segments. The segment S10 is formed by a single element, approximately trunk-conical. The posterior surface of S10 bears three short appendages: a middorsal epiproct and two lateral paraprocts. Each of these is shaped like a pyramid with a triangular base and an apex that is facing backwards. The posterior surface of S10 is almost entirely covered by these appendages. At about the fifth instar two cerci appear as short-pointed appendages between the epiproct and paraprocts. All these appendages are strongly sclerotized and together form the caudal or anal pyramid. In late instars of both Zygoptera and Anisoptera, the precursors of external genitalia appear. In female Zygoptera and in some families of Anisoptera (Aeshnidae, Cordulegastridae), the precursor of the ovipositor appears at S8 and S9. It is largest in the last larval instar. However, the ovipositor is invisible in female larvae of other Anisoptera. In late instars of male Zygoptera, two small spines appear at S9. In male Anisoptera, the precursors of the copulating organ can be seen at the sternite of S2 and the openings of the genital ducts at S9. However, in Anisoptera these are very small and at times hardly visible even in the last instar.

Overview of physiology

In the Zygoptera the respiratory organs are the caudal lamellae and the rectum, and possibly also the whole body surface. Anisopteran larvae have tracheal gills inside the rectum, which also contains muscles that allow the active circulation of water to irrigate the tracheal gills. Respiration in this suborder is confined to the rectum, thus allowing the larvae to burrow in mud, provided their anus protrudes into open water.

The respiratory systems are also used for locomotion, as larvae of almost all Zygoptera swim by oscillating their abdomen laterally and using the caudal lamellae as a tail fin, while the larvae of the Anisoptera can expel water forcefully from their rectum, thus propelling their bodies forward.

The caudal lamellae of the Zygoptera can also protect against predatory actions. In fact, several species have a preset fracture line at the point of attachment of the epiproct and paraprocts. Lamellar autotomy permits escape from predators, and individuals can survive the loss of all their lamellae. The lamellae and even the limbs, if lost totally or partially, are regenerated during subsequent molts. However, the shape and size of the regenerated parts do not resemble the normal form until after several molts.

The molting process between the various instars is usually fast. An individual stops eating for a short period of time, then the body swells, stretching the old cuticle which then ruptures. The larva slips out of the exuvia, and the cuticle hardens rapidly. The larva starts eating again after 1–2 h. The last molt, which results in the adult emerging, is more complex and takes longer. In this process: (A) the larva stops feeding several days before molting; (B) the wing sheaths become swollen and the mask empties out as the labium of the adult is withdrawn into the larval postmentum; and (C) eyes and ocelli of the adult become bigger and are more clearly visible under the larval skin. Many hours or sometimes days before the ecdysis, larvae move completely or partially out of the water and stay there for extended periods. Finally, they choose a suitable support (vertical for most of the species) near the water or somewhat away from it and start molting. The process is similar to the molts described above but takes more time. Once the adults are completely out of the larval skin, they remain on the exuviae, stretching and expanding wings and abdomen. During the time it takes for the tissues to harden, the colors of the adults intensify, which had already been visible under the larval skin. The exuviae rapidly dry out and remain in place, normally until some external force (such as a knock, strong wind, or rain) makes them fall to the ground.

Overview of biology

Egg stage

Fertilization is immediately followed by oviposition, which can be either endophytic (Zygoptera and aeshnid Anisoptera) or exophytic (the remaining families of Anisoptera). Endophytic oviposition can take place in parts of a plant which are either under or above water. In exophytic oviposition the eggs are commonly released individually or in clusters in the water or on soils that are temporary dry. Oviposition among Cordulegastridae and some Aeshnidae is similar, with eggs being inserted in mud banks. Endophytic eggs and those laid by species provided with a well-developed ovipositor are elongated and tapered at both ends, whereas the exophytic eggs are roughly spherical. Both types of eggs have a smooth and resistant chorion, while in some cases the exophytic eggs have an additional layer that produces a gelatinous coating upon contact with water. The development of the embryo begins with fertilization and in some species might be interrupted by a diapause. In those species with no diapause, the length of time necessary to complete development is highly variable, lasting 1–9 months. In the species with diapause, development takes several months longer. The duration of the diapause is regulated by genetic and environmental factors, resulting in synchronized hatching of the larvae in a favorable season, usually spring.

Larval stage

All odonates hatch from eggs as prolarvae, with all the appendages extended and positioned along the body, antennae, and mask included, and all lack apparent articulations. The duration of this form is extremely short, a few minutes at most, followed by the first molt, which results in the first larval instar.

The number of larval instars differs among species, varying from 8 to 18 (Corbet, 2002). The number of instars also varies within the same species, but by no more than a few stages, depending on environmental and perhaps genetic factors. The duration of the instars is also variable and normally increases as development progresses, from a minimum of about a week to a maximum of about 3 weeks. This time span is strongly influenced by environmental conditions, most importantly by temperature and food availability. Larval development can also be interrupted by a diapause which consists of cessation of molting and growth. Under natural conditions, the diapause often occurs in the penultimate or last instar, which delays the emergence of the terrestrial adult until the season is favorable. This results in a synchronization in the development of the individuals of the same population. The diapause can lengthen the duration of larval development by some months, and the aquatic phase of the life cycle can last from a few months to 5 years.

Life cycle

The duration of odonate life cycles is highly variable among species and in some cases even among different populations of the same species. Latitudinal gradients in life cycles among species have also been reported. However, it is possible to classify the types of life cycles of the various species into univoltine, semivoltine, and bivoltine. Univoltine life cycle is characterized by one generation per year with adults emerging in the year after the oviposition; this applies to most Zygoptera and to some Anisoptera of the Mediterranean fauna. In a semivoltine life cycle there is one generation every 2 years or more. Some species of Anisoptera have life cycles of 3 or more years, particularly under alpine conditions. Populations of some univoltine species may include a semivoltine fraction where individuals do not complete the development in time for the flight season but instead enter diapause, overwinter as larvae, and emerge early in the following year. In bivoltine life cycles, two generations are present every year. Populations of some species that are usually univoltine in colder climates may have a second or even third generation in the Mediterranean basin (*Ischnura elegans* is an example of the latter).

In general, the following classification regarding the flight period can be applied to Odonata. Among spring species (*sensu* Corbet, 1954), diapause occurs at the end of the last larval instar, with an early flight period that is short and synchronized within the same population. In contrast, among summer species (*sensu* Corbet, 1954), no diapause occurs or the odonate enters diapause before the last instar, with a later flight period that is longer and not synchronized within the same population. In some cases (univoltine species with semivoltine fraction or bivoltine species), there are an early and a late flight period which at times merge into one long flight period.

Ecology of larvae

General ecology

All larval Odonata are predators. They hunt by staying almost motionless and waiting for the prey to come within reach. However, at least some species actively search for a favorable hunting place and even drive away conspecifics. In the laboratory, we observed that some species actively moved towards an offered prey. The larvae locate the prey by sight and/or touch (with antennae and legs). Prey seems to be selected based on size relative to the size of the larva and its mask. However, the prey must move to elicit the larval predatory behavior. Within the limits indicated, the prey of larval Odonata include ciliates, rotifers, oligochaete worms, crustaceans, insects, snails, tadpoles, and small fishes. Feeding habits in nature are not well known, and little is known about cannibalism and predation among the different odonate species, especially between Anisoptera and Zygoptera.

The habitats hosting the highest biodiversity of odonate species are small- and medium-sized lakes, ponds, and bogs. Within large lakes, the shoreline is the habitat containing the highest number of species, particularly in areas with submerged and emergent plants. Some species can also reproduce in temporary ponds if the presence of water is compatible with the needs of embryonic and larval development. Thus, some species can breed in rice paddies or other artificial temporary environments, and some can even develop in brackish waters. Also bogs, including those at a high elevation, are colonized by various odonates. In all these peculiar environments population sizes can be considerable. In contrast, in large or small oligotrophic lakes and in springs, odonates are generally scarce. Flowing water habitats often host a considerable number of odonates. In fact, along watercourses it is often possible to find stretches with a weak current and abundant vegetation where species typical of standing waters can be found. Members of some families (Calopterygidae, Gomphidae, Cordulegastridae) and some species in other families (e.g., *Coenagrion caerulescens*) exclusively breed in running waters. However, river sections with very strong currents are normally not a suitable habitat for larvae. In each habitat that hosts odonates, it is possible to distinguish two groups of species which have different ecological characteristics: bottom species, which wander on the ground or burrow more or less deeply into the mud (Gomphidae, Cordulegastridae, and some Libellulidae and Corduliidae) versus species that live in the vegetation (Zygoptera, the remaining families and species of Anisoptera). All odonates can swim, but they are unable to complete their life cycle in open waters.

The environmental factors that most influence odonate populations are water movements; water permanence compatible with the presence of larva; content of salt and some organic substances; water temperature; abundance of prey; and presence of predators (mainly entomophagous fish).

For more information on the ecology of odonates, we recommend the book "Dragonflies: Behavior and Ecology of Odonata" (Corbet, 2004).

Importance as biological indicators

Odonates are considered good bioindicators of ecosystems as they meet many of the requirements which make a taxon useful for this purpose:

1. well-known and stable taxonomy and can be identified with relative ease;
2. biology/ecology is well known and shows low variability;
3. widely distributed, abundant, and easily surveyed;
4. limited mobility of the larvae;
5. relative long life cycle;
6. monitored changes responds in a predictable and gradual way to the intensity of the stressor;
7. of interest to the public and to policy makers.

Yet, some authors have identified features which render Odonata less useful as a bioindicator. For example, Chovanec & Raab (1997) stated that dragonflies are not very sensitive, especially in waters which are moderately polluted. However, large differences in sensitivity exist between different taxa and in relation to the various pollutants. For example, odonates in general seem to be more tolerant to metals compared to other aquatic invertebrates (Tollett et al., 2009). Larvae of Zygoptera, however, are in general sensitive to insecticides and pesticides, whereas Anisoptera were found to be relatively insensitive to this group of toxins. It is preferable to investigate Odonata assemblages rather than single species. For example, Samways (2003) showed in the tropical island Mayotte in the Indian Ocean that pollution by detergents was one of the main factors structuring the Odonata assemblages. Another approach to bioindication are sentinel organisms, which accumulate pollutants from their surrounding and/or food and thus they may provide time-integrated measures of pollution. For example, methylmercury (MeHg) bioaccumulates in aquatic larvae and its concentration increases with trophic level in the benthic food web. Predators, such as larval Odonata, consistently showed higher MeHg concentrations (Tremblay et al., 1996).

Broadly speaking, sampling larvae (or exuviae) is preferable for the biomonitoring of a specific waterbody, as adults might have arrived from elsewhere and do not necessarily reflect local conditions. Thus sampling of larvae and exuviae gives better insights into site-specific effects and might help to identify conditions which limit the survival of larvae. However, it is often not possible to determine all larvae (or exuviae) sampled, as some might be in the early instars and/or having missing lamellae. Generally, most of the larvae sampled can be determined in Europe and in North America, where keys are available, but in other parts of the world determination of larvae is often much more problematic.

Collection

The larvae of odonates can be collected rather easily with a hand dredge of fair size (for example 30 × 40 cm), having a straight base, if possible with teeth, and a bag that is long enough (e.g., twice the largest dimension of the opening). The handle should be rather long, very robust, and if possible extendable. The bag cloth should be made of nylon with a mesh not exceedingly fine (in fact the early instar larvae can be caught already with a mesh of 1 mm). When sampling sandy or muddy bottoms, the dredge should be passed 1–2 cm in the sediment and dragged swiftly, if possible. This should allow to catch stationary, burrowing larvae as well as larvae wandering on the substrate. In the vegetation, the dredge should be passed energetically, thereby combing the emergent vegetation from the base to the surface. At times it is useful to create a swirl by passing the dredge two or three times "empty" before pulling it out. The larvae can be found by searching directly in the bag of the dredge, but best results are obtained by pouring the contents into a white tray and then observing carefully. Often, the larvae that do not move are hardly visible on a natural background, even if rather large, and some species actively hide under leaves or in the mud. To collect larvae in large water bodies, dredges can also be dragged, but the results are generally inferior to a hand dredge, particularly if recovery is slow or has to stop. Sometimes an Ekman grab is used to sample larval Odonata. The simple collection of aquatic vegetation and which is then washed in a tray does generally not result in satisfactory numbers larvae. In fact, the species living in the vegetation are quite mobile and the larger individuals will leave the plants as these are pulled out of the water. For the same reason, artificial substrates do generally not work well.

After emergence of the adults, exuviae can be collected by hand. Simple emergence traps can be made, using coarse cloth supported by a frame to obtain quantitative data for a known area. The collection of larvae and exuviae provides evidence that a species reproduces in a certain waterbody. In contrast, the observations of an adult, which might have emerged at a different site, do not prove local reproduction. However, collecting adults is far easier than collecting larvae because they are conspicuous, and even a few isolated individuals will often not go undetected. Therefore species lists which are based on adult surveys generally contain more taxa than those obtained by investigating larvae or exuviae.

Fixation, conservation, preparation

The larvae, which are usually wet after collection, can be fixed in ethyl alcohol (80%). It is advisable to pierce the cuticle with a needle once the larvae are dead so that the alcohol can easily penetrate the entire body. For long-term conservation of larvae use 70% alcohol. It is also possible, principally for display purposes, to air-dry larvae in the desired position. The exuviae are normally air-dried and do not need any treatment. Exuviae can be stored dry or can be preserved in 70% alcohol (before preserving the exuvia in alcohol, soften it with a short immersion in water). For small and fragile exuviae (zygopteran molts preceding the last one), the latter method is more suitable. According to Gerken & Sternberg (1999) exuviae preserved wet may crumble, thus becoming impossible to study. However, in our collection, we have exuviae preserved in 70% alcohol that even after 40 years are still intact and can be studied. Also for exuviae, do not use denatured alcohol because it can deteriorate the thinner parts of the cuticle. As the abdominal segments are movable with respect to each other, larvae and exuviae might present a shorter or longer abdomen when compared to the living animal, and this needs to be taken in account when making length measurements.

For species determination, it might be necessary to observe the inner side of the palpi and the prementum. Even though this is not always necessary, it is generally useful to dissect and mount those parts. The procedure to follow is simple and consists in cutting of the postmentum at the joint with the prementum, and then cutting the latter slightly under the articulation with the labial palpus (it is not necessary to remove both labial palpi). Care should be taken not to detach the setae; however, even if that happens, their insertion points will still be visible. Prementum and palpus can be preserved in alcohol or dry, but preferably should be mounted on microscopic slides. It is usually not necessary to brighten them with sodium hydroxide, but while mounting the parts, suitable shims should be inserted to avoid deformation caused by pressure. Some details are visible at good magnification up to 40 × .

Rearing in captivity

It is relatively easy to rear the later larval instars until the adults emerge; however, completing the whole cycle in a laboratory is difficult. Eggs can often be collected easily. For species with endophytic oviposition, it is sufficient to collect the plants where oviposition had been observed. In some cases, it is possible to obtain oviposition in captivity in natural or artificial substrates, such as Styrofoam or paper (tissue paper and even toilet paper) soaked with water. In species with exophytic oviposition, females can be caught while laying eggs. They should be held by their wings; and when they are allowed to touch the water with the tip of the abdomen, they normally will continue to lay eggs in a bowl or test tube filled with water. In order to prevent the eggs from clumping together, it is better to slide the tip of the abdomen over a wet surface.

Larvae can be collected for rearing as mentioned above and transported in water-tight containers with no water but with plenty of wet vegetation, to which they cling. It is important not to expose the larvae to high temperatures or to let the plants dry out. If the larvae need to be kept for long periods of time, it is better to use an easily portable, and aerated aquarium. It is not advisable to feed the larvae during transport, but if fasting is too prolonged, the risk of cannibalism increases.

The larvae can be reared in groups, but it is better to keep them separate to avoid cannibalism; this will allow collection of successive exuviae of each individual separately. For example, larvae can be reared in shallow basins (15 cm), half filled with water in which are immersed cylindrical cages that are high enough to extend out of the water. These should be made with mosquito-netting (made of nylon or other types of plastic) and closed with a lid, preferably transparent. In these cages the individual larvae can be fed and food residues and feces can be removed from above; moreover, the exuviae can be collected without handling the larvae. These cages, made of mosquito netting, are also well suited for emergence, if sufficiently tall and large. The lids are necessary to prevent larvae from climbing out of the cage. For burrowing species, the bottom of the cages can be covered with sand; while for rheophilous species, an aerator can create a weak water current. In all cases it is advisable to keep the water clean with a mechanical filtration system that also circulates the water.

The larvae are very resistant to fasting; but to allow normal growth, it is important to provide abundant food. Only live prey is accepted, and therefore the cages should be checked frequently and prey fragments that were not eaten and dead individuals should be removed so as to preserve water quality. The early instar larvae feed on ciliates, rotifers, and other small metazoans, and they can also consume nauplii of brine shrimp (*Artemia salina*). When they grow, their diet extends to larger animals, until it includes for larger odonates small vertebrates such as tadpoles and small fishes.

At the time of the last molt, which can be detected by observing the above described indications, surveillance should be increased since the newly emerged adults or tenerals can fall into the water and die. Therefore it is better to remove the cage from the water as soon as the final molt begins.

Keeping the larvae in the laboratory can also be useful with individuals that are much covered with debris and periphyton. After the next molt, the larvae will be clean.

Taxonomic and distributional notes

For nomenclature and distributions, we followed Boudot et al. (2009) and Boudot and Kalkman (2015). However, the geographic area considered in these publications is much larger than the bioclimatic Mediterranean area. In fact, some areas of the Mediterranean countries are not included in this bioclimatic basin, such as most of northern Italy, a large part of central Italy, and the Nile valley and its delta. As a consequence, and following some personal communications with J.-P. Boudot on the present status of the odonate fauna in the bioclimatic Mediterranean Basin, the taxa to be considered here were reduced to 14 genera with 48 species for the Zygoptera and 27 genera with 85 species for the Anisoptera. Because the larvae of several of these species have not yet been described and some species cannot reliably be distinguished, we only included keys for families and genera. Keys that allow to arrive at the species level are available for some geographic areas (e.g., Gerken & Sternberg, 1999; Carchini, 2016; Conesa-Garcia, 2021). The following keys are mainly based on Carchini (2016), and the figures, unless otherwise noted, are originals or were taken from that book.

Keys

Odonata: Suborders

For larvae of any instar

1 Abdomen conical and terminating in three long lamellae, thread-like in the first instars, flattened laterally or inflated in the next instars (Fig. 11.1A) .... **Zygoptera**
1' Abdomen tapered and terminating in a pyramid formed by short and highly sclerotized appendages (Fig. 11.1B) .... **Anisoptera**

Zygoptera: Families

For larvae of any instar

1 Abdomen with lateral gills, at least on the central segments (Fig. 11.2) .... Euphaeidae

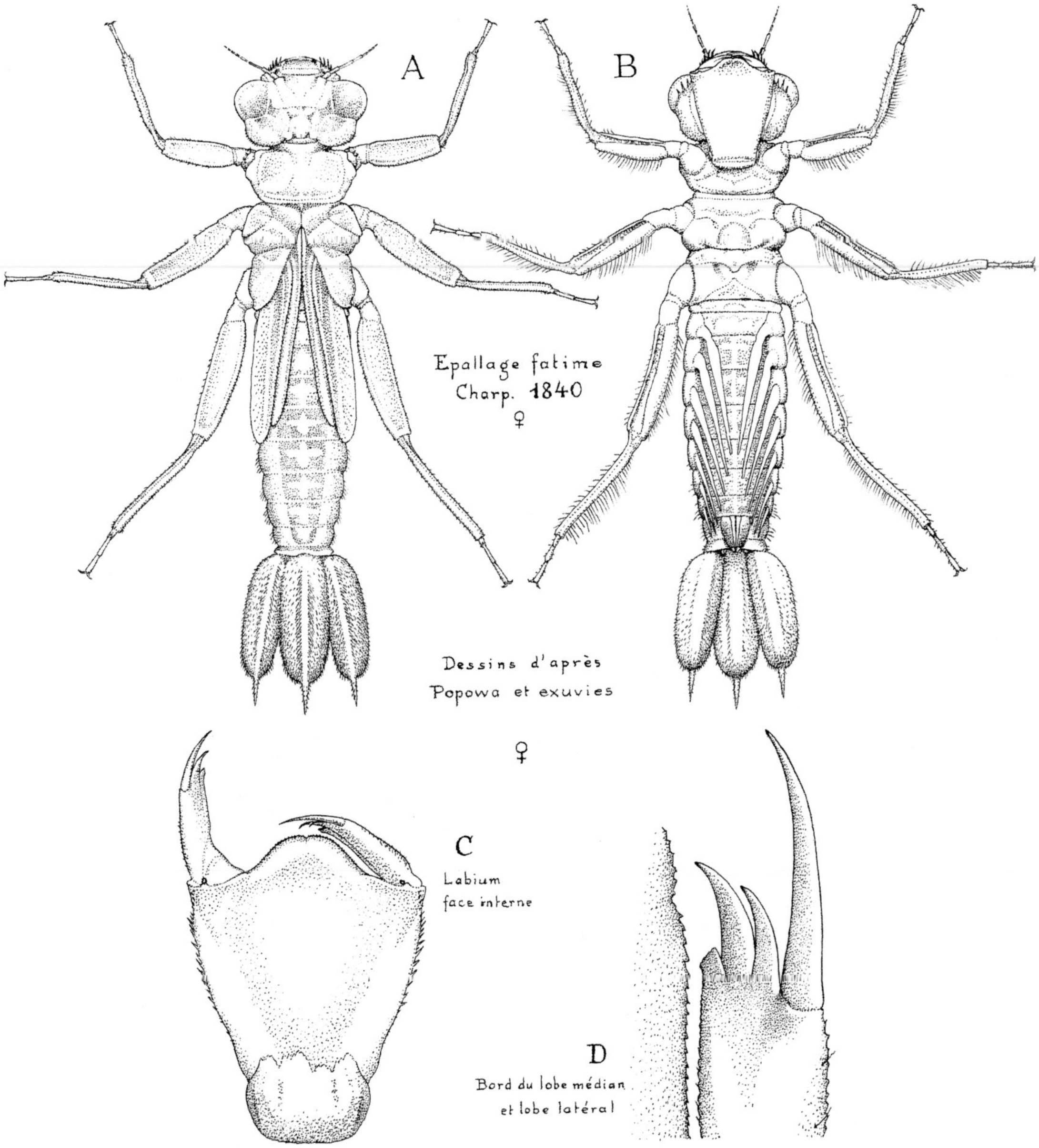

FIGURE 11.2 *Epallage fatime. Drawing by P.A. Robert from Brochard (2018).*

A single genus *Epallage* with one species: *E. fatime* (Fig. 11.3)
1' Abdomen without lateral gills 2
2(1) First segment of the antennae longer than the sum of the remaining segments (Fig. 11.4) .... **Calopterygidae**
A single genus *Calopteryx* with seven species (Fig. 11.5).
2' First segment of the antennae much shorter than the sum of the remaining segments (Fig. 11.1A) 3
3(2) Prementum with a short median slit-like incision at the distal margin (Fig. 11.6); movable hook with a row of long setae (Fig. 11.7) .... **Lestidae**
3' Prementum without median incision at the distal margin, movable hook with no long setae (Fig. 11.8) 4

FIGURE 11.3 *Epallage fatime. Foto by C. Brochard.*

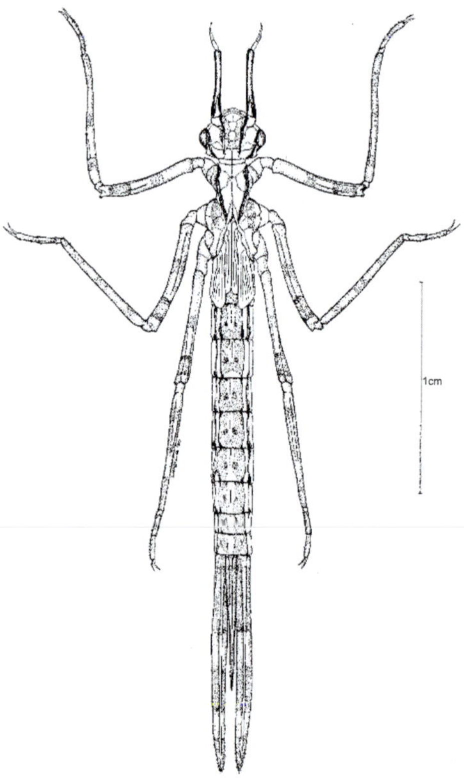

FIGURE 11.4 *Calopteryx splendens* a typical body shape for Calopterygidae.

FIGURE 11.5 *Calopteryx haemorroidalis. Foto by C. Brochard.*

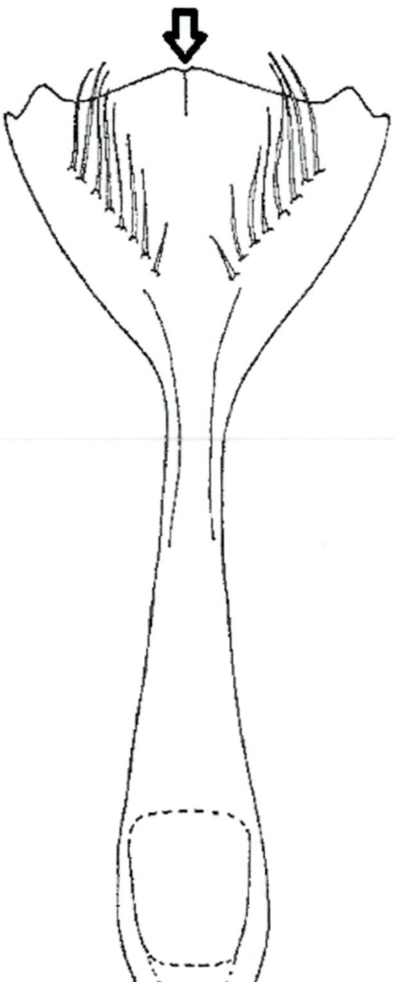

FIGURE 11.6 Prementum inner surface: *Lestes barbarus*. The arrow indicates the incision at the distal margin.

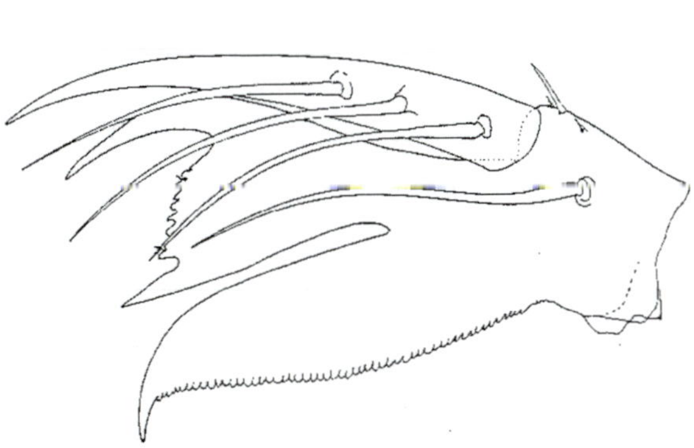

FIGURE 11.7 Palpus: *Lestes barbarous.*

4(3) Wing sheaths present (older larvae) 5
4' Wing sheaths absent (younger larvae) 6
5(4) Pronotum with tubercles (Fig. 11.9A); palpus with a row of long setae and a more marginal row of shorter setae (Fig. 11.10C) .... **Playcnemididae**
A single genus *Platycnemis* with five species (Figs. 11.11 and 11.12).
5' Pronotum with no tubercles (Fig. 11.9B); palpus with long setae, but without marginal setae (Fig. 11.10A and B) .... **Coenagrionidae**
6(4) Tarsi with at least two segments 7
6' Tarsi with one segment only 8
7(6) Dorsal tubercules on S1–S8 .... **Platycnemididae**
7' No dorsal tubercles on S1–S8 .... **Coenagrionidae**
8(6) Head (front view) with two dorsal protuberances .... **Platycnemididae**
8' Head (front view) with no dorsal protuberances .... **Coenagrionidae**

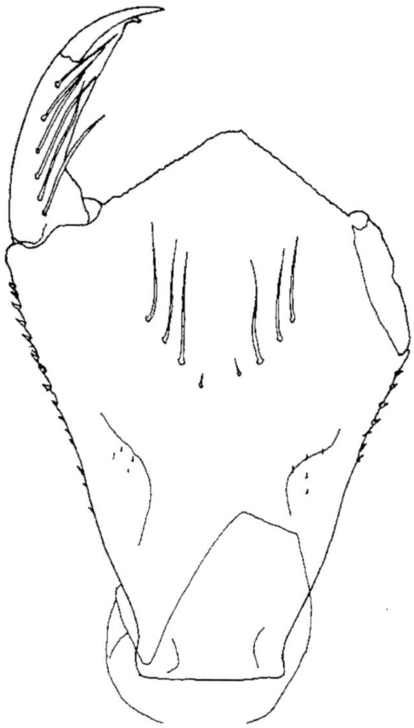

FIGURE 11.8 Prementum inner surface and palpus: *Pyrrhosoma nymphula*.

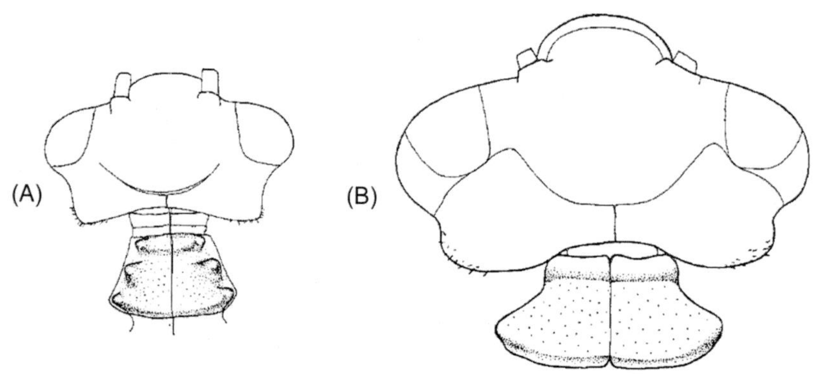

FIGURE 11.9 Head and pronotum in dorsal view: (A) *Platycnemis pennipes*; (B) *Pyrrhosoma nymphula*.

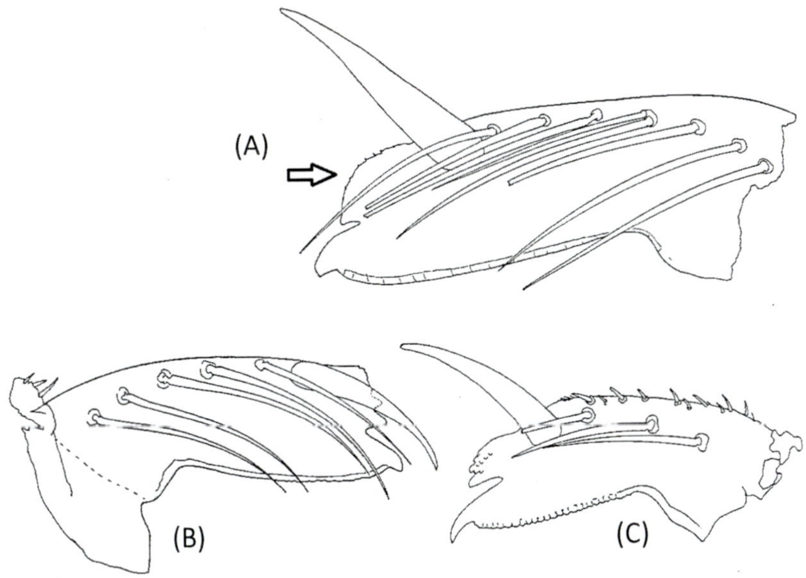

FIGURE 11.10 Palpus: (A) *Ceriagrion tenellum*; (B) *Ischnura elegans*; (C) *Platycnemis pennipes*. The arrow indicates the rounded distal margin of the palpus.

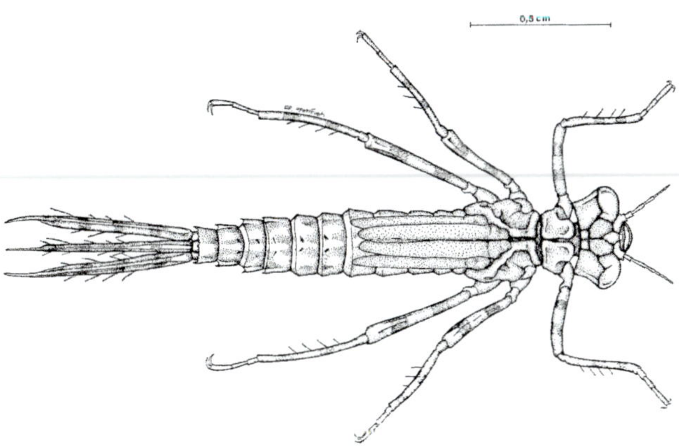

FIGURE 11.11 *Platycnemis pennipes* a typical body shape for Platycnemididae.

FIGURE 11.12 *Platycnemis pennipes. Foto by C. Brochard.*

Zygoptera: Genera and Species

The keys presented are only valid for larvae in which the posterior wing sheaths cover at least S3. In several genera the characters used in the keys are not consistent, so that the decision tree does not always lead to a group which includes all species of a genus. For this reason, in the following keys the same genus may appear at different endpoints of the decision tree.

Zygoptera: Lestidae: Genera and Species

Typical body shapes for the Lestidae are presented in Figs. 11.13 and 11.14.

1 Prementum racket-shaped (Fig. 11.15B and C) .... *Lestes* (six species)
1' Prementum trapezoidal 2
2(1) Aboral branch of the palpus with the distal margin having a series of nearly equal, more or less regularly spaced, crenations (Fig. 11.16A) .... *Chalcolestes* (two species)
2' Aboral branch of the palpus with a pointed tooth in the lateral distal corner (Fig. 11.16B) .... *Sympecma fusca*

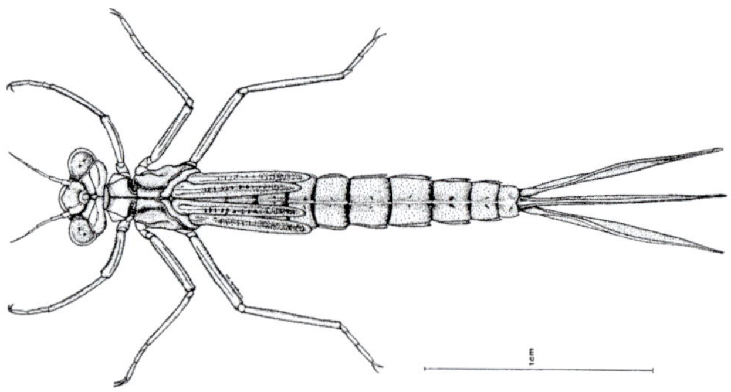

FIGURE 11.13 *Lestes barbarus*, a typical body shape for Lestidae.

FIGURE 11.14 *Chalcolestes parvidens. Foto by C. Brochard.*

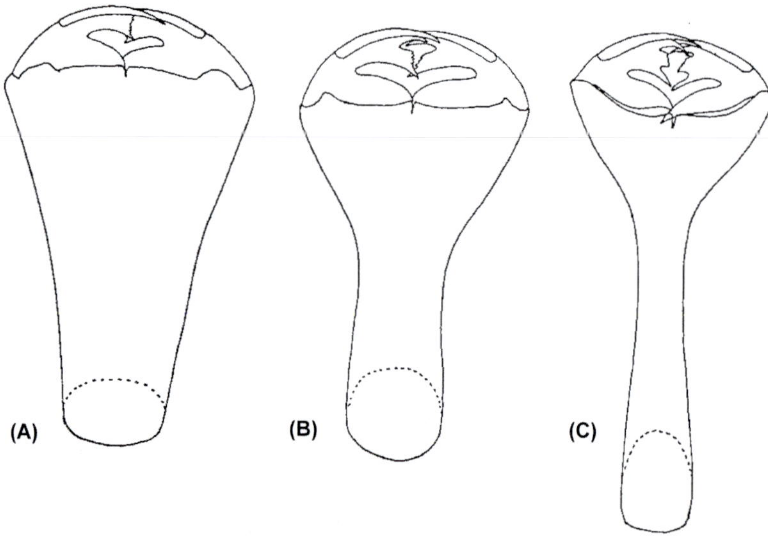

FIGURE 11.15 Labium in ventral view: (A) *Chalcolestes viridis*; (B) *Lestes macrostigma*; (C) *Lestes dryas*.

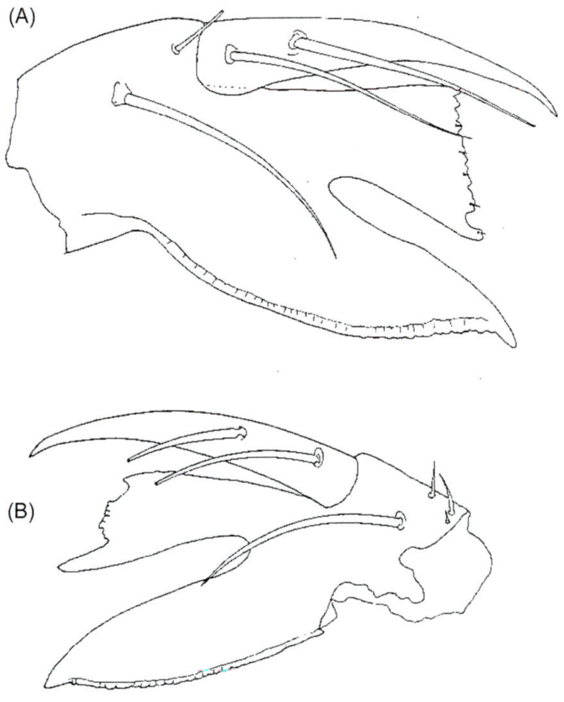

FIGURE 11.16 Palpus: (A) *Chalcolestes viridis*; (B) *Sympecma fusca*.

Zygoptera: Coenagrionidae: Genera and Species

Typical body shapes for the Coenagrionidae are presented in Figs. 11.17 and 11.18.

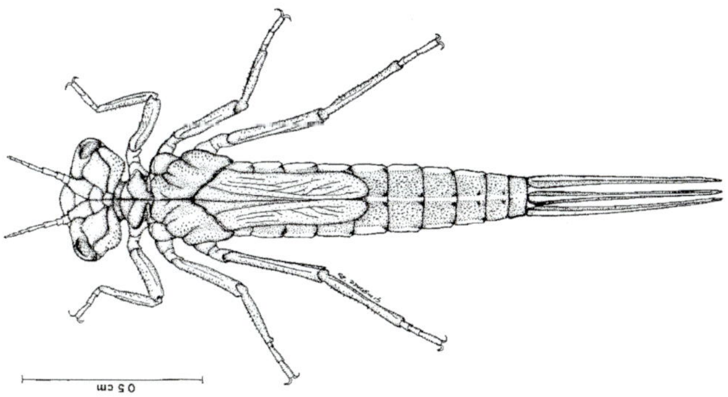

FIGURE 11.17 *Ischnura pumilio*, a typical body shape for Coenagrionidae.

FIGURE 11.18 *Ceriagrion georgifreyi. Foto by C. Brochard.*

1 Movable hook of palpus much shorter than the strongly developed internal end hook; this character is cited by Dumont (1991) but it is not re-detectable in several other congeneric species. Body of fully grown larvae not exceeding 1 cm, with lamellae measuring about 0.5 cm .... *Agriocnemis sania*
1' Movable hook of palpus much longer than internal end hook (Fig. 11.10A–C). Body of fully grown larvae longer than 1 cm 2
2(1) Tracheae of the central lamella with a distinct curve in correspondence of the nodal line (Fig. 11.19) .... *Pseudagrion* (three species)
2' Tracheae of the central lamella always clearly straight (Fig. 11.20A–C) 3
3(2) Ventral margin of the eye without thick spiniform setae (Fig. 11.21A) 4
3' Ventral margin of the eye with some (at times one) thick spiniform setae (Fig. 11.21B) 5
4(3) Prementum with 2 (rarely 4) long setae and between them a variable number of small spiniform setae not aligned with the long setae (Fig. 11.22A). Distal margin of the palpus rounded in the lateral distal corner (Fig. 11.10A) .... *Ceriagrion* (two species)

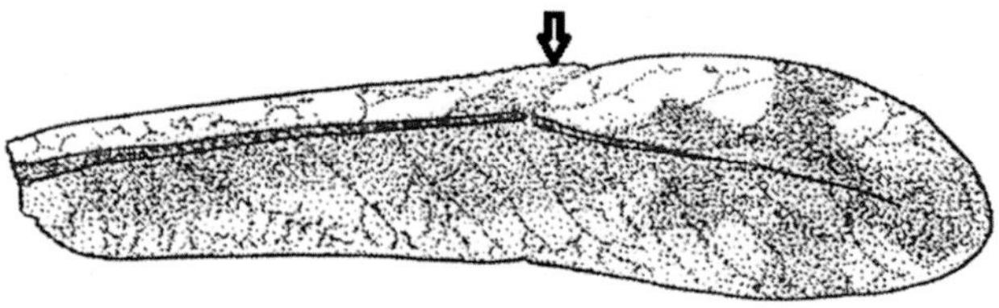

FIGURE 11.19 Caudal lamella of *Pseudagrion sublacteum*. The arrow indicates the curve of the central tracheae. *Drawing by Ole Müller from Suhling et al. (2014).*

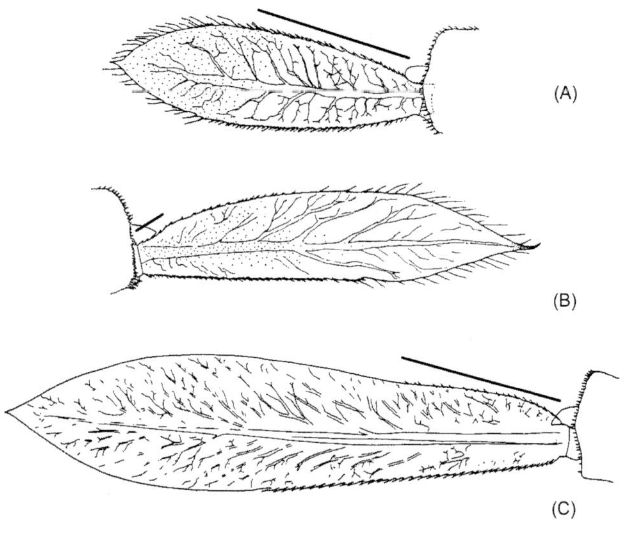

FIGURE 11.20 Lateral lamellae: (A) *Coenagrion caerulescens*; (B) *Coenagrion mercuriale*; (C) *Ischnura elegans*. Lines indicate the extent of groups of short, stout spiniform setae on dorsal margin of the lamellae.

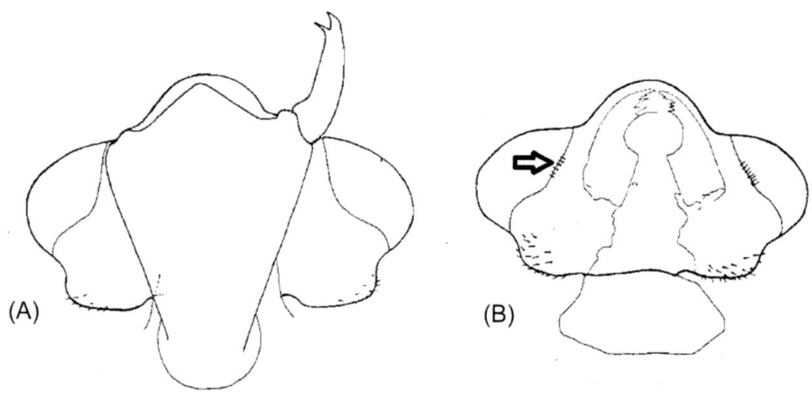

FIGURE 11.21 Head in ventral view: (A) *Pyrrhosoma nymphula*; (B) *Coenagrion castellani*. The arrow indicates a group of thick spiniform setae along the margin of the eye.

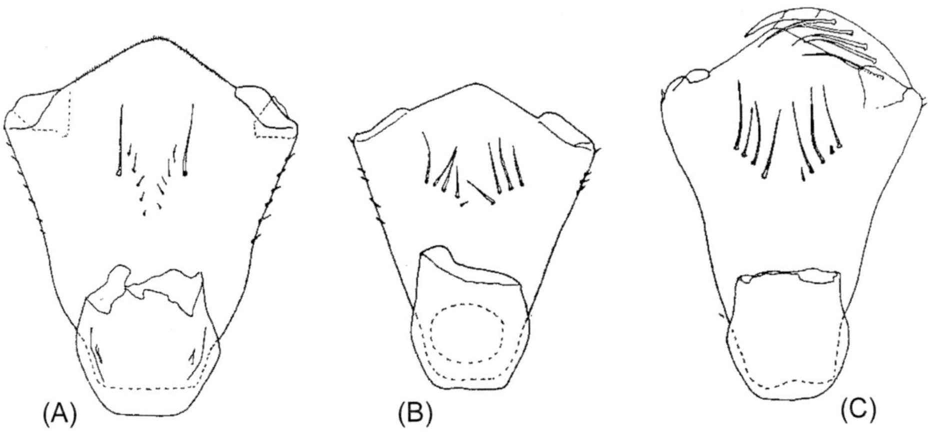

FIGURE 11.22 Prementum inner surface: (A) *Ceriagrion tenellum*; (B) *Coenagrion caerulescens*; (C) *Coenagrion pulchellum*.

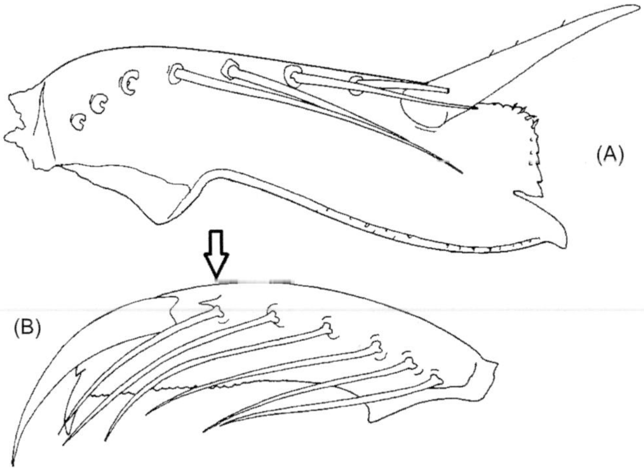

FIGURE 11.23 Palpus: (A) *Pyrrhosoma nymphula*; (B) *Enallagma cyathigerum*. The arrow indicates the small spine at the base of the most distal seta.

4' Prementum with more than 4 setae, more or less long, arranged in two rows and, between these no other small setae (Fig. 11.8). Distal margin of the palpus less rounded in the lateral distal corner (Fig. 11.23A) .... *Pyrrhosoma* (two species)
5(3) One or more rows of spiniform setae near to the posterior margin of the first sternite of the abdomen (Fig. 11.24A and B) .... *Erythromma* (three species)
5' No spiniform setae near to the posterior margin of the first sternite of the abdomen 6
6(5) Postocular lobes and occiput spots with a pigmentation made of small roundish patches (Fig. 11.25B) .... *Coenagrion* (group 1, four species)
6' Post-ocular lobes and occiput with no pigmentation or with a pigmentation made of irregular zones (Fig. 11.25A) 7
7(6) A very small spine at the base of the most distal seta of the palpus (Fig. 11.23B) .... *Enallagma* (two species)
7' No small spine at the base of the above-mentioned seta 8
8(7) Lateral carinae of S5–S8 with spiniform setae that are no bigger than those covering the rest of the abdominal segment (Fig. 11.26B) .... *Ischnura pumilio*
8' Lateral carinae of S5–S8 with spiniform setae that are bigger than those covering the rest of the abdominal segment (Fig. 11.26A and C) 9
9(8) Inner surface of prementum with two rows of long setae and two groups of smaller spiniform setae that are placed more proximally (Fig. 11.27) .... *Ischnura fountaineae*
9' Inner surface of prementum with only two rows of long setae, at most with a few and much smaller spiniform setae not grouped together (Fig. 11.28) 10

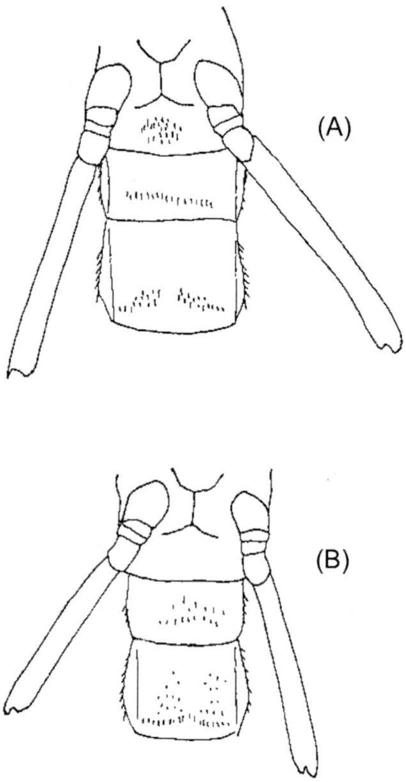

FIGURE 11.24 Metasternum and first abdominal segments: (A) *Erythromma najas*; (B) *Erythromma viridulum*.

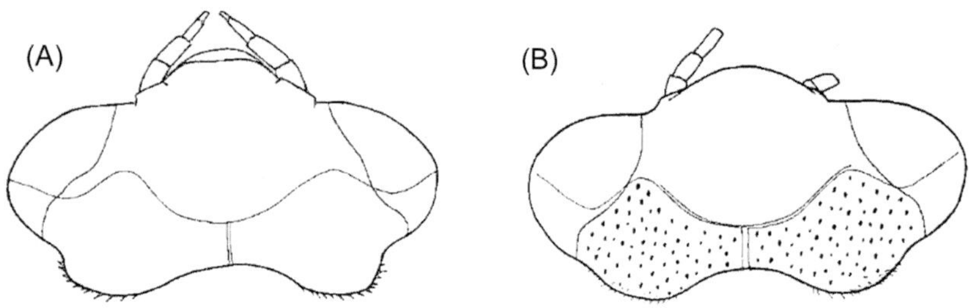

FIGURE 11.25 Head in dorsal view: (A) *Coenagrion castellani*; (B) *Coenagrin pulchellum*.

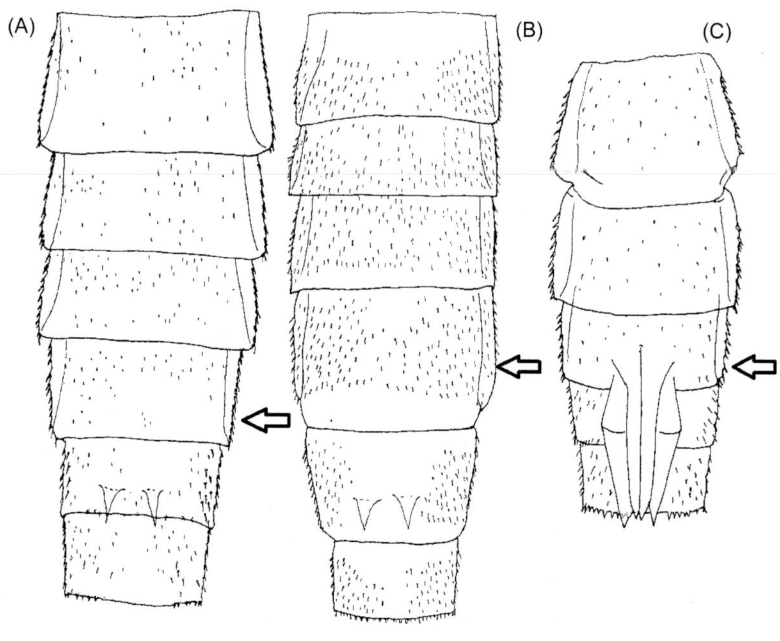

FIGURE 11.26 Last abdominal segments in ventral view: (A) *Ischnura elegans*; (B) *Ischnura pumilio*; (C) *Coenagrion caerulescens*. The arrows indicate the spiniform setae of the carinae of S8.

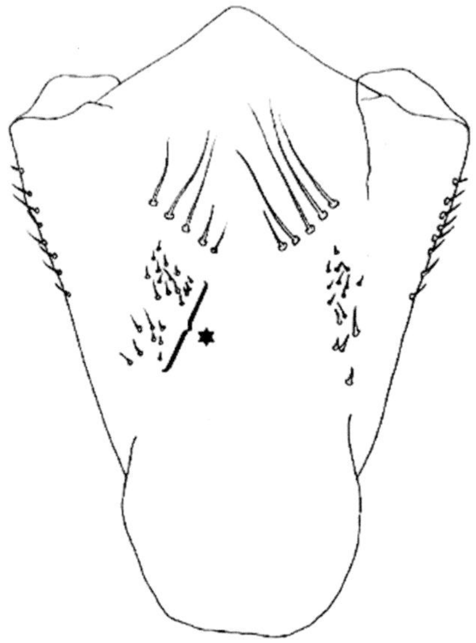

FIGURE 11.27 Prementum inner surface: *Ischnura fountainei* (* indicates the group of smaller spiniform setae that are placed more proximally).

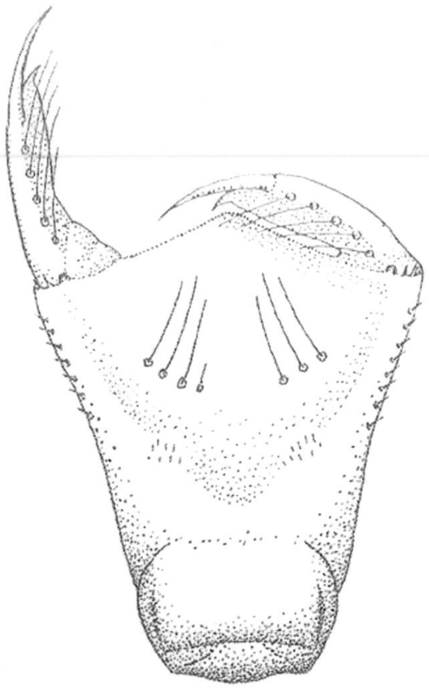

FIGURE 11.28 Prementum inner surface: *Coenagrion castellani*. *Drawing by P.A. Robert from Brochard (2018).*

10(9) Lateral lamellae with a row of short but stout spiniform setae almost exclusively on the ventral margin (Fig. 11.20B) .... *Coenagrion* (group 2, two species)
10' Lateral lamellae with rows of short but stout spiniform setae on the ventral and dorsal margins, the dorsal ones often being more sparse and shorter (Fig. 11.20A and C) 11
11(10) Dorsal surface of the anterior femurs with few and relatively long and pointed spiniform setae (Fig. 11.29E); lamellae with no pigmentation and main tracheas uniformly colored .... *Coenagrion scitulum*
11' Dorsal surface of the anterior femurs with numerous long as well as short spiniform setae, the long ones at times blunt, or with numerous small spines (Fig. 11.29A–D); lamellae with pigmented areas or tracheas with alternating dark and light tracts 12

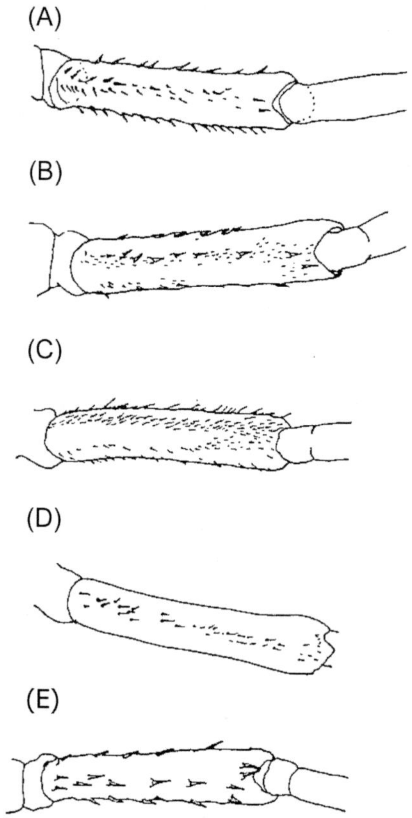

FIGURE 11.29 Dorsal surface of the anterior femurs: (A) *Coenagrion caerulescens*; (B) *Ischnura elegans*; (C) *Ischnura pumilio*; (D) *Coenagrion meruriale castellani*; (E) *Coenagrion scitulum*.

12(9) Caudal lamellae short, five times as long as S10, and with a strong pigmentation (Fig. 11.20A); setae on legs and head are blunt .... *Coenagrion caerulescens*

12' Caudal lamellae long, about seven times as long as S10, with no pigmentation but with tracheas having alternating light and dark parts (Fig. 11.20C); setae on legs and head are pointed .... *Ischnura* (group 1, three species)

Anisoptera: Families

For larvae of any instar

1 Labium flat, labial palps not covering the labrum (Fig. 11.30B) 2

1' Labium strongly concave, labial palps covering the labrum (Fig. 11.30A) 3

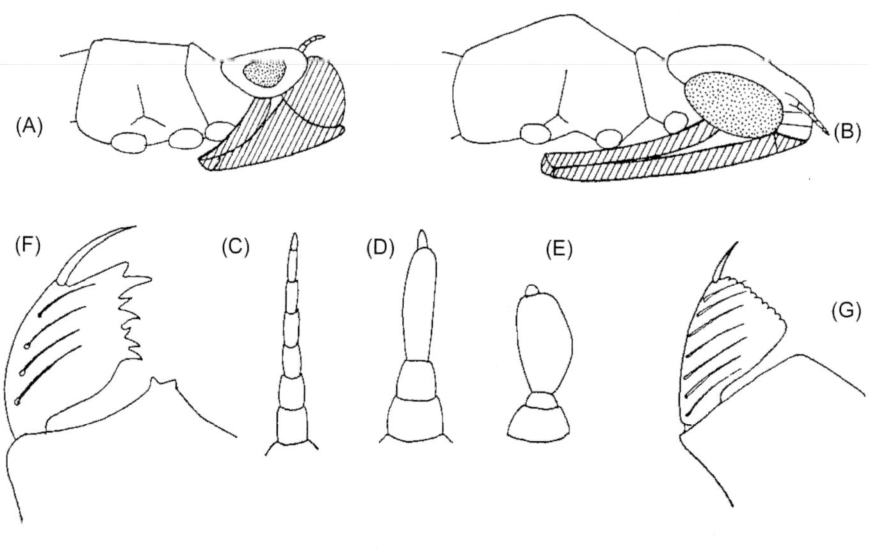

FIGURE 11.30 Labium (hatched areas) in side view: (A) concave type; (B) flat type. Antennae: (C) with subequal segments; (D, E) with third segment longer than the sum of the other together. Prementum: (F) distal margin of labial palpus deeply serrated and bifid protuberance on the median lobe of prementum (Cordulegastridae); (G) with labial palpus having subequal teeth and without bifid protuberance on the median lobe of prementum.

2(1) Antennae having four segments, with the third one longer than the sum of the others (Fig. 11.30D and E) .... **Gomphidae**

2' Antennae having four or more segments, with the third one shorter than the sum of the others (Fig. 11.30C) .... **Aeshnidae**

3(1) Frons of the head with a horn-like projection between antennae (Fig. 11.31A) .... **Macromiidae**
Only one genus *Macromia* (Fig. 11.32) and with one species: *M. splendens* (Fig. 11.33)

3' Frons of the head without or with two shallow projections between eyes (Fig. 11.31B) 4

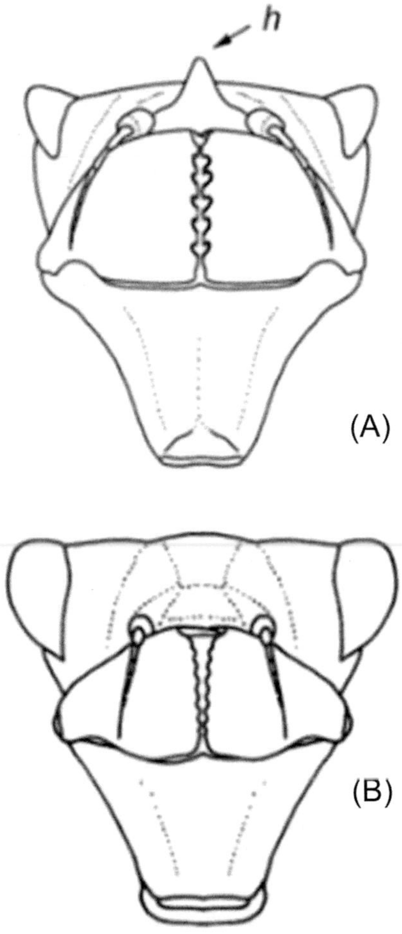

FIGURE 11.31 Head in frontal view: (A) with "horn" (h) between the antennae; (B) without "horn" between the antennae. *Drawing by Ole Müller from Suhling et al. (2014)*

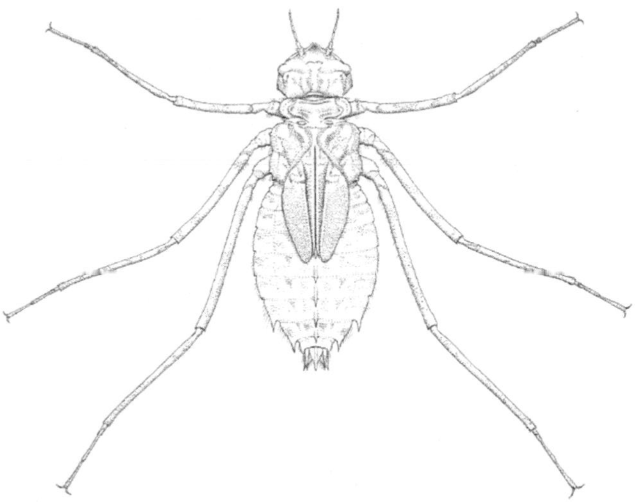

FIGURE 11.32 *Macromia splendens* (Macromiidae). *Drawing by P. A. Robert from Brochard (2018).*

FIGURE 11.33 Head in side view of *Macromia splendens*. Note the "horn" on the front.

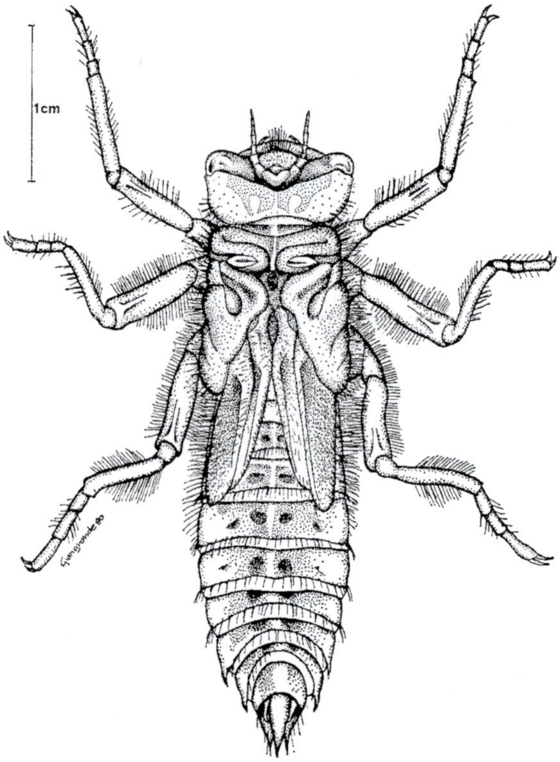

FIGURE 11.34 *Cordulegaster boltonii* a typical body shape of Cordulegasteridae.

FIGURE 11.35 *Cordulegaster trinacriae. Foto by C. Brochard.*

4(3) Median lobe of prementum bifid, the points ending in beak-like hooks; distal margin of labial palpus deeply serrated (Fig. 11.30F) .... Cordulegastridae
Only one genus *Cordulegaster* (Figs. 11.34 and 11.35) with seven species.
4' Median lobe of prementum not bifid; distal margin of labial palpus not deeply serrated (Fig. 11.30G) 5
5(4) Lower surface of the prementum with a median groove (Fig. 11.36A) .... **Corduliidae/Synthemistidae**
5' Lower surface of the prementum without median groove (Fig. 11.36B) .... **Libellulidae**

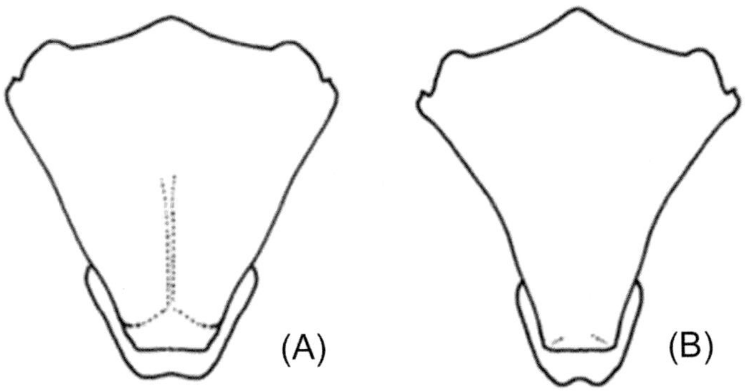

FIGURE 11.36 Prementum in ventral view: (A) with median groove; (B) without median groove. *Drawing by Ole Müller from Suhling et al. (2014).*

Anisoptera: Genera and Species

The keys presented are only valid for larvae in which the posterior wing sheaths cover at least S3. In some genera the characters of the larvae are not consistent, so that the decision tree does not always lead to a group which includes all species of a genus. For this reason, in the following keys, the same genus may appear at different endpoints of the decision tree.

Anisoptera: Gomphidae: Genera and Species

Typical body shapes for the Gomphidae are presented in Figs. 11.37 and 11.38.

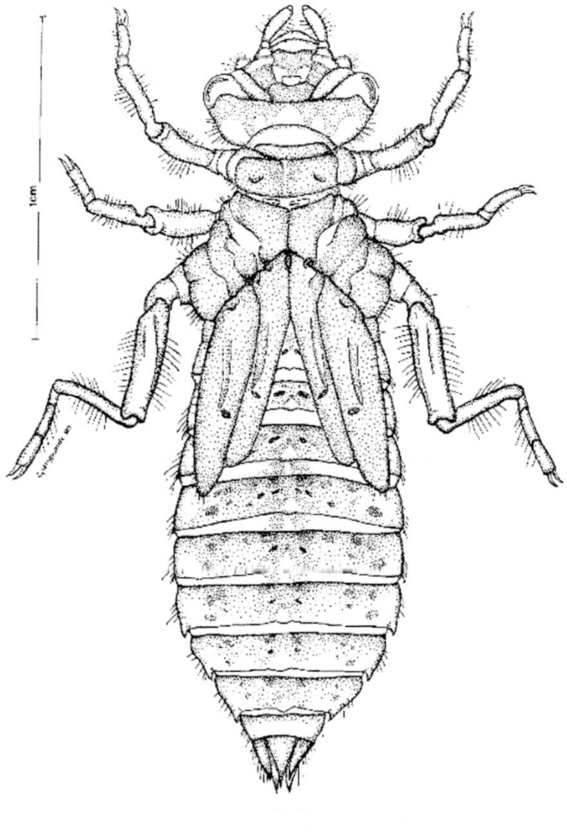

FIGURE 11.37 *Onychogomphus forcipatus unguiculatus*, a typical body shape for Gomphidae.

FIGURE 11.38 *Gomphus flavipes. Foto by C. Brochard.*

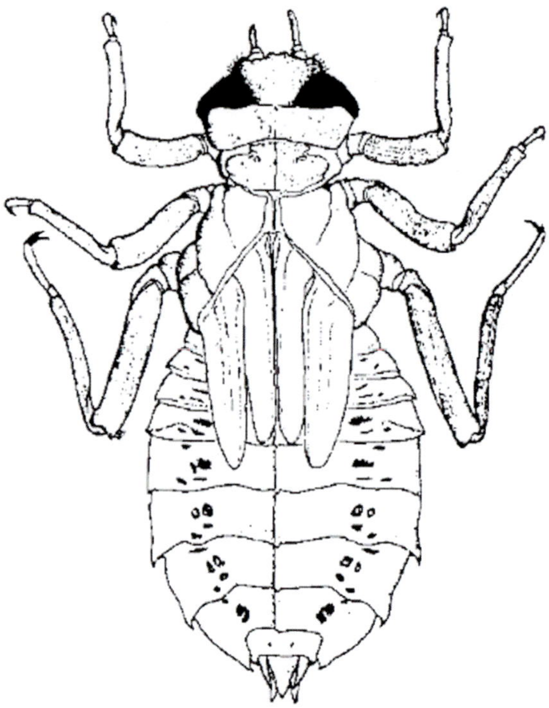

FIGURE 11.39 *Lindenia tetraphylla.*

1 Metatarsi with three segments, width of abdomen ½ its length or less (Fig. 11.37) 2
1' Metatarsi with two segments, width of abdomen more than ½ its length (Fig. 11.39) .... *Lindenia tetraphylla*
2(1) Fourth antennal segment long and falciform (Fig. 11.40A) .... Paragomphus (two species)
2' Fourth antennal segment extremely short and roundish (Fig. 11.40B and C) 3
3(2) First abdominal segments without middorsal spines (Fig. 11.41D); palpus with large end hook directed mesially (Fig. 11.42); third antennal segment cylindrical; head without a protuberance between eye and antenna; in larvae wing sheaths parallel and in contact dorsally .... *Gomphus* (nine species)
2' First abdominal segments with middorsal spines in form of obtuse projection (Fig. 11.41A–C); palpus with or without end hook (Fig. 11.43); third antennal segment strongly flattened dorso-ventrally; head with protuberance between eye and antenna (Fig. 11.40A–C); in larvae wing sheaths divergent (Fig. 11.37) .... *Onychogomphus* (nine species)

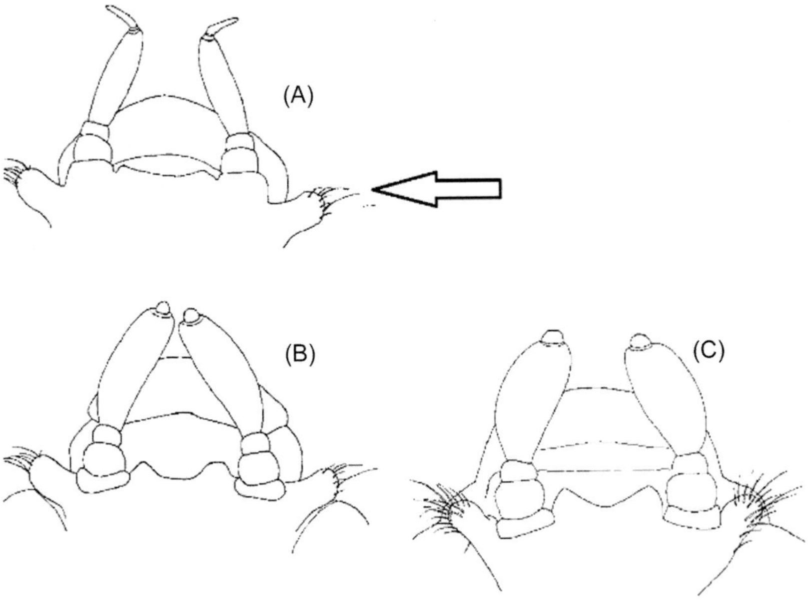

FIGURE 11.40 Antennae: (A) *Paragomphus genei*; (B) *Onychogomphus forcipatus unguiculatus*; (C) *O. uncatus*. Arrow indicates the protuberance between eye and antenna.

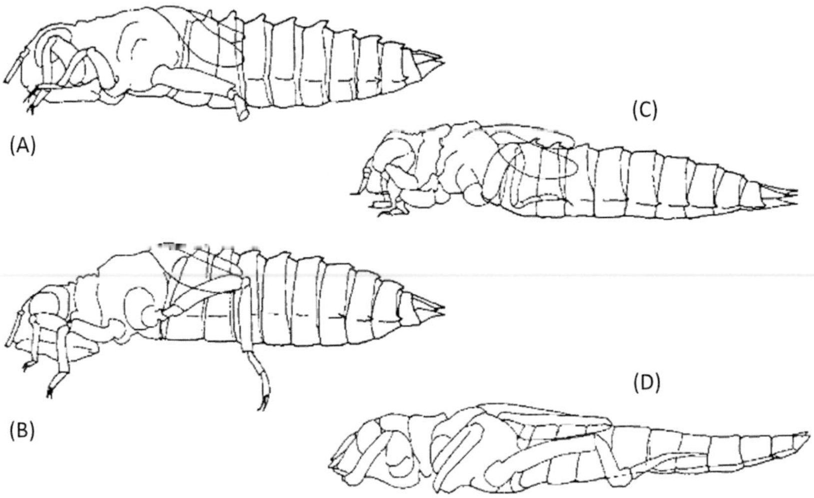

FIGURE 11.41 Abdomen in side view: (A) *Onychogomphus uncatus*; (B) *O. forcipatus unguiculatus*; (C) *Paragomphus genei*; (D) *Gomphus vulgatissimus*.

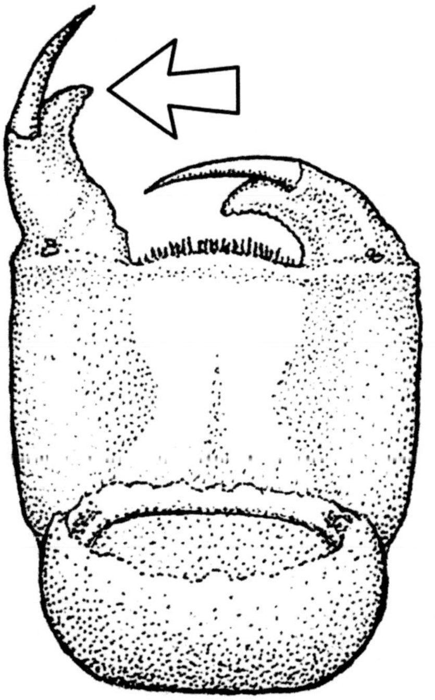

FIGURE 11.42 Inner surface of labium of *Gomphus pulchellus*. Arrow indicates the tip of the palpus with large end hook. *Drawing by P. A. Robert from Brochard (2018).*

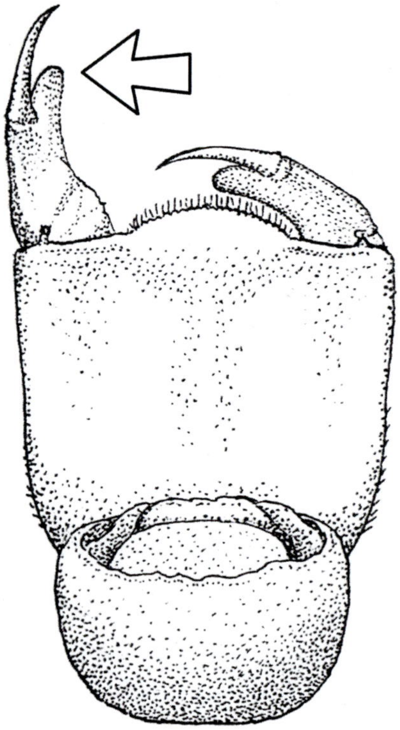

FIGURE 11.43 Inner surface of labium of *Onychogomphus uncatus*. Arrow indicates the tip of the palpus without end hook. *Drawing by P. A. Robert from Brochard (2018).*

Anisoptera: Aeshnidae: Genera and Species

Typical body shapes for the Aeshnidae are presented in Figs. 11.44 and 11.45.

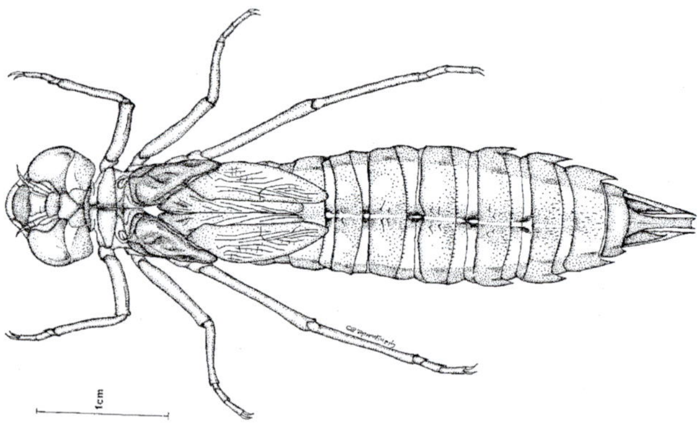

FIGURE 11.44 *Anax imperator*, a typical body shape for Aeshnidae.

FIGURE 11.45 *Caliaeshna microstigma. Foto by C. Brochard.*

1 Eyes small, in dorsal view, occupying at most a third of the lateral profile of the head (Fig. 11.46C); a blunt mid-dorsal spine on S9 (Fig. 11.47C), however, reduced in some individuals or sometimes present as vestigial also on S10 .... *Brachytron pratense*

1' Eyes big, occupying more than a third of the lateral profile of the head (Fig. 11.46A, B, and D); no middorsal spines or protuberances on the abdominal segments 2

2(1) In dorsal view lateral and posterior profiles of the head forming sharp angles (Fig. 11.46D) .... *Boieria* (two species)

2' In dorsal view lateral and posterior profiles of the head rounded, not forming sharp corners (Fig. 11.46A and B) 3

3(2) Epiproct terminating with only one apex; internal side of the paraproct with several short but strong spiniform setae (Fig. 11.48) .... *Caliaeshna microstigma*

3' Epiproct terminating with two apexes or more (Fig. 11.49A–C) 4

4 (3) Eyes elongated longitudinally, in dorsal view external profile of the eyes with uneven curvature, ratio of the width(A)/length (B) less than 1 (Fig. 11.46A); S6 with no lateral spines, not even vestigial (Fig. 11.47A) .... *Anax* (four species)

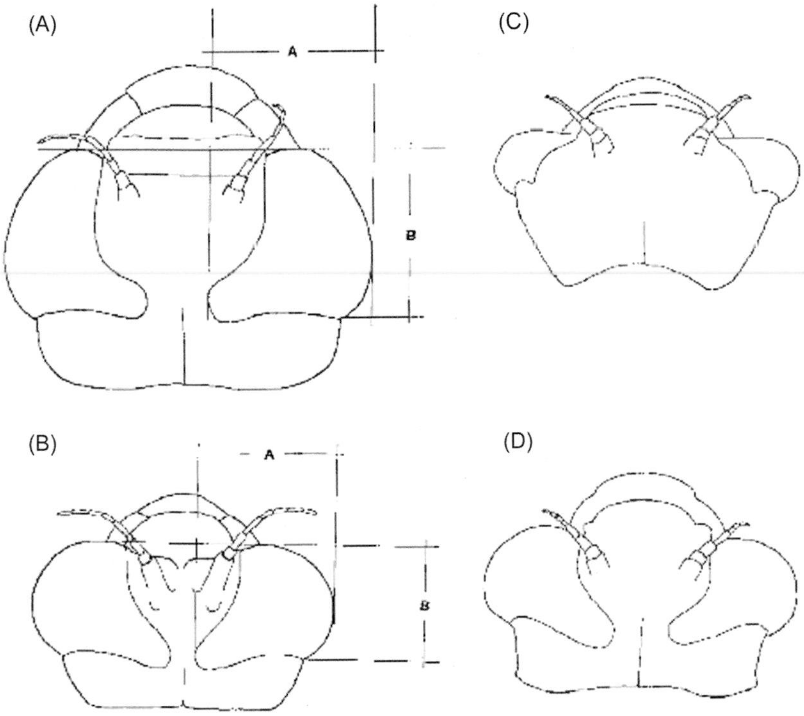

FIGURE 11.46 Head in dorsal view: (A) *Anax imperator*; (B) *Aeshna mixta*; (C) *Brachytron pratense*; (D) *Boyeria irene*.

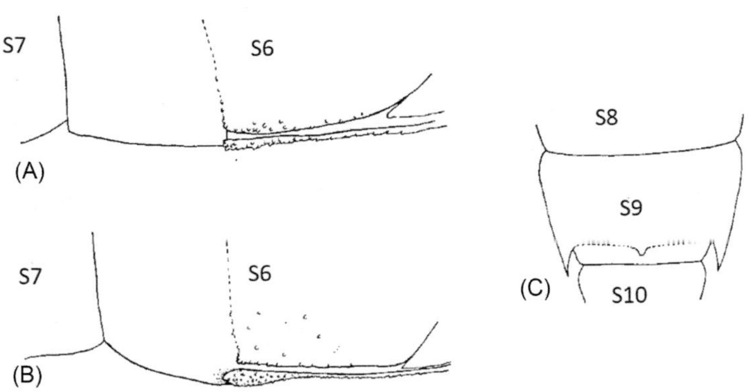

FIGURE 11.47 Lateral margin of S6 in ventral view: (A) *Anax imperator*; (B) *Aeshna juncea*. Last abdominal segments in dorsal view: (C) *Brachytron pratense*.

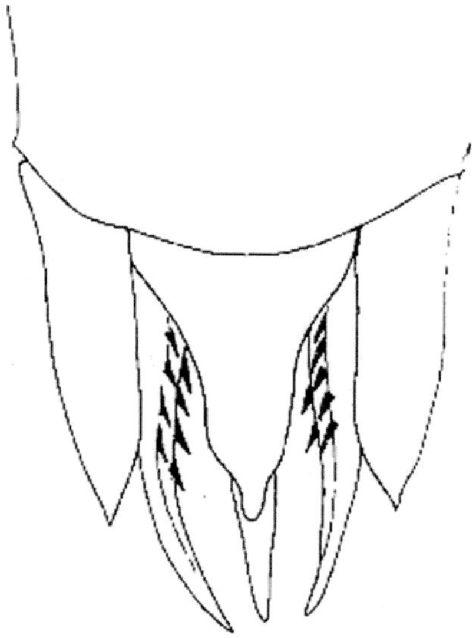

FIGURE 11.48 Male caudal pyramid in dorsal view: *Caliaeshna microstigma*.

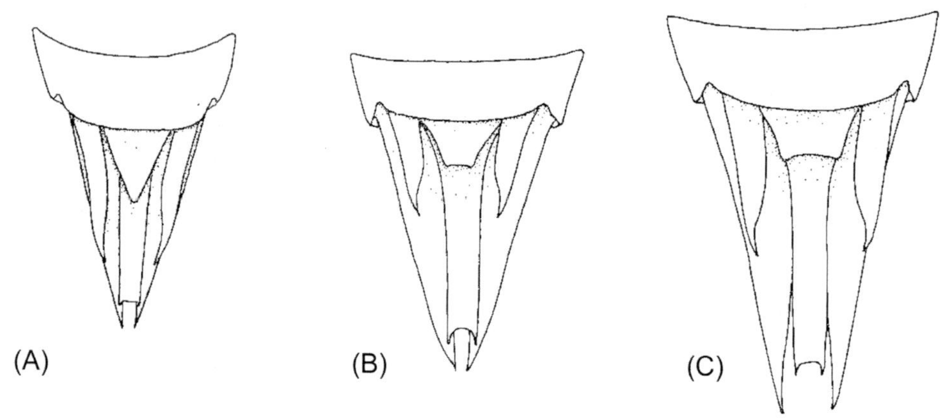

FIGURE 11.49 Male caudal pyramid in dorsal view: (A) *Aeshna isosceles*; (B) *Anax imperator*; (C) *A. parthenope*.

4' Eyes elongated transversally, in dorsal view external profile of the eyes with uniform curvature similar to an arc of a circle, ratio of the width (A)/length (B) bigger than 1 (Fig. 11.46B); S6 with lateral spines, at time vestigial or rarely absent (Fig. 11.47B) .... *Aeshna* (four species).

Anisoptera: Corduliidae: Genera and Species

Typical body shapes for the Corduliidae are presented in Figs. 11.50 and 11.51.

The key of the Corduliidae also includes the genus *Oxygastra*, although it is now considered to belong to the Synthemistidae. This genus is treated here because its larvae are morphologically similar to those of the family Corduliidae.

1 Abdominal segments with no middorsal spines, at most with vestigial spines on S5, S6, S7 (Fig. 11.52A and B) ... . *Oxygastra curtisii*

1' Abdominal segments with prominent middorsal spines, also on S8 and S9 (Fig. 11.53A–D) 2

2(1) S9 with a small dorsal spine, in lateral view it covers less than 1/5 of the remaining length of the segment (Fig. 11.53A and D); in living larvae a well-defined dark stripe is visible between the eyes (Fig. 11.50) .... *Cordulia aenea*

2' S9 with a large dorsal spine, in lateral view longer than 1/4 of the remaining length of the segment (Fig. 11.53B and C); pigmentation of the head different in life larvae .... *Somatochlora* (four species)

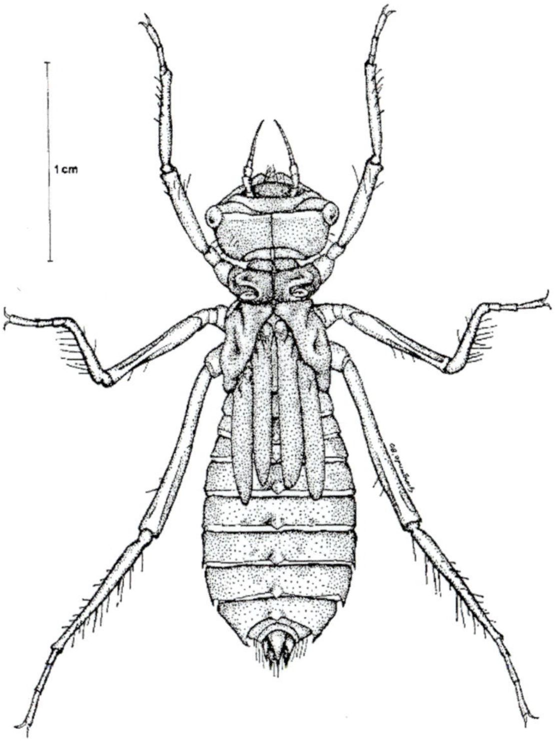

FIGURE 11.50 *Cordulia aenea*, a typical body shape of Corduliidae.

FIGURE 11.51 *Somatochlora meridionalis. Foto by C. Brochard.*

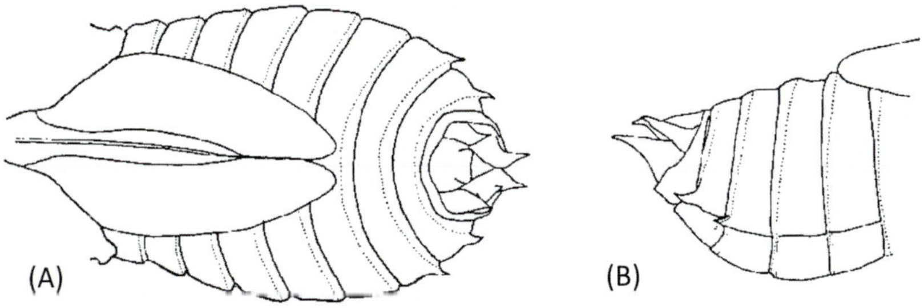

FIGURE 11.52 Last abdominal segment of *Oxygastra curtisii*: (A) dorsal view; (B) lateral view.

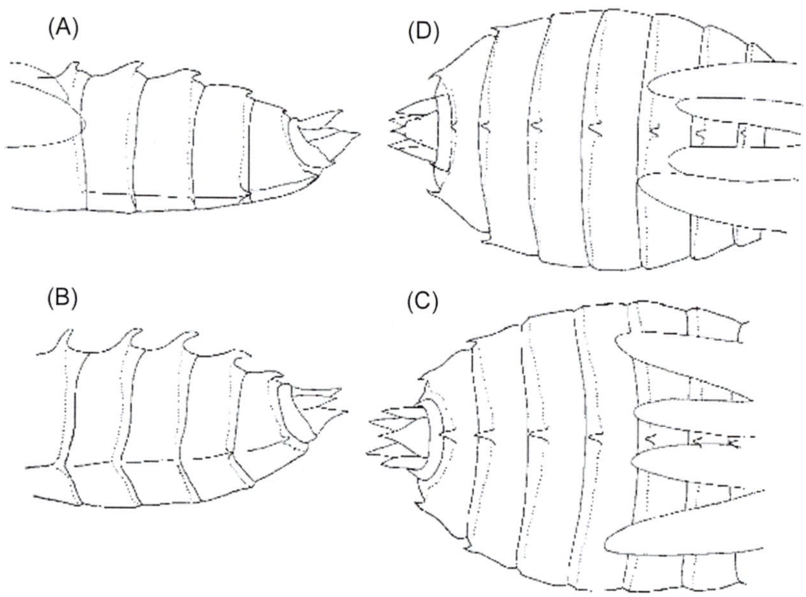

FIGURE 11.53 Last abdominal segments in (A, B) lateral and (D, C) dorsal view: (A, D) *Cordulia aenea*; (B, C) *Somatochlora metallica*.

Anisoptera: Libellulidae: Genera and Species

Typical body shapes for the Libellulidae are presented in Figs. 11.54–11.57.

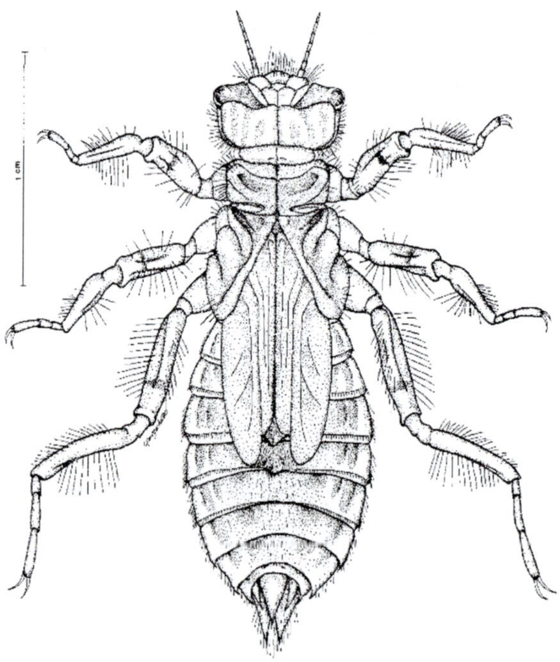

FIGURE 11.54 *Orthetrum cancellatum* (Libellulidae).

FIGURE 11.55 *Libellula quadrimaculata. Foto by C. Brochard.*

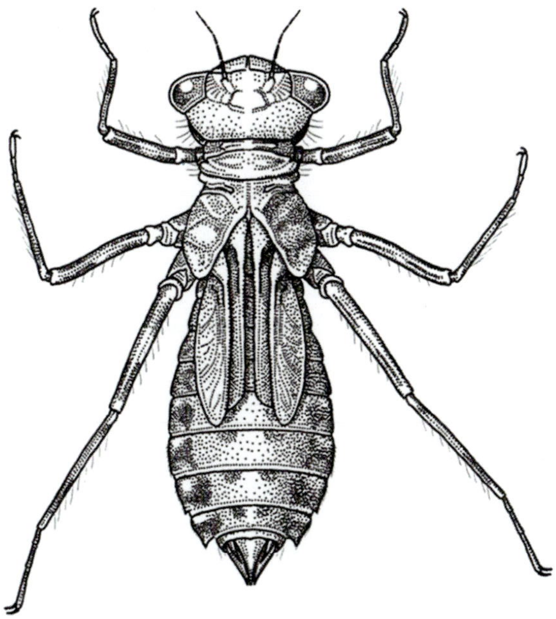

FIGURE 11.56 *Crocothemis erythraea* (Libellulidae). *Drawing by O. Müller from Suhling et al. (2014).*

FIGURE 11.57 *Trithemis kirby. Foto by C. Brochard.*

FIGURE 11.58 *Zygonyx torridus*, last exuvia in lateral view.

1 Abdomen with dorsal spines up to S9 and with a hint of a spine even on S10 (Fig. 11.58). In dorsal view eyes occupying about ½ of the lateral profile of the head. Postocular lobes tapering backwards and having posterior lateral angles acute. Prementum flattened and very short, not extending beyond the coxae of the first pair of legs (Fig. 11.59). Distal margin of palpus without undulations but with a line of short setae. Femora of all legs flattened .... *Zygonyx torridus*

1' Other characters 2

2(1) Eyes small; in dorsal view they occupy at most 1/3 of the lateral profile of the head; in frontal view the eyes do not reach the labium. In dorsal view margins of the head behind the eyes almost parallel. Thorax wide, particularly

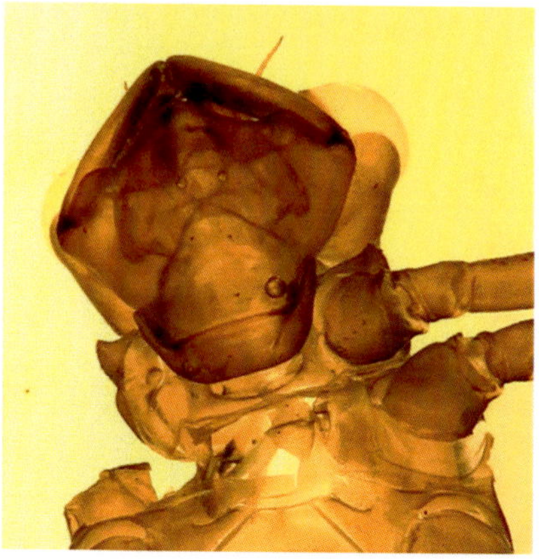

FIGURE 11.59 *Zygonyx torridus*, head and first two thoracic segments of last exuvia in ventral view.

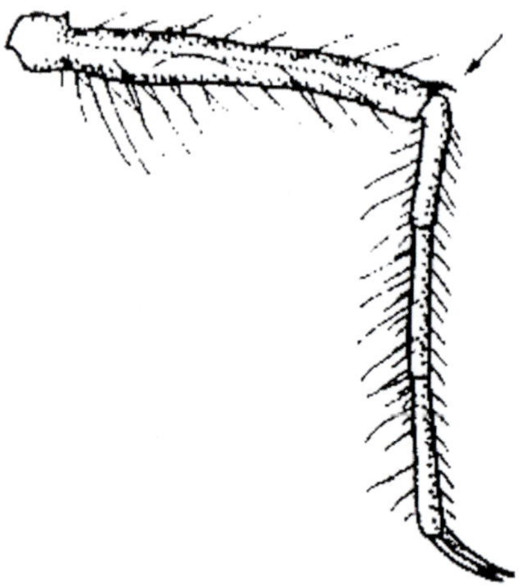

FIGURE 11.60 *Nesciothemis farinosa*; articulation between tibia and tarsus with a single large spine (arrow). *Drawing by O. Müller from Suhling et al. (2014).*

the pronotum, which in dorsal view is slightly narrower than the head. Legs robust, particularly the anterior femurs. Body and legs covered by dense piliform setae, in nature often encrusted with much debris (Fig. 11.54) 3

2' Eyes large, in dorsal view they occupy more than ½ of the lateral profile of the head; in frontal view the eyes reach the labium. In dorsal view margins of the head behind the eyes curved and convergent. Thorax narrow, particularly the pronotum, which is about half as wide as the head. Legs long and slender with straight femurs. Body and legs with fewer and more rigid setae, in nature often devoid of debris (Fig. 11.56) 5

3(2) Hind legs long and with a single large spine on each tibia near the articulation with the tarsus (Fig. 11.60) .... *Nesciothemis farinosa*

3' Hind legs short and no single large spine on each tibia near the articulation with the tarsus 4

4(3) S8 with no dorsal spine (Fig. 11.61A–C), at times tufts of piliform setae resembling a spine .... *Orthetrum* (ten species)

4' S8 with dorsal spine (Fig. 11.61D and E) .... *Libellula* (four species)

5(2) Abdominal segments without dorsal spines (Fig. 11.62A, F, and H) 6

5' Some abdominal segments with dorsal spines (Fig. 11.62B–E, G, I, and L) 11

6(5) Abdominal segments without lateral spines (Fig. 11.63) .... *Acisoma panorpoides*

6' Abdominal segments S8 and S9 with lateral spines (Fig. 11.64) 7

7(6) Lateral spines on S8 and S9 very long, the latter almost reaching the apex of the caudal pyramid (Fig. 11.64) .... *Pantala flavescens*

7' Lateral spines on S8 and S9 shorter, the latter not reaching the middle of the caudal pyramid (Fig. 11.62F and H) 8

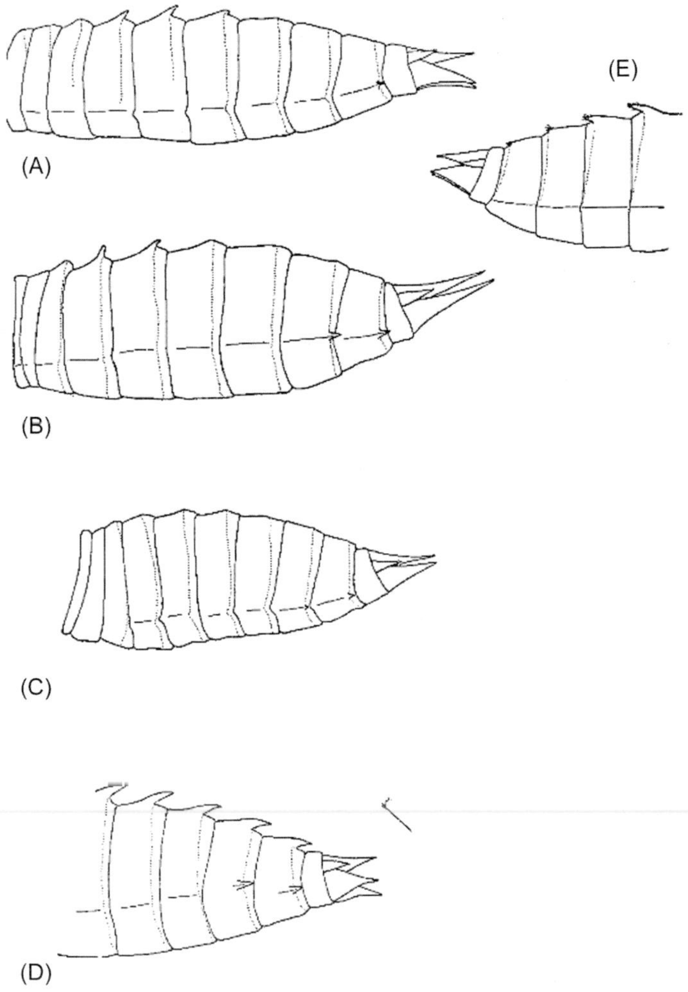

FIGURE 11.61 Abdominal segments in lateral view: (A) *Orthetrum cancellatum*; (B) *O. trinacria*; (C) *O. albistylum*; (D) *Libellula fulva*, (E) *L. depressa*.

8(7) Dorsal surface of the pronotum with a pair of long hairy setae symmetrically placed on both sides of the medial line (Fig. 11.65); legs short, articulation between tibia and tarsus of the posterior leg that do not surpass the end of anal pyramid .... *Diplacodes lefebvrii.*
8' Dorsal surface of the pronotum without a pair of long hairy setae, sometimes with groups of small spiniform or piliform setae; legs longer, articulation between tibia and tarsus of the posterior leg surpassing the end of anal pyramid 9
9(8) Posterior margin of the sternite of S8 without row of spiniform setae (Fig. 11.66A) ... *Sympetrum fosnscolombii*
9' Posterior margin of the sternite of S8 with a row of spiniform setae (Fig. 11.66B) 10
10(9) Lateral spines on S9 about twice as long as those of S8 .... *Sympetrum sinaiticum*
10' Lateral spines on S9 about as long as those of S8 or slightly longer (Fig. 11.62H) ..... *Crocothemis* (two species)
11(5) A big middorsal spine on S9 (Fig. 11.62B and D) 12
11' Middorsal spine absent from S9 13
12(11) The joint of prementum-postmentum reaches between the coxae of the first pair of legs. S3 without dorsal spine; lateral spines long, those of S9 reach about 1/2 of the caudal pyramid, caudal pyramid distinctly shorter than S9 and S10 together (Fig. 11.62B) .... *Brachythemis* (two species)
12' The joint of prementum-postmentum reaches the line separating the first and the second sternite. S3 with a dorsal spine; lateral spines short, those of S9 do not reach the posterior margin of S10; caudal pyramid distinctly longer than S9 and S10 together (Fig. 11.62D) .... *Trithemis* (four species)
13(11) Dorsal spine on S8 extends to the posterior margin of S9; lateral spines of S9 extend further than the anal pyramid (Fig. 11.67) .... *Urothemis edwardsii*

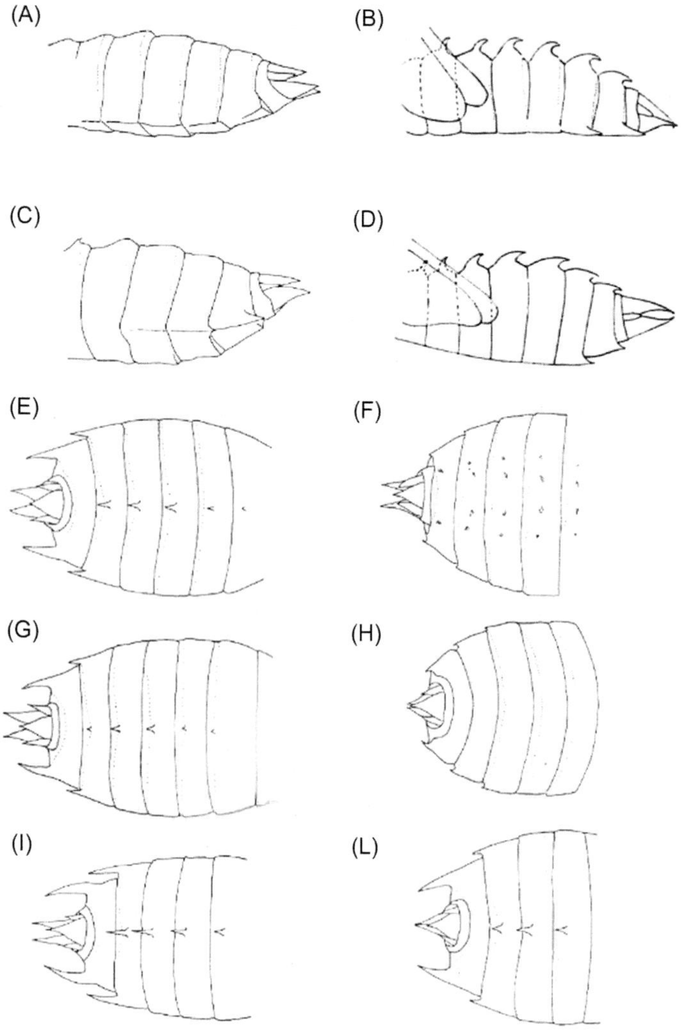

FIGURE 11.62 Last abdominal segments in lateral (A–D) and dorsal (E–L) view: (A, F) *Sympetrum fonscolombii*: (B) *Brachythemis impartita*; (C) *Leucorrhinia dubia*; (D) *Trithemis annulata*; (E) *S. striolatum*; (G) *S. meridionale*; (H) *Crocothemis erythraea*; (I) *S. depressiusculum*; (L) *S. vulgatum*.

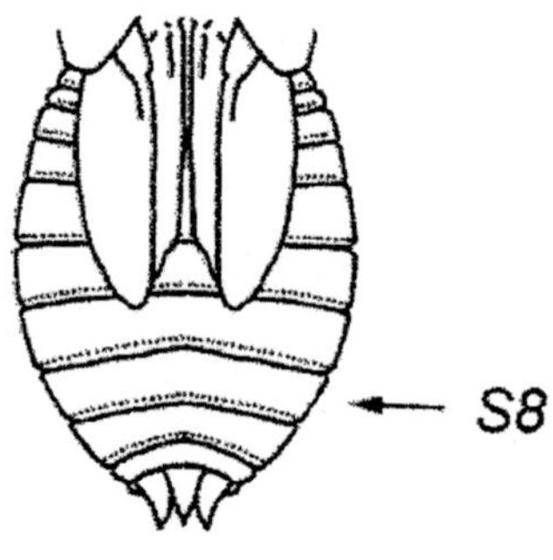

FIGURE 11.63 *Acisoma panorpoides*: last abdominal segments in dorsal view. *Drawing by O. Müller from Suhling et al. (2014).*

13' Dorsal spine on S8 and lateral spines of S9 clearly shorter 14

14(13) In dorsal view eyes elongated longitudinally, ratio between line A (connecting the anterior and posterior corners of the eye) and line B (perpendicular to line A) equals about 2 (Fig. 11.68A) .... *Selysiothemis nigra*

14' In dorsal view eyes less elongated longitudinally, ratio between line A (connecting the anterior and posterior corners of the eye) and line B (perpendicular to line A) less than 1.5 (Fig. 11.68B) .... *Sympetrum* group 3 (seven species)

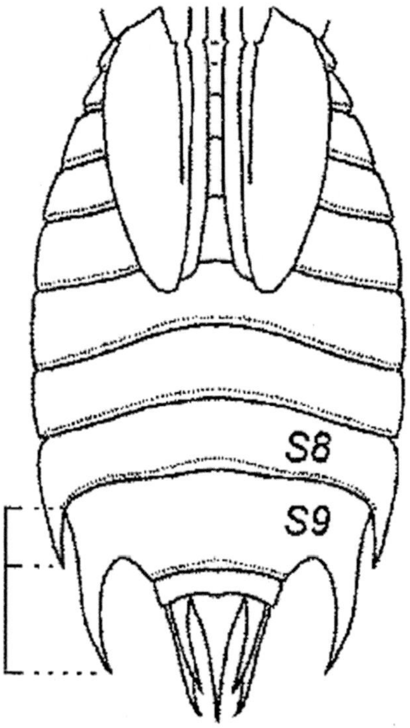

FIGURE 11.64 *Pantala flavescens*: abdomen in dorsal view. *Drawing by O. Müller from Suhling et al. (2014).*

FIGURE 11.65 *Diplacodes lefebvrii*: head and prothorax in dorsal view. Note the two long setae near the mid of prothorax.

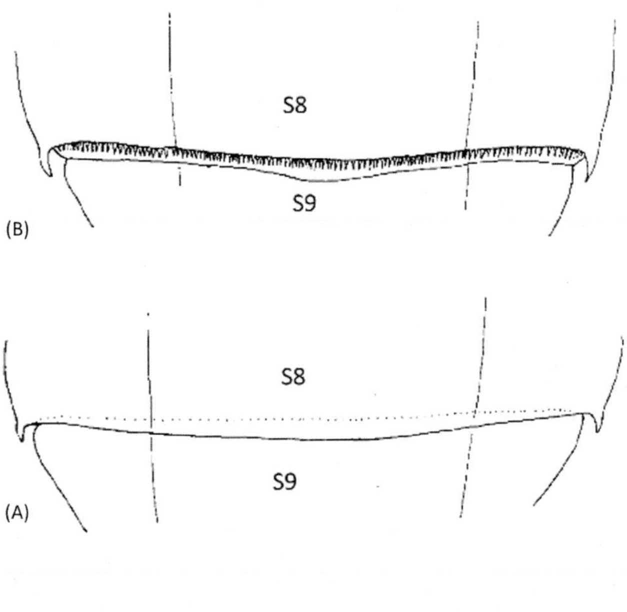

FIGURE 11.66 Abdominal segments S8 and S9 in ventral view: (A) *Sympetrum fonscolombii*; (B) *Crocothemis erythraea.*

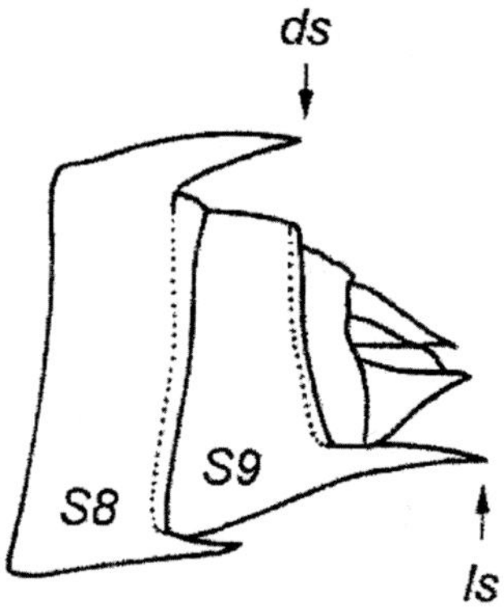

FIGURE 11.67 *Urothemis edwardsii*: last abdominal segments in lateral view; ds, dorsal spine; ls, lateral spine. *Drawing by O. Müller from Suhling et al. (2014).*

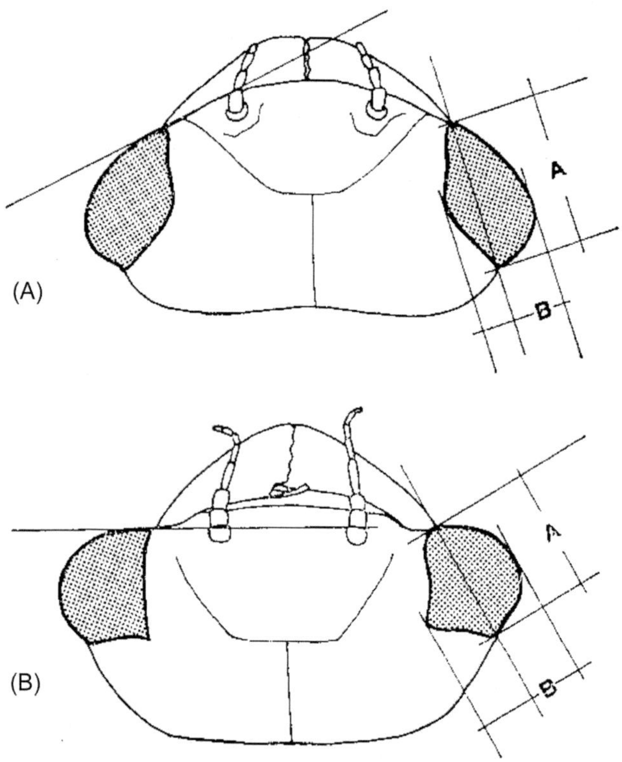

FIGURE 11.68 Head in dorsal view: (A) *Selysiothemis nigra*; (B) *Sympetrum* sp.

Acknowledgments

The authors thank J.-P. Boudot for useful information; O. Müller for the kind permission to reproduce drawings from the publication The dragonfly larvae of Namibia (Suhling F., Muller O. & Martens A. 2014); C. Brochard for the kind permission to reproduce drawings from the book Les larves de libellules de P. A. Robert (Brochard, 2018) and for the kind permission of reproduction of his original photos. A special thanks to M. Di Domenico for his original drawings, A. Cordero Rivera, S. Aguzzi, and F. Leandri for kindly supplying some exuviae.

References

Boudot, J.P. & V.J. Kalkman (eds.). 2015. Atlas of the European dragonflies and damselflies. KNNV Publishing, The Netherlands. 381 pp.

Boudot, J.P., V.J. Kalkman, M. Azpilicueta Amorín, T. Bogdanović, A. Cordero Rivera, G. Degabriele, J.L. Dommanget, S. Ferreira, B. Garrigós, M. Jović, M. Kotarac, W. Lopau, M. Marinov, N. Mihoković, E. Riservato, B. Samraoui. & W. Schneider. 2009. Atlas of the Odonata of the Mediterranean and North Africa. Libellula (Supplement) 9: 1−256.

Brochard, C. (ed.). 2018. Les Larves de Libellules de Paul-André Robert. KNNV Publishing, Zeist, 320 pp.

Carchini, G. 2016. Chiave per il riconoscimento delle larve delle specie italiane della Libellule (Odonata). Società italiana per lo studio e la conservazione delle Libellule, Carmagnola. 159 pp.

Chovanec, A. & R. Raab. 1997. Dragonflies (Insecta, Odonata) and the ecological status of newly created wetlands - Examples for long-term bioindication programmes. Limnologica 27: 381–392.

Conesa-Garcia M.A. 2021. Larvas de libélulas en la peninsula ibérica. Torres editores, Granada. 528 p.

Corbet, P.S. 1953. A terminology for the labium of larval Odonata. The Entomologist 86: 191–196.

Corbet, P.S. 1954. Seasonal regulation in British dragonflies. Nature 174: 655.

Corbet. P.S. 2002. Stadia and growth ratios of Odonata: a review. International Journal of Odonatology 5: 45–73.

Corbet, P.S. 2004. Dragonflies: Behaviour and ecology of Odonata. Harley Books, Martins, Great Horkesley. 829 + xii pp.

Dumont, H.J. 1991. Odonata of the Levant. Israel Academy of Sciences and Humanities, Jerusalem. 297 + ii pp.

Gardner, A.E. 1954. A key to the larvae of the British Odonata. Part I. Zygoptera. Entomological Gazette 5(3): 157-171. Part. II. Anisoptera. Ibid. 5 (4): 193–213.

Gerken, B. & K. Sternberg. 1999. Die Exuvien europäischer Libellen. Huxaria Druckerei GmbH, Höxter 264 pp.

Paulson, D. & Schorr, M. 2021. World Odonata List (Revision 28 June 2021). https://www2.pugetsound.edu/academics/academic-resources/slater-museum/biodiversity-resources/dragonflies/world-odonata-list2/.

Samways, M.J. 2003. Threats to the tropical island dragonfly fauna (Odonata) of Moyotte, Comoro archipelago. Biodiversity and Conservation 12: 1785–1792.

Suhling F., O. Müller & A. Martens. 2014. The dragonfly larvae of Namibia. Libellula (Supplement) 13: 5–106.

Tollett, V.D., E.L. Benvenutti, L.A. Deer, T.M. Rice. 2009. Differential toxicity to Cd, Pb, and Cu in dragonfly larvae (Insecta: Odonata). Archives of Environmental Contamination and Toxicology 56: 77–84.

Tremblay, A., M. Lucotte, M. Meili, L. Cloutier, P. Pichet. 1996. Total mercury and methylmercury contents of insects from boreal lakes: Ecological, spatial and temporal patterns. Water Quality Research Journal of Canada 31: 851–873.

Chapter 12

Order Hemiptera

Fabio Cianferoni[1,2]
[1]*Research Institute on Terrestrial Ecosystems (IRET), National Research Council of Italy (CNR), Florence, Italy,* [2]*Zoology, "La Specola," Natural History Museum of the University of Florence, Florence, Italy*

Introduction

The order Hemiptera Linnaeus, 1758, with more than 103,000 described extant species, is the fifth largest group of insects after Coleoptera, Lepidoptera, Diptera, and Hymenoptera and the most speciose among nonendopterygote insects (Forero, 2008; Zhang, 2013; Stork, 2018; Cianferoni, unpublished data). Hemipterans are currently divided into more than 300 families, the highest number among all insect orders (Drohojowska et al., 2020).

Within Hemiptera, the suborder Heteroptera Latreille, 1810, or "true bugs," has more than 45,000 described species (Henry, 2017) in over 90 families (Henry, 2017; cf. Schuh & Weirauch, 2020; Cianferoni, unpublished data), and it shows the most successful radiation of Hexapoda (Weirauch & Schuh, 2010). Heteroptera includes the only aquatic representatives of Hemiptera. No other insect suborder utilizes such a wide array of different habitats as does the Heteroptera: with the great majority occupying many terrestrial microhabitats. Others are found in aquatic environments like freshwaters and transition zones, in marine intertidal zones, and even a few living on the open ocean surface (the only ones among insects). A few others are ectoparasites of birds and mammals, are associated with nests of ants and termites, occupy webs of spiders where they have been found to steal their prey, or are adapted to life in caves. Heteropterans may be phytophagous, mycetophagous, zoophytophagous, predaceous, or hematophagous (Schuh & Slater, 1995; Schuh & Weirauch, 2020).

The immature stages of Heteropterans (sometimes called nymphs, Fig. 12.1A and B) typically resemble adults (Fig. 12.1C and D) and live in the same environments. Heteropterans have ordinarily five instars, with very few exceptions (see Štys & Davidová-Vilímová, 1989).

Four monophyletic infraorders of Heteroptera are more or less closely associated with water: Gerromorpha (Fig. 12.2A), Nepomorpha (Fig. 12.2B), Leptopodomorpha (Fig. 12.2C and D), and Dipsocoromorpha (Fig. 12.5A). Members of Leptopodomorpha inhabit damp areas adjacent to fresh- or saltwaters, some are intertidal and can remain submerged for a prolonged period, whilst others have no apparent association with water. Some members of Dipsocoromorpha live in the substrate (gravel-sand) or in mosses near freshwaters and can utilize plastron when submerged, but other taxa belonging to this group are not directly associated with water (Tamanini, 1979; Schuh & Slater, 1995; Schuh & Weirauch, 2020). However, the latter two infraorders do not live in or above water and are not included in this chapter.

Only the members of the infraorder Nepomorpha Popov, 1968 or "water bugs" have successfully adapted to a true aquatic existence even in their adult life stage, as is the case of only Coleoptera within insects (Schuh & Slater, 1995). Nepomorphans are known from definitive fossils of the Triassic (Schuh & Weirauch, 2020), although possibly they began to diversify by the late Permian (Ye et al., 2019), and the extant Nepomorpha has more than 2300 described species in 13 families (cf. Nieser, 2002; Polhemus & Polhemus, 2008; Ye et al., 2019; Wang et al., 2021). All Nepomorphans have secondarily shortened antennae, which can be hidden in a groove or concavity below the eyes (Fig. 12.3A). Their body is hydrodynamic, and many species possess swimming legs (Figs 12.3B and C and 12.4A; more or less modified) adapted to aquatic life. The members of Ochteroidea [i.e., Ochteridae (Fig. 12.1D) and Gelastocoridae] live out of the water, in transition zones between terrestrial and aquatic habitats (shores, beaches), and do not possess all these adaptative features to an aquatic life; however, they retained the shortened antennae. Nepoidea [i.e., Nepidae (Fig. 12.2B) and Belostomatidae (Fig. 12.3D)] possess a caudal respiratory siphon conducting air from

Identification and Ecology of Freshwater Arthropods in the Mediterranean Basin. DOI: https://doi.org/10.1016/B978-0-12-821844-0.00003-X

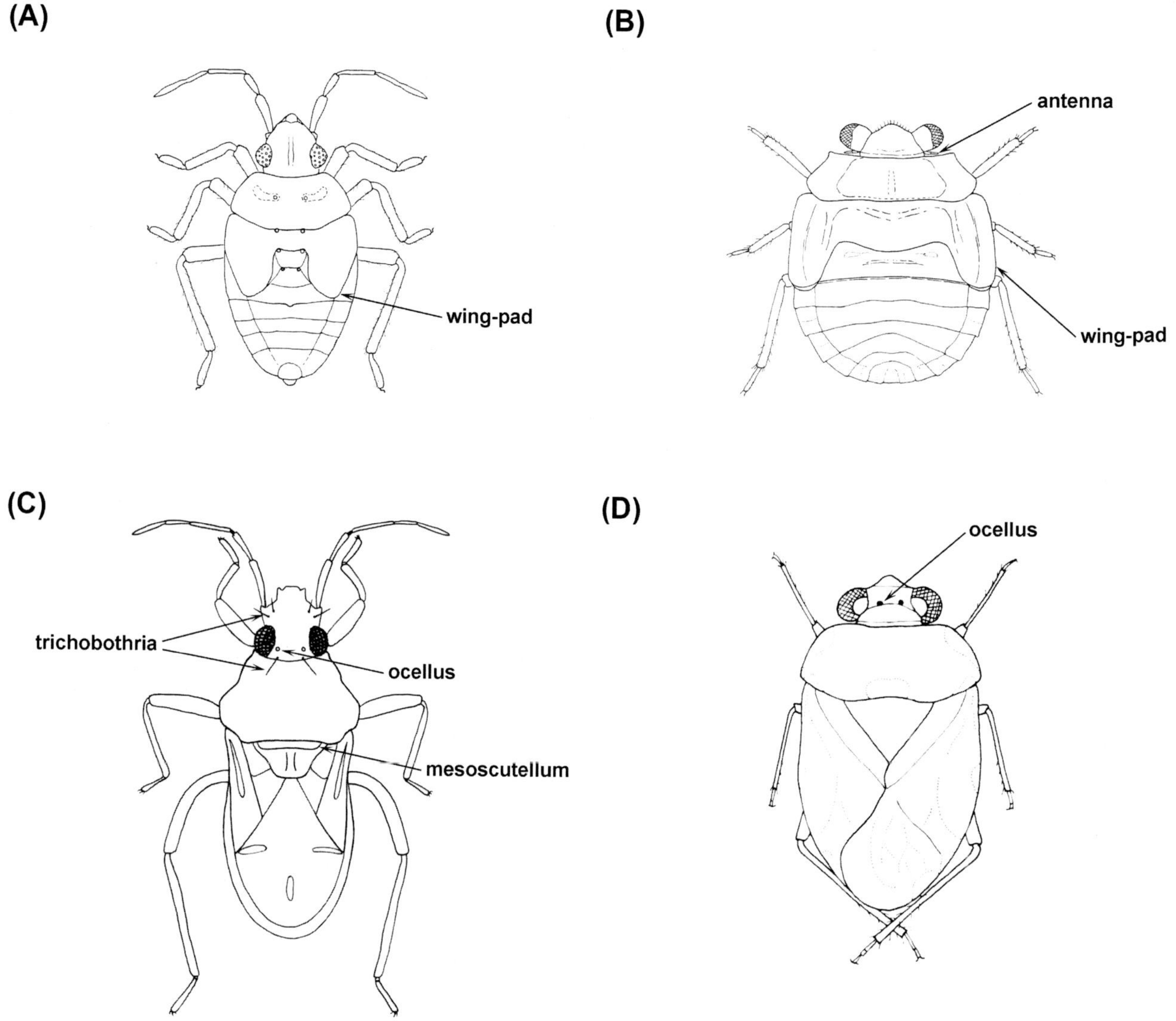

FIGURE 12.1 Habitus of (A) *Hebrus* sp. (Hebridae), V instar nymph; (B) *Ochterus marginatus* (Ochteridae), V instar nymph; (C) *Hebrus pusillus* (Hebridae), adult; (D) *Ochterus marginatus* (Ochteridae), adult. *Reproduced with modifications from (A, B, D) Tamanini (1979). (C) Original drawing.*

the atmosphere to the abdomen, but most of the species directly capture an air bubble on their venter (as some water beetles do). Others [Aphelocheiridae (Fig. 12.3C) and certain Naucoridae (Fig. 12.4A)] possess a true plastron (also present by evolutionary convergence in some aquatic beetles).

All nepormophans are predators of invertebrates and often possess forelegs strengthened and raptorial. The strongest and largest species feed on small vertebrates as well. Some of the Corixidae (Fig. 12.4B) may also digest and utilize plant material including algae. The feeding behavior of Micronectidae (Fig. 12.4C) is practically unknown (Hädicke et al., 2017). Several species have stridulatory structures used to produce sounds and attract partners. Nepomorphans' size ranges from very small (a few millimeters for some Helotrephidae and Pleidae, cf. Fig. 12.4D) to very large (more than 12 cm for some Belostomatidae, cf. Fig. 12.3D) (Tamanini, 1979; Schuh & Slater, 1995; Perez-Goodwyn, 2006; Schuh & Weirauch, 2020).

The Heteropterans of the infraorder Gerromorpha Popov, 1971 or "semiaquatic bugs" are water-surface dwellers and can walk or skate on the water surface of freshwater habitats, or in some cases even on the sea; however, marine species do not occur in the Mediterranean Basin. The oldest member of Gerromorpha (from bona fide fossils) is known from the Late Jurassic (Schuh & Weirauch, 2020), although the origin of this group can be dated back to the Triassic (Wang et al., 2017). Gerromorpha have generally been considered to be more basal than Nepomorpha, since the group

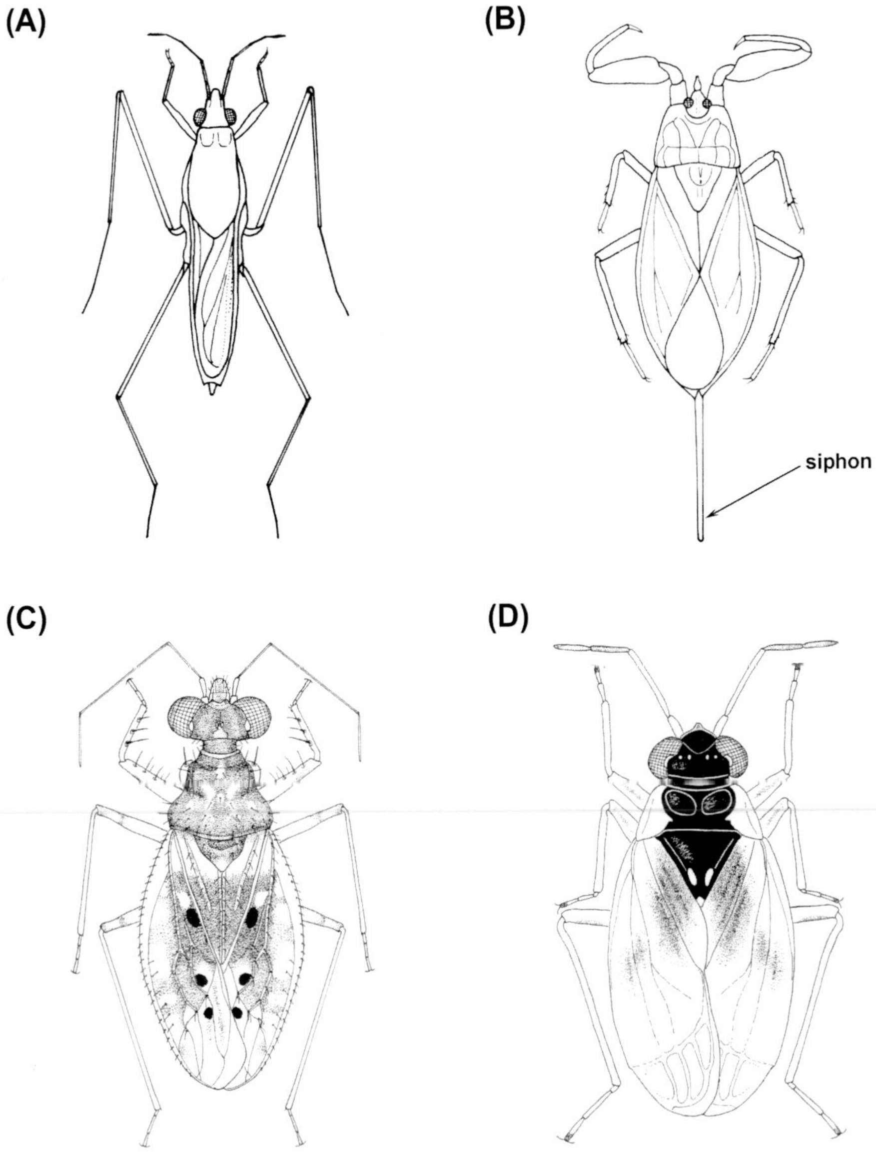

FIGURE 12.2 Habitus of (A) *Gerris* sp. (Gerridae), adult; (B) *Nepa cinerea* (Nepidae), adult; (C) *Leptopus* sp. (Leptopodidae), adult; (D) *Halosalda* sp. (Saldidae), adult. *Reproduced with modifications from Tamanini (1979).*

of Heteroptera to which they belong [with Dipsocoromorpha (Fig. 12.5A), and Enicocephalomorpha] began to diverge from the group that led to Panheteroptera (including Nepomorpha) already in the Permian (Wang et al., 2017; Johnson et al., 2018). Extant Gerromorpha include more than 2100 described species currently divided into eight families but their number, given the paraphyly of Veliidae, is probably bound to increase in the future (Polhemus & Polhemus, 2008; cf. Wang et al., 2017; Schuh & Weirauch, 2020).

Gerromorphans are the most abundant and widespread group among insects inhabiting the upper surface of natural waters (pleuston), for which they compete only with the beetles of the family Gyrinidae. Adaptations to this habitat include specialized pretarsi allowing the claws to be withdrawn into a cleft (Fig. 12.6A), elaborate pretarsal structures like swimming fans (Fig. 12.6B) that help them float in running waters, and the body surface covered by a double layer of cuticular expansions, called microtrichia and macrotrichia (Fig. 12.6C), ensuring a waterproof function. Most are predators and feed on insects and other arthropods, but they may even act as scavengers. No living species of Gerromorphans with stridulatory devices are found in the Mediterranean Basin. Gerromorpha range from very small (1 mm in some "Veliidae" s.l., cf. Fig. 12.5B, and Hebridae, cf. Fig. 12.1C) to medium-large as the Oriental gerrid *Gigantometra gigas* (China) with 3.6 cm of body length and a leg span that may exceed 20 cm (Andersen, 1977, 1982; Tamanini, 1979; Polhemus, 1994; Schuh & Slater, 1995; Hu et al., 2003; Perez-Goodwyn, 2009; Schuh & Weirauch, 2020).

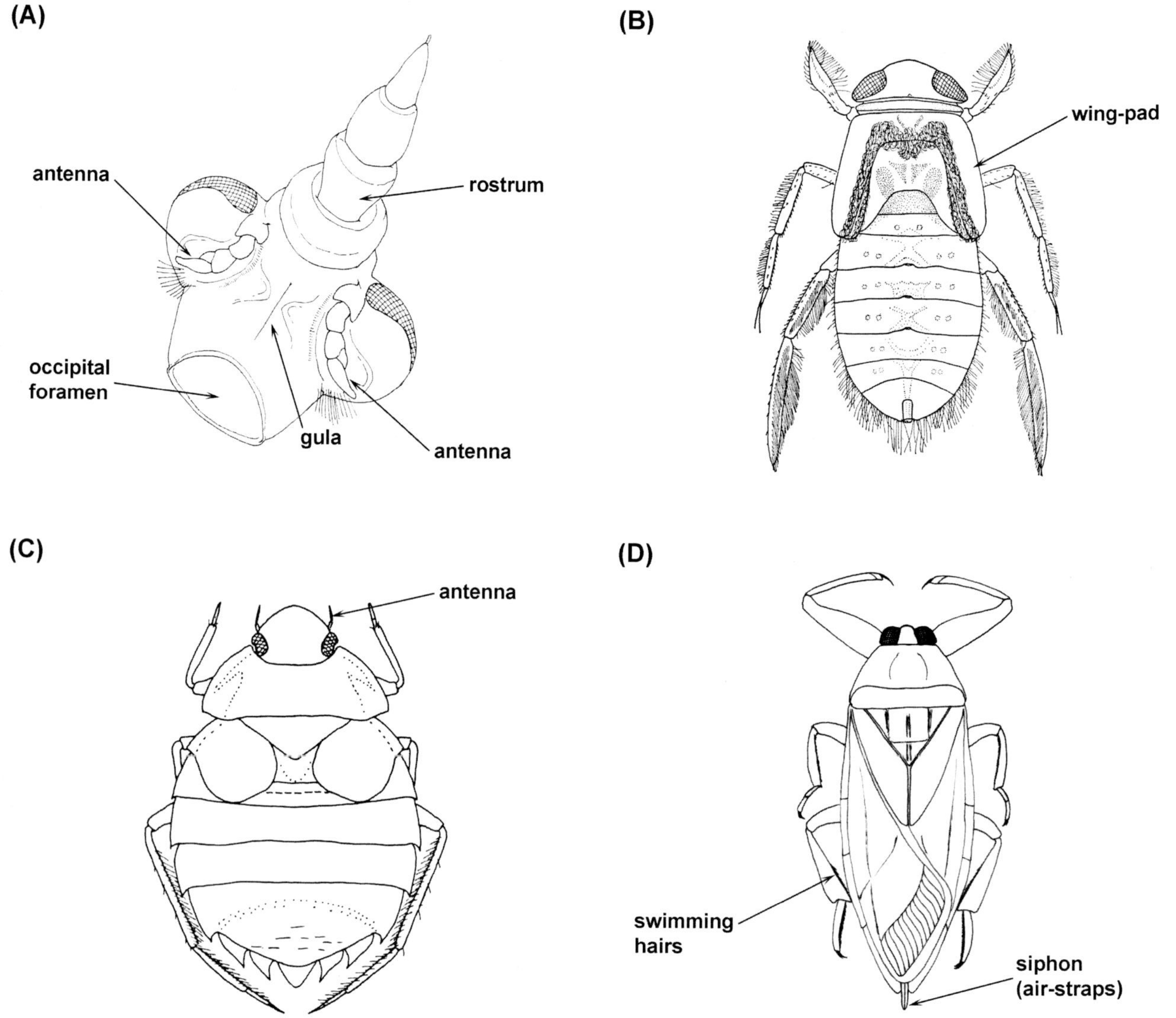

FIGURE 12.3 (A) Head of *Nepa cinerea* (Nepidae), seen from below. Habitus of (B) *Corixa punctata* (Corixidae), V instar nymph; (C) *Aphelocheirus aestivalis* (Aphelocheiridae), adult, brachypterous morph; (D) *Lethocerus patruelis* (Belostomatidae), adult. *Reproduced with modifications from (A, B, C) Tamanini (1979). (D) Original drawing.*

General ecology and distribution

Among the Nepomorpha living in the Mediterranean Basin, most of Nepidae (Fig. 12.2B) inhabit still waters of streams, rivers, springs, channels, marshes, etc. where they "crawl" (poor swimmers) or stay clung on vegetation waiting for their prey. They rarely fly.

Belostomatidae (Fig. 12.3D) live mainly in standing waters, but can also be found in slow-flowing portions of streams and rivers. They show male parental care; and at certain times, males can be found out of the water, on vegetation, guarding their eggs (in the Lethocerinae: *Lethocerus*) or exposing them to atmospheric air (in Belostomatinae, i.e., all the other genera found in the Mediterranean Basin, in which females glue their eggs on the back of males). All Belostomatidae are strong flyers and are particularly attracted to artificial lights (in fact they are known also as "electric light bugs").

Ochteridae (Fig. 12.1D) are the only Mediterranean nepomorphans not living underwater. The members belonging to the genus *Ochterus* (the only ones occurring in the considered area) can be found on the banks of streams and rivers. They have excellent eyesight and fly quickly, although not far when approached.

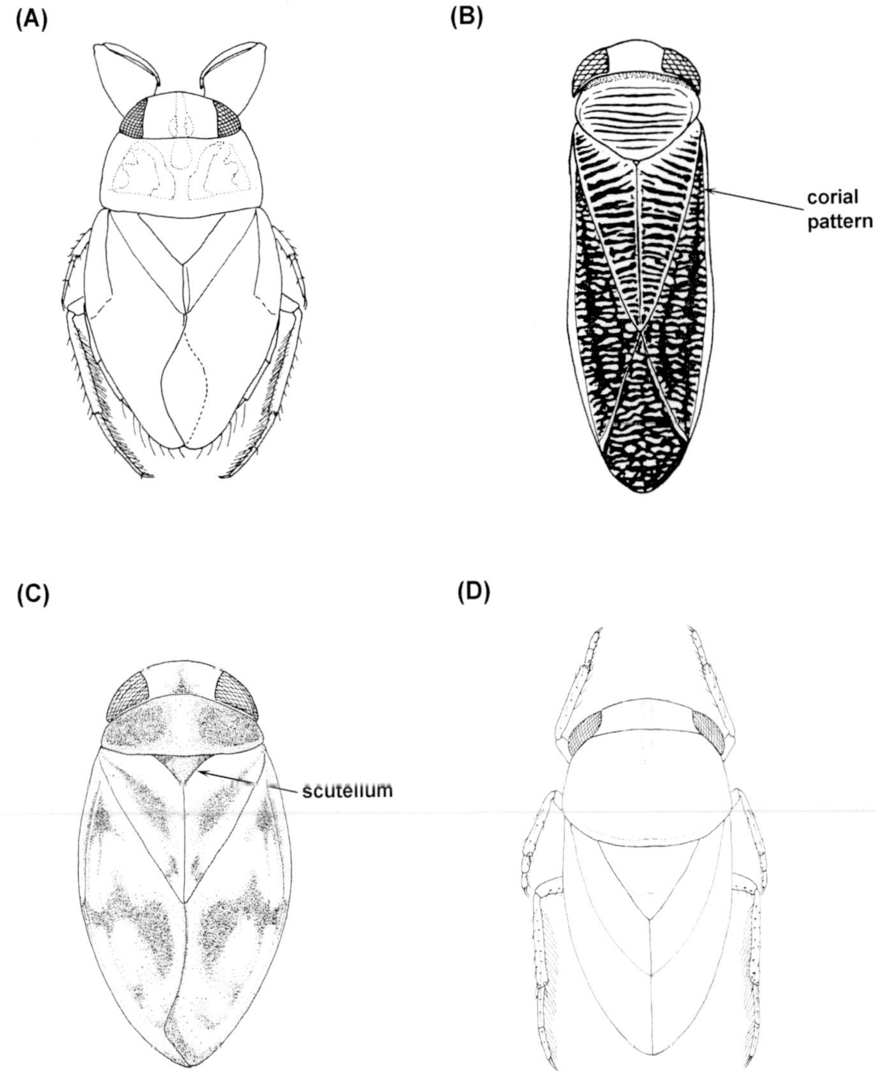

FIGURE 12.4 Habitus of (A) *Naucoris maculatus conspersus* (Naucoridae), adult; (B) *Sigara (Retrocorixa) limitata*, adult; (C) *Micronecta (Micronecta) poweri* (Micronectidae), adult, brachypterous morph; (D) *Plea minutissima* (Pleidae), adult. *Reproduced with modifications from Tamanini (1979).*

Corixidae (Fig. 12.4B) occupy a relatively wide range of habitats in the Mediterranean Basin: from seaside brackish waters to high mountain lakes. They live in both lentic and lotic waters (but in slow-flowing zones). Once they have captured a bubble of atmospheric air, they swim quickly toward the bottom or among the aquatic plants where they remain motionless.

Similarly, Micronectidae (Fig. 12.4C, Corixoidea) live in lakes, ponds, and in quiet stretches of rivers and streams.

Aphelocheiridae (Fig. 12.3C) are part of the benthos and inhabit the bottom of streams and rivers. Since they breathe through a plastron that allows them to extract oxygen directly from the water, they remain permanently submerged.

Naucoridae (Fig. 12.4A) live in still or very slow-flowing waters, rich in aquatic plants.

Likewise, Notonectidae (Fig. 12.7A–D) live in different aquatic environments characterized by still or calm waters in rivers, streams, lakes, and ponds, but they are also found in artificial habitats such as swimming pools and water reservoirs. Most species require vegetation since they lay their eggs in plant tissues; a species of the genus *Notonecta* (*N. maculata* Fabricius) is able to glue the eggs directly on the stone surface. Some species, like *N. viridis* Delcourt and *Anisops* spp., occur even in brackish waters. All notonectids swim on their backs (called "backswimmers"). Members of the subfamily Notonectinae (Fig. 12.7A, C, and D) cannot regulate their air storage and tend to float upwards when not actively swimming, and they have to come often to the surface to replenish their air reserves. Anisopinae (genus *Anisops*, Fig. 12.7B) have hemoglobin cells located anteriorly in their abdomens which enables them to store oxygen reserves during dives (Nieser, 2004). In contrast with the Notonectinae, they can regulate the amount of air in order to obtain neutral buoyancy (they are amongst the few truly planktonic insects).

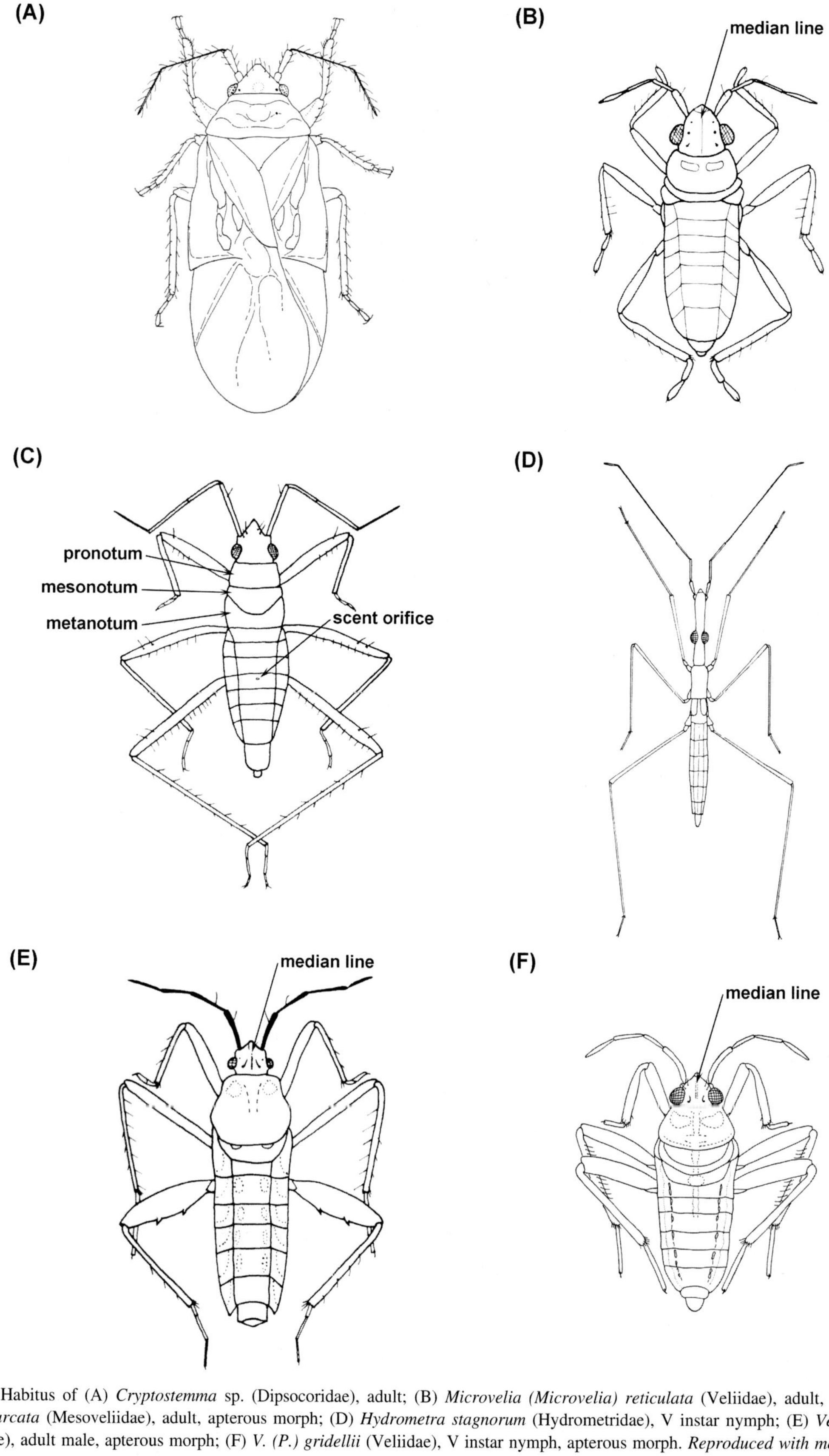

FIGURE 12.5 Habitus of (A) *Cryptostemma* sp. (Dipsocoridae), adult; (B) *Microvelia (Microvelia) reticulata* (Veliidae), adult, apterous morph; (C) *Mesovelia furcata* (Mesoveliidae), adult, apterous morph; (D) *Hydrometra stagnorum* (Hydrometridae), V instar nymph; (E) *Velia (Plesiovelia) currens* (Veliidae), adult male, apterous morph; (F) *V. (P.) gridellii* (Veliidae), V instar nymph, apterous morph. *Reproduced with modifications from Tamanini (1979).*

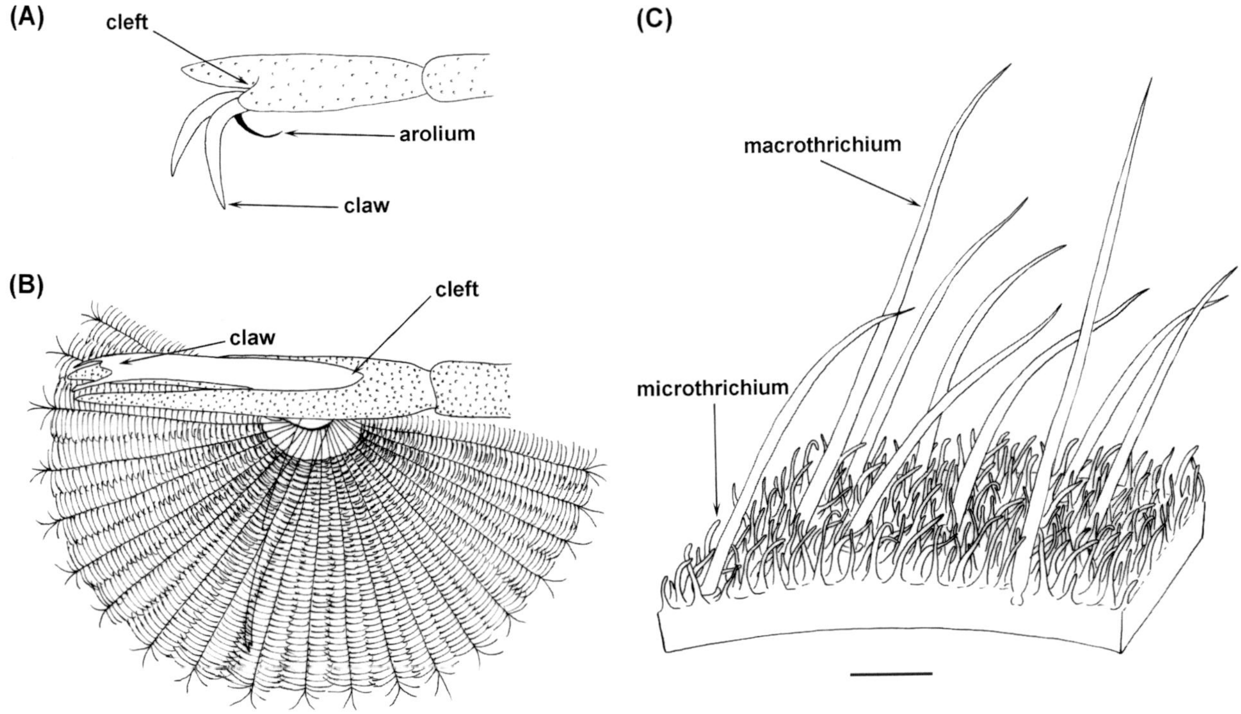

FIGURE 12.6 Tarsal structure of (A) *Microvelia (Microvelia) reticulata* (Veliidae) (B) *Rhagovelia* sp. (Veliidae). (C) Hair layer of *Gerris lacustris* (Gerridae), from mesosternum (Scale bar: 10 μm). *Redrawn with modifications from (A, B) Andersen (1982) and (C) Andersen (1977).*

Pleidae (Figs 12.4D and 12.14E) swim with their venters up like the Notonectidae. They require dense aquatic vegetation and quiet-water habitats (Poisson, 1957; Tamanini, 1979; Schuh & Slater, 1995; Chen et al., 2005; Schuh & Weirauch, 2020).

Members of the gerromorphan family Mesoveliidae (i.e., *Mesovelia* in the Mediterranean Basin; Fig. 12.5C) inhabit the surface of ponds, lakes, and slow-flowing portions of streams and rivers, rich in aquatic plants with emerging or floating parts, into the tissue of which the females insert their eggs.

Hebridae (i.e., *Hebrus* in the Mediterranean Basin; Fig. 12.1C) live around vegetated margins of ponds, marshes, lakes, peat bogs, and seepages, but also along the edges of streams and rivers. These small insects may be spotted walking on the surface of the water, often on floating vegetation, but frequently they stay on the shores hidden under plants, litter, and debris on the shores of water bodies, or deep in mats of moss and interstices. In these cases, they can be collected by sieving the substrate.

Hydrometridae (i.e., *Hydrometra* in the Mediterranean Basin; Fig. 12.5D) can be found on the surface or along the banks of still or running water bodies, walking on stones and riparian vegetation.

Members of "Veliidae" live on or near standing or flowing water. Most taxa of Microveliinae (Fig. 12.5B) live on the surface or the shores of ponds, lakes, marshes, and slow-flowing portions of streams and rivers (usually shaded). Members of the genus *Velia* (Fig. 12.5E and F; Veliinae) are more exclusive to streams, where they often stay in quiet loops, or pools where they skate on the surface or hide on the banks. Species of *Rhagovelia* species (Rhagoveliinae) are adapted to live on the surface of flowing water. They have swimming fans located on the pretarsi of the middle legs (Fig. 12.6B).

Gerridae (Fig. 12.2A) live on the surface of standing and flowing water. There are species occurring only in still waters, such as lakes, pools, swamps, and quiet portions of streams and rivers, and others living exclusively on running water where; however, they occupy the slow-flowing areas of torrents and streams (Tamanini, 1979; Schuh & Slater, 1995; Schuh & Weirauch, 2020; pers. observ.).

Morphological characteristics needed in identification

Morphological terms used for the identification of aquatic Heteroptera vary somewhat among families (peculiar characters are explained and illustrated directly in the dichotomous keys), but most of the terminology is used in common

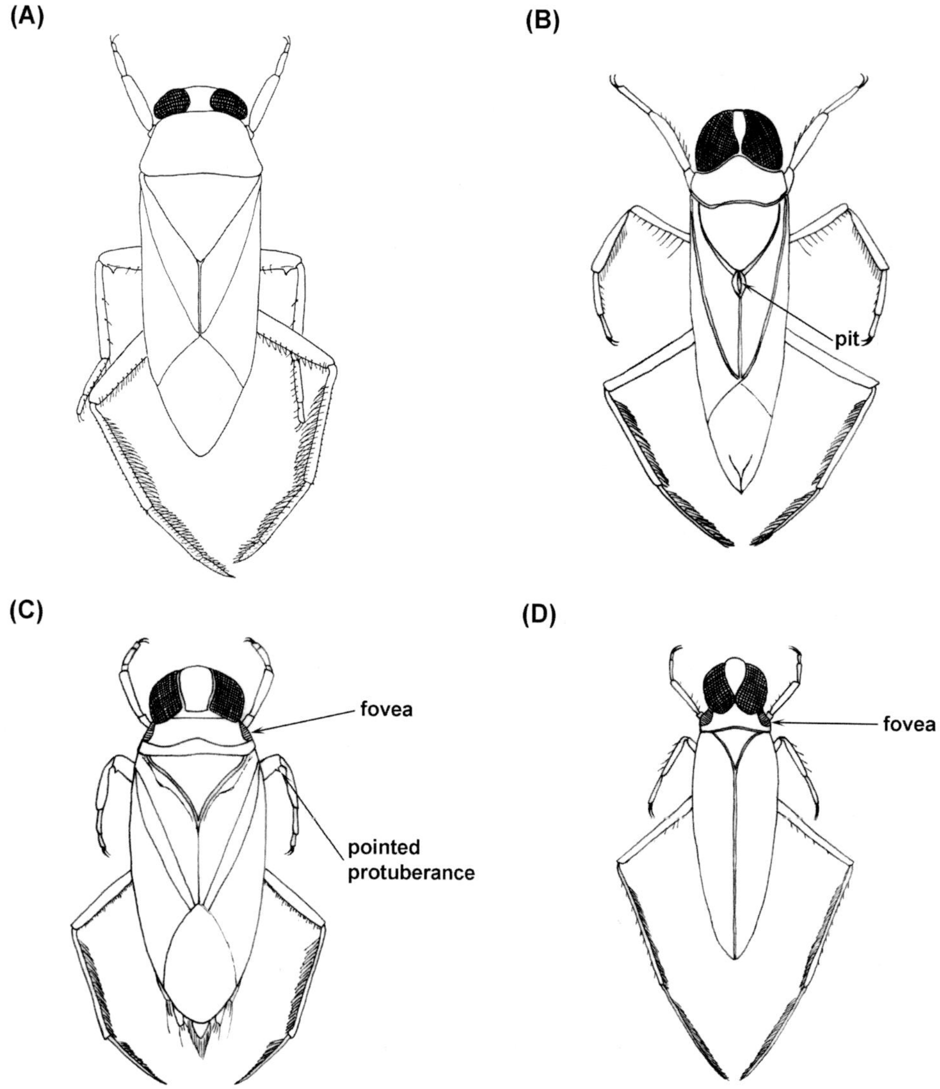

FIGURE 12.7 Habitus of (A) *Notonecta* sp. (Notonectidae, Notonectinae); (B) *Anisops* sp. (Notonectidae, Anisopinae); (C) *Enithares* sp. (Notonectidae, Notonectinae); (D) *Nychia* sp. (Notonectidae, Notonectinae). *Reproduced with modifications from (A) Tamanini (1979). (B–D) Redrawn with modifications from Chen et al. (2005).*

with other insects. Examples of the main morphological parts of Gerromorpha and Nepomorpha are presented in Fig. 12.8A and B.

The keys are valid for adults only. The immature stages (nymphs) are distinguished mainly by the lack of ocelli, genitalia, and wings (Fig. 12.1A–D). The latest instars (III to V) of the nymphs of macropterous morphs develop wing pads (Fig. 12.1A and B, not present in apterous morphs of Gerromorpha, Fig. 12.5F). The first instars (I–III) have no wing pads.

Literature and identification keys (especially for the Mediterranean Basin) for immatures aquatic Heteroptera are mainly limited to the last instars and restricted to some geographical areas: for example, on Gerridae (e.g., Brinkhurst, 1959 for Great Britain; Poisson, 1957 for France; Vepsäläinen & Krajewski, 1986 for Northern Europe) or some on Corixoidea (e.g., Cobben & Pillot, 1960 for the Netherlands; Jansson, 1969 for Northern Europe; Savage, 1999 for Great Britain).

Cobben (1968) provided a survey of heteropteran eggs, including examples for the aquatic groups (Fig. 12.9A and B).

An attempt was made in this chapter to make the keys usable for both sexes, but in some cases, a reliable recognition is possible only based on male specimens. In particular, in order to distinguish the genera of Corixidae (Fig. 12.4B) and the subgenera of *Micronecta* (Fig. 12.4C), *Microvelia* (Fig. 12.5B), and *Velia* (Fig. 12.5E), it is necessary to examine the male genitalia. In some cases, it is useful to extract the endophallus (Fig. 12.9D) from the genital capsule (Fig. 12.5C).

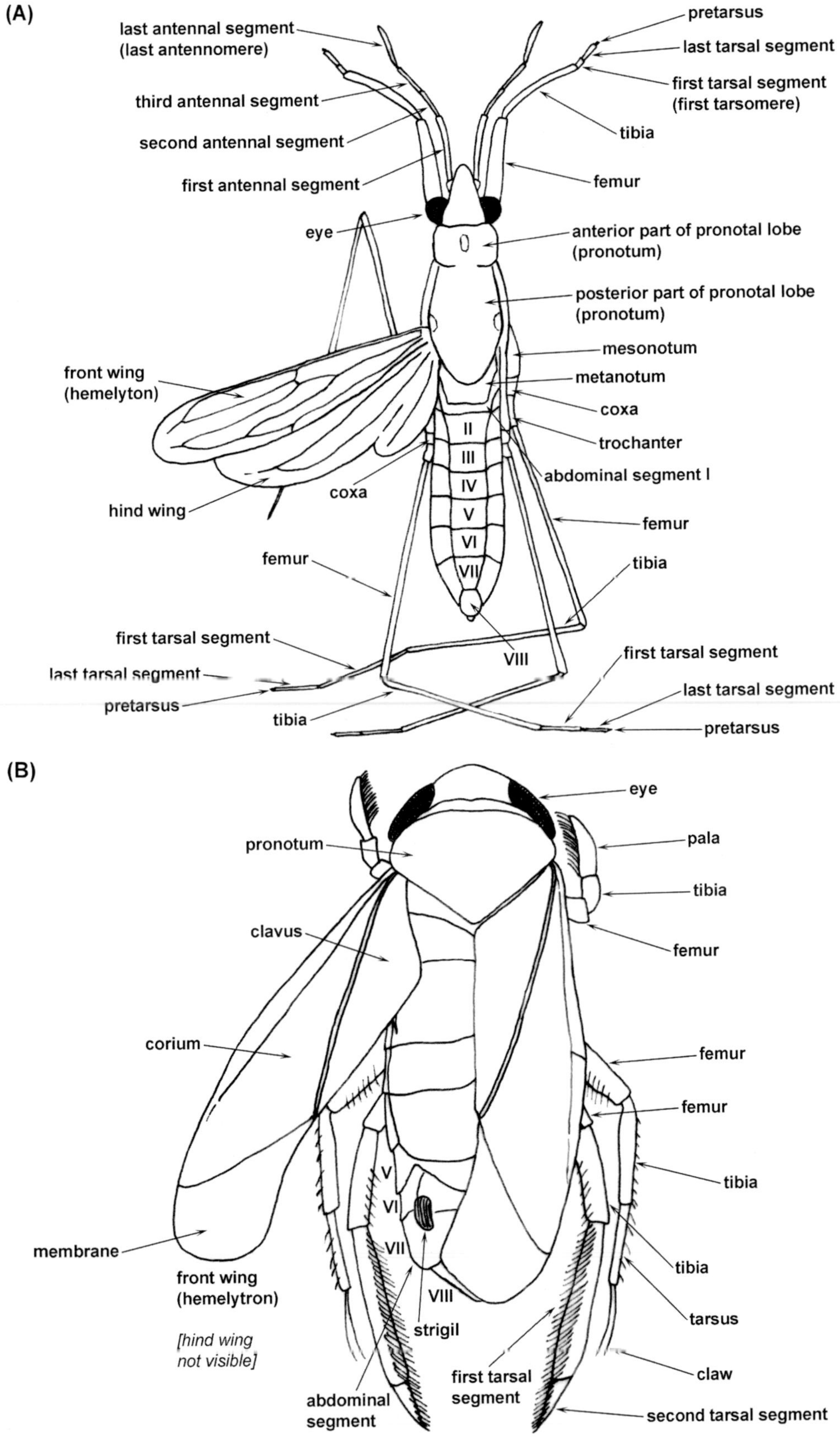

FIGURE 12.8 Dorsal view of a member of (A) Gerromorpha (*Gerris lacustris*, Gerridae); (B) Nepomorpha (*Corixa punctata*, Corixidae). *Redrawn with modifications from Tamanini (1979).*

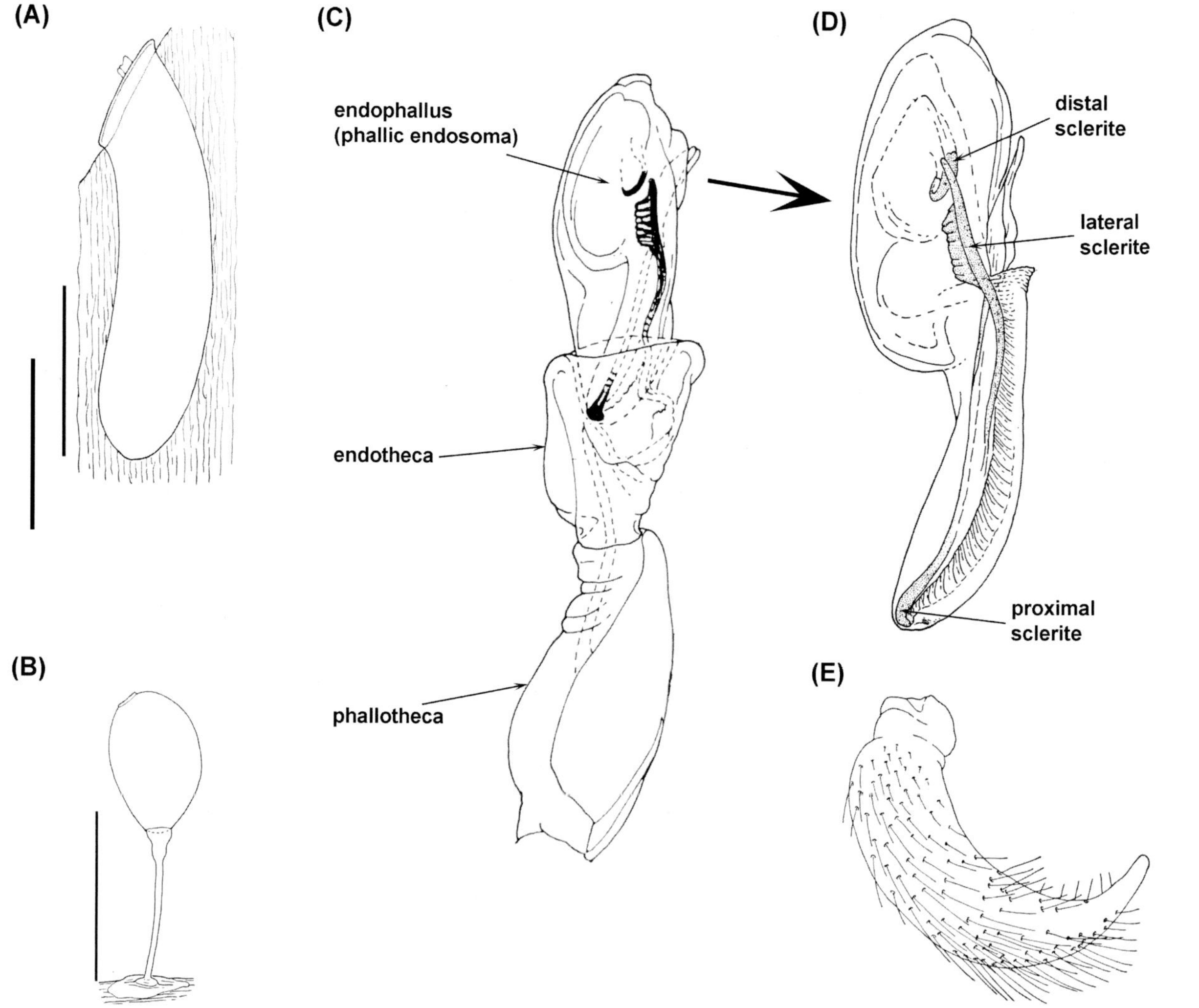

FIGURE 12.9 Egg of (A) *Ilyocoris cimicoides* (Naucoridae); (B) *Cymatia coleoptrata* (Corixidae). (C) Everted phallus and (D) endophallus of *Velia (Plesiovelia) currens* (Veliidae). (E) Paramere of *V. (P.) gridellii* (Veliidae). *Reproduced with modifications from Tamanini (1979) (A, B from original drawings by R. Cobben).*

The identification of isolated female specimens is often challenging and for certain groups only a few contributions based on female features are available. An example is the dichotomous key for the German Corixidae provided by Höregott & Jordan (1954).

In the keys, some exclusively montane taxa have also been included since they may fall within the bioclimatic limit of the Mediterranean Basin.

Material preparation and preservation

Preservation methods of aquatic Heteroptera depend on the intended use of the specimens. All stages (both adult and immature) can be directly collected and preserved in 70% ethanol. This is probably the most convenient method but in this way dissection of adult specimens (very often required for specific and sometimes also generic/subgeneric identification) can be more difficult due to the rigidity that the specimens acquire by remaining in alcohol. This often also prevents a satisfactory subsequent dry preparation of adults (just from a purely esthetic point of view, due often to the impossibility of mounting the appendages of the specimens symmetrically on entomological cards). For some groups [Corixoidea, i.e., Corixidae (Fig. 12.4B), and Micronectidae (Fig. 12.4C), in the considered area], adult specimens preserved in alcohol are not easy to subsequently be mounted dry in a satisfactory way, because the forewings (hemelytra) coarctate completely and it is, therefore, preferable to keep them in liquid. The nymphs must be stored in liquid to avoid coarctation, preferably in 70% ethanol.

Specimens collected and stored in pure ethanol—preferable for molecular sequencing and other kinds of analysis—can also be studied but with even greater difficulty due to the extreme hardness of the material.

For a detailed morphological study and optimal dry mounting, the specimens can be collected in containers with a few drops of ethyl acetate (or acetic ether) in shredded cork, wood, or absorbent paper. The paper allows small specimens to be found more quickly, but this volatile liquid tends to evaporate in a shorter time than when retained by cork and the risk of mold growth is higher. The ethyl acetate can however be added again before dry mounting of the specimens.

The dissection of specimens can be done with fine or dissection pins under a stereo microscope (or dissecting microscope). The dissected parts (genitalia, appendages, etc.) should be glued on the same card with the specimen or stored in small micro vials with a drop of glycerol attached to the same pin. However, in some cases, some morphological parts [parameres (Fig. 12.9E), endophallus (Fig. 12.9C and D), etc.] need to be examined under a classic compound microscope or biological microscope (capable of greater magnification) and require to be placed on glass slides, where they can be observed or fixed permanently.

Keys to Hemiptera

The systematic order is based on the phylogenies by Wang et al. (2017, 2021).

The keys are generally valid for adult specimens only (cf. Fig. 12.1C and D). The first subdivision (suborders) is also applicable to immature stages (cf. Fig. 12.1A and B). Any differences between males and females or the applicability of the key to only one sex is specified.

Hemiptera: suborders

This key is modified from Tamanini (1979), Dolling (1991), and Chen et al. (2005).

1 Ventral part of the head behind rostrum closed by a sclerotized part forming a distinct *gula*; rostrum originating from the anterior half of the head (Figs 12.3A and 12.10A and B) .... **Heteroptera**

1' Ventral part of the head behind rostrum not closed by a sclerotized part (*gula* absent); rostrum originating from the posterior half of the head (Fig. 12.10C) .... other terrestrial Hemiptera

Hemiptera: Heteroptera: infraorders

This key is modified from Andersen & Weir (2004) and Chen et al. (2005).

1 Antennae longer than head, inserted in front of eyes, with first segments clearly visible from above (Fig. 12.5B–F) 2

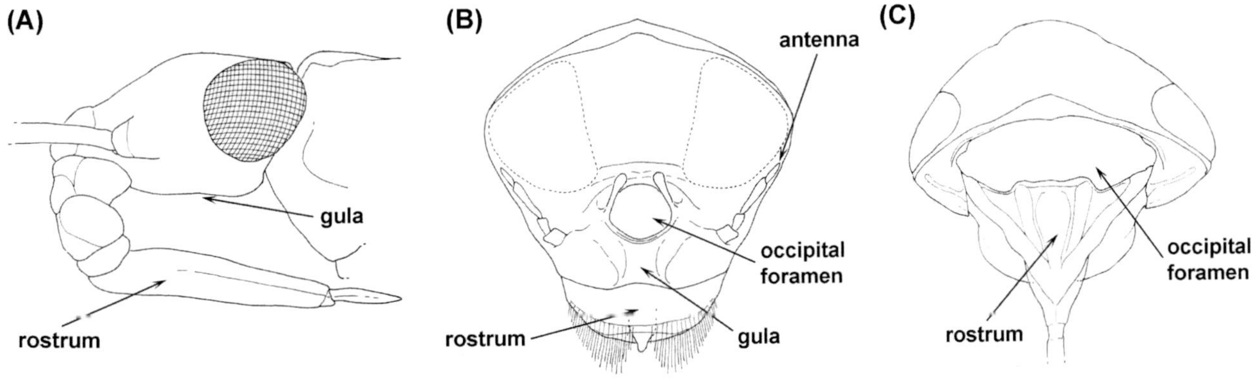

FIGURE 12.10 Head of a member of (A) Heteroptera Gerromorpha (*Gerris costae*, Gerridae), lateral view; (B) Heteroptera Nepomorpha (*Corixa punctata*, Corixidae), posterior view; (C) Hemiptera non-Heteroptera (Cicadomorpha, Aphrophoridae), posterior view. *Reproduced with modifications from Tamanini (1979).*

1' Antennae shorter than head and hidden (usually completely folded beneath eyes; Fig. 12.3A), in most cases not visible from above (Fig. 12.4D). In some families (Ochteridae and Aphelocheiridae) apex of antennae could be exposed beyond margin of large eyes; however, the first segments are not visible in dorsal view (Figs 12.1B and 12.3D) .... **Nepomorpha**
2(1) Head with at least three pairs of conspicuous trichobothria (appearing as long and slender setae inserted in deep pits) located near inner margin of eyes (Fig. 12.11A and B) .... **Gerromorpha**
2' Head without trichobothria, or if one or more pairs of setae (macrosetae) are present, these never inserted in deep pits (Figs 12.2C and D and 12.5A) .... remaining Hemiptera Heteroptera (including Dipsocoromorpha: minute litter bugs and Leptopodomorpha: shore bugs and allies).

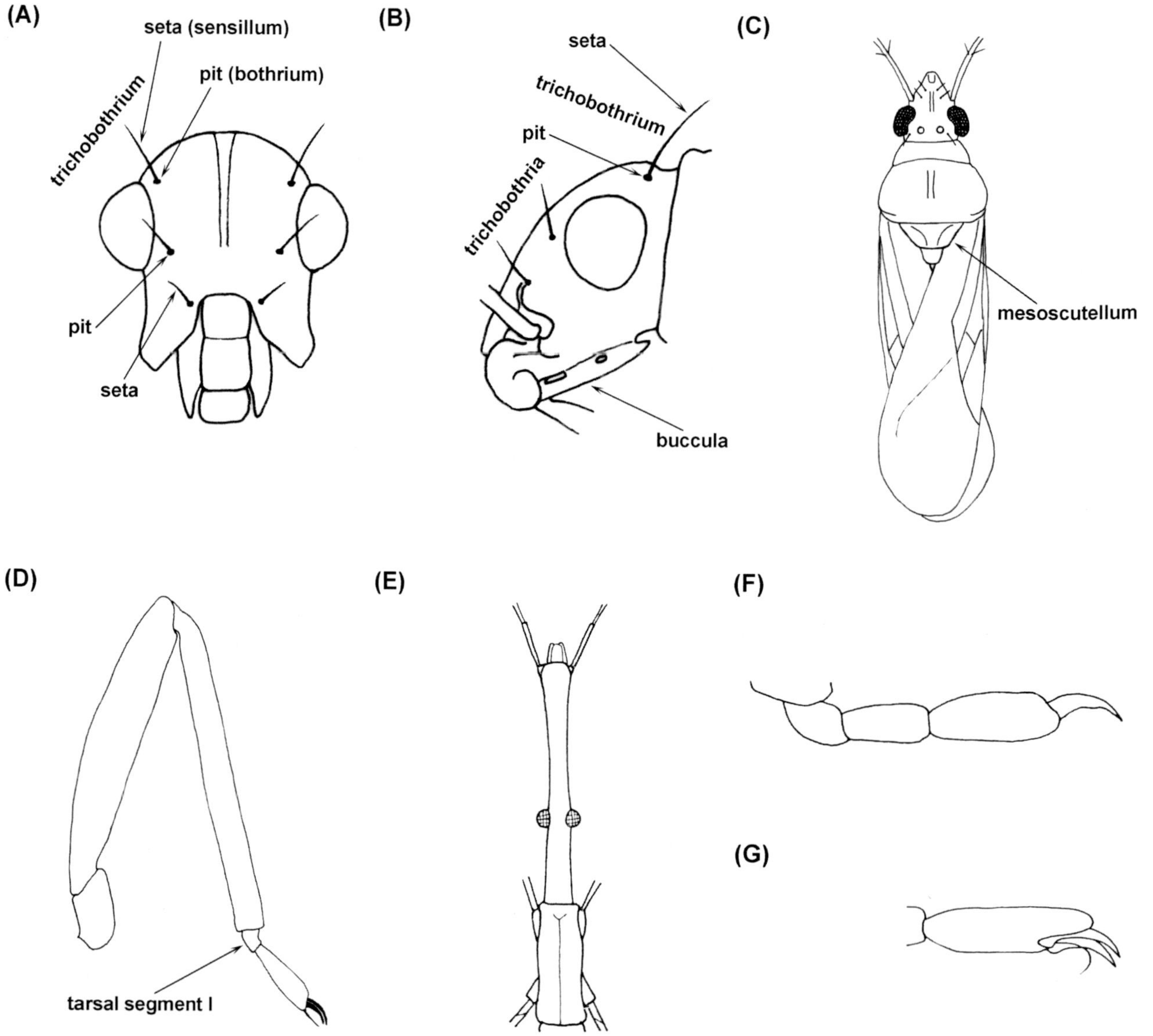

FIGURE 12.11 Head in (A) frontal and (B) lateral view of *Hebrus franzi* (Hebridae). (C) Habitus (legs and antennae partially removed) of *Mesovelia vittigera* (Mesoveliidae), macropterous female. (D) Hind leg of *Hebrus ruficeps* (Hebridae). (E) Head and pronotum of *Hydrometra stagnorum* (Hydrometridae). (F) Hind tarsus of *Mesovelia furcata* (Mesoveliidae). (G) Last tarsal segment of *Velia (Plesiovelia) currens* (Veliidae). *Redrawn with modifications from (A, B) Cianferoni et al. (2016) (from original drawings by H. Zettel); (C) Poisson (1933). (D–G) Reproduced with modifications from Tamanini (1979).*

Hemiptera: Heteroptera: Gerromorpha: families

This key is modified from Tamanini (1979), Andersen (1982), Andersen & Weir (2004), and Chen et al. (2005).

1 Macropterous (wings fully exposed; Figs 12.1C and 12.2A) 2
1' Apterous (wingless) or meiopterous (wings reduced; Fig. 12.5B, C, and E) 6
2(1) Mesoscutellum exposed, forming a subtriangular (Fig. 12.11C) to transverse (Fig. 12.1C) plate behind the pronotal lobe 3
2' Mesoscutellum not exposed, covered by the pronotal lobe posteriad (Fig. 12.5B and E) 4
3(2) Head ventrally with a pair of prominent, vertical plates (bucculae) covering the base of rostrum (Fig. 12.11B). Tarsi two-segmented, first segment very short (Fig. 12.11D). Habitus: cf. Fig. 12.1C .... Hebridae, one genus: *Hebrus*
3' Head ventrally lacking bucculae. Tarsi three-segmented. Habitus: cf. Fig. 12.11C .... Mesoveliidae, one genus: *Mesovelia*
4(2') Head distinctly prolonged behind the eyes, length of postocular part definitely longer than two times the diameter of one eye (Fig. 12.11E). Habitus: cf. Fig. 12.5D .... Hydrometridae, one genus: *Hydrometra*
4' Head not elongated behind the eyes, eyes situated close to the hind margin of head (Figs 12.2A and 12.5B). Habitus different (cf. Figs 12.2A and 12.5E and F) 5
5(4') Head with distinct longitudinal median impressed line on dorsal surface (Fig. 12.5B, E, and F). Habitus: cf. Fig. 12.5E and F .... **Veliidae**
5' Head without a median impressed line on dorsal surface. Habitus: cf. Fig. 12.2A .... **Gerridae**
6(1') Claws of middle and hind legs inserted apically on last tarsal segment (Fig. 12.11F). Abdominal tergum IV with a scent orifice (Fig. 12.5C) 7
6' Claws of middle and hind legs inserted distinctly before apex of last tarsal segment (Figs 12.11G and 12.6A and B). Abdominal tergum IV without a scent orifice (Fig. 12.5B and E) 9
7(6) Head distinctly prolonged behind the eyes, length of postocular part definitely longer than two times the diameter of one eye (Fig. 12.11E). Habitus: cf. Fig. 12.5D .... Hydrometridae, one genus: *Hydrometra*
7' Head not elongated behind the eyes, eyes situated at the hind margin of head (Fig. 12.5B) 8
8(7') Pronotum very reduced, exposing both meso- and metanotum (Fig. 12.5C). Tarsi three-segmented (Fig. 12.11F) .... Mesoveliidae, Habitus: cf. Fig. 12.5C, one genus: *Mesovelia*
8' Pronotum at least covering mesonotum (Fig. 12.1C). Head ventrally with a pair of prominent, vertical plates (bucculae) covering the base of rostrum (Fig. 12.11B). Tarsi two-segmented, first segment very short (Fig. 12.11D) .... Hebridae, one genus: *Hebrus*[1]
9(6') Head with distinct longitudinal median impressed line on dorsal surface (Fig. 12.5B, E, and F) .... **Veliidae**
9' Head without a median impressed line on dorsal surface (Fig. 12.2A) .... **Gerridae**

[1] The subgeneric division of the genus *Hebrus* Curtis, 1833 needs further study world-wide, using phylogenetic tools; thus, I do not consider a subgeneric assignment here (see also Kanyukova, 1997; Cianferoni et al., 2016; and Kment et al., 2016 for further information).

Hemiptera: Heteroptera: Gerromorpha: Veliidae: genera

This key is based on Linnavuori (1977) and Andersen (1982).

1 Middle tarsi deeply cleft, with plumose swimming fans arising from base of the cleft (easily detectable in liquid) (Fig. 12.6B) .... Rhagoveliinae: *Rhagovelia*[1] [Western, southern, and eastern Mediterranean Basin]
1' Middle tarsi not deeply cleft and without plumose swimming fans (Fig. 12.6A) 2
2(1') Middle tarsi two-segmented .... 3 (Microveliinae)
2' Middle tarsi three-segmented .... **Veliinae: *Velia*** [Meditterranean Basin]
3(2) Body length (of adults) more than 5 mm. First antennal segment slender and very long, at least 3/4 the width of head between eyes .... *Tenagovelia*, one species: *T. sjostedti* Kirkaldy, 1908 [Southern Mediterranean Basin: North Africa[2]]
3' Body length (of adults) less than 3 mm. First antennal segment shorter and stouter, less than 3/4 the width of head between eyes 4

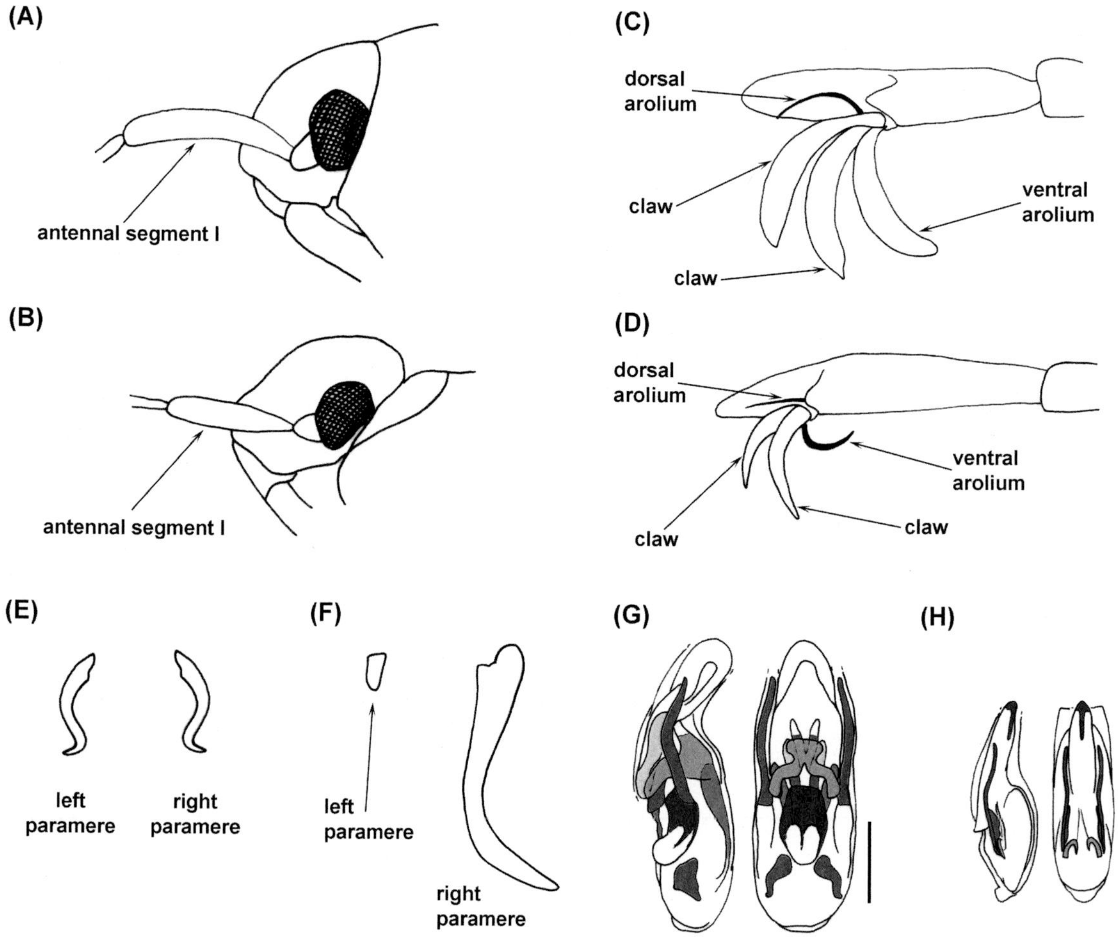

FIGURE 12.12 Head of (A) *Xiphoveloidea major* ssp. and (B) *Microvelia* sp. in lateral view. Tarsal structure of (C) *Xiphoveloidea major* ssp. and (D) *Microvelia (Microvelia) reticulata* in ventral view. Parameres of (E) *Microvelia* (s.str.) *reticulata* and (F) *M. (Picaultia) pygmaea*. Endophallus of (G) *Velia (Velia) rivulorum* and (H) *Velia (Plesiovelia) currens* in lateral and dorsal view. *Redrawn with modifications from (A, B) Linnavuori (1977); (C, D) Andersen (1982); (F) Poisson (1957); (G, H) Tamanini (1947). (E) Original drawing, based on Poisson (1957).*

4(3') First antennal segment incrassate, extending more than 2/3 of its length beyond apex of head (Fig. 12.12A). Last tarsal segment of middle tarsi with three leaflike/bladelike structures subapically (two claws and modified ventral arolium) (Fig. 12.12C) ....
...... *Xiphoveloidea*, one species, nominotypical subspecies: *X. major major* (Poisson, 1926) [Southern Mediterranean Basin: North Africa]

4' First antennal segment not incrassate, extending less than 2/3 of its length beyond apex of head (Fig. 12.12B). Last tarsal segment of middle tarsi without structures as above, with gracile claws and unmodified ventral arolium (two claws and filiform arolium) (Fig. 12.12D) .... ***Microvelia*** [Mediterranean Basin]

[1]All the species from the Mediterranean Basin belongs to the nominotypical subgenus.

[2]The occurrence of this taxon in the Mediterranean Basin *s.l.* remains to be verified.

Hemiptera: Heteroptera: Gerromorpha: Veliidae: *Microvelia*: subgenera

Part of this key is based on Andersen & Weir (2003).

1 Male genitalia symmetrical, parameres subequal in size (Fig. 12.12E) .... *Microvelia (Microvelia)*

1' Male genitalia asymmetrical, right paramere much larger than left paramere (Fig. 12.12F) ....
.... *Microvelia (Picaultia)*

Hemiptera: Heteroptera: Gerromorpha: Veliidae: *Velia*: subgenera

This key is based on Tamanini (1955), Andersen (1981), and Berchi et al. (2017).

1 Hind tibiae shorter than hind femora. Hind trochanter of male with at least one large tooth and several smaller teeth. Endophallus (see Fig. 12.9C and D) with more than 10 vesical sclerites (Fig. 12.12G). Macropterous or brachypterous morphs .... *Velia (Velia)*, one species: *V. (V.) rivulorum* (Fabricius, 1775)

1' Hind tibiae longer than hind femora. Hind trochanter of male with several small teeth, subequal in size. Endophallus (see Fig. 12.9C and D) with 4–9 vesical sclerites (Fig. 12.12H). Macropterous or apterous morphs .... *Velia (Plesiovelia)*

Hemiptera: Heteroptera: Gerromorpha: Gerridae: genera

This key is based on Poisson (1957), Tamanini (1979), and Andersen, (1982, 1990).

1 Middle femora shorter than middle tibiae .... *Naboandelus* [Southern and eastern Mediterranean Basin: North Africa and Middle East]

1' Middle femora longer than middle tibiae 2

2(1') Dorsal surface of the head dark with evident longitudinal light stripes or elongated spots extending also anteriad (Fig. 12.13A and B) .... *Limnogonus*, one species: *L. (Limnogonus) cereiventris* (Signoret, 1862) [Southern and eastern Mediterranean Basin: North Africa and Middle East]

2' Dorsal surface of the head almost uniformly dark, sometimes with light spots limited to the hind margin of the head (near pronotum) or close to the eyes (Fig. 12.13C) 3

3(2') Antennae longer than one-half body length. Hind femora longer than middle femora .... *Limnoporus* [Mediterranean Europe]

3' Antennae shorter than one-half body length. Hind femora subequal or shorter than middle femora 4

4(3') First antennal segment longer than second and third segment together (Fig. 12.13D). Fore femora uniformly dark. Abdominal segment VII ending with well-developed connexival spines (Fig. 12.13F and G) .... *Aquarius* [Mediterranean Basin]

4' First antennal segment subequal to or shorter than second and third segment together (Fig. 12.13E). Fore femora light with more or less extended dark stripes (Fig. 12.13J). Abdominal segment VII ending with a more or less acute angles but without spines (Fig. 12.13H and I) .... ***Gerris*** [Mediterranean Basin]

Hemiptera: Heteroptera: Gerromorpha: Veliidae: *Gerris*: subgenera

This key is modified from Andersen (1993).

1 Posterior part of pronotum (pronotal lobe) rufous (reddish-brown). Posterior margin of male abdominal sternum VII without median emargination (Fig. 12.13L) .... *Gerriselloides*

1' Posterior part of pronotum (pronotal lobe) dark brown, black, or yellowish brown. If yellowish brown (at least in the posterior part), then posterior margin of male abdominal sternum VII with angular or semicircular median emargination (Fig. 12.13K) .... *Gerris*

Hemiptera: Heteroptera: Nepomorpha: families

This key is modified from Poisson (1949, 1957), Tamanini (1979), Jansson (1986), Andersen & Weir (2004), and Chen et al. (2005).

1 Ocelli present (Fig. 12.1D) .... Ochteridae, one genus: *Ochterus*

1' Ocelli absent (Fig. 12.4A–D) 2

2(1') Rostrum short, broadly triangular, nonsegmented, with transverse sulcation (Fig. 12.14A; except Corixidae Cymatiainae having rostrum not ridged). Posterior margin of head covers anterior part of pronotum (Fig. 12.14B) 3

2' Rostrum more or less elongated, subcylindrical (more or less compressed dorsoventrally, Fig. 12.3A). Posterior margin of head not covering anterior part of pronotum (Fig. 12.14C) 4

3(2) Scutellum exposed (Fig. 12.4C) .... Micronectidae, one genus: ***Micronecta***

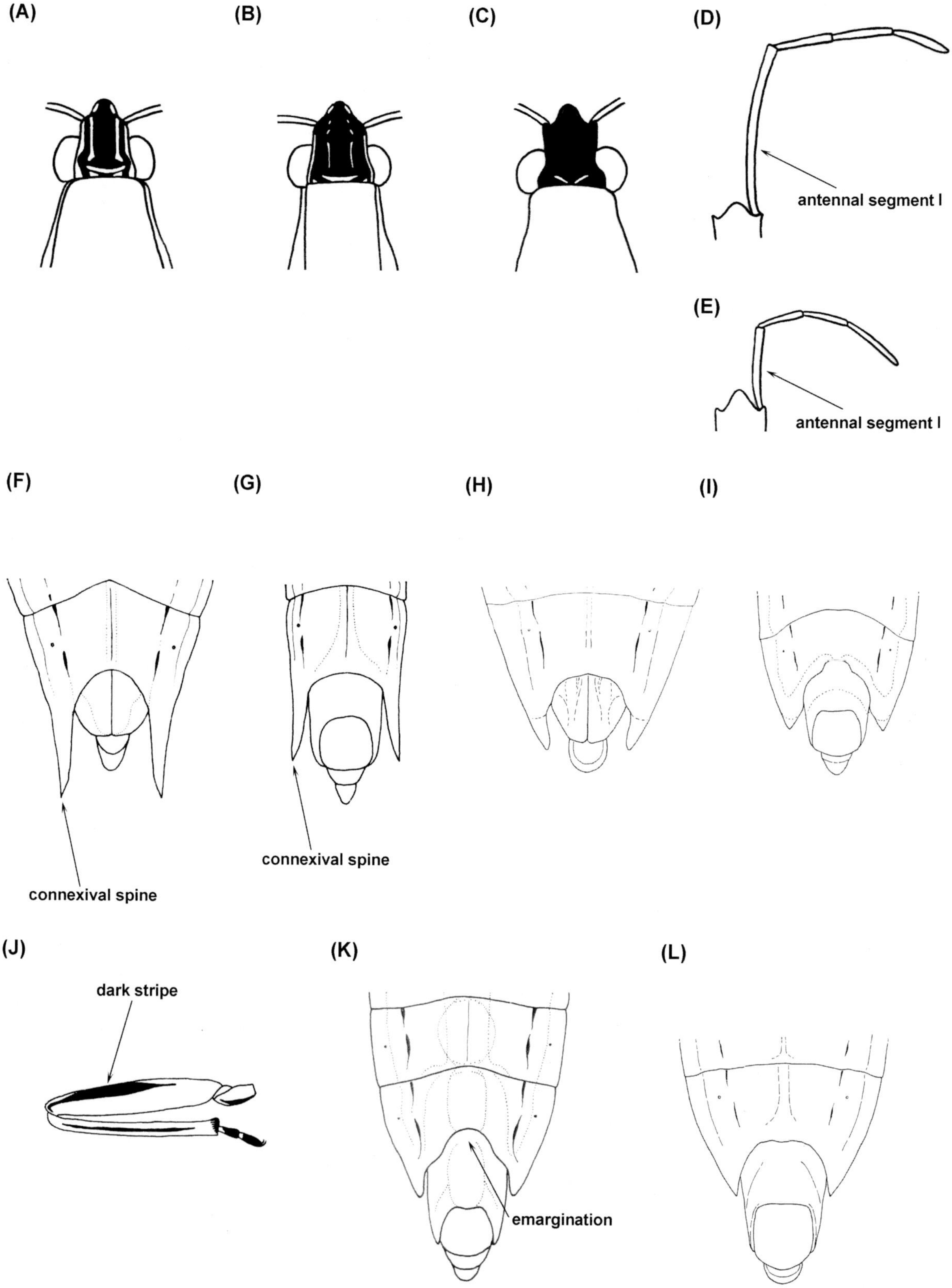

FIGURE 12.13 Head of (A, B) *Limnogonus cereiventris* and (C) *Gerris (Gerriselloides) lateralis*. Antenna of (D) *Aquarius* sp. and (E) *Gerris* sp. Last abdominal segments of (F) female *Aquarius paludum*, (G) male *A. najas*, (H) female *Gerris (Gerriselloides) asper*, (I) male *Gerris (Gerris) argentatus*, (K) male *Gerris (Gerris) costae*, (L) male *Gerris (Gerriselloides) lateralis*. (J) Front leg of *Gerris (Gerris) lacustris*. *Redrawn with modifications from (A, B) Poisson (1965a); (C) Andersen (1993). (D, E) Original drawings. (F–L) Reproduced with modifications from Tamanini (1979).*

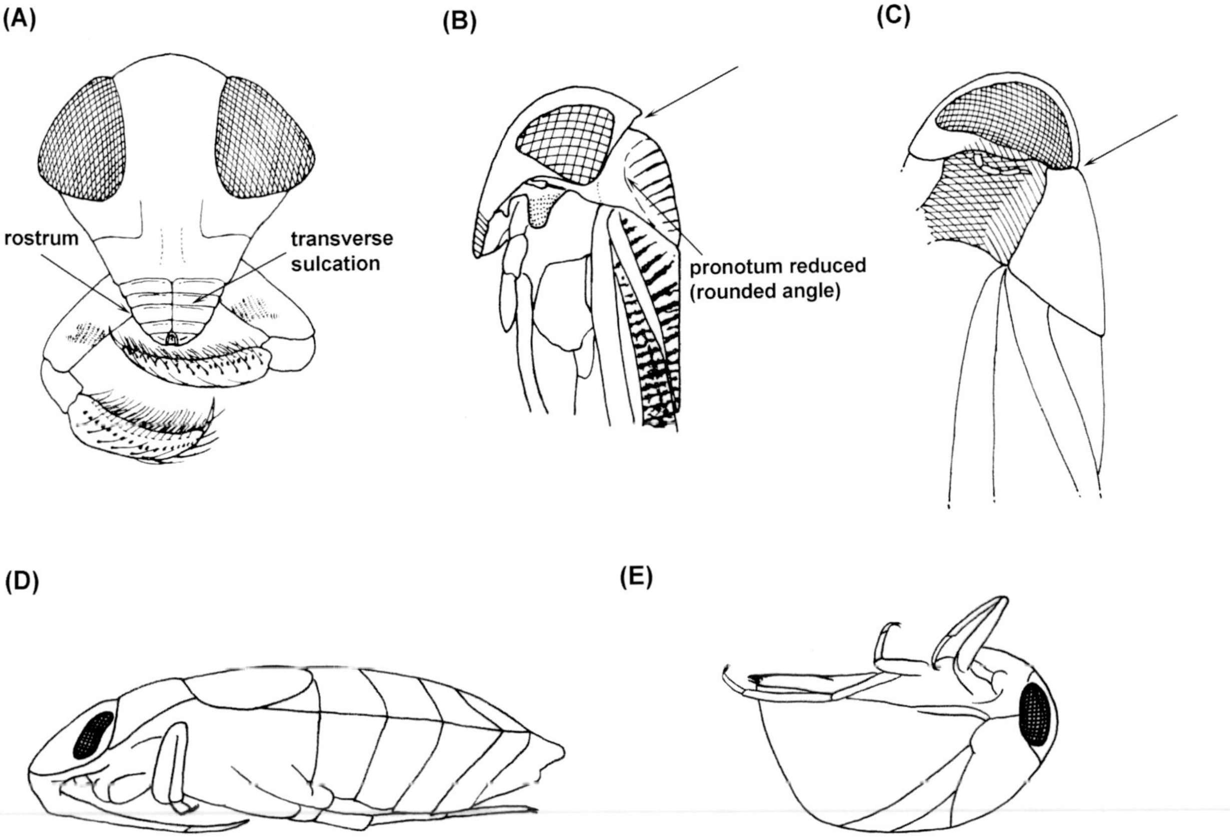

FIGURE 12.14 (A) Head and front legs of *Corixa* sp. (Corixidae). Head and pronotum of (B) *Sigara (Retrocorixa) limitata* and (C) *Notonecta* sp., in lateral view (free arrows indicate posterior margin of head covering or not pronotum). Shape of (D) *Aphelocheirus aestivalis* and (E) *Plea minutissima*, in lateral view. *Reproduced with modifications from (A–C) Tamanini (1979). (D, E) Original drawings.*

3' Scutellum covered by pronotum (Fig. 12.4B) .... **Corixidae**
4(2') Apex of abdomen with paired processes, either in the shape of a long siphon (Fig. 12.2B) or a pair of short air straps (Fig. 12.3D; attention: sometimes retracted) 5
4' Apex of abdomen without respiratory processes (Fig. 12.4B–D) 6
5(4) Respiratory siphon nonretractile, long, and tube-like (Fig. 12.2B; halves of the siphon are not entirely fused and the siphon may split, especially in dry specimens). All tarsi one-segmented. Hind tibiae without swimming hairs (Fig. 12.2B) .... **Nepidae**
5' Respiratory siphon retractile, flattened, and strap-like (Fig. 12.3D; often only apices visible if retracted). Tarsi two- or three-segmented (Fig. 12.20E, G, H, and I); in *Appasus urinator urinator* Dufour, 1863 the proximal segment is extremely reduced, appearing as one-segmented tarsus (Fig. 12.20F). Hind tibiae with swimming hairs (Fig. 12.3D) .... **Belostomatidae**
6(4') Body dorsoventrally flattened (Fig. 12.14D; up to moderately convex), oval to nearly circular in dorsal view (Figs 12.3C and 12.4A) 7
6' Body ventrally flat, dorsally strongly convex, inversely boat-shaped (swimming with venter upward) (Fig. 12.14E) 8
7(6) Fore tarsi one-segmented. Fore legs strongly raptorial with swollen femur (Fig. 12.4A). Rostrum reaching anterior coxae (not surpassing prosternum). Head only slightly produced in front of eyes (Fig. 12.4A) .... **Naucoridae**
7' Fore tarsi two-segmented. Fore legs not raptorial, femur rather slender (Fig. 12.3C). Rostrum reaching hind coxae. Head strongly produced in front of eyes (Fig. 12.3C) .... Aphelocheiridae, one genus: *Aphelocheirus*
8(6') Broadly oval, robustly built species (Figs 12.4D and 12.14E), under 3.5 mm long (adults). Antennae three-segmented. Hind wings coleopteriform and not bilobate, bearing numerous punctures .... **Pleidae**

8' Elongate, wedge-shaped species (Fig. 12.7A–D), over 4.5 mm long (adults). Antennae four-segmented. Hind wings with bilobate membrane, without punctures .... **Notonectidae**

Hemiptera: Heteroptera: Nepomorpha: Micronectidae: *Micronecta*: subgenera

This key is based on Hutchinson (1940), Jansson (1986), and unpublished data.

1 Abdominal sternite VII with two conspicuous bristle-like hairs[1] (Fig. 12.15B; cf. Fig. 12.15A) .... *Micronecta (Dichaetonecta)* [Mediterranean Basin]
1' Abdominal sternite VII with three to six conspicuous bristle-like hairs[1] (Fig. 12.15A) 2
2(1') Male genitalia: right paramere thin, styliform, more than 7.5 times as long as high (Fig. 12.15D and F); left paramere with small serrations (Fig. 12.15C and E) .... *Micronecta (Basileonecta)* [Southern and eastern Mediterranean: North Africa and Middle East]
2' Male genitalia: right paramere usually more robust, thick, and blade-like, less than 7.5 times as long as high (Fig. 12.15H and J); left paramere smooth, possibly with a denticle but without serrations (Fig. 12.15G and I) .... *Micronecta (Micronecta)* [Mediterranean Basin]
[1]When possible, check the hairs of more than one specimen.

Hemiptera: Heteroptera: Nepomorpha: Corixidae: genera

This key is based on Jaczewski (1926), Poisson (1957), Tamanini (1979), Jansson (1986), Andersen & Weir (2004), Chen et al. (2005), Cianferoni & Terzani (2013), and unpublished data.

1 Rostrum without transverse sulcations. Fore tarsus nearly cylindrical (similar in the two sexes), without palm; in males without palar pegs (Fig. 12.16A) .... Cymatiainae: *Cymatia* [Mediterranean Basin]
1' Rostrum with transverse sulcations (Fig. 12.14A). Fore tarsus modified ("pala"), more or less spoonshaped and flattened (especially in male; different in the two sexes: sexual dimorphism), with ventral palm (Fig. 12.16B); in males palar pegs present (Fig. 12.16C) .... 2 (Corixinae)
2(1') Abdomen of males (seen from above, with the wings apart) strigil lacking or vestigial (cf. Fig. 12.8B) 3
2' Abdomen of males (seen from above, with the wings apart) with strigil, although sometimes very reduced (cf. Fig. 12.8B) 5
3(2) Body length equal or less than 6 mm. Pala and right paramere of male as in Fig. 12.16D and E .... ***Sigara*** (partim): *S. (Vermicorixa) scripta* (Rambur, 1840) [Mediterranean Basin]
3' Body length equal or more than 6.5 mm. Pala and right paramere of male different (Fig. 12.16F–I) 4
4(3') In males: pala with upper margin not sinuous and palar pegs arranged in a single row (Fig. 12.16F). Paramere as in Fig. 12.16G .... *Paracorixa* (one species possibly occurring in the considered area, nominotypical subspecies: *P. concinna concinna*) (Fieber, 1848) [Northern Mediterranean: Europe]
4' In males: pala with upper margin sinuous and palar pegs arranged in two separate rows (Fig. 12.16H). Paramere as in Fig. 12.16I ....
Callicorixa (one species possibly occurring in the considered area, nominotypical subspecies: *C. praeusta praeusta*) (Fieber, 1848) [Northern Mediterranean: Europe]
5(2') Abdomen of males (seen from above, with the wings apart) with strigil on the left side (cf. Fig. 12.8B) 6
5' Abdomen of males (seen from above, with the wings apart) with strigil on the right side (cf. Fig. 12.8B) 8
6(5) In males: apex of fore tibia produced over base of pala (Fig. 12.16J) .... *Trichocorixa*, one species: *T. verticalis* (Fieber, 1851) [Nearctic species, introduced in the Mediterranean Basin, so far apparently confined to the western part[1]]
6' In males: apex of fore tibia not produced over base of pala (Fig. 12.17C and D) 7
7(6') Body length more than 8 mm. In males, pegs occurring along the entire length of the pala (Fig. 12.17C) .... *Corixa* [Mediterranean Basin]
7' Body length less than 6.5 mm. In males, pegs occurring approximately up to half the length of the pala (Fig. 12.17D) .... *Heliocorisa*, one species: *H. vermiculata* (Puton, 1874) [Mediterranean Basin]
8(5') Pronotum with well-developed median carina visible for its entire length (Fig. 12.17A) 9
8' Pronotum without developed median carina or with median carina visible only anteriorly, at most for one quarter of pronotum length (Fig. 12.17B) 10

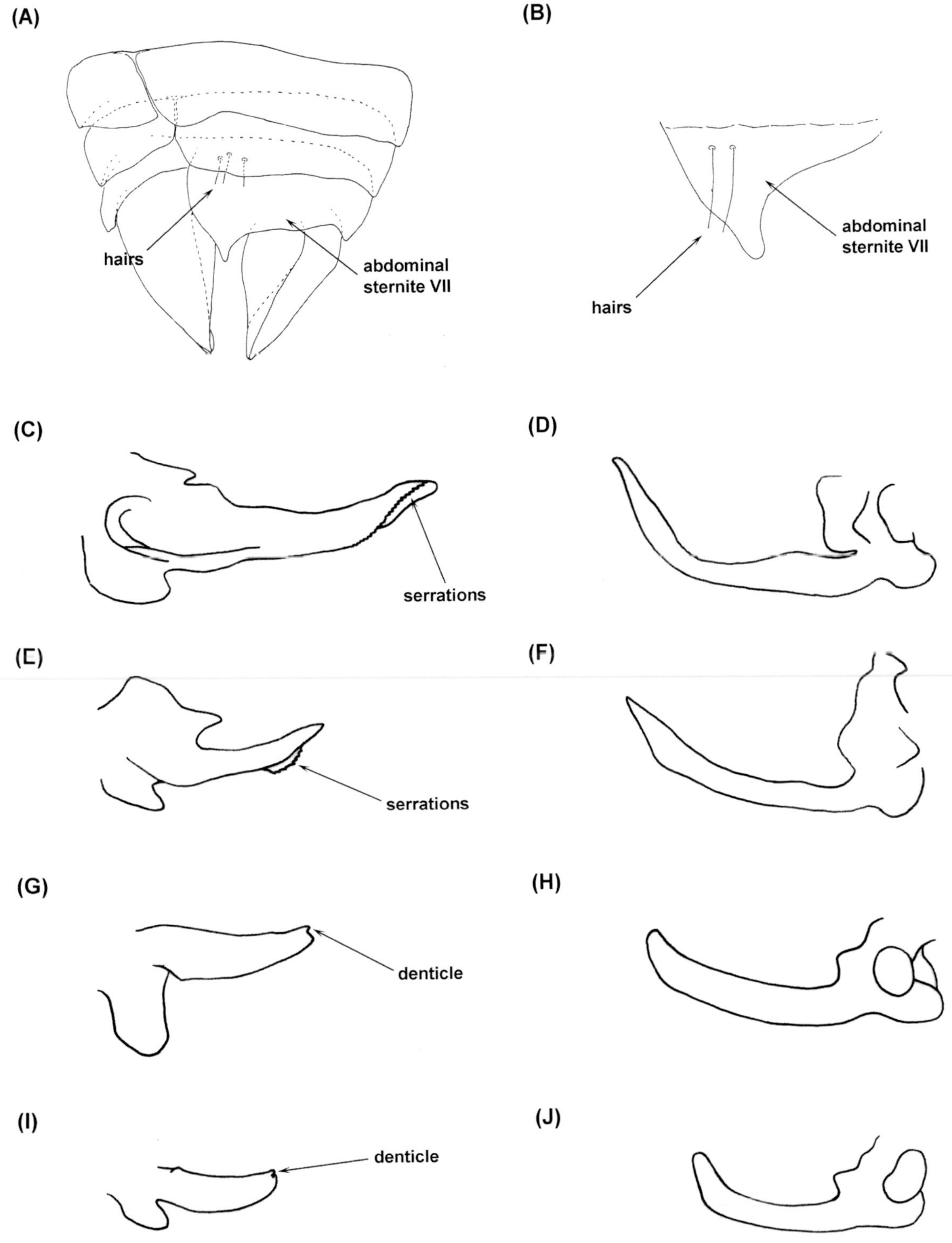

FIGURE 12.15 (A) Last abdominal segments of *Micronecta (Micronecta) minuscula*, seen from below. (B) VII abdominal sternite of *Micronecta (Dichaetonecta) scholtzi*. (C) Left and (D) right paramere of *Micronecta (Basileonecta) scutellaris*; the same for (E, F) *M. (B.) isis*, (G, H) *M. (Micronecta) leucocephala*, (I, J) *M. (Micronecta) anatolica*. *Reproduced with modifications from (A, B) Tamanini (1979). (C–F) Redrawn with modifications from Linnavuori (1964); (G–J) Jansson (1986).*

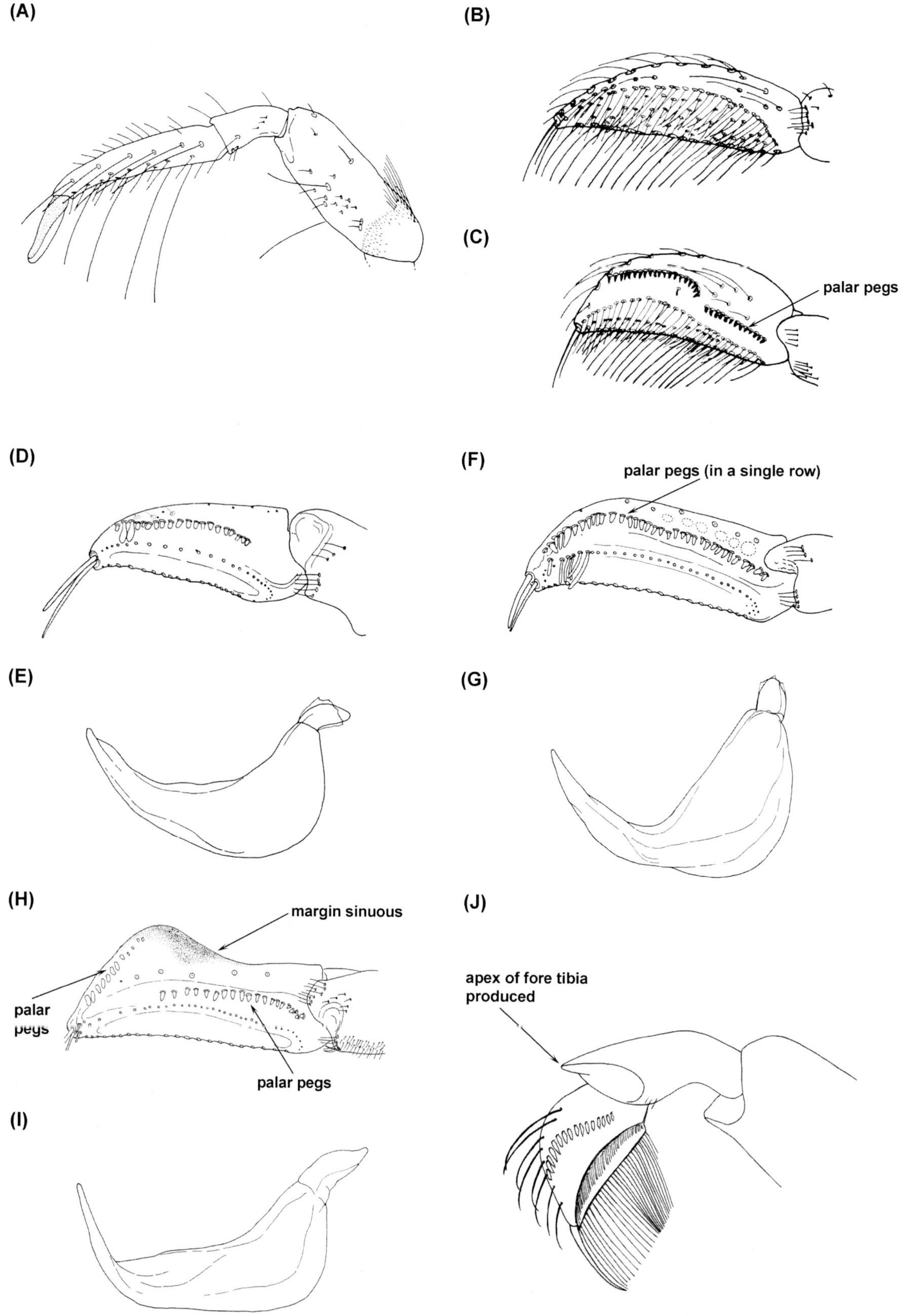

FIGURE 12.16 (A) Male front leg of *Cymatia coleoptrata*. Pala of (B) female and (C) male of *Sigara servadeii*. Pala and right paramere of (D, E) *Sigara (Vermicorixa) scripta*, (F, G) *Paracorixa concinna*, (H, I) *Callicorixa praeusta*. (J) Male front leg of *Trichocorixa verticalis*. *Reproduced with modifications from (A–I) Tamanini (1979). Redrawn with modifications from (J) Sailer (1948).*

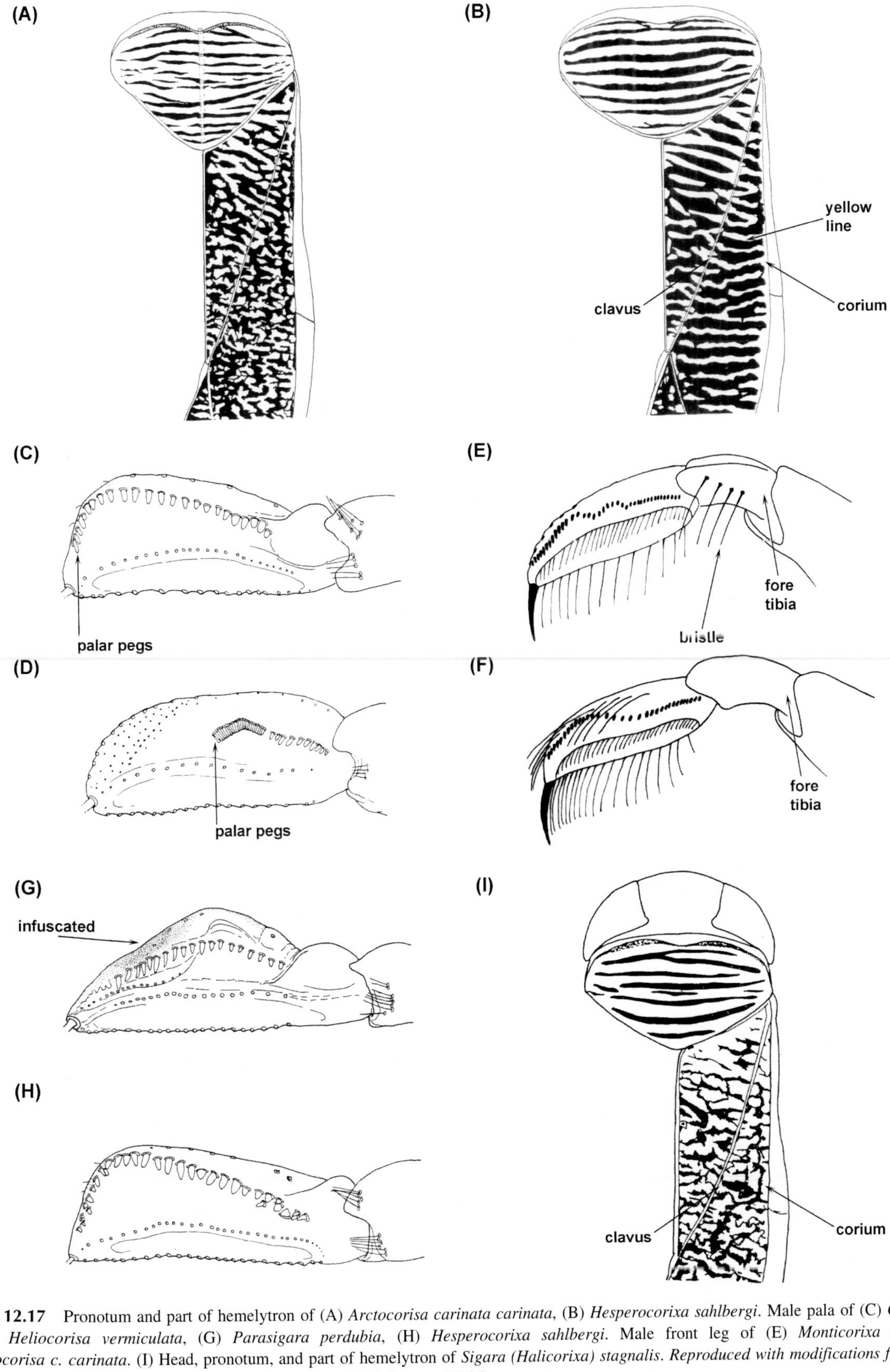

FIGURE 12.17 Pronotum and part of hemelytron of (A) *Arctocorisa carinata carinata*, (B) *Hesperocorixa sahlbergi*. Male pala of (C) *Corixa affinis*, (D) *Heliocorisa vermiculata*, (G) *Parasigara perdubia*, (H) *Hesperocorixa sahlbergi*. Male front leg of (E) *Monticorixa armeniaca*, (F) *Arctocorisa c. carinata*. (I) Head, pronotum, and part of hemelytron of *Sigara (Halicorixa) stagnalis*. *Reproduced with modifications from (A–D, G–I) Tamanini (1979). Redrawn with modifications from (E, F) Jansson (1986).*

9(8) Both sexes with three to four long bristles on fore tibia (Fig. 12.17E) .... *Monticorixa*, one species: *M. armeniaca* (Štys, 1975) [Anatolia, only at high altitudes in the mountains]
9' Males without any long bristles on fore tibia (Fig. 12.17F), females with two long bristles on fore tibia .... *Arctocorisa* [Southern montane populations in the Spanish mountain ranges, Pyrenees, Alps, Apennines, and Balkan mountains, only at high altitudes]
10(8') Pala apically infuscated in both sexes (Fig. 12.17G) .... *Parasigara* [Western Mediterranean: Italy, France, Iberian Peninsula, and Maghreb]
10' Pala apically not infuscated in either sex (Figs 12.16C and 12.17H) 11
11(10') Pronotum and hemelytra mainly concolorous .... *Agraptocorixa* [Possible occurrence in the southern part of the Mediterranean Basin: North Africa; *A. dakarica* Jaczewski, 1926 recorded from Egypt]
11' Pronotum and hemelytra dark with yellowish transverse bands (Fig. 12.4B) 12
12(11') Yellow lines of corium mostly unbroken and extending across the corium (Fig. 12.17B) .... *Hesperocorixa* [Mediterranean Basin]
12' Yellow lines of corium mostly broken, not extending across the corium (Fig. 12.17I) .... ***Sigara*** (partim) [Mediterranean Basin]
[1]Potentially able to invade the entire Mediterranean Basin (see Guareschi et al., 2013).

Hemiptera: Heteroptera: Nepomorpha: Corixidae: *Sigara*: subgenera

This key is based on Hutchinson (1940), Poisson (1957), Tamanini (1979), and Jansson (1986).

1 Pronotum with transverse lines interrupted by a longitudinal pale line .... *Sigara (Microsigara)*, one species: *S. (M.) hellensi* (C. Sahlberg, 1819) [Northern Mediterranean[1]]
1' Pronotal transverse lines medially uninterrupted (cf. Fig. 12.4B) 2
2(1') Male genitalia with right paramere apically clearly forked (Fig. 12.18A) .... *Sigara (Halicorixa)* [Mediterranean Basin]
2' Male genitalia with right paramere, apically unforked (Fig. 12.18B) or just slightly forked (Fig. 12.18C) 3
3(2') Second segment of hind tarsus infuscated (Fig. 12.18E and F). Hemelytral patterns vermiculate (Fig. 12.18G) .... *Sigara (Vermicorixa)* [Mediterranean Basin]
3' Second segment of hind tarsus not infuscated (check perfectly mature specimens). Hemelytral patterns variable, but not vermiculate (Fig. 12.18H) 4
4(3') Pronotum laterally slightly reduced, with rounded angles (Figs 12.4B, 12.14B, and 12.18I) 5
4' Pronotum laterally not reduced (Figs 12.17I and 12.18J) 6
5(4) Corial patterns transverse, sometimes somewhat fragmentary (Fig. 12.18H). Right paramere with no or weak sinuosity externally (Fig. 12.18D) .... *Sigara (Pseudovermicorixa)* [Mediterranean Basin]
5' Corial patterns finely divided and longitudinally coalescent in the median angle of corium (Fig. 12.4B). Right paramere with very pronounced sinuosity externally (cf. Fig. 12.18C) .... *Sigara (Retrocorixa)* [Northern Mediterranean: Europe, including Corsica and Sardinia[2]]
6(4') Lateral lobe of prothorax distally slightly broadened (Fig. 12.19A) .... *Sigara (Subsigara)* [Mediterranean Basin, except North Africa[3]]
6' Lateral lobe of prothorax as broad as or narrower than proximally (Fig. 12.19B) 7
7(6') Right paramere enlarged and crested apically (Fig. 12.19C and D) .... *Sigara (Tropocorixa)* [Southern and eastern Mediterranean: North Africa and Middle East]
7' Right paramere tapered (sometimes slightly forked, Fig. 12.19E) and not crested apically (Fig. 12.19E and F) .... *Sigara (Sigara)* [Northern and eastern Mediterranean: Europe and Middle East]
[1]Northern and Central European species, occurring in very few relict locations in the northern sector of the Mediterranean Basin.
[2]The subgenus *Sigara* (*Retrocorixa)* comprises mountain taxa that are beyond the scope of this book, but includes also a Sardo-Corsican species, *S. (R.) limitata remyi* (Poisson, 1954), collected in still waters at high altitude (see Poisson, 1957, Jansson, 1986).
[3]Since the occurrence of some species belonging to this subgenus just outside the Mediterranean Basin (i.e., Atlantic coast of Morocco) or in the Middle East (i.e., Israel), the presence in the western and eastern sectors of Mediterranean Africa cannot be excluded.

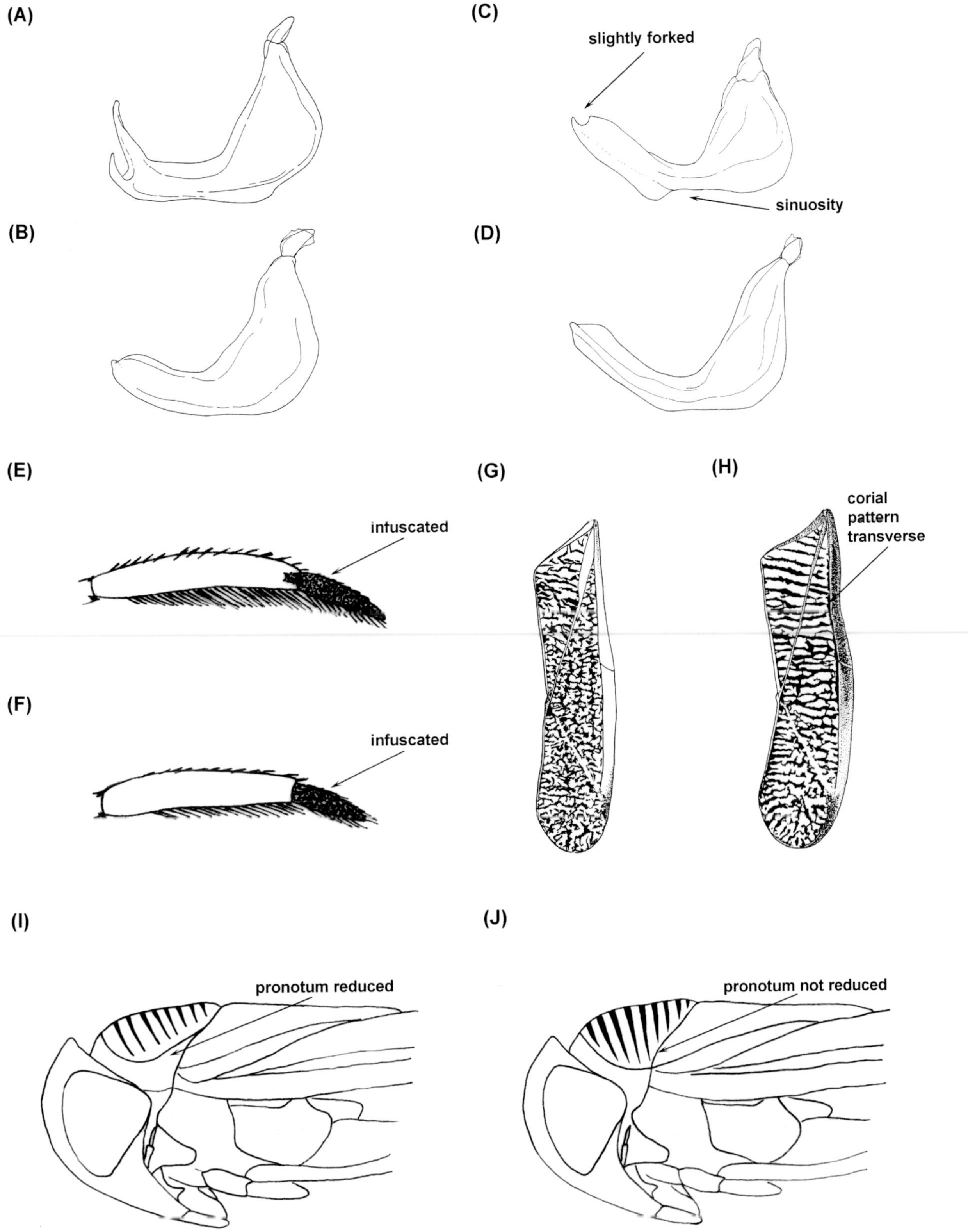

FIGURE 12.18 Right paramere of (A) *Sigara (Halicorixa) selecta*, (B) *S. (Vermicorixa) lateralis*, (C) *S. (Retrocorixa) limitata*, (D) *S. (Pseudovermicorixa) nigrolineata*. Hind tarsi of (E) *S. (Vermicorixa) lateralis* and *S. (Vermicorixa) scripta*. Hemelytron of (G) *S. (Vermicorixa) lateralis* and (H) *S. (Pseudovermicorixa) nigrolineata*. Anterior part of *Sigara* sp. in lateral view, with reduced (I) and not reduced (J) pronotum. *Reproduced with modifications from (A–D, G, H) Tamanini (1979). (E, F) Redrawn with modifications from Jansson (1986); (I, J) Tamanini (1979).*

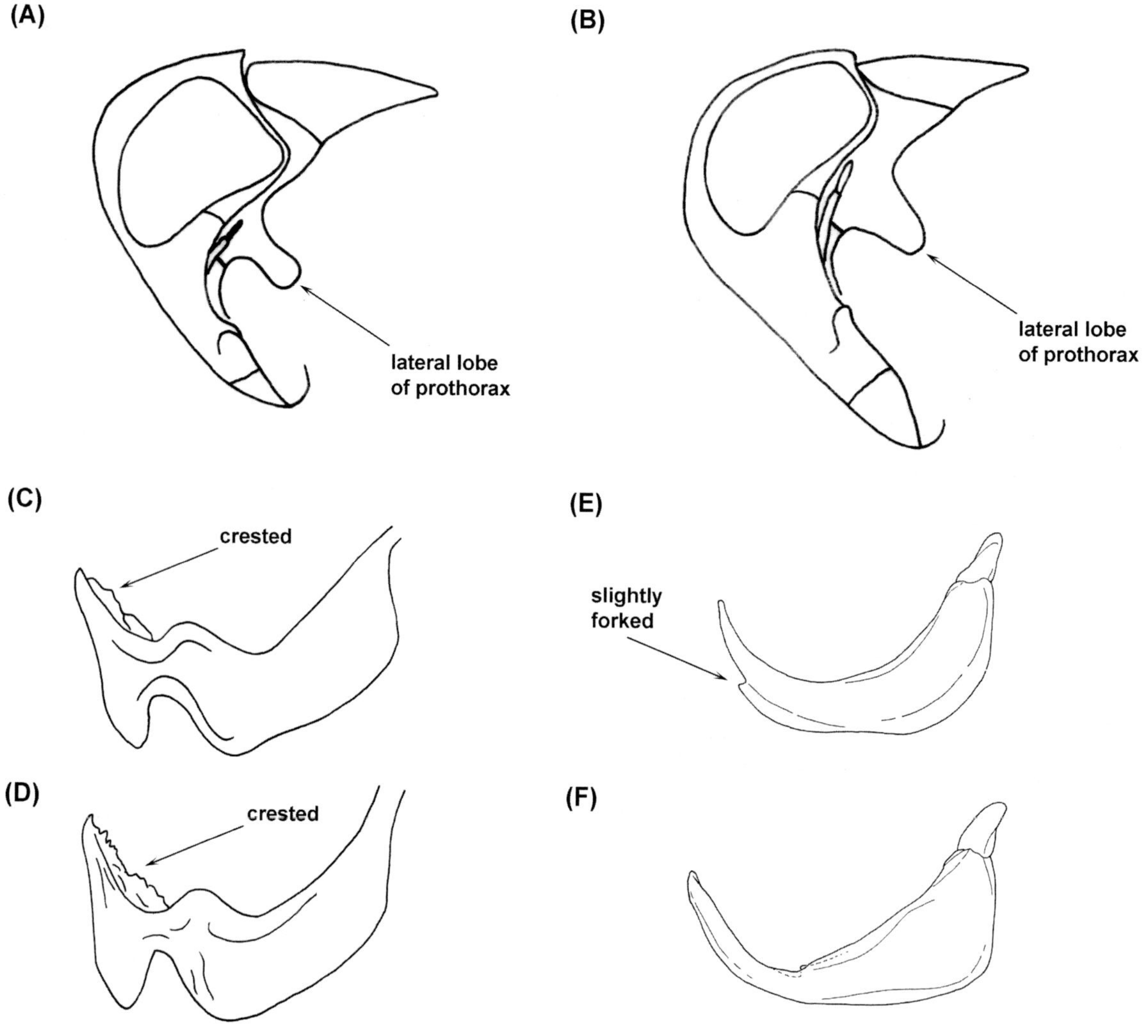

FIGURE 12.19 Lateral lobe of prothorax of (A) *Sigara (Subsigara)* sp. and (B) *Sigara (Sigara)* sp. Right paramere of (C, D) *Sigara (Tropocorixa) hoggarica*, (E) *S. (Sigara) servadeii*, (F) *S. (Sigara) striata*. *Redrawn with modifications from (A, B) Jansson (1986); (C) Poisson (1952); (D) Poisson (1936). (E, F) Reproduced with modifications from Tamanini (1979).*

Hemiptera: Heteroptera: Nepomorpha: Belostomatidae: genera

This key is in part modified from Poisson (1949), Lauck & Menke (1961), and Polhemus (1995).

1 Abdominal sternites V–VI subdivided laterally by a suture-like fold (Fig. 12.20A). Spiracles located on or adjacent to mesal margin of ventral laterotergites (Fig. 12.20A) .... Lethocerinae: *Lethocerus* [Eastern Mediterranean]

1' Abdominal sternites not subdivided by a suture (Fig. 12.20B). Spiracles well separated from the mesal margin of ventral laterotergites (Fig. 12.20B) .... 2 (Belostomatinae)

2(1') Front tarsi with only one developed claw (the other vestigial) (Fig. 12.20E) .... *Belostoma*, one species: *B. bifoveolatum* (Spinola, 1852) [Neotropical; introduced in Israel[1]]

2' Front tarsi with two developed claws (although one may be shorter) (Fig. 12.20F–I) 3

3(2') Pubescence of laterotergites IV not attaining external margin (Fig. 12.20C). Front tarsi appearing one- or two-segmented (Fig. 12.20F and G). Body length (adults) less than 30 mm .... *Appasus* [Southern and eastern Mediterranean: North Africa and Middle East]

3' Pubescence of laterotergites IV attaining entire external margin (Fig. 12.20D). Front tarsi appearing three-segmented (Fig. 12.20H and I). Body length (adults) more than 35 mm 4

4(3') Front tarsi with segments II and III subequal in length (Fig. 12.20H). Front tibiae and tarsi subcylindrical. Hind legs not modified for swimming .... *Limnogeton* [Southern and eastern Mediterranean: North Africa and Middle East]
4' Front tarsi with segment II shorter than III (Fig. 12.20I). Front tibiae and tarsi laterally flattened .... *Hydrocyrius*, one species, nominate subspecies: *H. colombiae colombiae* Spinola, 1850. Hind legs not modified for swimming [Southern and eastern Mediterranean: North Africa and Middle East]
[1]See Novoselsky et al. (2018) for further information.

Hemiptera: Heteroptera: Nepomorpha: Nepidae: genera

This key is in part modified from Andersen & Weir (2004), Chen et al. (2005), and in part based on Poisson (1965b).

1 Body broad and flat (Fig. 12.21C). Parasternites visible (Fig. 12.21A and C) 2
1' Body narrowly cylindrical (Fig. 12.21D). Parasternites concealed by the ventral parts of the laterotergites (Fig. 12.21B and D) .... *Ranatra*[1] [Mediterranean Basin]
2(1) Body length (of adults) less than 25 mm, length of respiratory siphon (in adults) less than 12 mm .... *Nepa* [Mediterranean Basin]
2' Body length (of adults) more than 35 mm, length of respiratory siphon (in adults) more than 60 mm (the latter can often be broken, especially in dry specimens) .... *Laccotrephes* [Southern and eastern Mediterranean: North Africa and Middle East]
[1]All the species from the Old World belong to the nominotypical subgenus.

Hemiptera: Heteroptera: Nepomorpha: Naucoridae: genera

This key is based on Poisson (1948, 1949), Popov (1970), Tamanini (1979), Chen et al. (2005), Mbogho & Sites (2013), and Polhemus & Polhemus (2013).

1 Apex of head folded down and backward, such that the rostrum arises at a point posterior to the anterior margin of the head when viewed laterally or ventrally (Fig. 12.22A and B). Foreleg pretarsus bearing two small claws. Males with well-developed tomentose patch on mesotibia (on females weakly developed) .... 2 (Laccocorinae)
1' Apex of head not folded down and backward, rostrum arising at anterior margin of head when viewed laterally (Fig. 12.22C and D). Foreleg pretarsus bearing one tiny claw. Absence of tomentose patch on mesotibia .... 3 (Naucorinae)
2(1) Labrum with bluntly rounded tip (Fig. 12.22E) .... *Heleocoris*, one species: *H. minusculus* (Walker, 1870) [Southern and eastern Mediterranean: North Africa and Middle East]
2' Labrum with sharp or narrowly rounded tip (Fig. 12.22F and G) .... *Laccocoris* [Southern Mediterranean: North Africa]
3(1') Males with tergites VII and VIII modified with asymmetrical medial lobes (Fig. 12.22H) .... *Macrocoris* [Southern Mediterranean: North Africa]
3' Males with tergites VII and VIII unmodified 4
4(3') Lateral margin of the pronotum with a distinct raised edge (Fig. 12.22I) .... *Ilyocoris* [Southern and eastern Mediterranean: Europe and Asia]
4' Lateral margin of the pronotum with indistinct edge (Fig. 12.22J) .... *Naucoris* [Mediterranean Basin]

Hemiptera: Heteroptera: Nepomorpha: Pleidae: genera

This key is based on Esaki & China (1928) and Andersen & Weir (2004).

1 Each of fore, middle, and hind tarsi three-segmented .... *Plea*, one species: *P. minutissima* Leach, 1817[1] [Mediterranean Basin]
1' Fore and middle tarsi two-segmented, only hind tarsi three-segmented .... *Paraplea*, one species: *P. pullula* (Stål, 1855) [North Africa]
[1]The need to use the nominotypical subspecies is due to *P. minutissima tassili* Poisson 1953, a taxon that still urges to be assessed.

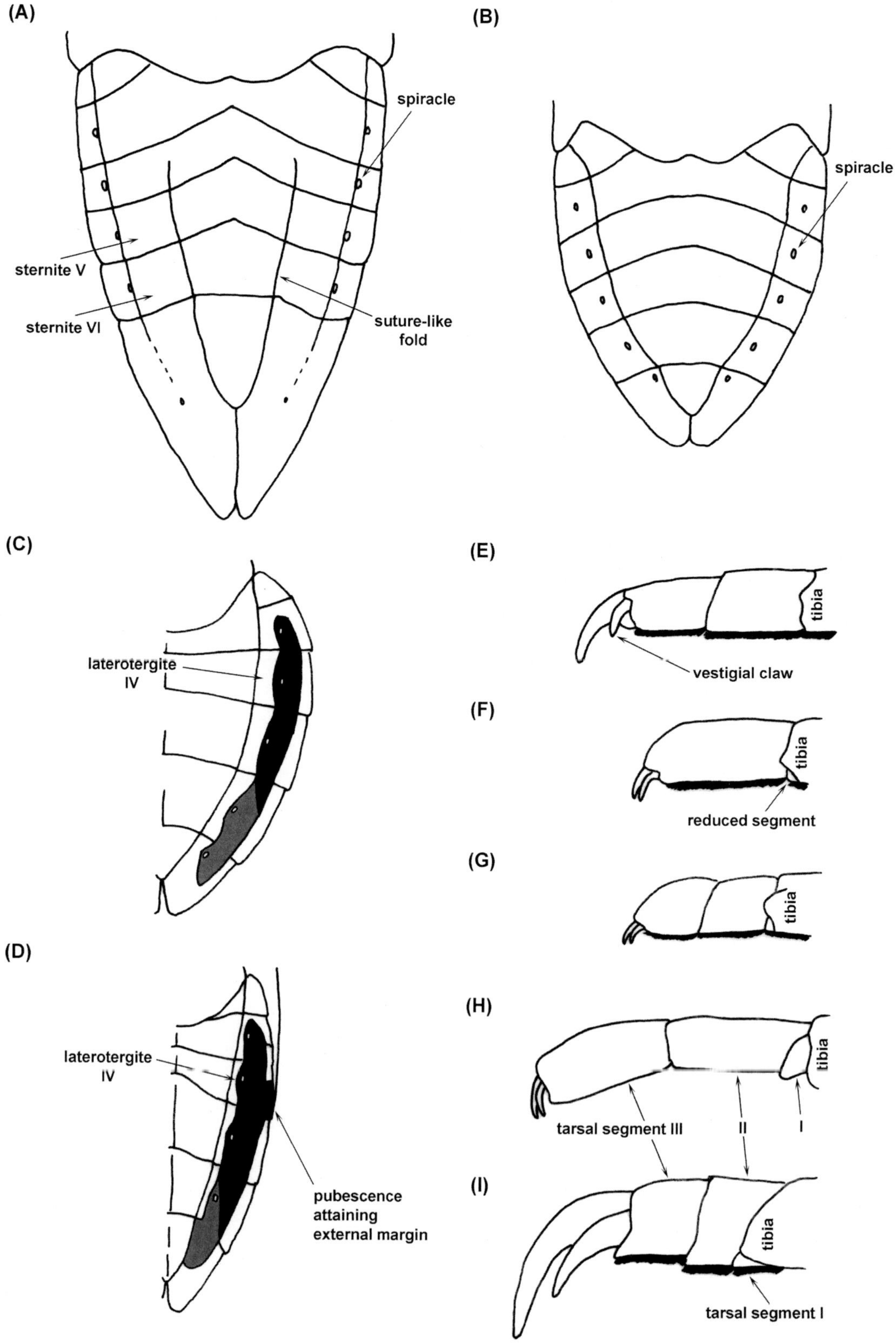

FIGURE 12.20 Abdomen in ventral view of (A) Lethocerinae and (B) Belostomatinae. Abdomen of Belostomatinae in ventral view, showing the pubescence of laterotergites IV not attaining external margin (C) and attaining entire external margin (D). Right front tarsi of (E) *Belostoma* sp., (F, G) *Appasus* sp., (H) *Limnogeton* sp., (I) *Hydrocyrius* sp. *Redrawn with modifications from (A, B) Chen et al. (2005); (C–I) Lauck & Menke (1961).*

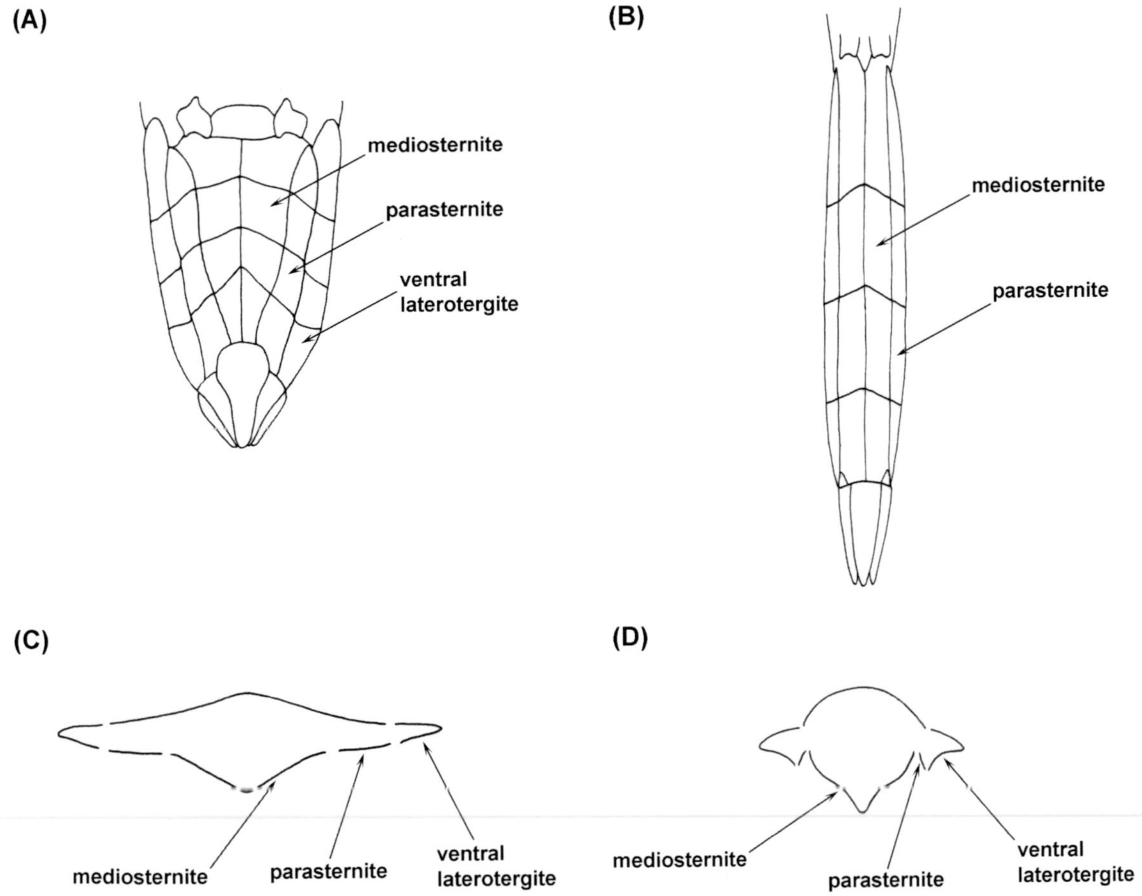

FIGURE 12.21 Abdomen in ventral view of (A) Nepinae and (B) Ranatrinae. Section of abdomen of (C) Nepinae and (D) Ranatrinae. *Redrawn with modifications from Chen et al. (2005).*

Hemiptera: Heteroptera: Nepomorpha: Notonectidae: genera

This key is modified from Andersen & Weir (2004) and Chen et al. (2005).

1 Hemelytral commissure with a prominent and well-defined hair-lined pit anteriorly, close to the apex of the scutellum (Fig. 12.7B) .... Anisopinae: *Anisops* [Mediterranean Basin]
1' Hemelytral commissure continuous, without a hair-lined pit anteriorly (Fig. 12.7A, C, and D) .... 2 (Notonectinae)
2(1') Anterolateral margins of prothorax foveate (Fig. 12.7C and D) 3
2' Anterolateral margins of prothorax not foveate (Fig. 12.7A) .... *Notonecta*[1] [Mediterranean Basin]
3(2) Middle femur with a pointed protuberance on ventral margin, before apex (Fig. 12.7C). Eyes dorsally widely spaced (Fig. 12.7C) .... *Enithares* [Southern and eastern Mediterranean: North Africa and Middle East]
3' Middle femur without subapical pointed protuberance (Fig. 12.7D). Eyes dorsally contiguous (Fig. 12.7D), forming an ocular commissure (appearing to be joined or overlapping) .... *Nychia*, one species: *N. marshalli* (Scott, 1872) [In the Mediterranean Basin recorded only from Corsica and North Africa]

[1]All the species from the Mediterranean Basin belongs to the nominotypical subgenus.

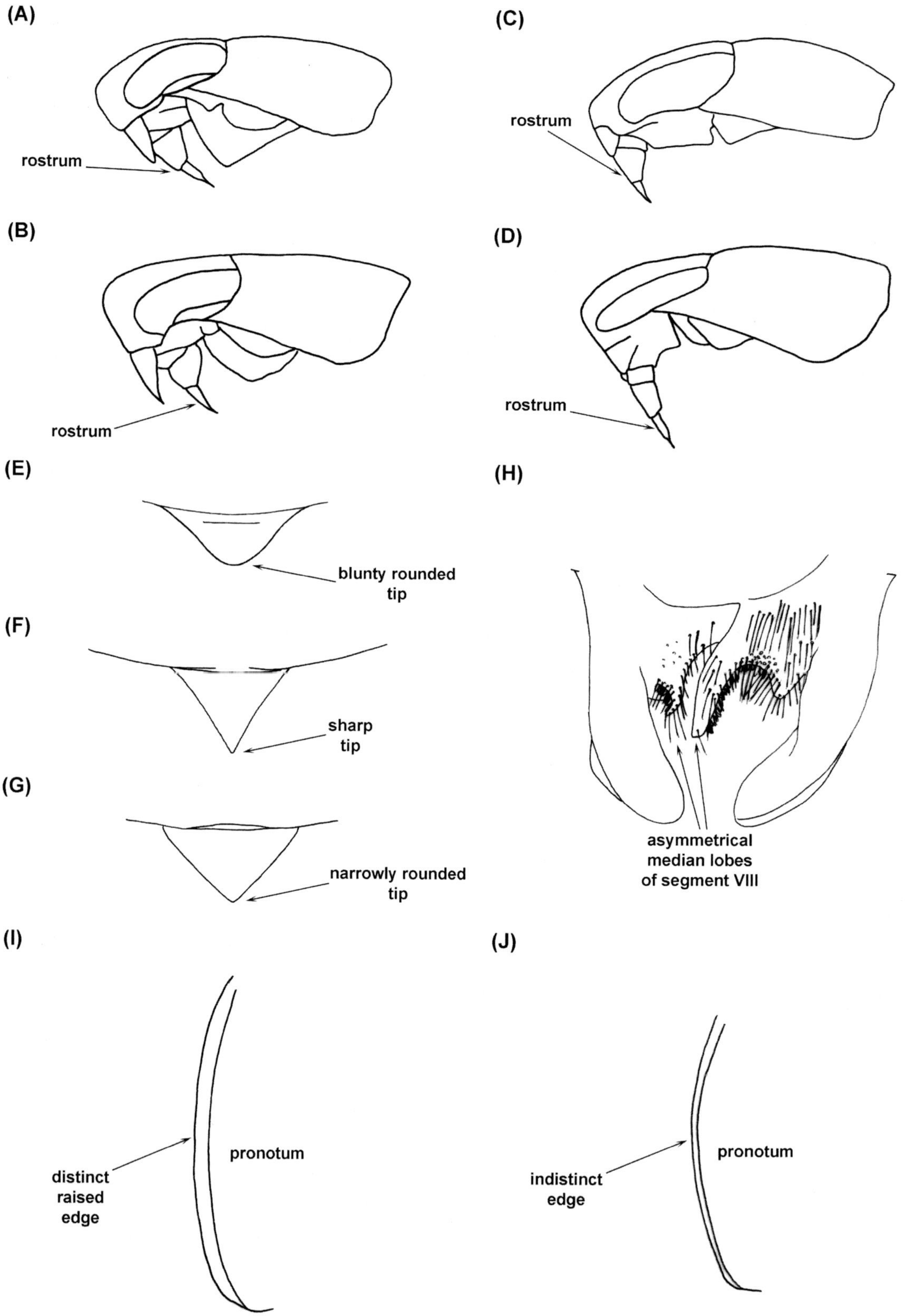

FIGURE 12.22 Head and prothorax in lateral view of (A) *Heleocoris* sp., (B) *Laccocoris* sp., (C) *Naucoris maculatus*, (D) *Ilyocoris cimicoides*. Labrum of (E) *Heleocoris* sp., (F, G) *Laccocoris* sp. (H) Median lobes of segment VIII of *Macrocoris flavicollis*. Lateral margin of the pronotum of (I) *Ilyocoris cimicoides* and (J) *Naucoris maculatus*. *Redrawn with modifications from (A–D) Popov (1970); (E–H) Poisson (1949); (I, J) Tamanini (1979).*

Acknowledgments

I am grateful to Petr Kment (National Museum, Prague, Czech Republic), Franco Faraci (Bardolino, Verona, Italy), Attilio Carapezza (Palermo, Italy), Francesca Graziani (Florence, Italy) for reading the manuscript and providing useful advice. This chapter is dedicated to Livio Tamanini (1907–1997) for his valuable contribution to the knowledge of aquatic heteropterans and for his wonderful drawings which I have the great pleasure of including part here.

References

Andersen, N.M. 1977. Fine structure of the body hair layers and morphology of the spiracles of semiaquatic bugs (Insecta, Hemiptera, Gerromorpha) in relation to life on the water surface. Videnskabelige Meddelelser fra Dansk naturhistorisk Forening 140: 7–37.

Andersen, N.M. 1981. A new genus of Veliinae and descriptions of new Oriental species of the subfamily (Hemiptera: Veliidae). Entomologica Scandinavica 12: 339–356.

Andersen, N.M. 1982. The semiaquatic bugs (Hemiptera, Gerromorpha). Phylogeny, adaptations, biogeography, and classification. Entomonograph, Scandinavian science press, Klampenborg, 3: 1–455.

Andersen, N.M. 1990. Phylogeny and taxonomy of water striders, genus *Aquarius* Schellenberg (Insecta, Hemiptera, Gerridae), with a new species from Australia. Steenstrupia 16 (4): 37–81.

Andersen, N.M. 1993. Classification, phylogeny, and zoogeography of the pond skater genus *Gerris* Fabricius (Hemiptera: Gerridae). Canadian Journal of Zoology 71 (12): 2473–2508.

Andersen, N.M. & T.A. Weir. 2003. The genus *Microvelia* Westwood in Australia (Hemiptera: Heteroptera: Veliidae). Invertebrate Systematics 17: 261–348.

Andersen, N.M. & T.A. Weir. 2004. Australian Water Bugs. Their Biology and Identification (Hemiptera-Heteroptera, Gerromorpha & Nepomorpha). Apollo Books, Stenstrup. Entomonograph 14, 344 pp.

Berchi, G.M., D. Copilaş-Ciocianu, P. Kment, F.M. Buzzetti, A. Petrusek, I. Rákosy, F. Cianferoni & J. Damgaard. 2017. Molecular phylogeny and biogeography of the West-Palaearctic *Velia* (Heteroptera: Gerromorpha: Veliidae). Systematic Entomology 43 (2) [2018]: 262–276.

Brinkhurst, R.O. 1959. A description of the nymphs of British *Gerris* species (Hemiptera-Heteroptera). Proceedings of the Royal Entomological Society of London, Series A 34: 130–136.

Chen, P.-P., N. Nieser & H. Zettel. 2005. The aquatic and semi-aquatic bugs (Heteroptera: Nepomorpha & Gerromorpha) of Malesia. Brill, Leiden. Fauna Malesiana Handbook 5, 546 pp.

Cianferoni, F., F.M. Buzzetti & H. Zettel. 2016. The "Italian hebrid" *Hebrus franzi* (Wagner, 1957): disentangling a halfcentury dilemma (Hemiptera: Heteroptera: Gerromorpha). Zootaxa 4132 (1): 127–134.

Cianferoni, F. & F. Terzani. 2013. Nuovi dati su Gerromorpha e Nepomorpha in Italia (Hemiptera Heteroptera). Bollettino della Società Entomologica Italiana 145 (2): 51–57.

Cobben, R.H. 1968. Evolutionary trends in Heteroptera. Part I. Eggs, architecture of the shell, gross embryology and eclosion. Centre for Agricultural Publishing and Documentation, Wageningen, 475 pp.

Cobben, R.H. & H.M. Pillot. 1960. The larvae of Corixidae and an attempt to key the last larval instar of the Dutch species (Hem., Heteroptera). Hydrobiologia 16: 323–356.

Drohojowska, J., J. Szwedo, D. Żyła, D.-Y. Huang & P. Müller. 2020. Fossils reshape the Sternorrhyncha evolutionary tree (Insecta, Hemiptera). Scientific Reports 10: 11390.

Dolling, W.R. 1991. The Hemiptera. Natural History Museum Publications. Oxford University Press, Oxford, 274 pp.

Esaki, T. & W.E., China. 1928. A monograph of the Helotrephidae, subfamily Helotrephinae (Hem. Heteroptera). EOS, Revista Española de Entomología 4 (2): 9–172.

Forero, D. 2008. The systematics of the Hemiptera. Revista Colombiana de Entomología 34 (1): 1–21.

Guareschi, S., C. Coccia, D. Sánchez-Fernández, J.A. Carbonell, J. Velasco, L. Boyero, A.J. Green & A. Millán. 2013. How far could the alien boatman *Trichocorixa verticalis verticalis* spread? Worldwide estimation of its current and future potential distribution. PLoS ONE 8 (3): e59757

Hädicke, C.W., D. Rédei & P. Kment. 2017. The diversity of feeding habits recorded for water boatmen (Heteroptera: Corixoidea) world-wide with implications for evaluating information on the diet of aquatic insects. European Journal of Entomology 114: 147–159.

Henry, T.J. 2017. Biodiversity of Heteroptera. Pages 279–335 *in*: Foottit, R.G. and P.H. Adler (eds.), Insect Biodiversity: Science and Society, Volume I, Second Edition. Wiley-Blackwell, Hoboken, NJ.

Höregott, H. & K.H.G. Jordan. 1954. Bestimmungstabelle der Weibchen deutscher Corixiden (Heteroptera: Corixidae). Beiträge zur Entomologie 4: 578–594.

Hu, D., B. Chan & J.W.M. Bush. 2003. The hydrodynamics of water strider locomotion. Nature 424: 663–666.

Hutchinson, G.E. 1940. A revision of the Corixidae of India and adjacent regions. Transactions of the Connecticut Academy of Arts and Sciences 33: 339–476.

Jansson, A. 1969. Identification of larval Corixidae (Heteroptera) of Northern Europe. Annales Zoologici Fennici 6 (3) 289–312.

Jansson, A. 1986. The Corixidae (Heteroptera) of Europe and some adjacent regions. Acta Entomologica Fennica 47: 1–94.

Jaczewski, T. 1926. Notes on some West-African Heteroptera. Annales Zoologici Musei Polonici 5 (2): 62–106, 3 pls.

Johnson, K.P., C.H. Dietrich, F. Friedrich, R.G. Beutel, B. Wipfler, R.S. Peters, J.M. Allen, M. Petersen, A. Donath, K.K.O. Walden, A.M. Kozlov, L. Podsiadlowski, C. Mayer, K. Meusemann, A. Vasilikopoulos, R.M. Waterhouse, S.L. Cameron, C. Weirauch, D.R. Swanson, D.M. Percy, N.B. Hardy, I. Terry, S. Liu, X. Zhou, B. Misof, H.M. Robertson & K. Yoshizawa. 2018. Phylogenomics and the evolution of hemipteroid insects. Proceedings of the National Academy of Sciences 115 (50): 12775–12780.

Kanyukova, E.V. 1997. Hebridae of Russia and adjacent countries (Heteroptera). Zoosystematica Rossica 6 (1/2): 223–236.

Kment, P., Z. Jindra & G.M. Berchi. 2016. Review of West-Palaearctic Hebridae with description of a new species and redescription of *Hebrus fulvinervis* (Hemiptera: Heteroptera). Zootaxa 4147 (3): 201–239.

Lauck, D.R. & A.S. Menke. 1961. The higher classification of the Belostomatidae (Hemiptera). Annals of the Entomological Society of America 54: 644–657.

Linnavuori, R. 1964. Hemiptera of Egypt, with remarks on some species of the adjacent Eremian region. Annales Zoologici Fennici 1: 306–356.

Linnavuori, R. 1977. On the taxonomy of the subfamily Microveliinae (Heteroptera, Veliidae) of West and Central Africa. Annales Entomologici Fennici 43 (2): 41–61.

Mbogho, A.Y. & R.W. Sites. 2013. Naucoridae Leach, 1815 (Hemiptera: Heteroptera) of Tanzania. African Invertebrates 54 (2): 513–542.

Nieser, N. 2002. Guide to Aquatic Heteroptera of Singapore and Peninsular Malaysia. IV. Corixoidea. The Raffles Bulletin of Zoology 50 (1): 263–274.

Nieser, N. 2004. Guide to Aquatic Heteroptera of Singapore and Peninsular Malaysia. III. Pleidae and Notonectidae. The Raffles Bulletin of Zoology 52 (1): 79–96.

Novoselsky, T., P.-P. Chen & N. Nieser. 2018. A review of the giant water bugs (Hemiptera: Heteroptera: Nepomorpha: Belostomatidae) of Israel. Israel Journal of Entomology 48 (1): 119–141.

Perez-Goodwyn, P.J. 2006. Taxonomic revision of the subfamily Lethocerinae Lauck & Menke (Heteroptera: Belostomatidae). Stuttgarter Beiträge zur Naturkunde, Serie A (Biologie) 695: 1–71.

Perez-Goodwyn, P.J. 2009. Chapter 4. Anti-Wetting Surfaces in Heteroptera (Insecta): Hairy Solutions to Any Problem. Pages 55–76 *in*: S.N. Gorb (ed.), Functional Surfaces in Biology. Little Structures with Big Effects. 1. Springer, Dordrecht.

Poisson, R. 1933. Note sur les *Mesovelia* de la faune française. [Hem. Mesoveliidae]. Bulletin de la Société entomologique de France 38 (12): 181–187.

Poisson, R. 1936. Hemiptera. I. Aquatica. Mission scientifique de l'Omo. Tome III. Fascicule 26. Mémoires du Muséum national d'Histoire naturelle, Nouvelle sèrie 4: 191–218.

Poisson, R. 1948. Sur quelques Naucoridae africains des Collections du Musée du Congo (Hémipt.—Hétéropt.). Revue de Zoologie et de Botanique Africaines 61 (2-3): 202–221.

Poisson, R. 1949. Hémiptères aquatiques. Exploration du Parc National Albert. Mission G. F. de Witte (1933-1935). Vol. 58. Institut des Parcs National du Congo Belge, Brussels, 94 pp.

Poisson, R. 1952. XIII. – Hydrocorises. La Réserve naturelle intégrale du Mt Nimba. Fascicule I. Mémoires de l'Institut Français d'Afrique Noire (IFAN) 19 (1): 277–287.

Poisson, R. 1957. Hétéroptères aquatiques. Faune de France. Vol. 61. Fédération Française des Sociétés de Sciences Naturelles, Paris, 263 pp.

Poisson, R. 1965a. Catalogue des Hétéroptères Gerridae Leach, 1807, africano-malgaches. Bulletin de l'Institut Français d'Afrique Noire (IFAN) 27 sér. A (1): 1466–1503.

Poisson, R. 1965b. Catalogue des Insectes Hétéroptères Hydrocorises africano-malgaches de la famille des Nepidae (Latreille) 1802. Bulletin de l'Institut Français d'Afrique Noire (IFAN) 27 sér. A (1): 229–269.

Polhemus, D.A. & J.T. Polhemus. 2013. Guide to the aquatic Heteroptera of Singapore and Peninsular Malaysia. XI. Infraorder Nepomorpha—Families Naucoridae and Aphelocheiridae. The Raffles Bulletin of Zoology 61 (2): 665–686.

Polhemus, J.T. 1994. Stridulatory mechanisms in aquatic and semiaquatic Heteroptera. Journal of the New York Entomological Society 102 (2): 270–274,

Polhemus, J.T. 1995. Nomenclatural and synonymical notes on the genera *Diplonychus* Laporte and *Appasus* Amyot and Serville (Heteroptera: Belostomatidae). Proceedings of the Entomological Society of Washington 97 (3): 649–653.

Polhemus, J.T. & D.A. Polhemus. 2008. Global diversity of true bugs (Heteroptera; Insecta) in freshwater. Hydrobiologia 595: 379–391.

Popov, Y.A. 1970. Notes on the classification of the recent Naucoridae (Heteroptera, Nepomorpha). Bulletin de l'Académie Polonaise des Sciences. Série des Sciences Biologiques 18 (2): 93–98.

Sailer, R.I. 1948. The Genus *Trichocorixa* (Corixidae, Hemiptera). Pages 289–407 *in*: H.B. Hungerford (ed.), The Corixidae of the Western Hemisphere (Hemiptera). The University of Kansas Science Bulletin 32: 1–827.

Savage, A.A. 1999. Key to the larvae of British Corixidae. Freshwater Biological Association. Scientific Publications 57, Ambleside, Cumbria, 56 pp.

Schuh, R.T. & J.A. Slater. 1995. True Bugs of the World (Hemiptera: Heteroptera). Classification and Natural History. Cornell University Press, Ithaca, New York, XII + 336 pp.

Schuh, R.T. & C. Weirauch. 2020. True Bugs of the World (Hemiptera: Heteroptera). Classification and Natural History (Second Edition). Siri Scientific Press. Monograph Series 8, Manchester, UK, 767 + (1) pp. + 32 pls.

Stork, N.E. 2018. How Many Species of Insects and Other Terrestrial Arthropods Are There on Earth? Annual Review of Entomology 63: 31–45.

Štys, P. & J. Davidová-Vilímová. 1989. Unusual numbers of instars in Heteroptera: a review. Acta Entomologica Bohemoslovaca 86: 1–32.

Tamanini, L. 1947. Contributo ad una revisione del genere *Velia* Latr. e descrizione di alcune specie nuove (Hemiptera, Heteroptera, Veliidae). Memorie della Società Entomologica Italiana 26: 17–74.

Tamanini, L. 1955. IV contributo allo studio del genere *Velia* Latr. con la descrizione di quattro nuove entità. Bollettino della Società Entomologica Italiana 85: 35–44.

Tamanini, L. 1979. Eterotteri acquatici (Heteroptera: Gerromorpha, Nepomorpha). Guide per il Riconoscimento delle Specie Animali delle Acque Interne Italiane. Consiglio Nazionale delle Ricerche AQ/1/43, 6: 1–106.

Vepsäläinen, K. & S. Krajewski. 1986. Identification of the waterstrider (Gerridae) nymphs of Northern Europe. Annales Entomologici Fennici 52: 63–77.

Wang, Y.-H., H.-Y. Wu, D. Rédei, Q. Xie, Y. Chen, P.-P. Chen, Z.-E. Dong, K. Dang, J. Damgaard, P. Štys, Y.-Z. Wu, J.-Y. Luo, X.-Y. Sun, V. Hartung, S.M. Kuechler, Y. Liu, H.-X. Liu & W. Bu. 2017. When did the ancestor of true bugs become stinky? Disentangling the phylogenomics of Hemiptera–Heteroptera. Cladistics 35 (1) [2019]: 42–66.

Wang, Y.-H., F.F.F. Moreira, D. Rédei, P.-P. Chen, S.M. Kuechler, J.-Y. Luo, Y. Men, H.-Y. Wu & Q. Xie. 2021. Diversification of true water bugs revealed by transcriptome-based phylogenomics. Systematic Entomology 46: 339–356.

Weirauch, C. & R.T. Schuh. 2010. Systematics and evolution of Heteroptera: 25 years of progress. Annual Review of Entomology 56: 487–510. [volume publication 2011].

Ye, Z., J. Damgaard, H. Yang, M.B. Hebsgaard, T. Weir & W. Bu. 2019. Phylogeny and diversification of the true water bugs (Insecta: Hemiptera: Heteroptera: Nepomorpha). Cladistics 36 (1) [2020]: 72–87.

Zhang, Z.-Q. 2013. Phylum Arthropoda. Pages 17–26 *in*: Z.-Q. Zhang (ed.), Animal Biodiversity: An Outline of Higher-level Classification and Survey of Taxonomic Richness (Addenda 2013). Zootaxa 3703: 1–82.

Chapter 13

Order Coleoptera

Andrés Millán[1], Antonio J. García-Meseguer[1], Félix Picazo[2], Pedro Abellán[3] and David Sánchez-Fernández[1]
[1]*Department of Ecology and Hydrology, Faculty of Biology, University of Murcia, Spain,* [2]*Department of Ecology/Research Unit Modeling Nature, Faculty of Sciences, University of Granada, Spain,* [3]*Department of Zoology, Faculty of Biology, University of Seville, Spain*

Introduction

What is a true water beetle?

Coleoptera (beetles) is by far the most speciose order of insects on earth, with approximately 400,000 species described, although some biodiversity experts estimate that millions of species may roam the earth (Ślipiński et al., 2011). The morphology of adult beetles is characterized by a strongly sclerotized exoskeleton, elytra that cover functional wings, and a generally compact body. The presence of elytra is indeed one of the features that underpin their evolutionary success, whose origins date back to the late Paleozoic (Zhang et al., 2018).

Even though only a small percentage (<5%) of Coleoptera species are aquatic, they represent one of the major groups of freshwater arthropods (likely surpassed only by Diptera; Yee & Kehl, 2015). The definition of "water beetle" varies depending on the context and author. As pointed out by Jäch & Balke (2008), the difficulties in this definition are mainly related to the amount of time spent in contact with water, the level of submergence, the degree of water dependence, and the motivation for getting into contact with water (food, refuge). These factors are highly variable and can be displayed in various combinations, showing an astonishing disparity between, and even within, beetle families (Fig. 13.1A–D). Furthermore, the habitat preference of a beetle (Fig. 13.2A–F) could differ between life-history stages, and so, some beetle families are predominantly aquatic as adults, larvae, or both. Additionally, some species of other predominantly terrestrial families can be considered aquatic species or associated with water during at least one life-history stage.

In this chapter, we mainly focus on "true water beetles" as defined by Jäch (1998). True water beetles ("water beetles" hereafter) are submerged for most of the time during their adult stage (larvae and pupae may be aquatic or terrestrial). These beetles are usually provided with conspicuous morphological adaptations to aquatic life: for example, swimming hairs on legs, divided eyes, plastron, large claws, streamlined body form, etc. We have only made the exception of including Scirtidae and Psephenidae larvae in the family keys because, despite the fact that the adults are terrestrial, their larvae are strictly aquatic. Thus those families or genera whose adults are always predominantly terrestrial (other than the two above-mentioned families) have not been considered, including: (1) phytophagous taxa that live and feed on aquatic plants and can therefore appear submerged for at least some time in any developmental stage (e.g., few genera belonging to the families Chrysomelidae and Curculionidae); (2) predominantly terrestrial beetle taxa that occasionally or regularly stay submerged for a limited period (for hunting, feeding, seeking refuge) in any of their life stages (e.g., some Carabidae and Staphylinidae species); (3) riparian species that live close to the water edge during all their developmental stages but which do not enter water voluntarily (e.g., some species of the families Carabidae, Limnichidae, Heteroceridae, Georissidae, and Staphylinidae); and (4) several genera of Hydrophilidae considered total or predominantly terrestrial, such as *Cercyon, Cryptopleurum, Dactylosternum, Megasternum, Pachysternum*, and *Sphaeridium*.

Diversity and distribution

Water beetles are distributed in all biogeographic regions and on nearly all continents (except Antarctica). The Palearctic, the Neotropical, and the Afrotropical regions harbor the highest diversity of water beetles (with a very similar number of species), followed by the Oriental, the Australian, and the Pacific regions, and with the Nearctic

Identification and Ecology of Freshwater Arthropods in the Mediterranean Basin. DOI: https://doi.org/10.1016/B978-0-12-821844-0.00015-6

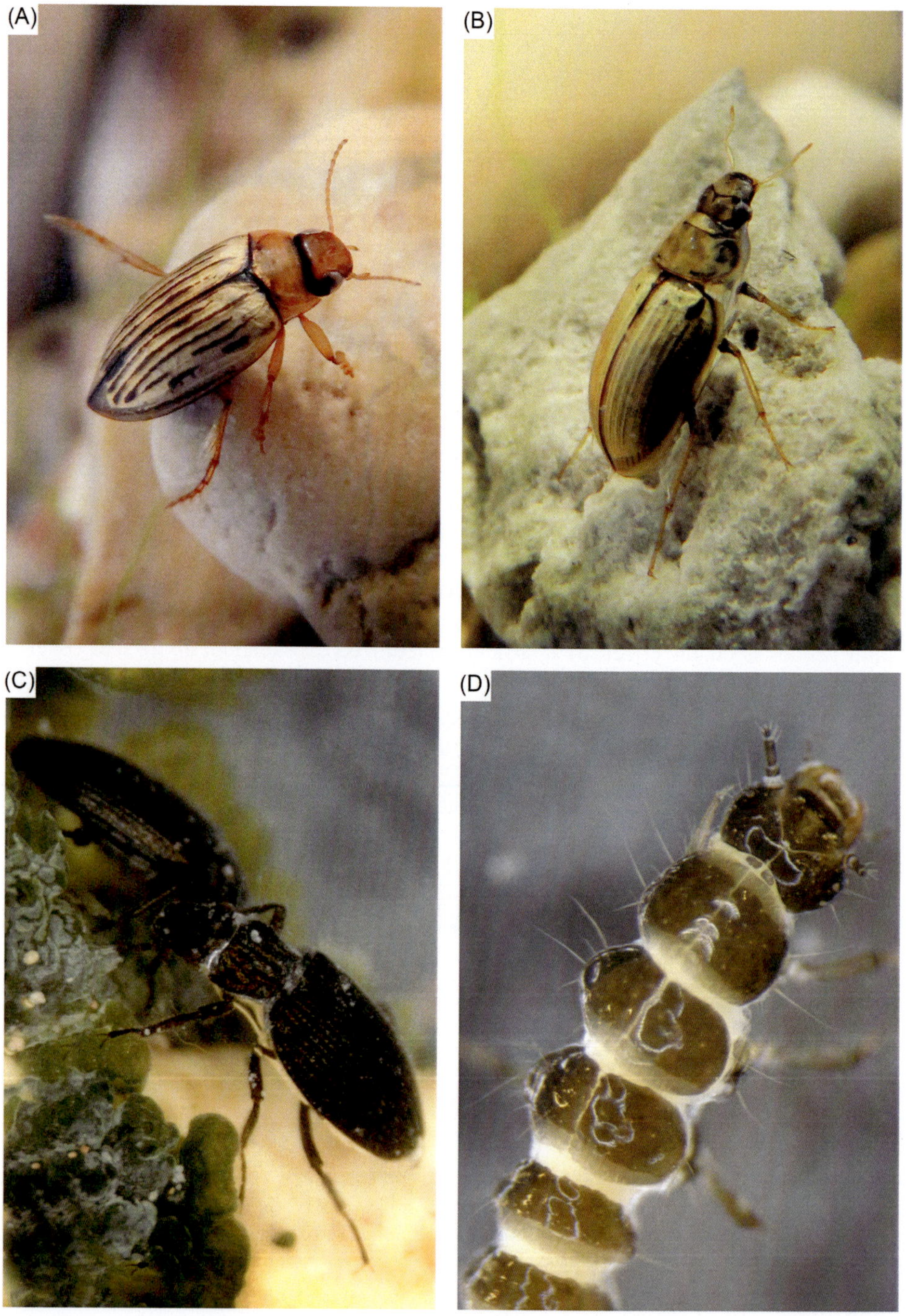

FIGURE 13.1 Some examples of species belonging to the most speciose water beetle families: (A) Dytiscidae, *Nebrioporus ceresyi* (Aubé, 1838); (B) Hydrophilidae, *Enochrus jesusarribasi* (C) Hydreanidae, *Ochthebius glaber* (D) Hydraenidae, *Ochthebius lejolisii* Mulsant & Rey, 1861 larva. *Aquatic Ecology group (University of Murcia, Spain).*

region being by far the poorest in terms of species diversity (see Jäch & Balke, 2008 for a detailed description of global diversity patterns of water beetles). Although there is a remarkable number of widespread Holarctic species, only very few species are found in more than two realms. Many species of most water beetle families display a high degree of local endemism. Water beetles are distributed across at least 20 families in three out of the four coleopteran suborders: Myxophaga, Adephaga, and Polyphaga. The other beetle suborder, Archostemata, lacks aquatic representatives. The most speciose families include Dytiscidae (c. 4000 species), Hydrophilidae (c. 2000 species), Hydraenidae (c. 1500 species), and Elmidae (c. 1500 species) (Jäch & Balke, 2008; Short, 2018).

FIGURE 13.2 Some examples of aquatic habitats occupied by water beetles: (A) Aguas Verdes lake (Sierra Nevada, Spain); (B) Estena River (Cabañeros, Spain); (C) Cala Reona supralittoral rockpools (Cabo de Palos, Spain); (D) Rambla Salada stream (Fortuna, Spain); (E) Khenifiss National Park (Morocco); (F) Traditional irrigation pond (Bagil, Spain). *Aquatic Ecology group (University of Murcia, Spain).*

Within the Palearctic region, the Mediterranean bioclimatic zone is a recognized biodiversity hotspot for water beetles, at least for certain families (Ribera, 2000; Millán et al., 2006, 2014). In this area, true water beetles are distributed across 14 families in the three coleopteran suborders previously mentioned, with a considerable variation in the number of water beetle genera and species among families [Table 13.1: Myxophaga (2 families; 2 genera), Adephaga (5 families; 59 genera), and Polyphaga (7 families; 38 genera)]. Mirroring the global patterns, Dytiscidae, Hydraenidae, Hydrophilidae, and Elmidae are the most speciose families across the Mediterranean Basin.

TABLE 13.1 Families of water beetles ("true water beetles" sensu Jäch, 1998) covered in this chapter. For each taxon, the common name (if used) is provided, along with their suborder, general geographic distribution, the number of aquatic genera in the Mediterranean region, the total number of known genera (approximated), and general habitat preferences.

Family (common name)	Suborder	Distribution	Aquatic genera (total)	Habitats
Hydroscaphidae (skiff beetles)	Myxophaga	Almost worldwide	1 (3)	Hygropetric habitats
Sphaeriusidae (minute bog beetles)	Myxophaga	Europe, North and Central America, Australia, Africa	1 (1)	Interstitial and marginal zones of running and/or standing water
Gyrinidae (whirligig beetles)	Adephaga	Worldwide except Antarctica	4 (13)	Surface of running and standing waters
Haliplidae (crawling water beetles)	Adephaga	Worldwide except Antarctica	3 (5)	Vegetated running and/or standing waters
Noteridae (burrowing water beetles)	Adephaga	Worldwide except Antarctica	4 (14)	Standing waters usually rich in organic matter
Hygrobiidae (squeak beetles)	Adephaga	Europe, Africa, China, Australia	1 (1)	Permanent and temporary standing waters
Dytiscidae (predaceous diving beetles)	Adephaga	Worldwide	47 (179)	A great variety of aquatic habitats
Dryopidae (long-toed water beetles)	Polyphaga	Nearly worldwide distribution	2 (33)	Running and standing waters
Elmidae (riffle beetles)	Polyphaga	Worldwide	9 (145)	Mainly running waters
Helophoridae	Polyphaga	Holarctic, Ethiopian	1 (1)	Edges of standing and running waters. Some species are terrestrial
Hydrochidae	Polyphaga	Worldwide	1 (1)	Edges of vegetated standing and running waters
Hydrophilidae (water scavenger beetles)	Polyphaga	Worldwide	21 (142)	Standing and running waters. Some genera are terrestrial (not considered in this study)
Spercheidae	Polyphaga	Almost worldwide distribution	1 (1)	Stagnant water with much organic matter
Hydraenidae (minute moss beetles)	Polyphaga	Cosmopolitan, except Antarctica	3 (40)	Interstitial zones of running and standing waters. Many are riparian and some terrestrial. Can colonize hypersaline waters, including marine rock pools. As far known, only larvae of *Ochthebius notabilis* Rosenhauer, 1856 can swim

General biology and ecology

Despite the great number of species, most aquatic coleopterans show clear habitat specificity. Indeed, the high diversity of life strategies displayed by water beetles allows them to be present in almost any kind of aquatic habitat (Jäch & Balke,

2008; Picazo et al., 2012; see Table 13.1). The broad spectrum of aquatic habitats occupied by water beetles ranges from cold and fast-moving water in mountain streams to brackish and stagnant waters of estuaries, lagoons, and saltmarshes, as well as lakes, ponds, reservoirs, puddles, phytotelmata, seepages, and groundwaters. They are even known to survive trapped in ice or in salt crystals (Gerdes et al., 1985; Millán et al., 2011). However, beetles do not inhabit the oceans, although numerous species live at their shores. For instance, they can be found in rockpools at the supralittoral strip, that is, the spray (or splash) area slightly above the intertidal zone (Villastrigo et al., 2020). One of the most readily observable habitat differences in aquatic Coleoptera (and macroinvertebrates in general) is whether they occupy running (lotic) or standing (lentic) waters. While some species can live in both habitats, most species are usually found in just one of them (Bilton et al., 2019; Table 13.1).

The distribution of water beetles varies within a specific ecosystem. These differences are intimately related to their morphological features, breathing strategies, and feeding behaviors (Yee & Kehl, 2015). Most water beetles dwell among rocks, detritus, or mud substrates. However, some species are neustonic and can glide on the water surface where they often congregate in large numbers (e.g., Gyrinidae adults). As distinctive features, the latter have the eyes completely divided into an upper and lower portion and bear a specialized organ on the antennal pedicellus that allows them to detect the small waves produced by any organism/object present in the water. A considerable number of water beetles are primarily pelagic as they can swim and dive (mainly adult Dytiscidae). The body of these species is streamlined and flattened (compared to other water beetles), and their hind legs are usually equipped with dense rows of swimming hairs. Other species (e.g., some species of Hydraenidae and Hydrophilidae) are not so hydrodynamic and can be found typically "walking" upside-down on the underside of the surface film, and even some of them exclusively appear on aquatic plants. Several families contain species that can be collected from vertical water surfaces (e.g., Hydraenidae and Elmidae), and some specimens can be found at depths of 10 m or more.

Respiration in water beetles conforms to three major modes: (1) reliance on self-contained air reserves (e.g., most Dytiscidae, Haliplidae, Hydrophilidae, Hydraenidae); (2) transcuticular respiration, with or without tracheal gills (larvae of most families); and (3) plastron respiration (e.g., adults of Elmidae and some genera of Dytiscidae, such as *Deronectes* or *Nebrioporus*, at least facultatively). Most species have an air reservoir that occupies the space beneath the elytra. These beetles must regularly return to the surface to renew their depleted air supplies. Beetles using plastron respiration mostly occupy fast-moving well-oxygenated water and are quite sensitive to pollutants. Many species, especially those that live in well-oxygenated running waters (most Elmidae adults), stay submerged for most of their lives and breathe by means of a microplastron (a very thin layer of air, held by a dense coating of hydrofuge setae).

The wide diversity of morphology and habitat preferences among beetle families is also reflected in the high variability of feeding modes which encompass predation (piercers, engulfers), herbivory (piercers, scrapers, shredders), and detritivory (collector-gathers). While the species from the suborder Myxophaga are largely scrapers of algae, the Adephaga are chiefly predaceous (except Haliplidae which have a mixed diet). They can engulf their prey or inject digestive enzymes through piercing mouthparts (especially larvae of Dytiscidae). Some large dytiscids can even attack small fish or tadpoles. However, scavenging, mainly in medium or bigger-sized Dytiscidae, may be much more common than currently recognized (Velasco & Millán, 1998). The species belonging to Polyphaga (e.g., Hydrophilidae, Elmidae) vary in feeding habits. Few families have been shown to have a unique food source, and it is common that herbivores also consume detritus and predators scavenge; thus, they may also consume plant material while feeding. Feeding mode could also vary with the life cycle. Some Hydrophilidae larvae are mainly carnivorous, while adults are mainly herbivores or even detritivores (White & Roughley, 2008).

In addition to preying on a range of species, many aquatic beetles are also food for other invertebrates (e.g., odonate larvae) and vertebrates (e.g., fishes). On land, dispersing terrestrial adults are presumably prey for various predators (e.g., spiders), and previous work suggests that larger beetles (Dytiscidae, Hydrophilidae) exhibit scars of encounters with predators (Peddle & Larson, 1999). Given the intensity of predation in aquatic systems, aquatic beetles have evolved various defense mechanisms, including chemicals, mechanical defense, and crypsis (Dettner, 2019).

Data on the life cycles and phenology of aquatic beetles are scarce and mainly come from a few studies using populations from typical temperate zones (Yee & Kehl, 2015), with few data including Mediterranean genera. Development times are often dependent on temperature; therefore, these are variable among different climatic zones, even for the same species. Thus, life cycles are often adapted to the climatic zones with hibernation during winter or summer droughts, an aspect that is especially important for water beetles from the Mediterranean Basin. Another aspect that further complicates the recognition of life cycle generalities is that larvae, pupae, and even eggs of many species are unknown. The lack of detailed information on oviposition modes and sites, ecology, habits, and phenology for the bulk of species makes a comprehensive summary of the life cycles of aquatic beetles rather problematic. Despite the high variety of life strategies exhibited by water beetles in terms of locomotion, respiration, feeding, or life cycles, among others, it is worth mentioning that most biological traits are generally available at the genus level (see, for instance, Picazo et al., 2012), so further studies refining them at the species level are required.

Systematic and phylogenetic relationships

Water beetles do not represent a single monophyletic clade, but they are an ecological guild (Hunt et al., 2007). There have been at least eight major independent transitions to the aquatic medium from terrestrial habitats throughout more than 300 Myr of evolutionary history of the Coleoptera, and the actual number may be double or more (Bilton et al., 2019). Additionally, several water beetle lineages have experienced one or more secondary transitions back to the terrestrial realm, or from freshwaters to saline waters (e.g., Hydrophilidae and Hydraenidae). Molecular studies propose the monophyly of each of the four suborders of Coleoptera. Although their evolutionary relationships are still debated, the most recent molecular data point out that Archostemata and Myxophaga are sister groups, Adephaga is sister to them, and Polyphaga is basal to all others (McKenna et al., 2015). Within Myxophaga, only Hydroscaphidae has been the subject of a recent comprehensive molecular phylogeny (Short et al., 2015). Molecular data suggest that the adephagan water beetle families (i.e., Hydradephaga) have invaded the aquatic environment only once (e.g., Ribera et al., 2002); however, recent studies (Baca et al., 2017, 2021; Vasilikopoulos et al., 2021) recover the idea that Adephaga are paraphyletic and that Gyrinidae are basal within Adephaga and probably colonized water first and separately to the remaining taxa, which agrees with their different morphology as well. Most species-rich families of Hydradephaga (e.g., Dytiscidae and Gyrinidae) have been the subjects of extensive phylogenetic studies in recent decades (e.g., Ribera et al., 2008; Miller & Bergsten, 2016; Baca et al., 2017), although the relationships within various genera, tribes, and subfamilies remain to be resolved (see Short, 2018 for a recent review).

Within the suborder Polyphaga, water beetles belong to the Staphyliniformia (e.g., superfamily Hydrophiloidea and family Hydraenidae) and Elateriformia (e.g., families Elmidae and Dryopidae). The monophyly of Hydrophiloidea and that of its six included families is not in dispute (Short, 2018). Except for Hydrophilidae (Short & Fikáček, 2013) and Helophoridae (Fikáček et al., 2012), the internal relationships of families within Hydrophiloidea have not been examined. The current Hydraenidae classification is largely based on the detailed morphological studies made by Perkins (1980, 1997), although the speciose genera *Hydraena* and *Ochthebius* have recently been the subject of several molecular phylogenies to explain their explosive diversity (e.g., Trizzino et al., 2013; Villastrigo et al., 2020). Finally, the relationships between and within the families of aquatic Elateriformia are the least understood among water beetle lineages (Short, 2018). Thus, no comprehensive phylogeny has been yet undertaken for the families Elmidae and Dryopidae, which are condensed in the superfamily Byrrhoidea.

Conservation and global change

Water beetles have a great potential for biodiversity and conservation assessment of Mediterranean aquatic ecosystems, showing some features that make them an excellent indicator group (see Sánchez-Fernández et al., 2006). Indeed, they have been used as a tool to address several conservation-related questions, such as setting conservation priorities for both species and areas and assessing the role of protected areas in their conservation (e.g., Abellán et al., 2007; Sánchez-Fernández et al., 2008), also taking into consideration the importance of conserving the evolutionary history of this group (Abellán et al., 2013). On the other hand, it is well known that global change is reducing the extent of suitable aquatic habitats in the Mediterranean mountains (Calosi et al., 2008), placing such taxa in double jeopardy. Indeed, global change involves multiple stressors operating synergistically, including increased temperatures and hypoxia (Verberk & Bilton, 2013; Gutiérrez-Cánovas et al., 2022). Furthermore, water beetles have been recently used to make important contributions to conservation biology and specifically to our knowledge of how aquatic insects could face global change (Arribas et al., 2012 a,b; Sánchez-Fernández et al., 2013; Pallarés et al., 2020).

Morphological characters needed for identification

To identify the families and genera of aquatic beetles, we need a good knowledge of the external morphology of both larvae and adults. In the case of the larvae, key morphological characters are the shape of the body, head, mandibles, number of segments in body, legs and antennae, the pilosity and length of the terminal cerci (urogomphi), and the presence of gills. Concerning adults, it is crucial to pay attention to the shape and body color, mainly in Dytiscidae.

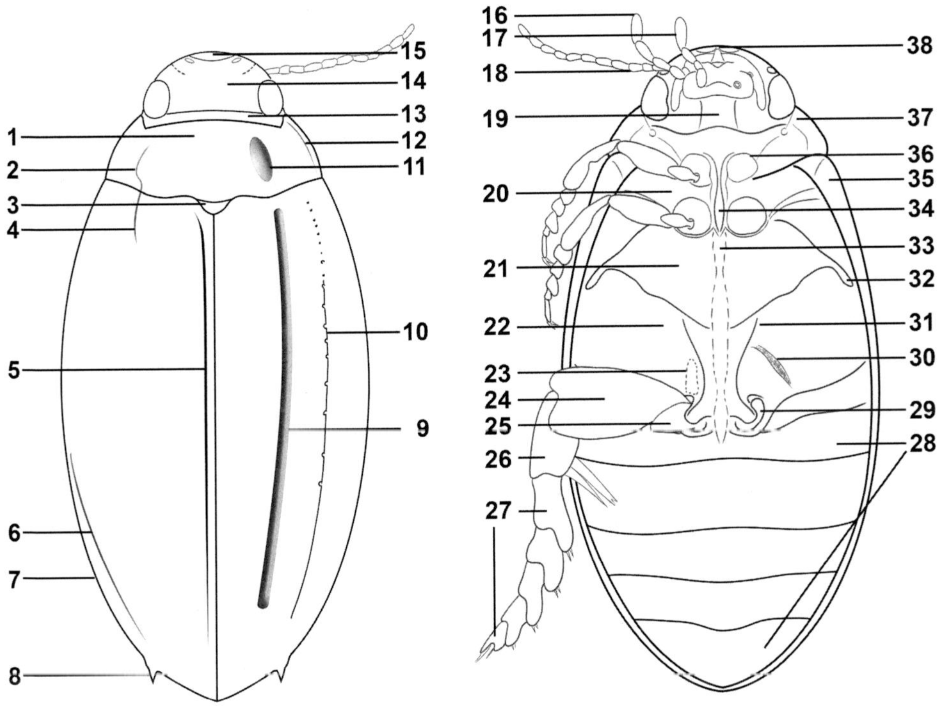

FIGURE 13.3 Main structures of an idealized water beetle (Larson et al., 2000). Dorsal and ventral view. (1) pronotum; (2) pronotal stria/carina/groove; (3) scutellum; (4) elytral stria; (5) elytral sutural line; (6) elytral lateral stria; (7) elytron; (8) elytral spine; (9) elytral carina; (10) elytral linear puncture row; (11) pronotal pit; (12) pronotal ridge; (13) occipital line; (14) head; (15) labrum; (16) maxillary palp; (17) labial palp; (18) antenna; (19) gula; (20) thoracic mesoventrite; (21) thoracic metaventrite; (22) thoracic metacoxal plate; (23) metaventrite plate; (24) metafemur; (25) metatrochanter; (26) metatibia; (27) metatarsomeres; (28) abdominal ventrites; (29) metacoxal process; (30) stridulatory file; (31) metacoxal line; (32) metaventral wing; (33) keel; (34) prosternal process (process of proventrite); (35) elytral epipleura; (36) procoxa; (37) prothoracic epipleura; (38) epistoma. *Modified from Larson et al. (2000).*

Also, other frequent characters are the shape and reticulation of the pronotum, elytra, and abdominal segments as well as the length, shape, and number of segments in the legs, maxillary palps, and antennae. There are many other external characteristics that could be observed to do a correct identification.

To facilitate the understanding of the keys, we presented an idealized adult water beetle (Fig. 13.3) showing most of the morphological structures used for its identification.

Sampling, preparation, and preservation

Aquatic Coleoptera are relatively easy to sample. As the name of the group denotes, they are mostly linked to aquatic habitats, with few exceptions. Such is the case of species showing high dispersal abilities that, during their flights, can be attracted towards different types of luminous reflections and thereby occasionally appear in terrestrial habitats (for instance, under lampposts at night or on shiny cars in the daytime).

As stated above, true water beetles can occupy a wide variety of aquatic habitats, from lentic to lotic, both at their margins and in areas of more open water. They can be found semiburied in the interstices of marginal sediments, amongst the aquatic vegetation, swimming or diving in open water, attached to bed substrata and stones in riffles and runs, or even deep in the hyporheic zone.

The most appropriate period to sample water beetles coincides with the climatic-linked activity peak of these organisms when they reach their highest abundances. In Mediterranean environments, this generally happens from mid-spring (in low altitudes, warmer areas) to late summer (at high altitudes). Nevertheless, they can be also collected in other seasons, given that most species have life cycles spanning more than a year. The sampling method will depend on specific objectives. To compare diversity among sites or dates, quantitative methods that allow the association of individuals and species to sampling efforts in terms of time, area, or volume are required. These methods are based on sampling tools with known dimensions, such as cylinders/cores (Fig. 13.4A) for lentic environments or surber samplers

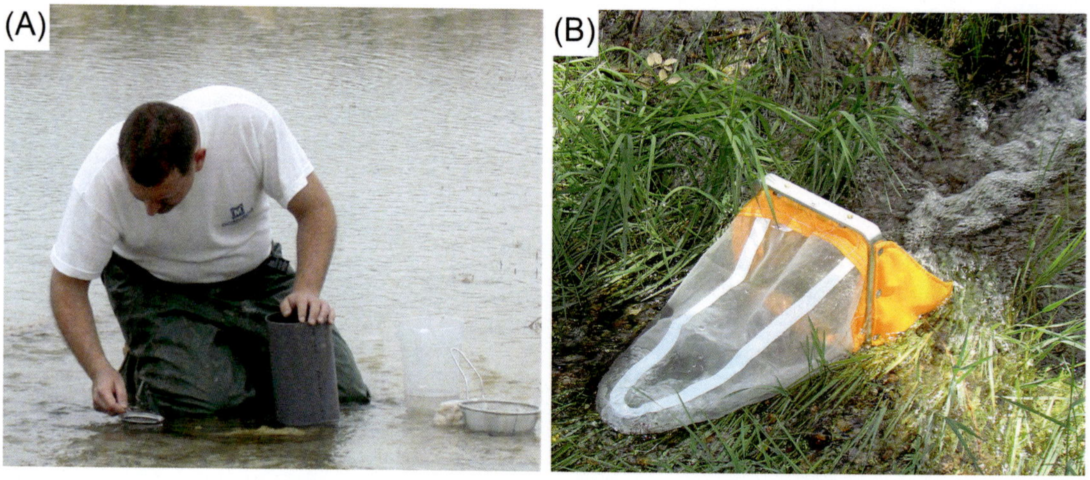

FIGURE 13.4 Sampling with core or cylinders (A) and surber (B). *Aquatic Ecology group (University of Murcia, Spain).*

FIGURE 13.5 Sampling with hand kick net accompanied by the use of small sieves, tweezers, brushes, and plastic white trays. *Aquatic Ecology group (University of Murcia, Spain).*

(Fig. 13.4B) for lotic environments. For extensive sampling aiming at elaborating species inventories and checklists—that is to maximize the number of recorded species richness per site and to obtain relative abundance in most areas—it is preferable to use aquatic kick nets with a D-shaped or triangular frame, with a 0.5–1.0 mm mesh and a 30–50 cm depth bag (Fig. 13.5). The use of the aquatic kick net should be ideally accompanied by the use of small sieves, tweezers, brushes, and white plastic trays.

Before beginning the sampling, the principal areas of the water body (pools, riffles, shores, with or without macrophytes/helophytes, different substrata, etc.) must be identified and roughly quantified so that the sampling effort will correspond to the proportion covered by each mesohabitat. One common method is to use "kick sampling"

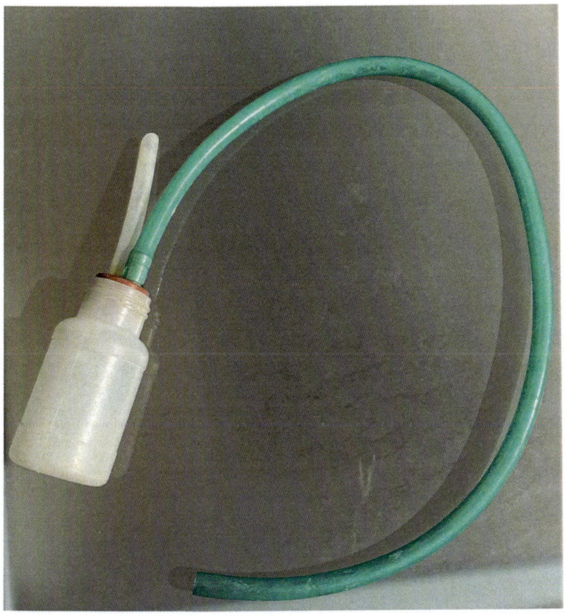

FIGURE 13.6 Mouth aspirator. *Aquatic Ecology group (University of Murcia, Spain).*

in a defined area by sweeping it with a commercial kick net of known mesh size. It is also highly recommendable to throw water onto the shore to wash off and capture edge-dwelling species with amphibious habits, which can sometimes be found floating on the water surface. After netting and draining the material collected, it can be placed on a plastic tray where any beetle can be removed using tweezers, a brush, or a mouth aspirator (Fig. 13.6). Beetles are usually fixed with a 70%–75% ethanol solution and preserved in plastic bottles. Alternatively, beetles can be efficiently killed using ethyl acetate using a small ball of saturated paper towel at the bottom of an empty bottle where the beetles are placed. If the aquatic sample contains a large amount of vegetation and/or detritus, it should be re-sieved the next day and more ethanol is added to ensure effective preservation of the invertebrates. If it is necessary to study DNA, the material should be preserved in 96% ethanol, changing it to 99% in the lab, and placing the sample in a −20°C freezer.

A key question in the field is when to stop the sampling. If a quantitative approach is desired, the standardization of the sampling effort by time and/or area is mandatory. Although no fixed rules are often established, a kick netting activity during 1–5 min per mesohabitat is adequate. However, if time and labor are not limited, it is desirable to extend the sampling until apparently no new species can be found. Under such an approach the preservation should be limited to just a few individuals per species for further identification in the lab. Once the individuals belonging to the most active species have been picked, it is recommendable to add some water into the tray and finish the sampling by collecting those species that float or swim. Also, in this case, it is ideal to standardize the sampling time to allow for cross-site richness comparisons. For instance, 30–60 min should be enough for sampling most sites. With a quantitative approach, all the material can be taken to the lab and carefully examined, sorted, and counted. It is also recommended to take notes describing the site (and pictures), including coordinates and elevation, as well as recording some basic environmental variables, such as water temperature, conductivity, salinity, and pH.

Although most genera can be distinguished just using external morphological characters, the way in which water beetles must be mounted and preserved for further identification to species level is similar to other beetles. In this regard, there are two main nonexclusive strategies: dry-mounting individuals or maintaining preservation in ethanol. One option is to dry mount one individual per species and locality or at least the most interesting or difficult species to identify and keep the rest in ethanol. The mounted individual should preferably be a male given that in most cases the distinction from closely related species requires the examination of male genitalia. Thus, the beetle and its genitalia will be placed separately on tiny rectangular cards or transparent methacrylate plates. The beetle can be properly attached by using a water-soluble gum, while the genitalia can be placed in a drop of DMHF (dimethyl hydantoin formaldehyde) on the same card/plate. DMHF is a water-soluble medium that allows removal for examination whenever needed. Once mounted, the card or plate will be pinned together with a paper label indicating the species, locality, date, and collector. A stereomicroscope will be needed for species identification for both dry- and wet-mounted species. In some cases, such as those involving small genitalia, the material must be examined under a compound microscope. After mounting dry specimens, it is necessary to place them in an appropriate box with some moth-proofing treatment to prevent pests. Finally, it is recommended to create a database containing all the information regarding the sample sites and any additional information of likely interest to future scientists examining those specimens and the local water beetle assemblages.

Keys to Adults and Larvae

The present work includes 3 suborders, 16 families, and 99 genera regularly found in the Mediterranean Basin. We provide keys to families for larvae and keys to families and genera for adults. We also provide a complete taxonomic list in the Appendix with an indication of the habitat preference for the different genera (i.e., lentic or lotic; and within lotic, the tendency for crenon, rhithron or potamon fluvial zones). The keys have been prepared based on field and laboratory experience and supported by the following literature: Nilsson (1996), Komarek (2003), Tachet et al. (2010), Foster & Friday (2011), Lawrence et al. (2011), Short et al. (2013, 2015, 2017); Foster et al. (2014), Biström et al. (2015), Thorp & Rogers (2015), Miller & Bergsten (2016), Nasserzadeh & Komarek (2017), Clarkson et al. (2018), and Rogers & Thorp (2019).

Key to Families

Coleoptera: Families

1. Elytra present (Fig. 13.7A) .... 2 (adults)
1'. Elytra absent (Fig. 13.7B) .... 16 (larvae)

Coleoptera: Families (Adults)

2. Metacoxae freely mobile and never exceeding the first abdominal segment (Fig. 13.7C, Polyphaga and Mixophaga in part) 3
2'. Metacoxae fused to venter and exceeding the first abdominal segment (Fig. 13.7D, Adephaga and Mixophaga in part) 11
3. Elytra truncated, exposing some abdominal segments (Fig. 13.7E) 4
3'. Elytra not truncated, covering all abdominal segments (Fig. 13.7F) 5
4. Maxillary palps shorter than antennae (Fig. 13.7G) .... Hydroscaphidae (one genus: *Hydroscapha*).
4'. Maxillary palps longer than antennae (Fig. 13.7H) .... **Hydraenidae** in part (one genus: *Limnebius*).
5(3'). Distinctly long and filiform antennae (Fig. 13.7J) .... **Elmidae**
5'. Short and nonfiliform antennae (Fig. 13.7K) 6
6. Maxillary palps shorter than antennae (Fig. 13.7G), body densely covered by hairs (Fig. 13.7L) and second segment of antennae expanded (Fig. 13.7N) .... **Dryopidae**
6'. Maxillary palps longer or equal than antennae (Fig. 13.7H and I), body glabrous or generally scarcely covered by hairs (Fig. 13.7M); second segment of antennae not expanded (Fig. 13.7O) 7
7. Maximum width of pronotum at its base (Fig. 13.8A); body shape oval and very convex (Fig. 13.8D and E) .... **Hydrophilidae**
7'. Maximum width of pronotum in the middle or apically (Fig. 13.8B and C); body shape elongated and flattened (Fig. 13.8F and G) except Spercheidae (see Fig. 13.8L) 8
8. Pronotum with five longitudinal grooves (Fig. 13.8H) .... Helophoridae (one genus: *Helophorus*).
8'. Pronotum without five longitudinal grooves (Fig. 13.8I–K) 9
9. Epistoma strongly indented along anterior margin; pronotum smooth, without pits (Fig. 13.8L) .... Spercheidae (one species: *Spercheus emarginatus* (Schaller, 1783)).
9'. Epistoma straight (Fig. 13.8M); pronotum with or without pits (Fig. 13.8I–K) 10
10. Prominent eyes (Fig. 13.8N); five pits always present on the pronotum (Fig. 13.8I) .... Hydrochidae (one genus: *Hydrochus*).
10'. Normal eyes (Fig. 13.8O); variable presence of pits or without pits (Fig. 13.8J, K) .... **Hydraenidae** in part
11(2'). Very small beetles, less than 1.5 mm long; spherical shape and antennae clubbed (Fig. 13.9A) .... Sphaeriusidae (one genus: *Sphaerius*).

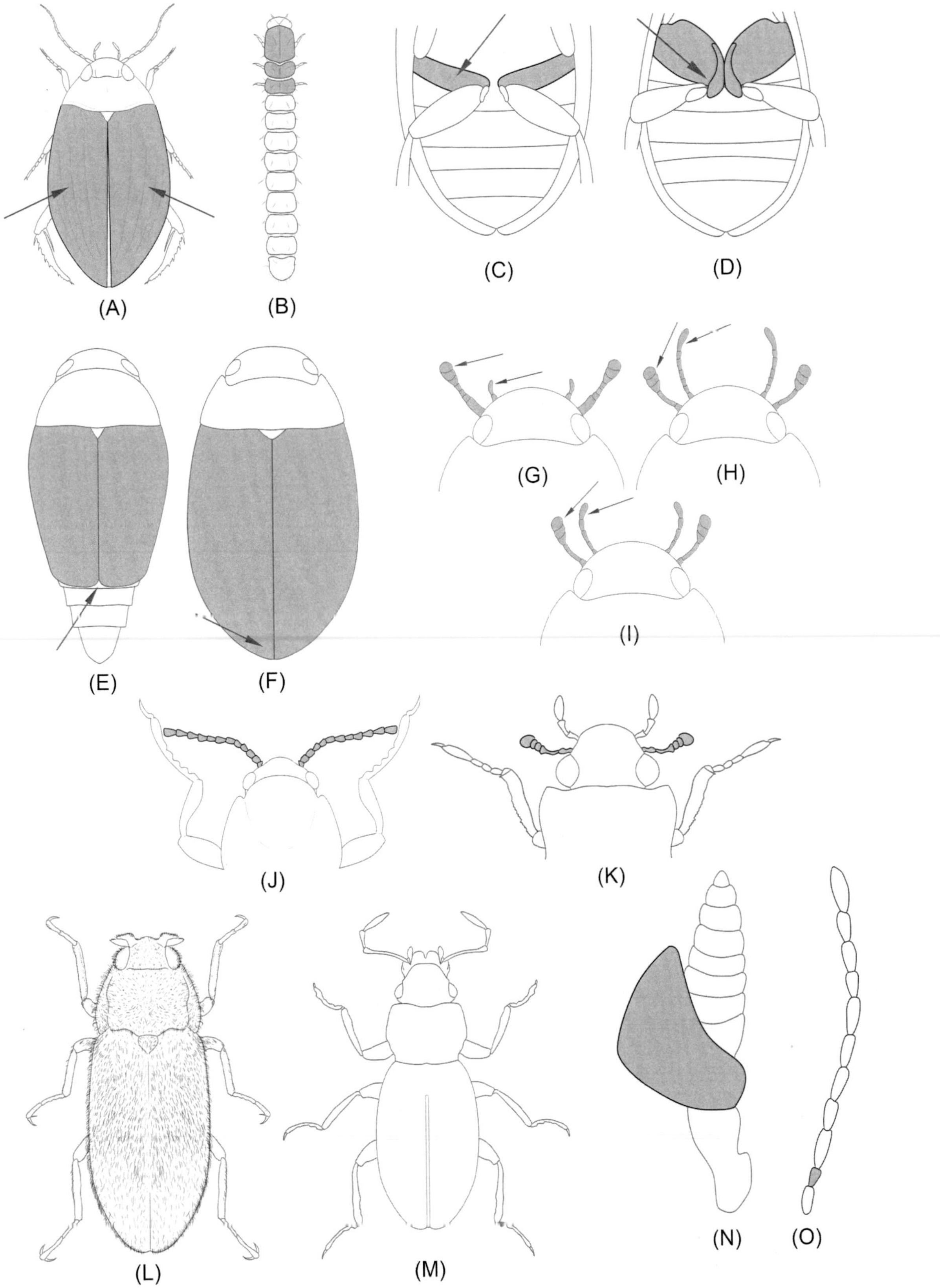

FIGURE 13.7 (A) Elytra in Dytiscidae adult; (B) Dryopidae larvae showing thoracic segments; (C) Hydrophilidae metacoxae; (D) Dytiscidae metacoxae; (E) *Limnebius* with truncated elytra showing abdominal segments; (F) *Agabus* elytra covering abdominal segments; (G) maxilar palps shorter than antennae in an idealized beetle; (H) maxilar palps longer than antennae in an idealized beetle; (I) maxilar palps equal than antennae in an idealized beetle; (J) Elmidae antennae; (K) *Ochthebius* antennae; (L) *Dryops* habitus; (M) *Hydraena* habitus; (N) second segment of antennae in Dryopidae; (O) second segment of antennae in Hydrophilidae.

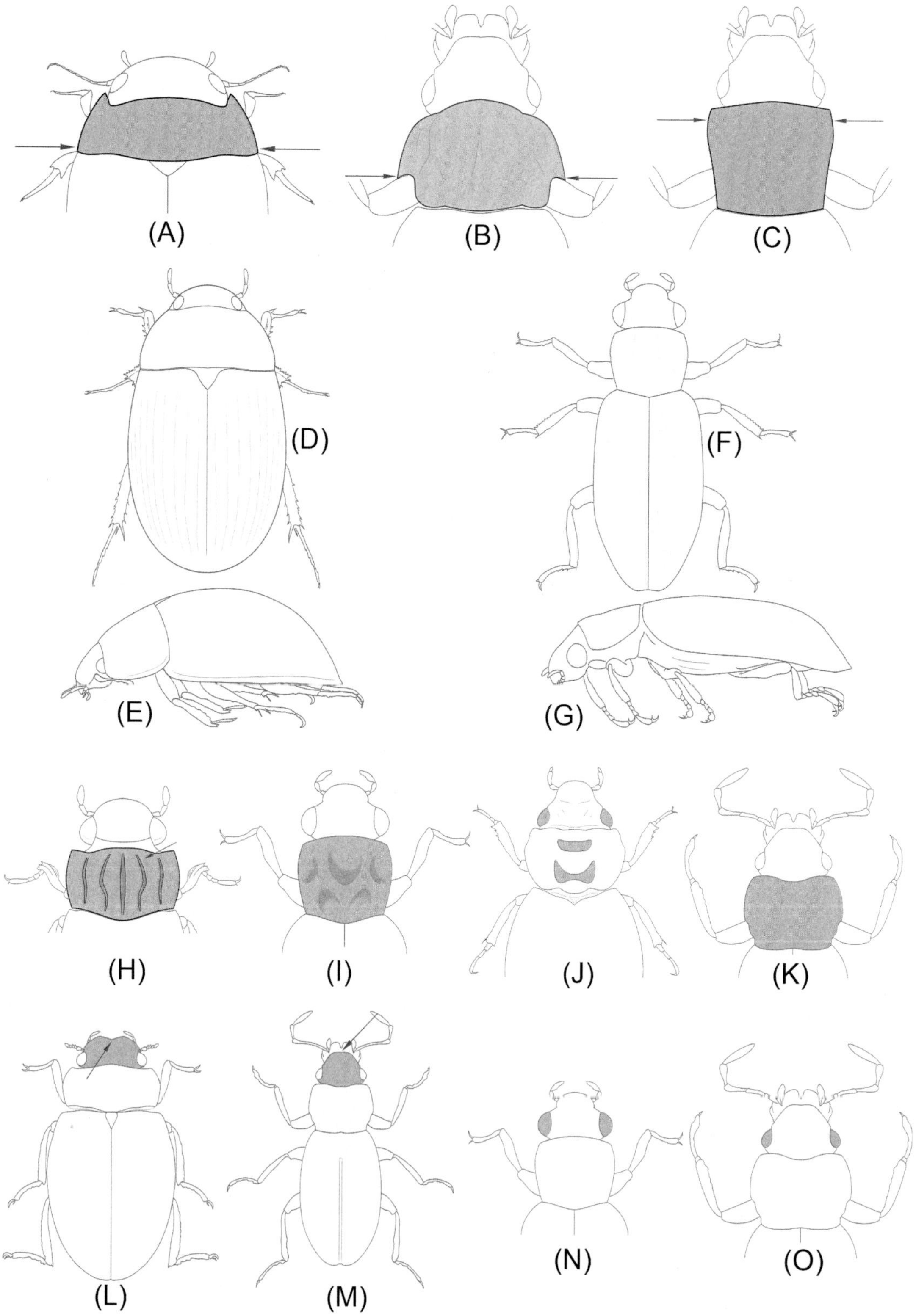

FIGURE 13.8 (A) Hydrophilidae pronotum; (B) *Ochthebius* pronotum; (C) Hydraena pronotum; (D) Hydrophilidae habitus, dorsal view; (E) Hydrophilidae lateral view; (F) *Hydrochus* habitus, dorsal view; (G) *Hydrochus* habitus, lateral view; (H) *Helophorus* pronotum; (I) *Hydrochus* pronotum; (J) *Ochthebius* pronotum; (K) *Hydraena* pronotum; (L) *Spercheus* habitus with detail of epistoma indentation; (M) *Hydraena* habitus with detail of epistome shape; (N) *Hydrochus* eyes; (O) Hydraena eyes.

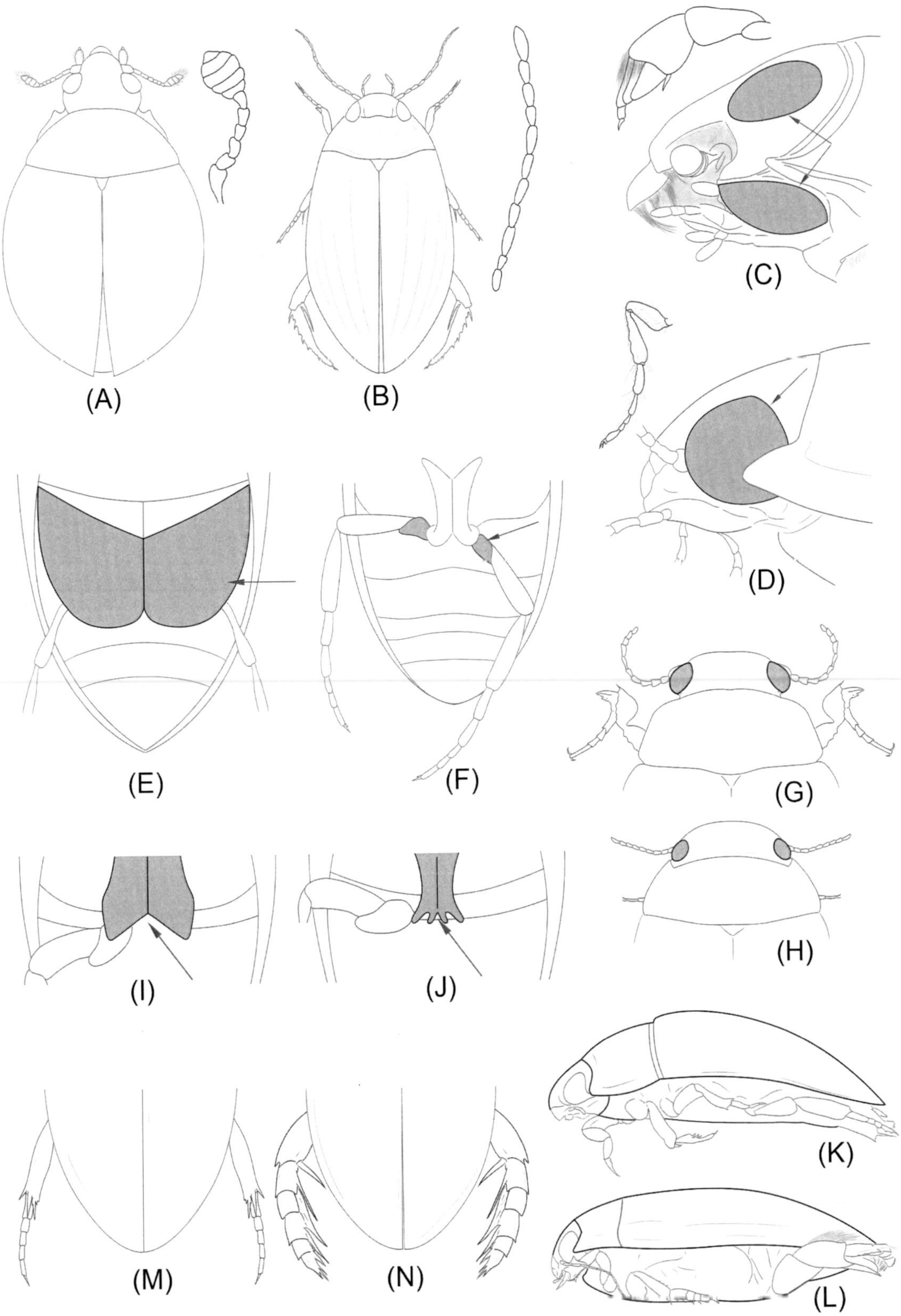

FIGURE 13.9 (A) *Sphaerius* habitus showing antennal shape; (B) Dytiscidae habitus showing antennal shape; (C) Gyrinidae head in lateral view showing middle leg detail; (D) Dytiscidae head in lateral view showing middle leg detail; (E) *Haliplus* hind coxae; (F) Dytiscidae hind coxae; (G) *Hygrobia* eyes; (H) Dytiscidae eyes; (I) Noteridae metacoxal process; (J) Dytiscidae metacoxal process; (K) Noteridae lateral view; (L) Dytiscidae lateral view; (M) Noteridae hind legs; (N) Dytiscidae hind legs.

11'. Variable body size, generally bigger than 1.5 mm; frequently oval shape and antennae never clubbed (Fig. 13.9B) 12
12. Eyes divided horizontally, giving the appearance of four eyes, and middle and hind legs shorter and broader than the front legs (Fig. 13.9C) .... **Gyrinidae**
12'. Eyes undivided and middle and hind legs as longer and broader as front legs (Fig. 13.9D) 13
13. Hind coxae with large, rounded plates, which cover half of the abdomen and the basal half of the hind femora (Fig. 13.9E) .... **Haliplidae**
13'. Hind coxae normal, not expanded and covering part of the abdomen (Fig. 13.9F) 14
14. Eyes strongly protuberant from the head (Fig. 13.9G) .... Hygrobiidae (one species: *Hygrobia hermanni* (Fabricius, 1775)).
14'. Eyes not so protuberant from head (Fig. 13.9H) 15
15. Metacoxal process wider than long, with posterior margin V-shaped (Fig. 13.9I); body highly convex dorsally and flattened ventrally (Fig. 13.9K); hind legs without swimming hairs (Fig. 13.9M) .... **Noteridae**
15'. Metacoxal processes longer than wide (Fig. 13.9J); body convex, both dorsally and ventrally (Fig. 13.9L); swimming hairs on hind legs (Fig. 13.9N) .... **Dytiscidae**

Coleoptera: Families (Larvae)

16(1'). Legs composed of four articles (Fig. 13.10A) 17
16'. Legs composed of five articles (Fig. 13.10B) 25
17. Eight abdominal segments (Fig. 13.10C) 18
17'. More than eight abdominal segments (Fig. 13.10D and E) 19
18. Small antennae, shorter or of similar length than head (Fig. 13.10F) .... Hydrophilidae
18'. Long antennae, clearly longer than head (Fig. 13.10G) .... Scirtidae (terrestrial adults).
19(17'). Head always visible dorsally (Fig. 13.10H) 20
19'. Head hidden dorsally (Fig. 13.10I) .... Psephenidae (terrestrial adults, one species: *Eubria palustris* (Germar, 1818)).
20. Antennae with two segments (Fig. 13.10J) 21
20'. Antennae with more than two segments (Fig. 13.10K) 22
21. Abdominal segments 1–8 with lateral digitiform lobes (Fig. 13.10L) .... Spaheriusidae
21'. Only abdominal segments 1 and 8 have lateral digitiform lobes (Fig. 13.10M) .... Hydroscaphidae
22(20'). Nine abdominal segments (Fig. 13.10N) 23
22'. Ten abdominal segments (Fig. 13.10O) 24
23. Gills present in the last abdominal segment (Fig. 13.10P) .... Elmidae
23'. Gills absent in the last abdominal segment (Fig. 13.10Q) .... Dryopidae
24(22'). Body shape elongated, with articulated urogomphi on abdominal segment nine (Fig. 13.11A) .... Hydraenidae
24'. Body shape globular, without articulated urogomphi (Fig. 13.11B) .... Spercheidae
25 (16'). Gills present in the abdominal segments (Fig. 13.11C) 26
25'. Gills absent in the abdominal segments (Fig. 13.11D) 27
26. Abdomen with 10 segments (Fig. 13.11E) and conspicuous hooks in distal abdominal segment (Fig. 13.11F) .... Gyrinidae
26'. Abdomen with 8 segments; without hooks in distal abdominal segment (Fig. 13.11G) .... Hygrobidae
27(25'). Legs with only one claw (Fig. 13.11H) .... Haliplidae
27'. Legs with two claws (Fig. 13.11I) 28
28. Head elongated (Fig. 13.11J) or rounded (Fig. 13.11K); with slender, channeled mandibles (Fig. 13.11L); .... Dytiscidae
28'. Always with rounded head (Fig. 13.11K); robust, denticulate and nonchanneled mandibles (Fig. 13.11M) .... Noteridae

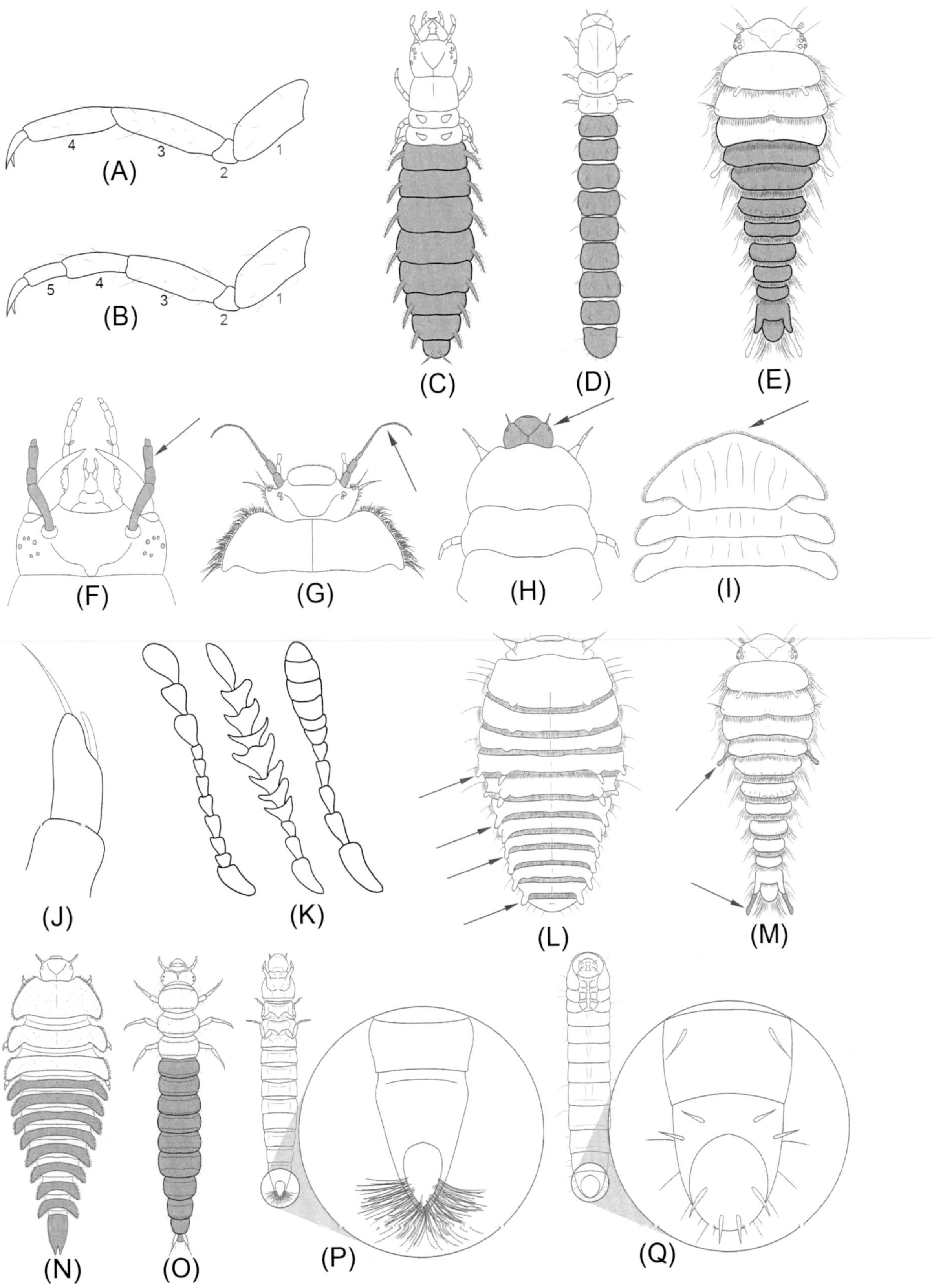

FIGURE 13.10 (A) Hydrophilidae posterior leg; (B) Elmidae posterior leg; (C) Hydrophilidae larvae; (D) Dryopidae larvae; (E) *Hydroscapha* larvae; (F) head in dorsal view of Hydrophilidae larvae; (G) head and first thoracic segment of Scirtidae larvae; (H) head and three first segments of Elmidae in dorsal view; (I) three first segments of *Eubria* larvae in dorsal view; (J) Mixophaga antennae; (K) different types of antennae in Polyphaga; (L) *Sphaerius* larvae; (M) *Hydroscapa* larvae; (N) Elmidae larvae; (O) Hydreaenidae larvae; (P) detail of last segment in ventral view of Elmidae larvae; (Q) detail of last segment in ventral view of Dryopidae larvae.

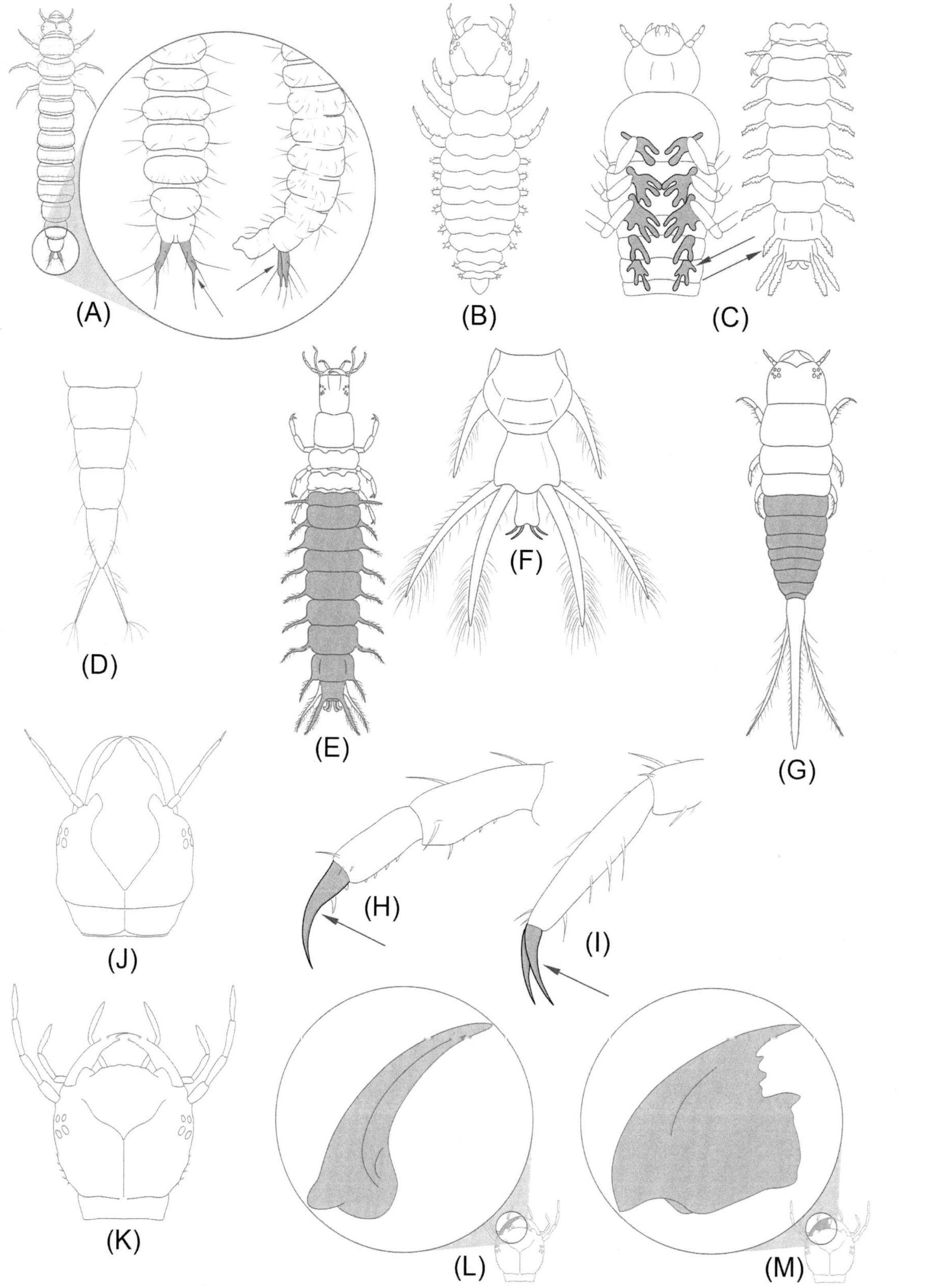

FIGURE 13.11 (A) Detail of last segments in Hydraenidae larvae; (B) body shape of *Spercheus* larvae; (C) branchial detail in Gyrinidae larvae; (D) last segments of Dytiscidae larvae; (E) larvae of Gyrinidae; (F) detail of hooks in the last segment of Gyrinidae larvae; (G) larvae of Hygrobiidae; (H) detail of fore leg claws in Haliplidae; (K) detail of fore leg claws in Dytiscidae; (J) Dytiscidae with elongated head; (K) Dytiscidae with rounded head; (L) mandible slender and channeled in Dytiscidae; (M) mandible robust and not channeled in Noteridae.

Keys to Genera (Adults)

Coleoptera: Dryopidae: Genera

1. Pronotum with a clear, entire sublateral carina on each side (Fig. 13.12A) .... *Dryops*
1'. Pronotum without sublateral carinae (Fig. 13.12B) .... *Pomatinus*
(one species: *P. substriatus* Müller, 1806)

Coleoptera: Elmidae: Genera

1. Body length >5.5 mm .... *Potamophilus*
(one species: *P. acuminatus*, Fabricius, 1792)
1'. Body length <5.5 mm 2
2. Pronotum smooth, without carinae, grooves or pits (Fig. 13.12C) .... *Riolus*
2'. Pronotum with carinae, grooves, pits, or tubercles and U-shaped carina (Fig. 13.12D, E, F, G) 3
3. Pronotum with a longitudinal carina on each side (Fig. 13.12D) 4
3'. Pronotum with grooves, pits, or tubercles and U-shaped or oblique carinae (Fig. 13.12E, F, G) 7
4. Scutellum conspicuous, large and wide (Fig. 13.12H) 5
4'. Scutellum very narrow, barely visible (Fig. 13.12I) 6
5. Body length clearly > 2.5 mm .... *Dupophilus*
(one species: *D. brevis* Mulsant & Rey, 1872)
5'. Body length clearly <2.5 mm .... *Oulimnius*
6(4'). Elytra without sublateral carina (Fig. 13.12J) .... *Limnius*
6'. Elytra with a sublateral longitudinal carina on each side (Fig.13.12K) .... *Esolus*
7(3'). Pronotum with a U-shaped carina, surrounding a hollow and elytra without carina (Fig. 13.12L) .... *Elmis*
7'. Pronotum and elytra with grooves and/or tubercles and carinae (Fig. 13.12M, N) 8
8. Legs not as long as body; pronotum with a longitudinal central pit surrounded by weak oblique carina; elytra with marked longitudinal carinae (Fig. 13.12M) .... *Stenelmis*
8'. Legs as long or even longer than the entire body; two shiny and hairy tubercles on the pronotum and the shoulders of the elytra (Fig. 13.12N) .... *Macronychus*
(one species: *M. quadrituberculatus* Müller, 1806).

Coleoptera: Gyrinidae: Genera

1. Large body size (9–15 mm); scutellum concealed (Fig. 13.13A, B) .... *Dineutus*
1'. Small body size (4–8 mm), scutellum exposed (Fig. 13.13C, D) 2
2. Body covered with hairs (Fig. 13.13E) .... *Orectochilus*
2'. Body without pilosity (Fig. 13.13F, G) 3
3. Pronotum with a transverse furrow/groove/sulcus and without a lateral yellow line (Fig. 13.13F) .... *Gyrinus*
3'. Pronotum smooth and with a lateral yellow line that continues on the elytra (Fig. 13.13G) .... *Aulonogyrus*

Coleoptera: Haliplidae: Genera

1. Pronotum with nearly parallel sides (Fig. 13.13H) .... *Brychius*
1'. Pronotum clearly wider in basal part (Fig. 13.13I) 2
2. Most of the head occupied by the eyes (Fig. 13.13J); pointed metacoxal plates (Fig. 13.13L) .... *Peltodytes*
2'. Eyes occupying a minor part of the head (Fig. 13.13K); not pointed metacoxal plates (Fig. 13.13M) .... *Haliplus*

Coleoptera: Noteridae: Genera

1. Metacoxal processes notably notched (Fig. 13.14A) .... *Neohydrocoptus*
(one species: *N. jaechi* (Wewalka, 1989)).
1'. Metacoxal processes without notches (Fig. 13.14B) 2
2. Metaventral plate, prosternal and metacoxal processes covered by dense hairs and antennae with uniform segments in both sexes (Fig. 13.14C) .... *Canthydrus*

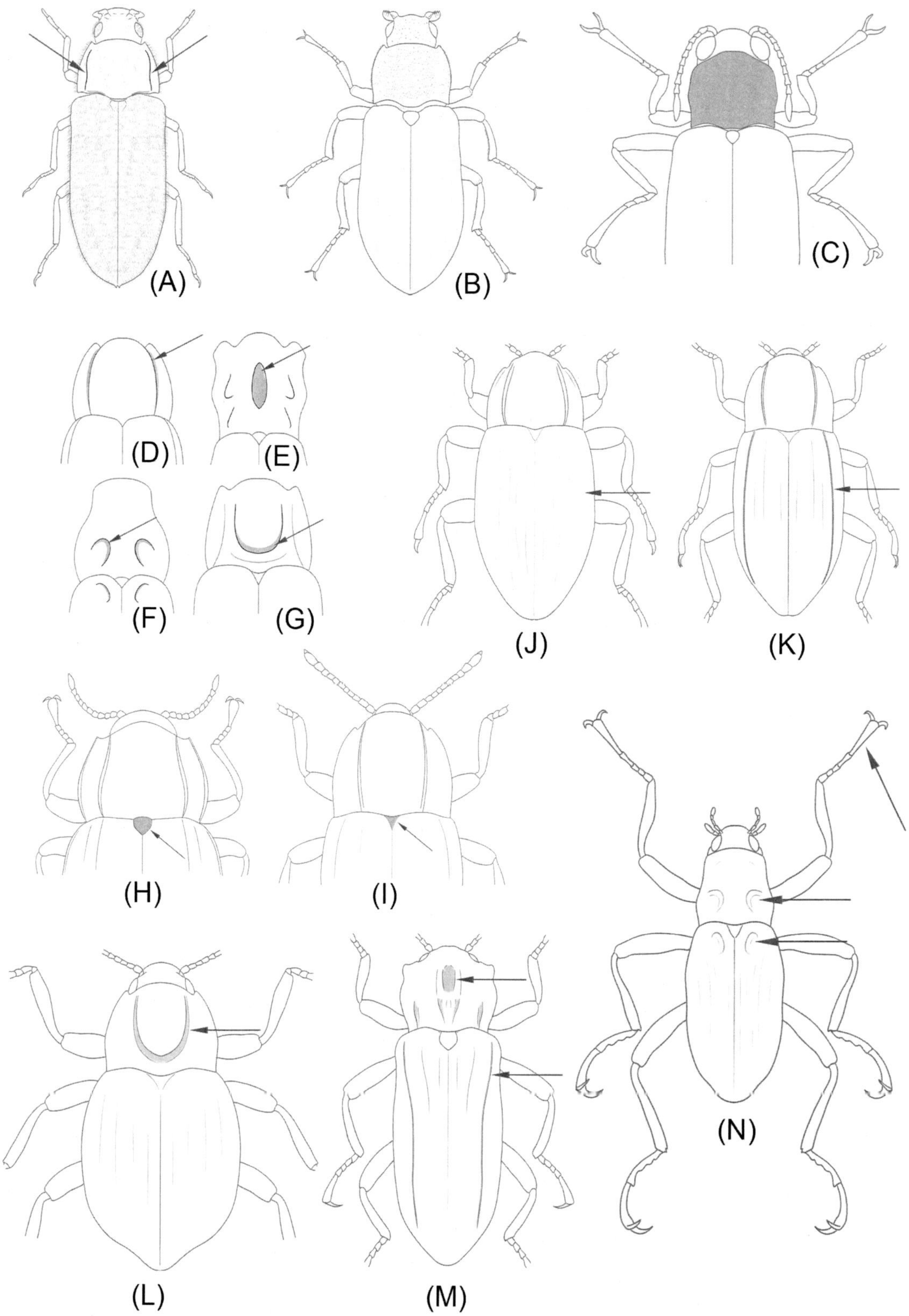

FIGURE 13.12 (A) *Dryops* habitus; (B) *Pomatinus* habitus; (C) *Riolus* pronotum; (D) *Esolus* pronotum; (E) *Stenelmis* pronotum; (F) *Macronychus* pronotum; (G) *Elmis* pronotum; (H) *Oulimnius* scutellum; (I) *Esolus* scutellum; (J) *Limnius* habitus; (K) *Esolus* habitus; (L) *Elmis* habitus; (M) *Stenelmis* habitus; (N) *Macronychus* habitus.

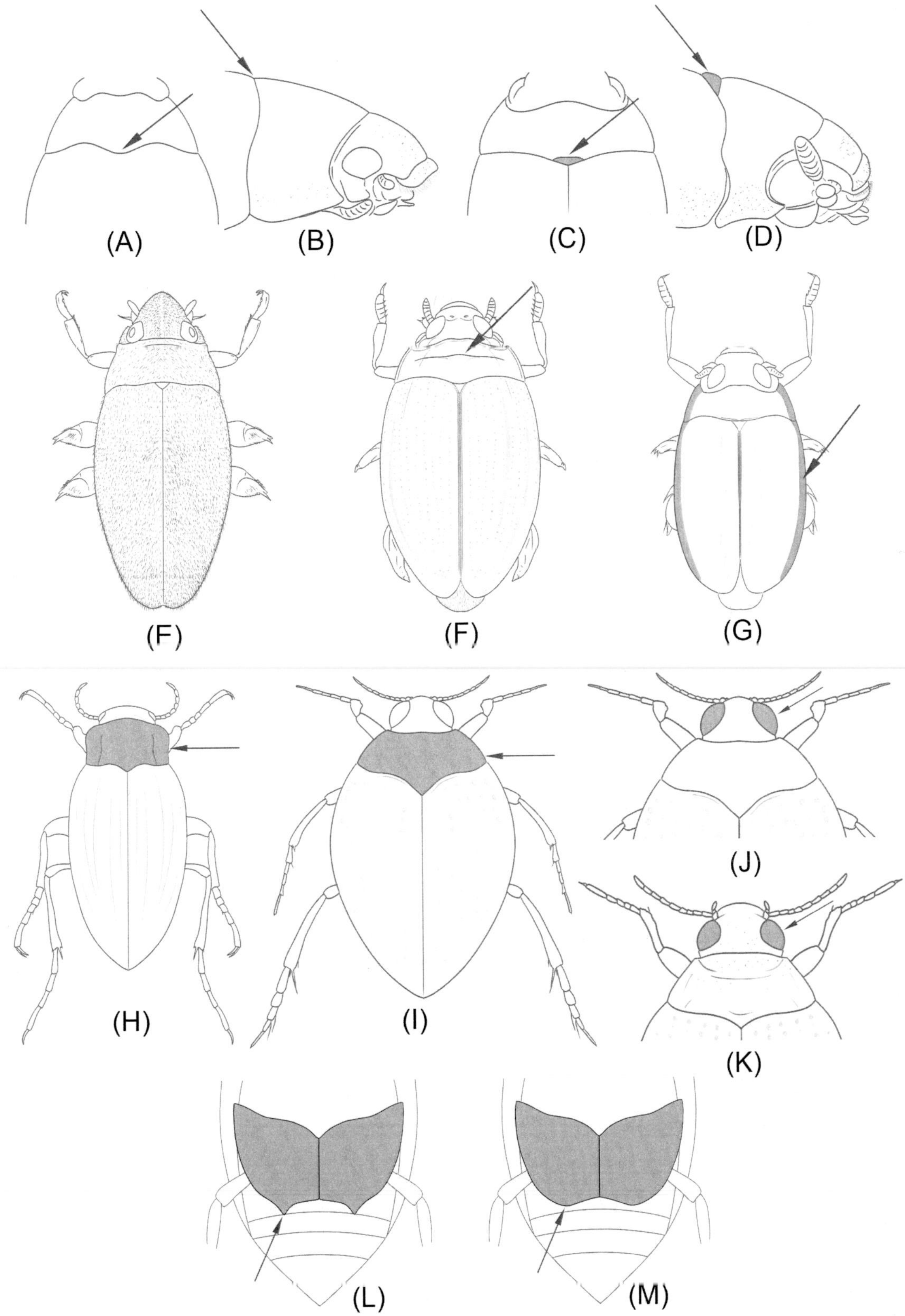

FIGURE 13.13 (A) *Dineutus* showing absence of the scutellum in dorsal view; (B) *Dineutus* showing absence of the scutellum in lateral view; (C) *Gyrinus* showing the scutellum in dorsal view; (D) *Gyrinus* showing the scutellum in lateral view; (E) *Orectochilus* habitus; (F) *Gyrinus* habitus showing detail of transverse lateral furrow; (G) *Aulonogyrus* habitus showing detail of lateral yellow line; (H) *Brychius* habitus showing detail of pronotum; (I) *Peltodytes* habitus showing detail of pronotum; (J) detail of *Peltodytes* eyes; (K) detail of *Haliplus* eyes; (L) *Peltodytes* metacoxal plates; (M) *Haliplus* metacoxal plates.

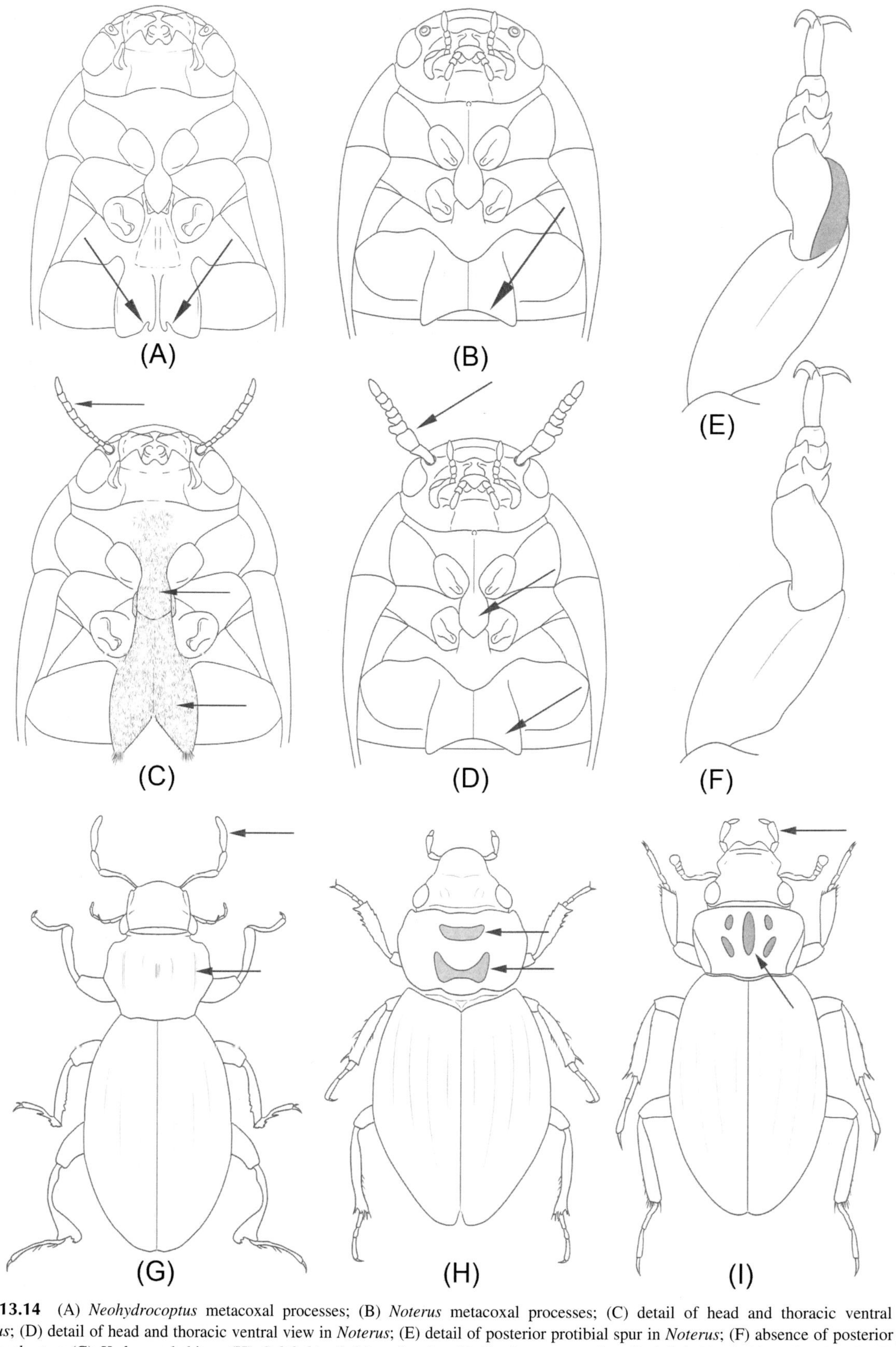

FIGURE 13.14 (A) *Neohydrocoptus* metacoxal processes; (B) *Noterus* metacoxal processes; (C) detail of head and thoracic ventral view in *Canthydrus*; (D) detail of head and thoracic ventral view in *Noterus*; (E) detail of posterior protibial spur in *Noterus*; (F) absence of posterior protibial spur in *Synchortus*; (G) *Hydraena* habitus; (H) *Ochthebius* habitus showing detail of pronotum pits; (I) *Ochthebius* habitus showing details of palpi and pronotum central sulcum.

2'. Metaventral plate, prosternal and metacoxal processes without hairs and males with some distal antennal segments enlarged (Fig. 13.14D) 3
3. Posterior protibial spur present (Fig. 13.14E) .... *Noterus*
3'. Posterior protibial spur absent (Fig. 13.14F) .... *Synchortus*
(one species: *S. imbricatus* (Klug, 1853))

Coleoptera: Hydraenidae: Genera

(except genus *Limnebius*, see 4' in Family keys for adults)
1. Maxillary palps much longer than antennae and pronotum with a longitudinal shallow depression on each side (Fig. 13.14G) .... *Hydraena*
1'. Maxillary palps equal to or shorter than antennae and pronotum with two transverse hollows/pits or a central longitudinal sulcus/furrow (Fig. 13.14H, I) .... *Ochthebius*

Coleoptera: Dytiscidae: Genera

1. Eyes absent (Fig. 13.15A); cuticle depigmented 2
1'. Eyes present (Fig. 13.15B); cuticle pigmented 4
2. Sublateral pronotal stria absent, and constriction present at the junction of pronotum and elytra (Fig. 13.15A) .... *Iberoporus* in part
2'. Sublateral pronotal stria present, and no constriction at the junction of pronotum and elytra (Fig. 13.15C) 3
3. Head regularly rounded (Fig. 13.15D); gap present between the prosternal process and the anteromedial metaventral process (Fig. 13.15E) .... *Siettitia*
3'. Head subrectangular (Fig. 13.15C); prosternal and anteromedial metaventral processes somewhat overlapped (Fig. 13.15F) .... *Etruscodytes*
(one species: *E. nethuns* Mazza, Cianferoni & Rocchi, 2013).
4(1'). Scutellum visible (Fig. 13.15G) 5
4'. Scutellum not visible (Fig. 13.15H) 19
5. Eyes anteriorly rounded, not emarginated (Fig. 13.15I and J) 6
5'. Eyes anteriorly emarginated (Fig. 13.15K and L) 13
6. Metacoxal lines absent or very closely approximated (Fig. 13.15M and N) 7
6'. Metacoxal lines distinctive, broadly separated or subparallel (Fig. 13.15O) 8
7. Metacoxal line absent (Fig. 13.15M) .... *Rhantaticus*
(one species: *R. congestus* Klug, 1833)
7'. Metacoxal line very closely approximated (Fig. 13.15N) .... *Liopterus*
8(6)'. Metatibia very short, with spurs different in size and shape (Fig. 13.15P) .... *Cybister*
8'. Metatibia normal, with spurs similar in size and shape (Fig. 13.15Q) 9
9. Apices of metatibial spurs bifid (Fig. 13.15 P) 10
9'. Apices of metatibial spurs simple (Fig. 13.15 Q) 11
10. Surfaces of pronotum, elytra, metaventrite, metacoxae, and abdominal segments with dense macropunctation (Fig. 13.15R) .... *Acilius*
10'. Surfaces of pronotum, elytra, metaventrite, metacoxae, and abdominal segments with fine and sparse punctation (Fig. 13.15S) .... *Graphoderus*
11(9'). Anterolateral margin of metaventral wing straight (Fig. 13.16A) .... *Hydaticus*
11'. Anterolateral margin of metaventral wing curved (Fig. 13.16B) 12
12. Posterolateral margin of elytra with a series of short spines (Fig. 13.16C); body length clearly smaller than 19 mm .... *Eretes*
12'. Posterolateral margin of elytra without spines (Fig. 13.16D); body length clearly longer than 19 mm .... *Dytiscus*
13(5'). Metafemur with a distinct brush of setae at the anteroapical angle (Fig. 13.16E) 14
13'. Metafemur without brush of setae at the anteroapical angle (Fig. 13.16F) 16
14. Prosternal process with lateral bead expanded (Fig. 13.16G) .... *Platambus*
14'. Prosternal process with lateral bead not expanded (Fig. 13.16H) 15
15. Anterior clypeal margin with bead (marginal groove) continuous (Fig. 13.16I) .... *Agabus*
15'. Anterior clypeal margin with bead clearly interrupted in the middle (Fig. 13.16J) .... *Ilybius*

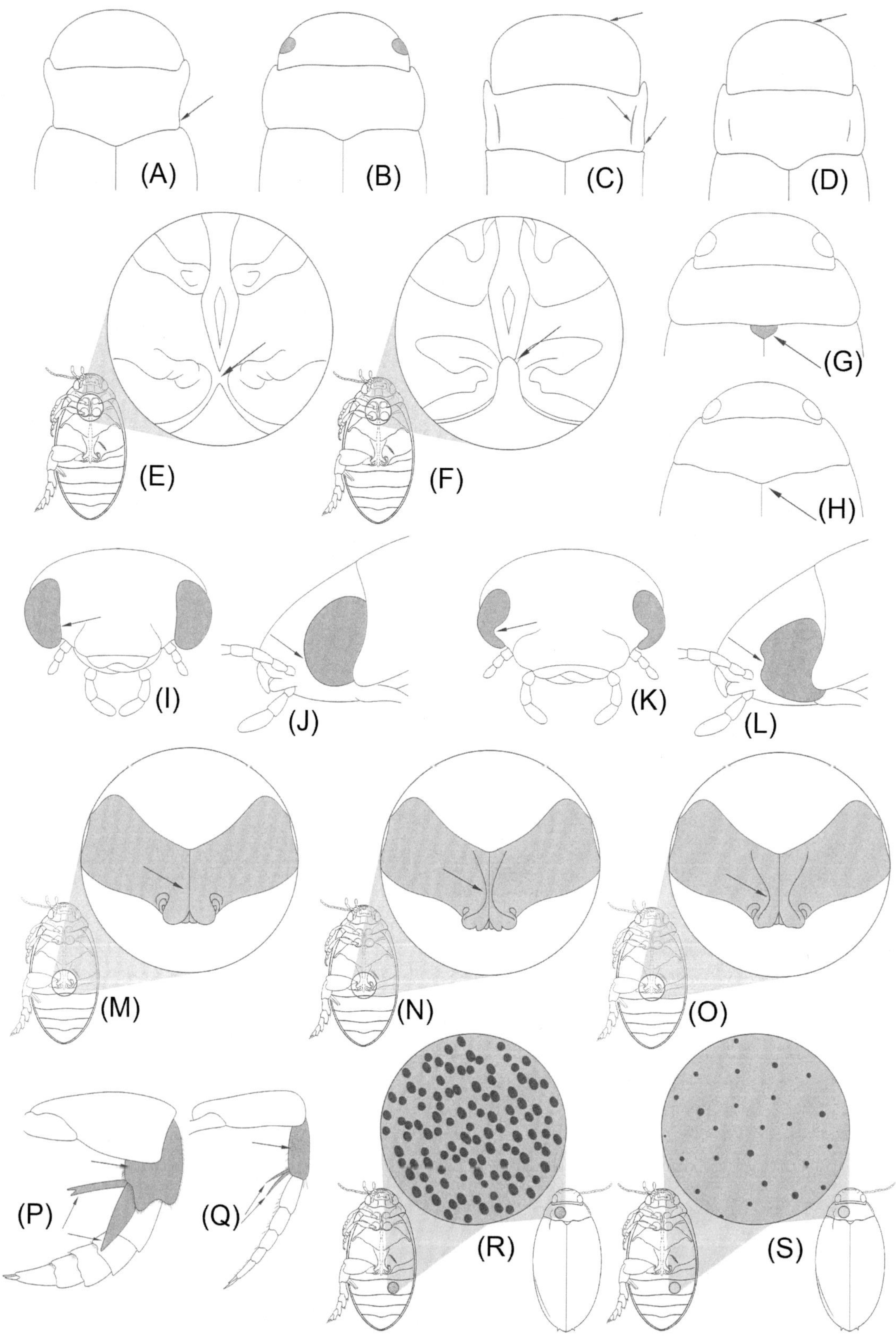

FIGURE 13.15 (A) *Iberoporus* head showing constriction between pronotum and elytra; (B) *Graptodytes* head shape; (C) *Etruscodytes* head showing pronotal striae and absence of constriction between pronotum and elytra; (D) *Siettitia* head shape; (E) detail of prosternal and anteromedial metaventral processes in *Siettitia*; (F) detail of prosternal and anteromedial metaventral processes in *Etruscodytes*; (G) *Liopterus* showing scutellum; (H) scutellum not visible in *Laccophilus*; (I) idealized Dytiscidae in front view showing eye shape; (J) idealized Dytiscidae in lateral view showing eye shape; (K) idealized Dytiscidae in front view showing eye emargination; (L) idealized Dytiscidae in lateral view showing eye emargination; (M) absence of metacoxal lines in *Rhantaticus*; (N) arrangement of metacoxal lines in *Liopterus*; (O) arrangement of metacoxal lines in *Agabus*; (P) *Cybister* metatibia showing spur; (Q) *Agabus* metatibia showing spur; (R) detail of ventral surface punctation in *Acilius*; (S) detail of ventral surface punctation in *Graphoderus*.

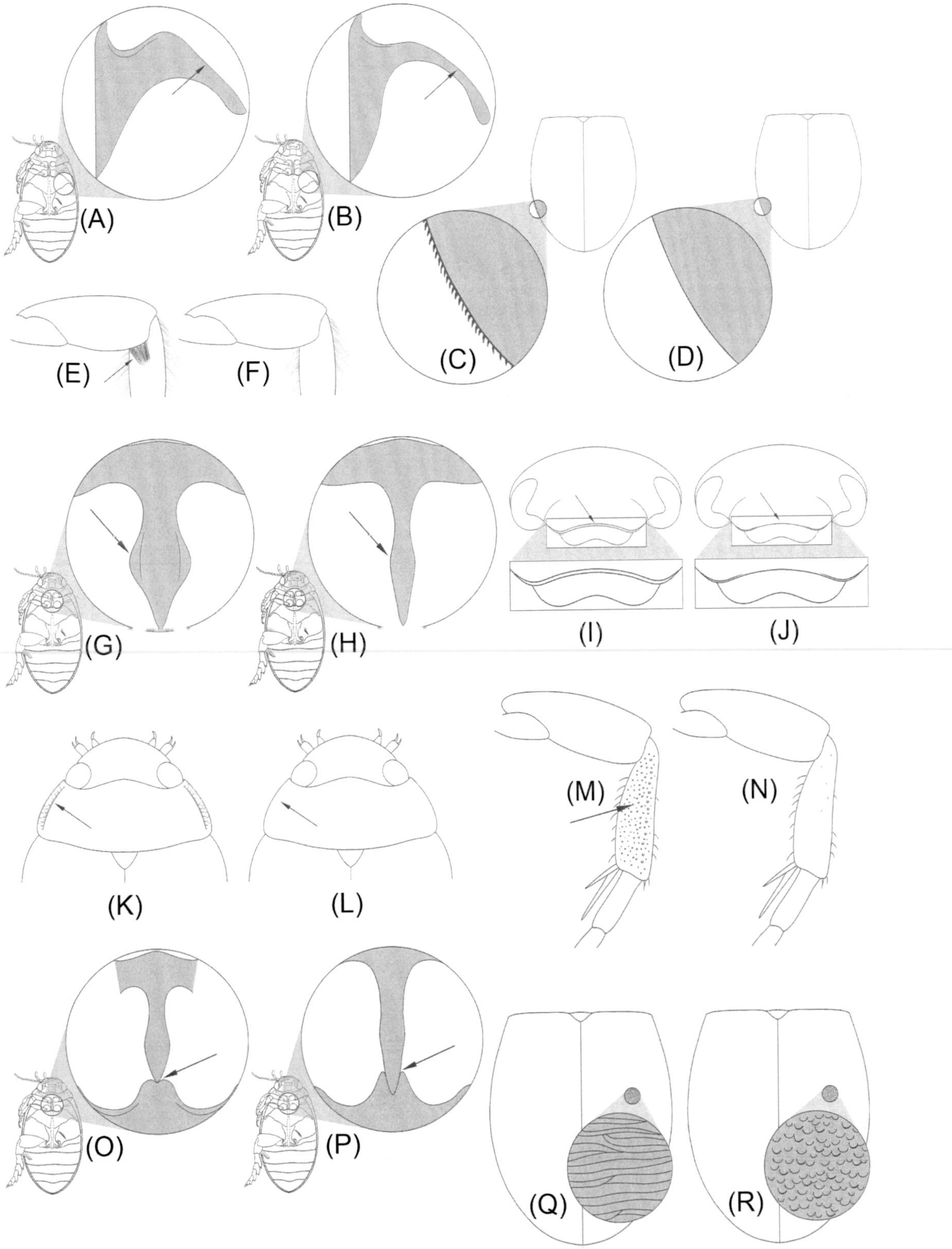

FIGURE 13.16 (A) *Hydaticus* metaventral wing shape; (B) *Eretes* metaventral wing shape; (C) posterolateral elytra margin in *Eretes*; (D) posterolateral elytra margin in *Dytiscus*; (E) *Agabus* metafemur; (F) *Rhantus* metafemur; (G) *Platambus* prosternal process; (H) *Agabus* prosternal process; (I) *Agabus* anterior clypeal margin detail; (J) *Ilybius* anterior clypeal marging detail; (K) *Rhantus* pronotum; (L) *Colymbetes* pronotum; (M) *Melanodytes* metatibia; (N) *Rhantus* metatibia; (O) detail of *Colymbetes* prosternal process incision in the metaventrite; (P) detail of *Meladema* prosternal process incision in the metaventrite; (Q) detail of *Colymbetes* elytra exculture; (R) detail of *Meladema* elytra exculture.

16 (13'). Pronotum with lateral stria (Fig. 13.16K) 17
16'. Pronotum without lateral stria (Fig. 13.16L) 18
17. Anterior surface of metatibia covered with setigerous punctures (Fig. 13.16M) .... *Melanodytes* (one species: *M. pustulatus* (Rosi, 1792))
17'. Anterior surface of metatibia without setigerous punctures (Fig. 13.16N) .... *Rhantus*
18(16') Anterior margin of metaventrite weakly incised for reception of the prosternal process (Fig. 13.16O); elytral sculpturing composed of dense, transverse, smoothly curved and parallel cells (Fig. 13.16Q) .... *Colymbetes*
18'. Anterior margin of metaventrite deeply incised for reception of the prosternal process (Fig. 13.16P); elytral sculpturing composed of semi-circular punctures (Fig. 13.16R) .... *Meladema*
19(4'). Protarsi clearly pentamerous (Fig. 13.17A) 20
19'. Protarsi pseudotetramerous, with tarsomere IV concealed within ventral lobes of tarsomere III (Fig. 13.17B) 22
20. Metatibial spurs (at least anterior spur) bifid (Fig. 13.17C) .... *Laccophilus*
20'. Metatibial spurs simple (see Fig. 13.17D) 21
21. Prosternal process apically trifid (Fig. 13.17E) .... *Neptosternus* (one species: *N. ornatus* Sharp, 1882).
21'. Proternal process normal (Fig. 13.17F) .... *Philodytes* (one species: *P. umbrinus* (Motschulsky, 1855))
22(19'). Metatarsal claws clearly unequal in length (Fig. 13.17G) 23
22'. Metatarsal claws equal or subequal (Fig. 13.17H) 24
23. Prosternal process apically narrow and metaventral wing medially narrow (Fig. 13.17I) .... *Hyphydrus*
23'. Prosternal process apically broad and metaventral wing medially broad (Fig. 13.17J) .... *Heterhydrus* (one species: *H. senegalensis* (Laporte, 1835)).
24(22'). Prosternal process short and broadly truncate (Fig. 13.17K) .... *Hydrovatus*
24'. Prosternal process elongated, apically pointed or rounded (Fig. 13.17L) 25
25. Abdomen and elytra apically acuminated (Fig. 13.17M) .... *Methles*
25'. Abdomen and elytra apically rounded or pointed, but never acuminated (Fig. 13.17N and O) 26
26. Metatibia elongated, gradually expanded apically and metacoxal lobes very small (Fig.13.17P); stria present at each side of the pronotum (Fig. 13.17R, S) 27
26'. Metatibia variable, but generally never expanded apically and metacoxal lobes larger (Fig. 13.17Q); most genera without striae on the pronotum (Fig. 13.17T) 31
27. Occipital line absent (Fig. 13.17T) .... *Hydroglyphus*
27'. Occipital line present (Fig. 13.17R) 28
28. Longitudinal carina present on the elytral disk (Fig. 13.18A) 29
28'. No longitudinal carina on the elytral disk (Fig. 13.18B) 30
29. Elytral punctures not organized in longitudinal series (Fig. 13.18C) .... *Yola*
29'. Elytral punctures organized in longitudinal series (Fig. 13.18D) .... *Yolina* (one species: *Y. insignis* (Sharp, 1882)).
30(28'). Elytral sutural line well developed (Fig. 13.18E) .... *Bidessus*
30'. Elytral sutural line absent (Fig. 13.18F) .... *Clypeodytes* (one species: *C. cribrosus* (Schaum, 1864)).
31(26'). Elytral epipleuron with transverse carina at humeral angle (Fig. 13.18G) 32
31'. Elytral epipleuron without transverse carina at humeral angle (Fig. 13.18H) 34
32. Stria present on each side of the pronotum (Fig. 13.17R and S) 33
32'. No stria on each side of the pronotum (Fig. 13.17T) .... *Hygrotus*
33. Body shape subparallel and elytra concolor (Fig. 13.18I) .... *Iberoporus* in part
33'. Body shape oval and elytra vittate (Fig. 13.18J) .... *Rhithrodytes*
34(31'). Stria present on each side of the pronotum (Fig.13.17R and S) 35
34'. No stria on each side of the pronotum (Fig. 13.17T) 39
35. Apical labial palpomere with distinct split (Fig. 13.18K) 36
35'. Apical labial palpomere without split (Fig. 13.18L) 38
36. Ventral surface densely shagreened, matt, opaque, with fine punctures in some cases (Fig. 13.18M) .... *Stictonectes*
36'. Ventral surface shiny, with coarse punctures (Fig. 13.18N) or coarse punctures and reticulation in some areas (Fig. 13.18O) 37

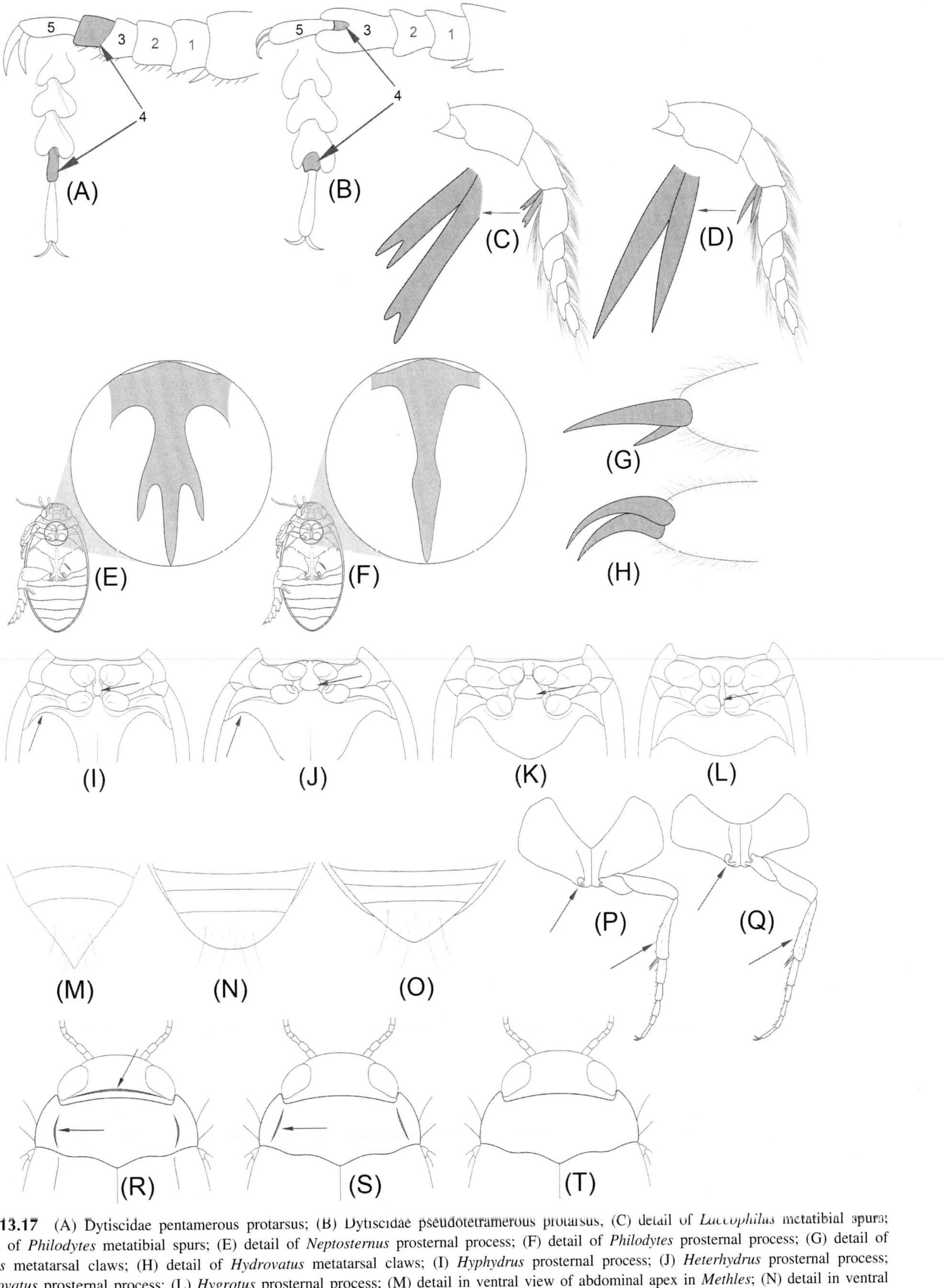

FIGURE 13.17 (A) Dytiscidae pentamerous protarsus; (B) Dytiscidae pseudotetramerous protarsus; (C) detail of *Laccophilus* metatibial spurs; (D) detail of *Philodytes* metatibial spurs; (E) detail of *Neptosternus* prosternal process; (F) detail of *Philodytes* prosternal process; (G) detail of *Hyphydrus* metatarsal claws; (H) detail of *Hydrovatus* metatarsal claws; (I) *Hyphydrus* prosternal process; (J) *Heterhydrus* prosternal process; (K) *Hydrovatus* prosternal process; (L) *Hygrotus* prosternal process; (M) detail in ventral view of abdominal apex in *Methles*; (N) detail in ventral view of abdominal apex in *Hygrotus*; (O) detail of abdominal apex in *Oreodytes*; (P) details of metacoxae and fore leg in *Hydroglyphus*; (Q) detail in ventral view of metacoxae and fore leg in *Hydroporus*; (R) detail of prontotum striae and occipital line in *Bidessus*; (S) detail of pronotum striae in *Graptodytes*; (T) absence of pronotum striae in *Hygrotus*.

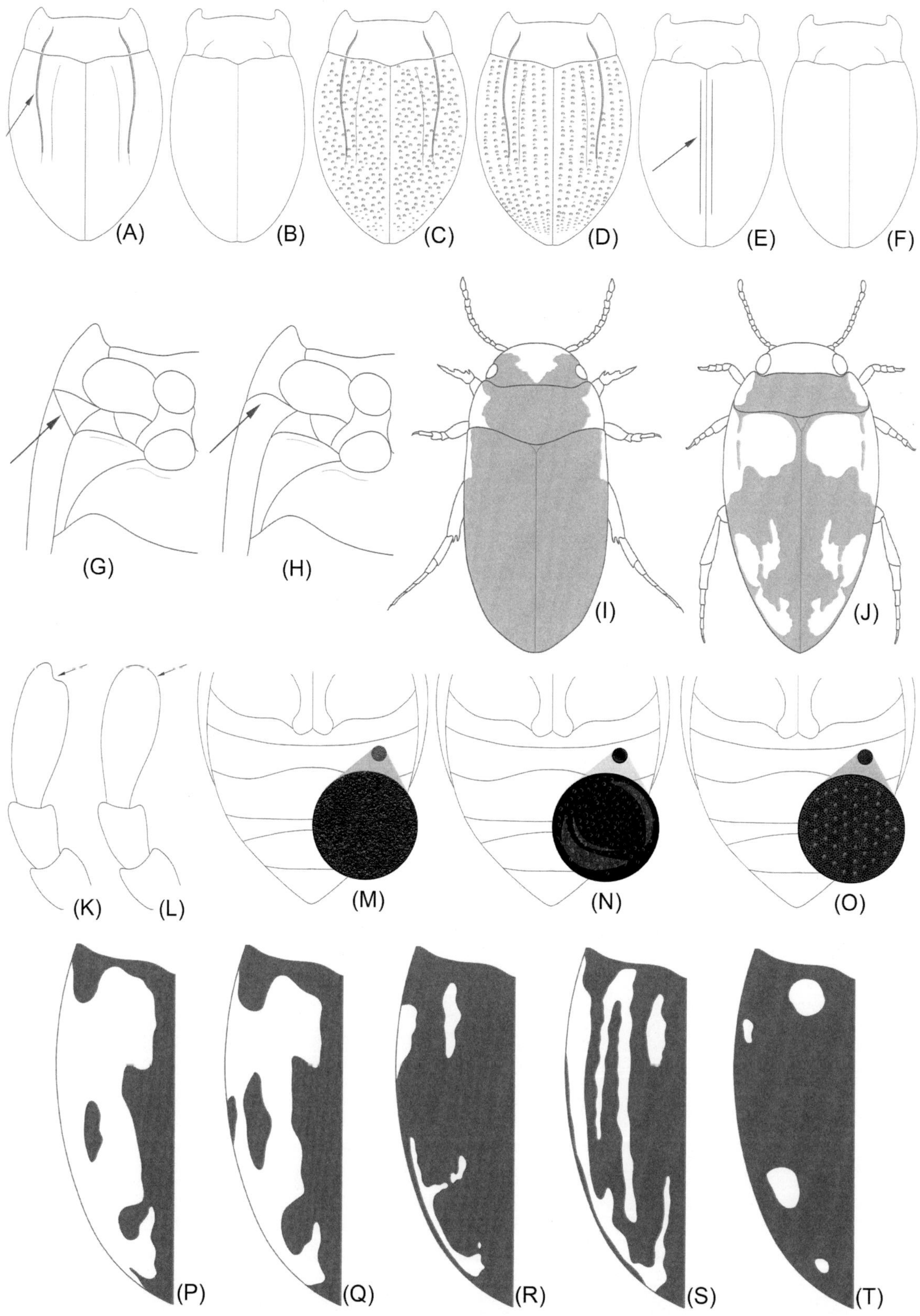

FIGURE 13.18 (A) *Yola* elytra showing longitudinal carina; (B) *Bidessus* elytra without carina; (C) *Yola* elytra punctures pattern; (D) *Yolina* elytra puncture pattern; (E) *Bidessus* elytra suture line; (F) Absence of suture line in *Clypeodytes*; (G) elytra epipleurum with transverse carina in *Hygrotus*; (H) elytra epipleurum showing absence of carina in *Stictonectes*; (I) *Iberoporus* habitus; (J) *Rhithrodytes* habitus; (K) *Stictonectes* apical palpomere; (L) *Oreodytes* apical palpomere; (M) *Stictonectes* ventral surface; (N) *Graptodytes* ventral surface; (O) *Metaporus* ventral surface; (P) *Graptodytes* elytron; (Q) *Graptodytes* elytron; (R) *Graptodytes* elytron; (S) *Graptodytes* elytron; (T) *Tassilodytes* elytron.

37. Elytra vittate (Fig. 13.18P–S) .... *Graptodytes*
37'. Elytra with dots, not vittate (Fig. 13.18T) .... *Tassilodytes*
(one species: *T. parisii* Gridelli, 1939).
38(35'). Body outline in dorsal view more or less continuous, without or with only a weak discontinuity at the base of the pronotum (Fig. 13.19A); prosternal process almost rectangular (Fig. 13.19C) .... *Oreodytes*
38'. Body outline in dorsal view clearly discontinuous at the base of the pronotum (Fig. 13.19B); prosternal process clearly elongated (Fig. 13.19D) .... *Nectoporus*
39(34'). Metacoxal process without any indentation or V-shaped hollow, as in Fig. 13.19E–H .... *Hydroporus*
39'. Metacoxal process with an indentation or V-shaped hollow, as in Fig. 13.19I 40
40. Ventral surface microreticulated (Fig. 13.18O) 41
40'. Ventral surface densely shagreened, matt, and opaque (Fig. 13.18M) 42
41. Dorsal surface microreticulated (Fig. 13.19J); body size <3 mm .... *Metaporus*
41'. Dorsal surface not microreticulated (Fig. 13.19K), body size >3 mm .... *Porhydrus*
42(40'). Elytral surface unicolored, from reddish-brown to black, or bicolored (Fig. 13.19L); metacoxal processes with interlaminar bridge exposed (Fig. 13.19O); metatibia with anterior surface covered with dense punctures (Fig. 13.19Q); metatarsomere V about twice the length of metatarsomere IV (Fig. 13.19S) .... *Deronectes*
42'. Elytral surface vittated or maculated (Fig. 13.19M and N); metacoxal processes with interlaminar bridge concealed in most species (Fig. 13.19P); metatibia with anterior surface not covered with dense punctures (Fig. 13.19R); metatarsomere V about one and a half the length of metatarsomere IV (Fig. 13.19T) 43
43. Elytra with or without subapical spine (Fig. 13.20A and B); ventral surface matt (Fig. 13.18M) or shiny (Fig. 13.18N) 44
43'. Elytra without subapical spine (Fig. 13.20B); ventral surface matt, never shiny (Fig. 13.18M) 45
44. Body with ventral surface matt, with dense and fine punctation (Fig. 13.18M); elytron with subapical spine (Fig. 13.20A) except in some species (*ceresyi* group, most species exclusive from hypersaline water) ... *Nebrioporus*
44'. Body with ventral surface shiny, with sparse and coarse punctation (Fig. 13.18N); elytron without subapical spine (Fig. 13.20B) .... *Scarodytes*
45(43'). Metatibia densely covered with nonsetiferous punctures (Fig. 13.19Q) 46
45'. Metatibia with sparse nonsetiferous punctures and a line of setiferous punctures (Fig. 13.19R) 47
46. Habitus as in Fig. 13.20C, elytra with distinct longitudinal grooves .... *Iberonectes*
(one species: *I. bertrandi* (Legros, 1956)).
46'. Habitus as in Fig. 13.20D, elytra without longitudinal grooves .... *Stictotarsus*
47(45'). Habitus as in Fig. 13.20E, body outline in dorsal view continuous .... *Boreonectes*
47'. Habitus as in Fig. 13.20F, body outline in dorsal view discontinuous at the base of the pronotum .... *Trichonectes*
(one species: *T. otini* (Guignot, 1941)).

Coleoptera: Hydrophilidae: Genera

1. Eyes completely divided into dorsal and ventral faces (Fig. 13.21A) .... *Amphiops*
1'. Eyes not divided into dorsal and ventral faces 2
2. Tarsi composed of five segments, with the first segment of middle and hind tarsi longer than second (Fig. 13.21B); maxillary palps clearly shorter than antennae (Fig. 13.7G); body shape clearly rounded .... *Coelostoma*
2'. Tarsi composed of four or five segments (Fig. 13.21C and D) but in the latter case the first segment of middle and hind tarsi is clearly shorter than the second; maxillary palps longer or equal than antennae (Fig. 13.7H and I) 3
3. Tarsi with four segments (Fig. 13.21C) .... *Cymbiodyta*
(one species: *C. marginella* (Fabricius, 1792)).
3'. Tarsi with five segments (Fig. 13.21D) 4
4. Fused ventral keel present along the center of the meso and metaventrites, which is prolonged into a spine over at least the base of the first abdominal ventrite (Fig. 13.21E); body length generally much bigger than 10 mm 5
4'. No ventral keel on meso and metaventrites, or if present, never fused and prolonged into a spine (Fig. 13.21F); body length generally less than 10 mm 9

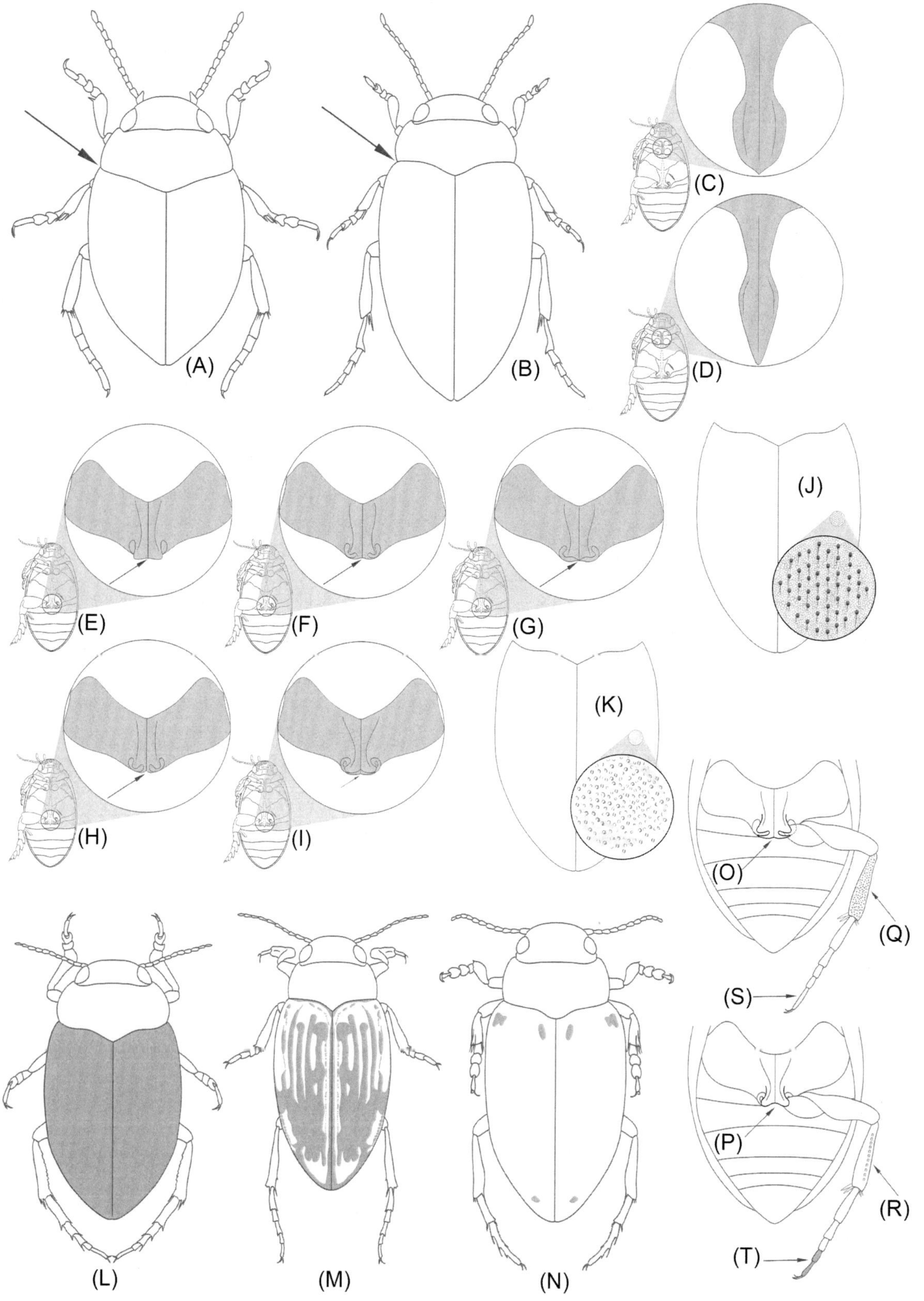

FIGURE 13.19 (A) *Oreodytes* body shape; (B) *Nectoporus* body shape; (C) *Oreodytes* prosternal process; (D) *Nectoporus* prosternal process; (E–H) Hydroporini metacoxal processes, being E) *Hydroporus*; (I) *Deronectini* metacoxal process; (J) *Metaporus* dorsal surface; (K) *Porhydrus* dorsal surface; (L) *Deronectes* elytral habitus; (M) *Nebrioporus* elytral habitus; (N) *Nebrioporus luctuosus* (Aubé, 1838) elytral habitus (inverted colors); (O) *Deronectes* metacoxal process; (P) *Nebrioporus* metacoxal process; (Q) *Deronectes* metatibia; (R) *Nebrioporus* metatibia; (S) *Deronectes* metatarsomeres; (T) *Nebrioporus* metatarsomeres.

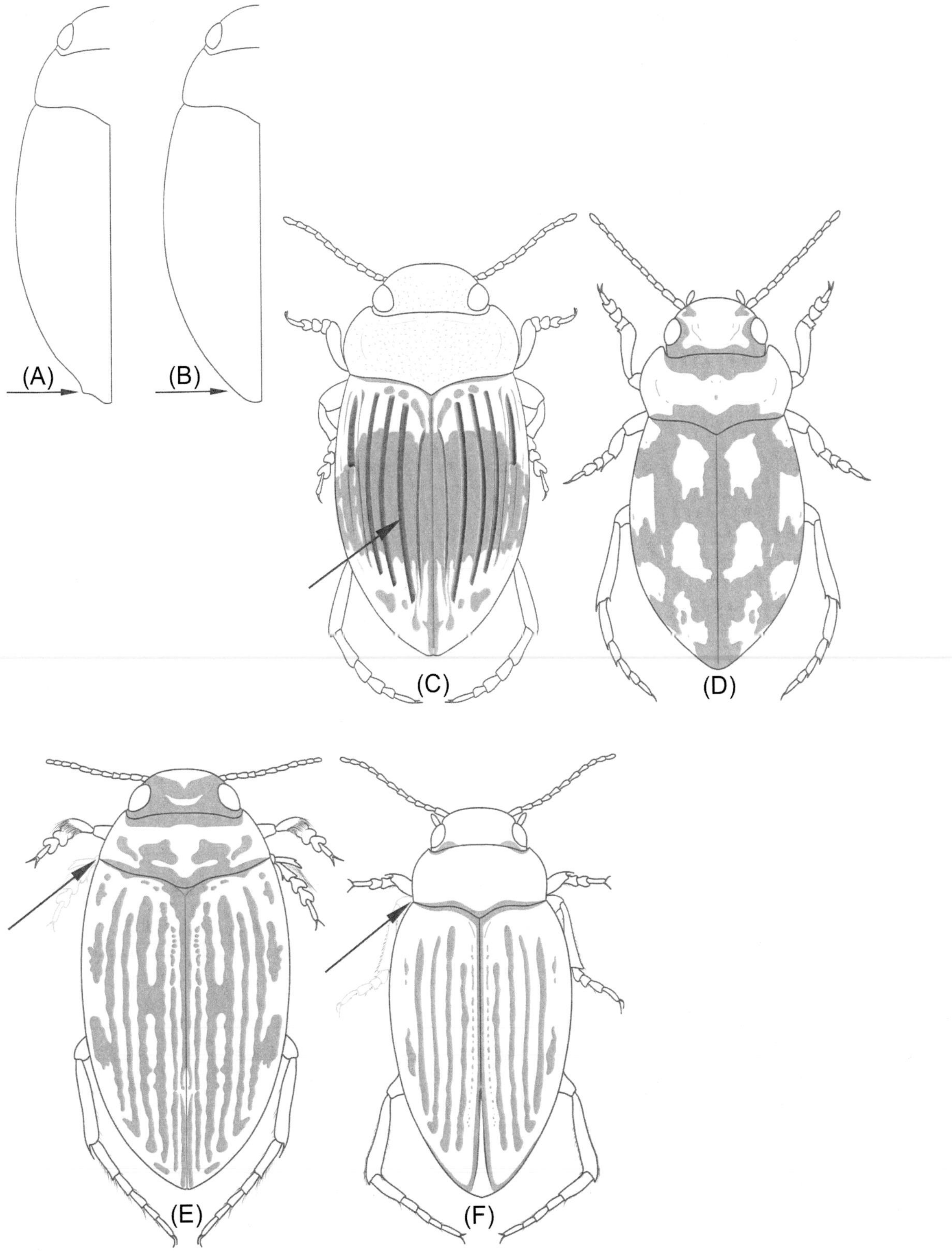

FIGURE 13.20 (A) *Nebrioporus* lateral outline showing subapical spine; (B) *Scarodytes* lateral outline; (C) *Iberonectes* habitus; (D) *Stictotarsus* habitus; (E) *Boreonectes* habitus; (F) *Trichonectes* habitus.

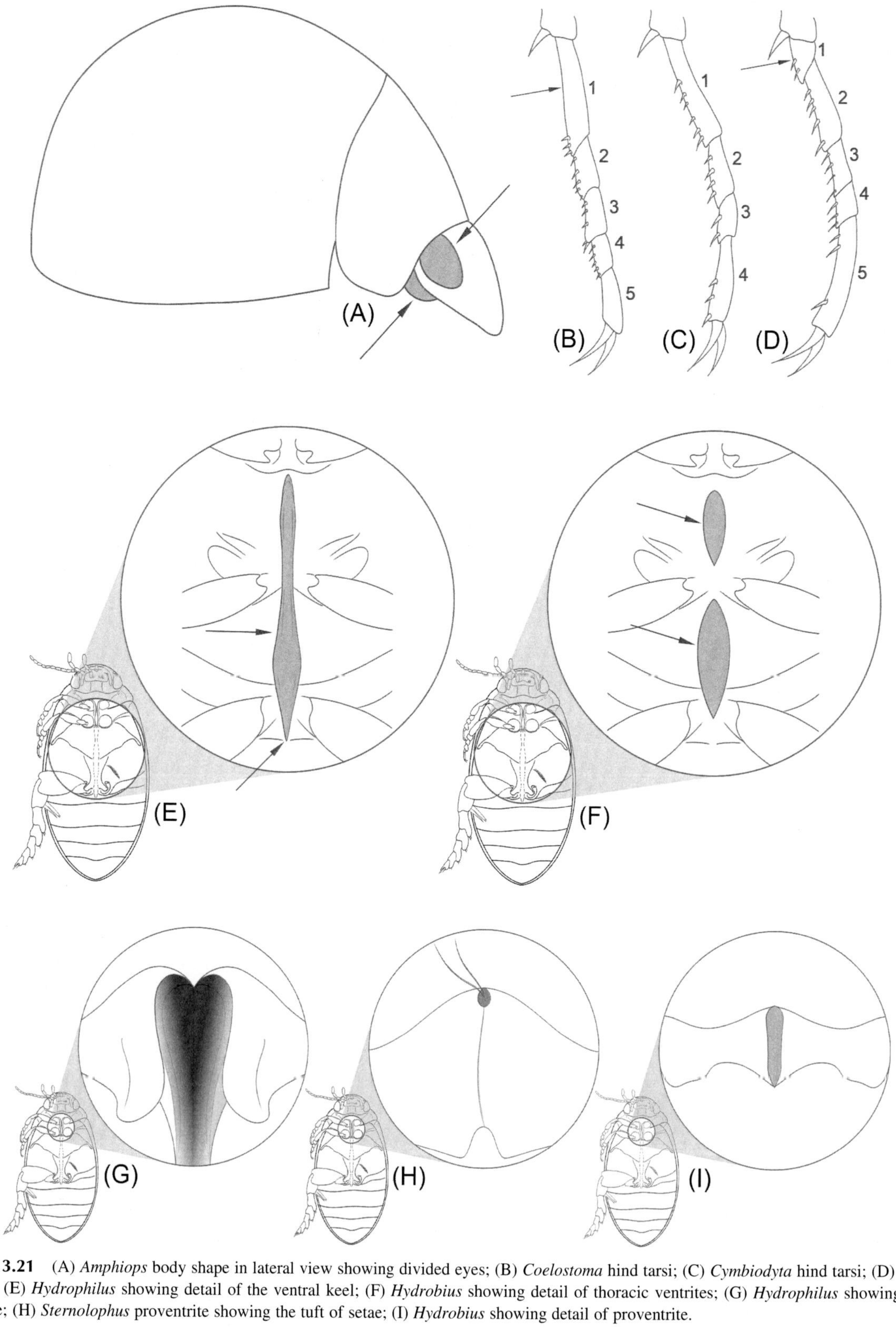

FIGURE 13.21 (A) *Amphiops* body shape in lateral view showing divided eyes; (B) *Coelostoma* hind tarsi; (C) *Cymbiodyta* hind tarsi; (D) *Enochrus* hind tarsi; (E) *Hydrophilus* showing detail of the ventral keel; (F) *Hydrobius* showing detail of thoracic ventrites; (G) *Hydrophilus* showing detail of proventrite; (H) *Sternolophus* proventrite showing the tuft of setae; (I) *Hydrobius* showing detail of proventrite.

5. Proventrite with median carina deeply emarginated or completely divided into two lobes for reception of anterior portion of mesosternal keel (Fig. 13.21G); body length 20–50 mm, frequently surpassing 30 mm .... *Hydrophilus*
5'. Proventrite with median carina very weakly or not emarginated or divided (Fig. 13.21H and I); body length 9–21 mm 6
6. Femora always with basally hydrofuge pubescence (Fig. 13.22A); anterior apex of median carina with a small tuft of setae (Fig. 13.21H); body length 9–12 mm .... *Sternolophus*
(one species: *S. solieri* Castelnau, 1840).
6' Middle and hind femora without hydrofuge pubescence basally (Fig. 13.22B); anterior apex of median carina without small tuft of setae (Fig. 13.21I); body length usually greater than 12 mm 7
7. Anterior margin of clypeus straight (Fig. 13.22C); segments of antennal club slightly asymmetrical (Fig. 13.22E); body length 10–20 mm .... *Hydrochara*
7. Anterior margin of clypeus broadly emarginated (Fig. 13.22D); first segment of antennal club deeply divided into two very asymmetrical lobes (Fig. 13.22F); body length 18–28 mm 8
8. First segment of antennal club set with long setae (Fig. 13.22G); aedeagus with the median lobe laterally compressed and expanded, and the basal piece strongly extended into a membranous lobe or a strongly acute spine (Fig. 13.22I) .... *Hydrobiomorpha*
8'. First segment of antennal club without setae (Fig. 13.22H); aedeagus simple, with the median lobe not laterally compressed, and the basal piece not extended (Fig. 13.22J) .... *Brownephilus*
9(4'). Scutellum clearly longer than wide (Fig. 13.22K); meso and metatibiae with swimming hairs (Fig. 13.22M) .... *Berosus*
9'. Scutellum as long as wide (Fig. 13.22L); meso and metatibiae without swimming hairs (Fig. 13.22N), but if visible, confined to middle and hind tarsi 10
10. Length of the tarsomeres about half the length of the tibia (Fig. 13.22O); basal second abdominal ventrite with a common excavation on each side that is covered with a bilobed ciliated plate (Fig. 13.22Q) .... *Chaetarthria*
10'. Length of the tarsomeres approximately similar to the length of the tibia (Fig. 13.22P); basal second abdominal ventrite without an excavation and ciliate plate on each side (Fig. 13.22R) 11
11. Third segment of maxillary palpi longer than the second (Fig. 13.23A) 12
11'. Third segment of maxillary palpi equal to or shorter than the second (Fig. 13.23B) 14
12. Elytra with sharp sutural striae at least in their posterior half (Fig. 13.23C); second segment of maxillary palpi concave or convex (Fig. 13.23E and F) 13
12'. Elytra without sutural striae (Fig. 13.23D); second segment of maxillary palpi always concave (Fig. 13.23E) .... *Helochares*
13. Second segment of maxillary palpi concave (Fig. 13.23E) .... *Chasmogenus*
(one species: *C. livornicus* (Kuwert, 1890)).
13'. Second segment of maxillary palpi convex (Fig. 13.23F) .... *Enochrus*
14(11'). Elytra without sutural striae (Fig. 13.23D), and around 20 highly regular linear series of punctures (Fig. 13.23G); abdomen with six distinct ventrites (Fig. 13.23K) .... *Laccobius*
14'. Elytra with sutural striae (Fig. 13.23C), and without regular series of punctures (Fig. 13.23H) or if any, much less than 20 (Fig. 13.23I); abdomen with five ventrites (Fig. 13.23L) 15
15. Elytra with linear rows (Fig. 13.23G and I) 16
15'. Elytra without linear rows (Fig. 13.23H) 20
16. Body length greater than 6 mm 17
16'. Body length less than 6 mm 19
17. Mesoventrite in lateral view strongly raised in the middle and forming a keel (Fig. 13.23M); elytral puncture rows not in grooved striae, except for the rear half of the rows either side of the suture, which form distinct sutural striae (Fig. 13.23J) .... *Limnoxenus*
17'. Mesoventrite in lateral view without a keel (Fig. 13.23N); each elytra with 10 punctured striae most strongly grooved at the rear, with weak series of punctures towards the front where intercalary series of punctures are also visible (Fig. 13.23I) 18
18. Process of the mesoventrite forming a weak to strongly elevated transverse ridge (Fig. 13.24A); aedeagus as in Fig. 13.24C; body moderately dorsoventrally compressed .... *Hydrobius*
18'. Process of the mesoventrite forming either a high longitudinal keel or a strongly elevated, broad tubercle (Fig. 13.24B); aedeagus as in Fig. 13.24D; body tending to be convex .... *Limnohydrobius*

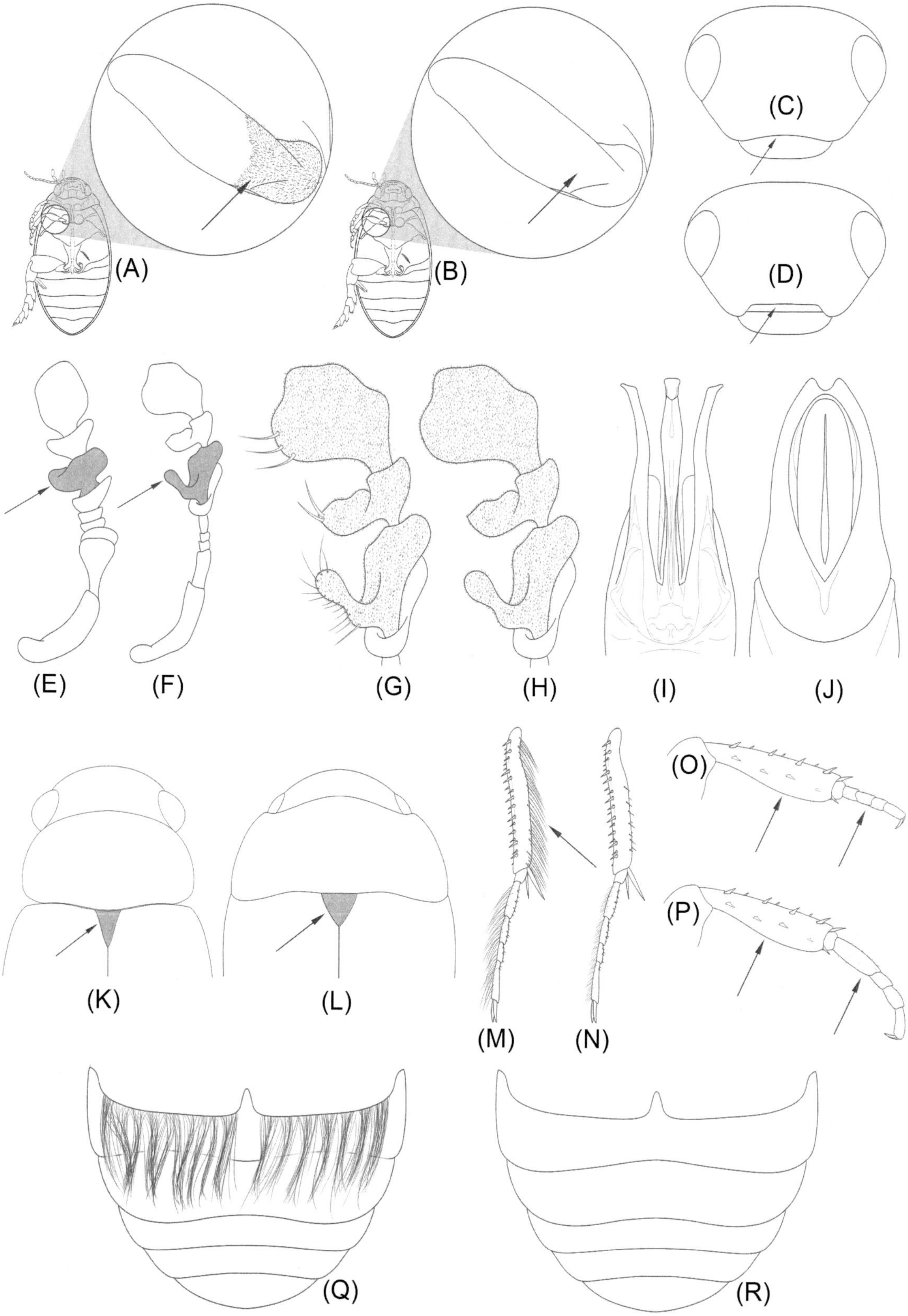

FIGURE 13.22 (A) *Sternolophus* hind femur; (B) *Hydrochara* hind femor; (C) *Hydrochara* clypeus; (D) *Hydrobiomorpha* clypeus; (E) *Hydrochara* antennae; (F) *Hydrobiomorpha* antennae; (G) detail of last antennae segments in *Hydrobiomorpha*; (H) detail of last antennae segments in *Brownephilus*; (I) *Hydrobiomorpha* aedeagus; (J) *Brownephilus* aedeagus; (K) *Berosus* scutellum; (L) *Chaethartria* scutellum; (M) *Berosus* metatibiae; (N) *Hydrobius* metatibiae; (O) *Chaetarthria* protarsomeres; (P) *Helochares* protarsomeres; (Q) *Chaetarthria* abdominal segments; (R) Paracymus abdominal segment.

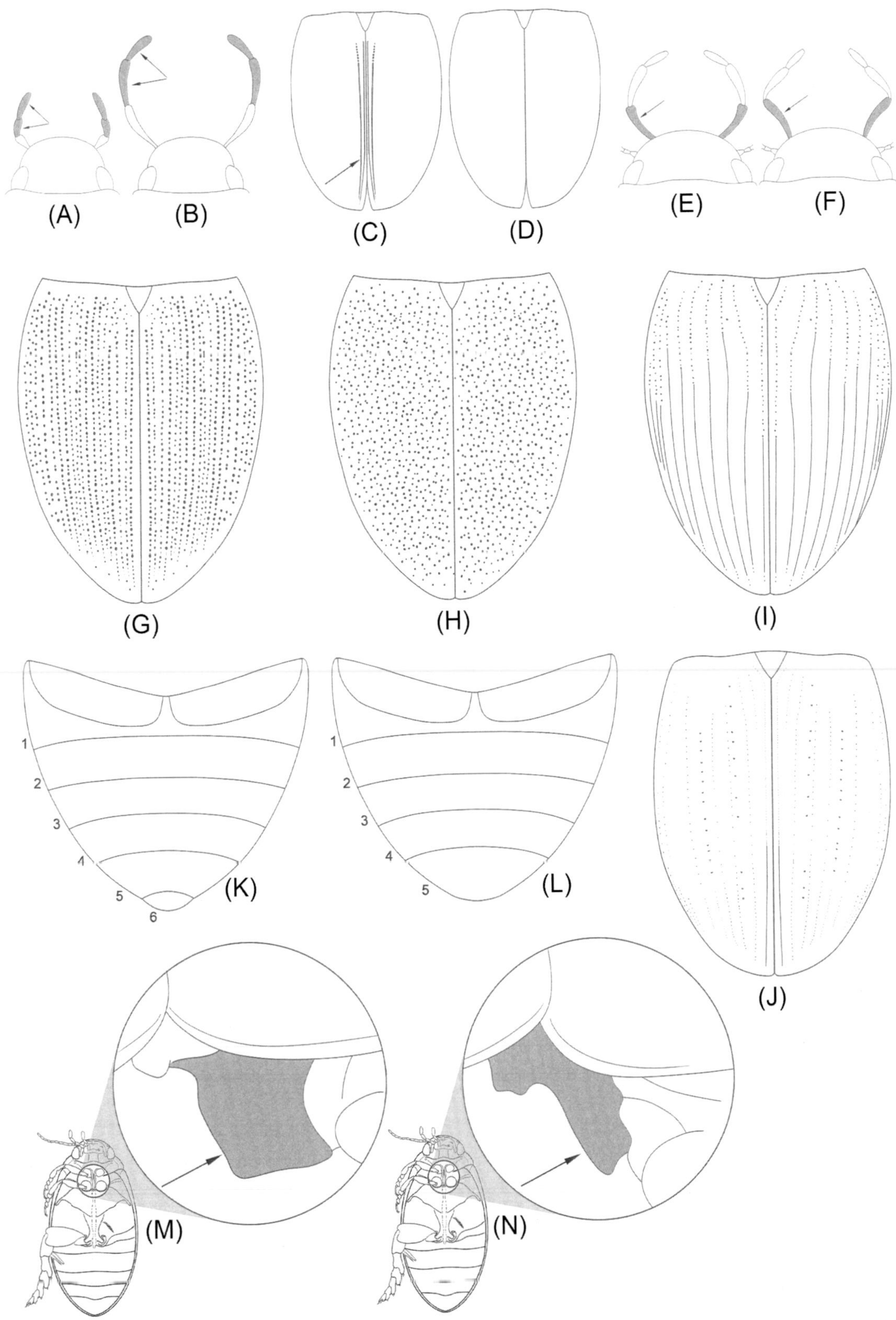

FIGURE 13.23 (A) *Paracymus* maxilar palps; (B) *Helochares* maxilar palps; (C) *Enochrus* sutural striae; (D) *Helochares* elytra; (E) *Chasmogenus* maxilar palp; (F) *Enochrus* maxilar palp; (G) *Laccobius* elytra; (H) *Anacaena* elytra; (I) *Hydrobius* elytra; (J) *Limnoxenus* elytra; (K) *Laccobius* abdominal segments; (L) *Anacaena* abdominal segments; (M) *Limnoxenus* mesoventrite in lateral view; (N) *Limnohydrobius* mesoventrite in lateral view.

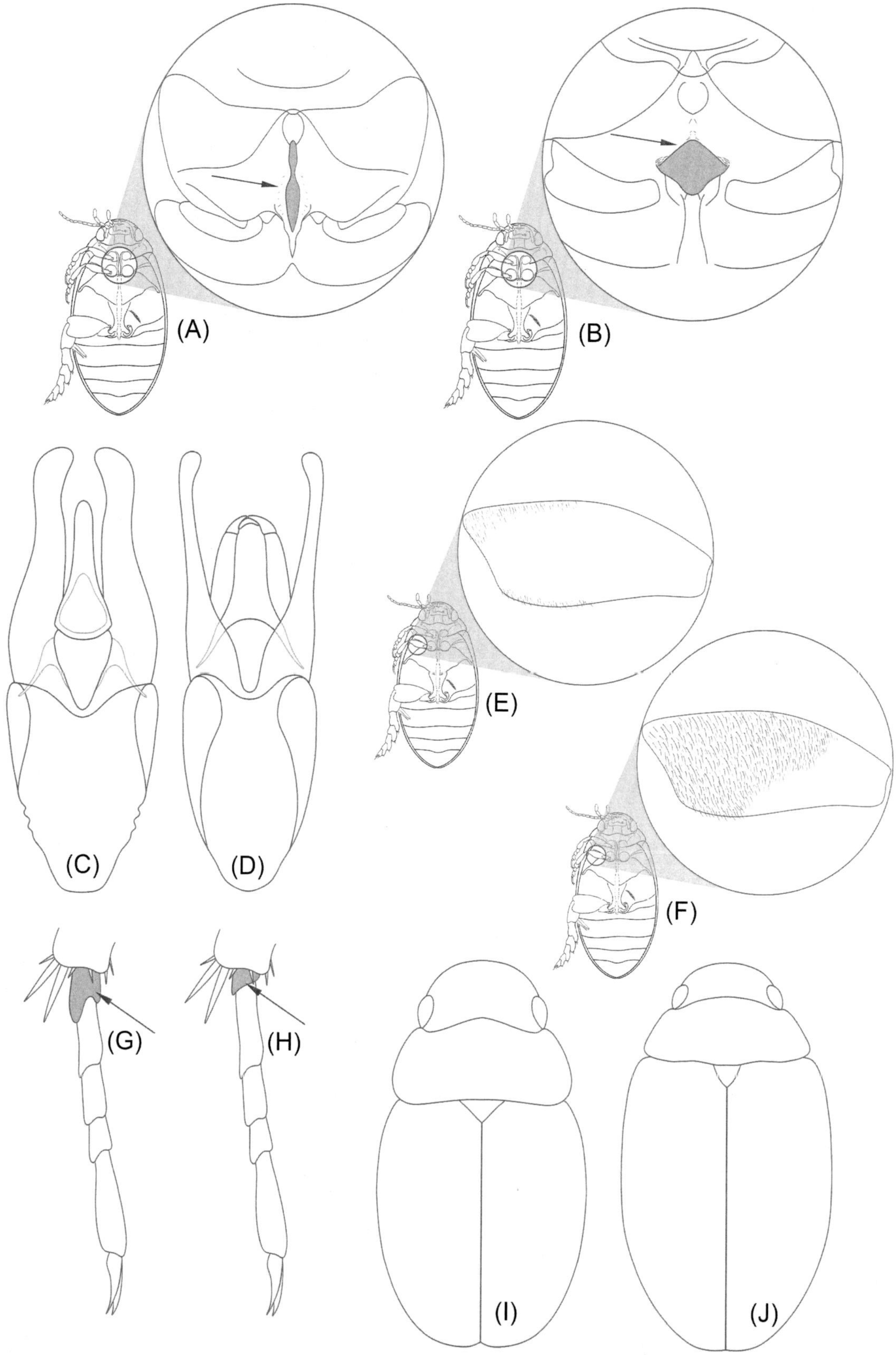

FIGURE 13.24 (A) Detail of mesoventrite process in *Hydrobius*; (B) detail of mesoventrite process in *Lymnohydrobius*; (C) *Hydrobius* aedeagus; (D) *Lymnohydrobius* aedeagus; (E) detail of *Paracymus* metafemur; (F) detail of *Anacaena* metafemur; (G) detail of metatarsal segments in *Paracymus*; (H) detail of metatarsal segments in *Anacaena*; (I) *Hemisphaera* body outline; (J) *Crenitis* body outline.

19(16'). Metafemur entirely smooth and glabrous (Fig. 13.24E); length of metatarsal segment 1 equal or greater than half the length of segment 2 (Fig. 13.24G); dorsal surface with fine metallic hue .... *Paracymus*
19'. Metafemur with its basal portion densely punctate and pubescent (Fig. 13.24F); length of metatarsal segment 1 less than half the length of segment 2 (Fig. 13.24H); dorsal surface not metallic .... *Anacaena*
20(15'). Very small, less than 1.5 mm in length; body shape clearly rounded at sides (Fig. 13.24I) ...... *Hemisphaera*
20'. Body length equal to or greater than 3 mm; body shape more elongated and parallel (Fig. 13.24J) .... *Crenitis* (one species: *C. punctatostriata* (Letzner, 1840)).

Acknowledgments

We want to thank Dr. David T. Bilton for reviewing an earlier version of the manuscript and checking the English writing. This work is dedicated to our colleague and friend, Ignacio Ribera, who passed away on April 15, 2020. The memory of his advice and friendship will remain with us forever.

Appendix

See Table 13.1.

APPENDIX: Taxonomic list of "true aquatic beetles" from the Mediterranean Basin with indicative number of Palaearctic and Mediterranean species/subspecies and habitat preferences. PS: Palaearctic species richness; MS: Mediterranean species richness; Habitat: Lotic (Lo), lentic (Le), subterranean (Sb). When there is a clear lotic preference, then the tendency to occupy crenon (C), rhithron (R) or potamon (P) is also specified.

Suborder	Family	Subfamily	Tribe	Genus	PS	MS	Habitat
MIXOPHAGA	HYDROSCAPHIDAE	Hydroscaphidae		*Hydroscapha*	9	3	Lo
MIXOPHAGA	SPHAERIUSIDAE	Sphaeriusidae		*Sphaerius*	8	2	Lo
ADEPHAGA	GYRINIDAE	Gyrininae	Gyrinini	*Aulonogyrus*	3	2	Lo-P
ADEPHAGA	Gyrinidae	Gyrininae	Gyrinini	*Dineutus**	13	3	Le
ADEPHAGA	Gyrinidae	Gyrininae	Gyrinini	*Gyrinus*	31	12	Lo/Le
ADEPHAGA	Gyrinidae	Gyrininae	Orectochilini	*Orectochilus*	25	4	Lo-P
ADEPHAGA	HALIPLIDAE			*Brychius*	2	2	Lo-R
ADEPHAGA	Haliplidae			*Haliplus*	57	22	Lo/Le
ADEPHAGA	Haliplidae			*Peltodytes*	7	2	Le
ADEPHAGA	NOTERIDAE	Noterinae	Hydrocanthini	*Canthydrus*	12	1	Le
ADEPHAGA	Noteridae	Noterinae	Neohydrocoptini	*Neohydrocoptus**	3	1	Le
ADEPHAGA	Noteridae	Noterinae	Noterini	*Noterus*	7	3	Le
ADEPHAGA	Noteridae	Noterinae	Noterini	*Synchortus**	1	1	Le
ADEPHAGA	HYGROBIIDAE			*Hygrobia*	2	1	Le
ADEPHAGA	DYTISCIDAE	Dytiscinae	Acilini	*Acilius*	6	3	Le
ADEPHAGA	Dytiscidae	Agabinae	Agabini	*Agabus*	116	34	Lo/Le
ADEPHAGA	Dytiscidae	Hydroporinae	Bidessini	*Bidessus*	22	20	Lo/Le
ADEPHAGA	Dytiscidae	Hydroporinae	Hydroporini	*Boreonectes*	7	6	Le
ADEPHAGA	Dytiscidae	Colymbetinae	Colymbetini	*Colymbetes*	16	4	Le
ADEPHAGA	Dytiscidae	Hydroporinae	Bidessini	*Clypeodytes**	3	1	Le
ADEPHAGA	Dytiscidae	Dytiscinae	Cybistrini	*Cybister*	31	7	Le
ADEPHAGA	Dytiscidae	Hydroporinae	Hydroporini	*Deronectes*	57	41	Lo-C

(Continued)

APPENDIX: (Continued)

Suborder	Family	Subfamily	Tribe	Genus	PS	MS	Habitat
ADEPHAGA	Dytiscidae	Dytiscinae	Dytiscini	*Dytiscus*	20	11	Lo/Le
ADEPHAGA	Dytiscidae	Hydroporinae	Hydroporini	*Etruscodytes*	1	1	Sb
ADEPHAGA	Dytiscidae	Dytiscinae	Acilini	*Graphoderus*	8	5	Le
ADEPHAGA	Dytiscidae	Hydroporinae	Hydroporini	*Graptodytes*	22	20	Lo/Le
ADEPHAGA	Dytiscidae	Hydroporinae	Pachydrini	*Heterhydrus**	1	1	Le
ADEPHAGA	Dytiscidae	Dytiscinae	Hydaticini	*Hydaticus*	31	9	Le
ADEPHAGA	Dytiscidae	Hydroporinae	Bidessini	*Hydroglyphus*	20	7	Le
ADEPHAGA	Dytiscidae	Hydroporinae	Hydroporini	*Hydroporus*	122	72	Lo/Le
ADEPHAGA	Dytiscidae	Hydroporinae	Hydrovatini	*Hydrovatus*	21	8	Le
ADEPHAGA	Dytiscidae	Hydroporinae	Hygrotini	*Hygrotus*[1]	36	17	Le
ADEPHAGA	Dytiscidae	Hydroporinae	Hyphydrini	*Hyphydrus*	25	6	Le
ADEPHAGA	Dytiscidae	Hydroporinae	Hydroporini	*Iberonectes*	1	1	Lo
ADEPHAGA	Dytiscidae	Hydroporinae	Hydroporini	*Iberoporus*	2	2	Sb
ADEPHAGA	Dytiscidae	Agabinae	Agabini	*Ilybius*	46	21	Lo/Le
ADEPHAGA	Dytiscidae	Laccophilinae	Laccophilini	*Laccophilus*	35	8	Lo/Le
ADEPHAGA	Dytiscidae	Copelatinae	Copelatini	*Liopterus*	2	2	Le
ADEPHAGA	Dytiscidae	Colymbetinae	Colymbetini	*Meladema*	3	2	Lo/Le
ADEPHAGA	Dytiscidae	Colymbetinae	Colymbetini	*Melanodytes*	1	1	Le
ADEPHAGA	Dytiscidae	Hydroporinae	Hydroporini	*Metaporus*	2	1	Le
ADEPHAGA	Dytiscidae	Hydroporinae	Methlini	*Methles*	3	2	Le
ADEPHAGA	Dytiscidae	Hydroporinae	Hydroporini	*Nebrioporus*	53	28	Lo-R
ADEPHAGA	Dytiscidae	Hydroporinae	Hydroporini	*Nectoporus*	3	2	Lo-C
ADEPHAGA	Dytiscidae	Laccophilinae	Laccophilini	*Neptosternus**	11	1	Lo-P
ADEPHAGA	Dytiscidae	Hydroporinae	Hydroporini	*Oreodytes*	7	3	Lo-C
ADEPHAGA	Dytiscidae	Laccophilinae	Laccophilini	*Philodytes**	1	1	Le
ADEPHAGA	Dytiscidae	Agabinae	Agabini	*Platambus*	33	2	Lo
ADEPHAGA	Dytiscidae	Hydroporinae	Hydroporini	*Porhydrus*	4	4	Le
ADEPHAGA	Dytiscidae	Dytiscinae	Acilini	*Rhantaticus**	1	1	Le
ADEPHAGA	Dytiscidae	Colymbetinae	Colymbetini	*Rhantus*	25	9	Le
ADEPHAGA	Dytiscidae	Hydroporinae	Hydroporini	*Rhithrodytes*	6	6	Lo
ADEPHAGA	Dytiscidae	Hydroporinae	Hydroporini	*Tassilodytes*	1	1	Lo
ADEPHAGA	Dytiscidae	Hydroporinae	Hydroporini	*Scarodytes*	9	8	Lo-R
ADEPHAGA	Dytiscidae	Hydroporinae	Hydroporini	*Siettitia*	2	2	Sb
ADEPHAGA	Dytiscidae	Hydroporinae	Hydroporini	*Stictonectes*	12	11	Lo-P
ADEPHAGA	Dytiscidae	Hydroporinae	Hydroporini	*Stictotarsus*	3	3	Lo-P
ADEPHAGA	Dytiscidae	Hydroporinae	Hydroporini	*Trichonectes*	1	1	Lo
ADEPHAGA	Dytiscidae	Hydroporinae	Bidessini	*Yola*	9	6	Lo
ADEPHAGA	Dytiscidae	Hydroporinae	Bidessini	*Yolina**	1	1	Le

(Continued)

APPENDIX: (Continued)

Suborder	Family	Subfamily	Tribe	Genus	PS	MS	Habitat
POLYPHAGA	DRYOPIDAE			*Dryops*	34	22	Lo
POLYPHAGA	Dryopidae			*Pomatinus*	1	1	Lo
POLYPHAGA	ELMIDAE	Elminae		*Dupophilus*	1	1	Lo-R
POLYPHAGA	Elmidae	Elminae		*Elmis*	18	13	Lo-C
POLYPHAGA	Elmidae	Elminae		*Esolus*	11	9	Lo-R
POLYPHAGA	Elmidae	Elminae		*Limnius*	19	12	Lo-R
POLYPHAGA	Elmidae	Elminae		*Oulimnius*	15	15	Lo
POLYPHAGA	Elmidae	Larinae		*Potamophilus*[2]	1	1	Lo-R
POLYPHAGA	Elmidae	Elminae		*Riolus*[3]	11	9	Lo-R
POLYPHAGA	Elmidae	Elminae		*Stenelmis*	74	3	Lo-R
POLYPHAGA	Elmidae	Elminae		*Macronychus*	7	1	Lo-R
POLYPHAGA	HELOPHORIDAE			*Helophorus*	151	63	Le
POLYPHAGA	HYDROCHIDAE			*Hydrochus*	29	14	Lo
POLYPHAGA	HYDROPHILIDAE	Hydrophilinae	Amphiopini	*Amphiops**	6	2	Lo-P
POLYPHAGA	Hydrophilidae	Hydrophilinae	Anacaenini	*Anacaena*	25	8	Lo
POLYPHAGA	Hydrophilidae	Hydrophilinae	Berosini	*Berosus*	38	16	Lo/Le
POLYPHAGA	Hydrophilidae	Hydrophilinae	Hydrophilini	*Brownephilus*	2	2	Le
POLYPHAGA	Hydrophilidae	Hydrophilinae	Chaetarthriini	*Chaetarthria*	5	2	Lo
POLYPHAGA	Hydrophilidae	Acidocerinae		*Chasmogenus*	3	1	Le
POLYPHAGA	Hydrophilidae	Sphaeridiinae	Coelostomatini	*Coelostoma*	15	3	Lo/Le
POLYPHAGA	Hydrophilidae	Hydrophilinae	Anacaenini	*Crenitis*	16	1	Lo
POLYPHAGA	Hydrophilidae	Enochrinae		*Cymbiodyta*	1	1	Le
POLYPHAGA	Hydrophilidae	Enochrinae		*Enochrus*	57	28	Lo/Le
POLYPHAGA	Hydrophilidae	Acidocerinae		*Helochares*	21	8	Lo/Le
POLYPHAGA	Hydrophilidae	Hydrophilinae	Chaetarthriini	*Hemisphaera*	3	3	Lo
POLYPHAGA	Hydrophilidae	Hydrophilinae	Hydrophilini	*Hydrobiomorpha**	4	2	Le
POLYPHAGA	Hydrophilidae	Hydrophilinae	Hydrobiusini	*Hydrobius*	5	2	Le
POLYPHAGA	Hydrophilidae	Hydrophilinae	Hydrophilini	*Hydrochara*	9	3	Le
POLYPHAGA	Hydrophilidae	Hydrophilinae	Hydrophilini	*Hydrophilus*	16	5	Lo/Le
POLYPHAGA	Hydrophilidae	Hydrophilinae	Laccobiini	*Laccobius*	118	35	Lo
POLYPHAGA	Hydrophilidae	Hydrophilinae	Hydrobiusini	*Limnoxenus*	2	2	Le
POLYPHAGA	Hydrophilidae	Hydrophilinae	Hydrobiusini	*Limnohydrobius*	2	1	Le
POLYPHAGA	Hydrophilidae	Hydrophilinae	Laccobiini	*Paracymus*	10	5	Le
POLYPHAGA	Hydrophilidae	Hydrophilinae	Hydrophilini	*Sternolophus**	6	1	Lo/Le
POLYPHAGA	SPERCHEIDAE	Spercheinae		*Spercheus*	6	2	Le
POLYPHAGA	HYDRAENIDAE*	Hydraeninae	Hydraenini	*Hydraena*	342	253	Lo
POLYPHAGA	Hydraenidae	Hydraeninae	Limnebiini	*Limnebius*	86	67	Lo
POLYPHAGA	Hydraenidae	Ochthebinae	Ochthebiini	*Ochthebius*[4]	288	193	Lo

**Only sporadic occurrence in the Mediterranean Basin. (1) We have not included the genus* Hyphoporus *as it is currently considered a subgenus of* Hygrotus *(Villastrigo et al., 2017); (2)* Potamophilus *has terrestrial habits as an adult; (3)* Normandia *has been recently synonymized with* Riolus *(Jäch et al., 2016); (4) Following Vilastrigo et al. (2019),* Aulacochthebius *and* Micragasma *are considered subgenus of* Ochthebius*.*

References

Abellán P., D. Sánchez-Fernández, J. Velasco & A. Millán. 2007. Effectiveness of protected area networks in representing freshwater biodiversity: the case of a Mediterranean river basin (SE Spain). Aquatic Conservation: Marine and Freshwater Ecosystems 17: 361–374.

Abellán P., D. Sánchez-Fernández, F. Picazo, A. Millán, J.M. Lobo & I. Ribera. 2013. Preserving the evolutionary history of freshwater biota in Iberian National Parks. Biological Conservation 162: 116–126.

Arribas P., P. Abellán, J. Velasco, D.T. Bilton, A. Millán & D. Sánchez-Fernández. 2012a. Evaluating drivers of vulnerability to climate change: a guide for insect conservation strategies. Global Change Biology 18: 2135–2146.

Arribas P., J. Velasco, P. Abellán, D. Sánchez-Fernández, C. Andújar, et al. 2012b. Dispersal ability rather than ecological tolerance drives differences in range size between lentic and lotic water beetles (Coleoptera: Hydrophilidae). Journal of Biogeography 39: 984–994.

Baca S.M., A. Alexander, G.T. Gustafson & A.E.Z. Short. 2017. Ultraconserved elements show utility in phylogenetic inference of Adephaga (Coleoptera) and suggest paraphyly of 'Hydradephaga'. Systematic Entomology 42: 786–795.

Baca S.M., G.T. Gustafson, A. Alexander, H. Gough & E.A. Toussain. 2021. Integrative phylogenomics reveals a Permian origin of Adephaga beetles. Systematic Entomology 46: 968–990.

Bilton D.T., I. Ribera & A.E.Z. Short. 2019. Water beetles as models in ecology and evolution. Annual Review of Entomology 64: 359–377.

Biström O., A.N. Nilsson & J. Bergsten. 2015. Taxonomic revision of Afrotropical *Laccophilus* Leach, 1815 (Coleoptera, Dytiscidae). ZooKeys 542: 1–379.

Calosi P., D.T. Bilton & J.I. Spicer. 2008. Thermal tolerance, acclimatory capacity and vulnerability to global climate change. Biology Letters 4: 99–102.

Clarkson B., M. Archangelsky, P.L.M. Torres & A.E.Z. Short. 2018. Chapter 15.4: Family Hydrophilidae. Pages 561-576 in: Thorp and Covich's Freshwater Invertebrates (Fourth Edition). Volume 3: Keys to Neotropica Hexapoda.

Dettner K. 2019. Defenses of water insects. Pages 191-262 *in*: Aquatic Insects. Springer, Cham.

Fikáček M., A.A. Prokin, R.B. Angus, A.G. Ponomarenko, Y. Yue, et al. 2012. Phylogeny and the fossil record of the Helophoridae reveal Jurassic origin of modern hydrophiloid lineages (Coleoptera: Polyphaga). Systematic Entomology 37: 420–447.

Foster G.N. & L.E. Friday. 2011. Key to adults of the water beetles of Britain and Ireland (Part 1). Handbooks for the Identification of British Insects. Vol. 4 Part 5 (2nd Ed.). The Field Studies Council, Telford.

Foster G.N., D.T. Bilton & L.E. Friday, 2014. Key to adults of the water beetles of Britain and Ireland (Part 2). Handbooks for the Identification of British Insects. Vol. 4 Part 5b. The Field Studies Council, Telford.

Gerdes G., J. Spira & C. Dimentman. 1985. 15. The fauna of the Gavish Sabkha and the Solar Lake – a comparative study. Pages 322-345 in: Friedman, G.M. & W.E. Krumbein (eds), Ecological Studies. Vol. 53: Hypersaline Ecosystems. Springer, Verlag.

Gutiérrez-Cánovas C., R. Arias-Real, D. Bruno, Marco J. Cabrerizo, J.M. González-Olalla, et al. 2022. Multiple-stressors effects on Iberian freshwaters: A review of current knowledge and future research priorities. Limnetica 41: 245–268.

Hunt T., J. Bergsten, Z. Levkanicoa, A. Papadopoulou, O. St John et al. 2007. A comprehensive phylogeny of beetles reveals the evolution origins of a superradiation. Science 318: 1913–1916.

Jäch M.A. 1998. Annotated check list of aquatic and riparian/littoral beetle families of the world (Coleoptera). Pages 25–42 in: Jäch, M.A. & L. Ji (eds.), Water Beetles of China, Vol. II. Zoologisch-Botanische Gesellschaft in Österreich and Wiener Coleopterologenverein, Wien.

Jäch M.A. & M. Balke. 2008. Global diversity of water beetles (Coleoptera) in freshwater. Hydrobiologia 595: 419–442.

Jäch, M., J. Kodada, M. Brojer, W. Shepard & F. Ciampor. 2016. World Catalogue of Insects: Coleoptera: Elmidae and Protelmidae (pp. 1–318). Leiden: Brill, XXI fl.

Komarek A. 2003. Hydrophilidae: I. Check list and key to Palearctic and Oriental genera of aquatic Hydrophilidae (Coleoptera). Pages 383–395 in: Jäch, M.A. & L. Ji (eds.), Water beetles of China. Vol. III. Zoologisch-Botanische Gesellschaft & Wiener Coleopterologenverein, Wien.

Larson D.J., Alarie Y., Roughley R.E. 2000. Predaceous diving beetles (Coleoptera: Dytiscidae) of the Nearctic region, with emphasis on the fauna of Canada and Alaska. NRC Research Press, Ottawa.

Lawrence J.F., A. Ślipinski, A.E. Seago, M.K. Thayer, A.F. Newton et al. 2011. Phylogeny of the coleóptera on morphological characters of adults and larvae. Annales Zoologici 61: 1–217.

Millán A., P. Abellán, I. Ribera, D. Sánchez-Fernández & J. Velasco. 2006. The Hydradephaga of the Segura basin (SE Spain): twenty five years studying water beetles. Memorie della Società Entomologica Italiana 85: 137–158.

Millán A., J. Velasco, C. Gutiérrez-Cánovas, P. Arribas, F. Picazo, et al. 2011. Mediterranean saline streams in southeast Spain: What do we know? Journal of Arid Environments 75: 1352–1359.

Millán A., D. Sánchez-Fernández, P. Abellán, F. Picazo, J.A. Carbonell, et al. 2014. Atlas de los Coleópteros Acuáticos de España Peninsular. MAGRAMA, Madrid.

Miller K. & R. Bergsten. 2016. Diving beetles of the world. Systematica and Biology of the Dytiscidae. Johns Hopkins University Press, Baltimore.

McKenna D.D., A.L. Wild, K. Kanda, C.L. Bellamy, R.G. Beutel, et al. 2015. The beetle tree of life reveals that Coleoptera survived end-Permian mass extinction to diversify during the Cretaceous terrestrial revolution. Systematic Entomology 40: 835–880.

Nasserzadeh H. & A. Komarek. 2017. Taxonomic revision of the water scavenger beetle genus *Sternolophus* Solier, 1834 (Coleoptera: Hydrophilidae). Zootaxa 4282: 201-254.

Nilsson A. (Ed.) 1996. Aquatic Insects of North Europe. A Taxonomic Handbook. Apollo Book, Stenstrup, Denmark.

Pallarés S., A. Millán, J.M. Mirón, J. Velasco, D. Sánchez-Fernández, et al. 2020. Assessing the capacity of endemic alpine water beetles to face climate change. Insect Conservation and Diversity 13: 271–282.

Peddle S.M. & D.J. Larson. 1999. Cuticular Evidence of Traumatic Experiences of Water Beetles (Coleoptera: Dytiscidae, Hydrophilidae). The Coleopterists Bulletin 53: 42–51.

Perkins P.D. 1980. Aquatic beetles of the family Hydraenidae in the Western Hemisphere: classification, biogeography and inferred phylogeny (Insecta, Coleoptera). Quaestiones Entomologicae 16: 3–554.

Perkins P.D. 1997. Life on the effective bubble: exocrine secretion delivery systems (ESDS) and the evolution and classification of beetles in the family Hydraenidae (Insecta: Coleoptera). Annals of Carnegie Museum 66: 89–207.

Picazo F., A. Millán & S. Doledec. 2012. Are patterns in the taxonomic, biological and ecological traits of water beetles congruent in Mediterranean ecosystems? Freshwater Biology 57: 2192–2210.

Ribera I. 2000. Biogeography and conservation of Iberian water beetles. Biological Conservation 92: 131–150.

Ribera I., J.E. Hogan & A.P. Vogler. 2002. Phylogeny of Hydradephagan water beetles inferred from 18S rDNA sequences. Molecular Phylogenetics and Evolution 23: 4–62.

Ribera I., A.P. Vogler & M. Balke. 2008. Phylogeny and diversification of diving beetles (Coleoptera: Dytiscidae). Cladistics 24: 563–590.

Rogers D.C. & J.H. Thorp. 2019. Keys to Palaearctic fauna: Thorp and Covich's freshwater invertebrates, vol 4. Academic Press/Elsevier, London.

Sánchez-Fernández D., P. Abellán, A. Mellado, J. Velasco & A. Millán. 2006. Are water beetles good indicators of biodiversity in Mediterranean aquatic ecosystems? The case of the Segura river basin (SE Spain). Biodiversity and Conservation 15: 4507–4520.

Sánchez-Fernández D., P. Abellán, F. Picazo, A. Millán, I. Ribera et al. 2013. Do protected areas represent species' optimal climatic conditions. A test using Iberian water beetles. Diversity and Distributions 19: 1407–1417.

Sánchez-Fernández D., D.T. Bilton, P. Abellán, I. Ribera, J. Velasco et al. 2008. Are the endemic water beetles of the Iberian Peninsula and the Balearic Islands effectively protected? Biological Conservation 141: 1612–1627.

Short A.E.Z. 2018. Systematics of aquatic beetles (Coleoptera): current state and future directions. Systematic Entomology 43: 1–18.

Short A.E.Z. & M. Fikáček. 2013. Molecular phylogeny, evolution, and classification of the Hydrophilidae (Coleoptera). Systematic Entomology 38: 723–752.

Short A.E.Z., J.L. Joly, M. García, A. Wild, D.D. Bloom et al. 2015. Molecular phylogeny of the Hydroscaphidae (Coleoptera: Myxophaga) with description of a remarkable new lineage from the Guiana shield. Systematic Entomology 40: 214–229.

Short A.E.Z., J. Cole & E.F.A. Toussaint. 2017. Phylogeny, classification and evolution of the water scavenger beetle tribe Hydrobiusini inferred from morphology and molecules (Coleoptera: Hydrophilidae: Hydrophilinae). Systematic Entomology 42: 677–691.

Ślipiński S.A., R.A.B. Leschen & J.F. Lawrence. 2011. Order Coleoptera Linnaeus, 1758. Pages 203–208 in: Zhang, Z.-Q. (ed.), Animal biodiversity: an outline of higher-level classification and survey of taxonomic richness Vol. 3148. Magnolia Press, Auckland.

Tachet H., P. Richoux, M. Bournard & P. Usseglio-Polatera. 2010. Invertébrés d'eau douce: Systématique, biologie, écologie. CNRS éditions, Paris.

Trizzino M., M.A. Jäch, P. Audisio, R. Alonso & I. Ribera. 2013. A molecular phylogeny of the cosmopolitan hyperdiverse genus Hydraena Kugelann (Coleoptera, Hydreanidae). Systematic Entomology 38: 192–208.

Thorp J.H. & D.C. Rogers. 2015. Ecology and general biology: Thorp and Covich's freshwater invertebrates, vol 1, 4th edn. Academic Press/Elsevier, Amsterdam.

Vasilikopoulos A., M. Balke, S. Kukowa, J. Pflug, S. Martin, et al. 2021. Phylogenomic analyses clarify the pattern of evolution of Adephaga (Coleoptera) and highlight phylogenetic artefacts due to model misspecification and excessive data trimming. Systematic Entomology 46: 991–1118.

Velasco J. & A. Millán. 1998. Feeding Habits of Two Large Insects from a Desert Stream: Abedus herberti (Hemiptera: Belostomatidae) and Thermonectus marmoratus (Coleoptera: Dytiscidae). Aquatic Insects 20: 85–96.

Verberk W.C.E.P. & D.T. Bilton. 2013. Respiratory control in aquatic insects dictates their vulnerability toglobal warming. Biology Letters 9: 20130473

Villastrigo A., C. Hernando, A. Millán & I. Ribera. 2020. The neglected diversity of the Ochthebius fauna from Eastern Atlantic and Central and Western Mediterranean coastal rockpools (Coleoptera, Hydraenidae). Organisms Diversity & Evolution 20: 785–801.

Villastrigo, A., M. Jäch, A. Cardoso, L.F. Valladares & I. Ribera. 2019. A molecular phylogeny of the tribe Ochthebiini (Coleoptera, Hydraenidae, Ochthebiinae). Systematic Entomology 44: 273–288.

Villastrigo, A., I. Ribera, M. Manuel, A. Millán & H. Fery. 2017. A new classification of the tribe Hygrotini Portevin, 1929 (Coleoptera: Dytiscidae: Hydroporinae). Zootaxa 4317(3): 499–529.

White D.S. & R.E. Roughley. 2008. Aquatic Coleoptera. Pages 571–671 *in*: Merritt, R.W., K.W. Cummins & M.B. Berg (eds.), An introduction to the aquatic insects of North America, Fourth. (Eds), Kendall/Hunt. Publishing Company, Iowa.

Yee D.A. & S. Kehl. 2015. Order Coleoptera. Pages 1003-1042 in: Thorp and Covich's Freshwater Invertebrates (4th Edition). Elsevier, London.

Zhang S.Q., L.H. Che, Y. Li, D. Liang, H. Pang, et al. 2018. Evolutionary history of Coleoptera revealed by extensive sampling of genes and species. Nature Communications 9: 205.

Chapter 14

Order Trichoptera ☆

Ioannis Karaouzas[1], Carmen Zamora-Muñoz[2], Marta Sáinz Bariáin[3], Johann Waringer[4] and Ralph W. Holzenthal[5]

[1]Institute of Marine Biological Resources and Inland Waters, Hellenic Centre for Marine Research, Anavyssos, Attica, Greece, [2]Department of Zoology, University of Granada, Granada, Spain, [3]Spanish Institute of Oceanography, Oceanographic Research Center of Santander, Santander, Cantabria, Spain, [4]Department of Functional and Evolutionary Ecology, University of Vienna, Vienna, Austria, [5]Department of Entomology, University of Minnesota, St. Paul, MN, United States

Introduction

The order Trichoptera Kirby, 1813, or caddisflies, is a monophyletic group of holometabolous insects (i.e., having a pupal stage in their developmental cycle) most closely related to the megadiverse order Lepidoptera (moths and butterflies). Together they form the monophyletic superorder Amphiesmenoptera. Basal diversification of existing lineages of Trichoptera dates back from Middle and Late Triassic times (De Moor & Ivanov, 2008). With presently 16,267 extant species worldwide, distributed into 52 families and 632 genera (Thomas et al., 2020; Morse, 2020), Trichoptera are among the most successful and diverse animals in freshwater ecosystems. Containing more species than any of the other fully aquatic insect orders combined (Ephemeroptera, Odonata, Plecoptera, Megaloptera), they are the seventh most species-rich order among Insecta (Holzenthal et al., 2007; Morse et al., 2019; Thomas et al., 2020) and the second most diverse clade of obligate freshwater inhabiting organisms (Malm et al., 2013). They inhabit a diverse range of aquatic habitats and occur on almost every continent except Antarctica (Holzenthal et al., 2011). Nevertheless, they are distributed unevenly with the highest number occurring in the Oriental Biogeographic Region (Southern Asia) and the lowest in the Afrotropical Region (Morse et al., 2019). Immature caddisflies can inhabit practically every available substrate in freshwater ecosystems where they play a fundamental role in energy flow, food webs, biological monitoring of ecological quality, as food for fish and other predators, and as engineers of substrate stability (Morse et al., 2019).

Unlike most moths and butterflies, the vast majority of caddisflies have aquatic egg, larval, and pupal stages except for a very few considered terrestrial or semiterrestrial and brackish water species, and the family Chathamiidae whose members are exclusively marine (Holzenthal et al., 2015). Adults are aerial and usually fly not far from the larval aquatic habitats. They are largely associated with riparian vegetation and have crepuscular or nocturnal activity and behavior. Most species have one generation per year (i.e., univoltine) and lay eggs (i.e., oviparous). Females lay 30–10,000 eggs in masses that form plaques attached to substrates, strings and loops, or gelatinous spheres (Lancaster & Downes, 2013). Eggs hatch in a few days or weeks, and the individuals spend most of their lives as larvae. Larval development typically involves five instars and may be rapid. Caddisfly larvae are free living and mobile or construct diverse and elaborate cases or retreats with the aid of silk from modified labial salivary glands. Based on larval morphology and construction behavior, caddisflies are traditionally split into two taxonomic subdivisions: the "retreat-making" Annulipalpia and the "case-making" Integripalpia including the "cocoon making" Spicipalpia, the latter now considered a paraphyletic group (Wiggins, 2004; Kjer et al., 2016; Holzenthal et al., 2007; Thomas et al., 2020). Annulipalpians construct retreats of detritus or mineral fragments, or both, held together with silk that are fixed to the substrate, from which silken nets are spun and used to filter food particles (Holzenthal et al., 2007; Morse et al., 2019). The majority of integripalpian families, now included in the infraorder Phryganides, create transportable, tubular cases made from a wide variety of materials, including sand, small stones, leaf and wood fragments, and even small snail shells. These act primarily as a protective device or camouflage against predators and also have an important function in increasing respiratory efficiency (Wiggins, 1996; Holzenthal et al., 2007). Finally, the cocoon making integripalpian families of the former "Spicipalpia" (Rhyacophilidae, Glossosomatidae, Hydroptilidae, Ptilocolepidae) include

☆. This chapter is dedicated to the memory of Carmen Zamora-Muñoz.

Identification and Ecology of Freshwater Arthropods in the Mediterranean Basin. DOI: https://doi.org/10.1016/B978-0-12-821844-0.00009-0

free-living larvae and "tortoise" and "purse" case-making larvae. When mature, larvae pupate by constructing stationary shelters or modifying their portable cases into stationary shelters, and then molting to the pupal form (Holzenthal et al., 2007; Morse et al., 2019). Pupation is completed after a few weeks, with adult emergence occurring often at specific times of day or night and at predictable times of year, cued by daylight changes, temperature, or other environmental conditions. Adult Trichoptera are short-lived, active flyers with reduced mouthparts, and mate in flight, on the ground or in riparian vegetation. The peak emergence time of adults varies according to their geographical distribution.

There are 2349 described species belonging to 229 genera and 28 families in the Palearctic Region (de Moor & Ivanov, 2008; Morse, 2020). The West Palearctic Region contains 1841 species and is the fourth highest region of Trichoptera diversity worldwide followed by the Nearctic (1581), East Palearctic (1426), and Afrotropical (1251) (Morse et al., 2019). The complex history (paleogeographical and paleoclimatic) and environmental features of Mediterranean Basin have contributed to the high biodiversity and endemism currently found in these areas (Tierno de Figueroa et al., 2013). To date, the Mediterranean Basin contains 25 families, 110 genera, and 1147 species of caddisflies (Table 14.1), but with the increasing use of molecular techniques in taxonomy and systematics, many new species, including cryptic species, continue to be described each year from several areas. Of the 1147 species present in the Mediterranean Basin, 403 are endemic, or 35% of the total number of species recorded (Table 14.1). However, the immature stages of many Mediterranean species remain, even today, largely unknown due to the lack of taxonomic studies to associate immature stages of rare and endemic species with adults.

General ecology and distribution

Trichoptera diversity and evolutionary success is attributed greatly to the various ways in which silk is used among the different groups, allowing larvae to exploit different kinds of microhabitats and food resources (Mackay & Wiggins, 1979; Wiggins, 2004). The ability of the larvae to spin silk and to use it for the construction of cases and retreats, as well as for other functions such as camouflage, supporting aquatic respiration, resistance to fast currents, or to capture food, are undoubtedly responsible for the diversity and evolutionary success of the order (Mackay & Wiggins, 1979; Wiggins, 2004). Trichoptera larvae live in a diverse range of aquatic habitats from springs, streams, and rivers to temporary pools, ponds, wetlands, lakes, and reservoirs, and even brackish or, rarely, sea water (Mackay & Wiggins, 1979). However, the habitat preferences of the individual species are usually quite restricted (Graf et al., 2008) and are further subdivided according to the physical and chemical factors of their living environment. Most caddisfly species are benthic and live underneath or on top or sides of rocks and stones, in stacks of fallen leaves and twigs or wood debris, burrowed in sand or silt on the bottom of water bodies, or attached to aquatic plants (e.g., emergent and/or submergent macrophytes, mosses, etc.). Some can even swim with their cases, such as *Triaenodes lefkas* Malicky, 1974 and *Setodes tineiformis* Curtis, 1834 (Leptoceridae). Only a few are semi-aquatic and live in moist terrestrial environments such as the limnephilids *Limnephilus centralis* Curtis, 1834 and *Enoicyla pusilla* Burmeister, 1839. Caddisfly larvae have exceptionally diverse feeding habits and diet, they can be herbivorous, detritivorous, carnivorous, or omnivorous (Wiggins, 2004; Wells, 2005; Morse & Lenat, 2005). The ecological characteristics and distributions of each family present in the Mediterranean Basin are briefly described as follows:

Rhyacophilidae: Larvae of the "free-living green caddisflies" are campodeiform (slender and with tapered ends) and medium-to-large in size (8–30 mm). As their common name indicates, they construct no larval structure. However, their last larval instar constructs a fixed, dome-like shelter of stones and silk, under which it spins a semipermeable silken cocoon and pupates within. Pupal shelters and cocoons of Rhyacophilidae resemble those of Glossosomatidae. Larvae are primarily predators and hunt for other invertebrates, but some species are omnivores or change their diet as they mature. They generally prefer cold, well-oxygenated, fast flowing springs and mountainous rivers, where they live underneath or on top or sides of rocks and stones. Most species are univoltine and rarely semivoltine (Graf et al., 2008). The family occurs primarily in the Northern Hemisphere, extending from North America and Europe, to India and the tropical areas of southeastern Asia. In the Mediterranean Basin, the family is represented by a single genus, *Rhyacophila* Pictet, with currently 100 species, 37 being endemic (Table 14.1).

Glossosomatidae: Larvae of "saddle or turtle-case caddisflies" are small (3–10 mm), slow-moving cyphosomatic species. Most construct a dome-like case of small stones and silk that is oval and has a transverse strap ventrally connecting the two longer sides (Wiggins, 2004). During pupation, the final larval instar removes the transverse ventral strap, cements the dome to the rock substrate with silk, spins a semipermeable cocoon, and pupates inside it. Larval habitat selection in this family is quite diverse, ranging from cold springs to warm large rivers and wave-swept lake shores, where larvae live on rocks, stones, wood, and leaves. Larvae are dominant grazers of

TABLE 14.1 List of families and genera of Trichoptera Kirby 1913 occuring in the Mediterranean Basin with the number of recorded species and their distribution. (Neu et al., 2018; Fauna Europaea, Trichoptera World Checklist). Distributional areas, following Tierno de Figueroa et al. (2013), are divided into four different areas: NW (North-West; Mediterranean Europe from Portugal to Italy, both including the Balearic Islands, Corsica, Sardinia, Sicily, and Malta, as all of the small West Mediterranean islands); NE (North-East; from the former Yugoslavian countries to North-West Turkey, including all of the Greek and West Turkish islands); SE (South- East; South Turkey, Cyprus, Mediterranean Near East, and East Libya); and SW (South-West; from Morocco to West Libya).

Taxon	Number of species/(endemics)	Distribution	Notes
Suborder Annulipalpia Martynov 1924			
Family Philopotamidae Stephens 1829			
Chimarra Stephens 1829	1	NW, SW	*C. marginata*
Philopotamus Stephens 1829	7 (2)	NW, NE, SW, SE	7 species and 15 subspecies. Endemics: *P. corsicanus* (Sardinia & Corsica), *P. ketama* (Morocco)
Wormaldia McLachlan 1865	40 (16)	NW, NE, SW, SE	40 species and 8 subspecies
Family Ecnomidae Ulmer 1903			
Ecnomus McLachlan 1864	6 (3)	NW, NE, SW, SE	Endemics: *E. galilaeus* (Palestine), *E. kurui* (Turkey), *E. relictus* (Algeria)
Family Polycentropodidae Ulmer 1903			
Cyrnus Stephens 1836	6	NW, NE, SW	
Holocentropus McLachlan 1878	3	NW, NE	*H. dubius, H. picicornis, H. stagnalis*
Neureclipsis McLachlan 1864	1	NW, NE	*N. bimaculata*
Nyctiophylax Brauer 1865	1 (1)	NW	Endemic: *N. gaditana* (Spain)
Plectrocnemia Stephens 1836	19 (3)	NW, NE, SW	19 species and 5 subspecies: Endemics: *P. alicatai* (Italy), *P. diakosensis* and *P. kydon* (Greece)
Polycentropus Curtis 1835	25 (10)	NW, NE, SW, SE	25 species, 10 subspecies within *P. ierapetra* and *P. flavomaculatus*
Family Pseudoneureclipsidae Ulmer 1951			
Pseudoneureclipsis Ulmer 1913	5 (2)	NE, SW, SE	Endemics: *P. maroccanus* (Morocco), *P. palmonii* (Israel)
Family Psychomyiidae Walker 1852			
Lype McLachlan 1878	3	NW, NE, SE	*L. auripillis, L. phaeopa, L. reducta*
Metalype Klapalek 1898	2	NW, NE	*M. fragilis, M. klapaleki*
Psychomyia Latreille 1829	4 (3)	NW, NE, SE	Endemics: *P. ctenophora* (Iberian P.),*P. dadayensis* and *P. mengensis* (Turkey)
Paduniella Ulmer 1913	1	NW, SW	*P. vandeli*
Tinodes Curtis 1834	75 (48)	NW, NE, SW, SE	65% endemism

(Continued)

TABLE 14.1 (Continued)

Taxon	Number of species/(endemics)	Distribution	Notes
Family Hydropsychidae Curtis 1835			
Cheumatopsyche Wallengren 1891	2	NW, NE, SW, SE	*C. capitella, C. lepida*
Diplectrona Westwood 1839	6 (2)	NW, NE	Endemics: *D. meridionalis* (Corsica), *D. ripollensis* (Spain)
Hydropsyche Pictet 1836	119 (49)	NW, NE, SW, SE	High endemism in Turkey, Greece and Iberian P.
Suborder Integripalpia Martynov 1924			
Family Glossosomatidae Wallengren 1891			
Agapetus Curtis 1834	28 (14)	NW, NE, SE, SW	
Catagapetus MacLachlan 1884	2 (1)	NW	Endemic: *C. maclachlani* (Iberian P.)
Glossosoma Curtis 1834	19 (8)	NW, NE, SE	
Synagapetus McLachlan 1879	22 (12)	NW, NE	
Family Hydroptilidae Stephens 1836			
Agraylea Curtis 1834	2	NW, NE	*A. multipunctata, A. sexmaculata*
Allotrichia McLachlan 1880	7 (2)	NW, NE, SE	Both Greek endemics: *A. laerma, A. millitsa*
Hydroptila Dalman 1819	66 (25)	NW, NE, SE, SW	
Ithytrichia Eaton 1873	4 (1)	NW, NE, SE, SW	Endemic: *I. aquila* (Spain)
Microptila Ris 1897	1	NE	*M. minutissima*
Orthotrichia Eaton 1873	6	NW, NE, SE, SW	
Oxyethira Eaton 1873	11 (4)	NW, NE, SE, SW	Endemics: *O. archaica* and *O. iglesiasi* (Porugal), *O. hartigi* (Sardinia & Corsica), *O. mithi* (Crete)
Stactobia McLachlan 1880	26 (17)	NW, NE, SE, SW	
Stactobiella Martynov 1924	2 (1)	NW, NE	Endemic: *S. celtikci* (Turkey)
Tricholeiochiton Kloet & Hincks 1944	1	NW, NE	*T. fagesi*
Family Ptilocolepidae Martynov 1913			
Ptilocolepus Kolenati 1848	4(1)	NW, NE	Endemic: *P. extensus* (Iberian P.)
Family Rhyacophilidae Stephens 1836			
Rhyacophila Pictet 1834	100 (37)	NW, NE, SE, SW	
Infraorder Phryganides			
Suborder Brevitentoria			
Family Calamoceratidae Ulmer 1905			

(*Continued*)

TABLE 14.1 (Continued)

Taxon	Number of species/(endemics)	Distribution	Notes
Calamoceras Brauer 1865	2	NW, NE	*C. illiesi* (NE), *C. marsupus* (NW)
Family Leptoceridae Leach 1815			
Adicella McLachlan 1877	15 (7)	NW, NE, SE	
Athripsodes Billberg 1820	23 (13)	NW, NE, SE, SW	
Ceraclea Stephens 1829	10 (2)	NW, NE, SE	Endemics: *C. litania* (Lebanon), *C. macronemoides* (Portugal)
Erotesis McLachlan 1877	3 (1)	NW	Endemic: *E. schachti* (Iberian P.)
Leptocerus Leach 1815	5 (1)	NW, NE, SE	Endemic: *L. aksu* (Turkey)
Mystacides Berthold 1827	3	NW, NE	*M. azureus, M. longicornis M. niger*
Oecetis McLachlan 1877	10 (3)	NW, NE, SW, SE	*O. brignolii* (Turkey), *O. grazalemae* (Spain), *O. uyulala* (Algeria)
Parasetodes McLachlan 1880	1	NW, NE	*P. respersellus*
Setodes Rambur 1842	13 (5)	NW, NE, SW, SE	Endemics: *S. dehensurae* and, *S. muglaensis* (Turkey), *S. holocercus* (Iberian P.), *S. kugleri* (Israel), *S. zerrouki* (Morocco)
Triaenodes McLachlan 1865	11 (2)	NW, NE, SW, SE	Endemics: *T. albicornis, T. laami* (both Morocco)
Family Molannidae Wallengren 1891			
Molanna Curtis 1834	2*	NW, NE	Europe but not in the Mediterranean (*2 species in proximities; Alps)
Molannodes McLachlan 1866	1	NW, NE	*M. tinctus* (Europe but not in the Mediterranean)
Family Odontoceridae Wallengren 1891			
Odontocerum Leach 1815	3 (1)	NW, NE	Endemic: *O. lusitanicum* (Iberian P.)
Family Beraeidae Wallengren 1891			
Beraea Stephens 1836	20 (14)	NW, NE	
Beraeamyia Mosely 1936	11 (8)	NW, NE	Almost all endemics from Greece and Turkey
Beraeodes Eaton 1867	1	NW, NE	*B. minutus*
Beraeodina Mosely 1931	1 (1)	NW	*B. palpalis* – Corsica & Sardinia endemic
Ernodes Wallengren 1891	8 (2)	NW, NE, SE	Endemics: *E. kakofonix* (Greece), *E. nigroauratus* (Italy)
Family Helicopsychidae Ulmer 1906			
Helicopsyche Siebold 1856	5 (2)	NW, NE	Endemics: *H. lusitanica* (Iberian P.), *H. megalochari* (Greece)
Family Sericostomatidae Stephens 1836			
Notidobia Stephens 1829	6 (2)	NW, NE	Endemics: *N. sagarrai*, (Spain), *N. salihli* (Turkey)

(Continued)

TABLE 14.1 (Continued)

Taxon	Number of species/(endemics)	Distribution	Notes
Oecismus McLachlan 1876	5 (2)	NW, NE	Endemics: *O. tjederi* (Greece), *O. turcicus* (Turkey)
Schizopelex McLachlan 1876	7 (5)	NW, NE	
Sericostoma Latreille 1825	19 (12)	NW, NE	
Suborder Plenitentoria			
Family Apataniidae Wallengren 1886			
Apatania Kolenati 1848	12 (2)	NW, NE, SE	Endemics: *A. theischingerorum* (Spain), *A. volscorum* (Italy)
Apataniana Mosely 1936	3 (3)	NE	All Greek endemics; *A. hellenica, A. strpones, A. vardusia*
Family Goeridae Ulmer 1903			
Goera Stephens 1829	1	NW, NE	*G. pilosa*
Larcasia Navas 1917	2	NW	*L. ligurica, L. partita*
Lithax McLachlan 1876	3	NW, NE, SE	*L. musaka, L. niger, L. obscurus*
Silo Curtis 1830	7 (3)	NW, NE	Endemics: *S. chrisiammos* (Greece), *S. mediterraneus* (Italy), *S. rufescens* (Corsica & Sardinia)
Silonella Fischer 1966	1	NW, SW	2 subspecies: *S. aurata aurata* and *S. a.ronda* (Spain)
Family Limnephilidae Kolenati 1848			
Subfamily Drusinae Banks 1916			
Anomalopterygella Stein 1867	1	NW	*A. chauviniana* (Iberian P. & France)
Cryptothrix McLachlan 1880	1	NW	*C. nebulicola* (Italy)
Drusus Stephens 1833	56(19)	NW, NE	Many endemics especially in the Dinaric region
Ecclisopteryx Kolenati 1848	8	NW, NE	
Hadimina Sipahiler 2002	1	NE	*H. torosensis* (Turkey)
Leptodrusus Ulmer 1913	1	NW	*L. budtzi*
Metanoea McLachlan 1880	3	NW, NE	
Monocentra Rambur 1842	1	NW	*M. lepidoptera* (Italy)
Subfamily Limnephilinae Kolenati 1848			
Tribe Chaetopterygini Hagen 1858			
Annitella Klapalek 1907	10 (8)	NW, NE	6 Iberian endemics
Chaetopteroides Kumanski 1987	2	NE	(in Med proximities)
Chaetopterygopsis Stein 1874	3	NW, NE	*C. maclachlani, C. sisestii, C. siveci*
Chaetopteryx Stephens 1829	19 (14)	NW, NE	Many Adriatic, Italian and Iberian endemics
Psilopteryx Stein 1874	1	NE	*P. montanus*

(Continued)

TABLE 14.1 (Continued)

Taxon	Number of species/(endemics)	Distribution	Notes
Tribe Limnephilini Kolenati 1848			
Anabolia Stephens 1837	4	NW, NE	*A. anatolia, A. fuscata, A. lombarda, A. nervosa*
Glyphotaelius Stephens 1833	1	NW, NE	*G. pellucidus*
Grammotaulius Kolenati 1848	3	NW, NE	*G. nigropunctatus, G. nitidus, G. submaculatus*
Limnephilus Leach 1815	38	NW, NE, SW, SE	
Rhadicoleptus Wallengren 1891	1	NW, NE	*R. alpestris* (3 subspecies)
Tribe Stenophylacini Schmid 1955			
Acrophylax Brauer 1867	1	NW	*A. zerberus*
Allogamus Schmid 1955	16 (6)	NW, NE	Majority of endemics from Iberian P.
Alpopsyche Botosaneanu & Giudicelli 2004	1	NW	*A. ucenorum*
Anisogamus McLachlan 1874	2 (1)	NW	Endemic: *A. waringeri* (France)
Consorophylax Schmid 1955	3	NW	*C. consors, C. corvo, C. demastroi*
Enoicyla Rambur 1842	3	NW, NE	Larvae terrestrial
Halesus Stephens 1836	7 (3)	NW, NE	Endemics: *H. appenninus, H. calabrus* and *H. nurag* (Italy)
Hydatophylax Wallengren 1891	1	NW, NW (outer Med limits)	*H. infumatus*
Melampophylax Schmid 1955	5 (3)	NW, NE	Endemics: *M. orientalopyrenaeus* (Pyrenees), *M. scalercioi* and *M. vestinorum* (Italy)
Mesophylax McLachlan 1882	4 (2)	NW, NE, SW, SE	Endemics: *M. morettii* and *M. sardous* (Sardinia & Corsica)
Micropterna Stein 1873	18	NW, NE, SW, SE	
Parachiona (Pictet 1834)	1	NE	*P. picicornis*
Potamophylax Wallengren 1891	16	NW, NE	
Stenophylax Kolenati 1848	10 (2)	NW, NE, SW, SE	Endemics: *S. bischofi* (Italy), *S. minoicus* (Greece)
Tribe Agaphylacini Oláh 2019			
Agaphylax Oláh, Kovács & Ibrahimi 2018	1	NE (in Med proximities)	*A. balcanicus*
Family Thremmatidae Martynov 1915			
Thremma McLachlan 1876	4 (2)	NW, SW, NE	Endemics: *T. sardoum* (Sardinia & Corsica), *T. tellae* (Iberian P.)
Family Brachycentridae Ulmer 1903			
Brachycentrus Curtis 1834	3	NW, NE, SE	*B. maculatus, B. montanus, B. subnubilus*
Micrasema McLachlan 1876	15 (6)	NW, NE	

(Continued)

TABLE 14.1 (Continued)

Taxon	Number of species/(endemics)	Distribution	Notes
Family Lepidostomatidae Ulmer 1903			
Crunoecia McLachlan 1876	5 (1)	NW, NE, SE	Endemic: *C. fortuna* (Italy)
Lepidostoma Rambur 1842	10 (6)	NW, SW, NE, SE	All endemics in Turkey
Family Phryganeidae Leach 1815			
Agrypnia Curtis 1835	3	NW, NE, SW	*A. obsoleta, A. pagetana, A. varia*
Hagenella Martynov 1924	1	NE: Croatia, Slovenia	*H. clathrate*
Oligostomis Kolenati 1848	1	NW, NE	*O. reticulata*
Oligotricha Rambur 1842	1	NW, NE	*O. striata*
Phryganea Linnaeus 1758	2	NW, NE	*P. bipunctata, P. grandis* (3 subspecies)
Trichostegia Kolenati 1848	1	NW, NE	*T. minor*

Notes: Many species from Turkey are currently considered endemic. Their status might change when the caddisfly fauna of adjacent unexplored areas is investigated.
Species from the Macaronesian Islands are not included in the list as these islands have variable climatic conditions (e.g. subtropical, maritime temperate, arid, etc.).

epilithic periphyton from large stones or other substrate and can depress the biomass of stream periphyton. Glossosomatidae have univoltine cycles; although, depending on environmental conditions, some species can show a more flexible life cycle (Graf et al., 2008). The family is cosmopolitan occupying all biogeographic regions; and in the Mediterranean Basin, it is represented by four genera (*Agapetus, Glossosoma, Synagapetus*, and *Catagapetus*) with 71 species of which 35 are endemic (Table 14.1).

Hydroptilidae: Larvae of the "microcaddisflies" are campodeiform and among the smallest in the order (2–6 mm). Larvae are free-living in their first four instars and build cases only in their final instars (Wiggins, 2004). Cases consist of two lateral valves made from silk, sand or plant materials and may be fixed or portable, joined on the long edges and open at the ends. The cases are very characteristic of each genus. For example, *Hydroptila* and *Allotrichia* have flat cases, those of *Oxyethira* are bottle-shaped, the cases of *Ithytrichia* resemble pumpkin seeds, and those of *Orthotrichia* and *Stactobia* oat seeds (Wiggins, 2008; Holzenthal et al., 2015). Pupation occurs between these valves where the cocoon spun by the last instar larva in preparation for pupation is semipermeable (Morse et al., 2019). They feed by gathering deposited fine particulate organic matter (FPOM), scraping algae and periphyton, or by piercing and sucking contents from individual cells of filamentous algae (Wiggins, 2004). They occupy all stream zonations with different flow requirements, from very cold springs to moderately warm estuaries and lakes. Hydroptilidae have unibi-voltine life cycles (Graf et al., 2008). Hydroptilidae are the most diverse family of the order, with 37 genera and more than 2000 described species found on every habitable continent (Holzenthal et al., 2007; Holzenthal & Calor, 2017). In the Mediterranean Basin, 10 genera are found with more than 125 species, of which 50 are considered endemic (Table 14.1).

Ptilocolepidae: Larvae of Ptilocolepidae are very similar to Hydroptilidae (collectively "microcaddisflies") except for a porous, rather than a semipermeable cocoon in the pupation stage (Wiggins, 2004). Larvae live in cold spring streams, usually in association with liverworts, which they eat and use also in constructing portable cases (Wiggins, 2004). They have univoltine life cycles (Graf et al., 2008). Ptilocolepidae is a small family with two genera *Palaeagapetus* Ulmer 1912 and *Ptilocolepus* Kolenati 1848. *Ptilocolepus* occupies Europe, India, and Southeast Asia with four species in the Mediterranean Basin (including Turkey).

Philopotamidae: Larvae of the "finger-net caddisflies" are caseless, campodeiform and can reach up to 18 mm in length. Larvae construct and live in silken tubes or sack-like nets, usually on the bottom of rocks in fast or slowly flowing waters, in cold to warm streams and rivers of all sizes. Their unmarked head with an unsclerotized

T-shaped labrum is unique and easily separates them from other net-spinning larvae that also have unsclerotized meso- and metanota. Larvae are omnivorous and feed on algae and detritus that they capture in their silken nets (Wiggins, 2004). Philopotamidae are cosmopolitan with 26 extant genera containing 1508 species (Morse et al., 2019). *Chimarra* is the most species rich genus in the family, followed by *Wormaldia* (Holzenthal et al., 2007). In the Mediterranean Basin, apart from the former two genera, *Philopotamus* is also present; the three genera totaling 48 described species, including 18 endemics that are mostly found within the genus *Wormaldia* (Table 14.1).

Polycentropodidae: Larvae of the "trumpet-net caddisflies" are caseless, campodeiform and can reach up to 20 mm in length. While polycentropodids predominantly live in clear, fast-flowing streams, they also inhabit a wide range of standing water habitats including swamps, pools, and littoral zones of lakes and temporary ponds. They build silken tubular or trumpet-shaped retreats with various shaped capturing nets (Lepneva, 1970; Chamorro & Holzenthal, 2011). *Neureclipsis* and *Polycentropus* spin bag-like filtering nets in slowly flowing water to filter FPOM and to capture small invertebrate prey. *Holocentropus* and *Plectrocnemia* are predators with silken tubular retreats suspended on aquatic plants and with many irregularly radiating silken threads to form capture nets (Lepneva, 1970; Wiggins, 1996). The pupae are generally sheltered in a domelike case of silk and sand. They have mainly univoltine life cycles with an emergence period that goes from spring to autumn (Graf et al., 2008). Polycentropodidae are cosmopolitan with a global diversity of 861 species in 18 genera (Morse et al., 2019). The Mediterranean Basin hosts 54 species belonging to six genera with 14 endemics (Table 14.1).

Pseudoneureclipsidae: Current taxonomy includes this clade as a family distinct from Polycentropodidae, where it was previously considered a subfamily (Chamorro & Holzenthal, 2011). *Pseudoneureclipsis* larvae are grazers and their retreats resemble the branching tube-dwellings buried in loose substrate constructed by Dipseudopsidae or unbranched tubes, attached to solid substrate as those of Psychomyiidae (Vieira-Lanero 2000; Tachet et al., 2001). Larvae are found in small and clear water streams in backwater areas, large rivers, and lake shores (Vieira-Lanero, 2000; Tachet et al., 2001; Holzenthal & Calor, 2017). *Pseudoneureclipsis* is the only genus present in the Mediterranean Basin with five species, including two endemics from Turkey (Table 14.1).

Ecnomidae: Larvae of Ecnomidae are small (5–10 mm) and inhabit small streams, permanent ponds and lakes, or slowly flowing waters. They live under rocks or wood, where they construct elongate, flimsy tubes of silk covered by sand and silk (Wiggins, 1996). Some are predatory, feeding on small invertebrates, but most eat algae and detritus (Holzenthal & Calor, 2017). They have univoltine life cycles with an emergence period spanning spring to autumn (Graf et al., 2008). Ecnomidae contain 12 genera distributed in all biogeographic regions (Morse et al., 2019). In the West Palearctic only the genus *Ecnomus* is present and is represented in the Mediterranean Basin by six species (Table 14.1), three considered endemics (*Ecnomus galilaeus*, *E. kurui*, and *E. relictus*; Darılmaz & Salur, 2015; Neu et al., 2018), while *E. tenellus* is the most common and widely distributed.

Psychomyiidae: Larvae of the "net-tube caddisflies" are small in size (up to 11 mm) and live in springs, streams, small to midsized rivers and wave-washed lake shores. They frequently live on hygropetric surfaces, where they build elongate silken galleries of irregular form, covered with sand or debris. They are also found on or in wood and rocks (Holzenthal & Calor, 2017), while *Tinodes* larvae may even inhabit springs or the splash zones of waterfalls or cascades. Psychomyiid larvae are sedentary grazers, feeding on algae, detritus and fungi. Some *Tinodes* species (e.g., *Tinodes waeneri*) are known to cultivate algae or diatoms in the silken mesh of their tubes, eating the old portions while constructing new ones to serve as silken gardens (Ings et al., 2010). Most Psychomyiidae have univoltine life cycles, although depending on environmental conditions, some species can show a more flexible life cycle (Graf et al., 2008). Ten genera are recognized with 600 extant species (Morse et al., 2019) that are widespread in all biogeographic regions apart from the Neotropics. In the Mediterranean Basin, Psychomyiidae are represented by 85 species in five genera (Table 14.1). *Tinodes* is one of the richest genera in the east Mediterranean Basin (75 species) with a remarkable rate of endemism. In Greece, 33 species are known, where 16 are endemic, mostly found on the Aegean Islands (Malicky, 2005). In Turkey, 32 species are known, 16 of which are considered endemic (Sipahiler, 2016).

Hydropsychidae: Larvae of the "net-spinning caddisflies" are medium to relatively large in size (10–25 mm) and are easily recognized by the numerous, branched filamentous gills on the ventrolateral and ventral surfaces of their thoracic and abdominal segments. They are found in streams and rivers of all sizes, currents, and types (i.e., high to low altitudes, very fast to very low water flow) where they construct fixed retreats to the sides of rocks, wood, or on leaf packs. Larvae are omnivorous; at the anterior upstream entrance of the retreat, hydropsychids spin a net or sieve made of fine silk situated against the flow to capture detritus, algae, diatoms, or smaller invertebrates (Wiggins, 2004). Net size, structure, and function vary considerably among genera depending on the food type. Hydropsychidae larvae are notable for being able to produce sound underwater, a rare accomplishment among immature insects, with a head/fore femur mechanism and use the sound as part of aggressive behavior for defense of their feeding nets (Aiken, 1985).

The pupal case is dome-shaped, made from small stones and mineral particles, sometimes incorporating snail shells and lined with silk (Karaouzas, 2018). Hydropsychidae species usually have univoltine life cycles, but some species can also have more than one generation per year. The emergence period usually is long and goes from spring to autumn, and some species are also able to emerge in winter (Malicky, 2005; Graf et al., 2008). Hydropsychidae are the third most diverse family of the order with 41 genera and 1982 described species and are present in all continents expect Antarctica (Holzenthal et al., 2007; Morse et al., 2019). In the Mediterranean Basin, Hydropsychidae are represented by three genera: *Diplectrona, Cheumatopsyche*, and *Hydropsyche*; the latter is the most diverse with 119 species including 49 endemics, about half found mainly in the Aegean Islands (Malicky, 2005; Karaouzas, 2018).

Phryganeidae: Larvae of the "giant case caddisflies" are large in size (15–45 mm) and construct cases out of plant fragments of longitudinal or spiral forms, sometimes incorporating tiny pebbles. They generally live in clean, cold waters, both lentic and gently flowing (e.g., ponds, marshes, pools) among aquatic plants and roots (Wiggins, 1998). Larvae are primarily shredders, consuming mostly the same kinds of leaves or wood with which they construct their cases. Some species of *Oligostomis, Oligotricha*, and *Phryganea* may also eat animal tissue by shredding and ingesting living insects (Wiggins, 1998). Unlike other case building caddisflies, they easily abandon the case if threatened, and frequently use empty cases of other individuals. Phryganeidae have univoltine life cycles with short emergence periods, mainly in spring and summer (Graf et al., 2008). The family lives only in the Northern Hemisphere in Asia, Europe, and North America (Holzenthal et al., 2007) with 15 genera and 81 species (Morse et al., 2019). In the Mediterranean Basin, six genera are recorded with only nine species (Table 14.1).

Brachycentridae: The "humpless case caddisflies" are distributed throughout the Northern Hemisphere, in both Holarctic and Oriental regions (Holzenthal et al., 2007). Larvae are small (6–14 mm) and inhabit running waters of various sizes and types, including calm rivers and rivulets, and slow-flowing marshy channels. They construct square or round cases made of plant and/or rock material; some genera, for example *Brachycentrus*, build tetrahedral cases. *Brachycentrus* and *Micrasema* are both widespread across the Mediterranean Basin. *Brachycentrus* larvae feed on microorganisms and detritus carried by the current which they capture with spinules and setae on their legs (Lepneva, 1970). Larvae of *Micrasema* are exclusively grazers and are associated with mosses. Brachycentridae species have univoltine, short life cycles. The emergence period goes from spring to summer, but on occasions they extend it to autumn and winter (Graf et al., 2008). Currently, 18 species are known from the Mediterranean Basin, including 15 species of *Micrasema*, with most of them occurring in the Iberian Peninsula (González & Menéndez, 2011).

Lepidostomatidae: Larvae of the "scaly-mouth caddisflies" are small sized (9–11 mm) and are most commonly found in small mountainous streams and springs, but may also be found along the shores of large rivers or lakes. They are primarily shredding detritivores, feeding on decaying plant material (Lepneva, 1970; Wiggins, 2004). Larvae usually construct quadrate cases made of plant material and rarely of sand grains, conical or straight (as in *Lepidostoma basale*). Lepidostomatidae species are univoltine, with short or long life cycles depending on geographical distribution (Graf et al., 2008). Lepidostomatidae are nearly cosmopolitan, known from all habitable continents except South America, with 519 species in seven genera; *Lepidostoma* is the most diverse genus (Morse et al., 2019). In the Mediterranean Basin, the family is represented by two genera, *Crunoecia* and *Lepidostoma* with 5 and 10 species, respectively. *Crunoecia irrorata, Lepidostoma hirtum* and *L. basale* are common, while the remaining are either endemics or range restricted (e.g., *Lepidostoma belkisae, L. holzschuhi, L. olimpense*—Turkey; *Crunoecia fortuna*—Italy).

Thremmatidae: Larvae of the "little northeastern caddisflies" are very small (4–6 mm) and construct fine-grained, flattened, cases resembling the freshwater limpet *Ancylus* (Waringer et al., 2020); the case structure minimizes hydraulic stress and also protects the larva against predators. They live in small cascading brooks and springs, and all are scrapers grazing on periphyton and FPOM. Until recently, the genus *Thremma*, the only representative of the family in the Mediterranean Basin, was considered as member of Uenoidae (Vshivkova et al., 2007; Holzenthal et al., 2007, 2011; Waringer et al., 2020). *Thremma* species have uni- or bivoltine life cycles (Décamps, 1967; Lavandier, 1979; Malicky, 2005). Seven taxa are known from the Mediterranean Basin: *T. anomalum* (Albania, Greece, Montenegro, Serbia and Turkey), *T. gallicum* (Iberian Peninsula, France), *T. g. arvernense* (French Massif Central, endemic), *T. martynovi* (Asia Minor), *T. sardoum* (Corsica and Sardinia, endemic), *T. s. africanum* (Algeria and Tunisia), and *T. tellae* (Iberian Peninsula).

Goeridae: This is a widely distributed family, found in all biogeographic regions except the Neotropical (Morse et al., 2019). Larvae of the "weighted-case caddisflies" are small (8–15 mm) and usually live in small to medium sized rivers and streams, brooks and rivulets with slow current, or open lakes shores (Lepneva, 1970). Cases are slightly curved and made entirely of rock fragments; *Goera* and *Silo* larvae incorporate larger rock fragments laterally. Goeridae are grazers feeding on periphyton and detritus (Holzenthal et al., 2007). Goeridae have univoltine life cycles and the emergence period goes from spring to summer, although some species extend the flight period to autumn (Graf et al., 2008). Eleven genera are known worldwide with a total of 189 species (Morse et al., 2019). In the Mediterranean Basin, five genera occur with 14 species, including three endemics (*Silo chrisiammos, S. mediterraneus, Larcasia ligurica*; Waringer et al., 2017).

Apataniidae: Larvae of the "mountain case caddisflies" (previously subfamily of Limnephilidae) are relatively small (e.g., *Apatania* 5–12 mm; *Apataniana* 5–7 mm) and construct papoose-shaped cases of fine sand grains on tops of rocks. They are northern hemisphere inhabits (North America, Asia, and Europe), and usually live in high altitude springs, streams, brooks, and rivulets, where they graze on periphyton from rocks with scraper mandibles (Lepneva, 1970; Holzenthal et al., 2007). European species of Apataniidae have semi-, uni-, or bivoltine life cycles (Graf et al., 2008). In the Mediterranean Basin, two genera are found, *Apatania* and *Apataniana* with 12 and 3 species, respectively, most of them microendemics. *Apataniana hellenica, A. stropones*, and *A. vardusia* occur only in the mountains of Ossa, Dirfis, and Vardousia in Greece, respectively (Malicky, 2005; Waringer & Malicky 2016); their closest relative from the Mediterranean Basin, *A. borcka*, is reported from Turkey (Sipahiler, 1996).

Limnephilidae: The "northern caddisflies" are the fourth largest family of the order and the most ecologically diverse (Vshivkova et al. 2007; Morse et al., 2019). Although they are distributed in all biogeographic regions apart from the Afrotropics, highest species diversity is usually found in high temperate latitudes of the northern hemisphere (Holzenthal et al., 2007). Larval lengths range considerably between genera and species from small to large sized (8–33 mm) and they inhabit practically all freshwater habitats, from springs, rivulets, and streams, to lakes, pools, temporary ponds, and marshes. They live on and between rocks and stones, on sand-detritus bottoms or among vegetation. Limnephilid larvae have very diverse cases differing in both materials and architecture, which can often facilitate identification of the genus or species. Larvae may build cases from plant materials (e.g., *Grammotaulius, Glyphotaelius, Limnephilus*), rock materials (e.g., *Drusus, Allogamus, Potamophylax, Stenophylax, Micropterna*), or a mixture of plants, rocks, and snail shells (e.g., *Limnephilus rombicus*). Usually those living in cold, fast, mountainous streams use rock materials for cases, while those in warmer, slow flowing, larger rivers use plant material. Most are shredding or scraping herbivores or detritivores (Wiggins, 2004), feeding on bryophytes (e.g., *Chaetopterygopsis, Limnephilus*), algae, epilithic diatoms (e.g., *Drusus*,) and leaf detritus (e.g., *Hydatophylax, Potamophylax*). Members of the European genera *Allogamus* and *Drusus* exhibit filtering bristles on their legs and occasionally on thoracic sterna as morphological adaptations to filter-feeding (Vitecek et al., 2015). Most European species have univoltine life cycles, but some species are semivoltine (e.g., *Drusus discolor*; Lavandier, 1979) or even multivoltine (e.g., *Glyphotaelius pellucidus*; Graf et al., 2008). Globally, 97 genera are known with 1037 species (Morse et al., 2019) and this number constantly increases with the aid of molecular analysis (e.g., Kučinić et al., 2013; Previšić et al., 2014; Oláh et al., 2018). In the Mediterranean Basin, 33 genera are currently known with 243 species (Table 14.1).

Leptoceridae: Larvae of the "long-horned caddisflies" are easily recognized by their long, conspicuous antennae (Malm & Johanson, 2011), as compared to most larval caddisfly antennae, which are barely visible as small nubs (Waringer & Graf, 2011). Larvae are relatively small (7–15 mm) and are found in virtually all freshwater environments, from high altitude springs, brooks, and streams to large lowland rivers, lakes, pools, and temporary ponds; some are semiterrestrial and may be found on splash areas of waterfalls (Holzenthal et al., 2007). Larval members of the family create a diverse range of cases, both in terms of structure and materials; tubular cases of rock or sand grains, plant material arranged spirally or transversely or made entirely of silk (Lepneva, 1970). *Mystacides* larvae use long conifer needles or leaf stems that fade towards the end of the case, while some *Ceraclea* construct flat, limpet-like cases made entirely of sand grains, while those that feed on freshwater sponges utilize sponge pieces and spicules in their cases (Whitlock & Morse, 1994). Larvae are mainly leaf detritus shredders and periphyton scrapers, but some are predators, preying on freshwater sponges, snails, and other insects (Morse & Lenat, 2005). They have univoltine life cycles (Graf et al., 2008). Leptoceridae are the second largest Trichoptera family and are found in all biogeographic regions with 49 genera and 2235 species (Morse et al., 2019). In the Mediterranean Basin, 94 species are currently known (Table 14.1), belonging to 10 genera (including the terrestrial genus *Homilia*).

Helicopsychidae: The name of this family refers to the helix-shaped larval cases ("snail-case caddisflies") made of sand and other mineral fragments. Larvae are small in size (2–7 mm) and are normally found in small mountainous to lowland medium sized streams with riffles or with moderate to slow water currents. They can live temporarily outside of the water in hygropetric environments (Vieira-Lanero et al., 2001). They shred and consume detritus, mosses, and liverworts and possibly also graze on attached periphyton (Wiggins, 2004). The European *Helicopsyche* species seem to be univoltine and on the wing mainly in spring and summer, but some species are collected also in autumn (Graf et al., 2008; Waringer et al., 2017). The family currently contains 270 species (Holzenthal et al., 2011) distributed in two genera: the monotypic *Rakiura*, from New Zealand, and the cosmopolitan *Helicopsyche* Siebold 1856, which was divided into six subgenera by Johanson (1998). In the Mediterranean Basin, *Helicopsyche* is represented by five species (Table 14.1), two of them being microendemics (*Helicopsyche megalochari*—Greece, *Helicopsyche lusitanica*—Iberian Peninsula).

Beraeidae: The greatest diversity of this small family (seven genera, 59 species; Morse et al, 2019) is found within the West Palearctic region, where five genera (*Beraea*, *Beraeamyia*, *Beraeodes*, *Beraeodina*, and *Ernodes*) occur in the Mediterranean Basin with 40 species (Table 14.1). Larvae are very small (5–7 mm) and occur in springs and small streams usually among mosses, leaf litter deposits, roots of emergent plants, and dense marginal vegetation (Holzenthal et al., 2007). They feed on microphytobenthos, especially mycelium of fungi. Frequently, *Ernodes* and *Beraea* larvae may be found outside the water, near springs among humid leaves and moss (Lepneva, 1970). Larvae build slightly to strongly tapered cases made entirely from sand grains. They usually have a short and semivoltine life cycle, with an emergence period that goes from spring to summer. Some species have also been observed on the wing during the autumn (Graf et al., 2008). In the Mediterranean Basin, Beraeidae are represented by 5 genera accounting for a total of 41 species with 25 endemics (Table 14.1).

Sericostomatidae: Larvae of the "bush-tailed caddisflies" build tubular, slightly curved, and tapered cases made entirely from sand grains. Larvae are medium sized (11–17 mm) and live in brooks and rivulets, streams, and rivers with rapid or slow currents depending on the genus or species; they often burrow in sandy deposits (Lepneva, 1970; Wiggins, 2004; Holzenthal et al., 2007). They are primarily leaf litter detritivorous and feed on plant remains and algae. The life cycle is variable, being semi- or univoltine. The emergence period goes from spring to summer, extending to autumn in some species (Graf et al., 2008). Sericostomatidae occur in every biogeographic region, except the Australasian with 103 species belonging to 16 genera (Morse et al., 2019). In the Mediterranean Basin, four genera are represented, *Notidobia, Oecismus, Schizopelex*, and *Sericostoma*, with 37 species, 21 of them being endemics (Table 14.1).

Odontoceridae: Larvae of the "mortar-joint case caddisflies" are eruciform and their cases are cylindrical (tubular) made of sand grains or larger mineral fragments. Cases are very resistant to crushing due to reinforcing silken mortar applied by the larva between the sand grains. Larvae are moderate in size (8–18 mm) and are usually found in slow to moderate running water of small to medium-sized streams to large rivers, normally associated with bottoms of gravel or sand, where they are sometimes found buried in the substrate. Larvae are omnivorous, feeding on organic detritus, vascular plants, moss, and algae as well as aquatic arthropods (Wiggins, 1996). *Odontocerum* species have uni- or bivoltine life cycles (Malicky, 2005; Graf et al., 2008). In the West Palearctic, the family is represented by the type genus only, *Odontocerum* Leach 1815, with three species, which are also present in the Mediterranean Basin (Table 14.1).

Calamoceratidae: Larvae of the "comb-lipped case caddisflies" are well known for their flattened cases made of large pieces of excised leaves that completely camouflage the larva from above. Larvae are small to medium in size (8–12 mm) and are frequently found in lateral pools and depositional areas of small streams and rivers and backwater areas, where they feed as shredders of leaf litter and other plant detritus. Adult emergence is short and takes place during spring and summer (Graf et al., 2008). Calamoceratidae with currently 190 described species are widely distributed in all biogeographic regions (Morse et al., 2019). *Calamoceras* is the only genus of Calamoceratidae in the West Palearctic Region with two species present in the Mediterranean Basin: *Calamoceras marsupus* Brauer 1865, which occurs in France, the Iberian Peninsula and Morocco and *Calamoceras illiesi* Kumanski & Malicky 1974, which occurs in Bulgaria, Croatia, Greece, and Turkey (Coppa & Tachet, 2010; Sipahiler, 2013; Evtimova & Kenderov, 2016).

Trichoptera adaptations to the Mediterranean Basin

Seasonal droughts are regular and frequent natural events in Mediterranean lotic ecosystems to which many invertebrate species, including caddisflies, have developed resilience through physiological and life-history adaptations (Williams, 1996; Hoppeler et al., 2016). *Limnephilus bipunctatus* distributed in the Mediterranean Basin is a typical inhabitant of small rivers and streams which can dry up in summer. Other species survive summer drought by emerging in the spring, passing the summer in an ovarial diapause and laying desiccation-resistant egg masses on dry pond basins and river beds in early autumn (Wiggins, 1973; Botosaneanu, 1974). *Mesophylax aspersus*, as do other subtroglophile species of the genus *Stenophylax*, estivate as adults in caves (e.g., Bouvet, 1975), and has evolved physiological and behavioral strategies to survive seasonal drought and stream drying in the Mediterranean Basin (Salavert et al., 2008, 2011). Caddisflies are only able to disperse among water bodies as flying adults. Some species, such as *Stenophylax* or *Mesophylax*, are strong fliers which are known to fly distances up to 5 km (Malicky, 1987). However, a recent review of reports concerning insect declines, especially in well-studied parts of the world (e.g., Europe and North America) revealed that Trichoptera species are being lost at a greater rate than other freshwater insect orders. Approximately 74% of the Trichoptera species cited in these reports are in decline or are extinct (Sánchez-Bayo & Wyckhuys, 2019), but these results are based on limited studies and knowledge of life history and habitat requirements. The potential main causes negatively affecting caddisfly populations include agrochemical pollution (fertilizers and pesticides), land use changes (from natural to agricultural and urban uses), climate change, pathogens, and alien species. However, not all species and European ecoregions are equally vulnerable to climate change. The highest fraction of potentially endangered species is found in the Mediterranean Basin, mainly in the Iberic-Macaronesian region (Hering et al., 2009). Endemic species appear to be more vulnerable to global change as a consequence of their low dispersal capacity as well as for their strict requirements for very particular ecological conditions (Hering et al., 2009; Múrria et al., 2020). These species will also be threatened by more generalist species that can expand their distribution range to higher altitudes, and therefore compete for resources (Domisch et al., 2013; Sáinz-Bariáin et al., 2016).

Material preparation and preservation

Trichoptera larvae are best identified by fully mature, fifth instar individuals, well preserved in an appropriate solution. Currently, high concentration ethyl alcohol (EtOH, 80%–95%) is recommended, especially if specimens are to be used for molecular studies. If the concentration becomes diluted with the introduction of water from wet specimens in the field, the alcohol can be decanted and replaced with fresh alcohol in the laboratory. Specimens permanently stored in 95% EtOH become brittle, so after DNA extraction is concluded it is best to permanently store specimens in 80% EtOH. Solutions containing formalin should be avoided, not only for human health reasons, but because of the chemical's destructive effect on DNA (Frandsen and Thomson, 2016). There seem to be as many devices and methods of collecting Trichoptera and other aquatic immature insects as there are taxa and microhabitats. These have been presented in detail by Jackson et al. (2019).

Except for the free-living families (Rhyacophilidae in the Mediterranean Basin) most larvae will be collected from some kind of retreat or case. While the retreat-makers are generally pulled out of their retreats when collected, the structure of the retreat should be noted or photographed. Case makers are usually collected case and all, but these generally need to be removed from the case for identification. A pair of fine tipped forceps can be slipped between the body of the larva and the case to gently pull the individual out, with no or minimal damage to the case or larva. After identification, the case and its larva should be preserved together; never discard the case as it is of diagnostic value. All caddisflies pupate in some kind of pupal structure or modified case. These can be collected in their entirety and later dissected to observe pupal characters and obtain larval sclerites from metamorphotype association (Wiggins, 1996). For additional methods of collecting, preserving, and assembling specimens for identification, see Holzenthal et al. (2015).

Morphological characters needed in identification

Morphological identification of caddisfly larvae uses characteristics of all body regions (head, thorax, and abdomen) and their appendages (Figs. 14.1 and 14.2). The overall body shape is elongate cylindrical. The case-making Integripalpia (taxon Phryganides) (*sensu* Thomas et al., 2020) have caterpillar-like eruciform or suberuciform larvae, with the head oriented more or less vertically and the mouth parts directed ventrally. In the other Integripalpia and in Annulipalpia the larvae are campodeiform with the head oriented more horizontally and the mouthparts directed forward.

The head is complete, rounded, or somewhat elongate, and composed of a well-sclerotized, usually pigmented cranium or head capsule; however, the ventral portion of the cranium is usually less well sclerotized and not as highly pigmented. Pigmentation on the head can include dark spots, patches, or stripes. "Muscle scars" marking the attachment of internal muscles are usually visible (Fig. 14.1a, b, g). A distinctive dorsal landmark on the cranium is the large, triangular frontoclypeal apotome, defined by the frontoclypeal sutures (Fig. 14.1g). Ventrally, lies the ventral apotome, which can be divided into anterior and posterior parts by the ventral ecdysial suture (Fig. 14.1f). Other typical features of the head include the genae and the postgenae below and behind it, of the large, paired parietal sclerites separated dorsally by the coronal suture (Fig. 14.1b, g). Laterally on the parietal sclerites lie the larval eyes, composed of several individual eye facets (stemmata) (Fig. 14.1b). The dorsum of the head is usually smooth, but is often textured, bears protuberances, patches of hairs, scales, ridges, or carinae, especially laterally. Also present on the head are primary and

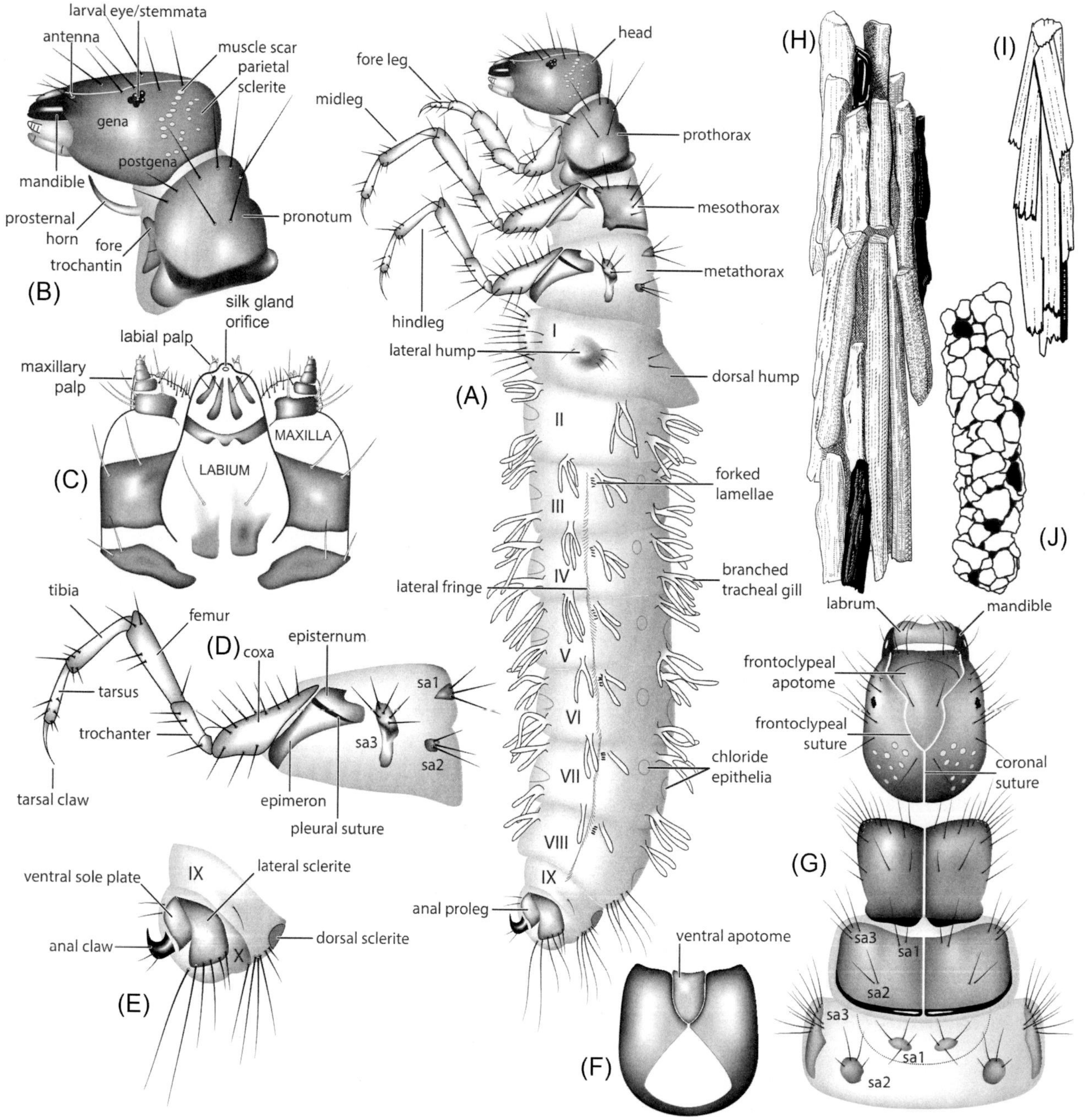

FIGURE 14.1 Trichoptera larval morphology (Limnephilidae); (a) body, lateral; (b) head and prothorax, lateral; (c) maxillolabial complex, ventral; (d) metathorax, lateral; (e) abdominal segments 9, 10, and anal proleg, lateral; (f) head, ventral; (g) head and thorax, dorsal; (h, i, j) larval cases, lateral. (Abbreviations: I–X, abdominal segments 1–10; sa, setal area.)

secondary setae that usually occur in a typical pattern and follow a more or less standard nomenclature (Williams & Wiggins, 1981).

Appendages on the head include a pair of antennae and the mouthparts. The antennae occur laterally on the head capsule between the eye and the base of the mandible. They are very short, about as long as wide, and one-segmented (e.g., Fig. 14.1b), but in some families, especially most Leptoceridae genera, they can be several times longer than wide and conspicuous (e.g., Fig. 14.24). As in almost all holometabolous larvae the mouthparts are mandibulate, composed of a simple labrum, a pair of mandibles, a pair of maxillae, and a labium (Fig. 14.1c). Mandibles and their dentition vary with the larval feeding method and can have blunt apical teeth and a molar region in shredders, sharp apical teeth

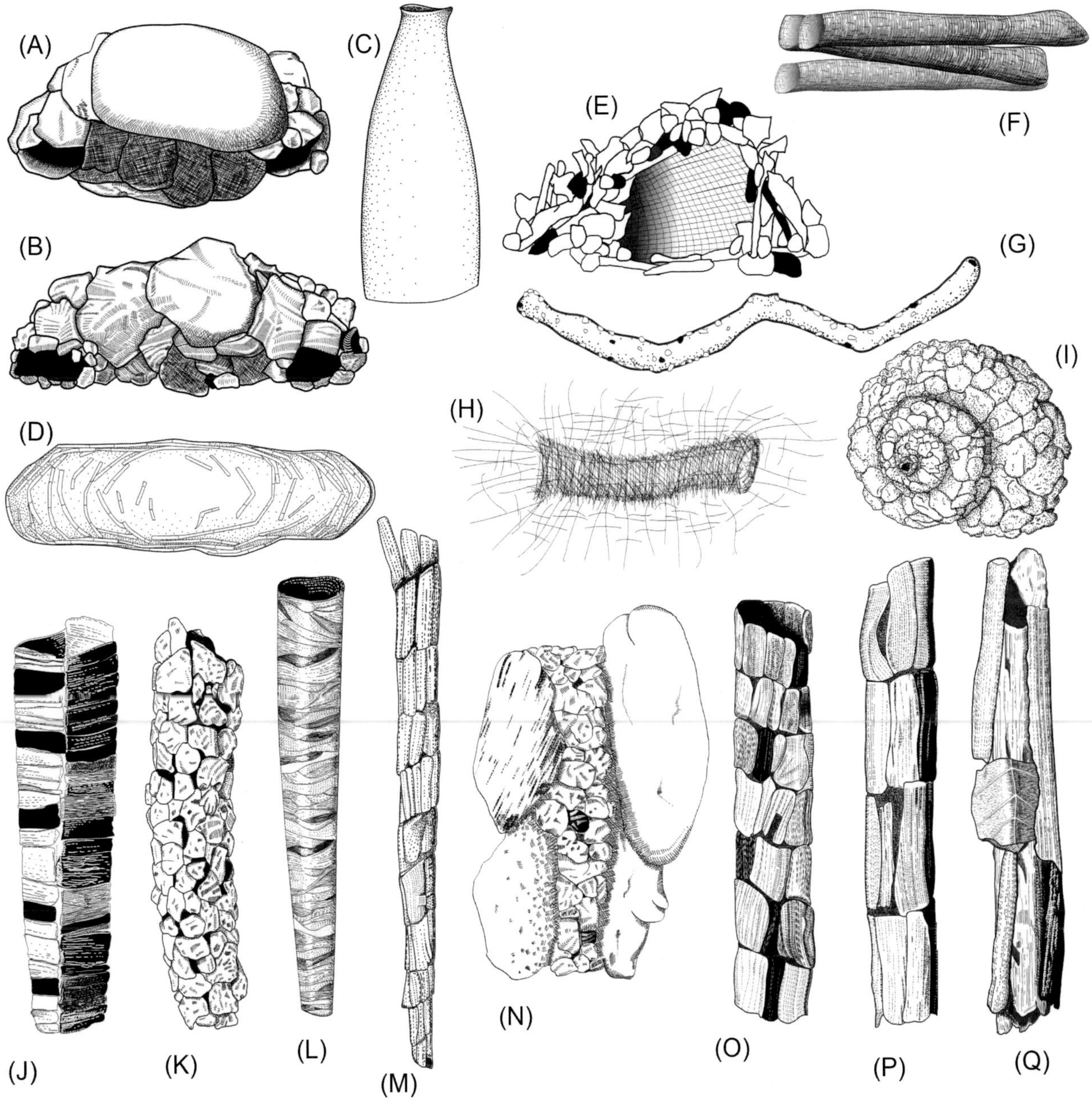

FIGURE 14.2 Larval cases and retreats; (a) Glossosmatidae, *Glossosoma* case, lateral; (b) Glossosomatidae, *Agapetus* case, lateral; (c) Hydroptilidae, *Oxyethira* case, lateral; (d) Hydroptilidae, *Agraylea* case, lateral; (e) Hydropsyche, *Hydropsyche* retreat, lateral; (f) Philopotamidae, *Chimarra* retreat, lateral; (g) Ecnomidae, *Ecnomus* retreat, dorsal; (h) Polycentropodidae, *Holocentropus* retreat, dorsal; (i) Helicopsychidae, *Helicopsyche* case, dorsal; (j) Brachycentridae, *Brachycentrus* case, lateral; (k) Odontoceridae, *Odontocerum* case, lateral; (l) Brachycentridae, *Micrasema* case, lateral; (m) Leptoceridae, *Triaenodes* case, lateral; (n) Goeridae, *Goera* dorsal, lateral; (o) Phryganeidae, *Phryganea* case, lateral; (p) Phryganeidae, *Agrypnia* case, lateral; (q) Calamoceratidae; *Calamocerus* case, lateral.

in predators, or smooth dorsal and ventral edges in scrapers, for example. The short maxillae are in close association with the labium, together forming the "maxillo-labial complex" bearing short maxillary and very reduced labial palps. Apically the labium bears the orifice (external opening) of the duct of the internal silk glands (Fig. 14.1c).

The prothorax, mesothorax, and metathorax are distinct, and each bears a pair of six-segmented legs, consisting of a coxa, trochanter (usually subdivided), femur, tibia (sometimes constricted in the middle), unsegmented tarsus, and pretarsus bearing a single terminal tarsal claw (Fig. 14.1a, d). The forelegs are usually the shortest and the hindlegs the longest, but in some taxa, they are all of similar length (Figs. 14.1a, 14.4, and 14.19). In most taxa the legs are "unmodified" and used from clinging to or

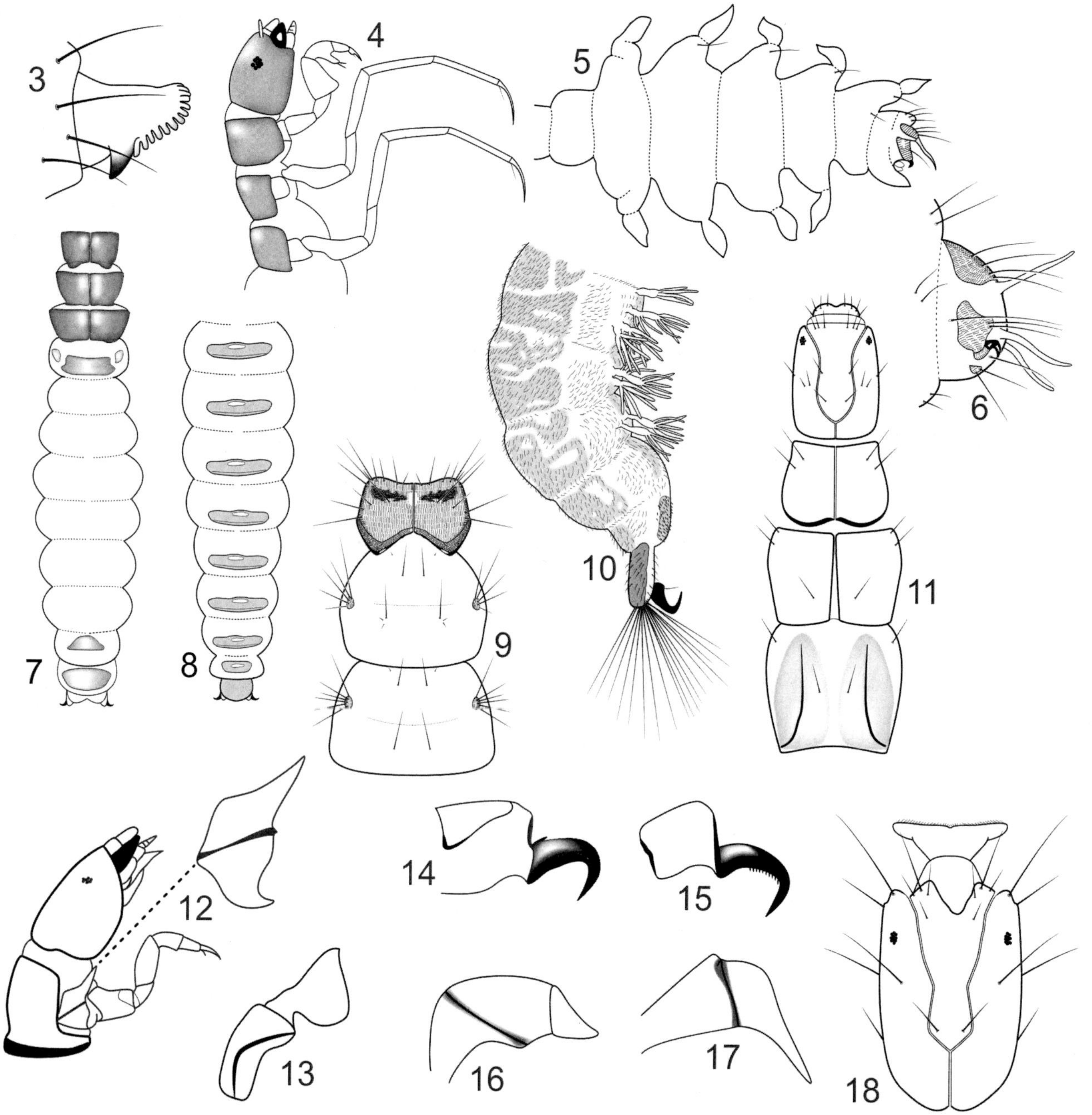

FIGURES 14.3–14.18 (3) Helicopsychidae, *Helicopsyche* anal claw, lateral; (4) Hydroptilidae, *Oxyethira* head and thorax, lateral; (5) Hydroptilidae, *Ithytrichia* abdomen, lateral; (6) Hydroptilidae, *Hydroptila*; abdominal segments 9, 10, and anal proleg, lateral; (7) Ptilocolepidae, *Ptilocolepus*, thorax and abdomen, dorsal; (8) Hydroptilidae, *Stactobia*, abdomen, dorsal; (9) Phryganeidae, *Agrypnia*, thorax, dorsal; (10) Hydropsychidae, *Hydropsyche* terminal abdominal segments, lateral; (11) Pseudoneureclipsidae, *Pseudoneureclpisis* head and thorax, dorsal; (12) Polycentropodidae, *Polycentropus* head and prothorax, inset foretrochantin, lateral; (13) Psychomyiidae, *Psychomyia* foretrochantin, lateral; (14) Pseudoneureclipsidae, *Pseudoneureclpisis* anal proleg and claw, lateral; (15) Ecnomidae, *Ecnomus* anal proleg and claw, lateral; (16) Pseudoneureclipsidae, *Pseudoneureclpisis* foretrochantin, lateral; (17) Ecnomidae, *Ecnomus* foretrochantin, lateral; (18) Philopotamidae, *Chimarra* head and labrum, dorsal.

crawling on the substrate, but is some families they are modified for predation (semi-raptorial or chelate), for stridulation, for swimming, or for filtering or snagging food, including detritus and invertebrate prey, suspended in flowing water.

Ventrally, the thoracic segments are membranous or bear only small sclerites or setae, but dorsally they often bear conspicuous and taxonomically diagnostic sclerites on one or all three nota (e.g., Figs. 14.1g, 14.19, and 14.36–14.40). The dorsal prothorax (pronotum) always has a pair of dorsal pronotal sclerites, separated along the dorsal mid-line by an ecdysial suture (Fig. 14.1g). Below the pronotal sclerites laterally on each side lies a propleural sclerite, the trochantin; the pleural

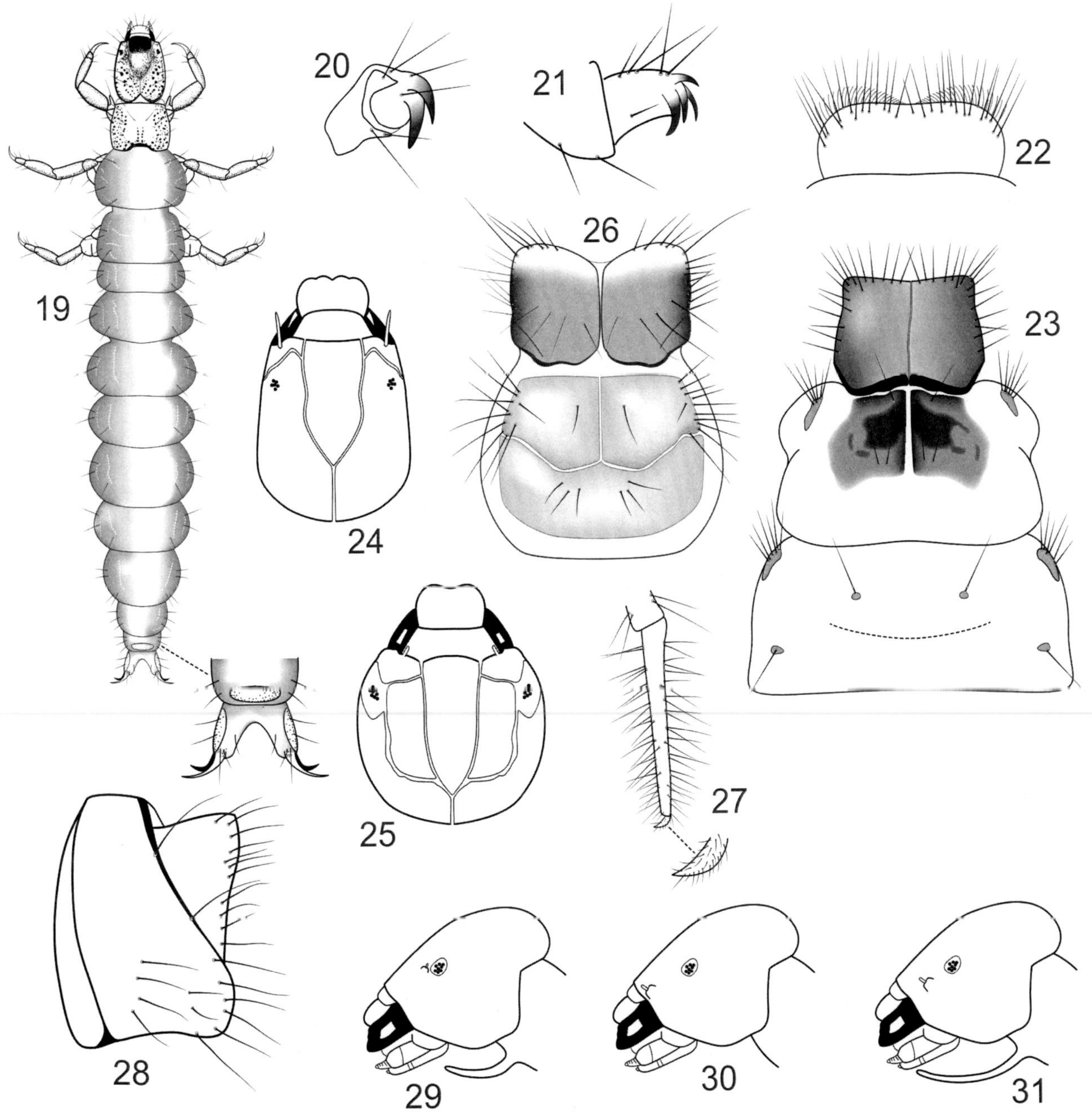

FIGURES 14.19–14.31 (19) Rhyacophilidae, *Rhyacophila* body, inset anal proleg, dorsal; (20) Glossosmatidae, *Glossosoma* anal claw, lateral; (21) Glossosmatidae, *Agapetus* anal claw, lateral; (22) Calamoceratidae, *Calamocerus* labrum, dorsal; (23) Calamoceratidae, *Calamocerus* thorax, dorsal; (24) Leptoceridae, *Oecetis* head, dorsal; (25) Leptoceridae, *Ceraclea* head, dorsal; (26) Molannidae, *Molanna* pro- and mesonota, dorsal; (27) Molannidae, *Molanna* hind tarsus and tarsal claw, lateral; (28) Brachycentridae, *Micrasema* pronotum, lateral; (29) Lepidostomatidae head, lateral; (30) Leptoceridae head, lateral; (31) Limnephilidae head, lateral.

sclerites of the meso- and metathorax are divided by a pleural suture into an episternum and an epimeron (Fig. 14.1d). The pronotum, depending on the family or genus, may also bear anterolateral extensions and/or carinae (Figs. 14.33 and 14.39). In several families, the prothorax ventrally has a median prothoracic horn, a structure bearing a gland at its tip believed to secrete a lipid to aid silk spinning (Figs. 14.1b and 14.31). In most annulipalpian families and in the integripalpian Rhyacophilidae and Glossosomatidae in the Mediterranean Basin the meso- and metathoracic nota are membranous or nearly so (Fig. 14.19). In the Hydropsychidae and Ecnomidae of Annulipalpia, there is a complete pair of notal sclerites on the meso- and metathorax (Figs. 14.44a and 14.48c). Likewise, in the integripalpian families Hydroptilidae and Ptilocolepidae, the meso- and metanota are fully sclerotized (Fig. 14.4). In all other Integripalpia, the mesonotum may be fully or partially

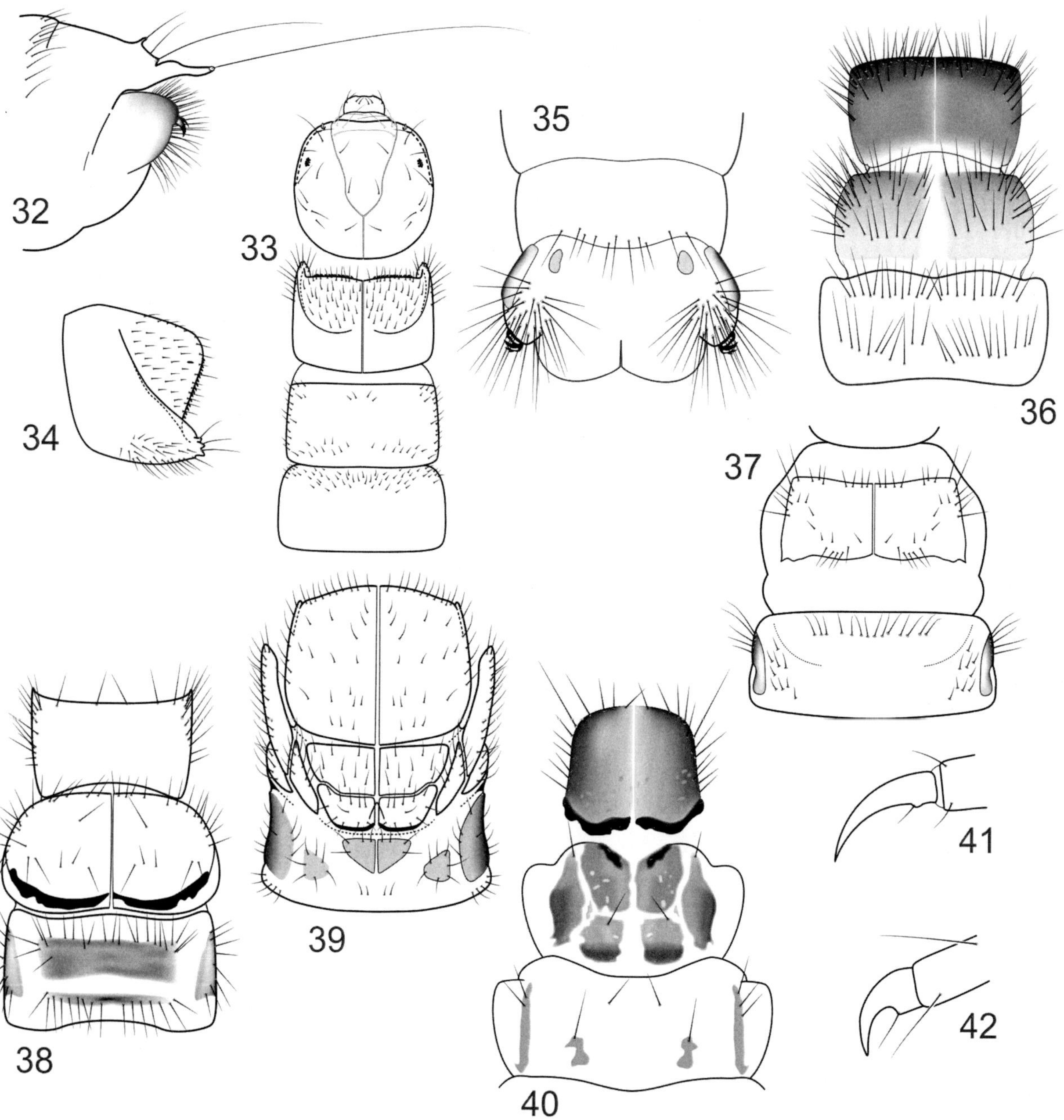

FIGURES 14.32–14.42 (32) Beraeidae, *Beraea* terminal abdominal segments, lateral; (33) Beraeidae, *Beraea* thorax and abdomen, dorsal; (34) Beraeidae, *Beraea* pronotum, lateral; (35) Sericostomatidae, *Sericostoma* terminal abdominal segments, dorsal; (36) Sericostomatidae, *Sericostoma* thorax, dorsal; (37) Apataniidae, *Apatania* meso- and metathorax, dorsal; (38) Odontoceridae, *Odontocerum* thorax, dorsal; (39) Goeridae, *Goera* thorax, dorsal; (40) Thremmatidae, *Thremma* thorax, dorsal; (41) Limnephilidae tarsal claw, lateral; (42) Apataniidae tarsal claw, lateral.

sclerotized, while the metanotum may be membranous or bear one to three pairs of small or medium-sized sclerites (e.g., Fig. 14.1g). Whether these sclerites are present or absent, their size and amount of division, and the distribution of setae on them are diagnostically important in discriminating among the various families and genera (generally clustered into three "setal areas"—sa1, sa2, sa3; Fig. 14.1g). As on the head, the pronotum especially may have anterolateral extensions, carinae, or ridges, and the setae present and their position on all nota follow a standard nomenclature.

Ten segments comprise the larval abdomen with the terminal segment only bearing a pair of prolegs (Fig. 14.1a). These anal prolegs each bear a single, terminal, anal claw (Fig. 14.1e). There are no spiracles as respiration is cuticular. In many taxa respiration is aided by filamentous tracheal gills, often only on various abdominal segments, but also on the thorax in some taxa. Commonly abdominal gills, when present, occur on abdominal segments 2 (rarely 1) to

7 (rarely 8) as either single filaments or in basal tufts of up to five or more filaments, or as numerous branches along a central stalk (Figs. 14.1a and 14.10). Gills can occur in dorsolateral, lateral, and ventrolateral rows along the abdomen and either anteriorly or posteriorly or both on an individual segment. The distribution of gills on the abdominal segments constitutes the "gill formula" and is usually depicted as a schematic illustration in larval descriptions. Except for Hydropsychidae, Annulipalpia lack gills, while most Integripalpia: Phryganides have some complement of gills. Also, in the phryganidean case-making families, with few exceptions, abdominal segment I bears a median dorsal and a pair of lateral "humps," fleshy, retractile protuberances providing a space between the body wall and case allowing oxygenated water to flow over the body and gills. In the Limnephilidae, the lateral humps bear small sclerites which are often taxonomically important in identifying the various genera (Fig. 14.1a). Other than the aforementioned humps, gills, and terminal prolegs, the abdomen does not have any other processes in all families except Hydroptilidae and some Ptilocolepidae. In these families, mature case-building larvae commonly have distended, laterally compressed abdomens with dorsal and ventral lobes or lateral, tube-like protuberances (Figs. 14.5 and 14.6). In the Phryganides, there is an abdominal lateral line system consisting of a lateral fringe, a row of short, fine, bifid filaments of the cuticle, and a series of fewer forked lamellae usually occurring from abdominal segments 2 (or 3) to 7 (or 8). This fringe aids and directs water flow over the abdomen, while the lamellae are believed to be sensory in function (Kerr & Wiggins 1995). While not as setose as the head and thorax, the abdomen does bear some primary setae that offer diagnostic value. However, in the Hydropsychidae, all abdominal segments are densely covered with short secondary hair-like setae or modified scale-hairs or both. In some Hydroptilidae and several Phryganides families, but especially Limnephilidae, elongate oval to circular areas of chloride epithelia occur; while often difficult to discern they are demarcated by a thin sclerotized line and specialized for ion transport. These demarcated areas of specialized cuticle commonly occur ventrally on segments 2–7, but may also occur dorsally and laterally. In the Annulipalpia, and also associated with osmoregulatory ion absorption, are membranous, retractile, anal papillae.

The larval abdomen terminates in the closely associated 9th and 10th segments (Fig. 14.1a). Segment 9 bears a dorsal sclerite in many families (Figs. 14.7 and 14.8) and in Hydropsychidae there are ventral sclerites on segment 9 as well as 8, these being especially diagnostic among the genera in this family (Fig. 14.10). The anal prolegs are borne on segment 10, each with a single anal claw (Fig. 14.1e). In Annulipalpia and non-phryganidean Integripalpia, the anal prolegs are usually elongate and separate from the body of the segment, with long anal claws (Figs. 14.14 and 14.15). In phryganidean Integripalpia, the anal prolegs are shorter, stouter, and more closely fused with segment 10 and the anal claw is shorter and stouter. In some families, the anal claw may have one (most Integripalpia) to several (Glossosomatidae) dorsal accessory hooks (Figs. 14.20 and 14.21). In *Helicopsyche*, the anal claw is pectinate (comb-shaped) along its margin (Fig. 14.3). There are various lateral plates (lateral sclerites, ventral sole plate) and primary setae associated with segment 10 and the anal proleg and these can be of diagnostic importance. Additional details of Trichoptera larval morphology can be found in the works of Wiggins (1996) and Holzenthal et al. (2015); the latter work also included descriptions of pupal and adult morphology.

Key to families

The keys below are to final instar larvae only.

1 Anal claw with comb-like dorsal process (Figs. 14.3 and 14.58c, arrow); head and pronotum without carina (Fig. 14.58b); larva with portable case shaped like snail's shell, constructed of sand grains or small rock fragments (Figs. 14.2I and 14.58a); .... Helicopsychidae, *Helicopsyche* (only genus in the Mediterranean Basin)

1' Anal claw hook-shaped, without comb-like dorsal process (Figs. 14.14, 14.20, and 14.21); larvae with portable cases of different shape (not like a snail's shell), fixed retreats, silken tubes, or free living without portable case (Figs. 14.1h, i, j, 14.2, and 14.19) 2

2(1') Dorsum of all three thoracic segments (pro-, meso-, and metanota) largely covered with sclerites (Fig. 14.4) 3

2' Pronotum covered with sclerites; mesonotum covered with large or small sclerites, or entirely membranous; metanotum entirely membranous or with small sclerites only (Figs. 14.11, 14.23, 14.36, and 14.40) 7

3(2) Abdominal segments and last two thoracic segments with ventrolateral rows of branched gills (Fig. 14.10); anal prolegs with dense brush of long terminal setae (Fig. 14.10); larvae in fixed retreats with associated silken net, on or under substrate in fast-flowing water (Fig. 14.2e) .... Hydropsychidae

3' Abdomen and thoracic segments without ventrolateral branched gills (Fig. 14.7); anal proleg without brush of terminal setae (Fig. 14.6); cases/retreats/habitats variable 4

4(3') Abdominal segment 9 with dorsal sclerite; tracheal gills, if present, only on abdominal segment 9 and at base of anal claws (Fig. 14.6); larval case (final instar only) of sand, algae, or silk, portable or fixed, usually laterally compressed or dorsoventrally flattened (Fig. 14.2c, d) 5

4' Abdominal segment 9 without dorsal sclerite; larvae in fixed silken and detritus tubes (Fig. 14.2g) 6
5(4) Dorsum of abdominal segment I with large sclerite (Figs. 14.7 and 14.51a); larval case only in final instar, purse-like case made of roundly cut plant material (Fig. 14.51b) ....
...... Ptilocolepidae, *Ptilocolepus* (only genus in the Mediterranean Basin) or *Microptila minutissima* (Hydroptilidae)
Note: Head capsule width (0.15 mm), tibial process of foreleg with pinnate spine and case of silk material of M. minutissima distinguish this species from Ptilocolepidae.
5' All abdominal segments I–VIII with dorsal sclerites (Fig. 14.8) or none (Fig. 14.5) .... Hydroptilidae
6(4') Metanotal sclerites each with narrow black stripe longitudinally in middle (Figs. 14.11 and 14.46a, black arrows); mesonotum and metanotum widely separated mesally by longitudinal membrane (Fig. 14.46a, white arrows); foretrochantin short, conical, separated from the second coxal pleurite (episternum) by a suture (Fig. 14.16); metathoracic claws clearly shorter than those of pro- and mesothoracic legs (Fig. 14.46b); anal claws strongly arched (Fig. 14.46b), each without teeth on inner margin (Fig. 14.14) ....
.... Pseudoneureclipsidae, *Pseudoneureclipsis* (only genus in the Mediterranean Basin)
6' Metanotal sclerite each without black stripes (Fig. 14.44a); fore trochantin long, pointed, fused with the second coxal pleurite (episternum; Fig. 14.17); lateral fringe of bristles on abdominal segments (Fig. 14.44b, arrow); anal claws right–angled, each with row of tiny teeth on inner margin (Fig. 14.15) ....
.... Ecnomidae, *Ecnomus* (only genus in the Mediterranean Basin)
7(2') Mesonotum without sclerites (mostly membranous) (Fig. 14.19) or with only few small sclerotized plates covering less than half of the notum (Fig. 14.9) 8
7' Mesonotum with sclerites (although sometimes weakly sclerotized) 13
8(7) Abdominal segment 9 membranous dorsally, without dorsal sclerite; larvae in fixed retreats or silken tubes, not building portable cases 9
8' Abdominal segment 9 with dorsal sclerite (Fig. 14.19, inset) 11
Note: Sometimes this sclerite is difficult to see and is only recognizable by its shiny surface.
9(8) Labrum membranous and T-shaped (Fig. 14.18); larvae construct fixed sac-shaped nets of silk (Fig. 14.2f) ....
.... Philopotamidae
Note: The labrum is often retracted under the frontoclypeus in preserved specimens.
9' Labrum sclerotized and rounded 10
10(9') Foretrochantin broad, hatchet-shaped apically (Fig. 14.13); larvae construct long tubular galleries on rocks ....
.... Psychomyiidae
10' Foretrochantin pointed apically (Fig. 14.12); larvae construct funnel-shaped or tubular filter nets or flattened retreats of silk (Fig. 14.2h) .... Polycentropodidae
11(8') Abdominal segment 1 with dorsal hump (Fig. 14.1a); prosternal horn present (Fig. 14.1b); larval case of plant, rarely mineral, particles arranged spirally or in rings; occasionally a piece of grass stem is used instead of a case (Fig. 14.2o, p) .... Phryganeidae
11'Abdominal segment 1 without dorsal hump; prosternal horn absent 12
12(11') Anal prolegs not joined to abdominal segment 9 (Fig. 14.19); anal claws without dorsal accessory hook, although a secondary lateral claw (sword process) may be present (Fig. 14.19, inset); larvae free-living, without cases ....
.... Rhyacophilidae
12' Half of anal prolegs joined to abdominal segment 9; anal claws with at least one dorsal accessory hook (Fig. 14.20); anal claw with accessory teeth similar to the claw (Fig. 14.21); tortoise-like portable cases with anterior and posterior openings directed ventrad (Fig. 14.2a, b) .... Glossosomatidae
13(7')Antenna long and prominent (length at least six times longer than width at its widest part) (Fig. 14.24) and/or hind legs longer than others .... Leptoceridae or *Beraeodes minutus* (Beraeidae)
*Note: Larvae of genus Ceraclea have short antennae protected by a lobe (*Fig. 14.25*). The mosaic pattern on the pronotum and the numerous setae on the frontoclypeus of B. minutus distinguish this species from Leptoceridae.*
13' Antenna short (no more than three times as long as wide); hindlegs no longer than others 14
14(13') Labrum with transverse row of at least 12 setae across central part (Fig. 14.22); anterolateral corners of pronotum extended and pointed (Fig. 14.53a, arrow); metanotum with 6 or 2 sclerites, 3 or 1 on each side; in the first case, the lateral sclerite is oblong, the median sclerites small, dot-shaped (Fig. 14.23); larval cases of two leaf pieces, dorsal piece usually larger than ventral, or larval case with hollowed out twig (Figs. 14.2q and 14.53b) ....
.... Calamoceratidae, *Calamoceras* (only genus in the Mediterranean Basin)
14' Labrum without transverse row of setae or with few setae 15

15(14') Tarsal claw of hindleg smaller than tarsal claws on fore and midlegs; hind tarsal claw stout, covered in setae (Figs. 14.27 and 14.55), or thin and elongated; mesonotum sclerotized with W-like transverse structure (Fig. 14.26); larval case flattened, shield-like, with an anterior hood, constructed of sand grains or small detritus particles .... Molannidae, *Molanna* and *Molannodes* (not present in the Mediterranean Basin; present in central and north Europe)
15' Tarsal claws on hindlegs similar to tarsal claws on fore—and midlegs 16
16(15') Abdominal segment 1 without lateral humps; pronotum divided transversely by a crease (Fig. 14.28); sometimes restricted to lateral sides only; elongated portable cases with four sides (most common) or rounded cases of mineral fragments, only silk, or both materials (Fig. 14.2j, l) .... Brachycentridae
16' Abdominal segment 1 with lateral humps (Fig. 14.1a) 17
17(16') Abdominal segment 1 without median dorsal hump; antenna situated close to anterior margin of eye (Fig. 14.29); cases of various materials and arrangement, usually tetrahedral .... Lepidostomatidae
Note: Genus Anisogamus (Limnephilidae) without median dorsal hump; in contrast to Lepidostomatidae, the antennae are located about halfway between eye and anterior border of the head.
17' Abdominal segment 1 with median dorsal hump; antenna situated close to the anterior margin of the head capsule (Fig. 14.30) or midway between anterior margin and the eye (Fig. 14.31) 18
18(17') Antenna situated near anterior margin of head capsule; prosternal horn absent (Fig. 14.30) 19
18' Antenna situated midway between eye and anterior margin of head capsule; prosternal horn present (Fig. 14.31) 21
Note: The same antennal position applies for the limnephilid Drusus chrysotus; however, a prosternal horn is missing in this species which is easily identified by the deep median incision of its headcapsule.
19(18) Anal proleg with a ventral brush of setae and with a prominent dorsal process bearing setae, one longer than others (Fig. 14.32); pronotum usually with transverse carina extended as rounded anterolateral lobes (Figs. 14.33 and 14.34); tubular cases of sand grains .... Beraeidae except *Beraeodes minutus* (see couplet 13, above)
19' Anal proleg without a ventral brush of setae or dorsal process; pronotum without transverse carina 20
20(19') Dorsum of anal proleg with about 30 long setae (Fig. 14.35); mesonotum membranous or sclerotized only anteriorly; metanotum with numerous setae (Fig. 14.36); abdominal segment 1 with three fleshy humps (Fig. 14.1a); abdominal segment 9 without dorsal sclerites (Fig. 14.35); larval case of small rock fragments ...... Sericostomatidae
20' Dorsum of anal proleg with no more than 3–5 long setae; mesonotum sclerotized (Fig. 14.56a); metanotum with four sclerites, with the two central ones arranged parallel to each other (Fig. 14.38); all tibiae with two apical spurs (Fig. 14.56c, arrow); abdominal segment 1 with three fleshy humps (Fig. 14.56a, arrows; Fig. 14.56b, arrows); abdominal segment 9 with a dorsal sclerite; larval case of coarse rock fragments (Figs. 14.2k and 14.56d) .... Odontoceridae, *Odontocerum* (only genus in the Mediterranean Basin)
21(18') Each half of mesonotum divided into two or three separate plates (Figs. 14.39 and 14.40) 22
Note: Some Brachycentridae larvae of genus Micrasema also share these features, but first abdominal segment lacks dorsal and lateral humps.
21' Each half of mesonotum not divided into separate plates (Fig. 14.1g) 23
22(21) Each half of mesonotum divided into two or three separate plates; metanotum with six or eight sclerites (Fig. 14.39); mesopleurites usually with prominent processes pointed anteriorly (Fig. 14.39); tubular case of coarse sand grains and with much larger stones or flat stones on each side (Fig. 14.2n) .... Goeridae
22 Each half of mesonotum divided into three separate plates; metanotum with four sclerites (Figs. 14.40 and 14.63a, numbers); shield-shaped case of sand grains (Fig. 14.63b) .... Thremmatidae, *Thremma* (only genus in the Mediterranean Basin)
23(21') Anterodorsal plates absent from metanotum, replaced by a transverse row of setae (Fig. 14.37); basal setae of each tarsal claw subequal to or as long as claw (Fig. 14.42); larval case of coarse sand grains .... Apataniidae
23' Metanotum with six small sclerites, including anterodorsal pair (Fig. 14.1g); the latter fused or replaced by setal groups in some taxa; basal setae of each tarsal claw shorter than claw (Fig. 14.41); larval cases vary considerably in size, shape, and composition (mineral or plant material or both, sometimes with snail shells incorporated; Fig. 14.1h, i, j) .... Limnephilidae

Keys to genera

The keys below are to final instar larvae only (the order of the key follows Table 14.1).

Trichoptera: Philopotamidae: Genera

1 Anterior margin of frontoclypeus convex without notch or with shallow asymmetrical notch (Fig. 14.43a); distal seta of forecoxa without long process (Fig. 14.43b, arrow) 2

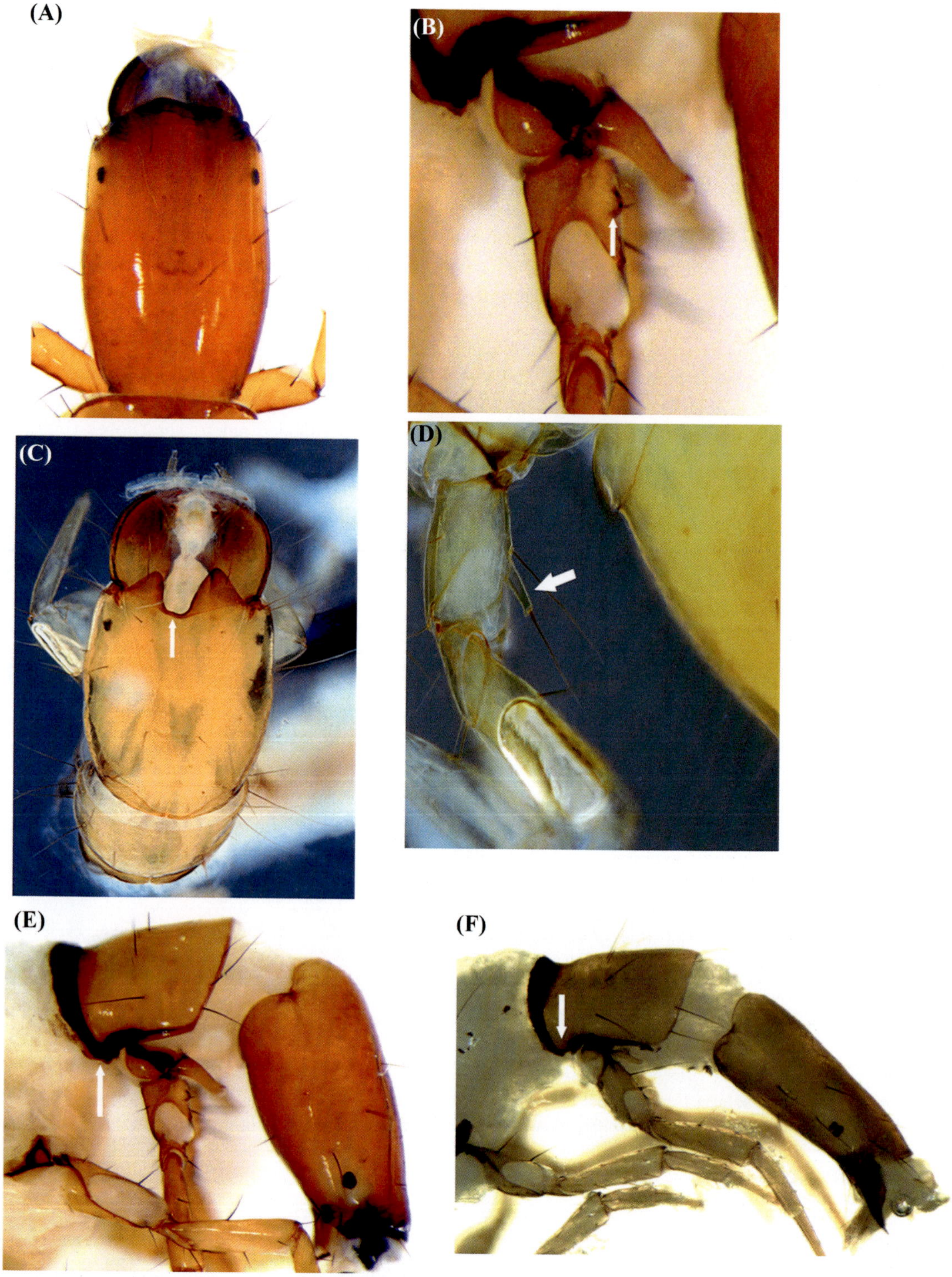

FIGURE 14.43 Philopotamidae; a, b, e: *Philopotamus achemenus*, (a) head, dorsal view, (b) forecoxa, lateral view, (e) head and pronotum lateral view; (c, d) *Chimarra marginata* (c) head, dorsal view, (d) forecoxa, lateral view; (f) *Wormaldia asterusia*, head and pronotum, lateral view.

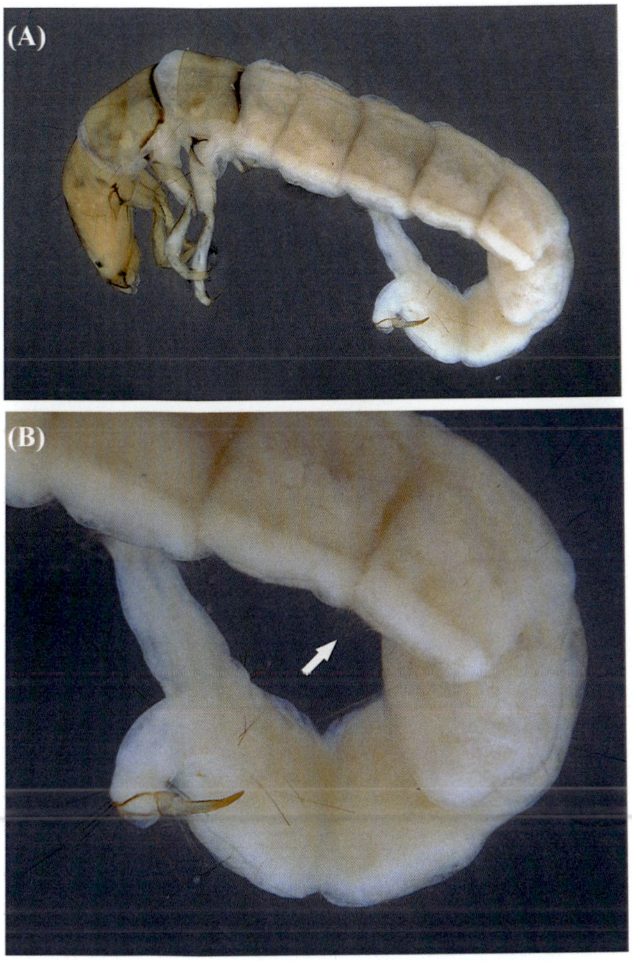

FIGURE 14.44 *Ecnomus* sp.; (a) larva, dorsolateral view, (b) abdomen and anal claw, dorsolateral view; white arrow indicating lateral fringe of fine bristles.

1' Anterior margin of frontoclypeus deeply notched asymmetrically (Fig. 14.43c, arrow); distal seta of fore coxa with long process arising near distal end (Fig. 14.43d, arrow) .... *Chimarra*
2(1) Lateral black stripe on each side of the pronotum continuous until posterior margin (Fig. 14.43e, arrow) .... *Philopotamus*
2' Lateral black stripe on each side of the pronotum interrupted by light area (Fig. 14.43f, arrow) .... *Wormaldia*

Trichoptera: Polycentropodidae: Genera

1 Basal segment of anal proleg with numerous bristles (Fig. 14.45a); abdominal segment 9 without ventral spines 2
1'Basal segment of anal prolegs without bristles (Fig. 14.45b, bs); abdominal segment 9 with two ventral short spines (Fig. 14.45b, vs) .... *Neureclipsis* (only *Neureclipsis bimaculata* (Linnaeus 1758))
2(1) Anal claw without teeth on inner apex (Fig. 14.45c) 3
2' Anal claw with four blunt teeth on inner apex (Fig. 14.45d) .... *Cyrnus*
3(2) Foretarsus at least 2/3 as long as tibia (Fig. 14.45e); anal claw obtusely curved (Fig. 14.45a) or right angled 4
3' Foretarsus half as long as tibia (Fig. 14.45f); anal claw obtusely curved (Fig. 14.45a); head width of final instar larvae < 1.65 mm .... *Polycentropus*
Note: In final instar larvae of Polycentropus species known so far, foretarsus is less than half length of tibia except in P. corniger McLachlan, 1884, P. schmidi Novák & Botosaneanu, 1965 (Urbanič 2006) and P. ierapetra Malicky, 1972 (Karaouzas & Waringer 2017), where foretarsus is more than half length of tibia.
4(3) Anal claw obtusely curved (Fig. 14.45a) .... *Plectrocnemia*
4' Anal claw right angled (Fig. 14.45c) .... *Holocentropus*

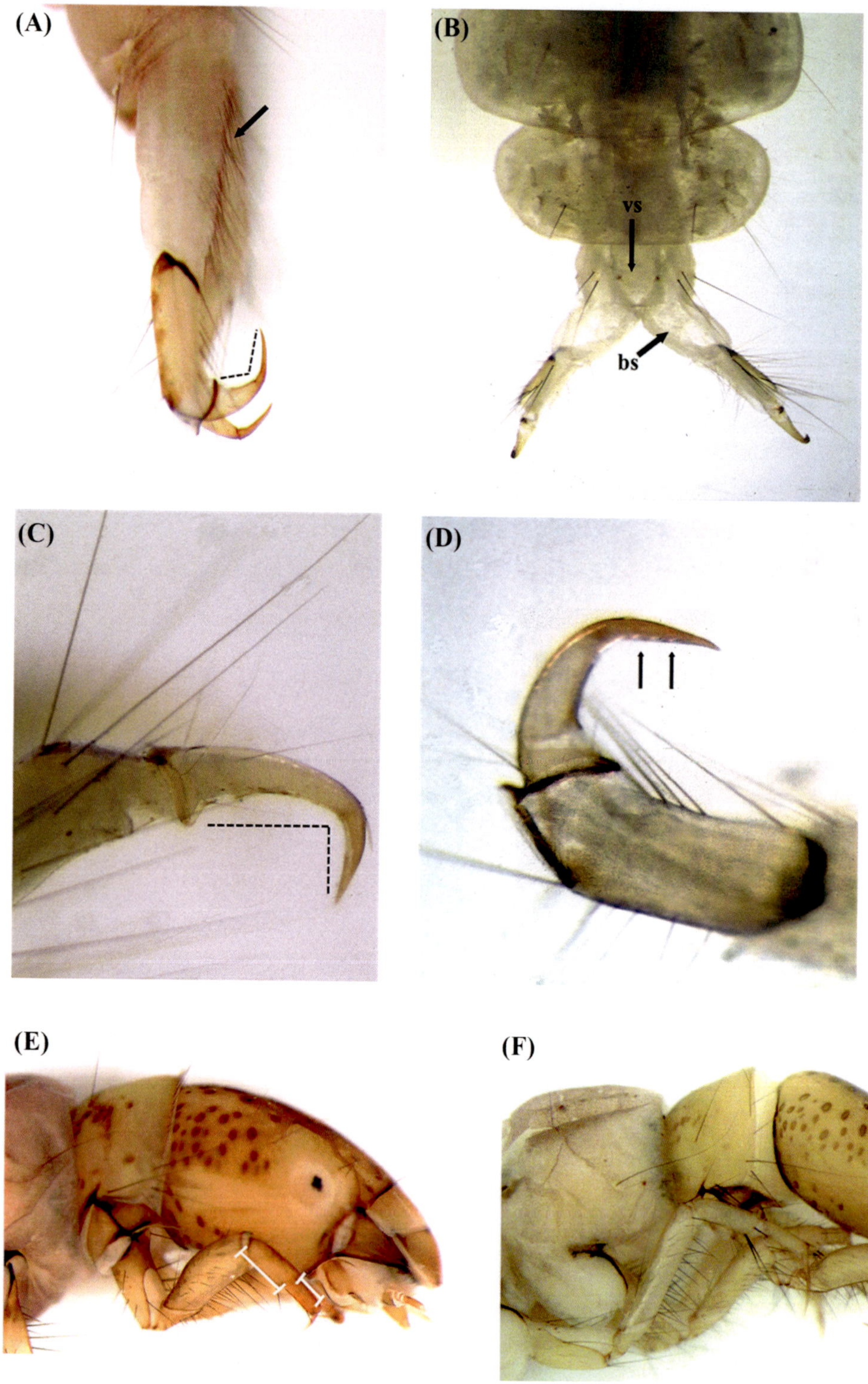

FIGURE 14.45 Polycentropodidae; (a, f) *Polycentropus ierapetra* (a) anal proleg and anal claw, lateral view, (f) head, pronotum and foreleg, lateral view; (b) *Neureclipsis bimaculata* abdominal segment 9, ventral spines (vs) and basal segment (bs) of anal prolegs, ventral view; (c) *Holocentropus* sp., anal claw, lateral view; (d) *Cyrnus trimaculatus*, anal claw, lateral view; (e) *Plectrocnemia* sp., head, pronotum and foreleg, lateral view.

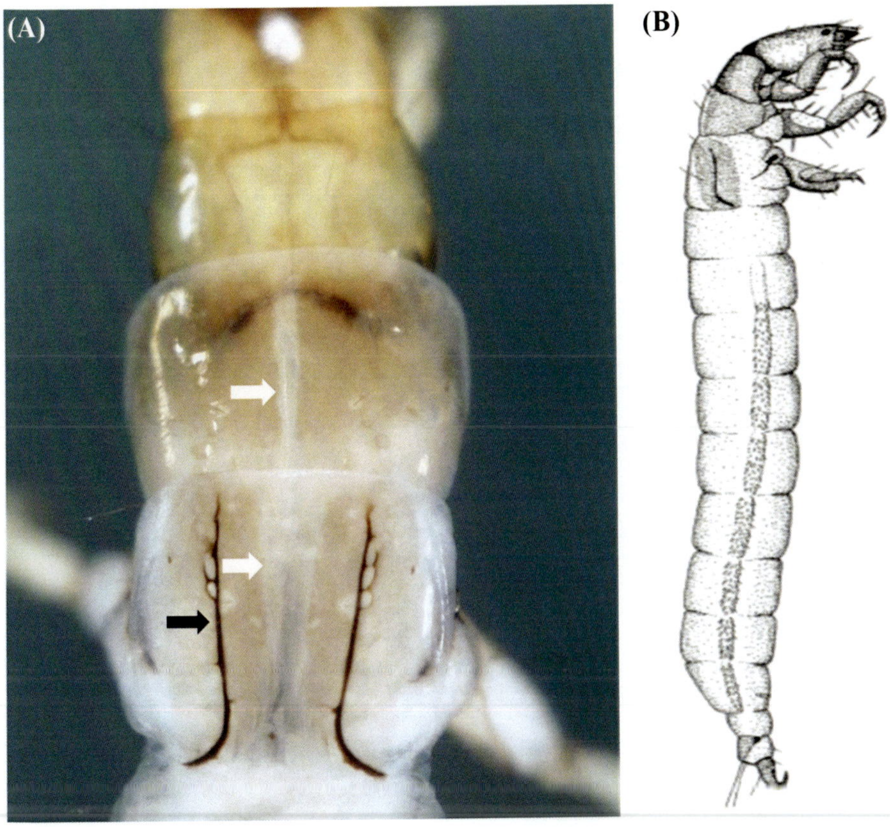

FIGURE 14.46 *Pseudoneureclipsis* sp.; (a) Pronotum, mesonotum and metanotum, dorsal view. White arrows indicate longitudinal membrane in meso- and metanotum, black arrow indicates black longitudinal stripe along the metanotum (photo by Rufino Vieira-Lanero); (b) larva, right lateral view (reproduced from Tachet et al., 2001; with permission from Taylor & Francis).

Note: European larva of genus Nyctiophylax (only Nyctiophylax (Paranyctiophylax) gaditana, is known from Europe) is still unknown.

Trichoptera: Psychomyiidae: Genera

1 Pronotum with a fold in posterolateral position (Fig. 14.47a, arrow) 2
1' Pronotum without such fold (Fig. 14.47b, arrow) 3
2(1) Submental sclerites longer than wide, black and heavily ornamented (Fig. 14.47c, arrow) .... *Psychomyia*
2' Submental sclerites wider than long, brown with smooth surface (Fig. 14.47d, arrow) 4
3(1') Anterior part of coxopleurite of fore leg with two vertical black bars (Fig. 14.47b, arrows) .... *Tinodes*
3' Anterior part of coxopleurite of fore leg with one vertical black bar only (Fig. 14.47f, arrow) .... *Lype*
4(2') Pair of prosternites present behind prothoracic coxae, blackish, narrow, diagonally transverse (Fig. 14.47g); posterolateral black thickening on the pronotum arched (Fig. 14.47h) .... *Metalype*
4' Prosternites absent (Fig. 14.47i); posterolateral black thickening on the pronotum straight (Fig. 14.47j) .... *Paduniella*

Trichoptera: Hydropsychidae: Genera

1 Frontoclypeus unconstricted at eye level (Fig. 14.48a); foretrochantin usually forked (Fig. 14.48b, arrow); meso- and metanotum without transverse sutures (Fig. 14.48c) 2

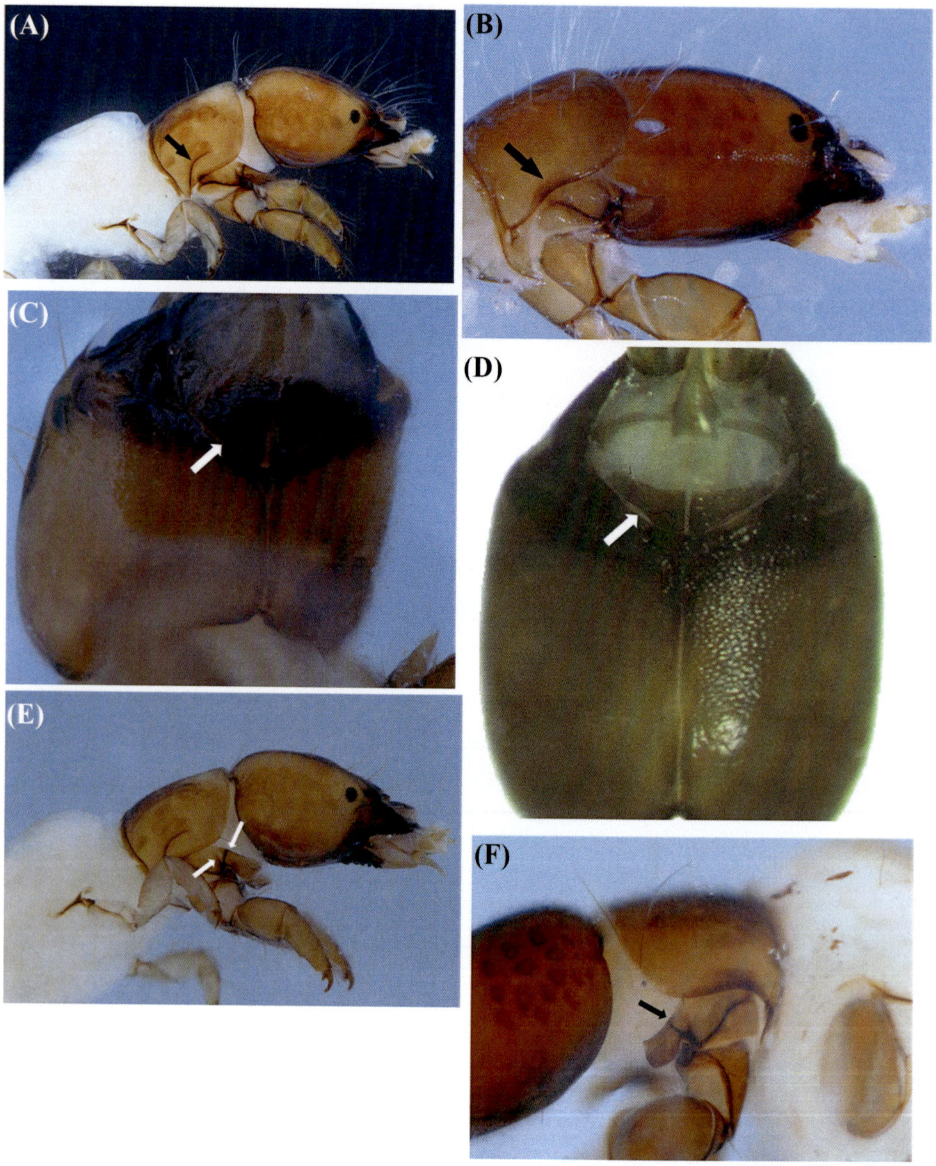

FIGURE 14.47 Psychomyidae; (a) *Psychomyia* sp., head and pronotum with black thickening in posterolateral position, lateral view; (b) *Tinodes* sp., head and pronotum, lateral view; (c) *Psychomyia* sp., submental sclerites, ventral view; (d) *Metalype fragilis*, submental sclerites, ventral view; (e) *Psychomyia* sp., coxopleurite of fore leg, lateral view; (f) *Lype* sp., coxopleurite of fore leg, lateral view; (g) *Metalype fragilis*, prosternites, ventral view; (h) *Metalype fragilis*, pronotum with posterolateral black thickening, lateral view; (i) *Paduniella vandeli*, prosternites, ventral view (photo by Henry Tachet); (j) *Paduniella vandeli*, pronotum with posterolateral black thickening, lateral view (photo by Henry Tachet).

1' Frontocypleus constricted at eye level (Fig. 14.48d); foretrochantin single and pointed (Fig. 14.48e, arrow); meso- and metanotum with transverse suture at posterior third of the sclerites (Fig. 14.48f, arrows) ...... *Diplectrona*

2(1) Posterior prosternites lacking or reduced to small sclerotized spots (Fig. 14.48g); anterolateral corner of pronotum and dorsal area of the head densely covered with short, tapered *hair-like* (*hs*) setae (Fig. 14.48h) .... *Cheumatopsyche*

2' Posterior prosternites well developed (Fig. 14.48i, arrows); dorsal area of head and pronotum without short tapered setae (Fig. 14.48j) .... *Hydropsyche*

Note: In some species, for example, Hydrospyche modesta, H. contubernalis, prosternites may be weakly sclerotized or have a light color.

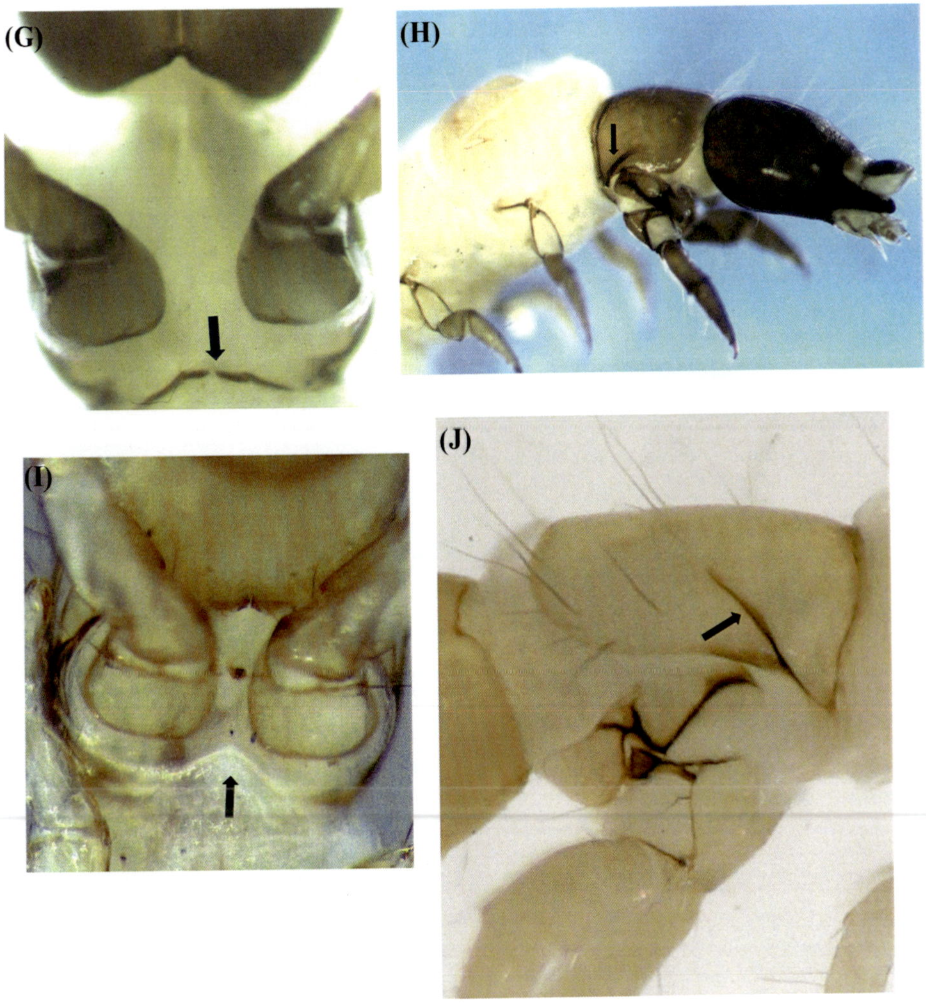

FIGURE 14.47 (Continued)

Trichoptera: Glossosomatidae: Genera

1 Pronotum fully sclerotized, meso- and metanota completely membranous (Fig. 14.49a) .... *Glossosoma*
1' Meso- and metanota incompletely sclerotized, each with two sclerites (Fig. 14.49b) .... 2 (*Agapetinae*)
2(1') Mesonotal sclerites large, in contact with anterolateral sa3 setae (Fig. 14.49e, arrow); basal spine of tarsal claw ending in a filiform prolongation that reaches the perpendicular tip of the claw (Fig. 14.49e, arrow); larval case saddle shaped (Fig. 14.49h) .... *Catagapetus*
2' Mesonotal sclerites small, clearly separated from anterolateral sa3 setae (Fig. 14.49b, c, d); larval case rounded (Fig. 14.49i, k) 3
3 (2') Each side of abdominal tergum 3 with one lateral seta (ls, Fig. 14.49f); basal spine of tarsal claw ending in a filiform prolongation that reaches the perpendicular tip of the claw (Fig. 14.49c); .... *Synagapetus*
3' Each side of abdominal tergum 3 without lateral setae (except *Agapetus fuscipes*) (Fig. 14.49g); basal spine of tarsal claw ending in a filiform extension that does not reach the perpendicular tip of the claw (Fig. 14.49d); .... *Agapetus*

Trichoptera: Hydroptilidae: Genera

Note: In the Hydroptilidae cases are built only by fifth instar larvae. In larval instars 1–4, dorsal sclerites are present on all the abdominal segments.

FIGURE 14.48 Hydropsychidae; (a–c) *Hydropsyche* spp., (a) head and frontoclypeal apotome, dorsal view; (b) foretrochantin, lateral view, arrow; (c) head, pro-, meso-, and metanotal sclerites, dorsal view; (d–f) *Diplectrona* sp.; (d) head and pronotum, dorsal view; (e) foretrochantin, lateral view, arrow; (f) head, pro-, meso-, and metanotal sclerites, dorsal view; (g, h) *Cheumatopsyche* sp.; (g) posterior prosternites, ventral view; (h) head and pronotum, left lateral view; (i, j) *Hydropsyche* sp., posterior prosternites, ventral view, arrows; (j) larva, left lateral view.

1 Abdominal segments 3–8 with dorsal and ventral ovoid processes (Fig. 14.50a); case oval with a narrow anterior opening made of secretion (Fig. 14.50a) .... *Ithytrichia*
1' Abdominal segments without dorsal and ventral ovoid processes 2
2(1') All or some abdominal segments 1–8 with dorsal sclerites (Fig. 14.50b) 3
2' Abdominal segments 1–8 without dorsal sclerites (Fig. 14.50e) 4
3(2) Dorsal sclerites present at all abdominal segments (Fig. 14.50b); case equal in width at anterior and posterior end, made of fine grains of sand (Fig. 14.50c) .... *Stactobia*

FIGURE 14.48 (Continued)

3' Dorsal sclerites present at first and last abdominal segment only; tibial process of foreleg (Fig. 14.50d, lower arrow t) with pinnate spine (Fig. 14.50d, upper arrow s); case flat made of silk with minor additions of inorganic particles and diatoms; strictly hygropetric .... *Microptila*

4(2') Mid- and hindlegs at least twice as long as forelegs (Fig. 14.50f); case flat made of silk only (Fig. 14.50g) 5

4' All legs approximately of similar length (Fig. 14.50h, k, n); case not flat, made of silk or silk with sand grains, detrital or filamentous algae (Fig. 14.50h, k, n) 6

5(4) Mid- and hindlegs four to five times longer than forelegs (Fig. 14.50e); case flat, purse-shaped, with both openings similar in width (Fig. 14.50e) .... *Tricholeiochiton*

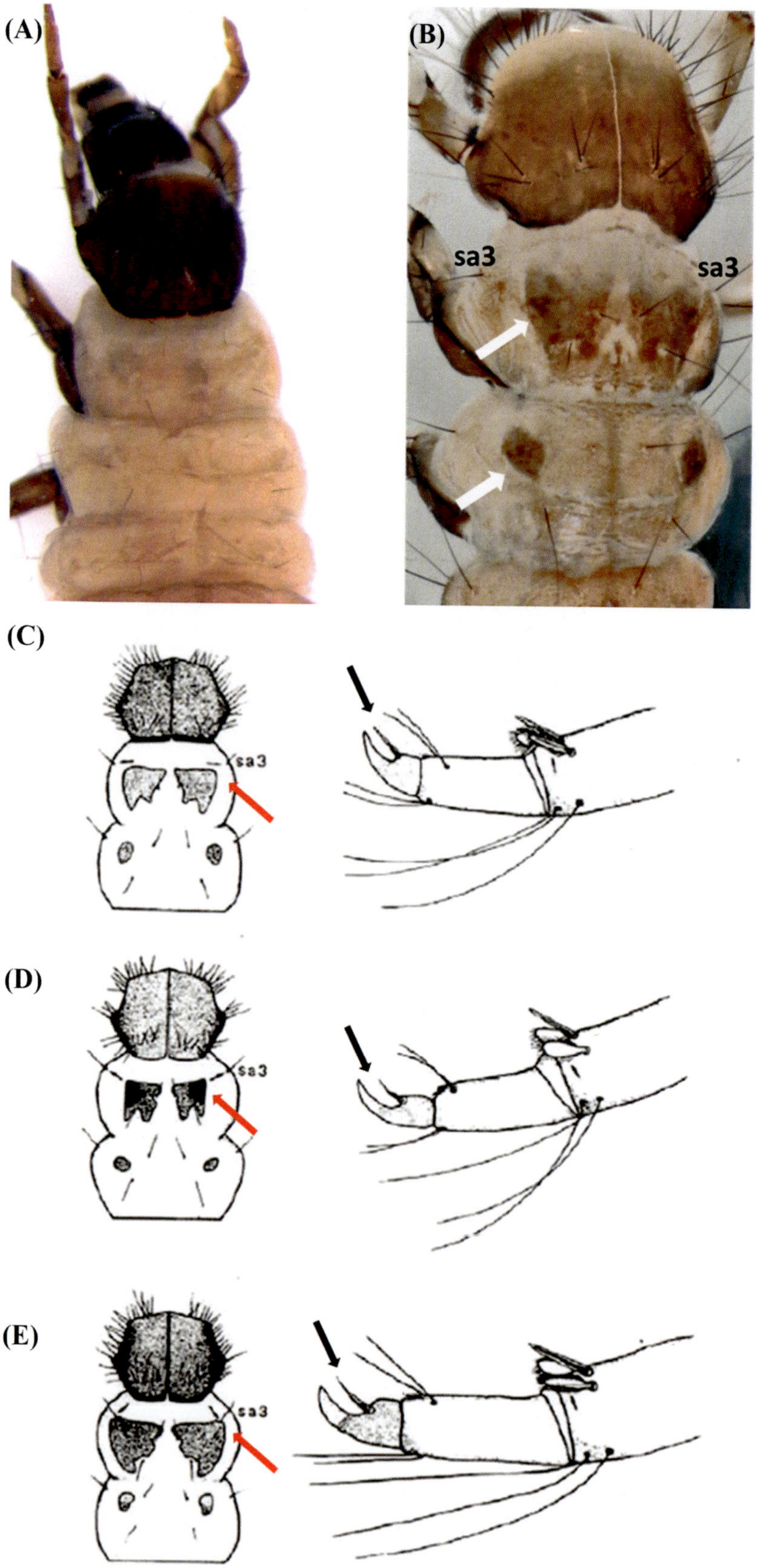

FIGURE 14.49 Glossosomatidae; (a) *Glossosoma* sp.; pro-, meso- and metanotum, dorsal view; (b) *Agapetus episkopi*; pro-, meso-, and metanotum, dorsal view; white arrows indicate meso- and metanotal sclerites; (c–k) subfamily *Agapetinae*; (c–e) from top to bottom *Synagapetus*, *Agapetus*, and *Catagapetus* spp.; pro-, meso-, and metanotum, dorsal view; tibia and tarsus, lateral view (figure reproduced with permission from J.A. Camargo & D. García de Jalón); (f) *Synagapetus* sp.; abdominal segments 1–4 with lateral setae (*ls*); (g) *Agapetus episkopi*; abdominal segments 1–2 with lateral (*ls*) and ventral setae (*vs*), segments 3–4 with only ventral setae (*vs*); (h, i, j) larva cases of *Catagapetus*, *Synagapetus*, and *Agapetus*, respectively.

FIGURE 14.49 (Continued)

5' Mid- and hindlegs two to three times longer than forelegs (Fig. 14.50f); case bottle-shaped, with an anterior opening much smaller than the posterior (Fig. 14.50g) .... *Oxyethira*

6(4') Case seed-like with longitudinal ridges (Fig. 14.50h); abdominal segment 2 with lateral protuberances; labrum asymmetrical, with beak-like projection (Fig. 14.50i, arrow) .... *Orthotrichia*

6' Case not seed-like (Fig. 14.50j); abdominal segment 2 without lateral protuberances; labrum symmetrical 7

7(6') Tarsal claw abruptly curved, with thick basal spur (Fig. 14.50k, arrow); case with two valves, made of silk (Fig. 14.50l) .... *Stactobiella*

7' Tarsal claw curved, with thin basal spur (Fig. 14.50m) 8

8(7') Case covered with sand grains, detrital particles or diatoms (Fig. 14.50n); anal proleg claw with accessory hooks (Fig. 14.50o, arrow) .... *Hydroptila*

FIGURE 14.50 Hydroptilidae; (a) *Ithytrichia lamellaris*, larva and case; (b) *Stactobia* sp., larva, dorsal view; (c) *Stactobia* sp., case, ventral view; (d) *Microptila minutissima*, left foreleg, posterior face. s: tibial process spine, t: tibial process (reproduced with permission from Graf et al., 2004, Taylor & Francis); (e) *Tricholeiochiton fagesii*, larva and case, lateral view, respectively; (f) *Oxyethira* sp., head and thoracic legs, lateral view; (g) *Oxyethira* sp., larva and case, lateral view; (h) *Orthotrichia* sp., larva and case, dorsal (left) and lateral (right) view (photo by Rufino Vieira-Lanero); (i) *Orthotrichia* sp., labrum with beak-like projection (arrow) (photo by Rufino Vieira-Lanero); (j) *Agraylea multipunctata*, larva and case, lateral view; (k) *Stactobiella risi*, head, pro- and mesonotum, and tarsal claw (arrow) lateral view; detail of tarsal claw of fore leg (bottom right corner, photo by Wolfram Graf); (l) *Stactobiella risi*, larva and case, lateral view (photo by Wolfram Graf); (m) *Hydroptila vectis*, tibial process of fore leg (drawing by Wolfram Graf); (n) *Hydroptila* sp., larva and case, lateral view; (o) *Hydroptila* sp., anal proleg claw, ventrolateral view (photo by Rufino Vieira-Lanero); (p) *Allotrichia pallicornis*, larva and case, lateral; (q) *Agraylea sexmaculata*, head and thorax, dorsal view; (r) *Allotrichia* sp., head and thorax, dorsal view (photo by Rufino Vieira-Lanero); (s) *Allotrichia* sp., tibiae of fore-, mid- and hindlegs, lateral views (photos by Rufino Vieira-Lanero).

FIGURE 14.50 (Continued)

8' Case made of silk, usually with incorporated concentric algal filaments (Fig. 14.50p); anal proleg claw without accessory hooks 9

9(8') Meso- and metanotum with dark markings (Fig. 14.50q); tibiae of mid- and hindlegs without a ventral prominence .... *Agraylea*

9' Meso- and metanotum without dark markings (Fig. 14.50r); tibiae of fore-, mid-, and hindlegs with a ventral prominence (Fig. 14.50s, arrow) .... *Allotrichia*

Trichoptera: Rhyacophilidae: Genera

The family Rhyacophilidae is represented by the single genus *Rhyacophila* in the Mediterranean Basin. Another genus is known from the Palaearctic Region and is represented by a single species; *Philocrena trialetica* Lepneva, 1956

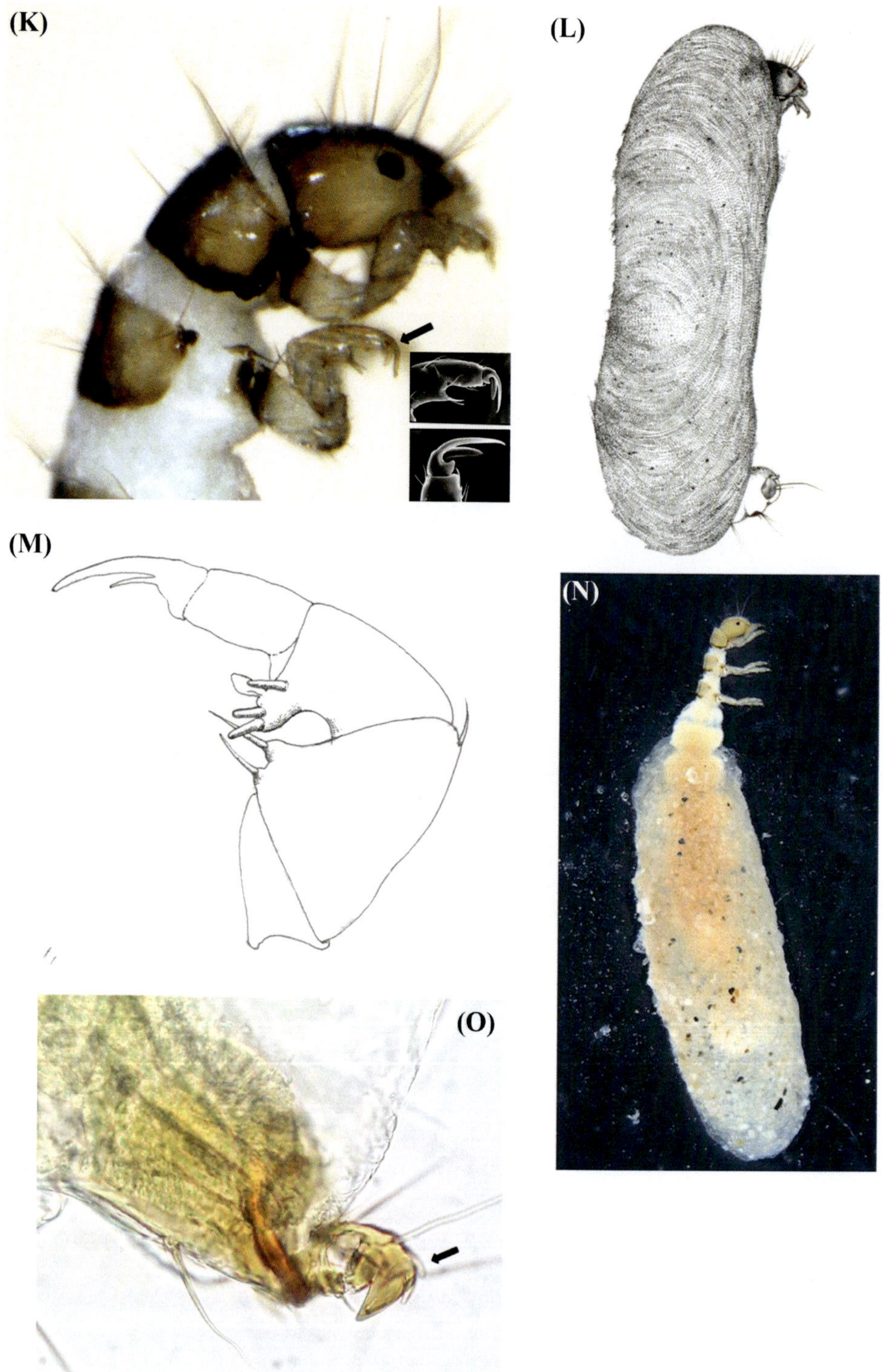

FIGURE 14.50 (Continued)

FIGURE 14.50 (Continued)

(Caucasus). European species of the genus *Rhyacophila* have been positioned by Döhler (1950) into six subgenera based on the morphology and number of larval gills: *Hyperrhyacophila, Pararhyacophila, Hyporhyacophila, Metarhyacophila, Prosrhyacophila,* and *Rhyacophila s.str.* Although this concept is not reflected in adult morphology (Schmid 1970), Döhler's classification is commonly used to characterise *Rhyacophila* larval morphology and has been used ever since in many Trichoptera larval keys. In the Mediterranean Basin, the larvae of five paraphyletic Döhler groups have been recorded so far: *Hyporhyacophila, Prosrhyacophila, Pararhyacophila, Hyperrhyacophila,* and *Rhyacophila s. str.*

1 Abdominal gills lacking (Fig. 14.52a) .... *Hyporhyacophila*
1' Abdominal gills present 2
2(1') Single filament gill at each side of abdomen (Fig. 14.52b) .... *Prosrhyacophila*
2' More than one gill filament at each side of abdomen 3
3(2') Two to four gill filaments at each side of abdomen (Fig. 14.52c) .... *Pararhyacophila*
3' Four to six fringed-lobe gill filaments at each side of abdomen or comb-like or tufted gills at each side of abdomen 4
4(3') Four to six fringed-lobe gill filaments at each side of abdomen (see Waringer & Graf 2011, p. 50, Figs. 19, 20) .... *Metarhyacophila*
Note: Members of this group (e.g., Rhyacophila bonaparti) are not present in the Mediterranean Basin.
4' Comb-like or tufted gills at each side of abdomen (Fig. 14.52d, e) 5
5(4') Comb-like gills at each side of abdomen (Fig. 14.52d) .... *Hyperrhyacophila*
5' Tufted gills with a short stalk and numerous filaments (Fig. 14.52e) .... *Rhyacophila s. str.*

Trichoptera: Leptoceridae: Genera

1 Tarsal claw of midlegs hook-shaped; tarsus curved (Fig. 14.54a, arrow) .... *Leptocerus*
1' Tarsal claw of midlegs normal, slightly curved; tarsus straight (Fig. 14.54b, white arrows) 2
2(1') Mandibles elongated, about three times as long as the width at the base and with only one cutting edge (Fig. 14.54b, c, black arrows) .... *Oecetis*
Note: Oecetis strucki, a non-Mediterranean species, has short mandibles with two cutting edges (Waringer & Graf 2014).
2' Mandible not elongated and with two cutting edges (Fig. 14.54d, white arrow) 3
3(2') Mesonotum with two posterolateral projections, usually very dark in color (they may be hidden in the intersegmental fold) (Fig. 14.54e, arrows); gills of several filaments (Fig. 14.54f, arrows) 4
3' Mesonotum without posterolateral projections (Fig. 14.54n, v, x); if present, gills of a single filament 6
4(3) Submentum triangular or subtriangular (Fig. 14.54g, arrow); case made of sand or plant pieces without an overhanging dorsal lip (Fig. 14.54h) 5
4' Submentum quadrangular or polygonal (Fig. 14.54i, arrow); case made of sand grains, secretion or a mixture of both, with an overhanging dorsal lip (Fig. 14.54k) .... *Ceraclea*

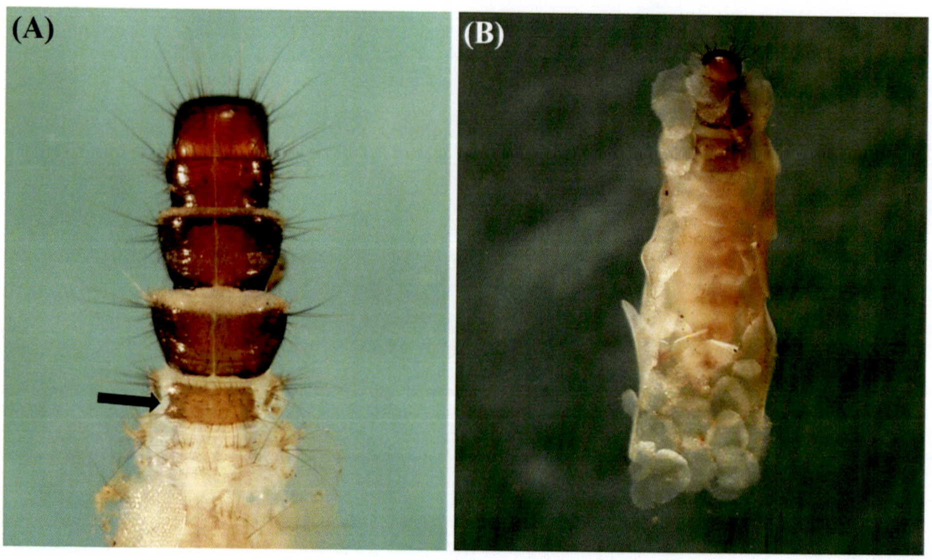

FIGURE 14.51 *Ptilocolepus* sp.; (a) head, thorax and abdominal segment 1, dorsal view; (b) case of instar V, dorsal view (photos by Rufino Vieira-Lanero).

FIGURE 14.52 Rhyacophilidae; (a) *Hyporhyacophila*; (b) *Prosrhyacophila* (small photo: ventral view of abdominal gills); (c) *Pararhyacophila*; (d) *Hyperrhyacophila*; (e) *Rhyacophila s. str.*

(A)

(B)

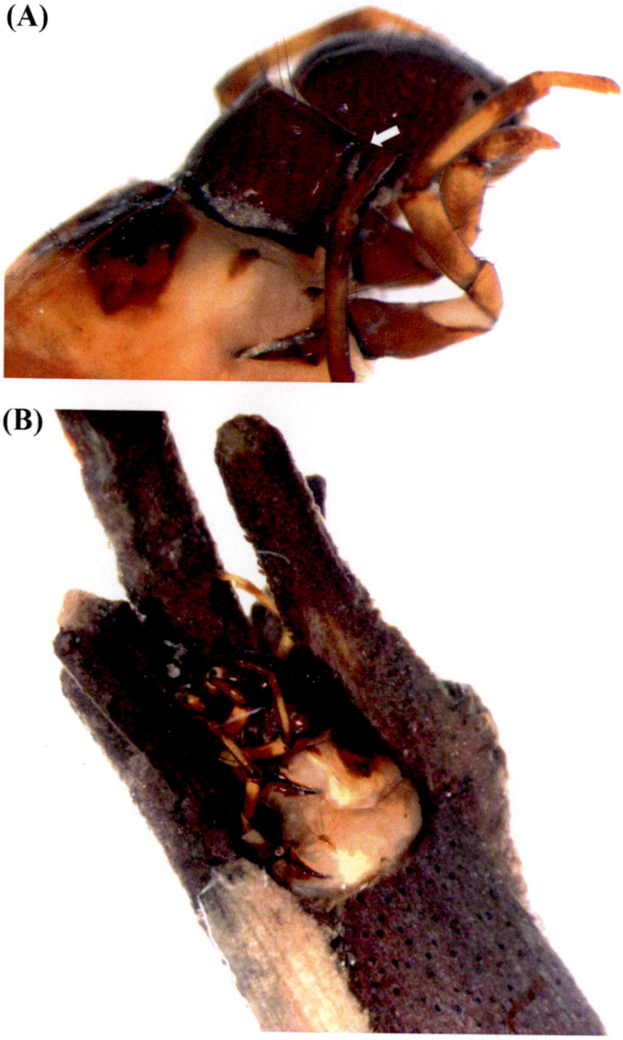

FIGURE 14.53 *Calamoceras illiesi*; (a) head and pronotum, lateral view; (b) larval case, lateral view.

5(4) Submentum triangular, more or less pointed at its posterior apex that reaches or nearly the posterior margin of the head (Fig. 14.54g, arrow); abdominal segments 4–8 without gills (Fig. 14.54f); case curved made of sand grains (Fig. 14.54h) .... *Athripsodes*
5' Submentum triangular, pointed at its posterior apex, but not reaching the posterior margin of the head (Fig. 14.54l, arrow); abdominal segments 4–8 with gills (Fig. 14.54j); case straight made of plant pieces with additional stem fragments attached longitudinally and irregularly (Fig. 14.54m) .... *Parasetodes*
6(3') Anal region surrounded by tooth edge plates or posterior part of anal proleg with two rows of strong spines directed posteriorly (Fig. 14.54o, p, arrows); case straight or slightly curved made of sand grains (Fig. 14.54q) .... *Setodes*
6' Anal region without that characteristics, there may be spinules and soft spines around the anal slit (Fig. 14.54r) 7
7(6') Lateral sclerite at abdominal segment 1 with a posterior projection bearing a dark bar (Fig. 14.54s) 8
7' Lateral sclerite at abdominal segment 1 without a dark bar on the posterior projection (Fig. 14.54x) 9
8(7) Dark bar of posterior projection of lateral sclerite at abdominal segment 1 thin and curved (Fig. 14.54s, arrow); tibia and tarsus of hindleg with a median constriction, usually marked with a pale transversal band (Fig. 14.54t, arrows); case straight or slightly curved made of sand grains and usually with plant pieces (Fig. 14.54u) *Mystacides*

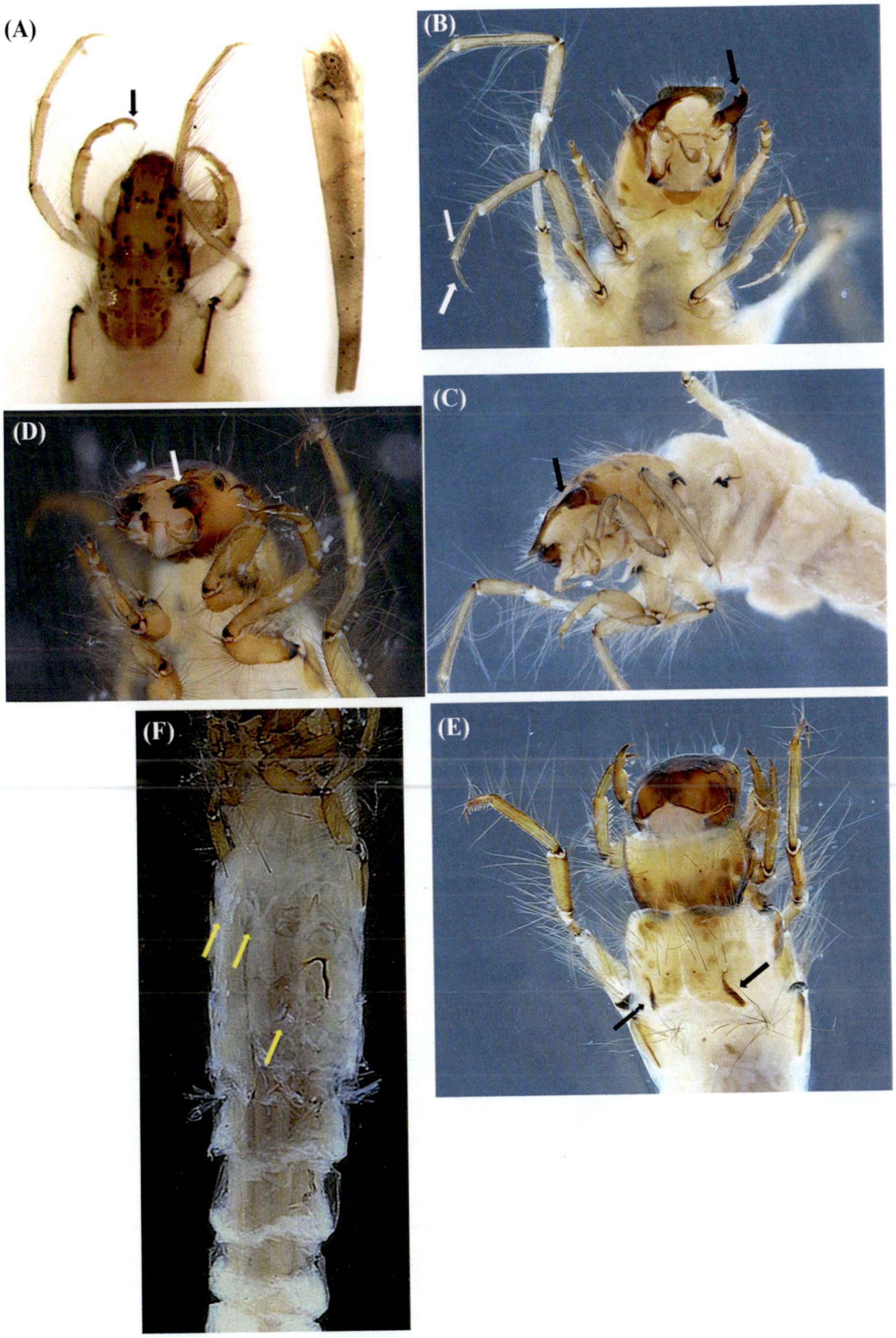

FIGURE 14.54 Leptoceridae; (a) *Leptocerus interruptus*, head, thorax, and thoracic legs, dorsal view; larva and case; (b) *Oecetis* sp., head and thoracic legs, ventral view; (c) *Oecetis* sp., head and thoracic legs, lateral view; (d) *Athripsodes* sp., head, ventral view, mandibles; (e) *Athripsodes* sp., head and thorax, ventral view, postero-lateral projections; (f) *Athripsodes* sp., abdominal gills, lateral view; (g) *Athripsodes* sp., head, ventral view, submentum (photo by Rufino Vieira-Lanero); (h) *Athripsodes* sp., larval case, lateral view; (i) *Ceraclea* sp., head, ventral view (photo by Rufino Vieira-Lanero); (j) *Parasetodes respersellus*, left lateral side of abdominal segments 1–4 showing gills on segments 2–4; d: dorsal, v- ventral, vl: ventrolateral (Móra et al., 2014, reproduced with permission from copyright holder); (k) *Ceraclea* sp., larval case, lateral view (photo by Rufino Vieira-Lanero); (l) *Parasetodes respersellus*, head, ventral, submentum (Mora *et al.*, 2014, reproduced with permission from copyright holder); (m) *Parasetodes respersellus*, larval case (Móra et al., 2014, reproduced with permission from copyright holder); (n) *Setodes* sp., thorax, dorsal view; (o) *Setodes* sp., anal prolegs, dorsal view; (p) *Setodes* sp., anal prolegs, ventral view; (q) *Setodes* sp., larval case, lateral view; (r) *Adicella* sp., anal prolegs, ventral view; (s) *Mystacides* sp., abdomen segment 1 with lateral protuberance, lateral view (photo by Rufino Vieira-Lanero); (t) *Mystacides* sp., head and thoracic legs, lateral view, tibia and tarsus of hindlegs; (u) *Mystacides* sp., larval case; lateral view; (v) *Adicella* sp., head and thoracic legs, lateral view, tibia and tarsus of hindlegs; (w) *Adicella* sp., larval case; lateral view; (x) *Triaenodes* sp., head and thoracic legs, lateral view, tibia and tarsus of hindlegs; (y) larval cases of *Triaenodes bicolor* (left) and *Erotesis baltica* (right, photo by Edyta Buczyńska); (z) *E. baltica*, head, thorax and abdominal segments, dorsal view (left), hindleg (right) (photos by Edyta Buczyńska)

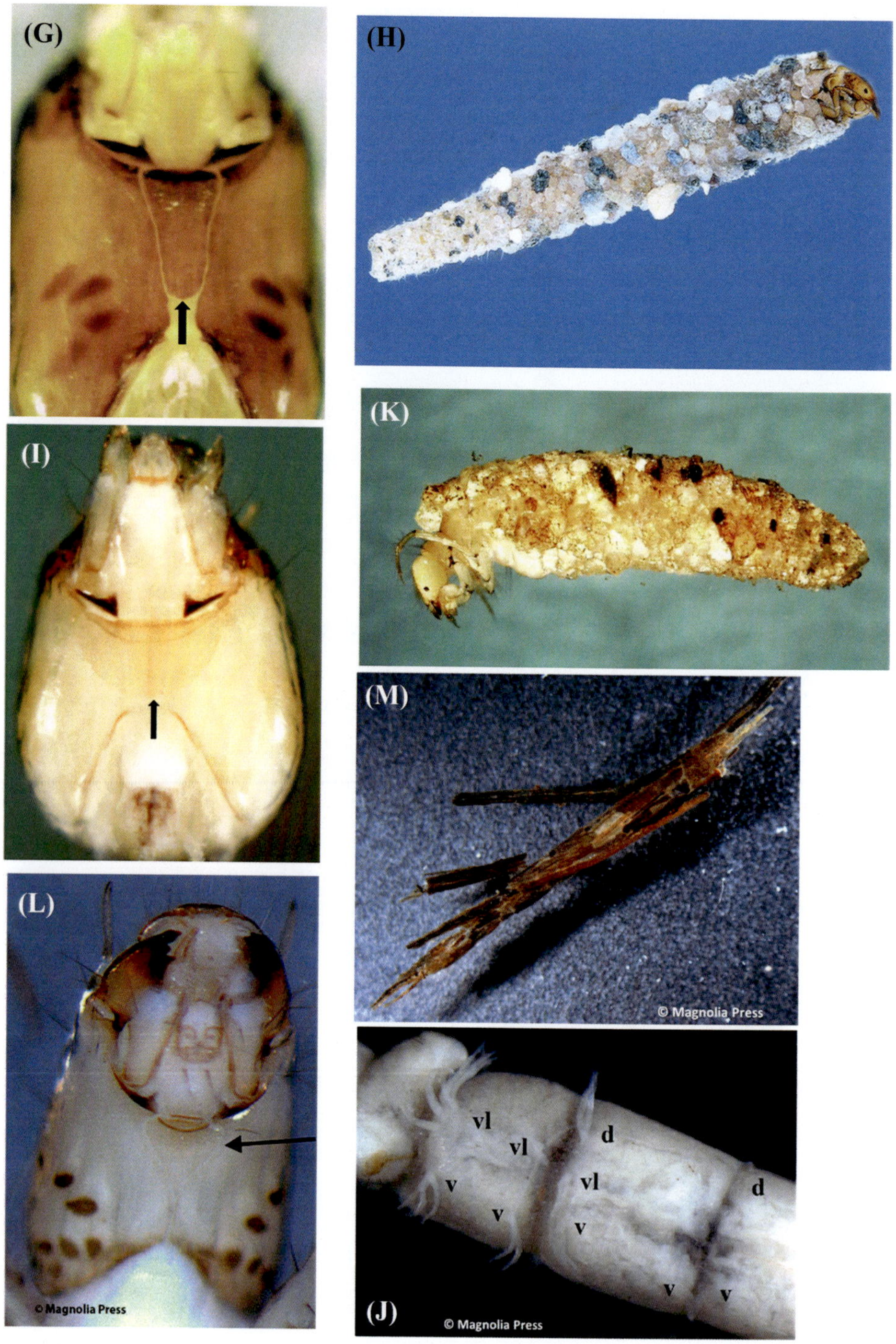

FIGURE 14.54 (Continued)

8' Dark bar of posterior projection of lateral sclerite at abdominal segment 1 broad and straight (Fig. 14.54v, arrow); tibia and tarsus of hindleg without a median constriction or pale transversal band (Fig. 14.54v, arrows); case either strongly curved made of sand grains or straight made of plant pieces, placed in spiral (Fig. 14.54w) .... *Adicella*
9(7') Hindleg with one or two long setal fringes (Fig. 14.54x, arrows); case straight made of plant material placed in spiral (Fig. 14.54y, left) .... *Triaenodes*

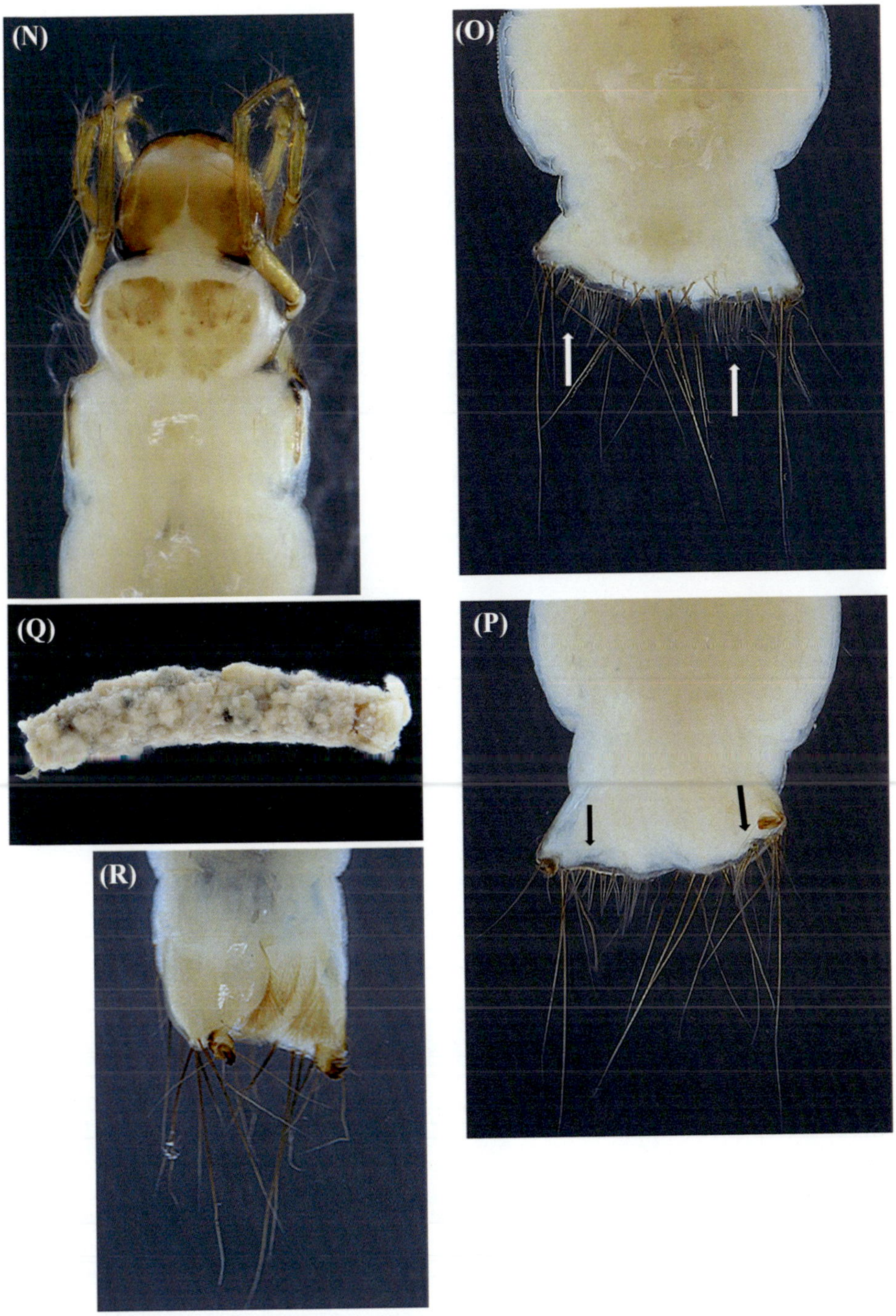

FIGURE 14.54 (Continued)

Note: Although genus Ylodes is considered as a subgenus of Triaenodes (Holzenthal & Andersen 2004), it is maintained in some keys as a separate genus. They can be separated by the number of setal fringes on hindlegs: Ylodes with one row of setal fringes and Triaenodes with two rows.

9' Hindleg without long setal fringes (Fig. 14.54z); case curved when early instars but straight in V instar, made of plant material arranged in two opposing spirals, one dextral and the other sinistral (Fig. 14.54y, right) .... *Erotesis*

FIGURE 14.54 (Continued)

Trichoptera: Beraeidae: Genera

1 Frontoclypeus with dense setal cover (Fig. 14.57a, white arrow) 2
1' Frontoclypeus without dense setal cover 3
2(1) Head with extensive black areas (Fig. 14.57a); pronotum with a typical mosaic of dark spots (Fig. 14.57a, black arrow) .... *Beraeodes*

FIGURE 14.54 (Continued)

FIGURE 14.55 *Molanna angustata*; claw of hindlegs, dorsal view.

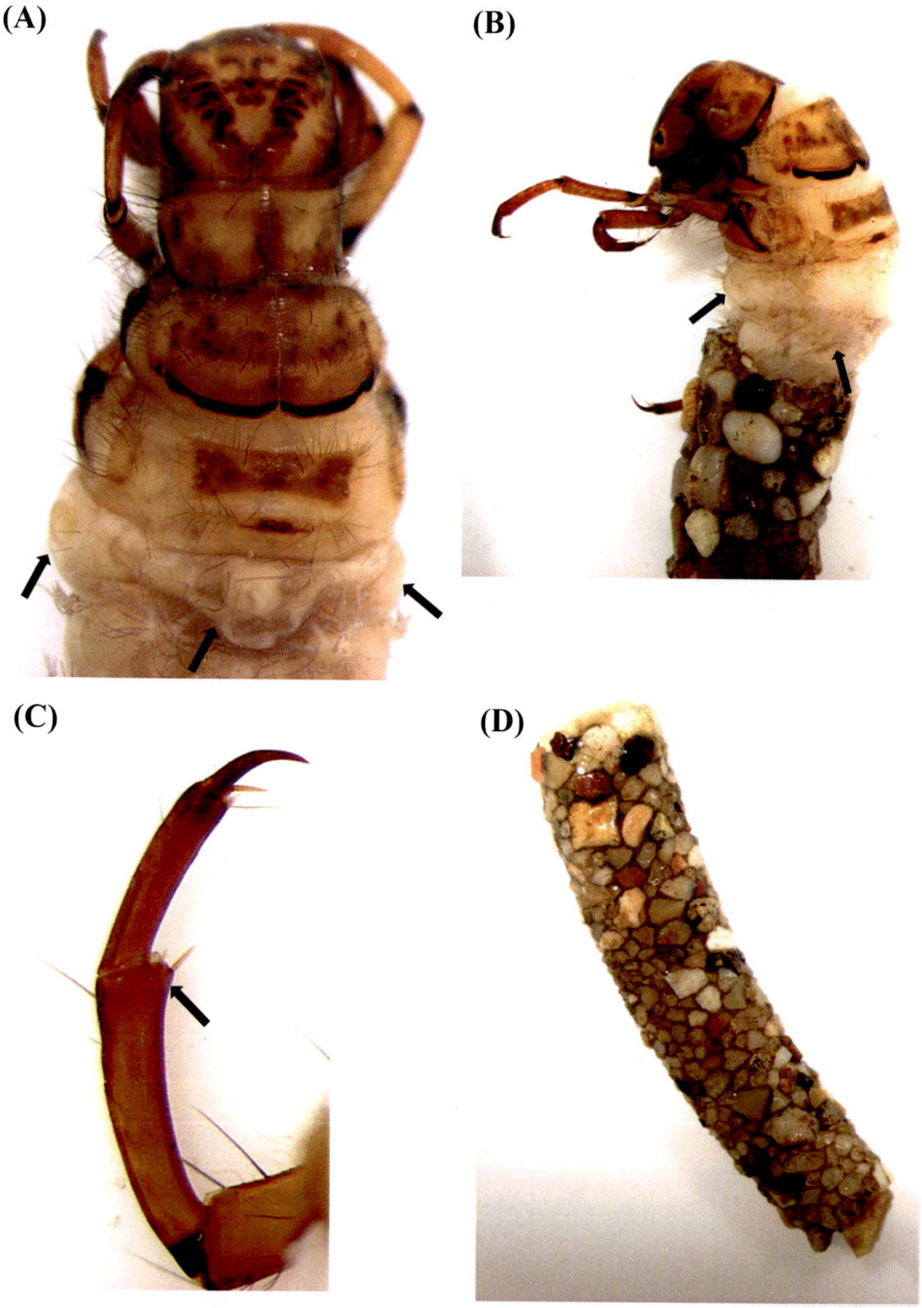

FIGURE 14.56 *Odontocerum albicorne*; (a) head, thorax, and abdominal segment 1, dorsal view; (b) head, thorax, and abdominal segment 1, lateral view; (c) tibiae of midleg with two apical spurs, dorsal view; (d) larval case.

2' Head and pronotum not as above .... *Beraeodina*

3(1') The longest dorsal seta of the anal proleg arises from a long conical base (Fig. 14.57b); pronotum with a bluntly pointed anteriorlateral corner (Fig. 14.57c, arrow); posterior tip of case with a hemispherical prominence on which the opening is situated ventrally (Fig. 14.57f, left) .... *Ernodes*

3' The longest dorsal seta of the anal proleg arises from a smaller tubercle or sclerotized plate (Fig. 14.57d, arrow); pronotum with a broadly rounded anteriorlateral corner (Fig. 14.57e, arrow); posterior opening without a hemispherical prominence (Fig. 14.57f, middle and right) 4

4(3') Posterior third and tip of frontoclypeus rounded (Fig. 14.57g, arrow); head with two carinae (Fig. 14.57g, white arrows; Fig. 14.57h, black arrows) .... *Beraeamyia*

4' Posterior third and tip of frontoclypeus broadly triangular; head capsule with only one carina across the eye (Fig. 14.57i) .... *Beraea*

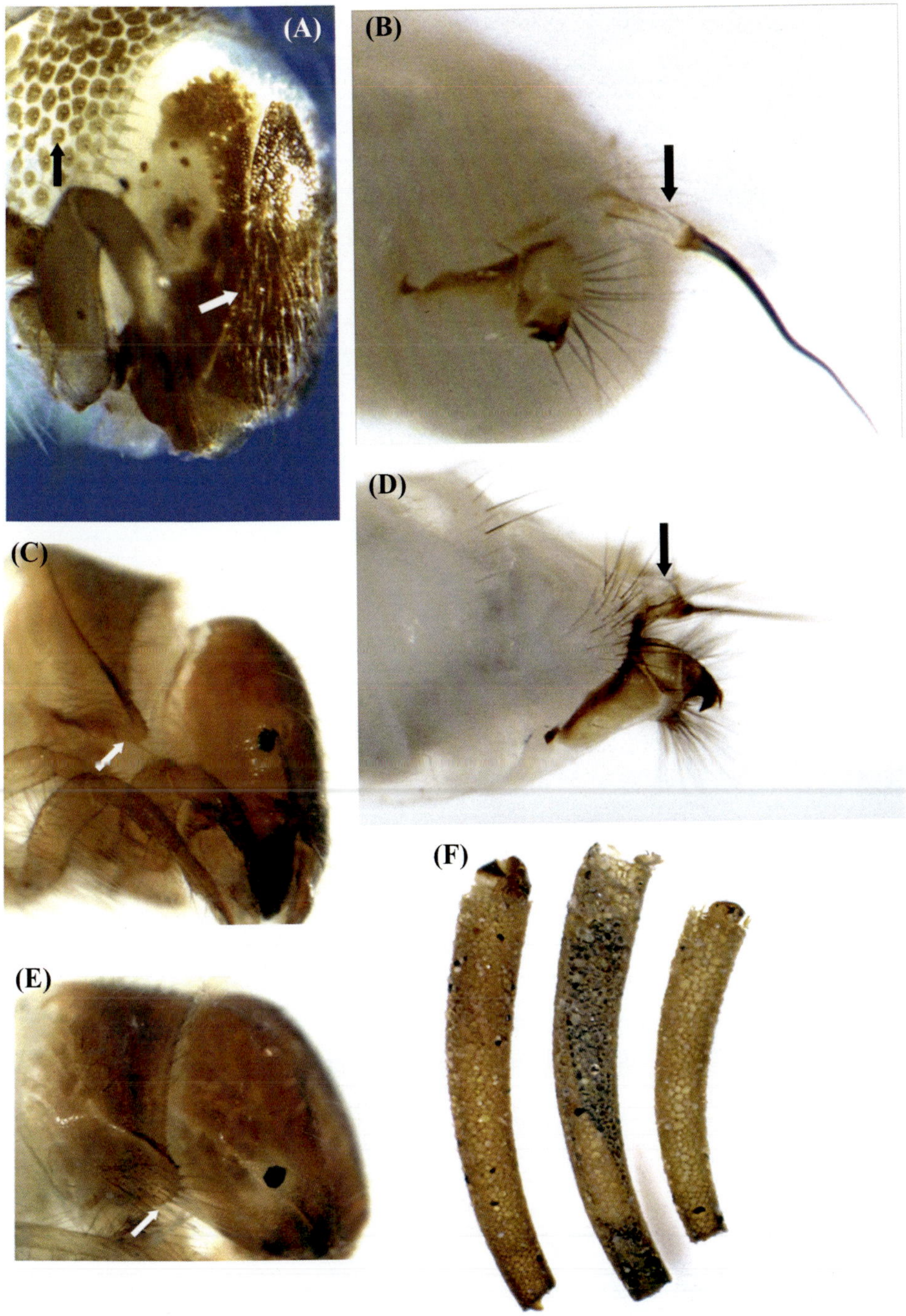

FIGURE 14.57 Beraeidae; (a) *Beraeodes minutus*, head and pronotum, lateral view; (b) *Ernodes vicinus*, tip of abdomen, lateral view; (c) *Ernodes vicinus*, head and pronotum, lateral view; (d) *Beraeamyia hrabei*, tip of abdomen, lateral view; (e) *Beraea dira*, head and pronotum, lateral view; (f) larval cases of *Ernodes articularis* (left), *Beraeamyia hrabei* (middle) and *Beraea maurus* (right); (g) *Beraeamyia schmidi*, head and pronotum, dorsal view; (h) *Beraeamyia schmidi*, head and pronotum, lateral view; (i) *Beraea malahiguerra*, head, dorsal view (photo by Rufino Vieira-Lanero).

Note: Some endemic Iberian species of Beraea cannot be separated from Beraeamyia using these characters: Beraea terrai Malicky, 1975 and B. alva Malicky, 1975 have two carinae, and B. malatebrera Schmid, 1952 and B. malahiguerra Schmid, 1952 have the posterior third and tip of frontoclypeus rounded (Vieira-Lanero et al. 2002; Fig. 14.57i*).*

Trichoptera: Sericostomatidae: Genera

1 Anterolateral corners of pronotum rounded (Fig. 14.59a, arrow); parietal ridge on head dorsally meets coronal suture (Fig. 14.59b, arrow) .... *Notidobia*

FIGURE 14.57 (Continued)

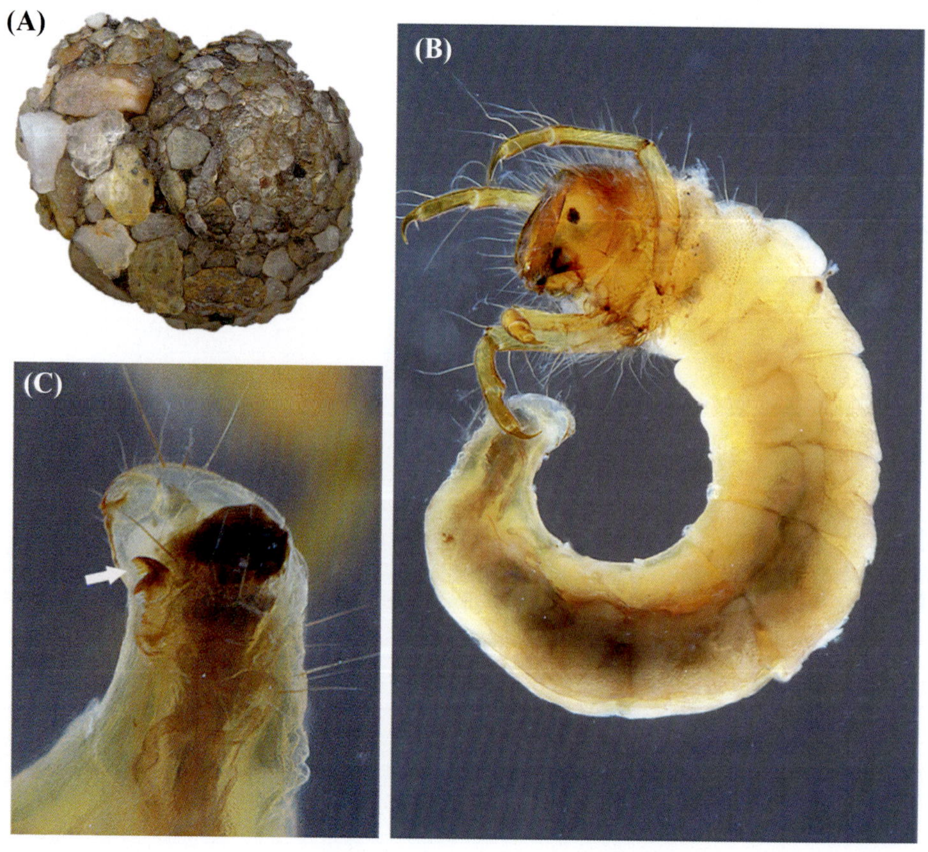

FIGURE 14.58 *Helicopsyche* sp.; (a) larval case, dorsal view; (b) larva, lateral view, (c) anal claw, dorsal view.

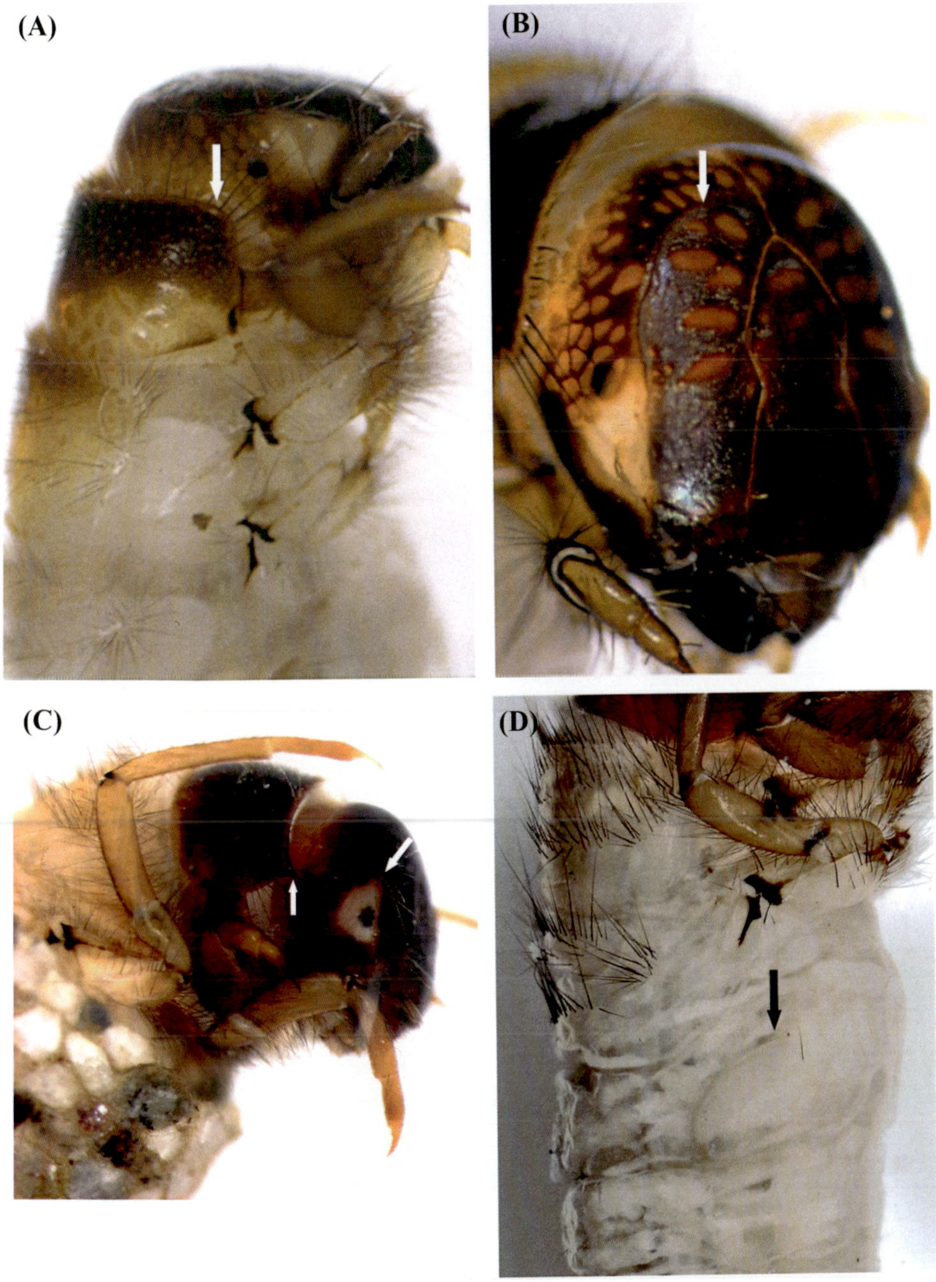

FIGURE 14.59 Sericostomatidae; (a) *Notidobia ciliaris*, head and pronotum, lateral view; (b) *Notidobia ciliaris*, head, dorsal view; (c) *Sericostoma* sp., head and pronotum, lateral view; (d) *Schizopelex huettingeri*, meso-, metanota, and abdominal segments 1 and 2, right lateral view; (e) *Sericostoma* sp., meso-, metanota, and abdominal segments 1 and 2, left lateral view; (f) *Schizopelex huettingeri*, head and pronotum, lateral view; (g) *Schizopelex huettingeri*, coxopleurite, right lateral view; (h) *Oecismus monedula*, head and pronotum, lateral view; (i) *Oecismus monedula*, tip of abdomen, dorsal view.

1' Anterolateral corners of pronotum conically prolonged and pointed or short and knob-like (Fig. 14.59c, f, h); parietal ridge on head short, exists only dorsally of the eye (Fig. 14.59c, arrow) 2

2(1') Lateral protuberance on abdominal segment 1 without comma-shaped black marking (Fig. 14.59d, arrow) 3

2' Lateral protuberance on abdominal segment 1 with comma-shaped black marking (Fig. 14.59e, arrow); abdominal dorsum 9 with 18–41 setae .... *Sericostoma*

3(2') Pronotum with straight ventrolateral border (Fig. 14.59f, line); anterior part of coxopleurite long and corniform (Fig. 14.59g, arrow); abdominal dorsum 9 with 18–41 setae .... *Schizopelex*

3' Pronotum with convex ventral border (Fig. 14.59h, line); coxopleurite without anterior process (Fig. 14.59h, arrow); abdominal dorsum 9 with 48–74 setae (Fig. 14.59i) .... *Oecismus*

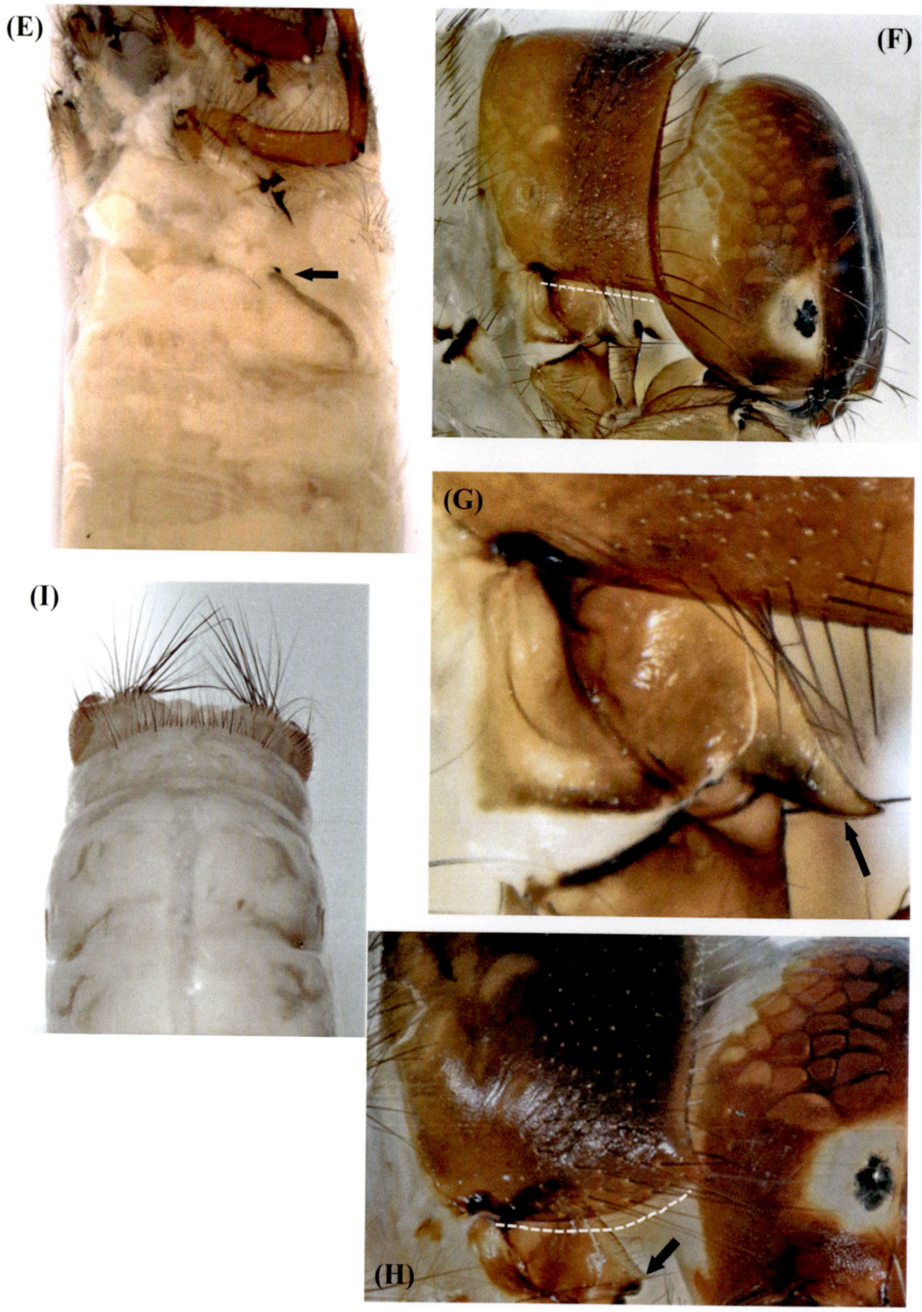

FIGURE 14.59 (Continued)

Trichoptera: Apataniidae: Genera

1 Setae at the anterior edge of pronotum short and dagger-shaped (Fig. 14.60a, arrow) .... *Apatania (partim)*
1' Setae at the anterior edge of pronotum long, tapering and with flexuous tips (Fig. 14.60b, arrow) 2
2 (1') Species from Hellenic Western Balkan (Greece) .... *Apataniana*
2' Species from other Mediterranean areas .... *Apatania (partim)*

Trichoptera: Goeridae: Genera

1 Pronotum almost as long as wide (Fig. 14.61a); mesepisternum without anterior process (Fig. 14.61a, arrow) .... *Larcasia*

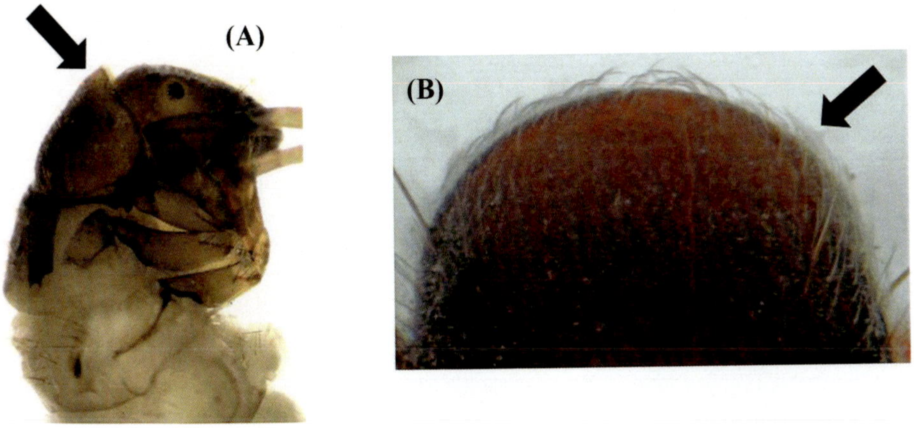

FIGURE 14.60 Apataniidae; (a) *Apatania* sp., head and thorax, lateral view; (b) *Apataniana*, pronotum, dorsal view.

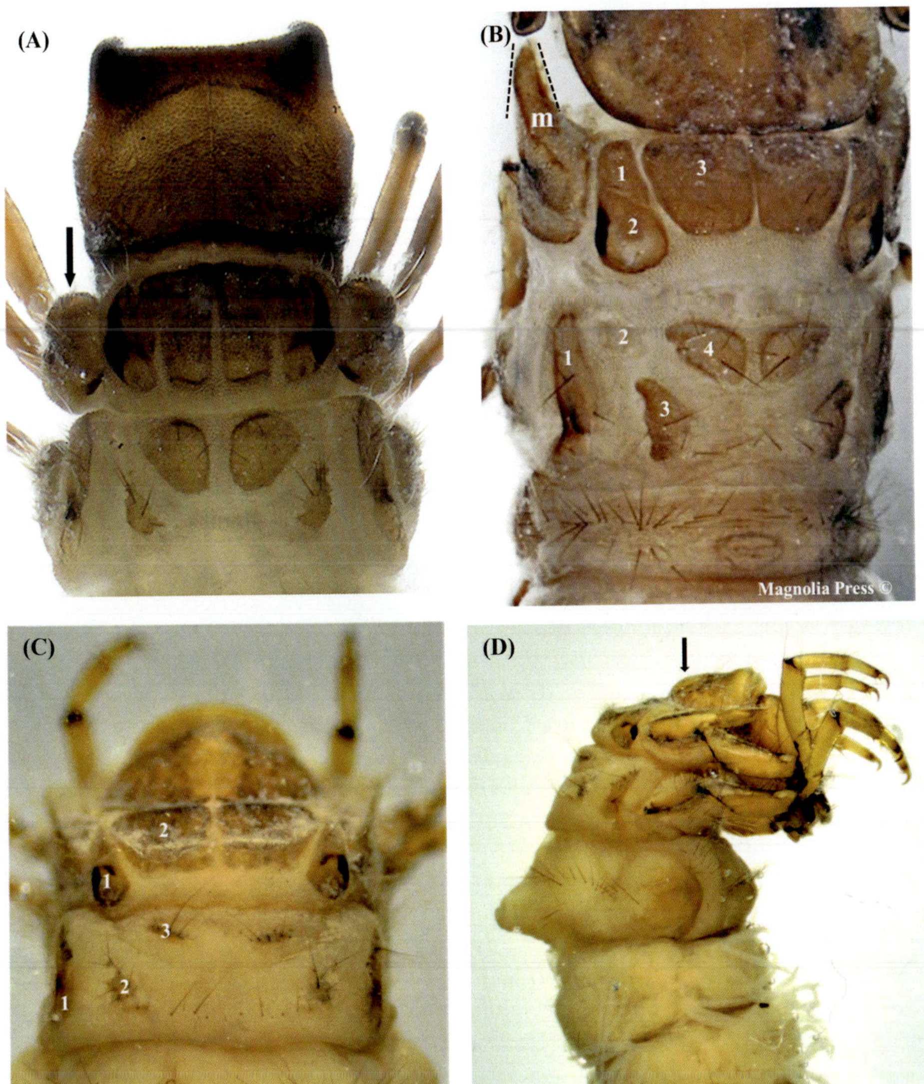

FIGURE 14.61 Goeridae; (a) *Larcasia partita*, pro-, meso and metanotum, dorsal view, mesepisternum (arrow); (b) *Silo chrisiammos*, pro-, meso and metanotum, dorsal view, mesepisternum (m); (c) *Goera pilosa*, meso- and metanotum, dorsal view; (d) *G. pilosa*, larva, right lateral view, pronotum hump (arrow); (e) *Lithax obscurus*, pro-, meso and metanotum, dorsal view; mesepisternum (arrow); (f) *Silonella aurata*, pronotal hump, lateral view; (g) *S. aurata*, pro-, meso and metanotum, dorsal view, mesepisternum (black dots).

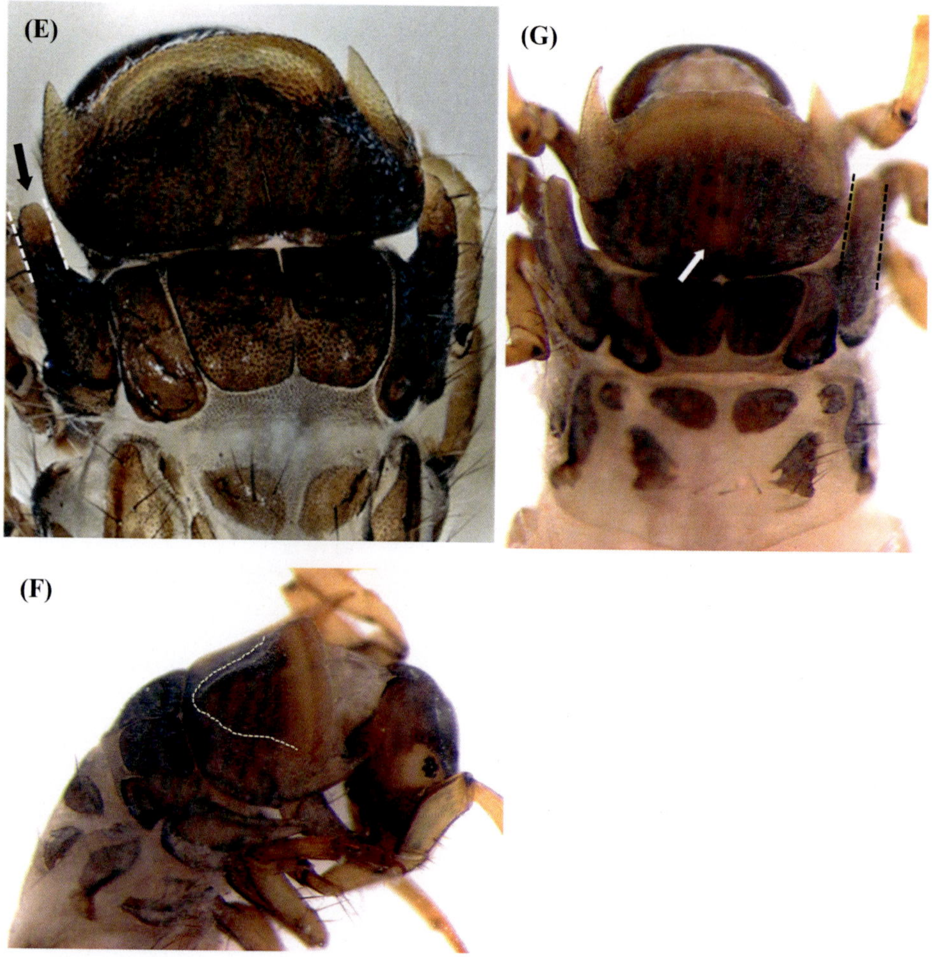

FIGURE 14.61 (Continued)

1' Pronotum wider than long; mesepisternum (m) long, pointed or projecting anteriorly (Fig. 14.61b) 2

Note: Larvae of Larcasia share the same meso- and metanotum sclerites features as Goera (mesonotum with two pairs of sclerites; metanotum with three pairs of sclerites). In addition, mesepisternum, epipleura and mesopleurites are terms used for the same structure throughout the literature.

2(1) Mesonotum with two pairs of sclerites (Fig. 14.61c); metanotum with three pairs of smaller sclerites (Fig. 14.61c); pronotum with one low, pale-colored hump in median part covered with spines (Fig. 14.61d, arrow) .. .. *Goera*

2' Mesonotum with three pairs of sclerites (Fig. 14.61b); metanotum with four pairs of sclerites (Fig. 14.61b); pronotum without central hump or with large central hump 3

3(2') Mesepisternum in dorsal view pointed anteriorly (Fig. 14.61b, dotted lines) .... *Silo* and *Lithax musaca* (Anatolia, European part of Turkey, Bulgaria)

3' Mesepisternum in dorsal view parallel-sided and apically rounded (Fig. 14.61e, dotted lines) 4

4(3) Pronotum with a broad triangular hump (Fig. 14.61f, dotted line); acute point of the triangle projecting towards the posterior part of pronotum (Fig. 14.61g, arrow) .... *Silonella*

4' Pronotum lacking dorsal humps or with a pronotal hump ridge-like, wider anteriorly and gradually descending to the posterior border of the pronotum (Fig. 14.61e) .... *Lithax*

Trichoptera: Limnephilidae: Genera

1 Larvae terrestrial; gills are lacking in all instars .... *Enoicyla*

1' Larvae aquatic; final instars with gills (Fig. 14.62a, d) 2

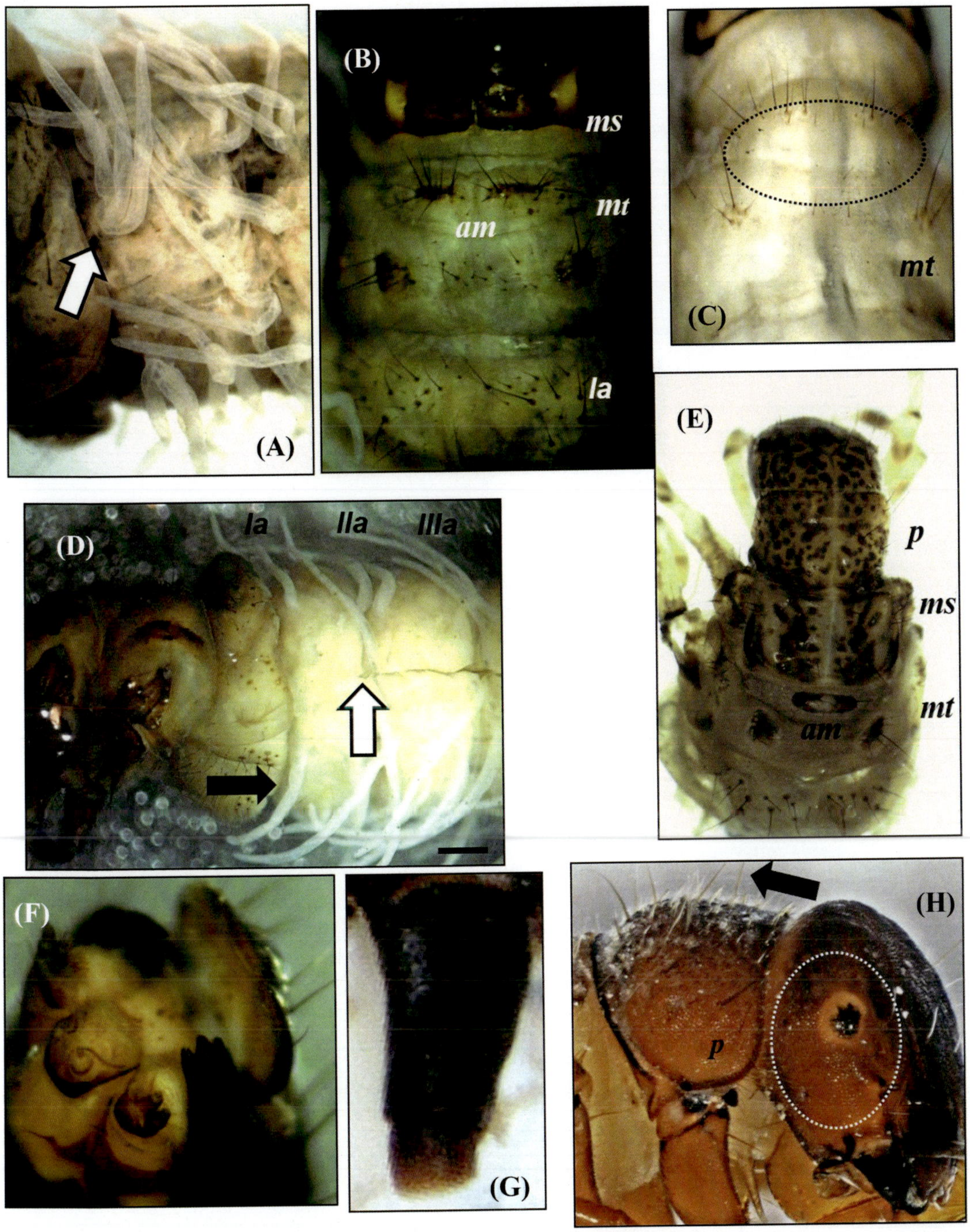

FIGURE 14.62 Limnephilidae; (a) *Grammotaulius nigropunctatus*, left lateral side of abdominal segment 1 showing gills; (b) *Stenophylax nycterobius*, thorax and abdominal segment 1, dorsal view; (c) *Limnephilus extricatus*, head and thorax, dorsal view; (d) *Stenophylax nycterobius*, thorax and abdominal segments 1–3, left lateral view; (e) *Hydatophylax infumatus*, head and thorax, dorsal view; (f) *Stenophylax nycterobius*, mandibles with teeth; (g) *Drusus* sp., mandibles without teeth; (h) *Drusus trifidus*, head and pronotum, lateral view; (i) *Metanoea* sp., first abdominal sternum, ventral view; (j) *Anomalopterygella chauviniana*, pronotum, lateral view; (k) *Drusus chrysotus*, metanotum, dorsal view; (l) *Stenophylax nycterobius*, femur of mesothoracic leg, anterior face; (m) *Drusus melanchaetes*, head, frontal view; (n) *Drusus camerinus*, head, frontal view; (o) *Drusus adustus*, hindtibia; (p) *Drusus biguttatus*, hindtibia; (q) *Drusus biguttatus*, abdominal segments 1–3, left lateral view; (r) *Drusus franzressli*, metanotum and abdominal segments 1–2, right lateral view; (s) *Annitella iglesiasi*, abdominal segments 1–2, right lateral view; (t) *Stenophylax nycterobius*, abdominal segment 1, left lateral view; (u) hind femora, lateral view; modified from Waringer & Graf (2013, p127); (v) *Drusus discolor*, head and pronotum, lateral view; (w) *Cryptothrix nebulicola*, head, frontal view; (x) *Stenophylax permistus*, head; (y) *Annitella iglesiasi*, head, right lateral view; (z) *Stenophylax nycterobius*, tip of abdomen, ventral view; (aa) *Allogamus ligonifer*, pronotum, dorsal view; (ab) *Potamophylax rotundipennis*, pronotum, dorsal view; (ac) *Acrophylax zerberus*, metanotum, dorsal view; (ad) *Consorophylax styriacus*, pronotum, dorsal view; (ae) *Chaetopterygopsis maclachlani*, abdominal sternum 1, ventral view; (af) *Annitella iglesiasi*, head and thorax, dorsal view; (ag) *Stenophylax nycterobius*, tergite IX, dorsal view; (ah) *Mesophylax* sp; (ai) *Limnephilus lunatus;* (aj) *Anabolia furcata*, head, dorsal view; (ak) *Drusus monticola*, head, dorsal view; (al) *Limnephilus rhombicus*, head, dorsal view; (am) *Anabolia nervosa*, larval case; (an) *Anabolia brevipennis*, forefemur; (ao) *Glyphotaelius pellucidus*, forefemur; (ap) *Anabolia brevipennis*, larval case; (aq) *Glyphotaelius pellucidus*, larval case; (ar) (1) *Limnephilus vittatus*, (2) *Rhadicoleptus alpestris*, larval cases*;* (as) *Rhadicoleptus alpestris*, tip of abdomen; black arrow: stout spines.

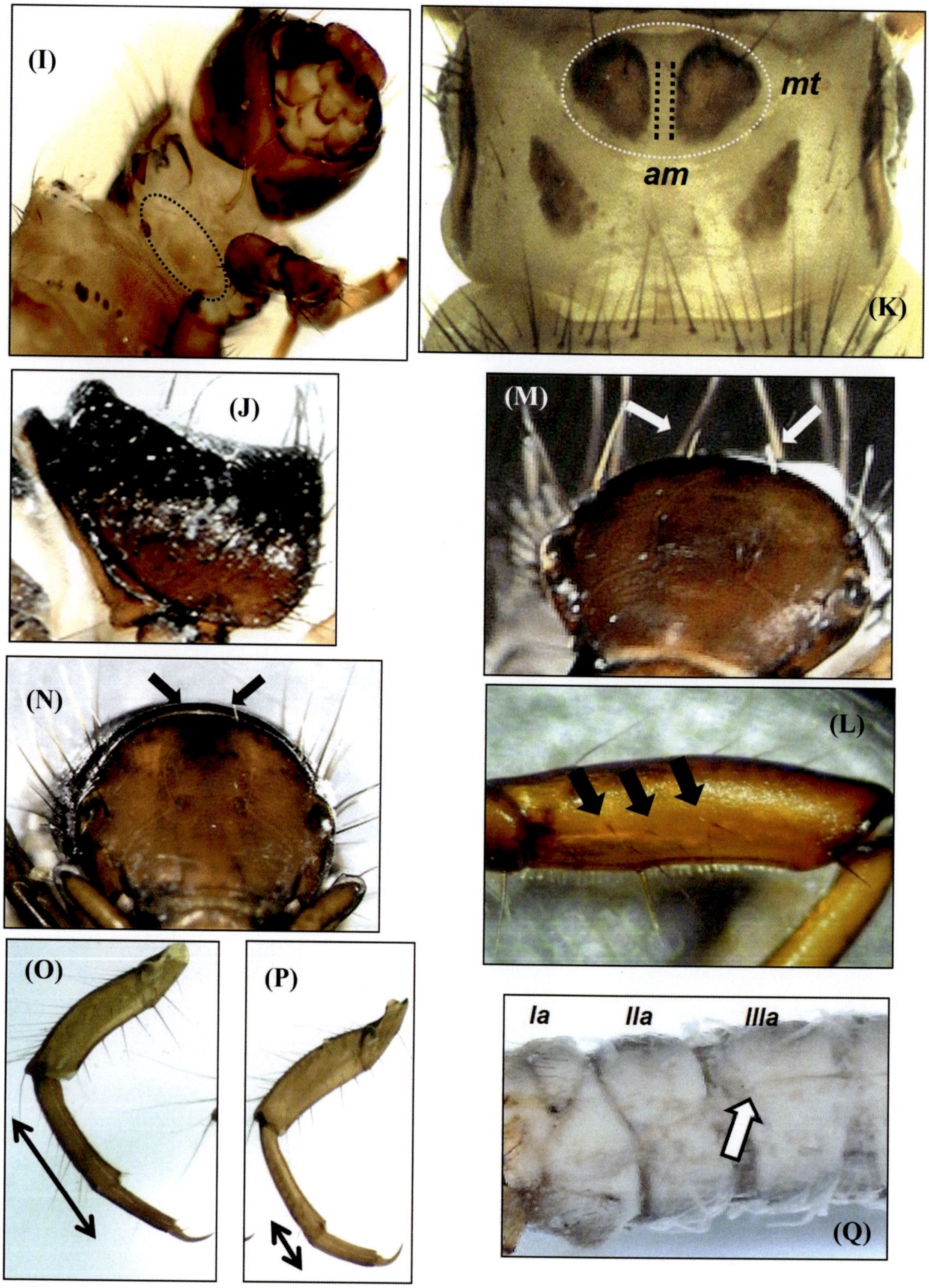

FIGURE 14.62 (Continued)

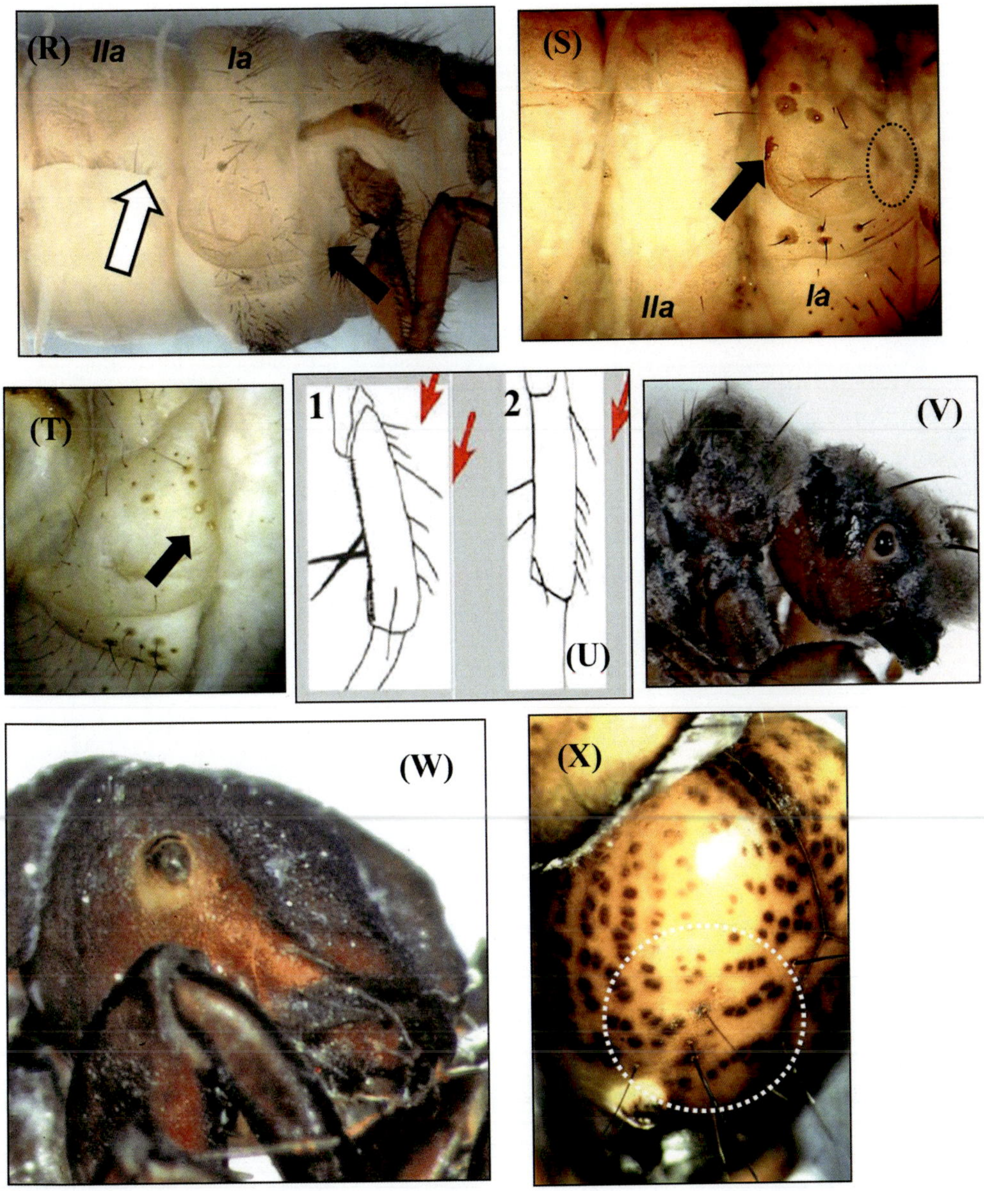

FIGURE 14.62 (Continued)

FIGURE 14.62 (Continued)

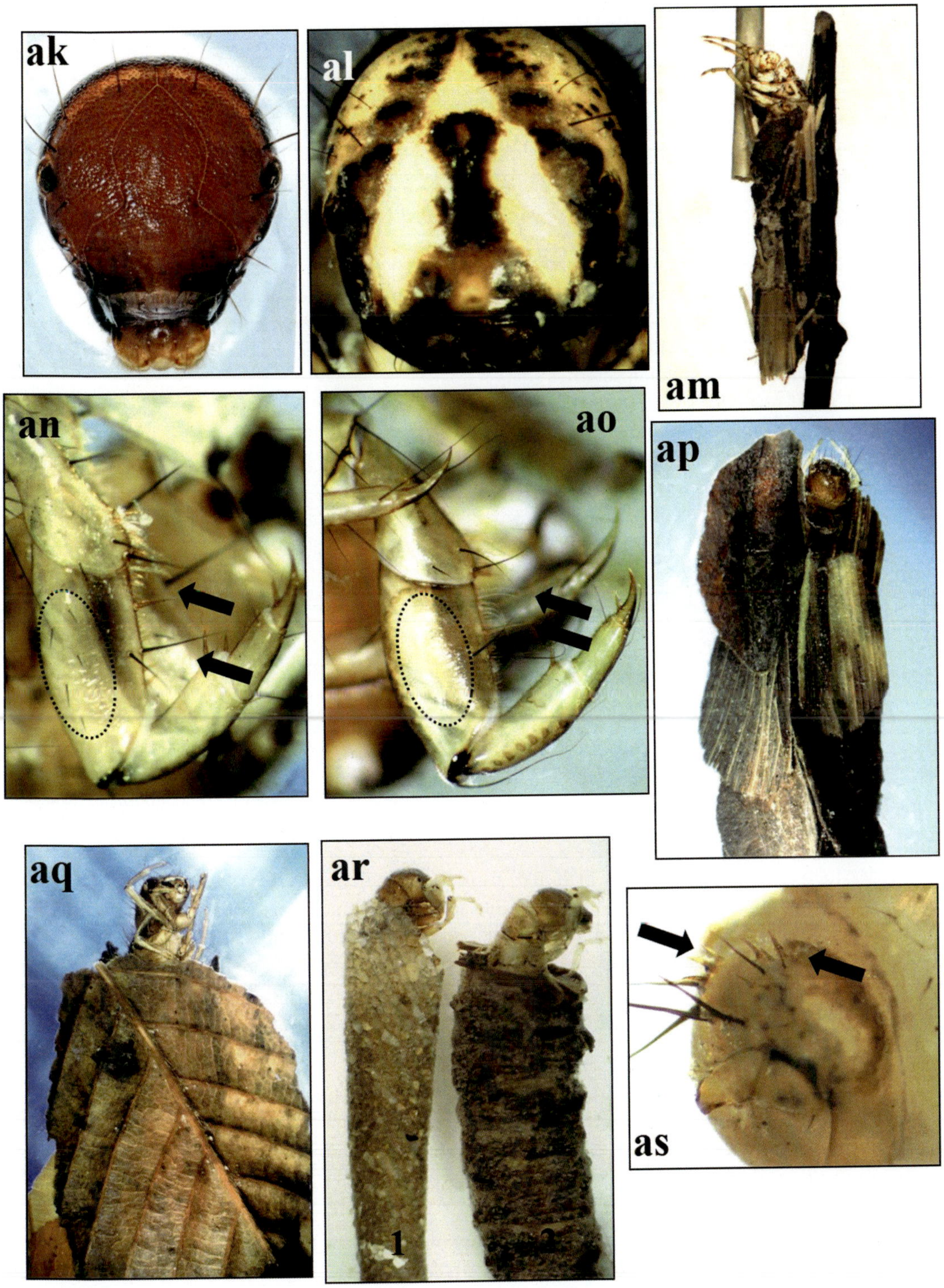

FIGURE 14.62 (Continued)

2(1') Metanotum covered by five or six small sclerites (Fig. 14.62b, *mt*) 4
2' Anterior and sometimes posterior metanotal sclerites replaced by setal groups (Fig. 14.62c, area surrounded by ring and arrows) 3
3(2') All gills consisting of single filaments (Fig. 14.62d, black arrow) .... *Parachiona*
Note: One species is known for the Mediterranean Basin, Parachiona picicornis.
3' Gills consisting of two to four filaments and additional single filaments (Fig. 14.62a, arrow) .... *Limnephilus (partim)*
Note: In *Limnephilus species keyed out here each posteromedian metanotal sclerite is replaced by more than two setae.*
4(2) Metanotum (Fig. 14.62e, *mt*) with one pair of lateral sclerites, one pair of posteromedian sclerites, and a single anteromedian sclerite (*am*); first abdominal sternum with two large sclerites bearing setae .... *Hydatophylax*
Note: One species is known for the Mediterranean Basin, Hydatophylax infumatus.
4' Metanotum (Fig. 14.62b, *mt*) with three pairs of sclerites: one pair of lateral sclerites, one pair of posteromedian sclerites, and one pair of anteromedian sclerites (*am*) 5
5(4') Cutting edge of mandibles with teeth (Fig. 14.62f) 18
5' Cutting edge of mandibles without teeth (Fig. 14.62g) 6
6(5') In addition to basal setation, spines or bristles are present at the parietalia around the eyes .... *Ecclisopteryx (partim), Drusus (partim)*
6' Without such additional spines or bristles; very small spinules (0.03 mm long) may be present 7
7(6') First abdominal sternum with a large homogeneous sclerotized patch (Fig. 14.62i, area surrounded by ring); pronotum finely granulated .... *Metanoea*
7' Without large sclerotized patch at the first abdominal sternum, but a number of fused setal bases may be present; pronotum coarsely granulated 8
8(7') Pronotum with a distinct dorsal ridge (Fig. 14.62j) 9
8' Pronotum rounded and without a distinct sharp ridge (Fig. 14.62h, *p*) 10
9(8) Pronotum with sharp ridge, which does not extend to the anterior pronotal margin .... *Drusus (partim); Ecclisopteryx (partim)*
9' Pronotum with ridge extending to the anterior pronotal margin (Fig. 14.62j) .... *Drusus (partim)*; *Anomalopterygella*
Note: One species is known for the Mediterranean Basin, Anomalopterygella chauviniana from the Iberian Peninsula and France.
10(8') Anterior metanotal sclerites shorter than their median separation (Fig. 14.62b, *am*) .... *Micropterna (partim)*
10' Anterior metanotal sclerites longer than their median separation (Fig. 14.62k, *mt*) 11
11(10') Setae are present at both faces of mid- and hindfemur (Fig. 14.62l, arrows) 12
11' Setae are lacking at both faces of mid- and hindfemur 17
12(11) With bristles near the middle of the anterior pronotal margin (Fig. 14.62m, arrows) 13
12' Without bristles near the middle of the anterior pronotal margin (Fig. 14.62n, arrows) .... *Drusus (partim)*
13(12) Dorsal-edge setae at the entire length of mid- and hindtibia (Fig. 14.62o, arrow) .... *Drusus (partim)*
13' Dorsal-edge setae only at the distal third of mid- and hindtibia (Fig. 14.62p, arrow) 14
14(13') Lateral fringe starting at abdominal segment 3 (Fig. 14.62q, white arrow) 15
14' Lateral fringe starting at abdominal segment 2 (Fig. 14.62d, r, white arrows) 16
15(14) Lateral fringe starting at the posterior third of abdominal segment 3 .... *Ecclisopteryx (partim)*
15' Lateral fringe starting at the anterior margin of abdominal segment 3 (Fig. 14.62q, white arrow) .... *Drusus (partim)*
16(14') Lateral fringe starting at the anterior third of abdominal segment 2 (Fig. 14.62r, white arrow) .... *Hadimina*
Note: One species is known for the Mediterranean Basin, Hadimina torosensis, endemic of Turkey.
16' Lateral fringe starting at the posterior third of abdominal segment 2 (Fig. 14.62d, white arrow) .... *Drusus (partim), Leptodrusus*
Note: Leptodrusus budtzi is endemic of Sardinia, Corsica, and the Balearic Isles. It has a distinct rim at the end of the anterior third of the pronotum, which is lacking in Drusus species keyed here.
17(11') Anterior section of lateral fleshy hump of the abdominal segment 1 with a setal band (Fig. 14.60r, black arrow) .... *Allogamus (partim)*
17' Anterior section of lateral fleshy hump without setal band (Fig. 14.62s, area surrounded by a ring) .... *Annitella (partim), Melampophylax (partim)*

18(5) All gills consisting of single filaments (Fig. 14.62d, black arrow) 19
18' Gills consisting of two to four filaments and additional single filaments (Fig. 14.62a, white arrow) 35
19(18) Posterior section of lateral hump of the abdominal segment 1 with 0–3 sclerites, without setae and with central hole (Fig. 14.62t, black arrow) 20
19' Posterior section of lateral hump with one large sclerite without setae and with one to three holes (Fig. 14.62s, black arrow) 24
20(19) Hindfemur with more than one proximodorsal seta (Fig. 14.62u1) 21
20' Hindfemur with one proximodorsal seta (Fig. 14.62u2) 22
21(20) Head flattened or head and pronotum with a dense layer of woolly hairs (Fig. 14.62v) .... *Drusus (partim)*
21' Head rounded; head and pronotum without a dense layer of woolly hairs; frontoclypeal suture running in a rim-like depression (Fig. 14.62w) .... *Cryptothrix*
Note: One species is known for the Mediterranean Basin, Cryptothrix nebulicola from Italy.
22(20') Anterior face setae present at the ventral third of the mid- and hindfemur 23
22' Without additional setae at the ventral third of the mid- and hindfemur .... *Potamophylax (partim), Stenophylax (partim)*
Note: One species is known for the Mediterranean Basin, Stenophylax vibex.
23(22) Head with spinules restricted to two small areas behind the eyes (Fig. 14.62x, area surrounded by a ring) .... *Stenophylax (partim)*
Note: Stenophylax permistus, S. mitis
23' Head spinules covering large areas of parietalia, behind eyes and frontoclypeus (Fig. 14.62y, area surrounded by a ring) .... *Micropterna (partim), Stenophylax (partim)*
Note: Stenophylax crossotus (Head spinules covering large areas of parietalia behind and above each eye), S. espanioli, S. nycterobius, S. sequax, and S. fissus (Head spinules covering large areas of parietalia and frontoclypeus).
24(19') Hindfemur with more than one proximodorsal seta .... *Halesus (partim)*
24' Hindfemur with one proximodorsal seta 25
25(24') Abdominal segment 9 with more than one posterolateral setae 26
25' Abdominal segment 9 with only one posterolateral setae (Fig. 14.62z, black arrow). 28
26(25) Anterior third of pronotum lighter than posterior area (Fig. 14.62aa) .... *Allogamus (partim)*
26' Pronotum concolorous (Fig. 14.62ab) 27
27(26') Anterior metanotal sclerites longer than their median separation (Fig. 14.62ac, *am*) .... *Acrophylax*
Note: One species is known for the Mediterranean Basin, Acrophylax zerberus.
27' Anterior metanotal sclerites shorter than their median separation (Fig. 14.62b, *am*) .... *Potamophylax (partim)*
28(25') Anterior third of pronotum darker than posterior area (Fig. 14.62ad) .... *Consorophylax*
28' Pronotum concolorous (Fig. 14.62ab) 29
29(28') Anterior section of lateral fleshy protuberance of the abdominal segment I with a setal band (Fig. 14.62r, black arrow) .... *Allogamus (partim)*
9' Setal band lacking at this position (Fig. 14.62s, area surrounded by ring) 30
30(29') First abdominal sternum with large and often fused setal bases (Fig. 14.62ae, area surrounded by ring); case made of watermoss leaves .... *Chaetopterygopsis*
30' First abdominal sternum with small and isolated setal bases 31
31(30') Anterior metanotal sclerites longer than their median separation (Fig. 14.62ac, *am*) 32
31' Anterior metanotal sclerites shorter than their median separation (Fig. 14.62b, *am*) 34
32(31) Inner margins of anteromedian sclerites divergent; sclerite margins irregular (Fig. 14.62af, *am*) .... *Annitella (partim)*
32' Inner margins of anteromedian sclerites almost parallel (Fig. 14.62k, *am*); sclerite margins smooth 33
33(32') Posterior section of dorsal fleshy protuberance of the abdominal segment I with setae .... *Allogamus (partim), Alpopsyche*
Note: One species is known for the Mediterranean Basin, Alpopsyche ucenorum, restricted to Italy.
33' Setae lacking at this position .... *Melampophylax (partim)*
34(31') Dorsal sclerite of abdominal segment 9 with one central seta .... *Chaetopteryx*
34' Dorsal sclerite of abdominal segment 9 with two central setae (Fig. 14.62ag) .... *Halesus (partim)*
35(18') Additional setae present at least at one face of mid- and hindfemur 36
35' Without additional face setae on mid and hindfemur 42
36(35) With three or more strong ventral edge setae on midfemur (Fig. 14.62ah) .... *Mesophylax*

Note: Species keyed out here, tibia with only one apical pair of stout setae. M. aspersus (forefemur with three ventral edge setae) and M. impunctatus (forefemur with only two ventral edge setae). Mesophylax impunctatus aduncus (Navás 1923) (Turkey and southern Balkan peninsula) lacks terminal teeth on mandibles.

36' With two strong ventral edge setae on midfemur (Fig. 14.62ai); additional slender setae may be present 37

37(36') Frontoclypeus with dark mushroom pattern (Fig. 14.62aj); case tubular with long twigs (Fig. 14.62am) .... *Anabolia (partim)*

Note: These features are known for Anabolia furcata and A. nervosa.

37' Frontoclypeus with different pattern (Fig. 14.62ak, al) 38

38(37') Ventral edge setae of fore femur contrasting in color (pale/dark) (Fig. 14.62an, arrows) 39

38' Both ventral edge setae pale (Fig. 14.62ao, arrows) 40

39(38) First femur with additional face setae (Fig. 14.62an, area surrounded by a ring); case cross-section triangular (Fig. 14.62ap) .... *Anabolia (partim)*

Note: These features are known only for Anabolia brevipennis.

39' First femur without additional face setae (Fig. 14.62ao, area surrounded by a ring); case flattened, made from large circular leaf pieces (Fig. 14.62aq) .... *Glyphotaelius*

Note: One species is known for the Mediterranean Basin, Glyphotaelius pellucidus.

40(38') Anterior third of pronotum darker than remaining pronotal surface (Fig. 14.62ad) .... *Limnephilus (partim)*

40' Pronotum concolorous (Fig. 14.62ab); transverse rim may be darker (Fig. 14.62ab, arrow) 41

41(40') Case straight and with triangular cross section (Fig. 14.62, ap) .... *Anabolia (partim)*

41' Case curved or straight with circular cross section (Fig. 14.62ar) .... *Limnephilus (partim)*

42(35') With squat, yellow setae on lateral sclerite of anal proleg (Fig. 14.62as, black arrows) .... *Rhadicoleptus*

Note: One species is known for the Mediterranean Basin, Rhadicoleptus alpestris.

42' Without squat setae on lateral sclerite 43

43(42') Ventral setae at mid- and hindfemur contrasting in color (pale/dark) 44

43' Both ventral setae at mid- and hindfemur dark 45

44(43) Case curved (Fig. 14.62ar2) .... *Limnephilus (partim)*

44' Case straight (Fig. 14.62ar1) .... *Limnephilus (partim), Grammotaulius (partim)*

Note: One species is known for the Mediterranean Basin, Grammotaulius nitidus.

45(43') Head with parietal and frontoclypeal bands [Fig. 14.62al .... *Limnephilus (partim)*]

45' Head without distinct bands (Fig. 14.62ak); small pale areas near the frontoclypeal suture may be present 46

46(45') Lateral fringe start on abdominal segment 2 (Fig. 14.62r) .... *Limnephilus (partim)*

46' Lateral fringe start on abdominal segment 3 (Fig. 14.62q) .... *Grammotaulius (partim); Limnephilus (partim)*

Note: In *Grammotaulius nigropunctatus and G. submaculatus*, the head width in last instars is wider than 1.9 mm.

Trichoptera: Brachycentridae: Genera

1 Mid and hind femur twice as long as the tibia (Fig. 14.64a); apex of tibia of mid and hind legs with a ventral prolongation bearing spinous setae (Fig. 14.64a, arrows); abdominal gills present dorsally and ventrolaterally; case made of sand grains, detritus, tetrahedral, or made of secretion, brown to dark brown, with irregular tetrahedral, rarely round cross section (Fig. 14.64c, left) .... *Brachycentrus*

1' Mid and hind femur as long as the tibia (Fig. 14.64b); tibia of mid and hind legs without a ventral prolongation (Fig. 14.64b); abdominal gills absent, rarely small and single; case smooth, conical, made of fine sand grains, plant fragments, rarely of secretion (as in *Micrasema longulum*) (Fig. 14.64c, right) .... *Micrasema*

Trichoptera: Lepidostomatidae: Genera

1 Anteromedian and posteromedian sclerites of metanotum each with one seta (Fig. 14.65a, arrows) .... *Lepidostoma*

1' Anteromedian sclerites of metanotum with one seta each; posteromedian sclerites with more than one seta each (Fig. 14.65b, arrows) .... *Crunoecia*

Trichoptera: Phryganeidae: Genera

1 Posterior section of frontoclypeus with dark, horseshoe-shaped pigmentation (Fig. 14.66a, arrow) ...... *Oligostomis*

1' Without such pattern on frontoclypeus 2

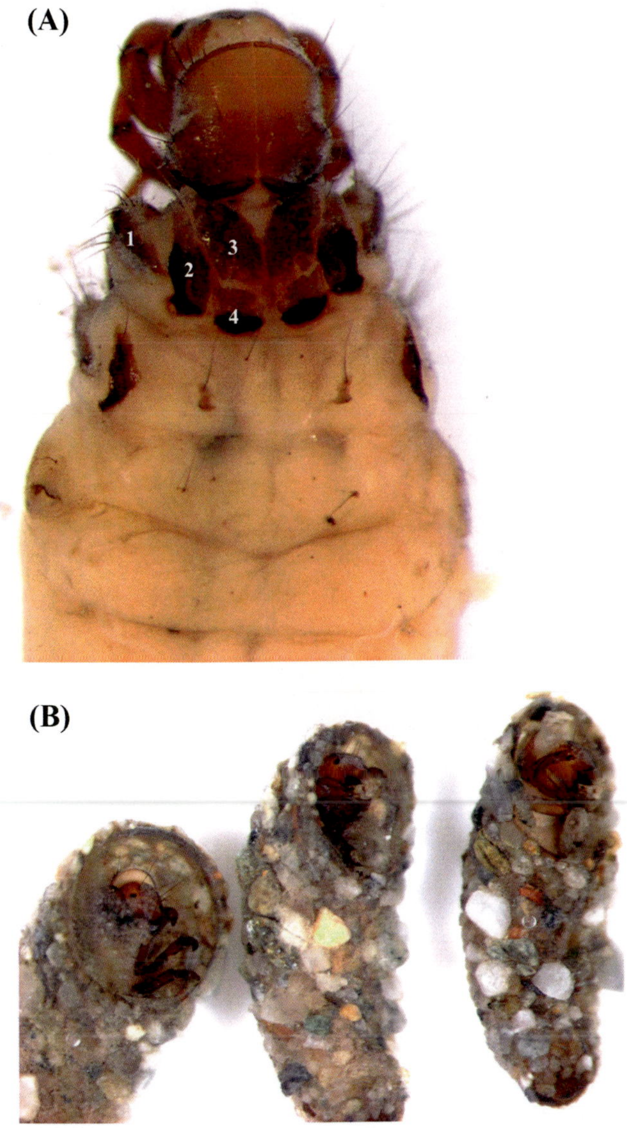

FIGURE 14.63 *Thremma anomalum*; (a) head and pronotum, dorsal view; (b) larval cases, ventrolateral view.

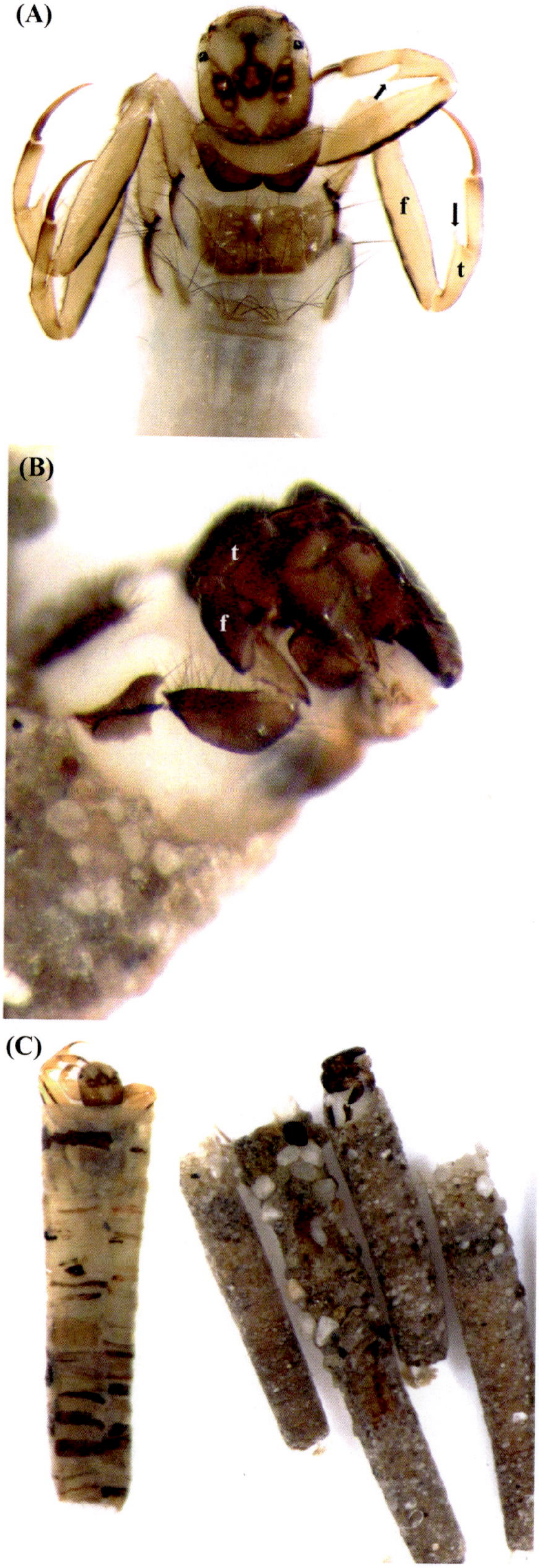

FIGURE 14.64 Brachycentridae; (a) *Brachycentrus subnubilus*, head, mid-, and hindlegs, dorsal view; f: femur, t: tibia; (b) *Micrasema* sp., head, mid-, and hindlegs, lateral view; (c) larval cases of *Brachycentrus subnubilus* (left), *Micrasema* sp. (right).

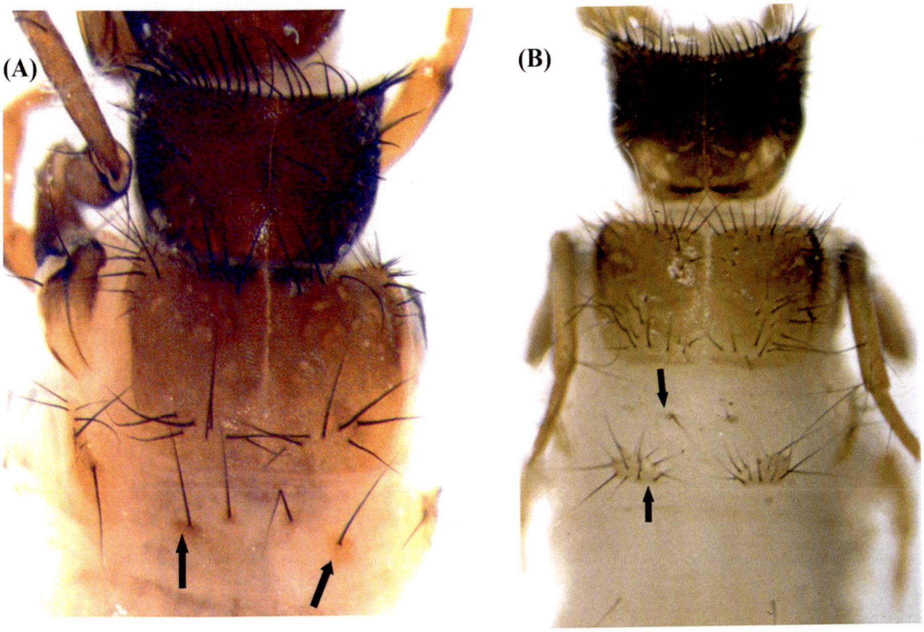

FIGURE 14.65 Lepidostomatidae; (a) *Lepidostoma doehleri*, head and thorax, dorsal view; (b) *Cruenocia irrorata*, head and thorax, dorsal view.

FIGURE 14.66 Phryganeidae; (a) *Oligostomis reticulata*, head, dorsal view, horseshoe pigmentation; (b) *Oligotricha striata*, head, thorax, and abdominal segments, dorsal view; (c) *Hagenella clathrata*, head, dorsal view; (d) *Agrypnia varia*, head and pronotum dorsal view; (e) *Trichostegia minor*, head, thorax, and abdominal segments, lateral view, lateral protuberance, arrow; (f) *A. varia*, head, thorax, and abdominal segments, lateral view, dorsal and lateral protuberances dotted circles; (g) *A. varia*, head and thorax, ventral view, prosternum sclerite, arrow; (h) *Phryganea grandis*, head and thorax, ventral view.

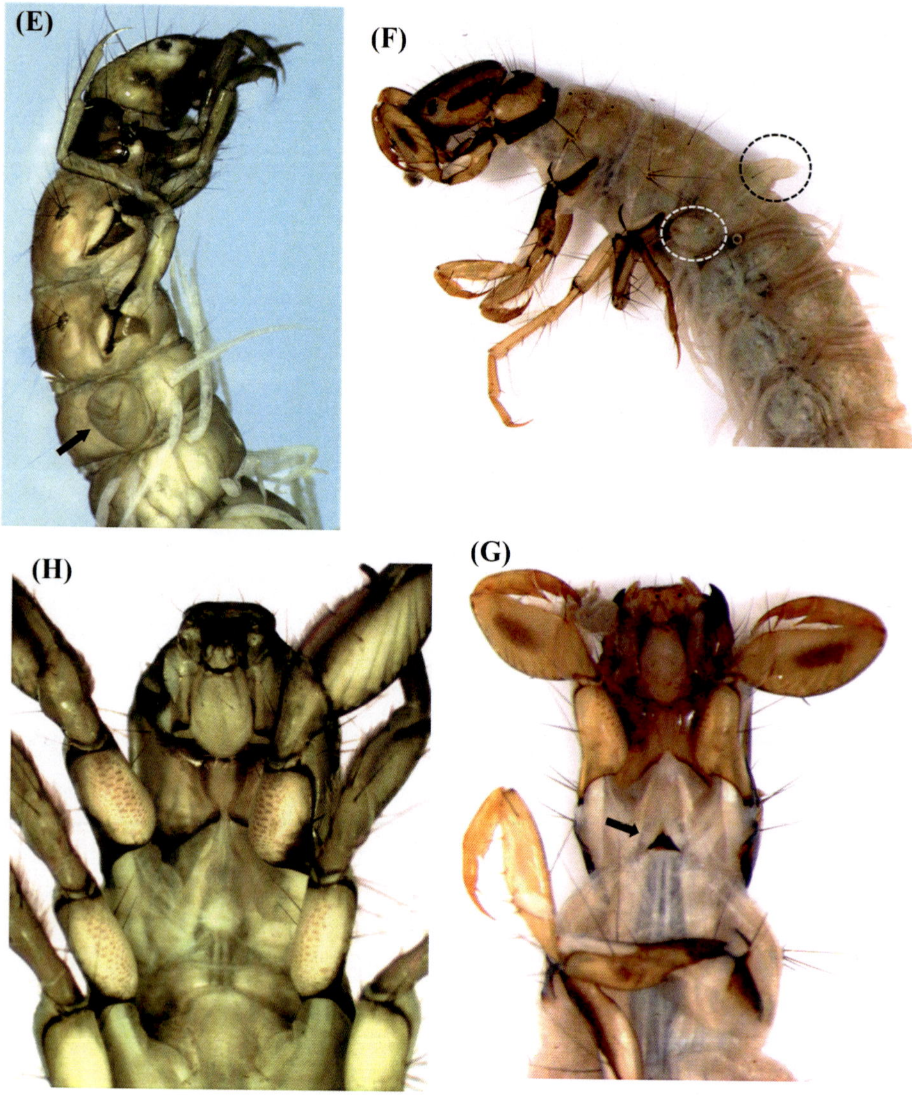

FIGURE 14.66 (Continued)

2(1') Dorsal surface of head, thorax and abdominal segment I with two dark (purple color) longitudinal bands (Fig. 14.66b) .... *Oligotricha*
2' Without dark longitudinal bands on thorax and abdomen 3
3(2') Dorsal surface of head plain brown without dark bands (Fig. 14.66c) .... *Hagenella*
3' Dorsal surface of head yellow or brown with conspicuous dark bands (Fig. 14.66d) 4
4(3') Abdominal segment 1 with only lateral protuberances (Fig. 14.66e, arrow) .... *Trichostegia*
4' Abdominal segment 1 with dorsal and lateral protuberances (Fig. 14.66f, dotted circles) 5
5(4') Prosternum with a small sclerite (may be hidden in the intersegmental fold) (Fig. 15.66g, arrow) .... *Agrypnia*
5' Prosternum without sclerite (Fig. 15.66h) .... *Phryganea*

Acknowledgments

We are grateful to Rufino Vieira-Lanero, Maxence Forcellini, Henri Tachet, Edyta Buczyńska, and Wolfram Graf for providing larval photos for several species and to Pedro Sandoval for taking stereoscopic photos for some species. We benefited from the stereoscopic system for high-precision stacked photography provided by the Spanish Ministerio de Ciencia, Innovación y Universidades funds (Project EQC2018-004655-P). We also thank the editors of this book, Alain Maasri and James H. Thorp, for their valuable comments and suggestions.

Uncited references

Holzenthal (2009); Johanson, Espeland (2010).

References

Aiken, R.B. 1985. Sound production by aquatic insects. Biological Reviews 60(2): 163–211.

Botosaneanu, L. 1974. Notes descriptives, faunistiques, écologiques, sur quelques trichoptères du ''trio subtroglophile'' (Insecta: Trichoptera). Travaux de l'Institut de Speologie ''Emile Racovitza'' 13: 61–75.

Bouvet, Y. 1975. Les Trichoptères du groupe de Stenophylax: conditions de vie et réactions aux variations des facteurs du milieu. Annales de Spéléologie 30: 207–229.

Décamps, H. 1967. Introduction à létude écologique des Trichoptères des Pyrénées. Annales de Limnologie 3: 101–176.

De Moor, F.C. & V.D. Ivanov. 2008. Global diversity of caddisflies (Trichoptera: Insecta) in freshwater. Hydrobiologia 595: 393–407.

Domisch, S., M.B. Araújo, N. Bonada, S.U. Pauls, S.C. Jähnig, & P. Haase. 2013. Modelling distribution in European stream macroinvertebrates under future climates. Global Change Biology 19: 752–762.

Chamorro, M.L. & R.W. Holzenthal. 2011. Phylogeny of Polycentropodidae Ulmer, 1903 (Trichoptera: Annulipalpia: Psychomyioidea) inferred from larval, pupal and adult characters. Invertebrate Systematics 25(3): 219–253.

Coppa, G. & H. Tachet. 2010. *Calamoceras marsupus* Brauer 1865 (Trichoptera, Calamoceratidae) in France: a rediscovered species. Denisia 29: 105–113.

Darılmaz, M.C. & A. Salur. 2015. Annotated catalogue of the Turkish caddisflies (Insecta: Trichoptera). Munis Entomology & Zoology 10: 521–734.

Evtimova, V.V. & L.A. Kenderov. 2016. First confirmed record of the larva of *Calamoceras illiesi* Malicky and Kumanski, 1974 (Trichoptera) from Europe, with notes on its morphology. Aquatic Insects 37(4): 317–326.

Graf, W.J. Murphy, J. Dahl, C. Zamora-Muñoz & M.J. López-Rodríguez. 2008. Distribution and ecological preferences of European freshwater organisms. Vol. 1. Trichoptera. A. Schmidt-Kloiber and D. Hering (Eds). Pensoft Publishers, Sofia-Moscow, 388 pp.

Hering, D., A. Schmidt-kloiber, J. Murphy, S. Lücke, C. Zamora-Muñoz, M. López-Rodríguez, T. Huber & W. Graf. 2009. Potential impact of climate change on aquatic insects: A sensitivity analysis for European caddisflies (Trichoptera) based on distribution patterns and ecological preferences. Aquatic Sciences 71: 3–14.

Frandsen, P.B. & R.E. Thomson. 2016. Collection and preservation of Trichoptera for use in DNA research. Zoosymposia 10: 200–202.

Holzenthal, R.W. 2009. Trichoptera (Caddisflies). pp 456–467 in: G.E. Likens (ed.), Encyclopedia of Inland Waters, Vol. 2, Elsevier, Oxford.

Holzenthal, R.W., & T. Andersen. 2004. The caddisfly genus *Trianeodes* in the Neotropics (Trichoptera: Leptoceridae). Zootaxa, 511: 1–80.

Holzenthal, R.W. & A.R. Calor. 2017. Catalog of the Neotropical Trichoptera (Caddisflies). ZooKeys 654: 1–566.

Holzenthal, R.W., R.J. Blahnik, A.L. Prather, K.M. Kjer. 2007. Order Trichoptera Kirby, 1813 (Insecta), Caddisflies. Zootaxa 1668: 639–698.

Holzenthal, R.W., J.C. Morse, K.M. Kjer. 2011. Order Trichoptera Kirby, 1813. Zootaxa 3148: 209–211.

Holzenthal, R.W., R.E. Thomson & B. Ríos-Touma. 2015. Order Trichoptera. Ecology and General Biology, Vol. I. pp 965–1002 in: J.H. Thorp & D.C. Rogers (eds) Thorp and Covich's Freshwater Invertebrates, 4th edn, Academic Press, Cambridge, Massachusetts.

Hoppeler, F., B. Rotter, N. Krezdorn & S.U. Pauls. 2016. A larval transcriptome of the limnephilid caddisfly *Micropterna lateralis* (Stephens, 1837) (Trichoptera: Limnephilidae). Aquatic Insects 37(3): 253–257.

Jackson, J.K., V.H. Resh, D.P. Batzer, R.W. Merritt & K.W. Cummins. 2019. Sampling aquatic insects: collection devices, statistical considerations, and rearing procedures. pp. 17–42 in: R.W. Merritt, K.W. Cummins, M.B. Berg (eds), An Introduction to the Aquatic Insects of North America, 5th edition. Kendall Hunt, Dubuque, Iowa.

Johanson, K.A. & M. Espeland. 2010. Phylogeny of the Ecnomidae (Insecta: Trichoptera). Cladistics 26: 36–48.

Johanson, K.A. 1998. Phylogenetic and biogeographic analysis of the family Helicopsychidae (Insecta: Trichoptera). Entomologica Scandinavica, Supplement 53: 1–172.

Ings, N.L., A.G. Hildrew & J. Grey. 2010. Gardening by the psychomyiid caddisfly *Tinodes waeneri*: Evidence from stable isotopes. Oecologia 163: 127–139.

Karaouzas, I. 2018. The larvae of three Greek species of *Hydropsyche* (Trichoptera: Hydropsychidae) and key for larvae of known Aegean *Hydropsyche* species. Zootaxa 4382 (2): 381–392.

Karaouzas, I. & J. Waringer. 2017. The larva of *Polycentropus ierapetra* Malicky 1972 (Trichoptera: Polycentropodidae) including a key to the larvae of genus *Polycentropus* (Curtis 1835) in the Hellenic western Balkan region. Zootaxa 4294 (5): 586–592.

Kjer, K.M., J.A. Thomas, X. Zhou, P.B. Frandsen, E. Predini & R.W. Holzenthal. 2016. Progress on the phylogeny of caddisflies (Trichoptera). Zoosymposia 10: 248–256.

Lancaster, J. & B.J. Downes. 2013. Aquatic entomology. Oxford Univ. Press, Oxford. UK.

Lavandier, P. 1979. Ecologie d'un torrent pyrénéen de haute montagne: l'Estaragne. Ph.D. Thése, Université Paul Sabatier, Toulouse. 532 pp.

Lepneva, S.G. 1970. Fauna of the USSR, Trichoptera (Vol. 2, No. 1). Larvae and Pupae of the Annulipalpia, Zoological Institute of the Academy of Science of the USSR, New Series, 88, Jerusalem: Israel Program for Scientific Translations.

Mackay, R.J. & G.B. Wiggins. 1979. Ecological diversity in Trichoptera. Annual review of entomology 24(1): 185–208.

Malicky, H. 1987. Anflugdistanz und Fallenfangbarkeit von Köcherfliegen (Trichoptera) bei Lichtfallen. Jahresberichte der biologischen Station Lunz 10: 140–157.

Malicky, H. 2005. Die Köcherfliegen Griechenlands (Trichoptera). Denisia 17: 1–240.

Malm, T. & K.A. Johanson. 2011. A new classification of the long-horned caddisflies (Trichoptera: Leptoceridae) based on molecular data. BMC evolutionary Biology 11(1): 10.

Malm, T., K.A. Johanson & N. Wahlberg. 2013. The evolutionary history of Trichoptera (Insecta): A case of successful adaptation to life in freshwater. Systematic Entomology 38: 459–473.

Morse, J.C. (ed.) 2020. Trichoptera World Checklist. https://entweb.sites.clemson.edu/database/trichopt/ (accessed on 23 March 2020).

Morse, J.C., P.B. Frandsen, W. Graf & J.A. Thomas. 2019. Diversity and ecosystem services of Trichoptera. Insects 10(5): 125.

Morse, J.C. & D.R. Lenat. 2005. A new species of *Ceraclea* (Trichoptera: Leptoceridae) preying on snails. Journal of the North American Benthological Society 24: 872–879.

Múrria, C., M. Sáinz-Bariáin, A.P. Vogler, A. Viza, M. González & C. Zamora-Muñoz. 2020. Vulnerability to climate change for two endemic high-elevation, low-dispersive *Annitella species* (Trichoptera) in Sierra Nevada, the southernmost high mountain in Europe. Insect Conservation and Diversity 13(3): 283–295.

Neu, P.J., H. Malicky, W. Graf & A. Schmidt-Kloiber. 2018. Distribution Atlas of European Trichoptera. Series: Die Tierwelt Deutschlands Volume: 84 (Ed). ConchBooks, Harxheim, Germany, 891 pp.

Salavert, V., C. Zamora-Muñoz, M. Ruiz-Rodríguez, A. Fernandez-Cortés & J.J. Soler. 2008. Climatic conditions, diapause and migration in a troglophile caddisfly. Freshwater Biology 53: 1606–1617.

Salavert, V., C. Zamora-Muñoz, M. Ruiz-Rodríguez & J.J. Soler. 2011. Female-biased size dimorphism in a diapausing caddisfly, *Mesophylax aspersus*: effect of fecundity and natural and sexual selection. Ecological Entomology 36: 389–395.

Sáinz-Bariáin, M., C. Zamora-Muñoz, J.J. Soler, N. Bonada, C.E. Sáinz-Cantero & J. Alba-Tercedor. 2016. Changes in Mediterranean high mountain Trichoptera communities after a 20-year period. Aquatic Sciences 78: 669–682.

Sánchez-Bayo, F. & K.A. Wyckhuys. 2019. Worldwide decline of the entomofauna: A review of its drivers. Biological conservation 232: 8–27.

Sipahiler, F. 1996. Two new isolated species of Limnephilidae (Trichoptera) from northern Turkey. Aquatic Insects 18(2): 117–127.

Sipahiler, F. 2013. The Larva of *Calamoceras illiesi* Malicky & Kumanski, 1974 (Trichoptera, Calamoceratidae). Nova Acta Científica Compostelana 20: 21–26.

Sipahiler, F. 2016. Two new species of Trichoptera (Psychomyiidae, Beraeidae) from Turkey. Nova Acta Científica Compostelana 23: 61–64.

Tachet, H., J.C. Morse & A. Berly. 2001. The Larva and Pupa of *Pseudoneureclipsis lusitanicus* Malicky, 1980 (Trichoptera: Hydropsychoidea): Description, Ecological Data and Taxonomical Considerations. Aquatic Insects 23(2): 93–106.

Thomas, J.A., P.B. Frandsen, E. Prendini, X. Zhou & R.W. Holzenthal. 2020. A multigene phylogeny and timeline for Trichoptera (Insecta). Systematic Entomology 45(3): 670–686.

Tierno de Figueroa, J.M.T., M.J. López-Rodríguez, S. Fenoglio, P. Sánchez-Castillo & R. Fochetti. 2013. Freshwater biodiversity in the rivers of the Mediterranean Basin. Hydrobiologia 719(1): 137–186.

Urbanič, G. 2006. Description of the larva of *Polycentropus schmidi* Novak & Botosaneanu, 1965 (Trichoptera: Polycentropodidae) with some notes on its ecology. Aquatic Insects 28(4): 257–262.

Vieira-Lanero, R. 2000. Las larvas de los Tricópteros de Galicia (Insecta:Trichoptera). Ph.D. Thesis. Department of Animal Biology, University of Santiago de Compostela, Santiago, 612 pp.

Vieira-Lanero, R., M.A. González & F..Cobo. 2001. Descripción de la larva de *Helicopsyche helicifex* (Allan, 1857) (Trichoptera, Helicopsychidae). Nova Acta Científica Compostelana (Bioloxía) 11: 215–223.

Vieira-Lanero, R., M.A. Gonzalez & F. Cobo. 2002. Descripción de las larvas de cuatro endemismos ibericos del genero Beraea (Trichoptera, Beraeidae). Nouvelle Revue d'Entomologie 19: 11–37.

Vshivkova, T.S., J.C. Morse & D. Ruiter. 2007. Phylogeny of Limnephilidae and composition of the genus *Limnephilus* (Limnephilidae: Limnephilinae, Limnephilini). Pages 309–319 in: J. Bueno-Soria, R. Barba-Álvarez & B.J. Armitage (eds.), Proceedings of the 12th International Symposium on Trichoptera. The Caddis Press, Columbus, Ohio.

Waringer, J. & W. Graf. 2011. Atlas of Central European Trichoptera larvae. Atlas der mitteleuropaeischen Koecherfliegenlarven. Erik Mauch Verlag, Dinkelscherben. 468 pp.

Waringer, J. & W. Graf. 2013. Key and bibliography of the genera of European Trichoptera larvae. Zootaxa 3640: 101–151.

Waringer, J., W. Graf & F. Sipahiler. 2017. The larvae of *Silo chrisiammos* Malicky 1984 and *Lithax musaca* Malicky 1972 (Trichoptera: Goeridae), including a key to the Goeridae larvae of the Eastern and Hellenic western Balkan regions. Zootaxa 4306(1): 67–80.

Waringer, J., H. Malicky, I. Živić & H. Vicentini. 2017. The larvae of the European *Helicopsyche* species (Trichoptera: Helicopsychidae). Zootaxa 4277: 561–572.

Waringer, J., M.A. González & H. Malicky. 2020. Discriminatory matrix for the larvae of the European *Thremma* species (Trichoptera: Thremmatidae). Zootaxa 4718(4): 451–469.

Wells, A. 2005. Parasitism by hydroptilid caddisflies (Trichoptera) and seven new species of Hydroptilidae from northern Queensland. Australian Journal of Entomology 44(4): 385–391.

Whitlock, H.N. & J.C. Morse. 1994. *Ceraclea enodis*, a new species of sponge-feeding caddisfly (Trichoptera: Leptoceridae) previously misidentified. Journal of the North American Benthological Society 13(4): 580–591.

Wiggins, G.B. 1973. A contribution to the biology of the caddisflies (Trichoptera) in temporary pools. Royal Ontario Museum Life Sciences Contributions 88: 1–28.

Wiggins, G.B. 1996. Larvae of the North American caddisfly genera (Trichoptera). 2nd ed. University of Toronto Press, Toronto, Canada. 457 pp.

Wiggins, G.B. 1998. The caddisfly family Phryganeidae (Trichoptera). University of Toronto Press, Toronto. ix + 306 pp.

Wiggins, G.B. 2004. Caddisflies: the underwater architects. University of Toronto Press, Toronto, Canada. 292 pp.

Williams, N.E. & G.B. Wiggins. 1981. A proposed setal nomenclature and homology for larval Trichoptera. Pages 421–429 in: G.P. Moretti (ed.), Proceedings of the Third International Symposium on Trichoptera Dr. W. Junk, The Hague.

Chapter 15

Order Diptera

Valeria Lencioni[1], Peter H. Adler[2] and Gregory W. Courtney[3]
[1]MUSE-Science Museum, Research and Museum Collections Office, Climate and Ecology Unit, Corso del Lavoro e della Scienza, Trento, Italy,
[2]Department of Plant and Environmental Sciences, Clemson University, Clemson, SC, United States, [3]Department of Plant Pathology, Entomology, and Microbiology, Iowa State University, Ames, IA, United States

Diversity, distribution, and ecology of Diptera

Midges, mosquitoes, and other flies are members of the order Diptera, one of the least-loved insect orders, popularly perceived as injurious to humans. Flies have particular medical and veterinary importance as vectors of disease agents. Many species are pests of plants and animals and are associated with cadavers, garbage, and manure. Biting flies cause annoyance that affects tourism, recreation, and agricultural and industrial production, and their effects on livestock can reduce egg, meat, and milk production. Adults of many aquatic families, such as Culicidae, Simuliidae, and Tabanidae, feed on the blood of humans, livestock, and wild animals, and some are vectors of organisms that cause diseases. Mosquitoes, for instance, can transmit the agents of diseases such as malaria, yellow fever, dengue, chikungunya, and West Nile encephalitis (Foster & Walker, 2019; Pichler et al., 2021). On the other hand, many species are economically beneficial as plant pollinators (e.g., some species of Syrphidae, Tipulidae, and even Culicidae), which visit flowers to imbibe nectar, and as parasitoids and predators (e.g., Empididae and Sciomyzidae) of arthropod and molluscan pests. The larvae of some flies (e.g., Chironomidae) are used in aquaculture as food for fish. Flies are many, and they are everywhere—and we have all had to deal with them. The scientific community is challenged with balancing the benefits and costs of Diptera and harnessing their services for human benefit while maintaining sustainable populations as more species face extinction (Adler & Courtney, 2019).

The order Diptera includes the true flies and is among the most species-rich insect orders, with a worldwide distribution. With nearly 160,000 species in approximately 10,000 genera and 158 families (Pape et al., 2011; Courtney & Cranston, 2015; Courtney et al., 2017), Diptera forms one of the most diverse groups of insects—nearly 15% of the world's known insect species (Adler & Foottit, 2017). The true number of fly species is probably many times greater (Hebert et al., 2016).

Dipterans are easy to distinguish from other insects. Dipteran adults have a single pair of functional wings, the anterior mesothoracic pair (Fig. 15.1). This characteristic gives the order its name, derived from the Greek δι- di- "two," and πτερόν pteron "wing." The posterior (metathoracic) wings typically found in other insects have evolved into mechanosensory gyroscopic organs known as "halteres." These halteres have a balancing function during flight, allowing dipterans to perform advanced aerobatics. The larvae have a distinct habitus, with no jointed thoracic legs and often with a maggot-like appearance.

Much has been written about Diptera in hundreds of chapters in books and encyclopedias to tens of thousands of papers on individual families, covering topics such as their taxonomy and systematics, ecology, and economic and medical importance. Notwithstanding our vast knowledge of flies, much remains to be discovered.

Dipteran diversity in terms of species richness, structure, life habits, and economic significance is associated with remarkable habitat use in terrestrial and aquatic environments, including freshwater and marine habitats (Figs. 15.2–15.3). Flies are ubiquitous, having conquered all continents. They are the only free-living insects to have colonized Antarctica, with three species of Chironomidae: *Belgica antarctica* Jacobs, 1900, *Eretmoptera murphyi* (Schaeffer, 1914) (subfamily Orthocladiinae), and *Parochlus steinenii* (Gercke, 1889) (subfamily Podonominae) (Allegrucci et al., 2006). Diptera are found in all aquatic habitats: microthin trickles to torrential waterfalls, tree holes to open oceans, and glacial meltwaters to hot springs (Adler & Courtney, 2019), from sea level up to 5600 m in the Himalayas, at depths of more than 1000 m in Lake Baikal, and at latitudes above 70°N (Ferrington & Berg, 2019).

Identification and Ecology of Freshwater Arthropods in the Mediterranean Basin. DOI: https://doi.org/10.1016/B978-0-12-821844-0.00001-6

FIGURE 15.1 Adults of Diptera. (A) Chironomidae (*Diamesa*), male. (B) Limoniidae (*Hexatoma*). (C) Stratiomyiidae (*Odontomyia*). (D) Tabanidae (*Tabanus*), male. (E) Sciomyzidae (*Dictya*). (F) Muscidae (*Muscina*). *Figures A, F © V. Lencioni; Figures B–E © G.W. Courtney.*

Diptera are holometabolous insects. A typical life cycle consists of an egg stage of a few days to years (typically depending on whether diapause or drought occur), a larval stage with three to seven or more instars spanning a combined total of about a week to more than a year, a pupal stage of varying length, and an adult stage lasting from less than 2 h (Deuterophlebiidae) to several months (Courtney, 2019).

Many keys for the identification of Diptera have been published, most of them for the adult stage. For larvae and pupae, taxonomic and ecological literature is uneven among families. Species in families such as the Culicidae and Simuliidae are well known in the immature forms, but for many other species the immature stages are not known. To date, almost nothing collectively has been published about freshwater Diptera of the Mediterranean Basin. In this chapter, we discuss Diptera associated with freshwaters, with only passing mention of Diptera associated with inland saline habitats.

Traditionally, on the basis of adult morphology, Diptera have been divided into two suborders: Nematocera (e.g., craneflies, gnats, midges, and mosquitoes) with slender antennae of multiple flagellomeres, and the typically heavier-built Brachycera (e.g., hover flies, and horse flies) with shorter, stouter antennae of only one flagellomere, although often with annulations. Nematocera is now considered a paraphyletic group, and the name has fallen out of favor; these flies now are often referred to informally as the "Lower Diptera." The classification within the Lower Diptera and Brachycera has been summarized by Courtney et al. (2017).

Diptera not only have a great number of species—perhaps more than any other order (Hebert et al., 2016)—but they also include a wide variety of adaptations that reflect their tremendous morphological and ecological diversity. Dipteran

FIGURE 15.2 Aquatic habitats of the Mediterranean Basin. (A) Maroglio River, estuary, Caltanisetta Province, Sicily, southern Italy (by L. Latella). (B) Agriokalami River, near Lempa, Cipro (by D. Spitale). (C) Amlay swamp (*Ranunculus aquatilis*, macrophytes) with herbaceous cover; Tetouan Province, Morocco (by K. Targuisti). (D) Río Hozgarganta, Cádiz, Spain (by N. Bonada). (E) Fifi Lake, surrounded by *Quercus canariensis*, Chefchaouen Province, Morocco (by K. Kettani). (F) Acheron River, the source, Glyki, Epirus, Greece (by B. Maiolini). (G) Río Genal, intermittent stream; Serranía de Ronda, Málaga, Spain (by M.S. Extremera). (H) "Fiumara" of Montebello, Reggio Calabria, Aspromonte Mt., southern Italy (by L. Latella).

larvae form a crucial component of many terrestrial and freshwater ecosystems in the extant biosphere. They play a central role in water purification, matter and energy transfer in riparian ecosystems, carbon cycling in lakes and forests, and decomposition and recycling of organic matter. Groups such as the Chironomidae, Culicidae, and Simuliidae occur in large numbers as larvae and adults and provide a major prey base for many other invertebrates as well as vertebrates such as amphibians, bats, birds, and fish. In turn, several families contain predators and parasitoids as larvae and adults, including the Dolichopodidae, Empididae, Syrphidae, and Tachinidae (Avesani & Lencioni, 2011).

FIGURE 15.3 Aquatic habitats of the Mediterranean Basin. (A) Alcantara Gorges, vulcanic canyon, Sicily, southern Italy (by V. Lencioni). (B) Lergue River, Hérault Department, southern France (by J. Moubayed-Breil). (C) Bracciano Lake, volcanic, shoreline, Rome Province, central Italy (by I. Davanzo). (D) Chellal Akchour, large pond with 40-m high waterfall, Morocco (by K. Kettani). (E) Large limnocrene spring, edge of Lake Skadar (near Poseljani), Montenegro (by M. Grabowski). (F) Charca del Lirio, Charca de Suárez, Granada, southern Spain (by C. Zamora-Muñoz). (G) Acheron River, the mouth, Ionian Sea coast, Ammoudia, Epirus, Greece (by B. Maiolini).

Aquatic Diptera

Diptera that spend most of their life in water as larvae and pupae are considered aquatic (Figs. 15.4–15.9). Almost no dipteran passes its entire life cycle in water; minimally, the adult stage is terrestrial. If "water" means wet habitats, most dipterans should be considered aquatic, because the vast majority of larval dipterans inhabit wet habitats, from those entirely immersed in free water to those in damp substrates or endoparasitic in other organisms (Adler & Courtney, 2019). The unique larval habitat—crude petroleum pools and seeps—of the ephydrid *Helaeomyia petrolei* (Coquillett, 1899) also complicates the definition of an aquatic insect, although the species has typically been considered aquatic (Adler & Courtney, 2019).

We consider the free-living dipterans as aquatic that inhabit a body of water or the saturated margins of the water body in at least the larval stage. Water bodies include lentic systems (i.e., standing waters) such as lakes, ponds and marshes, small pools, plant-held waters (i.e., phytotelmata, such as plant bracts and tree holes), stagnant ground-pools, and artificial containers, and lotic systems (i.e., flowing waters) such as rivers, streams, cold and hot springs, seepages

FIGURE 15.4 Larvae of Lower Diptera. (A) Simuliidae (*Simulium*). (B) Simuliidae (*Simulium*). (C) Ceratopogonidae: Forcipomyiinae (*Forcipomyia*). (D) Ceratopogonidae: Ceratopogoninae (*Palpomyia*). (E) Thaumaleidae, dorsal view. (F) Chironomidae: Chironominae (*Lauterborniella*) in its tube. (G) Chironomidae: Diamesinae (*Diamesa*). (H) Chironomidae Orthocladinae: (*Cricotopus*), head and thorax, lateral view. *Figures C, E, F, and H © G.W. Courtney; Figures A, B, D, and G © S.A. Marshall.*

and groundwater zones, trickles, and madicolous habitats (i.e., films of flowing water <2 mm deep). The full range of aquatic to semiaquatic species is found in groups such as the Chironomidae and Tipuloidea (Cylindrotomidae, Limoniidae, Pediciidae, and Tipulidae), which also include terrestrial species (Pinder 1995; De Jong et al. 2008; Ferrington & Berg 2019; Gelhaus & Podeniene 2019).

At least 41 (26%) of the 158 families of Diptera have definitively aquatic representatives including nearly 46,000 species worldwide—about 30% of all formally described species of Diptera (Adler & Courtney, 2019). Some of the more familiar aquatic groups are black flies, horse flies, midges, and mosquitoes. Diptera have more aquatic representatives than any other order of insects—at least three times more than the Coleoptera and Trichoptera (Adler & Courtney, 2019). Aquatic representation by Diptera is even greater if a wider definition of "aquatic" is used. For example, 10 additional

FIGURE 15.5 Larvae of Lower Diptera. (A) Culicidae (*Toxorhynchites*), lateral view. (B) Culicidae (*Aedes*), dorsolateral view. (C) Culicidae (*Aedes*), thorax and head, dorsal view. (D) Dixidae (*Dixella*), head and anterior part of body, ventrolateral view. (E–G) Chaoboridae (*Chaoborus*), lateral views: (E) Habitus. (F) Head and thorax. (G) Anal division showing abdominal air sac. *All figures © G.W. Courtney.*

families have been considered by various workers to have aquatic representatives: Anisopodidae, Calliphoridae *sensu lato*, Cecidomyiidae, Mycetophilidae, Pachyneuridae, Periscelididae, Rhagionidae, Sciaridae, Tachnidae, and Xylophagidae (Campaioli et al., 1999; Dobson, 2013; Fusari et al., 2018). Some families are unquestionably aquatic. The Chaoboridae, for example, are found almost exclusively in standing water bodies, whereas the Blephariceridae and Simuliidae live only in streams and rivers. The Chironomidae colonize any freshwater habitats, peat bogs, and the marine environment (intertidal zone). Several families (e.g., Bibionidae and Trichoceridae), which are occasionally encountered in stream samples, might be accidental because they occur in wet soils, and depending on the authors, these families have at times been considered aquatic (Dobson, 2013). Thus it is difficult to categorize the members of numerous dipteran families as aquatic or "water dependent" (Wagner et al., 2008). A key problem for most groups is the lack of knowledge on larval ecology and morphology of many taxa on the one hand and the great ecological plasticity on

FIGURE 15.6 Larvae of Lower Diptera. (A) Ptychopteridae (*Ptychoptera*). (B) Psychodidae (*Thornburghiella*), dorsal view. (C) Blephariceridae, dorsal view, (D) Blephariceridae, ventral view. (E) Cylindrotomidae (*Phalacrocera*). (F) Limoniidae (*Antocha*). (G) Pediciidae (*Pedicia*). (H) Tipulidae (*Tipula*). *Figures B and F © S.A. Marshall; Figures A, C–E, G, and H © G.W. Courtney.*

the other. The enormous genus *Dicranomyia* Stephens, 1829 in the family Limoniidae, for example, includes more than 1000 species but the immatures are described for only 20 species, and these include both terrestrial and aquatic species (Podeniene, pers. comm.). In general, many water-dependent larvae and pupae live in moist to wet ground (providing food and shelter) near lakes, ponds, rivers, springs, and streams, or in wetlands where they occupy a multitude of spatially and temporally variable habitats. Dipteran families with marine intertidal representatives include the Canacidae, Chironomidae, Coelopidae, Dryomyzidae, Helcomyzidae, and Heterocheilidae.

The services that aquatic Diptera can provide for humanity are limited only by the creativity of the investigator. Aquatic Diptera have played important roles, for example, in bioassessment of water quality and ecological and climate change, biological control, and forensic investigations; and although their value to bioengineering and the pharmaceutical industry has been recognized, their real potential has yet to be realized (Adler & Courtney, 2019). Some

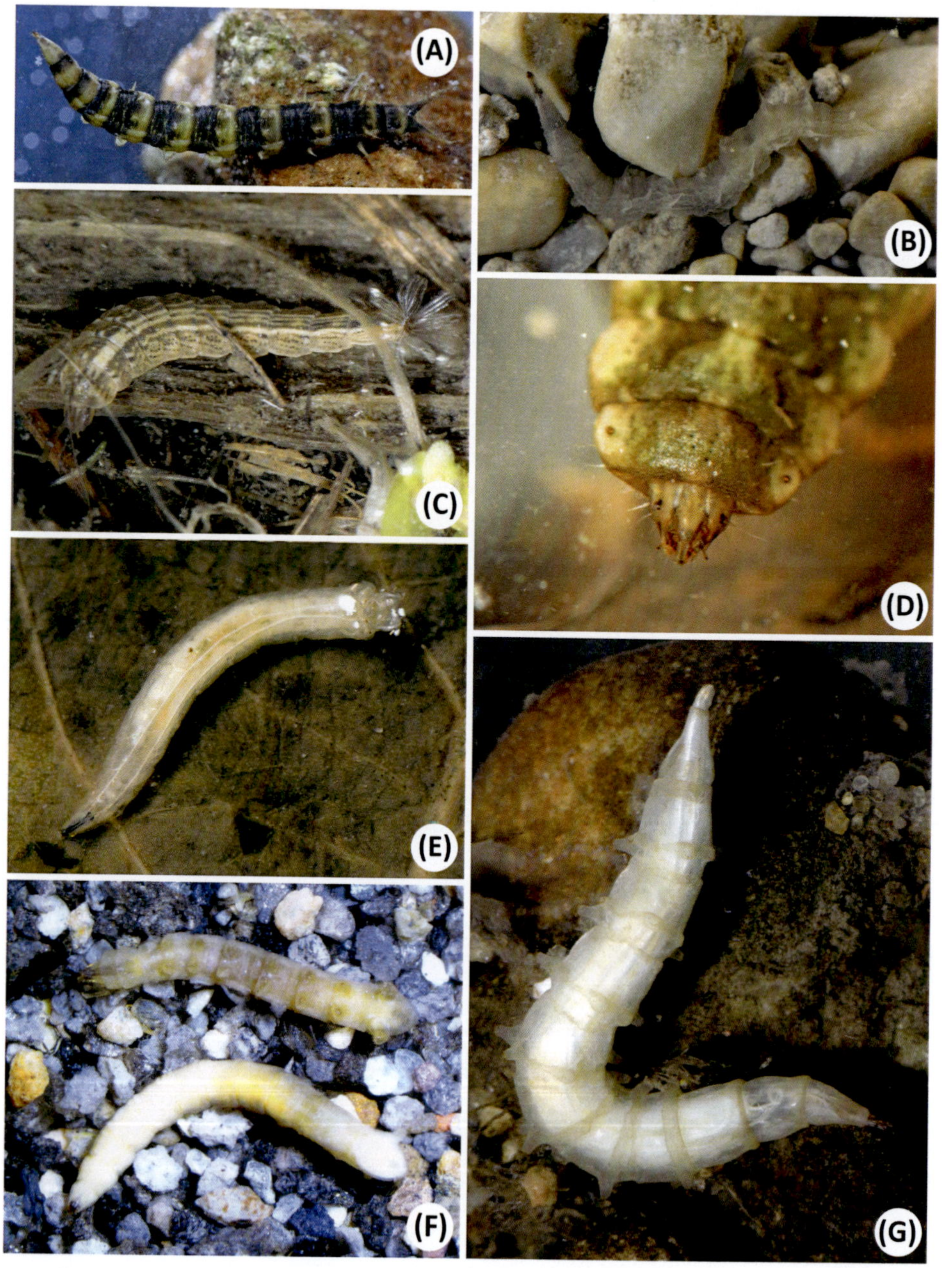

FIGURE 15.7 Larvae of Brachycera. (A, B) Athericidae (*Atherix*): (A) Dorsal view. (B) Lateral view. (C, D) Stratiomyiidae. (C) Dorsal view. (D) Head and thorax, dorsal view. (E) Dolichopodidae, dorsal view. (F) Empididae (Hemerodromiinae), dorsal view. (G) Tabanidae (*Tabanus*), dorsolateral view. *All Figures © G.W. Courtney.*

families of aquatic Diptera are important in water-quality and bioassessment studies to classify the degree of pollution in a water body. For example, larvae belonging to the nonbiting midge genus *Chironomus* Meigen, 1803 (Chironomidae, tribe Chironomini), which can tolerate low oygen levels in organically polluted habitats, have been referred to as "blood worms" because of the hemoglobin in their blood. These species and others, such as "rat-tailed maggots" (Syrphidae: *Eristalis* Latreille, 1804), are often used as indicators of polluted water or water low in oxygen. Other chironomids are highly sensitive to pollution and are indicators of clean, cold waters (e.g., subfamily Diamesinae; Lencioni, 2018).

Some aquatic Diptera have been the subject of ecotoxicological and molecular studies. For example, the nonbiting midge *Chironomus riparius* Meigen, 1804 (syn. *C. thummi* (Kieffer, 1911)) has been used extensively for testing pollutant toxicity in sediments and freshwater environments (EPA US 2000; OECD, 2018), with approaches ranging from

FIGURE 15.8 Larvae of Brachycera. (A) Syrphidae (*Eristalis*), dorsal view. (B) Sciomyzidae (*Sepedon*), dorsolateral view. (C) Sciomyzidae, dorsal view. (D) Ephydridae (*Setacera*), lateral view. (E) Ephydridae (*Ephydra*), lateral view. (F) Scathophagidae (*Acanthonema*) dorsal view. (G) Muscidae (*Lispe*), dorsal view. (H) Muscidae (*Limnophora*), lateral view. *Figures A–D and F–H © G.W. Courtney; Figure E © S.A. Marshall.*

ecotoxicological and genotoxic tests (Fisher et al., 2003; Martínez-Paz et al., 2013; Grazioli et al., 2016; Bernabò et al., 2017) to analyses of larval mouthpart deformities (Bisthoven et al., 1998) and structural and functional aberrations of giant polytene chromosomes from the larval salivary gland cells (Michailova et al., 2006).

Larval morphology of aquatic Diptera

Diptera originated in wet environments, and their morphology and life histories reflect that origin. The larvae of most species are vermiform and legless (Figs. 15.10–15.17). They develop mainly in moist or wet habitats, such as soil and decaying organic matter and in plant or animal tissues. The majority of species are liquid-feeders or microphages. Immature stages of

FIGURE 15.9 Pupae of Diptera: (A) Cylindrotomidae (*Phalacrocera*), dorsal view. (B) Ptychopteridae, lateral view. (C) Culicidae (*Aedes*), lateral view. (D) Blephariceridae, dorsal view. (E) Chaoboridae (*Chaoborus*), lateral view. (F) Empididae (*Hemerodromia*), lateral view. (G) Tabanidae, lateral view. (H) Sciomyzidae (*Dictya*), dorsal view. *All figures © G.W. Courtney.*

flies can be found in virtually every type of aquatic habitat, such as on or in the substrate, on or in aquatic plants, swimming or floating in open water, or associated with the air–water interface. Among aquatic insects, dipteran larvae are easily recognized by the absence of articulated thoracic appendages. Apodous larvae (those lacking legs) also occur in other orders, such as Siphonaptera and some Hymenoptera and Coleoptera, but few or no species of such groups inhabit aquatic or semiaquatic environments. Aquatic dipteran larvae have locomotory devices that include **prolegs** (e.g., Chironomidae and Simuliidae, Fig. 15.10A, B, K), **suctorial disks** (Blephariceridae, Fig. 15.6D), **friction pads** (some Psychodidae), **creeping welts**

FIGURE 15.10 Larvae of Lower Diptera (Chironomoidea). (A–D) Chironomidae: (A) Chironominae (*Chironomus riparius* Meigen, 1804), lateral view. (B) Diamesinae (*Diamesa zernyi* Edwards, 1933), lateral view. (C) Orthocladiinae [*Heterotanytarsus apicalis* (Kieffer, 1921)], head capsule, ventral view. (D) Diamesinae [*Prodiamesa olivacea* (Meigen, 1818)], mentum and mandibles, ventral view. (E–G) Ceratopogonidae (Forcipomyiinae): (E) Dorsal view. (F) Ventral view. (G) Lateral view. (H) Ceratopogonidae (Ceratopogoninae), dorsal view. (I–L) Simuliidae (*Prosimulium*): (I) Head capsule, dorsal view. (J) Head capsule, ventral view. (K) Habitus, lateral view. (L) Abdomen, posteroventral view. (M) Thaumaleidae, lateral view. *All figures © F. Pupin, except Figures C, D, I, J (C) V. Lencioni.*

FIGURE 15.11 Larvae of Lower Diptera. (A, B) Culicidae [*Aedes albopictus* (Skuse, 1894)]: (A) Dorsal view. (B) Lateral view. (C) Culicidae (*Anopheles*), head capsule, ventral view. (D, E) Dixidae, lateral view: (D) habitus, (E) posterior segment. (F, G) Blephariceridae: (F) Dorsal view. (G) Ventral view. (H–J) Psychodidae: (H) Ventral view. (I) Dorsal view. (J) Lateral view. (K, L) Trichoceridae: (K) Ventral view. (L) Anterior half of body, ventrolateral view. *All figures © F. Pupin, except Figure C © G.W. Courtney.*

(i.e., swollen ridges on the body) (e.g., some Limoniidae, Fig. 15.15C), and various fleshy projections (e.g., some Sciomyzidae, Fig. 15.15A). These protrusions work as supports and mechanisms for gaining traction that, together with peristaltic motions of the body, facilitate larval locomotion. Mandibles can aid this process by serving as additional anchoring points.

Dipteran larvae range from a few millimeters to 5 cm long. The general body shape can be eel-like (e.g., Ceratopogonidae, Fig. 15.4D), fusiform (e.g., Tabanidae, Fig. 15.7G), cylindrical (e.g., Chironomidae and

FIGURE 15.12 Larvae of Lower Diptera (Tipuloidae). (A) Tipulidae, lateral view. (B) Tipulidae (*Tipula*), head capsule, ventral view. (C) Tipulidae, spiracular disk, posterior view. (D–F) Limoniidae: (D) Spiracular disk, posterior view. (E, F) Lateral view. (G) Limoniidae (*Antocha*), ventral view with head exposed. (H) Pedicidae, lateral view, with head exposed. *All figures © F. Pupin, except Figure B © G.W. Courtney.*

FIGURE 15.13 Larvae of Brachycera. (A–D) Stratiomyidae: (A) Dorsal view. (B) Ventral view. (C) Lateral view. (D) Spiracular disk, dorsal view. (E, F) Tabanidae: (E) Dorsal view. (F) Ventral view. (G, H) Athericidae: (G) Lateral view. (H) Posterior end, ventral view. (I, J) Empididae: (I) Lateral view. (J) Ventral view. (K, L) Muscidae (*Lispe*): (K) Lateral view. (L) Posterior abdominal segments, lateral view. (M, N) Muscidae (*Limnophora*): (M) Lateral view. (N) Posterior abdominal segments, lateral view. *All figures © F. Pupin, except Figure D © G.W. Courtney.*

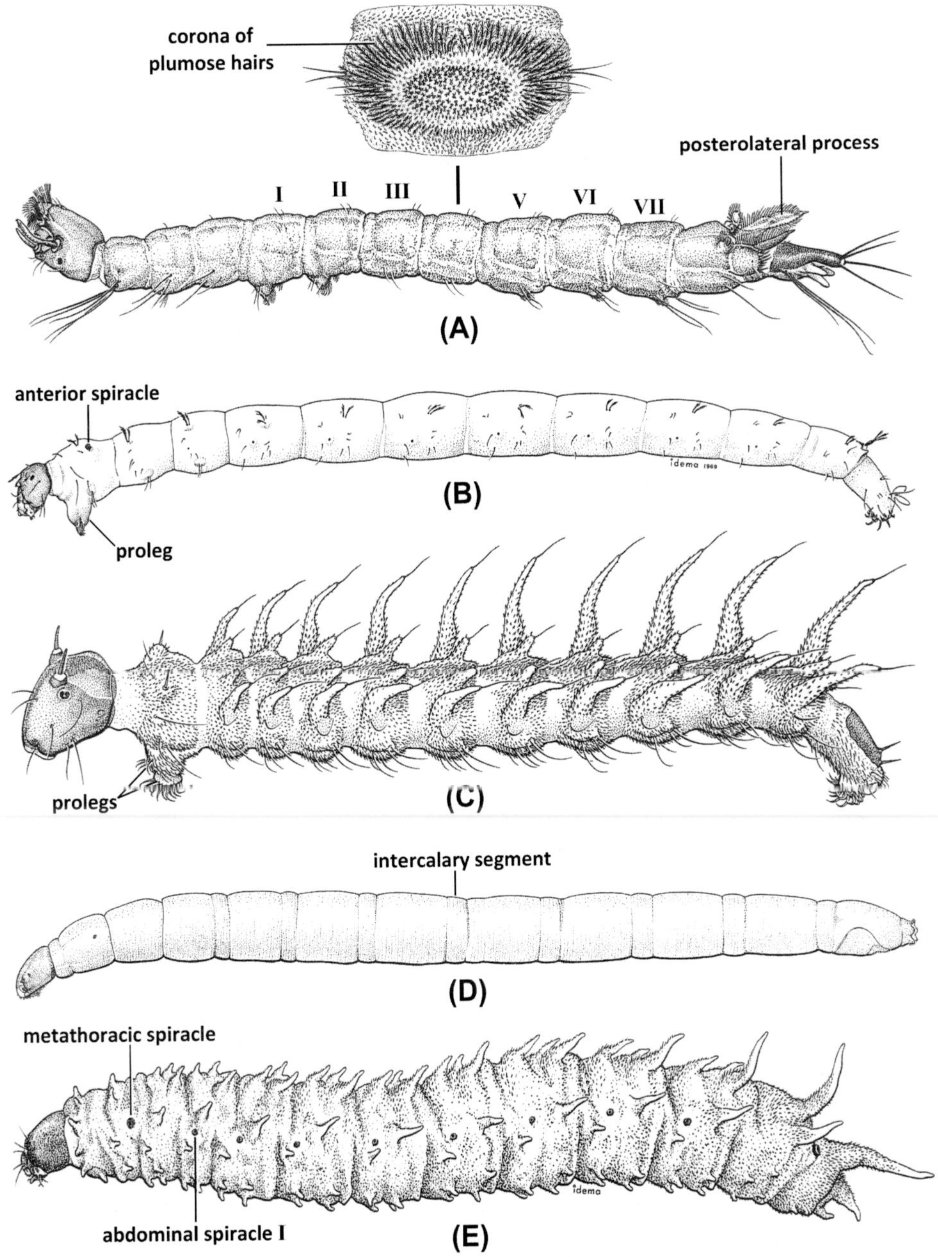

FIGURE 15.14 Larval Diptera (lateral view). (A) Dixidae, with dorsum of an abdominal segment. (B) Thaumaleidae. (C) Ceratopogonidae: Forcipomyiinae (*Atrichopogon*). (D) Anisopodidae (*Sylvicola*). (E) Bibionidae. *From* Manual of Nearctic Diptera, *Volume 1 (1981) by permission.*

Empididae, Figs. 15.4G and 15.7F), or dorsoventrally flattened (e.g., Stratiomyidae, Fig. 15.7C). Shape variants include an enlarged thorax (e.g., Chaoboridae and Culicidae, Fig. 15.5B, C, E, F) or an enlarged abdomen giving a distinctive flask or pear shape (e.g., Simuliidae (Fig. 15.4A, B)). Some syrphid larvae have a rat-shaped body with a truncated cephalic region and a long posterior respiratory siphon (Fig. 15.8A).

The most common body segmentation pattern has 12 segments, of which 3 are thoracic and 9 are abdominal. Numerous variations on this basic plan can be found. For example, the three thoracic segments can be fused and expanded (Chaoboridae and Culicidae, Fig. 15.11A, B) or fused with the first abdominal segment (Blephariceridae, Fig. 15.11F). Pseudosegmentation, in which each segment has two or three subdivisions (annuli), is found in families such as the Anisopodidae, Psychodidae, and Trichoceridae (Figs. 15.11H, 15.14D, and 15.16A). Brachycera frequently have only 11 obvious segments (Fig. 15.17E).

Aquatic dipteran larvae typically have a weakly sclerotized, rather membranous integument, varying from thick and opaque to transparent, with the internal organs clearly visible. The integument can be reinforced by sclerotized plates or

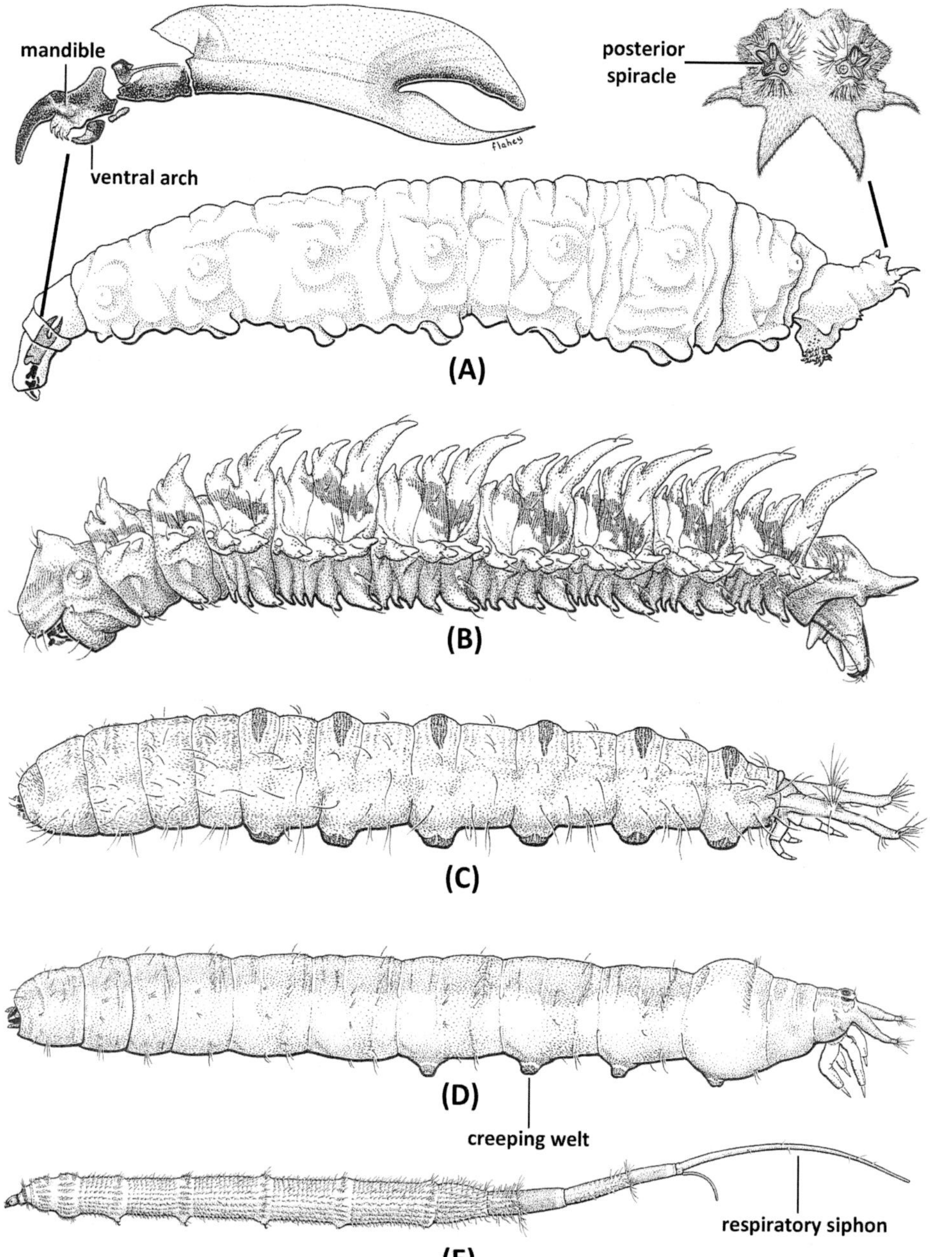

FIGURE 15.15 Larval Diptera (lateral view). (A) Sciomyzidae (*Sepedon*), with cephalopharyngeal skeleton and terminal view of posterior spiracular disk. (B) Cylindrotomidae. (C) Limoniidae (*Antocha*). (D) Pediciidae (*Pedicia*). (E) Ptychopteridae (*Ptychoptera*). *From* Manual of Nearctic Diptera, *Volume 1 (1981) by permission.*

calcareous deposits, or it can be lifted in conical evaginations. Body color in live larvae varies among species and can be blackish, brown, bluish, red, purple, whitish, or yellowish; there is no lack of variety.

The larval **head capsule** can be complete, sclerotized, and exposed, with biting and chewing mouthparts (Fig. 15.18). Or, it can be variously reduced, from partially undeveloped to completely absent, such as in maggot forms that consist only of the sclerotized mandibles ("**mouth hooks**") and supporting structures. Based on the development of the head, the larvae are characterized as eucephalic, hemicephalic, or acephalic, summarized as follows:

Eucephalic: The head is totally exposed, well-developed, and well-sclerotized (Fig. 15.10A–K). Most Lower Diptera are eucephalic, although some exceptions include the hemicephalic Tipulidae, in which the head is

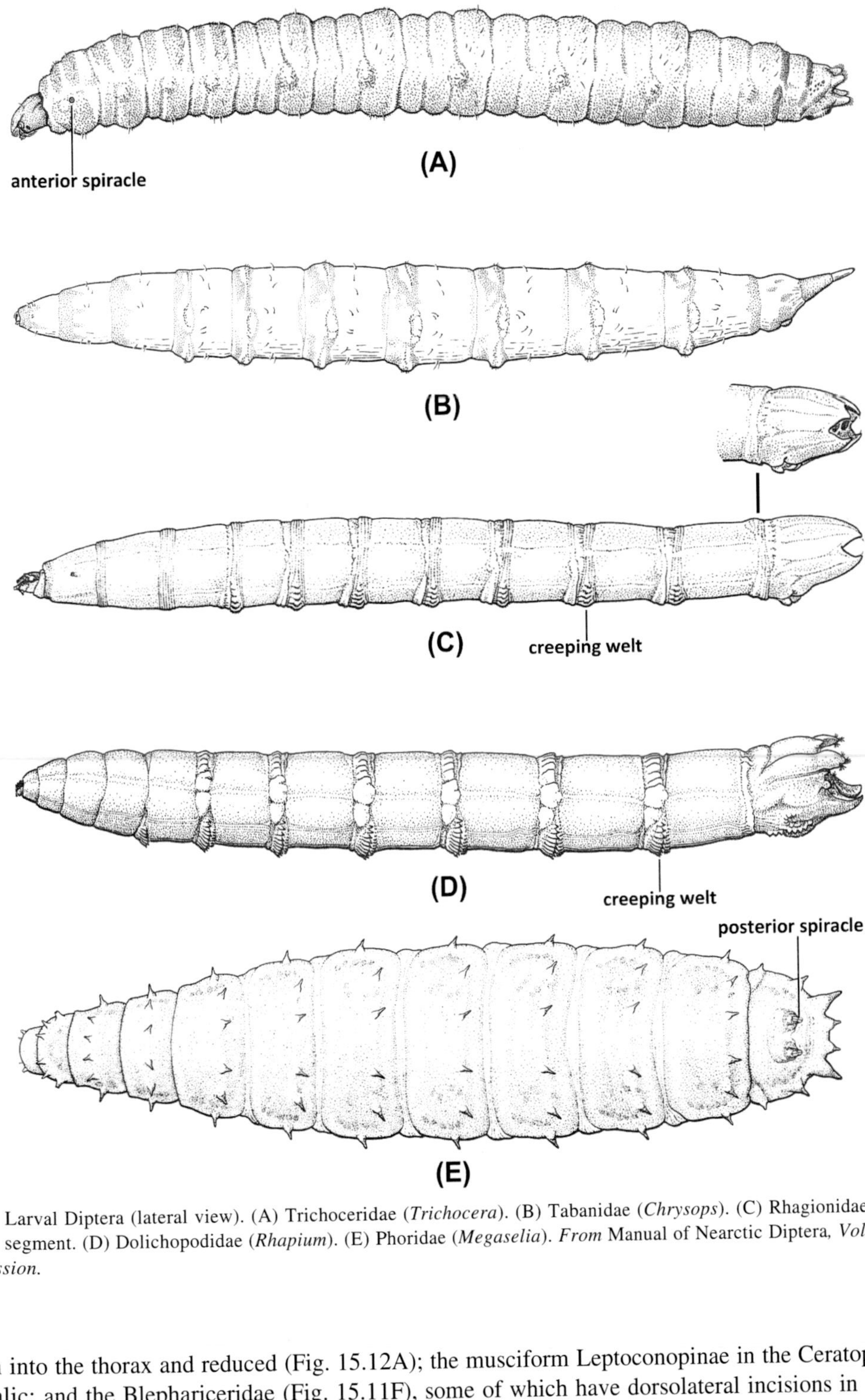

FIGURE 15.16 Larval Diptera (lateral view). (A) Trichoceridae (*Trichocera*). (B) Tabanidae (*Chrysops*). (C) Rhagionidae, with posterolateral view of terminal segment. (D) Dolichopodidae (*Rhapium*). (E) Phoridae (*Megaselia*). *From* Manual of Nearctic Diptera, *Volumes 1 (1981) and 2 (1987) by permission.*

withdrawn into the thorax and reduced (Fig. 15.12A); the musciform Leptoconopinae in the Ceratopogonidae, which are acephalic; and the Blephariceridae (Fig. 15.11F), some of which have dorsolateral incisions in the head capsule. Eucephalic larvae typically have opposable mandibles of the chewing type, which move in a pincer-like horizontal or oblique plane (Fig. 15.10C, D). Only the Stratiomyidae within the Brachycera are eucephalic (Fig. 15.7D), although their mandibles move in a vertical plane.

Hemicephalic: The head is relatively reduced posteriorly and partially retracted into the thorax, with sickle-shaped mandibles or mouth hooks that operate in a vertical plane (e.g., Empididae and Tabanidae) (Fig. 15.18J, K).

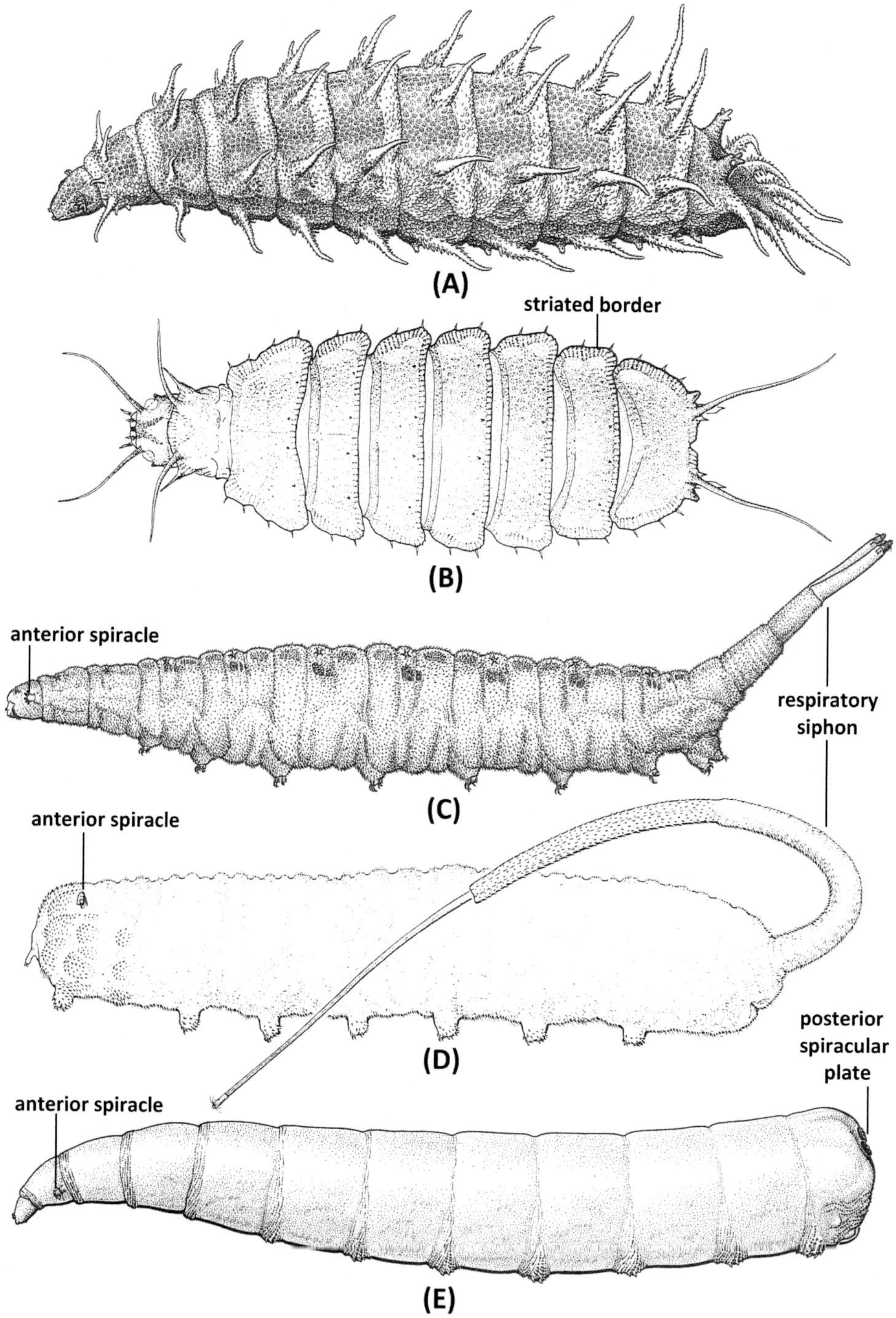

FIGURE 15.17 Larval Diptera. (A) Fanniidae [*Fannia canicularis* (L., 1761)], dorsolateral view. (B) Lonchopteridae (*Lonchoptera*), dorsal view. (C) Ephydridae (*Ephydra*), lateral view. (D) Syrphidae (*Eristalis tenax* (L., 1758)), lateral view. (E) Muscidae, lateral view. *From* Manual of Nearctic Diptera, *Volume 2 (1987) by permission.*

Acephalic: The head shows additional reduction, with desclerotization of all external elements of the head capsule, retraction of the cephalic region into the thorax, and development of an internal cephalopharyngeal skeleton that supports the musculature of the sucking pump and articulates with a pair of vertically oriented, sickle-shaped mandibles (i.e., mouth hooks) (e.g., Sciomyzidae and Muscidae) (Figs. 15.15A and 15.18L).

Larvae show a variety of respiratory adaptations. Oxygen may be acquired from the surrounding water (cuticular respiration), from the atmosphere (e.g., Dixidae and most Culicidae), or from plant tissues (e.g., some Culicidae).

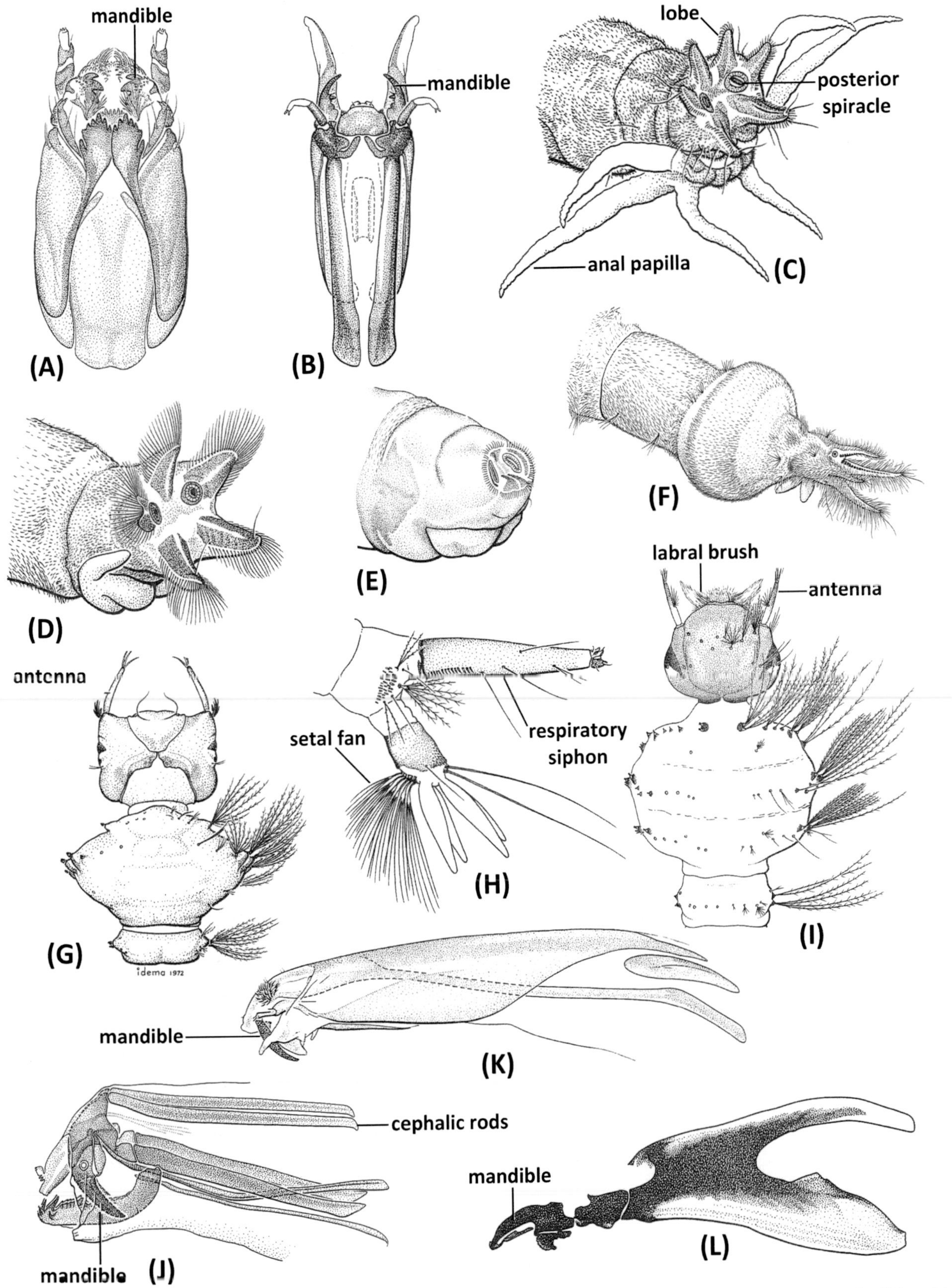

FIGURE 15.18 Structures of larval Diptera. (A) Head capsule (ventral) of Limoniidae (*Pseudolimnophila*). (B) Head capsule (dorsal) of Limoniidae (*Limnophila*). (C) Terminal segments (posterolateral view) of Tipulidae [*Tipula* (*Yamatotipula*)]. (D) Terminal segments (posterolateral view) of Limoniidae (*Ormosia*). (E) Terminal segments (posterolateral view) of Limoniidae (*Limnophila*). (F) Terminal segment (posterolateral view) of Limoniidae (*Limonia*). (G) Head, thorax, and abdominal segment I (dorsal view) of Chaoboridae. (H) Terminal segments (lateral view) of Culicidae (*Culex*). (I) Head and thorax (dorsal view) of Culicidae (*Culiseta*). (J) Sclerotized cephalic region (lateral view) of Empididae. (K) Head capsule (lateral view) of Tabanidae (*Tabanus*). (L) Cephalopharyngeal skeleton and mandibles (lateral view) of Muscidae. *From* Manual of Nearctic Diptera, *Volumes 1 (1981) and 2 (1987) by permission.*

Aquatic dipteran larvae can be classified according to the number and distribution of functional respiratory apertures (**spiracles**) as follows:

1. **Apneustic:** Although 10 pairs of spiracles are typically present, they are not functional, and air enters the tracheal system via diffusion across the body surface or especially in dedicated areas (e.g., branchiae or gills). Examples include the Blephariceridae, Chironomidae, and Simuliidae.
2. **Amphipneustic:** Two functional pairs of spiracles are present—one mesothoracic pair and one abdominal pair. Examples include the Psychodidae, Syrphidae, and some Muscidae (Fig. 15.17C–E).
3. **Holopneustic:** Ten pairs of functional spiracles are present, for example, in the Bibionidae (Fig. 15.14E).
4. **Metapneustic:** One functional pair of spiracles is present, located on the posterior of the abdomen. Examples include the Culicidae and Tipulidae (Figs. 15.18C–F and 15.18H).
5. **Peripneustic:** Nine pairs of functional spiracles are present, usually a thoracic pair and eight abdominal pairs. This condition is typical of terrestrial larvae. Among marginally aquatic Diptera, an example is the Scatopsidae.

In some families, the respiratory system changes with larval development. In the Ephydridae, for example, the first instars are metapneustic, whereas the second and third instars are amphipneustic. In particular families, such as the Ephydridae, Ptychopteridae, and Syrphidae, the spiracles are often at the end of a retractile **respiratory siphon** (respiratory tube or breathing tube) (Figs. 15.8A, 15.15E, and 15.17C, D). For example, this special morphological adaptation allows sapropagous aquatic larvae of *Eristalis* (Syrphidae) to live in slushy or putrid waters (e.g., sewage ponds) while acquiring oxygen from the air.

The structure of the spiracular plate and the lobes surrounding the posterior spiracles provide diagnostic characters at the family, genus, and species levels (Fig. 15.18C–F).

A number of families of aquatic Diptera, such as the Chironomidae, Culicidae, and Simuliidae, have pupae that are propneustic (Fig. 15.9C–E). That is, they have one functional (mesothoracic) pair of spiracles, each surrounded by a cuticular extension, the spiracular gill, or respiratory organ. In some of these groups (e.g., Simuliidae), the structure of the gills is of great taxonomic importance.

Further information on the general morphology and bionomics of aquatic Diptera is provided by Teskey (1981), Courtney et al. (2000), Fusari et al. (2018), Adler and Courtney (2019), Courtney (2019), and Merritt et al. (2019).

Sampling, identification, and preservation of larvae

Diptera are generally the most abundant macroinvertebrate group in benthic samples. The collection protocols vary according to habitat type, from the use of grabs for lake bottoms to handheld nets for streams and rivers. After sampling, larvae can be sorted in the field or transferred to the laboratory and typically fixed in 70%–95% ethanol, depending on the study aims (high percentages of ethanol, for example, are typically used in molecular analyses such as DNA barcoding). Alcohol-fixed larvae should be held in tightly sealed containers and stored in the dark at cool temperatures (15°C is generally sufficient) to avoid volatilization of the fixative and bleaching of the specimens. Depending on the purpose of the study, other fixatives may be necessary. For example, chromosomal studies (e.g., of Chironomidae or Simuliidae) require organisms to be fixed directly in Carnoy's solution (three-parts absolute ethanol to one-part glacial acetic acid) (Martin, 1979; Adler et al., 2016).

To identify dipteran larvae and pupae to the family level, a stereomicroscope usually is sufficient. Slide preparation might be needed to identify species and in some cases genera (also families if specimens are small). Slide preparation requires dissolving muscle tissue (i.e., clarification of the body), which can be done using solutions such as hot 85% lactic acid or cold or hot 10% potassium hydroxide (KOH) or sodium hydroxide (NaOH). Clearing time will depend on the specimen size and degree of sclerotization. Specimens subsequently must be dehydrated through an ethanol series (70%, 80%, 90%, and 100%), or in acetic acid and butanol, before being mounted in a more permanent medium, such as Euparal or Canada Balsam (Wirth & Marston, 1968; Lencioni et al., 2021). Analysis of temporary slides can be performed by mounting the specimen in an excavated slide with glycerin or a 1:1 mixture of glycerin and alcohol, and then covering the preparation with a coverslip. The specimen subsequently can be stored in a plastic or glass microtube with glycerin.

Aquatic and semiaquatic Diptera families in the Mediterranean Basin

Overall, 32 families of aquatic and semiaquatic Diptera live in freshwaters of the Mediterranean Basin (Table 15.1). Eight dipteran families consisting exclusively or partly of aquatic species are not found in the Mediterranean Basin: Axymyiidae, Corethrellidae, Deuterophlebiidae, Nymphomyiidae, Oreoleptidae, Pelecorhynchidae, Sarcophagidae, and Tanyderidae. Five additional families consist entirely or partly of species that are nearly exclusive to brackish or saline waters, including maritime coastal environments; they are not treated here: Canacidae, Coelopidae, Dryomyzidae,

TABLE 15.1 Estimated numbers of known extant species of Diptera with aquatic representatives in the Mediterranean basin, and their ecological characteristics.

Family	Total species in the world[a]	Aquatic species in the world[b]	Aquatic species in the Mediterranean Basin[c]	Predominant trophic group; habitat[d]
Ceratopogonidae	6206	4550	300	Collectors and predators; diverse lentic and lotic
Chaoboridae	51	51	4	Predators; lentic
Chironomidae	7290	7090	850	All trophic groups; all aquatic habitats
Culicidae	3725	3725	108	Collectors and some predators; lentic
Dixidae	197	197	27	Collectors; lentic and lotic (surfaces)
Simuliidae	2384	2384	105	Collectors; lotic
Thaumaleidae	183	183	25	Scrapers; madicolous
Ptychopteridae	80	80	4	Collectors; springs (mud)
Blephariceridae	330	330	56	Scrapers; lotic (rocks)
Bibionidae	1102	*[e]	*	Collectors?
Scatopsidae	407	5	**[f]	Collectors; tree holes
Psychodidae	3026	1988	200	Collectors and scrapers; lentic and lotic
Cylindrotomidae	71	2	2	Shredders; lentic and lotic
Limoniidae	10 813	8850	350–413	Collectors, predators, and shredders; lentic and lotic
Pediciidae	506	439	45	Predators; lentic and lotic
Tipulidae	4413	1550	100–127	Shredders; lentic and lotic
Anisopodidae	196	***[g]	***[g]	Collectors; lentic
Trichoceridae	157	*	*	Collectors; lentic and lotic (shores)
Stratiomyidae	2690	928	75	Collectors; lentic, madicolous, thermal springs
Athericidae	133	133	8	Predators; lotic
Rhagionidae	756	75	5	Predators; lentic and lotic (shores)
Tabanidae	4434	4434	218	Predators; lentic and lotic
Dolichopodidae	7910	3182	400–500	Predators; lentic and lotic
Empididae	3142	671	207	Predators; lotic
Lonchopteridae	65	2	**[f]	Collectors; lentic and lotic (shores)
Phoridae	4202	18	3	Collectors and predators; lentic (phytotelmata) and lotic (filter beds)
Syrphidae	6107	1341	60	Collectors; lentic and lotic margins (saturated wood)
Sciomyzidae	618	194	140	Predators; wetlands
Ephydridae	1994	1251	230–240	Collectors and shredders; lentic, lotic (margins) and marine intertidal
Muscidae	5218	701	59	Predators; lentic and lotic

(Continued)

TABLE 15.1 (Continued)

Family	Total species in the world[a]	Aquatic species in the world[b]	Aquatic species in the Mediterranean Basin[c]	Predominant trophic group; habitat[d]
Fanniidae	359	****[h]	****[h]	Collectors; lentic and lotic
Scathophagidae	419	150	6	Shredders and predators; lentic
Total	79 184	44 504	3587–3787	–

[a] *Pape et al. (2011), Grichanov (2016), Courtney et al. (2017), Adler (2020), and Borkent & Dominiak (2020).*
[b] *Borkent (2014), Morse (2017), and Adler & Courtney (2019); V. Podeniene (pers. comm. For families of Tipuloidae).*
[c] *Ceratopogonidae (A. Borkent, pers. comm.), Chaoboridae (Borkent, 2014); Culicidae (Robert et al., 2019); Chironomidae (Laville & Reiss, 1992; J. Moubayed-Breil, pers. comm.); Dixidae and Empididae (M. Ivković, pers. comm.); Simuliidae (Adler, 2020); Thaumaleidae (Wagner, 2013); Ptychopteridae (Campaioli, 1999, Fasbender, 2013); Blephariceridae (G. Courtney, unpublished); Psychodidae (R. Wagner, pers. comm.); Cylindrotomidae, Limoniidae, Pediciidae, and Tipulidae (V. Podeniene, pers. comm.); Stratiomyidae (M. Hauser and F. Mason, pers. comm.); Athericidae (Rozkošný, 2013); Rhagionidae (based on a rough estimate of 10% aquatic or semiaquatic species out of about 55 total species of Rhagionidae in the region); Tabanidae (Kiliç, 2006; Chvála, 2013; El Haouari et al., 2014); Dolichopodidae (I. Grichanov, pers. comm.); Phoridae (Disney, 1991, 1999); Syrphidae (D. Sommaggio, pers. comm.); Sciomyzidae (Vala, 1989); Ephydridae (W. Mathis and T. Zatwarnicki, pers. comm.); Muscidae (A. Pont, pers. comm.); Fanniidae (Rozkošný et al., 1997); Scathophagidae (De Jong, 2013).*
[d] *Merritt et al. (2019).*
[e] ** Occurrence of Bibionidae and Trichoceridae in aquatic environments is probably accidental; an estimated number of species is, therefore, not given.*
[f] *** Scatopsidae and Lonchopteridae are found in the Mediterranean Basin, but no aquatic records are known for the basin.*
[g] **** Four species of Anisopodidae are known for the Mediterranean Basin (De Jong, 2013), but no estimates of the world number of aquatic species are available and no aquatic records are known for the Mediterranean Basin.*
[h] ***** About 30 species of Fanniidae are found in the Mediterranean Basin, but no estimates of the world number of aquatic species are available and no aquatic records are known for the Mediterranean Basin.*

Helcomyzidae [with the Mediterranean species *Helcomyza mediterranea* (Loew)], and Heterocheilidae. In recognizing aquatic species in the Mediterranean Basin, we err on the liberal side and include families not typically considered to have aquatic representatives. These include families with species that inhabit saturated soil and might accidentally enter adjacent bodies of water (e.g., Bibionidae and Trichoceridae), treeholes that sometimes hold water (e.g., Anisopodidae and Faniidae), and other marginally aquatic habitats such as waterlogged leaves (e.g., Scatopsidae). The following three taxa (one superfamily and two families) with truly aquatic representatives are independently treated in detail elsewhere in this book (Chapters 15.1–15.3) and are not discussed below, although they are included in the identification key:

Chapter 15.1—Tipuloidae (crane flies)
Chapter 15.2—Chironomidae (nonbiting midges)
Chapter 15.3—Simuliidae (black flies)

Lower Diptera

Ceratopogonidae (biting midges)

The family Ceratopogonidae (Figs. 15.4C, D, 15.10E–H, and 15.14C) is large and diverse, with 112 genera and 6206 living species (Borkent & Dominiak, 2020). The family includes terrestrial, semiaquatic, and aquatic species. About 4550 species can be considered aquatic, with about 300 aquatic species in the Mediterranean Basin (Table 15.1).

Larvae range from 3 to 18 mm long. The larval body shape varies within the family, and four main groups can be distinguished (Campaioli, 1999):

1. Genuine (subfamily Forcipomyiinae; Figs. 15.4C, 15.10E–G, and 15.14C): Larvae are hypognathous and have paired prothoracic and anal prolegs with dark apical spines. The body is covered with conspicuous bristles and setae or fleshy processes. They live in damp soil, ponds, weakly flowing waters, and wastewater.
2. Intermediate (subfamily Dasyheleinae): Larvae are hypognathous and have one unpaired retractile proleg with hooks on the last anal segment only; the body is slender and without conspicuous setae. They are marginally aquatic, living in sluggish rivers and brackish water.
3. Maggot-like (subfamily Ceratopogoninae, tribe Palpomyiini; Figs. 15.4D and 15.10H): Larvae are prognathous and have a small head, thread-shaped body with or without pigmentation, evident segmentation, no prolegs, and the last segment with a crown of hooks. Larvae live in ponds, weakly flowing waters, and swamps rich in vegetation.
4. Musciform (subfamily Leptoconopinae): Larvae have a reduced, unsclerotized head capsule and thin body with supernumerary apparent segments (>20). They are widespread in wetlands with sandy or silty-clay soils and in estuaries, lagoons, and saline sandy soil on coastal and inland beaches.

Larvae of the Ceratopogonidae are characteristic of damp, semiaquatic habitats such as soil, mud, decaying vegetation, and sand in arid areas, coastal or ocean-side beach areas (subfamily Leptoconopinae), and flowing tree sap (subfamily Dasyheleinae). In addition, some genera develop in lentic systems such as tree holes (*Dasyhelea* Kieffer, 1913), rain pools, marshes, ponds, lakes, and algal mats in hot springs (*Bezzia* Kieffer, 1899). Other genera are fully aquatic and free swimming in the benthos of large lakes and streams. Larvae of aquatic members of the Ceratopogonidae have a unique swimming motion, gliding or undulating in a snake-like or eel-like fashion. The females of most ceratopogonids acquire protein by feeding on insect hemolymph, but the following four genera have representatives that take blood from humans and other vertebrates: *Culicoides* Latreille, 1809, *Leptoconops* Skuse, 1889, *Forcipomyia* Meigen, 1818 (subgenus *Lasiohelea* Kieffer, 1921), and *Austroconops* Wirth & Lee, 1958 (only present in the Australian region). The genus *Culicoides* is the most important in medical and veterinary entomology.

Chaoboridae (phantom midges)

The family Chaoboridae (Figs. 15.5E–G, 15.9E, and 15.18G) includes 51 extant species in six genera, of which four species are known in the Mediterranean Basin: *Chaoborus crystallinus* (De Geer, 1776), *C. flavicans* (Meigen, 1830), *C. pallidus* (Fabricius, 1794), and *Mochlonyx velutinus* (Ruthe, 1831) (Borkent, 2014). Commonly known as phantom midges or glassworms, they are common to abundant midges with a cosmopolitan distribution.

The larvae are of medium size (9–14 mm) with a large head and the thoracic segments fused and variously expanded (Figs. 15.5F and 15.18G). They are nearly transparent, sometimes with a slight yellowish cast, and have two air sacs (hydrostatic vesicles), one in the thorax and one in the abdomen (Fig. 15.5E–G).

Larvae and pupae live in lakes and ponds. The larvae are unique in their feeding method: the antennae are modified into grasping organs somewhat resembling the raptorial arms of a mantis, with which they capture prey. They feed primarily on small insects (e.g., mosquito larvae) and crustaceans (e.g., cladocerans). The antennae impale or crush the prey, and then bring them to the mouth. Although similar to the Culicidae, the Chaoboridae differ by having raptorial antennae and thoracic and abdominal air sacs visible through the integument. The larvae sometimes form large aggregations that move vertically at night in search of food (Campaioli, 1999). The adults are delicate flies resembling the Chironomidae and Culicidae and are not hematophagous.

Culicidae (mosquitoes)

The family Culicidae (Figs. 15.5A–C, 15.9C, 15.11A–C, and 15.18H, I) includes more than 3700 recognized species in about 40 genera and two or three subfamilies (Anophelinae, Culicinae, and, in some classifications, Toxorhynchitinae). Most species remaining to be discovered probably inhabit tropical rain forests (Foster & Walker, 2019). The Mediterranean Basin has seven genera and about 108 species (Robert et al., 2019).

Larvae are small to medium sized (5–10 mm), somewhat hairy, and eucephalic with a large head and the thoracic segments fused and expanded (Fig. 15.5B, C). The abdominal segments have long setae, and the last segment has a more or less developed setal fan or fringe of swimming hairs (Fig. 15.18H). A respiratory siphon is present in the Culicinae and Toxorhynchitinae (Fig. 15.5A, B), but absent in the Anophelinae.

Mosquitoes inhabit a wide variety of lentic systems and are generally not found in lotic systems except in littoral areas of slow-moving streams and rivers. Anthropogenically created larval mosquito habitats include barrels, bottles, cans, cemetery vases, discarded tires, plant pots, and rain gutters. The trophic or functional feeding relationships of the larvae involve mostly filter-feedering (e.g., many *Aedes* Meigen, 1818, *Anopheles* Meigen, 1818, and *Culex* L., 1758); however, some species scrape algae from rocks. Some taxa, such as the large toxorhynchitines, are predators on other mosquito larvae and invertebrates. The females of most mosquitoes are blood feeders, although some (e.g., toxorhynchitines) do not have cutting mouthparts.

Dixidae (meniscus midges)

Of the 197 world species in the family Dixidae (Figs. 15.5D, 15.11D, E, and 15.14A), all are aquatic in the immature stages, and 27 species are found in the Mediterranean Basin (Table 15.1).

The characteristic larvae are unmistakable among aquatic Diptera. They are eucephalic (Fig. 15.5D), slender and elongate, and less than 10 mm long. The abdomen has a pair of prolegs on the first one or two segments, often a corona of plumose hairs dorsally on the anterior segments, and a posterior set of processes with a few elongate setae and a fringe of hydrophobic hairs (Figs. 15.11D, E and 15.14A).

The larvae assume a U-shape while resting at the air–water interface or on damp surfaces in lotic and lentic habitats associated with aquatic vegetation. They feed on microorganisms and detritus either by scraping submerged substrates or by using their labral brushes to bring food to the mouth. Pupation typically occurs just above the waterline on emergent vegetation and stones. The adults are slender, superficially nondescript flies found near the larval habitat. They evidently have the ability to take liquids, although feeding habits are poorly studied.

Thaumaleidae (solitary midges or trickle midges)

Midges of the family Thaumaleidae (Figs. 15.4E, 15.10M, and 15.14B) are small and exclusively aquatic, with about 200 species in eight genera and worldwide distribution (Haubrock et al., 2017). About 25 species have been found in the Mediterranean Basin (Table 15.1).

Larvae are of small to medium length (6–15 mm), slender, and with one prothoracic and one anal proleg (Fig. 15.10M); the thorax and abdomen have a large dorsal plate on each segment. The larvae are hypognathous with small anterior bumps on the head. The anterior spiracles are on short stalks (Fig. 15.14B), and the posterior ones are united in a wide dorsal opening on the eighth segment from which two sclerotized procerci with long hairs protrude.

Thaumaleid larvae and pupae typically inhabit madicolous (hygropetric) habitats in vertical, thin water films alongside springs, streams, and waterfalls. They are associated with cool temperatures and are most frequent in shaded areas. Larvae are found on rocks in films rarely deeper than 2 mm, where they graze on diatoms and move across the substrate in a characteristic sidewinding fashion. The nonfeeding adults are usually found on foliage along the same streams in which the larvae are found (Wagner, 2002).

Ptychopteridae (phantom crane flies)

The family Ptychopteridae (Figs. 15.6A, 15.9B, and 15.15E) has 78 species in the northern hemisphere and the Afrotropical Region, and includes three genera: *Bittacomorpha* Westwood, 1835, *Bittacomorphella* Alexander, 1916, and *Ptychoptera* Meigen, 1803 (Fasbender, 2013). Four named *Ptychoptera* species inhabit the Mediterranean Basin: *Ptychoptera helena* (Peus, 1958) in Greece (Fasbender, 2013) and *P. albimana* (Fabricius, 1787), *P. contaminata* (L., 1758), and *P. paludosa* Meigen, 1804 in Italy (Campaioli, 1999).

Larvae are 15–35 mm long and caudally elongated, making them recognizable by the long telescopic respiratory siphon for breathing atmospheric oxygen (Figs. 15.6A and 15.15E). To cope with respiration after pupation, the right thoracic breathing horn is greatly elongated, up to twice the length of the pupa, whereas the other horn is atrophied (Fig. 15.9B). Adults superficially resemble those of the Tipulidae.

Larvae are associated with saturated mud or soil near water (Brindle, 1962) and shallow lentic waters with abundant vegetation and decaying organic matter, such as in marshes, ponds, springs, and even sulfurous pools (Campaioli, 1999).

Blephariceridae (net-winged midges)

The family Blephariceridae (Figs. 15.6C, D, 15.9D, and 15.11F, G) is found worldwide, with more than 300 exclusively aquatic species in about 40 genera (Pape et al., 2011). They are one of the most distinctive and specialized insect families. The Mediterranean Basin harbors 56 species in the following five genera: *Apistomyia* Bigot, 1862; *Blepharicera* Macquart, 1843; *Dioptopsis* Enderlein, 1937; *Hapalothrix* Loew, 1876; and *Liponeura* Loew, 1844.

Larvae are 5–12 mm long, with a distinct aspect—a cephalothorax (formed by fusion of the head, thorax, and first abdominal segment) and six ventral suctorial disks for adhering to rocks in torrential waters (Figs. 15.6C, D and 15.11F, G). This body shape makes them well adapted to live in the cascades, rapids, and waterfalls of mountain streams.

Blephariceridlarvae are grazers (scrapers), using their highly specialized mouthparts to feed on films of algae, bacteria, and other organic matter (i.e., periphyton) on submerged rocks. The dorsoventrally compressed and streamlined pupae are also well adapted to life in high current velocities (Fig. 15.9D). In some streams, densities of immature stages can exceed $1000/m^2$, making blephariceridsnot only the dominant grazer, but also one of the most abundant insects in these situations (Courtney, 2000).

Bibionidae (march flies)

The family Bibionidae (Fig. 15.14E) includes more than 1100 described species in 12 genera (Pape et al., 2011). A few species have been reported from aquatic environments; perhaps most are accidental in truly aquatic habitats (Wagner et al. 2008; Dobson 2013). We include the family here for the sake of completeness.

Larvae are up to 25 mm long. The body is cylindrical, without abdominal prolegs, but with rows of fleshy extensions that are longest on the posterior of the abdomen (Fig. 15.14E). Ten spiracles are on each side of the body.

The larvae are terrestrial or semiaquatic in decaying plants, damp detritus, and marshes, and occasionally along the banks of streams. They feed primarily on decaying organic matter and subterranean parts of plants. Adults can be abundant in the spring, giving rise to the common name "march flies."

Scatopsidae (minute black scavenger flies)

The family Scatopsidae is cosmopolitan, with more than 400 described species in 36 genera (Amorim, 2016). Only a little more than 1% of the species are considered aquatic (Adler & Courtney, 2019). Although no species are known to be aquatic in the Mediterranean Basin, a few might be expected.

Up to 8 mm long, the larvae have short extensions supporting spiracles at the end of the abdomen or a semicircular extension. Nine spiracles are visible along the abdomen, but no spiracles are on the meso- or metathorax. The abdomen lacks prolegs.

Larvae live in decaying plant or animal material, humid wood holes, and excrement. A small number of species have aquatic larvae that live under the surface of a water film among waterlogged leaves, and a few other species have larvae that are dendrolimnobiontic (e.g., inhabit phytotelmata). Pupae are sometimes found in the exuviae of the last larval instar (Wagner et al., 2008).

Psychodidae (moth flies, owl flies, and sand flies)

The family Psychodidae (Figs. 15.6B and 15.11H–J) is cosmopolitan, with more than 3000 described species in more than 140 genera (Pape et al., 2011; Courtney et al., 2017). Probably more than 65% of the species are aquatic (Morse, 2017; Adler & Courtney, 2019). Of the aquatic subfamilies, only the Psychodinae and the Sycoracinae (few species) are found in the Mediterranean Basin. The Mediterranean Basin is home to roughly 200 species (Table 15.1).

The larvae of most species are 4–5 mm long, usually elongated (Fig. 15.6B) but more oval in some species, and typically have the abdominal segments divided into two or three annuli (pseudosegments) (Fig. 15.11H–J). They are amphipneustic and have a well-sclerotized external head capsule.

Many species develop in muddy soils, mosses, tree holes, and detritus at the edges of ponds and streams. Some species develop in fast-flowing streams and madicolous habitats, such as on rocks in the splash zone adjacent to waterfalls. The adults are primarily nocturnal. Members of the subfamily Phlebotominae, of which the females feed on vertebrate blood, are terrestrial as larvae, at least those that are known.

Anisopodidae (wood gnats or window gnats)

The family Anisopodidae (Fig. 15.14D) has about 200 described species in 24 genera (Pape et al., 2011). About four named species inhabit the Mediterranean Basin, all in the genus *Sylvicola* Harris, 1780 (De Jong, 2013a, 2013b). The members of this family are primarily terrestrial, although some species have been found in freshwaters. No aquatic records are available from the Mediterranean Basin, but we include the family for completeness.

The larvae are 10–13 mm long, with a complete head capsule, an amphipneustic respiratory system, no prolegs, a rather bare body, and the abdomen with secondary annulations and a short conical endpiece or five short, fleshy lobes (Fig. 15.14D). The posterior spiracles are crescent-shaped with multiple openings.

The saprophagous larvae have been recorded from wet or damp areas with decaying organic matter, such as tree holes, moldering plants, sewage filter beds, between moist leaves, and in dung.

Trichoceridae (winter crane flies)

The family Trichoceridae (Figs. 15.11K, L, and 15.16A) includes about 160 described, extant species found predominantly in cool to moderate climates, with about seven known species extending into North Africa (Driauach et al., 2015). This family is typically considered terrestrial; however, the larvae are occasionally found in stream samples, probably entering accidentally from surrounding wet soils. We include the family for completeness.

Larvae are usually 6–8 mm long, and have a complete, nonretractable head capsule (Fig. 15.11L) with a characteristic beard-like fringe of hairs ventrally. Each body segment is secondarily annulated, prolegs are absent, and the body is covered with fine pubescence (Figs. 15.11K, L and 15.16A). The respiratory system is amphipneustic, and the spiracular disk is surrounded by four pigmented or unpigmented, equal-sized lobes.

The larvae are scavengers in terrestrial to semiaquatic habitats, such as moist soil, rotting plants, dung, and mushrooms.

Brachycera

Stratiomyidae (soldier flies)

The family Stratiomyidae (Figs. 15.1C, 15.7C, D, and 15.13A–D) is distributed worldwide, with roughly 2700 described species in 385 genera (Pape et al., 2011). Probably about one-third of all species are aquatic (Morse, 2017). About 75 aquatic species in 10 genera are known in the Mediterranean Basin (Table 15.1).

Mature larvae are typically flattened and about 10–55 mm long (Figs. 15.7C and 15.13A, C). The head capsule is well developed (Fig. 15.7D). Prolegs are absent, although some species have ventral hooks. Many aquatic species have a distinctive ring of hairs around the breathing tube at the posterior of the abdomen (Fig. 15.13D). The cuticular microsculpture, which has hexagonal calcium carbonate crystals, is characteristic.

The family includes species with terrestrial to fully aquatic habits. Aquatic and marginally aquatic species live in wet soils, shallow water often in association with emergent vegetation, and hygropetric areas. They have been found in thermal springs at some of the highest temperatures recorded for macroinvertebrates (Adler & Courtney, 2019). The larvae of aquatic species are primarily algal feeders and sarcophages. The adults often have colorful abdominal patterns (Fig. 15.1C).

Athericidae (water snipe flies)

The family Athericidae (Fig. 15.7A, B and 15.13G, H) includes 133 described species worldwide, presumably all of which are aquatic, in 12 genera (Pape et al., 2011). About eight species in four genera are found in the Mediterranean Basin (Table 15.1). Among the most common Mediterranean species (Campaioli, 1999) are *Atherix ibis* (Fabricius, 1798), *Atrichops crassipes* (Meigen, 1820), and *Ibisia marginata* (Fabricius, 1781).

Mature larvae are 16–30 mm long, tapered, and slender toward the front (Fig. 15.7A, B), and hemicephalic. Abdominal prolegs are present and the last abdominal segment has two often-feathered processes (Fig. 15.13G, H).

Larvae live in sandy sediments of streams and in rocky or gravelly riffle areas (Campaioli, 1999; Woodley, 2009). They are predaceous, feeding on larvae of other aquatic insects (Rozkošný & Nagatomi, 1997). Pupae are terrestrial or found in mosses or soil along stream margins.

Rhagonidae (snipe flies)

The family Rhagionidae (Fig. 15.16C) includes about 755 species in 47 genera worldwide (Pape et al., 2011). The larvae of most species are terrestrial, with some aquatic and semiaquatic species in three genera (Campaioli, 1999): *Chrysopilus* Macquart, 1826, *Rhagio* (Fabricius, 1775), and *Symphoromyia* Frauenfeld, 1867. Within these three genera, however, aquatic representation has rarely been addressed, although a rough estimate might be 10% of the species. In the Mediterranean Basin, about 55 species in these three genera have been reported (Majer, 2013), with perhaps 10% being aquatic or semiaquatic, including the stream inhabitant *Chrysopilus erythrophthalmus* Loew, 1840, known from France (Corsica) and the Italian mainland.

Mature larvae are up to 22 mm long, slightly tapered apically, hemicephalic, and with abdominal creeping welts (Fig. 15.16C). The last abdominal segment has a V-shaped transverse recess, in which small conical extensions protrude and the spiracular openings are located.

Aquatic larvae inhabit fine sediments under stones, lake and river shores, areas where rock substrates are covered with a thin layer of water, and aquatic bryophytes. Larvae and pupae of terrestrial species can be found in damp soils (Oscoz et al., 2011), under bark, and in mosses and rotting wood.

Tabanidae (horse flies)

The family Tabanidae (Figs. 15.1D, 15.7G, 15.9G, 15.13E, F, 15.16B, and 15.18K) has a global distribution, with approximately 4500 species in 156 genera (Pape et al., 2011). In the Mediterranean Basin, about 218 species in 14 genera have been recorded (Table 15.1).

Mature larvae are about 10 to more than 20 mm long and fusiform (Chrysopsinae and Tabaninae, Figs. 15.7G, 15.13E, and 15.16B) or broadened at the thorax (Pangoniinae). They are various colors, such as white, green, or dark brown, and can have longitudinal streaks. They have 11 body segments with a crown of three or four short prolegs along the anterior edge of each segment (Fig. 15.13F). The larvae are hemicephalic (Fig. 15.18K) with a completely retractile head capsule and partially retractile, short respiratory siphon on the last abdominal segment.

The larvae of most species are semiaquatic, inhabiting damp soil, swamps, and the banks of rivers and ponds. A few species are restricted to sand and gravel. Many larvae inhabit specialized niches, such as tree holes, bromeliads, or submerged aquatic vegetation. Members of *Chrysops* Meigen, 1803 and *Hybomitra* Enderlein, 1922 develop in ponds and streams. The larvae of most species are predators of small organisms, such as crustaceans, earthworms, insects, and snails, and sometimes practice cannibalism (Mullens, 2019), but some species are detritivores. Pupae tend to be terrestrial (Fig. 15.9G). Most females are hematophagous on vertebrates, primarily mammals. Males (Fig. 15.1D), like the females, feed on nectar and other plant sugars, but do not take blood.

Dolichopodidae (long-legged flies)

The family Dolichopodidae (Figs. 15.7E and 15.16D) is highly diverse and includes more than 7900 described species in 250 genera worldwide, with more than 40% considered aquatic (Yang et al., 2006; Grichanov, 2016; Adler & Courtney, 2019). About 1630 are known from the Palearctic Region (Grichanov, 2016). The Mediterranean Basin has an estimated 400–500 aquatic species (Table 15.1).

Mature larvae are 9–15 mm long, usually whitish, cylindrical, and tapered anteriorly, with an extremely reduced head capsule and abdominal creeping welts (Fig. 15.7E). The last abdominal segment bears four or more posteriorly directed lobes (Fig. 15.16D), producing a truncated, or less frequently, a rounded appearance.

Larvae are predominantly predaceous, although some are phytophagous. They are found in mud, decomposing wood, soil litter, and under bark. Semiaquatic species are typically found along the margins of lentic and lotic habitats, with some in salt marshes and seashore habitats. Adults prey on smaller insects and are often observed on foliage, tree trunks, and stones, especially along the shores of lentic and lotic habitats.

Empididae (balloon flies)

The family Empididae (Figs. 15.7F, 15.9F, 15.13I, J, and 15.18J) consists of 3142 species (Pape et al., 2011). About 671 species, or 21%, are considered aquatic, of which 207 are recorded from the Mediterranean Basin (Table 15.1).

Mature larvae are hemicephalic (Fig. 15.18J) and less than 20 mm long. They are apneustic except in a few aquatic genera that are metapneustic. The abdomen has seven or eight pairs of prolegs with hooks or crochets, and the terminal segment bears elongate lobes or small tubercles with apical setae (Figs. 15.7F and 15.13I, J).

The larvae of aquatic species are found in flowing water and along the shores of lentic habitats including bogs and marshes. They are predaceous on invertebrates, particularly dipteran larvae, killing the prey by piercing them and then sucking the contents. The adults are typically slender with a long or short, dagger-like proboscis. They are common in a wide variety of habitats where they prey on arthropods.

Lonchopteridae (spear-winged flies)

The family Lonchopteridae (Fig. 15.17B) includes about 65 species worldwide (Pape et al., 2011), mostly in a single genus, *Lonchoptera* Meigen, 1803. In Europe, at least two species are considered aquatic: *Lonchoptera nigrociliata* Duda, 1927 and *L. lutea* Panzer, 1809 (Wagner et al., 2008; Vaillant, 2002). No aquatic representatives are known from the Mediterranean Basin, but the family is included here for completeness.

Mature larvae are up to 5 mm long, rather flattened dorsoventrally with several long anterior and posterior bristle-like processes (Fig. 15.17B). They are amphipneustic, with the anterior and posterior spiracular pairs on short stalks. The striated margins of the tergites are diagnostic.

The larvae live in wet microenvironments, such as decaying organic matter, among dead leaves, in plant debris, in gravel bars, and under stones on shores of watercourses. They feed on decaying organic matter and microorganisms.

Phoridae (scuttle flies)

The family Phoridae (Fig. 15.16E) is found worldwide, with more than 4200 species (Pape et al., 2011), of which 18 (mostly in Southeast Asia) at least sometimes inhabit aquatic habitats (Disney, 1991, 2004). The Mediterranean Basin has three cosmopolitan species that occasionally are found in aquatic habitats: *Dohrniphora cornuta* (Bigot, 1857), *Megaselia rufipes* (Meigen, 1804), and *M. scalaris* (Loew, 1866) (Disney, 1991).

Phorid larvae that live in aquatic habitats are typically 5 mm long or less when mature, tapered at both ends or somewhat flattened, and with short, spine-like or plumose projections arising from the body (Fig. 15.16E). They are amphipneustic with unbranched anterior spiracles and posterior spiracles on cone-shaped projections, each with four spiracular openings.

Larval phorids are claimed to exploit a wider range of habitats than any other family of insects (Disney, 1999). Those that inhabit aquatic environments typically do so opportunistically, exploiting sewage-filter systems and phytotelmata, where the larvae scavenge dead invertebrates, fungi, pollen, films of microorganisms, and a range of organic detritus (Disney, 1999). Adult phorids, although capable fliers, often run rapidly, giving rise to the common name "scuttle flies."

Syrphidae (flower flies or hover flies)

The family Syrphidae (Figs. 15.8A and 15.17D) has a worldwide distribution with more than 6100 described species in about 210 genera (Pape et al., 2011). Roughly 22% of the species are aquatic (Morse, 2017; Adler & Courtney, 2019). In the Mediterranean Basin, about 60 of the 476 species are aquatic (Table 15.1).

Larvae are up to 20 mm long, not including the respiratory siphon that in the so called "rat-tailed maggots" can extend to several times the body length (Fig. 15.8A). The body form is highly varied but often maggot-like or with a wrinkled appearance (Fig. 15.17D). No external trace of a head capsule is visible.

Larvae inhabit a wide variety of habitats from truly aquatic to truly terrestrial. Aquatic habitats include standing waters with plenty of decaying vegetation, detritus-filled pools, and tree holes. Most aquatic species feed on decaying organic matter, although some are predaceous. The adults are common visitors to flowers from which they feed on the nectar and pollen.

Sciomyzidae (marsh flies or snail-killing flies)

The family Sciomyzidae (Figs. 15.1E, 15.8B, C, 15.9H, and 15.15A) is cosmopolitan with more than 600 described species in 66 genera (Pape et al., 2011). About a third of the species are deemed aquatic (Morse, 2017; Adler & Courtney, 2019), and about 140 species have been recorded in the Mediterranean Basin (Table 15.1).

Larvae are typically small (2–7 mm long, rarely up to 20 mm), heavily wrinkled, acephalic, and markedly tapered and retractile anteriorly (Fig. 15.8B, C). Prolegs are absent, although creeping welts are present in some aquatic forms. The last abdominal segment is truncated, with a flat spiracular plate surrounded by a variable number of lobes; the spiracles are often surrounded by short, water-repellent hairs (Fig. 15.15A). Pupae are enclosed in a rigid, somewhat barrel- or flask-shaped puparium that can float in aquatic forms (Fig. 15.9H).

Larvae live in standing waters in freshwater wetlands with emergent vegetation on which the adults (Fig. 15.1E) often rest head down. Larvae of most species are parasitoids or predators of aquatic or terrestrial mollusks (Gastropoda and Bivalvia). These feeding habits make the larvae ideal for biological control of agriculturally important slugs and snails and of vectors of parasitic organisms (snail-inhabiting trematodes) that cause diseases such as fascioliasis and schistosomiasis (Murphy et al., 2012).

Ephydridae (shore flies)

The family Ephydridae (Figs. 15.8D, E and 15.17C) is widespread and has nearly 2000 described species worldwide, the majority of which are aquatic, in 128 genera (Pape et al., 2011; Adler & Courtney, 2019). More than 230 named species are found in the Mediterranean Basin (Table 15.1).

Larvae are 6–13 mm long and acephalic, typically with a fusiform body, and with or without prolegs or creeping welts (Fig. 15.17C). The posterior spiracles are sometimes on elongate respiratory tubes (Figs. 15.8D, E and 15.17C). The integument is often transparent, sometimes thickened or with piliform processes.

Larvae of aquatic forms live in tree holes, stagnant waters, and alkaline lakes, along the margins of flowing and standing water and seashores, and even in pools of crude petroleum. They have a wide range of feeding habits, but many species graze on algae and other aquatic plants and, consequently, some species can be pests of crops such as rice.

Muscidae (house flies and relatives)

The family Muscidae (Figs. 15.1F, 15.8G, H, 15.13K–N, 15.17E, and 15.18L) is cosmopolitan and almost universally abundant, with about 5200 species of which about 14% are aquatic (Morse, 2017; Adler & Courtney, 2019). About 60 species with aquatic larvae in the following four genera are known in the Mediterranean Basin: *Limnophora*, *Lispe*, *Lispocephala*, and *Spilogona* (Table 15.1).

Mature larvae are 8–17 mm long, typically cylindrical and rather bare, acephalic, and tapered anteriorly (Figs. 15.8G, H and 15.13K, M). The abdomen is furnished with creeping welts and occasionally prolegs on segment 8 and is typically rounded or truncated posteriorly, with spiracles of three openings each (Fig. 15.17E). The cephalopharyngeal skeleton (Fig. 15.18L) is often visible through the integument.

Larvae of aquatic muscids can be found in running and standing waters (e.g., *Limnophora* Robineau-Desvoidy, 1830 and *Lipse* Latreille, 1796). They are generally collected in decaying organic matter and algal mats, presumably where they find their prey. Muscid larvae are coprophagous, saprophagous, or predaceous primarily on other Diptera. Adults (Fig. 15.1F) have extremely diverse habits including predation and bloodsucking (the latter not in aquatic species). Most, however, feed on decomposing animal or vegetable matter.

Fanniidae (little house flies)

The family Fanniidae (Fig. 15.17A) is widely distributed, although it is best represented in the Holarctic Region. At one time considered a subfamily of the Muscidae, it includes about 360 described species in four genera (Pape et al., 2011). About 30 nominal species are found in the Mediterranean Basin, of which a few sometimes might be found in aquatic habitats (Rozkošný et al., 1997).

Larvae are small (6–8 mm long), acephalic, and without prolegs. The body is dorsoventrally flattened and clearly segmented, with long, branched or unbranched, plumose dorsal and lateral processes (Fig. 15.17A).

Larvae live in decaying matter, bird's nests, cesspools, dung, fungi, latrines, nests of social Hymenoptera, rotting tree holes, and even sluggish streams, where they scavenge decaying organic matter, often scraping microorganisms from surfaces (Rozkošný et al., 1997; Faasch, 2015).

Scathophagidae (dung flies)

The family Scathophagidae (Fig. 15.8F) has more than 400 described species worldwide, of which more than a third are aquatic (Courtney et al., 2017; Adler & Courtney, 2019). About six species are recorded from the Mediterranean Basin (Table 15.1).

Larvae are 8–17 mm long, rather cylindrical, anteriorly tapered, and acephalic, with creeping welts (Fig. 15.8F). The anterior spiracles are two-branched and have up to 60 papillae. The terminal segment has four to eight pairs of tubercles surrounding a flattened area, and the posterior spiracles, each with three slits, are on short tubes. The pupae are barrel-shaped.

Most larvae are terrestrial or semiaquatic. The aquatic species are often found in detritus at the margins of water bodies. Selected examples of aquatic species include the leafminer *Hydromyza livens* (Fabricius, 1794) associated with water lilies (Nymphaeaceae), *Spaziphora hydromyzina* (Fallén, 1819) often in wastewater treatment plants, and *Acanthocnema glaucescens* (Loew, 1864) in the riparian zone of rivers (Campaioli, 1999). Larvae exhibit a range of feeding habits, from carnivory and coprophagy to phytophagy.

Key to larvae of aquatic and semiaquatic families of Diptera

Identification of larval Diptera is problematic, even to the family level. Examination of some diagnostic characters is not easy under a stereomicroscope, especially of the head capsule when it is retracted into the thorax. In some cases, it is necessary to dissect the head or other parts of the body for analysis at higher magnifications. This procedure can be time-consuming and requires taxonomic expertise. Many species are still unknown in the larval stage or are not included in commonly used general keys. Furthermore, many identification keys are incomplete or regionally restricted and often are written in different languages. Simpler keys for less taxonomically oriented users have been published, but simplification often comes at the cost of taxonomic detail, and consequently mistakes in determinations are common (Sundermann et al., 2007).

For a few families and geographical regions, comprehensive keys are available to species, but for most regions, keys are out of date, incomplete, or simply absent. In Europe, more than 10,000 species in 132 families of Diptera have been recorded (Oosterbroek, 2006). These numbers change, however, depending on the authors. For example, a figure of about 19,400 native species in 125 families was given by Skuhrava et al. (2010), emphasizing how much the knowledge of Diptera is in progress. The Mediterranean Basin is home to roughly 3600–3800 aquatic species of flies (Table 15.1). The following key pertains only to families with freshwater representatives that have been, or are likely to be, recorded in the Mediterranean Basin.

Diptera: Families

1 Mandibles moving against one another in horizontal or oblique plane (may require dissection if head is retracted into thorax) (Fig. 15.10C, D). Head capsule completely sclerotized (eucephalic, Figs. 15.4H, 15.5C, 15.10I, J, and 15.11L), or with short to long lengthwise incisions or slender rods in some tipuloids (Figs. 15.12B and 15.18A, B); fully exposed (Fig. 15.4H) to fully retracted (Fig. 15.12A) within thorax ("Lower Diptera") 2

1' Mandibles (mouth hooks) moving parallel to one another in vertical plane (may require dissection) (Figs. 15.15A and 15.18J–L), or absent (Syrphidae). Head capsule typically reduced, at least partially (hemicephalic) to completely unsclerotized (acephalic, but with pigmented cephalopharyngeal skeleton visible through integument) (Figs. 15.8H and 15.15A), partly to fully retracted within thorax (Figs. 15.13E–G, I–K, 15.16B–E, and 15.17A–E) (Brachycera) 22

2(1) Head capsule partially or fully retracted into thorax (Figs. 15.6E–H, 15.12A, E–H, and 15.15B–D) (**Tipuloidea**—Chapter 15.1) 3

2' Head capsule completely exposed (Figs. 15.4H, 15.5C, 15.10K, and 15.11L) 7

3(2) Body with longitudinal rows of thread-like or cone-like, fleshy processes (Figs. 15.6E and 15.15B) .... Cylindrotomidae

3' Body without thread-like or cone-like, fleshy processes, or if projections present, these shorter than their basal diameter 4

4(3) Abdomen with posterior spiracular disk surrounded by 6 lobes (4 in tiny first instar), some of which can be short and blunt (Figs. 15.12A, C, and 15.18C) .... Tipulidae

4' Abdomen with posterior spiracular disk surrounded by fewer than 6 lobes 5

5(4) Abdomen without posterior spiracles or spiracular disk (Figs. 15.12G and 15.15C) .... Limoniidae (in part; *Antocha*)

5' Abdomen with posterior spiracles on spiracular disk (Figs. 15.12D–F, 15.15D, and 15.18D–F) 6

6(5) Abdomen with posterior spiracular disk associated with 2 lobes and with prolegs or creeping welts (Figs. 15.6G and 15.15D); head capsule nearly complete (Fig. 15.12H) .... Pediciidae

6' Abdomen with posterior spiracular disk surrounded by 2, 4, or 5 lobes (Figs. 15.12D and 15.18D–F); prolegs absent; creeping welts present or absent; head capsule complete or reduced (Fig. 15.12E, F) .... Limoniidae (in part)

7(2) Body with 6 divisions (first division consisting of fused head, thorax, and first abdominal segment); each division with ventral suctorial disk (Figs. 15.6C, D and 15.11F, G) .... Blephariceridae

7' Body with more than 6 divisions, or segmentation unclear; ventral suctorial disks absent 8
8(7) Abdomen with elongate terminal, telescopic respiratory siphon typically ¼ or more length of body (Figs. 15.6A and 15.15E) .... Ptychopteridae
8' Abdomen without respiratory siphon, or with short terminal or subterminal, nonteloscopic respiratory siphon less than 1/5 length of body 9
9(8) Abdomen subterminally with pair of dorsal, flattened lobes (posterolateral processes) fringed with setae (Fig. 15.11D, E). Abdominal segments II–VII dorsally with or without corona of plumose setae (Fig. 15.14A) .... **Dixidae**
9' Abdomen subterminally without dorsal, flattened lobes fringed with setae (Figs. 15.10A, B, E–H, K, 15.14B–E, and 15.16A). Abdominal segments dorsally without corona of plumose setae 10
10(9) Thorax ventrally with single proleg or pair of prolegs (sometimes short or partially retracted) (Figs. 15.10A, B, G, K, M, and 15.14B, C) 11
10' Thorax without prolegs, although abdominal prolegs often present (Figs. 15.11B, J, 15.14D, E, and 15.16A) 14
11(10) Head capsule with pair of labral fans (sometimes adducted so that only fan stem is apparent) (Figs. 15.4A, B and 15.10I–K). Abdomen expanded posteriorly before narrowing to terminal posterior proleg with circlet of many rows of tiny hooks (Fig. 15.10K, L) .... **Simuliidae**—Chapter 15.3
11' Head capsule without labral fans. Abdomen not conspicuously expanded posteriorly nor with terminal ring of tiny hooks (although crochets may be present) 12
12(11) Thorax with anterior spiracles on short stalks (Fig. 15.14B). Abdominal segment VIII with transverse cleft into which posterior spiracles open. Thoracic proleg unpaired (Fig. 15.10M) .... Thaumaleidae
12' Thorax and abdomen without spiracles (apneustic). Thoracic proleg typically paired, although sometimes fused (Figs. 15.10A, B, G, and 15.14C) 13
13(12) Thorax and abdomen with conspicuous setae, and often with dorsal tubercles or processes (Figs. 15.4C, 15.10G, and 15.14C) .... Ceratopogonidae (in part, Forcipomyiinae)
13' Thorax and abdomen with inconspicuous or no setae, and without tubercles (Figs. 15.4G and 15.10A, B) .... **Chironomidae**—Chapter 15.2
14(10) Thorax consisting of single, swollen segment wider than abdominal segments (Figs. 15.11A, B and 15.18G, I). Abdomen posteriorly with ventral setal fan (Figs. 15.5G and 15.18H) 15
14' Thorax typically consisting of distinct segments as wide as or narrower than abdominal segments (Figs. 15.4D, 15.10H, 15.11H–K, and 15.14D, E). Abdomen without setal fan 16
15(14) Antennae in form of prehensile (grasping) organs with long apical setae (Fig. 15.5F). Head without labral brushes (Fig. 15.18G) .... Chaoboridae
15' Antennae not prehensile and with only short apical setae (Fig. 15.18I). Head with labral brushes of many fine or stout setae (Figs. 15.11C and 15.18I) .... Culicidae
16(14) Body slender, cylindrical, with smooth, shiny integument and few, small, inconspicuous setae, except sometimes on terminal segment (Figs. 15.4D, 15.10H, and 15.14D) 17
16' Body more robust, cylindrical or somewhat dorsolaterally flattened, with integument bearing numerous processes, fine hairs, setae, spicules, or spines (Figs. 15.6B, 15.11H–K, and 15.14E) 19
17(16) Abdomen with each of 8 segments preceded by shorter intercalary segment (Fig. 15.14D) .... Anisopodidae
17' Abdomen without intercalary segments (Figs. 15.4D and 15.10H) (Ceratopogonidae, in part) 18
18(17) Head capsule well-developed, sclerotized (Figs. 15.4D and 15.10H)...Ceratopogoninae and Dasyheleinae
18' Head capsule reduced, unsclerotized .... Leptoconopinae
19(16) Abdomen with all or most segments bearing sclerotized plates dorsally, and much of integument often with tiny dark spots (sclerotized platelets) (Fig. 15.11H–J) .... Psychodidae
19' Abdomen without dorsal sclerotized plates or platelets (Figs. 15.14E and 15.16A) 20
20(19) Body dorsoventrally rather flattened, with 9 pairs of spiracles (peripneustic) .... Scatopsidae
20' Body cylindrical, with 2 or 10 pairs of spiracles (i.e., amphipneustic or holopneustic, respectively) (Figs. 15.14E and 15.16A) 21
21(20) Body with 10 pairs of spiracles (holopneustic); segments not secondarily divided (although cuticle can be wrinkled into folds); terminal segment unlobed or with slender, fleshy processes (Fig. 15.14E) .... Bibionidae
21' Body with 2 pairs of spiracles (amphipneustic); segments secondarily divided; terminal segment ending in 4 blunt lobes (Fig. 15.16A) .... Trichoceridae

22(1) Cephalic region with sclerotized portions of head capsule and mandibles present and often at least partially exposed, although capable of complete retraction in some species (Figs. 15.7A–G, 15.13A–C, E–G, I, J, 15.16B–D, and 15.18K), sometimes reduced to slender internal rods (Fig. 15.18J) 23
22' Cephalic region with sclerotized portion reduced to internal pigmented cephalopharyngeal skeleton (acephalic condition) and usually mandibles (mouth hooks) (Fig. 15.18L), at least the tips of which sometimes can be seen externally (Figs. 15.8C–H, 15.13K, M, 15.15A, and 15.17A–E), although mandibles absent in Syrphidae 28
23(22) Body dorsoventrally rather flattened (Figs. 15.7C and 15.13A–C); integument roughened by calcium carbonate deposits in reticulate pattern. Head capsule typically with lateral eye prominences .... Stratiomyidae
23' Body typically not conspicuously flattened (Figs. 15.7E–G, 15.13E–G, I, J, and 15.16B–D); integument not roughened by calcium carbonate deposits in reticulate pattern. Head capsule without lateral eye prominences 24
24(23) Head capsule closed dorsally and at least partially ventrally (may require dissection) (Fig. 15.18K) 25
24' Head capsule open dorsally and ventrally, reduced to slender, elongate rods (may require dissection) (Fig. 15.18J) 27
25(24) Abdomen posteriorly with single tapered lobe or process (Figs. 15.7G, 15.13E, F, and 15.16B) .... Tabanidae
25' Abdomen posteriorly with 4 short lobes or 2 feathery processes or multiple long, fleshy, thread-like processes (Figs. 15.7A, B, 15.13G, H, and 15.16C) 26
26(25) Abdomen with prolegs and terminal segment with 2 feathery processes or long, thread-like processes (Fig. 15.13G, H) .... Athericidae
26' Abdomen without prolegs (although with creeping welts) and terminal segment with 4 short lobes (Fig. 15.16C) .... Rhagionidae
27(24) Abdomen typically with prolegs bearing apical hooks; if prolegs absent, then abdomen terminating in 1 lobe (Figs. 15.7F and 15.13I, J) .... Empididae
27' Abdomen without prolegs (although with creeping welts) and terminating in 4 lobes (Figs. 15.7E and 15.16D) .... Dolichopodidae
28(22) Thorax with anterior spiracles inconspicuous, unbranched, often slightly elevated on conical prominences, each with single opening. Body with short spine-like and/or plumose processes (Fig. 15.16E) .... Phoridae
28' Thorax with anterior spiracles either absent or branched, stalked, and bearing multiple openings. Body with or without processes, excluding spiracles that may be on processes (Figs. 15.8A–H, 15.13K, M, 15.15A, and 15.17A–E) 29
29(22) Body dorsoventrally flattened, with 1–3 pairs of bristle-like processes on anterior and posterior segments or with multiple slender, spiculate processes on thoracic and abdominal segments (Fig. 15.17A, B) 30
29' Body cylindrical, or if dorsoventrally flattened, then without anterior and posterior bristle-like processes or without multiple slender, spiculate processes on thoracic and abdominal segments (Figs. 15.8A–H, 15.13K, M, 15.15A, and 15.17C–E) 31
30(29) Body with more than 9 visible segments, each with slender, spiculate processes; without striated border along tergal plates (Fig. 15.17A) .... Fanniidae
30' Body with 8 or 9 visible segments, 1–3 pairs of bristle-like anterior and posterior processes, and striated border along all tergal plates (Fig. 15.17B) .... Lonchopteridae
31(29) Abdomen with posterior spiracles on pointed spines. Associated with plants as leaf miners or to obtain oxygen via spiracle-bearing posterior spines 32
31' Abdomen with posterior spiracles sessile or on telescopic respiratory tube, but not on pointed spines (Figs. 15.8A–H, 15.13K, M, 15.15A, and 15.17C–E). Free living 33
32(31) Mandibles (mouth hooks) present; filter chamber absent .... Ephydridae (in part)
32' Mandibles (mouth hooks) absent; ribbed filter chamber present instead .... Syrphidae (in part)
33(31) Abdomen with posterior spiracular plates fused or nearly so along midline, typically on distal end of telescopic respiratory tube (Figs. 15.8A and 15.17D). Mandibles (mouth hooks) absent; ribbed filter chamber present instead .... Syrphidae (in part)
33' Abdomen with posterior spiracular plates distinctly separated whether or not on distal end of telescopic respiratory tube (Figs. 15.8B–H, 15.13K, M, 15.15A, and 15.17C, E). Mandibles (mouth hooks) present; filter chamber absent (Figs. 15.15A and 15.18L) 34
34(33) Mandibles (mouth hooks) basally with sclerotized ventral arch of various shapes (may require dissection) (Fig. 15.15A) .... Sciomyzidae
34' Mandibles (mouth hooks) basally without sclerotized ventral arch (may require dissection) (Fig. 15.18L) 35

35(34) Abdomen with posterior segments tapered, typically ending with elongate respiratory tube (often forked), and with integument bearing setae, spinules, or setose tubercles (Figs. 15.8D, E and 15.17C) .... Ephydridae
35' Abdomen squared or truncated posteriorly, with or without short respiratory tube or concial spiracular processes (if on longer processes, then pair of ventral processes also present), and integument without setae or these restricted to intersegmental areas (Figs. 15.8F–H, 15.13K, M, and 15.17E) 36
36(35) Abdomen with posterior spiracular plate not bearing lobes, spines, or dorsal ridges, although spiracles may be on separate short or long processes (Figs. 15.13L, 15.13N, and 15.17E) .... Muscidae
36' Abdomen with posterior spiracular plate bearing marginal lobes, spines, or dorsal ridge (Fig. 15.8F) ....
.... Scathophagidae

Acknowledgments

We thank Fabio Pupin for photographs of larvae on a white background and Steve Marshall for some of the photographs of live larvae. The following workers provided habitat photos, for which we express our gratitude: N. Bonada, I. Davanzo, M. S. Extremera, M. Grabowski, K. Kettani, L. Latella, B. Maiolini, J. Moubayed-Breil, D. Spitale, K. Targuisti, and C. Zamora-Muñoz. We acknowledge the use of all line drawings by permission from Agriculture and Agri-Food Canada. We thank our friends and colleagues for providing estimates of the number of species of aquatic Diptera: Art Borkent (Ceratopogonidae), Joel Moubayed-Breil (Chironomidae), Igor Grichanov (Dolichopodidae), Martin Hauser and Franco Mason (Stratiomyidae), Marija Ivković (Dixidae and Empididae), Wayne Mathis and Tadeusz Zatwarnicki (Ephydridae), Virginija Podeniene (Cylindrotomidae, Limoniidae, Pediciidae, and Tipulidae), Adrian Pont (Muscidae), Daniele Sommaggio (Syrphidae), and Rüdiger Wagner (Psychodidae).

Subchapter 15.1

Superfamily Tipuloidea

Virginija Podeniene
Institute of Biosciences, Life Sciences Center, Vilnius University, Vilnius, Lithuania

Introduction

The superfamily Tipuloidea, or crane flies, consists of four families, Cylindrotomidae, Limoniidae, Pedicidae, and Tipulidae, with nearly 15,600 species worldwide and about 1100 species in the Mediterranean Basin (Oosterbroek, 2022). The family Cylindrotomidae includes 69 species in nine genera worldwide (Oosterbroek, 2022). At least two aquatic species are found in the Mediterranean Basin (Table 15.1). The family Limoniidae—the largest family of Diptera—is composed of more than 10,700 described species worldwide (Oosterbroek, 2022). In the Mediterranean Basin, 350 to more than 400 aquatic species are known (Table 15.1). The family Pediciidae includes 498 species worldwide in 10 genera, and the Mediterranean Basin has 45 aquatic species in five genera (Oosterbroek, 2022; Table 15.1). The family Tipulidae includes more than 4300 world species and more than 100 aquatic species in the Mediterranean Basin (Oosterbroek, 2022; Table 15.1).

General ecology

A typical life cycle of crane flies consists of an egg stage of a few days to several months, a larval stage with four instars (all combined lasts from about several months to several years), a pupal stage of a few weeks, and an adult stage lasting from a few days to about a week (Hadley 1971; Rogers 1927; MacLean 1973; Pritchard 1983).

Adults are nonfeeding (with a few exceptions), slowly flying insects usually hanging close to larval habitats (Alexander, 1920; Rogers, 1926). Adults of crane flies are easily recognized by their long legs which are easily broken and a V-shape suture on the thorax. Adults of different families of this superfamily are separated by wing venation, length of last palpal segment, length of the rostrum, number of antennal segments, and presence or absence of hairs over the eyes (Oosterbroek, 2006).

Larvae of terrestrial species pupate in the same habitat when aquatic larvae migrate to nearby terrestrial habitat to became a pupa (with a few exceptions). Larvae of Tipuloidea are terrestrial, aquatic, or semiaquatic and colonize a wide range of habitats. Specific habitats of larvae include the following (modified from Gelhaus & Podeniene, 2019):

1. bottom sediments in rapidly flowing streams: *Arctoconopa, Dicranota, Eloephila, Hexatoma, Scleroprocta, Hoplolabis, Rhabdomastix*
2. rocks in rapidly flowing streams: *Antocha, Dicranota*
3. leaf packs in smaller streams: *Dicranota, Tipula* (subgenera *Emodotipula, Yamatotipula*), *Molophilus, Gonempeda, Ormosia*
4. seepages or organically rich small streams: *Conosia, Dicranophragma, Pedicia, Tricyphona, Molophilus, Hoplolabis, Limnophila, Neolimnomyia, Paradelphomyia, Pilaria, Pseudolimnophila, Tipula (Acutipula)*
5. vegetated ponds and lake margins: *Prionocera*
6. aquatic to semiaquatic muddy or vegetated edge areas of streams and lakes: *Dicranomyia, Eutonia, Euphylidorea, Phylidorea, Helius, Pseudolimnophila, Tipula (Yamatotipula)*, and many other Limnophilinae and Chioneinae
7. open marshes: *Dicranomyia, Eutonia, Euphylidorea, Phylidorea, Helius, Idioptera, Pseudolimnophila, Paradelphomyia*
8. anaerobic marsh sediments with emergent macrophytes: *Erioptera (Erioptera)*
9. sandy, gravelly, or loamy soils with moderate humus, as found along stream borders: *Ellipteroides, Erioconopa, Erioptera, Gonomyia, Hoplolabis, Idiocera, Molophilus, Pilaria, Rhabdomastix, Symplecta*, also *Tipula* (subgenera *Beringotipula, Yamatotipula*), *Nigrotipula*
10. steep or vertical cliff faces kept wet by a film of water supporting algal growth: some species of *Dicranomyia, Elliptera, Geranomyia, Dactylolabis, Orimarga, Pedicia*
11. moist to wet cushions of mosses or liverworts, sometimes in streams: *Dicranomyia, Dolichopeza, Phalacrocera, Triogma, Tipula* (subgenera *Savtshenkia, Yamatotipula*)

12. saturated decaying wood in water: Lipsothrix
13. rich organic earth or mud: numerous genera and species, including, *Cheilotrichia, Erioptera, Gonomyia, Molophilus, Ormosia, Rhypholophus, Tricyphona, Tipula* (subgenera *Platytipula, Tipula, Schummelia, Yamatotipula*).

The larvae of the family Tipulidae usually are saprofagous, and only a few species are phytophagous which can become serious pests (Chiswell, 1956; Gelhaus, 1986; Blackshaw, 2009; Blackshaw & Hicks, 2012). The larvae of the family Cylindrotomidae are herbivorous and feed on plants and mosses (Peus, 1952: Imada, 2020). Larvae of all aquatic or semiaquatic Pediciidae are predators when the larvae of the family Limoniidae can be predators, saprofagous, detritofagous, or herbivorous (Lindner, 1959; Savchenko, 1986; Oosterbroek & Theowald, 1991; Ujvarosi et al., 2010).

Larval morphology and characteristics needed in identification

The larvae of crane flies are metapneustic (except larvae of genus *Antocha*) and hemicephalic. The body is elongate, subcylindrical (except in a few genera), and consists of eleven obvious segments: three thoracic and eight abdominal segments (Fig. 15.1.1A–F). The spiracular field is considered the ninth abdominal segment, and the anal papillae surrounding the anus represent the tenth segment (Gelhaus & Podeniene, 2019). The full-grown larvae range in length from 6 mm (family Limoniidae) to almost 80 mm (family Tipulidae). Thoracic segments are usually wider than long, whereas the length of the abdominal segments significantly exceeds their width. The body of crane fly larvae is covered with several types of hair-like structures, including macrosetae (with distinct basal sockets), and macroscopic and microscopic hairs. In some genera, abundant microscopic and macroscopic hairs that may be arranged in patterned lines can be an important taxonomic character (Fig. 15.1.1B). Body color in live larvae varies and can be brown, whitish, or yellowish. Abdominal locomotory structures are common among crane fly larvae and their presence is often used in the key. Creeping welts (ventral or both ventral and dorsal) on all or just the last few abdominal segments usually can be found in the family Limoniidae (Fig. 15.1.1C), whereas the family Pediciidae can possess not only creeping welts but also more prominent pseudopods (Fig. 15.1.1D). All known larvae of the family Cylindrotomidae possess well-developed processes on all body segments, which is one of the best distinguishing characteristics (Fig. 15.1.1E). The larvae of the family Tipulidae (those living in the Mediterranean Basin) usually are without any locomotory structures (Fig. 15.1.1F), but some genera bear lateral subconical lobes on abdominal segment VIII (Fig. 15.1.1G).

The heads of larval crane flies are hemicephalic and almost entirely retracted into the central thoracic region. Reduction of the head varies significantly among the different groups of crane flies and is one of the best distinguishing characteristics in this group. The head capsule can be slightly reduced, with short dorsal and coronal sutures and strongly sclerotized (Fig. 15.1.2A, B, E), or strongly reduced overall and consisting of only several pairs (two or three) of sclerotized plates or rods separated by deep incisions (Fig. 15.1.2C, D, F–J). The dorsal side of the head consists of several plates such as the labrum, clypeus, frons, and genae (internal and external lateralia). The shape of the labrum is more or less typical for families or subfamilies of crane fly larvae. The labrum can be fused with the clypeus or have additional structures (Fig. 15.1.2C, D, F). The clypeus can be separated from the labrum and frons (Fig. 15.1.2A) or be fused with the labrum or frons. The reduction of the frons is one of the most important distinguishing characteristics of crane fly larvae. The frons can be well developed and fused with the genae (Fig. 15.1.2B) or separated by frontal sutures or grooves (Fig. 15.1.10B). In the most reduced head types, the frons is completely reduced (Fig. 15.1.2H). The ventral side of the head capsule is more or less reduced in all crane fly larvae. In less reduced head capsule types, the strongly sclerotized genae are widely separated and diverging posteriorly (Fig. 15.1.2D). The genae in strongly reduced head capsules are represented by narrow rods. In genera such as *Hexatoma* and *Pilaria*, the entire ventral side of the genae is reduced. The ventral side of the head capsule is connected by the hypostomium or hypopharyngeal bar (Fig. 15.1.2I). The hypostomium can be a continuous plate, consists of two plates, or be interrupted by a membranous area (Fig. 15.1.2J). The mandible of crane flies is one-segmented except in the genus *Pilaria*, which has two-segmented mandibles. The mandibles are strongly sclerotized with a strong apical tooth and a few basal teeth and are conical or sickle-shaped. The maxilla, with the cardo at the base, usually consists of two short lobes (inner and outer). The lobes can be fused, in which case the maxillae are strongly elongated (Fig. 15.1.2F) and protrude from the first thoracic segment.

The main identification characteristics include the spiracular disk and the anal papillae. The only pair of spiracles (spiracles absent in genus *Antocha*) is in the area of the terminal segment, which is called the spiracular disk. It is surrounded by a variable number of lobes which are sclerotized to varying degrees according to the genus. The architecture of the larval spiracular disk is the most important character and is extensively used in this key. Larvae of the family Tipulidae have six lobes (Fig. 15.1.3A–C) surrounding the disk [except first-instar larvae, which have four lobes (Fig. 15.1.3D)]. Larvae of the family Limoniidae (in the Mediterranean Basin) have five, four, or two lobes, or the lobes

FIGURE 15.1.1 Larvae of Tipuloidea. (A) *Nigrotipula nigra* (Tipulidae), habitus of larva, lateral view. (B) *Dactylolabis* (Limoniidae), habitus of larva, dorsal view. (C) *Helius* (Limoniidae), habitus of larva, lateral view. (D) *Dicranota* (Pediciidae), habitus of larva, lateral view. (E) *Phalacrocera replicata* (Cylindrotomidae), habitus of larva, lateral view. (F) *Tipula* (*Beringotipula*) *unca* (Tipulidae), habitus of larva, lateral view. (G) *Dolichopeza* (*Oropeza*) (Tipulidae), terminal segment, dorsal view. *Figure A from Podeniene (2003), Figures B – G © V. Podeniene.*

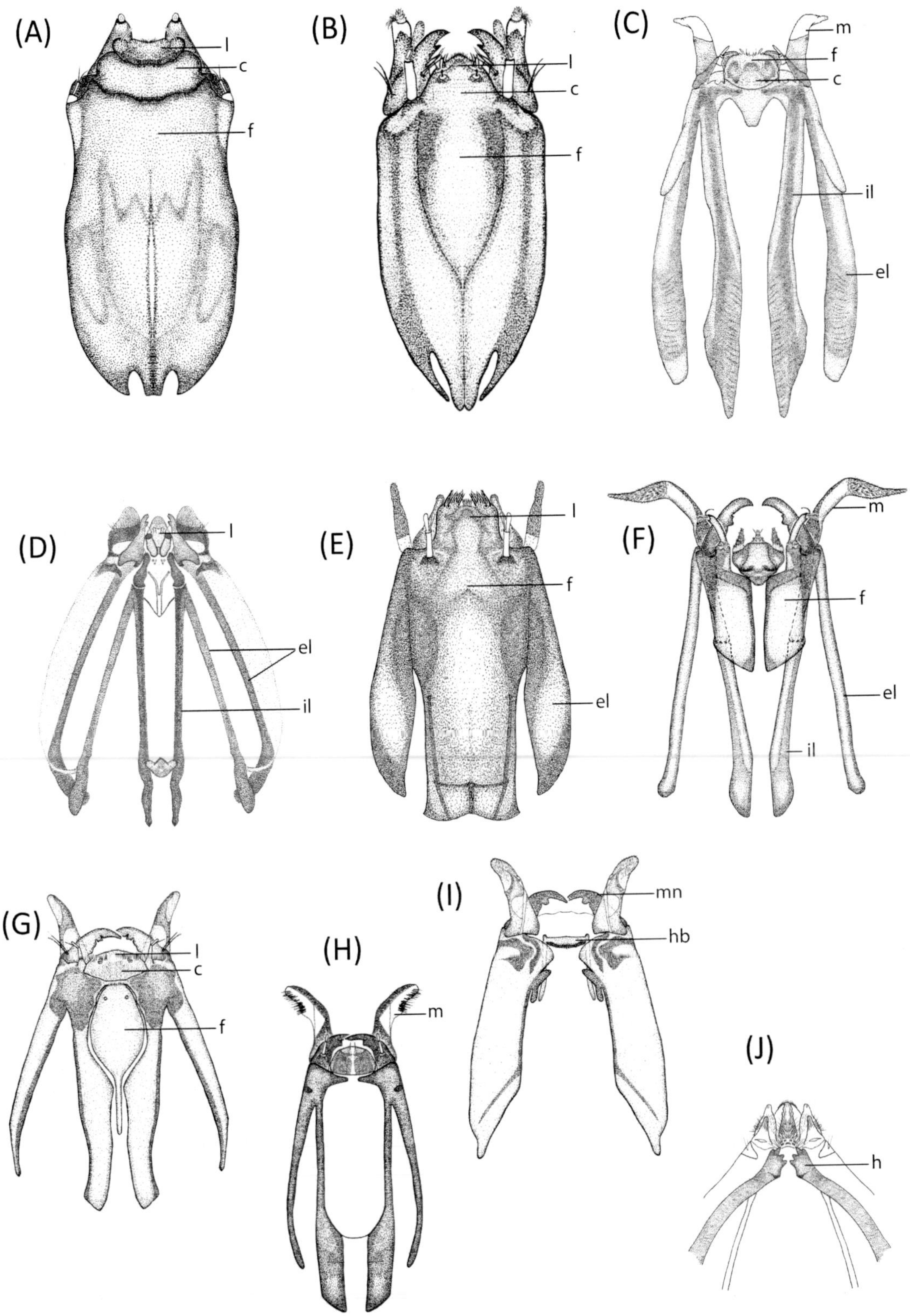

FIGURE 15.1.2 Head capsules of Tipuloidea. (A) *Pedicia rivosa* (Pediciidae), general view of head capsule, dorsal aspect, l—labrum, c—clypeus. f—frons. (B) *Paradelphomyia (Oxyrhiza) fuscula* (Limoniidae), general view of head capsule, dorsal aspect, abbreviations as in panel A. (C) *Idioptera* (Limoniidae), general view of head capsule, dorsal aspect, l—labrum, c—clypeus. m—maxillae, il—internolateralia, el—externolaterlia. (D) *Molophilus (Molophilus) crassipygus* (Limoniidae), general view of head capsule, dorsal aspect, abbreviations as in panel C. (E) *Pseudolimnophila (Pseudolimnophila) sepium* (Limoniidae), general view of head capsule, dorsal aspect, abbreviations as in panel C. (F) *Hexatoma (Hexatoma) vittata* (Limoniidae), general view of head capsule, dorsal aspect, abbreviations as in panel C. (G) *Limnophila (Limnophila) pictipennis* (Limoniidae) general view of head capsule, dorsal aspect, abbreviations as in panel C. (H) *Phylidorea* (Limoniidae) general view of head capsule, dorsal aspect, abbreviations as in panel C. (I) *Limnophila (Limnophila) pictipennis* (Limoniidae) general view of head capsule, ventral aspect, mn—mandible, hb—hypopharyngeal bar. (J) *Molophilus* (Limoniidae), general view of head capsule, ventral aspect, h—hypostomium. *Figures B, E, F, G, I from Podeniene (2002), Figures A, H from Podeniene (2003), Figure D from Podeniene (2009). Figure C © V. Podeniene.*

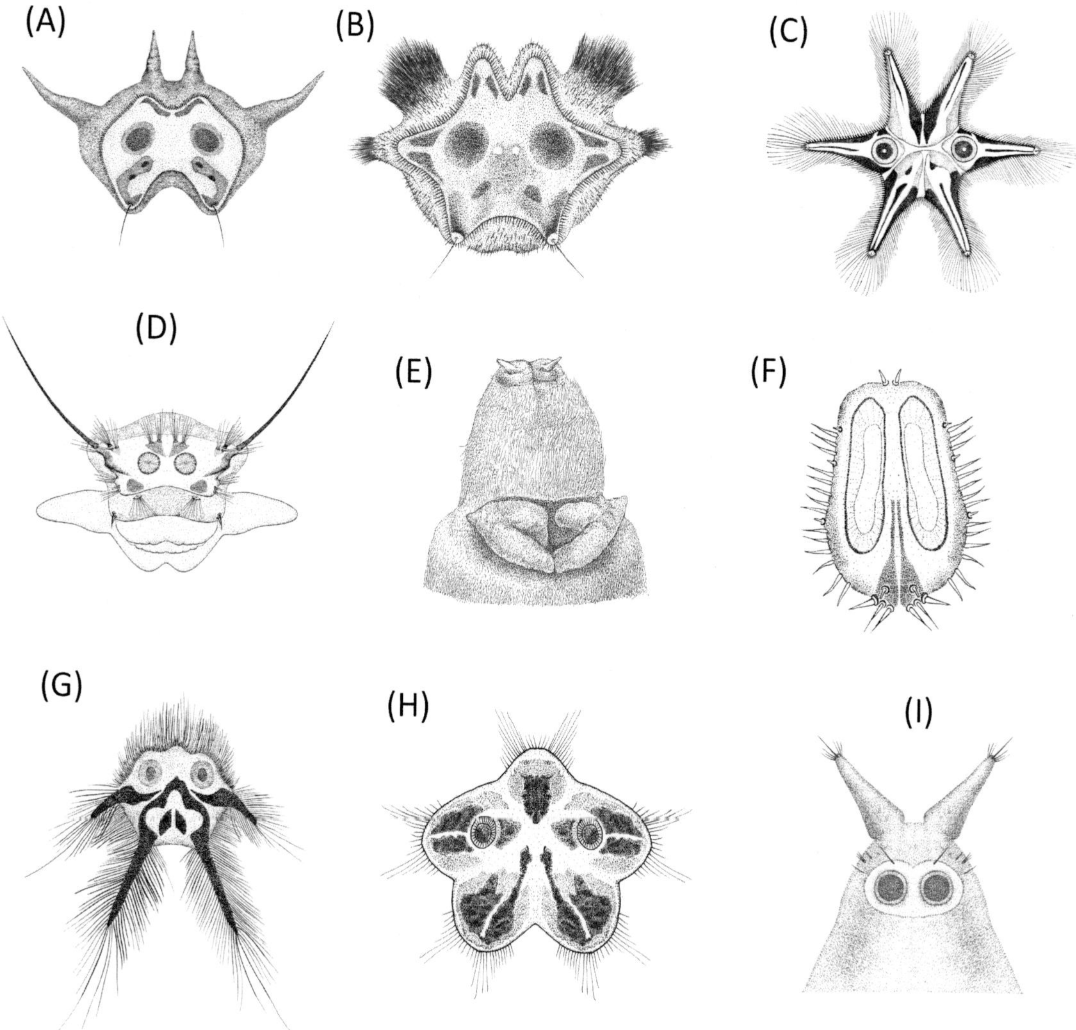

FIGURE 15.1.3 Larvae of Tipuloidea. (A) *Nigrotipula nigra* (Tipulidae), general view of spiracular field. (B) *Tipula* (*Yamatotipula*) *pruinosa* (Tipulidae), general view of spiracular field. (C) *Prionocera subserricornis* (Tipulidae), general view of spiracular field. (D) *Tipula* (*Beringotipula*) *unca* (Tipulidae), general view of spiracular field of egg–larva. (E) *Erioptera (Erioptera) straminea* (Limoniidae), terminal segment, ventral view. (F) *Dicranomyia* (Limoniidae) general view of spiracular field. (G) *Pseudolimnophila (Pseudolimnophila) sepium* (Limoniidae), general view of spiracular field. (H) *Erioconopa* (Limoniidae) general view of spiracular field. (I) *Pedicia* (*Pedicia*) *rivosa* (Pediciidae), general view of spiracular field. *Figures A–C, I from Podeniene (2003), Figure D–from Podeniene et al. (2019) by permision; Figure E from Gelhaus & Podeniene (2019) by permision; Figure G from Podeniene (2002). Figures F, H © V. Podeniene.*

can be reduced (Fig. 15.1.3E–H). Larvae of aquatic Pediciidae have only two (ventral) lobes (Fig. 15.1.3I). The lobes usually are bordered by a row of setae, sometimes very long on the apices. In most genera, spiracular lobes are covered by sclerites that form a great variety of patterns.

Anal papillae are located on the ventral side of the anal division surrounding the anus. These structures are important taxonomically, especially in the family Tipulidae. Crane fly larvae usually have one pair (Fig. 15.1.3E) or two pairs of lobes (Fig. 15.1.4A), although three (Fig. 15.1.4B) or four (Fig. 15.1.4C) pairs of lobes are found in some larvae of the family Tipulidae. In some genera, all pairs of anal papillae are reduced (Fig. 15.1.4D).

Identification key to the larvae of crane flies

Diptera: Tipulomorpha: Tipuloidea: Cylindrotomidae, Limoniidae, Pediciidae, Tipulidae: Genera

1 Thoracic and abdominal segments with dorsal and lateral rows of conspicuous, elongate fleshy projections (Fig. 15.1.1E) .... 57 (Family Cylindrotomidae)

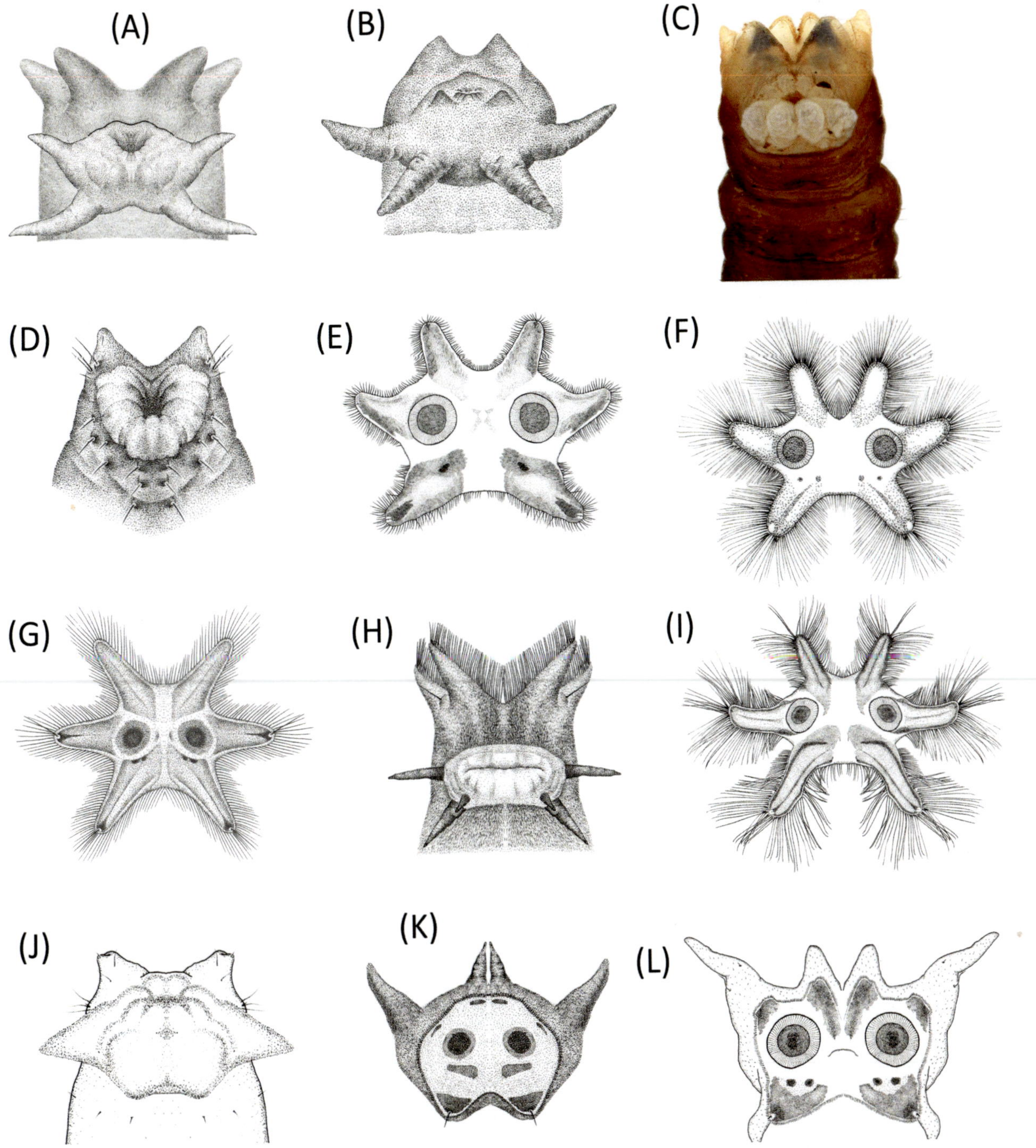

FIGURE 15.1.4 Larvae of Tipulidae. (A) *Tipula (Acutipula) luna*, terminal segment, ventral view. (B) *Tipula (Yamatotipula) pruinosa*, terminal segment, ventral view. (C) *Tipula (Savchenkia)*, terminal segment, ventral view. (D) *Nigrotipula nigra*: terminal segment, ventral view. (E) *Tipula (Acutipula) maxima*: general view of spiracular field. (F) *Tipula (Savchenkia) cheethami*, general view of spiracular field. (G) *Tipula (Platytipula) autumnalis*: general view of spiracular field. (H) *Tipula (Platytipula)*, terminal segment, ventral view. (I) *Tipula (Emodotipula) saginata*, general view of spiracular field. (J) *Tipula (Beringotipula) unca*, terminal segment, ventral view. (K) *Dolichopeza*, general view of spiracular field. (L) *Tipula (Beringotipula) unca*, general view of spiracular field. *Figures B, D, G, H from Podeniene (2003), Figures F, J redrawn from Chiswell (1956), Figures A, E, I, K, L © V. Podeniene.*

1' Thoracic and abdominal segments without dorsal longitudinal rows of conspicuous projections 2

2(1) Spiracular disk bordered by six lobes, orientation with two dorsal, two lateral, and two ventral (Fig. 15.1.3A–C); first instar larvae have 4 lobes (dorsal pair absent) (Fig. 15.1.3D) .... 3 (Family Tipulidae)

2' Spiracular disk bordered by 5, 4, or 2 lobes (Fig. 15.1.3G–I); lobes in some genera reduced and scarcely evident (Fig. 15.1.3F) 15
.... (Families Limoniidae and Pediciidae)
3(2) Anal papillae reduced (Fig. 15.1.4D); border of short setae only around central area of spiracular disk, missing on margins of lobes; posterior surface of ventral lobe with L-shaped sclerite (Fig. 15.1.3A); terrestrial and semiaquatic .... *Nigrotipula*
3' At least two pairs of anal papillae; sclerotization of spiracular field is different 4
4(3) Four pairs of anal papillae (Fig. 15.1.4C) 5
4' Three or two pairs of anal papillae (Fig. 15.1.4A, B) 6
5(6) Pairs of anal papillae different in length; spiracular field bordered with short setae (Fig. 15.1.4E); aquatic and semiaquatic .... *Tipula* (*Acutipula*), in part
5' All anal papillae short, similar in length (Fig. 15.1.4C); spiracular field bordered with long setae (Fig. 15.1.4F); develop on aquatic, semiaquatic and terrestrial mosses .... *Tipula* (*Savchenkia*)
6(4) Three pairs of anal papillae (Fig. 15.1.4B) 7
6' Two pairs of anal papillae (Fig. 15.1.4A) 10
7(6) Spiracular lobes long and slender, each 2–3 × as long as width at base; spiracular field bordered with long setae (Fig. 15.1.4G) 8
7' Spiracular lobes short, each just about 1.5 × as long as width at base; spiracular field bordered with short setae (Fig. 15.1.3B); aquatic, semiaquatic .... *Tipula* (*Yamatotipula*), in part.
8(7) Lobes of spiracular disk darkened along entire margin (Fig. 15.1.3C); anal papillae directed lateral, ventral, and dorsal 9
8' Lobes of spiracular field without darkened margins (Fig. 15.1.4G); anal papillae directed lateral or ventral, not dorsal (Fig. 15.1.4H); semiaquatic .... *Tipula* (*Platytipula*)
9(8) Spiracular lobes with a thin dark submedian line; two dark dots (as pair) at the bases of ventral lobes (Fig. 15.1.3C); aquatic, semiaquatic .... *Prionocera*
9' Spiracular lobes with a thin pale submedian line; dark dots at the bases of ventral lobes absent (Fig. 15.1.4I); semiaquatic .... *Tipula* (*Emodotipula*)
10(6) Anterior pair of anal papillae reduced, bump-shaped; posterior pair elongated, conus-shaped (Fig. 15.1.4J) or both pairs of anal papillae reduced, bump-shaped 11
10' Anterior and posterior pairs of anal papillae well developed, elongated (Fig. 15.1.4A) 13
11(10) Both pairs of anal papillae reduced, bump-shaped; dorsal lobes of spiracular disk closely appressed to one another (subgenus *Dolichopeza*) (Fig. 15.1.4K) or abdominal segment VIII bearing a subconical lobe at each side below and before dorsolateral lobe of spiracular disk (subgenus *Oropeza*) (Fig. 15.1.1G); develop in mosses and liverworts in wetlands and edge of aquatic habitats .... *Dolichopeza*
11' Anterior pair of anal papillae reduced to bumps, posterior pair elongated (Fig. 15.1.4J); dorsal lobes of spiracular disk not closely appressed; abdominal segment VIII without lateral subconical lobes 12
12(11) Lateral spiracular lobe significantly longer than dorsal lobe; ventral spiracular lobe with finger-like, subapical projection (Fig. 15.1.4L); terrestrial and semiaquatic .... *Tipula* (*Beringotipula*)
12' Lateral and dorsal spiracular lobes similar in length (Fig. 15.1.5A); ventral lobes without subapical projection; terrestrial and semiaquatic .... *Tipula* (*Tipula*)
13(10) Abdominal segment VIII with dorsal row of macroscopic hairs surrounding semicircular pilose area and pair of broad lateral swellings (Fig. 15.1.5B); posterior surface of each dorsal lobe with extensive dark sclerite (Fig. 15.1.5C); semiaquatic .... *Tipula* (*Schummelia*)
13' Abdominal segment VIII without semicircular row of hairs; sclerotization of dorsal lobe is different (Fig. 15.1.5D) 14
14(13) Dorsal and lateral lobes without sclerotization; ventral lobes with a thin submedian line (Fig. 15.1.5D); abdomen with hairs of greatly differing lengths, with distinct hair clusters or tufts near most lateral setae of dorsum and venter; dorsum patterned; aquatic and semiaquatic .... *Tipula* (*Yamatotipula*), in part
14' Sclerotization of spiracular field different (Fig. 15.1.5E); all abdominal hairs similar in length; dorsum unpaterned; aquatic and semiaquatic .... *Tipula* (*Acutipula*), in part
15(2) Terminal segment with two lobes (ventral) (Fig. 15.1.3I) 16
15' Terminal segment with four, five spiracular lobes or lobes are entirely reduced (Fig. 15.1.3F–H) 20
16(15) Paired ventral lobes of terminal segment short, spine-shaped (Fig. 15.1.3E); one pair of anal papillae; no creeping welts or pseudopods; in marshy sediments .... *Erioptera*, in part

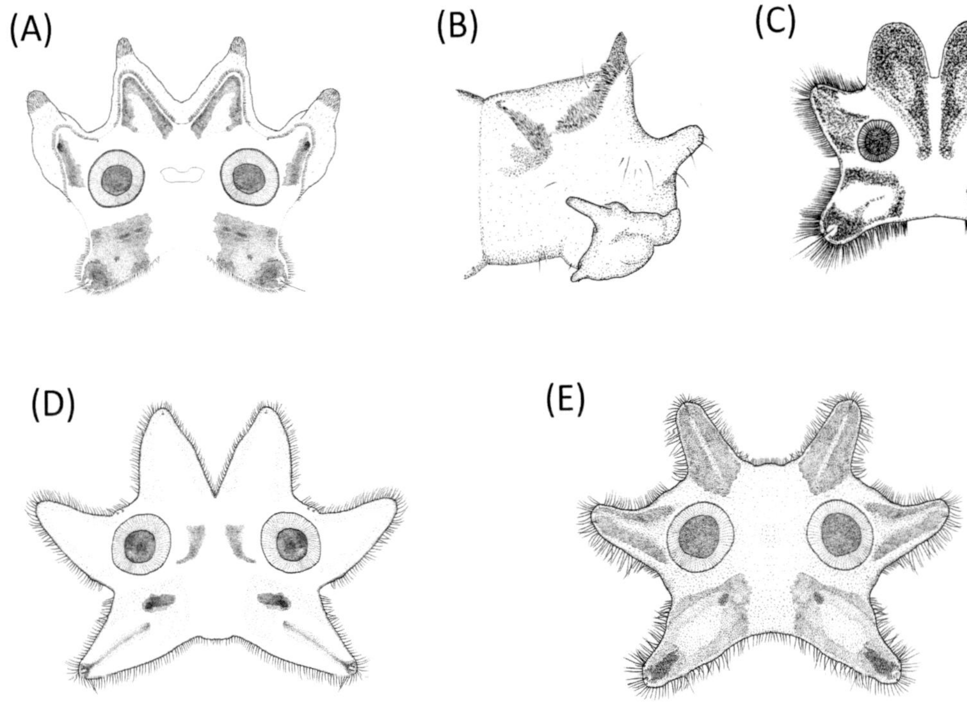

FIGURE 15.1.5 Larvae of Tipulidae. (A) *Tipula* (*Tipula*) *paludosa*, general view of spiracular field. (B) *Tipula* (*Schummelia*) *variicornis*, terminal segment, lateral view. (C) *Tipula* (*Schummelia*) *variicornis*, general view of spiracular field. (D) *Tipula* (*Yamatotipula*) *pierrei*: general view of spiracular field. (E) *Tipula* (*Acutipula*) *luna*, general view of spiracular field. *Figures B–C redrawn from Chiswell (1956), Figures A, D – E © V. Podeniene.*

16' Ventral lobes of terminal segment conical; two pairs of anal papillae; abdomen with creeping welts or pseudopods 17

17(16) Spiracles absent; ventral lobes of terminal segment elongate with a few tufts of long hairs; dorsal and ventral creeping welts conspicuous on abdominal segments II–VII (Fig. 15.1.6A); aquatic, on rock surfaces in streams .... *Antocha*

17' Spiracles present; ventral lobes with very short hairs and single setae; only ventral locomotory structures on abdominal segments III–VII or IV–VII 18

18(17) Spiracular field almost entirely covered with short hairs, except small areas around spiracles (Fig. 15.1.6B); venter of abdominal segments III–VII with paired pseudopods bearing apical hooks (Fig. 15.1.1D) .... *Dicranota*

18' Spiracular field not covered with short hairs; margins of spiracular field fringed with short setae; venter of abdominal segments IV–VII with creeping welts (Fig. 15.1.6C) 19

19(18) Bases of ventral lobes adjacent, forming V-shape between lobes (Fig. 15.1.3I); aquatic and semiaquatic .... *Pedicia*

19' Space between bases of ventral lobes broader, bases not adjacent, forming U-shape between lobes (Fig. 15.1.6D); aquatic and semiaquatic .... *Tricyphona*

20(15) Terminal segment extends into a single lobe, elongate, tapering to apex (Fig. 15.1.6E); semiaquatic .... *Neolimnomyia*

20' Terminal segment surrounded by four, five spiracular lobes or lobes indistinct 21

21(20) Spiracular disk without distinct lobes (Fig. 15.1.3F) 22

21' Spiracular disk with five or four lobes (Fig. 15.1.3G, H) 24

22(21) Abdominal segments without dorsal and ventral creeping welts; spiracular disk broadly oriented dorsally (Fig. 15.1.6F); thoracic segments hairy; larva in flattened, oblong, hardened case (Fig. 15.1.6G); semiaquatic .... *Thaumastoptera*

22' Abdominal segments II–VII with dorsal and ventral creeping welts on basal rings (Fig. 15.1.6H); thoracic segments not hairy; spiracular disk oriented posteriorly, not dorsally; larva not in hardened case, but may be in silken tube 23

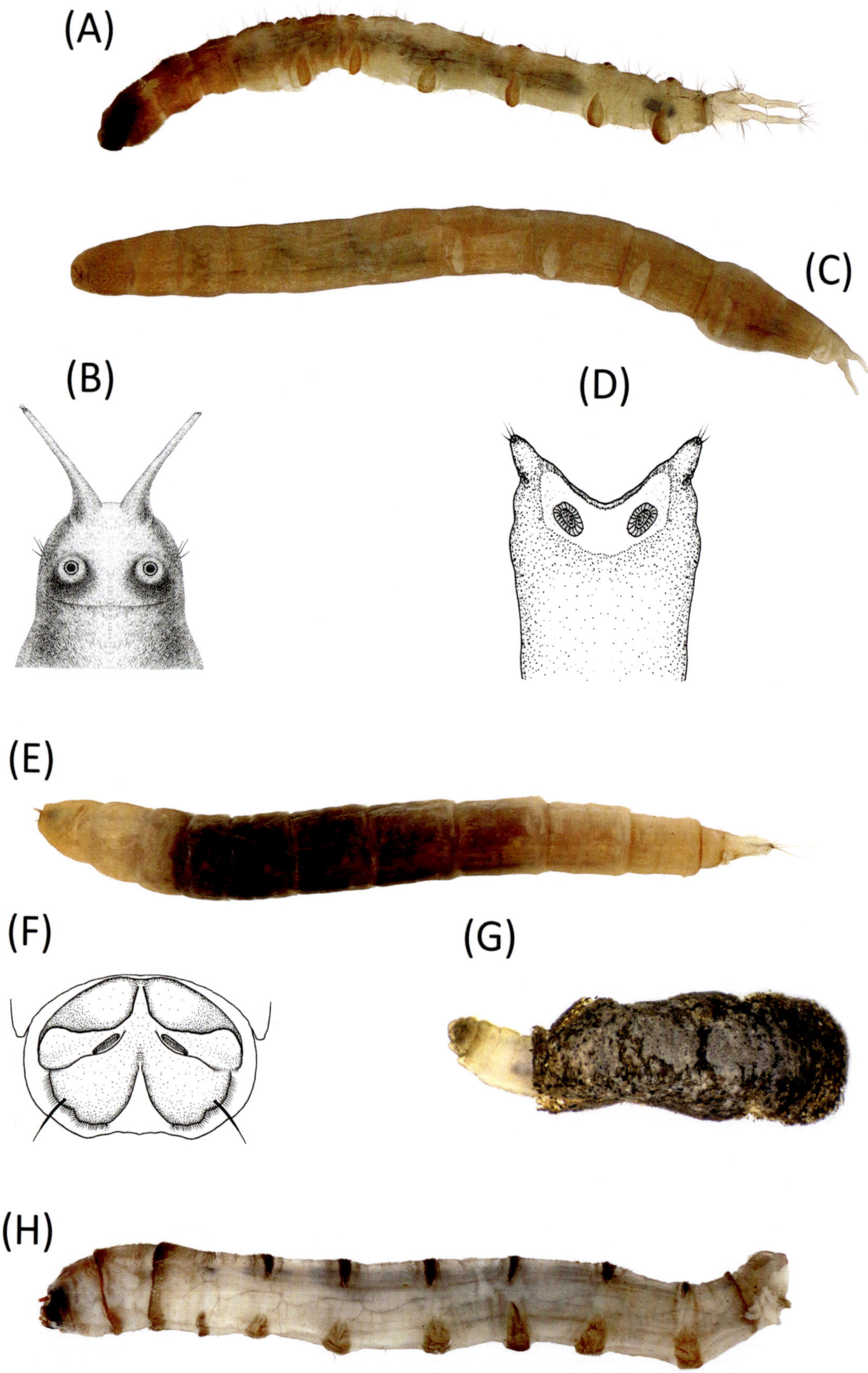

FIGURE 15.1.6 Larvae of Tipuloidea. (A) *Antocha* (Limoniidae), habitus of larva, lateral view. (B) *Dicranota* (Pediciidae) general view of spiracular field. (C) *Pedicia* (Pediciidae), habitus of larva, lateral view. (D) *Tricyphona* (*Tricyphona*) *immaculata* (Pediciidae), general view of spiracular field. (E) *Neolimnomyia batava* (Limoniidae), habitus of larva, lateral view. (F) *Thaumastoptera* (*Thaumastoptera*) *calceata* (Limoniidae) general view of spiracular field. (G) *Thaumastoptera* (*Thaumastoptera*) *calceata*, larva in case. (H) *Dicranomyia* (Limoniidae), habitus of larva, lateral view. *Figures B, from Podeniene (2003), Figure D redrawn from Gelhaus & Podeniene (2019), Figure F redrawn from Mauch (2017), Figure G © E. Mauch; Figures A, C, E, H © V. Podeniene.*

23(22) Spiracular disk roughly circular or broadly oval to transversely subrectangular; spiracles often large, oval, may be inclined together dorsally or not inclined; usually two ventral sclerites but rarely two ventral and two dorsal on spiracular disk or just one ventral; dorsal margin of spiracular field with two setae, lateral margin with three setae on each side (should be viewed under high magnification) (Figs. 15.1.3F and 15.1.7A); aquatic, semiaquatic, terrestrial .... *Dicranomyia*

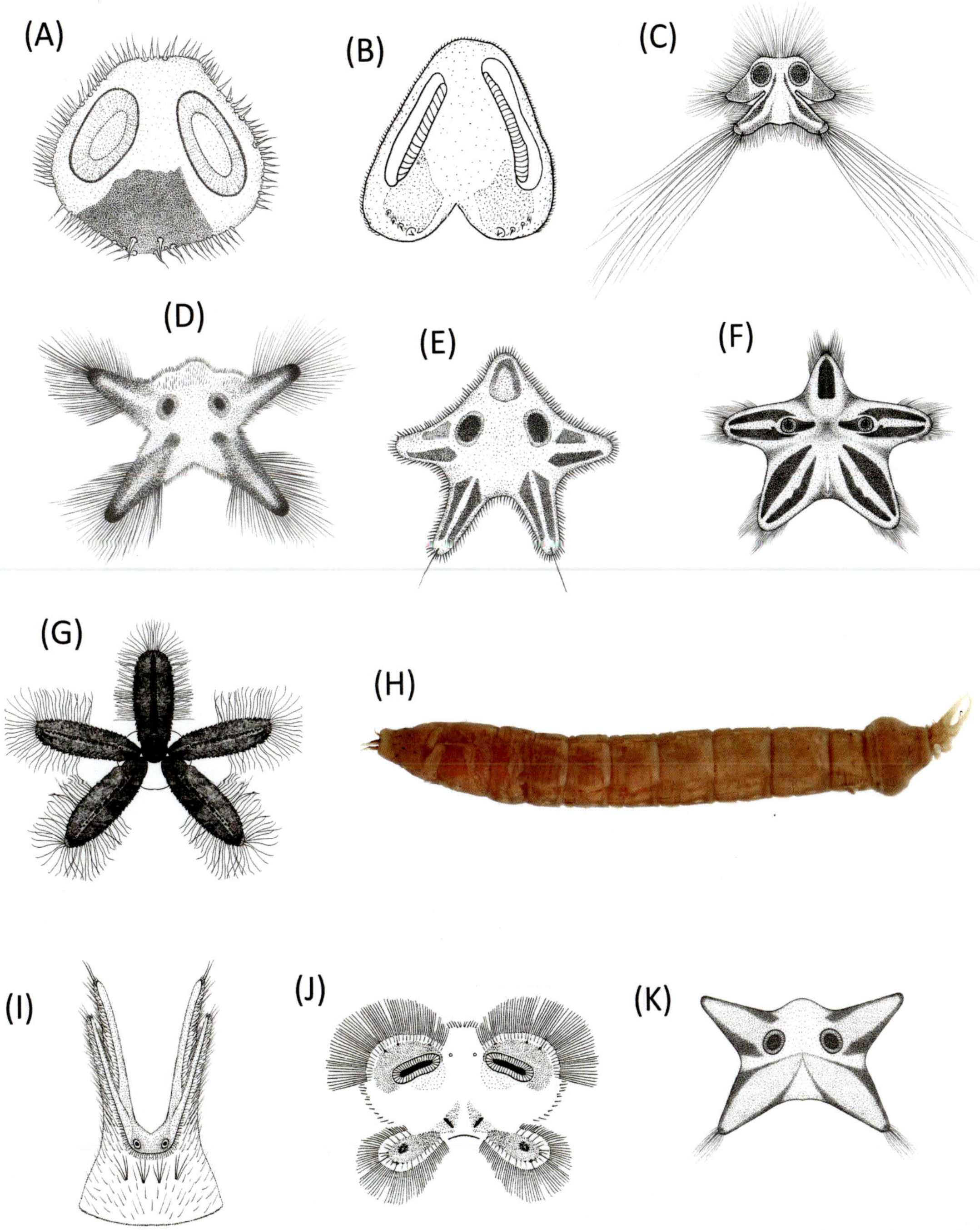

FIGURE 15.1.7 Larvae of Limoniidae. (A) *Dicranomyia (Dicranomyia) frontalis*, general view of spiracular field. (B) *Geranomyia*, general view of spiracular field. (C) *Paradelphomyia (Oxyrhiza) fuscula*, general view of spiracular field. (D) *Limnophila (Limnophila) pictipennis*, general view of spiracular field. (E) *Helius pallirostris*, general view of spiracular field. (F) *Molophilus (Molophilus) propinquus*, general view of spiracular field. (G) *Scleroprocta*, general view of spiracular field. (H) *Eloeophila*, habitus of larva, lateral view. (I) *Eloeophila*, terminal segment, dorsal view. (J) *Orimarga (Orimarga) attenuata*: general view of spiracular field. (K) *Dactylolabis*, general view of spiracular field. *Figure A from Gelhaus & Podeniene (2019) by permision; Figures C–E from Podeniene (2002), Figure F from Podeniene (2009), Figure J redrawn from Vaillant (1951), Figures B, G, H, I, K © V. Podeniene.*

23’ Spiracular field very small, lateral part of it inclined and infolded; two ventral sclerites only; spiracles large, oval, always inclined dorsally; dorsal and lateral margin of spiracular field without setae (should be viewed under high magnification) (Fig. 15.1.7B); aquatic .... *Geranomyia*
24(21) Dorsal lobe of spiracular disk greatly reduced in size or absent (Fig. 15.1.7C, D) 25
24’ Dorsal lobe of spiracular disk similar in size to other lobes (Fig. 15.1.7E–G) 40
25(24) Spiracular disk with lateral and ventral lobes forming long cylindrical processes (Fig. 15.1.7H); spiracular lobes could be pigmented or not pigmented; dorsal surface of spiracular lobes entirely covered with long hairs (Fig. 15.1.7I); aquatic .... *Eloeophila*
25’ Spiracular disk with lateral and ventral lobes flattened 26
26(25) Spiracular lobes wider than long (Fig. 15.1.7J) 27
26’ Spiracular lobes more elongate, longer than wide (Fig. 15.1.7K) 28
27(26) Spiracular lobes pigmented (Fig. 15.1.7J); aquatic .... *Orimarga*
27’ Spiracular lobes nonpigmented (Fig. 15.1.8A); aquatic, semiaquatic .... *Rhabdomastix*, in part
28(26) Body wide, flattened, dorsum of all segments with transverse bands or patches of dense pilosity (Fig. 15.1.1B); ventral pair of spiracular lobes bears only apical hairs (Fig. 15.1.7K); fauna hygropetrica .... *Dactylolabis*
28’ Body elongate, nearly cylindrical, dorsal side of body unpatterned 29
29(28) Abdominal segments with creeping welts 30
29’ Abdominal segments without distinct creeping welts 34
30(29) Abdominal segments with dorsal and ventral creeping welts (Fig. 15.1.8B) 31
30’ Abdominal segments with ventral creeping welts only (Fig. 15.1.8C) 32
31(30) Abdominal segments II–VII with dorsal and ventral creeping welts (Fig. 15.1.8B); both pairs of spiracular lobes fringed with long hairs; spiracle elongated dorsoventrally (Fig. 15.1.8D); fauna hygropetrica .... *Elliptera*
31’ Abdominal segments I–VII with dorsal and ventral creeping welts (Fig. 15.1.8E); marginal hairs of spiracular disk short (Fig. 15.1.8F); spiracle circular; larvae developing in rotten wood submerged in water .... *Lipsothrix*
32(30) Abdominal segments IV–VII with ventral creeping welts (Fig. 15.1.8G); apical hairs of ventral lobe very long, several times longer than lobe itself and longer than on lateral lobe (Fig. 15.1.7C); head capsule almost complete, with very short dorsal sutures (Fig. 15.1.2B); semiaquatic .... *Paradelphomyia*
32’ Abdominal segments V–VII or VI–VII with ventral creeping welts; marginal hairs of spiracular field short; length of apical hairs on ventral lobe are similar to that of lateral lobe (Fig. 15.1.8H); head capsule reduced (Fig. 15.1.2G) 33
33(32) Abdominal segments VI–VII with ventral creeping welts (Fig. 15.1.8I); marginal hairs of spiracular field very short, much shorter than lobes themselves (Fig. 15.1.8H); aquatic, semiaquatic .... *Dicranophragma*
33’ Abdominal segments V–VII with ventral creeping welts (Fig. 15.1.8C); marginal hairs on spiracular lobes longer, they almost equal to the lobe ‘s length (Fig. 15.1.7D); aquatic, semiaquatic .... *Limnophila*
34(29) Ventral lobes more than twice as long as lateral lobes (Fig. 15.1.9A) 35
34’ Ventral lobes subequal to lateral lobes or slightly longer (Fig. 15.1.9E) 37
35(34) Ventral and lateral lobes darkly pigmented (Fig. 15.1.3G) 36
35’ Ventral and lateral lobes bear narrow pale sclerites (Fig. 15.1.9A); semiaquatic .... *Conosia*
36(35) Ventral and lateral lobes entirely pigmented, elongate and narrow (Fig. 15.1.3G); head capsule completely sclerotized, with extended dorsal sutures (Fig. 15.1.2E); aquatic, semiaquatic .... *Pseudolimnophila*
36’ Pigmentation of ventral and lateral lobes discontinuous, with transverse striations, more continuous coloration toward apex or reduced to narrow line, lobes elongate but wide (Fig. 15.1.9C); head capsule reduced to rods, dorsal plate of head fused into spatulate plate, widest posteriorly (Fig. 15.1.9D); aquatic, semiaquatic .... *Pilaria*
37(34) Apical hairs on ventral lobes long, more than twice as long as lobe itself (Fig. 15.1.9B, E) 38
37’ Apical hairs on ventral lobes shorter (Fig. 15.1.9F) 39
38(37) Outer margin of maxilla with long dense hairs (Fig. 15.1.2H); aquatic, semiaquatic *Euphylidorea Phylidorea*
38’ Outer margin of maxilla without hairs (Fig. 15.1.2C); semiaquatic .... *Idioptera*
39(37) Ventral and lateral lobes almost entirely darkly pigmented (Fig. 15.1.9F); ventral part of head capsule connected with hypopharyngeal bar; frons reduced into small triangular plate (Fig. 15.1.9G); semiaquatic .... *Eutonia*

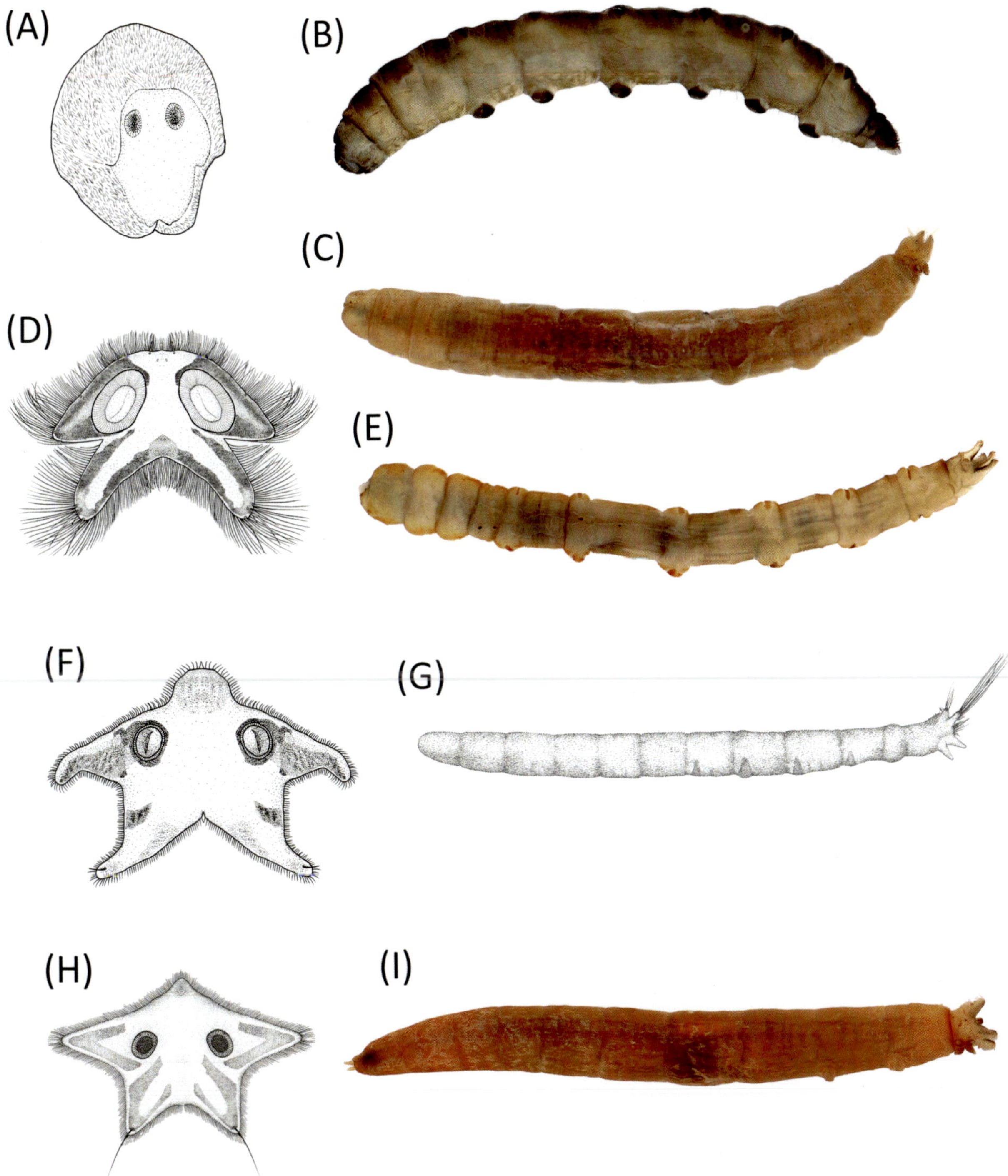

FIGURE 15.1.8 Larvae of Limoniidae. (A) *Rbabdomastix*, general view of spiracular field. (B) *Elliptera mongolica*, habitus of larva, lateral view. (C) *Limnophila*, habitus of larva, lateral view. (D) *Elliptera jacoti*: general view of spiracular field. (E) *Lipsothrix*, habitus of larva, lateral view. (F) *Lipsothrix*, general view of spiracular field. (G) *Paradelphomyia (Oxyrhiza) fuscula*, habitus of larva, lateral view. (H) *Dicranophragma* (*Brachylimnophila*) *nemorale*, general view of spiracular field. (I) *Dicranophragma* (*Brachylimnophila*) *nemorale*, habitus of larva, lateral view. *Figure B, D from Podeniene et al. (2021), Figures G–H from Podeniene (2003), Figures A, C, E–F, I © V. Podeniene.*

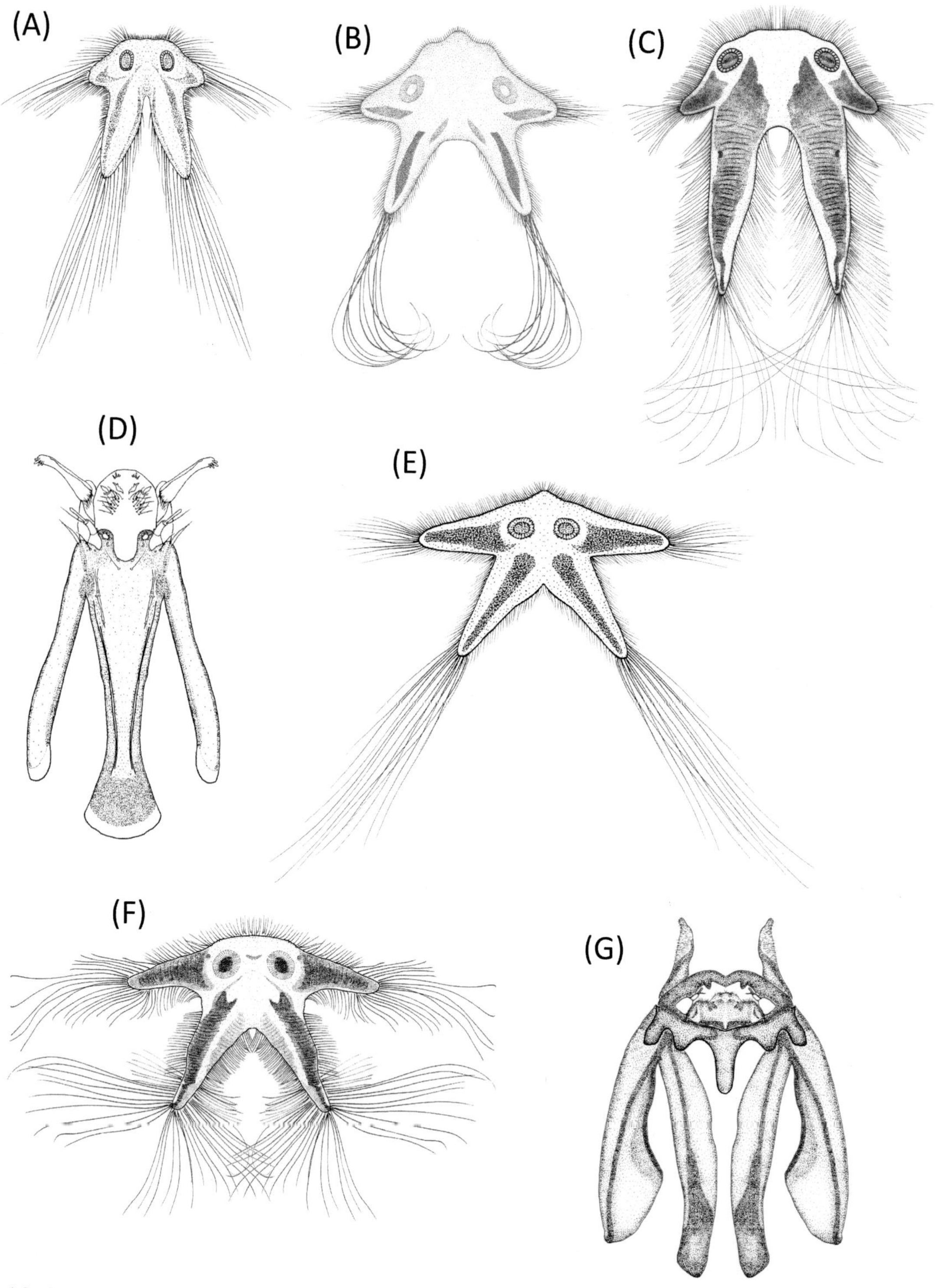

FIGURE 15.1.9 Larvae of Limoniidae. (A) *Conosia*, general view of spiracular field. (B) *Idioptera*, general view of spiracular field. (C) *Pilaria*, general view of spiracular field. (D) *Pilaria*, general view of head capsule, dorsal aspect. (E) *Phylidorea*, general view of spiracular field. (F) *Eutonia barbipes*, general view of spiracular field. (G) *Eutonia barbipes*, general view of head capsule, dorsal aspect. *Figure A redrawn from Wood (1952), Figure B from Podeniene (2004), Figure E from Podeniene (2003), Figures C–D, F–G © V. Podeniene.*

39' Lateral lobe of spiracular disk bears narrow, stripe-shaped sclerite; ventral lobe pigmented differently (Fig. 15.1.10A); ventral part of head capsule completely reduced and open, hypopharyngeal bar absent; frons divided into two large rectangular lateral plates (Fig. 15.1.2F); aquatic .... *Hexatoma*

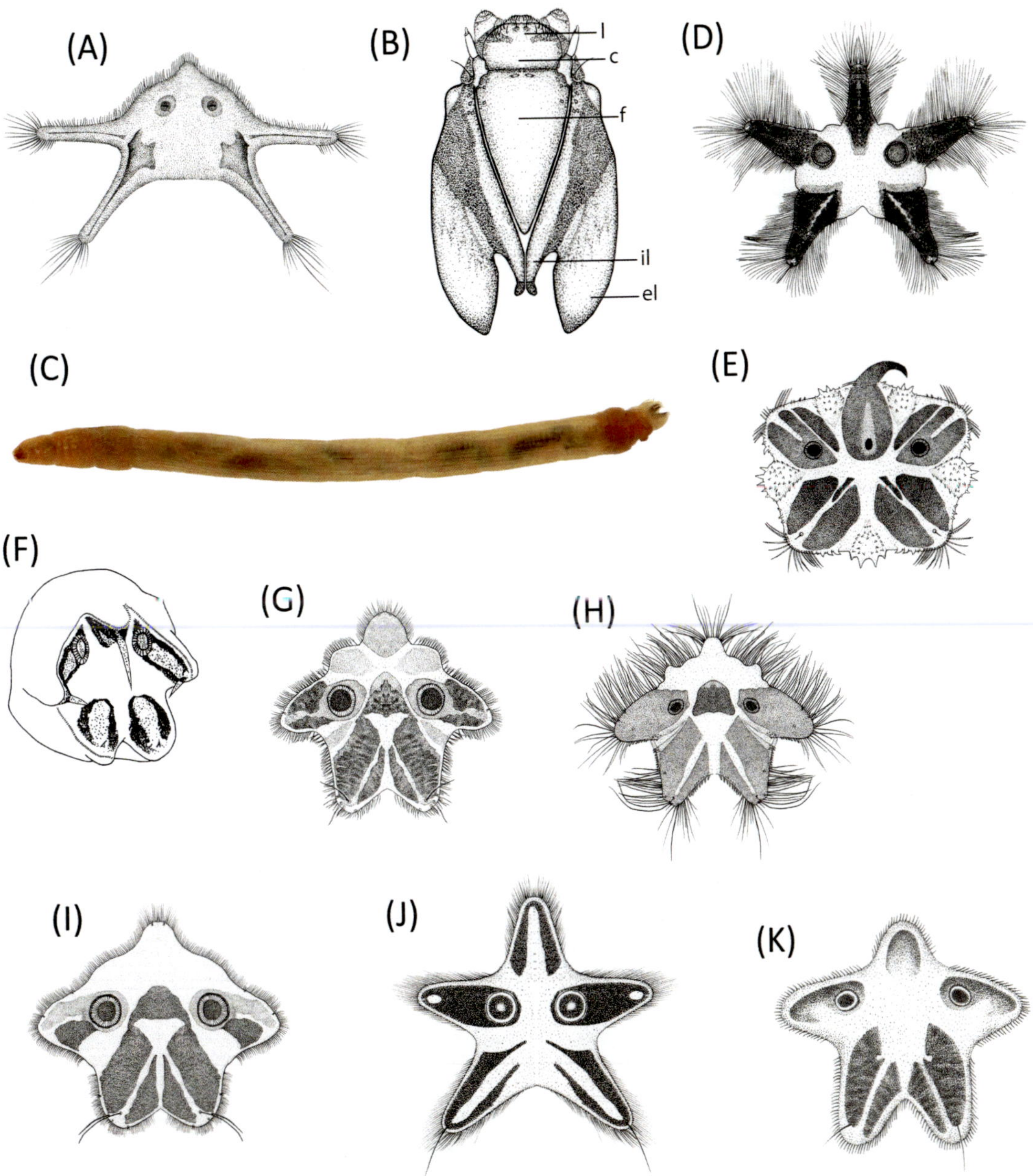

FIGURE 15.1.10 Larvae of Limoniidae. (A) *Hexatoma vittata*, general view of spiracular field. (B) *Helius (Helius) pallirostris*, general view of head capsule, dorsal aspect, abbreviations as in Fig. 15.1.2C. (C) *Molophilus*, habitus of larva, lateral view. (D) *Rhypholophus*, general view of spiracular field. (E) *Arctoconopa*, general view of spiracular field. (F) *Rhabdomastix (Sacandaga) laeta*, general view of spiracular field. (G) *Gonomyia*, general view of spiracular field. (H) *Ellipteroides*, general view of spiracular field. (I) *Idiocera*, general view of spiracular field. (J) *Cheilotrichia (Empeda) cinerascens*, general view of spiracular field. (K) *Gonempeda flava*, general view of spiracular field. *Figure A–B from Podeniene (2002), Figure F from Podeniene (2001), Figure J from Podeniene (2003), Figure K from Podeniene & Gelhaus (2002), Figures D–E, G–I © V. Podeniene.*

40(24) Abdominal segments II–VII with ventral creeping welts (Fig. 15.1.1C); head capsule almost complete (Fig. 15.1.10B); spiracles elongated (Fig. 15.1.7E); aquatic, semiaquatic .... *Helius*
40' Abdominal segments without creeping welts (Fig. 15.1.10C); head capsule reduced to elongate rods, Chioneinae-type (Fig. 15.1.2D); spiracle circular (Fig. 15.1.7F) 41
41(40) Spiracular disk surrounded by five large equal lobes, each almost entirely sclerotized (Fig. 15.1.10D) 42
41' Spiracular disk surrounded by five lobes which could be equal or different in size and shape, sclerotization of spiracular field is different (Fig. 15.1.3H) 43
42(41) Each spiracular lobe is a black spatulate plate with finely toothed margins (Fig. 15.1.7G); aquatic .... *Scleroprocta*
42' Spiracular lobes subconical in form, without fine teeth along margin (Fig. 15.1.10D); semiaquatic .... *Rhypholophus*
43(41) At least one spiracular lobe bears horn-like projection (Fig. 15.1.10E) 44
43' Lobes of spiracular field without projections 45
44(43) Median dorsal lobe of spiracular disk bearing a heavily sclerotized, horn-like projection with apex bent downward over disk; spiny wedge-shaped spots at periphery of disk between ventral lobes, between ventral and lateral lobes and between lateral and dorsal lobes (Fig. 15.1.10E); semiaquatic .... *Arctoconopa*
44' All five lobes with sclerotized projections, which differ in size, ventral projections are the smallest. Each lobe covered with large pigmented spots divided by a pale line (Fig. 15.1.10F); aquatic, semiaquatic .... *Rhabdomastix*
45(43) Anal field surrounded by two pairs of anal papillae; lateral lobes closer to ventral lobes; center of spiracular disk with single large solid triangular in shape sclerite (Fig. 15.1.10G–I); aquatic, semiaquatic
.... *Gonomyia*
.... *Ellipteroides*
.... *Idiocera*
45' Anal field surrounded by one pair of anal papillae; spiracular lobes more or less equidistant from each other; two sclerites between spiracles or they are absent (Fig. 15.1.10J) 46
46(45) Lateral lobe bears single sclerite (Fig. 15.1.10J) 47
46' Lateral lobe bears two sclerites (Fig. 15.1.3H) 48
47(46) Sclerites on lateral lobes very dark, each enclosing a small, pale subapical spot (Fig. 15.1.10J); semiaquatic .... *Cheilotrichia*
47' Pigmentation of lateral lobe pale, sclerites of lateral lobes extending as narrow band encircling spiracles (Fig. 15.1.10K); semiaquatic .... *Gonempeda*
48(46) Dorsal lobe with single sclerite (Fig. 15.1.7F) 49
48' Dorsal lobe with two sclerites (Fig. 15.1.11B) 50
49(48) Hypostomal prolongations expanded to form paired sclerotized plates with toothed anterior margins (Fig. 15.1.2J); pigmentation of spiracular disk black (Fig. 15.1.7F); aquatic, semiaquatic .... *Molophilus*
49' Hypostomal prolongations, if expanded, either not sclerotized or not toothed on anterior margins; inner part of spiracular lobes black, outer part light brown (Fig. 15.1.3H); semiaquatic .... *Erioconopa*
50(48) Central area of spiracular disk with pigmented spots extending to spiracles (Fig. 15.1.11C) 51
50' No pigmented spots between spiracles (Fig. 15.1.11E) 54
51(50) All spots on spiracular lobes of the same pigmentation (Fig. 15.1.11B) 52
51' Pigmentation of spots between spiracles much lighter than the darkest pigmentation of spiracular lobes (Fig. 15.1.11D) 53
52(51) Pigmentation of spiracular disk brown to light brown (Fig. 15.1.11A); aquatic, semiaquatic *Erioptera*, in part
52' Pigmentation of spiracular disk black (Fig. 15.1.11B); aquatic, semiaquatic .... *Hoplolabis*
53(51) Sclerites of spiracular disk entirely black; prominent pale setae near apex of each ventral spiracular lobe (Fig. 15.1.11D); semiaquatic .... *Ilisia*
53' Inner part of sclerites of spiracular lobes black, outer part light brown; no prominent pale setae near apex of each ventral spiracular lobe (Fig. 15.1.11C); aquatic, semiaquatic .... *Erioptera*, in part
54(50) Pigmentation of spiracular disk black (Fig. 15.1.11E) 55
54' Pigmentation of spiracular field pale (Fig. 15.1.11H) 56

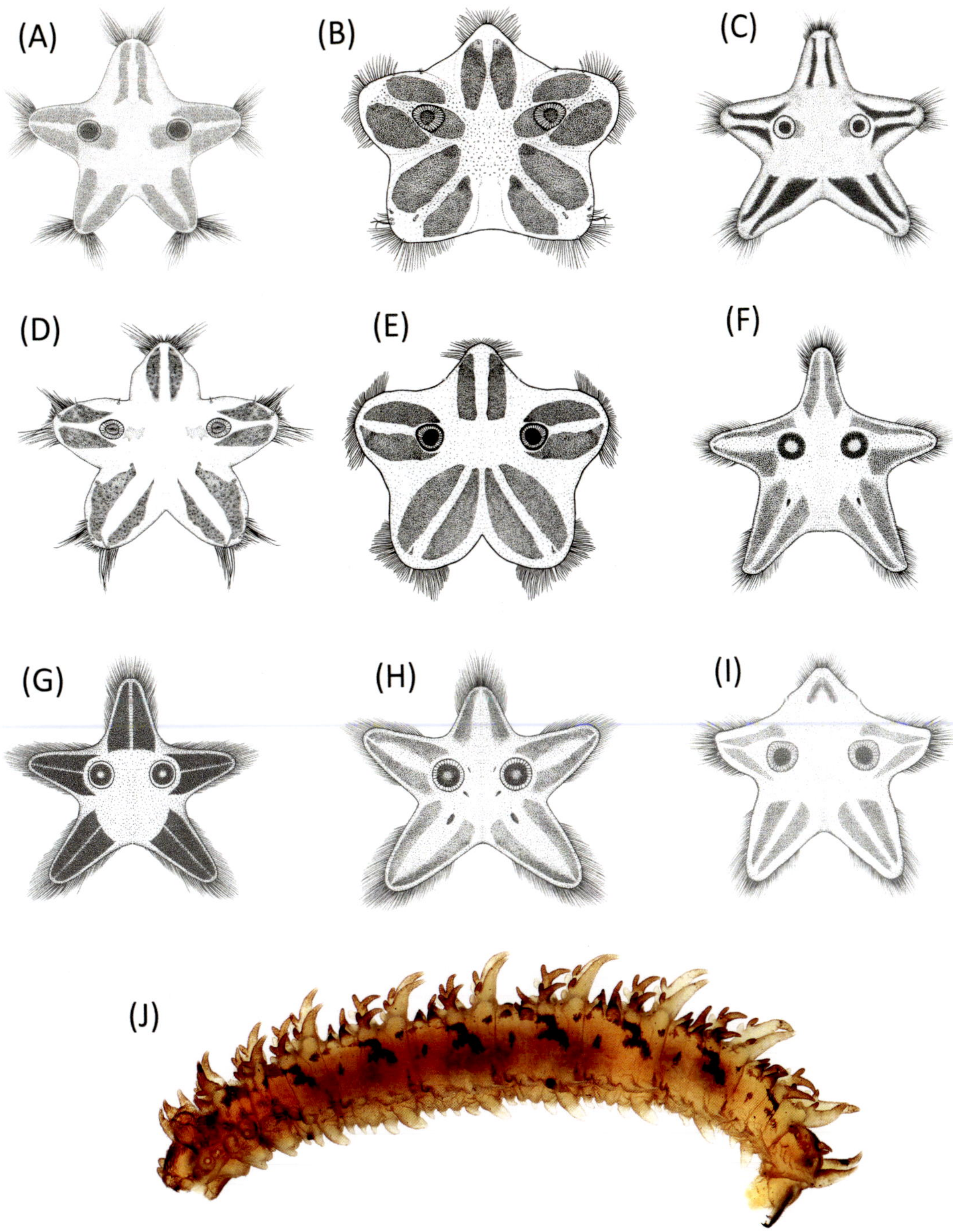

FIGURE 15.1.11 Larvae of Tipuloidea. (A) *Erioptera (Erioptera) lutea* (Limoniidae) general view of spiracular field. (B) *Hoplolabis (Parilisia)* (Limoniidae) general view of spiracular field. (C) *Erioptera (Mesocyphona) bivittata* (Limoniidae) general view of spiracular field. (D) *Ilisia venusta* (Limoniidae) general view of spiracular field. (E) *Ormosia Ormosia lineata* (Limoniidae), general view of spiracular field. (F) *Symplecta (Psiloconopa) stictica* (Limoniidae) general view of spiracular field. (G) *Trimicra pilipes* (Limoniidae), general view of spiracular field. (H) *Symplecta (Symplecta) hybrida* (Limoniidae), general view of spiracular field. (I) *Erioptera (Erioptera) divisa* (Limoniidae), general view of spiracular field. (J) *Triogma trisulcata* (Cylindrotomidae), habitus of larva, lateral view. *Figures A, E–H from Podeniene (2003), Figure C from Podeniene (2009), Figures I from Podeniene (2002), Figures B, D © V. Podeniene.*

55(54) Spiracular lobes only with apical setae; sclerites on spiracular lobes widely separated. (Fig. 15.1.11E) .... *Ormosia*

55' Spiracular lobes fringed with setae all around; sclerites on spiracular lobes close to each other (Fig. 15.1.11G) .... *Trimicra*, in part

56(54) Two to six small dots (as pairs) between and below spiracles (Fig. 15.1.11F, H); aquatic, semiaquatic .... *Symplecta*

56' No spots between and below spiracles (Fig. 15.1.11I); aquatic, semiaquatic .... *Erioptera*, in part

57(1) Dorsal projections of body mostly long, up to 10 times as long as basal diameter, posterior ones on abdominal segments often deeply bifurcate (Fig. 15.1.1E); in aquatic or semiaquatic mosses .... *Phalacrocera*

57' Dorsal projections shorter, length one to three times basal diameter; posterior part of dorsal projections on abdominal segments with three to four slender serrations, but not deeply divided (Fig. 15.1.11J); in aquatic or semiaquatic mosses .... *Triogm.*

Acknowledgments

I am very grateful for Dr Peter Adler for his correction of the text and for Erich Mauch for photographs of *Thaumastoptera* larvae.

Subchapter 15.2

Family Chironomidae

Valeria Lencioni[1], **Joel Moubayed**[2], **Peter H. Langton**[3]
[1]*MUSE-Science Museum, Research ad Museum Collections Office, Climate and Ecology Unit, Corso del Lavoro e della Scienza, Trento, Italy*
[2]*Freshwater and Marine biology, 10 rue des Fenouils, 34070 Montpellier, France*
[3]*16 Irish Society Court, Coleraine, Co. Londonderry Northern Ireland, BT52 1GX*

Introduction

The Chironomidae, popularly known as "nonbiting midges," are the freshwater insect family which comprises the highest number of species, both in lentic and lotic habitats (Cranston, 1995). This family is cosmopolitan and occurs in all zoogeographical regions of the world, including Antarctica (Ashe et al., 1987). There are very few areas where chironomids are absent.

Chironomids have colonized all freshwater habitats worldwide. They can be found in inland natural or man-made aquatic ecosystems, occupying extremely diverse habitats, and often constituting the most biodiverse and abundant taxa in habitats such as glacier-fed streams and springs (Lindegaard 1995; Rossaro et al. 2012; Lencioni 2018) (Fig. 15.2.1). Most larval chironomids reside in freshwaters, with only a few species occurring in marine waters (Neumann, 2003) or in moist ground and decaying matter (these are considered semiaquatic). Larvae and pupae can be found across the watershed, from fast-flowing headwaters to wider river sections and slow-moving lowland rivers. They are found also across floodplains, in wetlands and coastal lagoons, and in lakes of all types and shapes (Armitage et al., 1995).

Chironomids have attracted the attention of many scientists interested in studying the giant chromosomes in the larval salivary glands and the hemoglobin and morphological aberrations generally in relation to specific stressors (e.g., thermal and chemical) (Armitage et al., 1995). Their taxonomic diversity and worldwide distribution have encouraged many systematists to study Chironomidae phylogeny, and the family is considered interesting for biogeographic studies since the last century (e.g., Brundin, 1966; Serra-Tosio, 1973) to the present (e.g., Cornette et al., 2015; Krosch et al., 2017; Lin et al., 2018).

Furthermore, chironomid larvae are insects of commercial interest. They are recognized as an important food for many fishes and cultured crustaceans and are very popular in the aquarium fish trade (Das et al., 2012). Larvae are an excellent source of protein, lipid, vitamins, and minerals because of their high energy content and digestibility (Armitage, 1995).

Adults are not hematophagous and do not transmit any disease to humans. However, they are a source of disturbance, as is reported often in Florida, the United Kingdom, Sudan, Japan, and Italy. These flies can become a serious problem in residential areas near warm, polluted, eutrophic lakes, and lagoons (see Failla et al., 2015 for a review). "Globally, nearly 100 of the 4000 known chironomid species are documented as pestiferous" (Ali, 1996) and can be considered nuisances, public health risks (e.g., asthma and allergies), or economic pests (plaguing tourists).

Starting from this presentation, it is easy to understand why freshwater ecologists, histologists, molecular biologists, and systematists have been interested in these flies. The online Chironomid Worker Directory (http://www.chironomidae.net) provides a list of over 550 scientists working on chironomids and a forum to engage in exchange and discussions. The current bibliography of Chironomidae literature (http://literature.vm.ntnu.no/Chironomidae/ edited by Aagaard et al. 2011) contains more than 30,300 literature entries dating back to the 18th century (as of 13 October 2022) including articles, books, identification manuals, academic theses, project reports, and other items of "gray" literature. It is important to note that the number of papers per year increased significantly since 1965, reaching the highest values between 2009 and 2012 (Lencioni et al., 2018).

The knowledge of the Mediterranean chironomid fauna is relatively recent. The first list of species was published for the region only in 1977 (Reiss, 1977). Since then, knowledge has increased with the existence of faunal data from almost all the Mediterranean countries: Southern France and Corsica, Spain, Portugal, Italy, Albania, Croatia, Montenegro, Greece, Turkey, Syria, Lebanon, Israel, Tunisia, Algeria, and Morocco (Reiss, 1986; Boumaiza & Laville, 1988, Caspers & Reiss, 1989; Laville & Reiss, 1992; Langton & Casas, 1999; Moubayed & Langton, 2000; Laville & Langton, 2002; Moubayed-Breil & Dia, 2007, 2007; Moubayed-Breil et al., 2007; Mora & Csabai, 2008; Moubayed-Breil & Ashe, 2012; Moubayed-Breil et al., 2013; Płóciennik et al., 2014; Boulaaba et al., 2014; Moubayed-Breil & Ashe, 2016; Kettani & Moubayed-Breil, 2018; Zerguine et al., 2018; Bitušík & Trnková, 2019; Moubayed-Breil & Dominici, 2019;

FIGURE 15.2.1 Aquatic habitats of the Mediterranean Basin. (A) Oubeira Lake, El Kala, El Taref Province, Algeria (by A. Abdallah). (B) Drift nets collecting pupal exuviae in the Farda River, Tanger-Tetouan-Al Hoceima, Chefchaouen Province, Morocco (by K. Kettani). (C) Río Chíllar, Málaga, Spain (by N. Bonada). (D) Coastal spring, W-Corsica, France (by J. Moubayed). (E) Barouk spring, Lebanon (by J. Moubayed). (F) Lese spring, Cosenza Province, Sila Mt., southern Italy (by V. Lencioni).

Moubayed-Breil, 2020; Kettani et al., 2022; Aydın, 2022; Ozkan et al., 2010; Ozkan & Camur-Elipek, 2006; Moubayed & Lounaci, 1992).

Despite this progress, we are still far from having an accurate distribution for each of the known species of the Mediterranean Basin, and not all instars (larva, pupa, male and female adult) are known for all species.

Compiling the list of Mediterranean Basin species for some countries is challenging. The reason is that for some countries the Mediterranean Basin occupies only part of the country, for example, north-eastern and eastern Spain,

South-Eastern France, southern Asiatic Turkey, etc. According to our estimate, the species of the Mediterranean Basin represent about 11% of all known species of chironomids in the world. In fact, to date, 7290 species have been described (Lencioni et al., Chapter 15—Diptera, present volume), of which 174 genera and 850 species are reported for the Mediterranean Basin according to published works (see references above) and unpublished data from the private collections of Langton P. and Moubayed J., and the Diptera Collection of the MUSE-Science Museum of Trento, Italy (Lencioni, V. unpublished).

Considerably more species are expected in the region, particularly as cryptic species continue to be discovered. Laville & Reiss (1992) reported 703 species of chironomids for the Mediterranean countries: Italy, France (southern France and Corsica), Spain, Portugal, Morocco, Algeria, Tunisia, Lebanon, Syria, Turkey, Greece, and the former Yugoslavia. Of these, 97 of the species were exclusive to the Mediterranean Basin. The great majority of the other 570 species (81%) most likely have a Palearctic distribution, 29 species with an Afrotropical distribution, and 7 with a Panpalaeotropical distribution reaching into various parts of the Mediterranean Basin (Laville & Reiss, 1992). However, accurate distribution data for many of these species are currently unavailable. This holds in particular for species with an assumed South-Palearctic distribution. Laville and Reiss (1992) suggested that for chironomids of the Mediterranean fauna a differentiation into Circum- and West-Mediterranean subgroups is not yet possible. The Afrotropical species reach the Mediterranean Basin primarily through the Nile valley. Nevertheless, some apparently relict Afrotropic species occurring in the South of Morocco may suggest a West-African progression towards the Mediterranean. The Syrian-East African rift valley may also be considered as a migration path for West Palearctic and especially Oriental chironomids into the Afrotropical Region (Laville & Reiss, 1992; Moubayed & Langton, 2000).

Various identification keys to genera and species are available for pupal exuviae (e.g., Wilson & Ruse, 2005; Jacobsen, 2008; Prat et al., 2014); however, no comprehensive treatment is available for the entire Mediterranean fauna. We propose a new key to genera to identify pupal exuviae of chironomids from this region to supplement Langton (1991) and Langton & Visser (2003).

Ecology and distribution

Chironomids are diverse and exhibit a wide variety of ecological preferences that can be used to establish reference conditions for the bioassessment of freshwater ecosystems (Ferrington, 2008). However, due to taxonomic difficulties and poor knowledge of traits, chironomids are often neglected in biomonitoring programmes (Rossaro et al., 2012). This is especially true when bioassessment is based on larvae, whose sorting is very time consuming and identification is rarely possible at the species level. A more effective and less time-consuming method than traditional benthic sampling approaches was proposed in the 1970s (Wilson & McGill, 1977). It involves the collection and identification of pupal exuviae of chironomids, the so-called "Chironomid Pupal Exuvial Technique (CPET)" or "Chironomidae surface-floating pupal exuviae (SFPE)" (Ruse, 2002, 2010; Raunio & Muotka, 2005; Raunio et al., 2007, 2009; Kranzfelder et al., 2015; Prat et al., 2016). This is an efficient protocol for rapid bioassessment, to monitor and assess water quality and carry out studies on phenology.

Surface floating pupal exuviae are easily collected, processed, and identified especially when compared to the larvae (Ferrington et al., 1991). In addition to being more easily collected, SFPE can often be identified to lower taxonomic resolution (species level in many cases) (Ferrington et al., 1991). Immature stages (larvae and pupae) typically occur in aquatic habitats and adults emerge from the water, leaving their pupal skins, or exuviae, floating on the water's surface. Exuviae often accumulate along banks or behind obstructions pushed by the wind or water current where they can be collected to assess the chironomid communities (diversity and richness). Chironomids can be used as important biological indicators since some species are more tolerant to pollution than others. Therefore the relative abundance and species composition of collected SFPE reflect changes in water quality. For this reason, we propose a key to identify chironomids from the Mediterranean Basin as pupal exuviae useful to train new researchers. The key will also be of interest to environmental agencies measuring or managing the water quality of freshwater ecosystems. The collections of SFPE are easily standardized using timed-effort approaches, show high statistical precision, and require approximately one-third the time to process compared to benthic samples (Kranzfelder et al., 2015). Keying to genera and subgenera is complicated by the sharing of characters across the family, but in different arrangements. Further, increased taxonomic resolution can generally be achieved with pupal exuviae as most exuviae can be identified to species level (Langton & Visser, 2003).

In recent years, molecular-based approaches (e.g., DNA barcoding studies) were successfully adopted to delimit species of chironomids (e.g., Allegrucci et al., 2012; Carew & Hoffmann, 2015; Ekrem et al., 2010; Failla et al., 2016) despite concerns for some genera (e.g., Lencioni et al., 2021; Michailova et al., 2021).

FIGURE 15.2.2 Diptera Chironomidae, pupal habit. (A) *Zavrelimyia*, lateral view (by V. Lencioni). (B) *Chironomus*, lateral view (by F. Paoli). (C) *Pseudodiamesa*, lateral view (by V. Lencioni). (D) *Diamesa*, lateral view (by V. Lencioni). (E) *Diamesa*, pharate adult, lateral view (by V. Lencioni). *All figures (C) V. Lencioni, except Figure B (C) F. Paoli.*

Biology, morphology, and phenology

Most chironomid species have two generations per year; however, the range is between one generation per seven years to seven generations per year, depending on geographic latitude and altitude (Sæther 1968; Butler 1982). Chironomidae are holometabolous insects that have three development stages: larva, pupa (Fig. 15.2.2), and imago (=adult). During the larval stage, chironomids molt (= shed their exoskeleton =skin and head capsule) four times when they outgrow their exoskeleton and the different larval stages are called "instars." After molting, the more sclerotized parts, especially the head capsules. of the exoskeleton are deposited and preserved in the sediments.

After mating in aerial swarms or on the ground, adult female chironomids lay batches of eggs on a water surface. Each batch is a gelatinous mass containing from 20–30 to more than 2000 eggs, depending on the species. Hatching can occur within several hours or can take up to several days or weeks, and generally only a small number of eggs hatch (~2%). Subsequently, the larvae disperse through the water column looking for their favored habitat.

When the chironomid adult emerges, it leaves behind on the water surface its pupal exuviae (Figs. 15.2.3–15.2.4). Wind (even a mild breeze) and water movements result in exuviae accumulating at the lee shore, in bankside bays, and

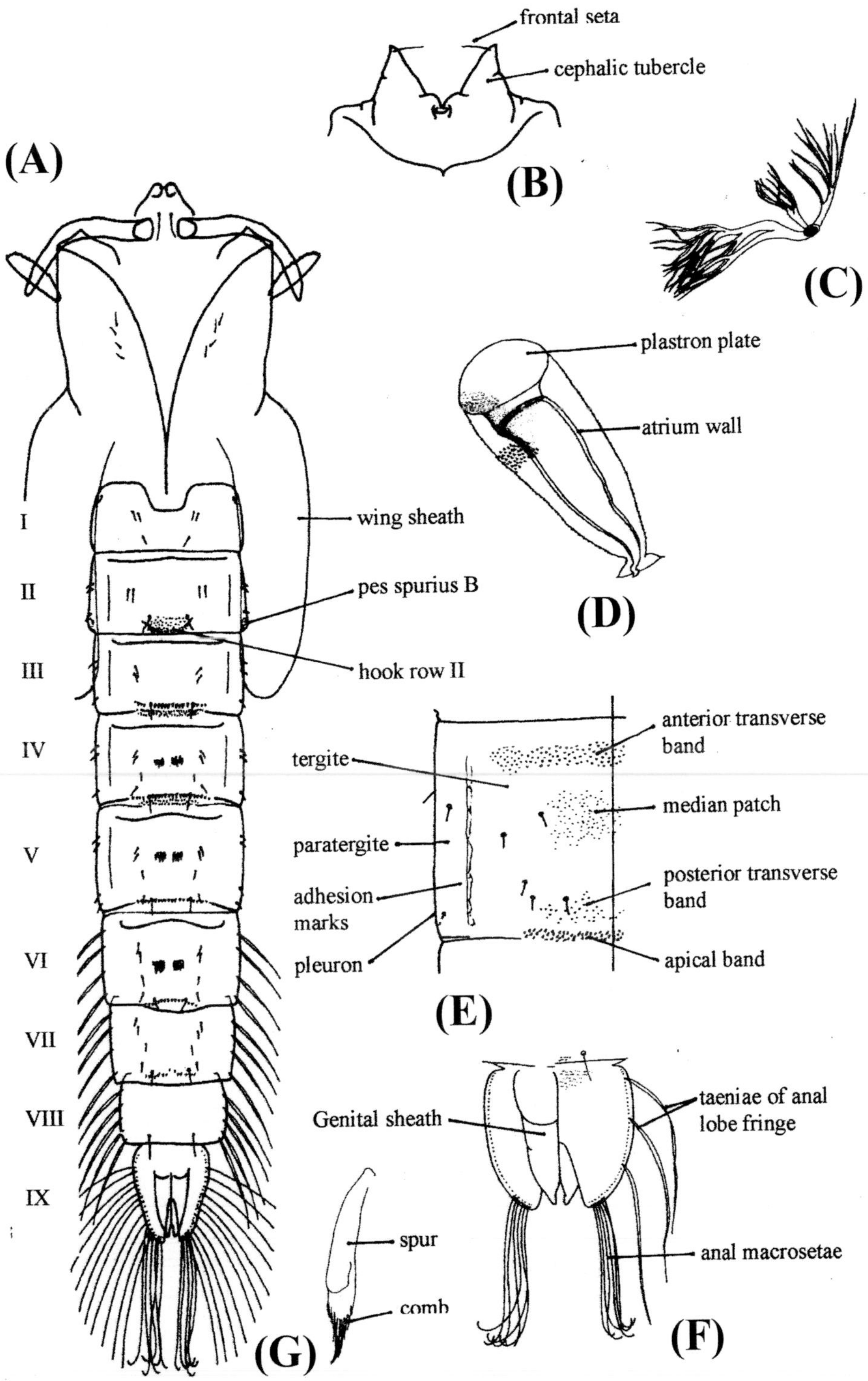

FIGURE 15.2.3 Pupal exuviae. (A) Orthocladiinae, *Psectrocladius* (s.str.) *barbimanus*. (B) Chironominae: *Paracladopelma nigritula*, frontal apotome. (C) *Kloosia pusilla*, thoracic horn. Tanypodinae: (D) *Procladius* (*Psilotanypus*) sp., thoracic horn. Chironominae: (E) *Polypedilum* (s.str.) sp., segment IV dorsal. (F) Orthocladiinae: *Psectrocladius* (s.str.) *barbimanus*, anal segment. (G) Chironominae: *Chironomus* (s.str.) sp., spur of segment VIII.

against debris dams where they are easily collected with a drift net, hand net, or sieve. Pupal exuviae float for about 2 days before sinking and disintegrating. Their collection provides, therefore a comprehensive sample of the chironomid species that recently emerged in the waterway (Bouchard and Ferrington, 2008).

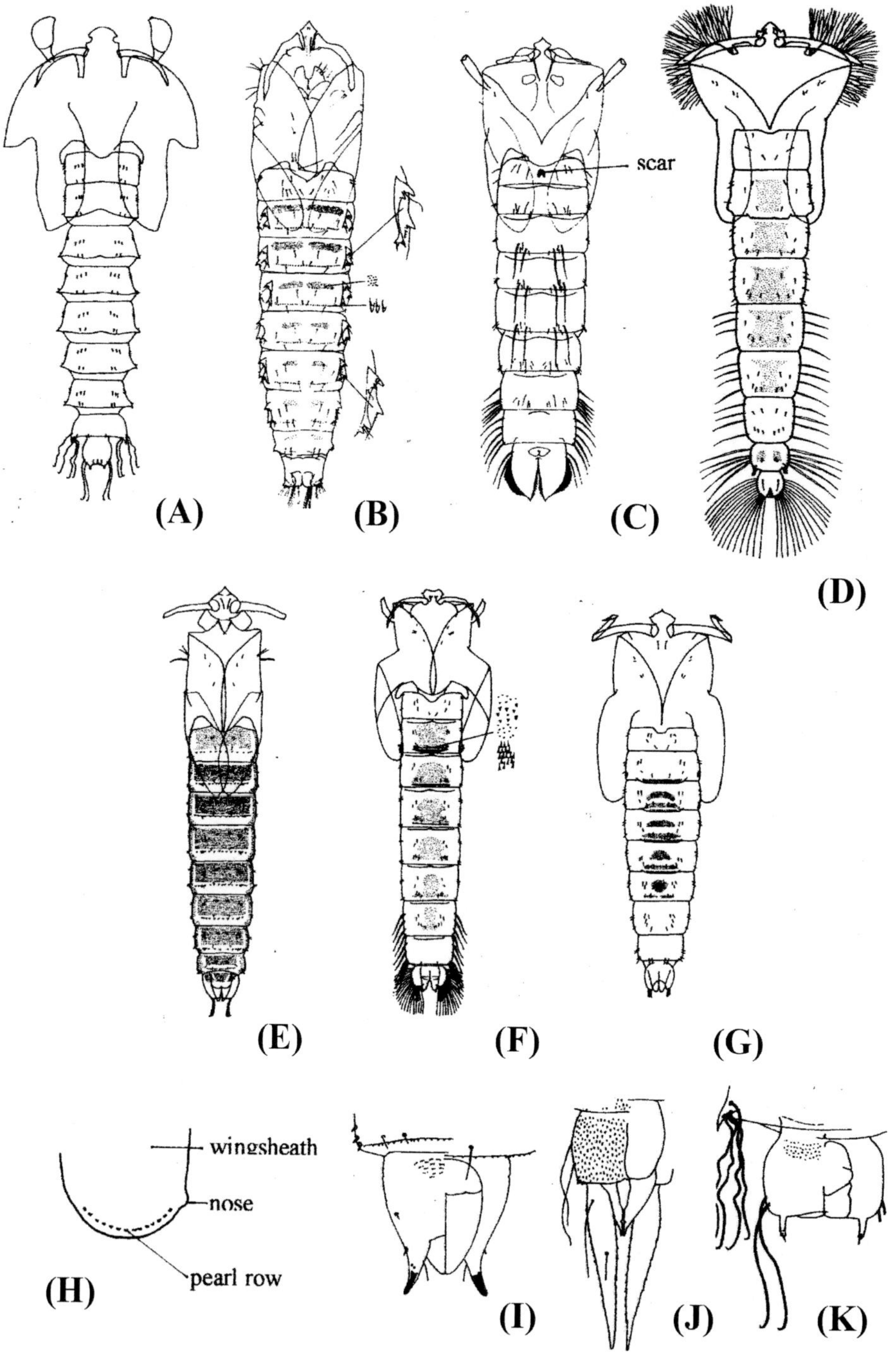

FIGURE 15.2.4 Pupal exuviae. (A) Podonominae, *Parochlus kiefferi*. (B) Buchonomyinae, *Buchonomyia thienemanni*. (C) Tanypodinae, *Psectrotanypus varius*. (D) Chironominae, *Chironomus* sp. (E) Diamesinae, *Diamesa* sp. (F) Prodiamesinae, *Prodiamesa olivacea*. (G) Orthocladiinae, *Cricotopus* (s.str.) *tremulus*; (H) *Paraphaenocladius* sp., wing sheath apex. (I) Podonominae: *Paraboreochlus minutissimus*, anal segment; (J) *Lasiodiamesa* ***gracilis***, anal segment. (K) *Parochlus kiefferi*, anal segment.

The larvae (which are recognized because they usually have anterior and posterior pairs of prolegs) are diverse in form and size. Among the species most peculiar in color are the larvae known as "bloodworms" that colonize anoxic waters. Their red color is due to the accumulation in the hemolymph of hemoglobin that increases the affinity to oxygen (Ferrarese & Rossaro 1981). When a critical temperature is reached (possibly in combination with a critical day length), the larvae pupate. Free pupae float or swim to the water surface, where the imagos emerge. The male adults form swarms, generally monospecific, but also scattered specimens from other species may join. Most species have a specific time and place of flying; the preferred place varies on local factors. A female enters this swarm for mating, and after mating, she deposits eggs on a water surface (Downes, 1969) as described above.

Depending on the life stage and habitat, the larvae are opportunistic in diet and ingest a wide variety of food items, from algae, detritus and associated microorganisms, to macrophytes, wood debris, and invertebrates (Berg, 1995) without being restricted to a single feeding mode (Henriques-Oliveira et al., 2003). However, the diet depends on the food quality available in the habitat (e.g., more algae on epiphyton or sand, more detritus in organic sediment). Most chironomid larvae are bottom-dwellers, where they live in tubes, filtering food from the water column. Others scrape or collect food from the lake's and river's sediments or from rocks. Some are predators, feeding on other chironomids [e.g., Tanypodinae and some Diamesinae such as *Pseudodiamesa branickii* (Nowicki)].

Larvae and pupae are an extremely important part of aquatic food webs, serving as prey for many aquatic organisms in general and more specifically for many species of carnivorous invertebrates (e.g., stoneflies) and fishes (Tokeshi, 1995). A more comprehensive review of the ecology of chironomids can be found in Armitage et al. (1995).

Morphological characters needed for pupal exuviae identification

The pupal exuviae (Fig. 15.2.1A) have two main sections: the cephalothorax and the abdomen. The most anterior part of the cephalothorax is the **frontal apotome** (Fig. 15.2.3B) between the bases of the antennal sheaths. The cephalothorax may bear a pair of **cephalic tubercles**. Commonly there is a pair of **frontal setae**; when cephalic tubercles are present, these are situated near to, or at the apex. In addition to the cephalic tubercles, there may be a pair of **frontal warts** (Fig. 15.2.14H), and the longitudinal split edge of the thorax is the **suture**. Anterolaterally the thorax usually has a respiratory structure, the **thoracic horn**, preceded by one to three **precorneal setae**. The thoracic horn may be multibranched (Fig. 15.2.3C) or simple (Fig. 15.2.3D). The thoracic horn is usually closed; however, in two subfamilies there may be an apical aeropyle: the **plastron plate** (sometimes surrounded by a clear rim, the **corona**) attached directly to an inner **atrium** by a **neck**, which may be long and contorted or absent (Fig. 15.2.7A). The **thorax comb**, found in Tanypodinae, is a vertical row or narrow band of enlarged granules or tubercles arising from near the base of the thoracic horn. The wing sheath may have a rounded projection, the **nose**, with or without a row or narrow band of small tubercles just within the apical margin, the **pearl row** (Fig. 15.2.4H).

The abdomen consists of nine segments (I–IX, Fig. 15.2.3A). I–VIII (Fig. 15.2.3E) are open-ended boxes with a **tergite** flanked by **paratergites** above, a **sternite** with **parasternites** below, and joined laterally by a thin membrane, the **pleuron**. Between the tergites and paratergites there may be a row of **adhesion marks**. Other adhesion marks occur on the tergites, each demarcated by a thin rim. Segment IX (Fig. 15.2.3F) is closed posteriorly apart from the anus, for, although the pupa has no mouth, it has an anus (the cast cuticle of the posterior gut remains attached to the anus and may be seen within the exuviae). The anal segment (segment IX) dorsally may be extended laterally by a flattened lobe, the **anal lobe**, which may bear a lateral **fringe** of setae, flattened setae (**taeniae**), and/or a number of **anal macrosetae**. Ventrally segment IX is closed off by the **genital sheaths**.

The abdominal segments are armed with **points** (resembling the point of a nail), **spines**, or **teeth** (enlarged, robust points). The **shagreen** is composed of minute points usually arranged in transverse or oblique rows, occasionally producing a fish scale effect (**scale shagreen**). The basic arrangement of the tergites' armament is an **anterior transverse band**, a **median patch**, a **posterior transverse band**, and an **apical band** (which may be displaced posteriad onto the inter-segmental conjunctiva). These armament areas may be fused as a general covering of the tergite, medially divided, or in part absent. The sternites are less extensively armed and frequently smooth. On segment II commonly, and on more posterior segments less frequently, the apical band is composed of **hooks** (hooked spines), for example, **hook row II**. Posterolaterally on segment II, there may be a swelling, the **pes spurius B** (plural: **pedes spurii**). These may also occur on segments I and III. Each segment has an array of small setae, generally, five on the tergite, numbered from anterior to posterior; the lateral setae on the parasclerites may be greatly elongated and taeniate. Segment VIII may be armed on the posterolateral corner with a group of strong points, teeth, or spines, the **comb of segment VIII;** there may be a projection posteriad of the posterolateral corner, the **spur**, with the comb at its apex (Fig. 15.2.3G).

In figures provided in this chapter, only the left side of the abdominal segments is depicted. Unless otherwise stated, only the dorsal side is shown.

Material preparation and preservation

Pupae and pupal exuviae can be collected easily using drift nets with a mesh size of 100–500 μm floating on the water surface and generally preserved in plastic bottles in 70%–75% ethanol.

Identifications to subfamilies or tribes (and in some cases to genus level) may be done at lower magnification (50 ×) under a stereomicroscope, while species determination requires higher magnification (up to 1000 ×) under a compound microscope. The pupae are identified by the size and color of the body, the relative length of the thoracic horn, and the abdominal armament pattern.

Microscope slide preparation is much more time-consuming, especially to mount pupae and pharate adults to separate the pupal exuviae. Before mounting in the medium (commonly Canada Balsam, or Euparal), the pupae are heated in a KOH 10% solution to macerate the organic matter (for 2 min) then dehydrated in increasing concentration of ethanol up to 100% or in glacial acetic acid (for 2 min) then butanol (for 2 min). For dissection, Tungsten needles are generally used. Pupae and pupal exuviae are dissected into head, thorax, and abdomen. The abdomen is mounted with the dorsal surface upwards and the cephalic capsule and thorax are mounted with the lateral surface upwards.

Key to subfamilies

1 Hook row of segment II absent 2

1' Hook row of segment II present (Fig. 15.2.3A) 6

2(1) Thoracic horn with plastron plate (Fig. 15.2.3D) 3

2' Thoracic horn, when present, without plastron plate 4

3(2) Segments VII–IX without lateral taeniae AND tergite I without scar (Fig. 15.2.4A) .... **Podonomiae**

3' At least one of segments VII–IX with taeniae (though they may be short) OR tergite I with scar (Fig. 15.2.4C) 5

4(2') Abdominal pleura II–VII dorsally with an anterior and posterior pair of strong, curved teeth (the anterior of each pair the larger), and ventrally with a similar tooth posteriorly, behind which is a row of about four smaller teeth. On segments III–V the anterior tooth of the posterior dorsal pair is jointed at its base, whereas on VI and VII it is the anterior tooth of the anterior dorsal pair. On segment VII the anterior dorsal pair of teeth is displaced medially, the others are missing (Fig. 15.2.4B) .... Buchonomyinae (one genus: *Buchonomyia)*

4' Abdominal pleura II–VII without lateral pairs of curved teeth 5

5(3',4') With one or more of the following characters: thoracic horn with plastron plate, thorax comb, scar on tergite I, tergites II–VIII with a dense mat of imbricating simple or forked spinules, anal lobes with two lateral taeniae with or without a setal fringe (Fig. 15.2.4C) .... **Tanypodinae**

5' Thoracic horn, if present, without a plastron plate. Thorax comb and scar of tergite I absent. Tergites II–VIII without a dense mat of spinules. Anal lobes very rarely with two lateral taeniae and then without an additional fringe 6

6(1',5') Thoracic horn, when present, branched or simple. Anal lobes without anal macrosetae, usually with a taeniate fringe; if fringe absent, either thoracic horn branched or wingsheaths with nose and tergite II with hook row. Hook row II usually present; if absent, thoracic horn branched (Fig. 15.2.4D) .... **Chironominae**

6' Thoracic horn, when present, simple. When anal lobes fringed, usually also with setaceous or taeniate macrosetae; if without additional macrosetae, hook row II absent and/or lateral margins of at least segments V and VI densely fringed with short setae. (Hook row II present or absent.)........

7(6') EITHER segments II–VIII with the lateral margin ending posteriorly in a right-angled corner or projecting tooth, the posterior margins of tergites and sternites (I,II)III–VIII armed with a row of strong teeth, OR anal lobes margined with about 10 setae, the most posterior spine-like and set in conspicuous sockets, OR thoracic horn small, triangular, pubescent and anal lobes small with three short, strong hooked macrosetae, OR anal lobes with three anal macrosetae and fringe of short hair-like setae, thoracic horn absent, OR anal lobes with a posterior triangular projection, toothed or smooth and with tergites II–VIII covered with small points and thoracic horn absent, OR tergites IV–VIII with a posterior row of strong teeth and thorax with a conspicuous tubercle above the insertion of the thoracic horn (Fig. 15.2.4E) .... **Diamesinae**

7' Without any of the above combinations of characters 8

8(7') Anal lobes fringed with taeniae, with four or five macrosetae or macrosetae absent; when macrosetae absent abdominal segments fringed with dense, short hairs. Tergite armament continuous, not divided into point patches (Fig. 15.2.4F) .... **Prodiamesinae**

8' If anal lobes fringed with taeniae and macrosetae (which can be taeniate), then macrosetae usually three in number; if five to nine, then tergite armament is divided into paired point patches. Segment VIII never with a lateral hair fringe (Fig. 15.2.4G) .... **Orthocladiinae**

Podonominae: Genera

1 Segment IX drawn out posteriorly into a toothed spur on each side with two widely separated, small spine or hair-like setae on each side and a fine hair-like seta on each apical spur internally (Fig. 15.2.4I) .... *Paraboreochlus*
1' Segment IX with long lateral setae 2
2 (1') Apical toothed spurs of segment IX very long, at least twice as long as the basal part of the segment and hinged to it (Fig. 15.2.4J) .... *Lasiodiamesa*
2' Anal segment with peg-like apical spurs (Fig. 15.2.4K) .... *Parochlus*

Tanypodinae: Genera and Subgenera

1. Anal lobes short (length: breadth 0.7–1.7:1) (Fig. 15.2.5B), rounded, the margins smooth, shorter than the median length of segment VIII (0.5–0.83:1); segment VIII extended posteriad on each side for one-fifth or more of its median length. Lateral taeniae EITHER forming a complete lateral fringe on segments (II)III–VIII, OR segment VII with 6(7) taeniae and VIII with 5(4). Thoracic horn (Fig. 15.2.5A). 645–1010 μm long, D-shaped, covered in small points, without plastron plate but with a terminal tubercle. (Tergite armament of small isolated points which may be grouped in short rows. No dark scar on segment I.) .... *Tanypus*
1' Anal lobes either narrowed to a point or broadly rounded (when they are margined with a tooth- or hair-fringe). Posterior edge of segment VIII usually straight or nearly so. When thoracic horn similar to the *Tanypus*-shape, it is EITHER less than 600 μm long, OR has an obvious plastron plate, OR the anal lobes are large and have a toothed margin 2
2(1') Segment VIII extended on each side posteriad for over one-fifth its median length. Anal lobes EITHER broadly rounded externally and armed with saw-like teeth, the inner margin straight and unarmed OR more or less evenly rounded apically and the spinous fringe extended onto the inner margin. Thoracic horn either D-shaped as in *Tanypus* or cylindrical (Fig. 15.2.3D), with a transversely oval to circular plastron plate filling the horn apex, which is joined to the broad, brown respiratory atrium by a broad, but short, neck 3
2' EITHER segment VIII not extended posteriad on each side by over one-fifth its median length (usually the posterior margin is more or less straight), OR anal lobes fringed with taeniae. When anal lobes fringed with saw-like teeth, the lobes are tapered to apex OR the teeth are also present on the inner margin as well 4
3(2) Respiratory atrium nearly as wide as the horn (Fig. 15.2.5C), or horn D-shaped. Anal lobes (Fig. 15.2.5D) with an internal projecting angle (Fig. 15.2.5D) .... *Procladius* (*Holotanypus*)
3' Respiratory atrium narrower than horn (Fig. 15.2.5E), never D-shaped. Anal lobes (Fig. 15.2.5F) rounded internally .... *Procladius* (*Psilotanypus*)
4(2') Anal lobes (Fig. 15.2.5H) somewhat *Procladius*-like, saw-toothed externally, but with teeth present on inner margin as well. Thoracic horn (Fig. 15.2.5G) flattish, 2–2½ times as long as broad with the transversely oval plastron plate occupying the whole width of the apex; atrium a straight and narrow median tube surrounded by dense irregular vesicles; neck broad, funnel-shaped .... *Anatopynia*
4' Anal lobes and thoracic horn otherwise 5
5(4') Anal lobes (Fig. 15.2.5J) with a rounded internal posterior projection, the margin fringed with taeniate setae as long as those on segment VIII .... *Clinotanypus*
5' Anal lobes narrowed to apex; if fringed, the setae hair-like 6
6(5') Anal lobes fringed with hair-like teeth. Thorax with well-spaced, regular, transverse grooves from suture. No thorax comb. Segment VII with 5 or 6 lateral taeniae, VIII with 5(4) 7
6' Anal lobes not fringed with hair-like teeth. Thorax, if so marked, with comb. Segment VII never with 6 lateral taeniae; when 5, then 6 taeniae on segments III–VI as well 10
7(6) Anal lobes (Fig. 15.2.5L) fringed externally only (but may be toothed internally). Intersegmentally shagreened II/III–VI/VII .... *Macropelopia*
7' Anal lobes fringed externally and internally. Intersegmentally not shagreened 8

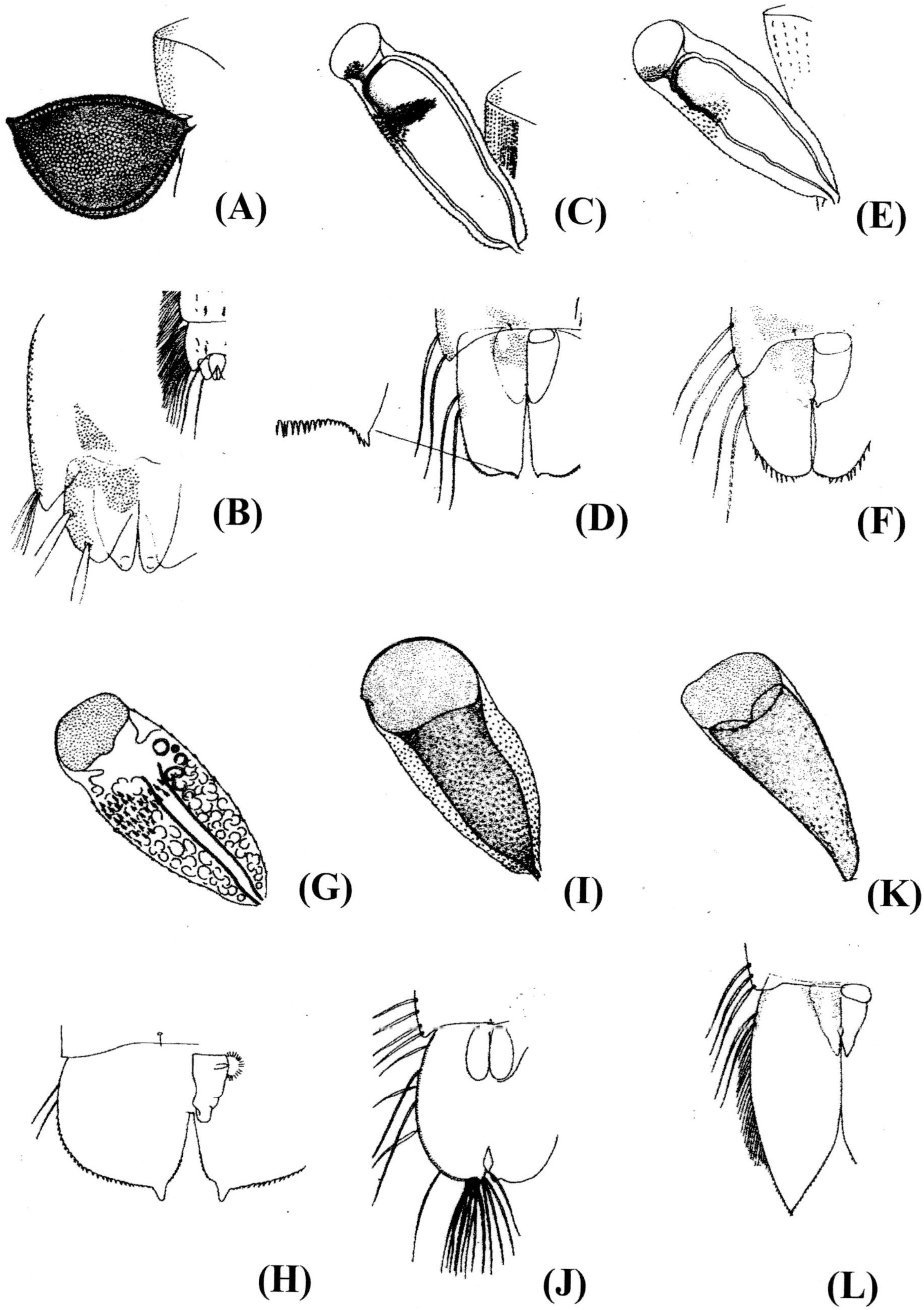

FIGURE 15.2.5 Pupal exuviae, Tanypodinae. (A) *Tanypus punctipennis*, thoracic horn; (B) posterior segments; *Procladius* (*Holotanypus*) *signatus*, (C) thoracic horn, (D) anal segments; (E) *Procladius* (*Psilotanypus*) *lugens*, thoracic horn, (F) anal segments; (G) *Anatopynia plumipes* thoracic horn, (H) segment IX; (I) *Clinotanypus nervosus*, thoracic horn, (J) segment IX; (K) *Macropelopia goetghebueri*, thoracic horn, (L) segment IX.

8(7') Plastron plate a narrow band across the apex of the thoracic horn (Fig. 15.2.6A). Segment VII anterior lateral taenia at 0.20–0.29 segment length. Anal lobes longer, about 2 ½ times or more the length of segment VIII (Fig. 15.2.6B) .... *Apsectrotanypus*
8' Plastron plate at least half as long as broad. Segment VII anterior lateral taenia at 0.44–0.55 length segment VIII. Anal lobes shorter, at most hardly exceeding twice the length of segment VIII 9
9(8') Thoracic horn atrium wall with struts broad and obvious even when the horn is flat (Fig. 15.2.6C) .... *Derotanypus*
9' Horn atrium wall with struts very narrow, showing up in surface view as dots, best seen in side view (Fig. 15.2.6E) .... *Psectrotanypus*
10(6') Segments II–VI with 6 lateral taeniae, VII and VIII with 5 (Fig. 15.2.6H) .... *Thienemannimyia*
10' Segments II–VI with 1 or 0 lateral taeniae 11
11(10') Thoracic horn (Fig. 15.2.6I) without plastron plate, toothed, more or less parallel-sided, rounded at apex and usually with a conspicuous lateral swelling near base .... *Arctopelopia*
11' IF thoracic horn without plastron plate and toothed, it is swollen either medially or apically 12
12(11') Abdominal tergite armament, at least in part, of a dense mat of simple and forked spinules 13
12' Tergite armament otherwise 16
13(12') Thoracic horn without plastron plate 14
13' Plastron plate present, wider than neck of horn atrium 15
14(13) Thoracic horn (Fig. 15.2.6K) with corona. Scar of tergite I distinct .... *Hayesomyia*
14' Corona absent (Fig. 15.2.6L). Scar of tergite I indistinct .... *Rheopelopia*
15(13') Thoracic horn (Fig. 15.2.7A) with plastron plate about 0.2 horn length, set in a wide corona about twice as wide as the plastron plate; horn atrium occupying the bulk of the width of the horn and attached to the plastron plate by a broad neck. A well-developed thorax comb present, of strong triangular teeth .... *Telopelopia*
15' Thoracic horn with corona (Fig. 15.2.7C) or without (Fig. 15.2.7D); if plastron plate small and set in a wide corona, then the atrium is for most of its length a narrow tube. Thorax comb reduced to large rounded or pointed tubercles .... *Conchapelopia*
16(12') Abdominal armament of fine scale shagreen at least posterolaterally on middle tergites. Thoracic horn (Fig. 15.2.7F) swollen, rounded apically, with a small offset apical plastron plate .... *Ablabesmyia*
16' If abdominal armament of scale shagreen then thoracic horn with a conspicuous plastron plate. If thoracic horn swollen and with a small offset plastron plate, it is flattened apically 17
17(16') Segments I–VII each with one long lateral taeniate seta and a very short seta inserted close to its base (Fig. 15.2.7I) .... *Natarsia*
17' Lateral setation of segments I–VII otherwise 18
18(17') Each anal lobe (Fig. 15.2.8B) about 1.7 times as long as broad, its apex smooth, bluntly pointed and curved outwards, with a conspicuous tubercle on its inner edge. Segment VII with three lateral taeniae .... *Krenopelopia*
18' Each anal lobe over twice as long as broad, or buckled inwards apically. Segment VII with 1, 3, or 4 lateral taeniae 19
19(18') Anal lobes (Fig. 15.2.8D) tapered, smooth, and buckled inwards apically producing a variable and low length:breadth ratio. Tergites II–VII with the points becoming larger posteriorly to form an obvious transverse band of granules in three or four rows; tergite VIII with a transverse row of teeth on its posterior margin. Segment VII with one lateral taenia, VIII with five. Thorax shagreened. Exuviae 2.4–2.9 mm long. Thoracic horn (Fig. 15.2.8C) with large plastron plate, offset apically, occupying over half the horn's length, without corona .... *Nilotanypus*
19' Anal lobes otherwise. Tergites without obvious posterior transverse bands of points or granules. Segment VII with 0, 3, or 4 lateral taeniae. Thorax wrinkled, granulate, or smooth. Plastron plate small 20
20(19') Anal lobes (Fig. 15.2.8F) over four times as long as broad, toothed externally beyond the posterior lateral taenia. Scar on segment I absent. Points of tergites isolated. Thoracic horn (Fig. 15.2.8E) broad, widened to apex, where it is broadly truncate; plastron plate minute, set on a transparent offset tubercle, joined to the respiratory atrium by a narrow sinuous neck .... *Labrundinia*
20' Anal lobes not four times as long as broad. Tergite I with a median mark. Thoracic horn very different in form and structure 21
21(20') Respiratory atrium (Fig. 15.2.8G) thin walled and much diverticulated .... *Larsia*
21' Respiratory atrium at most undulate, not diverticulated 22

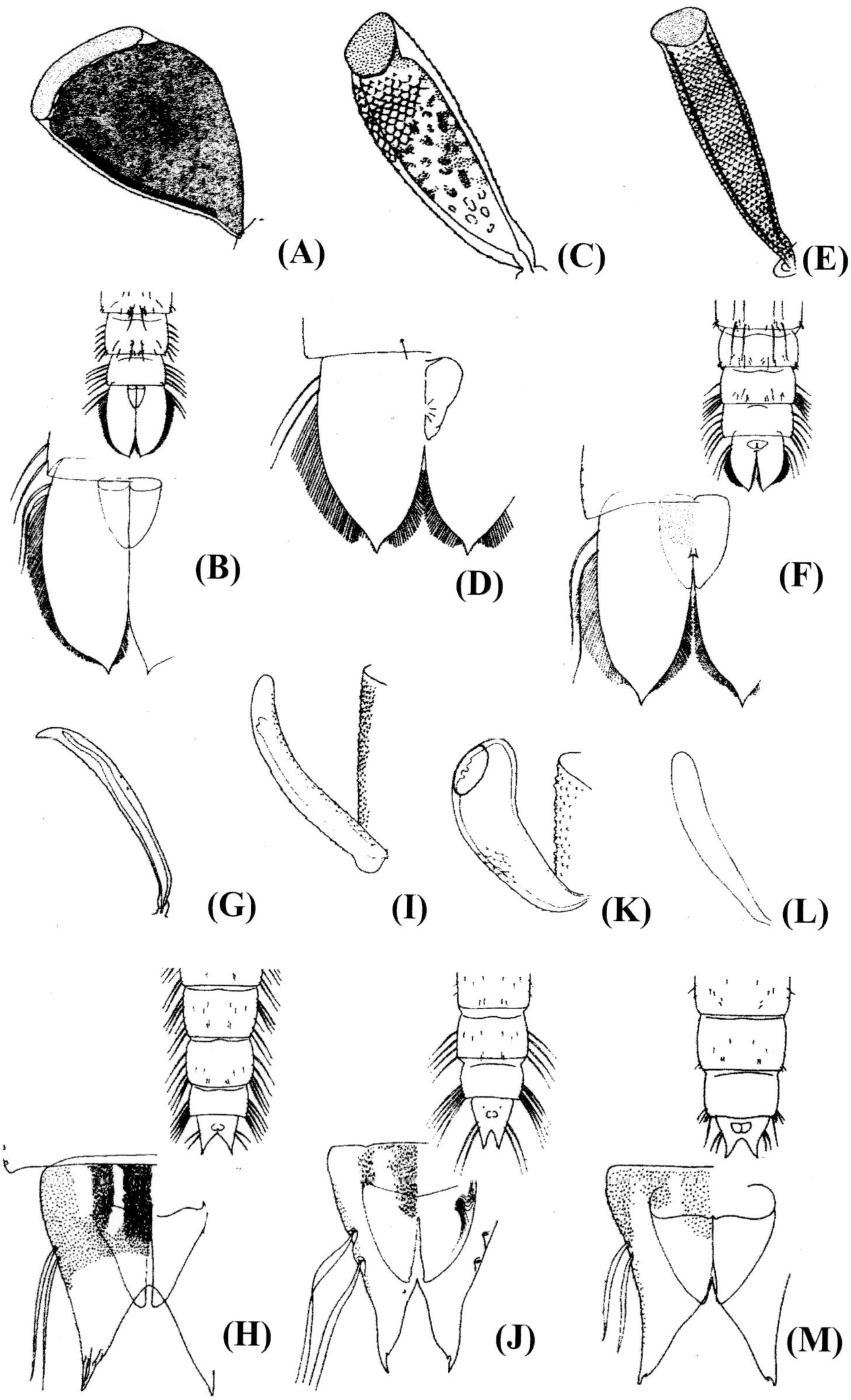

FIGURE 15.2.6 Pupal exuviae, Tanypodinae. (A) *Apsectrotanypus trifascipennis*, thoracic horn, (B) posterior segments; (C) *Derotanypus* sp., thoracic horn, (D) segment IX; (E) *Psectrotanypus varius*, thoracic horn, (F) posterior segments; (G) *Thienemannimyia lentiginosa*, thoracic horn, (H) posterior segments; (I) *Arctopelopia melanosoma*, thoracic horn, (J) posterior segments; (K) *Hayesomyia tripunctata*, thoracic horn; (L) *Rheopelopia ornata*, thoracic horn, (M) posterior segments.

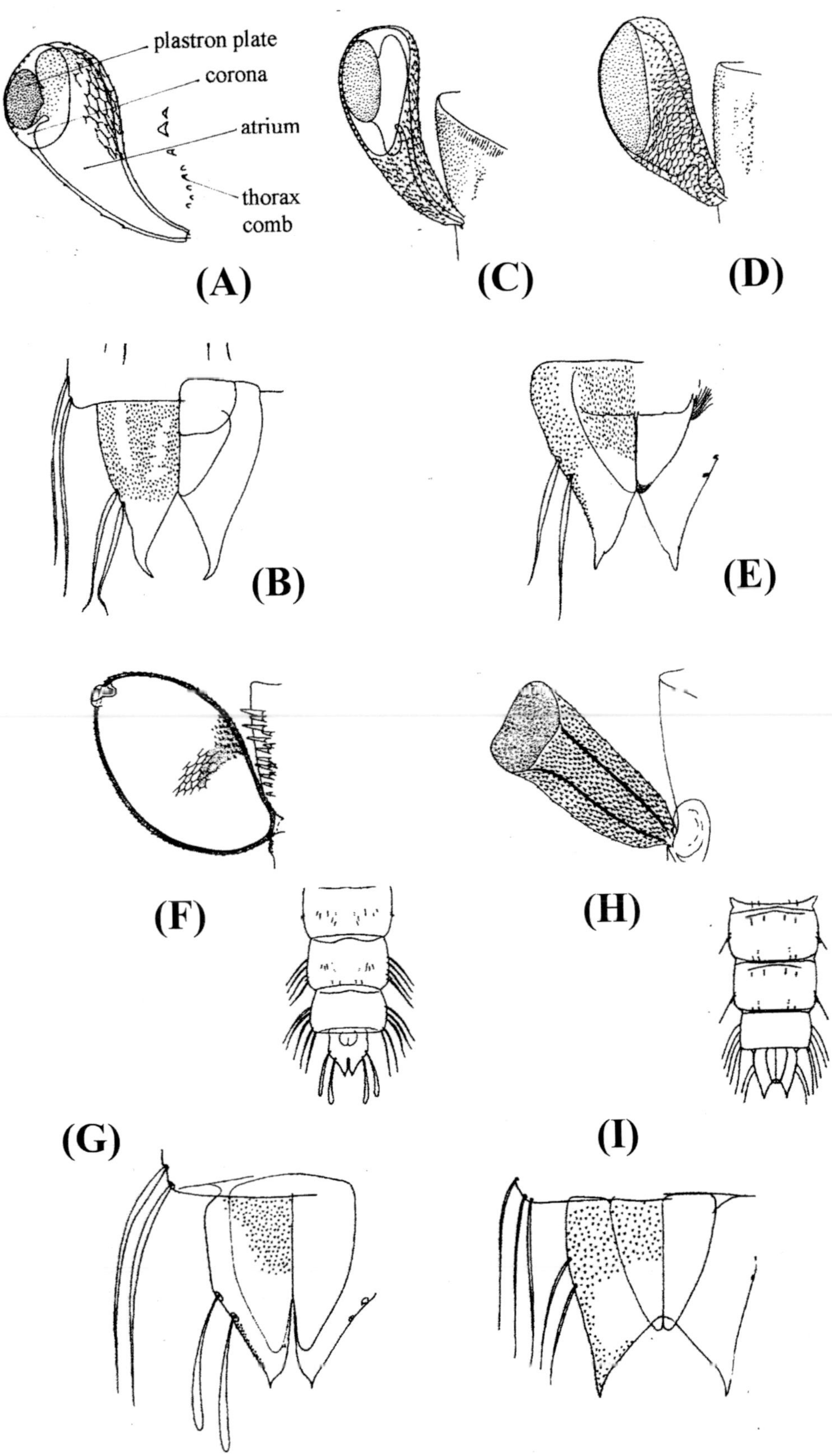

FIGURE 15.2.7 Pupal exuviae, Tanypodinae. (A) *Telopelopia fascigera*, thoracic horn, (B) segment IX; (C) *Conchapelopia viator*, thoracic horn; (D) *Conchapelopia melanops* thoracic horn, (E) segment IX; (F) *Ablabesmyia longistyla*, thoracic horn, (G) posterior segments; (H) *Natarsia nugax*, thoracic horn, (I) posterior segments.

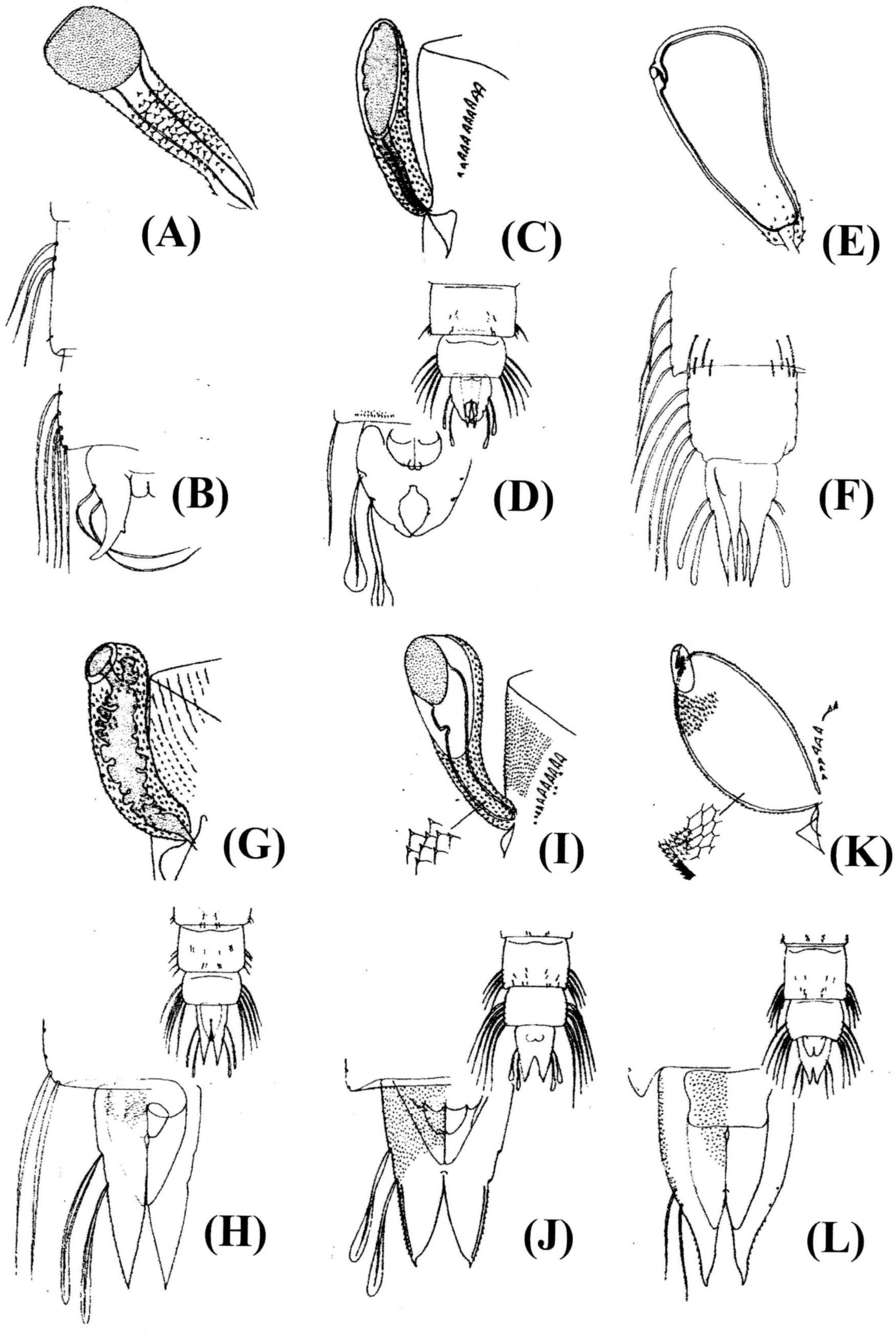

FIGURE 15.2.8 Pupal exuviae, Tanypodinae. (A) *Krenopelopia nigropunctata*, thoracic horn, (B) posterior segments; (C) *Nilotanypus dubius*, thoracic horn, (D) posterior segments; (E) *Labrundinia longipalpis*, thoracic horn, (F) posterior segments; (G) *Larsia curticalcar*, thoracic horn, (H) posterior segments; (I) *Trissopelopia longimana*, thoracic horn, (J) posterior segments; (K) *Guttipelopia guttipennis*, thoracic horn, (L) posterior segments.

22(21') Anal lobes (Fig. 15.2.8J) with external edge thickened from posterior lateral taenia to apex, toothed externally. Tergite armament of points grouped in short transverse rows. Thorax granulate anteriorly .... *Trissopelopia*
22' Anal lobes without a thickened lateral edge. If armament of abdominal tergites of group shagreen, thorax not granulate 23
23(22') Anal lobes (Fig. 15.2.8L) with anterior taenia inserted beyond 0.45 the lobe length; toothed externally from posterior taenia to apex. Segment VII usually nearly square. Thorax comb (Fig. 15.2.8K) of 6–11 hollow conical tubercles; basal tubercle of thoracic horn similar but larger .... *Guttipelopia*
23' Insertion of anterior anal taenia not beyond 0.45 lobe length. Segment VII transversely rectangular. Thorax comb of more or less parallel-sided teeth, or poorly developed, or absent. Thoracic horn basal tubercle rounded 24
24(23') Thoracic horn (Fig. 15.2.9A) flattish, toothed, elongate, gradually narrowed to the rounded apex; plastron plate occupies the whole of the horn apex, over one-third horn length, connected by a short neck to a thick-walled, unbranched, straight atrium, which occupies half to two-thirds the width of the horn lumen. Thorax comb absent. Many dorsal setae of the abdomen branched .... *Monopelopia*
24' Plastron plate only one-fifth thoracic horn length OR neck of respiratory atrium narrow OR atrial wall thin. Thorax comb present even if poorly developed. Dorsal setae not branched 25
25(24') Thorax comb (Fig. 15.2.9C) poorly developed, composed of a narrow band of 10–15 small granules on the otherwise smooth thorax .... *Telmatopelopia*
25' Thorax comb well developed, composed of strong, more or less parallel-sided teeth 26
26(25') Anal lobes toothed externally only 27
26' Anal lobes toothed strongly both externally beyond the lateral taeniae and internally, both margins convexly narrowed to apex 29
27(26) Thoracic horn atrium (Fig. 15.2.9E) narrow and with exceedingly thickened wall, joined to the plastron plate by a twisted neck .... *Xenopelopia*
27' Atrium of thoracic horn thin-walled, joined to the plastron plate by a short straight neck 28
28(27') Thoracic horn (Fig. 15.2.9G) about three times as long as broad, the atrium nearly filling the horn lumen; plastron plate without corona. Thorax smooth .... *Schineriella*
28' Thoracic horn (Fig. 15.2.9I) generally more than four times as long as broad, but, if less, the atrium is about two-thirds the width of the horn lumen; plastron plate with a smooth corona. Thorax granulate .... *Paramerina*

Paramerina and *Zavrelimyia* are extremely similar in their pupal structure, even to the form and arrangement of the dorsal and ventral shagreen. In general, the thoracic horn of *Paramerina* is widest near apex where it is nearly parallel-sided and truncate, the plastron plate lateral, whereas in *Zavrelimyia* the thoracic horn is widest below the apex to which it then narrows, the plastron plate being terminal, though offset. Also, in general, the male genital sheaths are much longer than the anal lobes in *Paramerina*, but shorter in *Zavrelimyia* and *Pentaneurella*.

29(26') Tergites and sternites I–VI armed at most with minute points, on III–VI contrasting with the parasternites which are densely armed with stronger points. Thoracic horn (Fig. 15.2.9J) with neck of plastron plate very short, about 6 μm long .... *Pentaneurella*
29' Tergites and sternites II–VIII armed conspicuously with spinulate points, which are longer than the points of the parasternites. Neck of plastron plate longer (Fig. 15.2.9L) .... *Zavrelimyia*

See comments on *Paramerina* and *Zavrelimyia* above.

Chironominae: Tribes

1 Thoracic horn with two to many branches. (Some of the branches may be apically small toothed or minutely setulated.) Tergites II–VI in general with an anterior and posterior transverse point band and a median patch of points, each developed to varying degrees; absent on some segments in some species, completely fused and covering the tergites in others. (In *Zavreliella* Kieffer, *Omisus* Townes and *Lauterborniella* Bause tergites II–V or II–VI bear a pair of anterior point patches and so resemble tanytarsine exuviae.) Wing sheaths without pearl row and, except in *Paralauterborniella*, nose 2
1' Thoracic horn unbranched (smooth, setulated or with long setae), or absent. Tergites (II)III–VI in general with a pair of anteromedian patches of points, spines, or spinules which may be extended posteriorly to produce longitudinal bands on each side of the mid-line; anteriorly the patches may be joined to produce a transverse band. Wing sheaths usually with nose and/or pearl row .... **Tanytarsini**
2(1) Sternite I (Fig. 15.2.13B) with two anterolateral mounds on each side, the outer always beset with points, the inner occasionally smooth. Tergite VII (Fig. 15.2.13A) with a continuous broad transverse band of small points

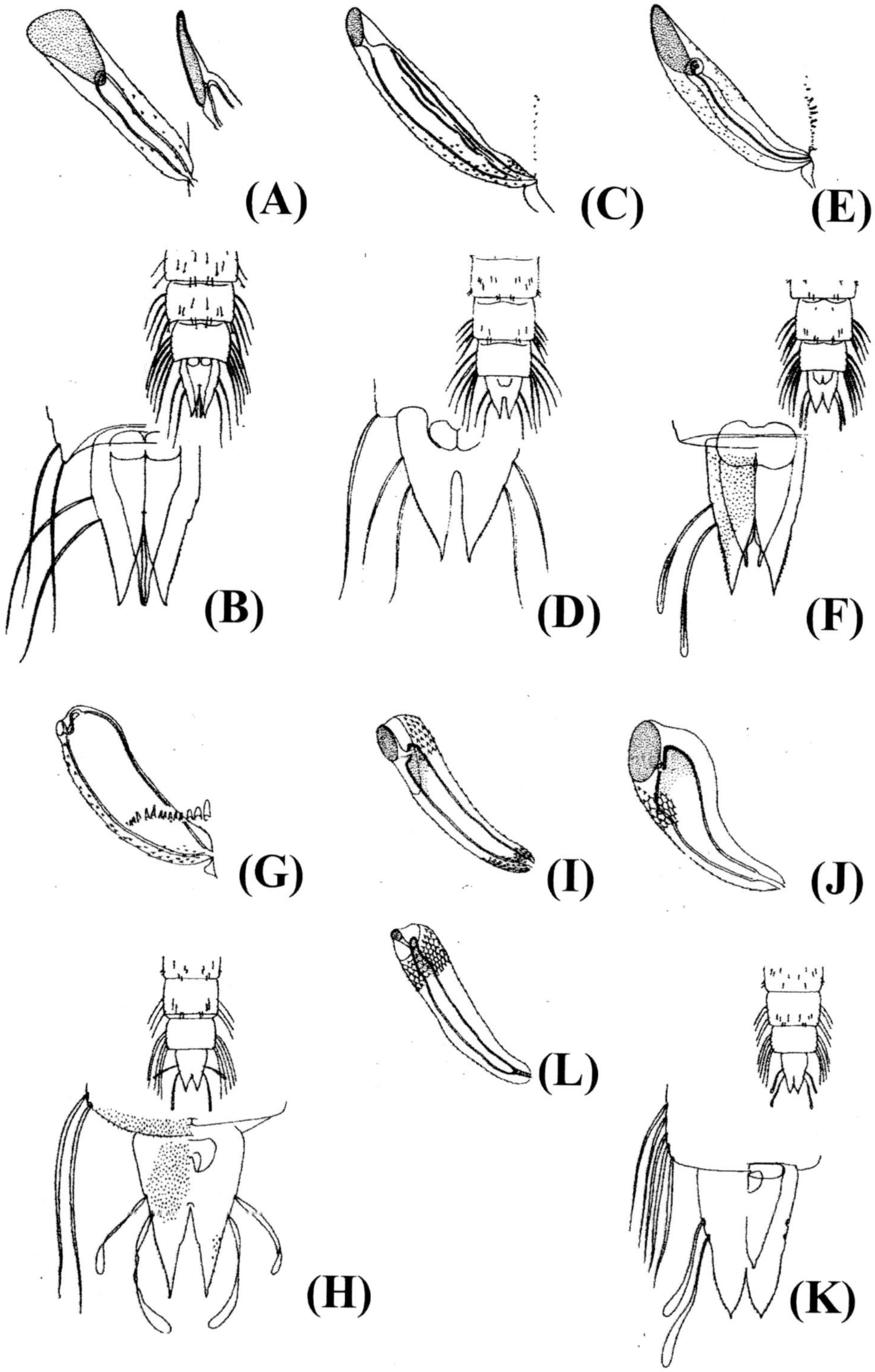

FIGURE 15.2.9 Pupal exuviae, Tanypodinae. (A) *Monopelopia tenuicalcar*, thoracic horn, (B) posterior segments; (C) *Telmatopelopia nemorum*, thoracic horn, (D) posterior segments; (E) *Xenopelopia nigricans*, thoracic horn, (F) posterior segments; (G) *Schineriella schineri*, thoracic horn, (H) posterior segments; (I) *Paramerina* sp., thoracic horn; (J) *Pentaneurella katterjokki*, thoracic horn, (K) posterior segments; (L) *Zavrelimyia berberi*, thoracic horn.

anteriorly. Thoracic horn two-branched, the branches swollen and rounded at the tip (Fig. 15.2.13A) .... Pseudochironomini, one genus: *Pseudochironomus* (Fig. 15.2.13A–C)
2' Without this combination of characters .... **Chironomini**

Chironominae: Tanytarsini: Genera

1 Anal lobes (Fig. 15.2.10C) without trace of a fringe or even an apical seta. Segment VIII without a posterolateral comb or spur. Tergites (Fig. 15.2.8B) II–V or II–VII with an anterior transverse band of dense small points; on II to IV also with a posteromedian pair of point patches which are much less dense and joined to the anterior transverse band to a greater or lesser degree on each side of the mid-line. Thoracic horn (Figs. 15.2.8 and 15.2.10A) setulated from about middle to tip, the longest setae about as long as the horn is maximally wide .... *Lithotanytarsus*
1' IF anal lobes without trace of a fringe and apical seta, then postero-lateral corners of segment VIII with a comb or spur(s). Tergite armament not as above 2
2(1') Segment VIII with posterolateral corners bearing a large spur (occasionally double and there may be additional lateral spurs on the segment) AND tergite IV with the armament extending into the posterior half 3
2' IF the posterolateral corners of segment VIII possess a single spur, tergite IV has the armament restricted to an anterior pair of point patches (though there may be fine shagreen scattered over the rest of the tergite) 7
3(2) Tergites III–VI with a narrow longitudinal band of points on each side of the mid-line, posteriorly widened or even extended laterally for a short distance, best developed on tergite V 4
3' Tergites III–VI with a median and posterior patch of points on each side of the midline, which may be sufficiently extensive to form a single large continuous patch 5
4(3) Frontal setae (Fig. 15.2.10D) thorn-like. Thoracic horn with base swollen, narrowing rapidly in the apical half. Thorax anteriorly by suture thickly set with small spinous teeth. Tergite V (Fig. 15.2.10E) with longitudinal point bands posteriorly extended laterad, the points of the extension enlarged and directed laterad. Segment VIII (Fig. 15.2.10F) with two to six pale golden spurs on each side and with 2 lateral taenlae .... *Constempellina*
4' Frontal setae elongate, flexible, narrow taeniate. Segment VIII with 3 or 4 lateral taeniae (Fig. 15.2.10I) .... *Stempellina*
5(3') Thoracic horn (Fig. 15.2.10J) with long setae forming a broad fringe from near base to tip. Tergite II (Fig. 15.2.10K) armed as in *Zavrelia* but pedes spurii B II absent .... *Neostempellina*
5' Thoracic horn at most short-toothed... Small pedes spurii II present 6
6(5') Tergite II (Fig. 15.2.10N) armament almost as extensive as on tergite III, extending beyond seta D1anteriorly .... *Zavrelia*
6' Armament of tergite II less extensive than on III (Fig. 15.2.10P), at most just reaching seta D1 anteriorly .... *Stempellinella*
7(2') Each anal lobe (Fig. 15.2.11C) strongly darkened, bearing dorsally on its posterior third dark points and 6–13 long brown taeniae. Tergites (Fig. 15.2.11B) II–VI with an anterior pair of longitudinally oval point patches. Very small exuviae (1.8–2.5 mm long) with very delicate, transparent thoracic horns (Fig. 15.2.11A), minutely toothed apically. Abdominal segments without lateral taeniae .... *Neozavrelia*
7' Without this combination of characters 8
8(7') Segment VIII (Fig. 15.2.11F) posterolateral corner with a simple, cleft or toothed spur, or three spurs, of which the outermost is the strongest and bent as a hook. Tergites (Fig. 15.2.11E) II–IV, II–V, or II–VI armed with paired anterior point patches .... *Rheotanytarsus*
8' Segment VIII posterolateral corner bears a comb of small teeth. If tergite II with anterior point patches, then point patches are also present on tergites III–VI 9
9(8') Wing sheath apex usually with a well-developed pearl row (Fig. 15.2.2H). Tergite armament varied (e.g., Fig. 15.2.11H, K, L, M); tergite IV usually with an anterior single, round to transversely elongate point patch. If tergite IV with a pair of anterior point patches, wing sheaths with a pearl row and tergite III with a pair of curved posterior spine bands. Thoracic horn (Fig. 15.2.11G, J) with or without setulae .... *Paratanytarsus*
9' Wing sheaths without a pearl row 10

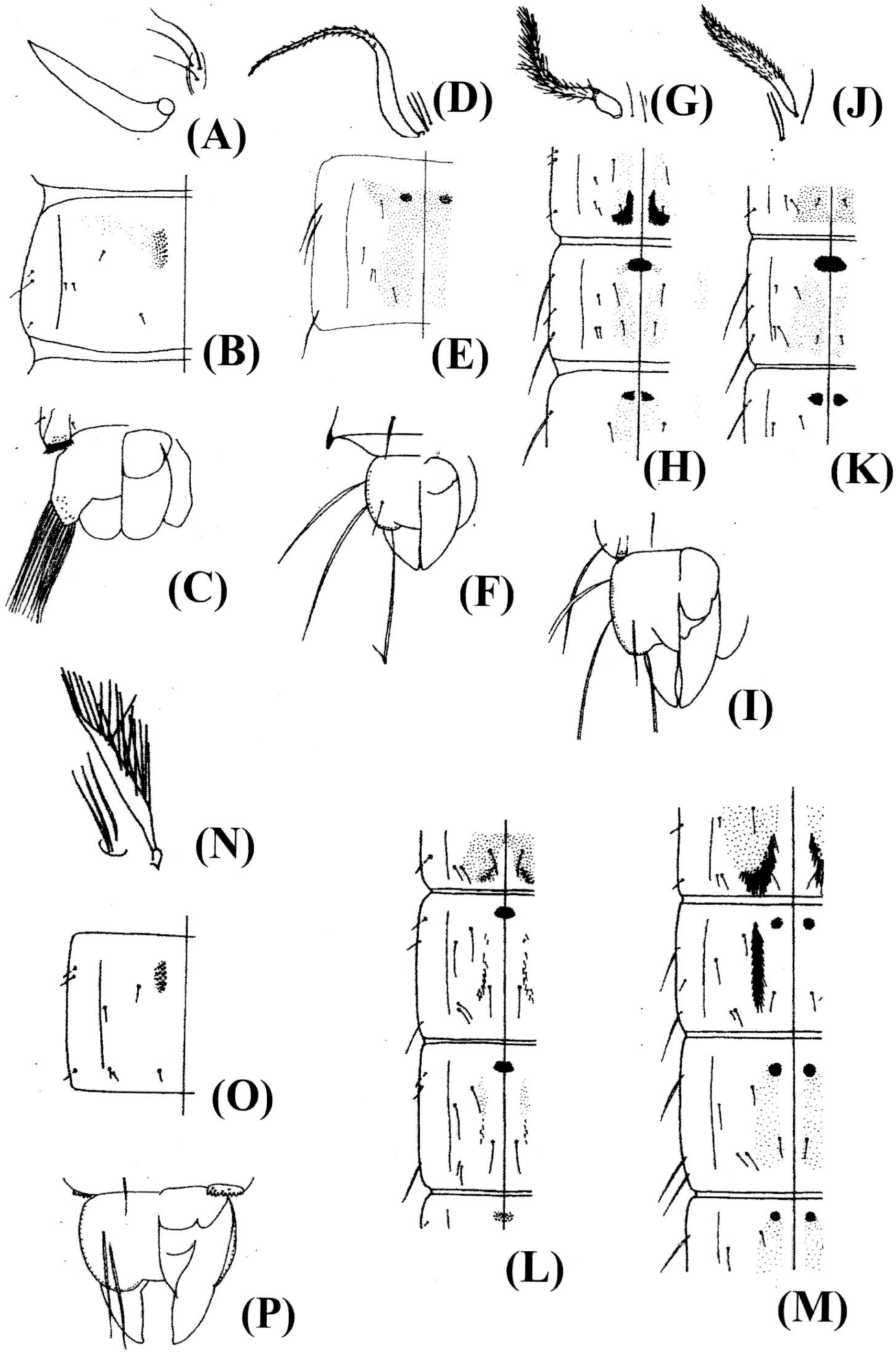

FIGURE 15.2.10 Pupal exuviae, Chironominae, Tanytarsini. (A) *Lithotanytarsus emarginatus*, thoracic horn and precorneal setae, (B) segment III and posterior II, (C) anal segment; (D) *Constempellina brevicosta*, frontal apotome, cephalic tubercles and frontal setae, (E) segment V; (G) *Stempellina bausei*, thoracic horn and precorneal setae, (H) segment IV, (I) anal segment and lateral VIII; (J) *Neostempellina thienemanni*, thoracic horn, (K) segment II, (L) anal segment and lateral VIII; (M) *Zavrelia pentatoma*, thoracic horn and precorneal setae, (N) segment II; (O) *Stempellinella minor*, thoracic horn and precorneal setae, (P) segment III and posterior II, (Q) anal segment and comb of segment VIII.

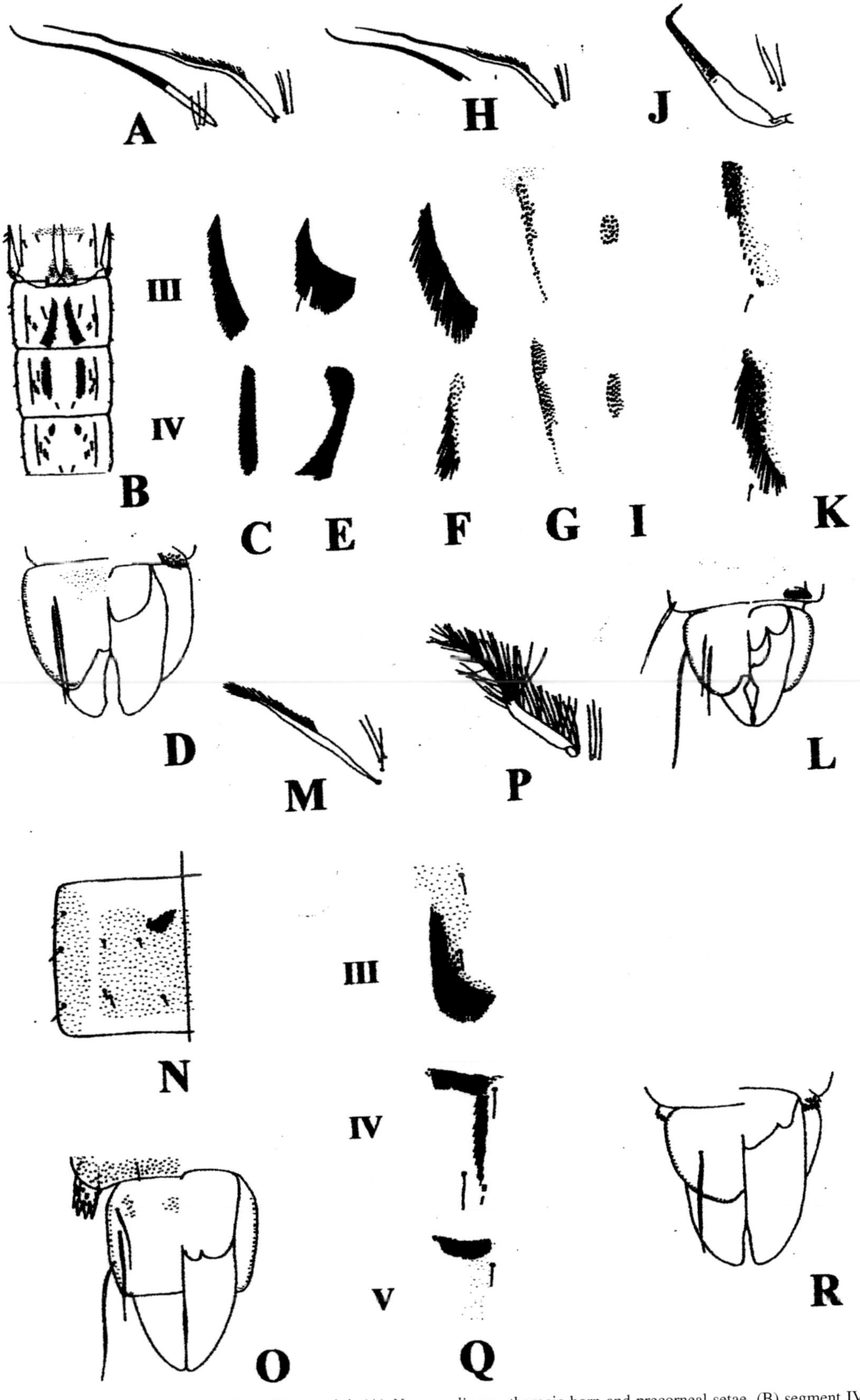

FIGURE 15.2.11 Pupal exuviae, Chironominae, Tanytarsini. (A) *Neozavrelia* sp., thoracic horn and precorneal setae, (B) segment IV, (C) anal segment; (D) *Rheotanytarsus* sp., thoracic horn and precorneal setae, (E) segment VI, (F) anal segment; (G) *Paratanytarsus austriacus*, thoracic horn and precorneal setae, (H) segments III–VI, (I) anal segment; (J) *Paratanytarsus inopertus*, thoracic horn and precorneal setae, (K) segments III–V: (L) *Paratanytarsus tenuis*, segments III–V; (M) *Paratanytarsus penicillatus*, segments III–V; (N) *Cladotanytarsus atridorsum*, thoracic horn and precorneal setae, (O) segment IV, (P) anal segment.

10(9') Exuviae usually very transparent, sometimes a little infuscated. Tergites (Fig. 15.2.11O) II–V each with an anterior pair of point patches. Segment VIII with 5 lateral taeniae. Thoracic horn (Fig. 15.2.11N) less than 400 μm long .... *Cladotanytarsus*
10' Exuviae usually considerably darkened. If tergite II with an anterior pair of point patches and segment VIII with 5 lateral taeniae, then the thoracic horn is longer than 500 μm 11
11(10') Tergites IV and V with the armament arranged in longitudinal bands 12
11' Tergites IV and V (Fig. 15.2.12N, Q) with an anterior pair of transverse, slightly oblique point patches, sometimes narrowly joined medially, frequently grading into spines laterally and posteriorly 13
12(11) Segment VIII with four or five lateral taeniae OR the comb small and with few marginal teeth. Tergite bands (e.g., Fig. 15.2.12C–G, I) of spines, points or a mixture of the two. Thoracic horn (e.g., Fig. 15.2.12A, H) armed with small points or setulae, or unarmed .... *Tanytarsus* (s.str.).
12' Segment VIII with three lateral taeniae AND the comb (Fig. 15.2.12L) with the posterior edge very straight and bearing many small teeth. Tergite IV point bands (Fig. 15.2.12K) armed with spines internally; tergite III may be similarly armed, but with fewer and shorter spines .... *Tanytarsus (Virgatanytarsus)*.
13 (11') Abdominal tergites II–VIII (e.g., Fig. 15.2.12N tergite IV) with extensive armament OR tergite II with a pair of anterior point patches in addition to those on tergites III–VI .... *Parapsectra*
13' Abdominal tergites VII and VIII with very restricted armament, or smooth. Tergite II without anterior point patches .... *Micropsectra*

Chironominae: Chironomini: Genera and Subgenera

1 On tergites (II)III–VI(VII) (e.g., Fig. 15.2.13D, G) dorsal abdominal armament restricted to large points set on the apices of a single, or a pair of posterior transverse mounds 2
1' Abdominal tergite armament not restricted to posterior mounds 3
2(1) Anal lobes (Fig. 15.2.13E) without fringe. Abdominal sternites without toothed mounds. Comb of segment VIII absent (Fig. 15.2.13E). Thoracic horn without an exceptionally long branch .... *Rheomus*
2' Anal lobes (Fig. 15.2.13H) fringed. At least some abdominal sternites armed with toothed mounds posteriorly. Comb of segment VIII (Fig. 15.2.13H) thorn-like, set close behind the fourth lateral taenia. Thoracic horn (Fig. 15.2.13F) with one branch very long and stiff, extending forwards for over half the exuvial length and not itself branched for its proximal half, the other initial branches small and clustered around its base .... *Cryptotendipes*
3(1') At least segments IV and V each with an anterior median dorsal spinous plate ('epaulet') distinct from the general armament of the tergite 4
3' Tergites without such spinous plates 7
4(3) Epaulets (Fig. 15.2.13I) small, restricted to tergites IV–VI .... *Demeijerea*
4' Epaulets larger, present on segments II/III–VI 5
5(4') Epaulets (Fig. 15.2.13J, K) racket-shaped, present on tergites II(III)–VI .... *Glyptotendipes* (s.str.)
5' Epaulets (Fig. 15.2.13L) foot-print shaped, with short, usually broad anterior limb; restricted to tergites III–VI 6
6(5') Thorax with extensive small granulation. Comb of segment VIII (Fig. 15.2.13M) composed of 4–18 colorless, elongate, basally curved spines. Segments V–VIII with 4, 4, 4, 5 lateral taeniae .... *Glyptotendipes (Heynotendipes)*
6' Thorax with a few granules near suture only. Comb of segment VIII composed of 1–8 dark teeth. Segments V–VIII with 4, 4, 4, 4(5) lateral taeniae .... *Glyptotendipes* (*Caulochironomus*)
7(3') Tergites II–VI (Fig. 15.2.14A) with small points arranged as an anterior and posterior transverse band joined by a median longitudinal band of smaller points, the lateral median space filled with scattered, strong, dark points set on small dark gray patches which coalesce to form an irregular pattern on a transparent background .... *Xenochironomus*
7' Tergites II–VI with armament not of this characteristic form (e.g., Fig. 15.2.14C, F, G) 8
8(7') Comb of segment VIII (Fig. 15.2.14D) of 4–6 equally spaced, strong, dark, narrow, curved teeth, situated on the lateral margin at the apical corner of the segment, which is itself very dark, and 0–2 smaller teeth internal to the apical tooth. Conjunctives III/IV, IV/V, and V/VI covered with dense spinules extending to the width of the tergite armament .... *Kiefferulus*
8' Comb of segment VIII of different form or absent. If conjunctives III/IV, IV/V, and V/VI armed, the armament is composed of minute shagreen or sturdy points 9

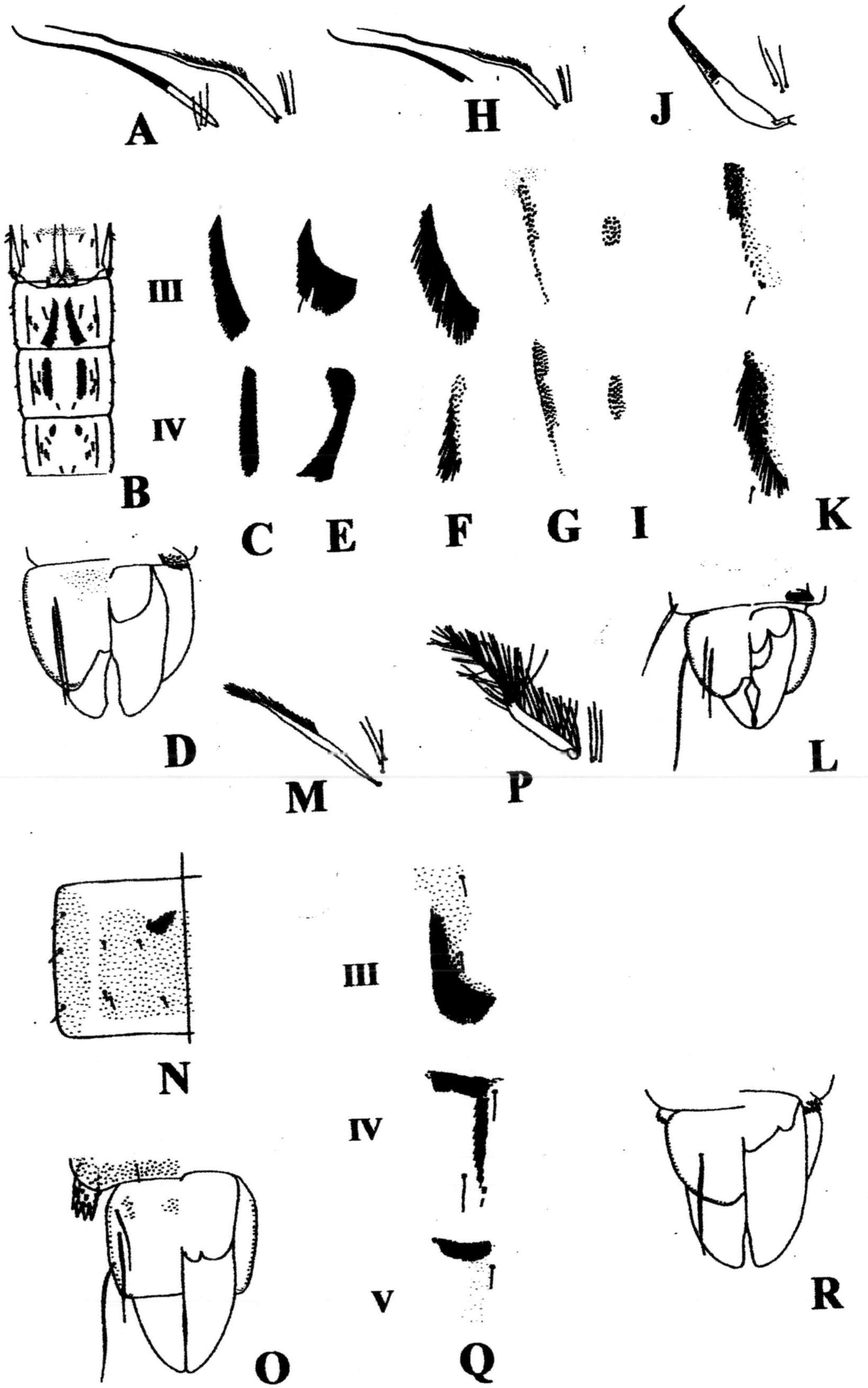

FIGURE 15.2.12 Pupal exuviae, Chironominae, Tanytarsini. (A) *Tanytarsus gregarius*, thoracic horn and precorneal setae, (B) segments II–V, (C) longitudinal bands tergites III and IV, (D) anal segment; (E) *Tanytarsus ejuncidus*, longitudinal bands tergites III and IV; (F) *Tanytarsus striatulus* longitudinal bands tergites III and IV; (G) *Tanytarsus gracilentus*, longitudinal bands tergites III and IV; (H) *Tanytarsus mendax*, thoracic horn and precorneal setae, (I) longitudinal bands tergites III and IV; (J) *Virgatanytarsus triangularis*, thoracic horn and precorneal setae, (K) longitudinal bands tergites III and IV, (L) anal segment; (M) *Parapsectra nana*, thoracic horn and precorneal setae, (N) segment IV, (O) anal segment; (P) *Micropsectra apposita*, thoracic horn and precorneal setae, (Q) tergite armament III–V, R. anal segment.

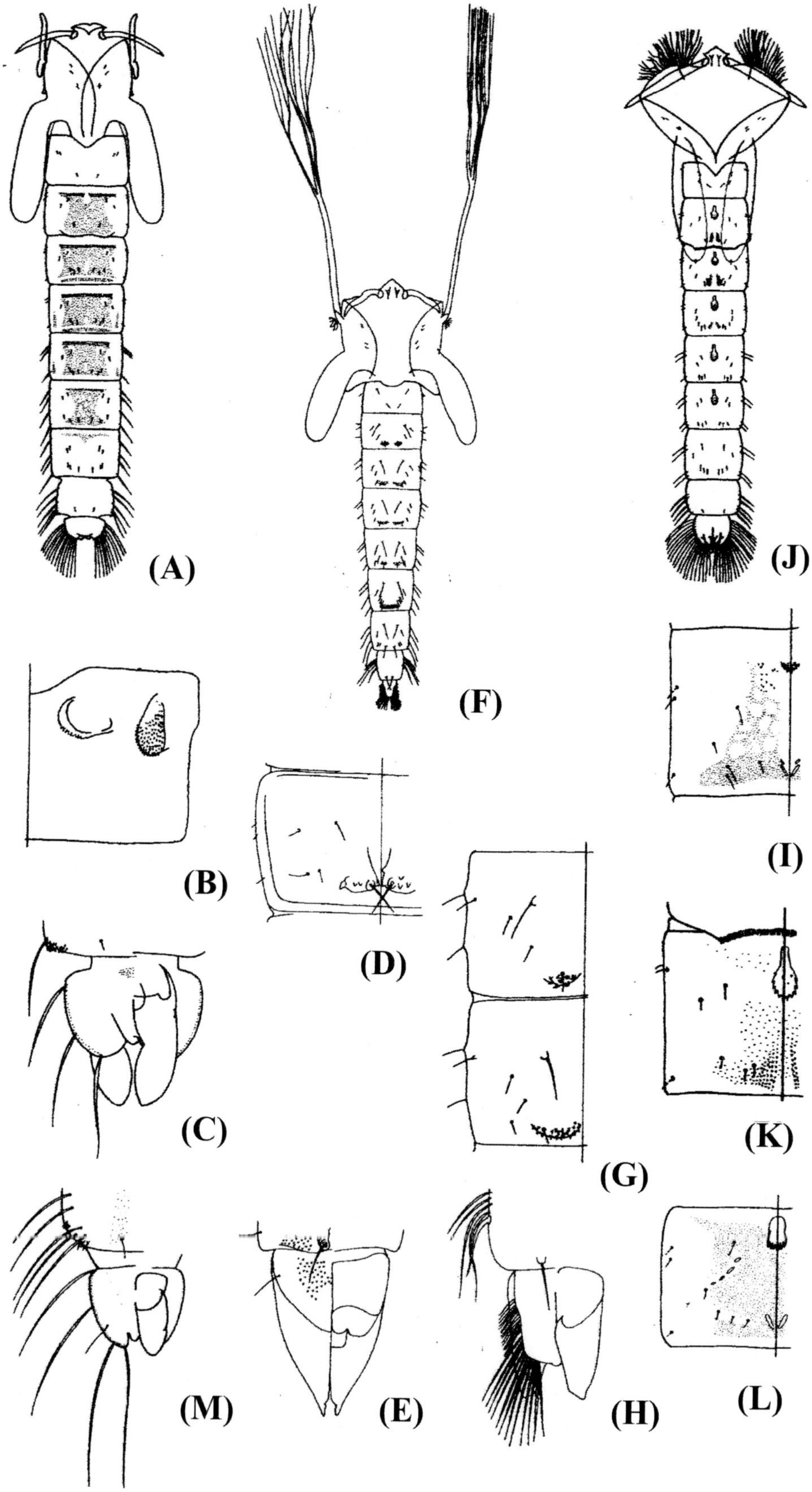

FIGURE 15.2.13 Pupal exuviae, Chironominae, Pseudochironomini. (A) *Pseudochironomus prasinatus* exuviae, (B) sternite I, (C) anal segment; Chironominae, Chironomini. (D) *Rheomus yahiae*, segment III, (E) anal segment; (F) *Cryptotendipes holsatus* exuviae, (G) segments II and III, (H) anal segment; (I) *Demeijerea rufipes*, segment IV; (J) *Glyptotendipes* (s.str.) *paripes* exuviae, (K) segment III; (L) *Glyptotendipes* (*Heynotendipes*) *signatus*, segment IV, (M) anal segment.

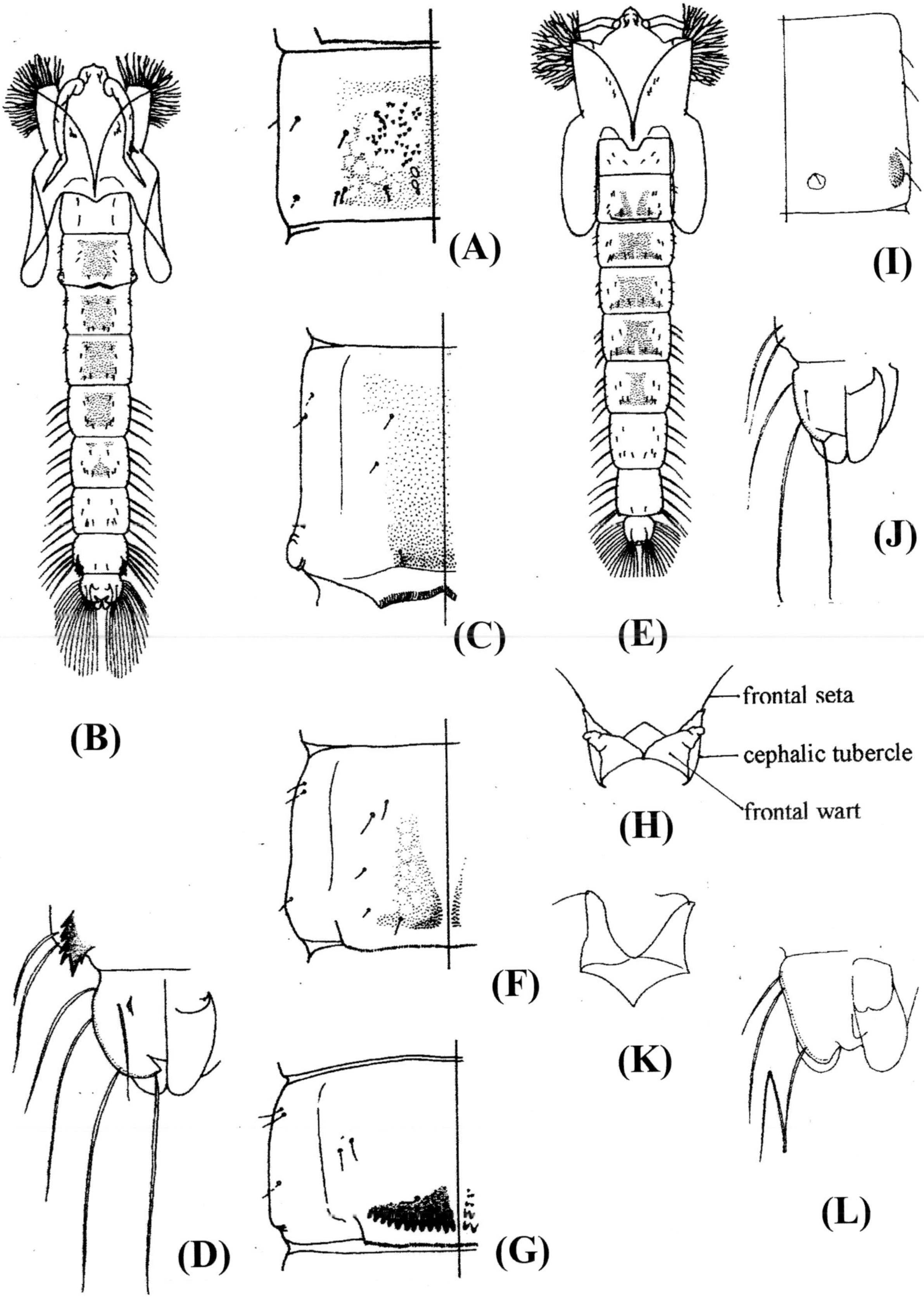

FIGURE 15.2.14 Pupal exuviae, Chironominae, Chironomini. (A) *Xenochironomus xenolabis* segment III; (B) *Kiefferulus tendipediformis* exuviae, (C) segment II, (D) anal segment and comb of segment VIII; (E) *Demicryptochironomus* (s.str.) *vulneratus* exuviae, (F) segment II; (G) *Demicryptochironomus* (*Irmakia*) sp. segment II; (H) *Benthalia dissidens*, frontal apotome, (I) segment IV ventral, (J) anal segment; (K) *Fleuria lacustris*, frontal apotome, (L) anal segment.

9(8') Tergite II (Fig. 15.2.14E, F) posteriorly with a transverse band of points medially broken by a narrow point-free space; from the posterior band points may spread forwards either as diverging marmorate patches or as narrowly separated longitudinal bands joined anteriorly. Tergites III–V (Fig. 15.2.14E) with the longitudinal patches more extensive, but always separate posteriorly 10

9' If the tergite armament is paired posteriorly, then the tergites also have conspicuously strong points anteriorly 11

10(9) Tergite II (Fig. 15.2.14F) and III with posterior points colorless .... *Demicryptochironomus* (s.str.)

10' Tergites (Fig. 15.2.14G) and III with posterior points large, crowded and dark, producing obvious transverse brown bands on the exuviae .... *Demicryptochironomus* (*Irmakia*)

11(9') Intersegmental membranes III/IV, IV/V, and V/VI armed with points. Comb of segment VIII absent (Fig. 15.2.14J) 12

11' If intersegmental membrane V/VI armed with points, comb of segment VIII present 13

12(11) Frontal apotome (Fig. 15.2.14H) with frontal warts in addition to the cephalic tubercles; in males they are of similar size, in females the frontal warts are narrower. Sternite IV (Fig. 15.2.14I) posteriorly with a circular or elongate mound on each side bearing a stiff seta. Sternites V and VI posteriorly with a pair of smaller hemispherical mounds, each bearing a stiff seta .... *Benthalia*

12' Cephalic tubercles (Fig. 15.2.14K) single. Sternites IV, V, and VI without mounds and stiff setae .... *Fleuria*

13(11') Tergites II–VI with armament in the form of an undivided, usually extensive, patch of strong points, the points in each segment increasing in size posteriorly. The point patches increase in extent from tergite II to V, but on VI the patch is more or less reduced. On at least tergites III and IV marmoration of the patches is peripheral or absent. Comb of segment VIII present 14

13' If the armament of tergite III is in the form of an undivided median point patch, the points are of much the same size OR anteriorly larger, OR segment VIII is posterolaterally unarmed 19

14(13) Comb of segment VIII (Fig. 15.2.15A) of 2 to 7 small teeth set on the apicolateral corner of the segment just inside the fifth lateral taenia and hardly, or not, projecting .... *Einfeldia*

14' Comb of segment VIII not as above (e.g., Figs. 15.2.3G and 15.2.15C) 15

15(14') Hook row II medially broken (Fig. 15.2.15B) .... *Chironomus* (*Lobochironomus*).

15' Hook row II entire 16

16(15') Comb of segment VIII (Fig. 15.2.3G) on the end of an elongate ventral cuticular mound which exceeds appreciably the apicolateral corner of the segment. (This mound is demarcated anteriorly in the flattened exuvial mount by a series of curved creases) 17

16' Comb of segment VIII usually an elongate thorn, or a few thorns, inserted directly on the posterolateral edge of the segment; if set on the end of an apicolateral extension of the segment, the ventral mound is not developed and the anterior creases are absent 19

17(16) Large: 12–17 mm long; brown. Sternites III and IV with shagreen forming an anterior transverse band, which may be widely broken medially and which may spread a little posteriorly on each side. Thoracic granulation strong and extensive .... *Chironomus* (*Camptochironomus*)

17' IF shagreen of sternites III and IV forms an anterior transverse band, then thorax nearly smooth OR exuvial length 10 mm or less 18

18(17') 10–11 mm long. Cephalic tubercles (Fig. 15.2.15D) with a short (50–75 μm) and narrow (30–45 μm) apex on a bulbous base. Comb of segment VIII usually with a single apical tooth. Thoracic granulation distinct and extensive .... *Chironomus* (*Chaetolabis*)

18' 5–17 mm long. If falling within the size range for *Chaetolabis*, differing in the characters above (cephalic tubercles (Fig. 15.2.15E) short or long conical, comb of segment VIII usually with many more than 1 apical tooth) .... *Chironomus* (s.str.)

19(13',16') Sternites I and II or I–III with transverse rows of backwardly directed colorless spines. Tergites II–VI (Fig. 15.2.15F) nearly covered with strong points which are elongated posteriorly to produce a transverse row of long spines. Hook row II broken medially. Exuvial length 2.9–3.5 mm .... *Kloosia*

19' If any sternites I–III with transverse rows of backwardly directed colorless spines, tergites without a posterior row of long spines 20

20(19') Sternites I and II (Fig. 15.2.15G) or I–III with or without posteriorly directed colorless spines. Tergite VI (Fig. 15.2.15H, I) with greatly enlarged points posteriorly, in some species medially as well. The median transverse pair of point patches may be separate or fused, in which case a longitudinally oval bare space is left posteriorly; the long points on each side of this may spread inwards posteriorly to enclose the space. Comb of segment VIII a few

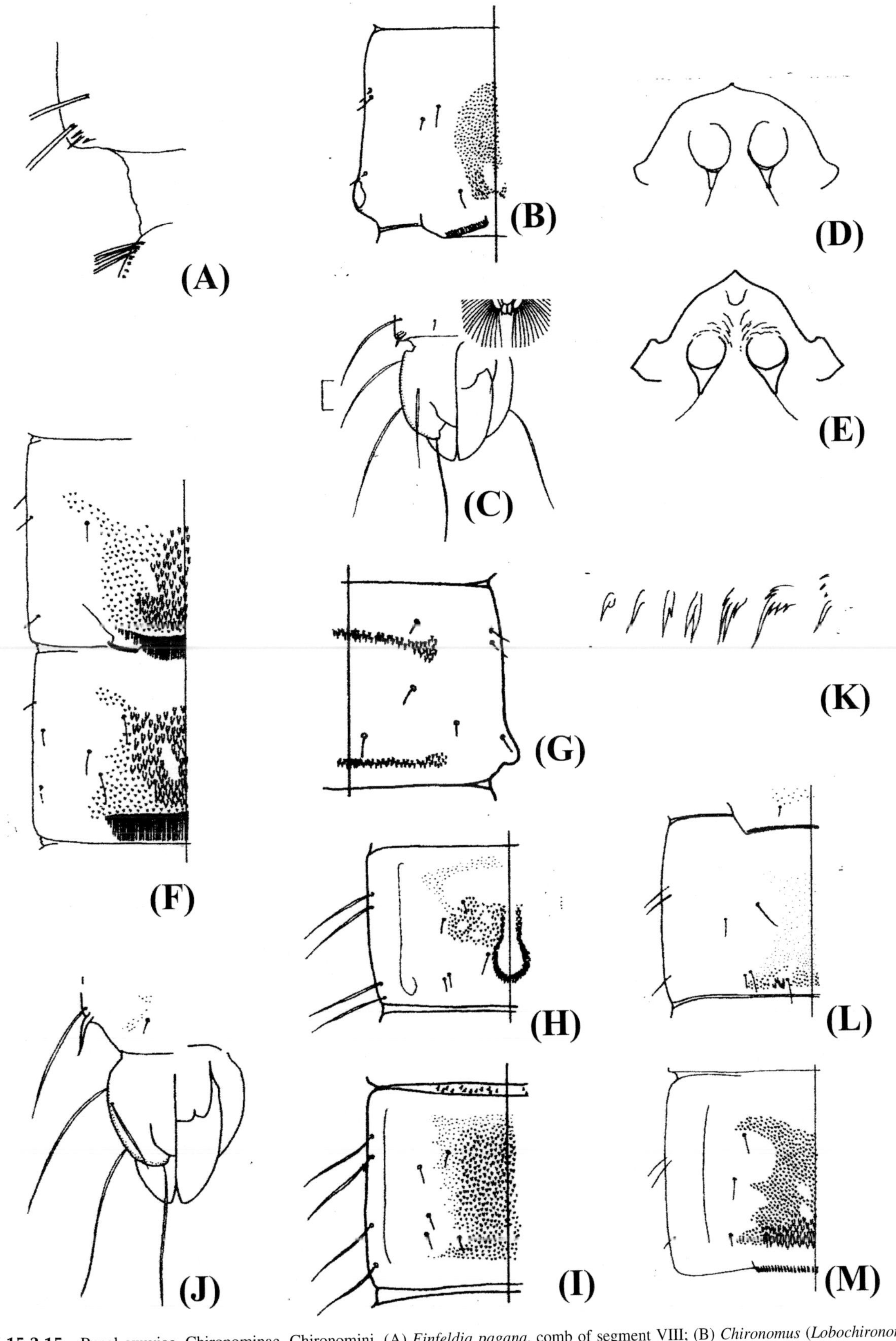

FIGURE 15.2.15 Pupal exuviae, Chironominae, Chironomini. (A) *Einfeldia pagana*, comb of segment VIII; (B) *Chironomus* (*Lobochironomus*) sp. segment II, (C) anal segment; (D) *Chironomus* (*Chaetolabis*) *macani*, frontal apotome; (E) *Chironomus* (s.str.) *aprilinus*, frontal apotome; (F) *Kloosia pusilla*, segments II, III; (G) *Dicrotendipes lobiger*, segment II ventral; (H) *Dicrotendipes pallidicornis*, segment VI; (I) *Dicrotendipes notatus*, segment VI, (J) anal segment; (K) *Paracladopelma nigritulum*, comb of segment VIII variation; (L) *Paracladopelma camptolabis*, segment III; (M) *Paracladopelma mikiana*, segment II.

small teeth on the posterolateral corner, a strong tooth on the end of a projecting spur or an elongated thorn (Fig. 15.2.15J) .... *Dicrotendipes*.

20' Sternites I–III without colorless spines or armament of tergite VI not of this characteristic form 21

21(20') Segment II without pedes spurii B. Comb of segment VIII (Fig. 15.2.15K) composed of a few (2–6) small, elongate, sinuous, brown teeth, usually joined towards their bases to form a short spur .... *Paracladopelma nigritulum* (Goetghebuer)

21' Segment II with or without pedes spurii B. Comb not as in *P. nigritulum*, or absent 22

22(21') Tergites (Fig. 15.2.15L) III–V(VI) with a posterior transverse band of points, on the same level as the setae D5, with a few larger yellowish or brownish mammillate points (swollen basally with a short conical or rounded-conical apex) near setae D5. Segment II without pedes spurii B .... *Paracladopelma camptolabis* group

22' Armament of tergites III–VI not in the form of a posterior transverse band with a few stronger mammillate points near setae D5. Segment II with pedes spurii B 23

23(22') Tergites II–V with median point patch and posterior band fused to form a nearly rectangular patch of strong points; on tergite VI the median patch and posterior band are separate or similar to that on V, and there is a reduced patch on VII. Comb of segment VIII absent 24

23' If tergite armament similar, then comb of segment VIII present 26

24(23) Cephalic tubercles conical with apical seta. Abdominal segments (Fig. 15.2.15M) without reticulation .... *Paracladopelma mikianum* (Goetghebuer)

24' Cephalic tubercles absent; frontal setae present or absent. At least posterior segments with extensive reticulation (Fig. 15.2.16B) 25

25(24') Segment II (Fig. 15.2.16A) with hook row widely broken medially and with pedes spurii B armed with colorless spinules. Frontal setae present. Sternites I–III without spines .... *Beckidia*

25' Hook row II continuous. Segment II without pedes spurii B. Frontal setae absent. Sternites I–(II)III with transverse spine rows, anteriorly and/or posteriorly .... *Robackia*

26(23') Tergite II only with an anterior transverse band of strong points (Fig. 15.2.16C). Anal lobes with the taeniate fringe continued around the inner margin as bristle-like setae .... *Nilodorum*

26' If tergite II has an anterior transverse band of strong points, so have tergites III–VI. If the fringe of the anal lobes is continued around the inner margin, it remains taeniate 27

27(26') Comb of segment VIII (Fig. 15.2.16E) a row of narrow elongate teeth set on the edge of a semicircular pad on the apico-lateral corner of the segment. (The teeth may be bluntly rounded at the apex, and an occasional tooth may occur further forward between the bases of lateral taeniae 3 and 4). Tergites (Fig. 15.2.16D) without an anterior transverse band of points, but III–VI with a posterior transverse band of strong points .... *Cyphomella*

27' Comb of segment VIII, if present, not as above 28

28(27') Hook row II continuous 29

28' Hook row II interrupted in the middle 30

29(28) Tergite II without armament (though occasionally represented by a short posterior band of small points). Thoracic horn much branched. Comb of segment VIII an elongate sinuous thorn, occasionally 2, or 1–9 small transparent spines in the apicolateral corner near the bases of the fourth and fifth lateral taeniae, or absent. Exuvial length 3.8–10.0 mm 32

29' Tergite II with median and/or anterior transverse point patches well developed, or exuviae about 3 mm long and thoracic horn of less than 10 branches 36

30(28') Tergites II–V without a distinct unbroken posterior row or band of strong points. Comb of segment VIII present, one or two elongate, narrow posterolateral thorns (Fig. 15.2.16H, K) or a group of small transparent spines on the apicolateral corner 31

30' Tergites II–V with a complete posterior transverse row or narrow band of strong points. Comb of segment VIII absent 33

31(30) Cephalic tubercles (Fig. 15.2.16F) conical, generally with small points apically. Tergite VI (Fig. 15.2.16G) with a small oval swelling bearing noticeably larger points in the middle of the posterior transverse band, with setae D3 set under its lateral edges .... *Cladopelma*

31' Cephalic tubercles (Fig. 15.2.16I) without small points apically. Tergite VI (Fig. 15.2.16J) without a median posterior oval swelling .... *Microchironomus*

32(29) Hook row of segment II (Fig. 15.2.16L) on the posterior edge of a projecting flap of the tergite; tergites III–VI with the points increasing in size posteriorly in each segment, the extent of the median patch and the size of

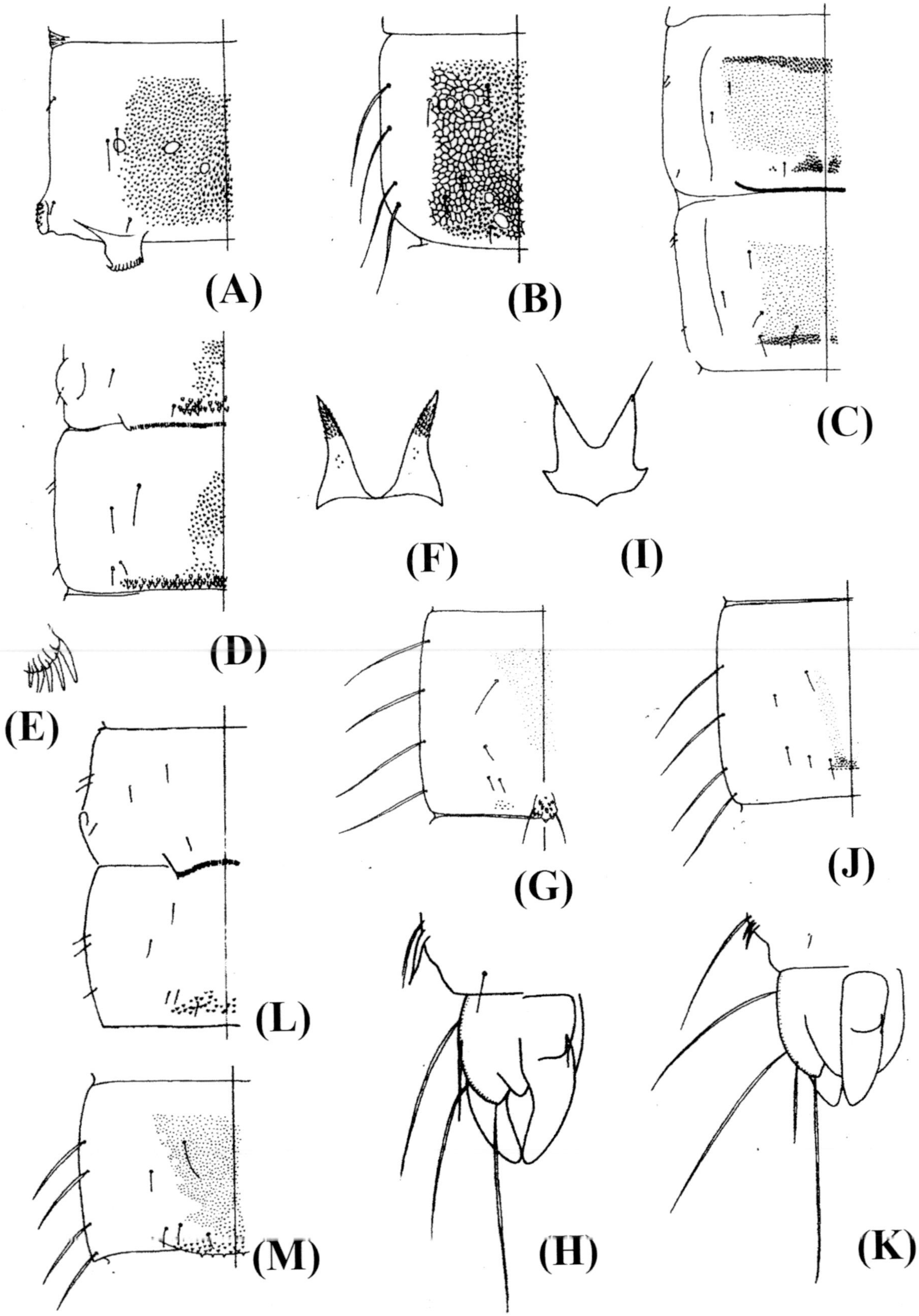

FIGURE 15.2.16 Pupal exuviae, Chironominae, Chironomini. (A) *Beckidia* sp. segment II; (B) *Robackia* sp. segment VI; (C) *Nilodorum brevibucca* segments II and III; (D) *Cyphomella cornea* segments II and III, (E) comb of segment VIII; (F) *Cladopelma virescens* cephalic tubercles, (G) segment VI, (H) anal segment and comb of segment VIII; (I) *Microchironomus tener* cephalic tubercles, (J) segment VI, (K) anal segment and comb of segment VIII; (L) *Parachironomus vitiosus* segments II and III, (M) segment VI.

the points increasing on succeeding segments; posterior points of tergite VI (Fig. 15.2.16M) set on a projecting flap and conspicuously larger than the preceding points .... *Parachironomus*

32' Tergite armament otherwise 35

33(30') Tergites II–V (Fig. 15.2.17A, C) with a posterior transverse row of strong points; hook row of segment II medially interrupted. (Comb of segment VIII absent.) 34

33' Tergites III–V (Fig. 15.2.17E) with a complete, unbroken posterior transverse band of sturdy points set on cuticle darker (more golden or browner) than the rest of the tergites .... *Saetheria reissi*

34(33) Tergite VII without a posterior transverse point row. Anal segment (Fig. 15.2.17B) without forked processes. Exuviae 4.2–6.7 mm long .... *Harnischia*

34' Tergite VII also with a posterior transverse point row. Forked anal processes present (Fig. 15.2.17D) Exuviae 6.0–15.1 mm long .... *Cryptochironomus*

35(32'') Tergites II–V (Fig. 15.2.17F) with a trapezoidal point patch, widest posteriorly (reduced on VI to a posterior transverse band), the points of the posterior three or four rows enlarged and overlapping. Segment VIII with four lateral taeniae and unarmed posterolaterally. (A pupal morphotype not specifically associated.) .... *Saetheria* sp.

35' Without the above combination of characters 36

36(35') Anal lobes (Fig. 15.2.17H) with a tuft of short broad darker taeniae posteriorly differing markedly from those of the rest of the fringe .... *Endochironomus*

36' Fringe of anal lobes normal (Fig. 15.2.17J), without a terminal bunch of conspicuously different taeniae 37

37(36') Cephalic tubercles broad shallow humps; frontal setae with small points around their base. Lateral taeniae of segments V–VIII 3,3,4,4 .... *Synendotendipes*

37' If the cephalic tubercles bear small points apically, the tubercles are long and curved, OR segment VIII with five lateral taeniae 38

38(37') Vortex absent 39

38' Vortex present 42

39(38) Comb of segment VIII a dark brown spur. Tergite VII nearly covered with points. [The following two morphotypes from the Mediterranean Basin have yet to be associated with a genus] 40

39' Comb of segment VIII not a dark spur. Tergite VII with restricted armament 41

40(39) Spur of segment VIII (Fig. 15.2.17L) with the sharp apical tooth extending about one-third the length of the anal lobes posteriad. Tergites II–VII (Fig. 15.2.17K) with anterior, median, and posterior point patches/bands fused to cover the tergite with even-sized points (smaller and less extensive on VII); tergite VIII with a median patch of small points; apical bands present on III/IV–V/VI, also in contact and of similar points, thus indistinct. Hook row II ofstrong hooks. Lateral setae of segments V–VIII taeniate. Anal lobes fringed for apical two-thirds .... Chironomini gen.? sp.? Pe 3. [Langton, 1991] [Tunisia].

40' Spur of segment VIII (Fig. 15.2.18B) rounded apically and fringed with 8 short teeth. Tergites (Fig. 15.2.18A) II–V with a distinct anterior transverse band of strong points contiguous across its breadth with the small points of the median patch plus posterior band, which cover the tergite except for a narrow point-free band posteriorly; tergites VI and VII have a similar patch but the anterior transverse band is not differentiated from the points of the remainder of the armament which are smaller in size and less extensive than on preceding segments; tergite VIII has a median patch of strong points. Hook row II is represented by a long band of reduced hooks similar to those of the apical bands of segments III and IV. Lateral setae of segments V–VIII becoming somewhat taeniate on VII and VIII. Anal lobes fringed for posterior half only .... Chironomini gen.? sp.? Pe 4. (Langton, 1991) [Mediterranean streams]

41(39') Thoracic horn much branched. Tergites armed as in Fig. 15.2.18C. Comb of segment VIII (Fig. 15.2.18D) with two to seven short, broadly triangular marginal teeth .... *Stenochironomus*

41' Thoracic horn with few branches 46

42(38') Tergites (Fig. 15.2.18E) II–VI with a pair of small longitudinally oval anterior point patches. Thoracic horn four-branched, two of the branches with points or spinules at the tip. Lateral taeniae of segments V–VIII: 3, 4, 4, 4. Comb of segment VIII (Fig. 15.2.18F) very small .... *Zavreliella*

42' If tergites II–VI possess a pair of anterior point patches, they are transversely elongate (Fig. 15.2.18G) 43

43(42') Tergites II–V or II–VI with a pair of anterior point patches .... *Lauterborniella.*

43' Armament of tergites (e.g., Fig. 15.2.18J) II–VI not of paired anterior point patches 44

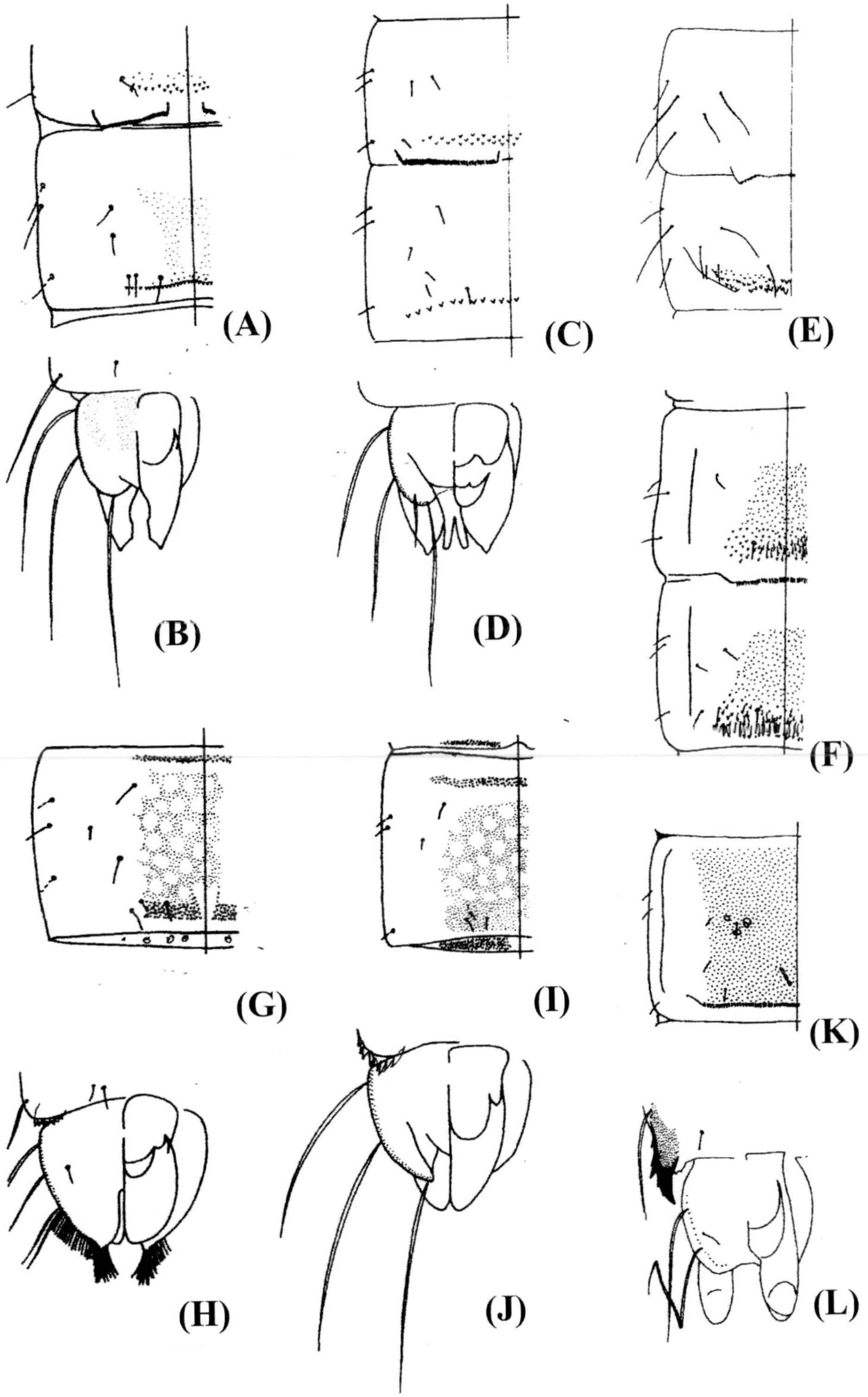

FIGURE 15.2.17 Pupal exuviae, Chironominae, Chironomini. (A) *Harnischia curtilamellatus*, segments II and III, (B) anal segment; (C) *Cryptochironomus supplicans*, segments II and III, (D) anal segment; (E) *Saetheria reissi*, segments II and III; (F) *Saetheria* sp., segments II and III; (G) *Endochironomus tendens*, segment IV, (H) anal segment and comb of segment VIII; (I) *Synendotendipes* sp., segment IV, (J) anal segment and comb of segment VIII; (K) Chironomini gen? sp? segment II, (L) anal segment and comb of segment VIII.

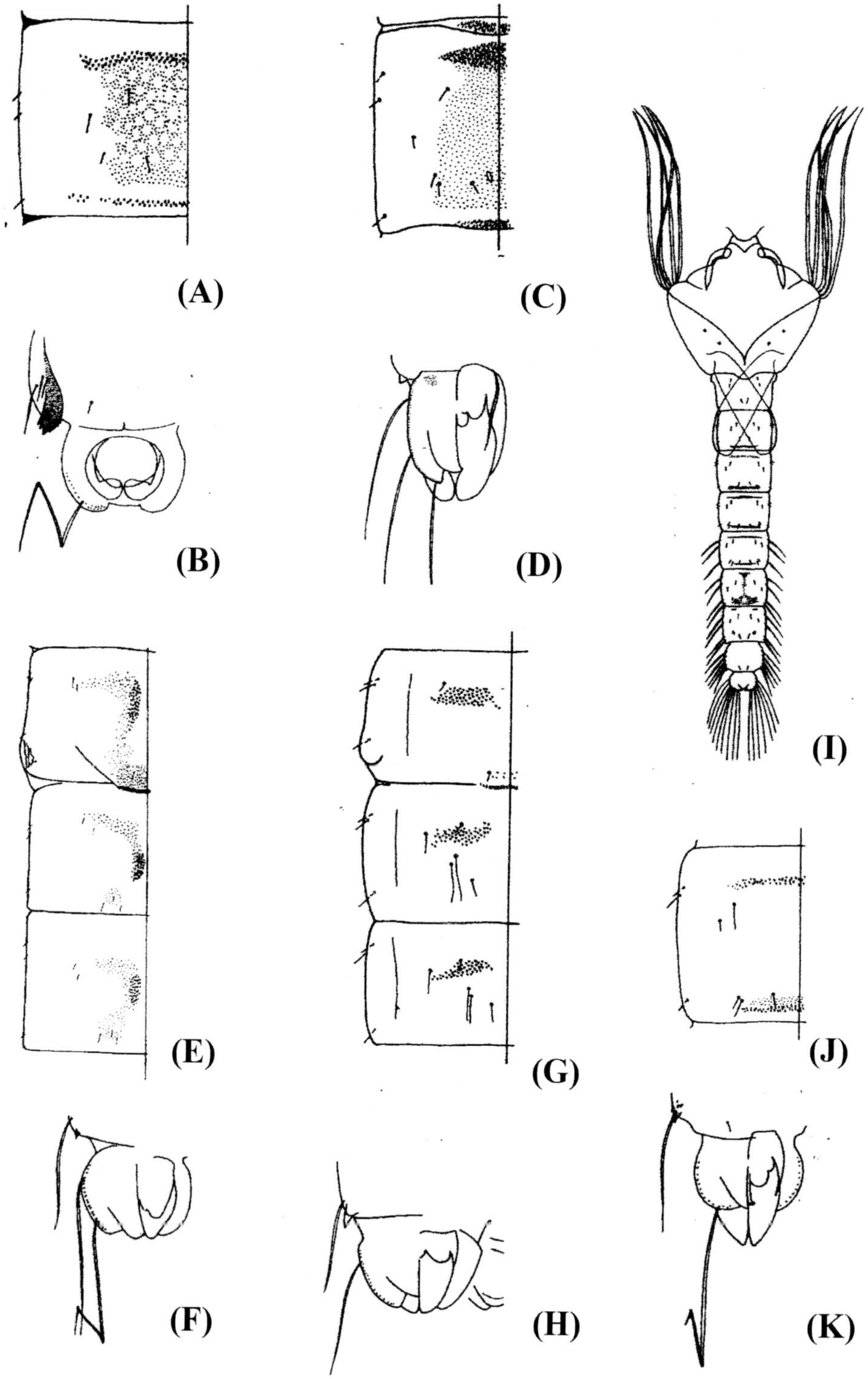

FIGURE 15.2.18 Pupal exuviae, Chironominae, Chironomini. (A) Chironomini gen? sp? Pe4, segment II, (B) anal segment and comb of segment VIII; (C) *Stenochironomus gibbus*, segment IV, (D) anal segment and comb of segment VIII; (E) *Zavreliella marmorata*, segments II–IV, (F) anal segment and comb of segment VIII; (G) *Lauterborniella agrayloides*, segments II–IV, (H) anal segment and comb of segment VIII; (I) *Pagastiella orophila* exuviae, (J) segment IV, (K) anal segment and comb of segment VIII.

44(43') Tergite II unarmed. Thoracic horn (Fig. 15.2.18I) branches very long, over half the length of the abdomen, but often broken off. Comb of segment VIII (Fig. 15.2.18K) a few small pale golden teeth .... *Pagastiella orophila* (Edwards)

44' Tergite II (Fig. 15.2.19B) with at least one of the point bands represented. Branches of thoracic horn much shorter 45

45(44') Thoracic horn many branched AND anterior transverse point band of tergites II–V separate from the median patch by a bare space or narrowly joined to it medially 46

45' If thoracic horn many branched, anterior transverse point band of tergites II–V broadly joined to the median patch 47

46(45) Cephalic tubercles (Fig. 15.2.19A) with swollen base, small conical apex and short frontal seta. (If cephalic tubercles swollen, without conical apex and with frontal setae surrounded at base by small points, see *Synendotendipes*, couplet 38). Comb of segment VIII (Fig. 15.2.19C) composed of isolated narrow thorns spreading anteriad along edge segment margin .... *Tribelos*

46' Cephalic tubercles (Fig. 15.2.19D) conical, with or without points; frontal setae long. Comb of segment VIII (Fig. 15.2.19E, F) composed of many teeth, which may be short or long, the most posterior joined at their bases .... *Stictochironomus*

47(45') Cephalic tubercles (Fig. 15.2.19I) elongate conical with an apical seta 48

47' Frontal setae absent, or cephalic tubercles at most rounded conical 49

48(47) Segments III and IV with four lateral setae. Anal lobes (Fig. 15.2.19H) with a dorsal taenia. Wing sheaths with a weak nose. Vortex absent .... *Paralauterborniella nigrohalteralis* (Malloch)

48' Segments II–IV (Fig. 15.2.19J) with three lateral setae. Anal lobes (Fig. 15.2.19K) with a dorsal seta. Wing sheaths without a nose. Vortex present .... *Paratendipes*

49(47') Pedes spurii B of segment II absent. Segments III and IV (Fig. 15.2.19N) with apical point bands usually present, but may be greatly reduced, or even absent. Three or four lateral taenia on segments V and VI. Frontal setae absent; cephalic tubercles (Fig. 15.2.19L, M) short or long conical, or strongly domed. Comb of segment VIII (Fig. 15.2.19O) composed of a few strong teeth on the lateral margin of the segment increasing in size towards the posterolateral corner .... *Microtendipes*

49' Pedes spurii B of segment II usually present; if absent, either apical point bands present on segments III, IV, and V or lateral taeniae absent on segments V and VI. (Frontal setae usually present; if absent, cephalic tubercles very weakly rounded.) 50

50(49') Cephalic tubercles more or less cylindrical, truncate and capped with fine spinules (often inverted in mounts). (Usually both segments III and IV with an apical band of points; sometimes only IV. Thoracic horn with 6–10 branches.) 51

50' Cephalic tubercles long conical, rounded or absent 52

51(50) Cephalic tubercles (Fig. 15.2.20A) short, usually less than 50 μm long. Thorax with a short patch of granules near suture, which may be extended along the suture as a narrow band. Conjunctives with a black mark laterally (Fig. 15.2.20B) .... *Phaenopsectra*

51' Cephalic tubercles (Fig. 15.2.20C) longer, usually about 100 μm long. Thorax with a narrow band of small granules along the suture. Conjunctives unmarked laterally .... *Sergentia*

52(50') Lateral taeniae IV-VIII: 1, 4, 4, 4, 4. Tergite and sternite VIII (Fig. 15.2.20D) with a median patch of points. (The anterior, median, and posterior point bands of tergites II–V completely fused, the points of the anterior band not appreciably larger than the median points. Anal lobes with a dorsal taenia situated close to the fringe in the posterior half of the lobes.) .... *Nilothauma brayi* (Goetghebuer)

52' Segment V with no more than three lateral taeniae. (Posterior tergites and sternites with or without point patches.) 53

53(52') Cephalic tubercles (Fig. 15.2.20E) fused medially to form a large conical projection from the apotome. Tergites II–VI (Fig. 15.2.20F) almost covered with small points. Comb of segment VIII (Fig. 15.2.20G) a short spur with about 6 short marginal teeth and with many surface points .... *Polypedilum* (*Cerobregma*)

53' Cephalic tubercles, if well developed, not fused together to form a single structure .... *Polypedilum* (s.str.)/(*Pentapedilum*)/(*Tripodura*)/(*Uresipedilum*). (This species-rich genus has individually recognizable species that do not separate satisfactorily into the adult-based subgenera (Fig. 15.2.20H–Q) show examples of each subgenus, but are not to be taken as definitive for the subgenus, but as an example of the range of armament and comb of segment VIII to be found in the genus as a whole). Further resolution of the taxa collected can be obtained from Langton (1991) or Langton & Visser (2003)

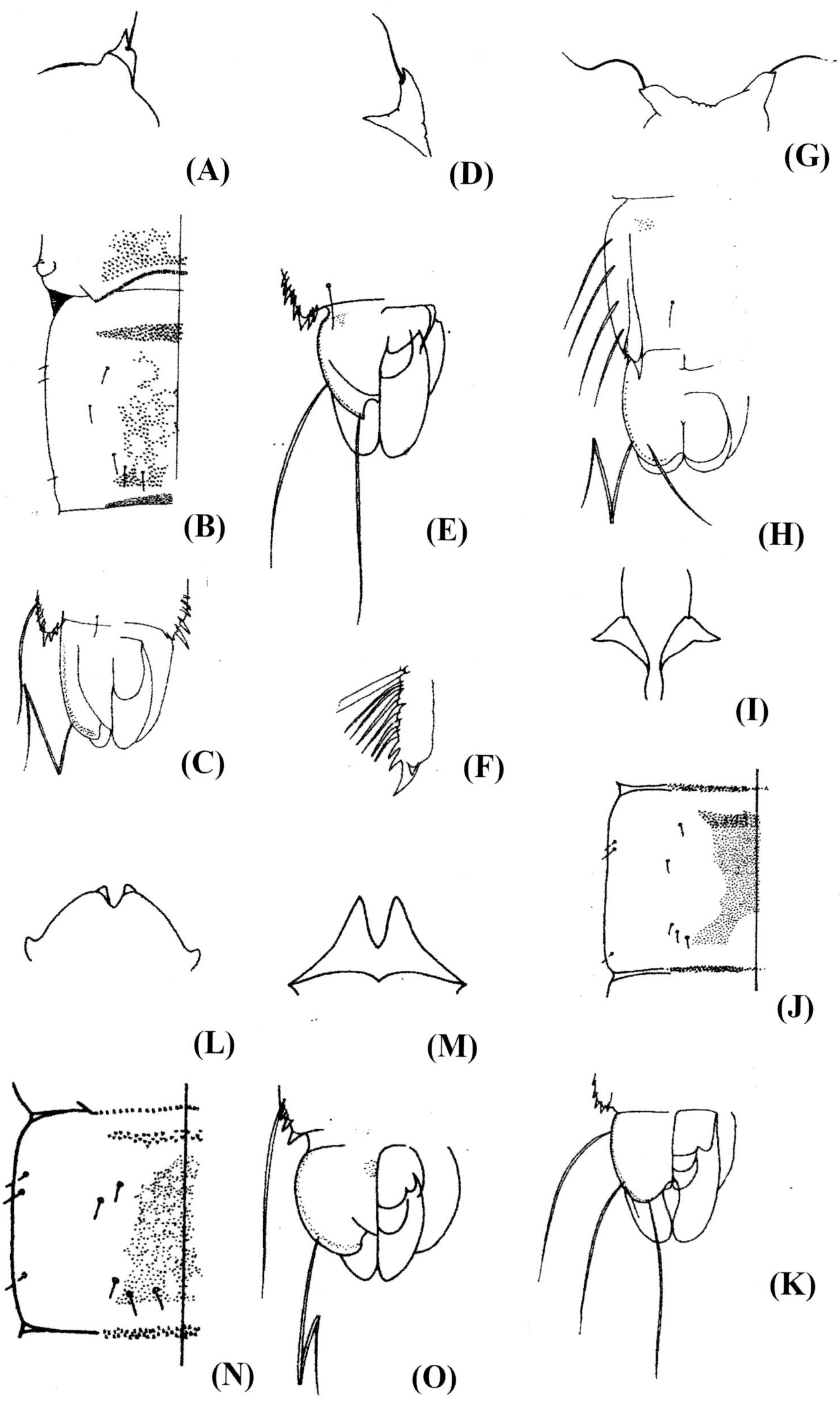

FIGURE 15.2.19 Pupal exuviae, Chironominae, Chironomini. (A) *Tribelos intextus*, cephalic tubercle, (B) segment III and posterior II, (C) anal segment and comb of segment VIII; (D) *Stictochironomus* sp., cephalic tubercle, (E) anal segment and comb of segment VIII; (F) *Stictochironomus caffrarius*, comb of segment VIII; (G) *Paralauterborniella nigrohalteralis*, cephalic tubercles, (H) anal segment and comb of segment VIII; (I) *Paratendipes albimanus*, cephalic tubercles, (J) segment III, (K) anal segment and comb of segment VIII; (L) *Microtendipes britteni*, cephalic tubercles; (M) *Microtendipes chloris*, cephalic tubercles, (N) segment III, (O) anal segment and comb of segment VIII.

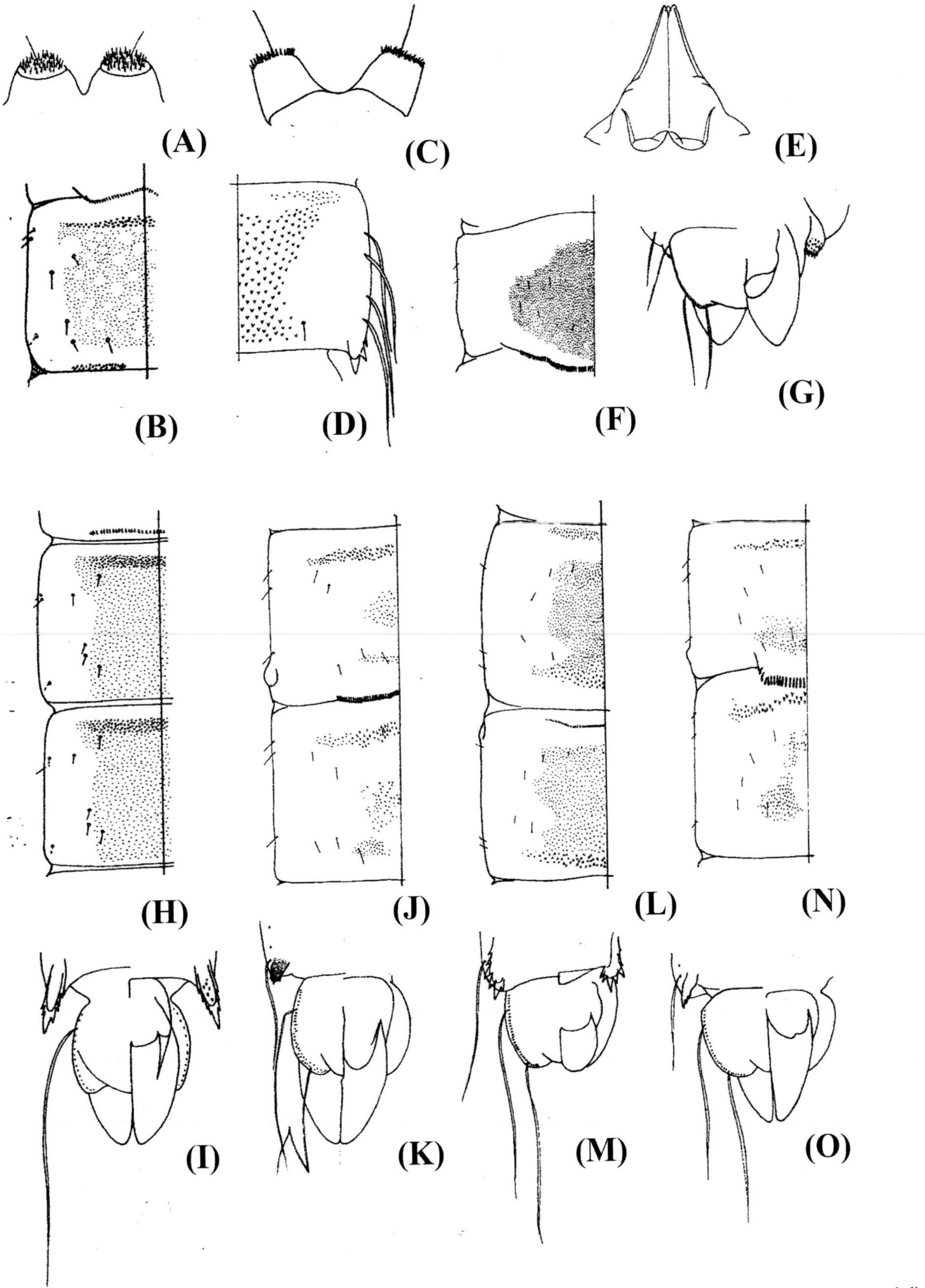

FIGURE 15.2.20 Pupal exuviae, Chironominae, Chironomini. (A) *Phaenopsectra* sp., cephalic tubercles (B) segment III; (C) *Sergentia* sp., cephalic tubercles; (D) *Nilothauma brayi* segment VIII ventral; (E) *Polypedilum* (*Cerobregma*) sp., cephalic tubercles, (F) segment II, (G) anal segment and comb of segment VIII; (H) *Polypedilum* (s.str.) *arundineti*, segments III and IV, (I) anal segment and comb of segment VIII; (J) *Polypedilum* (*Pentapedilum*) *sordens*, segments III and IV, (K) anal segment and comb of segment VIII; (L) *Polypedilum* (*Tripodura*) *bicrenatum*, segments III and IV, (M) anal segment and comb of segment VIII; (N) *Polypedilum* (*Uresipedilum*) *convictum*, segments III and IV, (O) anal segment and comb of segment VIII.

Diamesinae: Genera

1 Each anal lobe (Fig. 15.2.21A) bears about 10 marginal setae, the posterior of which are spine-like and inserted in conspicuous sockets (these setae are often broken off, but their sockets remain.) .... *Protanypus*
1' Anal lobes without a series of spine-like setae set in conspicuous sockets 2
2(1') Anal lobes (Fig. 15.2.21B) in addition to the three long apical macrosetae with a fringe of short hair-like setae. Abdominal segments with most of the lateral setae branched. Thoracic horn absent .... *Sympotthastia*
2' Without this combination of characters 3
3(2') Anal lobes (Fig. 15.2.21D) small, shorter than segment VIII, with three strong, hooked, macrosetae. Thoracic horn (Fig. 15.2.21C) small, brown, triangular. Frontal setae absent. Hook row II absent .... *Boreoheptagyia*
3' Without this combination of characters 4
4(3') Abdominal segments (Fig. 15.2.21E, F) II–VIII posterolaterally with a small projecting spur, progressively larger on succeeding segments .... *Pseudokiefferiella*
4' Abdominal segments without posterolateral spurs 5
5(4') Anal lobes posterior to three marginal setae with a short, broad triangular projection, which may be armed with small points or scales. Thoracic horn absent 6
5' If the anal lobes are drawn out posteriorly into a short tooth or spur, then a thoracic horn is present 7
6(5) Exuviae less than 8mm long. Apical tubercle of anal lobes (Fig. 15.2.21G) usually minutely toothed or scaled; if smooth, anterior dorsal seta of thorax many branched .... *Potthastia*
6' Exuviae usually more than 8 mm long. Apical tubercle of anal lobes (Fig. 15.2.21I) smooth, at most the lobe itself scaled near the tubercle. Dorsal setae of thorax simple .... *Pseudodiamesa*
7(6') Anal lobes (Fig. 15.2.21H) with an apical triangular projection. Thorax above the insertion of the thoracic horn with a conspicuous tubercle .... *Syndiamesa*
7' Anal lobes (Fig. 15.2.21J) without an apical projection .... *Diamesa*

Prodiamesinae: Genera

1 Anal lobes (Fig. 15.2.22B) without macrosetae. Abdominal segments (Fig. 15.2.22A) laterally with a dense fringe of short, fine, hairs; pedes spurii on segment II capped with spines .... *Monodiamesa*
1' Anal lobes with macrosetae in addition to the fringe of taeniae. Abdominal segments without a lateral fringe of hairs. Pedes spurii not with spines 2
2(1') Segment VIII (Fig. 15.2.22C) laterally strongly projecting posteriad. Anal lobes elongate, narrowed to apex, with five anal macrosetae .... *Odontomesa*
2' Segment VIII (Fig. 15.2.22D) not laterally projecting posteriad. Anal lobes rounded, with four or five macrosetae .... *Prodiamesa*

Orthocladiinae: Genera and Subgenera

1 Anal lobes fringed with taeniate setae 2
1' Anal lobes without fringe, or fringe of hair-like or spine-like setae 21
2(1) Tergite II without an apical transverse band of hooks or pad of hooked spines 3
2' Tergite II with either an apical band of hooks or pad of hooked spines (these may be overlooked if the posterior points of the tergite are large and dark) 6
3(2) Segments III–VIII with 1–4 lateral taeniae. Exuvial length less than 3 mm. Thoracic horn absent 4
3' Segments III–VI without lateral taeniae. Exuvial length greater than 3 mm 6
4(3) Anal macrosetae (Fig. 15.2.22E) taeniate. Pearl row present (see Fig. 15.2.4H) .... *Corynoneura*
4' Anal macrosetae (Fig. 15.2.22F) cylindrical, hooked at tip. Pearl row absent 5
5(4') Sternite II (Fig. 15.2.22G) with long dense spinules medially at least arranged in short transverse rows .... *Corynoneurella*
5' Sternite II without spinules; if armed, then spinules smaller, less dense, isolated .... *Thienemanniella*
6(3') Segment VII and VIII (Fig. 15.2.22H) with lateral setae only, though 3 of the 5 setae on VIII are conspicuously long .... *Parametriocnemus* (part)
6' Segment VII with at least one lateral seta and all those on segment VIII (Fig. 15.2.22K) taeniate 7

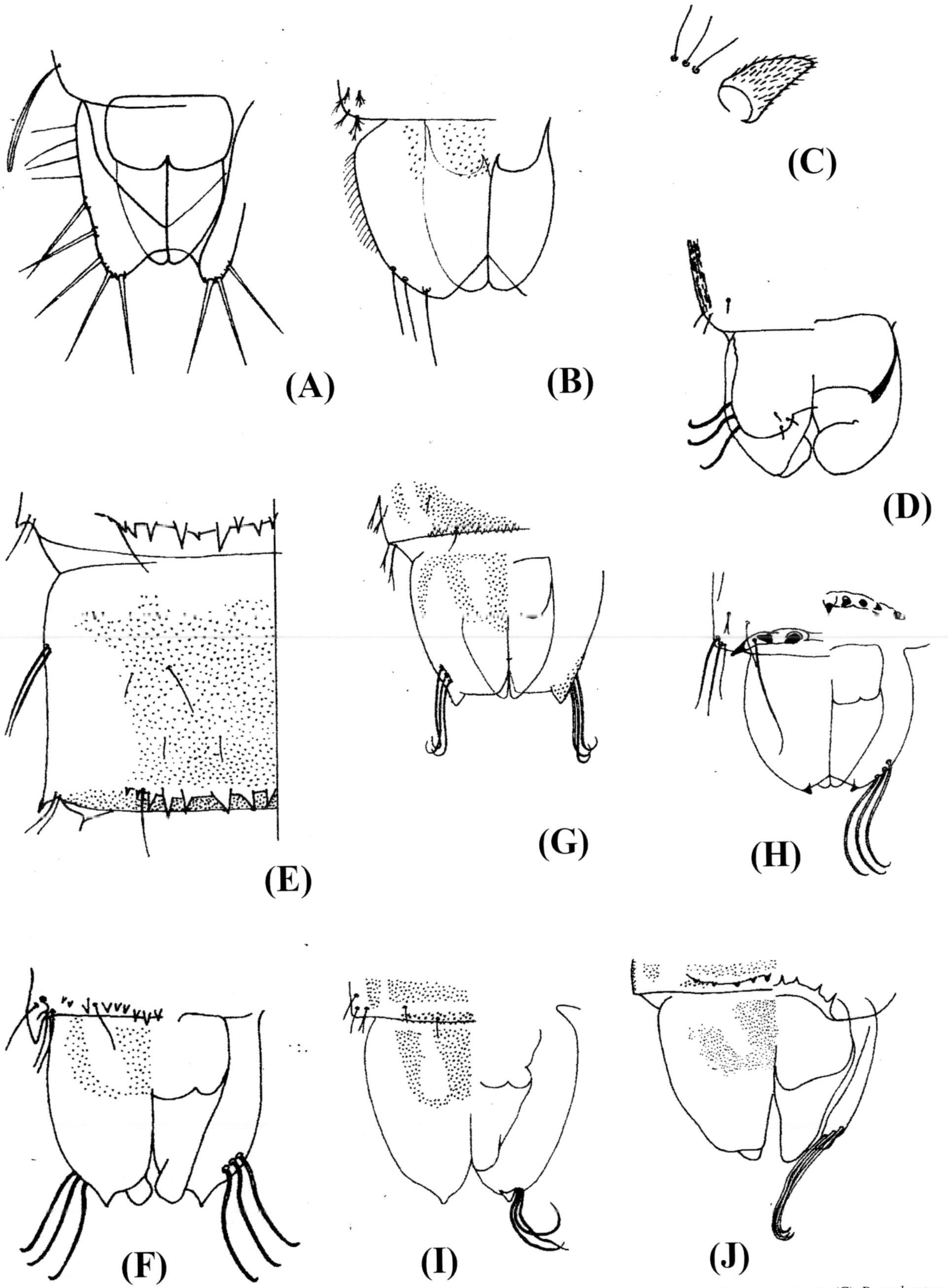

FIGURE 15.2.21 Pupal exuviae, Diamesinae. (A) *Protanypus morio*, anal segment; (B) *Sympotthastia zavreli*, anal segment; (C) *Boreoheptagyia monticola*, thoracic horn and precorneal setae, (D) anal segment; (E) *Pseudokiefferiella parva*, segment VII, (F) anal segment; (G) *Potthastia gaedii*, anal segment; (H) *Syndiamesa edwardsi*, anal segment; (I) *Pseudodiamesa branickii*, anal segment; (J) *Diamesa modesta*, anal segment.

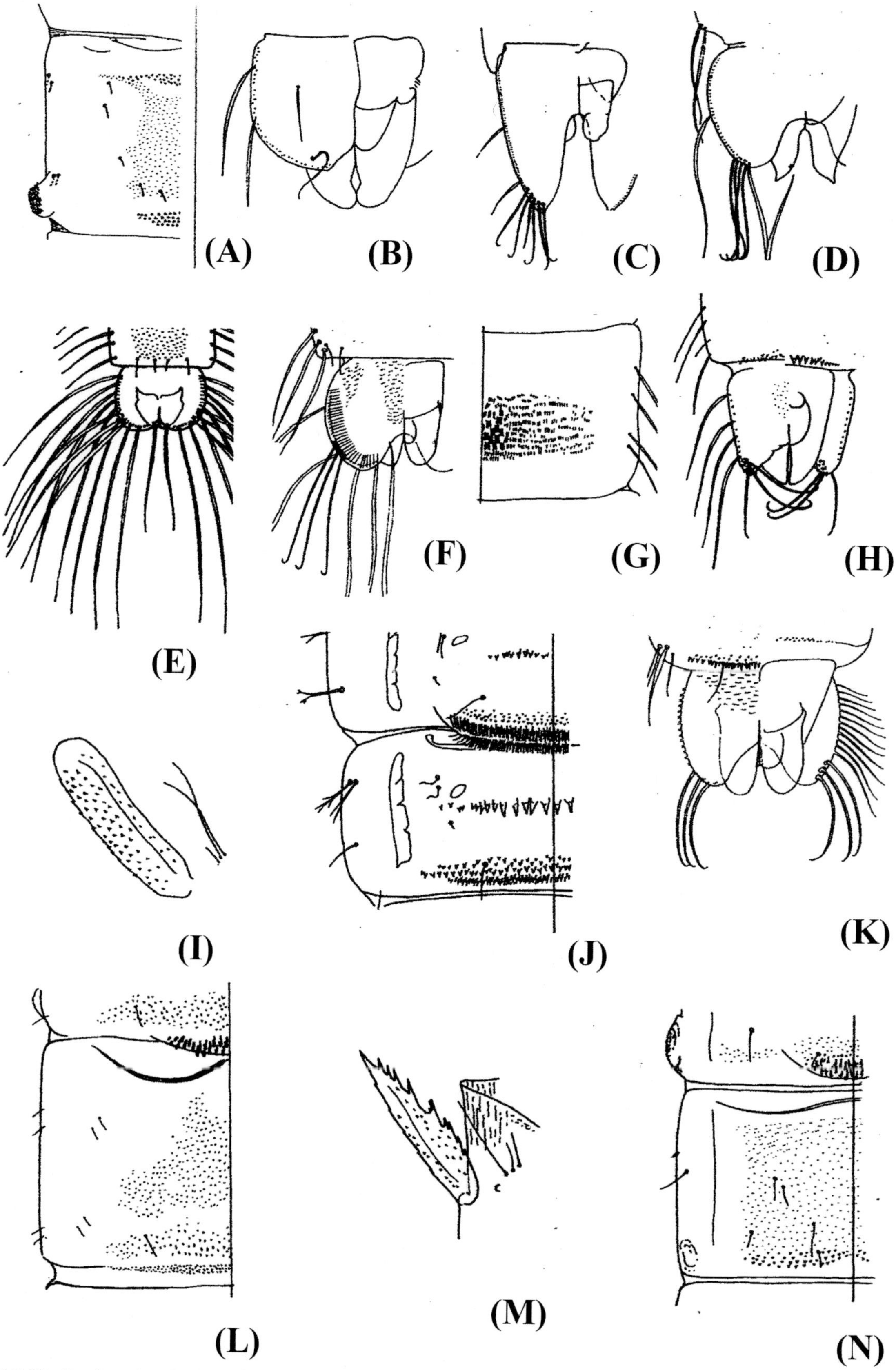

FIGURE 15.2.22 Pupal exuviae, Prodiamesinae and Orthocladiinae. (A) *Monodiamesa* sp., segment II, (B) anal segment; (C) *Odontomesa fulva*, anal segment; (D) *Prodiamesa olivacea*, anal segment; (E) *Corynoneura carriana*, anal segment; (F) *Thienemanniella* sp., anal segment; (G) *Corynoneurella paludosa*, segment II ventral; (H) *Parametriocnemus stylatus*, anal segment; (I) *Paracricotopus niger*, thoracic horn and precorneal setae, (J) segments II and III, (K) anal segment; (L) *Brillia bifida*, segment III; (M) *Heterotanytarsus apicalis*, thoracic horn and precorneal setae, (N) segment III.

7(6') Abdominal tergites (Fig. 15.2.22J) II–VI with a median transverse row or narrow band of strong points. Pedes spurii B absent. Precorneal setae (Fig. 15.2.22I) thin, not set on tubercles or mounds .... *Paracricotopus*
7' Tergites III–VI may have short transverse bands of points in midsegment, but then pedes spurii B on segment II conspicuous and precorneal setae large and set on individual tubercles or a common mound 8
8(7') Intersegmental membranes III/IV–V/VI (Fig. 15.2.22L) armed with an extensive band of points similar in form to those on the tergites; segments III–V with small points more or less covering the tergites, the posterior points in each segment a little stronger .... *Brillia*
8' Intersegmental membranes, if armed, with elongate spinules different in form to those on the tergites, or the tergites possess a posterior transverse band of conspicuously larger teeth 9
9(8') Segments II–VI (Fig. 15.2.22N) with at most 3 lateral setae, the anterior and posterior being reduced to minute thorns or absent. Thoracic horn (Fig. 15.2.22M) elongate, pointed at tip and coarsely toothed along one side .... *Heterotanytarsus*
9' Segments II–VI with three or four lateral setae; if very small, thin and hair-like. Thoracic horn small-toothed 10
10(9') Sternite VIII (Fig. 15.2.23B) of male armed with a posterior transverse row or band of conspicuous points which may be medially interrupted; of female, either with a row of similar points on a pair of projecting flaps, or with a pair of triangular projections each stiffened with an embedded spine 11
10' Sternite VIII not as above 12
11(10) Armed tergites (Fig. 15.2.23A) with small points which are gradually a little larger posteriorly on each segment. Sternite VIII (Fig. 15.2.23B) of male armed with a posterior band of conspicuous points along the evenly curved posterior margin; of female with a pair of triangular projections, each stiffened with an embedded spine .... *Heterotrissocladius*
11' Armed tergites (Fig. 15.2.23C) with a conspicuous posterior transverse band or row of large points. Sternite VIII of male armed with a row of strong points on a pair of posterior swellings; of female (Fig. 15.2.23D) with these points elevated on strongly projecting flaps .... *Euryhapsis*
12(10') Anal lobe fringe taeniae short AND tergites III–V (Fig. 15.2.23H) between the dorsal setae with more or less evenly distributed small points, which are gradually a little larger posteriorly on each segment 13
12' IF taeniae of anal lobes short, then tergites with conspicuous groups of large points 14
13(12) Each anal lobe (Fig. 15.2.23F) with more than five anal macrosetae. Thoracic horn (Fig. 15.2.23E) broad to the truncate apex, with a strong tooth to one side of apex .... *Propsilocerus*
13' Three anal macrosetae (Fig. 15.2.23I) on each anal lobe. Thoracic horn (Fig. 15.2.23G) narrowed to the pointed apex .... *Hydrobaenus*
14(12') Tergites II/III–VIII extensively shagreened. Many species have an embedded dark spine in the posterolateral corner of segment VIII 15
14' Tergites III–VIII with restricted shagreen, at least on VII and VIII, with conspicuous large points medially and/or posteriorly. Never with a dark embedded spine in the posterolateral corner of segment VIII 16
15(14) Tergites (Fig. 15.2.23J) III–VIII without posterior swellings; tergites III–VI posteriorly without larger points. Pedes spurii B on segment II well developed. Segment VIII (Fig. 15.2.23K) without an embedded spine in the posterolateral corners .... *Trissocladius*
15' Tergites (Fig. 15.2.23L) II/III–V(VI) with a swelling posteriorly covered with strong points. Pedes spurii B of segment II absent. Posterolateral corners of segment VIII (Fig. 15.2.23M) may have a dark embedded spine .... *Zalutschia*
16(14') Tergites without point patches medially, OR, if median patches are present on segments III–V, then the anterior anal macroseta is situated in the anterior half of the anal lobes 17
16' Tergites with median point patches, single or paired. The three anal macrosetae situated apically on the anal lobes 19
17(16) Tergites (Fig. 15.2.23O) III–VI with extensive shagreen; hook row II absent, its position occupied by a transverse band of anteriorly directed spinules, which may, however, be folded under the posterior edge of the tergite 18
17' If tergites III–VI have extensive shagreen, segment II has a posteromedian pad of strong hooks 21
18(17) Anal lobes with five to ten anal macrosetae (Fig. 15.2.23N) .... *Stygocladius*
18' Anal lobes with three anal macrosetae .... *Paratrissocladius*
19(16') Points of point patches small, the longest less than 30 μm long, usually shorter than the longest tooth of the posterior transverse rows. Anal lobes not drawn out posteriorly into a point 20

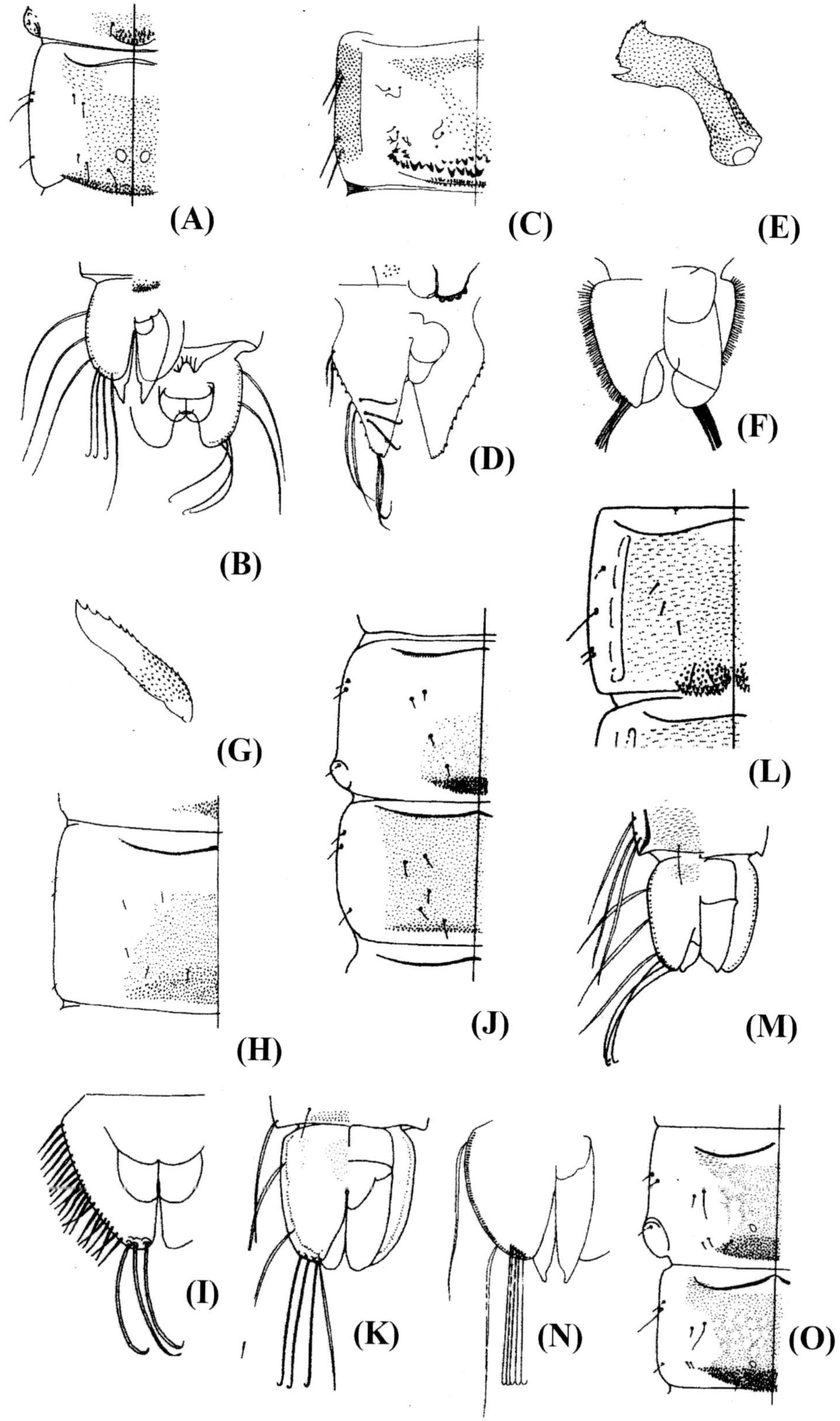

FIGURE 15.2.23 Pupal exuviae, Orthocladiinae. (A) *Heterotrissocladius marcidus*, segment III, (B) anal segment (male, left; female, right); (C) *Euryhapsis* sp., segment II, (D) anal segment; (E) *Propsilocerus paradoxus*, thoracic horn, (F) anal segment; (G) *Hydrobaenus distylus*, thoracic horn, (H) segment III, (I) anal segment; (J) *Trissocladius brevipalpis*, segments II and III, (K) anal segment; (L) *Zalutschia humphriesiae*, segment V, (M) anal segment; (N) *Paratrissocladius exerptus*, segments II and III.

19' Points of point patches large, the longest greater than 35 μm long, usually obviously exceeding the length of the longest tooth of the posterior transverse rows, OR the anal lobes drawn out posteriorly into a toothed point 24

20(19) Abdominal tergites (Fig. 15.2.24A) with point patches transversely elongate (occasionally split medially to produce paired point patches) or restricted to tergites VI and VII; pedes spurii B on segment II conspicuously large, usually elongate conical in shape .... *Nanocladius*

20' Tergites (Fig. 15.2.24B) IV–VI with point patches more or less circular; pedes spurii B on segment II in the form of rounded humps or absent .... *Rheocricotopus* (s.str.)

21(17') Exuvial length less than 5 mm .... *Rheocricotopus* (*Psilocricotopus*)

21' Exuvial length greater than 5 mm 23

22(1') Segments II–VII with three strong lateral setae (Fig. 15.2.24C, D), two close together at, or before, midsegment, the other well separated from them in the posterior half of the segment. Tergites II–VIII with a posterior transverse band of sturdy teeth (Fig. 15.2.24D); tergite II in addition with a small median pad of hooks (Fig. 15.2.24C). Anal lobes (Fig. 15.2.24D) with three long, curved anal macrosetae .... *Rheocricotopus* (*Psilocricotopus*) *tirolus* Lehmann

22' Without this combination of characters 25

23(21') Segments IV–VI (Fig. 15.2.24E) with no lateral taeniae, VII with 4, and VIII 5. Anal macrosetae (Fig. 15.2.24F) marginal, all three close together on the posterior extremity of the anal lobe .... *Psectrocladius* (*Mesopsectrocladius*)

23' Segments IV–VII (Fig. 15.2.24G) with 3 or 4 lateral taeniae, VIII with 5. Anal macrosetae (Fig. 15.2.24H) dorsal, at least the anterior and posterior displaced well within the bases of the marginal taeniae; the anterior seta situated in the anterior half of the lobe .... *Psectrocladius* (*Allopsectrocladius*)

24(19') Anal lobes (Fig. 15.2.24J) drawn out apically into a toothed point. Point patches single, present on tergites IV(V)–VII (Fig. 15.2.24I) .... *Psectrocladius* (*Monopsectrocladius*)

24' Anal lobes (Fig. 15.2.24L) rounded apically, without teeth. Abdominal tergites with median point patches single or double (Fig. 15.2.24K), if single, then absent on tergite VII .... *Psectrocladius* (s.str.)

25(22') Tergites III–V, III and IV, or V alone, with a transverse row of dark hooks at the posterior edge of the segment which may be widely broken medially. Tergite II without hook row, or rarely with one or two hooks 26

25' If hook rows are present on tergites III, IV, and/or V, then there is also a hook row on II, or on VI and VII 28

26(25) Abdominal segments (Fig. 15.2.24M) II–VIII dorsally nearly covered with small points except for the adhesion marks, AND thoracic horn absent .... *Tokunagaia*

26' IF abdominal segments approaching this level of armament, then thoracic horn present (Fig. 15.2.25A). IF thoracic horn absent, then armament of tergites much reduced, or, if extensive, composed of minute shagreen points 27

27(26') Pearl row absent. Anal lobes (Fig. 15.2.25C) usually without a fourth apical seta .... *Eukiefferiella*

27' Pearl row present. Anal lobes (Fig. 15.2.25D) with an additional fourth seta internally, well separated from the other three .... *Tvetenia*

28(25') Thoracic horn (Fig. 15.2.25E) of *Eukiefferiella* form, with a swollen base and long apical filament. (Pearl row present. Posterior transverse tooth row present on tergites (I)II–VII(VIII).) .... *Dratnalia*

28' Thoracic horn, if present, of different form 29

29(28') Tergite II (Fig. 15.2.25G) with a narrow posterior median projection crowned with a small pad of hooks; pedes spurii B large. Lateral setae of segments VII and VIII long, simple, forked or branched; rarely all simple (Fig. 15.2.25H). Thoracic horn (Fig. 15.2.25F) elongate egg-shaped, broadest near base, or elongate oval; minutely spinulate on apical half. Anal lobes (Fig. 15.2.25H) end in a point posteriorly .... *Acamptocladius*

29' Without this combination of characters 30

30(29') Anal lobes (Fig. 15.2.25K, M) drawn out posteriad into a point, the three apical setae, if present, displaced laterally. Small exuviae, less than 4 mm long, OR thorax anteriorly with dense elongate teeth. Tergites may have either an anterior or posterior transverse band of strong points, but not both 31

30' If anal lobes drawn out posteriad into a point, exuvial length 4.0 mm or more and thorax not so strongly armed OR tergites with both anterior and posterior transverse bands of strong points 35

31(30) Thorax anteriorly densely armed with elongate teeth (Fig. 15.2.25I). Abdominal tergites (Fig. 15.2.25J) IV–VI with an anteromedian circular point patch .... *Abiskomyia*

31' Thorax not so strongly armed. Abdominal tergites IV–VI without a distinct circular anteromedian point patch 32

32(31') Abdominal segments (Fig. 15.2.25L) with lateral margins hairy, segments II–VI with two setae amongst the hairs; tergites II–VII dorsally with an anterior and two posterior groups of small points .... *Epoicocladius*

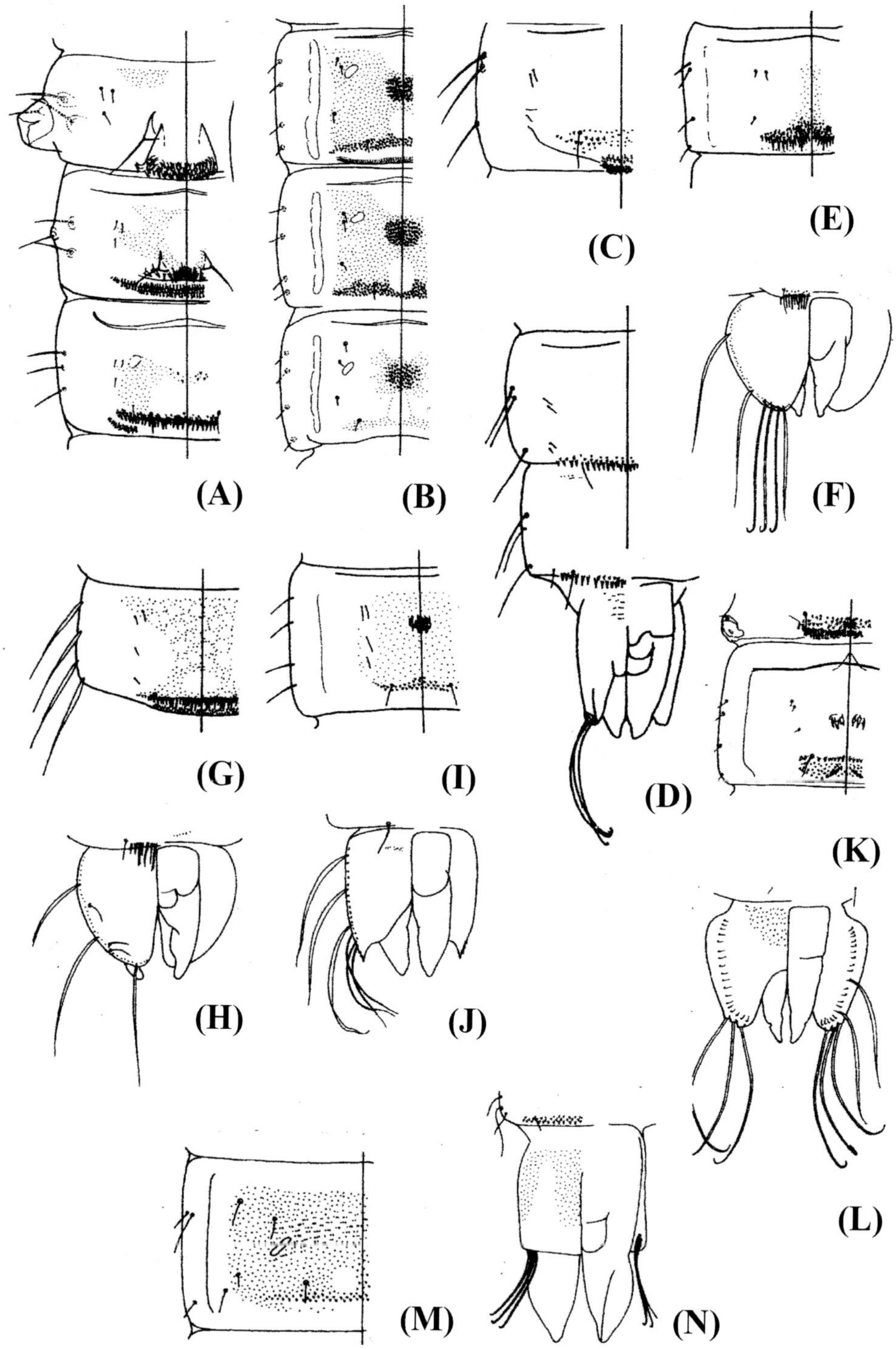

FIGURE 15.2.24 Pupal exuviae, Orthocladiinae. (A) *Nanocladius rectinervis*, segments II–IV; (B) *Rheocricotopus effusus*, segments II–IV; (C) *Rheocricotopus* (*Psilocricotopus*) *tirolus*, segment II, (D) segments VII–IX; (E) *Psectrocladius* (*Mesopsectrocladius*) *barbatipes*, segment VI, (F) anal segment; (G) *Psectrocladius* (*Allopsectrocladius*) *obvius*, segment VI, (H) anal segment; (I) *Psectrocladius* (*Monopsectrocladius*) *calcaratus*, segment VI, (J) anal segment; (K) *Psectrocladius* (s.str.) *octomaculatus*, segment III, (L) anal segment; (M) *Tokunagaia tonollii*, segment II, (E) anal segment.

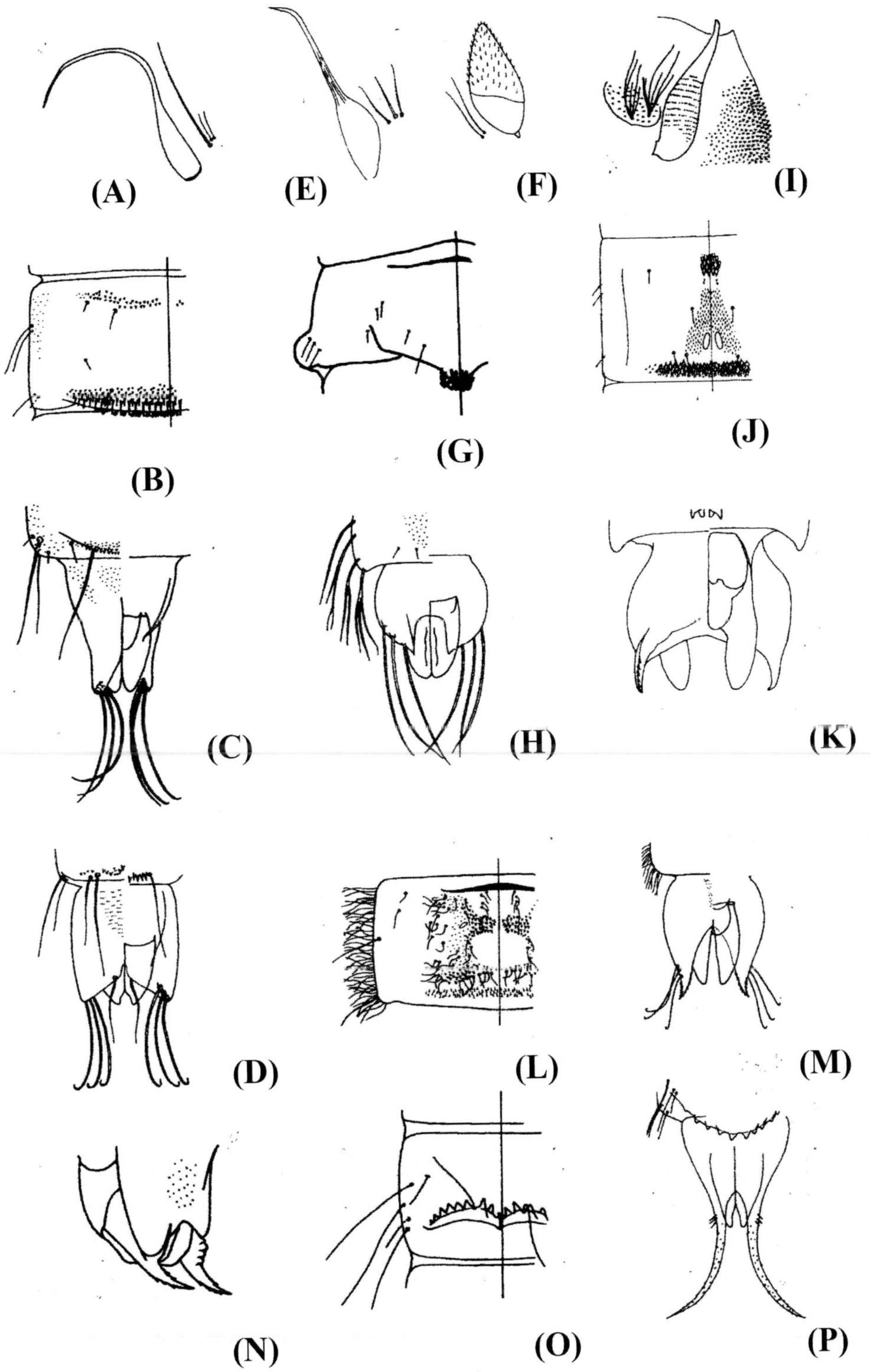

FIGURE 15.2.25 Pupal exuviae, Orthocladiinae. (A) *Eukiefferiella gracei*, thoracic horn and precorneal setae, (B) segment III, (C) anal segment; (D) *Tvetenia bavarica*, anal segment; (E) *Dratnalia potamophylaxi*, thoracic horn and precorneal setae; (F) *Acamptocladius reissi*, thoracic horn and precorneal setae, (G) segment II, (H) anal segment and lateral setation on segment VIII; (I) *Abiskomyia paravirgo*, anterior thorax, thoracic horn, and precorneal setae, (J) segment IV, (K) anal segment; (L) *Epoicocladius flavens*, segment IV, (M) anal segment; (N) *Rheosmittia spinicornis*, anal segment; (O) *Krenosmittia camptophleps*, segment IV, (P) anal segment.

32' Lateral margins of abdominal segments without a hair fringe 33

33(32') Anal lobes (Fig. 15.2.25N) without setae. Thoracic horn absent. (Sternites V and VI with posterior transverse point rows. Very small (less than 2.2 mm long), brown or brownish exuviae.) .... *Rheosmittia*

33' Anal lobes with setae. Thoracic horn present or absent 34

34(33') Abdominal tergites (Fig. 15.2.25O) (II)III–VIII with a posterior transverse row of large teeth elevated on a ridge of cuticle, which may be broken medially. Anal lobe projections may be very long (Fig. 15.2.25P) .... *Krenosmittia*

34' Abdominal tergites (Fig. 15.2.26A) with or without a posterior transverse band of points .... *Parakiefferiella*

35(30') Each anal lobe extended posteriorly into a long cylindrical projection that curves upwards to the tip, which bears small tubercles around the bases of the anal macrosetae (Fig. 15.2.26B). Sternites I–III unarmed, except for the posterior row of teeth on III. (Thoracic horn elongate, pointed at tip and toothed.) .... Orthocladiinae gen? sp? Pe1 (Langton, 1991). Described by Wülker (1957) as the possible pupal exuviae of *Krenosmittia hispanica* Wülker, because the exuviae and adults were collected at the same site in north-east Spain. (He recorded also an exuviae from southern Spain.) However, it more resembles that of *Heleniella* without the spines on sternites II and III, or the Nearctic *Lopescladius*. It has yet to be positively identified by association with an adult male.

35' If anal lobes somewhat similarly formed, sternites I–III are armed with points or spines 36

36(35') Tergites I(II)–VIII with at least the posterior margin armed with long narrow spines (Fig. 15.2.26C) 37

36' Posterior margins of tergites without long narrow spines 40

37(36) Anal macrosetae of anal lobes long, thin and usually curved (Fig. 15.2.26D). Anal lobes smooth. Thoracic horn reduced to a small tubercle 38

37' Anal macrosetae short, thorn-like, branched, or reduced to strong points. Anal lobes armed with points. Thoracic horn absent 39

38(37) Precorneal setae long (longest seta about 150 μm long) .... *Paralimnophyes*

38' Precorneal setae small (longest seta about 100 μm long) .... *Limnophyes*

39(37') Each anal lobe (Fig. 15.2.27F) with the macrosetae reduced to strong points on the posterior margin. Tergites (Fig. 15.2.27E) II–VIII covered with even-sized points except for the adhesion marks anterior to the posterior transverse row of spines .... *Molleriella*

39' Anal macrosetae (Fig. 15.2.27H) stout, the inner two thorn-like, the outer stout and usually branched. Tergites (Fig. 15.2.27G) II–VII medially with a transverse band of dark points, more or less divided into four sections .... *Cardiocladius*

40(36') Abdominal tergites II–VII or II–VIII each with a posterior transverse row of teeth which may be flattened and contiguous, without a similar anterior transverse row (Fig. 15.2.26K). Anal lobes flat, without strong teeth along the lateral margins and not extended posteriorly into a downwardly curved point 41

40' If abdominal tergites II–VIII each with a posterior transverse row of teeth, then the lateral margin of the anal lobes are bent upwards (the upturned edge often crested with strong points) and the apex may be extended posteriorly as a downwardly curved point; anterior transverse rows of strong points may also be present 46

41(40) Exuviae large, about 10 mm long. Abdominal tergites (Fig. 15.2.26E) with the posterior transverse tooth row set on the apices of longitudinal cuticular ridges. Anal lobes (Fig. 15.2.26F) with 4/5 apical taeniae ..... *Eurycnemus*

41' Exuviae smaller, less than 6 mm long. Abdominal tergites without longitudinal ridges posteriorly. Anal lobes (Fig. 15.2.26I) without taeniae; with three apical setae which may be very reduced and difficult to see 42

42(41') Sternites II (Fig. 15.2.26H) and III armed with strong spines anteriorly. Thoracic horn elongate, pointed at tip and toothed (Fig. 15.2.26G) .... *Heleniella*

43' Sternites II and III without spines anteriorly. Thoracic horn generally absent, but if present and elongate, smooth 43

43(42') Tergites VI- or VII–VIII with a median more or less circular patch of spinules (Fig. 15.2.26J) best developed on segment VIII .... *Thienemannia*

43' Tergites VI–VIII without a median circular patch of spinules 44

44(43') Sternites (III)IV–VII(VIII) with a posterior row of points similar to those on tergites (I)II–VIII 45

44' Sternites not armed posteriorly with teeth. Anal lobes (Fig. 15.2.26L) at most extended into a small posterolateral point .... *Metriocnemus*

45(44) Anal lobes (Fig. 15.2.26M) extended posteriorly in a broad claw-like projection .... *Parachaetocladius*

45' Anal lobes truncate posteriorly, where the anal macrosetae are inserted .... *Psilometriocnemus*

46(40') Anal lobes (Fig. 15.2.27B) rounded with the apical setae short and strong, thorn- or spine-like, when three, one is situated in the proximal half of the lobe, the other two in the distal half (but one may be missing). Abdominal

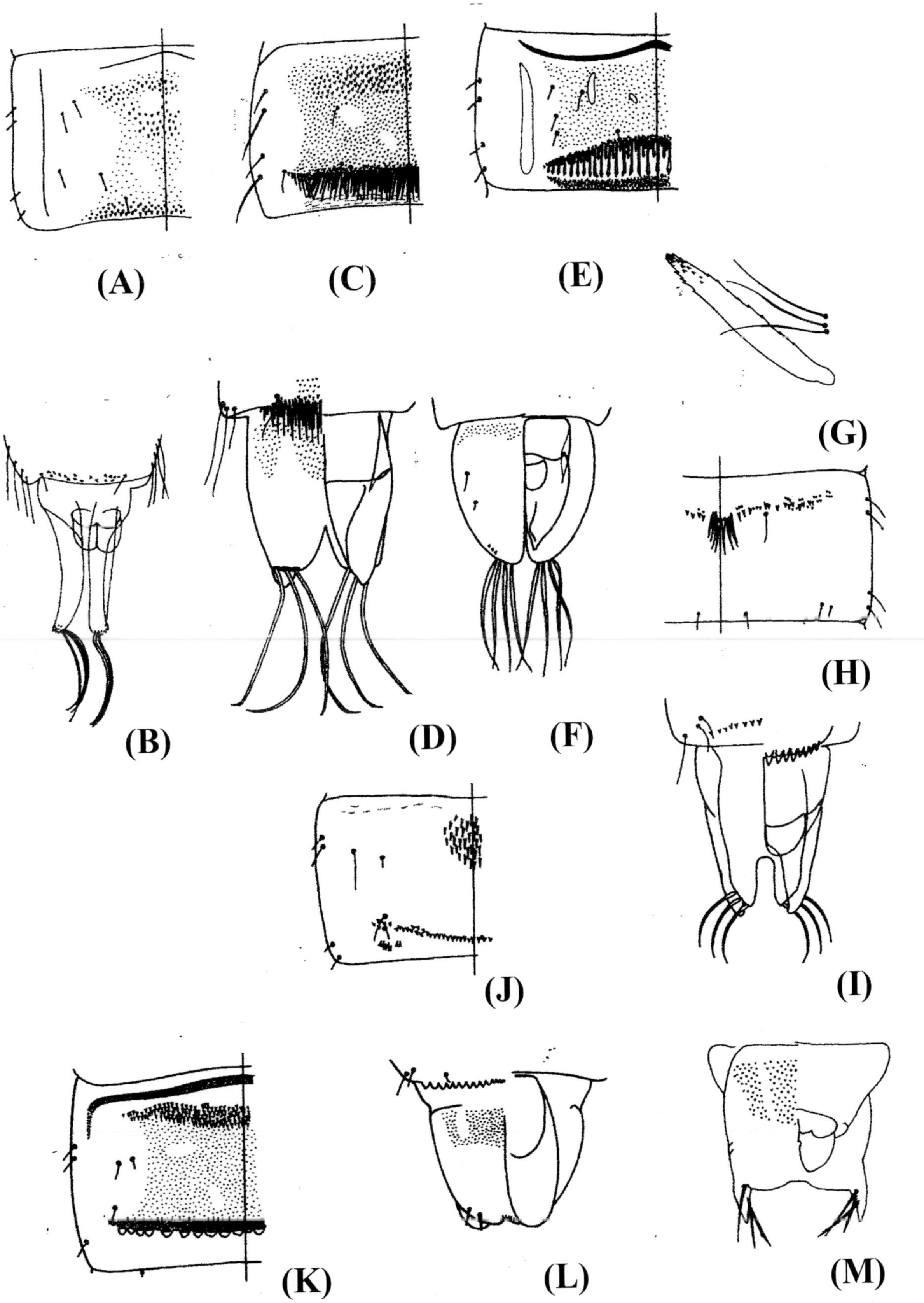

FIGURE 15.2.26 Pupal exuviae, Orthocladiinae. (A) *Parakiefferiella scandica*, segment IV; (B) Orthocladiinae gen? sp? Pe1, anal segment; (C) *Limnophyes edwardsi*, segment II, (D) anal segment; (E) *Eurycnemus crassipes*, segment II, (F) anal segment; (G) *Heleniella ornaticollis*, thoracic horn and precorneal setae, (H) segment II ventral, (I) anal segment; (J) *Thienemannia gracilis*, segment VII; (K) *Metriocnemus eurynotus*, segment VII, (L) anal segment; (M) *Parachaetocladius abnobaeus*, anal segment.

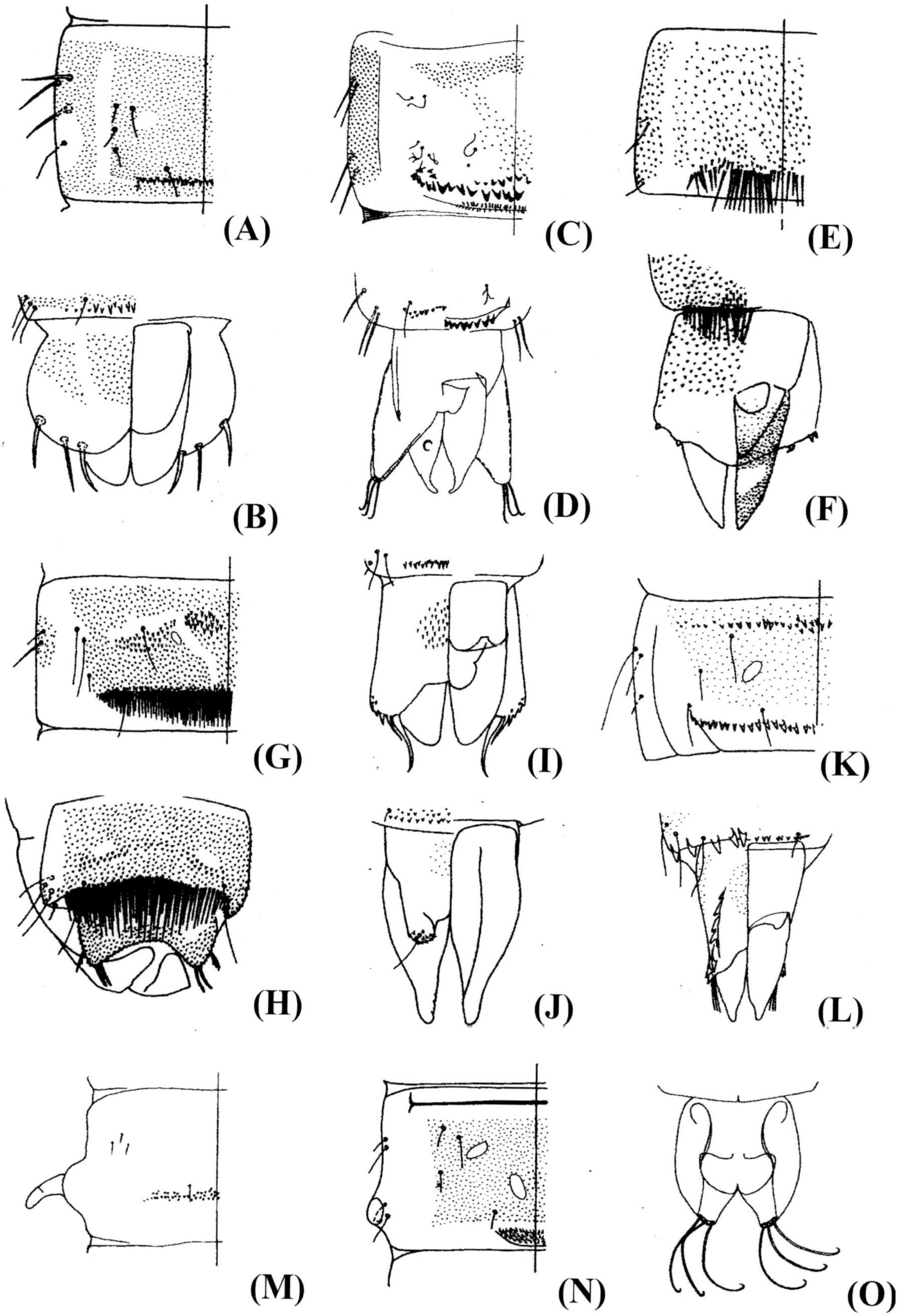

FIGURE 15.2.27 Pupal exuviae, Orthocladiinae. (A) *Chaetocladius insolitus*, segment II, (B) anal segment; (C) *Euryhapsis* sp., segment II, (D) anal segment; (E) *Molleriella calcarata*, segment VI, (F) anal segment; (G) *Cardiocladius capucinus*, segment II, (H) anal segment; (I) *Synorthocladius semivirens*, anal segment; (J) *Paraphaenocladius impensus*, anal segment; (K) *Pseudorthocladius filiformis*, segment II, (L) anal segment; (M) *Parametriocnemus boreoalpinus*, segment II; (N) *Diplocladius cultriger*, segment II, (O) anal segment.

tergites (Fig. 15.2.27A) II–VIII posteriorly with a weak to well-developed narrow transverse band of teeth, the remainder of the tergites covered with small points. (Thoracic horn, if present, elongate, small-toothed, pointed at apex.) .... *Chaetocladius*
46' IF anal macrosetae thorn- or spine-like, abdominal tergites with armament restricted to patches or bands, the remainder smooth 47
47(46') Tergites (Fig. 15.2.27C) II–VIII with posterior transverse bands of brown or black teeth. Sternites IV–V- or VI–VIII also with dark teeth similar to those on the tergites, but becoming progressively stronger on succeeding segments, at least on VIII medially divided and each half set on a rounded mound. Anal lobes (Fig. 15.2.27D) taper to the bluntly rounded apex which bears the three anal macrosetae .... *Euryhapsis*
47' Not as above 48
48(47') Anal lobes distally truncate (Fig. 15.2.27I) with one strong, straight spinous anal macroseta and the other two external to it progressively reduced in size, and with 2–6 additional teeth posterolaterally .... *Synorthocladius*
48' Anal lobes and apical setae not as above 49
49(48') IF pedes spurii of segment II (Fig. 15.2.27M) very large, elongate conical, wing sheaths with pearl row .... *Parametriocnemus boreoalpinus* Gowin
49' IF pedes spurii B of segment II very large, then pearl row absent 50
50(49') Anal lobes (Fig. 15.2.27J) usually folded inwards and so appearing parallel-sided and apically truncate; frequently apically toothed and with only one or two anal macrosetae, though there may be a small seta situated more anteriorly on the lobes. Wingsheaths with or without pearl row and nose .... *Paraphaenocladius*
50' Anal lobes at most folded upwards, in which case they are usually margined with strong points forming a dorsal "crest". Wing sheaths without pearl row and nose 51
51(50') Tergites (Fig. 15.2.27K) II–VIII with a posterior transverse row or narrow band of strong points; an anterior transverse row or band may be present, of strong points, small points or shagreen. Each anal lobe (Fig. 15.2.27L) folded upwards, the elevated margin often with a row of strong teeth from near base to the anal macrosetae, but this crest may be reduced to a short row of points little stronger than the general armament of tergite IX, or absent .... *Pseudorthocladius*
51' Tergites without armament of this main pattern. Anal lobes without strong lateral teeth 52
52(51') Anal lobes (Fig. 15.2.27O) dorsoventrally flattened laterally, rounded in outline, with three long, curved, hooked anal macrosetae. Tergite II (Fig. 15.2.27N) with a median pad of hooks. Tergites III–VIII generally covered with small points which are a little larger medially and posteriorly; tergites III–V with an additional apical transverse band of spinules. Segment VII with four conspicuous lateral setae, VIII with four or five .... *Diplocladius*
52' Without this combination of characters 53
53(52') Tergites (Fig. 15.2.28A) III–VIII extensively covered with small points which are a little larger posteriorly on each segment; hook row II and apical bands absent. Thorax with some strong, points near suture. Thoracic horn small, thin-walled, generally invisible. Pedes spurii B absent. Three well-developed anal macrosetae present (Fig. 15.2.28B) .... *Parorthocladius*
53' Without this combination of characters 54
54(53') Tergites (Fig. 15.2.28C) III–VIII armed with points which form an urn-shaped patch widest anteriorly, narrowing past the anterior adhesion marks towards setae D5, there expanding to the width of the tergite; on VII and VIII greatly reduced in extent and size of points; apical point bands present on tergites III and IV. On segment II hook row and pedes spurii B present. Anal macrosetae (Fig. 15.2.28D) long. Thoracic horn elongate, weakly toothed .... *Stilocladius*
54' Without this combination of characters 55
55(54') Anal lobes smoothly rounded, flared out laterad posteriorly (Fig. 15.2.28H), or reduced, with or without one or two small setae. If the hook row of segment II is present, the abdominal tergite armament is reduced to bands and patches of points 56
55' Anal lobes normal or drawn out posteriorly into a hooked lobe, with three macrosetae, or, if reduced, the hook row of segment II is present and the abdominal tergites are covered with shagreen 59
56(55) Tergites II–VIII, apart from small bare patches, covered with more or less evenly sized points (Fig. 15.2.28G). Thoracic horn usually absent 57
56' Abdominal tergites with the armament restricted to a posterior band of points, with or without an anterior band and/or median point patch. Thoracic horn usually present 60
57(56) Intersegmentally armed both dorsally and ventrally (Fig. 15.2.28E) .... *Pseudosmittia*

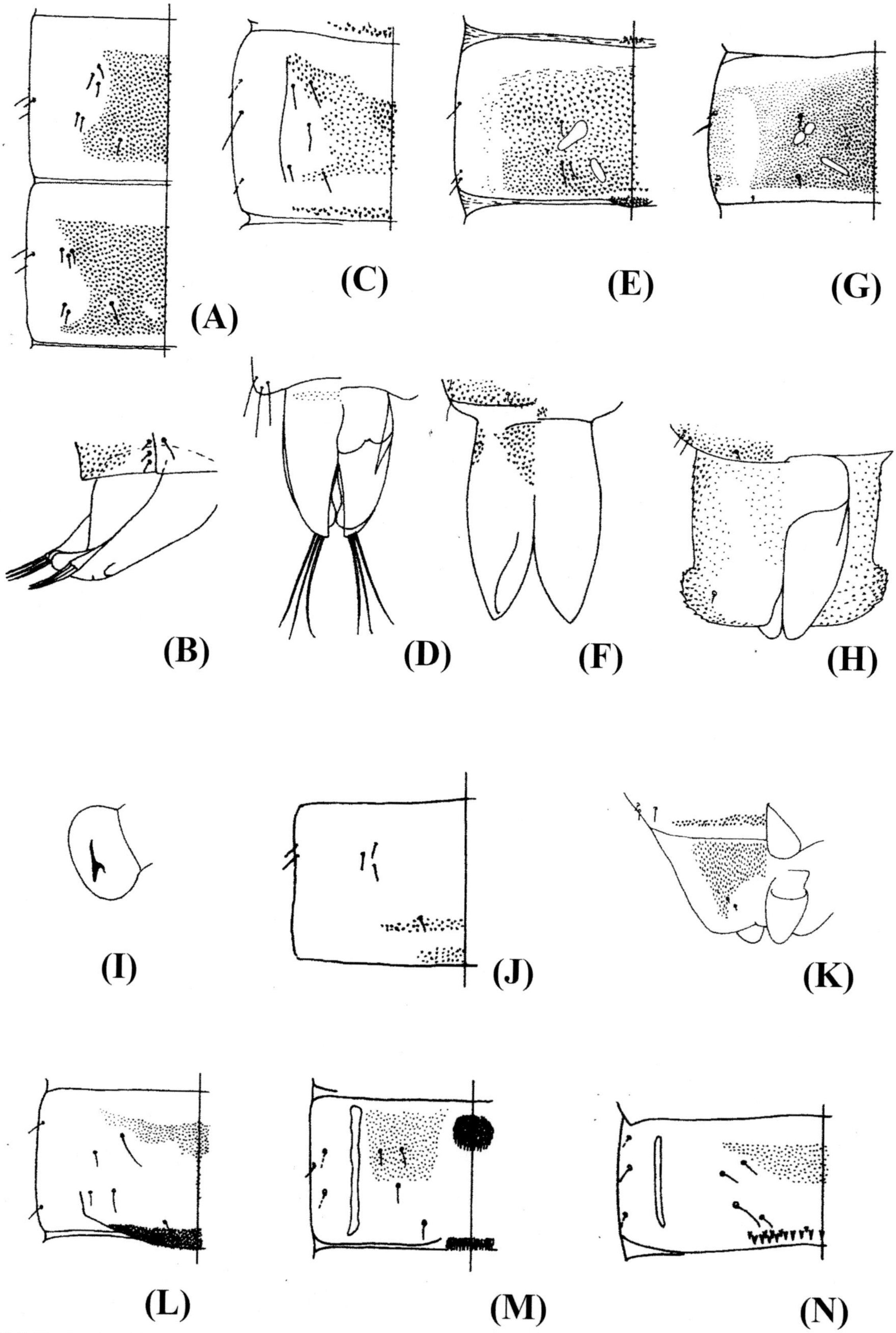

FIGURE 15.2.28 Pupal exuviae, Orthocladiinae. (A) *Parorthocladius nudipennis*, segments II and III, (B) anal segment; (C) *Stilocladius montanus*, segment III, (D) anal segment; (E) *Pseudosmittia recta*, segment III, (F) anal segment; (G) *Bryophaenocladius subvernalis*, segment IV, (H) anal segment; (I) *Symbiocladius rhithrogenae*, base of antennal sheath, (J) segment III, (K) anal segment; (L) *Orthocladius* (*Euorthocladius*) *calvus*, segment IV; (M) *Orthocladius* (*Euorthocladius*) *rivulorum* segment IV; (N) *Orthocladius* (*Euorthocladius*) *ashei*, segment IV.

57' Intersegmentally unarmed 58
58(57') Precorneals and antepronotals thorn-like, spine-like or very elongate, in which case they are about 3 μm wide at base .... *Bryophaenocladius*
58' Precorneals and antepronotals hair-like, narrow, about 1 μm across or less .... *Smittia*
59(55') Antennal sheath base with a sharp brown tooth (Fig. 15.2.28I). Tergites (Fig. 15.2.28J) III–VIII with a posterior transverse band of small points; III–V also with an apical band of small hook-like points; posterior tergites with a variably developed anteromedian patch of row shagreen. Anal lobes (Fig. 15.2.28K) with two very small anal macrosetae inserted dorsally .... *Symbiocladius*
59' Antennal sheath base without a sharp tooth. EITHER tergites II–V with apical bands of small, anteriorly directed hooks or spines; some tergites with posterior transverse bands of backwardly directed points; with or without an anterior point patch, OR tergites without apical bands or anterior point patches; some tergites with a posterior transverse band of points (e.g., Fig. 15.2.28L, M, N) .... *Orthocladius* (*Euorthocladius*)
60(56') Either hook row of segment II short, less than 0.34 the breadth of segment II, usually of three or more transverse rows of hooks, or absent (its position occupied by a band of anteriorly directed spinules). Anal lobes often with additional teeth laterally or apically. Four dorsocentral setae, the first, third and fourth thick, the second conspicuously thinner 61
60' Hook row II present, longer, more than 0.35 the breadth of segment II; if short, of two rows of hooks with the third represented at most by one or two hooks. Anal lobes very rarely with additional teeth laterally or apically. Dorsocentral setae without great differences in thickness 64
61(60) Hook row of segment II absent, or represented by a few hooks in the middle of the apical spinule band 62
61' Hook row II a discrete pad or band of hooks (Fig. 15.2.29F) 63
62(61) Thoracic horn (Fig. 15.2.29A) leaf-like, with a serrated edge. Tergites (Fig. 15.2.29B) III–VII with a median transverse pair of more or less circular point patches, sometimes indicated on tergite VIII, set in a field of shagreen that extends laterally to the tergite margins and posteriorly to the posterior transverse point bands. Tergite II shagreened, without median point patches, but with a transverse band of points in front of the apical spinule band. Tergites VIII and IX (Fig. 15.2.29C) covered with shagreen .... *Orthocladius* (*Pogonocladius*)
62' Thoracic horn (Fig. 15.2.29D) elongate oval, smooth, or nearly so. Tergites (Fig. 15.2.29E) II/III–VI/VII with transverse median point patches. Abdominal tergites with limited shagreen .... *Orthocladius* (*Eudactylocladius*)
63(61') Anal lobes (Fig. 15.2.29G) drawn out posteriorly into an inwardly curved point; anal macrosetae long, flexible and tapering to their tips .... *Orthocladius* (*Symposiocladius*)
63' Anal lobes (Fig. 15.2.29I) not drawn out into a point, but may be furnished apically with one or more teeth; anal macrosetae less flexible, usually hooked at apex. Segment II (Fig. 15.2.29H) may have pedes spurii B or not, the tergite armament may be divided into three transverse bands or fused to form a complete covering within the dorsal setae and the anal lobes may or may not have a fringe of setae .... *Orthocladius* (s.str.)
64(60') Tergites (Fig. 15.2.29K) (II)III–VI with a median transverse pair of point patches, often joined by small points. Thoracic horn (Fig. 15.2.29J) elongate, toothed (more densely towards apex), and narrowed to a pointed tip. Tergites VII and VIII without armament. Anal macrosetae (Fig. 15.2.29L) of equal length .... *Acricotopus*
64' Mid-tergite armament not obviously paired, or if somewhat paired, present on tergites III–VIII and the inner anal lobe seta smaller than the other two 65
65(64') Thorax (Fig. 15.2.29M) dorsally with dense short toothlets; more laterally granulate. Segment II (Fig. 15.2.29N) with pedes spurii B and hook row; tergite III–V armed with a trapezoidal patch with median and posterior points larger, less developed on VI. Anal segment (Fig. 15.2.29O) waisted .... *Paracladius*
65' Thorax at most strongly granulate. Anal segment (Fig. 15.2.30B) not narrowed medially. Segment II has a hook row, but pedes spurii may or may not be present 66
66(65') Segments (Fig. 15.2.30A) II–VIII with conspicuous armament covering all or much of the tergites. Thoracic horn and frontal setae absent .... *Cricotopus* (*Paratrichocladius*) *skirwithensis*-group
66' If tergites II–VIII covered with conspicuous armament, then frontal setae and thoracic horn present, or the anal lobes are reduced and lacking anal setae 67
67(66') Tergites (Fig. 15.2.30D) III–V with the median transverse point band medially arched (convex anteriorly, concave posteriorly); the median part of this transverse band on tergite III is situated in the second quarter of the tergite leaving the third quarter between it and the posterior band free of points; the median patch of tergite VI is often not concave posteriorly. Tergite II with a posterior transverse row or narrow band of points immediately in front of the hook row. Thoracic horn (Fig. 15.2.30C) toothed or nearly smooth, generally somewhat club-shaped with the

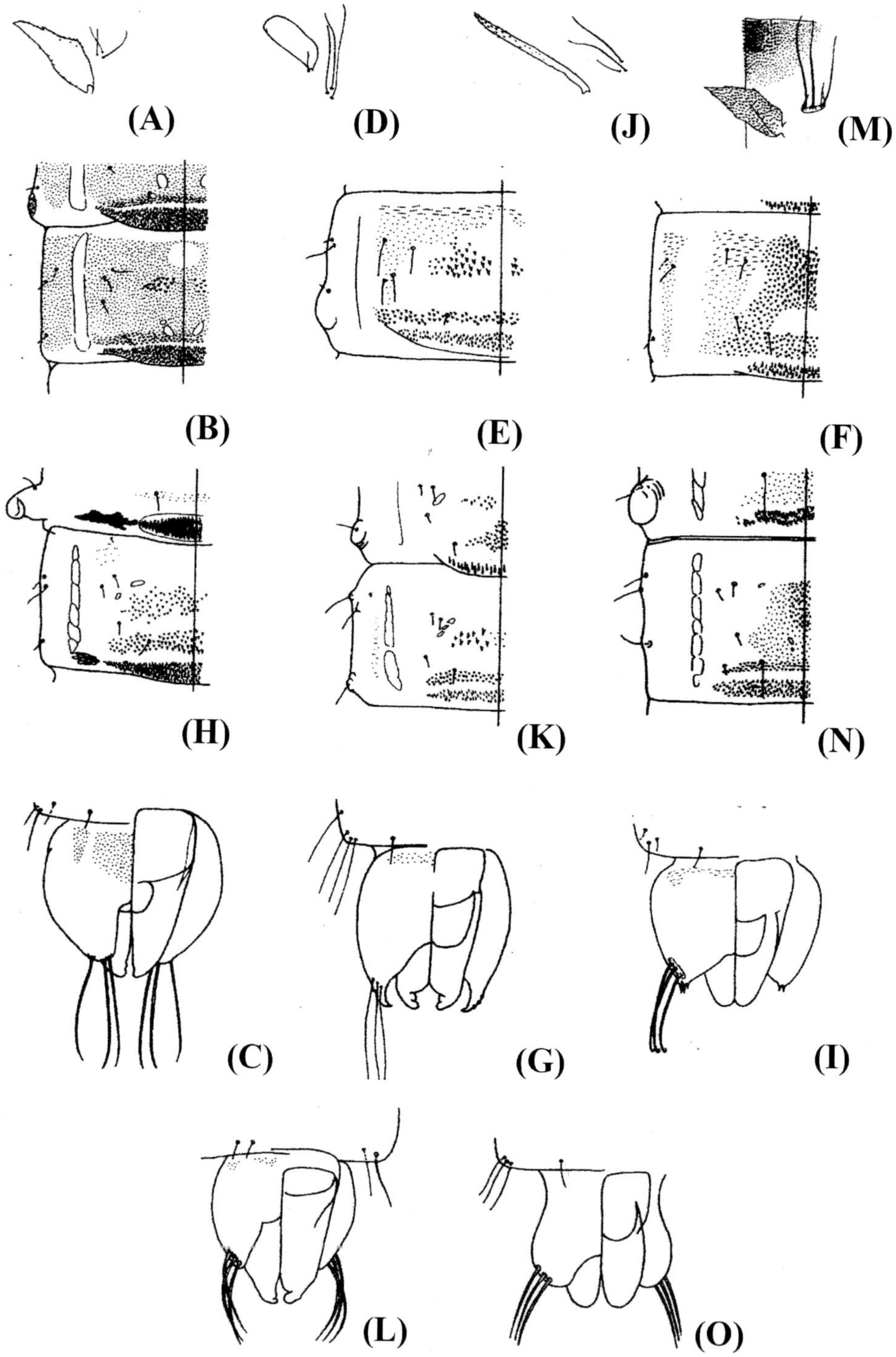

FIGURE 15.2.29 Pupal exuviae, Orthocladiinae. (A) *Orthocladius* (*Pogonocladius*) *consobrinus*, thoracic horn and precorneal setae, (B) segments II and III, (C) anal segment; (D) *Orthocladius* (*Eudactylocladius*) *olivaceus*, thoracic horn and precorneal setae, (E) segment II; (F) *Orthocladius* (*Symposiocladius*) *lignicola*, segment III, (G) anal segment; (H) *Orthocladius* (s.str.) *rubicundus*, segments II and III, (I) anal segment; (J) *Acricotopus lucens*, thoracic horn and precorneal setae, (K) segments II and III, (L) anal segment; (M) *Paracladius conversus*, anterior thorax, thoracic horn, and precorneal setae, (N) segments II and III, (O) anal segment.

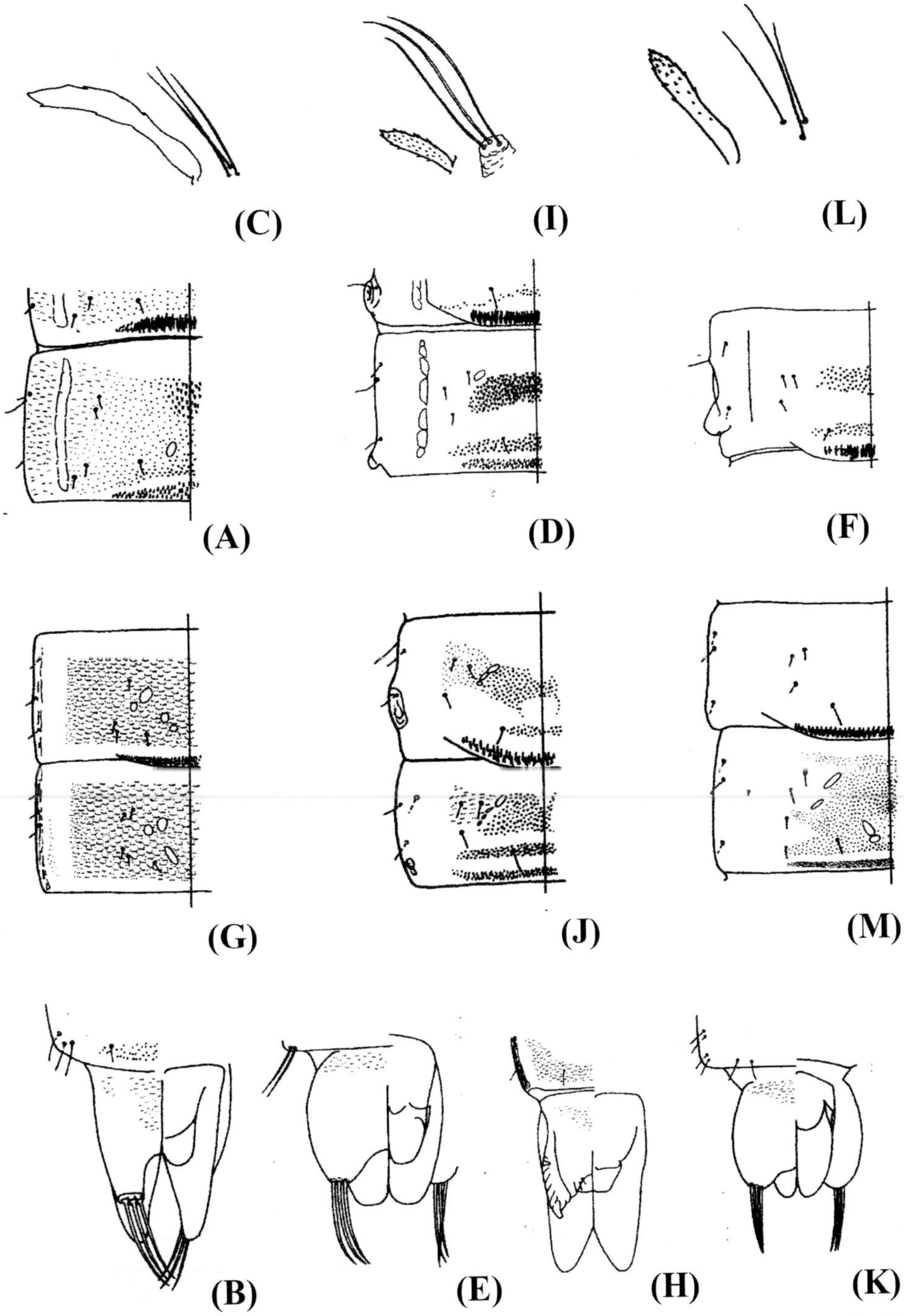

FIGURE 15.2.30 Pupal exuviae, Orthocladiinae. (A) *Cricotopus* (*Paratrichocladius*) *skirwithensis*, segments II and III, (B) anal segment; (C) *Cricotopus* (*Paratrichocladius*) *rufiventris*, thoracic horn and precorneal setae, (D) segments II and III, (E) anal segment; (F) *Cricotopus* (*Paratrichocladius*) *micans*, segment II; (G) *Cricotopus* (*Nostococladius*) *lygropis*, segments II and III, (H) anal segment; (I) *Cricotopus* (*Isocladius*) *intersectus*, thoracic horn and precorneal setae, (J) segments II and III, (K) anal segment; (L) *Cricotopus* (s.str.) *pilosellus*, thoracic horn and precorneal setae, (M) segments II and III.

apical half expanded. Segment VIII with five lateral setae, the fourth and fifth more or less equal in size (Fig. 15.2.30E). Frontal setae absent .... *Cricotopus* (*Paratrichocladius*) *rufiventris* (Meigen).
67' If tergites III–V with the median point band evenly arched forwards, then nearly the whole band is situated in the third quarter of the tergite. Four or five lateral setae on segment VIII; if five, and the fourth and fifth more or less equal in size, then frontal setae present 68
68(67') Tergite II (Fig. 15.2.30F) with a distinct median transverse point patch, widely separated from the posterior transverse band or joined to it on each side of midline; hook row II very short, about three-tenths the width of the segment .... *Cricotopus* (*Paratrichocladius*) *micans* (Kieffer).
68' If tergite II with a distinct median point patch, the hook row of segment II is much wider 69
69(68') Tergites (Fig. 15.2.30G) II–VIII covered with group shagreen forming a reticulate pattern. Anal lobes (Fig. 15.2.30H) reduced, transversely wrinkled .... *Cricotopus* (*Nostococladius*)
69' Abdominal tergites armed with patches and bands of points. Anal lobes not reduced 70
70(69') Tergites (Fig. 15.2.30J) III and IV anteriorly armed for at least five-sixths the breadth of the tergite (CAUTION: Sternite armament may confuse here), and on II extending over the posterior half of the tergite; or, if not so extensive on segment II, the precorneal setae (Fig. 15.2.30I) are set on a conspicuous mound. Frontal setae on frontal apotome or absent. Segment VIII (Fig. 15.2.30K) with four lateral setae AND thoracic horn present .... *Cricotopus* (*Isocladius*)
70' EITHER Tergites (Fig. 15.2.30M) III and IV anteriorly armed for less than three quarters the breadth of the tergite; OR, if greater than five-sixths tergite breadth, then on segment II not exceeding the posterior quarter of the tergite and the precorneal setae (Fig. 15.2.30L) are not set on conspicuous mounds; OR frontal setae on praefrons. If segment VIII with four lateral setae, then thoracic horn absent .... *Cricotopus* (s.str.)

Acknowledgments

We thank Aouadi Abdallah (Algeria), Núria Bonada (Spain), Kawtar Kettani (Morocco), and Francesca Paoli (Italy) for providing photographs of habitat and pupae.

Subchapter 15.3

Family Simuliidae

Peter H. Adler
Department of Plant and Environmental Sciences, Clemson University, Clemson, South Carolina, United States

Introduction

The Simuliidae, commonly called "black flies," are best known for the blood-feeding habits of the females, but they also are among the most abundant and functionally important macroinvertebrates in flowing water. They represent one of about 20 families of Diptera with exclusively aquatic larvae. All species of black flies develop in flowing water, regardless of the current velocity, whether barely perceptible or approaching the upper limit for naturally flowing water. The larvae and pupae inhabit virtually all freshwater flows worldwide, except in Antarctica and on a limited number of oceanic islands. Despite conventional claims that black flies are found only in clean water, some species can achieve large populations in polluted streams and rivers, possibly because they are often unfettered by competition or predation.

To date, 99 formally named species and 6 undescribed species in 7 genera have been found within the limits of the Mediterranean Basin, of which about 45% are endemic. Considerably more species are expected to be found in the basin, particularly as cryptic species continue to be discovered. Not all species known from a particular Mediterranean country are found within the Mediterranean Basin, especially in countries with high mountains, such as France, Italy, Morocco, and Slovenia; hence, the rich, and often unique, mountain simuliid fauna is excluded from the following treatment. The species of the Mediterranean Basin represent about 4.1% of the 2398 known species of black flies in the world (Adler, 2022).

Larvae and pupae are known for a remarkable 97% of the species recorded from the Mediterranean Basin, comparable to that for the Nearctic Region (Adler et al., 2004). Larvae and pupae are unknown for *Simulium flavipes* Austen, 1921, from Palestine/Jordan Valley (Jericho); *S. pulchripes* Austen, 1925, from Turkey (Çanakkale); and *S. rivosecchii* Rubtsov, 1964, from Italy (Lazio). Pupae are unknown for the undescribed species *Simulium* "IL-8" (Adler et al., 2016a) from Portugal. Considerable taxonomic work is still needed, particularly to resolve species complexes and species groups, some of which include an abundance of formal names, including numerous synonyms. The widely distributed and ubiquitous *S. ornatum* species group, for example, includes 80 formal names applied to 24 species in the Palearctic Region (Adler, 2022).

Various identification keys to species are available for selected countries, or parts thereof, in the Mediterranean Basin, such as Cyprus (Crosskey, 2004), Greece (Crosskey & Malicky, 2001), Italy (Rivosecchi, 1978), Morocco (Belqat & Dakki, 2004), Spain (Crosskey, 1991), and Turkey (Başören & Kazanci, 2015, 2016). However, no comprehensive treatment is available for the entire Mediterranean simuliid fauna.

Ecology and distribution

Black flies are typically represented in more than 90% of watercourses in any given area of the world, and can be found from below sea level to elevations of 5000 m above sea level. Any flowing freshwater provides suitable habitat, including large rivers, waterfalls, seepages, sulfur springs, hot springs, glacial meltwaters, slow-flowing swamps, impoundment outflows, intermittent streams, and subterranean flows. Their superabundance in streams and rivers often places black flies first in secondary production among benthic macroinvertebrates. A broader elevational range of sampling provides more habitat types and, therefore more simuliid diversity. Although more than one species typically inhabits a given site at any time of the year, species diversity is greatest in the spring. The most common taxa in the Mediterranean Basin are members of the *Simulium ornatum* species group and the subgenera *Eusimulium* and *Wilhelmia*. Species of *Prosimulium* are locally common in the spring.

Female black flies deposit their eggs in or near running water, either during flight or while landed on a wetted stone or trailing vegetation. The eggs generally cannot withstand desiccation, yet those of many species persist in dry streambeds, presumably protected in moist sediments. Depending on species and environmental conditions, the eggs hatch in a few days to more than 6 months if an obligatory diapause is involved. Larvae escape from the egg with the aid of an egg burster (a small cuticular tubercle) on the head of the first instar.

Larvae anchor themselves to any stable substrate in the current, including stones, trailing grasses, and rubbish such as metal and plastic. They can relocate either by looping inchworm-style over the substrate or by drifting downstream on a line of silk that they produce from their silk (i.e., salivary) glands. Larval black flies typically pass through seven instars, although more can occur if a larva is parasitized or otherwise environmentally stressed. Larval microdistribution patterns can exhibit even, random, or clumped spacing, and often provide optimal water velocities for the principal mode of feeding: filtering particulate matter from the water column (Eymann, 1991).

To filter-feed, larvae open their labral fans, twist their bodies 90–180 degrees, and lean in the direction of the current while the hooklets of their posterior proleg are enmeshed in a silk pad secured to the substrate (Chance & Craig, 1986). Any particle up to about 350 μm in diameter can be ingested, along with surface films and flocculated organic matter. Sometimes referred to as "ecosystem engineers," the larvae transform filtered particulate matter into larger pellets that are egested and then used as food by the benthic microbial and invertebrate communities (Wotton et al., 1998). Larvae can be reared to adults in aerated containers with a small amount of fish food.

Water temperature is a driving force in development and seasonality. Larvae and pupae of most species do well in cool water, but few species persist as water temperatures approach 30°C. Mediterranean species overwinter as eggs or larvae. Species that can tolerate higher temperatures—all of which are members of the genus *Simulium*—can complete multiple generations each year (multivoltinism). Species of all other Mediterranean genera are univoltine, the larvae typically hatching in late fall or winter and developing through the winter, and eventually emerging as adults in the spring. Transformation to the pupal stage takes place within a few weeks to half a year after hatching from the egg, depending on water temperature and species. The pharate pupa (i.e., the developing pupa within the larval cuticle) locates a suitable pupation site, scrapes it clean, spins a cocoon, and sheds the larval cuticle. Male development is typically more rapid than that of females; consequently, males emerge first, usually within a few days to a few weeks after initial pupation.

Larval black flies become prey for many invertebrates and vertebrates, such as predatory insects, birds, and fish. They also serve as hosts of numerous symbiotic organisms, including bacteria, fungi, helicosporidia, ichthyosporeans, mermithid nematodes, microsporidia, nematomorphs, protists, stramenopiles, and viruses (McCreadie et al., 2011). Pupae are sometimes found with ectoparasitic water mites in the cocoons, awaiting emergence of the adult flies from which they will feed on hemolymph (Gledhill et al., 1982).

Adults generally emerge from the pupa in the morning and are active in the day. Their principal activities include mating, sugar feeding, and for females, blood feeding and ovipositing. Most adults probably live less than a month in nature. Males of nearly all Mediterranean species form aerial swarms above or beside a landmark (e.g., a hilltop, branch, or stone). Females enter the swarm, coupling occurs, and the mating pair quickly leaves the swarm. Males and females feed on water and carbohydrates such as nectar from small flowers and honeydew from aphids and planthoppers (Burgin & Hunter, 1997).

Only female black flies feed on blood and only the blood from birds and mammals, driving their annoying, sometimes economically devastating, and even deadly consequences (Adler & McCreadie, 2019). In the Mediterranean Basin covered by this treatment, roughly 40% are primarily ornithophilic and 60% mammalophilic. The vast majority of species in the basin are anautogenous, requiring blood to complete all gonotrophic cycles. Some species might be facultatively autogenous (i.e., capable of completing the first gonotrophic cycle without blood). Transmission of parasites and pathogens often occurs during blood feeding, and in the Mediterranean Basin these agents can provoke diseases such as bovine onchocerciasis and avian leucocytozoonosis and trypanosomiasis. Black flies do not transmit the agents of any human disease in the Mediterranean Basin.

After converting a blood meal to eggs, females seek oviposition sites, generally under low illumination, especially toward the end of the day. Most species mature about 150–600 eggs per ovarian cycle, and could potentially undergo about six cycles of blood feeding and egg production if they live sufficiently long.

Comprehensive reviews of the ecology of black flies have been given by Crosskey (1990) and Adler et al. (2004).

Morphological characters needed in identification

Larvae

The fundamental organization of a larval black fly involves an external, well-sclerotized head capsule and an elongated body of three thoracic segments and nine apparent abdominal segments that somewhat enlarge posteriorly (Fig. 15.3.1A, B). Mature larvae of Mediterranean species are about 5–12 mm long. The body always bears nine pairs of nonfunctional spiracles (apneustic), a single ventral prothoracic proleg, and a terminal posterior proleg.

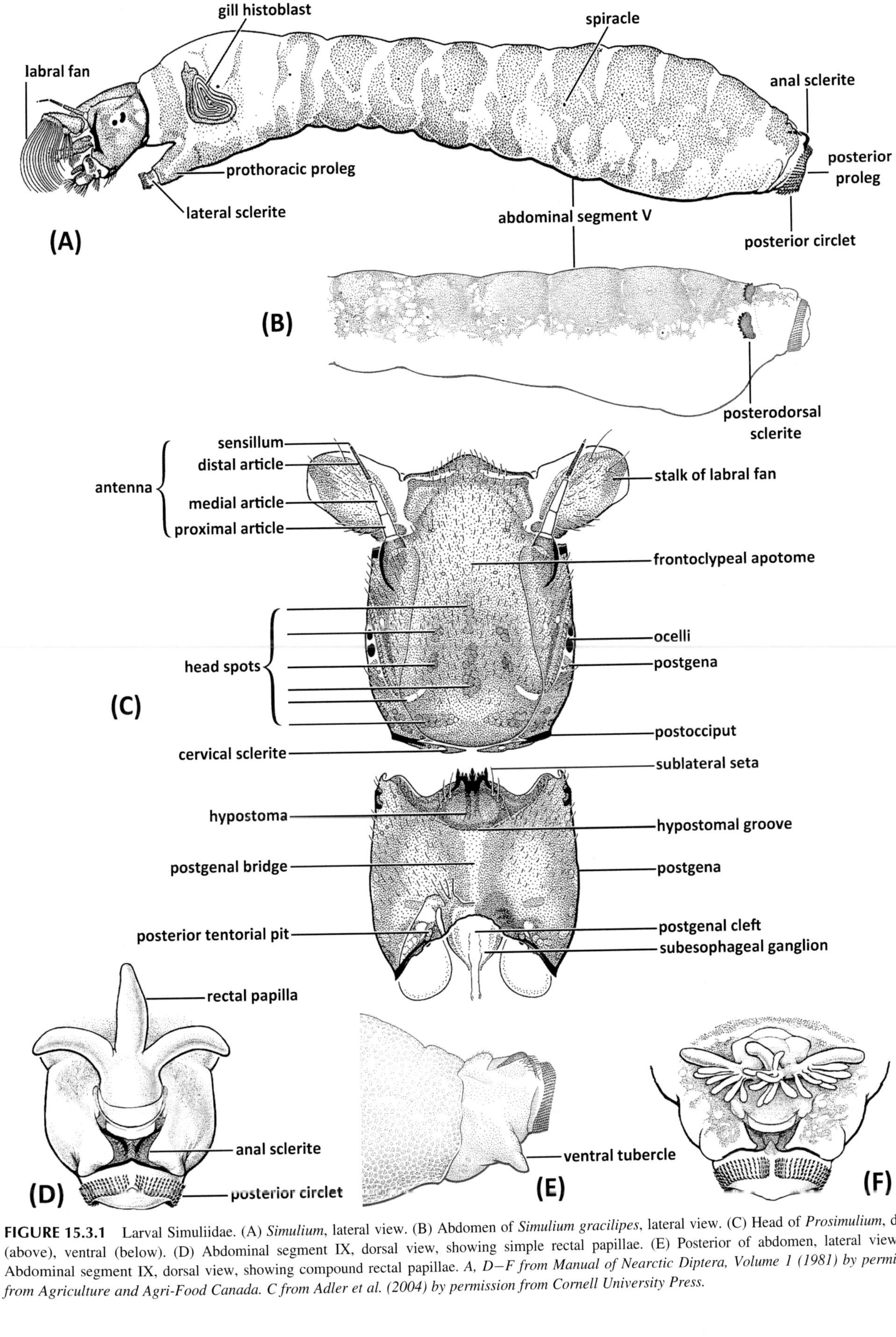

FIGURE 15.3.1 Larval Simuliidae. (A) *Simulium*, lateral view. (B) Abdomen of *Simulium gracilipes*, lateral view. (C) Head of *Prosimulium*, dorsal (above), ventral (below). (D) Abdominal segment IX, dorsal view, showing simple rectal papillae. (E) Posterior of abdomen, lateral view. (F) Abdominal segment IX, dorsal view, showing compound rectal papillae. *A, D–F from Manual of Nearctic Diptera, Volume 1 (1981) by permission from Agriculture and Agri-Food Canada. C from Adler et al. (2004) by permission from Cornell University Press.*

The most important features for identification of larvae are on the head (Fig. 15.3.1C) and in the configuration of the dissected mature **gill histoblast** on either side of the thorax (Fig. 15.3.1A). The gill histoblasts are the future pupal gills, and they darken in mature larvae (final instars). Features of secondary importance include the shape and pigmentation patterns of the body and a few characters related to the prothoracic and posterior prolegs (Fig. 15.3.1D–F).

The head capsule is composed of a dorsal plate, the **frontoclypeal apotome**, demarcated by the ecdysial line, and the lateral and ventral portions of the capsule, the **postgenae** (Fig. 15.3.1C). A thin sclerotized rim, the **postocciput**, defines the posterior border of the head and either encloses a pair of **cervical sclerites** or has a gap that leaves the sclerites free. The pattern of **head spots** on the frontoclypeal apotome represents the attachment points of internal muscles and is important for species identification. Pigmented eyespots (ocelli) lie on either side of the head. The ventral region of the head capsule has two taxonomically important features: the hypostoma and the postgenal cleft. The **hypostoma** is an anteriorly toothed plate, set off from the postgena posteriorly by a **hypostomal groove**. The hypostoma consists of a single median tooth and three sets of teeth on either side: sublateral, lateral, and paralateral teeth, the last-mentioned typically absent in the Mediterranean prosimuliines (*Helodon*, *Levitinia*, *Prosimulium*, and *Urosimulium*) (Fig. 15.3.2). In addition to the teeth, the margins typically bear lateral serrations and one or two rows of **sublateral setae**. The **postgenal cleft** is an area of thin, unpigmented cuticle of various shapes and depths, depending on species. In many (but not all) species, the **subesophageal ganglion** is ensheathed in pigment and visible in the postgenal cleft. A pair of **posterior tentorial pits** are situated laterally at the base of the postgenal cleft.

Articulated with the head capsule are the antennae and mouthparts. Each **antenna** consists of a proximal, medial, and distal article, the latter bearing an apical, conical sensory structure (Fig. 15.3.1C). Antennal pigmentation patterns can be useful in identification. The most conspicuous features of the larval head are the **labral fans**. Each fan consists of three sets of rays, the largest being the primary rays, about 20–80 per side in mature larvae, except in the Mediterranean genus *Levitinia* that lacks fans entirely. The labral fans filter particulate matter from the water column. The number of primary rays and the length of the labral fan stalk are environmentally influenced (Zhang & Malmqvist, 1997) and must be used cautiously as diagnostic characters. The mandibles have numerous teeth and brushes that

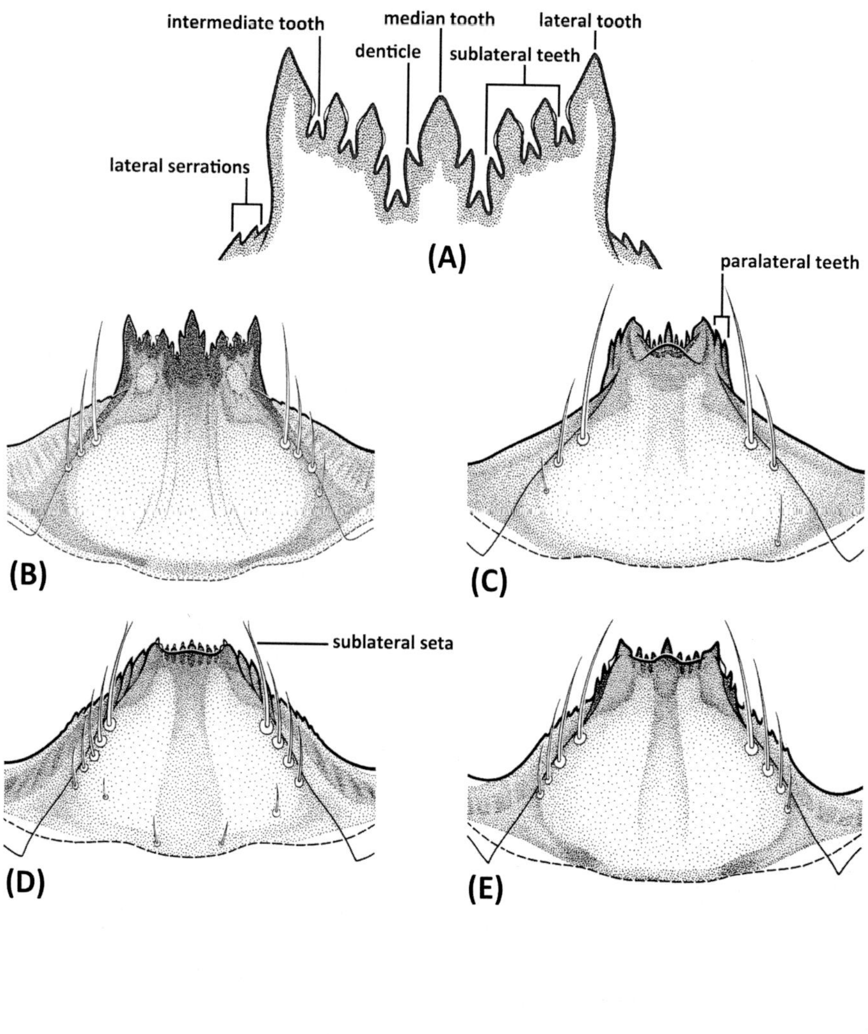

FIGURE 15.3.2 Hypostomata of larval Simuliidae, ventral view. (A) *Prosimulium*, anterior margin. (B) *Prosimulium*. (C) *Greniera*. (D) *Metacnephia*. (E) *Simulium*. *All from Adler et al. (2004) by permission from Cornell University Press.*

remove filtered matter from the labral fans, scrape the substrate for food or provide a clean attachment site, and in concert with the hypostomal teeth, cut lines of silk. Although the mandibles have been used in some identification keys, they require removal and slide mounting, and therefore are not used in the following key. Silk is extruded from the opening of the silk glands, large internal organs that extend deep into the abdomen and double back on themselves. The banding patterns of the giant chromosomes in the silk glands have played a vital role in taxonomy and identification of the Simuliidae (Adler et al., 2015).

Body pigmentation ranges from white or yellow to various shades of black, brown, green, or red. Colors and patterns can be useful for species identification, and are most easily discerned if collected larvae are placed in 95% ethanol or modified Carnoy's solution (1 part glacial acetic acid: 3 parts 95% or 100% ethanol). However, body color and intensity of the head spots in some species are influenced by the substrate to which the larvae attach: dark in larvae attached to stones, but paler in larvae attached to vegetation (Zettler et al., 1998). Color and head spot intensity, therefore, must be used cautiously in identification.

The most conspicuous extension from the body is the short **prothoracic proleg** that consists of two articles. The distal article, which is sometimes retracted within the basal article, has a pair of **lateral sclerites** and an apical ring of hooklets.

Abdominal shape comes in two forms: gradually expanded, which is typically associated with larvae inhabiting swift running water, and abruptly expanded at the fifth abdominal segment. At the end of the abdomen are several structures with variable taxonomic utility (Fig. 15.3.1D–F). The body terminates in a **posterior proleg** encircled by rows of minute hooks (**posterior circlet**) for engaging the silk pad spun from the silk glands onto a substrate. Anterior and dorsal to the posterior proleg is a sclerotized **anal sclerite**, X-shaped in all Mediterranean genera, except star-shaped in the genus *Levitinia*. A pair of conical **ventral tubercles** issues from the last segment of many species (Fig. 15.3.1E), particularly those found in small, slow-moving streams. Unpigmented, eversible **rectal papillae**, which function in osmoregulation, are anterior to the anal sclerite. They are often withdrawn in fixed larvae (Fig. 15.3.1E), but when everted they appear as three simple lobes (Fig. 15.3.1D) or three compound lobes with smaller lobules (Fig. 15.3.1F). They can be of taxonomic use, with the caveat that they are not always everted and can be subject to some environmental influence that determines whether lobules are present on the three principal lobes.

Pupae

Pupae are typically the favored life stage for identification of black flies because most identifications can be made, without dissection, based on two structures, the cocoon and the thoracic spiracular gills (i.e., respiratory organs). The gills link the pupa with the mature larva in which each developing gill (i.e., the gill histoblast) manifests as a dark spot on the side of the thorax.

The pupa is shaped like the adult, with an arched or domed thorax and nine visible abdominal segments (Fig. 15.3.3A). In female pupae, the antennal sheaths reach the posterior margin of the head, but in male pupae, they extend only half to three-quarters of the distance to the posterior margin. The microsculpture of the cuticle covering the dorsum of the head and thorax has taxonomic value. It can be smooth, finely wrinkled, or heavily rugose and can bear different densities of rounded, pointed, or elongated **microtubercles**, often with their own finer granules. Various **trichomes** (i.e., fine sensory hairs) arise from the head and thorax—typically a few pairs on the head and four to seven pairs on the thorax.

The paired spiracular **gills** arise from the thorax and are spectacular in their variability among species (Fig. 15.3.3). They consist of one or more filamentous or inflated branches of various lengths and surface sculpturing. The number of branches among Mediterranean species varies from 1 or 2 to more than 70 per gill, although the typical number is 4–16. Variation in number within a species is typically minimal except in species with more than 16 filaments.

The pupal abdomen bears numerous hooks, spines, setae, and combs for holding the pupa in its cocoon. In most species, a pair of **terminal spines** (Fig. 15.3.3A) arises from the ninth abdominal segment, and each is long and slender (e.g., most prosimuliines) or short (e.g., most species of *Simulium*). Setae of the pleural region of the eighth and ninth segments are typically unbranched, but are anchor- or grapnel-shaped in the genus *Metacnephia* and some members of the subgenus *Hellichiella* (Fig. 15.3.3C). Other sets of hooks and setae tend to be rather uniform across taxa. An additional abdominal character of taxonomic value is the striate pleural membrane that contains large **pleurites** on the fourth and fifth segments of the genera *Helodon*, *Prosimulium*, and *Urosimulium* (Fig. 15.3.3A). A large pleurite is present on the fifth (but not fourth) segment of the genus *Levitinia*. Large pleurites are absent in other taxa of the Mediterranean Basin.

A silk **cocoon** is characteristic of all species of black flies and provides a wealth of taxonomic characters. Prosimuliines and some simuliines (*Greniera*) have a shapeless, saclike cocoon covering all or part of the pupa (Fig. 15.3.3B), whereas the remaining Mediterranean genera (*Metacnephia* and *Simulium*) have a well-formed cocoon with somewhat rigid walls and either fine or coarse weaving. Well-formed cocoons come in three shapes: slipper

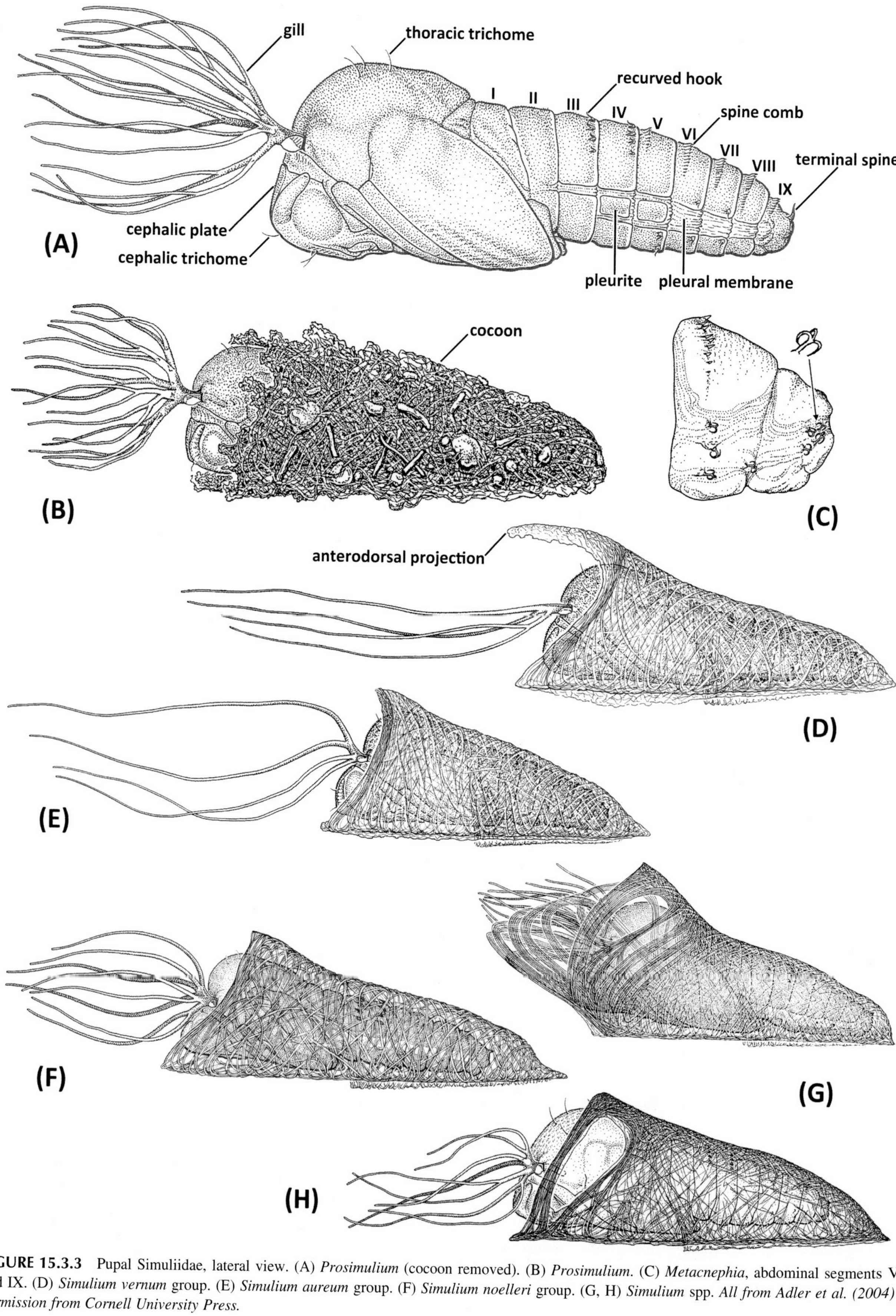

FIGURE 15.3.3 Pupal Simuliidae, lateral view. (A) *Prosimulium* (cocoon removed). (B) *Prosimulium*. (C) *Metacnephia*, abdominal segments VIII and IX. (D) *Simulium vernum* group. (E) *Simulium aureum* group. (F) *Simulium noelleri* group. (G, H) *Simulium* spp. *All from Adler et al. (2004) by permission from Cornell University Press.*

(Fig. 15.3.3D–F, H), shoe (i.e., with a short collar), and boot (i.e., with a high collar) (Fig. 15.3.3G). Additional features of well-formed cocoons include an anterodorsal projection (Fig. 15.3.3D) of various lengths, lateral fenestrations (windows) of various sizes (Fig. 15.3.3G, H), and dorsal incisions.

Material preparation and preservation

The intent of a particular study will determine the sampling and fixation procedures. To provide the most comprehensive picture of the simuliid fauna in a habitat, forceps can be used to hand pick larvae and pupae from the range of available substrates (e.g., leaves, rocks, trailing grasses, and discarded plastic) in different current velocities. Efforts should be made to collect both larvae and pupae to enhance the probability of accurate species identification. Studies designed to collect a broader range of aquatic insects often rely on techniques, such as kick nets, that are likely to yield few simuliid pupae.

When species-level identifications are critical, specific collecting procedures might be needed (Adler et al., 2004). Fixing larvae and pupae in modified Carnoy's solution provides the best specimens for morphological identification and simultaneously fixes larvae for chromosomal analysis of the giant polytene chromosomes. Chromosomal analyses require a subsequent staining procedure (Adler et al., 2016b). If molecular analyses are desired, placing larvae and pupae directly into 95% or 100% ethanol is best. Adults can be reared from pupae to provide additional character sources for identification. The procedure is straightforward. Pupae are lifted from the substrate by using fine forceps to grasp the margin of the cocoon. They are then held at room temperature in tubes with a damp absorbent substrate or in Petri dishes with moistened filter paper. Adults typically emerge within a week and are held in the dark for about 24 h before being dispatched in ethanol or by freezing.

The following identification keys rely on minute structures, particularly of the larval head and pupal thorax (Figs. 15.3.4–15.3.15), but they are written to minimize slide mounting of structures. Thus most identifications can be made with a stereomicroscope and a good light source. Nonetheless, some characters, such as those of the larval

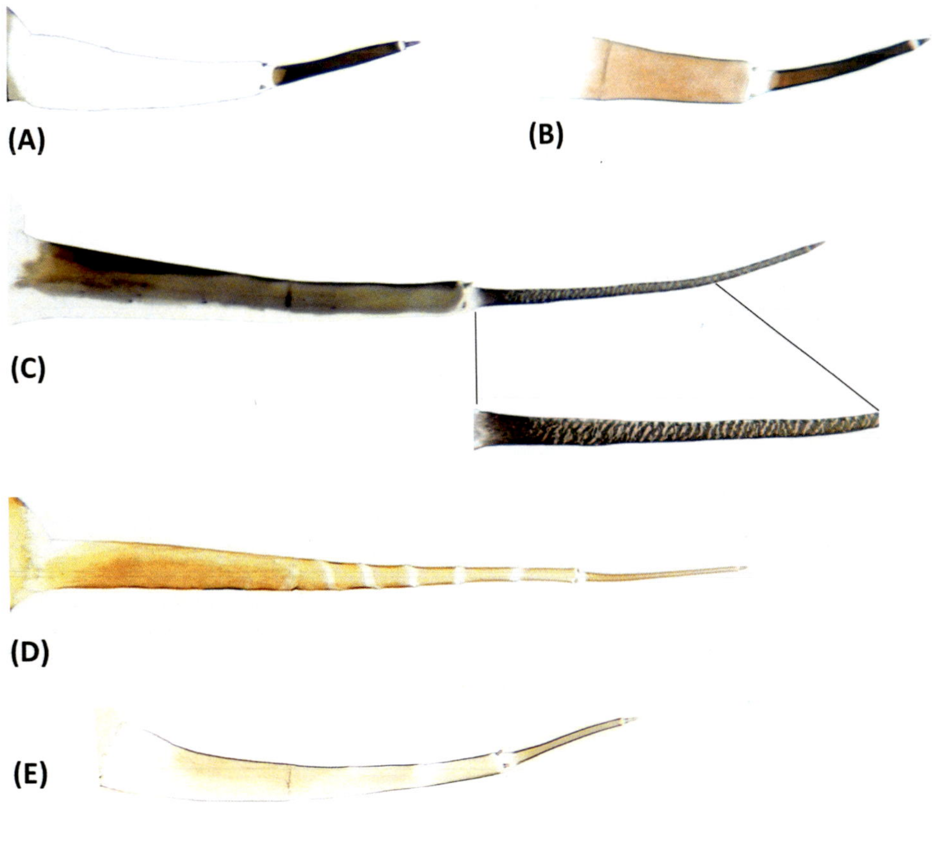

FIGURE 15.3.4 Antennae of larval Simuliidae. (A) *Prosimulium.* (B) *Urosimulium.* (C) *Greniera fabri.* (D) *Simulium* (*Hellichiella*). (E) *Simulium ornatum* complex.

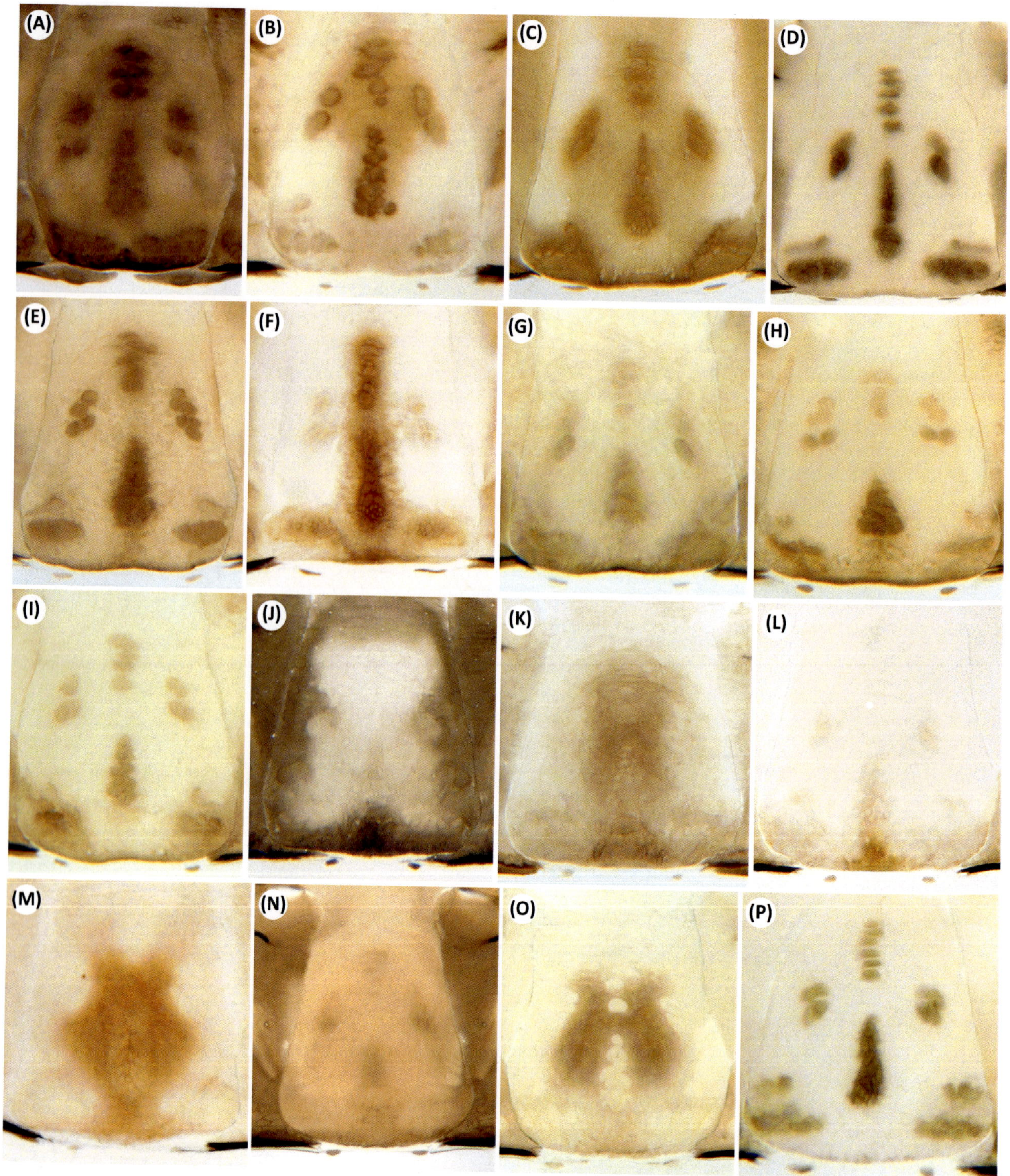

FIGURE 15.3.5 Frontoclypeal apotomes of larval Simuliidae, dorsal view. (A) *Prosimulium rachiliense*. (B) *Simulium erythrocephalum*. (C) *Simulium petricolum*. (D) *Simulium rubzovianum*. (E) *Simulium lundstromi*. (F) *Simulium ruficorne* complex. (G) *Simulium armoricanum*. (H) *Simulium costatum*. (I) *Simulium vernum* complex. (J) *Simulium aureofulgens*. (K) *Simulium noelleri* complex. (L) *Simulium reptantoides* complex. (M) *Simulium reptans* complex. (N) *Simulium monticola*. (O) *Simulium posticatum*. (P) *Simulium equinum*.

hypostoma might be easier to visualize if slide-mounted. For larvae, the identification key is based on final (mature) instars, although identifications of earlier instars can often be accomplished, either with the key or by association with mature larvae. Mature gill histoblasts can be useful for larval identification and typically require simple

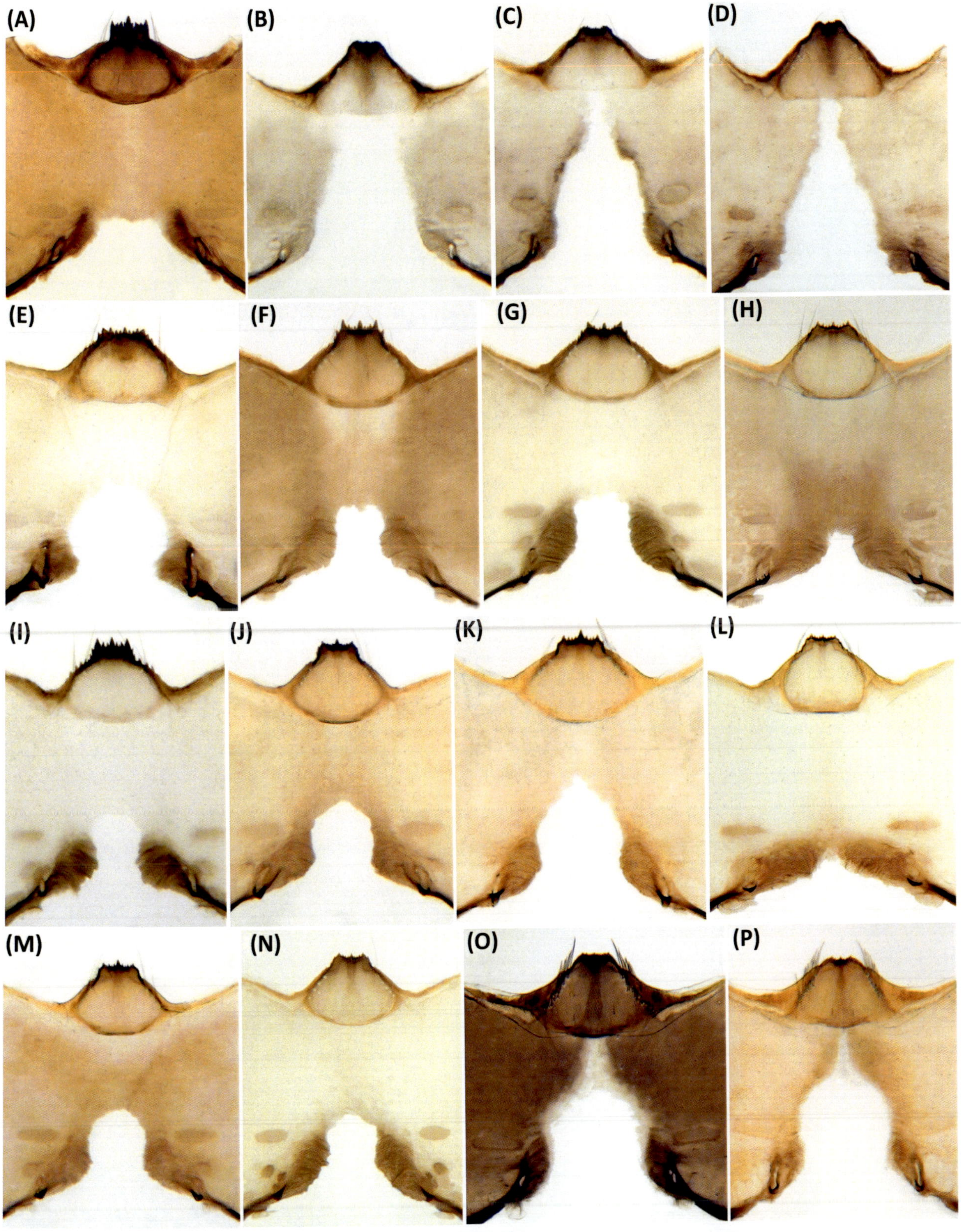

FIGURE 15.3.6 Head capsules of larval Simuliidae, ventral view; subesophageal ganglia not shown. (A) *Prosimulium tomosvaryi*. (B) *Metacnephia blanci*. (C) *Metacnephia nuragica*. (D) *Metacnephia phrygiensis*. (E) *Simulium erythrocephalum*. (F) *Simulium petricolum*. (G) *Simulium rubzovianum*. (H) *Simulium lundstromi*. (I) *Simulium ruficorne* complex. (J) *Simulium armoricanum*. (K) *Simulium carthusiense*. (L) *Simulium costatum*. (M) *Simulium cryophilum* complex. (N) *Simulium vernum* complex. (O) *Simulium aureofulgens*. (P) *Simulium hispaniola*.

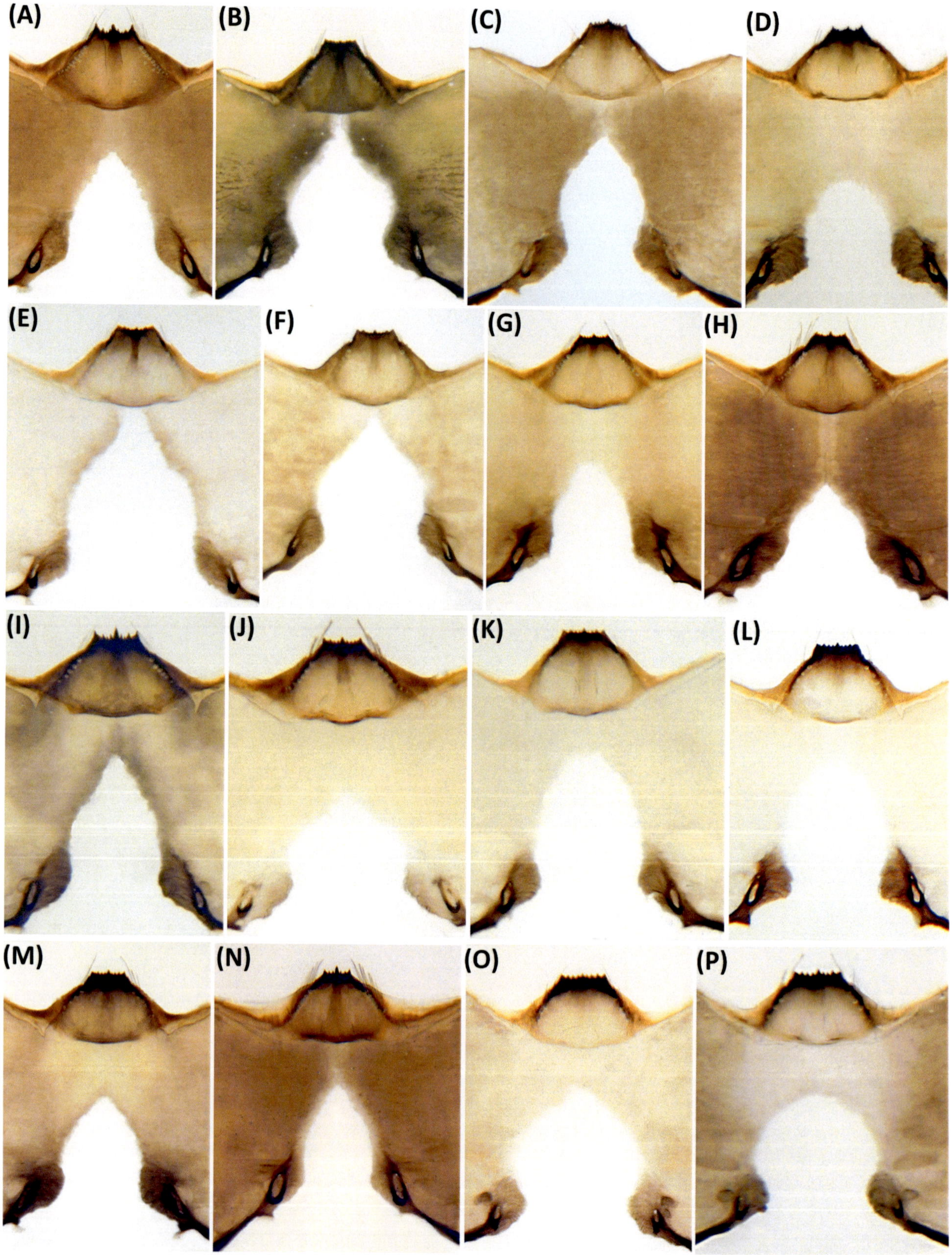

FIGURE 15.3.7 Head capsules of larval Simuliidae, ventral view; subesophageal ganglia not shown. (A) *Simulium bezzii* complex. (B) *Simulium bukovskii*. (C) *Simulium noelleri* complex. (D) *Simulium ornatum* complex. (E) *Simulium colombaschense*. (F) *Simulium reptantoides* complex. (G) *Simulium argyreatum*. (H) *Simulium monticola*. (I) *Simulium variegatum*. (J) *Simulium xanthinum*. (K) *Simulium bergi*. (L) *Simulium posticatum*. (M) *Simulium* indet. (N) *Simulium galloprovinciale*. (O) *Simulium equinum*. (P) *Simulium pseudequinum*.

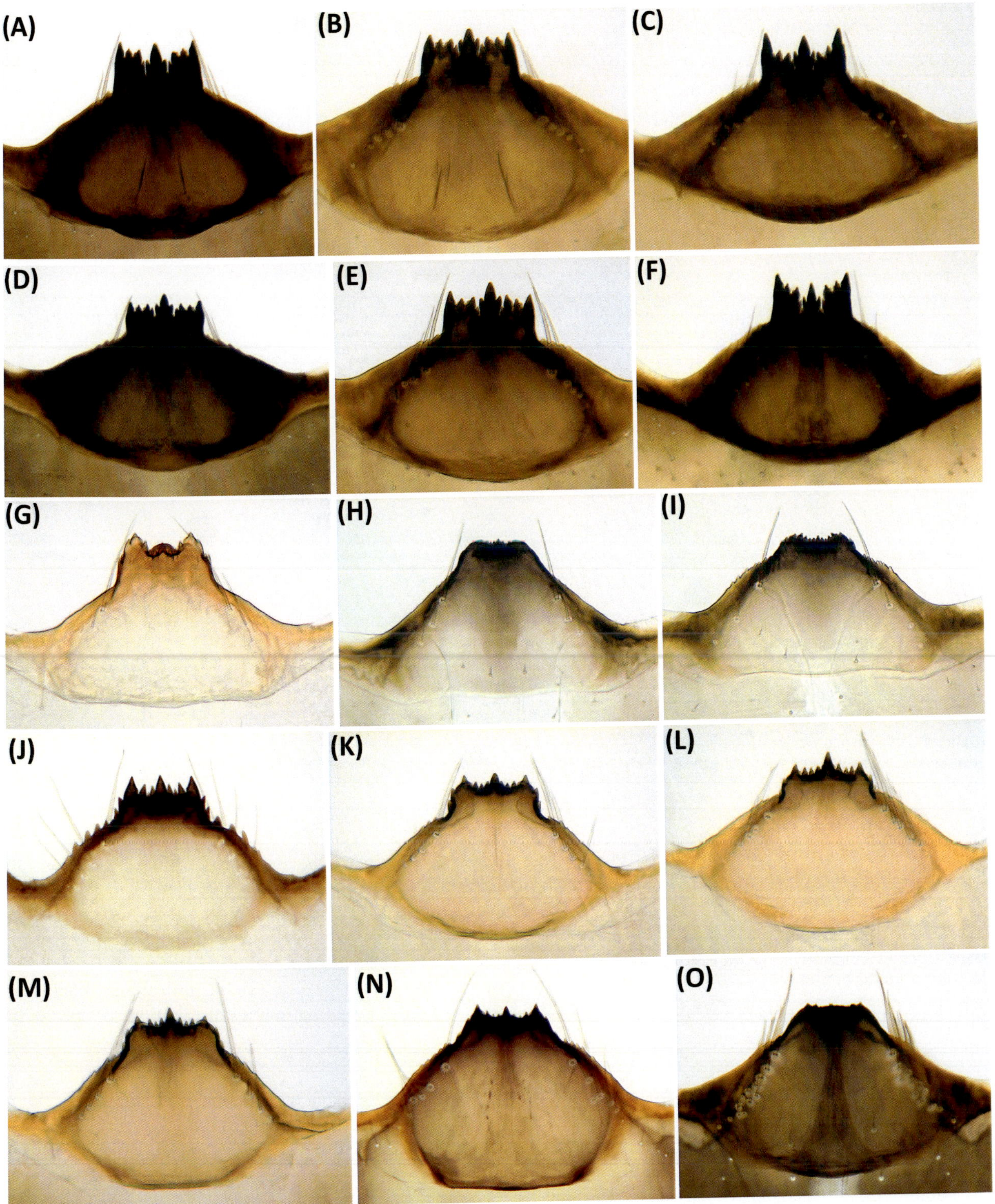

FIGURE 15.3.8 Hypostomata of larval Simuliidae, ventral view. (A) *Prosimulium anatoliense*. (B) *Prosimulium italicum*. (C) *Prosimulium latimucro*. (D) *Prosimulium rachiliense*. (E) *Prosimulium tomosvaryi*. (F) *Urosimulium faurei*. (G) *Greniera fabri*. (H) *Metacnephia blanci*. (I) *Metacnephia nuragica*. (J) *Simulium ruficorne* complex. (K) *Simulium armoricanum*. (L) *Simulium carthusiense*. (M) *Simulium cryophilum* complex. (N) *Simulium djafarovi*. (O) *Simulium aureofulgens*.

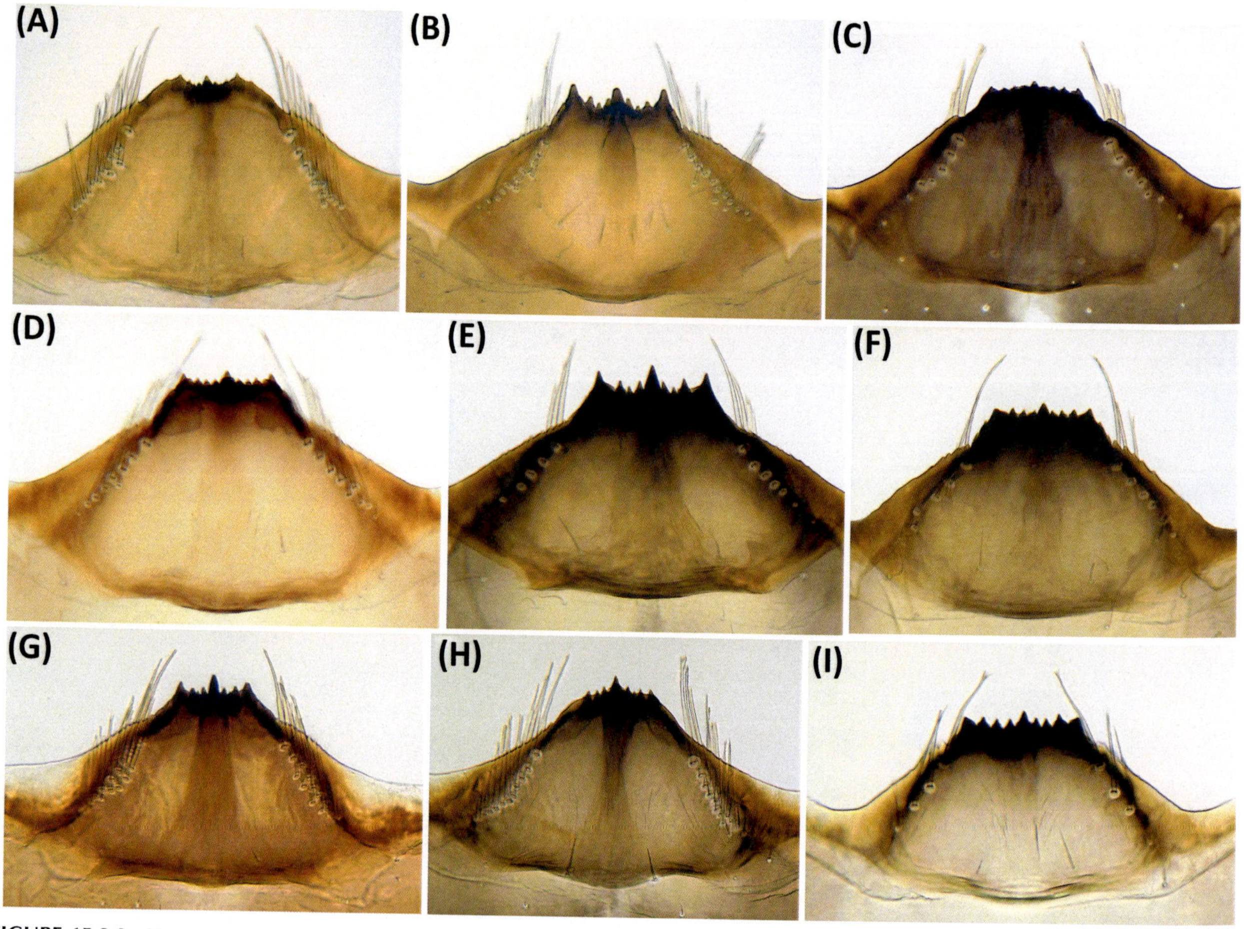

FIGURE 15.3.9 Hypostomata of larval Simuliidae, ventral view. (A) *Simulium hispaniola.* (B) *Simulium bezzii* complex. (C) *Simulium bukovskii.* (D) *Simulium argyreatum.* (E) *Simulium variegatum.* (F) *Simulium* indet. (G) *Simulium galloprovinciale.* (H) *Simulium albellum* group. (I) *Simulium pseudequinum.*

dissection. The clear overlying cuticle is cut with fine needles and the dark gill histoblast is lifted by its base and placed in a drop of 50% acetic acid to facilitate uncurling. Uncurling is particularly effective if the larvae have been fixed in Carnoy's solution.

The following keys are conservative; if a species is not reliably diagnosable as a larva or pupa across its entire Mediterranean range, it is not keyed beyond the most confident level. In these cases, additional characters from other life stages, larval chromosome patterns, or molecular sequences are needed for species-level identification. Nonetheless, geographic distributions can be useful guides for identification, with the caveat that not all distributions of species are fully known.

Keys to larvae and pupae of Simuliidae

Simuliidae: Genera (Larvae)

1 Labral fans absent. Head tapered anteriorly .... *Levitinia freidbergi* Beaucournu-Saguez & Braverman, 1987 [Golan]

1' Labral fans present (Fig. 15.3.1A), although often adducted. Head nearly parallel-sided, not tapered anteriorly (Fig. 15.3.1C) 2

2(1) Antenna with proximal and medial articles entirely free of pigment, contrasting with dark brown distal article (Figs. 15.3.1 and 15.3.4A), and postgenal cleft extended less than half distance from posterior tentorial pits to

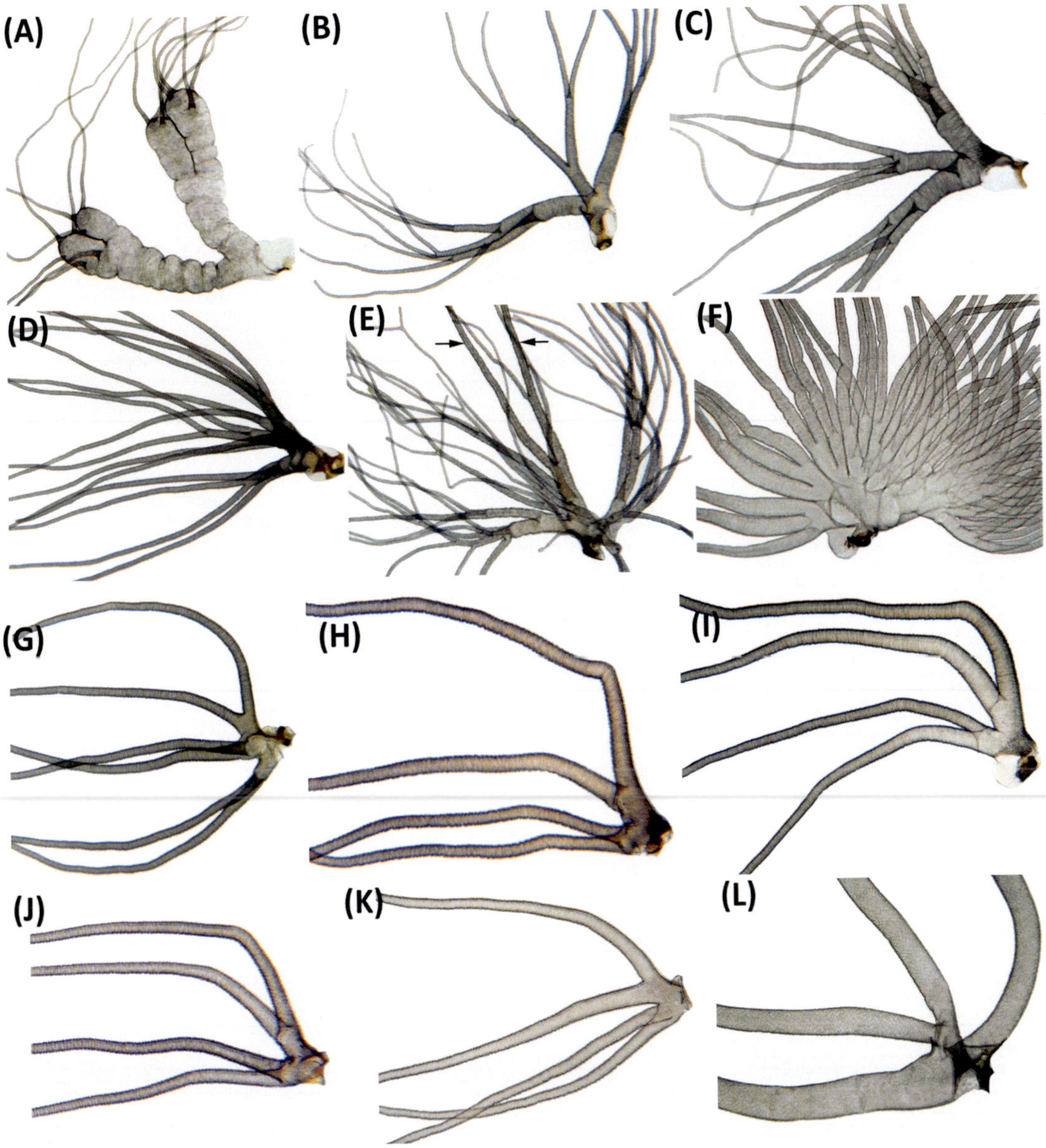

FIGURE 15.3.10 Gills of pupal Simuliidae (truncated distally), lateral view. (A) *Prosimulium albense*. (B) *Prosimulium anatoliense*. (C) *Prosimulium latimucro*. (D) *Prosimulium rachiliense*. (E) *Metacnephia phrygiensis*, arrows indicate diagnostic thick filaments. (F) *Metacnephia subalpina*. (G) *Simulium erythrocephalum*. (H) *Simulium angustipes*. (I) *Simulium rubzovianum*. (J) *Simulium angustitarse*. (K) *Simulium lundstromi*. (L) *Simulium ruficorne* complex.

hypostomal groove (Figs. 15.3.1C and 15.3.6A). Cervical sclerites encompassed by postocciput (Figs. 15.3.1C and 15.3.5A) 3

2' Antenna with proximal and medial article variously pigmented (Fig. 15.3.4B–E); if both articles entirely free of pigment and distal article dark brown, then postgenal cleft extended to hypostomal groove (Fig. 15.3.6B–D). Cervical sclerites either free (Fig. 15.3.5B–P) or encompassed by postocciput 4

3(2) Gill histoblast of 3 elongated, swollen, annulated tubes with few short apical thread-like filaments .... *Helodon laamii* (Beaucournu Saguez & Bailly-Choumara, 1981) [Morocco]

3' Gill histoblast of 14 to 27 stout or thread-like filaments (Fig. 15.3.10B–D), rarely arising from 2 elongated, swollen tubes (Fig. 15.3.10A) .... ***Prosimulium***

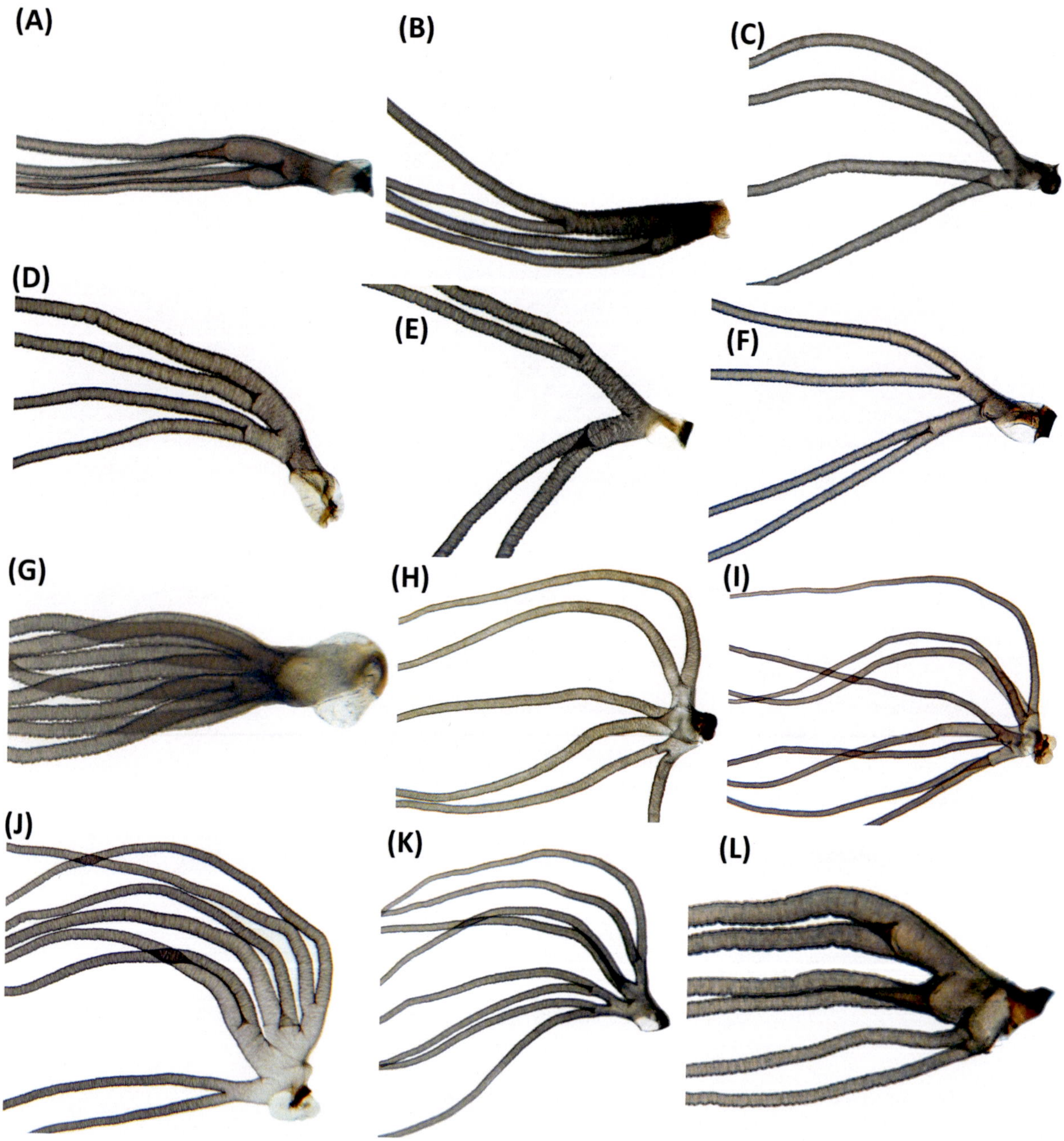

FIGURE 15.3.11 Gills of pupal Simuliidae (truncated distally), lateral view. (A) *Simulium armoricanum*. (B) *Simulium carthusiense*. (C) *Simulium costatum*. (D) *Simulium cryophilum* complex. (E) *Simulium djafarovi*. (F) *Simulium vernum* complex. (G) *Simulium aureofulgens*. (H) *Simulium bezzii* complex. (I) *Simulium noelleri* complex. (J) *Simulium ornatum* complex. (K) *Simulium reptans* complex. (L) *Simulium argyreatum*.

4(2) Hypostoma with all teeth uniformly minute and fewer than 7 sublateral setae (Figs. 15.3.2D and 15.3.8H, I). Postgenal cleft extended to hypostomal groove (Fig. 15.3.6B–D) .... ***Metacnephia***

4' Hypostoma with median and/or lateral teeth markedly larger than, and typically extended beyond, other teeth (Figs. 15.3.2E, 15.3.8F, G, J–N, and 15.3.9B–I); if in doubt, then more than 7 sublateral setae (Figs. 15.3.8O and 15.3.9A). Postgenal cleft extended various distances to hypostomal groove (Figs. 15.3.6E–P and 15.3.7) 5

5(4) Cervical sclerites encompassed by postocciput (as in Figs. 15.3.1C and 15.3.5A). Hypostoma with lateral teeth extended well beyond median tooth; paralateral teeth absent (Fig. 15.3.8F) .... ***Urosimulium***

FIGURE 15.3.12 Gills of pupal Simuliidae (truncated distally), lateral view. (A) *Simulium monticola*. (B) *Simulium variegatum*. (C) *Simulium xanthinum*. (D) *Simulium* indet. (E) *Simulium bergi*. (F) *Simulium posticatum*. (G) *Simulium margaritae*. (H) *Simulium balcanicum*, arrow indicates common petiole. (I) *Simulium equinum*. (J) *Simulium lineatum*. (K) *Simulium pseudequinum*. (L) *Simulium sergenti*.

5' Cervical sclerites not encompassed by postocciput (Fig. 15.3.5B–P). Hypostoma with lateral teeth extended posterior to or approximately to same level as median tooth; paralateral teeth present (Figs. 15.3.2C, E, 15.3.8G, J–O, and 15.3.9) 6

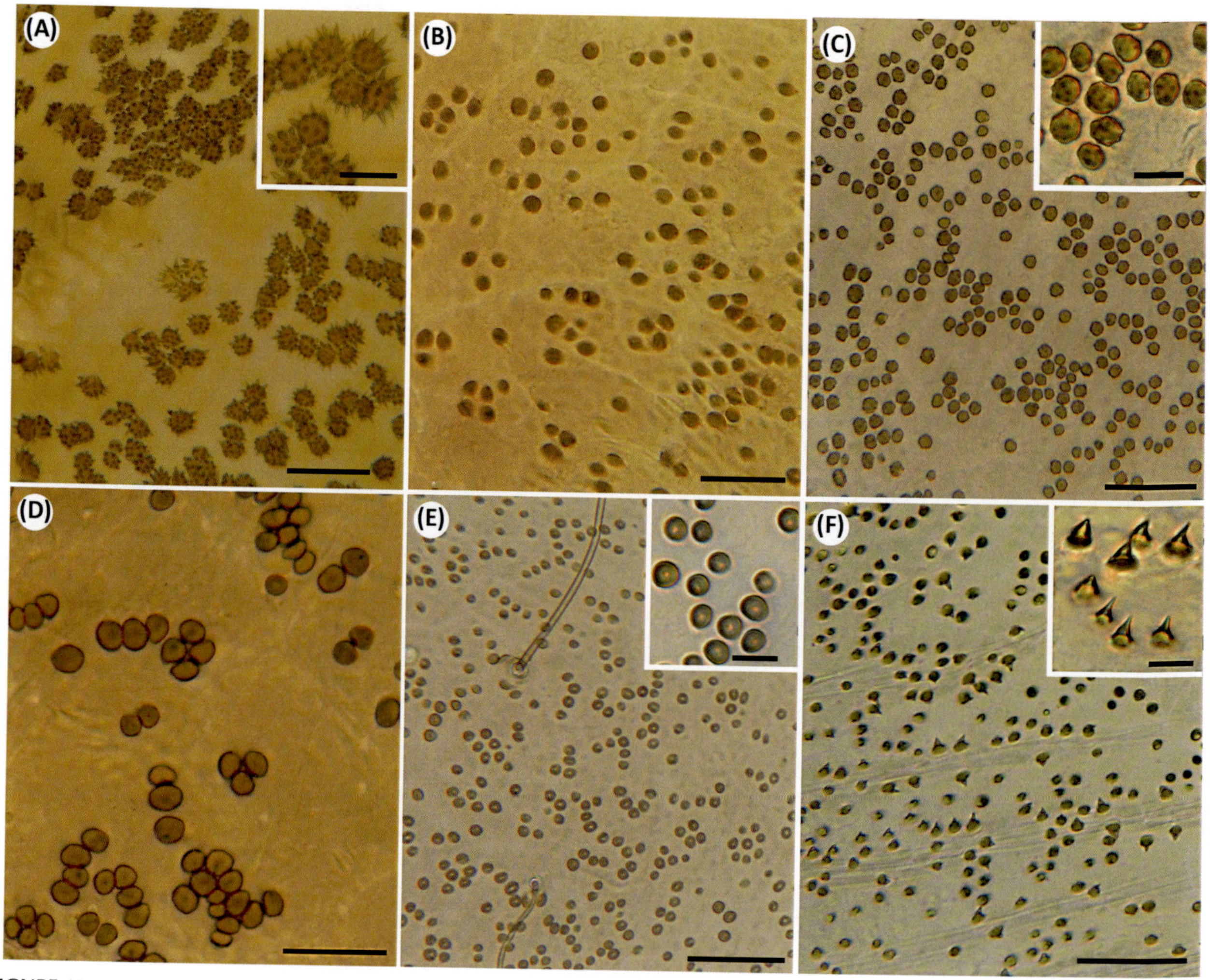

FIGURE 15.3.13 Thoracic surface sculpture of pupal Simuliidae, dorsal view of anterolateral region; large scale bars = 50 μm; small scale bars = 10 μm. (A) *Simulium sicanum*. (B) *Simulium kiritshenkoi*. (C) *Simulium carthusiense*. (D) *Simulium monticola*. (E) *Simulium urbanum*. (F) *Simulium trifasciatum*.

6(5) Hypostoma with teeth on 3 prominent lobes (i.e., 1 median tooth and 2 lateral teeth), with each lateral tooth bearing at least 1 sublateral and 1 paralateral tooth (Figs. 15.3.2C and 15.3.8G). Antenna (slide-mounted) with distal article often bearing fine, spiral banding (Fig. 15.3.4C). Rare .... ***Greniera***

6' Hypostoma with teeth not on 3 prominent lobes, although median and lateral teeth often prominent (Figs. 15.3.2E, 15.3.8J–O, and 15.3.9). Antenna with distal article often banded or spotted, but not bearing fine, spiral banding (Fig. 15.3.4D, E). Abundant .... ***Simulium***

Simuliidae: *Prosimulium*: Species (Mature larvae)

1 Hypostoma with lateral teeth extended well beyond median tooth (Fig. 15.3.2A) .... *Prosimulium petrosum* Rubtsov, 1955 [Lebanon]

1' Hypostoma with lateral teeth extended to about same level as, or posterior to, median tooth (Figs. 15.3.2B and 15.3.8A–E) 2

2(1) Hypostoma with median tooth extended beyond lateral teeth (Fig. 15.3.8B, D, E) 3

2' Hypostoma with median tooth extended to about same level as, or slightly posterior to, lateral teeth (Fig. 15.3.8A, C) 5

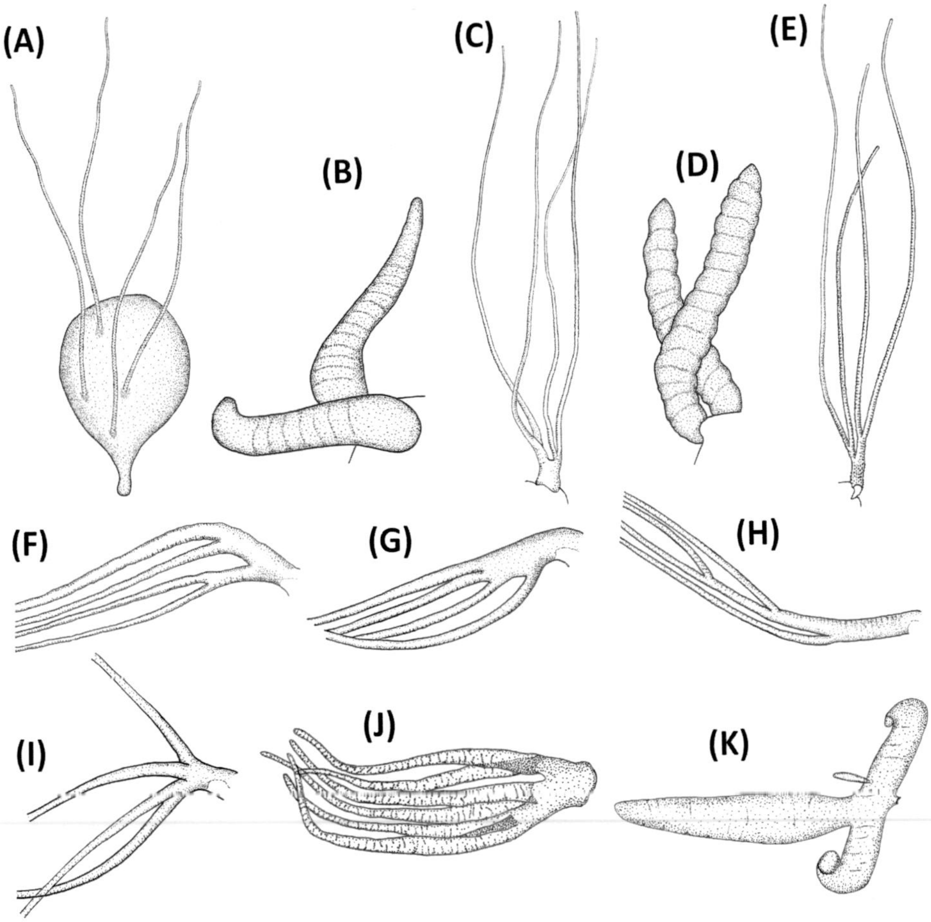

FIGURE 15.3.14 Gills of pupal Simuliidae. (A) *Simulium timondavidi*, ventral view. (B) *Simulium knidiri*, dorsal view of left gill. (C) *Simulium flexibranchium*, dorsal view of right gill. (D) *Simulium lamachi*, dorsal view of left gill. (E) *Simulium marsicanum*, dorsal view. (F) *Simulium brevidens*, lateral view of base. (G) *Simulium ichnusae*, lateral view of base. (H) *Simulium codreanui*, lateral view of base. (I) *Simulium fucense*, lateral view of base. (J) *Simulium continii*, lateral view. (K) *Simulium quadrifila*, lateral view.

3(2) Gill histoblast of 20 to 27 filaments .... *Prosimulium tomosvaryi* (Enderlein, 1921)
[Bosnia and Herzegovina. Croatia. France. Greece. Italy. Morocco. Portugal. Spain. Turkey]
3' Gill histoblast of 16 filaments 4
4(3) Hypostoma with median tooth extended well beyond other teeth (Fig. 15.3.8D) .... *Prosimulium calabrum* (Rivosecchi, 1966)
[Italy]
Prosimulium rachiliense (Djafarov, 1954) (complex)
[Greece. Morocco. Turkey]
4' Hypostoma with median tooth extended slightly beyond other teeth, giving anterior margin a rather straight appearance (Fig. 15.3.8B) .... *Prosimulium hirtipes* (Fries, 1824)
[Bosnia and Herzegovina. France. Italy (northern). Spain]
Prosimulium italicum (Rivosecchi, 1967)
[Italy]
5(2) Hypostoma with sublateral teeth extended to about same level with one another and slightly posterior to lateral teeth, not forming diagonal arrangement on each side; sublateral teeth extended well beyond denticles of median tooth (Fig. 15.3.8A) .... *Prosimulium anatoliense* (Adler & Şirin, 2015)
[Turkey]
5' Hypostoma with sublateral teeth extended to different levels and well posterior to lateral teeth, often forming diagonal arrangement on each side; sublateral teeth extended to about same level as denticles of median tooth (Fig. 15.3.8C) 6
6(5) Gill histoblast of 2 stout trunks, each giving rise to 8 apical or subapical thread-like filaments (Fig. 15.3.10A) .... *Prosimulium albense* (Rivosecchi, 1961)
[Algeria. Italy]

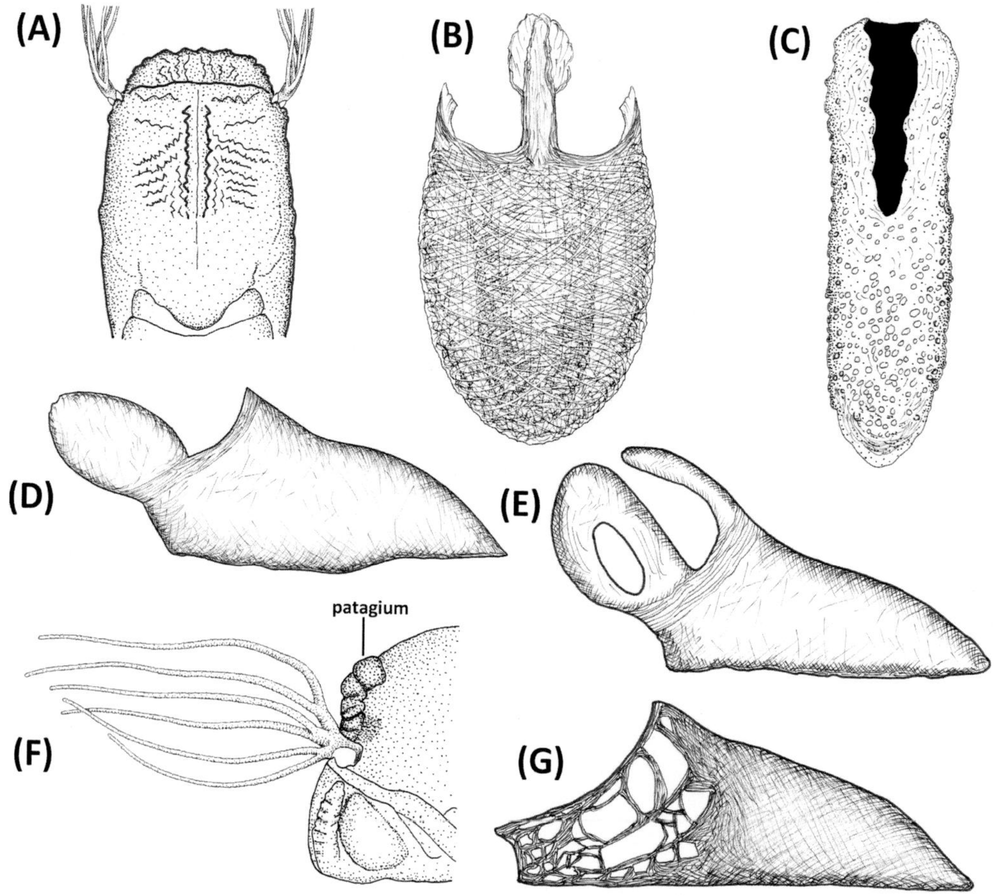

FIGURE 15.3.15 Pupae and cocoons of Simuliidae. (A) *Simulium gracilipes*, pupal head and thorax, dorsal view (gills truncated). (B) *Simulium bertrandi*, cocoon, dorsal view. (C) *Simulium ibericum*, cocoon dorsal view; after Crosskey & Santos Grácios (1985). (D) *Simulium aureofulgens*, cocoon, lateral view. (E) *Simulium argenteostriatum*, cocoon, lateral view. (F) *Simulium variegatum*, anterior portion of pupa, lateral view. (G) *Simulium bezzii* complex, cocoon, lateral view.

6' Gill histoblast of 16 thread-like filaments arising from 3 or more short basal trunks (Fig. 15.3.10C) .... *Prosimulium latimucro* (Enderlein, 1925) (complex)
[Bosnia and Herzegovina. France. Italy. Montenegro. Morocco. Portugal. Spain]

Simuliidae: *Urosimulium*: Species (Mature larvae)

1 Hypostoma with median tooth typically extended slightly beyond sublateral teeth (Fig. 15.3.8F) .... *Urosimulium faurei* (Bernard, Grenier & Bailly-Choumara, 1972)
[Algeria. Morocco. Spain]
1' Hypostoma with median tooth extended to about same level as, or posterior to, sublateral teeth 2
2(1) Hypostoma with 5 sublateral setae .... *Urosimulium aculeatum* (Rivosecchi, 1963)
[Balearic Islands: Majorca. Italy: Sicily, Sardinia]
2' Hypostoma with 3 sublateral setae .... *Urosimulium juccii* Contini, 1966
[Algeria(?). Italy: Sardinia. Tunisia(?)]

Simuliidae: *Greniera*: Species (Mature larvae)

1 Antenna (slide-mounted) with distal article bearing fine, pigmented spiral bands (Fig. 15.3.4C) .... *Greniera fabri* (Doby & David, 1959)
[Algeria. France. Italy. Morocco. Spain]

1' Antenna (slide-mounted) with distal article not bearing pigmented spiral bands .... *Greniera dobyi* (Beaucournu-Saguez & Braverman, 1987)
[Golan]

Simuliidae: *Metacnephia*: Species (Mature larvae)

1 Gill histoblast of 65 or more filaments .... *Metacnephia* "Crete" Procunier, 1982
[Greece: Crete]
1' Gill histoblast of fewer than 65 filaments 2
2(1) Gill histoblast of 18–28 filaments .... *Metacnephia persica* (Rubtsov, 1940)
[Lebanon]
2' Gill histoblast of fewer than 18 or more than 29 filaments 3
3(2) Postgenal cleft meeting hypostoma broadly, with apical width about equal to or greater than distance between lateral teeth (Fig. 15.3.6B) .... *Metacnephia blanci* (Grenier & Theodorides, 1953)
[Algeria. France. Italy: Sicily. Morocco. Portugal. Spain. Tunisia]
3' Postgenal cleft meeting hypostoma narrowly, with apical width less than distance between lateral teeth (Fig. 15.3.6C, D) 4
4(3) Postgenal cleft evenly tapered .... *Metacnephia sardoa* (Rivosecchi & Contini, 1965)
[Italy: Sardinia]
4' Postgenal cleft abruptly narrowed at about ½ to ¾ distance from posterior tentorial pits to hypostomal groove (Fig. 15.3.6C, D) 5
5(4) Gill histoblast of fewer than 18 filaments .... *Metacnephia nuragica* (Rivosecchi, Raastad & Contini, 1975)
[Italy: Sardinia. Morocco(?). Portugal. Spain]
5' Gill histoblast of 30 or more filaments 6
6(5) Gill histoblast with ventral filaments inflated, about twice diameter of other filaments (Fig. 15.3.10F) .... *Metacnephia subalpina* (Rubtsov, 1956)
[Turkey]
6' Gill histoblast with dorsal and ventral filaments of nearly same thickness (Fig. 15.3.10E) 7
7(6) Gill histoblast of 30 to 45 filaments; all filaments uniformly fine .... *Metacnephia nigra* (Rubtsov, 1940)
[Turkey]
7' Gill histoblast of more than 45 filaments; pair of filaments in middle group thicker than all others (Fig. 15.3.10E) .... *Metacnephia phrygiensis* (Şirin & Adler, 2015)
[Turkey]

Simuliidae: *Simulium*: Species (Mature larvae)

1 Abdomen, in lateral view, with ninth segment bearing pair of conical ventral tubercles (Fig. 15.3.1E) 2
1' Abdomen, in lateral view, with ninth segment bearing little more than wrinkled cuticle, without pair of conical ventral tubercles (Fig. 15.3.1A, B) 14
2(1) Antenna with medial article bearing 4 or more pale annulations (Fig. 15.3.4D). Gill histoblast of 10 slender filaments .... *Simulium saccai* (Rivosecchi, 1967)
[Italy]
2' Antenna with medial article bearing at most 1 or 2 pale markings (Fig. 15.3.4E). Gill histoblast of 2 or 4 slender filaments or inflated tubes 3
3(2) Postgenal cleft minute, extended less than 1/4 distance from posterior tentorial pits to hypostomal groove (Fig. 15.3.6H, L) 4
3' Postgenal cleft deeper, extended more than 1/4 distance from posterior tentorial pits to hypostomal groove (Fig. 15.3.6E–G, I–K, M, N) 7
4(3) Gill histoblast of 2 inflated tubes of roughly same diameter (subgenus *Rubzovia*) (Fig. 15.3.14B, D) .... *Simulium knidirii* (Giudicelli & Thiery, 1985)
[Morocco]
Simulium lamachi (Doby & David, 1960)
[Morocco. Spain]
4' Gill histoblast of 4 slender or inflated filaments (Figs. 15.3.10J, K and 15.3.11C) 5

5(4) Frontoclypeal apotome with posteromedian head spot about as long as wide, in form of equilateral triangle (Fig. 15.3.5H) .... *Simulium costatum* (Friederichs, 1920)
[Algeria. Bosnia and Herzegovina. Croatia. France. Italy. Montenegro. Morocco. Spain. Turkey]
5' Frontoclypeal apotome with posteromedian head spot longer than wide, in form of isosceles triangle (Fig. 15.3.5E) 6
6(5) Body length 10 mm or greater .... *Simulium fucense* (Rivosecchi, 1962)
[Italy: Abruzzo, Lazio, Umbria]
6' Body length 9 mm or less (*Simulium ruficorne* group, in part) .... *Simulium angustitarse* (Lundström, 1911)
[Algeria. Bosnia and Herzegovina. Croatia. France. Greece. Italy. Montenegro. Morocco. Portugal. Slovenia. Spain. Turkey]
Simulium ibleum (Rivosecchi, 1966) [Algeria. Cyprus. Greece: Crete. Lebanon. Libya. Morocco. Spain(?). Turkey. Tunisia]
Simulium lundstromi (Enderlein, 1921) [Algeria. Croatia. France. Italy. Lebanon. Morocco. Portugal. Spain. Turkey]
Simulium pinhaoi Santos Grácio, 1985
[Portugal]
7(3) Gill histoblast of 6 filaments (Fig. 15.3.10G) .... *Simulium erythrocephalum* (De Geer, 1776) (in part)
[Croatia. France. Greece. Italy. Portugal. Slovenia. Spain. Tunisia(?). Turkey]
7' Gill histoblast of 4 filaments (Figs. 15.3.10H, L and 15.3.11A, B, D–F), rarely arising from disk-like base (Fig. 15.3.14A) 8
8(7) Frontoclypeal apotome with anterolateral head spots faint or absent; anteromedian and posteromedian spots separated by gap less than length of anteromedian spot, forming thick or thin median line of spots (Fig. 15.3.5F). Hypostoma with lateral serrations prominent (Fig. 15.3.8J). Gill histoblast with filaments inflated at least basally (Fig. 15.3.10L) .... *Simulium ruficorne* (Macquart, 1838) (complex)
[Algeria. Egypt. Israel. Jordan. Lebanon. Libya. Malta. Morocco. Portugal. Spain. Syria. Tunisia]
8' Frontoclypeal apotome with anterolateral head spots distinct; anteromedian and posteromedian spots separated by gap of various lengths (Fig. 15.3.5C, D, G, I). Hypostoma with lateral serrations minute (Fig. 15.3.8K–N). Gill histoblast with filaments thread-like (Figs. 15.3.3D, E, 15.3.10H, I, and 15.3.11A, B, D–F) or arising from short, round, disk-like base (Fig. 15.3.14A) 9
9(8) Frontoclypeal apotome with posteromedian head spot elongate; anteromedian and posteromedian spots separated by gap less than length of anteromedian spot (Fig. 15.3.5C, D) (*Simulium aureum* group) 10
9' Frontoclypeal apotome with posteromedian head spot subtriangular; anteromedian and posteromedian spots separated by gap about equal to or greater than length of anteromedian spot (Fig. 15.3.5G, I) (*Simulium vernum* group, in part) 11
10(9) Head capsule infused with brown pigment in addition to typical spots (Fig. 15.3.5C) .... *Simulium krymense* (Rubtsov, 1956)
[Greece]
Simulium mellah (Giudicelli & Bouzidi, 2000) (in part)
[Algeria. Morocco]
Simulium petricolum (Rivosecchi, 1963)
[Algeria. Bosnia and Herzegovina. Cyprus. France. Greece. Italy. Libya. Morocco. Portugal. Spain]
10' Head capsule typically yellowish brown or pale yellowish, contrasting with typical spots; without infuscation (Fig. 15.3.5D) .... *Simulium angustipes* (Edwards, 1915)
[Algeria. Bosnia and Herzegovina. Croatia. France. Italy. Lebanon. Morocco. Portugal. Spain. Tunisia. Turkey]
Simulium aureum (Fries, 1824)
[France. Italy. Portugal. Spain]
Simulium flexibranchium (Crosskey, 2001)
[Greece: Rhodes, Crete, Kithira, Naxos]
Simulium mellah (Giudicelli & Bouzidi, 2000) (in part)
[Algeria. Morocco]
Simulium rubzovianum (Sherban, 1961)
[Algeria. Balearic Islands: Majorca, Minorca. Bosnia and Herzegovina. Croatia. Cyprus. France. Greece. Israel. Italy. Jordan. Libya. Malta. Montenegro. Morocco. Portugal. Slovenia. Spain. Tunisia. Turkey]
Simulium velutinum (Santos Abreu, 1922) (complex, including 3 undescribed species: "K" Leonhardt, 1985; "3" Adler, Cherairia, Arigue, Samraoui & Belqat, 2015; and "5" Adler, Cherairia, Arigue, Samraoui & Belqat, 2015)
[Algeria. Morocco. Spain]

11(9) Gill histoblast of 1 short, disk-like, round club with 4 slender filaments (Fig. 15.3.14A) .... *Simulium timondavidi* (Giudicelli, 1961)

[France: Corsica]

11' Gill histoblast of 4 slender filaments arising from 1 or 2 petioles or stalks (Figs. 15.3.3D and 15.3.11A, B, D–F) 12

12(11) Hypostoma with median tooth extended well beyond lateral teeth (Fig. 15.3.8L, M) 13

12' Hypostoma with median tooth extended only slightly beyond lateral teeth or to level of, or slightly posterior to, lateral teeth (Fig. 15.3.8K, N) .... *Simulium armoricanum* (Doby & David, 1961)

[Portugal. Spain]

Simulium brevidens (Rubtsov, 1956)

[Algeria(?). Bosnia and Herzegovina. France. Italy. Montenegro. Morocco(?). Spain]

Simulium cryophilum (Rubtsov, 1959) (complex) (in part)

[Algeria. Balearic Islands: Majorca. Bosnia and Herzegovina. Croatia. Cyprus. France. Greece. Italy. Lebanon. Montenegro. Morocco. Portugal. Slovenia. Spain. Tunisia. Turkey]

Simulium djafarovi (Rubtsov, 1962)

[Turkey]

Simulium ichnusae (Rivosecchi & Contini, 1994)

[Italy: Sardinia]

Simulium marsicanum (Rivosecchi, 1962)

[Italy]

Simulium urbanum (Davies, 1966)

[Spain]

Simulium vernum (Macquart, 1826) (complex)

[Bosnia and Herzegovina. Croatia. France. Italy. Morocco. Portugal. Slovenia. Spain. Tunisia. Turkey]

Simulium "IL-8" Adler, Belqat, González, Pérez & Seitz, 2016

[Portugal]

13(12) Gill histoblast with 1 pair of filaments on long stalk from which 2 other filaments arise independently (Fig. 15.3.14H) .... *Simulium codreanui* (Sherban, 1958)

[Bosnia and Herzegovina. Turkey]

13' Gill histoblast with both pairs of filaments arising on independent stalks (Fig. 15.3.11B, D) .... *Simulium bertrandi* (Grenier & Dorier, 1959)

[Bosnia and Herzegovina. France. Italy. Montenegro. Spain]

Simulium carthusiense (Grenier & Dorier, 1959)

[France. Italy. Morocco. Spain]

Simulium cryophilum (Rubtsov, 1959) (complex) (in part)

[Algeria. Balearic Islands: Majorca. Bosnia and Herzegovina. Croatia. Cyprus. France. Greece. Italy. Lebanon. Montenegro. Morocco. Portugal. Slovenia. Spain. Tunisia. Turkey]

14(1) Gill histoblast of short, wrinkled, finger-like tubes (Fig. 15.3.12H–L). Postgenal cleft strongly bowed outward, sometimes nearly circular or with small apical point (Fig. 15.3.7O, P), although sometimes difficult to discern if head capsule pale and subesophageal ganglion pigmented (subgenus *Wilhelmia*) 15

14' Gill histoblast of slender filaments (Figs. 15.3.11G–L and 15.3.12A–G). Postgenal cleft of various shapes, but often not strongly bowed outward or circular (Fig. 15.3.7A–N) 20

15(14) Gill histoblast of 4 or 6 tubes (including 2 basal tubes that lie close to pupa) 16

15' Gill histoblast of 8 tubes (including 2 basal tubes that lie close to pupa) 17

16(15) Gill histoblast of 4 tubes, 3 of which are inflated and robust and 1 of which is small and thin (Fig. 15.3.14K) .... *Simulium quadrifila* (Grenier, Faure & Laurent, 1957)

[Algeria. Morocco. Spain]

16' Gill histoblast of 6 tubes, 4 of which are inflated and robust and 2 of which are small and thin (Fig. 15.3.12L) .... *Simulium sergenti* (Edwards, 1923)

[Algeria. Morocco. Portugal. Spain. Tunisia]

17(15) Gill histoblast with tubes narrowest at their attachment points to inflated basal tubes (Fig. 15.3.12K) .... *Simulium bravermani* (Beaucournu-Saguez, 1986)

[Israel]

Simulium paraequinum Puri, 1933

[Croatia. Greece. Israel. Italy. Jordan. Lebanon. Turkey]
Simulium pseudequinum (Séguy, 1921)
[Algeria. Bosnia and Herzegovina. Croatia. Cyprus. France. Greece. Israel. Italy. Jordan. Lebanon. Libya. Montenegro. Morocco. Portugal. Slovenia. Spain. Tunisia. Turkey]
17' Gill histoblast with 6 filaments widest at their attachment points to inflated basal tubes (Fig. 15.3.12H–J) 18
18(17) Gill histoblast with 6 upright, strongly inflated tubes about 1/3 to more than ½ width of each of 2 inflated basal tubes (Fig. 15.3.12I) .... *Simulium equinum* (L., 1758)
[Bosnia and Herzegovina. Croatia. France. Italy. Montenegro. Morocco. Portugal. Slovenia. Spain. Turkey]
18' Gill histoblast with 6 upright tubes less than 1/3 width of each of 2 inflated basal tubes (Fig. 15.3.12H, J) 19
19(18) Gill histoblast with 1 pair of tubes arising from base on short common petiole (Fig. 15.3.12H) .... *Simulium balcanicum* (Enderlein, 1924)
[Albania. Bosnia and Herzegovina. Croatia. Greece. Italy. Montenegro. Slovenia. Turkey]
19' Gill histoblast with all tubes arising independently from base (Fig. 15.3.12J) .... *Simulium golani* (Beaucournu-Saguez & Braverman, 1977)
[Golan]
Simulium lineatum (Meigen, 1804)
[Bosnia and Herzegovina. Croatia. France. Italy. Montenegro. Portugal. Slovenia. Spain]
Simulium turgaicum (Rubtsov, 1940)
[Bosnia and Herzegovina. Lebanon. Slovenia. Turkey]
20(14) Abdomen with 4 dark, posterodorsal sclerites (Fig. 15.3.1B) (subgenus *Crosskeyellum*) .... *Simulium gracilipes* (Edwards, 1921)
[Algeria. Morocco]
20' Abdomen without posterodorsal sclerites (Fig. 15.3.1A) 21
21(20) Frontoclypeal apotome with pigmentation forming short or long "H" or inverted "U" pattern, sometimes faint (Fig. 15.3.5K, O) 22
21' Frontoclypeal apotome without pigmentation forming short or long "H" or inverted "U" pattern (Fig. 15.3.5B, J, L–N) 23
22(21) Gill histoblast of 8 filaments. Frontoclypeal apotome with pigmentation forming elongated "H" pattern (Fig. 15.3.5K). Postgenal cleft with anterior margin well defined (Fig. 15.3.7C) .... *Simulium noelleri* (Friederichs, 1920) (complex)
[Slovenia. Turkey]
22' Gill histoblast of 6 filaments. Frontoclypeal apotome with pigmentation forming short "H" or inverted "U" pattern (Fig. 15.3.5O). Postgenal cleft with anterior margin often poorly defined (Fig. 15.3.7L) .... *Simulium posticatum* (Meigen, 1838)
[Slovenia. Turkey].
23(21) Postgenal cleft extended about half distance from posterior tentorial pits to hypostomal groove; anterior margin broadly rounded (Figs. 15.3.6E and 15.3.7D) 24
23' Postgenal cleft typically extended more than half distance from posterior tentorial pits to hypostomal groove; sides converging to narrowed apex (Figs. 15.3.6O, P and 15.3.7A, B, E–K, M, N) 26
24(23) Frontoclypeal apotome with head spots positive and laterally compressed such that anterolateral spots closely approach anteromedian and posteromedian spots (Fig. 15.3.5B). Gill histoblast of 6 filaments .... *Simulium erythrocephalum* (De Geer, 1776) (in part)
[Croatia. France. Greece. Italy. Portugal. Slovenia. Spain. Tunisia(?). Turkey]
24' Frontoclypeal apotome with head spots positive or negative, either rather evenly spaced or with some absent (as in Fig. 15.3.5N). Gill histoblast of 8 or 10 filaments (*Simulium ornatum* group) 25
25(24) Gill histoblast of 8 filaments .... *Simulium baracorne* (Smart, 1944)
[Greece. Turkey]
Simulium egregium (Séguy, 1930)
[Morocco]
Simulium fontanum (Terteryan, 1952)
[Turkey]
Simulium intermedium (Roubaud, 1906) (complex)
[Algeria. Balearic Islands: Majorca, Minorca. France. Italy. Malta. Montenegro. Morocco. Portugal. Spain. Tunisia]
Simulium kiritshenkoi (Rubtsov, 1940)

[Cyprus. Turkey]
Simulium ornatum (Meigen, 1818) (complex)
[Algeria. Bosnia and Herzegovina. Croatia. France. Greece. Israel. Italy. Jordan. Lebanon. Montenegro. Morocco. Portugal. Spain. Tunisia. Turkey]
Simulium trifasciatum (Curtis, 1839)
[Algeria(?). Croatia. France. Greece. Italy. Morocco(?). Portugal. Spain. Turkey]
25' Gill histoblast of 10 filaments .... *Simulium pontinum* (Rivosecchi, 1960)
[Italy]
26(23) Gill histoblast of 8 or more filaments. Postgenal cleft often arrowhead-shaped or miter-shaped, and extended ¾ or more from posterior tentorial pits to hypostomal groove, sometimes reaching hypostomal groove as narrowed elongation (Figs. 15.3.6O, P and 15.3.7B, E, F, K) 27
26' Gill histoblast of 6 filaments. Postgenal cleft often subtriangular, and extended ¾ or less from posterior tentorial pits to hypostomal groove, not reaching hypostomal groove as narrowed elongation (Fig. 15.3.7A, G–J, M, N) 34
27(26) Gill histoblast of 8 to 16 filaments. Body pale grayish, greenish, or yellowish 28
27' Gill histoblast of 8 or more than 24 filaments. Body dark to medium brown or gray 31
28(27) Gill histoblast of 10–16 filaments. Postgenal cleft typically reaching hypostomal groove as narrowed elongation (Fig. 15.3.7E) .... *Simulium colombaschense* (Scopoli, 1780) (complex)
[Bosnia and Herzegovina. Croatia. Greece. Italy]
Simulium liriense (Rivosecchi, 1961) [Italy: Abruzzo, Lazio (extinct?)].
28' Gill histoblast of 8 filaments. Postgenal cleft typically not reaching hypostomal groove (Fig. 15.3.7F, K) 29
29(28) Postgenal cleft bowed slightly outward (Fig. 15.3.7K) .... *Simulium bergi* (Rubtsov, 1956)
[Turkey]
29' Postgenal cleft bowed strongly outward (Fig. 15.3.7F) 30
30(29) Frontoclypeal apotome with dark central area largely obscuring head spots (Fig. 15.3.5M) .... *Simulium reptans* (L., 1758) (complex)
[Bosnia and Herzegovina. Croatia. France. Greece. Italy. Portugal. Spain. Tunisia. Turkey]
30' Frontoclypeal apotome without dark central area; head spots absent or individually defined (Fig. 15.3.5L) .... *Simulium reptantoides* (Carlsson, 1962) (complex)
[Bosnia and Herzegovina. Croatia. Montenegro]
31(27) Gill histoblast of more than 24 filaments. Hypostoma with 4–7 sublateral setae per side (Fig. 15.3.9C) (*Simulium bukovskii* group) 32
31' Gill histoblast of 8 filaments. Hypostoma with 8–12 sublateral setae per side (Figs. 15.3.8O, 15.3.9A) (*Simulium argenteostriatum* group) 33
32(31) Gill histoblast of fewer than 30 filaments .... *Simulium bukovskii* (Rubtsov, 1940)
[Turkey].
32' Gill histoblast of 30 or more filaments .... *Simulium degrangei* (Dorier & Grenier, 1960)
[Bosnia and Herzegovina. France. Greece. Italy. Montenegro]
33(31) Frontoclypeal apotome with all head spots distinct against pale background .... *Simulium hispaniola* (Grenier & Bertrand, 1954)
[Algeria. France. Italy. Portugal. Spain. Turkey].
33' Frontoclypeal apotome with lateral head spots indistinct within dark pigmentation (Fig. 15.3.5J) .... *Simulium argenteostriatum* (Strobl, 1898)
[Algeria. Bosnia and Herzegovina. France. Italy. Montenegro. Slovenia. Spain. Tunisia]
Simulium aureofulgens (Terteryan, 1949)
[Turkey]
34(26) Abdomen in dorsal and lateral views gradually expanded posteriorly until narrowing just anterior to posterior circlet (as in Fig. 15.3.1B). Posterior proleg (slide-mounted) with more than 100 rows of hooklets (*Simulium albellum* group) 35
34' Abdomen in dorsal and lateral views rather abruptly expanded at segment V until narrowing just anterior to posterior circlet (Fig. 15.3.1A). Posterior proleg (slide-mounted) with fewer than 100 rows of hooklets 36
35(34) Hypostoma with sublateral setae in 2 subparallel rows on each side .... *Simulium brevifile* (Rubtsov, 1956)
[Italy]
Simulium continii (Rivosecchi & Cardinali, 1975)
[Italy: Sardinia].

35' Hypostoma with sublateral setae in 1 row on each side (Fig. 15.3.9G) .... *Simulium auricoma* (Meigen, 1818)
[Croatia. Cyprus. France. Greece. Italy. Lebanon. Morocco. Portugal. Spain. Turkey]
Simulium galloprovinciale (Giudicelli, 1963)
[Algeria. France. Italy. Morocco. Spain]
Simulium ibericum Crosskey & Santos (Grácio, 1985)
[Portugal. Spain]
Simulium margaritae (Rubtsov, 1956)
[Turkey]
Simulium marocanum (Bouzidi & Giudicelli, 1988)
[Algeria. Morocco]
36(34) Body pale yellowish white. Postgenal cleft bowed strongly outward, abruptly tapered after midpoint (Fig. 15.3.7J) .... *Simulium xanthinum* (Edwards, 1933)
[Algeria. France. Morocco. Spain].
36' Body brownish or grayish. Postgenal cleft bowed slightly outward, rather evenly tapered toward apex (Fig. 15.3.7A, G–I, M) 37
37(36) Head capsule yellowish, without distinct head spots (except along posterior border) .... *Simulium niha* (Giudicelli & Dia, 1986)
[Lebanon. Turkey]
Simulium sicanum (Rivosecchi, 1963) (in part)
[Italy: Calabria, Sicily]
37' Head capsule suffused with brown, with or without distinct head spots (Fig. 15.3.5N) 38
38(37) Subesophageal ganglion pigmented, often obscuring outline of postgenal cleft .... 39.
38' Subesophageal ganglion unpigmented; outline of postgenal cleft clearly demarcated (Fig. 15.3.7H) .... *Simulium monticola* (Friederichs, 1920)
[Algeria. Bosnia and Herzegovina. Croatia. France. Italy. Portugal. Spain]
Simulium sicanum (Rivosecchi, 1963) (in part)
[Italy: Calabria, Sicily]
39(38) Hypostoma with lateral and median teeth small, projected only slight beyond other teeth (Fig. 15.3.9F) .... *Simulium* "indet." Belqat, Adler & Dakki, 2001
[Morocco]
39' Hypostoma with lateral and median teeth prominent, projected well beyond other teeth (Fig. 15.3.9B, D, E) 40
40(39) Rectal papillae of 3 lobes (not always everted), each with 0–3 short secondary lobules (Fig. 15.3.1D) .... *Simulium bezzii* (Corti, 1914) (complex)
[Algeria. Croatia. Cyprus. France. Greece. Israel. Italy. Lebanon. Montenegro. Morocco. Spain. Turkey]
40' Rectal papillae of 3 lobes (not always everted), each with 2 or more long secondary lobules (Fig. 15.3.1F) 41
41(40) Gill histoblast with dorsalmost filament more than twice as thick basally as ventralmost filament (Fig. 15.3.11L) .... *Simulium argyreatum* (Meigen, 1838)
[Bosnia and Herzegovina. Croatia. France. Greece. Italy. Montenegro. Portugal. Spain. Turkey]
41' Gill histoblast with dorsalmost filament less than twice as thick basally as ventralmost filament (Fig. 15.3.12B) .... *Simulium variegatum* (Meigen, 1818)
[Algeria. Bosnia and Herzegovina. Croatia. France. Greece. Italy. Lebanon. Montenegro. Morocco. Portugal. Slovenia. Spain. Turkey]

Simuliidae: Genera (Pupae)

1 Cocoon absent or shapeless and saclike (Fig. 15.3.3B), covering various portions of pupa .... 2.
1' Cocoon well formed, slipper-, shoe-, or boot-shaped, typically covering thorax and abdomen, but sometimes loosely woven and bearing perforations and windows (Fig. 15.3.3D–H) 6
2(1) Gill of 3 elongated, swollen tubes with few short apical thread-like filaments .... *Helodon laamii* (Beaucournu-Saguez & Bailly-Choumara, 1981)
[Morocco]
2' Gill of 8 or more stout or thread-like filaments (Fig. 15.3.10B–D), rarely arising from 2 elongated, swollen lobes (Fig. 15.3.10A) 3

3(2) Gill of 8 filaments .... *Levitinia freidbergi* (Beaucournu-Saguez & Braverman, 1987)
[Golan].
3' Gill of 12 to 27 filaments 4
4(3) Gill of 14 to 27 filaments 5
4' Gill of 12 filaments (*Urosimulium*) .... *Urosimulium aculeatum* (Rivosecchi, 1963)
[Balearic Islands: Majorca. Italy: Sicily, Sardinia]
Urosimulium faurei (Bernard, Grenier & Bailly-Choumara, 1972)
[Algeria. Morocco. Spain]
Urosimulium juccii (Contini, 1966)
[Algeria(?). Italy: Sardinia. Tunisia(?)]
5(4) Abdominal segments IV and V without large sclerites in pleural membrane. Rare .... ***Greniera***
5' Abdominal segments IV and V with large sclerites in pleural membrane (Fig. 15.3.3A). Common .... ***Prosimulium***
6(1) Pleural region of abdominal segments VIII and IX with anchor- or grapnel-shaped setae (Fig. 15.3.3C). Gill of 12 to more than 60 filaments .... ***Metacnephia***
6' Pleural region of abdominal segments VIII and IX with unbranched setae, or if with anchor- or grapnel-shaped setae, then gill of 10 filaments. Gill of 2 to 30 filaments or inflated tubes .... ***Simulium***

Simuliidae: *Prosimulium*: Species (Pupae)

1 Gill of 20 to 27 filaments .... *Prosimulium tomosvaryi* (Enderlein, 1921)
[Bosnia and Herzegovina. Croatia. France. Greece. Italy. Morocco. Portugal. Spain. Turkey]
1' Gill of 14 to 16 filaments (Fig. 15.3.10A–D) 2
2(1) Gill of 2 swollen trunks, each giving rise to apical or subapical thread-like filaments (Fig. 15.3.10A) .... *Prosimulium albense* (Rivosecchi, 1961)
[Algeria. Italy]
2' Gill of 3 or more slender to moderately stout basal trunks with thread-like filaments (Fig. 15.3.10B–D) 3
3(2) Gill of 14 filaments .... *Prosimulium calabrum* (Rivosecchi, 1966)
[Italy]
3' Gill of 16 filaments 4
4(3) Gill filaments splayed in 3 well-spaced groups, each with at least 1 trunk 3 or more times as long as its width (Fig. 15.3.10B, C) .... *Prosimulium anatoliense* (Adler & Şirin, 2015)
[Turkey]
Prosimulium latimucro (Enderlein, 1925) (complex)
[Bosnia and Herzegovina. France. Italy. Montenegro. Morocco. Portugal. Spain]
Prosimulium petrosum (Rubtsov, 1955)
[Lebanon]
4' Gill filaments clustered in bundle, arising from 3 short trunks, each no more than twice as long as its width (Fig. 15.3.10D) .... *Prosimulium hirtipes* (Fries, 1824)
[Bosnia and Herzegovina. France. Italy (northern). Spain]
Prosimulium italicum (Rivosecchi, 1967)
[Italy]
Prosimulium rachiliense (Djafarov, 1954) (complex)
[Greece. Morocco. Turkey]

Simuliidae: *Greniera*: Species (Pupae)

1 Gill of 14 filaments .... *Greniera fabri* (Doby & David, 1959)
[France. Algeria. Italy. Morocco. Spain]
1' Gill of 18 filaments .... *Greniera dobyi* (Beaucournu-Saguez & Braverman, 1987)
[Golan]

Simuliidae: *Metacnephia*: Species (Pupae)

1 Gill of 13 to 25 filaments 2
1' Gill of more than 25 filaments 3
2(1) Gill of 13 to 18 filaments .... *Metacnephia nuragica* (Rivosecchi, Raastad & Contini, 1975)
[Italy: Sardinia. Morocco(?). Portugal. Spain]
2' Gill of 20 to 25 filaments .... *Metacnephia persica* (Rubtsov, 1940)
[Lebanon]
3(1) Gill of 65 or more filaments .... *Metacnephia* "Crete" Procunier, 1982
[Greece: Crete]
3' Gill of fewer than 60 filaments 4
4(3) Gill with ventral filaments basally inflated, about twice diameter of other filaments (Fig. 15.3.10F) .... *Metacnephia subalpina* (Rubtsov, 1956)
[Turkey]
4' Gill with dorsal and ventral filaments of nearly same basal thickness (Fig. 15.3.10E) 5
5(4) Cocoon with numerous perforations and apertures. Gill with 1 or 2 filaments thicker than others (Fig. 15.3.10E) .... *Metacnephia phrygiensis* (Şirin & Adler, 2015)
[Turkey]
5' Cocoon tightly woven, without perforations or apertures. Gill with all filaments of about same diameter 6
6(5) Western half of Mediterranean .... *Metacnephia nigra* (Rubtsov, 1940)
[Turkey]
6' Eastern half of Mediterranean .... *Metacnephia blanci* (Grenier & Theodorides, 1953)
[Algeria. France. Italy: Sicily. Morocco. Portugal. Spain. Tunisia]
Metacnephia sardoa (Rivosecchi & Contini, 1965)
[Italy: Sardinia]

Simuliidae: *Simulium*: Species (Pupae)

1 Gill of more than 20 slender filaments (*Simulium bukovskii* group) 2
1' Gill of fewer than 20 slender filaments and/or inflated tubes 3
2(1) Gill of 25–29 filaments .... *Simulium bukovskii* (Rubtsov, 1940)
[Turkey]
2' Gill of 30–40 filaments .... *Simulium degrangei* (Dorier & Grenier, 1960)
[Bosnia and Herzegovina. France. Greece. Italy. Montenegro]
3(1) Gill of 1 short, round club with 4 slender filaments arising from ventral surface (Fig. 15.3.14A) .... *Simulium timondavidi* (Giudicelli, 1961)
[France: Corsica]
3' Gill of 2–16 slender filaments and/or inflated tubes 4
4(3) Gill of 2 inflated tubes of about same diameter (Fig. 15.3.14B, D) (subgenus *Rubzovia*) .... 5.
4' Gill of 4–16 slender filaments (Figs. 15.3.10G–K, 15.3.11A–L, and 15.3.12A–G) or 4–8 variously inflated tubes (Figs. 15.3.10L and 15.3.12H–L) 6
5(4) Gill in dorsal view with one inflated tube projected anteriorly and one projected laterally (Fig. 15.3.14B). Cocoon excavated anteriorly, without anterodorsal projection .... *Simulium knidirii* (Giudicelli & Thiery, 1985)
[Morocco]
5' Gill in dorsal view with both inflated tubes projected anteriorly (Fig. 15.3.14D). Cocoon with short anteromedian projection .... *Simulium lamachi* (Doby & David, 1960)
[Morocco. Spain]
6(4) Gill of 10–16 slender filaments 7
6' Gill of 4–8 slender filaments or inflated tubes 9
7(6) Cocoon with anterodorsal projection; apertures and perforations absent (as in Fig. 15.3.3D) (subgenus *Hellichiella*) .... *Simulium saccai* (Rivosecchi, 1967)
[Italy]
7' Cocoon without anterodorsal projection; apertures and perforations present or absent (Fig. 15.3.3G) 8

8(7) Cocoon shoe- or boot-shaped, with apertures and perforations, especially anterolaterally (Fig. 15.3.3G) .... *Simulium colombaschense* (Scopoli, 1780) (complex)

[Bosnia and Herzegovina. Croatia. Greece. Italy]
Simulium liriense Rivosecchi, 1961
[Italy: Abruzzo, Lazio; extinct?].
8' Cocoon slipper-shaped, with at most some small perforations from loose weaving .... *Simulium pontinum* (Rivosecchi, 1960)

[Italy]
9(6) Gill with 2 inflated tubes lying against and encircling pupal cephalothorax, plus 2 to 6 additional, variously inflated tubes (Figs. 15.3.12H–L and 15.3.14K) (subgenus *Wilhelmia*) 10
9' Gill without 2 inflated tubes lying against pupal cephalothorax; all 4 to 8 filaments slender (Figs. 15.3.10G–K, 15.3.11A–L, and 15.3.12A–G), or if inflated, then only 4 filaments, with at most 1 lying close to pupa (Fig. 15.3.10L) 15
10(9) Gill of 4 or 6 total tubes (Figs. 15.3.12L and 15.3.14K) 11
10' Gill of 8 total tubes (Fig. 15.3.12H–K) 12
11(10) Gill of 4 tubes, 3 of which are inflated and robust and 1 of which is much smaller and thinner (Fig. 15.3.14K) .... *Simulium quadrifila* (Grenier, Faure & Laurent, 1957)
[Algeria. Morocco. Spain]
11' Gill of 6 tubes, 4 of which are inflated and robust and 2 of which are smaller and thinner (Fig. 15.3.12L) .... *Simulium sergenti* (Edwards, 1923)

[Algeria. Morocco. Portugal. Spain. Tunisia]
12(10) Gill with filaments or tubes narrowest at their attachment points to inflated basal tubes (Fig. 15.3.12K) .... *Simulium bravermani* (Beaucournu-Saguez, 1986)

[Israel]
Simulium paraequinum (Puri, 1933)
[Croatia. Greece. Israel. Italy. Jordan. Lebanon. Turkey]
Simulium pseudequinum (Séguy, 1921)
[Algeria. Bosnia and Herzegovina. Croatia. Cyprus. France. Greece. Israel. Italy. Jordan. Lebanon. Libya. Montenegro. Morocco. Portugal. Slovenia. Spain. Tunisia. Turkey]
12' Gill with filaments or tubes widest at their attachment points to inflated basal tubes (Fig. 15.3.12H–J) 13
13(12) Gill with 6 inflated tubes about 1/3 to more than ½ width of 2 inflated basal tubes (Fig. 15.3.12I) .... *Simulium equinum* (L., 1758)

[Bosnia and Herzegovina. Croatia. France. Italy. Montenegro. Morocco. Portugal. Slovenia. Spain. Turkey]
13' Gill with 6 slender tubes less than 1/3 width of 2 inflated basal tubes (Fig. 15.3.12H, J) .... 14.
14(13) Gill with 1 pair of filaments arising from base on short common petiole (Fig. 15.3.12H) .... *Simulium balcanicum* (Enderlein, 1924)

[Albania. Bosnia and Herzegovina. Croatia. Greece. Italy. Montenegro. Slovenia. Turkey]
14' Gill with all filaments arising independently from base (Fig. 15.3.12J) .... *Simulium golani* (Beaucournu-Saguez & Braverman, 1977)

[Golan]
Simulium lineatum (Meigen, 1804)
[Bosnia and Herzegovina. Croatia. France. Italy. Montenegro. Portugal. Slovenia. Spain]
Simulium turgaicum Rubtsov, 1940
[Bosnia and Herzegovina. Lebanon. Slovenia. Turkey]
15(9) Gill of 4 filaments 16
15' Gill of 6–8 filaments 33
16(15) Cocoon without anterodorsal projection (Fig. 15.3.3E), although minute bump or few silk loops sometimes present along anterior margin 17
16' Cocoon with distinct anterodorsal projection of various lengths (Fig. 15.3.3D) 26
17(16) Gill, in lateral view, with filaments in compact bundle directed ventrally and running along substrate, at least basally (Fig. 15.3.14G, H) 18
17' Gill, in lateral view, with filaments spaced apart, projected forward or dorsally, not running along substrate (Figs. 15.3.10H–J, 15.3.11C, E, and 15.3.14F, I) 19

18(17) Gill with 1 pair of filaments on long stalk from which 2 other filaments arise independently (Fig. 15.3.14H) .... *Simulium codreanui* (Sherban, 1958)
[Bosnia and Herzegovina. Turkey]
18' Gill with all filaments arising near base of common trunk (Fig. 15.3.14G) .... *Simulium ichnusae* (Rivosecchi & Contini, 1994)
[Italy: Sardinia]
19(17) Gill with dorsal 2 filaments sessile and ventral 2 filaments on short petiole; dorsalmost filament, in lateral view, strongly divergent from other 3 filaments (Fig. 15.3.14I) .... *Simulium fucense* (Rivosecchi, 1962)
[Italy: Abruzzo, Lazio, Umbria]
19' Gill with dorsal 2 filaments on petiole of various lengths and ventral 2 filaments either sessile or on petiole of various lengths; dorsalmost filament, in lateral view, variously divergent from other 3 filaments (Figs. 15.3.3E, 15.3.10H–J, and 15.3.11C, E) 20
20(19) Cocoon loosely woven, net-like; anterior border not reinforced with silk .... *Simulium angustitarse* (Lundström, 1911)
[Algeria. Bosnia and Herzegovina. Croatia. France. Greece. Italy. Montenegro. Morocco. Portugal. Slovenia. Spain. Turkey]
Simulium ibleum (Rivosecchi, 1966)
[Algeria. Cyprus. Greece: Crete. Lebanon. Libya. Morocco. Spain(?), Tunisia, Turkey]
20' Cocoon tightly woven, lacking perforations; anterior border typically reinforced with silk (Fig. 15.3.3E) 21
21(20) Gill, in lateral view, with dorsalmost filament angled near base, sometimes diverging from all other filaments (Figs. 15.3.3E and 15.3.10H, I) (subgenus *Eusimulium*) 22
21' Gill, in lateral view, with dorsalmost filament gently and smoothly curved near its base, not diverging from all other filaments (Fig. 15.3.11C, E) (*Simulium vernum* group, in part) 24
22(21) Cocoon with raised anteroventral collar .... *Simulium mellah* (Giudicelli & Bouzidi, 2000)
[Algeria. Morocco]
22' Cocoon without raised anteroventral collar (Fig. 15.3.3E) 23
23(22) Gill, in dorsal view, with upper pair of filaments angled inward so that all filaments not in same vertical plane (Fig. 15.3.14C) .... *Simulium flexibranchium* (Crosskey, 2001)
[Greece: Crete, Kithira, Naxos, Rhodes]
23' Gill, in dorsal view, with all filaments approximately in same vertical plane .... *Simulium angustipes* (Edwards, 1915)
[Algeria. Bosnia and Herzegovina. Croatia. France. Italy. Lebanon. Morocco. Portugal. Spain. Tunisia. Turkey]
Simulium aureum (Fries, 1824)
[France. Italy. Portugal. Spain]
Simulium krymense (Rubtsov, 1956)
[Greece]
Simulium petricolum (Rivosecchi, 1963)
[Algeria. Bosnia and Herzegovina. Cyprus. France. Greece. Italy. Libya. Morocco. Portugal. Spain]
Simulium rubzovianum (Sherban, 1961)
[Algeria. Balearic Islands: Majorca, Minorca. Bosnia and Herzegovina. Croatia, Cyprus, France. Greece. Israel. Italy. Jordan. Libya. Malta. Montenegro. Morocco, Portugal. Slovenia. Spain. Tunisia. Turkey]
Simulium velutinum (Santos Abreu, 1922) (complex, including 3 undescribed species: "K" Leonhardt, 1985; "3" Adler, Cherairia, Arigue, Samraoui & Belqat, 2015; and "5" Adler, Cherairia, Arigue, Samraoui & Belqat, 2015)
[Algeria. Morocco. Spain]
24(21) Gill, in lateral view, with dorsal and ventral pairs of filaments each grouped and strongly divergent in vertical plane (Fig. 15.3.11E) .... *Simulium djafarovi* (Rubtsov, 1962)
[Turkey]
24' Gill, in lateral view, with dorsal and ventral pairs of filaments not each strongly grouped and only weakly divergent in vertical plane (Figs. 15.3.11C and 15.3.14F) 25
25(24) Gill with dorsalmost filament thicker than other 3 filaments and dorsal petiole thicker than ventral petiole (Fig. 15.3.14F); gill surface with small black dots .... *Simulium brevidens* (Rubtsov, 1956)
[Algeria(?). Bosnia and Herzegovina. France. Italy. Montenegro. Morocco(?). Spain]

25’ Gill with all 4 filaments of subequal diameter on petioles of subequal length (Fig. 15.3.11C); gill surface without small black dots .... *Simulium costatum* (Friederichs, 1920)
[Algeria. Bosnia and Herzegovina. Croatia. France. Italy. Montenegro. Morocco. Spain. Turkey]
26(16) Gill filaments inflated at least basally (Fig. 15.3.10L) and, in dorsal view, not all in vertical plane .... *Simulium ruficorne* (Macquart, 1838) (complex)
[Algeria. Egypt. Israel. Jordan. Lebanon. Libya. Malta. Morocco. Portugal. Spain. Syria. Tunisia]
26’ Gill filaments slender, only slightly thickened basally (Figs. 15.3.3D, 15.3.10K, and 15.3.11A, D, F) and, in dorsal view, none (Fig. 15.3.14E) or all approximately in vertical plane .... 27.
27(26) Gill with dorsalmost filament divergent from other 3 filaments (Fig. 15.3.10K) .... *Simulium lundstromi* (Enderlein, 1921)
[Algeria. Croatia. France. Italy. Lebanon. Morocco. Portugal. Spain. Turkey]
27’ Gill with all filaments approximately equally spaced, dorsalmost filament not strongly divergent from other 3 filaments (Figs. 15.3.3D and 15.3.11A, B, D, F) 28
28(27) Cocoon with anterodorsal projection short, about as long as its basal width .... *Simulium pinhaoi* (Santos Grácio, 1985)
[Portugal]
28’ Cocoon with anterodorsal projection longer than its basal width (Fig. 15.3.3D) 29
29(28) Cocoon with anterodorsal projection expanded in distal half (Fig. 15.3.15B) Cephalothoracic integument with microtubercles longer than wide .... *Simulium bertrandi* (Grenier & Dorier, 1959)
[Bosnia and Herzegovina. France. Italy. Montenegro. Spain]
29’ Cocoon with anterodorsal projection tapered (Fig. 15.3.3D). Cephalothoracic integument with microtubercles about as long as wide (Fig. 15.3.13C, E) 30
30(29) Cocoon with some thicker areas of silk, especially along dorsal median line and on anterodorsal projection .... *Simulium armoricanum* (Doby & David, 1961)
[Portugal, Spain]
30’ Cocoon with thin, even surface, although anterior margin slightly thickened (Fig. 15.3.3D) 31
31(30) Gill with filaments nearly sessile and, in dorsal view, spaced fan-like in horizontal plane (Fig. 15.3.14E) .... *Simulium marsicanum* (Rivosecchi, 1962)
[Italy]
31’ Gill with filaments on petioles of various lengths (Fig. 15.3.11B, D, F) and, in dorsal view, grouped in vertical plane 32
32(31) Gill, in lateral view, with all filaments in compact bundle directed ventrally and running along substrate (Fig. 15.3.11B) .... *Simulium carthusiense* (Grenier & Dorier, 1959)
[France. Italy. Morocco. Spain]
32’ Gill, in lateral view, with filaments spaced apart, especially dorsal and ventral pairs (Fig. 15.3.11D, F) .... *Simulium cryophilum* (Rubtsov, 1959) (complex)
[Algeria. Balearic Islands: Majorca. Bosnia and Herzegovina. Croatia. Cyprus. France. Greece. Italy. Lebanon. Montenegro. Morocco. Portugal. Slovenia. Spain. Tunisia. Turkey]
Simulium urbanum (Davies, 1966)
[Spain]
Simulium vernum (Macquart, 1826) (complex)
[Bosnia and Herzegovina. Croatia. France. Italy. Morocco. Portugal. Slovenia. Spain. Tunisia. Turkey]
33(15) Gill of 6 filaments 34
33’ Gill of 8 filaments 49
34(33) Pupal thorax with median, rugose carina (Fig. 15.3.15A). Cocoon shoe-shaped, with entire opening surrounded by thickened silk rim (subgenus *Crosskeyellum*) .... *Simulium gracilipes* (Edwards, 1921)
[Algeria. Morocco]
34’ Pupal thorax smooth, rugose, or with microtubercles, but without median carina. Cocoon variously shaped, with opening not surrounded by thickened silk rim 35
35(34) Cocoon boot-shaped, without apertures or perforations; gills recessed within cocoon (*Simulium albellum* species group) 36
35’ Cocoon slipper- or shoe-shaped, with or without apertures or perforations; gills partially or entirely exposed (Fig. 15.3.3F, H) 40

36(35) Cocoon with dorsal longitudinal slit; surface warty (Fig. 15.3.15C) .... *Simulium ibericum* (Crosskey & Santos Grácio, 1985)
[Portugal. Spain].
36' Cocoon without any slits; surface not warty 37
37(36) Head and thorax with wrinkled cuticle 38
37' Head and thorax without wrinkled cuticle, except possibly on margins of head .... *Simulium auricoma* (Meigen, 1818)
[Croatia. Cyprus. France. Greece. Italy. Lebanon. Morocco. Portugal. Spain. Turkey]
Simulium brevifile (Rubtsov, 1956)
[Italy]
Simulium margaritae (Rubtsov, 1956)
[Turkey]
38(37) Gill filaments, in lateral view, in compact bundle (Fig. 15.3.14J) .... *Simulium continii* (Rivosecchi & Cardinali, 1975)
[Italy: Sardinia]
38' Gill filaments, in lateral view, well-spaced (as in Fig. 15.3.12G) 39
39(38) Head and thorax strongly wrinkled .... *Simulium galloprovinciale* (Giudicelli, 1963)
[Algeria. France. Italy. Morocco. Spain]
39' Head and thorax weakly wrinkled .... *Simulium marocanum* (Bouzidi & Giudicelli, 1988)
[Algeria. Morocco]
40(35) Thorax, in dorsal and lateral views, with rounded protuberance (patagium) near base of each gill (Fig. 15.3.15F) .... *Simulium variegatum* (Meigen, 1818)
[Algeria. Bosnia and Herzegovina. Croatia. France. Greece. Italy. Lebanon. Montenegro. Morocco. Portugal. Slovenia. Spain. Turkey]
40' Thorax, in dorsal and lateral views, without rounded protuberance near base of each gill 41
41(40) Cocoon with perforations and apertures (Fig. 15.3.15G) 42
41' Cocoon tightly woven, without perforations or apertures 43
42(41) Gill with brownish to pale gray, slender filaments (Fig. 15.3.11H); in dorsal view, with all filaments in vertical plane. Cocoon with multiple anterior apertures (Fig. 15.3.15G) .... *Simulium bezzii* (Corti, 1914) (complex)
[Algeria. Croatia. Cyprus. France. Greece. Israel. Italy. Lebanon. Montenegro. Morocco. Spain. Turkey]
42' Gill with dark gray to blackish filaments, somewhat swollen basally (Fig. 15.3.12C); in dorsal view, with middle filaments not in vertical plane. Cocoon with open weave of many apertures .... *Simulium xanthinum* (Edwards, 1933)
[Algeria. France. Morocco. Spain]
43(41) Gill, in lateral view, with filaments in tight bundle (Figs. 15.3.11L and 15.3.12A) 44
43' Gill, in lateral view, with filaments well-spaced (Figs. 15.3.10G and 15.3.12D, F) 47
44(43) Gill, in lateral view, with dorsalmost filament arched dorsally at base (Fig. 15.3.11L) .... *Simulium argyreatum* (Meigen, 1838)
[Bosnia and Herzegovina. Croatia. France. Greece. Italy. Montenegro. Portugal. Spain. Turkey]
44' Gill, in lateral view, with dorsalmost filament straight, often directed ventrally at base (Fig. 15.3.12A) 45
45(44) Thoracic integument (slide-mounted) with large rounded microtubercles bearing spines (Fig. 15.3.13A) .... *Simulium sicanum* (Rivosecchi, 1963)
[Italy: Calabria, Sicily]
45' Thoracic integument (slide-mounted) with smooth rounded microtubercles (Fig. 15.3.13D) 46
46(45) Thoracic integument with microtubercles scattered individually or in clumps (Fig. 15.3.13D) .... *Simulium monticola* (Friederichs, 1920)
[Algeria. Bosnia and Herzegovina. Croatia. France. Italy. Portugal. Spain]
46' Thoracic integument with microtubercles densely distributed .... *Simulium niha* (Giudicelli & Dia, 1986)
[Lebanon. Turkey]
47(43) Gill, in dorsal view, with middle filaments divergent in horizontal plane (Fig. 15.3.10G) .... *Simulium erythrocephalum* (De Geer, 1776)
[Croatia. France. Greece. Italy. Portugal. Slovenia. Spain. Tunisia(?). Turkey]
47' Gill, in dorsal view, with all filaments nearly in vertical plane 48

48(47) Gill, in lateral view, with all filaments of nearly same diameter (Fig. 15.3.12F) .... *Simulium posticatum* (Meigen, 1838)
[Slovenia. Turkey]
48' Gill, in lateral view, with diameter of dorsalmost filament about twice that of ventralmost filament (Fig. 15.3.12D) .... *Simulium* "indet." Belqat, Adler & Dakki, 2001
[Morocco]
49(33) Cocoon, in lateral view, shoe- or boot-shaped, with notch or anterodorsal projection (Fig. 15.3.15D, E) (*Simulium argenteostriatum* group) 50
49' Cocoon, in lateral view, slipper-shaped, without notch or anterodorsal projection, although lateral apertures can be present (Fig. 15.3.3F, H) 51
50(49) Cocoon, in lateral view, with large aperture on each side and with anterodorsal projection (Fig. 15.3.15E) .... *Simulium argenteostriatum* (Strobl, 1898)
[Algeria. Bosnia and Herzegovina. France. Italy. Montenegro. Slovenia. Spain. Tunisia]
50' Cocoon, in lateral view, without apertures or anterodorsal projection, but with lateral notch on each side (Fig. 15.3.15D) .... *Simulium aureofulgens* (Terteryan, 1949)
[Turkey]
Simulium hispaniola (Grenier & Bertrand, 1954)
[Algeria. France. Italy. Portugal. Spain. Turkey]
51(49) Cocoon without lateral apertures, although weave can be coarse (Fig. 15.3.3F) 52
51' Cocoon with lateral apertures (Fig. 15.3.3H) .... *Simulium reptans* (L., 1758) (complex)
[Bosnia and Herzegovina. Croatia. France. Greece. Italy. Portugal. Spain. Tunisia. Turkey]
Simulium reptantoides (Carlsson, 1962) (complex)
[Bosnia and Herzegovina. Croatia. Montenegro]
52(51) Gill with dorsalmost petiole bearing 3 filaments (Fig. 15.3.11I) .... *Simulium noelleri* (Friederichs, 1920) (complex)
[Slovenia. Turkey]
52' Gill with dorsalmost petiole bearing 2 filaments (Figs. 15.3.11J and 15.3.12E) 53
53(52) Gill with ventralmost petiole bearing 2 pairs of filaments (Fig. 15.3.12E) .... *Simulium bergi* (Rubtsov, 1956)
[Turkey]
53' Gill with ventralmost petiole bearing 1 pair of filaments (Fig. 15.3.11J) (*Simulium ornatum* group, in part) 54
54(53) Thoracic integument (slide-mounted) with conical microtubercles anterolaterally (Fig. 15.3.13F) .... *Simulium trifasciatum* (Curtis, 1839)
[Algeria(?). Croatia. France. Greece. Italy. Morocco(?). Portugal. Spain. Turkey]
54' Thoracic integument (slide-mounted) with rounded microtubercles anterolaterally (Fig. 15.3.13B) .... *Simulium baracorne* (Smart, 1944)
[Greece. Turkey]
Simulium egregium (Séguy, 1930)
[Morocco]
Simulium fontanum (Terteryan, 1952)
[Turkey]
Simulium intermedium (Roubaud, 1906) (complex)
[Algeria. Balearic Islands: Majorca, Minorca. France. Italy. Malta. Montenegro. Morocco. Portugal. Spain. Tunisia]
Simulium kiritshenkoi (Rubtsov, 1940)
[Cyprus. Turkey]
Simulium ornatum (Meigen, 1818) (complex)
[Algeria. Bosnia and Herzegovina. Croatia. France. Greece. Israel. Italy. Jordan. Lebanon. Montenegro. Morocco. Portugal. Spain. Tunisia. Turkey].

References

Adler, P.H. 2020. World blackflies (Diptera: Simuliidae): a comprehensive revision of the taxonomic and geographical inventory [2020]. 142 pp. http://biomia.sites.clemson.edu/pdfs/blackflyinventory.pdf Accessed: 30 January 2021.

Adler, P.H. 2022. World blackflies (Diptera: Simuliidae): a comprehensive revision of the taxonomic and geographical inventory [2022]. 145 pp. http://biomia.sites.clemson.edu/pdfs/blackflyinventory.pdf Accessed: 26 March 2022.

Adler P.H. & G.W. Courtney. 2019. Ecological and societal services of aquatic Diptera. Insects 10: 70.

Adler, P.H. & R.G. Foottit. 2017. Introduction. pp. 1-7 *in*: R.G. Foottit & P.H. Adler (eds.), Insect Biodiversity: Science and Society, Volume I, 2nd edition. John Wiley & Sons Ltd., Chichester, UK.

Adler, P.H., T. Kúdelová, M. Kúdela, G. Seitz & A. Ignjatović-Ćupina. 2016. Cryptic biodiversity and the origins of pest status revealed in the macrogenome of *Simulium colombaschense* (Diptera: Simuliidae), history's most destructive black fly. PLoS ONE 11: e0147673.

Adler, P.H. & J.W. McCreadie. 2019. Black flies (Simuliidae). pp. 237-259 *in*: G. R. Mullen & L. A. Durden (eds.), Medical and Veterinary Entomology. 3rd edition. Elsevier, San Diego, CA.

Adler, P.H., D.C. Currie & D.M. Wood. 2004. The Black Flies (Simuliidae) of North America. Cornell University Press, Ithaca, NY. xv + 941 pp. + 24 color plates.

Adler, P.H., M. Cherairia, S.F. Arigue, B. Samraoui & B. Belqat. 2015. Cryptic biodiversity in the cytogenome of bird-biting black flies in North Africa. Medical and Veterinary Entomology 29: 276–289.

Adler, P.H., B. Belqat, J. Garrido González, A.J. Pérez & G. Seitz. 2016a. Chromosomal relationships of *Simulium armoricanum* and its undescribed sister species in the *Simulium vernum* species group (Diptera: Simuliidae). Zootaxa 4137: 211-222.

Adler, P.H., T. Kúdelová, M. Kúdela, G. Seitz & A. Ignjatović-Ćupina. 2016b. Cryptic biodiversity and the origins of pest status revealed in the macrogenome of *Simulium colombaschense* (Diptera: Simuliidae), history's most destructive black fly. PLoS ONE 11 (1): e0147673.

Alexander, C.P. 1920. The Crane-flies of New York. Part II. Biology and Phylogeny. Memoirs, Cornell University Agricultural Experiment Station 38: 691–1133. Cornell University, Ithaca.

Ali, A. 1996. A concise review of chironomid midges (Diptera: Chironomidae) as pests and their management. Journal of Vector Ecology 21: 105–121.

Allegrucci, G., G. Carchini, P. Convey & V. Sbordoni. 2012. Evolutionary geographic relationships among orthocladine chironomid midges from maritime Antarctic and sub-Antarctic islands. Biological Journal of the Linnean Society 106, 258–274.

Allegrucci, G., G. Carchini, V. Todisco, P. Convey & V.A. Sbordoni 2006. A molecular phylogeny of Antarctic Chironomidae and its implications for biogeographical history. Polar Biology 29: 320–326.

Amorim, D.S. 2016. Family Scatopsidae. pp. 239-245 *in*: M. Wolff, S.S. Nihei & C.J.B. de Carvalho (eds.), Catalogue of Diptera of Colombia. Zootaxa 4122: 1–949.

Armitage, P., P.S. Cranston & L.C.V. Pinder. 1995. Chironomidae - Biology and ecology of non-biting midges. Chapman & Hall, London. 572 pp.

Armitage, P.D. 1995. Chironomidae as food. In: Armitage P.D., P.S. Cranston & L.C.V. Pinder (eds). The Chironomidae. Springer, London: 423–435.

Ashe, P., D.A. Murray & F. Reiss. 1987. The zoogeographical distribution of Chironomidae (Insecta: Diptera). Annales de Limnologie - International Journal of Limnology 23(1): 27–60.

Avesani D. & V. Lencioni 2011. Ditteri (Capitolo 4). Pages 37-167 *in*: M. Gobbi & L. Latella (eds.), La fauna dei prati – 1: tassonomia, ecologia e metodi di studio dei principali gruppi di invertebrati terrestri italiani. Quaderni del Museo delle Scienze di Trento 4/1: 1–176.

Aydın, G.B. 2022. Contributions to the faunistic knowledge of Chironomidae (Diptera) of Turkey based on the adult males collected around Hazar Lake (Elazığ). Journal of the Entomological Research Society 24(3): 375–387.

Başören, Ö. and N. Kazanci. 2015. The species key to the Simuliidae (Insecta, Diptera) larvae in running waters of Eastern Black Sea Region (Turkey). Review of Hydrobiology 8: 105–118.

Başören, Ö. and N. Kazanci. 2016. Identification key to the Simuliidae (Insecta, Diptera) pupae in running waters of the Eastern Black Sea Region (Turkey). Review of Hydrobiology 9: 123–134.

Belqat, B. & M. Dakki. 2004. Clés analytiques des simulies (Diptera) du Maroc. Zoologica Baetica 15: 77–137.

Bernabò, P., M. Gaglio, F. Bellamoli, G. Viero & V. Lencioni. 2017. DNA damage and translational response during detoxification from copper exposure in a wild population of *Chironomus riparius*. Chemosphere 173: 235–244.

Bisthoven, L.J. De, A. Vermeulen & F. Ollevier. 1998. Environmental contamination and toxicology experimental induction of morphological deformities in *Chironomus riparius* larvae by chronic exposure to copper and lead. Archives of Environmental Contamination and Toxicology 256: 249–256.

Bitušík, P. & K. Trnková. 2019. A preliminary checklist of Chironomidae (Diptera) from Albania with first records for the Balkan Peninsula. Zootaxa 4563 (2): 361–371.

Bouchard, R.W. & J.L.C. Jr, Ferrington. 2008. Identification guide and key to Chironomid pupal exuviae in Mongolian Lakes. Chironomid Research Group, 72.

Boulaaba, S., S. Zrelli, M. Płóciennik & M. Boumaiza. 2014. Diversity and distribution of Chironomidae (Insecta: Diptera) of protected areas in North Tunisia. Knowledge and Management of Aquatic Ecosystems 415, 06. https://doi.org/10.1051/kmae/2014031.

Boumaiza, M. & Laville, H. 1988. Premier inventaire faunistique (Diptera, Chironomidae) des eaux courantes de la Tunisie. Annales de Limnologie 24: 173–181.

Brundin, L., 1966. Transantarctic relationships and their significance, as evidenced by chironomid midges, with a monograph of the Subfamilies Podonominae and Aphroteniinae and the austral Heptagyiae. Kungliga Svenska Vetenskapsakademiens Handlingar 11: 1–472.

Blackshaw, R.P. 2009. A comparison of management options for leatherjacket populations in organic crop rotations using mathematical models. Agricultural and Forest Entomology 11: 197–203.

Blackshaw, R.P. & H. Hicks. 2012. Distribution of adult stages of soil insect pests across an agricultural landscape. Journal of Pest Sciences 86: 53–62.

Borkent, A. 2014. World catalogue of extant and fossil Chaoboridae (Diptera). Zootaxa 3796: 469–493.

Borkent A. & P. Dominiak. 2020. Catalog of the biting midges of the world (Diptera: Ceratopogonidae). Zootaxa 4787: 1–377.

Brindle, A. 1962. Taxonomic notes on the larvae of British Diptera, 9: the family Ptychopteridae. Entomologist 95: 212–216.

Burgin, S.G. & F.F. Hunter. 1997. Nectar versus honeydew as sources of sugar for male and female black flies (Diptera: Simuliidae). Journal of Medical Entomology 34: 605–608.

Campaioli S., 1999. Ditteri. Pages 403-443 *in*: S. Campaioli, P.F. Ghetti, A. Minelli & S. Ruffo (eds.), Manuale per il Riconoscimento dei Macroinvertebrati delle Acque Dolci Italiane. Volume II. Provincia Autonoma di Trento, Italy.

Carew, M.E. & A.A. Hoffmann. 2015. Delineating closely related species with DNA barcodes for routine biological monitoring. Freshwater Biology 60: 1545–1560.

Caspers, N. & Reiss, F. 1989. Die Chironomidae der Türkei. Teil I: Podonominae, Diamesinae, Prodiamesinae, Orthocladiinae (Diptera, Nematocera, Chironomidae). Entomofauna, Zeitschrift Für Entomologie 10(8/1): 105–160.

Chance, M.M. & D.A. Craig. 1986. Hydrodynamics and behaviour of Simuliidae larvae (Diptera). Canadian Journal of Zoology 64: 1295–1309.

Chiswell, J.R. 1956. A taxonomic account of the last instar larvae of some British Tipulinae (Diptera: Tipulidae). Transactions of the Royal Entomological Society of London 108: 409–484.

Chvála, M. 2013. Fauna Europaea: Tabanidae. *In*: T. Pape & P. Beul (coords.), Fauna Europaea: Diptera. Fauna Europaea version 2017.06. https://fauna-eu.org Accessed: 12 September 2021.

Cornette, R., O. Gusev, Y. Nakahara, S. Shimura, T. Kikawada & T. Okuda. 2015. Chironomid midges (Diptera, Chironomidae) show extremely small genome sizes. Zoological Science 32 Issue 3(1): 248–254.

Courtney, G.W. 2000. Family Blephariceridae. pp. 7–30 *in*: L. Papp & B. Darvas (eds.), Contributions to a Manual of Palaearctic Diptera. Appendix. Science Herald, Budapest, Hungary.

Courtney, G.W. 2019. Aquatic Diptera. pp. 925–1022 *in*: Merritt, R.W., K.W. Cummins & M.B. Berg (eds.), An Introduction to the Aquatic Insects of North America, 5th edition. Kendall/Hunt Publishing Co., Dubuque, IA.

Courtney, G.W. & P.S. Cranston. 2015. Order Diptera. pp. 1043-1058 *in*: Thorp, J.H. & D.C. Rogers (eds.), Thorp and Covich's Freshwater Invertebrates, Volume 1, 4th edition. Academic Press, London, UK.

Courtney, G.W., T. Pape, J.H. Skevington & B.J. Sinclair. 2017. Biodiversity of Diptera. pp. 229–278 *in*: R.G. Foottit & P.H. Adler (eds.), Insect Biodiversity: Science and Society, Volume I, 2nd edition. John Wiley & Sons Ltd., Chichester, UK.

Courtney, G.W., B.J. Sinclair & R. Meier. 2000. Morphology and terminology of Diptera larvae. pp. 85–161 *in*: L. Papp & B. Darvas (eds.), Contributions to a Manual of Palaearctic Diptera, Volume 1. Science Herald, Budapest, Hungary.

Cranston, P.S. 1995. Chironomids: from genes to ecosystems. CSIRO East Melbourne, Victoria 3002, Australia. pp. 482.

Crosskey, R.W. 1990. The Natural History of Blackflies. John Wiley & Sons Ltd., Chichester, England. 711 pp.

Crosskey, R.W. 1991. The blackfly fauna of Majorca and other Balearic islands (Diptera: Simuliidae). Journal of Natural History 25: 671–690.

Crosskey, R.W. 2004. The blackfly fauna of the island of Cyprus (Diptera: Simuliidae). Entomologist's Gazette 55: 49–66.

Crosskey, R.W. and H. Malicky. 2001. A first account of the blackflies (Diptera, Simuliidae) of the Greek Islands. Studia Dipterologica 8: 111–141.

Das, P., S.C. Mandal, S.K. Bhagabati, M.S. Akhtar & S.K. Singh. 2012. Important live food organisms and their role in aquaculture. In: M. Sukham (Ed.), Frontiers in Aquaculture, Narendra Publishing House, 69–86.

De Jong, H. 2013a. Fauna Europaea: Scathophagidae. *In*: T. Pape & P. Beul (coords.), Fauna Europaea: Diptera. Fauna Europaea version 2017.06. https://fauna-eu.org Accessed: 3 August 2021.

De Jong, H. 2013b. Fauna Europaea: Anisopodidae. *in*: T. Pape & P. Beul (coords.), Fauna Europaea: Diptera. Fauna Europaea version 2017.06. https://fauna-eu.org Accessed: 2 February 2020.

De Jong, H., P. Oosterbroek, J. Gelhaus, H. Reusch & C. Young 2008. Global diversity of craneflies (Insecta, Diptera: Tipulidea or Tipulidae sensu lato) in freshwater. Hydrobiologia 595: 457–467.

Disney, R.H.L. 1991. The aquatic Phoridae (Diptera). Entomologica Scandinavica 22: 171–191.

Disney, R.H.L. 1999. British Dixidae (Meniscus Midges) and Thaumaleidae (Trickle Midges): keys with ecological notes. Freshwater Biological Association Scientific Publication 56: 1–129.

Disney, R.H.L. 2004. Insecta: Diptera, Phoridae. Pages 818-825 *in*: C.M. Yule & H.S. Yong (eds.), Freshwater Invertebrates of the Malaysian Region. Academy of Sciences Malaysia, Kuala Lumpur, Malaysia.

Dobson M. 2013. Family-level keys to freshwater fly (Diptera) larvae: a brief review and a key to European families avoiding use of mouthpart characters. Freshwater Reviews 6: 1–32.

Downes, J.A. 1969. The swarming and mating flight of Diptera. Annual Review of Entomology 14(1): 271–298.

Driauach, O., E. Krzemińska & B. Belqat. 2015. Genus *Trichocera* in Morocco: first records from Africa and a new species (Diptera: Trichoceridae). Zootaxa 4059: 181–190.

El Haouari, H., K. Kawtar & M. Ghamizi 2014. Les Tabanidae (Insecta: Diptera) du Maroc. Bulletin de la Societe Zoologique de France 139: 91–105.

Ekrem, T., E. Willassen & E. Stur. 2010. Phylogenetic utility of five genes for dipteran phylogeny: a test case in the Chironomidae leads to generic synonymies. Molecular Phylogenetics and Evolution 57: 561–571.

EPA US. 2000. Methods for Measuring the Toxicity and Bioaccumulation of Sediment-associated Contaminants with Freshwater Invertebrates, 2nd edition. EPA 600/R-99/064, Environmental Protection Agency, Washington, DC. pp. 192.

Eymann, M. 1991. Dispersion patterns exhibited by larvae of the black flies *Cnephia dacotensis and Simulium rostratum* (Diptera: Simuliidae). Aquatic Insects 13: 99–106.

Faasch, H. 2015. Identification guide to aquatic and semi-aquatic Diptera larvae. Deutsche Gesellschaft für Limnologie–Arbeitshilfe 1-2015, Essen, Germany. 179 pp.

Failla, A.J., A.A. Vasquez, M. Fujimoto & J.L. Ram. 2015. The ecological, economic and public health impacts of nuisance chironomids and their potential as aquatic invaders. Aquatic Invasions 10: 1–15.

Failla, A.J., A.A. Vasquez, P. Hudson, M. Fujimoto & J.L. Ram. 2016. Morphological identification and COI barcodes of adult flies help determine species identities of chironomid larvae (Diptera, Chironomidae). Bulletin of Entomological Research, 106 (1): 34–46.

Fasbender, A. 2013. Phylogeny and diversity of the phantom crane flies (Diptera: Ptychopteridae). PhD. Thesis, Iowa State University, Ames, IA. pp. 855.

Ferrington, L.C., Jr. & M.B. Berg. 2019. Chironomidae. pp. 1119–1274 *in*: Merritt, R.W., Cummins, K.W. & M.B. Berg (eds.), An Introduction to the Aquatic Insects of North America, 5th edition. Kendall/Hunt Publishing Co., Dubuque, IA.

Ferrington, L.C. Jr. 2008. Global diversity of non-biting midges (Chironomidae; Insecta-Diptera) in freshwater. Hydrobiologia 595: 447–455.

Ferrington, L.C. Jr, M.A. Blackwood, C.A. Wright, N.H. Crisp, J.L. Kavanaugh & F.J. Schmidt. 1991. A protocol for using surface-floating pupal exuviae of Chironomidae for rapid bioassessment of changing water quality, pp. 181–190. In: Sediment and stream water quality in a changing environment: trends and explanations. Vol. 203. N.E. Peters & D.E. Walling (eds.). IAHS Press, Oxfordshire, UK.

Fisher, T., M. Crane & A. Callaghan. 2003. Induction of cytochrome P-450 activity in individual *Chironomus riparius* Meigen larvae exposed to xenobiotics. Ecotoxicology and Environmental Safety 54: 1–6.

Foster W.A. & E.D. Walker. 2019. Mosquitoes (Culicidae). pp. 261–325 *in*: G.R. Mullen & L.A. Durden (eds.), Medical and Veterinary Entomology, 3rd edition. Elsevier, San Diego, CA.

Fusari, L.M., G.P.S. Dantas & L.C. Pinho. 2018. Order Diptera. pp. 607–623 *in*: Hamada, N., J.H. Thorp & D. C. Rogers (eds.), Thorp and Covich's Freshwater Invertebrates: Keys to Neotropical Hexapoda, 4th edition. Elsevier, London, UK.

Gelhaus, J. & V. Podeniene. 2019. Tipuloidea. pp. 1023–1070 in: R.W. Merritt, M.B. Berg and K.W. Cummins (eds.), An Introduction to the Aquatic Insects of North America, 5th edition. Kendall/Hunt Publ. Co., Dubuque, IA.

Gelhaus, J.K. 1986. Larvae of the crane fly genus Tipula in North America (Diptera: Tipulidae). The University of Kansas Science Bulletin 53: 121–182.

Gelhaus, J. & V. Podeniene. 2019. Tipuloidea. pp. 1023–1070 in: R.W. Merritt, M.B. Berg and K.W. Cummins (eds.), An Introduction to the Aquatic Insects of North America, 5*th* edition. Kendall/Hunt Publ. Co., Dubuque, IA.

Gledhill, T., J. Cowley & R.J. M. Gunn. 1982. Some aspects of the host: parasite relationships between adult blackflies (Diptera; Simuliidae) and larvae of the water-mite *Sperchon setiger* (Acari; Hydrachnellae) in a small chalk stream in southern England. Freshwater Biology 12: 345–357.

Grazioli, V., B. Rossaro, P. Parenti, R. Giacchini & V. Lencioni. 2016. Hypoxia and anoxia effects on alcohol dehydrogenase activity and hemoglobin content in *Chironomus riparius* Meigen, 1804. Journal of Limnology 75: 347–354.

Grichanov, I. 2016. A checklist of species of the family Dolichopodidae (Diptera) of the World arranged by alphabetic list of generic names. http://dolicho.narod.ru/Genera3.htm Accessed: 1 February 2021.

Hadley, M. 1971. Aspects of the larval ecology and population dynamics of *Molophilus ater* Meigen (Diptera, Tipulidae) on Pennine Moorland. Journal of Animal Ecology 40: 445–466.

Haubrock, P.J., G.M. Kvifte, H. Langguth & R. Wagner. 2017. Glacial and postglacial species divergence and dispersal of European trickle midges (Diptera: Thaumaleidae). Arthropod Systematics & Phylogeny 75: 523–534.

Hebert, P.D.N., S. Ratnasingham, E.V. Zakharov, A.C. Telfer, V. Levesque-Beaudin, M.A. Milton, S. Pedersen, P. Jannetta & J.R. deWaard 2016. Counting animal species with DNA barcodes: Canadian insects. Philosophical Transactions of the Royal Society B 371: 20150333.

Henriques-Oliveira, A.L., J.L. Nessimian & L.F.M. Dorvillé. 2003. Feeding habits of chironomid larvae (Insecta: Diptera) from a stream in the Floresta da Tijuca, Rio de Janeiro, Brazil. Brazilian Journal of Biology 63: 269–281.

Imada, Y. 2020. Moss mimesis par excellence: integrating previous and new data on the life history and larval ecomorphology of long-bodied craneflies (Diptera: Cylindrotomidae: Cylindrotominae). Zoological Journal of the Linnean Society 193: 1156–1204.

Jacobsen, R.E. 2008. A key to the pupal exuviae of the midges (Diptera: Chironomidae) of everglades National Park, Florida (No. Scientific Investigations Report 2008-5082). Fort Lauderdale, FL: US Geological Survey.

Kettani, K., M.J. Ebejer, D.M. Ackland, G. Bächli, D. Barraclough, M. Barták, M. Carles-Tolrá, M. Černý, P. Cerretti, P. Chandler, M. Dakki, C. Daugeron, H. De Jong, J. Dils, H. Disney, B. Droz, N. Evenhuis, P. Gatt, G. Graciolli, I.Y. Grichanov, J.-P. Haenni, M. Hauser, O. Himmi, I. MacGowan, B. Mathieu, M. Mouna, L. Munari, E.P. Nartshuk, O.P. Negrobov, P. Oosterbroek, T. Pape, A.C. Pont, G.V. Popov, K. Rognes, M. Skuhravá, V. Skuhravý, M. Speight, G. Tomasovic, B. Trari, H.-P. Tschorsnig, J.-C. Vala, M. von Tschirnhaus, R. Wagner, D. Whitmore, A.J. Woźnica, T. Zatwarnick & P. Zwick. 2022. Catalogue of the Diptera (Insecta) of Morocco an annotated checklist, with distributions and a bibliography. ZooKeys 1094: 1–466.

Kettani, K. & J. Moubayed-Breil. 2018. Communities of Chironomidae (Diptera) from four ecological zones delimited by the Mediterranean coastal ecosystems of Morocco (Moroccan Rif). Updated list and faunal data from the last two decades: Chironomidae from coastal ecosystems of Morocco. Journal of Limnolology 77(sl): 141–144.

Kranzfelder, P., A.M. Anderson, A.T. Egan, J.E. Mazack, R.W. Bouchard Jr, M.M. Rufer & L.C. Ferrington Jr. 2015. Use of Chironomidae (Diptera) surface-floating pupal exuviae as a rapid bio-assessment protocol for water bodies. JoVE (Journal of Visualized Experiments), (101), e52558.

Krosch, M.N., P.S. Cranston, L.M. Bryant, F. Strutt & S.R. McCluen. 2017. Towards a dated molecular phylogeny of the Tanypodinae (Chironomidae, Diptera). Invertebrate Systematics 31(3): 302–316.

Kiliç, A.Y. 2006. New additions and errata to the checklist of Tabanidae (Insecta: Diptera) fauna of Turkey. Turkish Journal of Zoology 30: 335–343.
Langton, P.H. 1991 A key to the pupal exuviae of West Palaearctic Chironomidae. Privately published. Huntingdon, PE17 1YH, England. 386 pp.
Langton, P.H., & J. Casas. 1999. Changes in chironomid assemblage composition in two Mediterranean mountain streams over a period of extreme hydrological conditions. Hydrobiologia 390: 37–49.
Langton, P.H., & H. Visser (2003). Chironomidae exuviae: a key to pupal exuviae of the West Palaearctic Region. Expert Center for Taxonomic Identification, University of Amsterdam. Expert Center for Taxonomic Identification, 2003 (Ed.). World biodiversity database CD-ROM series.
Laville, H. & P.H. Langton. 2002. The lotic Chironomidae (Diptera) of Corsica (France). Annales de Limnologie 38(1): 63–64.
Laville, H. & F. Reiss. 1992. The Chironomid Fauna of the Mediterranean Region reviewed. Netherlands Journal of Aquatic Ecology 26(2-4): 239–245.
Lencioni, V. 2018. Glacial influence and stream macroinvertebrate biodiversity under climate change: Lessons from the Southern Alps. Science of the Total Environment 622-623: 563–575.
Lencioni, V., P.S. Cranston & E.A. Makarchenko. 2018. Recent advances in the study of Chironomidae: An overview. Journal of Limnology 77: 1–6.
Lencioni, V., A. Rodriguez-Prieto & G. Allegrucci. 2021. Congruence between molecular and morphological systematics of Alpine non-biting midges (Chironomidae, Diamesinae). Zoologica Scripta 50(4): 455–472.
Lin, X.L., E. Stur & T. Ekrem. 2018. Molecular phylogeny and temporal diversification of Tanytarsus van der Wulp (Diptera: Chironomidae) support generic synonymies, a new classification and center of origin. Systematic Entomology 43(4): 659–677.
Lindegaard, C. 1995. Chironomidae of European cold springs and factors influencing their distribution. Journal of Kansas Entomological Society 68 (2):108–131.
Lindner, E. 1959. Beitrage zur Kenntnis der Larven der Limoniidae (Diptera). Zeitschrift fur Morphologie und Okologie der Tiere 48: 209–319.
MacLean, S.F. 1973. Life cycle and growth energetics of the Arctic crane-fly *Pedicia hannai antennata*. Oikos 24: 434–443.
Majer, J. 2013. Fauna Europaea: Rhagionidae. *In*: T. Pape & P. Beul (coords.), Fauna Europaea: Diptera. Fauna Europaea version 2017.06. https://fauna-eu.org Accessed: 2 September 2021.
Martin, J. 1979. Chromosomes as tools in taxonomy and phylogeny of Chironomidae (Diptera). Entomologica Scandinavica Supplement 10: 67–74.
Martínez-Paz, P., M. Morales, J.L. Martínez-Guitarte & G. Morcillo. 2013. Genotoxic effects of environmental endocrine disruptors on the aquatic insect *Chironomus riparius* evaluated using the comet assay. Mutation Research–Genetic Toxicology and Environmental Mutagenesis 758: 41–47.
Mauch, E. 2017. Aquatische Diptera-Larven in Mittel- Nordwest- und Nordeuropa. Übersicht über die Formen und ihre Identifikation. Lauterbornia 83: 1–404.
McCreadie, J.W., P.H. Adler & C.E. Beard. 2011. Ecology of symbiotes of larval black flies (Diptera: Simuliidae): distribution, diversity, and scale. Environmental Entomology 40: 289–302.
Merritt, R.W., K.W. Cummins & M.B. Berg (Eds.) 2019. An Introduction to the Aquatic Insects of North America, 5th edition. Kendall/Hunt Publishing Company, Dubuque, IA. 1480 pp.
Michailova, P., N. Petrova, J. Ilkova, S. Bovero, S. Brunetti, K. White & G. Sella. 2006 Genotoxic effect of copper on salivary gland polytene chromosomes of *Chironomus riparius* Meigen 1804 (Diptera, Chironomidae). Environmental Pollution 144: 647–654.
Michailova P., Lencioni V., Nenov M., & S. Nikolov. (2021). Can DNA barcoding be used to identify closely related Clunio Haliday, 1855 species (Diptera: Chironomidae, Orthocladiinae)? Zootaxa 4927 (1): 001–008. https://doi.org/10.11646/zootaxa.4927.1.1
Mora, A., & Z. Csabai. (2008). First annotated checklist of Chronomidae of Rhodos, Greece (Insecta, Diptera). Spixiana, 31(2): 223–231.
Morse, J.C. 2017. Biodiversity of aquatic insects. pp. 205–227 *in*: R.G. Foottit & P.H. Adler (eds.), Insect Biodiversity: Science and Society, Volume I, 2nd edition. John Wiley & Sons, Chichester, UK.
Moubayed, J., Ait-Mouloud, S. & Lounaci, A. 1992. Les Chironomidae (Diptera) d'Agérie. I. Bassin de l'Oued Aissi (Grande Kabylie). Nachrichtenblatt der Bayerrischen Entomologen 41(1): 21–29.
Moubayed, J. & P.H. Langton. 2000. Community of Chironomidae (Diptera) from Lebanese estuaries. Faunal and biogeographic outline. In: Late 20th Century Research on Chironomidae: and Anthology from the 13th International Symposium on Chironomidae: 565–569.
Moubayed-Breil, J. 2020. Chironomidae from the Mediterranean ecosystem of continental France sensu lato. Faunal and biogeographic data over the last four decades (Diptera). Ephemera 21(1): 31–69.
Moubayed-Breil, J. & P. Ashe. 2012. An updated checklist of the Chironomidae of Corsica with an outline of their altitudinal and geographical distribution [Diptera, Chironomidae]. Ephemera 13(1): 13–39.
Moubayed-Breil, J. & P. Ashe. 2016. New records and additions to the database on the geo-graphical distribution of some threatened chironomid species from continental France [Diptera, Chironomidae]. Ephemera 16(2): 121–136.
Moubayed-Breil, J., & A. Dia. 2007a. Les Chironomidae de deux rivières côtières du Liban méridional: le Damour et l'Awwali (Diptera: Chironomidae). Entomofauna Zeitschrift Für Entomologie 14(14): 269–276.
Moubayed-Breil, J., & A. Dia. 2007b. New records of non-biting midges for Lebanon and the Near East (Diptera: Chironomidae). Ephemera 8(2): 101–107.
Moubayed Breil, J., & J.M., Dominici. 2019 Clunio boudouresquei sp. n. and Thalassosmittia ballestai sp. n., two Tyrrhenian marine species occurring in Scandola Nature Reserve, West Corsica (Diptera: Chironomidae). Chironomus Journal of Chironomidae Research. 32: 4–24.
Moubayed-Breil, J., A. Lounaci & D. Lounaci-Daoudi. 2007. Non-biting midges from Algeria, North Africa (Diptera: Chironomidae). Ephemera 8(2): 93–99.
Moubayed-Breil, J., M. Verlaque, J.M. Dominici & C.H. Bianconi. 2013. Estuarine zones of Corsica: Faunal, ecological and biogeographical data. Travaux de l'Institut Scientifique, Rabat, Série Zoologie 49: 43–58.

Mullens B.A. 2019. Horse flies and deer flies (Tabanidae). pp. 327–343 *in*: G.R. Mullen & L.A. Durden (eds.), Medical and Veterinary Entomology, 3rd edition. Elsevier, San Diego, CA.

Murphy, W.L., L.V. Knutson, E.G. Chapman, R.J. McDonnell, C.D. Williams, B.A. Foote & J.-C. Vala 2012. Key aspects of the biology of snail-killing Sciomyzidae flies. Annual Review of Entomology 57: 425–447.

Neumann D. 2003. Adaptations of Chironomids to Intertidal Environments. Annual Review of Entomology 21(1): 387–414.

OECD. 2018. Chironomid toxicity test using spiked sediment (OECD TG 218) or spiked water (OECD TG 219). pp. 241–251 *in*: Revised Guidance Document 150 on Standardised Test Guidelines for Evaluating Chemicals for Endocrine Disruption. OECD Publishing, Paris, France.

Oosterbroek, P. 2006. The European Families of the Diptera: Identification, Diagnosis, Biology. KNNV Uitgeverij, 2nd edition. 204 pp.

Oosterbroek, P. 2022. Catalogue of the Craneflies of the World (CCW). https://ccw.naturalis.nl/ Accessed 23 September 2022.

Oosterbroek, P. & B. Theowald 1991. Phylogeny of the Tipuloidea based on characters of larvae and pupae (Diptera, Nematocera) with an index to the literature except Tipulidae. Tijdschrift voor Entomologie 134: 211–267.

Oscoz, J., D. Galicia & R. Miranda 2011. Taxa description and biology. pp. 47–148 *in*: J. Oscoz, D. Galicia & R. Miranda (eds.), Identification Guide of Freshwater Macroinvertebrates of Spain. Springer, Dordrecht, The Netherlands.

Ozkan, N. & Camur-Elipek, B. 2006. The dynamics of Chironomidae larvae (Diptera) and the water quality in MeriÇ River (Edirne/ Turkey). Tiscia 35: 49–54.

Ozkan, N., Moubayed-Breil, J. & Çamur-Elipek, B. 2010. Ecological analysis of chironomid larvae (Diptera, Chironomidae) in Ergene River Basin (Turkish Thrace). Turkish Journal of Fisheries and Aquatic Sciences 10: 93–99.

Pape, T., V. Blagoderov & M.B. Mostovski. 2011. Order Diptera Linnaeus, 1758. pp. 222–229 *in*: Z.Q. Zhang (ed.), Animal biodiversity: an outline of higher-level classification and survey of taxonomic richness. Zootaxa 3148: 1–237.

Peus, F. 1952. 17. Cylindrotomidae. pp. 1–80 *in*: Lindner, E. (ed.), Die Fliegen der palaearktischen Region, 3(5)3, Lieferung 169.

Pichler V., E. Mancini, M. Micocci, M. Calzetta, D. Arnoldi, A. Rizzoli, V. Lencioni, F. Paoli, R. Bellini, R. Veronesi, S. Martini, A. Drago, C. De Liberato, A. Ermenegildi, J. Pinto, A. della Torre & B. Caputo 2021. A novel allele specific polymerase chain reaction (AS-PCR) assay to detect the V1016G knockdown resistance mutation confirms its widespread presence in *Aedes albopictus* populations from Italy. Insects 12: 79.

Pinder, L.C.V. 1995. The habitats of chironomid larvae. pp. 107–135 *in*: P.D. Armitage, P.S. Cranston & L.C.V. Pinder (eds.), The Chironomidae: Biology and Ecology of Non-Biting Midges. Springer (Chapman & Hall), Dordrecht, The Netherlands.

Płóciennik, M., P. Gadawski & J. Kazimierczak. 2014. New records of non-biting midges from coastal regions of Croatia and Montenegro. Spixiana 37 (1): 89–92.

Podeniene, V. 2001. Notes on the larvae of *Rhabdomastix* (*Sacandaga*) *laeta* (Loew, 1873) (Diptera, Limoniidae). Acta Zoologica Lituanica 11: 385–387.

Podeniene, V. 2002. Records on new and little-known larvae of the family Limoniidae (Diptera, Nematocera) from Lithuania. Acta Zoologica Lituanica 12: 294–308.

Podeniene, V. 2003. Morphology and ecology of the last instar larvae of the crane flies (Diptera, Tipulomorpha) of Lithuania. Doctoral dissertation, Vilnius University. 295 pp.

Podeniene, V. 2004. Records on little-known larvae of *Idioptera pulchella* (Meigen, 1830) (Diptera, Limoniidae, Limnophilinae). Acta Zoologica Lituanica 14: 37–41.

Podeniene, V. 2009. Lithuanian Chioneinae (Limoniidae, Diptera): larval habitat preferences and problems of identification, with description of last instar larvae of *Molophilus* (*Molophilus*) *crassipygus* de Meijere, 1918, *M.* (*M.*) *griseus* (Meigen, 1804), *M.* (*M.*) *ochraceus* (Meigen, 1818), *M.* (*M.*) *propinquus* (Egger, 1863). Lauterbornia 68: 135–145.

Podeniene, V. & J.K. Gelhaus 2002. The first description of the larva of the crane fly genus *Gonempeda* Alexander, 1924 (Limoniidae: Chioneinae), with new information for understanding the phylogenetic relationships of the genus. Proceedings of the Academy of Natural Sciences of Philadelphia 152: 67–73.

Podeniene, V., Podenas, S., Park, S.-J., Kim, A.-Y., J.A. Kim & J.K. Gelhaus. 2021. Review of East Palaearctic *Elliptera* (Diptera, Limoniidae) immatures with description of a new species. European Journal of Taxonomy 735: 110–132.

Podeniene, V., N. Naseviciene, & S. Podenas, 2019. Notes on first instar larvae of the genus *Tipula* (Diptera: Tipulidae). Zootaxa 4567: 90–110.

Prat, N., J.D. González-Trujillo & R. Ospina-Torres. 2014. Key to chironomid pupal exuviae (Diptera: Chironomidae) of tropical high Andean streams. Revista de Biologia Tropical 62(4); 1385–1406.

Prat, N., T. Puntí & M. Rieradevall. 2016. The use of larvae and pupal exuviae to study the biodiversity of Chironomidae in Mediterranean streams. Journal of Entomological and Acarological Research 48(1): 29–36.

Pritchard, G. 1983. Biology of Tipulidae. Annual Review of Entomology 28: 1–22.

Raunio, J. & T. Muotka. 2005. The use of chironomid pupal exuviae in river biomonitoring: the importance of sampling strategy. Archiv für Hydrobiologie 164(4): 529–545.

Raunio, J., L. Paasivirta & H. Hämäläinen. 2009. Assessing lake trophic status using spring-emerging chironomid pupal exuviae. Fundamental and applied limnology, 176(1), 61–73.

Raunio, J., Paavola, R. & T. Muotka. 2007. Effects of emergence phenology, taxa tolerances and taxonomic resolution on the use of the Chironomid Pupal Exuvial Technique in river biomonitoring. Freshwater Biology 52: 165–176.

Reiss, F. 1977. Verbreitungsmuster bei palaearktischen Chironomidenarten (Diptera, Chironomidae). Spixiana 1: 85–97.

Reiss, F. 1986. Ein Beitrag zur Chironomiden auna Syriens (Diptera, Chironomidae). Entomofauna Zeitschrift Für Entomologie 7(11): 153–166.

Rossaro B., A. Boggero, B. Lods-Crozet, G. Free, V. Lencioni, L. Marziali & G. Wolfram. 2012. A benthic quality index for European alpine lakes. Fauna Norvegica 31: 95–107.

Ruse, L. 2002. Chironomid pupal exuviae as indicators of lake status. Archiv für Hydrobiologie 153(3): 367–390.

Ruse, L. (2010). Classification of nutrient impact on lakes using the chironomid pupal exuvial technique. Ecological Indicators 10(3): 594–601.

Rivosecchi, L. 1978. Simuliidae: Diptera Nematocera. Fauna d'Italia 13: 1–533.

Robert, V., F. Günay, G. Le Goff, P. Boussès, T. Sulesco, A. Khalin, J.M. Medlock, H. Kampen, D. Petrić & F. Schaffne. 2019. Distribution chart for Euro-Mediterranean mosquitoes (western Palaearctic region). Journal of the European Mosquito Control Association 37: 1–28.

Rogers, J.S. 1926. Some notes on the feeding habits of adult crane-flies. Florida Entomologist 10: 5–8.

Rogers, J.S. 1927. Notes on the biology and immature stages of *Geronomyia* (Tipulidae, Diptera). 1. *Geranomyia rostrata* (Say). Florida Entomologist 11: 17–26.

Rozkošný, R. 2013. Fauna Europaea: Athericidae. *In*: T. Pape & P. Beul (coords.), Fauna Europaea: Diptera. Fauna Europaea version 2017.06. https://fauna-eu.org Accessed: 6 August 2021.

Rozkošný, R. & A. Nagatomi. 1997. Family Athericidae. pp. 439–446 *in*: L. Papp & B. Darvas (eds.), Manual of Palaerctic Diptera, Volume 2. Nematocera and Lower Brachycera. Science Herald, Budapest, Hungary.

Rozkošný, R., F. Gregor & A.C. Pont. 1997. The European Fanniidae (Diptera). Acta Scientiarum Naturalium Academiae Scientiarum Bohemicae Brno 31 (New Series): 1–80.

Savchenko, E.N. 1986. Limoniid–flies. General description, subfamilies Pediciinae and Hexatominae. Fauna Ukrainy 14(2): 1–380.

Serra-Tosio, B. 1973. Ecologie et biogéographie des Diamesini d'Europe (Diptera: Chironomidae). Travaux du Laboratoire d'Hydrobiologie et de Pisciculture de Grenoble 63: 5–175.

Skuhrava, M., M. Martinez & A. Roques. 2010. Diptera. Chapter 10 in: A. Roques, M. Kenis, D. Lees, C. Lopez-Vaamonde & W. Rabitsch (eds.), Alien Terrestrial Arthropods of Europe. BioRisk 4 (2): 553–602.

Sundermann A., S. Lohse, L.A. Beck & P. Haase. 2007. Key to the larval stages of aquatic true flies (Diptera), based on the operational taxa list for running waters in Germany. Annales de Limnologie – International Journal of Limnology 43: 61–74.

Teskey, H.J. 1981. Morphology and terminology-larvae. pp. 65–88 *in*: J.F. McAlpine, B.V. Peterson, G.E. Shewell, H.J. Teskey, J.R. Vockeroth & D. M. Wood (eds.), Manual of Nearctic Diptera, Volume 1. Monograph No. 27. Research Branch, Agriculture Canada, Ottawa, Ontario.

Tokeshi, M. 1995. Life cycles and population dynamics. pp. 225–268 in: P.D. Armitage, P.S. Cranston & L.C.V. Pinder (eds.), The Chironomidae: Biology and Ecology of Non-Biting Midges. Springer Netherlands, Dordrecht.

Ujvarosi, L.; Kolcsar, L.P.; Balint, M. & M. Ciprian. 2010. Pediciidae larvae (Insecta, Diptera) in the Carpathian Basin: Preliminary results and further perspectives. Acta Biologica Debrecina Oecologica Hungarica 21: 233–246.

Vaillant, F. 2002. Insecta: Diptera: Lonchopteridae. pp. 1–14 *in*: J. Schwoerbel & P. Zwick (eds.), Süßwasserfauna von Mitteleuropa [Freshwater Fauna of Europe], Volume 21, Insecta: Diptera, 22, 23, Lonchopteridae, Sciomyzidae. Springer Spektrum, Heidelberg, Germany

Vaillant, F. 1951. Sur *Orimarga hygropetrica* n.sp. Travaux de Laboratoire d'Hydrobiologie et Pisciculture de l'Universite de Grenoble 41/42: 43–47.

Vala, C. 1989. Diptères Sciomyzidae Euro-méditerranéens. Faune de France. France et Régions Limitrophes. N° 72. Fédération Française des Sociétés de Sciences Naturelles, Paris. 300 pp.

Wagner, R., 2002. Insecta: Diptera: Thaumaleidae. pp. 41–110 *in*: J. Schwoerbel & P. Zwick (eds.), Süßwasserfauna von Mitteleuropa [Freshwater Fauna of Europe], Volume 21, Insecta: Diptera: 10/11, Chaoboridae, Thaumaleidae. Spektrum Akademischer Verlag, Gustav Fischer, Heidelberg, Germany.

Wagner, R. 2013. Fauna Europaea: Thaumaleidae. *In*: T. Pape & P. Beul (coords.), Fauna Europaea: Diptera. Fauna Europaea version 2017.06. https://fauna-eu.org Accessed: 2 February 2020.

Wagner R., M. Barták, A. Borkent, G.W. Courtney, B. Goddeeris, J.-P. Haenni, L. Knutson, A. Pont, G.E. Rotheray, R. Rozkošný, B. Sinclair, N. Woodley, T. Zatwarnicki & P. Zwick. 2008. Global diversity of dipteran families (Insecta Diptera) in freshwater (excluding Simulidae, Culicidae, Chironomidae, Tipulidae and Tabanidae). Hydrobiologia 595: 489–519.

Wilson, R.S. & J.D. McGill. 1977. A new method of monitoring water quality in a stream receiving sewage effluent, using chironomid pupal exuviae. Water Research 11(11): 959-962.

Wilson, R.S. & L.P. Ruse. 2005. A guide to the identification of genera of chironomid pupal exuviae occurring in Britain and Northern Ireland (including common genera from Northern Europe) and their use in monitoring lotic and lentic fresh waters. Freshwater Biological Association. Special Publication 13, pp. 176.

Wirth, W.W. & N. Marston. 1968. A method for mounting small insects on microscope slides in Canada balsam. Annals of the Entomological Society of America 61: 783–784.

Woodley, N.E. 2009. Athericidae (athericid flies). pp. 491–493 *in*: B.V. Brown, A. Borkent, J.M. Cumming, D.M. Wood, N.E. Woodley & M.A. Zumbado (eds.), Manual of Central American Diptera, Volume 1. National Research Council Press, Ottawa, Ontario, Canada.

Wood, H.G. 1952. The crane-flies of the South-West Cape (Diptera, Tipuloidea). Annals of the South African Museum 39: 1–327.

Wotton, R.S., B. Malmqvist, T. Muotka & K. Larsson. 1998. Fecal pellets from a dense aggregation of suspension-feeders in a stream: an example of ecosystem engineering. Limnology and Oceanography 43: 719–725.

Yang, D., Y. Zhu, M. Wang & L. Zhang. 2006. World Catalog of Dolichopodidae (Insecta: Diptera). China Agricultural University Press. Beijing, China. 740 pp.

Zettler, J.A., P.H. Adler & J.W. McCreadie. 1998. Factors influencing larval color in the *Simulium vittatum* complex (Diptera: Simuliidae). Invertebrate Biology 117: 245–252.

Zhang, Y. & B. Malmqvist. 1997. Phenotypic plasticity in a suspension-feeding insect, *Simulium lundstromi* (Diptera: Simuliidae), in response to current velocity. Oikos 78: 503–510.

Zerguine, K., Z. Bensakhri, D. Bendjeddou & O. Khaladi. (2018). Diversity and distribution of Chironomidae (Insecta: Diptera) of the Oued Charef basin, North-Eastern Algeria. Annales de la Société entomologique de France (N.S.) 54(2): 141–155.

Index

Note: Page numbers followed by "*f*" and "*t*" refer to figures and tables, respectively.

D

H

I

J

K

L

N

O

Q

R

S